Osvaldo Gervasi   Marina L. Gavrilova
Vipin Kumar   Antonio Laganà
Heow Pueh Lee   Youngsong Mun
David Taniar   Chih Jeng Kenneth Tan (Eds.)

Computational Science and Its Applications – ICCSA 2005

International Conference
Singapore, May 9-12, 2005
Proceedings, Part II

Volume Editors

Osvaldo Gervasi
University of Perugia
E-mail: ogervasi@computer.org

Marina L. Gavrilova
University of Calgary
E-mail: marina@cpsc.ucalgary.ca

Vipin Kumar
University of Minnesota
E-mail: kumar@cs.umn.edu

Antonio Laganà
University of Perugia
E-mail: lag@dyn.unipg.it

Heow Pueh Lee
Institute of High Performance Computing, IHPC
E-mail: hplee@ihpc.a-star.edu.sg

Youngsong Mun
Soongsil University
E-mail: mun@computing.soongsil.ac.kr

David Taniar
Monash University
E-mail: David.Taniar@infotech.monash.edu.au

Chih Jeng Kenneth Tan
Queen's University Belfast
E-mail: cjtan@optimanumerics.com

Library of Congress Control Number: Applied for

CR Subject Classification (1998): D, F, G, H, I, J, C.2.3

ISSN   0302-9743
ISBN-10   3-540-25861-2 Springer Berlin Heidelberg New York
ISBN-13   978-3-540-25861-2 Springer Berlin Heidelberg New York

This work is subject to copyright. All rights are reserved, whether the whole or part of the material is concerned, specifically the rights of translation, reprinting, re-use of illustrations, recitation, broadcasting, reproduction on microfilms or in any other way, and storage in data banks. Duplication of this publication or parts thereof is permitted only under the provisions of the German Copyright Law of September 9, 1965, in its current version, and permission for use must always be obtained from Springer. Violations are liable to prosecution under the German Copyright Law.

Springer is a part of Springer Science+Business Media

springeronline.com

© Springer-Verlag Berlin Heidelberg 2005
Printed in Germany

Typesetting: Camera-ready by author, data conversion by Scientific Publishing Services, Chennai, India
Printed on acid-free paper   SPIN: 11424826   06/3142   5 4 3 2 1 0

Preface

The four-volume set assembled following *the 2005 International Conference on Computational Science and Its Applications*, ICCSA 2005, held in Suntec International Convention and Exhibition Centre, Singapore from 9 May 2005 till 12 May 2005, represents the fine collection of 540 refereed papers selected from nearly 2700 submissions.

Computational science has firmly established itself as a vital part of many scientific investigations, affecting researchers and practitioners in areas ranging from applications such as aerospace and automotive, to emerging technologies such as bioinformatics and nanotechnologies, to core disciplines such as mathematics, physics, and chemistry. Due to the sheer size of many challenges in computational science, the use of supercomputing, parallel processing, and sophisticated algorithms is inevitable and becomes a part of fundamental theoretical research as well as endeavors in emerging fields. Together, these far-reaching scientific areas contribute to shape this conference in the realms of state-of-the-art computational science research and applications, encompassing the facilitating theoretical foundations and the innovative applications of such results in other areas.

The topics of the refereed papers span all the traditional as well as emerging computational science realms, and are structured according to six main conference themes:

- Computational Methods and Applications
- High-Performance Computing, Networks and Optimization
- Information Systems and Information Technologies
- Scientific Visualization, Graphics and Image Processing
- Computational Science Education
- Advanced and Emerging Applications

In addition, papers from 27 workshops and technical sessions on specific topics of interest, including information security, mobile communication, grid computing, modeling, optimization, computational geometry, virtual reality, symbolic computations, molecular structures, Web systems and intelligence, spatial analysis, bioinformatics and geocomputation, to name a few, complete this comprehensive collection.

The warm response of the great number of researchers to the offer to present high-quality papers in ICCSA 2005 took the conference to record heights. The continuous support of computational science researchers has helped build ICCSA into a firmly established forum in this area. We look forward to building on this symbiotic relationship together to grow ICCSA further.

We recognize the contribution of the International Steering Committee and we deeply thank the International Program Committee for their tremendous support in putting this conference together, nearly 900 referees for their diligent work, and the Institute of High Performance Computing, Singapore for its generous assistance in hosting the event.

We also thank our sponsors for their continuous support without which this conference would not have been possible.

Finally, we thank all authors for their submissions and all invited speakers and conference attendees for making the ICCSA conference truly one of the premium events in the scientific community, facilitating the exchange of ideas, fostering new collaborations, and shaping the future of computational science.

May 2005

Marina L. Gavrilova
Osvaldo Gervasi

on behalf of the co-editors

Vipin Kumar
Antonio Laganà
Heow Pueh Lee
Youngsong Mun
David Taniar
Chih Jeng Kenneth Tan

Organization

ICCSA 2005 was organized by the Institute of High Performance Computing (Singapore), the University of Minnesota (Minneapolis, MN, USA), the University of Calgary (Canada) and the University of Perugia (Italy).

Conference Chairs

Vipin Kumar (Army High Performance Computing Center and University of Minnesota, USA), Honorary Chair
Marina L. Gavrilova (University of Calgary, Canada), Conference Co-chair, Scientific
Osvaldo Gervasi (University of Perugia, Italy), Conference Co-chair, Program
Jerry Lim (Institute of High Performance Computing, Singapore), Conference Co-chair, Organization

International Steering Committee

Alexander V. Bogdanov (Institute for High Performance Computing and Information Systems, Russia)
Marina L. Gavrilova (University of Calgary, Canada)
Osvaldo Gervasi (University of Perugia, Italy)
Kurichi Kumar (Institute of High Performance Computing, Singapore)
Vipin Kumar (Army High Performance Computing Center and University of Minnesota, USA)
Andres Iglesias (University de Cantabria, Spain)
Antonio Laganà (University of Perugia, Italy)
Heow Pueh Lee (Institute of High Performance Computing, Singapore)
Youngsong Mun (Soongsil University, Korea)
Chih Jeng Kenneth Tan (OptimaNumerics Ltd., and Queen's University Belfast, UK)
David Taniar (Monash University, Australia)

Local Organizing Committee

Kurichi Kumar (Institute of High Performance Computing, Singapore)
Heow Pueh Lee (Institute of High Performance Computing, Singapore)

Workshop Organizers

Approaches or Methods of Security Engineering
Haeng Kon Kim (Catholic University of Daegu, Korea)
Tai-hoon Kim (Korea Information Security Agency, Korea)

Authentication, Authorization and Accounting
Eui-Nam John Huh (Seoul Women's University, Korea)

Component-Based Software Engineering and Software Process Models
Haeng Kon Kim (Catholic University of Daegu, Korea)

Computational Geometry and Applications (CGA 2005)
Marina Gavrilova (University of Calgary, Canada)

Computer Graphics and Geometric Modeling (TSCG 2005)
Andres Iglesias (University of Cantabria, Spain)
Deok-Soo Kim (Hanyang University, Korea)

Computer Graphics and Rendering
Jiawan Zhang (Tianjin University, China)

Data Mining and Bioinformatics
Xiaohua Hu (Drexel University, USA)
David Taniar (Monash University, Australia)

Digital Device for Ubiquitous Computing
Hong Joo Lee (Daewoo Electronics Corp, Korea)

Grid Computing and Peer-to-Peer (P2P) Systems
Jemal H. Abawajy (Deakin University, Australia)
Maria S. Perez (Universitad Politecnica de Madrid, Spain)

Information and Communication Technology (ICT) Education
Woochun Jun (Seoul National University, Korea)

Information Security and Hiding, ISH 2005
Raphael C.W. Phan (Swinburne University of Technology, Sarawak, Malaysia)

Intelligent Multimedia Services and Synchronization in Mobile Multimedia Networks

Dong Chun Lee (Howon University, Korea)
Kuinam J. Kim (Kyonggi University, Korea)

Information Systems Information Technologies (ISIT)

Youngsong Mun (Soongsil University, Korea)

Internet Comunications Security (WICS)

Josè Sierra-Camara (University Carlos III of Madrid, Spain)
Julio Hernandez-Castro (University Carlos III of Madrid, Spain)
Antonio Izquierdo (University Carlos III of Madrid, Spain)
Joaquin Torres (University Carlos III of Madrid, Spain)

Methodology of Information Engineering

Sangkyun Kim (Somansa Ltd., Korea)

Mobile Communications

Hyunseung Choo (Sungkyunkwan University, Korea)

Modeling Complex Systems

Heather J. Ruskin (Dublin City University, Ireland)
Ruili Wang (Massey University, New Zealand)

Modeling of Location Management in Mobile Information Systems

Dong Chun Lee (Howon University, Korea)

Molecular Structures and Processes

Antonio Laganà (University of Perugia, Italy)

Optimization: Theories and Applications (OTA 2005)

In-Jae Jeong (Hanyang University, Korea)
Dong-Ho Lee (Hanyang University, Korea)
Deok-Soo Kim (Hanyang University, Korea)

Parallel and Distributed Computing

Jiawan Zhang (Tianjin University, China)

Pattern Recognition and Ubiquitous Computing

Woongjae Lee (Seoul Women's University, Korea)

Spatial Analysis and GIS: Local or Global?

Stefania Bertazzon (University of Calgary, Canada)
Borruso Giuseppe (University of Trieste, Italy)
Falk Huettmann (Institute of Arctic Biology, USA)

Specific Aspects of Computational Physics for Modeling Suddenly Emerging Phenomena

Paul E. Sterian (Politehnica University, Romania)
Cristian Toma (Titu Maiorescu University, Romania)

Symbolic Computation (SC 2005)

Andres Iglesias (University of Cantabria, Spain)
Akemi Galvez (University of Cantabria, Spain)

Ubiquitous Web Systems and Intelligence

David Taniar (Monash University, Australia)
Wenny Rahayu (La Trobe University, Australia)

Virtual Reality in Scientific Applications and Learning (VRSAL 2005)

Osvaldo Gervasi (University of Perugia, Italy)
Antonio Riganelli (University of Perugia, Italy)

Program Committee

Jemal Abawajy (Deakin University, Australia)
Kenny Adamson (EZ-DSP, UK)
Srinivas Aluru (Iowa State University, USA)
Frank Baetke (Hewlett-Packard, USA)
Mark Baker (Portsmouth University, UK)
Young-Cheol Bang (Korea Polytechnic University, Korea)
David Bell (Queen's University Belfast, UK)
Stefania Bertazzon (University of Calgary, Canada)
Sergei Bespamyatnikh (Duke University, USA)
J.A. Rod Blais (University of Calgary, Canada)
Alexander V. Bogdanov (Institute for High Performance Computing and
    Information Systems, Russia)
Richard P. Brent (University of Oxford, UK)
Peter Brezany (University of Vienna, Austria)
Herve Bronnimann (Polytechnic University, NY, USA)
John Brooke (University of Manchester, UK)
Martin Buecker (Aachen University, Germany)
Rajkumar Buyya (University of Melbourne, Australia)
YoungSik Choi (University of Missouri, USA)

Hyunseung Choo (Sungkyunkwan University, Korea)
Bastien Chopard (University of Geneva, Switzerland)
Min Young Chung (Sungkyunkwan University, Korea)
Toni Cortes (Universidad de Catalunya, Spain)
Yiannis Cotronis (University of Athens, Greece)
Danny Crookes (Queen's University Belfast, UK)
Josè C. Cunha (New University of Lisbon, Portugal)
Brian J. d'Auriol (University of Texas at El Paso, USA)
Alexander Degtyarev (Institute for High Performance Computing and Data Bases, Russia)
Frédéric Desprez (INRIA, France)
Tom Dhaene (University of Antwerp, Belgium)
Beniamino Di Martino (Second University of Naples, Italy)
Hassan Diab (American University of Beirut, Lebanon)
Ivan Dimov (Bulgarian Academy of Sciences, Bulgaria)
Iain Duff (Rutherford Appleton Laboratory, UK and CERFACS, France)
Thom Dunning (NCSA, USA)
Fabrizio Gagliardi (CERN, Switzerland)
Marina L. Gavrilova (University of Calgary, Canada)
Michael Gerndt (Technical University of Munich, Germany)
Osvaldo Gervasi (University of Perugia, Italy)
Bob Gingold (Australian National University, Australia)
James Glimm (SUNY Stony Brook, USA)
Christopher Gold (Hong Kong Polytechnic University, China)
Yuriy Gorbachev (Institute of High Performance Computing and Information Systems, Russia)
Andrzej Goscinski (Deakin University, Australia)
Jin Hai (Huazhong University of Science and Technology, China)
Ladislav Hlucky (Slovak Academy of Science, Slovakia)
Shen Hong (Japan Advanced Institute of Science and Technology, Japan)
Paul Hovland (Argonne National Laboratory, USA)
Xiaohua Hu (Drexel University, USA)
Eui-Nam John Huh (Seoul Women's University, Korea)
Terence Hung (Institute of High Performance Computing, Singapore)
Andres Iglesias (University of Cantabria, Spain)
In-Jae Jeong (Hanyang University, Korea)
Elisabeth Jessup (University of Colorado, USA)
Peter K. Jimack (University of Leeds, UK)
Christopher Johnson (University of Utah, USA)
Benjoe A. Juliano (California State University at Chico, USA)
Peter Kacsuk (MTA SZTAKI Research Institute, Hungary)
Kyung Woo Kang (KAIST, Korea)
Carl Kesselman (University of Southern California, USA)
Daniel Kidger (Quadrics, UK)
Deok-Soo Kim (Hanyang University, Korea)

Haeng Kon Kim (Catholic University of Daegu, Korea)
Jin Suk Kim (KAIST, Korea)
Tai-hoon Kim (Korea Information Security Agency, Korea)
Yoonhee Kim (Syracuse University, USA)
Mike Kirby (University of Utah, USA)
Jacek Kitowski (AGH University of Science and Technology, Poland)
Dieter Kranzlmueller (Johannes Kepler University Linz, Austria)
Kurichi Kumar (Institute of High Performance Computing, Singapore)
Vipin Kumar (Army High Performance Computing Center and University of Minnesota, USA)
Domenico Laforenza (Italian National Research Council, Italy)
Antonio Laganà (University of Perugia, Italy)
Joseph Landman (Scalable Informatics LLC, USA)
Francis Lau (University of Hong Kong, Hong Kong, China)
Bong Hwan Lee (Texas A&M University, USA)
Dong Chun Lee (Howon University, Korea)
Dong-Ho Lee (Hanyang University, Korea)
Heow Pueh Lee (Institute of High Performance Computing, Singapore)
Sang Yoon Lee (Georgia Institute of Technology, USA)
Tae Jin Lee (Sungkyunkwan University, Korea)
Bogdan Lesyng (ICM Warszawa, Poland)
Zhongze Li (Chinese Academy of Sciences, China)
Laurence Liew (Scalable Systems Pte., Singapore)
David Lombard (Intel Corporation, USA)
Emilio Luque (Universitat Autonoma of Barcelona, Spain)
Michael Mascagni (Florida State University, USA)
Graham Megson (University of Reading, UK)
John G. Michopoulos (US Naval Research Laboratory, USA)
Edward Moreno (Euripides Foundation of Marilia, Brazil)
Youngsong Mun (Soongsil University, Korea)
Jiri Nedoma (Academy of Sciences of the Czech Republic, Czech Republic)
Genri Norman (Russian Academy of Sciences, Russia)
Stephan Olariu (Old Dominion University, USA)
Salvatore Orlando (University of Venice, Italy)
Robert Panoff (Shodor Education Foundation, USA)
Marcin Paprzycki (Oklahoma State University, USA)
Gyung-Leen Park (University of Texas, USA)
Ron Perrott (Queen's University Belfast, UK)
Dimitri Plemenos (University of Limoges, France)
Richard Ramaroson (ONERA, France)
Rosemary Renaut (Arizona State University, USA)
Alexey S. Rodionov (Russian Academy of Science, Russia)
Paul Roe (Queensland University of Technology, Australia)
Reneé S. Renner (California State University at Chico, USA)
Heather J. Ruskin (Dublin City University, Ireland)

Ole Saastad (Scali, Norway)
Muhammad Sarfraz (King Fahd University of Petroleum and Minerals,
   Saudi Arabia)
Edward Seidel (Louisiana State University, USA and Albert Einstein Institute,
   Germany)
Josè Sierra-Camara (University Carlos III of Madrid, Spain)
Dale Shires (US Army Research Laboratory, USA)
Vaclav Skala (University of West Bohemia, Czech Republic)
Burton Smith (Cray, USA)
Masha Sosonkina (University of Minnesota, USA)
Alexei Sourin (Nanyang Technological University, Singapore)
Elena Stankova (Institute for High Performance Computing and Data Bases,
   Russia)
Gunther Stuer (University of Antwerp, Belgium)
Kokichi Sugihara (University of Tokyo, Japan)
Boleslaw Szymanski (Rensselaer Polytechnic Institute, USA)
Ryszard Tadeusiewicz (AGH University of Science and Technology, Poland)
Chih Jeng Kenneth Tan (OptimaNumerics, UK and Queen's University
   Belfast, UK)
David Taniar (Monash University, Australia)
John Taylor (Quadrics, UK)
Ruppa K. Thulasiram (University of Manitoba, Canada)
Pavel Tvrdik (Czech Technical University, Czech Republic)
Putchong Uthayopas (Kasetsart University, Thailand)
Mario Valle (Visualization Group, Swiss National Supercomputing Centre,
   Switzerland)
Marco Vanneschi (University of Pisa, Italy)
Piero Giorgio Verdini (University of Pisa and Istituto Nazionale di Fisica
   Nucleare, Italy)
Jesus Vigo-Aguiar (University of Salamanca, Spain)
Jens Volkert (University of Linz, Austria)
Koichi Wada (University of Tsukuba, Japan)
Kevin Wadleigh (Hewlett-Packard, USA)
Jerzy Wasniewski (Technical University of Denmark, Denmark)
Paul Watson (University of Newcastle upon Tyne)
Jan Weglarz (Poznan University of Technology, Poland)
Tim Wilkens (Advanced Micro Devices, USA)
Roman Wyrzykowski (Technical University of Czestochowa, Poland)
Jinchao Xu (Pennsylvania State University, USA)
Chee Yap (New York University, USA)
Osman Yasar (SUNY at Brockport, USA)
George Yee (National Research Council and Carleton University, Canada)
Yong Xue (Chinese Academy of Sciences, China)
Igor Zacharov (SGI Europe, Switzerland)
Xiaodong Zhang (College of William and Mary, USA)

Aledander Zhmakin (SoftImpact, Russia)
Krzysztof Zielinski (ICS UST/CYFRONET, Poland)
Albert Zomaya (University of Sydney, Australia)

Sponsoring Organizations

The Institute of High Performance Computing, Singapore
University of Perugia, Perugia, Italy
University of Calgary, Calgary, Canada
University of Minnesota, Minneapolis, USA
Queen's University Belfast, UK
Society for Industrial and Applied Mathematics, USA
The Institution of Electrical Engineers, UK
OptimaNumerics Ltd., UK
MASTER-UP, Italy

Table of Contents – Part II

Approaches or Methods of Security Engineering Workshop

Implementation of Short Message Service System to Be Based Mobile Wireless Internet
Hae-Sool Yang, Jung-Hun Hong, Seok-Hyung Hwang, Haeng-Kon Kim 1

Fuzzy Clustering for Documents Based on Optimization of Classifier Using the Genetic Algorithm
Ju-In Youn, He-Jue Eun, Yong-Sung Kim 10

P2P Protocol Analysis and Blocking Algorithm
Sun-Myung Hwang 21

Object Modeling of RDF Schema for Converting UML Class Diagram
Jin-Sung Kim, Chun-Sik Yoo, Mi-Kyung Lee, Yong-Sung Kim 31

A Framework for Security Assurance in Component Based Development
Gu-Beom Jeong, Guk-Boh Kim 42

Security Framework to Verify the Low Level Implementation Codes
Haeng-Kon Kim, Hae-Sool Yang 52

A Study on Evaluation of Component Metric Suites
Haeng-Kon Kim 62

The K-Means Clustering Architecture in the Multi-stage Data Mining Process
Bobby D. Gerardo, Jae-Wan Lee, Yeon-Sung Choi, Malrey Lee 71

A Privacy Protection Model in ID Management Using Access Control
Hyang-Chang Choi, Yong-Hoon Yi, Jae-Hyun Seo, Bong-Nam Noh, Hyung-Hyo Lee 82

A Time-Variant Risk Analysis and Damage Estimation for Large-Scale Network Systems
InJung Kim, YoonJung Chung, YoungGyo Lee, Dongho Won 92

Efficient Multi-bit Shifting Algorithm in Multiplicative Inversion
Problems
  Injoo Jang, Hyeong Seon Yoo 102

Modified Token-Update Scheme for Site Authentication
  Joungho Lee, Injoo Jang, Hyeong Seon Yoo 111

A Study on Secure SDP of RFID Using Bluetooth Communication
  Dae-Hee Seo, Im-Yeong Lee, Hee-Un Park 117

The Semantic Web Approach in Location Based Services
  *Jong-Woo Kim, Ju-Yeon Kim, Hyun-Suk Hwang, Sung-Seok Park,
  Chang-Soo Kim, Sung-gi Park* 127

SCTE: Software Component Testing Environments
  Haeng-Kon Kim, Oh-Hyun Kwon 137

Computer Security Management Model Using MAUT and SNMP
  Jongwoo Chae, Jungkyu Kwon, Mokdong Chung 147

Session and Connection Management for QoS-Guaranteed Multimedia
Service Provisioning on IP/MPLS Networks
  Young-Tak Kim, Hae-Sun Kim, Hyun-Ho Shin 157

A GQS-Based Adaptive Mobility Management Scheme Considering the
Gravity of Locality in Ad-Hoc Networks
  Ihn-Han Bae, Sun-Jin Oh 169

A Study on the E-Cash System with Anonymity and Divisibility
  Seo-Il Kang, Im-Yeong Lee 177

An Authenticated Key Exchange Mechanism Using One-Time Shared
Key
  Yonghwan Lee, Eunmi Choi, Dugki Min 187

Creation of Soccer Video Highlight Using the Caption Information
  Oh-Hyung Kang, Seong-Yoon Shin 195

The Information Search System Using Neural Network and Fuzzy
Clustering Based on Mobile Agent
  Jaeseon Ko, Bobby D. Gerardo, Jaewan Lee, Jae-Jeong Hwang 205

A Security Evaluation and Testing Methodology for Open Source
Software Embedded Information Security System
  Sung-ja Choi, Yeon-hee Kang, Gang-soo Lee 215

An Effective Method for Analyzing Intrusion Situation Through
IP-Based Classification
  *Minsoo Kim, Jae-Hyun Seo, Seung-Yong Lee, Bong-Nam Noh,
  Jung-Taek Seo, Eung-Ki Park, Choon-Sik Park* 225

A New Stream Cipher Using Two Nonlinear Functions
  Mi-Og Park, Dea-Woo Park 235

New Key Management Systems for Multilevel Security
  *Hwankoo Kim, Bongjoo Park, JaeCheol Ha, Byoungcheon Lee,
  DongGook Park* 245

Neural Network Techniques for Host Anomaly Intrusion Detection
Using Fixed Pattern Transformation
  ByungRae Cha, KyungWoo Park, JaeHyun Seo 254

The Role of Secret Sharing in the Distributed MARE Protocols
  Kyeongmo Park 264

Security Risk Vector for Quantitative Asset Assessment
  *Yoon Jung Chung, Injung Kim, NamHoon Lee, Taek Lee,
  Hoh Peter In* 274

A Remote Video Study Evaluation System Using a User Profile
  Seong-Yoon Shin, Oh-Hyung Kang 284

Performance Enhancement of Wireless LAN Based on Infrared
Communications Using Multiple-Subcarrier Modulation
  Hae Geun Kim 295

Modeling Virtual Network Collaboration in Supply Chain Management
  Ha Jin Hwang 304

SPA-Resistant Simultaneous Scalar Multiplication
  Mun-Kyu Lee 314

HSEP Design Using F2mHECC and ThreeB Symmetric Key Under
e-Commerce Environment
  Byung-kwan Lee, Am-Sok Oh, Eun-Hee Jeong 322

A Fault Distance Estimation Method Based on an Adaptive Data
Window for Power Network Security
  *Chang-Dae Yoon, Seung-Yeon Lee, Myong-Chul Shin,
  Ho-Sung Jung, Jae-Sang Cha* 332

Distribution Data Security System Based on Web Based Active Database
Sang-Yule Choi, Myong-Chul Shin, Nam-Young Hur, Jong-Boo Kim, Tai-hoon Kim, Jae-Sang Cha 341

Efficient DoS Resistant Multicast Authentication Schemes
JaeYong Jeong, Yongsu Park, Yookun Cho 353

Development System Security Process of ISO/IEC TR 15504 and Security Considerations for Software Process Improvement
Eun-ser Lee, Malrey Lee 363

Flexible ZCD-UWB with High QoS or High Capacity Using Variable ZCD Factor Code Sets
Jaesang Cha, Kyungsup Kwak, Changdae Yoon, Chonghyun Lee 373

Fine Grained Control of Security Capability and Forward Security in a Pairing Based Signature Scheme
Hak Soo Ju, Dae Youb Kim, Dong Hoon Lee, Jongin Lim, Kilsoo Chun 381

The Large Scale Electronic Voting Scheme Based on Undeniable Multi-signature Scheme
Sung-Hyun Yun, Hyung-Woo Lee 391

IPv6/IPsec Conformance Test Management System with Formal Description Technique
Hyung-Woo Lee, Sung-Hyun Yun, Jae-Sung Kim, Nam-Ho Oh, Do-Hyung Kim 401

Interference Cancellation Algorithm Development and Implementation for Digital Television
Chong Hyun Lee, Jae Sang Cha 411

Algorithm for ABR Traffic Control and Formation Feedback Information
Malrey Lee, Dong-Ju Im, Young Keun Lee, Jae-deuk Lee, Suwon Lee, Keun Kwang Lee, HeeJo Kang 420

Interference-Free ZCD-UWB for Wireless Home Network Applications
Jaesang Cha, Kyungsup Kwak, Sangyule Choi, Taihoon Kim, Changdae Yoon, Chonghyun Lee 429

Safe Authentication Method for Security Communication in Ubiquitous
Hoon Ko, Bangyong Sohn, Hayoung Park, Yongtae Shin 442

Pre/Post Rake Receiver Design for Maximum SINR in MIMO Communication System
   Chong Hyun Lee, Jae Sang Cha 449

SRS-Tool: A Security Functional Requirement Specification Development Tool for Application Information System of Organization
   Sang-soo Choi, Soo-young Chae, Gang-soo Lee 458

Design Procedure of IT Systems Security Countermeasures
   Tai-hoon Kim, Seung-youn Lee 468

Similarity Retrieval Based on Self-organizing Maps
   Dong-Ju Im, Malrey Lee, Young Keun Lee, Tae-Eun Kim,
   SuWon Lee, Jaewan Lee, Keun Kwang Lee, Kyung Dal Cho 474

An Expert System Development for Operating Procedure Monitoring of PWR Plants
   Malrey Lee, Eun-ser Lee, HeeJo Kang, HeeSook Kim 483

Security Evaluation Targets for Enhancement of IT Systems Assurance
   Tai-hoon Kim, Seung-youn Lee 491

Protection Profile for Software Development Site
   Seung-youn Lee, Myong-chul Shin 499

Information Security and Hiding (ISH 2005) Workshop

Improved RS Method for Detection of LSB Steganography
   Xiangyang Luo, Bin Liu, Fenlin Liu 508

Robust Undetectable Interference Watermarks
   Ryszard Grząślewicz, Jarosław Kutyłowski, Mirosław Kutyłowski,
   Wojciech Pietkiewicz 517

Equidistant Binary Fingerprinting Codes. Existence and Identification Algorithms
   Marcel Fernandez, Miguel Soriano, Josep Cotrina 527

Color Cube Analysis for Detection of LSB Steganography in RGB Color Images
   Kwangsoo Lee, Changho Jung, Sangjin Lee, Jongin Lim 537

Compact and Robust Image Hashing
   Sheng Tang, Jin-Tao Li, Yong-Dong Zhang 547

Watermarking for 3D Mesh Model Using Patch CEGIs
Suk-Hwan Lee, Ki-Ryong Kwon 557

Related-Key and Meet-in-the-Middle Attacks on Triple-DES and DES-EXE
Jaemin Choi, Jongsung Kim, Jaechul Sung, Sangjin Lee, Jongin Lim 567

Fault Attack on the DVB Common Scrambling Algorithm
Kai Wirt 577

HSEP Design Using F2mHECC and ThreeB Symmetric Key Under e-Commerce Envrionment
Byung-kwan Lee, Am-Sok Oh, Eun-Hee Jeong 585

Perturbed Hidden Matrix Cryptosystems
Zhiping Wu, Jintai Ding, Jason E. Gower, Dingfeng Ye 595

Identity-Based Identification Without Random Oracles
Kaoru Kurosawa, Swee-Huay Heng 603

Linkable Ring Signatures: Security Models and New Schemes
Joseph K. Liu, Duncan S. Wong 614

Practical Scenarios for the Van Trung-Martirosyan Codes
Marcel Fernandez, Miguel Soriano, Josep Cotrina 624

Obtaining True-Random Binary Numbers from a Weak Radioactive Source
Ammar Alkassar, Thomas Nicolay, Markus Rohe 634

Modified Sequential Normal Basis Multipliers for Type II Optimal Normal Bases
Dong Jin Yang, Chang Han Kim, Youngho Park, Yongtae Kim, Jongin Lim 647

A New Method of Building More Non-supersingular Elliptic Curves
Shi Cui, Pu Duan, Choong Wah Chan 657

Accelerating AES Using Instruction Set Extensions for Elliptic Curve Cryptography
Stefan Tillich, Johann Großschädl 665

Modeling of Location Management in Mobile Information Systems Workshop

Access Control Capable Integrated Network Management System for TCP/IP Networks
  Hyuncheol Kim, Seongjin Ahn, Younghwan Lim, Youngsong Mun 676

A Directional-Antenna Based MAC Protocol for Wireless Sensor Networks
  Shen Zhang, Amitava Datta 686

An Extended Framework for Proportional Differentiation: Performance Metrics and Evaluation Considerations
  Jahwan Koo, Seongjin Ahn 696

QoS Provisioning in an Enhanced FMIPv6 Architecture
  Zheng Wan, Xuezeng Pan, Lingdi Ping 704

Delay of the Slotted ALOHA Protocol with Binary Exponential Backoff Algorithm
  Sun Hur, Jeong Kee Kim, Dong Chun Lee 714

Design and Implementation of Frequency Offset Estimation, Symbol Timing and Sampling Clock Offset Control for an IEEE 802.11a Physical Layer
  Kwang-ho Chun, Seung-hyun Min, Myoung-ho Seong, Myoung-seob Lim 723

Automatic Subtraction Radiography Algorithm for Detection of Periodontal Disease in Internet Environment
  Yonghak Ahn, Oksam Chae 732

Improved Authentication Scheme in W-CDMA Networks
  Dong Chun Lee, Hyo Young Shin, Joung Chul Ahn, Jae Young Koh 741

Memory Reused Multiplication Implementation for Cryptography System
  Gi Yean Hwang, Jia Hou, Kwang Ho Chun, Moon Ho Lee 749

Scheme for the Information Sharing Between IDSs Using JXTA
  Jin Soh, Sung Man Jang, Geuk Lee 754

Workflow System Modeling in the Mobile Healthcare B2B Using
Semantic Information
  *Sang-Young Lee, Yung-Hyeon Lee, Jeom-Goo Kim,
  Dong Chun Lee* 762

Detecting Water Area During Flood Event from SAR Image
  Hong-Gyoo Sohn, Yeong-Sun Song, Gi-Hong Kim 771

Position Based Handover Control Method
  Jong chan Lee, Sok-Pal Cho, Hong-jin Kim 781

Improving Yellow Time Method of Left-Turning Traffic Flow at
Signalized Intersection Networks by ITS
  Hyung Jin Kim, Bongsoo Son, Soobeom Lee, Joowon Park 789

Intelligent Multimedia Services and Synchronization in Mobile Multimedia Networks Workshop

A Multimedia Database System Using Dependence Weight Values for a
Mobile Environment
  Kwang Hyoung Lee, Hee Sook Kim, Keun Wang Lee 798

A General Framework for Analyzing the Optimal Call Admission
Control in DS-CDMA Cellular Network
  Wen Chen, Feiyu Lei, Weinong Wang 806

Heuristic Algorithm for Traffic Condition Classification with Loop
Detector Data
  Sangsoo Lee, Sei-Chang Oh, Bongsoo Son 816

Spatial Data Channel in a Mobile Navigation System
  Yingwei Luo, Guomin Xiong, Xiaolin Wang, Zhuoqun Xu 822

A Video Retrieval System for Electrical Safety Education Based on a
Mobile Agent
  Hyeon Seob Cho, Keun Wang Lee 832

Fuzzy Multi-criteria Decision Making-Based Mobile Tracking
  Gi-Sung Lee 839

Evaluation of Network Blocking Algorithm based on ARP Spoofing
and Its Application
  Jahwan Koo, Seongjin Ahn, Younghwan Lim, Youngsong Mun 848

Design and Implementation of Mobile-Learning System for Environment Education
Keun Wang Lee, Jong Hee Lee 856

A Simulation Model of Congested Traffic in the Waiting Line
Bongsoo Son, Taewan Kim, Yongjae Lee 863

Core Technology Analysis and Development for the Virus and Hacking Prevention
Seung-Jae Yoo 870

Development of Traffic Accidents Prediction Model with Intelligent System Theory
SooBeom Lee, TaiSik Lee, Hyung Jin Kim, YoungKyun Lee 880

Prefetching Scheme Considering Mobile User's Preference in Mobile Networks
Jin Ah Yoo, In Seon Choi, Dong Chun Lee 889

System Development of Security Vulnerability Diagnosis in Wireless Internet Networks
Byoung-Muk Min, Sok-Pal Cho, Hong-jin Kim, Dong Chun Lee 896

An Active Node Management System for Secure Active Networks
Jin-Mook Kim, In-sung Han, Hwang-bin Ryou 904

Ubiquitous Web Systems and Intelligence Workshop

A Systematic Design Approach for XML-View Driven Web Document Warehouses
Vicky Nassis, Rajugan R., Tharam S. Dillon, Wenny Rahayu 914

Clustering and Retrieval of XML Documents by Structure
Jeong Hee Hwang, Keun Ho Ryu 925

A New Method for Mining Association Rules from a Collection of XML Documents
Juryon Paik, Hee Yong Youn, Ungmo Kim 936

Content-Based Recommendation in E-Commerce
Bing Xu, Mingmin Zhang, Zhigeng Pan, Hongwei Yang 946

A Personalized Multilingual Web Content Miner: PMWebMiner
Rowena Chau, Chung-Hsing Yeh, Kate A. Smith 956

Context-Based Recommendation Service in Ubiquitous Commerce
  Jeong Hee Hwang, Mi Sug Gu, Keun Ho Ryu 966

A New Continuous Nearest Neighbor Technique for Query Processing
on Mobile Environments
  Jeong Hee Chi, Sang Ho Kim, Keun Ho Ryu 977

Semantic Web Enabled Information Systems: Personalized Views on
Web Data
  *Robert Baumgartner, Christian Enzi, Nicola Henze, Marc Herrlich,
  Marcus Herzog, Matthias Kriesell, Kai Tomaschewski* 988

Design of Vehicle Information Management System for Effective
Retrieving of Vehicle Location
  Eung Jae Lee, Keun Ho Ryu 998

Context-Aware Workflow Language Based on Web Services for
Ubiquitous Computing
  Joohyun Han, Yongyun Cho, Jaeyoung Choi 1008

A Ubiquitous Approach for Visualizing Back Pain Data
  T. Serif, G. Ghinea, A.O. Frank 1018

Prototype Design of Mobile Emergency Telemedicine System
  *Sun K. Yoo, S.M. Jung, B.S. Kim, H.Y. Yun, S.R. Kim,
  D.K. Kim* 1028

An Intermediate Target for Quick-Relay of Remote Storage to Mobile
Devices
  Daegeun Kim, MinHwan Ok, Myong-soon Park 1035

Reflective Middleware for Location-Aware Application Adaptation
  *Uzair Ahmad, S.Y. Lee, Mahrin Iqbal, Uzma Nasir, A. Ali,
  Mudeem Iqbal* 1045

Efficient Approach for Interactively Mining Web Traversal Patterns
  Yue-Shi Lee, Min-Chi Hsieh, Show-Jane Yen 1055

Query Decomposition Using the XML Declarative Description Language
  Le Thi Thu Thuy, Doan Dai Duong 1066

On URL Normalization
  Sang Ho Lee, Sung Jin Kim, Seok Hoo Hong 1076

Clustering-Based Schema Matching of Web Data for Constructing
Digital Library
  Hui Song, Fanyuan Ma, Chen Wang 1086

Bringing Handhelds to the Grid Resourcefully: A Surrogate Middleware
Approach
  *Maria Riaz, Saad Liaquat Kiani, Anjum Shehzad,
  Sungyoung Lee* 1096

Mobile Mini-payment Scheme Using SMS-Credit
  Simon Fong, Edison Lai 1106

Context Summarization and Garbage Collecting Context
  Faraz Rasheed, Yong-Koo Lee, Sungyoung Lee 1115

EXtensible Web (xWeb): An XML-View Based Web Engineering
Methodology
  *Rajugan R., William Gardner, Elizabeth Chang,
  Tharam S. Dillon* 1125

A Web Services Framework for Integrated Geospatial Coverage Data
  Eunkyu Lee, Minsoo Kim, Mijeong Kim, Inhak Joo 1136

Open Location-Based Service Using Secure Middleware Infrastructure
in Web Services
  Namje Park, Howon Kim, Seungjoo Kim, Dongho Won 1146

Ubiquitous Systems and Petri Nets
  *David de Frutos Escrig, Olga Marroquín Alonso,
  Fernando Rosa Velardo* 1156

Virtual Lab Dashboard: Ubiquitous Monitoring and Control in a Smart
Bio-laboratory
  *XiaoMing Bao, See-Kiong Ng, Eng-Huat Chua,
  Wei-Khing For* 1167

On Discovering Concept Entities from Web Sites
  Ming Yin, Dion Hoe-Lian Goh, Ee-Peng Lim 1177

Modelling Complex Systems Workshop

Towards a Realistic Microscopic Traffic Simulation at an Unsignalised
Interscetion
  Mingzhe Liu, Ruili Wang, Ray Kemp 1187

Complex Systems: Particles, Chains, and Sheets
  R.B Pandey 1197

Discretization of Delayed Multi-input Nonlinear System via Taylor Series and Scaling and Squaring Technique
  Yuanliang Zhang, Hyung Jo Choi, Kil To Chong 1207

On the Scale-Free Intersection Graphs
  Xin Yao, Changshui Zhang, Jinwen Chen, Yanda Li 1217

A Stochastic Viewpoint on the Generation of Spatiotemporal Datasets
  MoonBae Song, KwangJin Park, Ki-Sik Kong, SangKeun Lee 1225

A Formal Approach to the Design of Distributed Data Warehouses
  Jane Zhao 1235

A Mathematical Model for Genetic Regulation of the Lactose Operon
  Tianhai Tian, Kevin Burrage 1245

Network Emergence in Immune System Shape Space
  Heather J. Ruskin, John Burns 1254

A Multi-agent System for Modelling Carbohydrate Oxidation in Cell
  Flavio Corradini, Emanuela Merelli, Marco Vita 1264

Characterizing Complex Behavior in (Self-organizing) Multi-agent Systems
  Bingcheng Hu, Jiming Liu 1274

Protein Structure Abstraction and Automatic Clustering Using Secondary Structure Element Sequences
  Sung Hee Park, Chan Yong Park, Dae Hee Kim, Seon Hee Park, Jeong Seop Sim 1284

A Neural Network Method for Induction Machine Fault Detection with Vibration Signal
  Hua Su, Kil To Chong, A.G. Parlos 1293

Author Index 1303

Table of Contents – Part III

Grid Computing and Peer-to-Peer (P2P) Systems Workshop

Resource and Service Discovery in the iGrid Information Service
Giovanni Aloisio, Massimo Cafaro, Italo Epicoco, Sandro Fiore, Daniele Lezzi, Maria Mirto, Silvia Mocavero 1

A Comparison of Spread Methods in Unstructured P2P Networks
Zhaoqing Jia, Bingzhen Pei, Minglu Li, Jinyuan You 10

A New Service Discovery Scheme Adapting to User Behavior for Ubiquitous Computing
Yeo Bong Yoon, Hee Yong Youn 19

The Design and Prototype of RUDA, a Distributed Grid Accounting System
M.L. Chen, A. Geist, D.E. Bernholdt, K. Chanchio, D.L. Million ... 29

An Adaptive Routing Mechanism for Efficient Resource Discovery in Unstructured P2P Networks
Luca Gatani, Giuseppe Lo Re, Salvatore Gaglio 39

Enhancing UDDI for Grid Service Discovery by Using Dynamic Parameters
Brett Sinclair, Andrzej Goscinski, Robert Dew 49

A New Approach for Efficiently Achieving High Availability in Mobile Computing
M. Mat Deris, J.H. Abawajy, M. Omar 60

A Flexible Communication Scheme to Support Grid Service Emergence
Lei Gao, Yongsheng Ding 69

A Kernel-Level RTP for Efficient Support of Multimedia Service on Embedded Systems
Dong Guk Sun, Sung Jo Kim 79

Group-Based Scheduling Scheme for Result Checking in Global Computing Systems
HongSoo Kim, SungJin Choi, MaengSoon Baik, KwonWoo Yang, HeonChang Yu, Chong-Sun Hwang 89

Service Discovery Supporting Open Scalability Using FIPA-Compliant
Agent Platform for Ubiquitous Networks
Kee-Hyun Choi, Ho-Jin Shin, Dong-Ryeol Shin 99

A Mathematical Predictive Model for an Autonomic System to Grid
Environments
Alberto Sánchez, María S. Pérez 109

Spatial Analysis and GIS: Local or Global? Workshop

Spatial Analysis: Science or Art?
Stefania Bertazzon 118

Network Density Estimation: Analysis of Point Patterns over a Network
Giuseppe Borruso 126

Linking Global Climate Grid Surfaces with Local Long-Term Migration
Monitoring Data: Spatial Computations for the Pied Flycatcher to
Assess Climate-Related Population Dynamics on a Continental Scale
Nikita Chernetsov, Falk Huettmann 133

Classifying Internet Traffic Using Linear Regression
Troy D. Mackay, Robert G.V. Baker 143

Modeling Sage Grouse: Progressive Computational Methods for Linking
a Complex Set of Local, Digital Biodiversity and Habitat Data Towards
Global Conservation Statements and Decision-Making Systems
Anthonia Onyeahialam, Falk Huettmann, Stefania Bertazzon 152

Local Analysis of Spatial Relationships: A Comparison of GWR and
the Expansion Method
Antonio Páez 162

Middleware Development for Remote Sensing Data Sharing and Image
Processing on HIT-SIP System
*Jianqin Wang, Yong Xue, Chaolin Wu, Yanguang Wang,
Yincui Hu, Ying Luo, Yanning Guan, Shaobo Zhong, Jiakui Tang,
Guoyin Cai* 173

A New and Efficient K-Medoid Algorithm for Spatial Clustering
Qiaoping Zhang, Isabelle Couloigner 181

Computer Graphics and Rendering Workshop

Security Management for Internet-Based Virtual Presentation of Home Textile Product
   Lie Shi, Mingmin Zhang, Li Li, Lu Ye, Zhigeng Pan 190

An Efficient Approach for Surface Creation
   L.H. You, Jian J. Zhang 197

Interactive Visualization for OLAP
   Kesaraporn Techapichetvanich, Amitava Datta 206

Interactive 3D Editing on Tiled Display Wall
   Xiuhui Wang, Wei Hua, Hujun Bao 215

A Toolkit for Automatically Modeling and Simulating 3D Multi-articulation Entity in Distributed Virtual Environment
   Xiaohui Liang, Chuanpeng Wang, Yinghui Che, Jiangying Yu, Na Qu 225

Footprint Analysis and Motion Synthesis
   Qinping Zhao, Xiaoyan Hu 235

An Adaptive and Efficient Algorithm for Polygonization of Implicit Surfaces
   Mingyong Pang, Zhigeng Pan, Mingmin Zhang, Fuyan Zhang 245

A Framework of Web GIS Based Unified Public Health Information Visualization Platform
   Xiaolin Lu 256

An Improved Colored-Marker Based Registration Method for AR Applications
   Xiaowei Li, Yue Liu, Yongtian Wang, Dayuan Yan, Dongdong Weng, Tao Yang 266

Non-photorealistic Tour into Panorama
   Yang Zhao, Ya-Ping Zhang, Dan Xu 274

Image Space Silhouette Extraction Using Graphics Hardware
   Jiening Wang, Jizhou Sun, Ming Che, Qi Zhai, Weifang Nie 284

Adaptive Fuzzy Weighted Average Filter for Synthesized Image
   Qing Xu, Liang Ma, Weifang Nie, Peng Li, Jiawan Zhang, Jizhou Sun 292

Data Mining and Bioinformatics Workshop

The Binary Multi-SVM Voting System for Protein Subcellular Localization Prediction
Bo Jin, Yuchun Tang, Yan-Qing Zhang, Chung-Dar Lu, Irene Weber 299

Gene Network Prediction from Microarray Data by Association Rule and Dynamic Bayesian Network
Hei-Chia Wang, Yi-Shiun Lee 309

Protein Interaction Prediction Using Inferred Domain Interactions and Biologically-Significant Negative Dataset
Xiao-Li Li, Soon-Heng Tan, See-Kiong Ng 318

Semantic Annotation of Biomedical Literature Using Google
Rune Sætre, Amund Tveit, Tonje Stroemmen Steigedal, Astrid Lægreid 327

Fast Parallel Algorithms for the Longest Common Subsequence Problem Using an Optical Bus
Xiaohua Xu, Ling Chen, Yi Pan, Ping He 338

Estimating Gene Networks from Expression Data and Binding Location Data via Boolean Networks
Osamu Hirose, Naoki Nariai, Yoshinori Tamada, Hideo Bannai, Seiya Imoto, Satoru Miyano 349

Efficient Matching and Retrieval of Gene Expression Time Series Data Based on Spectral Information
Hong Yan 357

SVM Classification to Predict Two Stranded Anti-parallel Coiled Coils Based on Protein Sequence Data
Zhong Huang, Yun Li, Xiaohua Hu 374

Estimating Gene Networks with cDNA Microarray Data Using State-Space Models
Rui Yamaguchi, Satoru Yamashita, Tomoyuki Higuchi 381

A Penalized Likelihood Estimation on Transcriptional Module-Based Clustering
Ryo Yoshida, Seiya Imoto, Tomoyuki Higuchi 389

Conceptual Modeling of Genetic Studies and Pharmacogenetics
Xiaohua Zhou, Il-Yeol Song 402

Parallel and Distributed Computing Workshop

A Dynamic Parallel Volume Rendering Computation Mode Based on Cluster
Weifang Nie, Jizhou Sun, Jing Jin, Xiaotu Li, Jie Yang, Jiawan Zhang 416

Dynamic Replication of Web Servers Using Rent-a-Servers
Young-Chul Shim, Jun-Won Lee, Hyun-Ah Kim 426

Survey of Parallel and Distributed Volume Rendering: Revisited
Jiawan Zhang, Jizhou Sun, Zhou Jin, Yi Zhang, Qi Zhai 435

Scheduling Pipelined Multiprocessor Tasks: An Experimental Study with Vision Architecture
M. Fikret Ercan 445

Universal Properties Verification of Parameterized Parallel Systems
Cecilia E. Nugraheni 453

Symbolic Computation, SC 2005 Workshop

2d Polynomial Interpolation: A Symbolic Approach with Mathematica
Ali Yazici, Irfan Altas, Tanil Ergenc 463

Analyzing the Synchronization of Chaotic Dynamical Systems with Mathematica: Part I
Andres Iglesias, Akemi Gálvez 472

Analyzing the Synchronization of Chaotic Dynamical Systems with Mathematica: Part II
Andres Iglesias, Akemi Gálvez 482

A Mathematica Package for Computing and Visualizing the Gauss Map of Surfaces
Ruben Ipanaqué, Andres Iglesias 492

Numerical-Symbolic *Matlab* Toolbox for Computer Graphics and Differential Geometry
Akemi Gálvez, Andrés Iglesias 502

A LiE Subroutine for Computing Prehomogeneous Spaces Associated with Real Nilpotent Orbits
Steven Glenn Jackson, Alfred G. Noël 512

Applications of Graph Coloring
  Ünal Ufuktepe, Goksen Bacak 522

Mathematica Applications on Time Scales
  Ahmet Yantır, Ünal Ufuktepe 529

A Discrete Mathematics Package for Computer Science and Engineering Students
  Mustafa Murat Inceoglu 538

Circle Inversion of Two-Dimensional Objects with Mathematica
  Ruben T. Urbina, Andres Iglesias 547

Specific Aspects of Computational Physics for Modeling Suddenly-Emerging Phenomena Workshop

Specific Aspects of Training IT Students for Modeling Pulses in Physics
  Adrian Podoleanu, Cristian Toma, Cristian Morarescu,
  Alexandru Toma, Theodora Toma 556

Filtering Aspects of Practical Test-Functions and the Ergodic Hypothesis
  Flavia Doboga, Ghiocel Toma, Stefan Pusca, Mihaela Ghelmez,
  Cristian Morarescu 563

Definition of Wave-Corpuscle Interaction Suitable for Simulating Sequences of Physical Pulses
  Minas Simeonidis, Stefan Pusca, Ghiocel Toma, Alexandru Toma,
  Theodora Toma 569

Practical Test-Functions Generated by Computer Algorithms
  Ghiocel Toma 576

Possibilities for Obtaining the Derivative of a Received Signal Using Computer-Driven Second Order Oscillators
  Andreea Sterian, Ghiocel Toma 585

Simulating Laser Pulses by Practical Test Functions and Progressive Waves
  Rodica Sterian, Cristian Toma 592

Statistical Aspects of Acausal Pulses in Physics and Wavelets Applications
  Cristian Toma, Rodica Sterian 598

Wavelet Analysis of Solitary Wave Equation
Carlo Cattani 604

Numerical Analysis of Some Typical Finite Differences Simulations of the Waves Propagation Through Different Media
Dan Iordache, Stefan Pusca, Ghiocel Toma 614

B–Splines and Nonorthogonal Wavelets
Nikolay Strelkov 621

Optimal Wavelets
Nikolay Strelkov, Vladimir Dol'nikov 628

Dynamics of a Two-Level Medium Under the Action of Short Optical Pulses
Valerică Ninulescu, Andreea-Rodica Sterian 635

Nonlinear Phenomena in Erbium-Doped Lasers
Andreea Sterian, Valerică Ninulescu 643

Internet Communications Security (WICS) Workshop

An e-Lottery Scheme Using Verifiable Random Function
Sherman S.M. Chow, Lucas C.K. Hui, S.M. Yiu, K.P. Chow 651

Related-Mode Attacks on Block Cipher Modes of Operation
Raphael C.-W. Phan, Mohammad Umar Siddiqi 661

A Digital Cash Protocol Based on Additive Zero Knowledge
Amitabh Saxena, Ben Soh, Dimitri Zantidis 672

On the Security of Wireless Sensor Networks
Rodrigo Roman, Jianying Zhou, Javier Lopez 681

Dependable Transaction for Electronic Commerce
Hao Wang, Heqing Guo, Manshan Lin, Jianfei Yin, Qi He, Jun Zhang 691

On the Security of a Certified E-Mail Scheme with Temporal Authentication
Min-Hua Shao, Jianying Zhou, Guilin Wang 701

Security Flaws in Several Group Signatures Proposed by Popescu
Guilin Wang, Sihan Qing 711

A Simple Acceptance/Rejection Criterium for Sequence Generators in Symmetric Cryptography
Amparo Fúster-Sabater, Pino Caballero-Gil 719

Secure Electronic Payments in Heterogeneous Networking: New Authentication Protocols Approach
Joaquin Torres, Antonio Izquierdo, Arturo Ribagorda, Almudena Alcaide 729

Component Based Software Engineering and Software Process Model Workshop

Software Reliability Measurement Use Software Reliability Growth Model in Testing
Hye-Jung Jung, Hae-Sool Yang 739

Thesaurus Construction Using Class Inheritance
Gui-Jung Kim, Jung-Soo Han 748

An Object Structure Extraction Technique for Object Reusability Improvement Based on Legacy System Interface
Chang-Mog Lee, Cheol-Jung Yoo, Ok-Bae Chang 758

Automatic Translation Form Requirements Model into Use Cases Modeling on UML
Haeng-Kon Kim, Youn-Ky Chung 769

A Component Identification Technique from Object-Oriented Model
Mi-Sook Choi, Eun-Sook Cho 778

Retrieving and Exploring Ontology-Based Human Motion Sequences
Hyun-Sook Chung, Jung-Min Kim, Yung-Cheol Byun, Sang-Yong Byun 788

An Integrated Data Mining Model for Customer Credit Evaluation
Kap Sik Kim, Ha Jin Hwang 798

A Study on the Component Based Architecture for Workflow Rule Engine and Tool
Ho-Jun Shin, Kwang-Ki Kim, Bo-Yeon Shim 806

A Fragment-Driven Process Modeling Methodology
Kwang-Hoon Kim, Jae-Kang Won, Chang-Min Kim 817

A FCA-Based Ontology Construction for the Design of Class Hierarchy
  Suk-Hyung Hwang, Hong-Gee Kim, Hae-Sool Yang 827

Component Contract-Based Formal Specification Technique
  Ji-Hyun Lee, Hye-Min Noh, Cheol-Jung Yoo, Ok-Bae Chang 836

A Business Component Approach for Supporting the Variability of the Business Strategies and Rules
  Jeong Ah Kim, YoungTaek Jin, SunMyung Hwang 846

A CBD Application Integration Framework for High Productivity and Maintainability
  Yonghwan Lee, Eunmi Choi, Dugki Min 858

Integrated Meta-model Approach for Reengineering from Legacy into CBD
  Eun Sook Cho 868

Behavior Modeling Technique Based on EFSM for Interoperability Testing
  Hye-Min Noh, Ji-Hyen Lee, Cheol-Jung Yoo, Ok-Bae Chang 878

Automatic Connector Creation for Component Assembly
  Jung-Soo Han, Gui-Jung Kim, Young-Jae Song 886

MaRMI-RE: Systematic Componentization Process for Reengineering Legacy System
  Jung-Eun Cha, Chul-Hong Kim 896

A Study on the Mechanism for Mobile Embedded Agent Development Based on Product Line
  Haeng-Kon Kim 906

Frameworks for Model-Driven Software Architecture
  Soung Won Kim, Myoung Soo Kim, Haeng Kon Kim 916

Parallel and Distributed Components with Java
  Chang-Moon Hyun 927

CEB: Class Quality Evaluator for BlueJ
  Yu-Kyung Kang, Suk-Hyung Hwang, Hae-Sool Yang, Jung-Bae Lee, Hee-Chul Choi, Hyun-Wook Wee, Dong-Soon Kim 938

Workflow Modeling Based on Extended Activity Diagram Using ASM Semantics
  Eun-Jung Ko, Sang-Young Lee, Hye-Min Noh, Cheol-Jung Yoo, Ok-Bae Chang 945

Unification of XML DTD for XML Documents with Similar Structure
  Chun-Sik Yoo, Seon-Mi Woo, Yong-Sung Kim 954

Secure Payment Protocol for Healthcare Using USIM in Ubiquitous
  Jang-Mi Baek, In-Sik Hong 964

Verification of UML-Based Security Policy Model
  Sachoun Park, Gihwon Kwon 973

Computer Graphics and Geometric Modeling (TSCG 2005) Workshop

From a Small Formula to Cyberworlds
  Alexei Sourin 983

Visualization and Analysis of Protein Structures Using Euclidean Voronoi Diagram of Atoms
  Deok-Soo Kim, Donguk Kim, Youngsong Cho, Joonghyun Ryu, Cheol-Hyung Cho, Joon Young Park, Hyun Chan Lee 993

C^2 Continuous Spline Surfaces over Catmull-Clark Meshes
  Jin Jin Zheng, Jian J. Zhang, Hong Jun Zhou, Lianguan G. Shen .. 1003

Constructing Detailed Solid and Smooth Surfaces from Voxel Data for Neurosurgical Simulation
  Mayumi Shimizu, Yasuaki Nakamura 1013

Curvature Estimation of Point-Sampled Surfaces and Its Applications
  Yongwei Miao, Jieqing Feng, Qunsheng Peng 1023

The Delaunay Triangulation by Grid Subdivision
  Si Hyung Park, Seoung Soo Lee, Jong Hwa Kim 1033

Feature-Based Texture Synthesis
  Tong-Yee Lee, Chung-Ren Yan 1043

A Fast 2D Shape Interpolation Technique
  Ping-Hsien Lin, Tong-Yee Lee 1050

Triangular Prism Generation Algorithm for Polyhedron Decomposition
Jaeho Lee, JoonYoung Park, Deok-Soo Kim, HyunChan Lee 1060

Tweek: A Framework for Cross-Display Graphical User Interfaces
Patrick Hartling, Carolina Cruz-Neira 1070

Surface Simplification with Semantic Features Using Texture and Curvature Maps
Soo-Kyun Kim, Jung Lee, Cheol-Su Lim, Chang-Hun Kim 1080

Development of a Machining Simulation System Using the Octree Algorithm
Y.H. Kim, S.L. Ko 1089

A Spherical Point Location Algorithm Based on Barycentric Coordinates
Yong Wu, Yuanjun He, Haishan Tian 1099

Realistic Skeleton Driven Skin Deformation
X.S. Yang, Jian J. Zhang 1109

Implementing Immersive Clustering with VR Juggler
Aron Bierbaum, Patrick Hartling, Pedro Morillo, Carolina Cruz-Neira 1119

Adaptive Space Carving with Texture Mapping
Yoo-Kil Yang, Jung Lee, Soo-Kyun Kim, Chang-Hun Kim 1129

User-Guided 3D Su-Muk Painting
Jung Lee, Joon-Yong Ji, Soo-Kyun Kim, Chang-Hun Kim 1139

Sports Equipment Based Motion Deformation
Jong-In Choi, Chang-Hun Kim, Cheol-Su Lim 1148

Designing an Action Selection Engine for Behavioral Animation of Intelligent Virtual Agents
Francisco Luengo, Andres Iglesias 1157

Interactive Transmission of Highly Detailed Surfaces
Junfeng Ji, Sheng Li, Enhua Wu, Xuehui Liu 1167

Contour-Based Terrain Model Reconstruction Using Distance Information
Byeong-Seok Shin, Hoe Sang Jung 1177

An Efficient Point Rendering Using Octree and Texture Lookup
  Yun-Mo Koo, Byeong-Seok Shin 1187

Faces Alive: Reconstruction of Animated 3D Human Faces
  Yu Zhang, Terence Sim, Chew Lim Tan 1197

Quasi-interpolants Based Multilevel B-Spline Surface Reconstruction from Scattered Data
  Byung-Gook Lee, Joon-Jae Lee, Ki-Ryoung Kwon 1209

Methodology of Information Engineering Workshop

Efficient Mapping Rule of IDEF for UMM Application
  Kitae Shin, Chankwon Park, Hyoung-Gon Lee, Jinwoo Park 1219

A Case Study on the Development of Employee Internet Management System
  Sangkyun Kim, Ilhoon Choi 1229

Cost-Benefit Analysis of Security Investments: Methodology and Case Study
  Sangkyun Kim, Hong Joo Lee 1239

A Modeling Framework of Business Transactions for Enterprise Integration
  Minsoo Kim, Dongsoo Kim, Yong Gu Ji, Hoontae Kim 1249

Process-Oriented Development of Job Manual System
  Seung-Hyun Rhee, Hoseong Song, Hyung Jun Won, Jaeyoung Ju, Minsoo Kim, Hyerim Bae 1259

An Information System Approach and Methodology for Enterprise Credit Rating
  Hakjoo Lee, Choon Seong Leem, Kyungup Cha 1269

Privacy Engineering in ubiComp
  Tae Joong Kim, Sang Won Lee, Eung Young Lee 1279

Development of a BSC-Based Evaluation Framework for e-Manufacturing Project
  Yongju Cho, Wooju Kim, Choon Seong Leem, Honzong Choi 1289

Design of a BPR-Based Information Strategy Planning (ISP) Framework
  Chiwoon Cho, Nam Wook Cho 1297

An Integrated Evaluation System for Personal Informatization Levels
and Their Maturity Measurement: Korean Motors Company Case
*Eun Jung Yu, Choon Seong Leem, Seoung Kyu Park,
Byung Wan Kim* 1306

Critical Attributes of Organizational Culture Promoting Successful KM
Implementation
Heejun Park 1316

Author Index 1327

Table of Contents – Part IV

Information and Communication Technology (ICT) Education Workshop

Exploring Constructivist Learning Theory and Course Visualization on Computer Graphics
Yiming Zhao, Mingming Zhang, Shu Wang, Yefang Chen 1

A Program Plagiarism Evaluation System
Young-Chul Kim, Jaeyoung Choi 10

Integrated Development Environment for Digital Image Computing and Configuration Management
Jeongheon Lee, YoungTak Cho, Hoon Heo, Oksam Chae 20

E-Learning Environment Based on Intelligent Synthetic Characters
Lu Ye, Jiejie Zhu, Mingming Zhang, Ruth Aylett, Lifeng Ren, Guilin Xu 30

SCO Control Net for the Process-Driven SCORM Content Aggregation Model
Kwang-Hoon Kim, Hyun-Ah Kim, Chang-Min Kim 38

Design and Implementation of a Web-Based Information Communication Ethics Education System for the Gifted Students in Computer
Woochun Jun, Sung-Keun Cho, Byeong Heui Kwak 48

International Standards Based Information Technology Courses: A Case Study from Turkey
Mustafa Murat Inceoglu 56

Design and Implementation of the KORI: Intelligent Teachable Agent and Its Application to Education
Sung-il Kim, Sung-Hyun Yun, Mi-sun Yoon, Yeon-hee So, Won-sik Kim, Myung-jin Lee, Dong-seong Choi, Hyung-Woo Lee ... 62

Digital Device for Ubiquitous Computing Workshop

A Space-Efficient Flash Memory Software for Mobile Devices
Yeonseung Ryu, Tae-sun Chung, Myungho Lee 72

Security Threats and Their Countermeasures of Mobile Portable
Computing Devices in Ubiquitous Computing Environments
  Sang ho Kim, Choon Seong Leem 79

A Business Model (BM) Development Methodology in Ubiquitous
Computing Environment
  Choon Seong Leem, Nam Joo Jeon, Jong Hwa Choi,
  Hyoun Gyu Shin 86

Developing Business Models in Ubiquitous Era: Exploring
Contradictions in Demand and Supply Perspectives
  Jungwoo Lee, Sunghwan Lee 96

Semantic Web Based Intelligent Product and Service Search Framework
for Location-Based Services
  Wooju Kim, SungKyu Lee, DeaWoo Choi 103

A Study on Value Chain in a Ubiquitous Computing Environment
  Hong Joo Lee, Choon Seong Leem 113

A Study on Authentication Mechanism Using Robot Vacuum Cleaner
  Hong Joo Lee, Hee Jun Park, Sangkyun Kim 122

Design of Inside Information Leakage Prevention System in Ubiquitous
Computing Environment
  Hangbae Chang, Kyung-kyu Kim 128

Design and Implementation of Home Media Server Using TV-Anytime
for Personalized Broadcasting Service
  Changho Hong, Jongtae Lim 138

Optimization: Theories and Applications (OTA) 2005 Workshop

Optimal Signal Control Using Adaptive Dynamic Programming
  Chang Ouk Kim, Yunsun Park, Jun-Geol Baek 148

Inverse Constrained Bottleneck Problems on Networks
  Xiucui Guan, Jianzhong Zhang 161

Dynamic Scheduling Problem of Batch Processing Machine in
Semiconductor Burn-in Operations
  Pei-Chann Chang, Yun-Shiow Chen, Hui-Mei Wang 172

Polynomial Algorithm for Parallel Machine Mean Flow Time Scheduling
Problem with Release Dates
    Peter Brucker, Svetlana A. Kravchenko 182

Differential Approximation of MIN SAT, MAX SAT and Related Problems
    Bruno Escoffier, Vangelis Th. Paschos 192

Probabilistic Coloring of Bipartite and Split Graphs
    Federico Della Croce, Bruno Escoffier, Cécile Murat,
    Vangelis Th. Paschos 202

Design Optimization Modeling for Customer-Driven Concurrent
Tolerance Allocation
    Young Jin Kim, Byung Rae Cho, Min Koo Lee,
    Hyuck Moo Kwon 212

Application of Data Mining for Improving Yield in Wafer Fabrication
System
    Dong-Hyun Baek, In-Jae Jeong, Chang-Hee Han 222

Determination of Optimum Target Values for a Production Process
Based on Two Surrogate Variables
    Min Koo Lee, Hyuck Moo Kwon, Young Jin Kim, Jongho Bae 232

An Evolution Algorithm for the Rectilinear Steiner Tree Problem
    Byounghak Yang 241

A Two-Stage Recourse Model for Production Planning with Stochastic
Demand
    K.K. Lai, Stephen C.H. Leung, Yue Wu 250

A Hybrid Primal-Dual Algorithm with Application to the Dual
Transportation Problems
    Gyunghyun Choi, Chulyeon Kim 261

Real-Coded Genetic Algorithms for Optimal Static Load Balancing in
Distributed Computing System with Communication Delays
    Venkataraman Mani, Sundaram Suresh, HyoungJoong Kim 269

Heterogeneity in and Determinants of Technical Efficiency in the Use
of Polluting Inputs
    Taeho Kim, Jae-Gon Kim 280

A Continuation Method for the Linear Second-Order Cone
Complementarity Problem
    Yu Xia, Jiming Peng 290

Fuzzy Multi-criteria Decision Making Approach for Transport Projects
Evaluation in Istanbul
    E. Ertugrul Karsak, S. Sebnem Ahiska 301

An Improved Group Setup Strategy for PCB Assembly
    V. Jorge Leon, In-Jae Jeong 312

A Mixed Integer Programming Model for Modifying a Block Layout to
Facilitate Smooth Material Flows
    Jae-Gon Kim, Marc Goetschalckx 322

An Economic Capacity Planning Model Considering Inventory and
Capital Time Value
    S.M. Wang, K.J. Wang, H.M. Wee, J.C. Chen 333

A Quantity-Time-Based Dispatching Policy for a VMI System
    Wai-Ki Ching, Allen H. Tai 342

An Exact Algorithm for Multi Depot and Multi Period Vehicle
Scheduling Problem
    Kyung Hwan Kang, Young Hoon Lee, Byung Ki Lee 350

Determining Multiple Attribute Weights Consistent with Pairwise
Preference Orders
    Byeong Seok Ahn, Chang Hee Han 360

A Pricing Model for a Service Inventory System When Demand Is Price
and Waiting Time Sensitive
    Peng-Sheng You 368

A Bi-population Based Genetic Algorithm for the Resource-Constrained
Project Scheduling Problem
    Dieter Debels, Mario Vanhoucke 378

Optimizing Product Mix in a Multi-bottleneck Environment Using
Group Decision-Making Approach
    Alireza Rashidi Komijan, Seyed Jafar Sadjadi 388

Using Bipartite and Multidimensional Matching to Select the Roots of
a System of Polynomial Equations
    Henk Bekker, Eelco P. Braad, Boris Goldengorin 397

Principles, Models, Methods, and Algorithms for the Structure
Dynamics Control in Complex Technical Systems
    B.V. Sokolov, R.M. Yusupov, E.M. Zaychik 407

Applying a Hybrid Ant Colony System to the Vehicle Routing Problem
  Chia-Ho Chen, Ching-Jung Ting, Pei-Chann Chang 417

A Coevolutionary Approach to Optimize Class Boundaries for
Multidimensional Classification Problems
  Ki-Kwang Lee 427

Analytical Modeling of Closed-Loop Conveyors with Load Recirculation
  Ying-Jiun Hsieh, Yavuz A. Bozer 437

A Multi-items Ordering Model with Mixed Parts Transportation
Problem in a Supply Chain
  Beumjun Ahn, Kwang-Kyu Seo 448

Artificial Neural Network Based Life Cycle Assessment Model for
Product Concepts Using Product Classification Method
  Kwang-Kyu Seo, Sung-Hwan Min, Hun-Woo Yoo 458

New Heuristics for No-Wait Flowshop Scheduling with Precedence
Constraints and Sequence Dependent Setup Time
  Young Hae Lee, Jung Woo Jung 467

Efficient Dual Methods for Nonlinearly Constrained Networks
  Eugenio Mijangos 477

A First-Order ε-Approximation Algorithm for Linear Programs
and a Second-Order Implementation
  *Ana Maria A.C. Rocha, Edite M.G.P. Fernandes,
  João L.C. Soares* 488

Inventory Allocation with Multi-echelon Service Level Considerations
  Jenn-Rong Lin, Linda K. Nozick, Mark A. Turnquist 499

A Queueing Model for Multi-product Production System
  Ho Woo Lee, Tae Hoon Kim 509

Discretization Approach and Nonparametric Modeling for Long-Term
HIV Dynamic Model
  Jianwei Chen, Jin-Ting Zhang, Hulin Wu 519

Performance Analysis and Optimization of an Improved Dynamic
Movement-Based Location Update Scheme in Mobile Cellular Networks
  Jang Hyun Baek, Jae Young Seo, Douglas C. Sicker 528

Capacitated Disassembly Scheduling: Minimizing the Number of
Products Disassembled
  *Jun-Gyu Kim, Hyong-Bae Jeon, Hwa-Joong Kim, Dong-Ho Lee,
  Paul Xirouchakis* 538

Ascent Phase Trajectory Optimization for a Hypersonic Vehicle Using
Nonlinear Programming
  *H.M. Prasanna, Debasish Ghose, M.S. Bhat,
  Chiranjib Bhattacharyya, J. Umakant* 548

Estimating Parameters in Repairable Systems Under Accelerated Stress
  Won Young Yun, Eun Suk Kim 558

Optimization Model for Remanufacturing System at Strategic and
Operational Level
  Kibum Kim, Bongju Jeong, Seung-Ju Jeong 566

A Novel Procedure to Identify the Minimized Overlap Boundary of
Two Groups by DEA Model
  Dong Shang Chang, Yi Chun Kuo 577

A Parallel Tabu Search Algorithm for Optimizing Multiobjective VLSI
Placement
  Mahmood R. Minhas, Sadiq M. Sait 587

A Coupled Gradient Network Approach for the Multi-machine Earliness
and Tardiness Scheduling Problem
  Derya Eren Akyol, G. Mirac Bayhan 596

An Analytic Model for Correlated Traffics in Computer-Communication
Networks
  Si-Yeong Lim, Sun Hur 606

Product Mix Decisions in the Process Industry
  Seung J. Noh, Suk-Chul Rim 615

On the Optimal Workloads Allocation of an FMS with Finite In-process
Buffers
  Soo-Tae Kwon 624

NEOS Server Usage in Wastewater Treatment Cost Minimization
  *Isabel A.C.P. Espoírito-Santo, Edite M.G.P Fernandes,
  Madalena M. Araújo, Eugenio C. Ferreira* 632

Branch and Price Algorithm for Content Allocation Problem in VOD
Network
    Jungman Hong, Seungkil Lim 642

Regrouping Service Sites: A Genetic Approach Using a Voronoi Diagram
    Jeong-Yeon Seo, Sang-Min Park, Seoung Soo Lee, Deok-Soo Kim ... 652

Profile Association Rule Mining Using Tests of Hypotheses Without
Support Threshold
    Kwang-Il Ahn, Jae-Yearn Kim 662

The Capacitated max-k-cut Problem
    Daya Ram Gaur, Ramesh Krishnamurti 670

A Cooperative Multi-Colony Ant Optimization Based Approach to
Efficiently Allocate Customers to Multiple Distribution Centers in a
Supply Chain Network
    Srinivas, Yogesh Dashora, Alok Kumar Choudhary,
    Jenny A. Harding, Manoj Kumar Tiwari 680

Experimentation System for Efficient Job Performing in Veterinary
Medicine Area
    Leszek Koszalka, Piotr Skworcow 692

An Anti-collision Algorithm Using Two-Functioned Estimation for
RFID Tags
    Jia Zhai, Gi-Nam Wang 702

A Proximal Solution for a Class of Extended Minimax Location Problem
    Oscar Cornejo, Christian Michelot 712

A Lagrangean Relaxation Approach for Capacitated Disassembly
Scheduling
    Hwa-Joong Kim, Dong-Ho Lee, Paul Xirouchakis 722

General Tracks

DNA-Based Algorithm for 0-1 Planning Problem
    Lei Wang, Zhiping P. Chen, Xinhua H. Jiang 733

Clustering for Image Retrieval via Improved Fuzzy-ART
    Sang-Sung Park, Hun-Woo Yoo, Man-Hee Lee, Jae-Yeon Kim,
    Dong-Sik Jang 743

Mining Schemas in Semi-structured Data Using Fuzzy Decision Trees
Sun Wei, Liu Da-xin 753

Parallel Seismic Propagation Simulation in Anisotropic Media by
Irregular Grids Finite Difference Method on PC Cluster
Weitao Sun, Jiwu Shu, Weimin Zheng 762

The Web Replica Allocation and Topology Assignment Problem in
Wide Area Networks: Algorithms and Computational Results
Marcin Markowski, Andrzej Kasprzak 772

Optimal Walking Pattern Generation for a Quadruped Robot Using
Genetic-Fuzzy Algorithm
Bo-Hee Lee, Jung-Shik Kong, Jin-Geol Kim 782

Modelling of Process of Electronic Signature with Petri Nets and
(Max, Plus) Algebra
Ahmed Nait-Sidi-Moh, Maxime Wack 792

Evolutionary Algorithm for Congestion Problem in Connection-Oriented
Networks
Michał Przewoźniczek, Krzysztof Walkowiak 802

Design and Development of File System for Storage Area Networks
Gyoung-Bae Kim, Myung-Joon Kim, Hae-Young Bae 812

Transaction Reordering for Epidemic Quorum in Replicated Databases
Huaizhong Lin, Zengwei Zheng, Chun Chen 826

Automatic Boundary Tumor Segmentation of a Liver
Kyung-Sik Seo, Tae-Woong Chung 836

Fast Algorithms for l1 Norm/Mixed l1 and l2 Norms for Image
Restoration
*Haoying Fu, Michael Kwok Ng, Mila Nikolova, Jesse Barlow,
Wai-Ki Ching* 843

Intelligent Semantic Information Retrieval in Medical Pattern Cognitive
Analysis
Marek R. Ogiela, Ryszard Tadeusiewicz, Lidia Ogiela 852

FSPN-Based Genetically Optimized Fuzzy Polynomial Neural Networks
Sung-Kwun Oh, Seok-Beom Rob, Daehee Park, Yong-Kah Kim 858

Unsupervised Color Image Segmentation Using Mean Shift and
Deterministic Annealing EM
    Wanhyun Cho, Jonghyun Park, Myungeun Lee, Soonyoung Park 867

Identity-Based Key Agreement Protocols in a Multiple PKG
Environment
    Hoonjung Lee, Donghyun Kim, Sangjin Kim, Heekuck Oh 877

Evolutionally Optimized Fuzzy Neural Networks Based on Evolutionary
Fuzzy Granulation
    Sung-Kwun Oh, Byoung-Jun Park, Witold Pedrycz,
    Hyun-Ki Kim 887

Multi-stage Detailed Placement Algorithm for Large-Scale Mixed-Mode
Layout Design
    Lijuan Luo, Qiang Zhou, Xianlong Hong, Hanbin Zhou 896

Adaptive Mesh Smoothing for Feature Preservation
    Weishi Li, Li Ping Goh, Terence Hung, Shuhong Xu 906

A Fuzzy Grouping-Based Load Balancing for Distributed Object
Computing Systems
    Hyo Cheol Ahn, Hee Yong Youn 916

DSP-Based ADI-PML Formulations for Truncating Linear Debye and
Lorentz Dispersive FDTD Domains
    Omar Ramadan 926

Mobile Agent Based Adaptive Scheduling Mechanism in Peer to Peer
Grid Computing
    SungJin Choi, MaengSoon Baik, ChongSun Hwang, JoonMin Gil,
    HeonChang Yu 936

Comparison of Global Optimization Methods for Drag Reduction in
the Automotive Industry
    Laurent Dumas, Vincent Herbert, Frédérique Muyl 948

Multiple Intervals Versus Smoothing of Boundaries in the Discretization
of Performance Indicators Used for Diagnosis in Cellular Networks
    Raquel Barco, Pedro Lázaro, Luis Díez, Volker Wille 958

Visual Interactive Clustering and Querying of Spatio-Temporal Data
    Olga Sourina, Dongquan Liu 968

Breakdown-Free ML(k)BiCGStab Algorithm for Non-Hermitian Linear Systems
  Kentaro Moriya, Takashi Nodera 978

On Algorithm for Efficiently Combining Two Independent Measures in Routing Paths
  Moonseong Kim, Young-Cheol Bang, Hyunseung Choo 989

Real Time Hand Tracking Based on Active Contour Model
  Jae Sik Chang, Eun Yi Kim, KeeChul Jung, Hang Joon Kim 999

Hardware Accelerator for Vector Quantization by Using Pruned Look-Up Table
  Pi-Chung Wang, Chun-Liang Lee, Hung-Yi Chang, Tung-Shou Chen 1007

Optimizations of Data Distribution Localities in Cluster Grid Environments
  Ching-Hsien Hsu, Shih-Chang Chen, Chao-Tung Yang, Kuan-Ching Li 1017

Abuse-Free Item Exchange
  Hao Wang, Heqing Guo, Jianfei Yin, Qi He, Manshan Lin, Jun Zhang 1028

Transcoding Pattern Generation for Adaptation of Digital Items Containing Multiple Media Streams in Ubiquitous Environment
  Maria Hong, DaeHyuck Park, YoungHwan Lim, YoungSong Mun, Seongjin Ahn 1036

Identity-Based Aggregate and Verifiably Encrypted Signatures from Bilinear Pairing
  Xiangguo Cheng, Jingmei Liu, Xinmei Wang 1046

Element-Size Independent Analysis of Elasto-Plastic Damage Behaviors of Framed Structures
  Yutaka Toi, Jeoung-Gwen Lee 1055

On the Rila-Mitchell Security Protocols for Biometrics-Based Cardholder Authentication in Smartcards
  Raphael C.-W. Phan, Bok-Min Goi 1065

On-line Fabric-Defects Detection Based on Wavelet Analysis
  Sungshin Kim, Hyeon Bae, Seong-Pyo Cheon, Kwang-Baek Kim 1075

Application of Time-Series Data Mining for Fault Diagnosis of Induction Motors
Hyeon Bae, Sungshin Kim, Yon Tae Kim, Sang-Hyuk Lee 1085

Distortion Measure for Binary Document Image Using Distance and Stroke
Guiyue Jin, Ki Dong Lee 1095

Region and Shape Prior Based Geodesic Active Contour and Application in Cardiac Valve Segmentation
Yanfeng Shang, Xin Yang, Ming Zhu, Biao Jin, Ming Liu 1102

Interactive Fluid Animation Using Particle Dynamics Simulation and Pre-integrated Volume Rendering
Jeongjin Lee, Helen Hong, Yeong Gil Shin 1111

Performance of Linear Algebra Code: Intel Xeon EM64T and ItaniumII Case Examples
Terry Moreland, Chih Jeng Kenneth Tan 1120

Dataset Filtering Based Association Rule Updating in Small-Sized Temporal Databases
Jason J. Jung, Geun-Sik Jo 1131

A Comparison of Model Selection Methods for Multi-class Support Vector Machines
Huaqing Li, Feihu Qi, Shaoyu Wang 1140

Fuzzy Category and Fuzzy Interest for Web User Understanding
SiHun Lee, Jee-Hyong Lee, Keon-Myung Lee, Hee Yong Youn 1149

Automatic License Plate Recognition System Based on Color Image Processing
Xifan Shi, Weizhong Zhao, Yonghang Shen 1159

Exploiting Locality Characteristics for Reducing Signaling Load in Hierarchical Mobile IPv6 Networks
Ki-Sik Kong, Sung-Ju Roh, Chong-Sun Hwang 1169

Parallel Feature-Preserving Mesh Smoothing
Xiangmin Jiao, Phillip J. Alexander 1180

On Multiparametric Sensitivity Analysis in Minimum Cost Network Flow Problem
Sanjeet Singh, Pankaj Gupta, Davinder Bhatia 1190

Mining Patterns of Mobile Users Through Mobile Devices and the
Musics They Listen
  John Goh, David Taniar 1203

Scheduling the Interactions of Multiple Parallel Jobs and Sequential
Jobs on a Non-dedicated Cluster
  Adel Ben Mnaouer 1212

Feature-Correlation Based Multi-view Detection
  Kuo Zhang, Jie Tang, JuanZi Li, KeHong Wang 1222

BEST: Buffer-Driven Efficient Streaming Protocol
  *Sunhun Lee, Jungmin Lee, Kwangsue Chung, WoongChul Choi,
  Seung Hyong Rhee* 1231

A New Neuro-Dominance Rule for Single Machine Tardiness Problem
  Tarık Çakar 1241

Sinogram Denoising of Cryo-Electron Microscopy Images
  Taneli Mielikäinen, Janne Ravantti 1251

Study of a Cluster-Based Parallel System Through Analytical Modeling
and Simulation
  Bahman Javadi, Siavash Khorsandi, Mohammad K. Akbari 1262

Robust Parallel Job Scheduling Infrastructure for Service-Oriented
Grid Computing Systems
  J.H. Abawajy 1272

SLA Management in a Service Oriented Architecture
  James Padgett, Mohammed Haji, Karim Djemame 1282

Attacks on Port Knocking Authentication Mechanism
  *Antonio Izquierdo Manzanares, Joaquín Torres Márquez,
  Juan M. Estevez-Tapiador, Julio César Hernández Castro* 1292

Marketing on Internet Communications Security for Online Bank
Transactions
  José M. Sierra, Julio C. Hernández, Eva Ponce, Jaime Manera ... 1301

A Formal Analysis of Fairness and Non-repudiation in the RSA-CEGD
Protocol
  *Almudena Alcaide, Juan M. Estévez-Tapiador, Antonio Izquierdo,
  José M. Sierra* 1309

Distribution Data Security System Based on Web Based Active Database
Sang-Yule Choi, Myong-Chul Shin, Nam-Young Hur, Jong-Boo Kim, Tai-Hoon Kim, Jae-Sang Cha 1319

Data Protection Based on Physical Separation: Concepts and Application Scenarios
Stefan Lindskog, Karl-Johan Grinnemo, Anna Brunstrom 1331

Some Results on a Class of Optimization Spaces
K.C. Sivakumar, J. Mercy Swarna 1341

Author Index 1349

Table of Contents – Part I

Information Systems and Information Technologies (ISIT) Workshop

The Technique of Test Case Design Based on the UML Sequence Diagram for the Development of Web Applications
Yongsun Cho, Woojin Lee, Kiwon Chong 1

Flexible Background-Texture Analysis for Coronary Artery Extraction Based on Digital Subtraction Angiography
Sung-Ho Park, Jeong-Hee Cha, Joong-Jae Lee, Gye-Young Kim 11

New Size-Reduced Visual Secret Sharing Schemes with Half Reduction of Shadow Size
Ching-Nung Yang, Tse-Shih Chen 19

An Automatic Resource Selection Scheme for Grid Computing Systems
Kyung-Woo Kang, Gyun Woo 29

Matching Colors with KANSEI Vocabulary Using Similarity Measure Based on WordNet
Sunkyoung Baek, Miyoung Cho, Pankoo Kim 37

A Systematic Process to Design Product Line Architecture
Soo Dong Kim, Soo Ho Chang, Hyun Jung La 46

Variability Design and Customization Mechanisms for COTS Components
Soo Dong Kim, Hyun Gi Min, Sung Yul Rhew 57

A Fast Lossless Multi-resolution Motion Estimation Algorithm Using Selective Matching Units
Jong-Nam Kim 67

Developing an XML Document Retrieval System for a Digital Museum
Jae-Woo Chang 77

WiCTP: A Token-Based Access Control Protocol for Wireless Networks
Raal Goff, Amitava Datta 87

An Optimized Internetworking Strategy of MANET and WLAN
Hyewon K. Lee, Youngsong Mun 97

An Internetworking Scheme for UMTS/WLAN Mobile Networks
Sangjoon Park, Youngchul Kim, Jongchan Lee 107

A Handover Scheme Based on HMIPv6 for B3G Networks
*Eunjoo Jeong, Sangjoon Park, Hyewon K. Lee, Kwan-Joong Kim,
Youngsong Mun, Byunggi Kim* 118

Collaborative Filtering for Recommendation Using Neural Networks
Myung Won Kim, Eun Ju Kim, Joung Woo Ryu 127

Dynamic Access Control Scheme for Service-Based Multi-netted
Asymmetric Virtual LAN
Wonwoo Choi, Hyuncheol Kim, Seongjin Ahn, Jinwook Chung 137

New Binding Update Method Using GDMHA in Hierarchical Mobile
IPv6
*Jong-Hyouk Lee, Young-Ju Han, Hyung-Jin Lim,
Tai-Myung Chung* 146

Security in Sensor Networks for Medical Systems Torso Architecture
Chaitanya Penubarthi, Myuhng-Joo Kim, Insup Lee 156

Multimedia: An SIMD – Based Efficient 4x4 2 D Transform Method
*Sang-Jun Yu, Chae-Bong Sohn, Seoung-Jun Oh,
Chang-Beom Ahn* 166

A Real-Time Cooperative Swim-Lane Business Process Modeler
Kwang-Hoon Kim, Jung-Hoon Lee, Chang-Min Kim 176

A Focused Crawling for the Web Resource Discovery Using a Modified
Proximal Support Vector Machines
YoungSik Choi, KiJoo Kim, MunSu Kang 186

A Performance Improvement Scheme of Stream Control Transmission
Protocol over Wireless Networks
*Kiwon Hong, Kugsang Jeong, Deokjai Choi,
Choongseon Hong* 195

Cache Management Protocols Based on Re-ordering for Distributed
Systems
SungHo Cho, Kyoung Yul Bae 204

DRC-BK: Mining Classification Rules by Using Boolean Kernels
Yang Zhang, Zhanhuai Li, Kebin Cui 214

General-Purpose Text Entry Rules for Devices with 4x3 Configurations of Buttons
Jaewoo Ahn, Myung Ho Kim 223

Dynamic Load Redistribution Approach Using Genetic Information in Distributed Computing
Seonghoon Lee, Dongwoo Lee, Donghee Shim, Dongyoung Cho 232

A Guided Search Method for Real Time Transcoding a MPEG2 P Frame into H.263 P Frame in a Compressed Domain
Euisun Kang, Maria Hong, Younghwan Lim, Youngsong Mun, Seongjin Ahn 242

Cooperative Security Management Enhancing Survivability Against DDoS Attacks
Sung Ki Kim, Byoung Joon Min, Jin Chul Jung, Seung Hwan Yoo 252

Marking Mechanism for Enhanced End-to-End QoS Guarantees in Multiple DiffServ Environment
Woojin Park, Kyuho Han, Sinam Woo, Sunshin An 261

An Efficient Handoff Mechanism with Web Proxy MAP in Hierarchical Mobile IPv6
Jonghyoun Choi, Youngsong Mun 271

A New Carried-Dependence Self-scheduling Algorithm
Hyun Cheol Kim 281

Improved Location Management Scheme Based on Autoconfigured Logical Topology in HMIPv6
Jongpil Jeong, Hyunsang Youn, Hyunseung Choo, Eunseok Lee 291

Ontological Model of Event for Integration of Inter-organization Applications
Wang Wenjun, Luo Yingwei, Liu Xinpeng, Wang Xiaolin, Xu Zhuoqun 301

Secure XML Aware Network Design and Performance Analysis
Eui-Nam Huh, Jong-Youl Jeong, Young-Shin Kim, Ki-Young Mun 311

A Probe Detection Model Using the Analysis of the Fuzzy Cognitive Maps
Se-Yul Lee, Yong-Soo Kim, Bong-Hwan Lee, Suk-Hoon Kang, Chan-Hyun Youn 320

Mobile Communications (Mobicomm) Workshop

QoS Provisioning in an Enhanced FMIPv6 Architecture
  Zheng Wan, Xuezeng Pan, Lingdi Ping 329

A Novel Hierarchical Routing Protocol for Wireless Sensor Networks
  Trong Thua Huynh, Choong Seon Hong 339

A Vertical Handoff Algorithm Based on Context Information in
CDMA-WLAN Integrated Networks
  Jang-Sub Kim, Min-Young Chung, Dong-Ryeol Shin 348

Scalable Hash Chain Traversal for Mobile Device
  Sung-Ryul Kim 359

A Rate Separation Mechanism for Performance Improvements of
Multi-rate WLANs
  Chae-Tae Im, Dong-Hee Kwon, Young-Joo Suh 368

Improved Handoff Scheme for Supporting Network Mobility in Nested
Mobile Networks
  *Han-Kyu Ryu, Do-Hyeon Kim, You-Ze Cho, Kang-Won Lee,
  Hee-Dong Park* 378

A *Prompt Retransmit* Technique to Improve TCP Performance for
Mobile Ad Hoc Networks
  Dongkyun Kim, Hanseok Bae 388

Enhanced Fast Handover for Mobile IPv6 Based on IEEE 802.11
Network
  *Seonggeun Ryu, Younghwan Lim, Seongjin Ahn,
  Youngsong Mun* 398

An Efficient Macro Mobility Scheme Supporting Fast Handover in
Hierarchical Mobile IPv6
  Kyunghye Lee, Youngsong Mun 408

Study on the Advanced MAC Scheduling Algorithm for the Infrared
Dedicated Short Range Communication
  Sujin Kwag, Jesang Park, Sangsun Lee 418

Design and Evaluation of a New Micro-mobility Protocol in Large
Mobile and Wireless Networks
  Young-Chul Shim, Hyun-Ah Kim, Ju-Il Lee 427

Performance Analysis of Transmission Probability Control Scheme in Slotted ALOHA CDMA Networks
  In-Taek Lim 438

RWA Based on Approximated Path Conflict Graphs in Optical Networks
  Zhanna Olmes, Kun Myon Choi, Min Young Chung, Tae-Jin Lee, Hyunseung Choo 448

Secure Routing in Sensor Networks: Security Problem Analysis and Countermeasures
  Youngsong Mun, Chungsoo Shin 459

Policy Based Handoff in MIPv6 Networks
  Jong-Hyouk Lee, Byungchul Park, Hyunseung Choo, Tai-Myung Chung 468

An Effective Location Management Strategy for Cellular Mobile Networks
  In-Hye Shin, Gyung-Leen Park, Kang Soo Tae 478

Authentication Authorization Accounting (AAA) Workshop

On the Rila-Mitchell Security Protocols for Biometrics-Based Cardholder Authentication in Smartcards
  Raphael C.-W. Phan, Bok-Min Goi 488

An Efficient Dynamic Group Key Agreement for Low-Power Mobile Devices
  Seokhyang Cho, Junghyun Nam, Seungjoo Kim, Dongho Won 498

Compact Linear Systolic Arrays for Multiplication Using a Trinomial Basis in $GF(2^m)$ for High Speed Cryptographic Processors
  Soonhak Kwon, Chang Hoon Kim, Chun Pyo Hong 508

A Secure User Authentication Protocol Based on One-Time-Password for Home Network
  Hea Suk Jo, Hee Yong Youn 519

On AAA with Extended IDK in Mobile IP Networks
  Hoseong Jeon, Min Young Chung, Hyunseung Choo 529

Secure Forwarding Scheme Based on Session Key Reuse Mechanism in HMIPv6 with AAA
  Kwang Chul Jeong, Hyunseung Choo, Sungchang Lee 540

A Hierarchical Authentication Scheme for MIPv6 Node with Local Movement Property
  Miyoung Kim, Misun Kim, Youngsong Mun 550

An Effective Authentication Scheme for Mobile Node with Fast Roaming Property
  Miyoung Kim, Misun Kim, Youngsong Mun 559

A Study on the Performance Improvement to AAA Authentication in Mobile IPv4 Using Low Latency Handoff
  Youngsong Mun, Sehoon Jang 569

Authenticated Key Agreement Without Subgroup Element Verification
  Taekyoung Kwon 577

Multi-modal Biometrics with PKIs for Border Control Applications
  Taekyoung Kwon, Hyeonjoon Moon 584

A Scalable Mutual Authentication and Key Distribution Mechanism in a NEMO Environment
  Mihui Kim, Eunah Kim, Kijoon Chae 591

Service-Oriented Home Network Middleware Based on OGSA
  Tae Dong Lee, Chang-Sung Jeong 601

Implementation of Streamlining PKI System for Web Services
  Namje Park, Kiyoung Moon, Jongsu Jang, Sungwon Sohn, Dongho Won 609

Efficient Authentication for Low-Cost RFID Systems
  Su Mi Lee, Young Ju Hwang, Dong Hoon Lee, Jong In Lim 619

An Efficient Performance Enhancement Scheme for Fast Mobility Service in MIPv6
  Seung-Yeon Lee, Eui-Nam Huh, Sang-Bok Kim, Young-Song Mun ... 628

Face Recognition by the LDA-Based Algorithm for a Video Surveillance System on DSP
  Jin Ok Kim, Jin Soo Kim, Chin Hyun Chung 638

Computational Geometry and Applications (CGA'05) Workshop

Weakly Cooperative Guards in Grids
  Michał Małafiejski, Paweł Żyliński 647

Mesh Generation for Symmetrical Geometries
  Krister Åhlander 657

A Certified Delaunay Graph Conflict Locator for Semi-algebraic Sets
  François Anton 669

The Offset to an Algebraic Curve and an Application to Conics
  *François Anton, Ioannis Emiris, Bernard Mourrain,
  Monique Teillaud* 683

Computing the Least Median of Squares Estimator in Time $O(n^d)$
  Thorsten Bernholt 697

Pocket Recognition on a Protein Using Euclidean Voronoi Diagram of Atoms
  *Deok-Soo Kim, Cheol-Hyung Cho, Youngsong Cho, Chung In Won,
  Dounguk Kim* 707

Region Expansion by Flipping Edges for Euclidean Voronoi Diagrams of 3D Spheres Based on a Radial Data Structure
  Donguk Kim, Youngsong Cho, Deok-Soo Kim 716

Analysis of the Nicholl-Lee-Nicholl Algorithm
  Frank Dévai 726

Flipping to Robustly Delete a Vertex in a Delaunay Tetrahedralization
  Hugo Ledoux, Christopher M. Gold, George Baciu 737

A Novel Topology-Based Matching Algorithm for Fingerprint Recognition in the Presence of Elastic Distortions
  Chengfeng Wang, Marina L. Gavrilova 748

Bilateral Estimation of Vertex Normal for Point-Sampled Models
  Guofei Hu, Jie Xu, Lanfang Miao, Qunsheng Peng 758

A Point Inclusion Test Algorithm for Simple Polygons
  Weishi Li, Eng Teo Ong, Shuhong Xu, Terence Hung 769

A Modified Nielson's Side-Vertex Triangular Mesh Interpolation Scheme
  Zhihong Mao, Lizhuang Ma, Wuzheng Tan 776

An Acceleration Technique for the Computation of Voronoi Diagrams Using Graphics Hardware
  Osami Yamamoto 786

On the Rectangular Subset Closure of Point Sets
Stefan Porschen 796

Computing Optimized Curves with NURBS Using Evolutionary Intelligence
Muhammad Sarfraz, Syed Arshad Raza, M. Humayun Baig 806

A Novel Delaunay Simplex Technique for Detection of Crystalline Nuclei in Dense Packings of Spheres
A.V. Anikeenko, M.L. Gavrilova, N.N. Medvedev 816

Recognition of Minimum Width Color-Spanning Corridor and Minimum Area Color-Spanning Rectangle
Sandip Das, Partha P. Goswami, Subhas C. Nandy 827

Volumetric Reconstruction of Unorganized Set of Points with Implicit Surfaces
Vincent Bénédet, Loïc Lamarque, Dominique Faudot 838

Virtual Reality in Scientific Applications and Learning (VRSAL 2005) Workshop

Guided Navigation Techniques for 3D Virtual Environment Based on Topic Map
Hak-Keun Kim, Teuk-Seob Song, Yoon-Chu Choy, Soon-Bum Lim 847

Image Sequence Augmentation Using Planar Structures
Juwan Kim, Dongkeun Kim 857

MultiPro: A Platform for PC Cluster Based Active Stereo Display System
Qingshu Yuan, Dongming Lu, Weidong Chen, Yunhe Pan 865

Two-Level 2D Projection Maps Based Horizontal Collision Detection Scheme for Avatar in Collaborative Virtual Environment
Yu Chunyan, Ye Dongyi, Wu Minghui, Pan Yunhe 875

A Molecular Modeling System Based on Dynamic Gestures
Sungjun Park, Jun Lee, Jee-In Kim 886

Face Modeling Using Grid Light and Feature Point Extraction
Lei Shi, Xin Yang, Hailang Pan 896

Virtual Chemical Laboratories and Their Management on the Web
  *Antonio Riganelli, Osvaldo Gervasi, Antonio Laganà,
  Johannes Froehlich* 905

Tangible Tele-meeting System with DV-ARPN (Augmented Reality
Peripheral Network)
  Yong-Moo Kwon, Jin-Woo Park 913

Integrating Learning and Assessment Using the Semantic Web
  *Osvaldo Gervasi, Riccardo Catanzani, Antonio Riganelli,
  Antonio Laganà* 921

The Implementation of Web-Based Score Processing System for WBI
  Young-Jun Seo, Hwa-Young Jeong, Young-Jae Song 928

ELCHEM: A Metalaboratory to Develop Grid e-Learning Technologies
and Services for Chemistry
  *A. Laganà, A. Riganelli, O. Gervasi, P. Yates, K. Wahala,
  R. Salzer, E. Varella, J. Froeklich* 938

Client Allocation for Enhancing Interactivity in Distributed Virtual
Environments
  Duong Nguyen Binh Ta, Suiping Zhou 947

IMNET: An Experimental Testbed for Extensible Multi-user Virtual
Environment Systems
  Tsai-Yen Li, Mao-Yung Liao, Pai-Cheng Tao 957

Application of MPEG-4 in Distributed Virtual Environment
  Qiong Zhang, Taiyi Chen, Jianzhong Mo 967

A New Approach to Area of Interest Management with
Layered-Structures in 2D Grid
  Yu Chunyan, Ye Dongyi, Wu Minghui, Pan Yunhe 974

Awareness Scheduling and Algorithm Implementation for Collaborative
Virtual Environment
  Yu Sheng, Dongming Lu, Yifeng Hu, Qingshu Yuan 985

M of N Features vs. Intrusion Detection
  Zhuowei Li, Amitabha Das 994

Molecular Structures and Processes Workshop

High-Level Quantum Chemical Methods for the Study of Photochemical Processes
  Hans Lischka, Adélia J.A. Aquino, Mario Barbatti, Mohammad Solimannejad 1004

Study of Predictive Abilities of the Kinetic Models of Multistep Chemical Reactions by the Method of Value Analysis
  Levon A. Tavadyan, Avet A. Khachoyan, Gagik A. Martoyan, Seyran H. Minasyan 1012

Lateral Interactions in O/Pt(111): Density-Functional Theory and Kinetic Monte Carlo
  A.P.J. Jansen, W.K. Offermans 1020

Intelligent Predictive Control with Locally Linear Based Model Identification and Evolutionary Programming Optimization with Application to Fossil Power Plants
  Mahdi Jalili-Kharaajoo 1030

Determination of Methanol and Ethanol Synchronously in Ternary Mixture by NIRS and PLS Regression
  Q.F. Meng, L.R. Teng, J.H. Lu, C.J. Jiang, C.H. Gao, T.B. Du, C.G. Wu, X.C. Guo, Y.C. Liang 1040

Ab Initio and Empirical Atom Bond Formulation of the Interaction of the Dimethylether-Ar System
  Alessandro Costantini, Antonio Laganà, Fernando Pirani, Assimo Maris, Walther Caminati 1046

A Parallel Framework for the Simulation of Emission, Transport, Transformation and Deposition of Atmospheric Mercury on a Regional Scale
  Giuseppe A. Trunfio, Ian M. Hedgecock, Nicola Pirrone 1054

A Cognitive Perspective for Choosing Groupware Tools and Elicitation Techniques in Virtual Teams
  Gabriela N. Aranda, Aurora Vizcaíno, Alejandra Cechich, Mario Piattini 1064

A Fast Method for Determination of Solvent-Exposed Atoms and Its Possible Applications for Implicit Solvent Models
  Anna Shumilina 1075

Thermal Rate Coefficients for the $N + N_2$ Reaction: Quasiclassical, Semiclassical and Quantum Calculations
Noelia Faginas Lago, Antonio Laganà, Ernesto Garcia, X. Gimenez 1083

A Molecular Dynamics Study of Ion Permeability Through Molecular Pores
Leonardo Arteconi, Antonio Laganà 1093

Theoretical Investigations of Atmospheric Species Relevant for the Search of High-Energy Density Materials
Marzio Rosi 1101

Pattern Recognition and Ubiquitous Computing Workshop

ID Face Detection Robust to Color Degradation and Facial Veiling
Dae Sung Kim, Nam Chul Kim 1111

Detection of Multiple Vehicles in Image Sequences for Driving Assistance System
SangHoon Han, EunYoung Ahn, NoYoon Kwak 1122

A Computational Model of Korean Mental Lexicon
Heui Seok Lim, Kichun Nam, Yumi Hwang 1129

A Realistic Human Face Modeling from Photographs by Use of Skin Color and Model Deformation
Kyongpil Min, Junchul Chun 1135

An Optimal and Dynamic Monitoring Interval for Grid Resource Information System
Angela Song-Ie Noh, Eui-Nam Huh, Ji-Yeun Sung, Pill-Woo Lee 1144

Real Time Face Detection and Recognition System Using Haar-Like Feature/HMM in Ubiquitous Network Environments
Kicheon Hong, Jihong Min, Wonchan Lee, Jungchul Kim 1154

A Hybrid Network Model for Intrusion Detection Based on Session Patterns and Rate of False Errors
Se-Yul Lee, Yong-Soo Kim, Woongjae Lee 1162

Energy-Efficiency Method for Cluster-Based Sensor Networks
Kyung-Won Nam, Jun Hwang, Cheol-Min Park, Young-Chan Kim 1170

A Study on an Efficient Sign Recognition Algorithm for a Ubiquitous
Traffic System on DSP
  Jong Woo Kim, Kwang Hoon Jung, Chung Chin Hyun 1177

Real-Time Implementation of Face Detection for a Ubiquitous
Computing
  Jin Ok Kim, Jin Soo Kim 1187

On Optimizing Feature Vectors from Efficient Iris Region Normalization
for a Ubiquitous Computing
  Bong Jo Joung, Woongjae Lee 1196

On the Face Detection with Adaptive Template Matching and Cascaded
Object Detection for Ubiquitous Computing Environment
  Chun Young Chang, Jun Hwang 1204

On Improvement for Normalizing Iris Region for a Ubiquitous
Computing
  *Bong Jo Joung, Chin Hyun Chung, Key Seo Lee, Wha Young Yim,
  Sang Hyo Lee* 1213

Author Index 1221

Implementation of Short Message Service System to Be Based Mobile Wireless Internet[†]

Hae-Sool Yang[1], Jung-Hun Hong[2], Seok-Hyung Hwang[3], and Haeng-Kon Kim[4]

[1] Department of Computer Science, HoSeo Venture University
hsyang@office.hoseo.ac.kr
[2] Department of Computer Science, HoSeo Venture University
jaspers74@naver.com
[3] Department of Computer Science, SunMoon University
hsyang@office.hoseo.ac.kr
[4] Department of Computer Information & Communication Eng.,
Catholic University of Daegu

Abstract. In our time, everything to be connected is Internet and Mobile. Most people using Internet and Mobile and we can't do many works without that. Alike this, it gives us an advantage and can get many useful information anywhere, anytime. Moreover it will be a symbol meaning the mobile wireless internet to will be lead the information society. In this research, as develop SMS service to be based mobile wireless internet without SMS server in each communication company, it can service each kind advertisement and various information transmission with an expenses reduction.

1 Introduction

It has meaning a short message on wireless internet to transmit various information to another user on Mobile or PDA. Usually, it used for right terminology, "Short Message Service". It can send a message, 80byte. We can send 80 characters of English or 40 characters of Korean to use 80 byte. As this service, many users can send about 40 characters Korean message without addition equipment in Korea. and it's growing up users more and more as to be spread a mobile. Specially, people of all ages and both sexes are using SMS service because anybody can use easily. In Korea, most credit card company send SMS message to users whenever they use card to prevent an accident. And it's spread to electronic settlement and electronic commerce. But after past few years, there have only there are serviced just similar case alike message advertisement and credit card using, etc.and there have not any improvement and expansion. Now, it need to pay 30won per message in all communication company. Though it's cheaper than SMS service in USA, Europe, We can't passing over the market in Korea. The present time, several company supplying SMS service for business advertisement to low price. but it only can use on the wire system(Desktop computer, Notebook, etc). so it's so restrictive. In this system, As I developed SMS messaging

---

[†] This research was supported by University IT Research Center Project.

service application to be connected wireless internet on mobile, it seek the maximum profit for efficiency and activity. And if it will be open wireless internet netting and will be activity, it will bring more effect. Moreover, in this research, I designed and developed the SMS messaging service to be connected wireless internet on mobile to different with existing SMS service. The first chapter in this research, it describe about the present condition of SMS service market, a charge system and a composition of to way to SMS sending way. And the body chapter in this research, it describe about a composition of developed system(FreeMsg), and mobile packet charge system, and then describe the function to send message, the function to save data and the function to send advertisement and it shows a display of this system. And the end chapter in this research, it describe a value of this system and a hereafter questions.

2 Reference Research

2.1 KVM

Many people interpret the 'K' meaning of KVM, but As representation of Sun Microsystems.Co.Ltd, there are meaning a Kilobyte of 10 units. KVM designed for Mobile Device to has a limitation and designed to fit by microprocessor and controller of 16/32-bit 의 RISC or CISC way. Presently, NTT DoCoMo.Co.Ltd presenting a service to be loaded KVM and also Palm pilot is to be loaded KVM. It's based on most spec of JVM.

2.2 CDDC / MIDP

Before introduce about CLDC/MIDP platform, try to investigate about followings. It can help your understanding about a term of program to be based on mobile device.

- Java Language, Virtual Machine, platform and Java 2 platform
- J2EE/J2SE/J2ME
- Configuration and Profile
- KVM/CLDC/CDC
- MID Profile / Personal Profile / Foundation Profile / RMI Profile

Many people still think that Java is just programming language. But it has a computing platform with programming language.

It's composed to two points. It's Java virtual machine to execute java action code and class library to can use to coding java program, standard API assemblage. And Java platform is divide to three parts of J2EE, J2SE, J2ME with progressing include marketing concept. This three parts divide Enterprise server market, Desktop market, Consumer/Embedded device market individually. It shows a platform division of Java 2 platform, (Figure 1).

2.3 Market of SMS Service

We can use SMS service to 30won. It's some different to each communication companies but We can use 80byte message(except Callback number) per SMS and mobile telecommunication companies(SK Telecom, KTF, LG019) offer 100-200 SMS mes

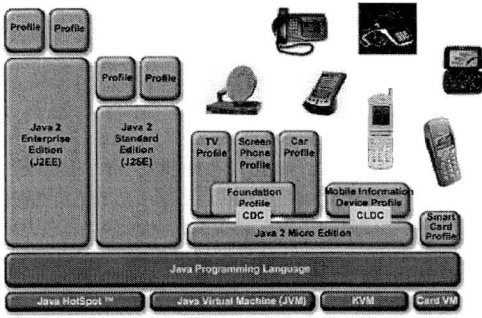

Fig. 1. Java Platform each device

sage to free per price system. As different with this, Some companies service for business. They offer a SMS service to be based on wire internet(Desktop computer, Notebook). So if you use this service, you can send SMS message to wireless device(PCS, Cellular, PDA, etc) on wire internet device. Through this service, users can get a message about their reservation(Hotel, Theater, etc) or the information about credit card. But there need a price to 20won per SMS.

2.4 Constitution of SMS Service

SMS(Short messaging service) is a favorite with everyone in this world. SMS is the acronym for Short Message Service. SMS allows up to 40 korean characters of text to be sent or received by a digital mobile phone. GSM(Global System for Mobile communication) researched about SMS to sent 190billions messages in 2002. It's so easy to use. Just type your message onto the keypad of your cellular phone. If your mobile is turn off, it's saved till your mobile can use. and then you can get a message.

The beginning, SMS is used by teenagers because it's cheaper than speech calling. But little by little, it's expanded for marketing and customer management. The persons concerned SMS Service emphasized that there are effective so much in the part of time and costs. If they employ a working student for calling to customers, it can't over 500 in a day. But if you use SMS service for marketing, you can send about hundred thousand message in 20minutes and can refer the result real time. But There can't find profitable SMS model to connected with price and system in SMS to service presently. It shows the composition of SMS to service presently in (Figure 2).

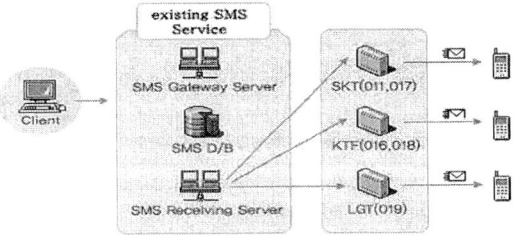

Fig. 2. The system of existing SMS service

3 Standard Construction of Proposal System

The purpose of development this system is to make the fewest cost SMS service to each telecommunication companies(011/017, 016/018, 019) to be based on JAVA Virtual Machine with mobile wireless internet. The basic formation of SMS service is constructed by following. Device mobile base station SMS Center Costs Center customer number Database Center Costs Center SMS Center Device. That is to say, it's passed by following. SMS service with mobile wireless internet is consist with Database server, SMS cooperation company for sending. I give a name of this system to "FreeMsg".

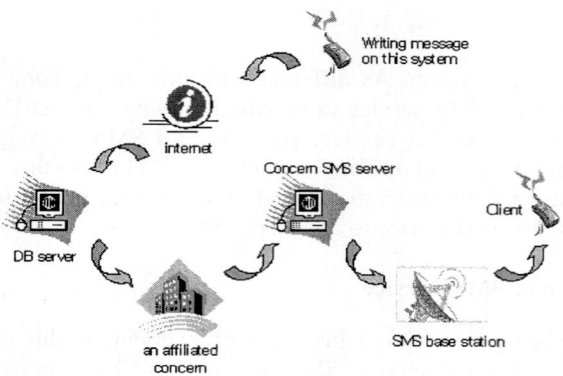

Fig. 3. Composition of FreeMsg system

It's consisted of following system. At first, user download this SMS program and then can send SMS message via this program on wireless internet. And this message is saved in SMS cooperation company Database server. and then this company send message data to each telecommunication companies. and the other person can get message. User don't need to pay for SMS(30won) because it use only mobile wireless internet. In other words, user just need to pay for wireless internet cost. It use a text service when it use this system, FreeMsg, and it impose about 6.5won per packet to each telecommunication companies. It means 512byte per 1packet. Suppose to use 40 Korean characters per SMS message to the highest degree, it means that has a 80byte data transmission. So if use this system, it can send message 6times per packet at a minimum.

4 Implementation and Function of System

FreeMsg System developed to be based on J2ME with following system.

- OS(operating system) : Windows 2000 Professional
- Language : Java SDK 1.3.1, MIDP, SK-VM
- Database : MySql 3.23
- Web Server : Apache Tomcat 4.1.12

It's divided into Message Sending Function, Message Save and Management, Advertisement Sending to Client.

4.1 Message Sending Function

It send message data to Database with wireless internet different with existing form. SMSMessage class to be sent message define the function of short message service with SMS, SMSListener. SMSMessage include a content of message and is 80byte to the maximum. Short message service type of n.TOP Wizard(name of wireless internet in SKTelecom) is divided into SHORT_MESSAGE, DOWNLOAD_ NOTIFICATION, APPLICATION_DATA. It can send any type in this three, but only can get APPLICATION_DATA in client device. It will be ask if user want to send message or not when it use send() method on device. n.TOP wizard shows a display for application data management before conduct the message. Programmer can get a filed data to wanted via getName(), getAppData() when arrive application data. This system save in Database after put user's number, client number, and message into StringBuffer. This message data to saved in database is transmitted to client by cooperation company for SMS service.

4.2 Message Save and Management Function

It's only proper a data saving for framing of a valuable application. It's not meaning to save data for a time while running application. It means that it can use anytime after finish running application. We call this function a permanency and heap memory for space to use object in Java Application. It can't save any data because all object disappear unless you don't use any device to have a permanency because there have not a permanence in heap memory. Contrary to this, it's called to RMS(Record management System) self-data saving space to be defined in MIDP. It create save file for message, "sendMsg" after bring recordStore via getRecordStore() method. User can decide a file name to it user like. And it put a message to created recordStore, and then save a recordStore, name and data(message) to use ByteArrayOutputStream and DataOutputStream.

4.3 Function of Advertisement Sending to Client

It's a new hybrid advertisement type to use circuit netting and packet netting to use MIDP. It shows in client device after sending SMS message. This system, FreeMsg that I developed can use onto SK Nate Wizard(SKTelecom) to be based on SK-VM. SK-VM is the J2ME(Java 2 Micro Edition) running environment to be based on clean room in SKTelecom. It's a part of reading advertisement image of this system in (Figure 4).

It's check a image size after put advertisement image into image vector(addElement). And then it's called on device after read image from buffer to put FileInputStream. t's a part of sending image to device after read an advertisement image by following.

It's not supported usual image format(BMP, JPG or GIF) on SK-VM for send imgae because it can't use voluminous file on mobile device. so usually it use LBM and PNG image format. If programmer use a voluminous file on mobile device, it will

Fig. 4. A part of saving advertisement image

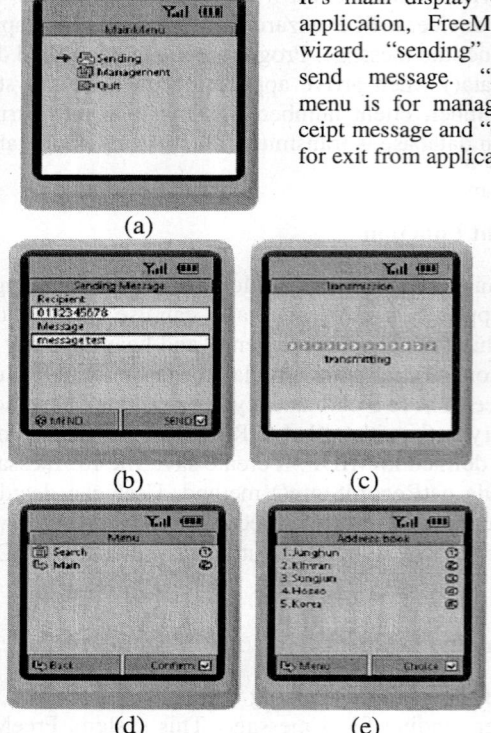

(a)

It's main display when running application, FreeMsg on n.Top wizard. "sending" menu is for send message. "Management" menu is for management of receipt message and "Quit" menu is for exit from application

(b) (c)

It's main display or send message in (b). In this display, user can input client's number data and message data. And then t will be sent when push "send" button alike (c)

(d) (e)

"search" in figure (d) menu is for search client number be saved in user mobile device. Figure (e) is a result display after run "search" menu.

(f)

It's a display for advertisement. It shows an advertisement image on the user's mobile device after success sending message.

Fig. 5. Simulation of this system

happen an OutOfMemoryError. It will be bigger if it change the PNG format to LBM format. But LBM format is not a compress format so it has an efficiency to make jar file. So it become same size format when it loaded PNG format and LBM format on memory.

4.4 Display of This System

It shows a display formation from start application till finish to send message in (Figure 5).

5 An Application and Evaluation of This System

The present time, there have several SMS service in Korea as BtoB, BtoC model. But there are a restriction for supplying this service.
So, it can bring several efficiency. It is a feature of this system as follows.

- curtailment of expenditure for marketing one to one : it's possible to send SMS message to each customer(example : stock information, recommended product in internet shopping mall, weather information, traffic information).
- curtailment of expenditure for marketing by time : it's possible a SMS AD to customer, and strategic AD by time(example : restaurant publicizing, communication of discount time in internet shopping mall, communication of anniversary).
- curtailment of expenditure for timing marketing : SMS alarm service whenever happen specific event(example: stock program alarm, deposit withdrawal, credit card settlement).

Table 1. Comparison with existing SMS service

	existing SMS service	FreeMsg SMS service
Sending system	Send via SMS server	send via wireless internet
Expenses	30won per sending	6.5won per 7-8 times sending
Using state	·using usually ·business advertisement ·send an information for credit card ·basic character advertisement	·using usually to inexpensive charge ·business advertisement ·send an information for credit card ·image advertisement ·possibility an LBS (Location Based Service) ad model ·customer marketing one to one ·possibility a various information

Table 1 shows a specific character after comparison and analysis between existing SMS service and this service, FreeMsg. You can know there are many difference via this table.
Not to conclude, it can define the value of this system by followings.

- An expenses reduction via business efficiency : it's possible to reduce a system building time, personnel expenses contrary to existing SMS service.
- Possibility an effective public information to inexpensive expenses.
- inducement a customer participation : communication participation via various useful information.
- An enterprise impression of ascension via SMS service.
- An increase of main users, teenagers.

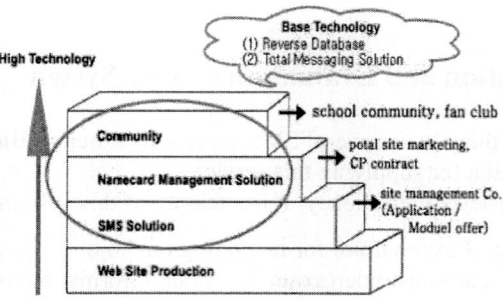

Fig. 6. Practical using via this service

6 A Conclusion and a Problem Awaiting Solution to Be Solved

For development this system, it used J2ME, MIDP and SK-VM. This system only can use on SKTelecom mobile device because it's developed to be based on SK-VM platform to use JAVA Language. And it's still postpone a opening of wireless netting moreover it's expensive to use wireless internet. Henceforward it will be achieved an environment to can use wireless internet anytime with a opening of wireless netting and an improvement of a charge system. If it will be achieved, there don't need to cooperate with SMS enterprise and it's possible to send on client's mobile device directly. Now, it's completed to this part. In the next version, I will try to develop a GPS AD model and discount coupon to be based on GPS(Global Positioning System) and will apply ringing sound and SIS(Simple Image Service) service. But there have some improvement matters. For this, it will be completed an extension of mobile device users and must develop an AD model to can appeal to users unlike static AD model. Telecommunication's must open and assist for using wireless internet freely. Moreover they must supply a better service and various information to be based on mobile wireless internet.

Acknowledgments

This research was supported by the MIC(Ministry of Information and Communication), Korea, under the ITRC(Information Technology Research Center) support program supervised by the IITA(Institute of Information Technology Assessment).

References

[1] J2ME, http://java.sun.com/j2me
[2] J2SE, http://java.sun.com/j2se
[3] J2EE, http://java.sun.com/j2ee
[4] Apach tomcat, http://httpd.apache.org
[5] MIDP, http://java.sun.com/products/midp
[6] KVM, http://java.sun.com/products/kvm
[7] SK-VM, http://developer.xce.co.kr
[8] J2ME for Moble Devices, Bill Day, sun Microsystems
[9] J2ME Wireless Toolkit, http://java.sun.com/products/j2mewtoolkit/down load.html
[10] RMI Trail of the Java Tutorial, http://java.sun.com/ docs/books/tutorial/ rmi/index.html
[11] Implementation a MIDP Sample to use mobile database, http://www. mobilejava.co.kr /bbs/temp/lecture/j2me/mdb1.html
[12] Qusay Mahmoud, "MIDP Database Programming using RMS: a Persistent Storage for MIDlets", 2000. 10
[13] Qusay Mahmoud, "MIDP Network Programming using HTTP and the Connection Framework", 2000. 11
[14] Qusay Mahmoud, "Advanced MIDP Networking, Accessing Using Sockets and RMI from MIDP-enabled Devices", 2002. 1
[15] MIDP Code Samples, http://wireless.java.sun.com/midp/samples
[16] Compile, beforehand verification, and running, http://www.mobilejava co.kr/bbs/temp/ lecture/midp/midp_compile.html, 2000. 11
[17] MIDP Style Guide, http://java.sun.com/j2me/docs/alt-html/midp-style-guide7/ index.html
[18] Eric Giguere, "Wireless Messaging API Basics", 2002. 4

Fuzzy Clustering for Documents Based on Optimization of Classifier Using the Genetic Algorithm

Ju-In Youn, He-Jue Eun, and Yong-Sung Kim

Chonbuk National University, Division of Electronics and Information Engineering,
664-14 1 ga, Duckjin-Dong Duckjin-Gu Jeonju, Republic of Korea
hjeun@chonbuk.ac.kr

Abstract. It is a problem that established document categorization method reflects the semantic relation inaccurately at feature expression of document. For the purpose of solving this problem, we propose a genetic algorithm and C-Means clustering algorithm for choosing an appropriate set of fuzzy clustering for classification problems of documents. The aim of the proposed method is to find a minimum set of fuzzy cluster that can correctly classify all training documents. The number of fuzzy pseudo-partition and the shapes of the fuzzy membership functions that we use the classification criteria are determined by the genetic algorithms. Then, the classifier decides using fuzzy c-means clustering algorithms for documents classification. A solution obtained by the genetic algorithm is a set of fuzzy clustering, and its fitness function is determined by fuzzy membership function.

1 Introduction

With the development of Internet, the volume of information has rapidly expanded and it is increasing at the rate we are able to predict. In the information Retrieval System, the main objective is to satisfy the user's need. To achieve this, it is essential to start with a phase of representing queries and documents. These representations will be used to achieve the second phase that determines the relevant documents according to the query by comparing them [1].

The Web search engine or user who uses the information retrieval system constructs a search query language which is most suitable in the search domain. In case of the retrieval system accomplishes a retrieval capability with the provided query language, and against the result of the feedback, or from the user it receives, to use the result which search a query language with the automatically modified query language and important degree [1, 2]. The existing research regarding the automatic classification of the document uses the methods such as probability method, statistical technique method, and vector similarity method. Bayesian probability method among these document classification methods shows generally high classification efficiency from each language.

Since Zadeh's formulation of fuzzy set theory, many fuzzy set-based approaches such as control, pattern recognition, decision-making, and clustering have been developed and applied to system with uncertainty. The basic idea of these approaches is to represent the uncertainty of the given system by means of fuzzy rules and their membership functions defined over appropriate discourses. One of the most prominent applications of it may be a fuzzy logic-based modeling by means of fuzzy clustering [2, 3].

Cluster analysis is to place elements into groups or clusters suggested by a given data set $X = \{x_1, x_2, \cdots, x_n\} \subset R^p$ which are n points in the p-dimensional space for summarizing data or finding "natural" or "real" substructures in the data set.

The Fuzzy C-Means (FCM) algorithm [1] and its derivatives based on the possibilistic approach for the cluster analysis have been the dominant approaches in both theory and practical applications of fuzzy techniques to unsupervised classification for the last two decades [2, 4]. As pointed out by Milligan, a cluster analysis will not only refer to clustering methods such as the FCM and the possibilistic approach but also to the overall sequence of steps such as clustering elements, clustering variables, variable standardization, measure of association, number of clusters, interpretation, testing, and replication [5, 6]. In recent years, many literatures have paid a great deal of attention to cluster validity issues, and may functional have been proposed for validation of partitions of data produced by the FCM algorithm.

Therefore, in this paper we consider average recall, which applies the fitness function in genetic algorithm. In Section 2, we describe the basic ideas of genetic algorithm and its operation. In Section 3, we describe the decision method of classifier using the fuzzy genetic algorithm and in Section 4, we propose the documents clustering method using fuzzy c-means clustering algorithm. In Section 5, we describe how to use the proposed algorithms.

2 Genetic Algorithm

Genetic algorithms are unorthodox search or optimization algorithms, which were first suggested by John Holland in his book Adaptation in Natural and Artificial Systems (Univ. of Michigan Press, Ann Arbor, 1975) [4]. As the name suggests, the processes observed in natural evolution inspired genetic algorithms. They attempt to mimic these processes and utilize them for solving a wide range of optimization problems. In general, genetic algorithms perform directed random searches through a given set of alternatives with the aim of finding the best alternative with respect to given criteria of goodness. These criteria are required to be expressed in terms of an objective function, which is usually referred to as a fitness function [1, 4, 5].

Genetic algorithms require that the set of alternatives to be searched through be finite. If we want to apply them to an optimization problem where this requirement is not satisfied, we have to discredit the set involved and select an appropriate finite subset. It is further required that the alternatives be coded in strings of some specific finite length which consist of symbols from some finite alphabet. These strings are called chromosomes; the symbols that form them are called genes, and their set is called a gene pool. Genetic algorithms search for the best alternative (in the sense of a given fitness function) through chromosomes evolution [4, 8].

2.1 Basic Idea of Genetic Algorithms

To describe a particular type of genetic algorithm in greater detail, let G denote the gene pool, and let n denote the length of strings of genes that form chromosomes.

That is, chromosomes are in *n-tuples* in G^n [4, 5]. The size of the population of chromosomes is usually kept constant during the execution of a genetic algorithm.

That is, when new members are added to the population, the corresponding numbers of old members are excluded. Let *m* denote this constant population size. Since each population may contain duplicates of chromosomes, we express populations by *n-tuples* whose elements re *n-tuples* from the set G^n [3, 4, 8]. Finally, let *f* denote the fitness function employed in the algorithm [4, 7].

Step 1: An initial population $p^{(k)}$ of a given size *m*, where *k=1* of chromosomes is randomly selected. This selection is made randomly from the set G^n. The choice of value *m* is important. If it is too large, the algorithm does not differ much from an exhaustive search, if it is too small, the algorithm may not reach the optimal solution.

Step 2: Each of the chromosomes in the population $p^{(k)}$ is evaluated in terms of fitness function. This is done by determining for each chromosome *x* in the population the value of the fitness function, *f(x)*.

Step 3: A new population $p^{(k)}$ of chromosomes is selected from the given population $p^{(k)}$ by giving a greater change to select chromosomes with high fitness. This is called natural selection. The new population may contain duplicates. If given stopping criteria are not met some specific, genetic-like operations are performed on chromosomes of the new population [8, 9]. These operations produce new chromosomes, called offspring. In the same steps of this process, evaluation and natural selection are then applied to chromosomes of the resulting population. The whole process is repeated until the given stopping criteria are met. The best chromosome in the final population expresses the solution. We describe only one possible procedure of natural selection, which is referred to as deterministic sampling. According to this procedure, we calculate the value *e(x)=mg(x)* for each *x* in $p^{(x)}$, where *g(x)* is a relative fitness defined by formula.

$$g(x) = \frac{f(x)}{\sum_{x \in p^{(k)}} f(x)}$$

Step 4: Then the number of copies of each chromosome x in $p^{(x)}$ that is chosen for $p_n^{(k)}$ is given by the integer part of *e(x)*. If the total number of chromosomes chosen in this way is smaller than m (the usual case), then we select the remaining chromosomes for $p_n^{(k)}$ by the fractional parts of *e(x)*, from the highest values down. In general, the purposes of this procedure are to eliminate chromosomes with low fitness and duplicate those with high fitness [4, 10].

Step 5: If stopping criteria are not met, go to Step 6; otherwise, stop.

Step 6: Produce a population of new chromosomes, $p^{(k+1)}$, by operating on chromosomes in population $p_n^{(k)}$.

2.2 Operation of Genetic Algorithm

They include some or all of the following four operations;
- **Simple crossover:** Given two chromosomes $x = \{x_1, x_2, \cdots, x_n\}$, $y = \{y_1, y_2, \cdots, y_n\}$ and an integer $i \in N_{n-1}$, which is called a crossover position, the operation of simple crossover applied to x replaces y these chromosomes with their offsprings, $x' = \{x_1, \cdots, x_i, y_{i+1}, \cdots, y_n\}$ $y' = \{y_1, \cdots, y_i, x_{i+1}, \cdots, x_n\}$ chromosomes x and y, to which this operation is applied, are called mates.
- **Double crossover:** Given the same chromosomes mate x, y as in the simple crossover and two crossover positions $i, j \in N_{n-1}(i < y)$, the operation of double crossover applied x and y to replaces these chromosomes with their offsprings,
$$y' = \{y_1, \cdots, y_i, x_{i+1}, \cdots, x_j, y_{j+1}, \cdots, y_n\}$$
$$x' = \{x_1, \cdots, x_i, y_{i+1}, \cdots, y_j, x_{j+1}, \cdots, x_n\}$$
- **Mutation:** Given a chromosome $x = \{x_1, x_2, \cdots, x_n\}$ and an integer $i \in N_n$, which is called a mutation position, the operation of mutation replaces x with $x' = \{x_1, \cdots, x_{i-1}, z, x_{i+1}, \cdots, x_n\}$, where z is a randomly chosen gene from the gene poll G [1, 4]. A crossover operation is employed in virtually all types of genetic algorithms, but the operations of mutation and inversion are sometimes omitted. Their role is to produce new chromosomes not on the basis of the fitness function, but for the purpose of avoiding a local minimum. This role is similar to the role of a disturbance employed in neural networks. If these operations are employed, they are usually chosen with small probabilities. The mates in the crossover operations and the crossover positions in the algorithm are selected randomly. When the algorithm terminates, the chromosome in $p^{(x)}$ with the highest fitness represents the solution.

2.3 Fuzzy System and Genetic Algorithm

The connection between fuzzy systems and genetic algorithms is bidirectional. In one direction, genetic algorithms are utilized to deal with various optimization problems involving fuzzy systems. One important problem for which genetic algorithms have proven very useful is the problem of optimizing fuzzy inference rules in fuzzy controllers. In the other direction, classical genetic algorithms can be fuzzified. The resulting fuzzy genetic algorithms tend to be more efficient and more suitable for some applications. In this paper, we discuss how classical genetic algorithms can be fuzzified; the use of genetic algorithms in the area of fuzzy systems is covered only by a few relevant references [4, 6]. There are basically two ways of fuzzifying classical genetic algorithms. One way is to fuzzify the gene pool and the associated coding of chromosomes; the other one is to fuzzify operations on chromosomes. In classical genetic algorithms, the set *{0, 1}* is often used as the gene pool, and binary numbers codes chromosomes. These algorithms can be fuzzified by extending their gene pool to the whole unit interval *[0, 1]*.

Consider chromosomes $x = \{x_1, x_2, \cdots, x_n\}$ and $y = \{y_1, y_2, \cdots, y_n\}$ are taken from a given gene pool. Then, the simple crossover with the crossover position $i \in N_{n-1}$ can be formulated in terms of a special **n-tuples** $t = \{t_j \mid t_j = 1, j \in N_i, t_j = 0, N_{i+1,n}\}$ referred to as a template, by the formulas $x' = (x \wedge t) \vee (y \wedge \overline{t})$, $y' = (y \wedge \overline{t}) \vee (y \wedge t)$, where $\wedge$ and $\vee$ are *min* and *max* operation on tuples and $\overline{t} = \{\overline{t_j} \mid \overline{t_j} = 1 - t_j\}$ [9].

We can see that the template *t* defines an abrupt change at the crossover position *i*. This is the characteristic of the usual crisp operation of simple crossover. The change can be made gradually by defining the crossover position approximately. This can be done by a fuzzy template.

$$f = \{f_i \mid i \in N_n, f_1 = 1, f_n = 0, i < j \Rightarrow f_i \geq f_j\}$$

Assume that chromosomes $x = \{x_1, x_2, \cdots, x_n\}$ and $y = \{y_1, y_2, \cdots, y_n\}$ are given, whose components are, in general, numbers in [0, 1]. Assume further that a fuzzy template $f = \{f_1, f_2, \cdots, f_n\}$ is given. Then, the operation of fuzzy simple crossover of mates *x* and *y* produces offsprings x' and y' defined by the formulas

$$x' = (x \wedge f) \vee (y \wedge \overline{f}), \; y' = (y \wedge \overline{f}) \vee (y \wedge f)$$

These formulas can be written, more specifically, as

$$x' = \max[\min(x_i, f_i), \min(y_i, \overline{f_i})] \mid i \in N_n$$
$$y' = \max[\min(x_i, \overline{f_i}), \min(y_i, f_i)] \mid i \in N_n$$

The operation of a double crossover as well as the other operations on chromosomes can be fuzzified in a similar way. Experience with fuzzy genetic algorithm seems to indicate that they are efficient, robust, and better attuned to some applications than their classical, crisp counterparts.

3 Classifier Decision Using the Fuzzy Genetic Algorithm

In this chapter, we apply the fuzzy membership function in order to construct information of the gene pool field, which of interest to user. The membership functions are represented by sigmoid function, which uses transfer function in neural network.

3.1 Construction Chromosome Using Sigmoid Function

In this chapter, we describe that the mapping form occurrence frequencies or counts to possibilistic membership degrees is thus a sigmoid function, with its steep part around the "critical" area of occurrence-the concrete values. Also, we consider with occurrence frequency of keyword related each occurrence location of document. Therefore we define different sigmoid functions, which have different critical value as shown in figure 1 [8, 13, 14].

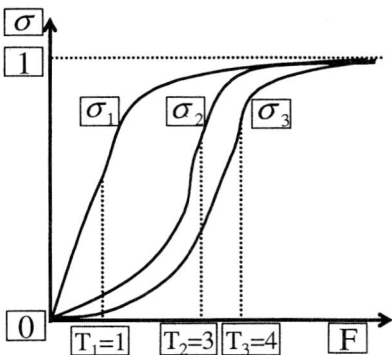

Fig. 1. Sigmoid functions

In practice σ is not necessarily continuously differentiable, but its characteristics should be nevertheless "S-shaped". Although occurrence frequencies are integer numbers, it is reasonable to introduce the sigmoid mapping over the whole positive half of the real lines [8, 9, 13].

3.2 Classifier Decision Algorithm

In this section we propose the classifier decision algorithm using operation of the genetic algorithm. The Genetic Algorithm optimizes knowledge base, which it composed of the association keyword extracted by a data mining technique. Knowledge base constructed by the exclude association keyword that the user does not prefer from the association keyword. The association keyword that the user prefers does the web document that the user prefers in standard. In this case, the web document that the user prefers is the document where the search frequency of the users is high from information retrieval engine.

Therefore, in this paper we apply the frequency of occurrence of significant words that extracted in web documents. It is connected with the importance of that word in the document. Also, it will be that pairs of words occurring frequently in the same document or the same part of a document might be connected in the meaning. The initial population with the web document, which the user prefers, shows the whole document, which it extracts and of the chromosome it composed. The initial population evolves through process of fitness calculation, reproduction, selection, crossover, mutation and fitness evaluation. The population that evolved through evaluation is decided whether it will continue an evolution with the next generation. The following pseudo-code shows the genetic algorithm through process of initialization, fitness calculation, reproduction, selection, crossover, mutation, and fitness evaluation, which decide the classifier of the clustering.

```
Procedure    fuzzy_GA(Doc[clustering.doc.n],    sigmoid_f[clustering,    keywordset,
fitness_threshold])
   // Extract keyword in web documents//
   Doc[clu_ID, doc_num, n] =Extract(Webdoc_set);
   // Initial population //
   for(doc_num=1; doc_num<n; doc_num++) do begin{
      for(word_num=1; word_num<t; word++)
       // calculate importance degree of keyword//
       if(word_loc ==title or keywordset)
         select(sigmoid_f[1]);
       else if(word_loc ==abstract or conclusion)
         select(sigmoid_f[2]);
       else(word_loc==text)
         select(sigmoid_f[3]);
   while(average_fitness < fitness_threshold) do begin{
      calculation_fitness( );
     //reproduction for simple section//
     for(doc_num=1; doc_num<n; doc_num ++) do begin{
       fitness_s[clustering.doc] = fitness[clu.doc]/sum_fitness;
     }
     //selection, crossover, mutation operation//
     while(doc<=n) do begin {
        selection( );
        crossover( );
        mutation( );
     }
   calculation_fitness( );
     if average(fitness>=fitness_threshold) then exit;
     fitness[clustering.doc] <= fitness_s[clustering.doc]
   end.
```

4 Document Clustering Using Fuzzy c-Means Algorithm

In this chapter, we describe the fuzzy clustering and used it to cluster documents. Clustering is one of the most fundamental issues in pattern recognition. It plays a key role in searching for structures in data. Given a finite set of data, X, the problem of clustering in X is to find several cluster centers that can properly characterize relevant classes of X. In classical cluster analysis, these classes are required to form a partition of X such that the degree of association is strong for data within blocks of the partition and weak for data in different blocks. However, this requirement is too strong in many practical applications, and it is thus desirable to replace it with a weaker requirement. When the requirement of a crisp partition of X is replaced with a weaker requirement of fuzzy partition or a fuzzy pseudo-partition on X, we refer to the emerging problem area as fuzzy clustering. Fuzzy pseudo-partitions are often called fuzzy c-partitions, where c designates the number of fuzzy classes in the partition [4, 11, 15].

We describe the feature selection of document, which extracted from search engine based on Web and how to transfer fuzzy value. To illustrate this issue let $X = \{x_1, x_2, \cdots, x_n\}$ be a set of given document data.

A fuzzy pseudo-partition or fuzzy c-partition of x is a family of fuzzy subsets of X, denoted by $P = \{A_1, A_2, \cdots, A_c\}$ which satisfies $\sum_{i=1}^{c} \mu_{ij} = 1$ for all $j \in N_n$ and $0 < \sum_{j=1}^{n} \mu_{ij} < n$ for all $i \in N_c$ where c is positive integer. Given a set of data $X = \{x_1, x_2, \cdots, x_n\}$, where x_k in general, is a vector $X = \{x_{k1}, x_{k2}, \cdots, x_{kp}\} \in R^p$ for all $k \in N_n$, the problem of fuzzy clustering is to find a fuzzy pseudo-partition and the associated cluster centers by which the structure of the data is represented as best as possible.

This requires some criterion expressing the general idea that associations (in the sense described by the criterion) be strong within clusters and weak between clusters [13, 15]. To solve the problem of fuzzy clustering, we need to formulate this criterion in terms of a performance index. Usually, the performance index is based upon cluster centers. Given a pseudo-partition $P = \{A_1, A_2, \cdots, A_c\}$, the c cluster centers, $\{v_1, v_2, \cdots, v_c\}$ associated with the partition are calculated by the formula

$$V_i = \frac{\sum_{j=1}^{n}(\mu_{ij})^m X_j}{\sum_{j=1}^{n}(\mu_{ij})^m}$$, for all $i \in N_c$, where $m>1$ is real number that governs the

influence of membership grades. Observe that the vector v_i calculated by above formula, which is viewed as the cluster center of the fuzzy class A_i, is actually the weighted average of data in A_i. The weight of a datum x_k is the mth power of the membership grade of x_k in the fuzzy set A_i.

In this paper, we use the fuzzy c-mean clustering method, which find a fuzzy pseudo-partition P that minimizes the object function. The performance index of a fuzzy pseudo-partition $P, J_m(P)$ is then defined in terms of the cluster centers by the object function such as following:

$$J_m(P) = \sum_{j=1}^{n}\sum_{i=1}^{c}(\mu_{ij})^m \| x_j - v_i \|^2$$, where $\| \cdot \|$ is some inner product-include norm in

space R^P and $\| x_j - v_i \|^2$ represents the distance between x_j and v_i. This performance index measures the weighted sum of distance the value of $J_m(P)$, the better the fuzzy pseudo-partition P. Therefore, the goal of the fuzzy c-means clustering method is to find a fuzzy pseudo-partition P that minimizes the performance index $J_m(P)$ [4, 8, 15]. The fuzzy c-means algorithm is based on the assumption that the desired number of clusters in given and, in addition, a particular distance, a real number $m \in (1, \infty)$, and a small positive number ε, serving as a stopping criterion, are chosen. Therefore, we propose the document-clustering algorithm that uses the genetic algorithm, which will produce clustering classifier and fuzzy c-means algorithm such as following:

***Step 1:* Select an initial fuzzy pseudo-partition**
```
Procedure document_clustering ( )
   extract_keyword(doc_num);
   apply_sigmoid function(keyword);
   random_pseudo-partition(doc_num );
   initial_partition(doc_num);
End
```

***Step 2:* Calculate the c cluster center**
```
Procedure clusterCenter_Calculation( )
   for(doc_num=1; doc_num< n; doc_num++)
   do begin{
      if(clucenter[doc_num] > clucenter[doc_num ++])
         Clucenter[doc_num]= clucenter[doc_num]/ Sum_clucenter[doc_num]; End
```

***Step 3*: update clustering**
```
Procedure update_clustering( )
   if(average_center>=classifier)
      Clusterceneter ++;
   else
      clusterCenter_Calculation( )
End
```

***Step 4*: Compare former fuzzy clustering and update clustering**
```
Procedure compare_clustering( )
   if(former_clustering- update_clustering <= ε )
      exit;
   else
      Cluster_num ++;
   return step 2;
End
```

5 Conclusion

In this paper, we suggest document-clustering method, which uses the genetic algorithm and fuzzy c-means algorithm. In general, user interest is reflected to extract keyword that stands for each document's content and clustering which constructing the same clustering with the document connected semantically. The suggested document clustering algorithm differs from previous the method in view of the use of keyword's occurrence frequency which extracted from document. We also define the fuzzy membership function of keyword's degree of importance and define fuzzy relation between keywords and documents. Also, different membership function was defined to give keyword's weight according to occurrence area extracted keyword and we tried to reflect the user interest. In this paper, we also use the genetic algorithm for the decision of the classifier of clustering and it uses to calculate the clustering center value. Document clustering method calculates similarity degree between documents. It also constructs the same clustering for semantically connected documents to classify document among higher similarity degree. Consequently, information retrieval method that applies genetic algorithm and fuzzy c-means algorithm, which were suggested in this paper offers document that reflected user's interests. Also, these algorithms offer appreciated documents more than the information retrieval method of simple keyword directly matching by document, which semantically connected, and classify the document set.

For further research, we will study automated indexing method in constructed document set using document categorization algorithm suggested in this paper. Also we will apply hierarchical classification for connected document based on content. Consequently, we will be improving the retrieval speed, precision and recall of similar documents using index, which stand for document set use. In addition, when arbitrary documents are stored in database, we will classify automatically the documents with the same category and connected semantically.

References

1. Su-Jeon Ko, "Bayesian Automatic Document Categorization Using Apriori-Genetic Algorithm", Vol. 8, No. 3, 2003.6.
2. Soon H, Kwon, "A Cluster Validity Index for Fuzzy Clustering" Electronics Letters, Vol. 34, No, 22, 2002 .
3. R. Baeza-ates, B. Ribeiro-Neto, "Modern Information Retrieval", p.230-255, 1998.
4. G. J. Klir, B. Yuan, "Fuzzy Sets and Fuzzy Logic Theory and Applications", 1998.
5. Soo-Jung Ko, "Optimization of Associative Word Knowledge Base Using Apriori-genetic algorithm" KISS , Vol 28, No. 8, 2003.
6. Keon-Myung Lee, "Classification Rule Mining from Fuzzy Data based on fuzzy decision Tree" KISS, Vol 28, No. 1, 2003.
7. K. Hyun-Jin, "Clustering Korean Nouns Based On Syntactic Relation and Corpus Data" Proceedings of the LASTED International Conference Artificial Intelligence and Soft Computing, 2003.
8. M. Gondon, "Probabilistic and genetic algorithms for document retrieval" Communication of the ACM, 31, 2000.
9. L. T. Koczy, "Information retrieval by fuzzy relations and hierarchical co-occurrence", 1997
10. P. Baranyi, T. D. Gedeon, L. T. Koczy, "Improved fuzzy and neural network algorithms for frequency prediction in document filtering", TR 97-02, 1997.
11. L. T. Koczy, T. D. Gedeon, J. A. Koczy, "The construction of fuzzy relational maps in information retrieval", IETR 98-01, 1998.
12. L. T. Koczy, T. Gedeon, "Information retrieval by fuzzy relations and hierarchical co-occurrence", Part I. TR99-01, Dept. of Info. Eng., School of Comp. Sci. & Eng., UNSW, 1999.

13. Seok-Woo Han, Hye-Jue Eun, Yong-Sung Kim, Laszlo T. Koczy, " A Document Classification Algorithm Using the Fuzzy Set Theory and Hierarchical Structure of Document", ICCSA(1) 2004, Page 122-133.
14. Eun, Hye-jue, "An Algorithm of Documents classification and Query Extension using fuzzy function", Journal of KISS : Software and applications Vol28, No. 2, 2001.
15. T.C.Chen, "A Fuzzy Network for the Document Clustering Based on the Measurement of Information Pattern", "Neural Networks", 4, 1991.

P2P Protocol Analysis and Blocking Algorithm

Sun-Myung Hwang

Department of Computer Engineering, Daejeon University, 96-3 Yongun-dong,
Dong-gu, Daejeon 300-716, South Korea
sunhwang@dju.ac.kr

Abstract. P2P (Peer to Peer) technology provides methods for overcoming many weak points of conventional client-server mechanism, and consequently, many efforts in many fields are made to apply it. On the other side of the coin, these strong points of the P2P technology have been used for bad purposes, causing many problems and concerns. This paper proposes a method to block P2P applications fundamentally in order to eliminate illegal data or files. We use Ethereal, a reliable network packet analysis tool, and analyze the packets receive and send when P2P applications run. Then, in this paper we examine the packet architecture and characteristics of each P2P application, and propose the algorithms that can block P2P applications. When being used for blocking up P2P applications, these proposed algorithms can play important roles in reducing excessive P2P traffic and illegal data sharing.

1 Introduction

Client-server based network solutions have been used most wide. This allows server to make up for client's performance lack, and this solution has been used most wide. For this client-server mechanism, a concept of file server storing clients' data has been used for data sharing among clients. This concept has caused server to experience overload. To solve this, a new network mechanism appeared. This concept is a P2P technology negating the existence of server. This technology has become rapidly popular.

Client-server based mechanism concentrates data into one system, and disables data sharing when server is unable to provide services. The P2P technology eliminates these problems by requesting data from connected clients (hereinafter "peers") so that data can be shared even if some peers are absent. Additionally, the P2P technology enables a part of data to be requested simultaneously from several peers in data downloading. This enables faxter data sharing.

The P2P technology provides methods for overcoming many weak points of conventional client-server mechanism, and consequently, many efforts in many fields are made to apply it. On the other side of the coin, these strong points of the P2P technology have been used for bad purposes, causing many problems and concerns.

Napster is a milestone for rise of P2P programs. Napster is a service designed to provide sharing of mp3 files. The advent of Napster caused explosive growth of many similar applications. Since Napspter allowed users to share music data, many music companies resisted the service, causing Napster to stop its service. Other applications were also made for the purpose of data sharing. Legal restrictions cannot be imposed on data sharing and websites providing data sharing cannot be closed forcefully. Consequently, many problems were caused by these P2P applications.

There has not been enough studies on blocking P2P applications. Moreover, there are not any commercially available applications for preventing use of P2P applications. Teenagers and even juniors use P2P applications to share data illegally. This has become a social issue which has not encountered appropriate solutions.

This paper is to analyze the protocols used by popular P2P applications, to propose the algorithms blocking the applications, to build the algorithms into blocking programs, and to ensure that the applications are blocked effectively by the blocking programs.

2 P2P Technologies: An Overview

For P2P model, a peer receives services via other peers and in turn, provides services for other peers. A peer has become able to perform more tasks than a conventional client on client-server network. This is because of changes in internet environment. Enhanced performance of PC and provision of sufficient bandwidth have made it inefficient for a host on the internet to play only the role of client. Moreover, users' demand on anonymity without server has increased.

2.1 P2P Models

Since the P2P model has no server, even if a peer is down or attacked, the whole network is not affected much. Moreover, P2P model allows anonymity because a peer cannot have information on all other peers. Each peer provides and accesses services by sharing its own storage space, memory, computing capacity or bandwidth.

The P2P model requires communication methods, which are different from those of server-client model, in order to provide peer-to-peer communication. For example, when peers provide file sharing services to each other, a peer should search for a file it needs. In the server-client model, files are searched for and downloaded from server. In the P2P model, a peer should know which peers have files it needs because the files are in other peers. The P2P model's file searching method has been more complex and many studies have been conducted on distributed search on the P2P network.

2.2 Classifications of P2P

P2P applications are divided into two groups, depending on the characteristics of each application.

A. Hybrid P2P programs
Hybrid P2P applications mean a concept of server added to original P2P model. Hybrid P2P applications search for data via server and the server delivers data lists to

each peer. However, the server does not involve in data downloading. Hybrid P2P programs enable server to manage peers efficiently, providing efficient data search. When the server is overcrowded with traffic or goes down, no search is possible.

B. Pure P2P programs

Pure P2P applications provide P2P services without server in order to fit to the original concept of P2P. The applications allow a peer to directly communicate with other peers. Therefore, no server is required and data is searched in the way of broadcast. Even if several peers are down, search or downloading can be done well. However, these applications provide slower file search than Hybrid P2P applications.

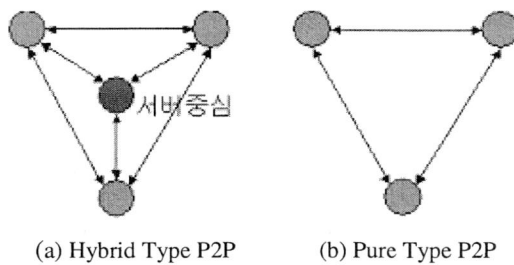

(a) Hybrid Type P2P   (b) Pure Type P2P

Fig. 1. Hybrid Type and Pure Type P2P

2.3 Types of P2P Applications

The table below shows the usage of P2P in Korea. Especially, Soribada and E-Donkey are most commonly used. Moreover, E-Donkey has taken increasingly higher share since many similar applications have appeared.

Table 1. P2P application trands in korea

P2P Application	Use rate (%)
Soribada	60.6
Gru-Gru	14.6
E-Donkey	11.9
Win-MX	6.4
V-Share	2.2
Etc	4.3

A. E-Donkey

This is the most popular P2P application. There are many similar applications of which protocols are modified versions of E-Donkey protocol(Emule, Overnet, etc.). Since it is very popular, the characteristics of its protocol are well known. E-Donkey is a typical hybrid P2P application that is dealt most intensively by this paper.

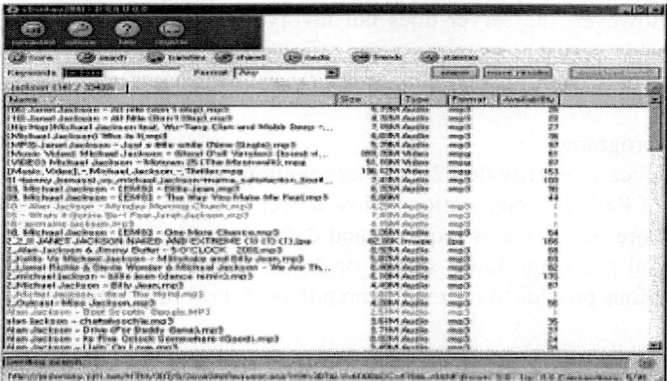

Fig. 2. An example of E-Donkey applicaton

B. Gnutella

Gnutella is a typical pure P2P application. There are many similar applications which use the same format of protocol as Gnutella protocol. The applications use modified versions of Gnutella protocol, making it difficult to analyze them.

C. Soribada

Soribada is a P2P application made in Korea, which takes a much different approach from conventional P2P applications. Since it has server, it is a hybrid applicaion. Unlike other P2P applications downloading a part of file from several peers in file download, Soribada chooses a peer from which file is to be downloaded and lets the whole file to be downloaded from the selected peer. Strictly, Soribada cannot be classified as a P2P application. Since it is used most wide at home and abroad, this paper will address it.

3 Analysis of P2P Protocols

This study analyzed P2P protocols by using Ethereal, which is a network packet analysis tool. This study tried to establish P2P connection with network, used Etherea to capture data packets on the network, and examined the characteristics of each P2P application that existed inside the packets.

3.1 E-Donkey

A. Architecture of E-Donkey packet
E-Donkey protocol has following architecture.

The first 1 byte is the protocol ID. It judges E-Donkey, Emule extension module and if compression is done. The following 4 bytes indicate the size of data area. The third 1 byte shows the signature indicating which command the present packet issues, and actual data is contained in the following data area.

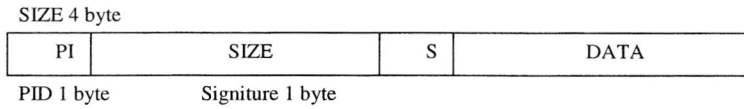

Fig. 3. Protocol structure of E-Donkey class

Table 2. Protocol id of E-Donkey class

protocol ID	description
0xe3	Orignal E-donkey protocol
0xc5	Extend protocol in E-mule
0xd4	Data compression protocol

Table 3. Signature in E-Donkey class

Instruction	Signature no. 당나귀	Signature no. Emule	Description
hello	0x01	0x01	서버의 존재여부 확인
hello answer	0x4c	0x02	서버 존재여부의 확인에 대한 응답
Req-Slist	0x14	0x14	서버리스트들을 해당 요청한다.
Rep-Slist	0x32	0x32	데이터에 서버 리스트 포함
Req-SState	0x96	0x96	서버의 상태를 요청한다.
Rep-SState	0x97	0x97	데이터에 유저수, 파일 개수, 최대사용가능 유저수가 포함(각각 4 바이트)
Req-Sinf	0xa2	0xa2	서버의 정보를 서버에 요청한다.
Rep-Sinf	0xa3	0xa3	서버 정보들을 보낸다.
Req-Search	0x16	0x16	데이터에 검색 문자열 길이 2바이트와 그 뒤에 검색 문자열을 포함
Rep-Search	0x33	0x33	검색에 대한 응답 E-mule은 메시지 압축
Req-Source	0x19	0x19	소스를 가지고 있는 피어들의 리스트를 요청한다.
Rep-Source	0x42	0x42	데이터에 소스를 가지고 있는 피어의 리스트 포함
Req-File	0x58	0x81	파일을 요청한다.
Rep-File	0x59	0x82	파일 요청에 대한 응답.
Req-Slot	0x54	0x54	파일을 받기위한 슬롯 할당을 요청한다.
Rep-Allslot	0x55	0x55	슬롯 할당에 대한 응답

B. Analysis of E-donkey protocols

E-Donkey protocols have their own specific ID's which are dividied into three: protocol ID only for E-Donkey, protocol ID used by Emule extension module and protocol ID used when data area is compressed.

Table 3 shows signatures which are commonly used and contain protocols used uniquely by Emule. For example, Emule uses E-Donkey's specific signature and Emule extension signature simultaneously.

C. Protection algorithm
```
for(i=0; i< protocolTableSize; i++){
         If(str(PacketString[0] == protocolIDTable[i])
           protocolIDFlag = true;
}
if(protocolIDFlag){
         for(j=0; j < DonkeySigSize; j++){
           if(strPacketString[SigSpace] == Donkey-Sig[j])
             return true;
}
}
return false;
```

Fig. 4. E-Donkey Protocol protection algorithm

3.2 Gnutella

A. Architecture of Gnutella packets
Gnutella protocol's packet architecture comprises 16-byte descriptor ID and 1-byte payloader descriptor ID. The 16-byte descriptor ID is a sole ID that a peer has on the network. Other peers use the 16-byte descriptor ID to establish connections with each other. Payloader descriptor ID means a signature showing the four basic commands used by Gnutella protocol.

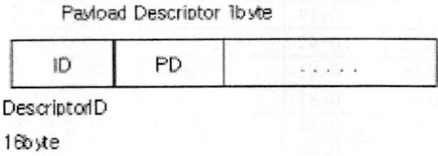

Fig. 5. Packet Structure of Gnutella Protocol

B. Analysis of protocol
Ping message: Ping message means that each peer (called servant in case of Gnutella protocol) broadcasts its own descriptor ID to the Gnutella network in order to log on to the Gnutella network.

Pong message: Pong message is the reply of each servant the ping message. The payload descriptor ID for pong message is 0x01.
Data query

Query hit: Payload descriptor ID has a value of 0x81. This is a list of query hits. When each servant's data has hits corresponding to queries, the list of hits are to the servants that requested it. Analysis of transferred messages shows that the list of data of the servants that contains actual query 'california' is included.

Actual data transfer: Data is transferred actually by TCP/IP stream. Data heads have slightly different values depending on the characteristics of relevant applications. Limewire, an application used for analysis, allows stream transfer as shown in Fig. 6.

```
GET /uri-res/N2R?urn:sha1:MLX6DEJZMUHSK4PITMQYIDOJRBGARAKI
HTTP/1.1
HOST: 70.64.173.156:6346
User-Agent: LimeWire/3.8.3
X-Queue: 0.1
X-Gnutella-Content-URN:
urn:sha1:MLX6DEJZMUHSK4PITMQYIDOJRBGARAKI
X-Alt: 155.97.202.30, 152.3.72.242:6348
Range: bytes=1499870-1599879
Chat: 203.237.140.179:6346


HTTP/1.1 206 Partial Content
Server: LimeWire/4.0.8
Content-Type: application/binary
Content-Length: 100010
Date: Thu, 23 Sep 2004 16:41:45 GMT
Content-Range: bytes 1499870-1599879/4202496
X-Gnutella-Content-URN:
urn:sha1:MLX6DEJZMUHSK4PITMQYIDOJRBGARAK
```

Fig. 6. Data Send/Receive in Limewire

C. Protection algorithm

```
for(i=0; i < payloaderDesTableSize; i++){
        if(strPacketString[GnuSigSpace] == payloaderDesTa-
ble[i])
            return true;
}
return false;
```

Fig. 7. Protection algorithm of Gnutella class application

3.3 Soribada3

A. Analysis of protocol

Access to Soribada server: When Soribada is run, access to the website www.soribada.com is established (http protocol). Then, a user enters his/her ID and password at soribada3.phtml?action=nick&id=XXXXX&pw= XXXXX and gets user authentification. This method of user authentification is used wide on ordinary websites. However, this authenthification is exposed to TCP stream so that this study could analyze the protocol easily.

Obtainment of the list of peers: After logging on to server, the application uses http://211.233.14.157/habor.html?action=list to get the list of peers.

Peer-to-peer receiving and sending: The application sends messages to the list of peers that is received from server and makes sure that peers can receive or send data to and from each other. The messages sent contain 0x1014 signature (UDP protocol). In reply to this, each peer exchanges data by transferring the messages containing 0x1015 signature to users.

Request for file search: Users use the messages that they have searched for by using their lists of peers, and sends request for data search to the peers. The signature used for this is 0x01.

Sending search results: When a peer received request for data search has data corresponding to search conditions, it sends the list of the data.

File receiving and sending: Since files are transferred by TCP stream as in case of Gnutella, transferred data is analyzed for stream comparison so that the application can be recognized to be Soribada.

B. Blocking algorithms

Since the signature held by Soribada is used only for query and search, the application can be recognized correctly by comparing this signature and the texts of GETMP3 and soribada3 existing in TCP stream. Recognition mechanism of Soribada application is a little complex.

```
for(i=0; i < SoribadaSig1byteTableSize; i++){
        if(strPacketString[0] == SoribadaSig[i])
           return true;
}
for(i=0; i < SoribadaSig2byteTableSize; i++){
        if(strPacketString[0] == Sori-
badaSig2byte[i][0] && strPacketString[1] ==
        SoribadaSig2byte[i][1])
           return true;
}
Return false;
```

Fig. 8. Soribada Protocol algorithm

4 P2P Blocking Programs

This study used a library called WinPCap to build a P2P blocking program which can block P2P programs by using P2P programs' packet architecture analyzed above and the proposed algorithms. Fig. 6 is shows the screen of loaded blocking program. When the blocking program starts to capture packets, all packets existing on the network are captured. Then the blocking program uses the proposed algorithms to make sure that there is any P2P packet and shows the list of present packets.

Additionally, a dialogue box to prompt the blocking of P2P packets is embedded in the blocking program so that the P2P packets can be selectively blocked. Fig. 7

This study shows that it is possible to efficiently block E-Donkey family (E-Donkey, Emule, Overnet, Pruna), Gnutella family (Morpheus, Limewire) and Soribada3, which are subjects of analysis, by using the proposed algorithms.

5 Conclusions

P2P protocols have some advantages that can be used very usefully. P2P model has lower dependence on server than conventional client-server mechanism. Instead, the model focuses on each peer. Because of this, when it appeared, it was expected that P2P model would be of great help in creating more innovative solutions than conventional ones. However, P2P solutions have been used for bad purposes, giving losses to many people.

Using P2P applications, users can easily find and download commercially available soft wares (games, operating systems, development tools etc.) and even download mp3 files, movie files and porn files.

For this, many users take it granted that they share data. They are not aware that their data sharing via P2P services bring losses to many other people. Software development companies undergo increasingly larger damages and losses because they do not have any measures and technologies to prevent their software products from being distributed free via the P2P applications. Additionally, software piracy has brought serious social problems.

This paper proposes a method to block P2P applications fundamentally in order to eliminate software piracy. This study used Ethereal, a reliable network packet analysis tool, and analyzed the packets received and sent when P2P applications run. Then, the study examined the packet architecture and characteristics of each P2P application, and proposed the algorithms that can block P2P applications.

When being used for blocking up P2P applications, these proposed algorithms can play important roles in reducing excessive P2P traffic and illegal data sharing.

However, the algorithms do not necessarily block up all P2P applications. Nobody knows which kind of P2P protocols using new methods will appear. It is not possible to assure that delicate matters excluded from the subjects of this study can be examined. Consequently, more studies on the blocking algorithms will provide more complete and more reliable P2P blocking mechanisms.

Moreover, current blocking algorithms use block up P2P applications, disallowing the P2P applications used for good purposes. If there are algorithms which can take a closer look at exchanged packets and determine what kind of data is exchanged, it will be possible to develop the methods that can filter only illegal data sharing through P2P applications.

References

[1] Ian D Graham and Join G Cleary, "Cell level measurements of ATM traffic," Proceedings of the Australian Telecommunications Networks and Applications Conference, pp.495-500, December 1996.
[2] Cisco, White Paper, "NetFlow Services and Applications,"http://www.cisco.com/warp/public/cc/pd/iosw/ioft/netflct/tech/napps_wp.htm.
[3] P.Phaal, S. Panchen, N. McKee, "InMon Corporation's sFlow: A Method for Monitoring Traffic in Switched and Routed Networks", IETF RFC 3176, September 2001.
[4] N. Brownlee, C. Mills, G. Ruth, "Traffic Flow Measurement: Architecture", IETF RFC 2722, October 1999.
[5] Argus, http://www.qosient.com/argus/.
[6] Se-Hee Han, Myung-Sup Kim, Hong-Taek Ju and James W. Hong, "The Architecture of NG-MON: A Passive Network Monitoring System", LNCS 2506, DSOM 2002, October, 2002, pp16-27.
[7] 소리바다, http://www.soribada.com/.
[8] Napster, http://www.napster.com/.
[9] Gnutella, http://gnutella.wego.com/.
[10] MSN Messenger, http://messenger.msn.co.kr/, Microsoft.
[11] Yahoo Messenger, http://kr.messenger.yahoo.com/, Yahoo.
[12] eDonkey2000, http://www.edonkey2000.com/.
[13] Matei Ripeanu, "Peer-to-Peer Architecture Case Study: Gnutella Network", Techreports TR-2001-26, University of Chicago, July, 2001.
[14] Subhabrata Sen, Jia Wang, "Analyzing Peer-to-Perr Traffic Across Large Networks", IMW2002 Workshop, 2002.
[15] AOL, http://www.aol.com/.
[16] Hun-Jeong Kang, Hong-Taek Ju, Myung-Sup Kim and James W. Hong, "Towards Streaming Media Traffic Monitoring and Analysis", APNOM2002, 2002, pp 503-504.
[17] Ethereal, http://www.ethereal.com/.

Object Modeling of RDF Schema for Converting UML Class Diagram

Jin-Sung Kim[1], Chun-Sik Yoo[1], Mi-Kyung Lee[2], and Yong-Sung Kim[1]

[1] Division of Electronic and Information Engineering, Chonbuk National University,
664-14 1ga Duckjin-Dong, Duckjin-Gu, Jeonju, 561-756,
Republic of Korea
[2] Seoul-Jungsoo Polytechnic College,
Department of Information & Data communication
kpjiju@chonbuk.ac.kr

Abstract. With increasing amounts of information on the web and the need to access it accurately, it is very important to standardize metadata and to store and manage the metadata system. The RDF (Resource Description Framework) is a framework for representing, exchanging, and reusing metadata. In this paper, we propose rules and an algorithm to convert the RDF schema into a UML (Unified Modeling Language) class diagram and formal models to represent an object-oriented schema for the RDF schema. The proposed rules and algorithm are useful for natural mapping, and the object modeling of RDF schema can be easily converted into the object-oriented schema.

1 Introduction

In order to find the exact document needed by the users on the Internet, we should know the exact information. The information is not only a description of document's content, but also the information of the document itself, which is called "metadata". These days, as the importance of metadata is emphasized, many researchers have made progress in the standardization work of metadata and construction of metadata for efficient and systematic management [1]. For the standardization of metadata, RDF based on the XML (eXtensible Markup Language) is able to cover all metadata on the web and furnish the standardized framework for the description of various resources scattered on the internet to save RDF schema defined metadata into a database and manage systematically, object modeling is needed. Therefore, this study shows that RDF schema converses into object modeling by utilizing UML being the standard of object-oriented method, and then, create object-oriented schema code. Because RDF schema and UML both have a common feature having a schema, and are used in other application purposes, they need mapping with each other. In the case of E-R diagram like UML, the model of entities and relations should be described. In the case of node and arc diagram like RDF schema, it is used to describe the node and arc model [2]. Therefore, this study suggests the rule of mapping RDF schema into the UML class diagram and algorithm, and then creates an object modeling by converting RDF schema. It also shows the basis of document management to construct documents in an object-oriented database by proposing formal models in the form of object-oriented database schema converted to a RDF schema.

2 Related Works

There is also a modeling method for XML documents using the Elm tree diagram [3] and UML class diagram [4]. The Elm tree diagram suggests a notation method based on the tree structure. But, in declaration of elements, there is a weak point on the bracket and link processing. As compared with it, the model using the UML class diagram supports the various cases with the XML link and so on. But, the modeling method of [4] is that the element of XML converts into UML class and, in the declaration of elements in the XML document, the parts of the name and content attribute have an aggregation relationship between elements. In the RDF schema, the element tag describes all relationships between classes. Therefore, a new modeling method for the RDF schema, not supported in the modeling of the general XML document, is needed. In this study, the rule and the algorithm in which the RDF schema is converted into the UML class diagram will be suggested.

3 RDF Schema and UML Class Diagram

In this section, we will examine the important components of the RDF schema and the UML class diagram.

3.1 RDF Schema

The metadata working group acting to establish a framework for the information related with web is the field of PICS and RDF. The RDF suggests interoperability between applications exchanging machine-readable information [5]. RDF uses XML as a common syntax for processing and exchanging metadata. RDF schema suggests interpreter information about statements provided by the RDF data model. The RDF schema mechanism suggests a type system for use in the RDF model [6]. In Figure 1, a class is described by a rounded rectangle; a resource is depicted by a large dot, and arrows are drawn from a resource to the class it defines.

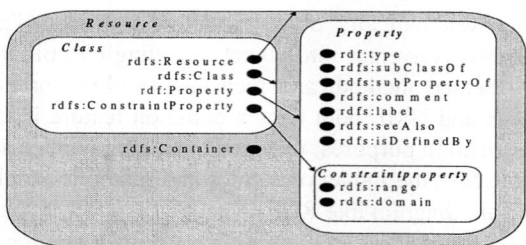

Fig. 1. Classes and Resources

3.2 UML Class Diagram

The UML class diagram consists of the classes, relationships, and constraints [11].

(1) Class
The class consists of a class name, an attribute list, and an operation list. Every class provides a type of a class.
(2) Aggregation Relationships
Aggregation relationship represents the relationship of 'part-of' between superclass and subclass, and is notated as a '◇'.
(3) Generalization Relationships
Generalization relationship represents the relationship of 'is-a' between superclass and subclass, and is notated as a '△'.
(4) Constraints
Several aggregation relationships connect with a dotted line, and give the constraints as 'or'. An 'ordered' constraint expresses the case that subclasses have an order.

4 Formal Models and Algorithm

This study describes the rules, modeling functions, formal models, algorithm, system and application results, and comparison analyses for converting RDF schema into a UML class diagram.

4.1 Conversion Rules

For converting the collection (set) of RDF schema into the collection of the UML class diagram, we define the rules as follows.

(1) Class and Attribute

> [Rule 1] The rdf:Description of the RDF schema becomes class in the UML class diagram, and rdf:type becomes the type of class.

Example 1) Class
<rdf:Description ID= "Resource"> <rdf:type resource="#Class"/> </rdf:Description>

Fig. 2. UML Class Diagram of Ex1

> [Rule2] The property and value, which represent attributes of the RDF schema, becomes a private attribute and the initial value of the attribute.

Example 2) Attribute
<rdf:Description about="http://www.w3.org/Home/Lassila">
    <s:Creator>Ora Lassila</s:Creator> </rdf:Description>

> [Rule 3] The constraints of RDF schema become a public attribute of class in the UML class diagram

Example 3) Constraints
<rdf:Property ID="result"> <rdf:domain rdf:resource="#SearchQuery"/>
<rdfs:range rdf:resource="#SearchResult"/> </rdf:Property>

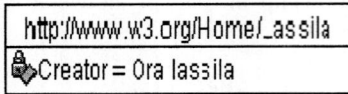
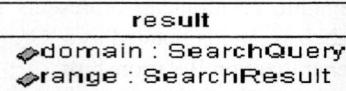

Fig. 3. UML Class Diagram of Ex2   **Fig. 4.** UML Class Diagram of Ex3

(2) Relationships

> [Rule 4] The rdfs:subClassOf RDF Schema becomes generalization relationships of the UML Class Diagram

Example 4) Subclass
<rdf:Description ID="Class">
<rdf:type resource="#Class"/> <rdfs:subClassOfrdf:resource="#Resource"/></rdf:Description>

> [Rule 5] The rdfs:Container of RDF schema properties become aggregation relationships of the UML class diagram. In rdfs:Container, rdf:Seq has an 'ordered' constraint, and rdf:Alt has an '{or}' constraint. When there's no subclass that becomes the attribute type of class, it is automatically generated.

Example 5) Properties
<rdf:Description about="http://mycollege.edu/courses/6.001">
 <s:Students> <rdf:Bag>
    <rdf:li resource="http://mycollege.edu/students/Amy"/>
   <rdf:li resource="http://mycollege.edu/students/Tim"/>
 </rdf:Bag> </s:students> </rdf:Description>

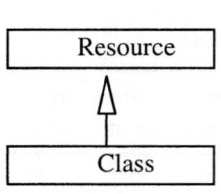
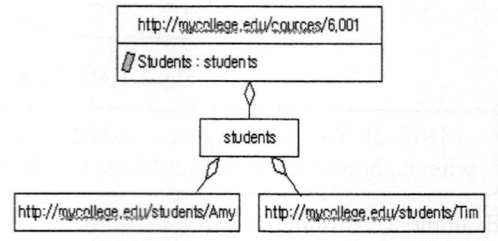

Fig. 5. UML Diagram of Ex4   **Fig. 6.** UML Class Diagram of Ex5

Example 6) Ordered Constraints
<rdf:Description about="http://www.foo.com/cool.html">
    <dc:Creator> <rdf:Seq ID="creatorSurname">
        <rdf:li>Mary Andrew</rdf:li>
        <rdf:li>Jacky Crystal</rdf:li>
    </rdf:Seq> </dc:Creator> </rdf:Description>

Example 7) Or Constraints
<rdf:Description about="http://x.org/pakages/x11">
  <s:DistributionSite> <rdf:Alt> <rdf:li>ftp.x.org</rdf:li>
        <tdf:li>ftp.cs.purdue.edu</rdf:li>
      </rdf:Alt> </s:DistributionSite> </rdf:Description>

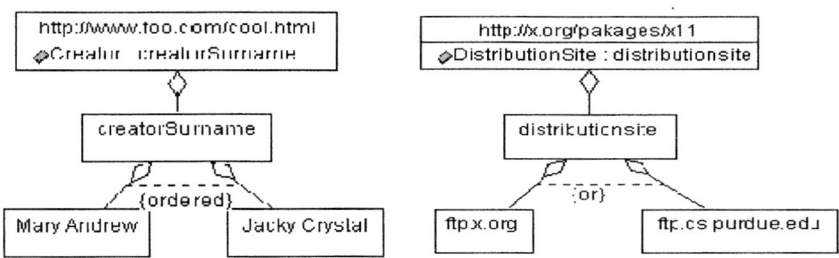

Fig. 7. UML Class Diagram of Ex6   **Fig. 8.** UML Class Diagram of Ex7

Attribute Students and Distributionsite of Example 5 & Example 7 are not specified as an attribute type, therefore, suitable class is automatically generated. Between classes and subclasses, they have aggregation relationships.

(3) Comments

> [Rule 6] The rdfs:comment of RDF schema becomes a note of the UML class diagram.

Example 8) Comments
<rdfs:Class ID="Resource" rdfs:label="Resource" rdfs:comment="The most general class/">

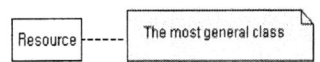

Fig. 9. UML Class Diagram of Ex 8

4.2 Formal Models

In order to convert the RDF schema into the object-oriented database schema, the form of schema is needed which is not supported by the present database. In this study, based on the O2 database, we extend a schema, and define the formal model of rdf:Container. By doing this, we give several definitions and hypothesis as follows.

[Hypothesis 1] The set of base types is dom, the set of attributes is $A = \{a_1, a_2, ..., a_n\}$, the set of objects is $O = \{o_1, o_2, ..., o_n\}$, and the set of classes is $C = \{c_1, c_2, ..., c_n\}$

[Hypothesis 2] The values of object set O are described as $V = \{v_1, ..., v_i, ..., v_n\}$. Where, an element of V, v_i becomes *nil* or dom or each elements of O. Therefor, the tuple $[a_1: v_1, ..., a_n: v_k]$, the set $\{v_1, ..., v_k\}$, and the list $[v_1, ..., v_k]$ are also to be the elements of V. [4].

[Hypothesis 3] The set of class, the type-set of each C elements is $\tau = \{\tau_1, ..., \tau_i, ..., \tau_n\}$, here the interpretation of it is $dom(\tau)$.

(1) Class

Based on the above hypothesis, the following are the definitions of rdf:Seq of rdf:Container, rdf:Bag of rdf:Container, rdf:Alt.

[Definition 1] Tuple $[a_1: \tau_1, ..., a_n: \tau_n]$ is Seq type, exactly, it is $dom([a_1: \tau_1, ..., a_n: \tau_n]) = \{[a_1: v_1, ..., a_n: v_n, ..., a_{n+l}: v_{n+l}] \mid vi \in dom(\tau_i), i = 1, ..., n; l \geq 0\}$[4].

(2) Relationships

[Definition 2] Tuple $[a_1: \tau_1 \& ... \& a_n: \tau_n]$ is Bag type. Exactly, it is $dom([a_1: \tau_1 \& ... \& a_n: \tau_n]) = \{[a_{i1}: v_{i1}, ..., a_{in}: v_{in}] \mid v_{im} \in dom(\tau_i), i = 1, ..., n\}$, where each $i_1, ..., i_m, ..., i_n$ is the permutation of $1, ..., n$.

[Definition 3] Set $\{a_1: \tau_1 + ... + a_n: \tau_n\}$ is Alt type. Exactly, it is $dom([a_1: \tau_1 + ... + a_n: \tau_n]) = \cup \{dom([a_i: v_i]) \mid 1 \leq i \leq n\}$.

4.3 Modeling Functions

In the UML class diagram, as inserted into the parts of operation lists of each class, it is easy to understand the structure of the class diagram. By using them, the modification and unification of the RDF schema is practicable. The hypotheses for defining the modeling functions are as follows.

[Hypothesis 4] The set of connector is Con = {Inherit, Seq, Bag, Alt}.

[Hypothesis 5] Class diagram consists of a pair of (S, p), where S is a set of class about $\forall t \in S$ and function p is the application S on to S.

In the above [Hypothesis 5], root class t of S is satisfying the features such as Table 1[4].

Table 1. The features of root class

functions	meaning
p(t) = t	the parents of t is t.
$\forall x \in S, \exists k \in N, p^k(x)=t$, where, k is a level of tree , N is a set of integers.	t is an ascendant of all x.

Table 2. RDF resource class modeling function

Function Name	Description
p(t)	Superclass of Class t
c(t)	Subclass of Class t
cons(t)	Connector between subclass of Class t
r(t)	Order of Subclass of Class t
m(t)	Defined attributes of Class t

Based on the above hypotheses, modeling functions are defined as follows.

(1) Superclass and Subclass

[Definition 4] $\forall t \in S$, p, c : S → p(S), where, p,c indicate each superclass and $t_2 = c(t_1)$ is $t_1 \neq t_2$.

(2) Connector of Subclass

That is, the relationship between the superclass and subclass is generated in rdf:subClassOf and rdf:Container of the RDF schema.

[Definition 5] $\forall t \in S$, cons : S →Con, where cons indicates a kind of a connector.

(3) Order of Subclass

That is, rdf:subClassOf of the RDF schema is cons(t) = inherit, and the cons(t) of rdf:Seq, rdf:Bag, rdf:Alt in rdf:Container has the value of Seq, Bag, Alt respectively.

[Definition 6] $\forall t \in S$, if r : S →Integer then cons(t) = Seq, where it expresses the procedure of the subclass.

(4) Attribute

That is, when rdf:Container of the RDF schema is rdf:Seq, the subclass has the value r(t) in the order of 1, 2,...

[Definition 7] $\forall t \in S$, m : S → A, where function m indicates all attributes belonged to the class.

That is, the constraints of the RDF schema, rdf:resource, rdf:range becomes the attributes.

4.4 Algorithm

From the mapping rule, the following is the algorithm, which converts the data model of the RDF schema into a UML class diagram.

```
Input  : Data model of RDF Schema
Output : UML Class Diagram.
begin {
  while ( RDF Data Model Element ) {
    if ( rdf:Description ) make_class()      // Class Generation
    else if ( rdf:comment ) make_note()      // Comment Attachment
  }
  while ( RDF Data Model Element) {
    if ( Property or Constraint ) {
      if ( rdf:type ) class_type()           // Definition of Class Type
      else if ( rdf:subClassOf ) make_inheritance()   //inheritance relationship
      else if ( rdf:Container ){
            if ( there is no attribute type) make_class()
            make_aggregation()               // aggregation relationship
            switch ( rdf:Container ) {
                case : rdf:Seq  order_constraint()   // sequence relationship
                case : rdf:Alt  or_constraint()
      } }
      else insert_attribute()
    }
  }
                                        }
                                       end;
```

4.5 System Architecture and Result of Application

This section shows the structure of system and the result of application.

4.5.1 System Structure and Functions

The Configuration of System is Fig.10.

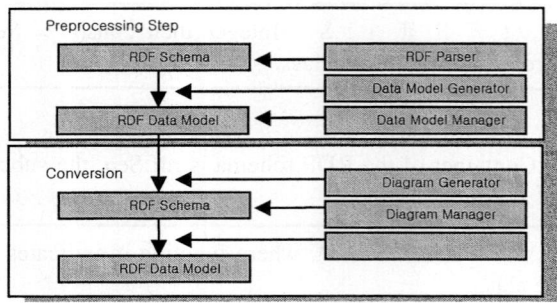

Fig. 10. System architecture

(1) Preprocessing Step
- RDF parser: Parsed an RDF schema using an XML, it examines whether there are some syntax errors.
- Data Model Generator: It converts a parsed RDF schema into an RDF data model.
- Data Model Manager: In the RDF data model, it saves and manages the information of nodes and arcs.

(2) Conversion Step
The configuration of the conversion step is shown Figure 11.

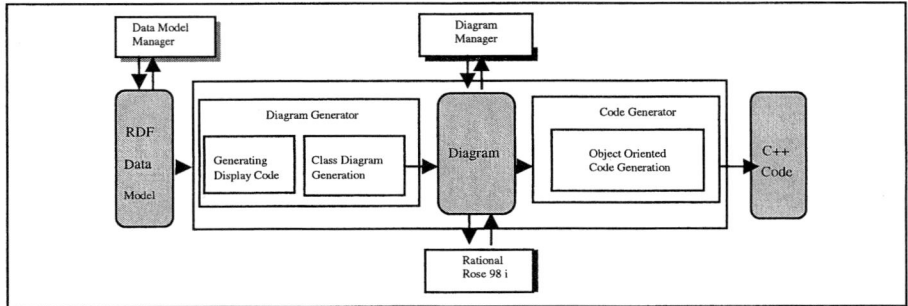

Fig. 11. The Configuration of the Conversion Step

Above Fig11, each component function and characteristic is the following.
- Diagram Generator: It counts a data a data model into a class diagram by mapping algorithm
- Diagram Manager: It manager and store location information and etc. of class diagram
- Code Generator: It generates the object code about the class diagram.

4.5.2 Result of Application
The following are the results, which generate a data model and class diagram by inputting a RDF schema in this system and then creating a C++ code.

(1) RDF Schema
The RDF schema, an input of this system, is in MortorVechicle's XML form.

```
                    <rdf:RDF xml:lang="en">
      xmlns:rdf=http://ww.w3.org/2004/02/22-rdf-syntax-
   ns#"xmlns:rdfs=http://www.w3.org/TR/2004/PR-rdf-schema-
                            20040303#>
              <rdf:Description ID="MortorVechicle">
          <rdf:type resource=http://www.w3.org./TR/2004/
                    PR-rdf-schema-20040303#Class/>
   <rdfs:subClassof ref:resource=http://www.w3.org/ TR/2004/PR-
                    rdf-schema-20040303#Resource"/>
                        </rdf:Description>
              <rdf:Description ID="PassengerVechicle">
    <rdf :type resource= "http://www.w3.org/TR/2004/ PR-rdf-schema-
```

```
                    20040303#Class"/>
       <rdfs:subClassOf rdf:resource="#Motor Vechicle"/>
                        </rdf:Description>
              <rdf :Description ID= "Truck" >
<rdf:typeresource=http://www.w3.org/TR/2004/PR-rdf-schema-
                        20040303#Class/>
        <rdfs:subClassOF rdf:resource="Motor Vechicle"/>
                        </rdf:Description>
               <rdf:Description   ID="Van">
  <rdf:type resource=http://www.w3.org./TR/2004/ PR -rdf-schema-
                        20040303#Class/>
        <rdfs:subClassOf   rdf:resource="MotorVechicle"/>
                        </rdf:Description>
               <rdf:Description ID="MiniVan">
 <rdf :type resource= "http ://www.w3.org/TR/2004/ PR-rdf-schema-
                        20040303#Van"/>
    <rdfs :subClassOf rdf :resource= "#Passenger Vechicle "/>
                        <.rdf:Description>
                          </rdf:RDF>
```

(2) Data Model

The following (Figure 12) is the data model of the RDF Schema which is easy to convert into a class diagram.

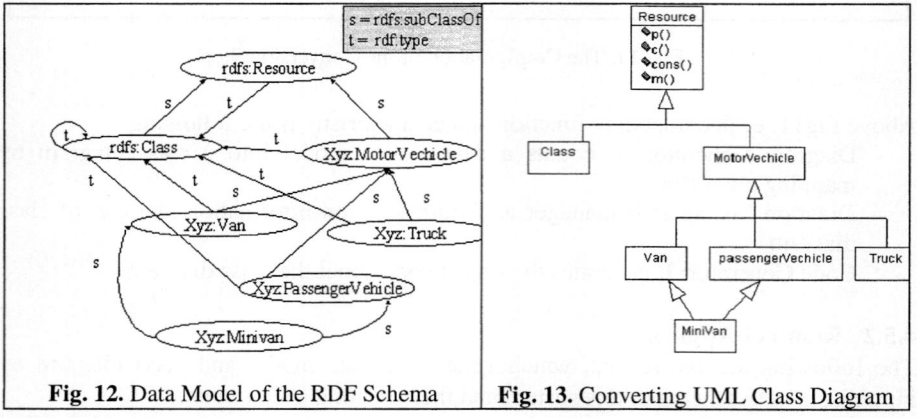

Fig. 12. Data Model of the RDF Schema   **Fig. 13.** Converting UML Class Diagram

(3) UML Class Diagram

By inputting a data model of the RDF schema, a UML class diagram is generated. The generated UML class diagram of (Figure 12) is (Figure 13).

4.6 Comparison Analysis

In order to evaluate the modeling method of this study, we would like to compare and analyze [3][4] research modeling RDF data model and an XML document. Table 2 is the comparison of the modeling method about each RDF resources.

Comparing with it, the modeling of this study expresses exactly the system structure between classes mapped into a class diagram of UML, according to the semantics of the RDF resources, and it has an advantage in which it converts easily into an object-oriented schema.

Table 3. The Comparison of RDF Modeling Methods

RDF resource		Elm tree diagram	UML class diagram	RDF data model	Model of this study		
rdf:Description				node	Class		
rdf:type				arc	Class type		
Property types:Description		Class or rectangle	Class	arc	private attribute		
rdfs:domain				arc	public attribute		
rdfs:range				arc	public attribute		
rdf:subClassOf				arc	generalization		
rdfs:Container	rdf:Seq			arc	aggregation	{order}	
	rdf:Alt			arc		{or}	
Rdf:comment				arc	comment		

5 Conclusion and Future Works

This study suggests the rules and the algorithms, which an RDF schema in the form of metadata in the web documents converts into an UML class diagram, object-oriented modeling method. In order to structure identification, conversion, and unification of the UML class diagram for the RDF schema, an operation is inserted in the operation list of each class, and the formal model is suggested in the form of the object-oriented database for the RDF schema. In accordance with it, this study will be the base to save and manage a RDF schema in an object-oriented database. In future research work, we suggest an algorithm for automatic generation of MCF (Meta Content Format), and WF (Warick Framework) into an object-oriented modeling and an object-oriented schema.

References

1. Ho-Taek Jung, Young-Jong Yang, Sun-Young Kim, Sang-Duk Lee, Youn-Chal Choi, "Suggestion of Metadata model and Development of Management system for Electron document on the Web", Korea Information Processing Society Journal, VOL.5, NO.4, pp.924-940, 1998. 4.
2. W3C, "A Discussion of the Relationship Between RDF-Schema and UML", August 1998, http:// www.w3.org /TR/1998/NOTE-rdf-uml-19980804
3. ArborText Inc. "Data modeling Report prepared for:W3C XML Specification DTD("XML spec")", September 1998, http://www.oasis-open.org/cover /xml-report-19980910.html.
4. Won-Seok Chae, Yan Ha, Yong-Sung Kim, "Structure Diagramming for XML documents Using UML Class Diagram", Korea Information Processing Society Journal, VOL.6, NO.3, pp. 2670-2679, 1999. 10.
5. W3C, "Resource Description Framework(RDF) Model and Syntax Specification", August 1998, http://www.w3.org/TR/ 1998/WD-rdf-syntax-19980819/
6. W3C, Resource Description Framework(RDF) Schema Specification, March 1999, http://www. w3.org/TR/Tr-rdf-schema.
7. Bonstrom, V.; Hinze, A.; Schweppe, H, "Storing RDF as a graph", Web Congress, 2003. Proceedings. First Latin American , NO. 10-12, pp. 27 – 36, Nov. 2003
8. Gasevic, D.; Djuric, D.; Devedzic, V.; Damjanovic, V., "From UML to ready-to-use OWL ontologies", Intelligent Systems, 2004. Proceedings. 2004 2nd International IEEE Conference , Vol. 2 , NO. 22-24, pp. 485-490, June. 2004.

A Framework for Security Assurance in Component Based Development

Gu-Beom Jeong[1] and Guk-Boh Kim[2]

[1] Dept. of Computer Engineering, Sangju National University, Sang Ju, South Korea
jgb@sangju.ac.kr
[2] Dept. of Computer Engineering Daejin University, Po Cheon, South Korea
kgb@road.daejin.ac.kr

Abstract. This paper will investigate the fundamental issues related to building and composing secure components. While all participants will closely cooperate, each will have primary responsibility in one area. The approach outlined in this paper develops a certification process for testing software components for security proper- ties. The anticipated results from this paper are a process, set of core white-box and black-box testing technologies to certify the security of software components and a framework for constructing compositional Component Security Assurance (CSA) based on the security property exposed by the atomic components. The manifestation of the product is a stamp of approval in the form of a digital signature.

1 Introduction

The e-commerce systems of today are composed of a number of components including: a commerce server, component transaction protocols, and client software from which transactions originate. While most of the attention in e-commerce security has been focused on encryption technology and protocols for securing the data transaction, it is critical to note that a weakness in any one of the components that comprise an e-commerce system may result in a security breach. For example, a flaw in the Web server software may allow a criminal access to the complete transaction records of an online bank without forcing the criminal to break any cipher text at all. Similarly, vulnerabilities in security models for mobile code may allow insecure behavior to originate from client-side software interaction. Until the security issues of software-component based commerce are adequately addressed, electronic commerce will not reach mass market acceptance[1.2.3]. Assurance is a vital component in any security consideration. In effect, assurance measures the trust that the consumer can put in the claims of the producer. While the software producer may claim that the product is secure, the consumer and ultimate user of that software typically does not simply believe the developers claims, but wants additional assurance that those claims are true. This additional assurance is needed whether the software is purchased from a software vendor or developed "in-house" by the same company. Assurance is usually demonstrated by means of testing, a formal proof, expert review, or relevant process controls during development. Components are comprised of a number of objects assembled into a unified, meaningful, single module. In this way they resemble an entire system. They have well defined inputs, outputs, and functions. We anticipate that a subset of Common Criteria specifications will apply to components. We also anticipate that some modifications and additions will have to be made to some specifications in order for them to apply. The anticipated results from this paper are a process, set of core white-box and black-box testing technologies to certify the security of software components and a framework for constructing compositional Component Security Assurance (CSA) based on the security property exposed by the atomic components. The manifestation of the product is a stamp of approval in the form of a digital signature.

2 Related Works

2.1 CBD Process

Development of e-business systems involves collaborative work of several different types of specialist with different areas of expertise; for example, business process consultants, software architects, legacy specialists, graphic designers and server engineers. We'll need a coordinating framework for dealing with these diverse skill sets and introduce a track-based pattern to help. It's also important to have a good idea of the kinds of deliverable that we can expect to produce. We describe a broad set of deliverables that work well on CBD projects. Techniques can then be applied in flexible fashion within our overall process framework of track-based pattern plus deliverables. e-Business process improvement provides the right business context for CBD, as shown in figure 1. Of particular importance for transitioning to e-business using CBD are the overall e-business improvement plan, which provides business direction for architecture planning and the business models, which focus on understanding specific processes requiring e-business solutions. While the business improvement plan ideally encompasses the entire enterprise, the overall vision may be developed incrementally, leading to a succession of more narrowly focused action plans[3,4,5,6].

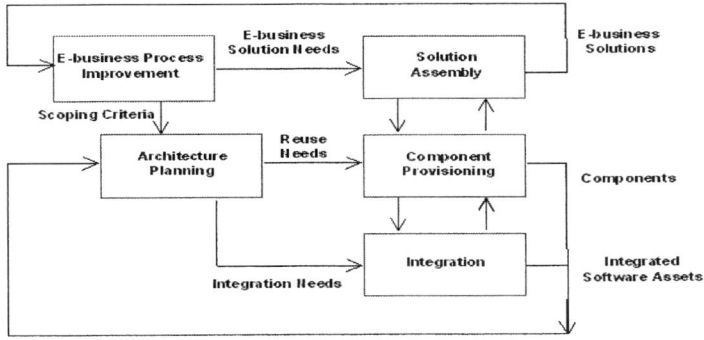

Fig. 1. Component Based Development Process

2.2 Connection and Adaptation

Once the decision to reuse a certain component is made, it will have to be configured within a Component configuration consists of connection and adaptation. Components have to be connected to each other so that they can cooperate. In the simplest case, the connector is just a link between a given required service and a given provided service. In other words, a connector establishes how a requirement is fulfilled. But connectors can be more complex; it is useful to have them encapsulate functionality that logically belongs within a shared infrastructure (for example, communication protocols in a distributed system) rather than to either of the two components that are being connected [13]. Adaptation increases the value of components. The more flexible and adaptable a component is, the more often it will be reused. Ideally, a component will provide ways for application composers to adapt it. However, a component manufacturer will not be able to foresee all adaptations that might be necessary. For this reason, there should be means to adapt a component externally without having to interact with it, for example wrapping.

2.3 Security of Information Technology Products and Systems

In component based software engineering, the security issue becomes more prevalent. Many components of a target software system to be assembled may be acquired from or delegated to third parties. The security properties of each component will be part of and impact on the target system's security. In such a scenario, we must know the security characteristics of the components to be able to evaluate the assembled system. Another equally important aspect that impacts on the target system's security is the system architecture that connects the components in a specific manner. In addressing the issue of security characterization of software components and component-based systems, we propose to identify and measure the security characteristics of a software component through the use, adaptation and formalization of the Common Criteria, and analysis and evaluate the security properties of a composed system in terms of the characteristics of its components and its system architecture. The Common Criteria identify the various security requirements for IT products and systems, and provide a good starting point for characterizing software components, i.e., with the components being regarded as IT products/systems. However, the Common Criteria do not directly address system composition, and therefore much investigation is required to evaluate a composed system based on the component characteristics and the system architecture[8,9,10].

3 Frameworks for Assuring Security in Components

The system architecture developed and used through this paper, CSA(Component Security Assurance) shown in figure 3. It provides a schematic representation of the some inter models such as component security requirements model, assuring security model and modeling component security assurance. It shows the relationship between e-business domain and security model-based framework. Component domain requirements, security requirements model and CBD workbench/security architecture model apply in this paper context in any of the delivery tracks. However, as we'll see a little later, de-scoping may occur at regular points through the lifecycle, resulting in hybrid projects. So, for example, a solution assembly project could branch into separate smaller assembly, provisioning and integration projects. Also software requirements techniques may also be used within a e-business domain services. They are especially useful in conjunction with prototyping, as a means of scouting ahead to explore different designs.

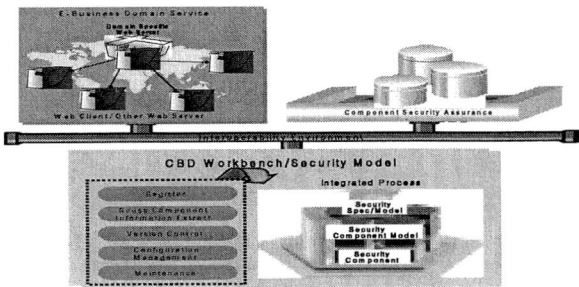

Fig. 2. Component Security Assurance System Architectures

3.1 Component Security Requirement Model

One of the most important aspects of any security architecture model is the ability to manage/maintain an accurate and consistent level of component security controls. An integrated risk management program is critical in securing business objectives requiring the enforcement of confidentiality, integrity, availability, and accountability. Confidentiality ensures the protection of component from unauthorized access throughout an organization's information architecture, which extends to all component directly associated with the architecture's applications, component stores, communication links and/or processes. Integrity ensures that component, services, and other controlled resources are not altered and/or destroyed in an unauthorized manner. Integrity based controls provide safeguards against accidental, unauthorized, or malicious actions that could result in the alteration of security protection mechanisms, security classification levels, addressing or routing information, and/or audit information. Because of increasing information sharing and the cost of securing information, it is important to classify information correctly. Under-classification of sensitive information can have serious consequences. The security requirements for each classification level are defined, based on the organization's particular application of the technology. The model includes categories for security, and the criteria by which the information is to be secured. Categories relate to both business processes (applications and component) and information technologies (hardware, software and services) that support that environment. The categories define a comprehensive structure for IT risk management issues and represent the components of the information systems environment.

3.2 Assuring Security Model in Component

The proposed approach for assuring security in components is illustrated by the Component Security Assurance (CSA) dimension as in figure 4. The CSA dimension is an architecture for providing security-oriented testing processes to a software component. The dimension consists of several processes including the construction of test plans, analysis using white- box testing techniques, black-box testing techniques, and the stamping with a digital signature of the relative security rating based on the metrics evaluated through the testing. The processes are broken out into sub-pipes of test plans, white-box testing, and black-box testing. The first stage to component certification is the development of a test plan. The application in which the JavaBean component will be used will influence the security policy, test suites, assertions, and fault perturbations used in both white-box and black-box testing processes. Based on the security policy, input generation will make use of test suites delivered from the applicant for certification as well as malicious inputs d signed to violate the security policy. The definition of the security policy is used to code security assertions that dynamically monitor component vulnerability during security analyses. Finally, perturbation classes are generated for white-box fault injection analysis according to the application in which the component will be used.

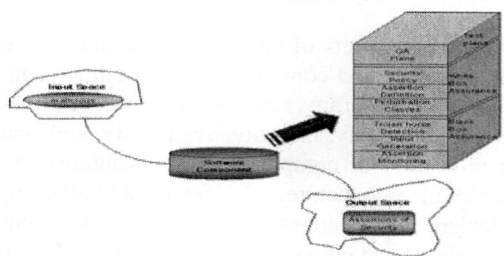

Fig. 3. Component Security Assuring Dimension

For the certification process, the property that is analyzed is the secure behavior for software components. One example of using assertions internally to the code is to determine if a user is granted access to a privileged resource when the use has not been authenticated. Assertions can also be employed external to the program to monitor system-wide properties. For example, an assertion can be used to determine if a portion of the file system has been accessed by an unprivileged process. And coverage analysis, property-based testing of the software component can be performed until a required degree of confidence is reached when the component will be certified. The degree of confidence necessary will be determined by the application in which the component will be employed.

3.3 System-Level Component Security Analysis

As stated earlier, composing secure components in a system does not guarantee secure system behavior. Conventional engineering of large systems follows the doctrine of "divide and conquer." That is, large systems are broken into smaller subsystems and each is individually engineered. Component-based software is aimed at building systems from the ground up from software components. One consequence of building large systems from components is the loss of system-wide robustness properties such as security due to the increase in the number of components that must be maintained and the number of interfaces that must be robust. A component designed and built for one application might behave remarkably different when employed in a different application. Even when component interfaces match (which is a difficult enough problem without universal acceptance of component standards); the system-wide behavior of components hooked together is as unpredictable as strange bedfellows. Unintended interactions between components can result in emergent system behavior that is unpredictable and possibly insecure.

3.5 Security Characteristics of Software Components

Since a software component can be regarded as an IT product or system, it is natural to use the Common Criteria in assessing its security properties. The Common Criteria provide a framework for evaluating IT systems, and enumerate the specific security requirements for such systems. The security requirements are divided into two categories: security functional requirements and security assurance requirement. The security functional requirements describe the desired security behavior or functions expected of an IT system to counter threats in the system's operating environment. These requirements are classified according to the security issues they address, and with varied levels of security strength. They include requirements in the following classes: security audit, communication, cryptographic support, user data protection, identification and authentication, security management, privacy, protection of system security functions (security meta-data), resource utilization, system access, and trusted path/channels. The security assurance requirements mainly concern the development and operating process of the IT system, with the view that a more defined and rigorous process delivers higher confidence in the system's security behavior and operation. These requirements are classified according to the process issues they address, and with varied levels of security strength. The process issues include: life cycle support, configuration management, development, tests, vulnerability assessment, guidance documents, delivery and operation, and assurance maintenance. The Common Criteria have also identified seven evaluation assurance levels by including assurance requirements of appropriate strength into each of these levels.

3.6 Security Properties of Component-Based Software Systems

The security characterization of a software system assembled from components should take a form similar to that of a component. After all, the composed system is an IT system and may be used as a component of another larger system. As such, the security characterization of the target system could be done in a way similar to that of an atomic component. Given that the security properties of the components used are already available. However, it is natural and advisable to use these component properties together with the system's composition architecture and process to arrive at the composed system's security characterization. It is even more so in cases where detailed analysis of third party components are not possible due to the lack of development information. As such, the security properties of a component-based system should be derived from those of the components used and the system architecture. Assuming the component properties are characterized and defined as outline above, we have to consider the ways of interaction between these components according to the system architecture and how these interactions impact on the components and the composed system. Therefore, we need a component-based and architecture-directed composition model for software security. While the Common Criteria do not directly address system composition issues, the security concerns they address do suggest that the software security composition model be based on the aspects such as the security properties of individual components, the system architecture of the target system, and the process of architecture design and system composition.

4 Modeling Components Security Assurance

Components are often under specified, which makes their proper reuse a risk in the development process. To remove this shortcoming a more precise specification is needed. Interfaces as we know them from object oriented programming provide a so called functional specification of the component. But there are non-functional issues which have to be specified. Prominent examples for non-functional aspects of a component are performance and security. For instance, a component customer may be interested in knowing the time complexity of a component's computation. A component user may want to know whether the component encrypts the component that it sends over the network. Along the same vein, knowledge of bandwidth and latency property of a component is an important issue for component deployment. We will make the requirements of a component explicit so that the developer can see as early as possible what it takes to get the shop offers an interface for ordering goods. But most likely the component needs some other component that offers a database interface so that the e-shop can store the customer component. The designer should be able to deduce easily from the component's specification whether the e-shop component has a dependency on the database component. Most currently used component models do not make these dependencies explicit. Component users have to search for this information in the written documentation (when this information is supplied at all). Good component architectures should not only expose and specify the contracts they offer but also make explicit the contracts they require from others.

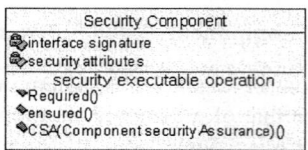

Fig. 4. Security Component Interface with UML

4.1 Compound Assurance

In more complex settings than those in the above example the relationships between required contracts and offered ones can become quite extensive, especially when level 4 (those with non functional properties) contracts are involved. When dealing with more complicated components we will discover that a set of contracts may have exclusive-or semantics. That means only one contract can be active at a certain time. These contracts often share common subparts. To make these modeling issues more explicit we introduced the concept of compound contracts. In our example the shopping contract is such a compound contract, which is a composition of other contracts.

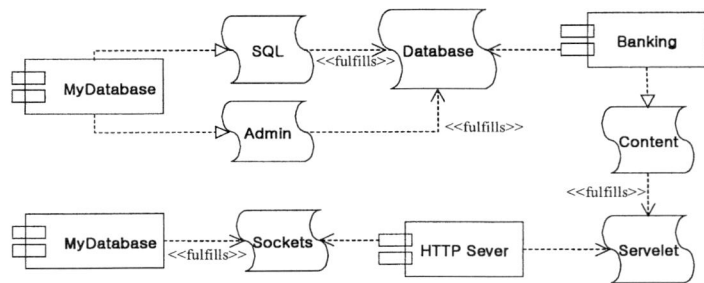

Fig. 5. A component assembly

A compound contract can play two different roles. It is either a useful is the combination of functional and non-functional contracts. For example some component demands a certain throughput for an SQL interface it is using. By grouping the interface and the non-functional contract in a compound contract we can express this relationship. The exclusive-or semantics can be expressed using UML stereotypes across several dashed lines. In UML this is already done with associations. For example you can model that an instance of some class may participate in one of several associations but not in two of them at the same time. We just applied this concept to contracts. Another reason for using compound contracts is that they enable very high level views of components and their contractual relationships. A real life component will offer a rich set of interfaces and non-functional contracts.

4.2 Contract Types and Contracts

In the previous sections we used the word "contract" with a somewhat vague meaning. However, when we want to model systems with components we need to be more precise. It is crucial to distinguish between contracts and contract types. Contracts and contract types are similar to the concept of objects and classes. Until now we have used the word "contract" where "contract type" would have been more appropriate. We use classes and contract types. Contracts are instantiations of contract types and exist at run time. Since the UML provides a means for modeling instances, this allows us to describe the over time. Therefore, we provide a means for modeling contracts in addition to contract types: contract entities coexist with component instances in deployment diagrams. As we will describe in more detail in the next section it is possible to parameterize contracts. While typical interfaces are not subject to parameterization during the design and implementation phase, quality of service contracts (level 4) are just templates that have to be filled at run time with concrete values. For example in a concrete application some component might not be able to ensure availability without knowing its run time environment and this environment's QoS. Another important issue in the development, distribution and application of software components is related to the integrity o the components, namely, how to ensure that any unauthorized modification of a software component, be it accidental or malicious, can be easily detected by a customer or another software component that depends on it. It applies to not only the implementation of the component functionality but also the characterization or interface definition of the component. The issue is especially important in dynamically configurable distributed software systems, where system components may be acquired or purchased on the Internet on a per-use basis. It has the same importance for software systems involving mobile agents.

5 Conclusion

Components are specifically designed to be combined into systems, and, in fact, it is these systems that ultimately need security assurance. Yet, composing security assurance specifications into specifications for larger systems is not only a non-trivial task, but considered one of the hard unsolved problems in computer security. when components are combined into systems, the specifications will be available and the characteristics of the combination systems can also be determined. This paper describes a new approach for certifying software components for security using both tried and true software engineering analysis techniques applied to security problems as well as novel security analysis techniques. The objective of this research is to invent a process methodology that can be used to certify the security of software components used in e-commerce applications. By providing a means for assessing the security of software components, the old practice of security through obscurity will no longer be a viable technique, and y-by-night software development organizations will not get away with selling supposedly secure products. Billions of dollars are spent on computer security, and most of it is wasted on insecure products. After all, weak security looks the same on the shelf as strong cryptography. Two e-mail encryption products may have almost the same user interface, yet one is secure while the other permits eavesdropping. A comparison chart may suggest that two programs have similar features, although one has gaping security holes that the other doesn't. An experienced cryptographer can tell the difference. This paper has demonstrated that an active interface can provide the basis for reasoning and assessing a component's suitability to meet certain security requirements of a particular application. From a security point of view, it is unrealistic to tell the component users or the system composers whether a software component is secure or not, rather it is much useful to expose what security properties are implemented. In a distributed environment, it would not be realistic to expect that all components would provide same degree of security to others. The proposed framework lets the human users and software components judge the trustworthy of a component by reasoning the security properties that it exposes. One of the secondary benefits of our framework is to separate the interface code from the application code of the component. This framework enforces a clear separation of concerns between the interface introspection and the application of the functionality. We conclude with our belief that a security characterization mechanism providing a full disclosure of security properties in both human and machine comprehensible terms could build a confidence and trust on a viable software component market.

References

[1] Jonathan Stephenson: "Web Services Architectures for Security," CBDi Journal, http://www.cbdiforum.com/, Feb. (2003)
[2] Mikio Aoyama: "New Age of Software Development : New Component-Based Software Engineering Changes the Way of Software Development," 1998 International Workshop on CBSE, ICSE, (1998) 124-128
[3] CBSE98. Proceedings of International Workshop on Component-Based software Engineering. www.sei.cmu.edu/cbs/ics98/, Kyoto Japan, April (1998)
[4] Peter Herzum, Oliver Sims: Business Component Factory : A Comprehensive Overview of CBD for the Enterprise, OMG press, December (1999)
[5] Clemens Szyperski: Component Software : Beyond Object-Oriented Programming, Addison-Wesley http://www.sei.cmu.edu/cbs/icse98/papers/p14.htm, January (1998)
[6] Desmond Francis D'Souza, Alan Cameron Wills: Objects, Components, and Frameworks With UML : The Catalysis Approach, Addison-Wesley Object, October (1998)
[7] "Information Technology-Software Life cycle Process, (ISO/IEC 12207)," http://standards.ieee.org/reading/ieee/std/, (1998)
[8] Monika Vetterling , Guido Wimmel and Alexander Wisspeintner: "Requirements analysis: Secure systems development based on the common criteria: the PalME project," Proceedings of the tenth ACM SIGSOFT symposium on Foundations of software engineering, Nov (2002) 129-138
[9] Robert C. Seacord: "Software Engineering Component Repository," Proceedings of 1999 International Workshop on CBSE, Los Angeles, at URL: http://www.sei.cmu.edu/ cbs/icse99/cbsewkshp.htm, (1999)
[10] Luqi, Jiang Guo: "Toward Automated Retrieval for a Software Component Repository," IEEE Conference and Workshop on Engineering of Computer-Based Systems, March (1999)
[11] Haeng-Kon Kim, Jung-Eun Cha, Ji-Young Kim, Eun-Ju Park: Identification of Design Patterns and Components for Network Management System_, SNPD '00 International Conference, Vol. 1, NO. 1, May (2000) 426-431
[12] D.D'Souza and A. Wills.: Objects, Components and Frameworks with UML:The Catalysis Approach. Addition-Wesley, (1998)

Security Framework to Verify the Low Level Implementation Codes

Haeng-Kon Kim[1] and Hae-Sool Yang[2]

[1] Department of Computer Information & Communication Engineering,
Catholic University of Daegu
hangkon@cu.ac.kr

[2] Graduate School of Venture, HoSeo Univ. Bae-Bang myon, A-San,
Chung-Nam,336-795, South Korea
hsyang@office.hoseo.ac.kr

Abstract. With the development of web-application, especially E-commerce, many software designers need to incorporate either low-level security functionalities into their programs. This involves the implementation of security features using Java Cryptography Architecture (JCA), Java Cryptography Extension (JCE) and Java Secure Socket Extension (JSSE) API provided by Sun Corporation [1]. Through our discovery, we find that many functional security related features in software systems are usually implemented by a few methods. The use of these methods results to some necessary structural patterns in reduced control flow graph of the program. In this papers, we present our way to recover the security features by recognizing these methods invocations automatically and transform the reduced control flow graph to state transition diagram through functional abstractions. We believe that it would not only facilitate the comprehension of the security framework implemented in the program, but also make the further verification of the security features possible.

Keywords: Slicing, Security features, Reduced Control Flow Graph, State Transition Diagram.

1 Introduction

Program slicing is a fundamental operation for many software engineering tools, including tools for program understanding, debugging, maintenance, testing, and integration [2,3]. Slicing was first defined by Mark Weiser,who gave algorithms for computing both intra- and interprocedural slices [4,5]. However, two aspects of Weiser's interprocedural-slicing algorithm can cause it to include "extra" program components in a slice: (1) A procedure call is treated like a multiple assignment statement "$v_1, v_2, \ldots, v_n := x_1, x_2, \ldots, x_m$", where the v_i are the set of variables that might be modified by the call, and the x_j are the set of variables that might be used by the call. Thus, the value of every v_i after the call is assumed to depend on the value of every x_j before the call. This may lead to an overly conservative slice. (2) Whenever a procedure P is included in a slice, *all* calls to P (as well as the computations of the actual parameters) are included in the slice. Design recovery plays very important role in the program comprehension. Along with the development of E-commerce applications, many software designers need to incorporate either low-level or high-level security functionalities into their programs, so it is useful to automatically recover and verify security features from program source code. In this paper, firstly we give the definitions of the Security-Related Variables and Security-Related methods. Then we describe our ways to correctly set the slicing criterions regarding two different groups of Security-Related variables. In order to include all the statements involving the invocation of Security-Related Methods, which are important to show the full scope of the security features in the program, we propose two ways to solve this problem, including pre-parsing process and modifications of the DEF set [2]. We slice the program to get the necessary statements, which directly or indirectly affects the implementation of Security features in the given program. And with the state's definition, locating and forming criterion provided, we turn the RCFG (Reduced-Control-Flow-Graph) to State transition diagram to represent the security features in the program [2,3,4].

2 Constructing Security-Based Reduced Control Flow Graph from Low Level Codes

Program slicing is widely used in program comprehension, debugging, testing, maintenance [5] and integration of program version (e.g., [2,3,4]). A program slice of a program P with respect to a slicing criterion <p,V>, for a p location in P and V a set of variables in P referenced at p, is the set of statements and predicates in P that might affect the value of variables in V at p[5]. In order to make the slicing applicable to large software systems, much research has been done to develop the interprocedural slicing techniques based on the system dependence graph (SDG).Various researchers have put off the way for constructing the system dependence graph (SDG) for a given Java program, this includes [6,7,8]. The Horwitz-Reps-Binkley algorithm operates on a program representation called the system dependence graph (SDG). The algorithm involves two steps: first, the SDG is augmented with summary edges, which represent transitive dependences due to procedure calls; second, one or more slices are computed using the augmented SDG. The two steps of the algorithm (as well as the construction of the SDG) require time polynomial in the size of the program. The cost of the first step—computing summary edges—dominates the cost of the second step.

2.1 Setting Security-Based Slicing Criterions

In order to set the correct slicing criterions, we'd like to give the definition of Security-Related Methods and Security-Related Variables first.

(1) Definition (Security-Related Methods)
Let Avar be list of parameters of method f, c is a class, $m_1, m_2....mn$ are packages defined in JCA ,JCE and JSSE, $C(m_1, m_2....mn)$ be set of classes defined in the $m_1, m_2....mn$.
　　Security-Related Methods in the set
　　Fsrm={ fl (c∈C($m_1, m_2....mn$) ∪ IsMethod(f(Aval, Avar),C($m_1, m_2....mn$)))}.

(2) Definition (Security-Related Variables)
Let Avar be list of parameters of method f, c is a class, m_1, m_2....mn are packages defined in JCA ,JCE and JSSE, $C(m_1, m_2....mn)$ be set of classes defined in the m_1, m_2....mn.

Let instance (v,c): variable v is the instance of class c. and IsMethod(f(Avar), $C(m_1, m_2....mn))$: f(Avar) is a method in a calss in $C(m_1, m_2....mn)$.

Security-Related Variables in the set
Varv={vl (c∈$C(m_1, m_2....mn)$ ∪ ((instance(v,c))) ∪ ((v∈Avar) ∪ IsMethod(f(Aval, Avar), $C(m_1, m_2....mn))))$.

As what is stated in the definition, the set of Security-Related Variables contains the variables that directly or indirectly affect the implementations of the security features in the programs. By saying that a variable directly affecting the implementation of security features in the program, we mean that this variable is the instance of the classes defined in JCA, JCE and JSSE [1], which are used as the basic computational unit for constructing the security framework. Thus, the statements related with these variables are considered as basic structure of the security framework implemented in the programs. The variables, which affect the implementation of security features indirectly, are the instances of the classes (exclude the classes defined in JCA, JCE and JSSE), used as the parameters of the Security-Related Methods. We need to treat these two groups of Security-Related Variables differently. Slicing criterion <p,V> comprises two parts, a p a location in P and V a set of variables. As for the first group of Security-Related Variables, V is the subset of Varv, which includes all the Security-Related Variables that belong to the first group and p is the end point of the program. However, for the other group of Security-Related Variables, the criteria are much more difficult to set. Firstly we need to locate all the statements, which involve the invocations of the Security-Related Methods. Then for each of these statements s, its program point p and the parameters set V used in methods invocations together form the slicing criterions for s, that is <p,V>.

2.2 Security-Based Slicing

2.2.1 Pre-processing

After we get the slicing criterions based on the Security-Related Variables, it is a set of criterions regarding all the Security-Related Variables in the program. According to the slicing algorithm given by Horwitz, we say that the final slices are the union of every slice based on the separate slicing criteria. Since the setting of our slicing criterions for second group of Security-Related Variables is based on the program points of Security-Related Methods invocation statements, more often than less, there might be more than one Security-Related Methods invocation statements concerning the same Security-Related Variable. Before applying the SDG-based slicing techniques on the given program, we need to find the Optimized-slicing criterion for all Security-Related Variables respectively in order to reduce the redundant work during the slicing phase.

(3) Definition (Optimized-Slicing Criteria)
Let s be the slicing criteria, S be a set of slicing criterions, predicate Is(s,<p,V>) means that s is the slicing criteria <p,V>.
if ($\exists s \in S$) ($\forall s' \in S$)(Is(s,<p1,V>))(Is(s',<p2,V>))(p2<=p1)
we say s is the Optimized-slicing criteria for V in the program.

Note here, by saying p2<=p1, we mean program point p2 appears later than p1 in sequence of the statements. In the example above, the slicing criterions for the first group of Security-Related Variables remain the same, while for the second group, <s33,alias> is the Optimized-slicing criteria for String variable alias, so <s32,alias> is excluded in the slicing criterions set. Thus after we find all the Optimized-slicing criterions for all the Security-Related Variables in the program, SDG-based slicing techniques is ready to be applied.

2.2.2 Slicing

As what we have described above, our recovery of Security features in the given program involves two aspects of concerns. Firstly, in the traditional static slicing view, the statements, which affect the value of Security-Related Variables in the program, will be picked out with the existing slicing techniques. Secondly, some of the statements which involve the invocations of the Security-Related methods through which the security functionalities are implemented, though itself may not have any impact on the value changing of Security-Related Variables. The other method is to expand the DEF sets of them to make them included after the SDG-based slicing process. Here, we'd like to describe them respectively.

(1) Locate and mark the statements related with Security-Related Methods invocation compared with the classical static slicing. This step is mainly a parsing process of the Java program. Firstly, we clearly set the phase-structure grammar rule for methods in vacation in Java language. Then according to this parsing rule, we go through the program to locate all the methods invocation statements and mark them. With the symbol table constructed for this program, we can pick out the methods invocation statements regarding the variables we care, which belong to the first group of Security-Related Variables. Here, besides the symbol table, Static Security-Related Methods table is also needed, for the reason that the static methods invocation is different from the ordinary methods invocation in the since that they can be called without constructing the specific instance of the class. Thus we cannot recognize them only with the symbol table information provided. In order to correctly find these statements, we build the Static Security-Related Methods table, which contains the name, parameter types and the return value type of the static methods defined in JCA, JCE and JSSE. If the methods invocation statements inside the program match the items in this table, then they are the Static Security-Related methods invocation statements, which should also be picked out. So the final marked statements set contains two kinds of Security-Related methods invocation statements.

(2) Expanding DEF set. The other approached is to expand the DEF set of the statements involved with Security-Related Methods invocation. This is done in the stage of constructing the SDG. Expanding rule here is to add the instance of the first group of Security-Related classes whose defined method is invoked in the statement.

For example, like statement s11, s.update(smsg.msg); its DEF(s11) is empty, so it is not included when we slice with the criteria <end,s>. With the expanding rule we have described above, since s is the instance of Class Signature, which belongs to the first group of Security-Related Variables, and its defined method update () is invoked in this statement, we add variable s to the DEF set of this statement. Then DEF (s11) would be {s}. With this new computation of DEF and REF schema, this statement will be kept in the final slice with the slicing criteria <end,s>. Expanding rule here is only for the purpose of program comprehension in this approach, and whether it is useful for some other approaches of recovering other features of a given program still remain unproved.

2.3 Security-Based Reduced Control Flow Graph

After we get the correct slice of the given program, Control Flow Graph (CFG) is used to represent it. A Control flow graph G=(V,vs,vf,A) is a directed graph where V is a finite set of vertices; vs ∈V is the start vertex; vf ∈V is the final vertex; and A is a finite set of arcs. A vertex represents a statement and an arc represent possible flow of control between statements. Thus the Reduced Control Flow Graph of the example program is shown as Figure 1. Here, we use the vertices to connect the calling and called statements to show the interprocedural relations in the programs. So the final graph shows the whole sequences of Security-Related statements that implemented in the program, as what you can see in Figure 2. After we get the RCFG of the program, every node in the CFG represents a single statement from the program; we need to give the more abstract and semantic representations for better program comprehension. With this purpose, we turn to the state transition diagram.

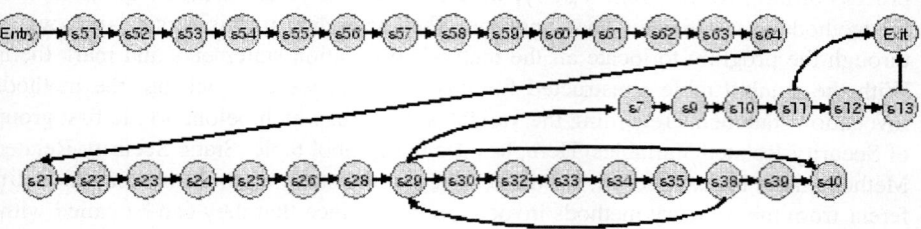

Fig. 1. RCFG of the example program

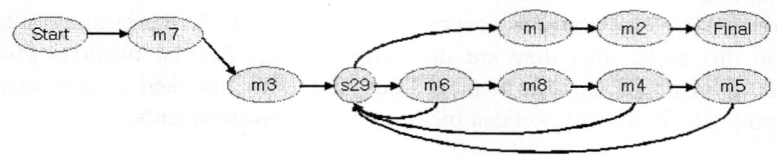

Fig. 2. State transition diagram representation

3 Transformation Security-Based RCFG into State Translation Diagram

3.1 Identifying and Constructing State

Here, we say a state is a group of adjacent statements in RCFG, which are used to fulfill the same functionality in the design recovery view. Thus every state here is labeled by Security-Related functionality type f and the Security-Related Variable v it targets to. In the appendix C table 1, we list some major types of the states used in our approach. For example, by saying that a state m1 is labeled {f=f1, v=p}, it means that state m1 is the type of " Setting Signature to be verified" and its target variable is p, the instance of Class Signature. In order to correctly form the group of adjacent statements into state respectively, the first step is how to locate the state in the RCFG. Since every state is labeled with f and v, f represents the type of the security functionality implemented in this state. Through our discovery, we find that these security functionalities are usually implemented by a few methods invocation statements. The use of these methods results to some necessary structural patterns in reduced control flow graph of the program. By recognizing these methods invocation statements automatically, we can locate the state in RCFG, and then based on the data dependence information to group the adjacent statements to form the state correctly. Here we call these methods invocation statements focus statements, for the reason that they are adequate to show the security functionality implemented in the state, and guide us to locate the state. In the table 2, focus statements of the states defined are also listed. As what shows in the appendix, the focus statements of most states is a single statement of the invocation of Security-Related method defined in JCA, JCE and JSSE. For example, as for the state type "Verify Signature", the focus statement is the invocation of method verify (byte signature) defined in Signature class, which is implemented to verify the pass-in signature. And the focus statements for some states can be sequence of Security-Related methods invocation statements in the sense that all these statements together form the Security functionality implemented in the state. For example, like the state type "Setting the Signature to be verified", the focus statements of it are initVerify(Publickey pub) and update(byte data) defined in Signature class. Because initVerify() is designed to initiate the object for verification and update() is to update the information to be verified, these two methods invocations together make the Signature instance ready to be verified. In appendix C, table 2 , we have listed the focus statements of the major types of state we use in this paper. After we locate the state by mapping the focus statements of it with the actual statements in the RCFG, the next step here is to form the correct state with the focus statements and the adjacent statements which both have the data dependence relations with them and are used to do the preliminary work before the invocation of the focus statement. For example, in the RCFG, as for the state M1{f=f1,v=IsInstance(Signature,s)}, firstly we locate it by s10 and s11 to match with the state type f1, "Setting Signature to be verified". Then based on the data dependence relations, s7 is used to define variable pubKey which acts as the parameters in s10, and the DEF set of s9 is {s}, the instance of class Signature, which is the target variable of state M1. Thus these two statements, along with the focus statements s10, s11, form the state M1. For state M3{f=f3,v=IsInstance(KeyStore ks)} in RCFG, where f3 is the state type "load KeyStore", since its focus statement is the invocation of the method load() defined in KeyStore class, we can locate the state with the statement s28. And the adjacent statements s21,s22,s24,s25,s26 are used to define the variable ks, ks_path and spass which are used as either the pass-in parameters or the target variable of the focus statement s28 of M3. So statement s21,s22,s24,s25,s26,s28, altogether form the state M3. With the methods described above, we can correctly locate and form the states in the RCFG. For the example program, the states and its formation statements are listed in table 1.

Table 1. States of example program

State	Description	Statements contained	State	Description	Statements contained
M1	{ f=f1,v=s}	s7,s9,s10,s11	M5	{f=f5,v=cert}	s39,s40
M2	{f=f2,v=s}	s12	M6	{f=f6,v=ks}	s30,s32
M3	{f=f3,v=ks}	s21,s22,s24, s25, s26,s28, s61,s62,s63, s64	M7	{f=f8,v=sslFact}	s51,s52,s53,s54,s55,s56 ,s57,s58,s59,s60
M4	{f=f4,v=x509cert}	s33,s34,s35,s38	M8	{f=f7,v=ks}	s33

Table 2. Focus statements of States

State Type	Focus statements	State Type	Focus Statements
Setting Signature to be verified—f1	s.initVerify() s.update()	Verify Signature—f2	s.verify()
Load KeyStore instance—f3	ks.load()	Find Certificate issuer—f4	P=x509cert.getSubjectDN() X509cert.getIssuer().equals(p)
Verify Certificate—f5	cert.verify()	Check certificate Entry—f6	ks.isCertificateEntry()
SSL Connection—f8	sslFact.CreateSocket()	Get Certificate—f7	ks.getCertificate()
Set Signature Before Signing—f9	s.initSign(); s.update();	Sign message—f10	s.sign()
Set KeypairGenerator—f11	KeyGen.initialize()	Generate Asymmetric Key—f12	KeyGen.generateKeyPair()
Set Message Digest—f13	Sha.digest()	Update Message digest input—f14	Sha.update()
Check Key Entry—f15	Ks.isKeyEntry()	Get Certificate Chain—f16	Ks.getCertificateChain()

3.2 Constructing STD (State Transition Diagram)

With all the states we get in the previous step, we use the state transition diagram to represent the state transitions of the Security-Related states of the program. And Kripke structure is used. Here, we use Kripke structure M=(Q,qint,L,R) where Q is a finite set of states; qint $\in$ Q is the initial state; L: Q→2AP is the function labeling each state with a subset of the set AP of atomic propositions; R$\subseteq$ Q X Q is the transition relation which is total, i.e., for every state m, there is a state m' such that (m,m')$\in$ R. And L(ms)={f=fs,v=∅}, L(mf)={f=ff,v=∅}, where fs is the state type "Start" and ff is the state type "Final" respectively. And L(mi)={f=fi,v=vi}for every mi ∈ M-{ms,mf}. The tuple (mf,mf) is necessary to guarantee that the transition relation be total. For the example program, the final state transition diagram representation is shown as Fig.3.

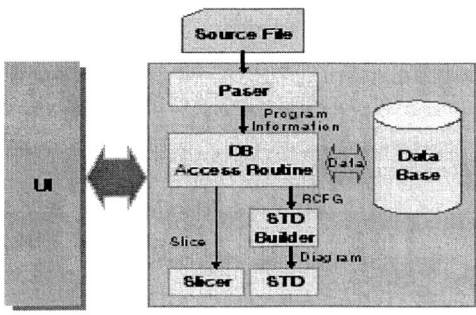

Fig. 3. System Architecture of JSR

4 Implementation

With the purpose of implementing the above approach, a prototype system Java SecurityRecover (JSR) is under development. Here, we'd like to give a brief description of the general architecture of it. We need to adopt a data-centered architecture. A database system will be the main part of this architecture, which is used to store the control flow and data flow information about the program, also can provide with the reuse information for the analyzing purpose. The system architecture is shown in Figure 3 and the system has the following subsystems:

Parser: Parse source code, gather basic analysis information about that source code, and store the information about all the statements in the database. It includes control-flow, local data-flow and symbol table information of a given program. This is done by using JavaCC (Java Compiler Compiler).

Slicer: Based on the information stored in the database, slice will perform the slicing actions based on the given slicing criterions, for our case, they are Security-related variables. The first step is to decide which variables are considered as Security-related ones. This can be achieved by getting the information about the symbol table and methods calling statements. After all the Security-related variables are found, it will do the slicing based on these variables, it needs to do the data-flow and control-dependence analysis. By extracting all the information about the Security-Related-Variables, the output will be a partial program, which is not necessarily to be executable but with all the security features inside the given program. As for the future process, the output result of the slicing will also be stored in the database. Currently, for this part, our work is mainly based on the Java Model Checker, Bandera [13]. Some modifications are carried out in order to include the statements involving the Security-Related methods invocation.

Information Database: This is the center part of the system, which is used to store the parsing and slicing result of the given program and also provide with the information needed for the analyzing. Database access routines are designed to provide the access to the Information Database.

Database Access Routines: it is used to manage the access to the database.

User Graphic Interface: It is used to provide the user with a friendlier interface for accepting the user input and showing the analysis result to the user.

STD Builder: It is used to build the security-related state transition diagram of the program. It will use the slice stored in Information Database to construct the diagram. And to find the basic states described above to find the general security structures the program implemented.

5 Conclusions

We have described our approach to recover the security features implemented in the given Java program. And proposed our way to construct the RCFG (Reduced Control Flow Graph) from Java source code for the statements that directly and indirectly affect the implementation of the security features. We also described our way to set the different slicing criterions for different groups of Security-Related Variables and how to include all the statements, which show the full-scope of security implementation. State transition diagram representation is applied in order to give the more compact and semantically meaningful means to show the security functionalities and its sequence implemented in the given program. Finally, we also briefly describe the prototype system under developed to implement the approaches described.

Based on the approaches above, two questions have been raised. They are, 1)"Can this approach be extended to check some properties?" Till now, the answer to this question seems to be not so clear in the sense that it depends on what kind of properties we are supposed to check and how important these are. For example, we have once figured out some empirical-based properties concerning the correct implementation of the security features in the java program. Like the property " Verify the certificate before verifying the signature of a signed message". In real application, this is a securer way, compared with purely verifying the signature. And this can be checked by querying on the state transition diagram obtained by our approach. Once enough properties are got, we will consider furthering the current approach to new stage. 2) "Can the current approach be applied in some other fields besides Security-Related one?"

Till now, we think the answer is yes. If only for the recovering concerns, as long as we can categorize certain packages, which are used for the same or related aspect, this approach is applicable. Further research will be done concerning these two questions.

Acknowledgments

This research was supported by the MIC(Ministry of Information and Communication), Korea, under the ITRC(Information Technology Research Center) support program supervised by the IITA(Institute of Information Technology Assessment).

References

[1] Sun Microsystems, "The Java Enterprise JavaBeans," 2nd Edition, Wisley, 2002
[2] G. Antoniol, R. Fiutem, G. Lutteri, P. Tonella, S. Zanfei, and E. Merlo. "Program understanding and maintenance with the CANTO environment," In International Conference on Software Maintenance, pp. 72–81, 1997.
[3] T. Ball and S. G. Eick, "Visualizing program slices," In IEEE Symposium on Visual Languages, pp. 288–295, 1994.
[4] F. Balmas. "Displaying dependence graphs: a hierarchical approach," In Proc. Eighth Working Conference on Reverse Engineering, pp. 261–270, 2001.
[5] Y. Deng, S. Kothari, and Y. Namara, "Program slice browser," In Ninth International Workshop on Program Comprehension (IWPC'01), pp. 50–59, 2001.
[6] J. Krinke, "Evaluating context-sensitive slicing and chopping," In International Conference on Software Maintenance, pp.22–31, 2002.
[7] Zhengqiang Chen and Baowen Xu, "Slicing Object-Oriented Java Programs." ACM SIGPLAN Notices, Vol.36 No.4, April 2001
[8] Corbett J, Dwyer M, Hatcliff J, Laubach S, Pasareanu C, Robby, Zheng H. Bandera, "Extracting finite-state models
from Java," In Proceedings of 22nd international conference on software Engineering. Limerick, Ireland, June 2000, ACM Press
[9] M. Weiser, "Program slicing," IEEE Transaction on Software Engineering, Vol.10 No.4, pp.352–357, July 1984.
[10] B.Joy, G. Steele, J.Godling, and G. Bracha, The Java Language Specification, Addison-Wesley, available from http://java.sun.com/docs/books/jls/index.html
[11] S.Horwitz, T.Reps, and D. Binkley, "Interprocedural slicing using dependence graphs," ACM Transaction on Programming Languages and Systems, Vol.12 No.1, pp.26-60, Jan., 1990
[12] L. Larsen and M. Harrold Slicing object oriented software. In 18th International Conference on Software Engineering, pages 495–505, March 1996.
[13] D. Liang and M. Harrold. Slicing objects using system dependence graphs. International Conference on Software Maintenance, pages 358–367, November 1998.
[14] J. Zhao. Applying program dependence analysis to java software. In Proc. Workshop on Software Engineering and Database Systems, pages 162–169, Taiwan, December 1998.
[15] P.Tonella, et al. Flow in-sensitive c++ pointers and polymorphism analysis and its application to slicing. In 19th International Conference on Software Engineering, pages 433-443, May 1997

A Study on Evaluation of Component Metric Suites

Haeng-Kon Kim

Department of Computer Information & Communication Engineering,
Catholic University of Daegu, Kyungbuk,
712-702, South Korea
hangkon@cu.ac.kr

Abstract. The requirement to improve software productivity and software quality has promoted the research on software metrics technology. Traditional metrics cannot be used for the object oriented paradigm because there are no metrics for the concepts like encapsulation, inheritance, and coupling so Various object-oriented metrics have been proposed by various researchers. Two of the widely accepted metrics are CK and MOOD Metrics. They have been analyzed according to their validation criteria and it has been observed that CK suite which was build on the validation criteria given by Weyukar fail to satisfy it completely. MOOD metrics on the other hand fail to satisfy the validation criteria given by the MOOD team themselves thus showing that MOOD Metrics is working with an inaccurate and imprecise understanding of the OO paradigm. Hence showing that the genesis of the metrics is controversial. Further many inconsistencies have been observed in CK and MOOD Metrics.

Keywords: Object Oriented Paradigm, CK Suite, MOOD Metrics, Coupling, Cohesion, Polymorphism, Component Based Development.

1 Introduction

Measurement is the process by which numbers or symbols are assigned to attribute of entities in the real world in such a way so as to describe them according to clearly defined rules. Measurement is very important in software industry because

- Software metrics can help to fully understand both the design and architecture information of the software system.
- Software design metrics can aid to discover the underlying errors in the software design at the early stage of software development life circle.
- Software metrics can evaluate the quality of the software and provide cost estimation of software project.

Object oriented design and development is becoming very popular in today's software development environment. So, to analyze object oriented system we need different set of metrics, which are different from traditional metrics because Object-oriented programming identifies the object types in the applications and structured programming models the applications as a set of functions. Also, there are no metrics for the concepts like encapsulation, inheritance etc. It is because of all these limitations a set of new software metrics adapted to the characteristics of object technology were purposed; CK and MOOD Metrics being one of them.

2 Chidamber and Kemerer Metrics

The best-known OO metrics are the suite of six proposed by MIT researchers. Various fundamental flaws and inconsistencies have been observed in them as shown under.

2.1 Validation Criteria for CK Metrics

The Metrics have been validated against Weyuker's axioms. Weyuker is used to evaluate software complexity measures. CK has evaluated its metrics according to Weyuker, which means that CK is considering its metrics to be complexity metrics and at the same time CK has not defined complexity and it seems that it has got different definition of complexity for each metrics, which is not even clear[12].

2.1.1 Weyuker's Properties

Of Weyuker's nine properties, three will be considered briefly here [9]. **Weyuker's second property, "granularity"**, only requires that there be a finite number of cases having the same metric value. Since the universe of discourse deals with at most a finite set of applications, each of which has a finite number of componentes, this property will be met by any metric measured at the component level.

The "**renaming property**" which is **Property eighth** requires that when the name of the measured entity changes, the metric should remain unchanged.

Weyuker's seventh property requires that permutation of elements within the item being measured can change the metric value. The intent is to ensure that metric values change due to permutation of program statements. The remaining six properties are mentioned below:

Property 1: Non-coarseness
Given a component P and a metric μ another component Q can always be found such that: $\mu(P) \neq \mu(Q)$. This implies that not every component can have the same value for a metric; otherwise it will lose its value as a measurement.

Property 2: Non-uniqueness (notion of equivalence)
There can exist distinct componentes P and Q such that $\mu(P) = \mu(Q)$. This implies that two componentes can have the same metric value, i.e., the two componentes are equally complex.

Property 3: Design details are important
Given two component designs, P and Q, which provide the same functionality, do not imply that $\mu(P) = \mu(Q)$. The specifics of the component must influence the metric value. The intuition behind Property 3 is that even though two component designs perform the same function, the details of the design matter, in determining the metric for the component.

Property 4: Monotonicity
For all componentes P and Q, the following must hold:
$\mu(P) \leq \mu(P+Q)$ and $\mu(Q) \leq \mu(P+Q)$ where P + Q implies combination of P and Q. This implies that the metric for the combination of two componentes can never be less than the metric for either of the component componentes.

Property 5: Non-equivalence of interaction
$\exists$ P, $\exists$ Q, $\exists$ R, such that
$\mu(P) = \mu(Q)$ does not imply that $\mu(P+R) = \mu(Q+R)$.
This suggests that interaction between P and R can be different than interaction between Q and R resulting in different complexity values for P+R and Q+R.

Property 6: Interaction increases complexity
$\exists$ P and $\exists$ Q such that:
$\mu(P) + \mu(Q) < \mu(P+Q)$
The principle behind this property is that when two componentes are combined, the interaction between componentes can increase the complexity metric value.

2.2 Inconsistencies in Chidamber and Kemerer Metrics

The CK metrics were validated by using six of Weyuker's nine axioms, and were found to be generally compliant with most of the properties [4]. None of the metrics was found to comply with either of properties 6 and 7 of Weyuker and DIT and LCOM fail to satisfy property four about monotonicity. Failure to meet the sixth property suggests that it is probably not applicable to OO systems, where interaction might in fact decrease complexity by rendering componentes closer to the abstractions they are supposed to portray. According to CK, failure to meet the seventh property suggests that permutation might not be significant in OO systems. Besides its non-compliance to Weyuker's axioms an inadequacy has also been observed in CK Suite that it has got no metrics to measure encapsulation and polymorphism and hence its inability to measure the entire object oriented properties.

2.2.1 Weighted Method per Component
WMC breaks an elementary rule of measurement theory that a measure should be concerned with a single attribute [9]. If all method complexities equal one then WMC gives a count of the methods. Once weighting, in the form of a complexity value is added, this count is lost. A component with one method that has a complexity value of twenty would give the same WMC value as another component with twenty methods, each having a complexity value of one. The viewpoints for WMC are largely concerned with method count, so the addition of a complexity value seems an unnecessary complication and will further make the word 'weighted' superfluous.

2.2.2 Depth of Inheritance Tree
Deeper inheritance trees are likely to increase the complexity, from the maintenance viewpoint, as the child componentes use the ancestral methods. At the same time, Chidamber and Kemerer gives contradicting statement and propose that it is better to have depth than breadth in the inheritance hierarchy [3]. CK's DIT metrics have other

ambiguities as well. The definition of DIT is ambiguous when multiple inheritance and multiple roots are present. Consider the component inheritance tree with multiple roots in Figure 1.

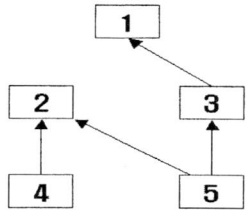

Fig. 1. A Component inheritance tree with multiple roots

The maximum length from component 5 becomes unclear. There are two roots in this design; the maximum length of component 5 from root 2 is (DIT(5)=1) and the maximum length of component 5 from root 1 is (DIT(5)=2) [3].

The other factors, which makes DIT ambiguous lies in the disagreeing goals stated in the definition and the theoretical basis for the DIT metric. The theoretical basis states that DIT is a measure of how many ancestor componentes can potentially affect this component. It is seen and understood to indicate that the DIT metric should measure the number of ancestor componentes of a component. However, the definition of DIT stated that it should measure the length of the path in the inheritance tree, which is the distance between two nodes in a graph; which conflicts with the measurement attribute declared in the theoretical basis. This ambiguity is only visible when multiple inheritance is present, where the distance between a component and the root component in the inheritance tree no longer yield the same number as the number of ancestor componentes for the component. This conflict is visualized in Figure 2. In this figure, according to the definition, componentes 4 and 5 have the same maximum length from the root of the tree to the nodes respectively; thus DIT(4)=DIT(5)=2.[3].

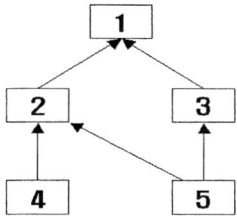

Fig. 2. A component Inheritance tree with single root

However, component 5 inherits from more componentes than 4 does so, these componentes should have different DIT values.

2.2.3 Number of Children

Just as DIT fails to count all the ancestor componentes, NOC fails to count all the descendants of a component because as per the definition it counts only the immediate children. Only counting immediate subcomponentes may give a distorted view of the system, as the following example illustrates [10]:

Example. A hierarchy is structured thus: Component A is the root component, component B extends A, component C extends B and componentes D, E, F, G and H extend C. Both A and B have a NOC value of one, but a total of seven componentes inherit A's properties and a total of six inherit B's properties. As the hierarchy grows bigger so, does the discrepancy.

2.2.4 Lack of Cohesion in Methods

LCOM is calculated as number of disjoint sets formed by the intersection of n sets. The idea behind this metric is that a high value suggests that the methods in the component are not really related to each other and vice versa. LCOM is given by

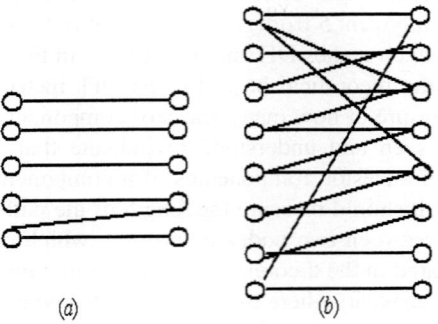

Fig. 3. Example to show LCOM

In the example shown above in Fig 3(a) the value of LCOM is 8 ($|P|= 9$, $|Q|= 1$). Whereas in (b) also LCOM is 8 ($|P|= 18$, $|Q| = 10$). The component (a) is less cohesive and component (b) is more cohesive but the value of LCOM for both the componentes is 18. In this sense, one of the basic axioms of measurement theory, that a measure should be able to distinguish two dissimilar entities, is violated by the LCOM definition. Furthermore, it is of serious concern that, whilst a high value of LCOM implies low similarity and therefore low cohesion, a value of LCOM = 0 does not imply the reverse [1]. If $|P| \leq |Q|$, LCOM = 0 and this can occur even for cases of obvious distinction.

3 MOOD Metrics

The MOOD metrics set was first proposed in 1994 by the MOOD project team, headed F. B. e Abreu. The metrics are designed to meet a particular set of criteria, also proposed by the team.

3.1 Validation Criteria for MOOD Metrics

The criteria are listed here [11]:
1. Metrics determination should be formally defined.
2. Non-size metrics should be system size independent.
3. Metrics should be dimensionless or expressed in some consistent unit system.
4. Metrics should be obtainable early in the life-cycle.
5. Metrics should be down-scaleable.
6. Metrics should be easily computable.
7. Metrics should be language independent.

The MOOD metrics are all system-wide measurements. Each metric is a quotient where the numerator is the use of a particular mechanism in the system being measured and the denominator is the maximum possible use of the same mechanism. The value for each metric will therefore be in the range 0-1 with 0 indicating no usage and 1 indicating maximum usage. The values for the metrics are expressed as percentages, 0-100%. The definition for each metric is given as a mathematical expression. It has been observed that majority of the MOOD Metrics are fundamentally flawed because they either fail to meet the MOOD team's own criteria or are founded on an imprecise or inaccurate, view of the OO paradigm.

3.2 Method Inheritance Factor

Definitions for MIF and AIF are inconsistent with the 0-1 scale as in [11]. Consider the following system with the hierarchical structure: A → B → C.

component A{public void x();public void y();}
component B extends A {//no methods defined}
component C extends B {//no methods defined}

B and C both inherit the two methods defined in component A and define no further methods. This is the maximum possible method inheritance in this system (i.e. all methods that can be inherited have been inherited, by all componentes that are able to inherit them). Intuitively, it seems that the MIF value for this system should be 100%, but in fact it is **66.6%**

$M_i(A) = 0$, $M_i(B) = 2$, $M_i(C) = 2$, Total = 4
$M_a(A) = 2$, $M_a(B) = 2$, $M_a(C) = 2$, Total = 6

This can be further illustrated. If component C in the above example had a new method added it should not change the MIF value for the system. This is consistent with our intuitive understanding of method inheritance. In fact we find that, with $M_a(C) = 3$ MIF becomes 4/7 = 0.57 (57%).

3.3 Method Hiding Factor

In it is recommended that MHF should not be lower than a particular (as yet undefined) value but suggest that there is no upper limit, thus implying that it is 'good' for all methods in a component to be hidden (private). However, the number of private methods in a component does not tell us anything about the degree of information hiding in the component. It may tell us that a particular method (or methods) has been broken down into a number of smaller methods to avoid duplication or for clarity of understanding. Such methods would only need to be visible to the containing component. But whether or not a method is broken down this way the containing component's implementation is still hidden. In the following example both componentes have equal 'information-hiding' levels:

```
component A { private int x; public int m0()
{ do_1; do_2; do_3; return x; }
}
   component B { private int x;
       public int m0() { m1(); m2(); m3(); return x;}
          private void m1() { do_1;}
          private void m2() { do_2;}
   private void m3() { do_3;}
   }
```

In component A all of method *m0*'s behavior is contained in the body of *m0*. In component *B* the behavior has been separated into three smaller methods which are called by *m0*. Both componentes have identical interfaces and their respective implementations are equally well hidden from client componentes. A count of the number of private methods in a component is not a particularly useful metric, and certainly does not contribute anything to our knowledge of a component's encapsulation level.

3.4 Polymorphism Factor

It is possible, indeed highly likely, that a sub-system will consist of a set of componentes that extend a framework. This may be a set of library componentes or a framework of low(er) level system componentes. When measuring the sub-system it should be only the componentes that belong to the sub-system that are measured; componentes outside of its boundaries should not be considered. In such cases the denominator for the POF measure may be less than the numerator, resulting in a value greater than 1. An example will make this clear. Sub-system "S" produces a value for POF which is outside the range 0-1 as shown in Figure 4.

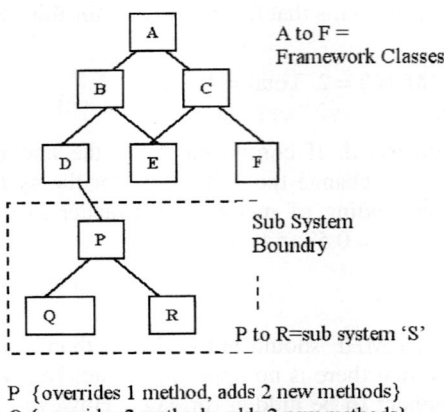

P {overrides 1 method, adds 2 new methods}
Q {overrides 2 methods, adds 2 new methods}
R {overrides 2 methods, adds 2 new methods}

Fig. 4. Example for Showing POF

$M_o(P) = 1$, $M_o(Q) = 2$, $M_o(R) = 2$
$(M_n(P)=2 * DC(P)=2) = 4$,
$(M_n(Q)=2 * DC(Q)=0) = 0$,
$(M_n(R)=2 * DC(R)=0) = 0$

Therefore, POF for sub-system "S" = (1+2+2) / (4+0+0) = 5/4 (and 5/4 > 1). This is especially likely in languages that are shipped with large component libraries, such as Java or Smalltalk. In such cases the whole system could produce POF>1.

3.5 Coupling Factor

There are basically two types of relationships inheritance and client supplier relationship i.e. one component uses the other component as an instance variable.

The fact here is that the relationship between any two componentes in a system is not confined to any of the one. As an example consider the two componentes, *Component* and *Container*, from the Java *java.awt* library package. *Component* is the super component of all graphical components and *Container* is one of its subcomponentes. Thus the two componentes are in an inheritance relationship. However, each component also contains an attribute of the other component type. *Container's* use of a set of *Components* has nothing to do with the fact that *Component* is its supercomponent, indeed if the hierarchy was redesigned the client supplier relationship would still hold. The question that the MOOD team does not adequately answer is whether a client-supplier relationship under these conditions is counted. There is probably no 'correct' way of dealing with this situation in terms of the COF metric but a decision needs to be made one way or the other and it needs to be explicit in the metric's definition.

4 Conclusion

A detailed investigation into the validity of the MOOD metrics and CK Metrics has led to conclude that, as far as *information hiding, inheritance, coupling,* and *cohesion* are concerned the metrics can be shown to the valid within the context of theoretical framework but the main problem occurred when the metrics were put into use practically taking real world examples. Here the research shows that the validation criteria, which CK and MOOD Metrics were required to fulfill initially, was not fulfilled. Instead CK and MOOD Metrics showed many inconsistencies. Analyses have also shown that MOOD Metrics operate at system level whereas CK Metrics operates at component level. It is because of this reason CK Metrics are useful to the designers and developers giving them the evaluation of the system at the component level and MOOD Metrics are useful to the project managers for the overall assessment of the system.

References

[1] B.Henderson – Sellers and L. L. Constantine. *"Coupling and Cohesion (towards a valid metrics suite for object oriented analysis and design."* Object Oriented Systems 3 (1996) 143-158.

[2] Fernando Brito e Abreu, Miguel Goulão, Rita Esteves, *"Toward the Design Quality Evaluation of Object-Oriented Software Systems"*. Proceedings of 5th International Conference on Software Quality, Austin, Texas, 23 to 26 October 1995.
[3] Frederick T. Sheldon, Kshamta Jerath, Hong Chung, *"Metrics for Maintainability of Component Inheritance Hierarchies"*. Journal of Software Maintenance, Issue 3 (May 2002) pp. 147 – 160, 2002.
[4] Joe Raymond Abounader, David Alex Lamb, *"Data model for Object Oriented System"*. Technical Report, Queen's University School of Computing 1997.
[5] Norman E. Fenton, Shari Lawrence Pfleeger *"Software Metrics – A Rigorous & Practical Approach"* 2nd Revision edition Brooks Cole.
[6] Rachel Harrison, Steve J. Counsell and Reuben V. Nithi. *"An Evaluation of the MOOD set of Object Oriented software metrics"* IEEE Transactions on Software Engineering VOL.24 No. 6 June
[7] Ralph D. Neal, Richard J Coppins, H. Roland Weistroffer, *"The Assignment of Scale to Object Oriented Software Measures."* Working Paper, Virginia Commonwealth University, Richmond, VA, 1997.
[8] Roger S. Pressman *"Software Engineering, A Practitioner's Approach"* Fifth Edition, New York: McGraw-Hill
[9] S. R. Chidamber and C. F. Kemerer, *"A Metric Suite for Object Oriented Design"*, IEEE Trans. Software Eng., Vol. 20, pp. 476-493, 1994.
[10] Tobias Mayer, Tracy Hall, *"Critical Analysis of Current OO Design Metrics."* Software Quality Journal, 8, 97-110, 1999.
[11] Tobias Mayer and Tracy Hall. *"Measuring OO Systems: A Critical analysis of the MOOD Metrics"* in TOOLS Proceedings, Nancy, France, July 1999, pp. 108-117.
[12] Weyuker, E., *"Evaluating Software Complexity Measures"*, IEEE Transactions on Software Engineering, Vol. 14, pp. 1357-1365, 1988.

The K-Means Clustering Architecture in the Multi-stage Data Mining Process

Bobby D. Gerardo[1], Jae-Wan Lee[1], Yeon-Sung Choi[1], and Malrey Lee[2]

[1] School of Electronic and Information Engineering, Kunsan National University
68 Miryong-dong, Kunsan, Chonbuk 573-701, South Korea
{bgerardo, jwlee, yschoi}@kunsan.ac.kr
[2] School of Electronic and Information Engineering, Chonbuk National University
664-14 Deokjin-dong, Jeonju, Chonbuk 561-756, South Korea
mrlee@mail.chonbuk.ac.kr

Abstract. In this paper, we used software engineering principles for the development of models and proposed the K-Means clustering architecture implemented on the multi-stage data mining process. We developed a modified architecture and expanded it by showing refinements on every process of the clustering and knowledge discovery stages. We used the mentioned hierarchical clustering model to partition the data into smaller groups of attributes so that we would determine the data structure before applying the data mining tools. The experiment shows that the model using the clustering resulted to an isolated but imperative association rules based on clustered data, which in return could be practically explained for decision making purposes. Shorter processing time had been observed in computing for smaller clusters implying faster and ideal processing period than dealing with the entire dataset.

1 Introduction

The impact of software on our civilization and culture is overwhelming. As its role expands, programmers continually attempt to develop technologies that will make it easier, faster, and less expensive to build high quality programs [1]. It is apparent that this software community had developed or innovated numerous productivity software used for a variety of applications such as in engineering, scientific experiments, entertainment and business.

The present trends in knowledge discovery process show that vendors of data management software are becoming aware of the need for integration of data mining capabilities into database engines, and some companies are already allowing for integration of database and the data mining software. Some of these tools were developed or innovated to address major applications in the academic, business or industrial purposes. And other tools were used for concept description, association analysis, classification, prediction, and cluster analysis.

Cluster analysis is utilized to group data into clusters so that the degree of association is strong between members of the same cluster and weak between members of different clusters, and thus, each cluster describes the class to which its members belong. Ideally, cluster analysis can reveal similarities in data which may have been otherwise impossible to find.

Some of the notable constraints that are observed in the data mining tasks are heterogeneity of database, computing speed, and reliability of the approach for computation. Dealing with dataset containing large tuples or attributes may lead to extra efforts in the processing for information. In addition, it is impractical to explain vast arrays of discovered knowledge, for instance, generating many rules or complex prediction models and having difficulty of explaining the results. Most of the time these are restraints that defeat typical and popular mining approach.

To address some of the restraints mentioned earlier, we propose a clustering architectural model using K-means model which will be implemented on the multi-stage data mining process. This study investigates the formulation of the cluster analysis as integrated component of the proposed model to partition the original data prior to the implementation of other data mining tools.

2 Related Works

2.1 Cluster Analysis

Cluster analysis is a statistical method used for partitioning a sample into homogeneous classes to create an operational classification. Such classification may help formulate hypotheses concerning the origin of the sample, describe a sample in terms of a typology, predict the future behavior of population types, optimize functional processes for business site locations or product design, assist in identification such as used in the medical sciences, and measure the different effects of treatments on classes within the population [4].

The cluster analysis process performs categorization of attributes like those used for consumer products; partition it into clusters or groups so that the degree of correlation is strong between members of the same cluster and weak between members of different clusters. Each group describes the class in terms of the data collected to which its members belong.

Usually, cluster analysis is also used as a tool for knowledge discovery. It may show structure and associations in data, although not previously evident, but are sensible and useful once discovered. The results of cluster analysis may contribute to the definition of a formal classification scheme, such as in taxonomy for related animals, insects or plants; suggest statistical models with which to describe populations; indicate rules for assigning new cases to classes for identification and diagnostic purposes; provide measures of definition, size and change in what previously were only broad concepts; and find patterns to represent classes [5].

2.2 Data Mining

There are many data mining algorithms that have been recently developed to facilitate the processing, discovery of knowledge, and interpretation of information generated from large databases. One example is the association rule algorithm, which discovers correlations between items. This data mining tool uses Apriori algorithm to find for candidate patterns and those candidates that receive sufficient support from the database are considered for transformation into a rule. Some limitations of association rule algorithms, such as the Apriori is that only database entries that exactly match the candidate patterns may contribute to the support of that candidate pattern.

Some research objectives are to develop association rule algorithms that accept partial support from data. In the past years, there were plenty of studies on faster, scalable, efficient and cost-effective way of mining a huge database in a heterogeneous environment. Most studies have shown modified approaches in data mining tasks which eventually made significant contributions in this field.

3 The Architectural Model of the Proposed System

Based on the literatures that had been reviewed, we developed the proposed architectural model as shown in Figure 1 and will be presented in refined view in the succeeding sections. The proposed architecture describes that integration of data is achieve by first performing data cleaning on the distributed database. The data cubes are generated as a result of the data aggregation by implementing extraction or transformation. The purpose of such cubes is to reduce the size of database by extracting dimensions that are relevant to the analysis. The final stage is utilization of the result for decision-making or strategic planning.

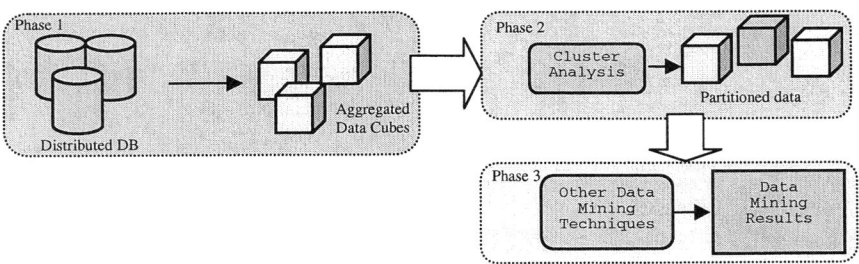

Fig. 1. The Proposed 3-Stage Data Mining Process

Figure 1 shows the proposed three phase implementation architecture for the data mining process. The first phase is the data cleaning process that performs data extraction, transformation, and loading. This will result to an aggregated data cubes as illustrated in the same figure. Figure 2 shows the implementation of the cluster analysis using the K-means partitioning algorithm while Figure 4 is the implementation of data mining techniques.

Figure 3 shows the refined process of the K-means partitioning method. A further refinement of process A depicts detail in the form of transforms A_1, A_2, and A_3. Small circle (bubble) A_1 is a transform which means that initialization of means from μ_1 until μ_k of k clusters shall be started. Bubble A_2 will perform randomization while A_3 performs random selection of objects.

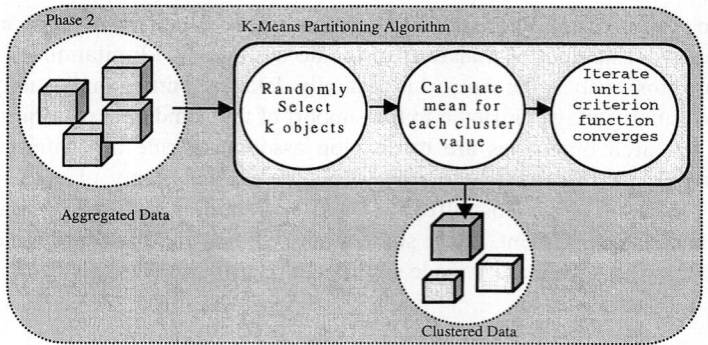

Fig. 2. The Cluster Analysis Model using K-means Partitioning

The next transform is shown by bubble B_1 which will calculate for the minimum Euclidean distance that will be used to determine the membership for the respective clusters. Process B_2 will determine the membership and assign each X to corresponding clusters. The initially partitioned objects are reflected in process B_3.

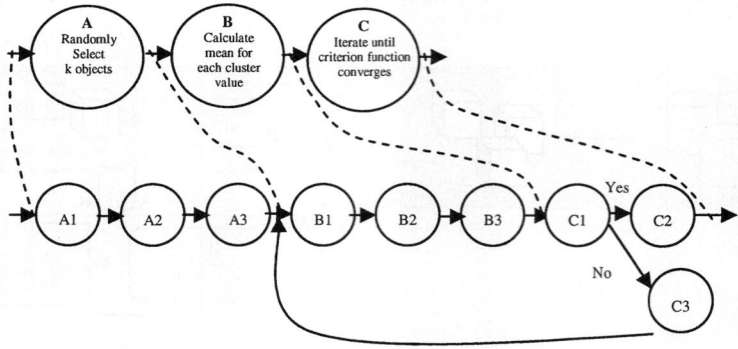

Fig. 3. Information flow refinement of K-means partitioning

Finally, process C will perform iteration until the criterion function converges. The closer view of the processes in C shows that a particular process C_1 re-computes μ_i until there is no change in the value of the mean. If there is no change then the clusters will be generated as reflected in C_2, otherwise, process C_3 will perform looping and return to the process B.

Figure 4 shows the refined view of Phase 3. In this illustration, association rule algorithm is used as part of the data mining process. The successions of transforms for association rule algorithm which are represented by bubbles are shown in the shaded rectangle.

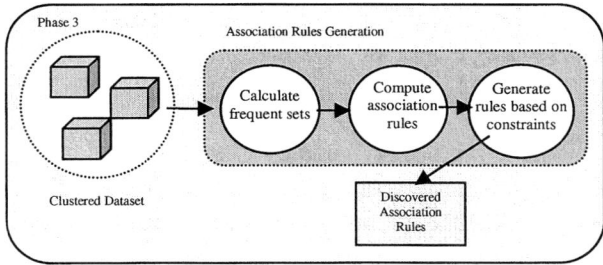

Fig. 4. Association rule discovery process

Phase 3 is the final stage in which the association rule algorithm will be employed to generate the association rules. The refinements are illustrated by the same model and it means that the first process shall calculate for the frequent itemsets then compute for the association rules (as extended data mining process) using algorithms for support and confidence. Finally, the discovered rules will be generated based on the assumptions on support and confidence threshold set by the researchers. The output is given by the last rectangle showing the discovered rules.

4 The Cluster Analysis Models

There are varieties of algorithms that are used in cluster analysis, these include hierarchical or non-hierarchical algorithm (also known as partitioning method), density-based, grid-based and model-based methods.

Some literature claimed that hierarchical clustering models are faster than non-hierarchical models in terms of data processing. This study will put more emphasis on the use of non-hierarchical clustering such as the K-means model.

4.1 Cluster Analysis and Its Models

4.1.1 Hierarchical Clustering Algorithms

The hierarchical method creates a hierarchical decomposition of the given set of data objects while the non-hierarchical method produces disjoint clusters and thus works well when a given set is composed of a number of distinct classes. Among the most common hierarchical clustering methods are Nearest-Neighbor, Farthest-Neighbor, and Minimal Spanning Tree.

4.1.2 Non-hierarchical Clustering Algorithms

The popular models of non- hierarchical methods include K-Means, Fuzzy K-Means, and Sequential K-Means. Non-hierarchical clustering algorithms produce disjoint clusters and thus work well when a given set is composed of a number of distinct classes or when the data description is flat [5].

4.1.2.1 K-Means. Assuming that we have an n feature vectors, $X_1, X_2, \ldots, X_n$, which belong to the same class C and suppose that we know that they belong to k clusters such that $k < n$. If clusters are well separated, then we use a minimum distance classifier to separate them. To do this, we first initialize the means $\mu_1 \ldots \mu_k$ of k clusters. One of the ways to do this is to assign random numbers to them. We then determine the membership of each X by taking the $|X - \mu_i|$. The minimum Euclidean distance determines X's membership in a respective cluster [5]. Below shows the K-Means algorithm:

1. **begin**
2.   **initialize** n; c; $\mu_1 \ldots \mu_c$
3.   **do**
4.   classify n samples according to nearest μ_i
5.   recompute μ_i
6.   **until** no change in μ_i
7.   **return** $\mu_1 \ldots \mu_c$
8. **end**

Figure 5 shows an example of K-means clustering and explain how the algorithm works when k=2. This is done by randomly selecting the values for μ_1 and μ_2, and the algorithm converges when the means have no longer changes. Assume that μ_i = mi, then the graph below is obtained:

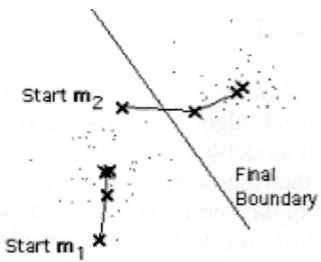

Fig. 5. K-Means graph for k=2

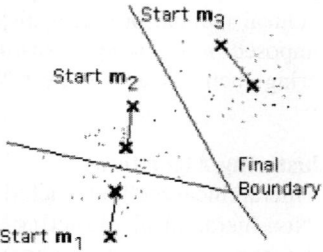

Fig. 6. K-Means graph for k=3

The results depend so much on the value of k, for instance the actual number of clusters. Another example is shown in Figure 6. The graph shown in this clustering has a k=3.

The graphs can be extended or we can further note its characteristics if we assume values until k=n. However, this procedure will give us the method known as the Nearest Neighbor Classifier which is one of the hierarchical clustering methods. It performs better if the number of feature vectors is large, but will require more computations [5]. Note that the K-means is more sensitive to noise and other outlier data points because small number of such data can affect the mean value [2].

4.1.2.2 Fuzzy K-Means. We can extend the preceding clustering method by assuming that each feature vector belongs to exactly one cluster for every iteration of the classical k-means procedure. This particular clustering uses the method called the Fuzzy K-means.

4.1.2.3 Sequential K-Means. If we assume that all the data points are not known, then we can use the sequential K-means by gradually obtaining such data points over a period of time. This procedure offers the flexibility of updating the means as new data points. The goal is to find such partitions of a set of n samples which maximize the criterion function like the distance. Due to overwhelming number of partitions of n elements into c subsets, it is difficult to assume every possibility. One approach is Iterative Optimization described by the basic iterative minimum-squared-error clustering which could be modified to a Sequential K-Means procedure. The main thought of this process is to come up with a reasonable initial partition and to move samples from one group to another if such move improves the value of the criterion function [5].

4.2 Implementation of the Desired Models

The output of the clustering method shall then be processed by implementing the extended data mining tool for association discovery. Note that other mining techniques such as concept description, classification, and prediction can also be implemented after Phase 2 of the proposed model. This multi-stage implementation of data mining tools is done on clustered dataset and we will anticipate more practical processing and discovery of knowledge. Tables showing the comparison of results of the original dataset, proposed model, and the discovered rules are presented in the next section.

5 Experimental Evaluations

The experimental evaluation was done on the dataset containing 1,200 tuples with 30 attributes and comprising of six (6) major dimensions of e-commerce and transactional types of data. These main dimensions are Electronics, Books, Entertainment, Gifts and Health, and these refer to main categories of items used as part of the transactional data. The evaluation platforms used in the study were IBM compatible PC with Window OS, C++, Python and application like SPSS.

The main goal of the clustering process is to partition the dataset into smaller clusters and then generate association rules from these smaller groups of attributes. We can then address some problems like determining how often consumers buy products and knowing the probability of the patterns of purchasing some items online. The results of our experiment will be presented in the subsequent sections.

5.1 The Results of Cluster Analysis

The result in Table 1 attempts to identify relatively homogeneous groups of cases based on selected characteristics. The same table shows the result using the K-Means cluster analysis algorithm and the membership of each case to a cluster with its corresponding distance.

Table 1. Cluster Analysis Result Using K-Means Algorithm

Cluster							
1		2		3		4	
Case No.	Distance	Case No.	Distance	Case No.	Distance	Case No.	Distance
6	0.765	1	1.111	19	0.513	5	0.192
9	0.651	2	0.849	55	0.513	7	0.192
15	0.765	3	1.439	84	0.513	14	0.192
18	0.651	4	1.111	120	0.513	16	0.192
24	0.765	8	1.111	156	0.513	23	0.192
27	0.651	10	1.165	185	0.513	25	0.858
33	0.765	11	0.849	221	0.513	32	0.192
36	0.651	12	1.439	257	0.513	34	0.192
42	0.765	13	2.669	286	0.513	41	0.192
45	0.651	17	1.111	322	0.513	43	1.889
...	...	...	...	...	...	...	...

A total of 4 clusters had been identified and the group membership of each case is partially indicated in the same table. The membership of each case is based on the minimum Euclidean distance that was calculated during the clustering process. The same table shows the minimum distance of each case and its corresponding membership to the cluster. Cluster 1 has a total of 260 cases (21.7%), cluster 2 has 660 cases (55.0%), cluster 3 has 41 (3.4%), and cluster 4 has 239 cases (19.9%).

5.2 The Data Mining Results After Clustering Using K-Means Method

Table 2 shows the membership of each attribute to the respective clusters. This explains that of a total of 30 attributes, 11 attributes belong to the first cluster, 14 for the second, 4 for the third and 1 attribute for the fourth cluster. In addition, it shows the discovered rules after the clustering process. For illustrative purposes, the two computations are shown on the same table showing their corresponding values, i.e. for complete dataset and clustered attributes.

Table 2. Attribute clustering membership, processing time and summary of the discovered rules with support ≥0.90

	Original Data	Clustered Attributes			
		1	2	3	4
Clusters→	(1200)	(11)	(14)	(4)	(1)
Member Attributes→	All	A1, A2, B1, B2, D2, D3, E1, E2, E3, E4, E5	A3, A4, A5, A6, B3, B4, C1, C2, C3, C4, D1, D5, D6, E6	B5, B6, C5, C6	D4
Number of Rules Generated→	610	2	2	2	0
Processing Time (sec.)	5.18	0.07	0.06	0.05	0.0

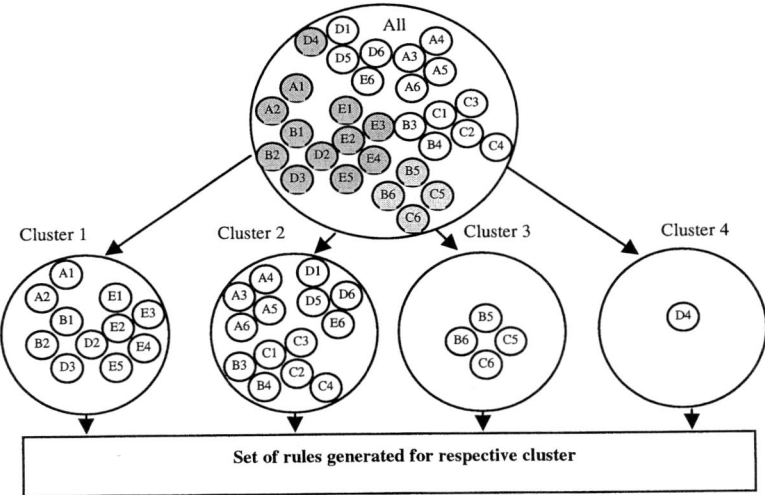

Fig. 7. Pictorial view of the clustered attributes

Figure 7 shows the graphical representation of clustered attributes. Each cluster is comprised of member attributes and each has a corresponding set of rules obtained.

It is imperative to learn that there would be an improvement in the processing time since the computation of rules was based on clustered attributes. It is also interesting to note the difference of computing time characteristics revealed by the processing time as presented in Table 2, which shows the comparison of the original and clustered dataset. Shorter processing time had been observed when computing for smaller clusters implying faster and ideal processing period than dealing with the entire dataset.

The result further implies that the blending of cluster analysis and association generation model specifically isolate groups of correlated cases using the K-means partitioning then using of the extended data mining process like the algorithm for association rule generation. This resulted to some clusters where we could conveniently analyze specific associations among clusters of attributes.

Table 3. Comparison of the discovered rules

Models	Discovered Rules (showing first 5 rules generated)	Supp.	Conf.
Original (610 rules)	D1=Buy -> D5=Buy	0.971	0.978
	D1=Buy -> E4=Always E5=Always	0.958	0.966
	D1=Buy -> E4=Always	0.958	0.966
	D1=Buy -> E5=Always	0.958	0.966
	D1=Buy -> D5=Buy E4=Always	0.937	0.944
Cluster Analysis 1 (2 rules)	E5=Always -> E4=Always	0.966	0.999
	E4=Always -> E5=Always	0.966	0.999
2 (2 rules)	D1=Buy -> D5=Buy	0.971	0.978
	D5=Buy -> D1=Buy	0.971	0.992
3 (2 rules)	D1=Buy -> D5=Buy	0.971	0.978
	D5=Buy -> D1=Buy	0.971	0.992
4 (0 rules)	No rules		

For the original dataset, we only show the first five rules generated as shown in Table 3. The support threshold that we set prior to the experiment was 0.90. On the same table, the rules with its corresponding support and confidence are presented. In the original dataset, those who buy D1 (Jewelry) will most likely buy D5 (Beauty Set) with support of 0.971 and confidence of 0.978 (97.8% probability of buying). The second rule means that those who buy D1 (Jewelry) will most likely buy E4 (Cosmetics) and E5 (supplements) with support of 0.958 and confidence of 0.966 (96.6%). The rest of the 608 rules can be explained in the same fashion.

We noted earlier in cluster 1 that it has 11 member attributes and with it generated 2 rules. The first rule for this cluster means that those who buy E5 (Supplements) will most likely buy E4 (Cosmetics) with support of 0.966 and confidence of 0.999 (99.9%). The same explanation can be made for its second rule. While in second cluster (with 2 rules), those who buy D1 (Jewelry) will most likely buy D5 (Beauty Set) with support of 0.971 and confidence of 0.978 (97.8%). The second rule for cluster 2 and other rules for cluster 3 can be explained in the same fashion.

6 Conclusions

The results in our experiment were implemented using the heuristic approach. The findings were obtained using the blending of the data mining algorithms and the proposed clustering models implemented on the synthetic data. Shorter processing time had been observed in computing for smaller clusters of attributes implying faster and ideal processing period than dealing with the entire dataset. The result further implies that the combination of cluster analysis and association discovery model specifically isolate groups of correlated cases or attributes using the K-means partitioning. We have provided examples, showed the result of clustered attributes, and generated rules but more rigorous treatment maybe needed if dealing with more complex and real world databases.

For future investigations, the researchers recommend the implementation of other clustering methods like agglomerative and divisive hierarchical clustering and other non-hierarchical clustering methods to diligently compare other interesting results in accordance to the premise of the models that we proposed.

References

1. Pressman, R. Software Engineering: a practitioner's approach, 5th Edition. USA: McGraw-Hill (2001).
2. Han J. & Kamber M.: Data Mining Concepts & Techniques. USA: Morgan Kaufmann (2001).
3. Chen, B., Haas, P., and Scheuermann, P.: A new two-phase sampling based algorithm for discovering association rules. Proceedings of ACM SIGKDD International Conference on Knowledge Discovery & Data Mining, 2002.
4. Cluster Analysis defined: Available at: http://www.clustan.com/ what_is_cluster_ analysis.html
5. Determining the Number of Clusters.: Available at: http://cgm.cs.mcgill.ca/~soss/ cs644/ projects/ siourbas/ cluster.html# kmeans.
6. Using Hierarchical Clustering in XLMiner.: Available at: http://www. resample.com/ xlminer/ help/ HClst/ HClst_intro.htm.
7. Agglomerative Hierarchical Clustering.: Available at http://www2.cs.uregina. ca/~hamilton/ courses/831/ notes/ clustering/ clustering.htm.
8. Ertz L., Steinbach M., & Kumar V. Finding Topics in Collections of Documents: A Shared Nearest Neighbor Approach. Text Mine '01, Workshop on Text Mining, First SIAM International Conference on Data Mining, Chicago, IL, (2001).

A Privacy Protection Model in ID Management Using Access Control[†]

Hyang-Chang Choi[1], Yong-Hoon Yi[1], Jae-Hyun Seo[2],
Bong-Nam Noh[1], and Hyung-Hyo Lee[3,*]

[1] Dept. of Information Security, Chonnam National University, Gwangju, Korea, 500-757
{hcchoi, yhyi, bongnam}@lsrc.jnu.ac.kr
[2] Div. of Information Engineering, Mokpo National University, Mokpo, Korea, 534-729
jhseo@mokpo.ac.kr
[3] Div. of Information and EC, Wonkwang University, Iksan, Korea, 570-749
hlee@wonkwang.ac.kr

Abstract. The problem of privacy of the Identity Management System (IMS) is the most pressing concern of ordinary users. Uncertainty about privacy keeps many users away from utilizing IMS. Most privacy-enhancing technologies such P3P, E-P3P and EPAL use purposes or policies to ensure privacy that is set by users. Access control is arguably the most fundamental and pervasive security mechanism in use. This paper proposes a privacy protection model using access control for IMS. The proposed model protects privacy using access control techniques with privacy policies in a single circle of trust. We address characteristics of components for the proposed model and describe access control procedures. After that, we show protection architecture and XML-based schema for privacy policies.

1 Introduction

Information society is a collection of huge amounts of information resources and serves the convenience of the users. Recently, Identity Management System (IMS) has been under the spotlight for its great convenience in information society. IMS describes the infrastructure in which identity management applications as components are coordinated and denotes an infrastructure within one or between several organizations that have agreed upon a mutual model of trust in managing and using identities[1]. Moreover, IMS can also denote an implementation of identity management encompassing a whole society.

On the other hand, the problem of privacy on the identity management system is the most pressing concern of ordinary users[2]. Uncertainty about privacy keeps many users away from utilizing IMS. Ordinarily, IMS manages personal information with

---

[†] This research was supported by the MIC (Ministry of Information and Communication), Korea, under the ITRC (Information Technology Research Center) support program supervised by the IITA (Institute of Information Technology Assessment).
[*] Correspondent author.

access control and provides personal information only to authorized users. Although privacy has to be protected for users, IMS is more focused on securing information whether its users are authorized or not. Therefore the existing access control approach can easily violate users' privacy. For example, IMS can not trade e-mail addresses to marketing companies without a user's permission even if IMS can[8, 9]. To solve the privacy problem, privacy-enhanced technologies such as E-P3P and EPAL are proposed by IBM[3, 4]. But they deal with an enterprise viewpoint rather than user concerns and are focused on user information in the enterprise environment.

Most privacy-enhancing technologies such as P3P, E-P3P and EPAL use purposes or policies to maintain privacy set by users. Basically, privacy policies are divided into two parts in IMS: *personal information permission policies* and *personal information request policies*. The personal information permission policies are set by users who offer their personal information, and the personal information request policies are given by service providers which want to use the personal information in IMS environments. We proposed a privacy protection model of IMS in which access to personal information is controlled by both personal information permission and request policies.

The paper is organized as follows. Section 2 illustrates related research projects such as RAPID[5] and PRIME[6]. Section 3 proposes a privacy protection model for identity management system and presents access control procedures and examples. Section 4 discusses the infrastructure of proposed models. Section 5 gives conclusions and research directions.

2 Related Work

2.1 ID Management System

Identity management is a broad administrative area that deals with identifying individuals in a system and controlling their access to resources within that system by associating user rights and restrictions with the established identity. Typical systems for identity management are Microsoft Passport[10], Liberty Alliance[11] and SourceID of PingIdentity[7]. Figure 1 shows a simple identity management system in a Circle of Trust.

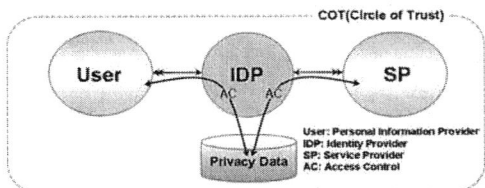

Fig. 1. A Simple Identity Management System

Even if typical identity management systems control stored personal information, they lack privacy concerns because they are just focused on functional aspects. They maintain personal information securely and provide Single Sign-On.

2.2 Project for Privacy and Identity Management

RAPID (Roadmap for Advanced research in Privacy and Identity management) is an EU project aimed at identifying research topics in the area of Privacy and Identity Management (PIM) and at building and strengthening the EU research community in this area. In order to support the goals of RAPID, five specific PIM themes were studied: privacy enhancing technologies in infrastructures, PIM in enterprise system, multiple and dependable identity management, legal PIM issues and socio-economic PIM issues. Moreover, for infrastructure research, RAPID defines address privacy, location privacy, service-level privacy and authorization privacy[5].

PRIME (Privacy and Identity Management for Europe) is a project leading experts from application and service providers, data protection authorities, academic and industrial research to join its reference group. PRIME addresses an integrative approach of the legal, social, economic and technical areas of concern to build synergies about the research, development and evaluation of solutions on privacy-enhancing identity management that focus on end-users. The PRIME framework aims to provide the basis for the widespread deployment of privacy-enhancing mechanisms and identity management[6].

Although RAPID and PRIME provide a big map of privacy-enhancing identity management for terminology, concepts, application scenarios and legal, social and technical options, they do not propose a specific privacy protection model in practical ways because they focus on overall frameworks for privacy-enhancing identity management.

3 Privacy Protection Model for ID Management System

3.1 Identity Management Privacy Protection (IDMP) Model

Privacy-enhancing technologies usually use purpose or policy to keep privacy that is set by users. For example, a P3P policy describes a website's privacy practices. When users and webservers want to maintain privacy, they automatically exchange policies in P3P. In the same way, the main purpose of privacy policies is to maintain privacy of IMS. In IMS, privacy policies are merely divided into information permission policy and information request policy. Figure 2 shows that only users can access objects (personal information) when information permission policy and information request policy are agreed on properly.

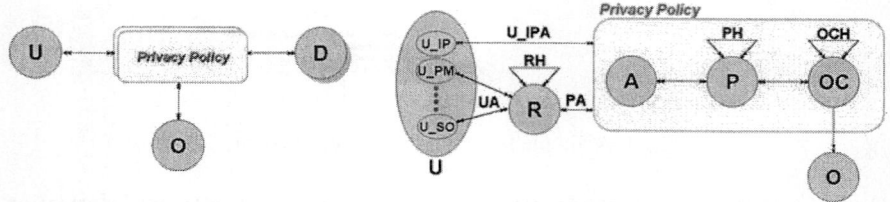

Fig. 2. A Basic Concept   **Fig. 3.** IDMP Model

U represents users for identity management systems, O illustrates privacy data and D indicates duty that is the same concept of obligation in EPAL[4]. In Figure 2, U can access O with duty or obligation when privacy policy is agreed on between permission policy and request policy. However, the formal definition and characteristics of duty are not included in this paper.

Figure 3 shows a detailed privacy protection model based on basic concepts. The IDMP model in Figure 3 is also defined with set relations in details as follows:

Definition 1. Identity Management Privacy Protection (IDMP) Model

- **User:** $U = U\_IP \cup U\_PM \cup U\_SM \cup U\_SU \cup U\_SO$
    - U_IP: A set of users who provide personal information to IDP or SP and define their own personal information permission policies.
    - U_PM: A set of privacy policy managers for administrating the privacy of personal information stored in either IDP or SP and creating a personal information request policy.
    - U_SM: A set of operators responsible for service management of IDP and SP.
    - U_SU: A set of users who belong to a single COT. It can also include all SP users managed by IDP.
    - U_SO: A set of users of special purposes such as law enforcement officers.

- **Object:** $O = O\_IP \cup O\_SM \cup O\_PP$
    - O_IP: A set of personal information data stored IDP or SP.
    - O_SM: A set of data newly generated and maintained by U_SM.
    - O_PP: A set of privacy policies generated by U_IP or U_PM.

- **Object Category:** *OC*, A set of categories for personal information. Examples of OC are personal information, notice information, financial information and medical information.

- **Object Category Hierarchy:** $OCH \subseteq OC \times OC$, Hierarchical relationship among object categories.

- **Object Tree:** $OT \subseteq OC \times 2^{Object}$, Relations between OC and objects.

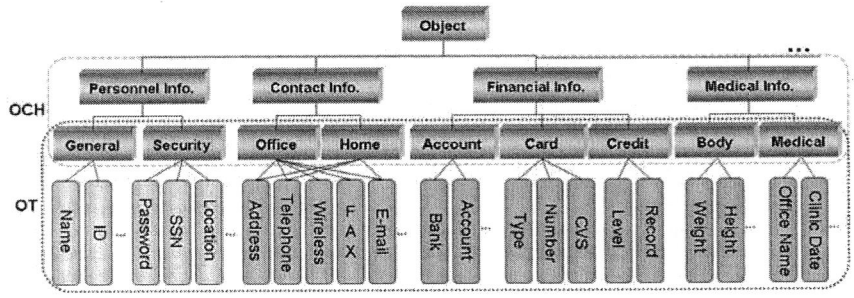

Fig. 4. An Example of OT and OCH

- **Role:** $R = R\_IP \cup R\_PM \cup R\_SM \cup R\_SU \cup R\_SO$
  R_IP: A set of roles to which users in U_IP can be assigned.
  R_PM: A set of privacy manager roles in IMS.
  R_SM: A set of service provider roles.
  R_SU: A set of roles to which users in U_SU can be assigned.
  R_SO: A set special purpose roles.

- **Role Hierarchy:** $RH \subseteq R \times R$, Partial order relation.

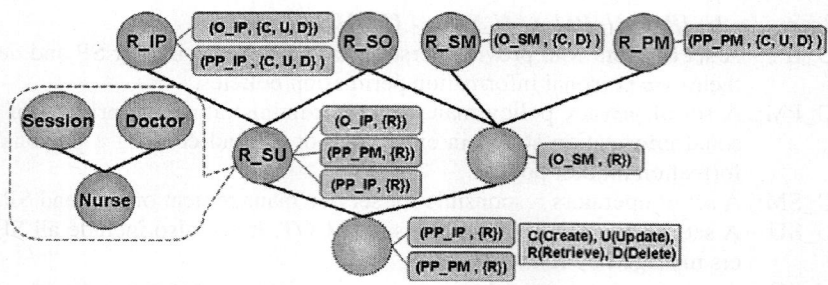

Fig. 5. An Example of Role Hierarchy

- **Purpose:** $P = P\_IP \cup P\_PM$
  P_IP : The purposes for permission of personal information.
  P_PM: The purposes for request of personal information by U_PM.

- **Purpose Hierarchy:** $PH \subseteq P \times P$, Partial order relation.

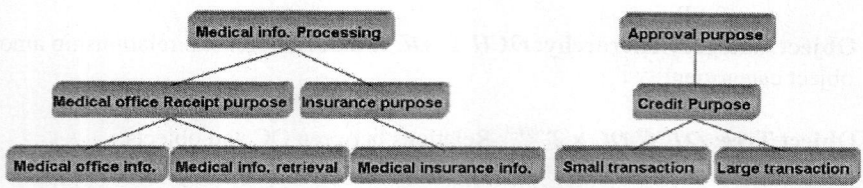

Fig. 6. An Example of Purpose Hierarchy

- **Privacy Policy:** $PP = PP\_IP \cup PP\_PM$
  $PP\_IP = P\_IP \times R \times A \times 2^{Object} \times OC$, Personal information permission policy. This privacy policy set by U_IP.
  $PP\_PM = P\_PM \times R \times A \times 2^{Object} \times OC$, PP_PM Personal information request policy. This policy is set by U_PM.

- **Access Mode:** $A = \{Create, Delete, Update, Retrieve\}$, A set of access operations.
- $U\_IPA \subseteq U\_IP \times PP\_IP$, A many-to-many mapping between U_IP and PP_IP.

- $UA \subseteq U \times R$, A many-to-many mapping between users and roles.
- $PA \subseteq R \times PP\_PM$, A many-to-many mapping between roles to PP_PM.

3.2 Characteristics of Components

IDMP model components are defined in due consideration of privacy in IMS. We describe characteristics of components for an understanding of the IDMP model.

- O_PP is composed of PP_IP and PP_PM. PP_IP is generated by U_IP and PP_PM is generated by U_PM. PP_IP are created in processing when U_IP offers O_IP. At this time, uesrs decide on their PP_IP such as P_IP, A and D policy of their personal information. Thus PP_IP can maintain a distinct policy with each user relevant to U_IP. PP_PM is a policy describing how to request O_IP when O_IP is needed or when U_SU uses service.
- O_IP is classifed into OCs with categories for flexibility of management. OC with categories can help U_IP and U_PM set up PP such as PP_IP and PP_PM. In setting PP, only OC needs to be considered, not the magnitude of O_IP. That is, there is a reduction effect from $R \times A \times 2^o \times O$ to $R \times A \times 2^o \times OC$ in relation size. On the other hand, because PP's subject is not O but OC for the protection of privacy, user's influence on privacy that is concerned with O can be reduced.
- A category set of Os is in OC and if OC has a hierarchy relation, it is expressed as OCH. OT is a special form of OCH.
- Introducing R such as R_IP, can reduce user's influence on privacy. However, it can improve the efficiency of personal information stored by IDP or SP in managing and processing. Thus U_PM and U_SM should decide the use of R_IP under the prior consultation with U_IP. When it is decided to use R_IP, U_PM and U_SM decides R_IP to the system management of IDP or SP. If, on the other hand, there is no R_IP, user's influence in privacy can be improved but the efficiency in managing personal information can be reduced. In this case, a unique PP_IP is generated and connected to a user's identity ID. Because R_IP and a user's identity ID are heterogeneous, they can be concurrently used by PP. So, user's identity ID can be used as the role of R_IP level.

3.3 Access Control Procedure

The proposed model uses access control procedure for access permission policy and access request policy to maintain privacy protection. Therefore, we define the access control procedure. Definition 2 shows the information structure used when U requests access.

Definition 2. Information system of access request
Access Request = (u, r, a, o, p) where $u \in U$, $r \in R$, $a \in A$, $o \in O$, $p \in P$

In Definition 3, the get_ip_policy and get_pm_policy functions return information permission policy and information request policy corresponding to input values u, r, o.

Definition 3. Functions for access control procedure of the proposed model
Table 1 descirbes access control functions needed in executing access requests.

Table 1. Denitions of functions for the access control procedure

Function Definition	Type
$get\_oc\_set(o) = \bigcup_{oc \subseteq OT} oc$ where $o \in oc$	$O \rightarrow 2^{OC}$
$get\_purpose\_set(r, oc\_set) = \bigcup_{oc \subseteq oc\_set}(oc)get\_oc\_purpose(r, oc)$	$R \times 2^{OC} \rightarrow 2^{P}$
$get\_oc\_purpose(r, oc) =$ $p \quad \exists (r, a, oc', p) \in PP, oc = oc'$ $\cup oc' \; get\_oc\_purpose(r, oc')$ where $(oc', oc) \in OCH$	$R \times OC$ $\rightarrow 2^{P}$
$get\_pm\_policy(u, r, oc\_set) =$ $\exists_{pp\_pm \in PP\_PM} pp\_pm(r_i, a_i, oc_i, p_i)$ where $(u_i=u, r_i=r, oc_i=oc)_{\forall 1 \leq i \leq n}$	$U \times R \times 2^{OC}$ $\rightarrow PP\_PM$
$get\_ip\_policy(u, r, oc\_set) =$ $\exists_{pp\_ip \in PP\_IP} pp\_ip(r_i, a_i, oc_i, p_i)$ where $(u_i=u, r_i=r, oc_i=oc)_{\forall 1 \leq i \leq n}$	$U \times R \times 2^{OC}$ $\rightarrow PP\_IP$
$inclusive\_purpose(p, p') = TRUE \; p \leq p'$ $FALSE$ otherwise	$P \times P$ $\rightarrow$ Boolean
$inclusive\_role(r, r') = TRUE \; r \geq r'$ $FALSE$ otherwise	$R \times R$ $\rightarrow$ Boolean

Definition 4. Authorization Rule

Authorization = inclusive_purpose(p of pp_pm, p of pp_ip)
$\wedge$ *inclusive_role(r of pp_pm, r of pp_ip)*

The authorization rule is a main part is a main function of access request and returns a Boolean value. That is, the function receives information permission policy and information request policy, processes *inclusive_role, inclusive_purpose* and determines whether it permits the requested information or not.

Scenario. Let us assume that Alice is registered in an ID management system and there are two medical sites, SP1 and SP2 in a COT. In the two sites, privacy is managed by PP_IP which is created by Alice's decision. PP_PM is created by U_PM and maintained in the ID management system. Personal information structures, roles and information request purposes in Figure 4, 5 and 6 are referred to in the scenario. Table 2 shows privacy policy examples of PP_IP and PP_PM. Alice wants her medical information for the purpose of a receipt from the SP2 site. A user of nurse role in SP2 requests access to Alice's medical information through the ID management system. Access procedures for the medical information follow the next steps and Figure 7 shows representation of the access control procedure.

Table 2. An example of PP_IP and PP_PM

Policy	R	A	OC	P
SP1's PP_PM	nurse	Retrieve	Medical record	Medical info. Retrieval
Alice's PP_IP	doctor	Retrieve	Medical record	Medical office Receipt

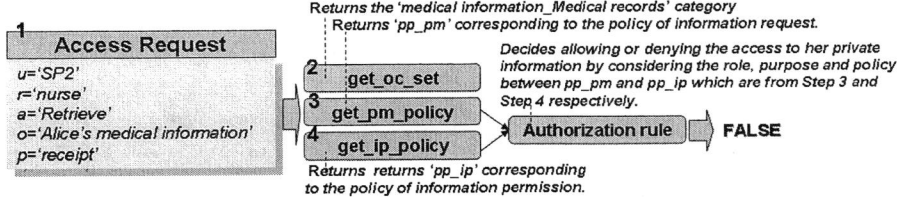

Fig. 7. The graphical representation of access control procedure

[Step 1] Access Request
→ ('SP2', 'nurse', 'Retrieve', 'Alice's medical information', 'receipt')
[Step 2] get_oc_set('Alice's medical information')
→ {medical information-Medical records}
[Step 3] get_pm_policy('SP1', 'nurse', {medical information_Medical records})
→ PP_PM(R: nurse, A: r, OC: Medical record, P_PM: Medical info Retrival)
[Step 4] get_ip_policy('Alice', 'doctor', {medical information-Medical records})
→ PP_IP(R: doctor, A: r, OC: Medical record, P_IP: Medical office Receipt)
[Step 5] Authorization
→ Step 3's purpose & role
    p: get_purpose_set('nurse', 'medical information-Medical records')
    = {'Medical info. Retrieval'}
    r: nurse
Step 4's purpose & role
    p´: get_purpose_set('doctor', 'medical information-Medical records')
    = {'Mdical office Receipt', 'Mdical office info., 'Medical info. Retrieval'}
    r´: doctor
→ inclusive_purpose({'Medical info. Retrieval'}, {'Mdical office Receipt', 'Mdical office info., 'Medical info. Retrieval'})= {'Medical info. Retrieval'} < {'Mdical office Receipt', 'Mdical office info., 'Medical info. Retrieval'})
    ∧ inclusive_role('nurse', 'doctor')= nurse < doctor
    = TRUE ∧ FALSE → FALSE

4 IDMP Architecture

4.1 The Overview of System Architecture

Figure 8 depicts the system architecture of IMS with access control based privacy protection model. Those are information provider group, information user group and user group. These groups are allowed to use all services provided by applications. Before service usage, they should be authorized by the *Privacy Policy Decision Engine* (PPDE). The PPDE is a core part of IDMP system architecture. The PPDE is responsible for protecting privacy by comparing PP_IP policy with PP_PM policy. To make a right decision, PPDE operates access control procedures defined in section 3.3 and authorizes an access request to personal information to O_IP, O_SM and O_PP. Authorized personal information can be used during service usage. After that, IDMP logs personal information usage and audit data.

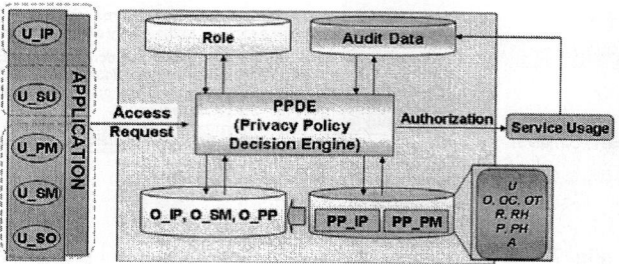

Fig. 8. IDMP System Architecture

4.2 Specifying Privacy Policy in ID Management System

Figure 9 shows the privacy policy schema. We illustrate the schema with Design View of XML SPY[12] for our model. When the policy for permission and request of personal information are created, illustrated schema are used. We assumed that user 'Alice' clicks the checkbox to permit the usage of an e-mail address for notice purposes. Then PP_IP policy for an e-mail address of 'Alice' is created with the privacy policy schema. Figure 10 shows the detailed privacy policy for Alice.

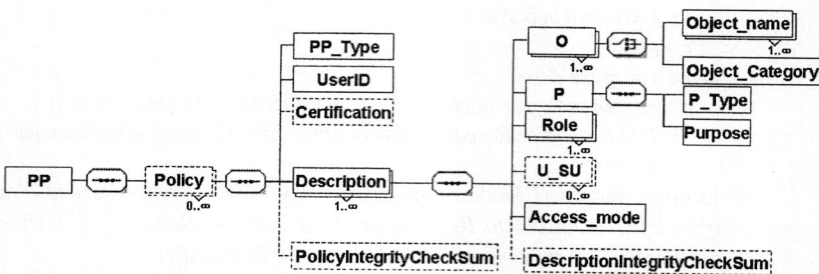

Fig. 9. A Privacy Policy Schema

```
<?xml version="1.0" encoding="UTF-8"?>
<PP xmlns:xsi="http://www.w3.org/2001/XMLSchema-instance
 " xsi:noNamespaceSchemaLocation="http://x.x.x.x/pixml/pp_schema.xsd">
 <Policy>
    <PP_Type>PP_IP</PP_Type>   <UserID>AliceID</UserID>   <Certification>0e0c95d2caca1d</Certification>
    <Description>
       <O> <Object_name>Name_CodeNo.00003</Object_name>   <Object_name>E-mail_CodeNo.00025</Object_name>   </O>
       <P> <P_Type>P_IP</P_Type> <Purpose>For notice purpose_CodeNo.P003</Purpose> </P>
       <Role>SP_NoticeRole_CodeNo.SPR023</Role>
       <U_SU>BobID</U_SU>
       <Access_mode>Retrieve</Access_mode>
       <DescriptionIntegrityCheckSum>c748f4df00d698ad2c98ac61b0ebc746f4c800dc98a12cb0fcc74af49500d398ff2cc
       </DescriptionIntegrityCheckSum>
    </Description>
    <PolicyIntegrityCheckSum>a36b0fcc71df49400d298a42c92ac31b0aac71ff494008298f12cc4ac35b0aec798f7
    </PolicyIntegrityCheckSum>
 </Policy>
</PP>
```

Fig. 10. An Example of PP_IP for Alice

5 Conclusions

In the paper, we propose a privacy protection model in ID management using access control. We also define the component of the model and describe the characteristics of the components. After that, we illustrate the system architecture of IDMP model and XML-based schema for privacy policies. In the future, the formal definition and role of the duty component in the access decision procedure need to be added to the model. Also, IDMP can be used in the framework of personal information protection which needs various techniques for privacy protection such as anonymity, password and policy-based personal information protection.

References

1. Identity Management Systems (IMS): Identification and Comparison Study. PRIME Project (2003), http://www.datenschutzzentrum.de/idmanage/study/ICPP_SNG_IMS-Study.pdf
2. Cranor, L.F.: Web Privacy with P3P. AT&T (2002)
3. Ashley, P., Hada, S., Karjoth, G., Schunter, M.: Privacy Policies and Privacy Authorization (E-P3P). WPES (2002)
4. Ashley, P., Hada, S., Karjoth, G., Powers, C., Schunter, M.: Enterprise Privacy Authorization Language. W3C (2003), http://www.w3.org/Submission/2003/SUBM-EPAL-20031110/
5. RAPID: Roadmap for Advanced Research in Privacy and Identity Management. RAPID Project (2001), http://www.ra-pid.org
6. PRIME: Privacy and Identity Management for Europe Date of preparation. PRIME Project (2004), http://www.prime- project.eu.org/
7. Sourceid: Open Source Federated Identity Management. Ping Identity (2004), http://www.sourceid.org/
8. Warren, A.D., Brandeis, L.D.: The Right to Privacy, Harvard Law Review (1980)
9. Magnuson, G., Reid, P.: Privacy and Identity Management Survey. IAPP Conference (2004)
10. Microsoft .NET Passport: Microsoft (2004), http://www.microsoft.com/net/services/passport/
11. Liberty Alliance: Introduction to the Liberty Alliance Identity Architecture, Liberty Alliance Project (2003)
12. XML SPY. Altova (2004), http://www.xml.com/pub/p/15

A Time-Variant Risk Analysis and Damage Estimation for Large-Scale Network Systems

InJung Kim[2], YoonJung Chung[1], YoungGyo Lee[2], and Dongho Won[2]

[1] Electronics and Telecommunication Research Institute,
Republic of Korea
{cipher, yjjung}@etri.re.kr
[2] ICSL, Sungkyunkwan University, Republic of Korea
{yglee, dhwon}@dosan.skku.ac.kr

Abstract. Risk analysis for preventing network intrusions and attacks and estimation of damages resulting from intrusions and attacks are routine exercises for large-scale network systems. However, previous methodologies for risk analysis and network administration techniques for controlling system failures have been limited to the offering of safeguards based on identification of assets and resources at risks, potential threats and system vulnerabilities. They fail to provide exact estimations as to the effect of eliminating threats and vulnerabilities, which may be done through real-time analysis, or to assess the scope of damage, in the event of an attack, incurred until the final recovery.

In this paper, we propose a time-variant risk analysis technique, which, based on previous risk analysis models for large-size networking systems and used in conjunction with the safeguards developed by these models, is able to identify real-time risk levels. Furthermore, to assess the scope of system damages resulting from a network intrusion, we propose a method for estimating the total cost incurred from the point of the occurrence of damage to that of recovery.

1 Introduction

Today's network systems are becoming growingly complex and also large in capacity. The rapidly-expanding information assets stored within networked environments are giving rise to new types of security accidents not encountered before. The latest pattern in network security accidents consists in the exploitation by hackers of system vulnerabilities in networks by breaking into a handful of system units to spread worms or to launch distributed denial of service (DDOS) attacks. This type of attack is able to paralyze the entire system of a large-scale network. Network attacks, by their very nature, cannot be entirely foreseen or prevented, as no system is totally infallible even with the seemingly most perfect defense architecture. Thus, a preventive measure is not the most effective means to guarantee security. Losses incurred in the event of a network security breach can be enormous especially for large-size network systems. Figure 1 below illustrates how an attack on one system can affect linked systems in remote locations to disable all functions related to the information assets within the network or to cause disruption of network-based services. To effectively tackle these problems, it is important that, in addition to resorting to preventive techniques based on risk analysis, a monitoring system be implemented to detect intrusions, as well as a system to shorten the steps in recovery procedures to more efficiently respond to network outages, should there be an attack [1-2]. However, existing risk analysis procedures are confined to identification of network assets, vulnerabilities and threat factors, and assessment of degree of exposure to damages and losses resulting from potential threats. Under these methods, it is far from easy to estimate asset values and the scope of damages and losses incurred.

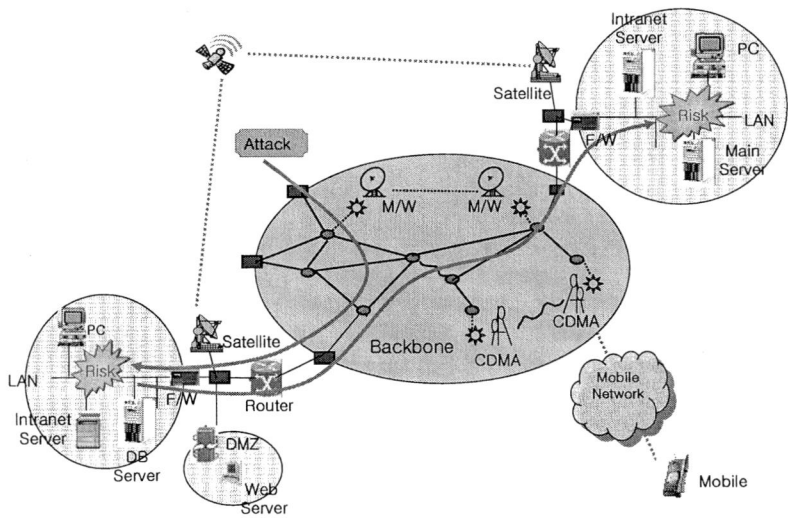

Fig. 1. Network Attack Damage Propagation Path

Accordingly, it is urgent to develop time-variant analysis techniques to move beyond these existing methods, and to compute varying risk levels over time on a real-time basis, and also techniques to estimate damages and losses sustained from the point of an attack to the final recovery. In other words, system administrators must implement preventive measures to minimize security accidents, and, once an accident occurs, must assess how much system damages have been reduced through the response procedures in place. Without an exact analysis of the effectiveness of measures in place, it is impossible to select adequate methods for protecting information. Hence, in this study, we propose a technique to calculate risk levels through a time-variant risk analysis and a method to estimate damages sustained in the event of a security accident, for large-size network systems.

2 Previous Risk Analysis Methods and Damage Estimation Models

2.1 Risk Analysis Methodologies

With the introduction of risk analysis and risk management techniques to the information security field, risk analysis standards and models such as ISO13335 [3], BS7799 Information Security System [4], System Security Engineering Capability Maturity Model [5], OCTAVE [6] and CSE[7] have been developed and are in use. Moreover, various automated risk analysis tools [8] are also developed and are popularly used. These techniques identify network assets and analyze vulnerabilities and threats specific to different assets to calculate risk levels. Risk levels are computed either by calculating threat probabilities corresponding to each loss over the total potential loss, or by multiplying the total asset value by the risk level and then dividing the result by the number of countermeasures in place. These methods, however, when applied to a large-size network system, reveal the following limitations:

- A system failure or exploit(e_i) affecting one asset(a_i) can and does affect the entire network(N).
- The linkage between assets(A), threats(T), vulnerabilities(V) and safeguards (S) does not necessarily correspond to the network reality.
- The information protection solutions are not measures designed for individual assets, threats or vulnerabilities, but are safeguards for the overall system.
- New threats and vulnerabilities continue to emerge during the risk analysis, and during the course of designing safeguards, these measures often are already irrelevant or outdated.
- As safeguards are established most often after the development and implementation of a system, the introduction of information protection solutions sometimes turns out to be impossible.

The extent to which one can anticipate new types of threats and vulnerabilities is highly limited. This is more so for systems using commercial software. As these applications are never entirely free of bugs, to radically eliminate the possibility of attacks by hackers targeting system bugs, one's only option is running a closed internal network. Even so, there are still possibilities for a system breakdown, whether by an internal member harboring resentment against the organization or through simple mistakes. A security program for real-time protection from system failure or intrusion therefore is a must in any case.

2.2 Damage Estimation Model

There have been numerous studies elaborating models for estimating costs of IT-related accidents. ICAMP [9], to calculate IT accident costs, took into account IT staff compensation cost, consultant compensation cost, user-side cost and equipment and software purchase cost. NPO Japan Network Security Association [10], in its damage calculation, distinguishes damage amounts into surfaced damage amount and potential damage amount. The amount of damage obtained through such a simple calculation formula, however, cannot reflect the variability of damage size depending on the circumstances. For example, an accident occurring during a weekend day or a holiday needs to be clearly distinguished from one taking place during peak business hours. Furthermore, the scope of damages can vary significantly depending on the type of security accident. Damages caused by email-distributed worms and virus, leaving relatively ample amounts of time to respond, are not comparable in scale to those resulting from network attacks exploiting system vulnerabilities, capable of wreaking havoc in a short time span by stealing, deleting or modifying critical information.

3 Time-Variant Risk Analysis for Large-Scale Networks

To begin with, circuit failures and equipment malfunctioning constitute a distinct area from system security, and two different staff members or specialists therefore must be assigned to these two duty areas. Repeated risk analysis procedures are unnecessary and simply burdensome for the concerned consultant or system administrator. Accordingly, for systems having already undergone a risk analysis, repeat analyses must be skipped, and analysis must be conducted only for necessary areas such as newly-generated assets, new threats and new vulnerabilities. The risk analysis concerning potential network attacks, while based on existing risk analyses [11-13], must include the process of identifying threats and vulnerabilities on a real-time basis. The safeguard must be formulated according to real-time assessment, and must be immediately implemented. This requires a sensor allowing immediate detection of threats and vulnerabilities. Large-scale network systems would need information protection solutions such as an early alert system, an ESM or a security monitoring system. The time-variant risk analysis presented in this paper assumes that these systems are already in place. These early detection tools are of critical importance, as a network assault by today's sophisticated hackers can disable an entire system in a matter of hours. For the purpose of calculating risk levels based on potential threats and vulnerabilities, in this paper, we assume the following:

Assumptions

- New threats/vulnerabilities occur periodically.
- Newly-detected threats/vulnerabilities are distinct from previous ones.
- Newly-detected threats/vulnerabilities, if provided with adequate safeguard, do not increase the level of risk deriving from the same type of threats/vulnerabilities.

When new threats and vulnerabilities continue to mount on a periodical basis, risk levels for large-scale network systems naturally increase. These risks can be reduced or brought to a standstill by implementing corresponding countermeasures. Figure 2 below illustrates the change in levels of threats/vulnerability and risk over time, when necessary countermeasures are adopted:

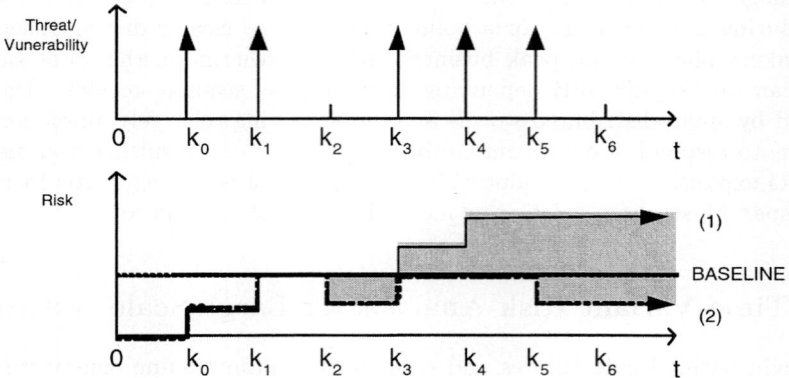

Fig. 2. Time-variant Risk Analysis

As can be noted in the figure above, when new threats/vulnerabilities occur over time, the risk level continues to rise (Curve 1). Here, we divided the risk levels into 5 different grades. When the risk level reaches grade 3 or above, the implementation of a countermeasure must be considered. Once the countermeasure is put into place, the risk level declines, as illustrated by Curve 2. As there is always a time lag between the detection of threats/vulnerabilities and the adoption of corresponding countermeasures, the risk can continue to increase in the meantime; timely response and implementation of countermeasures are of paramount importance. The area between Curve 1 and Curve 2 indicates the margin by which the risk level has been reduced. Figure 3 below is a block diagram illustrating the time-variant risk analysis.

- $N_{(}C, I, A)$: Network system(Confidentiality, Integrity, Availability), $N_{(C,I,A)} = \{1, 2, 3, 4, 5\}$
- $S(k)$: Security solution(Prevention,Detection,Recovery), $S_{(P,D,R)} = \{1, 2, 3, 4, 5\}$
- $r(k)$, $y(k)$: Risk rate, $r(k) = \{1, 2, 3, 4, 5\}$
- $e(k)$: failure, exploit, $e(k) = \{1, 2, 3, 4, 5\}$
- $t(k)$: Threat, $t(k) = \{0, 1\}$
- $v(k)$: Vulnerability, $v(k) = \{0, 1\}$
- k : time

The security risk level equation for the above-defined system is as follows:

$$e_k = y_k \bigoplus (v_k \bigodot t_k), N_{(C,I,A)}\{e_k\} = C\{e_k\} \bigotimes I\{e_k\} \bigotimes A\{e_k\}$$

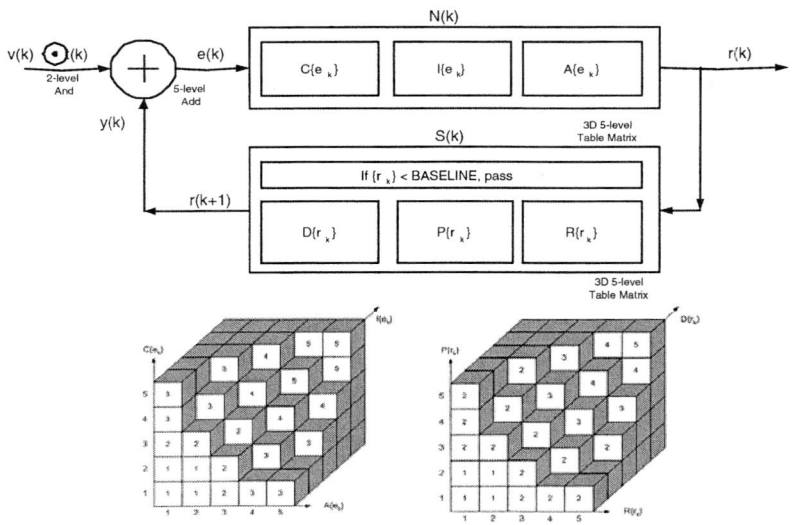

Fig. 3. Equation of State for a Large-scale Network System

$$C\{e_k\} = C\{e_1, e_2, \ldots, e_i\}, I\{e_k\} = I\{e_1, e_2, \ldots, e_i\}, A\{e_k\} = A\{e_1, e_2, \ldots, e_i\}$$

$$S_{(P,R,D)}\{e_k\} = P\{e_k\} \bigotimes R\{e_k\} \bigotimes D\{e_k\}$$

$$P\{e_k\} = P\{e_1, e_2, \ldots, e_i\}, R\{e_k\} = R\{e_1, e_2, \ldots, e_i\}, D\{e_k\} = D\{e_1, e_2, \ldots, e_i\}$$

$$r_k = \overline{N_{(C,I,A)}e_k}, r_{k+1} = y_k = S_{(P,R,D)}\{r_k\}$$

where, $\bigotimes$ is 3Dimension Table matrix.

4 Damage Estimation

Previously, the main concern for network systems was increasing usability. Issues at hand were therefore limited to controlling and maintaining the system to prevent network failures and downtimes and malfunctioning caused by software bugs. This is no longer the case with today's network environment, relentlessly challenged by intrusion attempts and hacker attacks. Thus, the top concern for system administrators is now safeguarding the network resources from external assaults and being able to provide real-time services without disruption or outage. However, units and personnel in charge of system security have thus far failed to adequately respond to attacks on a real-time basis. Unlike equipment or application malfunctions, which can be dealt with simply by replacing the problem-causing assets, responding to a network attack requires the elimination of root causes of security breach and involves measures to stop further propagation of damages sustained. When an intrusion occurs at the level of a single asset, the state of this asset is as follows: when an intrusion is made, the targeted asset experiences extra network loads. The increased traffic loads cause service disruptions and slowdown, leading eventually to a system halt. Figure 4 below are graphs illustrating this state. The related damages may consist of different types of functions.

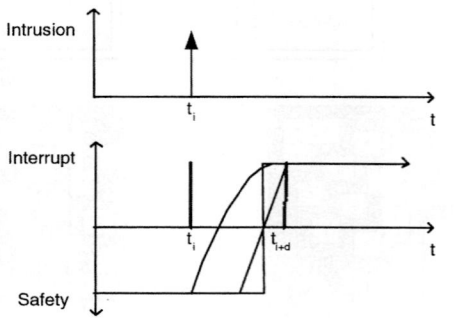

Fig. 4. A Damage Analysis on a Single Asset Unit

To simulate the characteristics of a single asset unit, including type of asset, type of intrusion, and whether or not security patches, upgrades and service packs have been applied, these variables can be expressed as follows:

```
class Result Asset (int I_Category, class A_Type)
   R.delay = td - ti;
   R.interrupt =  t(i+d);
   A.Security.Service = parameter;
   A.Security.Update = level;
   Case Virus : R.wave = Linear_Fn();
   Case Worm:R.wave=exponential_Fn();
   Case Hacking : R.wave = Pulse_Fn();
   Case Mistake  : R.wave = Log_Fn();
   Case end;
```

Estimation of intrusion-related damage sustained by a large-scale network system can be calculated by adding up the damage amounts corresponding to each single asset unit. An attack targeting a single asset triggers a chain reaction over the entire network. The speed of such propagation and its scope are illustrated in Figure 5 below.

In this graph where $f(t)$, the function of damage occurred, and $g(t)$, the function of recovery, the total damage amount is equal to the sum of the damage created during the intrusion and the cost incurred during the recovery efforts.

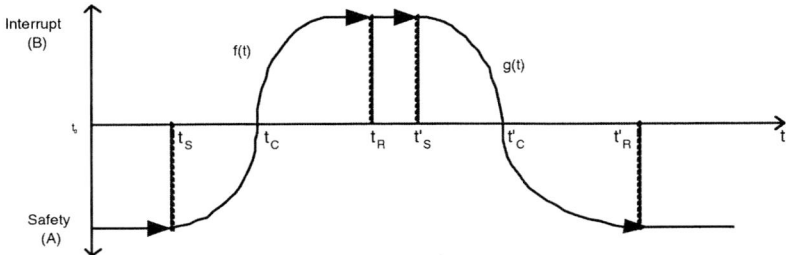

Fig. 5. Intrusion Damage and Recovery for a Large-scale Network

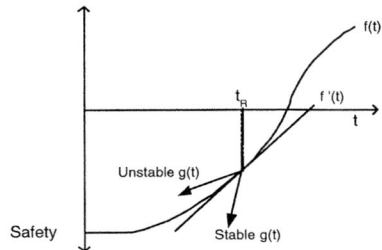

Fig. 6. Stable vs. Unstable State

$$R = A \times \{\{(t_R - t_S) \times (-100) - \int_{t_S}^{t_C} f(t)dt\} - 100 \times (t_S - t_R)$$

$$+\{(t'_R - t'_S) \times (-100) - \int_{t'_C}^{t'_R} g(t)dt\}\} + B \times \{\int_{t'_C}^{t'_R} f(t)dt + 100 \times (t_S - t_R) + \int_{t'_S}^{t'_C} g(t)dt\}$$

Here, damage amounts(A), in cases where system overloads occur, include administrative and operating costs, and, in cases(B) where the intrusion leads to a system outage, cost of service loss in addition to administrative and operating costs. Further, the calculation estimates the basic overhead costs, corresponding to monitoring and early alert system costs, which are classified into prevention, detection and recovery costs. Prevention cost corresponds to expenditures related to preventing network intrusions; and detection cost, to expenses incurred for monitoring and detecting intrusions. Finally, recovery cost is the cost of system recovery, when an attack has been sustained.

Next, in order to minimize the damage, system administrators must speedily respond to an attack, before it propagates throughout the network. A response must be made before the system comes to a halt, and the speed of response must exceed the rate of acceleration of the attack's damaging effects. Accordingly, using the slope of the functions $f(t)$ and $g(t)$, one can determine the level of response. If the inclination of $g(t)$ is higher than that of $f(t)$, this is an indication toward the adequacy of response. The inverse suggests the opposite, and the damage will continue to snowball.

- Stable Case :
$$\mid \frac{d}{dt}g(t) \mid >= \mid \frac{d}{dt}f(t) \mid$$

- Unstable Case :
$$\mid \frac{d}{dt}g(t) \mid < \mid \frac{d}{dt}f(t) \mid$$

5 Implementation

We developed an automated risk analysis tool and estimated intrusion damages. The automated tool is based on a server/client structure, and calculates, when a risk analysis consultant enters a specific potential threat/ vulnerability into the common item field and link the potential threat to a security measure, the risk level before the implementation of the countermeasure and the residual risk level after the implementation. Figure 7 below gives the results of a time-variant risk analysis performed using this tool. The example shows the risk levels for a large-size network system, when a threat against a specific router is entered, for the respective cases where a safeguard measure was applied to the router in question, and where no action was taken. The risk level, as can be observed in the figure, is reduced, when a safeguard is adopted.

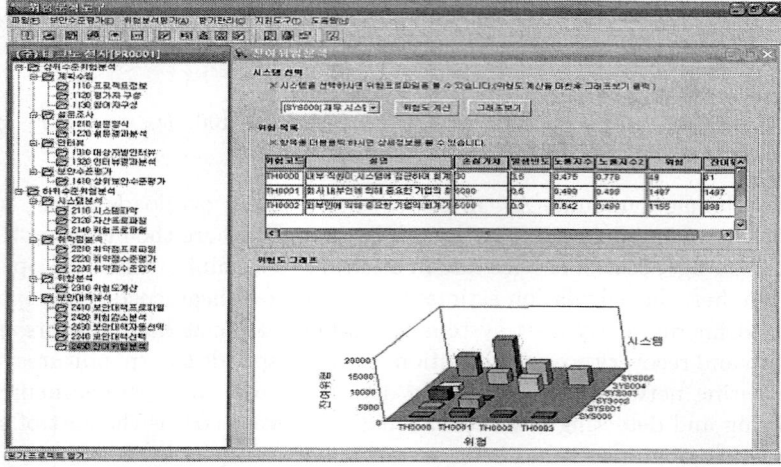

Fig. 7. Risk Analysis Tool and Damage Estimation result Developed by this Study

6 Conclusions

Large network systems run the inherent risk of seeing their entire system grinding to a halt in the event of an attack on a single system unit. Hence, in addition to preventive measures against security accidents, they must be able to assess the actual risk levels based on the scope of recovery following an accident. Previous works on network intrusion and related damage and recovery have been mostly centered on information assets at risk. Accordingly, there has been a lack of techniques for analyzing risks run by large-size network systems along a time axis and estimating potential damages to be sustained. In this study, we proposed a risk analysis model capable of estimating damages from network attacks for large-scale network systems and analyzing risk levels on a real-time basis. By introducing the time concept, this model notably enables a comprehensive assessment of the effectiveness of countermeasures in place in fending off risks and controlling damages, in addition to risk analysis.

References

1. Chandana Lala and Brajendra Panda, "Evalusating Damage From Cyber Attacks:A Model and Analysis," Systems, Man and Cybernetics, Part A, IEEE Transactions on, Volume:31 Issue:4, July 2001.
2. S. Jajodia et al., "Trusted recovery," Commun. ACM, vol. 42, no. 7, pp.71-75, July 1999.
3. ISO/IEC TR 13335, Information technology - Guidelines for the management of IT Security:GMITS, 1998.
4. Information Security Management, Part 2. Specification for Information Security Management System, British Standards Institution(BSI).
5. http://www.sse-cmm.org, System Secuity Engineering-Capability Maturity Model:SSE-CMM.
6. CMU, OCTAVE(Operationally Critical Threat, Assets and Vulnerability Evalustion), 2001. 12.
7. CSE(Canadian Secutiy Establishment), A Guide to Security Risk Management for IT Systems, Government of Canada, Communications Security Establishment(CSE), 1996.
8. http://www.cramm.com, CRAMM(CCTA Risk Analysis and Management Method).
9. A Report to the USENIX Association, Incident Cost Analysis and Modeling Project, I-CAMP II, 2000.
10. Janpena Security Association, Fiscal 2003-Information Security Incident Survey and Damage Calculation Model, March, 2004.
11. YoonJung Jung, Injung Kim, SeungHyun Kim, Hoh Pete In, "A Model-Based Security Risk Analysis Too," Proceeding of 2nd Second International Summer School Risks and Challenges of the Network Society by IFIP, 2003.
12. YoonJung Jung, Injung Kim, JoongGil Park, Dongho Won, "A Practical Security Risk Analysis Process of Information System," 4th APIS2005, 2005.
13. Young-Hwan Bang, YoonJung Jung, Injung Kim, Namhoon Lee, GangSoo Lee, "The Design and Development for Risk Analysis Automatic Tool," ICCSA 2004, LNCS 3043, pp.491-499, 2004.
14. Injung Kim, YoonJung Jung, JoongGil Park, Dongho Won, "A Study on Security Risk Modeling over Information and Communication Infrastructure," SAM04, pp. 249-253, 2004.

Efficient Multi-bit Shifting Algorithm in Multiplicative Inversion Problems

Injoo Jang and Hyeong Seon Yoo

School of Computer Science and Engineering, Inha University,
Incheon, 402-751, Korea
g2032076@inhavision.inha.ac.kr, hsyoo@inha.ac.kr

Abstract. This paper proposes an efficient inversion algorithm for Galois field $GF(2^n)$ by using a modified multi-bit shifting method. It is well known that the efficiency of arithmetic algorithms depends on the basis and many foregoing papers use either polynomial or optimal normal basis. An inversion algorithm, which modifies a multi-bit shifting based on the Montgomery algorithm, is studied. Trinomials and AOPs (all-one polynomials) are tested to calculate the inverse. It is shown that the suggested inversion algorithm reduces the computation time 1 ~ 26 % of the forgoing multi-bit shifting algorithm. The modified algorithm can be applied in various applications and is easy to implement.

1 Introduction

There has been an increasing attention in the design of fast arithmetic operations in the Galois field $GF(2^n)$ which have many industrial applications including cryptography. In these applications, efficient arithmetic algorithms and hardware structures are crucial factors for good performance. The efficiency of these basic operations is closely related in the way of presenting element bases. It is also related with the kind of irreducible polynomials for polynomial bases.

The most popular inversion algorithms are either based on Fermat's little theorem, on the Montgomery algorithm or related ones [1,2,3,4,5,6]. The Itoh and Tsujii inversion algorithm is developed for extension fields $GF(q^m)$ [2,3]. There are also many papers about Montgomery methods that are shown to be effective in modular arithmetic and multiplicative inversion problems that are prime importance in cryptographic applications [5,6]. And software impelmenting issue is covered in a book by Rosing [7]. In a series of papers, Koç et al published papers about Montgomery algorithm in which covers multi-bit shifting method [8,9]. They adapted multi-bit shifting method to eliminate the repetitive work, and the 3-bit was an optimal number for efficient implementation.

In this paper we are forcusing the multi-bit method in Montgomery algorithm. Our first idea is that if we modify the algorithm to adapt multi-bit case not just for 3-bit case but also for general multi-bit, we could reduce the number of repetitive works and save the computation time. Since the performance might be different irreducible polynomial types, the second one is to find a proper irreducible type for multi-bit implementations.

2 The Montgomery Inversion Algorithm

The Montgomery inversion algorithm is defined as in eq.(1),

$$b = a^{-1}2^n \pmod{p}, \quad p > a > 0 \tag{1}$$

where p is a prime number, and $n = \lceil \log_2 p \rceil$ [4,5,9].

The algorithm consists of two phases. The phase I computes the almost Montgomery inversion of a, integer r such that $r = a^{-1}2^k \pmod{p}$, where $n \leq k \leq 2n$.

$$(r,k) := AlmMonInv(a) = a^{-1}2^k \pmod{p} \tag{2}$$

The phase II is the correction step and calculates a slightly different Montgomery inverse, eq.(3) [9].

$$b = MonInv(a2^n) = a^{-1}2^n \pmod{p} \tag{3}$$

This algorithm can be modified for the binary extension field GF(2^n). Let $p(x)$ be an irreducible polynomials over GF(2) and $a(x)$ be an element of GF(2^n), eq.(4).

$$\begin{aligned}p(x) &= x^n + p_{n-1}x^{n-1} + p_{n-2}x^{n-2} + \cdots + p_1x + p_0 \\ a(x) &= a_{n-1}x^{n-1} + a_{n-2}x^{n-2} + \cdots + a_1x + a_0, \quad a_i \in \{0,1\}\end{aligned} \tag{4}$$

The arithmetic operations on the elements in GF(2^n) can be done as the conventional polynomial arithmetic. The Montgomery inversion algorithm is as following [9]:

Algorithm A
Phase I
Input : $a(x)$ and $p(x)$, where $\deg(a(x)) < \deg(p(x))$
Output : $s(x)$ and k, where $s = a(x)^{-1}x^k \pmod{p(x)}, \deg(s(x)) < \deg(p(x))$
1: $u(x) := p(x), v(x) := a(x), r(x) := 0$, and $s(x) := 1$
2: $k := 0$
3: while ($u(x) \neq 0$)
4:    if $u_0 = 0$ then $u(x) := u(x)/x, s(x) := xs(x)$
5:    else if $v_0 = 0$ then $v(x) := v(x)/x, r(x) := xr(x)$

6:     *else if* $\deg(u(x)) \geq \deg(v(x))$ *then*
$$u(x) := (u(x)+v(x))/x, \; r(x) := r(x)+s(x), \; s(x) := xs(x)$$
7:     *else* $v(x) := (v(x)+u(x))/x, \; s(x) := s(x)+r(x), \; r(x) := xr(x)$
8:    $k := k+1$
9: *if* $s_{n+1} = 1$ *then* $s(x) := s(x) + xp(x)$
10: *if* $s_n = 1$ *then* $s(x) := s(x) + p(x)$
11: *return* $s(x)$ *and* k

Phase II

Input: $s(x)$ and k *from*   Phase I
Output: $b(x)$ where $b(x) = a(x)^{-1} x^{2n} \pmod{p(x)}$
12:    *for* $i = 1$ to $2n - k$ *do*
13:        $s(x) := xs(x) + s_{n-1} p(x)$
14:    *return* $b(x) := s(x)$

The variant of multi-bit shifting Montgomery algorithm uses 3-bit shifting. The Three bits for shifting is rather an optimal number especially when considered the implementation by hardware. To check more than one bit at a time, it is needed a more complicated hardware. Therefore we must consider whether it will be worthwhile. In this point of view, three-bits is the most proper number. The 3-bit shifting algorithm modifies the step 4 and 5 in Algorithm A, [9]. They check and select the three least significant bits of $u(x)$ or $v(x)$ instead of routine checking. The variant method can be described as following:

Algorithm A-1

4:      *if* $u_2 u_1 u_0 = 000$ *then* $\{u = ShiftR(u,3); \; s = shiftL(s,3); \; k = k+2\}$
4.1:   *if* $u_2 u_1 u_0 = 100$ *then* $\{u = ShiftR(u,2); \; s = shiftL(s,2); \; k = k+1\}$
4.1:   *if* $u_2 u_1 u_0 = X10$ *then* $\{u = ShiftR(u,1); \; s = shiftL(s,1)\}$
5:      *if* $v_2 v_1 v_0 = 000$ *then* $\{v = ShiftR(v,3); \; r = shiftL(r,3); \; k = k+2\}$
4.1:   *if* $v_2 v_1 v_0 = 100$ *then* $\{v = ShiftR(v,2); \; r = shiftL(r,2); \; k = k+1\}$
4.1:   *if* $v_2 v_1 v_0 = X10$ *then* $\{v = ShiftR(v,1); \; r = shiftL(r,1)\}$

3   A New Approach

The Montgomery inversion algorithm, especially, performs repetitive work by the type of least significant digit. By considering hardware implementation property, the

3-bit shifting method was proposed in Algorithm A-1. The 3-bit was the optimal number since it increases the hardware complexity for large number of bits. Even in this case, we still reduce the computation time by employing multi-bit algorithm that can utilize many continuous zero terms during the arithmetic operations. Therefore, we might accelerate if we could handle more bits of zero. It is possible since the Montgomery inversion algorithm has repetitive work for bits of zero.

We modify the step 4 and 5 in order to check and adapt the continuous zero bits from the least significant digit of $u(x)$ or $v(x)$. The variable cZ counts the number of continuous zero-bit and determines the size of shifting. The modified multi-bit shifting algorithm is as following Algorithm B. In step 4.1, the variable cZ is updated while $u(x)$ has continuous zero bits. In the next step $u(x)$ and $s(x)$ are shifted with updated cZ. It follows the similar steps for $v(x)$ and $r(x)$.

Algorithm B

4: if $u_0 = 0$;

4.1:   then while $u_i = 0$; update cZ;

4.2 $u(x) := u(x) / x^{cZ}$;

     $s(x) := x^{cZ} s(x)$;

5: elseif $v_0 = 0$;

5.1:   then while $v_i = 0$; update cZ;

5.2: $v(x) := v(x) / x^{cZ}$;

     $r(x) := x^{cZ} r(x)$;

4 Examples

It is common that the number of bits in elliptic curve cryptography is between 150 and 250. In this range we choose several numbers such as {148, 156, 162, 178, 180, 196, 210, 228, 256} by considering irreducible polynomials with both trinomial and AOP types. We choose the four different sets of data, which has 3, 5, 9, and 13 non-zero terms. The points of non-zero terms are arbitrary selected on the entire region. The suggested multi-bit shifting Montgomery algorithm is tested on the 2.00G Hz Pentium machine.

The numerical results in Table 1 are for original Montgomery algorithm; Alg.A, 3-bit shifting algorithm; Alg.A-1, and the modified algorithm case; Alg.B. The points of non-zero term are (128, 64, 0).

Fig.1 shows the modified multi-bit shifting decreases the execution time to 10.77 ~ 26.65% for the trinomial prime and 0.97 ~ 20.15% for the AOP prime. The Tables 2, 3 and 4 give a little bit different numerical results but those also show the new method is more efficient than previous Montgomery methods. Using the 5 non-zero terms' polynomial, the modified multi-bit shifting algorithm decreases the execution time to 9.5 ~ 27.7% on the trinomial prime and 1 ~ 11.2% on the AOP.

Table 1. Inversion for the polynomial, which has 3 non-zero terms (m sec)

	Trinomial			AOP		
	Alg.A	Alg.A-1	Alg.B	Alg.A	Alg.A-1	Alg.B
148	3.64	3.13	2.98	3.97	3.84	3.83
156	3.55	2.97	2.84	3.91	3.61	3.44
162	4.17	3.56	3.41	4.42	4.16	4.14
178	3.91	3.30	3.13	4.78	4.69	4.70
180	4.06	3.67	3.55	4.81	4.80	4.77
196	4.34	3.95	3.86	5.77	5.00	4.69
210	4.64	4.23	4.14	6.19	5.41	5.08
228	4.28	3.39	3.14	6.05	5.22	4.83
256	5.95	5.45	5.31	6.02	5.27	4.91

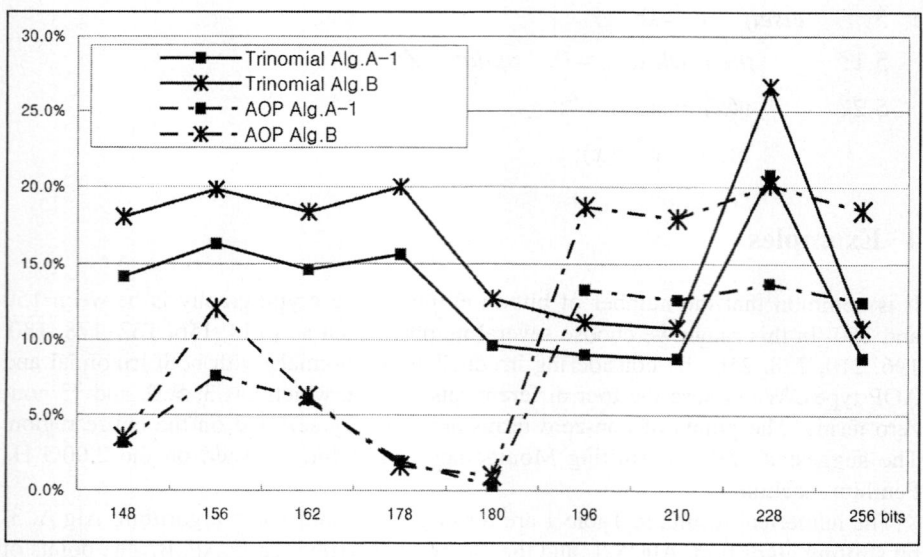

Fig. 1. The relative decrease of Alg.A-1 and Alg.B to Alg.A, for 3 non-zero terms

With the 9 non-zero terms' polynomial, the modified method decreases the execution time to 7.3 ~ 11.0% on the trinomial prime and 2.4 ~ 8.2% on the AOP. When we choose the non-zero points (160,140,120,100,80,60,40,20,0) in 178 bits, the modified algorithm decreases the consuming time 67.88% and for the point (200,175,150,125,100,75,50,25,0), About 71.53%, while on the other, 47.99% and 51.04% on 3-bit shifting operation.

With the 13 non-zero terms' polynomial, the modified method decreases the execution time to 7.7 ~ 12.2% on the trinomial prime and 6.2 ~ 11.1% on the AOP.

As we show in the follow graphs Fig.2, the case of the trinomial represented speed up about 7.14 ~ 26.65% for computing the Montgomery Inversion. In Fig.3, the case of AOP was showed speed up from 0.95% to 20.15%.

Table 2. Inversion for the polynomial, which has 5 non-zero terms(m sec)

	Trinomial			AOP		
	Alg.A	Alg.A-1	Alg.B	Alg.A	Alg.A-1	Alg.B
148	4.00	3.66	3.58	3.81	3.48	3.45
156	3.91	3.45	3.42	3.98	3.77	3.78
162	4.08	3.44	3.36	4.15	3.73	3.69
178	4.45	3.80	3.67	4.55	4.44	4.44
180	5.02	4.58	4.53	4.88	4.83	4.83
196	5.11	4.69	4.63	4.95	4.66	4.63
210	5.61	5.09	5.00	5.31	4.83	4.80
228	5.55	4.94	4.84	5.75	5.55	5.52
256	5.88	4.47	4.25	6.83	6.33	6.25

Table 3. Inversion for the polynomial, which has 9 non-zero terms(m sec)

	Trinomial			AOP		
	Alg.A	Alg.A-1	Alg.B	Alg.A	Alg.A-1	Alg.B
148	3.94	3.66	3.63	3.88	3.64	3.67
156	3.98	3.61	3.59	4.23	4.11	4.06
162	4.45	4.08	4.02	4.47	4.36	4.36
178	4.38	4.00	4.00	4.83	4.66	4.69
180	4.73	4.39	4.39	4.94	4.60	4.59
196	4.98	4.51	4.42	5.13	4.70	4.70
210	5.50	5.09	5.09	5.66	5.31	5.34
228	5.72	5.22	5.17	6.13	5.75	5.75
256	6.53	5.83	5.81	6.69	6.19	6.17

Table 4. Inversion for the polynomial, which has 13 non-zero terms(m sec)

	Trinomial			AOP		
	Alg.A	Alg.A-1	Alg.B	Alg.A	Alg.A-1	Alg.B
148	3.94	3.61	3.61	3.77	3.53	3.53
156	3.88	3.52	3.48	4.14	3.86	3.86
162	4.36	4.02	3.98	4.34	4.02	4.02
178	4.36	3.98	3.95	4.73	4.53	4.52
180	4.86	4.47	4.48	4.97	4.66	4.58
196	5.02	4.58	4.58	5.33	4.89	4.90
210	5.52	4.91	4.86	5.67	5.31	5.32
228	5.53	4.94	4.86	5.75	5.17	5.11
256	6.58	5.97	5.89	6.81	6.23	6.22

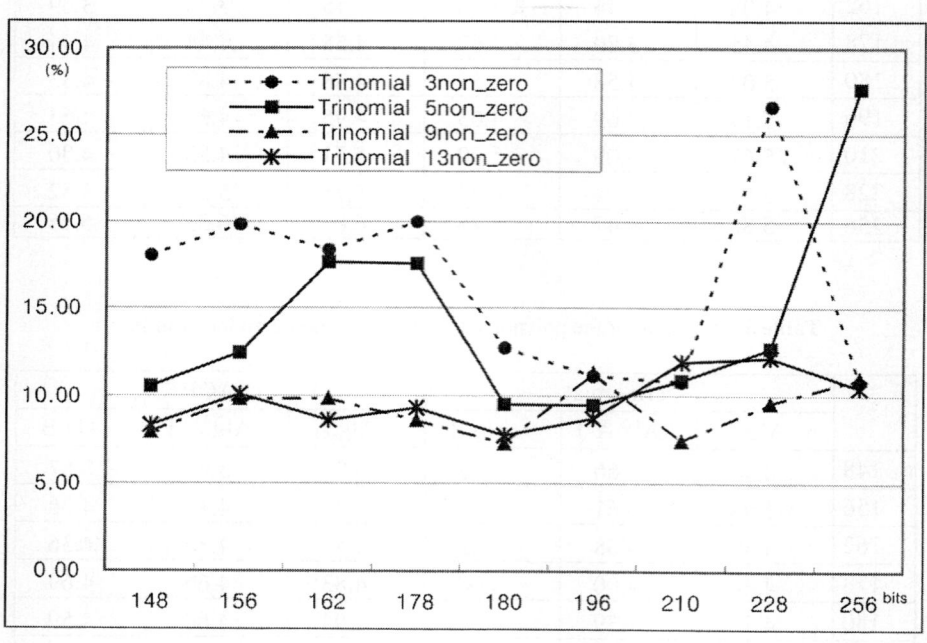

Fig. 2. The relative decrease of Alg.B to Alg.A, for trinomials

5 Conclusions

In this paper, we presented a modified multi-bit shifting Montgomery Inversion Algorithm allowing more than 3 zero-bits, and tested computation time for different number of bits and the prime polynomial types. The relative decreases for trinomial cases were about 7.14 ~ 26.65%. There were no correlations for different number of bits, but the relative decreases were relatively small for large non-zero input cases.

The AOP cases showed the same pattern as trinomials about from 0.95% to 20.15%, but were inferior to the trinomial cases in all test range. It is demonstrated that the suggested multi-bit Montgomery Inversion algorithm is easy to implement and does not take longer time than the previous Montgomery Inversion methods, even for the worst case.

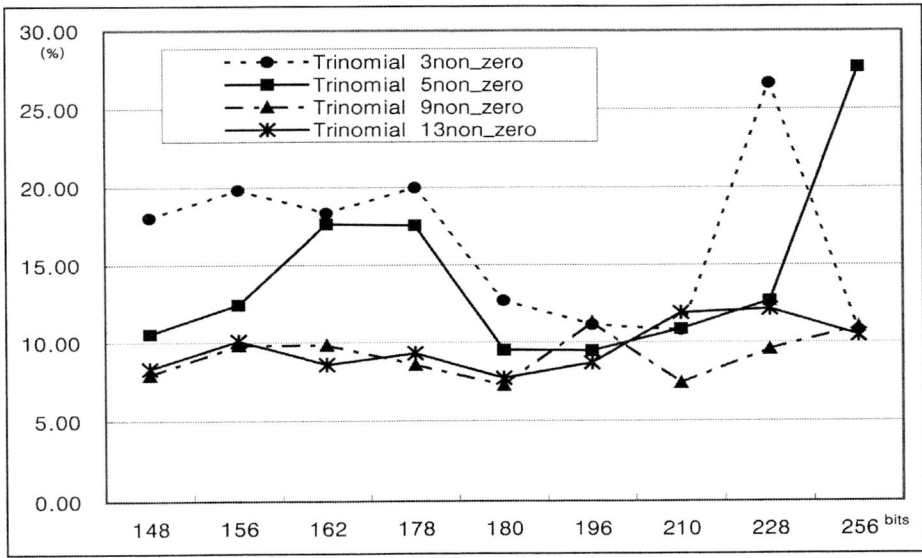

Fig. 3. The relative decrease of Alg.B to Alg.A , for AOPs

Acknowledgements

This research was supported by the MIC, Korea, under the ITRC support program supervised by the IITA.

References

1. T. Itoh and S. Tsujii, "A fast algorithm for computing multiplicative inverses in $GF(2^m)$ using normal bases," Information and Computation, 78, 171-177, 1988
2. J. Juajardo and and C. Paar, "Itoh-Tsujii inversion in standard basis and its application in cryptography and codes," 25, 207-216, 2002
3. D. V. Bailey and C. Paar, "Optimal extension fields for fast arithmetic in public-key algorithms," CRYPTO '98, LNCS 1462, 472-485, 1998

4. B.S. Kaliski Jr., "The Montgomery inverse and its applications," IEEE trans. On Computers, 44, 8, 1064-1065, 1995
5. C. K. Koç and T. Acar, "Montgomery multiplication in $GF(2^k)$," Design, Codes and Cryptography, 14, 1, 57-69, 1998
6. E. Savas and C. K. Koç, "The Montgomery modular inverse - Revisited," IEEE Trans. On Computers, 49, 7, 763-766, 2000
7. M. Rosing, *Implementing elliptic curve cryptography*, Manning Publ. Co., Greenwich, CT, 1999
8. E. Savas, A.F. Tenca, M.E. Ciftcibasi and C. K. Koç, "Multiplier architecture for GF(p) and $GF(2^n)$," IEE Proc. – Computers and Digital Tech., 151, 2, 2004
9. E. Savas, M. Naseer, A. A-A. Gutub, and C. K. Koç, "Efficient unified montgomery inversion with multibit shifting," Proceedings – Computers and Digital Tech.," to appear, 2004

Modified Token-Update Scheme for Site Authentication

Joungho Lee, Injoo Jang, and Hyeong Seon Yoo

School of Computer Science and Engineering,
Inha University, Incheon, 402-751, Korea
hsyoo@inha.ac.kr

Abstract. This paper proposes a new site authentication algorithm using a token-update. One-time password and smart card schemes are widely studied to protect servers from various malicious activities. Most papers discuss schemes with nounce and token updates. Since the LAN card' ID could be used to identify a PC, an authentication scheme with LAN card's Mac address might strengthen the conventional token update method. The suggested modified token-upgrade method algorithm is shown to be very effective in site authentication.

1 Introduction

It is common that multi-user systems require id and password to identify and permit logging onto the system. The user's id and password can be transferred either in plaintext or cipher-text. Since the transmitted data are always open to an attacker, the simple id and password scheme has weak points for authentication [1,2,3]. One time password system (OTP) is studied to prevent from repeated use of the logon data [3]. Using smart card systems that have memory and computing power can further strengthen OTP. In this system user selects a logon data and gets a smart card identifier (CID) through the key information center [KIC]. The CID is stored in a smart card and is used for mutual authentication [4,5]. The authentication scheme with smart card has some advantages such as it does not need a verification table and can authenticate each other. Users can choose their own passwords and it provides computation capability [5].

An authentication scheme without KIC is studied, which updates an authentication token with mutual random numbers [6,7]. The random numbers are generated by each identity and transferred to update the token. In this system, there is no need to have authentication table, but it transfers the data by plaintext type and likely to loose crucial data to an attacker by packet sniffing.

In this paper, we propose a modified token update method that uses PC's id numbers and encrypted data transmissions. Since the Mac address of LAN card could be used as an identifying number, it might replace the KIC. If we change and encrypt the login data, we can have secure authentication scheme.

2 Password Authentication Scheme

In the OTP system, server (S) selects nounce N and successively computes hash value N times $h(R)^{(N)}$ and sends N-1 and R to clients. The clients (C) calculate and return hash value $h(R)^{(N-1)}$. The server computer calculates hash with the transmitted value and compare with the stored hash value $h(R)^{(N)}$. It two values are equal the authentication step is finished [2,3]. In this system if N is constant, it is possible for an attacker to fool by stealing N. This process is described as following.

OTP Authentication:

S selects N, computes $h(R)^{(N)}$;

1. $S \rightarrow C$: $N-1, R$;
2. $C \rightarrow S$: $h(R)^{(N-1)}$
3. *S calculates* $h(h(R)^{(N-1)})$,

   accepts   if $h(R)^{(N)} = h(h(R)^{(N-1)})$

   rejects   otherwise

Peyravian and Zunic proposed a password transmission scheme for remote user applications [6]. User submits his user identification (id) and password (pw) to the remote user (client; C). The client generates and sends a random number (rc) with id to the server. The server returns a random number (rs). The client generates a hash of id and pw. He also generates and sends a onetime token with his id to the server. The server verifies the onetime token. In this scheme the server keeps the hash of id and pw, not pw itself.

1. $U \rightarrow C$: $\{id, pw\}$
2. $C \rightarrow S$: $\{id, rc\}$
3. $S \rightarrow C$: $\{rs\}$
4. C *calculates* *token* $= h(h(id, pw), rc, rs)$
5. $C \rightarrow S$: $\{id, token\}$
6. *S calculates   token**

   accepts   if   token* = token

   rejects   otherwise

In this scheme the onetime token is changing by random numbers for each logon, so it gives no information to logon. But if an attacker obtains every transmitted variable such as id, pw, rc and rc, then it is possible to get information. So it is necessary to encrypt or hashed them. Yang and Shieh published a paper that use nounce-bases password authentication scheme, but it needs a CID through the key information center [5].

3 A Modified Token-Upgrade Approach

If we use any PC's identifying number, there will be no need for adopting agency like KIC. So it is proposed to use Mac address of LAN card that is given as manufacturing. The Mac address is a physical address can be a site identifying number. The weak points by sending plaintext random numbers could be avoided by employing encrypting algorithms and hash functions. We restrict authentication problems to registered users (U) and adopt Mac address scheme with token update.

3.1 Registration Phase

Registration phase is described as following steps and Fig.1 is presented to illustrate the steps.

(R.1) U $construct$ $token1 = (id \parallel pw \parallel ma)$

A user starts by construct token1 with user identification (id), password (pw) and Mac address (ma).

(R.2) $U \rightarrow S : \{h(token1), \varepsilon_{ks}(token1)\}$

He computes hash of token1 h(token1) and encrypts token1 with server's public key ks. The user sends hash value of token1, and encrypted token1 to server.

(R.3.1) S $decrypts\ and\ reconstructs$ $h(token1)*$

$accepts$ $if\ h(token1) = h(token1)*, returns$ $\varepsilon_{ku}(h(rs))$

$rejects$ $otherwise.$

Server reconstructs hash value of token1 from the encrypted token1. If two hashes are identical, server accepts login phase. The server selects a random number (rs) and returns encrypted hash of random number rs to the user.

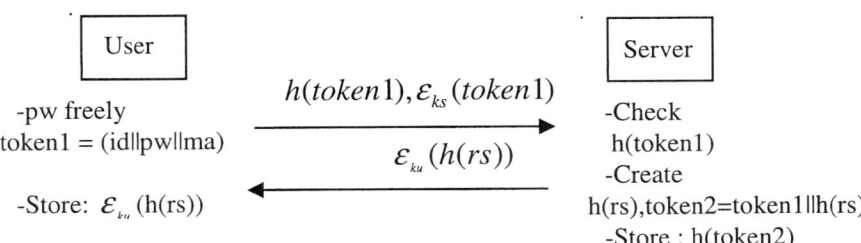

Fig. 1. The registration phase of the proposed scheme

(R.3.2) S $constructs$ $token2 = token1 \parallel h(rs),$ $h(token2)$

Server constructs token2 with token1 and h(rs), and keeps hash value of token2.

3.2 Login and Authentication Phase

In login and authentication phase, it is designed to update token2 by renewing the random number. The scheme is shown in Fig.2.

(A.1) U decrypt $\varepsilon_{ku}(h(rs))$
  construct token2 = token1 $\parallel$ h(rs)
  $\rightarrow S:$ {id, h(token2)}

User decrypts the transferred encryption data, and constructs topken2 with token1 and hash of random number h(rs). He sends to the server id and hash of token2 h(token2).

(A.2) S accepts if $h(token2) = h(token2)*$
  updates token2* with rs*, returns $\varepsilon_{ku}(h(rs*))$
  rejects otherwise.

Server checks the hash of token2 h(token2). If two hashes are identical, server accepts login and authentication phase. The server selects a new random number rs*, and returns the encrypted hash of rs* to the user. The updated hash of token2* is stored for another login phase.

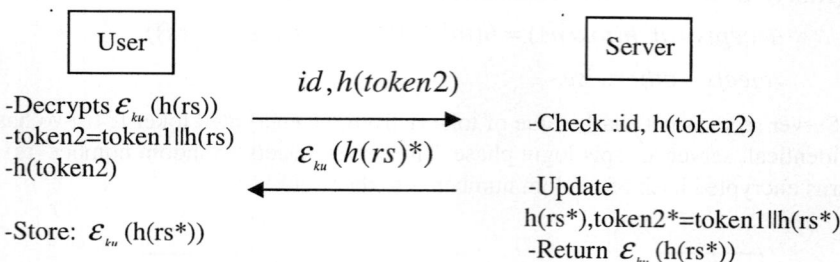

Fig. 2. The login and authentication phase

3.3 Password Recovery Phase

Password recovery phase is designed for recovering password, as in Fig.3.

(R.1) U constructs token3 = {id $\parallel$ ma $\parallel$ h(rs)}
  $\rightarrow S:$ {h(token3).ε_{ks}(token3)}

User starts by computing token3 combining id, ma and updated h(rs). He sends to the server id, hash of token3, and encrypted token3.

(R.2) S reconstructs $h(token3)*$
accepts if $h(token3) = h(token3)*$, returns $\mathcal{E}_{ku}(id, pw, h(rs))$
rejects otherwise.

Server reconstructs token3 and checks if the two hashes are identical. If they are identical, returns encryption of id, pw, and h(rs).

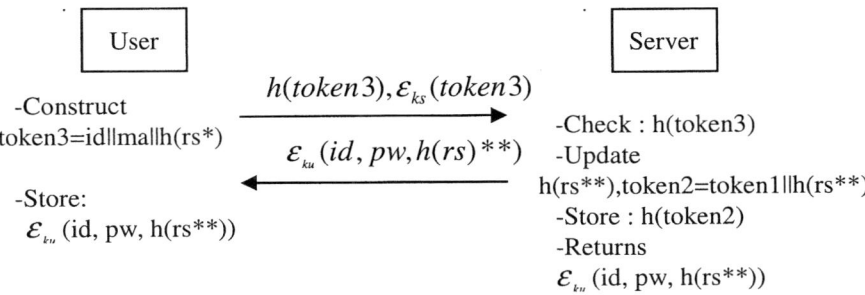

Fig. 3. The password recovery phase

4 Security Analysis

4.1 Password-Guessing Attack

Since every user would select easy password, it might be possible for an attacker to guess password in registration or login and authentication phase. But it could not be as there is no plaintext transmission in the suggested scheme. The crucial data such as id, pw, and ma are all transferred as hash or encrypted values. Even if there is a chance to get userid (id) and password (pw), an attacker still need Mac address (ma) to fool the server. The ma is not possible for a remote logger to obtain in Linux or Unix system.

4.2 Replay Attack

An attacker may try to login with id and old h(token2), but the token2 is updated every login, server will check the old, invalid hash. The token2 consists of registration data token1 and hash of one time random number (rs), so it is impossible to reuse the old hash of token2. And the random number rs is also hard since it is double hashed.

4.3 Packet Sniffing

An attacker may obtain login packet such as id, hash of token1 and encrypted hash, but it is impossible to obtain the token1 by decrypting the encrypted data. And it is also impossible to use the old packet of hashed token2, since it is updated every login time. In the password recovery phase, it is also forbidden to use the packet, as the transmission is by hash and encrypted data.

4.4 Forgery Attack

An attacker may try to register by using possible leaked id, pw, and a forged ma. But it is not possible since there is a registered user. If an attacker gets token1, it is impossible to construct token2 since he has no decryption key.

5 Conclusions

In this paper, we suggested a modified token-update scheme with Mac address that is hard to obtain for a possible remote attacker. The modified scheme adopts one-time random number; it is safe in various circumstances such as password-guessing attack, replay attack, packet sniffing, and forgery attack. In this scheme a user is free to select id, pw, but does not need to generate a random number. Only the registered user can authenticate since the necessary token2 is updated every login time. The suggested scheme can provide an efficient password recovery method by the Mac number and a random number by the server.

Acknowledgements

This research was supported by the MIC, Korea, under the ITRC support program supervised by the IITA.

References

1. M. Bishop, "Password Management," Proceedings of COMPCON, 167-169, 1991
2. B. Soh and A. Joy, "A Novel Web Security Evaluation Model for a One-Time-Password System," Proceedings of the IEEE/WIC, WI('03), IEEE Computer Society, 2003
3. J. Archer Harris, "OPA: A One-time Password System," ICPPW'02, IEEE Computer Society, 2002
4. W.S. Juang, "Efficient Password Authenticated Key Agreement Using Smart Cards," Computers & Security, 23, 167-173, 2004
5. W. Yang and S. Shieh, "Password Authentication Schemes with Smart Cards," Computers & Security, 18, 727-733, 1999
6. M. Peyravian and N. Zunic, "Methods for Protecting Password Transmission," Computers & Security, 19, 466-469, 2000
7. T.H. Chen, W.B. Lee and G. Horng, "Secure SAS-like password authentication schemes," Computer Standards & Interfaces, 27, 25-31, 2004

A Study on Secure SDP of RFID Using Bluetooth Communication

Dae-Hee Seo[1], Im-Yeong Lee[1], and Hee-Un Park[2]

[1] Division of Information Technology Engineering, SoonChunHyang University,
#646, Eupnae-ri, Shinchang-myun, Asan-si, Coogchungnam-Do,
336-745, Republic of Korea
{patima, imylee}@sch.ac.kr
http://sec-cse.sch.ac.kr
Phone +82-41-542-8819   Fax +82-41-530-1548
[2] Korea Information Security Agency (KISA), Republic of Korea
hupark@kisa.or.kr

Abstract. Recently, much research has been actively conducted for a new kind of network environment ubiquitous computing. This paper will define the essential technology called "ad hoc network" and the smart-tag technology required by a ubiquitous environment. We will describe how to apply smart-tag-related Radio Frequency Identification(RFID) research for Bluetooth, a local-area wireless-communication technology. In order to implement RFID technology for the ubiquitous-computing infrastructure, a number of important technology factors and structures should be considered. These include the realization of low-priced tags and the provision of technical service for the tag's security. For the passive RFID tags, the functions of each RFID tag and maintenance service should be considered to guarantee the price efficiency. As for the active RF tags' support of local-area wireless communication, one of the main issues is enhancing the level of security. This paper will present secured RF-tag service for the RF and Bluetooth modes for local-area wireless communication. A method of applying the generated service to the EPC code developed by MIT will also be suggested.

Keywords: RFID, Bluetooth, Service Discovery Protocol.

1 Introduction

The ubiquitous-network environment optimizes the user's connection to a network by intelligently capturing the user's circumstances or environment[1]. In addition, it arranges a network to utilize contents freely and securely. The technical requirements in assembling a ubiquitous network include a flexible broadband, teleportation, an agent, contents, an appliance, a platform, and a sensor network. Among these requirements, the ubiquitous-sensor network is most essential because it collects and

---

[1] This research was supported by the Program for the Training of Graduate Students in Regional Innovation which was conducted by the Ministry of Commerce Industry and Energy of the Korean Government.

manages information autonomously by communicating with peripheral equipment around users. More importantly, as an essential ubiquitous technology, RFID enables information exchange by recognizing and sensing information remotely through wireless communication. Since RFID is expected to replace the current offline bar-code system, more applications will be possible not only for individual consumers but also for entire industries. Not surprisingly, there has been vigorous research and development for RFID these days. The research on the RFID system mainly focuses on security issues for the RF reader and RFID tags. Due to its low price, RFID tags can be utilized not only by individual users but also by industrial users. Using RFID tags for most aspects of industrial operation, however, brings with it a major security issue. Chapter 2 presents a brief description of the RFID system and Bluetooth Service Discovery Protocol (SDP). Chapter 3 contains an analysis of the security weakness of the current RFID system. This paper also presents the additional security requirement to construct an RFID system using Bluetooth SDP. Chapter 4 suggests a secure and efficient RFID system satisfying the security requirements presented in Chapter 3. The suggested method is analyzed in Chapter 5, with conclusive results offered in Chapter 6.

2 Technology Overview

In This Chapter presents an introduction to the RFID system and Bluetooth SDP.

2.1 RFID System Overview

RFID is a wireless communication system composed of a RF reader having decoding capability and RF tags providing information. RFID is a wireless communication system consisting of an RF reader, which decodes signals, and RF tags, which provide information. It is an identification system in which unique information is embedded into objects such as machines and humans to distinguish one from the other. The embedded information can be decoded without contact, using wireless communication. Therefore, diverse offline applications could be automated with an RFID system. Its major characteristics are as follows:

- It is convenient to use since it is capable of simultaneous recognition of several tags.
- It can recognize fast-moving objects, paving the way for a more efficient usage of time.
- It is simple to install, depending on a system's characteristics or environment.
- It can be relatively inexpensive to maintain, and is very durable (free from wear) because the tag is read without direct contact.
- By programming tags with OTP, RFID provides a service secured against forgery or falsification.
- It is easy to upgrade.
- It can recognize signals in Mutual Authentication.

RFID is a system that functions as "The Internet of Things." Suggested by MIT Auto-ID Labs (formerly the Auto-ID Center), "The Internet of Things" is designed to sense items with wireless tags through the Internet (or a similar network) remotely and in real time. This system offers a new way to use the Internet. The service, however, will need billions of RFID tags and a much more efficient wireless network. It will also require new software and bar-code systems to handle numerous items and to support diverse applications.

2.2 Bluetooth SDP Overview

Nowadays the term Bluetooth is constantly used in electronics. The term dates back to the time of Viking Herald that brought Scandinavian countries under a single authority. Bluetooth is the name of a project aiming for a low-electric power and low-cost wireless interface-system. Ericsson in Sweden created this interface-system for wireless communication at a close range. A few companies that were interested in Bluetooth organized a group for developing a project for radio communication at a short distance, in May 1998. The name of the group was SIG (Special Internet Group).

Bluetooth can be in the bandwidth: in 2.4Ghz~ 2.5Ghz. SDP is a support protocol for Bluetooth devices, which can be divided into SDP clients and the SDP server. All Bluetooth devices can include the SDP server, SDP client, or both, depending on the required function. Since SDP has client and server structures, it is capable of requesting and responding, just like L2CAP. Since servers are designed to respond to any request, the SDP server has an internal database table containing all the service information provided by the Bluetooth device. If there is a request from a client, the server will reply to the request message based on the database.

3 Analysis and Research on RFID and Requirement for Security

In this chapter, the security issues of the RFID system will be analyzed based on the current research. The security requirement for the RFID system using Bluetooth SDP will also be discussed.

3.1 Analysis of the Current Research on the RFID System

Research work done on the tag-based RFID has been on the spotlight along with ubiquitous computing. A major method of the existing research is the formula suggested by MIT Auto-ID Center, in which a protocol that provides security service using a Hash function is used.

(A) Hash lock scheme, by MIT : This scheme is low cost, since all it requires is a hash function. Each tag verifies the reader as follows. The reader has key k for each tag, and each tag holds the result metaID, metaID = hash(k) of a hash function. A tag receives a request for ID access and sends metaID in response. The reader sends a key that is related to metaID received from the tag. The tag then calculates the hash function from the received key and checks whether the result of the hash function corresponds to the metaID held in the tag. Only if both data sets agree does the tag send its own ID to the reader. Although this scheme offers good reliability at low cost, since metaID is fixed, the adversary can track the tag via metaID. To avoid this, the metaID should be changed repeatedly, however, operating the system in a way that satisfies this requirement in practice is difficult.

(B) Randomized hash lock scheme, by MIT : This is an extension of the hash lock type scheme. It requires the tag to have a hash function and a pseudo-random generator. Each tag calculates the hash function based on the input from pseudo-random generated, r and id, i.e., c = hash(id|r). The tag then sends c and r to the reader. The reader sends the data to the back-end database. The back-end database calculates the hash function using the input as the received r and id for each ID stored in the back-end database. The back-end database then identifies the id that is related to the received and sends the id to the reader.

(C) Hash-chain Scheme by NTT : Suggested by Networks Takashi Miyamura (NTT), the method generates Hash chains using a secure Hash function. The initial value to generate the Hash-chain value is irrelevant to the RF tags. Based on this value, mutual authentication is performed by generating a secure Hash-chain value. The NTT method suggests an extended Electronic Product Code (EPC) in place of the existing EPC code, which is a way to satisfy both Perfect Forward Secrecy (PFS) and indistinguishability. This method, however, does not consider the partitioning of information in the database server and security from the point of confidentiality, and will only provide PFS for the transmitted data after the initial data transmission. Therefore, it can be said that the weakness of the NTT method lies in the lack of confidentiality in its transmission of the initial value to generate a hash chain and in the backward server's unsecure storing of the genuine ID and initial value of the corresponding secret information.

3.2 Analysis of the Security Requirement of the RFID System Based on the Bluetooth SDP

To build a Bluetooth SDP-based system with RFID, the following security requirements are suggested:

- ACIN of the existing communication: Bluetooth communication requires additional security service for the RFID system, aside from the security service provided by Bluetooth itself.
- Channel security: To generate and provide services though secure communication with tags and readers, there should be efficient for forward and backward security.
- Protocol security: For the RFID protocol, security service should be provided for secure communication with tags permitted by illegal query modification or changes.
- Security to generate and inquire on new services: To use Bluetooth SDP with RFID, a security service is required to maintain security in generating and inquiring about new services.
- Technology to acquire secure status: To assemble an RFID-based network, ensuring secure service for all kinds of RF-tag status is required.
- Extensibility: It should be possible to apply diverse kinds of applications based on the suggested RFID-based authorization.

4 Research on Secure SDP of RFID Using Bluetooth Communication

This paper presents a secure SDP protocol for RFID Bluetooth communication. When an RF-tag-based network is established for Bluetooth communication, the suggested method proposes a formula to generate and provide automatic services for the RF tags by performing a secure Service Discovery Protocol.

4.1 System Parameters

Each system factor of an entity is followed below.
* : T : RF Tag, R : Reader, H : Host Server
RFCode (State, Num_Cmd) : initial RF Code value
SR : Service Record
Cnt : Counter Number
RID, HID, RFID : *'s ID(Reader, Host server, Tag)
H(), h : Secure hash function, Secure hash value
E() : Encryption algorithm

4.2 Secure RF-Tag-Based SDP

When a certain network is constructed with RFID tags for Bluetooth communication, it follows the secure SDP processes described below. Our method suggests a network composed of a host server, a reader, and n RFID tags.

A) Assumption
The RFID tag stores SDP records for Bluetooth communication, whose structure is explained below.

① As a passive tag, an RFID tag stores the SDP record for Bluetooth communication. The characteristics of the tag and record are shown in the following:

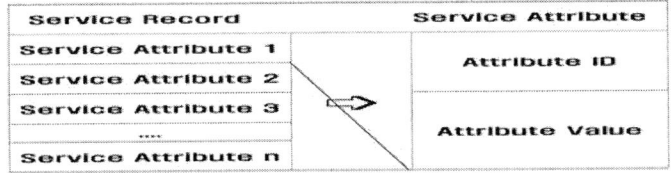

Fig .1. RFID Tag's Bluetooth Record Attribution

② It is composed of 32 bits uniquely categorized for the service records.
③ The characteristics of each service are composed independently.
④ To distribute the initial RFID tag, a secure registration process to a host is undertaken. The host stores the code information of RFID tags into the database. The RFID tag stores the HID of the host.

⑤ The communication between a host and the RF reader is done by a wired channel, in which certain TP keys are shared beforehand.
⑥ An RFID tag can perform a secure Hash function and XOR calculation. It has a counter generator similar to that of the host server.

B) Initial Service Registration and Authorization Process

In the initial service registration process, n RFID tags register each service to a certain host server.

① The RFID tag requests for service to the RF reader by transmitting a Service Request, SR[i], and Cnt_i

② The RF reader, which receives the Service Request, generates h_R based on SR[i] transmitted from the RFID tags. It then transmits RID, h_R and SR[i] to the host server.

$$h_R = H(SR[i] \| RID)$$

③ The host server confirms the integrity by verifying h_R after checking the transmitted RID from the RF reader. If the verification of the transmitted information is correct, the Service Attribute will be defined corresponding to the SR[i] of RFID tags. A random string P_{Cnt_i} corresponding to Cnt_i will be transmitted to the reader by creating the string P from a certain random string generator.

④ The RF reader will confirm HID to transmit SA[i], RID, and h_R to the RFID tags.

$$h_R = H(SA[i] \| P_{Cnt_i})$$

⑤ The RFID tag will check the integrity of SA[i] and P_{Cnt_i} at the transmitted h_R. In case it is correct, it will calculate RFID and h_T to transmit the result to the RF reader.

$$h_T = H(RF\ Code \| RFID)$$

⑥ The RF reader will transmit h_T and RFID transmitted from the RFID tags to the host server. The host server will store P_{Cnt_i} and RFID in the corresponding table.

Based on the process stated above, the RF tags perform the initial process to authorize after registering the service to the host server.

C) Inquiry for a New Service and Registration

Unlike the initial service registration, the inquiry and registration process for new kinds of services for mobile RF tags involve requesting for service after participating in the network. If the service is provided by a host, it will transfer SA[] to the RFID tags after the service inquiry. If the service is not provided, the RFID tags will perform the process to register a new service to the host.

① The RF reader transfers the query message to the RFID tags to provide the service.
② The tag participating in a certain network is an RFID tag requiring a new service. It transmits the current status message, service request message, and MAC_S to the RF reader.

$$MAC_S = H(Cnt_i, P_{Cnt_i} \oplus Service_{Request})$$

③ After receiving the service, the RF reader transfers the current status of the RFID tags with SAN[i] and RID to the host server.
④ After receiving SAN[i] and RID, the host server performs two kinds of detailed processes by following the SAN[i] from the RF reader.

(a) In case a service is already registered.
⑤ In case the service is already registered as SAN[i] in the database, the backward host will verify SAN[i] and MAC_S. After calculating $MAC_S' = H(Cnt_i', P_{Cnt_i} \oplus Service_{Request})$, and if the result is $MAC_S' \equiv MAC_S$, the host will transmit SRN[i] and $Service_{Response}$ to the RF reader.
⑥ The RF reader will confirm the service-permitting message $Service_{Response}$ at SRN[i] and $Service_{Response}$ to deliver the service to the RFID tags.

(b) In the case of a new service.
In the case of a new service, the inquiry message is transferred to the RF reader with SRN[i] and HID after registering the generated SRN[i] to the database.
⑤' The RF reader will confirm HID at inquiry, SRN[i], and HID transmitted from the host. It will transfer RID to the RFID tag that requested for a new service.
⑥' The RFID tag will store HID and RID temporarily to transmit the RF code (which is the initial code value of an RF tag) and Cnt_i to the RF reader.
⑦' The host stores the information about the new RFID tags to the database at step (1)-④. After this process, the host server transmits the following information to the RF reader by generating a random string at step (2).
⑧' After the process described above, the same process is undertaken following stage ④ to ⑥ of (2).

With the above protocol, authentication for the new RFID tags, the initiation process and registration process for a new service are then performed.

D) Status Acquisition for all RFID Tags on the RF Network
To acquire the current status of the RFID tags servicing on the RF network, the RF reader will store it in the host after the following process:
① The RF reader will broadcast the ID list of currently registered RFID tags.
② After receiving the broadcasted message, each RFID tag transmits the HID stored by the RFID tags, and the current status of RF Code, to the RF reader.

③ After extracting the RF Code[i] from the transmission (HID, RF Code[i]), the RF reader will transmit HID together with a certain random number r to the host.

④ After receiving (HID, r) from the RF reader, it will store V_{data} in the database after generating the session key K to store the Code for the RFID tags secretly based on a random number r.

E) Application of EPC Code to Auto-ID Center

Auto-ID Center was established at MIT as a national research institution in 1999. Auto-ID Center defines the RFID tags as the next-generation bar code, promoting its usage through a number of researches. EPC, on the other hand, is a scheme to connect the RFID system of each company, based on the international standard, through the Internet as an RFID-based global food-recognition network. The EPC network is based on the RFID system developed by Auto-ID Center at MIT. EPC Global, established jointly by European Article Number (EAN) and Uniform Commercial Code (UCC), is working to promote the EPC network. To apply the expanded EPC of the suggested method, a secure Hash-code value will be set as 1 by combining the existing EPC Manager space and Object Class space. The space allocated in this area will be 64 bits. In addition, if a secure Hash-code value is set at 2 for communication, the space would be 128 bits. In case the security service is required for confidentiality on the EPC applied system, the initial key storage space will be allocated as 56 bits, which can be used as an optional field. If the service does not require confidentiality service, it is possible to utilize it as the space for Time Stamp to transmit data, or to use it as a reserved field.

Therefore, the application of the definite EPC codes of the suggested method will be composed as follows:

① The reader transmits the expanded EPC Code to the ONS server.
② The ONS server transmits the MAC of the backward server and its corresponding reply message to the reader.
③ The reader transmits the expanded EPC Code to the backward server.
④ After hashing the factory code and the product code at the EPC code of Auto-ID, the backward server transmits the hashed result to the reader by storing the secure Hash-value field 1,.
⑤ The reader performs the suggested protocol.

5 Analysis of the Suggested Method

Through the analysis of the suggested security, this paper will explain the characteristics of security and its differences compared to the existing method.

① In this method, which is concerned with the application of the Bluetooth SDP protocol to the RFID tags, the security requirements mentioned in Chapter 3 satisfy the following:
- ACIN: In the suggested method, the Hash value is created by using a secure hash function H . The integrity service is provided by verifying this process. It presents the authentication and non-repudiation service between the host

server and RFID tags by generating P with Pseudo-Random Number and a secure hash value. As for confidentiality service, storing the initial information of RFID tags involves the secretly saved service of the backward server, which creates the session key K using a random number r and the previously shared TP key.
- Channel security: Forward channel security will be performed to maintain the security between a host and tags for a reader. Based on the information Cnt generated by the tags and P_{Cnt} by their host, security is provided by confirming the authentication of the MAC value.
- Protocol security: The suggested method is still vulnerable to eavesdropping and security attacks. To circumvent this weakness, an additional password algorithm is required.
- New service generation and security for inquiries: The suggested method is performed on the Bluetooth SDP to create and inquire about a new service. If the service is provided by the host, SA[]will be transmitted to the RFID tag after the service inquiry. In case the service is not provided, the RFID tags will register a new service.
- As for the backward server, in case of the registered service by the transmitted message, the initially registered service will be provided by verifying the process of $MAC_S' = H(Cnt_i', P_{Cnt_i} \oplus Service_{Request})$. If the service is not registered, the initial information of the RFID tags will be registered by transmitting RF Code and Cnt_i. Based on this, a new kind of service can be registered by receiving P_{Cnt_i} and SA[i] from the backward server.

② Technology to acquire security status: To acquire the status of all of the tags, the security of the stored data can be maintained by acquiring a security status and keeping it secretly by creating the session key K = H(TP ǁ r) based on the TP key shared beforehand. Compared with the existing method analyzed in Chapter 3, the suggested method has the following features, which make it efficient:
- Number of passes: MIT method 1 involves a handshake of number 8, and MIT method 2, number 6. The NTT method, on the other hand, proceeds as 4 passes. For the suggested method, the initial registration and authentication process will be performed by the number-6 handshake. Comparing the number of passes, this way is more efficient than that of MIT method 1. (It is still less efficient, however, than the NTT method.)
- Calculation: As for the suggested method, based on the RFID tags, the number-2 hash calculation is performed on the process for initial registration and authentication. Therefore, it can have the same level of efficiency as MIT method 1 and the NTT method. Compared to MIT method 2, however, the calculation would be more efficient if inefficient random numbers were not used.
- Usage algorithm: The usage algorithm produces one hash algorithm. It also provides a higher level of efficiency compared to the previous method. While the previous method was based on the low price tag in the range of the 900

MHz band, the suggested method considers the tags on the 2.4-GHz band range. The suggested method, however, can be constructed using one hash algorithm used for the protocol designed on the 900-MHz range. Therefore, it provides more advantages in terms of cost and efficiency.
- How to implement: The suggested method uses a counter-information instead of a random number generator, which makes it feasible to implement.

6 Conclusion

To fully utilize the next generation of IT-based environment for ubiquitous computing, this paper studied the aspects of secure SDP technology based on wireless RFID communication technology. Creating a user-friendly network environment, which is one of the key factors in ubiquitous computing, requires local-area wireless-communication technology. This study also delved into various security technology designed to protect the privacy of users.

The suggested method, however, was found vulnerable to attacks, both physical and through the network. Therefore, along with means to circumvent such a diverse security threat, more research on security-service protocols should be performed for the development of practical RFID tags.

References

1. Miyako Ohkubo, Koutarou Suzuki and Shingo Kinoshita, "Cryptographic Approach to "Privacy-Friendly" Tag" RFID Privacy Workshop@MIT, Nov, 2003
2. Sanjay E.Sarma, Stephen A. Weis and Daiel W. Engels, "Radio-Frequency Identification : Secure Risks and Challenges", RSA Laboratories Cryptobytes, vol. 6, no.1, pp.2-9. Spring 2003
3. Sanjay E.Sarma, Stephen A. Weis and Daiel W. Engels, "Radio-frequency identification systems", In Proceeding of CHES '02, pp454-469. Springer-Verlag, 2002. LNCS no. 2523.
4. Sanjay E.Sarma, "Towards the five-cent Tag", Technical Report MIT-AUTOID-WH-006, MIT Auto ID Center, 2001. Available from http://www.autoidcenter.org
5. Stephen A. Weis, "Security and Privacy in Radio-Frequency Identification Devices", Masters Thesis. MIT. May, 2003

The Semantic Web Approach in Location Based Services

Jong-Woo Kim[1], Ju-Yeon Kim[1], Hyun-Suk Hwang[2],
Sung-Seok Park[3], Chang-Soo Kim[3], and Sung-gi Park[4]

[1] Pukyong National University,
Interdisciplinary Program of Information Security, Korea
{jwkim73, jykim}@mail1.pknu.ac.kr
[2] Pukyong National University, Institute of Engineering Research, Korea
hhs@mail1.pknu.ac.kr
[3] Pukyong National University, Dept. of Computer Science, Korea
{sspark, cskim}@pknu.ac.kr
[4] JC System Inc., Korea
sgpark77@empal.com

Abstract. In recent years, there has been growing interest in Location Based Service (LBS). However, the existing LBS has some limitations in the management of dynamic location-dependent contents and interoperability between the different platforms and various domains. We have focused on how LBS applications obtain integrated, dynamic, and sensitive contents on the different domains. We approach the Semantic Web technologies to resolve these issues. In this paper, we design the architecture to combine the Semantic Web technologies with the LBS and implement the prototype to describe the Semantic LBS.

1 Introduction

The Location Based Services (LBS) provide context sensitive information based on mobile user's location. The LBS include services such as local maps, local weather, traffic condition, tour guide, and shopping guide. For example, when travelers visit a city first, they can search not only the location of hotels and ATM machines near by user's current location but also addition information. Therefore, LBS provides different results as the users' position even though users request the same services [10].

However, existing LBS systems have some limitations of implementation. First, it is difficult to manage data as a general data management method because LBS needs to manage dynamic information. Second, it is difficult to share their information because the location-based services are different in processing methods, information transfer methods, and platforms. In recent, the organizations that are involved in working the LBS standards - International Standard Organization (ISO), Mobile/Automotive Geographic Information Core Services Forum (MAGIC), Open Geospatial Consortium (OGC), etc - have researched LBS standards and the information management technologies. Nevertheless, there are some difficulties in sharing and integrating each of information between different domains to use different DBMS.

We focus on how LBS applications can obtain dynamic, context-sensitive, and integrated contents on the different LBS domains. Even though the Web provides such dynamic information, it is difficult to use contents extracted from the Web in the LBS because the Web has amounts of data and is not structured. Therefore, we combine the Semantic Web technologies to resolve these issues on the LBS. There have proceed a number of researches about the Semantic Web which is an extension of the current Web in which the dispersed information is searched semantically and generated new knowledge through inference as well as shared and integrated.

In this paper, we design the architecture to combine the Semantic Web to the LBS and describe implementation with an example. The system can not only solve the existing LBS problems but also provide better user-preferred search results. This paper is organized as follows. In section 2, we discuss architecture and limitations of the LBS and the applied fields of Semantic Web technologies. We propose the Location Base Service Architecture using the Semantic Web technologies in section 3. In section 4, we present the process of semantic approach in LBS with a simple example. The final section summarizes and presents the future work.

2 Location Based Services and Semantic Web

In this section we describe the LBS standards architecture and the Semantic Web technologies. We present necessity to use Semantic Web approach in LBS.

2.1 The Architecture and Limitations of Location Based Services

It is a problem that the existing Local Based Services cannot share their information and contents with other LBS because the systems developed on the different platforms and protocols. In order to solve the issue, the studies about LBS standards started. Recently, organizations, ISO (TC/211 19132[1], TC/211 19133[2], TC/211 19134[3].), OGC (OPenLS Initiative [9], OpenLS Core Service [7]), and etc., are involved in working the standards to service effective contents.

Particularly, OpenLS Core Services specification [7] in OGC defines the services specification, the information model, and the system architecture needed to design Location Services. The OpenLS architecture is organized as some compo-

---

[1] Location Based Services PossibleStandards,
   http://www.isotc211.org/scope.htm#19132.
[2] Location Based Services Tracking and Navigation,
   http://www.isotc211.org/scope.htm#19133.
[3] Multiple Location Based Services for Routing and Navigation,
   http://www.isotc211.org/scope.htm#19134.

nents which are Mobile Terminal, GMLC/MPC, Service Platform Portal, Geo-Mobility Server (GMS), 3rd-party Content Provider. GMLC/MPC collects and provides a user's location collection by the information in the components of CDMA or WCDMA network. It is not necessary to request the information of the GMLC/MPC when the mobile terminal uses a GPS device. The Service Platform Portal provides front-end functions needed for LBS such as session management, user authentication, request handling, billing, privacy management, and roaming. This component can also be a simple relay, transferring requests for services from the user to 3rd-party applications. GMS achieves to get information from contents, which consist of three services - OpenLS Applications, OpenLS Core Services, and Location Contents Database. OpenLS Applications and Core Services can also access other location contents databases hosted by 3rd-parties.

We focus on the way how GMS and 3rd-party Contents Providers represent contents for LBS. The GMS and 3rd-party Contents Providers provide contents, which are stored at Database Management Systems (DBMS) to search contents easily. However, it makes the LBS system to be a stovepipe system which is a system where all the components are hardwired to only work together. Therefore, it is very difficult to exchange and integrate information between the appropriate platforms and data structures based on DBMS. In other words, the DBMS-based system causes a poor content aggregation problem [2].

2.2 Using Contents from Semantic Web

In order to integrate contents between different databases and to provide dynamic information, we propose the Semantic Web approach in Location-Based Services.

The Semantic Web [1] is a technology to add information on Web to well-defined meaning, to enable computers as well as people to understand meaning of the documents easily, and to automate the works such as information search, interpretation, and integration. The automated agents and dedicate search engines can be a high level of automation and intelligence because the Semantic Web documents have a meaning that a computer can interpret, contrary to existing HTML-based Web documentations. The Semantic Web must be the special hierarchical structure to enable Web documents to intellectualize. The Semantic Web consists of Uniform Resource Identifier (URI), UNICODE, Resource Description Framework (RDF), RDF Schema, and Ontology hierarchically. The RDF is an XML-based language to describe resources like images, audio files, or concepts available via the Web. The RDF model is called as a triple because it has three parts, subject, predicate, and object. The subject and object means a resource, and the predicate defines their relation. The ontology defines the common words and concepts used to describe and represent an area of knowledge. The Semantic Web documents must be represented as the XML/RDF which defines common data of special domains based on an ontology which represents data relation. The contents on this structure can be searched, integrated, and inferred semantically.

The systems using the Semantic Web have been developed in practical fields, and its effectiveness was also presented. Tanen et al. [11] have developed the Courseware Watchdog which is a comprehensive approach for supporting the learning need of individual in the environment of fast change. It integrated the Semantic Web vision by using ontology and a peer-to-peer network of semantically annotated learning material. They showed how an ontology-based tool is allowed to make the most of the resources available on the web.

Lee et al. [6] presented an architecture how a service flow can be automatically composed by syntactic, semantic, and pragmatic knowledge. They demonstrated how heterogeneous Web services can be made interoperable within their framework. Their motivating examples are drawn from medical services, which should be called "Semantic Medical Services (SMS)."

Hunter et al. [3] presented a Webservices-based system which they have developed to enable organizations to semiautomatically preserve their digital collections by making preservation software modules available on Web services and describing them semantically as a machine-possible ontology (OWL-S).

We challenge to develop the Location-Based System using the Semantic Web technologies to provide more semantic, integrated, and intelligent information based on the location.

3 Architecture of Semantic Location Based Services

In this section, we design the architecture of Location-Based System using the Semantic Web technologies. We also present an example of ontologies needed for the Semantic Location Based Services (SemanticLBS).

The architecture of the SemanticLBS is shown in Fig. 1. Our system can provide the information considering users' preference from the users' current location with refined and integrated data from Web. The components of our system consist of Mobile Terminal, SLBS Portal, Personal Privacy Server (PPS), Semantic GeoMobility Server (SGMS), 3rd-party Content Providers. These components are based on the OpenLS Core Services architecture [7], and we changed SGMS internal structure and added PPS for supporting Semantic Web.

- Mobile Terminal obtains the location information through GPS(Global Position System) and GMLS/MPC like OpenLS architecture.
- SLBS Portal provides front-end common functions for Location-Based Services such as session management, user authentication, request handling, billing, privacy management, and roaming like OpenLS. This component supports user's interface to manage the personal information.
- Personal Private Server (PPS) provides RDF documents of the user's personal information based on the personal ontology. The documents are used in the SemanticLBS agent to provide user-preferred services.
- The Semantic GeoMobility Server (SGMS) have some subcomponents. The Core Services is to support map services, and the Internal Contents Services is to provide semantic contents. The SGMS needs Inference Component, Integration Component, and Discovery Component to search semantic infor-

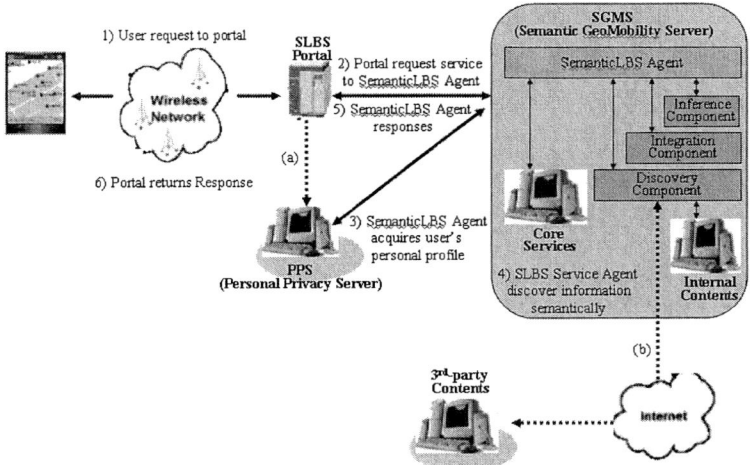

Fig. 1. The architecture of the Semantic Location Based Services

mation from contents. We need the SemanticLBS Agent to integrate these sub components and to provide better user-preferred information.
- The 3rd-party Providers have the contents of some special domains based on the Semantic Web. SGMS also use the information extracted from 3rd-party contents.

The general procedure of our system is in the following manner.

1. The Mobile user connects SLBS Portal through the wireless network.
2. The SLBS Portal gets the services requested by the user, and requests the requested information to SemanticLBS agent.
3. The SemanticLBS agent assures user's requirement and preference from PPS, which provides the user's information given through (a) as RDF type of documentation.
4. The SemanticLBS Agent analyzes the information searched, integrated and inferred semantically from Core Services and Internal Contents. Also, the SemanticLBS agent can access 3rd-party provider like (b) to obtain the contents.
5. The SemanticLBS agent sends the searched results to the SLBS Portal.
6. The Mobile user receives the user preferred results from SLBS Portal.

We need to define ontologies for the Location Based Service to support dynamic and intelligent services. The SemanticLBS ontologies can be defined by industrial organizations for standards or the government, and the Personal Ontology can be defined by LBS Providers. For example, the government establishes the ontology of traffic and weather information, and the industrial organizations for standards describe the ontology of hotels and restaurants information. The 3rd-party Contents Providers define classes of each domain from well-defined ontologies in the SemanticLBS, and create RDF documents for each of content. The Semantic LBS system can search, integrate, and infer more information as users' request based on the semantic contents, and it provides the user-preferred information through the well-defined Personal Ontology which stores users' characteristic information.

4 SemanticLBS Implementation

To describe the semantic Web technology on LBS, we introduce a simple SemanticLBS example and show the semantic search process. Because SemanticLBS project is an ongoing project, we implement some parts of the SemanticLBS functions described in Section 3. In this paper, we implement semantic information search in hotel domain and integration with GIS Services.

Example Scenario. Professor Kim visited a city Busan in Korea to attend a conference. He did not have a time to register a hotel even though he visited the place first. He arrived at the city at night, and he must go to the conference place in the next morning. Professor Kim wants search not only a hotel near the conference place but also the additional information such as prices and classification of hotels, kinds of rooms, kinds of the morning meals. In other words, he wants to choose a hotel considering his preference near his location.

4.1 The System Architecture

We design the simple system architecture to implement our simple example instead of the complete architecture of the proposed Fig. 1. The system is shown in Fig. 3, without a few components in the proposed SemanticLBS: SLBS Portal for providing functions like user authentication and billing, PPS for providing the semantic personal profile, and Inference and Integration Component in the SGMS. The system consists of the three major modules: the SemanticLBS Client to SemanticLBS Server, and LBS Contents Server. The main component of the SemanticLBS Client is MGIS Agent and LBS Agent. MGIS Agent for providing map services is implemented using our Mobile GIS model [5] developed in our previous work. The LBS Agent is responsible for the message process to exchange the query and the response with SemanticLBS Server through the Web Services. The LBS Application obtains the present location through GPS, processes user's input keyword, and provides the Location-Based Services for Hotel domains through the MGIS and LBS Agent. For further details, see Section 4.3. SemanticLBS Sever is responsible to semantically search information from LBS contents. The SemanticLBS Agent is a main process which controls the use of the other components. In order to search the semantic information, we have used Jena API and Joseki which are Semantic Web Toolkit. Query/Response Interface module converts client input keyword into the query for Jena and searched results into return values. For further details, see Section 4.2. SLBS Contents Server provides LBS contents, and executes real searches from the RDF documents using Jena API and Joseki.

4.2 Semantic Search with Ontologies and RDF Contents

We present a prototype of the proposed SemanticLBS system in figure 2. The user of the example scenario in section 4 wants to know hotel information near his location. For the semantic search, ontology of each domain is defined, and the SLBS Contents Provider refers the defined ontologies to make their contents. To create our ontology for hotel domains and RDF documents, we use Protege-2000 which is an ontology modeling tool. A modeling example of the ontology for Location Based Services is shown in Fig. 3(a). The classes are defined in the side of left, and the slots (attributes) of the hotel class are shown. We created

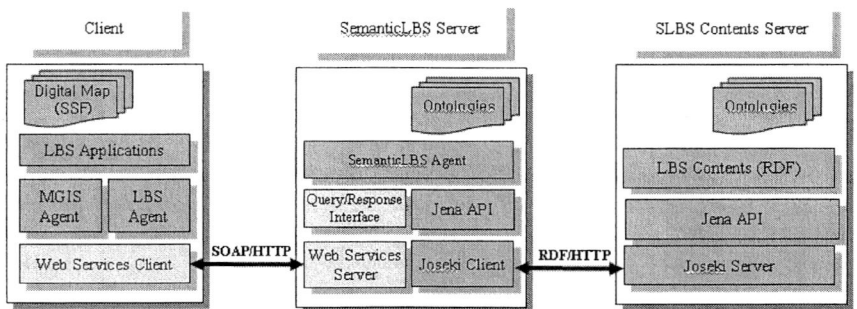

Fig. 2. Architecture of a SemanticLBS example

a RDF instance based on the defined ontology with contents of each hotel and motel and show it on the XML_writer editor Fig. 3(b). For searching, we used Jena API and Joseki tool in SemanticLBS Server and SLBS Contents Server. Jena API is a Java application programming, and creates and manipulates RDF documents, and Joseki[4] is a Java client and a server that implements the Jena network API over HTTP [4, 8]. We can semantically search the instances of RDF documents through RDQL[5] which is Jena's query language [4]. We present a RDF query (Fig. 4 (a)) and its results (Fig. 4 (b)) on the command screen to search the lodge information. The query is for searching the specific information considering room's type and meal as well as price in hotels near user's location. The results are actually processed by SemanticLBS Agent, and then these are sent to the SemanticLBS Client.

4.3 Integrating the Results of Semantic Search with Mobile GIS

We present input and output interface supported by Semantic LBS Client which provide map services and location based services on PDA. We use our efficient Mobile GIS model [5], which can present the map directly on PDA, not show

---

[4] Joseki Documentation, http://www.joseki.org/documentation.html.
[5] RDQL - A Qurey Language for RDF, http://www.w3.org/Submission/RDQL.

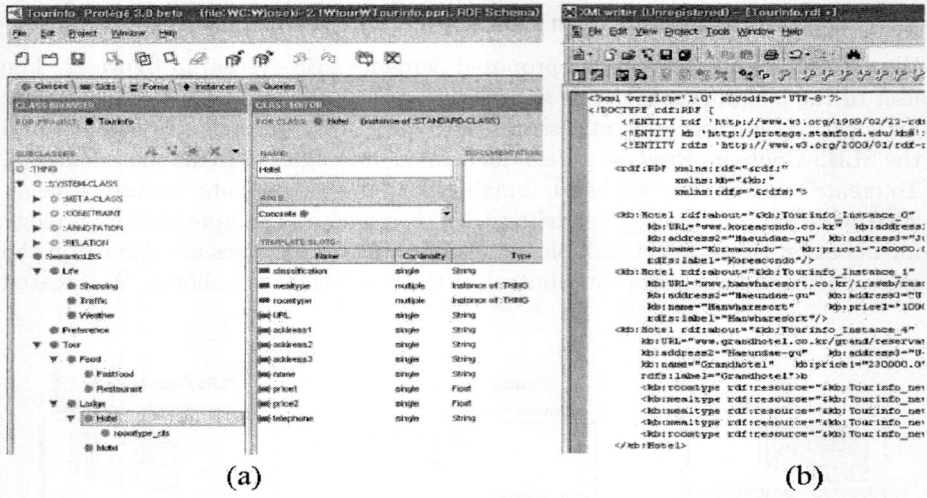

Fig. 3. Example of SemanticLBS ontology and an Instance of hotel domain

Fig. 4. A RDQL query and its semantic search results

the map received from a server. The Semantic LBS Server and Client exchanges the information through SOAP-based Web Services[6].

A user obtains the information from SemanticLBS by 4 steps: (1) A user requests a service with a location and the addition keyword on the input interface as the Fig. 5(a), (2) The user's request is sent to SemanticLBS Server through LBS Agent to execute the search, (3) LBS agent provides the results received from SemanticLBS Server on the PDA as the Fig. 5(b), and (4) user can get more detail information and the digital map of a hotel as the Fig. 5(c). The

---

[6] SOAP Version 1.2 Part 0: Primer, http://www.w3.org/TR/2003/REC-soap12-20030624/.

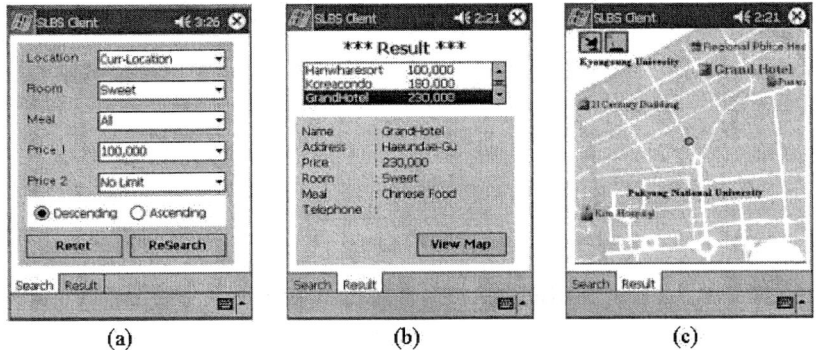

Fig. 5. Integrating the results of semantic search with Mobile GIS

information provided in SemanticLBS is the semantically searched information from semantically represented LBS contents, and user will be able to get user-preferred information through the inference.

5 Conclusion and Future Work

In this paper, we present the efficient Location Based Services with Semantic Web technologies for servicing more accurate and user-tailored information. We designed the architecture of Semantic Web LBS (SemanticLBS), which included the core ontologies for the LBS domains and represented location-dependent contents to the XML/RDF types of documents based on the ontologies.

It is need to provide more specific and exact information because LBS applications mostly work on mobile devices with wireless network. The advantages of this Semantic Web approach in LBS include (1) retrieving more exact information for LBS, (2) integrating the location-dependent contents of each domain, (3) providing interoperability between the different platforms through standard protocols, and (4) providing user-tailored information based on each personal preference and interest.

It is still not common to provide contents with XML/RDF for semantic Web. As Internet Services based on the semantic Web technologies are closer to general users, it is also convenient to implement LBS for supporting the semantic functions. However, the LBS to be incorporated with the semantic Web will be necessary to define more exact ontologies on various LBS domains and to implement the inference function based on LBS ontologies.

Acknowledgements. This research was supported by the Program for the Training Graduate Students in Regional Innovation which was conducted by the Ministry of Commerce, Industry and Energy of the Korean Government.

References

1. Berners-Lee, T., Hendler, J. and Lassila, O.: The Semantic Web. Scientific American (2001)
2. Daconta, M., Obrst, L. J., Smith, K.T.: The Semantic Web: A Guide to the Future of XML, Web Services, and Knowledge Management. WILEY (2003)
3. Hunter, J., Choudhury, S.: A Semi-Automated Digital Preservation System based on Semantic Web Services. JCDL'04, ACM (2004)
4. Jeremy J. Carroll, Lan Dickinson, and Chris Dollin.: Jena: Implementing the Semantic Web Recommendations. ACM (2004)
5. Kim, J. W. Park, S. S. Kim, C. S. Lee, Y.: The Efficient Web-Based Mobile GIS Service System through Reduction of Digital Map. Lecture Notes in Computer Science, Vol.3043. Springer-Verlag, Berlin Heidelberg (2004)
6. Lee, Y., Patel, C., Chun, S. A., Geller, J.: Compositional Knowledge Management for Medical Services on Semantic Web. WWW2004. ACM (2004)
7. Mabrouk, M. et. al.: OpenGIS Locaation Services(OpenLS): Core Services. OpenGIS Implementation Specification. OGC (2004)
8. McBride, B.: Jena: A Semantic Web Toolkit. IEEE INTERNET COMPUTING. IEEE (2002)
9. OpenLS Initiative.: A Request for Technology in Support of an Open Location Services Testbed. OGC (2000)
10. Shiowyang Wu, Kun-Ta Wu. : Dynamic Data Management for Location Based Services in Mobile Environments. Proc. 2003 Int. Conf. Database Engineering and Applications symposium. IEEE (2003)
11. Tane, J. Schmitz, C., Stumme, G.: Semantic Resource Management for the Web: An E-Learning Application. WWW2004, ACM (2004)

SCTE: Software Component Testing Environments

Haeng-Kon Kim[1] and Oh-Hyun Kwon[2]

[1] Department of Computer Information & Communication Eng,
Catholic University of Daegu, Korea 712-702
hangkon@cu.ac.kr
[2] Department of Computer Engineering, Tong Myong University of Information,
Technology Busan, Korea
ohkwon@tit.ac.kr

Abstract. There are various number of software development approaches that can be believed as component-based software. Setting up the rules for the prediction of properties in components based system, It is necessary to count the variety of approaches that components appear in systems. This paper describes automated software component testing environment to support the state based testing. This tool extracts information from two commercial CASE tools – Paradigm Plus and Select Enterprice. It supports test generation for C++ and Java programs to develop the specific components.

Keywords: Component Based Developments, CASE, CBD design, software testing, class testing, automated testing, CBD testing Environments.

1 Introduction

CBD applies the same manufacturing principle to software. Within the CBD framework, software parts are designed according to pre-defined specifications, so they can be assembled together to create entire applications. CBD systems are built using components that all conform to predefined specifications, changes and enhancements can happen more seamlessly[1]. The need to "retool" new components or the system to which they are being added is essentially eliminated. There may be a plethora of software development experience that is claimed to be component-based. The variety of CBD approaches can be viewed as space of instantiations derived from a common (meta)model. One portion of those approaches Somewhere in that space is package objects as advocated by D'Souza and Wills[2]. Elsewhere in that space may be an e-business strategy: e-Business improvement planning is an incremental and continuous process to recognize the situations. It is important to the success of the business strategy that are focused on the use of CBD but these may occupy quite different spaces with the view of long term involvement of component vendors represented by Morisio et al. [3,4,5]. If a software application is assembled from components, then it should be easy to reconfigure the components to support desired changes in the business process. The reliability of the application as a whole is a separate issue, but is clearly enhanced when the application is constructed from reliable components. Component-based development and software reuse places new demands on software testing and quality assurance. The user or purchaser of a software component wants to know: what evidence is there that this component is likely to work properly in my application? has the component been tested in a way that is relevant to my intended use? how much serious usage has this component already had, in areas similar to my intended use? what are the implications of using this component? performance/capacity, reliability/robustness and maintainabiliy / portability. A responsible developer of a software component wants to know: what evidence is there that this component is likely to work properly in real user applications? has the component been tested in a sufficient variety of situations? is the component designed for efficient performance in a reasonable range of contexts? A software component may be used for many different applications, in different business and technical environments, by different developers using different methods and tools for different users in different organizations. Development environments are available to support most activities in the software development process. In this paper we will concentrate on CBD testing environments. Developments environments to support analysis and design are widely used in industry. They can support a specific design and analysis method or constitute more general diagram editing systems augmented with knowledge of the most common methods. For some kinds of software systems the effort to test the software is half or even more of the total effort to develop software. Any activities decreasing the testing time give financial effects decreasing the cost and facilitate the software development process. The automated software testing environment, that are able to work with the CBD CASE tool are rare. We describes automated software component testing environment to support the state based testing. This tool extracts information from two commercial CASE tools and supports test generation for C++ and Java programs to develop the specific components.

2 Related Works

2.1 CBD Process

Development of e-business systems involves collaborative work of several different types of specialist with different areas of expertise; for example, business process consultants, software architects, legacy specialists, graphic designers and server engineers. We'll need some coordinating framework for dealing with these diverse skill sets and introduce a track-based pattern to help. It's also important to have a good idea of the kinds of deliverable that we can expect to produce. We describe a broad set of deliverables that work well on CBD projects. Techniques can then be applied in flexible fashion within our overall process framework of track-based pattern plus deliverables. e-Business process improvement provides the right business context for CBD, as shown in figure 1. While the business improvement plan ideally encompasses the entire enterprise, the overall vision may be developed incrementally, leading to a succession of more narrowly focused action plans.

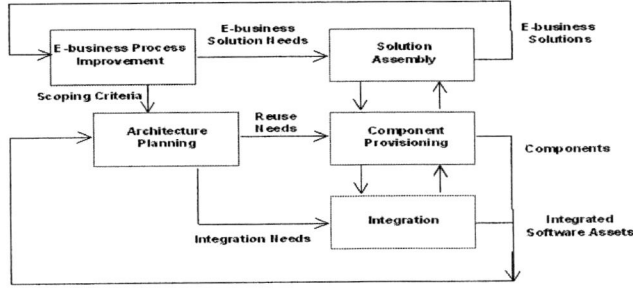

Fig. 1. Component Based Development Process

2.2 CBD Testing and Tools

Component-based development and software reuse places new demands on software testing and quality assurance. This is especially true if components are to be traded between organizations. In testing software components, two main approaches can be identified:

- functional testing:
the unit under test is treated as "black box", the unit's functions are tested.

- structural testing:
the internal structure of the unit under test is known, unit is seen as a "white box", and the tests are prepared to check some structure properties eg statement coverage.

Structural testing is a necessary part in the testing process, but it should not be the primary basis for testing. Even if a code based test suite achieves high coverage it can not prove the absence of faults, validate requirements, or test missing functions. An implementation-based test can only show that the code does what it does, but it is not necessarily what it is supposed to do (this could be checked by functional testing). Code-based test generation can be automated. Some testing tools are briefly described below. These tools are dedicated to object programming language C++ or Java.

The well known and used commercial testing tools are:

WhiteBox is a product of Reliable Software Technologies (www.rstcorp.com). This testing tool is integrated with MS Visual Studio and can support coverage testing of C++ and JAVA programs. Following types of code coverage can be tested:

- Branch (BC) – for each branch point checks if all branches can be reached
- Condition Decision (CD) - for each two branch points, like `if` and `while`, checks if all branches can be reached
- Multiple Condition (MCC) – used to test logical expressions
- Function (FC) – checking if all functions have been executed.

In WhiteBox testing, some metrics in the prediction of testing complexity be calculated. For example: number of attributes (private, protected, public), number of methods, depth of inheritance, number of inheritance hierarchies. The tested program has to be compiled with a special option, enabling the generation of a test report during the execution of program. The test report has to be viewed with WhiteBox viewer.

ParaSoft Jtest can support the test of classes, applets or even application written in Java. Blackbox and whitebox test can be executed In Jtest. A part of Jtest is CodeWizard, which can be used to statically analyse the code. CodeWizard contains some rules checking properties of the source code (eg. `for` statement with empty body). User can add his own rules to be checked. In Jtest some metrics can also be calculated (similar to metrics in WhiteBox). Jtest for C++ programs is also available.

Panorama++ (www.softwareautomation.com) supports testing of C++ programs. Automatic tests are not generated but some analysis during the execution of a program is made. Frequency of instruction's execution can be given, cyclomatic complexity and other metrics are calculated and function dependency graphs can be seen. Control flow of the program can be also observed.

Rational Suite TestStudio contains several programs supporting different testing domains eg. software planning, managing and monitoring, performing functional, efficiency and coverage test. TestStudio enables simulation of many users (Virtual Users component) and measuring response time of the program under different load (in GUI Users component). Purify, Quantify and PureCoverage programs enable some quality checks of the system under test. Rational Purify is used to find run-time errors and memory leaks. Program can be integrated with Visual C++ environment or can work independently. Rational Visual Quantify program can be used to find parts of the tested application that are executed slowly. PureCoverage program is developed to find different kinds of code coverage in the tested program.

CBD can be viewed as a set of cooperating agents. Each agent is responsible for its state. System behavior is the result of interacted individual behaviors. To develop and deliver trustworthy CBD systems a high level of confidence is needed. That means: each component will behave correctly, collective behavior is correct and no incorrect collective behavior will be produced. Although CBD design and programming support many kinds of fault prevention, testing is necessary. An effective approach for testing must enable effective tests of components and collections of components. CBD testing is well suited to state based testing because:

(1) In many CBD analysis and design methods eg. UML [1,2,3], the behaviour of a class is modeled by finite state machines. Methods of a class must be used sequentially. Some method's sequences may be prohibited by specification or may cause the implementation to fail.

(2) Model used for testing must help to find faults. State control is typically distributed over an entire system. Individual and collective behaviour faults are likely as a result of this complex implicit structure.

State based testing provides a straightforward means to develop test suites that will reveal these faults. State machines may be used to model the behaviour at any scope: class, cluster, component, subsystem or system. Details must decrease when state machines are developed for larger scope. The packaging of instance variables and methods into a class is fundamental to CBD programming. Although the number of message sequences and instance variable value combinations is infinite, a state machine can nevertheless provide a compact and predictable model of behaviour.

3 Design and Implementation of SCTE

3.1 Architectures of SCTE

The CBD approach becomes considerably more integrated with the incremental nature of the process, where the progressive stages of testing are planned and executed during each increment. It has great advantages over structured approaches where a large proportion of the testing effort is applied towards the end of the project and the deadline pressures are required to finalize the project. The iterative nature of CBD also has an impact on testing, a greater emphasis being placed on regression testing to ensure functional and non-functional stability across increments. Planning – identifying the requirements for test for each increment, types of test, and the overall test effort; Specification – specifying the test cases, created test scripts, ensures coverage is met; Execution – running and evaluating the tests, managing defects. We concentrate

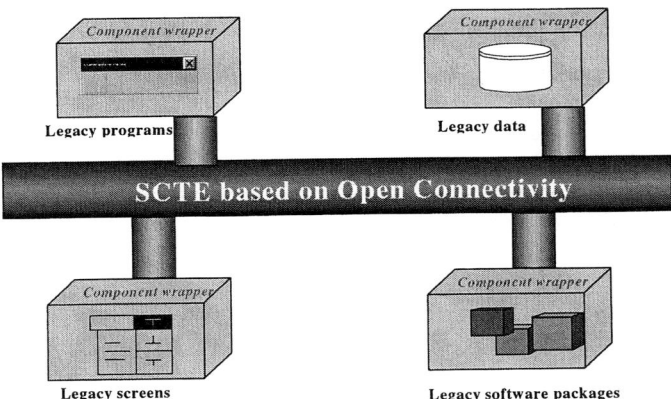

Fig. 2. Design and implementation of SCTE

here on the design and implementation of SCTE as Figure 2 and describe how the principles of CBD are leveraged to provide more rigorous testing techniques. Identifying requirements for test is key to the increment test planning activity; they provide a means of measuring the test coverage. The scope for a chosen increment is defined by a number of requirements and Use Cases, themselves identified by business priority, value, dependencies, etc; the duration of the increment is set by a time-box. Requirements for test are identified from the project artifacts, the main source of which is the model, containing key products: business process, requirements (use cases, rules, non-functional), component specifications, and component design. The model imposes trace-ability between these products, established via the Use Case, with backward trace-ability to the Business Processes and forward trace-ability to the architecture, design and code.

Apart from the first increment, CBD testing is essential to the success of the operational solution, ensuring stability as each increment is completed. Ideally each increment would re-run all the previous tests to verify the functional and non-functional integrity. However this may not be practical and a risk driven prioritized approach should be applied. Regression tests are conducted at Unit and System level, the encapsulation of the component means that it is relatively easy to run pre-built black box tests on a component. Regression testing at system level is a more common practice. Therefore, where a component's implementation has been affected by the current increment, its' Service Operations will undergo Unit regression tests. To identify System tests, it becomes a matter of identifying which Use Cases are fulfilled by the components Service. The traceability within the model can be used to instantly provide this information.

3.2 State Based Testing Method

The main goal of this method is to check, if the behaviour of the implemented class is in accordance with its state transition model. State transition models are present in all CBD design methods and are used to design the dynamic behaviour of objects. The state transition diagram consists of nodes representing states and edges showing the possible state transition caused by stimuli (eg. an operation call). Depending on the result of an operation call following types of operation can be given:

OP1. Changing component's state to a new one – according to the transition on the diagram to a new state

Op2. Remaining component's state unchanged - according to the loop transition on the diagram (the same input and output state)

OP3. Inappropriate for current state call – there is no transition from the current state mark with this operation

Op4. Changing component's state to an undefined one, erroneous – there are no nodes and edges on state transition diagram

Op5. Changing component's state to a defined one, but inappropriate - there is no such edge on state transition diagram

Op6. Lleaving component's state unchanged when it is supposed to change.

The state-based testing method assigns each operation call to one of the above groups, so one of the following defects in operation calls could be found:

(1) Inappropriate operation for current state (2) Operation forcing a component to change its state to an undefined one and (3) An inappropriate one.

The first defect is caused by inappropriate use of a component while the second and the third defect, could be the effect of errors in the operation implementation.

The component's state is determined by values of its data members: attributes and the current point in time. Attributes are used to store data and control information (used to trigger events). If the number of control attributes in a component increased The interaction between operations is more complicated in the state vector method. This affects the time required to design and generate state-based tests. Attribute values can be divided into tree groups:

AV1. Specific values: described in the design as elements of special significance (example: integer attribute 5 is a specific value)
AV2. General values: a group of attribute values considered in the same manner, a set of values (example: integer attribute values greater than 5 constitute general value)
AV3. Dynamic values: an attribute value will be established during program execution.

The state of component is the combined value of all its attributes at the current point in time. Unit white box tests are derived from the design and implementation of the component. Achieving full white box coverage is impractical, so testing the main decision-to-decision paths through code is considered a more practical solution. The design of each operation on a component's Service Interface is described by an Object Sequence Diagram (OSD), which depicts the paths through the code. Therefore the OSD is used to identify the decision points, whilst the test data is derived from the service operation (reused from the black box test cases) to ensure that each decision is evaluated. An emphasis on architecture will have helped minimize these dependencies, however the Unit tester will sometimes need to simulate the dependant components.

The example of a linked list written in C++ is from [6]:

```
class list
{
        public:
        // class interface
        protected:
        struct list_element {
                list_element *pNext;
                TYPE tItem;
        };
        // pTop - pointer to the first element of list
        list_element *pTop;
        // pCur - pointer to the current list element
        list_element **pCur;
};
```

Attribute pTop points to the first element in the linked list or is equals NULL if there is no list. Attribute pCur points to a pointer to the current element in the list. For the above shown class the attributes are of following types:

A. pTop attribute

pTop_1 - specific value equals NULL

pTop_2 - is not equal to NULL (general value)

B. pCur attribute

pCur_1 - specific value equals NULL

pCur_2 - dynamic value *pCur == pTop

pCur_3 - dynamic value *pCur != pTop

These states are created on the basis of the class design not the class code (as there may be errors in the code). In SCTE these states are created from state transition diagram. Exceptions are constructor and destructor. To enable state-based testing some modifications in the class code are necessary. The method code is enlarged with some instructions allowing state determination before and after the method execution. It is also necessary to determine if the change in state is correct and to report information about tested situation.

3.3 State-Based Testing Process

We proposed the state - based testing process consists of following steps:

Step1: For each attribute define its values
Step2: For each state define vector of attributes values
Step3: For each method define initial and final states
Step4: Generate code causing the component to be in the initial state.
Step5. Generate sequences of methods calls
Step6: Generate code causing the component to be in the final state.
Step7: Generate code necessary to finish the test
Step8: Run the program
Step9: Analyze test results.

State–based testing can be used in testing interactions between different classes. This involves creation of states, not only for the main component under test, but also for the components that will be passed as parameters. Integration testing of two components A and B comprises following steps:

Step1: Create component A, move it to state in the interactions with B takes place
Step2: Create component B, move it to state in the interactions with A takes place
Step3: Call the method of interaction (the parameter of method is B component)
Step4: Validate resultant state of component B
Step5: Validate resultant state of component A.

3.4 Assessment of the Method

State-based testing is useful for more than simply detecting the change in state of a component, it can be used to detect the correct construction of a more complex dynamic data structure (for example a linked list). State vector testing has particular classes on which it is more effective than on others. If a class is designed simply as a repository of information or as non-dynamic data structure then this method will have limited effect. For dynamic data structures it is essential to determine hen and which particular changes can occur to the structure. For dynamic classes state-based testing enables the detection of errors which are difficult to detect using other testing methods. It should be considered as a complimentary technique for testing interactions with a component's state.

4 Example of State Based Testing

As an example of state based testing in SCTE simple class Stack will be used. This class provides methods to push/ pop an integer value on/from the stack. During creation of a stack component, the initial size is given. In figure 3 the model of the class Stack is shown.

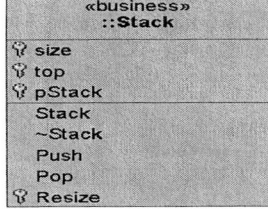

Fig. 3. Select model of Stack class

The declaration of the class Stack in C++ is given below.

```
#define STACK_SIZE_INCREASE 10
class Stack
{
protected:
    int size;
    int top;
    int * pStack;
public:
    Stack(int initial_size);
    ~Stack();
    void Push(const int value);
    int Pop();
protected:
    void Resize(const int new_size);
};
```

Attribute `pStack` points to an array of integer elements, the size of this array is in `size` attribute. Attribute `top` contains the index of the first free element in table `pStack`. The method `Resize` is changing the size of table containing stack elements. In figure 4 hierarchical state transition diagram describing the behaviour of `Stack` class is shown.

Table 1. Attribute value types

ATTRIBUTE	Value name	Value Type	Value
size	initialized	general	size>0
top	correct value	general	top>=0
	empty stack	specific	0
	full stack	dynamic	top==size
	not full stack	dynamic	top>0 && top<size
pStack	initialized	general	pStack!=NULL

According to the rules of state based testing types of attribute's values and state vectors for each state must be determined. In table 1 types of values and in table 2 state vectors for class Stack are given. For each state determined state vectors are shown in table 2.

States **initial** and **final** are specific and for these states the state vectors are not determined. Component is in **initial** state only at the moment constructor is called. State **final** is the ending state for destructor. After determination of state vectors the mapping between methods and source files is performed. Next the transitions on state transition diagram are linked with respective methods. At last the test cases are executed.

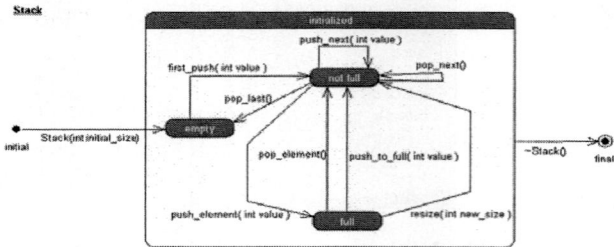

Fig. 4. State transition diagram for Stack class

Table 2. State vectors

State	State vector (attribute [value name])
initial	
initialised	size[initialised] top[correct value] pStack[initialised]
empty	size[initialised] top[empty stack] pStack[initialised]
full	size[initialised] top[full stack] pStack[initialised]
not full	size[initialised] top[not full stack] pStack[initialised]
final	

5 Conclusions

This paper describes automated software component testing environment to support the state based testing. This tool extracts information from two commercial CASE tools – Paradigm Plus and Select Enterprice and supports test generation for C++ and Java programs to develop the specific components. SCTE supports C++ and JAVA components testing but could be modified to support components integration testing as well in the future.

References

1. B. Henderson-Sellers, B. Unhelkar, OPEN modeling with UML, Addison-Wesley, 2000.
2. D.F. D'Souza, A.C. Wills, Objects, Components, and Frameworks wit UML, Addison-Wesley, 1998.
3. P. Allen, Realizing e-Business with Components, Addison-Wesley, 2001.
4. X.Cai, M.R. Lyu, K. Wong, Component-Based Soft-ware Engineering: Technologies, Development Frameworks, and Quality Assurance Schemes, Pro-ceedings APSEC 2000, Seventh Asia-Pacific Software Engineering Conference, Singapore, December 2000, pp 372-379
5. M. Morisio, C.B. Seaman, A.T. Parra, V.R. Basili, S.E. Kraft, S.E. Condon, Investigating and Improving COTS-Based Software Development Process, Proceedings ICSE 2000, 22nd International Conference on Software Engineering, Limerick, June 2000, pp 32-41.
6. C.D. Turner and D.J. Robson „The state-based testing of CBD programs" Proc. IEEE Conf. Software Maintenance, 1993, pp. 302-310.

Computer Security Management Model Using MAUT and SNMP[†]

Jongwoo Chae, Jungkyu Kwon, and Mokdong Chung

Dept. of Computer Engineering, Pukyong National University,
599-1 Daeyeon-3Dong, Nam-Gu, Busan,
608-737, Korea
madcow5980@hanmail.net, puker@puker.net,
mdchung@pknu.ac.kr

Abstract. We develop a computer security management model, which could dynamically adapt security policies. The proposed model consists of an adaptive security level algorithm based on MAUT and Simple Heuristics, and an adaptive security management system based on SNMP and mobile agent. The security level algorithm could adapt several security services according to the changes of network environment. And the security management system utilizes SNMP protocol and mobile agent to deal with diverse network environments. Therefore, the proposed model is expected to provide more flexible security management in the heterogeneous network environments.

Keywords: security management, adaptive security, MAUT (Multi-Attribute Utility Theory), Simple Heuristics, SNMP, mobile agent.

1 Introduction

The current computing environment is composed of the heterogeneous networks, where there are diverse characteristics of the network properties such as transmission media types, bandwidth, device types, and so on. And the properties of these heterogeneous networks are dynamically changing in accordance with the changes of the environment. The existing security models, however, could not provide an appropriate security management since they have only static security management policies.

Thus, we develop an adaptive security management model, which could dynamically adapt security policies according to the changes of dynamic network environments. We use MAUT (Multi-Attribute Utility Theory) and Simple Heuristics to introduce appropriate security policies, and use SNMP (Simple Network Management Protocol) to manage diverse network environment. Also we use mobile agent to achieve autonomous behaviors and to reduce excessive network traffic.

The structure of the paper is as follows. Section 2 discusses related work. Section 3 an adaptive security model for heterogeneous networks. Section 4 the evaluation of the proposed model. Section 5 concludes this paper with the future work.

[†] This work was supported by "Research Center for Logistics Information Technology (LIT)" hosted by the Ministry of Education & Human Resources Development in Korea.

2 Related Work

2.1 Multi-attribute Utility Theory and Simple Heuristics

The MAUT (Multi-Attribute Utility Theory) [1, 2] is a systematic method that identifies and analyzes multiple variables in order to provide a common basis for arriving at a decision. As a decision making tool to predict security levels depending on the security context, MAUT suggests how a decision maker should think systematically about identifying and structuring objectives, about vexing value tradeoffs, and about balancing various risks.

The Center for Adaptive Behavior and Cognition is an interdisciplinary research group founded in 1995 to study the psychology of bounded rationality and how good decisions can be made in an uncertain world. This group studies Simple Heuristics [3]. One of them is Take The Best which tries cues in order, searching for a cue that discriminates between the two objects. It serves as the basis for an inference, and all other cues are ignored.

2.2 SNMP (Simple Network Management Protocol)

The SNMP (Simple Network Management Protocol)[4], issued in 1988, was designed to provide an easily implemented, low-overhead foundation for multivendor network management of routers, servers, workstations, and other network resources. SNMP operates in the way of client/server model where NMS (Network Management System) requests for SNMP MIB (Management Information Base) [5] of the SNMP Agents in the other nodes. But this client/server model may cause excessive traffic and overload to the NMS. We, therefore, try to propose a different approach based on SNMP and mobile agent.

2.3 Network Security Management and Network Management

Kim and Chung [6] suggested an integrated security management, which considers the sensitivity of information asset to distribute different policies to each security system. Gavalas et al. [7] suggested a hierarchical network management with a scalable and dynamic mobile agent-based approach which introduced the concept of adaptive hierarchical management and provided a rationale for the use of MA (Mobile Agent) technology. Lee et al. [8] made a comparison between SNMP and MA for network management and introduced mixed mode. In the case of pure SNMP, the number of transactions between NMS and managed nodes is excessive, which means a serious drawback of SNMP approach. Additionally, MA also has limitation, that is, it carries accumulated data over network which may cause transmission delay. By the way, in the mixed mode, the agent doesn't accumulate data collected from nodes, but stores them in the temporary storage of the managed node, and moves to the next node. And then the managed node could send the data to the manager independently from MA.

2.4 Existing Static Security Management Model

The existing security management model usually considers static security policies. Thus, it will be quite difficult to deal with the changes of the environment promptly and appropriately. For instance, SSL/TLS (Secure Socket Layer/Transport Layer Security) [9, 10] selects the most secure Cipher Suite in the Cipher Suite List transmitted from the client in the Client Hello step. Unfortunately, this static policy could result in an excessive overload and long latency to the users who have low-end computing devices. However, if we adopted the adaptive security management policies, we could reduce the latency time to those users.

3 Adaptive Security Management Model

In this section, we suggest an adaptive security level algorithm using MAUT and Simple Heuristics. And we construct an adaptive security management model based on this algorithm, SNMP, and mobile agent.

3.1 Adaptive Security Level Algorithm

We present the security policy algorithm that dynamically adapts the security level according to the domain independent properties such as terminal types, and the domain dependent properties such as the sensitivity of information using MAUT and Simple Heuristics. The variables of the algorithms are as follows:

1. Domain independent variables $I = (i_1, i_2, ..., i_n)$: data size, computing power, network type, terminal type, and so on.
2. Domain dependent variables $X = (x_1, x_2, ..., x_n)$: user attributes, system attributes.
3. Security level $SL = (0, 1, 2, ..., 5)$: The larger the number is, the stronger the strength is. If SL is 0, we can not utilize the security system.

The equations of determining U and SL in our algorithm are as follows.

$$U = \sum_{i=1}^{n} k_i u_i(x_i), \ (0 \leq U \leq 1)$$
$$SL = \lceil U^*10 \rceil / 2, \ (SL = 0, 1, ..., 5)$$

$u_i(x_i)$ is converted utility value for the variable x_i, k_i is a scaling constant of each variable which is determined by security polices and security preference of user. U is the total utility value. SL is the final security level in the proposed model.

The overall algorithms for determining adaptive security level are as follows.

```
SecurityLevel(securityProblem)
// securityProblem: Determining security level
// Utilization of domain independent properties
   calculate SL by I end
   if SL = 0 then return SL // no use of security system
// Utilization of domain dependent properties
// Selection between MAUT and S. Heuristics
   if MAUT then SL = MAUT(X)
   if Simple Heuristics then SL = TakeTheBest(X);
   return SL; end;
```

```
MAUT(X)
// Determine total utility function by the interaction
// with the user according to MAUT
   u(x_1,x_2,...,x_n)=k_1u_1(x_1)+k_2u_2(x_2)+... +k_nu_n(x_n);
// k_i: set of positive scaling constants for all i
// x_i: domain dependent variable, where u_i(x*_i)=0,u_i(x*_i)=1
   ask the user's preference and decide k_i
   for i = 1 to n
      do u_i(x_i) = GetUtilFunction(x_i);
   end
   return u(x_1,x_2,...x_n); end;

GetUtilFunction(x_i)
// Determine utility function due to users' preferences
// x_i is one of domain dependent variables
   uRiskProne    : user is risk prone for x_i - convex
   uRiskNeutral  : user is risk neutral for x_i - linear
   uRiskAverse   : user is risk averse for x_i - concave
   x : arbitrary chosen from x_i
   h : arbitrary chosen amount
   <x+h, x-h>    : lottery from x+h to x-h
// where the lottery (x*, p, x°) yields a p chance at x*
// and a (1-p) chance at x°
   ask user to prefer <x+h, x-h> or x; // interaction
   if user prefer <x+h, x-h> then
      return uRiskProne;           // e.g. u = b(2^cx-1)
   elseif user prefer x then
      return uRiskAverse;          // e.g. u = blog_2(x+1)
   else return uRiskNeutral;  end; // e.g. u = bx
```

3.2 Security Policy and Service Policy

Security policy plays a role in determining security level, and service policy determines parameters for security services such as access control, authentication, confidentiality, digital signature, and so on. Table 1 is a typical example of security policy, where x_{att} is the strength of the cipher, x_{auth} is the authentication method, and x_{res} is the level of protection of the resource to which the user is trying to access.

Table 1. An example of security policy

A Security Policy for *Managed Resource A*	
Security Service	Reading
Utility Function	$u(x_{att}, x_{auth}, x_{res}) = k_{att} u(x_{att}) + k_{auth} u(x_{auth}) + k_{res} u(x_{res})$;
Security Contexts	$comp \geq 200$ MHz; $nType \geq 100$ Kbps; $tType$ = PC/PDA/Cell;
User's Preference	$uRiskProne=2^{2(x-1)}$; $uRiskNeutral=x$; $uRiskAverse=log_2(x+1)$;

Table 2 is an example of conversion table from the values of the security variables to the corresponding quantitative utility values. This table is used for utility analysis.

Table 2. Conversion table of each variable

variable \ utility	0.2	0.5	1.0
x_{att}	$\geq 10^{0.5}$	$\geq 10^5$	$\geq 10^{11}$
x_{auth}	One-Time Password [11]	OTP + Certificate	OTP + Biometrics
x_{res}	Low	Medium	High

Table 3 is an example of service policy for access control where reading or writing access right is given to the user. *SL* is the lower bound of security level. Any user cannot adopt *SL* lower than 3 for writing operation. If the user is administrator and *SL* is higher than 3, then he or she can have writing permission.

Table 3. An example of access control

A Service Policy for *Managed Resource A*
If ((SL ≥ 2) **and** ((Role = administrator) **or** ((Role = user) **and** (Date = Weekdays) **and** 8:00 < Time < 18:00)))) **Then** resource A can be read
If ((SL ≥ 3) **and** (Role = administrator)) **Then** resource A can be written

3.3 ASMIB (Adaptive Security MIB)

There are five kinds of information groups in ASMIB as follows:

- **SystemInfo** group: information about system, such as host name, system type, the capability of CPU, memory space, CPU usage, memory usage, and etc.
- **NetworkInfo** group: information about network, such as IP address, network type, network speed, the number of connection, network usage, and etc.
- **AdaptiveSecurity** group: domain-dependent variables and adaptive security level which is determined by the security level algorithm.
- **SecurityProperty** group: available security services and services' parameters.
- **Policy** group: security policies and service policies.

3.4 Architecture of the Proposed Model

To construct the security management model by client/server may cause excessive overload of NSMS (Network Security Management System). When traffic congestion occurs, centralization of management traffic would cause another problem. We, therefore, can construct our model in the distributed manner based on MA and SNMP.

3.4.1 Overall Architecture of the Proposed Model

Fig.1 shows the two important components of the proposed model. Context Engine consists of Security Module, Adaptive Security Level Algorithm, MA Generator and SNMP Manager.

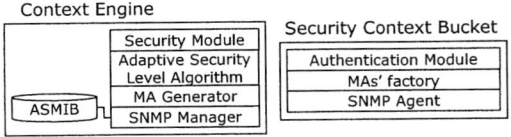

Fig. 1. Context Engine and Security Context Bucket

Context Engine creates MAs(Mobile Agents) using MA Generator to gather ASMIB information, and then MAs travel over several nodes as planned, collect

ASMIB information, and store this information in the temporary storages(Fig. 1). Then, SNMP Manager sends ASMIB object to NSMS by asynchronous Trap PDU (Protocol Data Unit) using SNMP function independently from MA. Adaptive Security Level Algorithm determines security level and parameters for security service according to the collected ASMIB. Security Module provides security services according to the security level and these parameters. MA of Security Context Bucket collects management information via ASMIB and Context Bucket transfers the collected ASMIB to the Context Engine via Trap PDU. We use mixed mode [8] with SNMP Trap to support interoperability of the legacy system which are not equipped with SNMP and/or MA. Fig. 2 shows the mixed mode with SNMP trap. We have eliminated SNMP polling-based collection to reduce excessive usage of network resources.

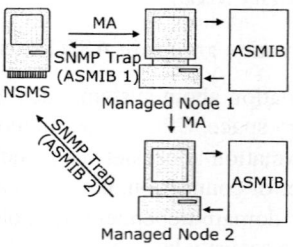

Fig. 2. Mixed mode with SNMP Trap

To achieve domain-dependent variables about user, e.g. user's security preference, MA should be intelligent, therefore it has Adaptive Security Level Algorithm. Fig.3 shows the overall architecture of the proposed model in this paper.

The role of NSMS is as follows:

- Create MAs, collect ASMIB through MAs, and get Trap PDU from managed nodes
- Establish Security/Service Policies and enforce them to the managed nodes

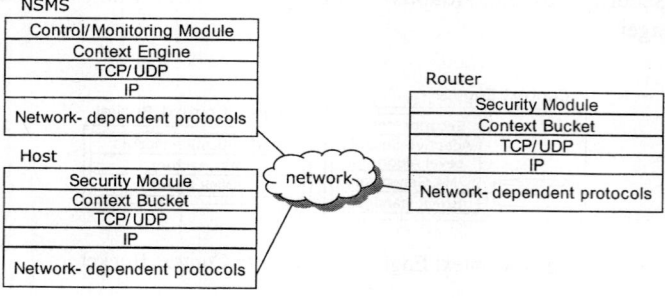

Fig. 3. Architecture of the proposed model

3.4.2 Hierarchical Decentralization of NSMS with Proxy

Our model uses MA-based collection of ASMIB with Trap protocol instead of polling-based collection. Therefore, our model may reduce the number of transactions to transfer management information. But the processes of control information such as Security Policy and Service Policy are still centralized. Therefore, we consider hierarchical decentralization of NSMS with Proxy.

The role of NSMS is extended as follows:

- Distribute Security Policies and Service Policies to Proxy (shown in Fig. 4).

The role of Proxy is as follows:

- Collects ASMIB through MAs and get Trap PDU from managed nodes
- Send the collected ASMIB to NSMS with asynchronous Trap
- Enforce Security Policies and Service Policies to the managed nodes
- Manage nodes which do not use SNMP.

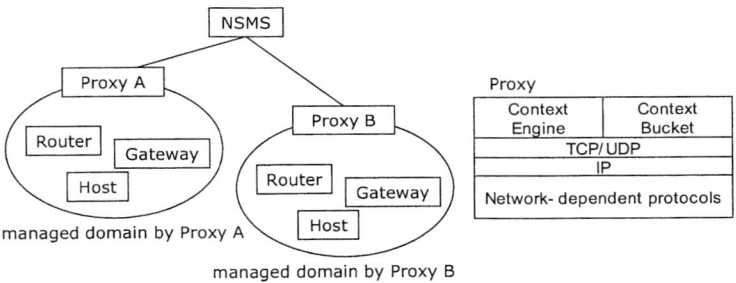

Fig. 4. Hierarchical decentralization of NSMS with Proxy

3.5 Dynamic Changes of Adaptive Security Level

When the security level is changed due to environmental issues such as user's preference and network's configuration, this type of change should be dynamically reflected to the current security services. Thus, we should have the mechanism to change the security services dynamically according to the modified security level.

3.5.1 Examples of Authentication Method and Symmetric Cipher

Table 4 shows One-Time Password-based authentication [11] and composite authentication methods, where multiple authentication methods are used at the same time. Using other authentication methods with OTP-based authentication could provide stronger authentication without changing the whole authentication scheme.

If stronger authentication is needed due to environmental changes such as user's preference and network's configuration, we can solve this problem by adding digital signature or biometric methods to the existing authentication mechanism. *Seed* is a random number and H_n is a Hash function.

Table 4. Computation of Authentication Data

OTP-Based	$S = seed\|passphrase$ $P_{(i)} = H_{n-i+1}(...H_2(H_1(S)))$ $P_{(i+1)} = H_{n-i}(...H_2(H_1(S)))$
OTP+Certificate-Based	$P_{(i)} = Sig(H_{n-i+1}(...H_2(H_1(S))))$ $P_{(i+1)} = Sig(H_{n-i}(...H_2(H_1(S))))$
OTP+Biometrics	$S = seed\|(\text{Biometrics data})$ $P_{(i)} = H_{n-i+1}(...H_2(H_1(S)))$ $P_{(i+1)} = H_{n-i}(...H_2(H_1(S)))$

When the stronger level of confidentiality is necessary due to environmental changes, we should adopt stronger symmetric cipher. This means that we should be able to change the symmetric cipher algorithms and the secret keys on the fly while we have confidential procedure.

4 Evaluation of the Proposed Model

4.1 Load Analysis for Changing Security Level

Security services may be changed according to the update of security level. To analyze the amount of computation in changing the security services, we measured the average response times of symmetric cipher and authentication methods. The result is shown in Table 5.

Table 5. Response time comparison of key generation and symmetric cipher

	Diffie-Hellman	Key generation	DES	3DES	AESLight	AESFast
Average response time	231	5	5	10	4	2

We used 1024bit Diffie-Hellman algorithm, and SHA1 where iteration count is 1000, and used 10 Kbytes for symmetric cipher encryption/decryption. Table 5 means that the increase of computation time is not so great in changing symmetric ciphers.

Table 6. Response time comparison of authentication methods

	authentication data size (byte)	average response time (millisecond)
OTP-Based	346	45
OTP+Certificate-based	371	67
note	SHA1, RSA1024 XML tag is included	Certificate verification time is excluded

Table 6 shows that the response time versus data size is increased slightly after digital signature is added to OTP-based authentication. Meanwhile the level of the strength of the composite authentication method is stronger than the single OTP-based authentication method.

4.2 Reduction of Management Traffic by Mobile Agent

Adaptive security level algorithm is rather complex because it requires many interactions between user and system. In the proposed algorithm *SecurityLevel()*, we assume the user selects MAUT, attribute $X = \{x_1, x_2, ..., x_i\}$ and the scaling constants $K = \{k_1, k_2, ..., k_i\}$. I denotes the required number of interactions to determine adaptive security level. Then, I is expressed as the following equation (1):

$$I = (|K| - 1) + |X| \approx 2|X|, \textit{ where attributes are additive and utility independent} \qquad (1)$$

To determine scaling constants according to the user's preference, at least $(|K| - 1)$ interactions are required. Simply, in the case of three attributes, the algorithm asks the user's preference k_1 to $k_2 + k_3$ and asks the user's preference k_2 to k_3. Thus the scaling constants are determined. After all, $(|K| - 1)$ is the least number of interactions to fix scaling constants. To decide utility function according to the user's risk-proneness, we need at least $|X|$ interactions. With the assumption that the variables are additive and utility independent, the number of scaling constants is equal to the number of variables.

Unfortunately, these interactions are performed over the network in the way of request and response, therefore the number of network transmissions would be two times of the interactions in equation(1), that is $4|X|$. To reduce the excessive number of interactions between user and system, MA may use adaptive security level algorithm. MA performs interaction with user locally. Therefore, we need only one asynchronous Trap after the end of MA's interaction. Thus, the proposed model greatly reduce the number of interactions in comparison with SNMP-based model (it may require $4|X|$).

Kim and Chung [6] compare their security management model with existing ISM (Integrated Security Management) model. Their model can remove unnecessary resource consumption, since it considers the sensitivity of resources. However, there are many other variables to consider in security as well as the sensitivity of resources, such as physical location, security preference, and etc. The proposed model deals with multiple variables instead of single variable in the existing approach. Therefore, our model could provide more flexible security management.

5 Conclusion

The existing security models could not provide an appropriate security management since they have only static security management policies. In this paper, we proposed a security management model, which could dynamically adapt policies according to the changes of diverse and dynamic network environments. The characteristics of the proposed model are as follows.

Firstly, the proposed model needs only one asynchronous Trap after the end of MA's interaction between user and system. Thus, the model reduces the number of interactions between user and system in comparison with the pure SNMP-based model.

Secondly, the proposed model deals with multiple variables instead of single variable in the existing approach. Therefore, our model could provide more flexible security management in the heterogeneous network environments.

Finally, traditional security management model usually works according to a static decision-making approach. On the other hand, the proposed model can reduce the waste of resources by adaptively applying appropriate cryptographic techniques and protocols according to the characteristics of the resources. Therefore, the proposed model increases efficiency and availability of the resources.

However, numerous interactions between MA and user could be problematic. We need to research on reducing these interactions in the future work.

References

[1] R.L. Keeney and H. Raiffa, *Decisions with Multiple Objectives: Preferences and Value Tradeoffs*, John Wiley & Sons, New York, NY, 1976.
[2] D. Winterfeld, von and W. Edwards, *Decision Analysis and Behavioral Research*, Cambridge, England: Cambridge University Press, 1986.
[3] L. Martignon and U. Hoffrage, *Why Does One-Reason Decision Making Work? In Simple Heuristics That Make Us Smart*, Oxford University Press, New York, 1999, pp. 119-140.
[4] W. Stallings, *SNMP, SNMPv2, and RMON: Practical Network Management*, 3rd ed., Reading, MA: Addison-Wesley, 1999.
[5] *Management Information Base for Network Management of TCP/IP-based internets: MIB-II*, http://www.ietf.org/rfc/rf c1213.txt
[6] D.S. Kim and T.M. Chung, "A Design of Preventive Integrated Security Management System Using Security Labels and a Brief Comparison with Existing Models," LNCS3043, 2004, pp. 183-190.
[7] D. Gavalas et al., "Hierarchical network management: a scalable and dynamic mobile agent-based approach," Computer Networks: The International Journal of Computer and Telecommunications Networking, Vol. 38, Issue 6, 2002, pp. 693-711.
[8] Jung-Woo Lee et al., "*Analytical Models and Performance Evaluations of SNMP and Mobile Agent*," Journal of Korean Institute of Communication Sciences, Vol. 28, No. 8B, 2003, pp. 716-729.
[9] *The SSL Protocol Version 3.0*, http://wp.netscape.com/eng/ssl3/draft302.txt.
[10] *The TLS Protocol Version 1.0*, http://www.ietf.org/rfc/rfc2246.txt.
[11] *A One-Time Password System*, http://www.ietf.org/rfc/rfc/2289.txt.

Session and Connection Management for QoS-Guaranteed Multimedia Service Provisioning on IP/MPLS Networks

Young-Tak Kim, Hae-Sun Kim, and Hyun-Ho Shin

Dept. of Info. & Comm. Eng., Graduate School, Yeungnam University,
Gyeongsan-Si, Gyeongbuk, 712-749, Korea
ytkim@yu.ac.kr, sunni@yumail.ac.kr, srobic@hanmail.net*

Abstract. In order to provide QoS-guaranteed realtime multimedia services on IP-based transport network, tightly coupled interactions of session & connection management and CAC (connection admission control) is essential. Also, for efficient QoS-guaranteed DiffServ provisioning across multiple AS domain networks, a scalable transit networking scheme must be provided so as to configure per-class-type QoS-guaranteed packet processing and to provide scalable connection admission control (CAC). In this paper, we propose a functional architecture of session & connection management with SIP, RSVP-TE, CAC and QoS-guaranteed virtual networking. We also propose a per-class-type virtual networking scheme for scalable QoS-guaranteed DiffServ provisioning across multiple autonomous system (AS) domain networks. The implementations of the proposed distributed session & connection management functions are based on Web Service architecture and XML/SOAP-based network management.

1 Introduction

In order to provide QoS-guaranteed realtime multimedia services on IP-based transport network, two signaling functions must be prepared: (i) end-to-end signaling to initialize a session, and (ii) UNI/NNI signaling to establish QoS & bandwidth-guaranteed virtual circuit for media packet flow. And these signalings must be tightly coupled with CAC (connection admission control) that controls the overall traffic flow over a traffic engineering transit tunnels at appropriate operational level for guaranteed QoS provisioning. For end-to-end session initialization and description, SIP (session initiation protocol) / SDP (session description protocol) can be used, while RSVP-TE can be implemented as the UNI and NNI signaling to establish QoS-guaranteed virtual circuit. Since current IP/MPLS implementations do not provide per-class-type packet processing

---

* This work has been supported by Yeungnam University IT Research Center (ITRC) Project.

and NNI signaling across multiple AS (autonomous system) domain networks, an efficient alternative transit networking must be provided so as to configure per-class-type QoS-guaranteed DiffServ packet processing, and enable scalable connection admission control (CAC) and QoS-guaranteed transit networking across multiple AS domain networks.

Session Initiation Protocol (SIP) is an application-layer control (signaling) protocol for creating, modifying, and terminating sessions with one or more participants[1,2], such as VoIP telephone calls, multimedia distribution & conferences. SDP is used for describing multimedia sessions for the purposes of session announcement, session invitation, and other forms of multimedia session initiation [2]. In order to organize a multimedia communication session, SDP/SIP messages must be exchanged among participants to check the availability of participant, capability of terminal, and determination of media type, transport protocol, format of media, and related network & port addresses.

After determination of parameters for a multimedia session, QoS-guaranteed connections for the session must be established among participant terminals. In IP/MPLS network, RSVP-TE (resource ReSerVation Protocol with extension of Traffic Engineering) can be used for UNI and intra-AS NNI signaling. MPLS NNI signaling standard for inter-AS domain networks, however, is not yet well standardized to establish per-class-type QoS-guaranteed connections across multiple AS domain networks. As a consequence, an efficient alternative transit networking scheme must be provided so as to configure per-class-type QoS-guaranteed DiffServ provisioning and to support scalable connection admission control (CAC).

In this paper, we propose a functional architecture of session & connection management with SIP, RSVP-TE, CAC and QoS-guaranteed virtual networking. We also propose a per-class-type virtual networking scheme for scalable QoS-guaranteed DiffServ provisioning across multiple autonomous system (AS) domain networks. The implementations of the proposed distributed session & connection management functions are based on Web Service architecture and XML/SOAP-based network management. Based on the proposed architecture, DiffServ-aware-MPLS traffic engineering can be efficiently implemented across multiple AS domain networks of different network operators.

The rest of this paper is organized as follows. In section 2, related works are briefly introduced. In section 3, QoS-guaranteed per-class-type virtual networking in intra-AS network and inter-AS networks, functional model of interactions among session and connection management protocols, QoS-routing, and CAC (connection admission control) functions are explained. In section 4, implementations of NMS (network management system), EMS (element management system) and CNM (customer network management) system for per-class-type virtual networking and end-to-end connection management are explained. Also, configuration of DiffServ-aware-MPLS TE at provider edge (PE) router is explained. Finally, we conclude this paper in section 5.

2 Related Work

2.1 Session Initiation Protocol (SIP) and Session Description Protocol (SDP)

SIP is an application -layer control (signaling) protocol for creating, modifying, and terminating sessions with one or more participants[1]. The session may be VoIP telephone calls, multimedia distribution & conferences. SIP supports five facets of establishing and terminating multimedia communications: user location, user availability, user capabilities, session setup, and session management. SIP makes use of proxy servers to help the routing request to the user's current location, to authenticate and authorize users for services, to implement provider call-routing policies, and to provide features to users. SIP also provides a registration function that allows users to register their current locations for use by proxy servers. SIP may run on top of several different transport protocols, such as UDP and TCP.

SIP invitations used to create sessions carry session descriptions that allow participants to agree on a set of compatible media types. SDP (session description protocol) [2] has been designed to convey session description information, such as session name and purpose, the time(s) when the session is active, the media comprising the session, information to receive those media (addresses, ports, format, etc), the type of media (video, audio, etc), the transport protocol (RTP/UDP/IP, H.320, etc), and the format of the media (H.261 video, MPEG video, etc). For a multicast session, multicast address and transport port for media are delivered; for a unicast session, remote address for media and transport port for contact address are delivered. SDP is not intended to support negotiation of session content or media encodings.

When a multimedia communication session is determined by SDP/SIP, the network addresses of participants and the related traffic & QoS parameters are determined. Based on these connection parameters, connection establishment is requested using UNI signaling function. The traffic parameters will include committed information rate (CIR), committed burst size (CBS), excess burst size (EBS), and peak information rate (PIR). The QoS parameters will include end-to-end transfer delay, delay variation tolerance, packet error rate / bit error rate, service availability and protection mode.

2.2 UNI Signaling with Traffic Engineering Extensions (RSVP-TE)

After determination of QoS & traffic parameters for a multimedia session, QoS-guaranteed per-class-type connections for the session must be established among the participant terminals. In IP/MPLS network, connection establishment is accomplished by UNI (user-network interface) and NNI (network node interface) signaling. For UNI signaling between user terminal and ingress edge router, RSVP-TE can be used to carry the connection request [3]. In order to support per-class-type DiffServ provisioning, RSVP-TE must provide traffic engineering extensions so as to deliver the traffic & QoS parameters.

The user agent in multimedia terminal must provide RSVP-TE client function, while the ingress edge IP/MPLS router must support the RSVP-TE server function. Since RSVP-TE establishes only unidirectional connection, two PATH-RESV message exchanges should be implemented to establish bi-directional path between user terminal and ingress router.

2.3 Differentiated Service Provisioning with Multiple Virtual Networks

The realtime multimedia service will require guaranteed QoS provisioning and traffic parameters according to its applications [4]. The highly interactive realtime transaction service, such as user-to-user signaling and highly interactive transaction service, will require less than 50 msec of end-to-end delay. Highly interactive realtime CBR(constant bit rate) conversation service, such as VoIP, will require end-to-end delay of 100 msec and jitter of 50 msec, and can be supported by EF (expedited forwarding) class-type in DiffServ standard. Highly interactive realtime VBR (variable bit rate) conversation service, such as multimedia phone, will require same QoS parameters, but may be provided by AF (assured forwarding) class-type that handles efficiently the variable bit rate (VBR) characteristics.

Multimedia conference will require extended end-to-end delay of 400 msec, and jitter of 50 msec. Interactive transaction data, such as telnet session, will require end-to-end delay of 400 msec without jitter constraints, while Web search or bulk data transfer will require much loose time constraints, such as more than 1 sec of end-to-end delay. Finally, the best effort service is a class type for legacy Internet service that does not guarantee time constraints nor traffic parameters.

As explained above, each class-type requires different QoS requirements; some service classes requires tight end-to-end delay of 100 or 400 msec and limited jitter, while the other non-realtime service classes do not require tight end-to-end delay. The different service class-type will require different fault restoration capability; highly interactive realtime transaction or conversational services will require fast restoration with 1+1, 1:1 or M:N backup path, while non-realtime services will require less strict fault restoration with 1:N or sometimes may allow no restoration.

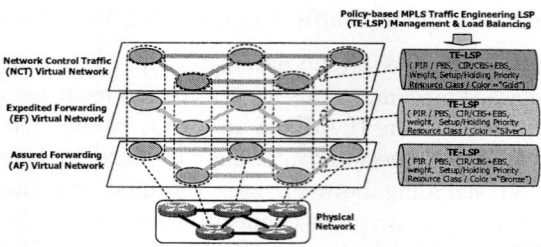

Fig. 1. Differentiated service provisioning with multiple virtual networking

In order to provide QoS-guaranteed service while maintaining the network in optimal resource utilization level, we need to configure the MPLS transit network to provide multiple per-class-type virtual networks with different operation mode. Fig. 1 shows the concept of multiple virtual network on MPLS network, where virtual networks for network control traffic (NCT), expedited forwarding (EF), and assured forwarding (AF) class-types are configured and managed separately [4].

3 Session and Connection Management for QoS-Guaranteed Multimedia Service Provisioning

3.1 QoS-Guaranteed per-Class-Type Virtual Networking in an Intra-AS Domain Network

Configuration of scalable per-class-type virtual networks in an intra-AS domain network is one of the key traffic engineering function in QoS-guaranteed DiffServ provisioning. As shown in Fig. 2, the NMS (network management system) in each AS domain network configures multiple virtual networks for each DiffServ class-type considering the QoS parameters. In a per-class-type virtual network, multiple QoS-guaranteed TE-LSPs are established among PE routers to configure connectivity of full mesh topology. The DiffServ-aware-MPLS AS domain network provides the network-view information of IP/MPLS routers, data links among routers, and CSPF (constraint-based shortest path first) routing function.

The QoS-guaranteed per-class-type virtual network function can be provided as the connectivity management API (application programming interface) of Parlay/OSA (Open Service Architecture) standard [5].

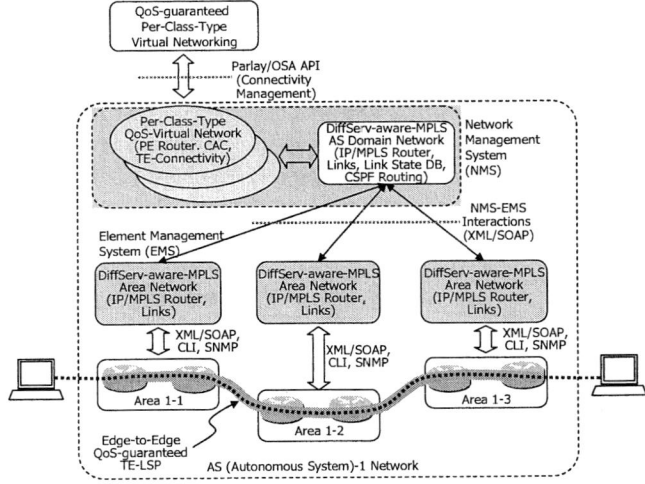

Fig. 2. QoS-guaranteed per-class-type virtual networking in an AS domain network

3.2 QoS-Guaranteed per-Class-Type Virtual Networking Among Inter-AS Domain Networks

The configuration of per-class-type virtual networks across multiple AS domain network is very important to support QoS-guaranteed DiffServ efficiently through multiple AS domain networks without scalability problem. In order to configure virtual networks for DiffServ class-types, the NMS establishes edge-to-edge TE-LSP in two-level hierarchy: (i) establishment of ASBR (autonomous system boundary router)-to-ASBR trunk TE-LSP as inter-AS transit tunnels, and (ii) establishment of edge-to-edge QoS-guaranteed TE-LSP through the transit tunnels among ASBRs.

In order to establish QoS-guaranteed ASBR-to-ASBR transit TE-LSP, the network resource availability and traffic engineering parameters of each AS domain network must be collected. Current BGP (border gateway protocol), unfortunately, only provides reachability & route information, and does not provide traffic & QoS information of the route. As an alternative solution, Web-service architecture can be used to implement the interactions among NMSes for AS domain networks [6].

As shown in Fig. 3, NMS of each AS domain network would register the available connectivity services among ASBR in the AS domain network through Web service registration. The ingress NMS queries the UDDI registry to get the URL of WSDL for the NMS of destination network to which the destination CPN (customer premises network) is attached. It then retrieves the information of neighbor NMS, recursively, until one of the neighbor of the intermediate NMS is itself. Based on the collected AS domain network connectivity information and the available transit networking attributes (i.e., available bandwidth, edge-to-edge transfer delay, etc.), the originating NMS can find the constraint-based shortest path between the ASBRs of the originating AS domain and the destination AS domain. Multiple ASBR-to-ASBR transit TE-LSPs may be configured with different route for the virtual transit networks according to DiffServ class-types.

The ingress NMS then configures intra-AS DiffServ-aware-LSP in the originating AS domain network with configuration of DiffServ-aware packet processing at ingress provider edge (PE) router, requests the destination NMS to setup the LSP at destination domain network, and finally completes the edge-to-edge DiffServ-aware-LSP through the transit trunk TE-LSP of transit virtual network for the requested DiffServ class-type.

3.3 Session and Connection Management

Fig. 4 shows the overall interactions among session management functions and connection management functions. The user terminal will firstly discover required service from service directory that provides service profiles. If an appropriate application service is selected and agreed via SLA (service level agreement), session setup & control for the multimedia application will be initiated. SIP and SDP will be used to find the location of destination, to determine the availability and capability of the terminal. The SIP proxy server may utilize location

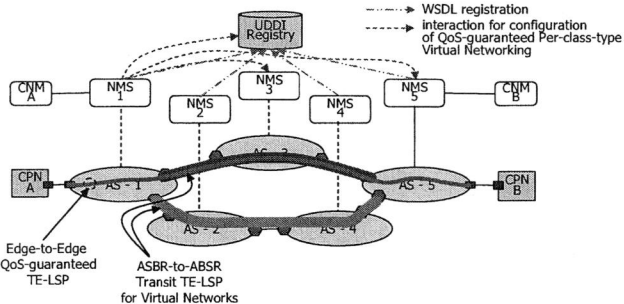

Fig. 3. Configuration of transit virtual networks

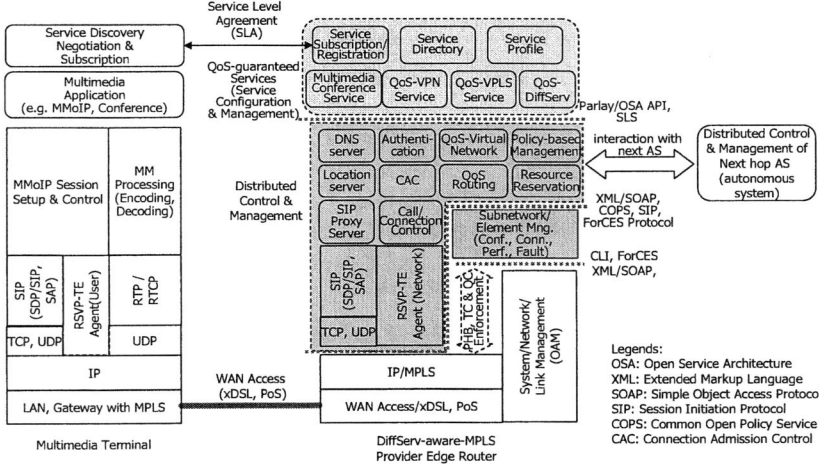

Fig. 4. Interaction among session and connection management functions

server to provide some value-added service based on presence and availability management functions.

Once session establishment is agreed by participants, QoS-guaranteed per-class-type end-to-end connection (or packet flow) establishment will be requested through UNI signaling. RSVP-TE may be used in IP/MPLS network. The edge IP/MPLS router of ingress provider network will contain connection control and management functions for on-demand connection establishments.

When the customer premises network (CPN) or participant's terminal does not support RSVP-TE signaling function, an end-to-end connection will not be established; instead, per-class-type packet flow must be registered and controlled by the connection management function of ingress PE router with CAC (connection admission control). The customer network management (CNM) system may support the procedure of per-class-type packet flow registration.

3.4 QoS-Routing and Connection Admission Control (CAC)

When a QoS-guaranteed connection setup (or per-class-type packet flow registration) request is arrived, the connection management module must check the available network resource, find appropriate route for the requested QoS & traffic parameters using CSPF (constraint-based shortest path first) QoS-routing function, and finally make decision on the admission of the requested connection establishment. The CSPF for the requested connection will consider bandwidth (CIR/CBS, EBS), end-to-end delay, jitter, bit/packet error rate and required protection mode. When a connection establishment request is accepted, the DiffServ-aware-MPLS connection control function will establish an end-to-end LSP for the requested connection.

The per-class-type virtual networking that has been explained in Section 2 will minimize the complexity of QoS-routing and scalability problem of CAC, and provides flexibility in the management of QoS-guaranteed transit network resources. Since near full-mesh topology can be configured among PE-PE pairs for each class-type, the user connection setup request can be handled easily based on the pre-estimated traffic between the ingress PE router and the egress PE router. Also, the per-class-type QoS-guaranteed virtual networking makes it easy to support the differentiated guaranteed bandwidth provisioning and controlled link utilization that manages the queuing delay and packet loss by buffer overflow. In the case of network link/node failure, it is much easier to reroute the TE-LSP for each class-type that contains multiple user traffic flows, instead of rerouting multiple LSPs for each user packet flow.

4 Implementations and Analysis

4.1 Implementation of NMS, EMS and CNM

In order to provide Web service based distributed connection management function, CNM (Customer Network Management) system and NMS (Network Management System) have been implemented based on JBuilder X with Apache AXIS toolkit, JAX-RPC for XML/SOAP messaging and DII (dynamic invocation interface). JWSDP with JAXR is used to configure private UDDI which provides the URL of neighbor NMS for neighbor AS domain network. Fig. 5 shows the architecture of NMS with Web service functions.

In order to support distributed connection management, each NMS registers the information of its client network addresses and related URLs of WSDL(Web Services Description Language) documents from which other NMSes can query for the availability to establish a connection with the requested QoS & traffic parameters to the destination. Each NMS also registers URLs of WSDL documents for the query of its neighbor AS domain networks with traffic & QoS parameters of the transit link (i.e., capacity, physical distance, supported DiffServ classes and their profiles).

In each AS domain network, an NMS may control multiple EMSes, where each EMS controls a sub-domain MPLS network. EMS provides basic func-

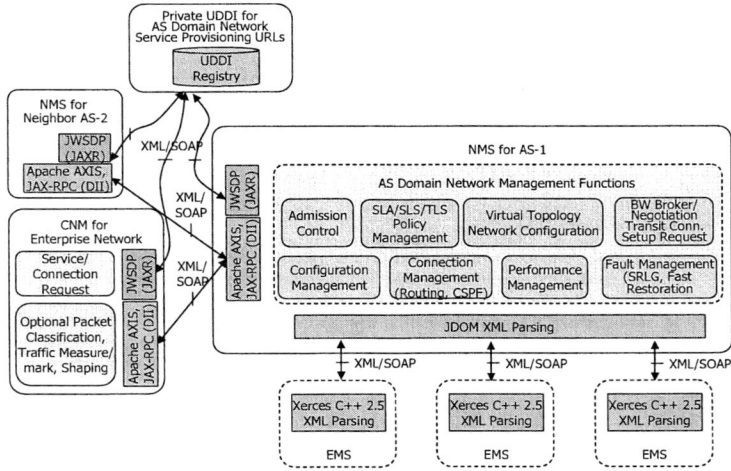

Fig. 5. Network Management System (NMS) with Web Service functions

tions of network configuration, traffic engineering tunnel (TE-LSP) establishment, node/link protection mode setup, and configuration of fault notification and fast restoration. The EMS is tightly related with the signaling module for on-demand dynamic connection establishment. In the initial stage of MPLS network installation when MPLS signaling function is not fully mature, the EMS may be used to provide traffic engineering tunnel establishments.

The EMS may be implemented with various programming language and platforms. Usually, the network element (NE, i.e. MPLS LSR) is developed together with its related EMS function for local management and test purpose. The legacy EMS modules (i.e., configuration management, connection management, performance & fault management) which have been implemented by C++ on UNIX platform must be provided with adaptation function for XML/SOAP access from NMS. Fig. 6 depicts the legacy EMS modules with adaptation functions.

4.2 Interactions Among NMS, EMS and NE

In the Web service architecture for distributed connection management across multiple AS networks, each AS domain network is assumed to be managed by an NMS which takes care of the overall AS domain network, and several EMSes (Element Management System) that take care of one or more areas within the AS. The QoS-guaranteed connection across different AS domains is established by interactions between the neighbor NMSes of the AS networks which include inter-AS negotiation for differentiated packet processing and bandwidth allocations.

Each intermediate NMS and the destination NMS must provide the information of reachability, available bandwidth and related QoS parameters between AS boundary edge points to the originating NMS that performs constraint-routed shortest path first (CSPF) route for the requested end-to-end connection with

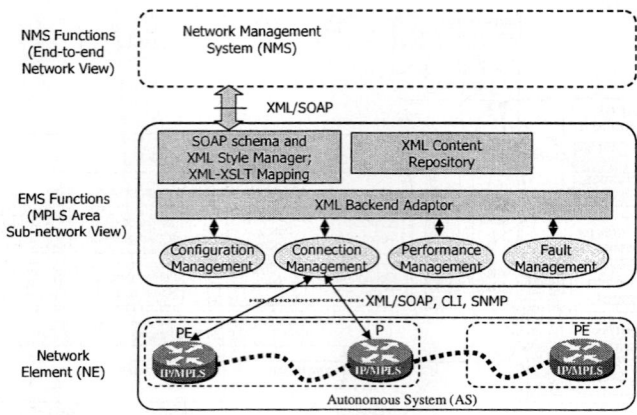

Fig. 6. Element Management System (EMS) for subnetwork management

specific traffic & QoS parameters. To provide the connectivity among AS networks and the edge points, the dedicated NMS of each AS network registers to the UDDI registry.

The interaction between NMS and EMS is implemented by XML/SOAP, and the interaction between EMS and network element nodes (i.e., MPLS router, host and switch) is implemented by XML/SOAP, SNMP or CLI (Command Line Interface). Currently, the SNMP interface is not fully supported in commercial MPLS routers. The CLI is usually proprietary standard which is specified by the equipment vendor; for specific equipments from a vendor, adaptation functional modules for CLI must be provided.

CNM (Customer Network Management) system is used to manage the customer's network, to configure differentiated services with appropriate SLA/SLS, and to generate end-to-end connection setup request for QoS-guaranteed service.

4.3 Configuration of DiffServ-Aware-MPLS TE at Provider Edge (PE) Router

Fig. 7 depicts the configuration of DiffServ-aware-MPLS TE at ingress provider edge router. DiffServ-aware-MPLS traffic engineering is configured firstly at the provider edge (PE) router where incoming packet flow is classified by pre-defined classification options, metered and marked for each class-type. If the packet exceeds the committed information rate (CIR) of the class-type, it may be remapped to lower priority class-type or dropped.

The packet flow of determined class-type is then enqueued in the specified output queue to be forwarded to the transit network where multiple virtual networks are configured for DiffServ class-types to the destination PE router.

In a test-bed network with Cisco 7200 series IP/MPLS routers, the DiffServ-aware-MPLS TE configuration has been implemented with Class Map, Route Map, Access List and Policy Map [7]. Class-map command and its subcommands are applied on per-interface basis to define packet classification, marking,

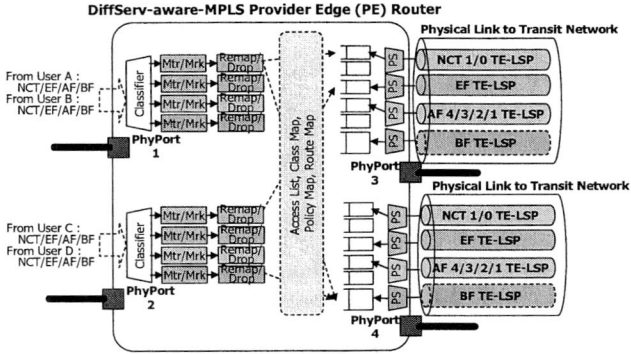

Fig. 7. Configuration of DiffServ-aware-MPLS TE at provider edge (PE) router

aggregate, and flow policing as part of a globally named service policy. Access lists perform packet filtering to control the flow of packets.

A policy map associates a traffic class with one or more QoS actions. The router supports QoS policing functions such as class-based weighted fair queuing (CBWFQ). Paths among AS domain networks are created by applying the route-map to PE. Filtering of source IP address of packet and specification of destination IP address can be possible, using the route-map.

5 Conclusion

Since current IP/MPLS implementations do not provide per-class-type packet processing and NNI signaling among different AS (autonomous system) domain networks, an efficient alternative transit networking must be provided so as to support efficient per-class-type QoS-guaranteed DiffServ provisioning, scalable connection admission control and QoS-guaranteed transit networking. In this paper, we proposed per-class-type virtual networking scheme for efficient QoS-guaranteed DiffServ provisioning across multiple AS domain networks. We also proposed a functional model of interactions among SIP/SDP, RSVP-TE, CAC and QoS-guaranteed virtual networking.

Currently, we are implementing NMS, EMS and CNM for virtual networking and distributed connection management to provision end-to-end QoS guaranteed realtime multimedia services across multiple AS domain networks. The interaction of NMSes among inter-AS domain networks is implemented based on Web Service architecture.

References

1. J. Rosenberg et. al., "SIP: Session Initiation Protocol," IETF RFC 3261, June 2002.
2. M. Handley and V. Jacobson, "SDP: Session Description Protocol," IETF RFC 2327, April 1998.

3. D. Awduche, et. al., "RSVP-TE: Extensions to RSVP for LSP Tunnels," IETF RFC 3209, December 2001.
4. RFC 3270, "Multiprotocol Label Switching (MPLS) support of Differentiated Services," April 2002.
5. Young-Tak Kim, Hae-Sun Kim, "Per-Class-type Virtual Networking for QoS-guaranteed DiffServ Provisioning on IP/MPLS Networks," submitted to ICC'2005.
6. Parlay version 4.0, Parlay Group, http://www.parlay.org.
7. Youngtak Kim, Hyun-Ho Shin, "Web Service based Inter-AS Connection Managements for QoS-guaranteed DiffServ-aware-MPLS Internetworking," Proc. of International Conference on Software Engineering Research, Management & Applications (SERA 2004), San Francisco, pp. 256-261.
8. Eric Osborne and Ajay Simha, Traffic Engineering with MPLS - Design, configure, and manage MPLS TE to optimize network performance, Cisco Press, 2003.

A GQS-Based Adaptive Mobility Management Scheme Considering the Gravity of Locality in Ad-Hoc Networks

Ihn-Han Bae[1] and Sun-Jin Oh[2]

[1] School of Computer and Information Communication Eng.,
Catholic University of Daegu, Gyeongbuk 712-702, Korea
ihbae@cu.ac.kr
[2] School of Computer and Information Science,
Semyung University, Chungbuk 390-711, Korea
sjoh@semyung.ac.kr

Abstract. Mobile Ad-Hoc Network (MANET) is a network of mobile nodes that do not have a fixed infrastructure. Recent research in this field addresses ways of solving existing problems in a MANET by using the node location information. However, maintaining the location information of the nodes is one of the essential challenges in mobile Ad-Hoc networks because the location of a node changes frequently. In this paper, we propose a Group Quorum System (GQS) based adaptive mobility management scheme considering the gravity of locality that is managing the location information for all nodes efficiently in a MANET. In a proposed scheme, the mobility database storing the location information of a mobile node can be selected adaptively from the GQS by considering the gravity of locality. The performance of a proposed scheme is to be evaluated by an analytical model, and compared with that of existing Uniform Quorum System (UQS) based mobility management scheme.

1 Introduction

The advancement in wireless communications and small-sized, lightweight, portable computing devices enables the mobile computing environment. One research issue that has attracted a lot of attention recently is the design of mobile Ad-Hoc network consisting of a set of mobile hosts that may communicate with one another and roam around at their will. Communication is accomplished through wireless links among mobile hosts, but there is no base station to support communication in a MANET. Therefore, excess the limitation of the transmission distance means that the mobile host may not be able to communicate with each other directly. Hence, a multi-hop scenario occurs, where a packet may need to be relayed by several hosts before reaching its final destination. This requires each mobile host in a MANET serve as a router [1].

Applications of MANETs occur in situation like battlefields, festival grounds, assemblies, outdoor activities, rescue actions, or major disaster areas, where networks need to be deployed immediately but base stations or fixed network infrastructures are not available. A scenario of a MANET is illustrated in Figure 1. Recent research in this field addresses ways of solving existing problems in MANETs by the use of node location

Fig. 1. Typical Architecture of a MANET

information. For example, some routing algorithms, including LAR (Location-Aided Routing) [2] and GRID [3] use information about the geographic location of nodes to optimize the routing process. Node location information may be used to provide various services such as location dependent query processing, navigation, geographic messaging and neighbor and service discovery [4].

In this paper, we design a GQS-based adaptive mobility management scheme considering the gravity of locality of a mobile node in a MANET. In a proposed scheme, the topology of the Ad-Hoc network is divided into several logical regions of 2-layered grid structure, and single home region is selected from each 2- dimensional region by using the mapping function. Then, the logically spread surface quorum system is composed from these selected N home regions, and the GQS is constructed from this system. If one mobile node updates its location, quorum is selected from the GQS by considering the gravity of locality of that mobile node, then the location information of a mobile node is stored to the nodes in a selected quorum.

The rest of the paper is organized as follows. Section 2 gives a brief description of related works for mobility management in mobile Ad-Hoc networks. Section 3 illustrates the proposed GQS-based adaptive mobility management scheme considering the gravity of locality. Section 4 presents the performance of the proposed scheme that is evaluated through an analytical model. Finally, Section 5 concludes the paper and describes our future works.

2 Related Work

In routing protocol of an Ad-Hoc network, the location service uses the location information of a node for packet routing. So, many researches for location management in an Ad-Hoc network were performed recently. In [5], the Ad-Hoc mobility management scheme was proposed and analyzed, which is using location databases forming virtual backbone that is dynamically constructed and distributed among network nodes. Such databases are serviced as a container for both location update and search. Routing is performed under the single network structure including all nodes in a network. These databases are composing the Uniform Quorum System (UQS). For location update or call delivery, the location information of a mobile node is written to or read from all databases in a quorum selected non-deterministic manner.

In [6], based on the virtual backbone structure, randomized database group method is proposed for Ad-Hoc mobility management. Similar to the UQS of Ad-Hoc mobility management, this method is doubly distributed at the point of view that both database allocation and database access are dynamic and non-deterministic. A location database allocation to a mobile host is flexible, and its constraints are stability of a network node, network traffic, and mobility pattern. During the location update of a mobile host or when the call is delivered to a mobile host, the location of a mobile host is recorded to or read from some group of K databases selected randomly.

In [4], two location management algorithms based on the probabilistic quorum system are developed. On the other hand, quorums are overlapped with some probability in probabilistic quorum system. In the first algorithm, a node selects size of k quorum by including itself and all reachable nodes for all operations, then transfers the location update or query messages to the selected quorum. In the second algorithm, a node selects one neighbor node randomly, and transfers the location update or query messages to the selected neighbor. Its neighbor node transfers that message to its neighbor node randomly that does not receive that message so on. If this message already visited k nodes or if there are no nodes that did not receive this message, then the message transfer is no longer performed.

In [7], many different query methods for randomized database group scheme are studied and their performances are compared. Especially, the optimal update-query size and query-group size are determined. Furthermore, it shows the probability of the first query being successful, and the average query latency searching for the location of a mobile host, then evaluates the cost for implementing randomized database group scheme with different number of database functions.

Quorum system can be used to implement shared object with single-writer and multiple-reader in which the read and write operations are executable. Read operation is composed of reads from all nodes in some quorum, and write operation is also composed of writes to all nodes in some quorum. Read operation can generate recent information in a quorum system because of its intersection property [4].

3 GQS-Based Adaptive Mobility Management Scheme

3.1 System Model

Given a square region of area A, GQS-based adaptive mobility management scheme divides the topography into G logical unit regions (referred to as Order-1 regions), where each node is aware of the size of the topography as well as the size of an Order-1 region. It then combines K^2 Order-1 regions to form Order-2 regions. Each node selects a home region in each Order-2 region via a function F that maps roughly the same number of nodes to each Order-1 region in an Order-2 region. Hence, every node has $O(\frac{A}{K^2})$ home regions in A [9]. The rest excepting home regions are considered far regions. Any node within every home region is selected randomly as a location server. At any given time, a relatively limited number of key locales in the network, where important events or activities are taking place, is referred to as focus or hot locales. The mobile nodes reside in this locales are called as focus or hot nodes. Otherwise, it is referred to as cold nodes respectively [10]. Figure 2 shows a sample square topography divided

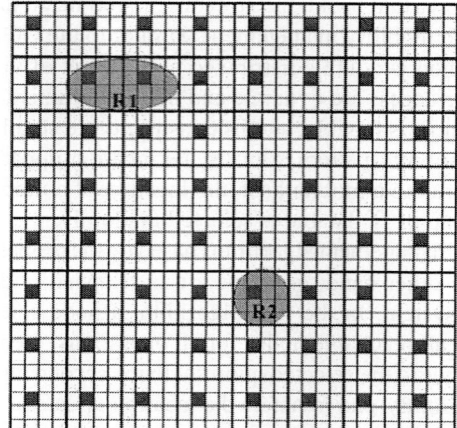

Fig. 2. The topolography of Ad-Hoc Network is divided into Order-2 regions

into Order-1 and Order-2 grids, where an Order-2 grid consists of 16 Order-1 grids. The shaded grid represents the home region in each Order-2 grid, and R_1 and R_2 represent the focus locale respectively.

The mobile Ad-Hoc network can be modeled as a set of N mobile nodes roaming freely in the predetermined 3-dimensional region. We assume that each node has a unique ID ranged between 0 and $N-1$, and each node can identify its location through a service like GPS. Furthermore, mobile node selected in each home region plays a role as a location server maintaining the location information of other nodes in a system. Each location data item is related to the timestamp representing the time obtained that data item.

3.2 GQS Construction

In a proposed adaptive mobility management scheme, the update or query operation for the location information of a mobile node is performed based on the GQS [8]. GQS can be constructed from the unfolded surface quorum system with home regions in an Ad-Hoc network as shown in Figure 3.

If we spread three sides of the cubic on a plane, the 3-group quorum system, $S_3 = (C_1, C_2, C_3)$ can be constructed as shown in Figure 3. Each quorum in C_1 corresponds to the vertical line crossing the right row of the square. Moreover, each quorum in C_2 corresponds to the horizontal line and vertical line of the lower two squares of the right row respectively. Finally, each quorum in C_3 corresponds to the horizontal line crossing lower two squares. In general, each C_i in S_m needs $m-1$ squares. By sharing more than one square of another $m-1$ cartel with each of these, lines corresponding to these two cartel intersect exactly one or more than one nodes on this square. On the whole, there are $\frac{m(m-1)}{2}$ squares.

Let k be the width of each square. Then, since each square is composed of k^2 nodes, the total number of nodes in $\frac{m(m-1)}{2}$ squares are $k^2 \frac{m(m-1)}{2}$. $k^2 \frac{m(m-1)}{2} = h$, where, h represents the total number of home regions composing the system. Therefore, $k = \sqrt{\frac{2h}{m(m-1)}}$, $m > 1$. The size of quorum is $q = (m-1)k = \sqrt{\frac{2h(m-1)}{m}}$.

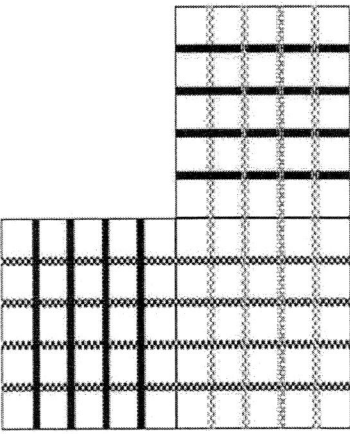

Fig. 3. Unfolded surface quorum system S_3

3.3 Location Update and Query Operation

In order to update the location information of a mobile node, we can use the following methods.

1. Call-origination update: When a mobile host initiates a call, it queries a quorum for the location of a destination and, at the same time, writes its current location into the queried quorum.
2. Location-change update: When a mobile host changes its location, it updates its new location in a quorum.
3. Periodic update: In order to avoid call loss during long time of lack of activity and immobility, a mobile host sends its location information to the quorum periodically.

In a proposed GQS-based adaptive mobility management scheme, location update operation adopts periodic update method. Therefore, the location update operation is initiated when the deadline of the location update period of a mobile node is due. We consider the gravity of locality while executing the location operation. If the mobile node performing the location update is a node in a focus local, the call connection for this mobile node is occurred more frequently than other mobile nodes. Therefore, the location information of the focus mobile node must be managed in much more home regions than other mobile nodes. If the mobile node is a focus node when the location update is occurred, two quorums are selected randomly from C_1-quorum group and C_3-quorum group respectively. Otherwise, one quorum is selected randomly in turns from C_1-quorum group and C_3-quorum group. After then, new location information is sent to all mobile nodes in a selected quorum.

In case of one mobile node searches for the location of other mobile nodes, the location information is returned if the location server of the home region that the querying mobile node resides, has the effective location information for the target mobile node. Otherwise, one quorum is selected from C_2-quorum group, and sends the location query messages to all home regions in this quorum. After then, selects the recent location information by examining the sequential number (timestamp) for the location information in response messages coming from the quorum nodes of that mobile node. The source mobile can always receive recent location information of a target mobile since the quorum of one group in GQS has the intersection property with the quorum of other group.

4   Performance Evaluation

The mobility management cost is evaluated by a total location management cost that adds the location update cost and the location query cost.

$$C_{total} = \{P_{h_1}C_{u-hot} + (1 - P_{h_1})C_{u-cold}\} + \frac{\lambda}{\mu}\{P_{h_2}C_{q-hot} + (1 - P_{h_2})C_{q-cold}\}$$

$$C_{u-hot} = (2q - 1) \cdot C_{u-cost}$$
$$C_{u-cold} = q \cdot C_{u-cost}$$
$$C_{q-hot} = P_{focus-home} \cdot C_{home-q-cost} + (1 - P_{hot-home}) \cdot q \cdot C_{far-q-cost} \quad (1)$$
$$P_{hot-home} = \frac{2q - 1}{h}$$
$$C_{q-cold} = P_{cold-home} \cdot C_{home-q-cost} + (1 - P_{cold-home}) \cdot q \cdot C_{far-q-cost}$$
$$P_{cold-home} = \frac{q}{h}$$

where, P_{h_1} is the probability that the mobile node updating its location is a focus node, P_{h_2} is the probability that the target mobile is a focus node, $P_{focus-home}$ and $P_{cold-home}$ is the probabilities that the location information of a target mobile will be founded from the home region in case of the target mobile focus or not respectively, C_{u-hot} and C_{u-cold} are the location update costs in case of the location update mobile node is focus or not respectively, C_{q-hot} and C_{q-cold} are the location query costs in case of the target mobile node is focus or not respectively, $\frac{\lambda}{\mu}$ represents the call-mobility ratio, C_{u-cost} represent the unit location update cost, and finally, $C_{home-q-cost}$ and $C_{far-q-cost}$ represent the unit location query costs according as home region or far region, respectively.

Figure 4 shows the total location management cost of the GQS-based adaptive mobility management (GQS-AMM) scheme proposed under the environment as in Table 1, and existing UQS-based mobility management (UQS-MM) scheme [5].

Figure 4 shows the total location management cost over call-mobility ratio. As shown in the figure, the total cost of the proposed scheme (GQS-AMM) is much lower than that of the existing UQS-based mobility management scheme (UQS-MM) regardless of the call-mobility ratio. Since the quorum size of the GQS is much smaller than that of the UQS, and the location information of the mobile is stored in home regions adaptively by considering the gravity of locality, we know that the performance of the GQS-AMM scheme is much superior than that of the UQS-MM scheme in higher call mobility ratio because of the locality in a location query operation. Furthermore, the number of control messages and electric power used in the mobility management are also reduced by

Table 1. Parameters used in the performance evaluation

Parameters	Value
G	2,304
K	4
h	144
m	3
q	14
P_{h_1}	0.25
P_{h_2}	0.75
C_{u-cost}	2
$C_{home-q-cost}$	1
$C_{far-q-cost}$	2

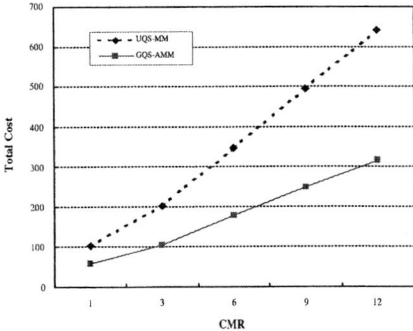

Fig. 4. Total location management cost over call-mobility ratio

using the adaptive location update considering the gravity of locality in location update operation and by performing the location query using small-sized quorum in location query operation. Moreover, the load balancing for the location query on the home regions in a system can be achieved because the number of quorums included in the arbitrary i-th home region are the same.

5 Conclusion

One of the main motive of adopting location management schemes in mobile ad hoc networks is to enable the geographic routing of packets because of scalability issues. In this paper, we propose a GQS-based adaptive mobility management scheme (GQS-AMM), and evaluate the performance of the proposed scheme by an analytical model. According to the results of the performance evaluation, we know that the performance of the proposed GQS-AMM scheme is much superior than that of the existing UQS-based mobility management scheme (UQS-MM). Our future work is to evaluate the performance of our GQS-AMM scheme by a simulation.

References

1. Yu-Chee, Shih-Lin Wu, Wen-Hwa Liao, Chih-Min Chao: Location Awareness in Ad Hoc Wireless Mobile Networks. IEEE Computer 34(6), (2001) 46-52
2. Y.-B. Ko and N. Vaidya: Location-Aided Routing(LAR) in Mobile Ad Hoc Networks. In Proc. 4th Annu. ACM/IEEE Int. Conf. Mobile Computing and Networking, (1998) 66-75
3. W.-H. Liao, Y.-C. Tseng, and J.-P. Sheu: GRID: A Fully Location-Aware Routing Protocol for Mobile Ad Hoc Networks. Telecommunication Systems 18(1), (2001) 37-60
4. S. Bhattacharya: Randomized Location Service in Mobile Ad Hoc Networks. Proceedings of the 6th international workshop on MSWIM'03, (2003) 66-73
5. Z. J. Haas and B. Liang: Ad-Hoc Mobility Management with Uniform Quorum Systems. IEEE/ACM Transaction on Networking 7(2), (1999) 228-240
6. Z. J. Haas and B. Liang: Ad-Hoc Mobility Management with Randomized Database Group. International Conference on Communication (ICC'99), (1999) 6-10
7. J. Li, Z. J. Haas, and B. Liang: Performance Analysis of Random Database Group Scheme for Mobility Management in Ad hoc Network. IEEE International Conference on Communications (ICC2003), (2003) 11-15
8. Yuh-Jzer Joung: Quorum-Based Algorithms for Group Mutual Exclusion. IEEE Transactions on Parallel and Distributed Systems 14(6), (2003) 463-476
9. S. J. Philip, C. Qiao: ELF : Efficient Forwarding on Ad hoc Networks. Proceedings of the IEEE Globecom, (2003)
10. S. Bhattacharya, H. Kim, S. Prabh, T. Abdelzaher: Energy-Conserving Data Placement and Asynchronous Multicast in Wireless Sensor Networks. The First International Conference on Mobile Systems, Applications, and Services (MobiSys), (2003)
11. Yuan Xue, Baochun Li, Klara Nahrstedt: A Scalable Location Management Scheme in Mobile Ad-hoc Networks. Proceedings of the 26th Annual IEEE Conference on Local Computer Networks, (2001) 102-111

A Study on the E-Cash System with Anonymity and Divisibility

Seo-Il Kang and Im-Yeong Lee

Division of Information Technology Engineering Soonchunhyang University,
#646,Eupnae-ri, Shinchang-myun, Asan-si, coogchungnam-Do,
336-745, Republic of Korea
Phone +82-41-542-8819, Fax +82-41-530-1548
{kop98, imylee}@sch.ac.kr
http://sec-cse.sch.ac.kr

Abstract. The micropayment system offers ways of payment in electronic commercial trades in accordance with the purchase of a commodity. The ways of payment include credit cards, on-line money transfer, electronic cash, and others. Electronic cash has been developed to complement the weaknesses of real currency for the sake of electronic commercial trades, and has many requirements and security measures to be met so as to be used like real currency. The author has analyzed the anonymity of electronic cash, among the many requirements and security measures, in order to come up with effective and safe ways to offer anonymity for the micropayment system.

Keywords: Electronic cash, Micropayment, Anonymity, Divisibility.

1 Introduction

Electronic cash has been developed due to the increase in electronic commercial trades. In the initial stage of electronic commercial trades, a bankbook was used to pay the price of a commodity bought from an Internet shopping mall. Using real money for electronic commercial trades, a buyer could pay the price by remitting money through a bank account. But, such practices incurred inconveniences, which called for the development of digitalized currency. Electronic cash, which functions as real currency in an electronic manner, has many of its kinds. First of all, in terms of hardware, there are an IC card type and a network type; in terms of money transmittance, an open type which can transmit the value and a closed type which cannot. Currently, in most cases, the electronic currency used is an IC card type, which can save the monetary value in an IC card. In the case of an IC card type, we need a place where an IC card can be recharged and a terminal wherewith to recharge the card. Such a kind is being used mainly for a transportation card which saves information both on the monetary value and the user. IC card is being utilized for the mobile payment system where payment can be made using a wireless terminal. As for relevant software, a user shall install an application program to conduct an e-commercial trade. The software comprises an electronic wallet or connects a user to a specific site to access the money of the user saves in the database. In order to use electronic money, a user should charge the electronic money with real money by depositing money in a bank or putting the value on a credit car beforehand. Such an electronic currency service is now being commercialized and utilized in our daily life. This thesis is aimed to suggest ways for a user to draw the electronic cash the user has deposited with a bank and utilize it in a store. In the process of such an e-commercial transaction, basic requirements for financial security should be met in an effective and anonymous way. In particular, a new way to the anonymity has been provided by employing the recipient-designated signature system, instead of the existing blind signature system.

2 Outline of Electronic Cash

Electronic cash is an expression of digital information and is being used to pay the price of a commodity in various sectors. Presently, electronic cash is being increasingly utilized by installing software, subscribing to a site so as to have access to the relevant database, or using an IC card so as to prevent credit cards from being used in an illegal way. An IC card has a memory and an operation device to support the saving and encipherment of data. Fabricating an IC card is harder than forging the existing magnetic cards, so that many European countries enjoy using IC cards. By the way, the following requirements and security measures are demanded in order that electronic cash may have an equal value with real currency.

2.1 Requirement of Electronic Cash

What real currency provides may be compared to what electronic cash provides, in the following way:

- Anonymity
  - ▸Real currency provides a user with privacy.
  - ▸If the digital data of electronic cash has or is connected to the information on a user, the cash cannot provide the user with privacy.
- Divisiveness
  - ▸Real currency can be divided or its changes can be offered because the currency has a basic unit.
  - ▸As for electronic currency, the issued data shall be divided.
- Transference
  - ▸Real currency can be transferred to a third party through offering the appropriate amount of money.
  - ▸Electronic currency can also be transferred to a third party through trans mitting data; but, the security should be kept as at the time of issuing the currency.
- Prevention of double use
  - ▸Real currency cannot be used for a second time unless it is faked.
  - ▸Electronic currency can be used for a second time if the saved information can be copied.

Electronic cash must meet all the above-mentioned requirements, but the current electronic cash cannot. Especially, electronic cash cannot provide transfer and division services due to an increase in its data and difficulty of maintaining the security.

3 Analysis of Electronic Cash

The weaknesses of electronic cash sue to its data increase, so that it cannot meet all the aforesaid requirements and security matters. Currently, electronic cash meets some of the requirements and provides some of the services users demand for the sake of safety and protection of money transactions. In this thesis, the author considers and analyzes electronic cash's anonymity, forgery, fabrication, double use, and certification. The anonymity of electronic cash is necessary to protect the privacy of users. Electronic currency contains information on users. So, which user has used the currency may be revealed and his privacy infringed upon. To protect the privacy, electronic currency must have its anonymity. At the same time, it should be possible to verify that data have been transmitted from a qualified user. To do this, blind signatures have been mainly used in addition to zero knowledge, virtual ID, and virtual certificates. The most frequently used is the blind signature system. However, in this system, a signatory does not know the content of the message when providing his/her signature. Stefan Brands suggested a way to provide the anonymity and prevent the double use with the help of discrete algebra on the Internet. But it also has its own weakness in that illegal money laundry can be committed because it provides unconditional anonymity.[1] After that, research has been actively conducted to come up with methods to manage the anonymity, centering on trust organizations. The methods include Yaug BO. The former is to issue a license to a group of users, and to verify the license when an illegal act is committed by one of the users. In the latter's case, a user has a bank provide an identifying value which will be signed by a trust organization. When an illegal act of the user is found, the anonymity can be restored through the trust organization.[2] In Korea, researches have suggested ways to manage blind signatures, including how to provide a blind signature to a value containing an identifier and, in the case of an user's illegal act or double use, to extract the identifier from the electronic cash used.

3.1 New Off-line Electronic Currency Using Hash Chain

A new off-line electronic currency using hash chain where electronic cash is composed of hash chain and utilizes zero knowledge and blind signature to provide the anonymity.[6] The flow chart of its withdrawal protocol is in Fig 1.

The new off-line electronic currency using hash chain utilizes zero knowledge and blind signature to provide the anonymity in the process of withdrawal. The zero knowledge certificate demonstrates that a message has been generated from a user in a rightful way so as to receive a blind signature. Then, upon receiving the blind signature, the anonymity is provided to the user. In this method, it is necessary that the confidence of a bank in the user should be really high, because all the relevant information is provided by the user and the bank provides only certifications.

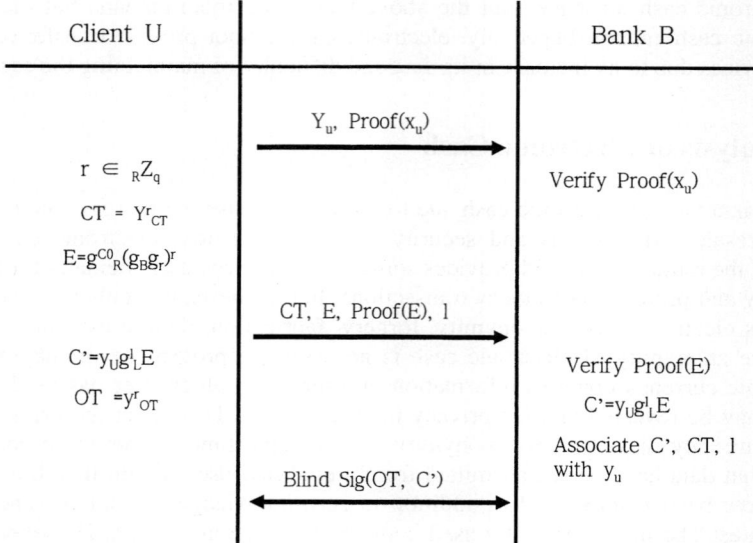

Fig. 1. New off-line electronic currency using hash chain withdrawal protocol flowchart

3.2 Safe Micropayment System Providing Effective Anonymity

The thesis of Payeras-Capella Magdalena has presented a way to the anonymity of a micropayment system and to enhance the effectiveness of the system. Hash function shall be used to issue electronic cash so that the original value may not be revealed and the payment of money shall be verified by the used value applicable to the hash

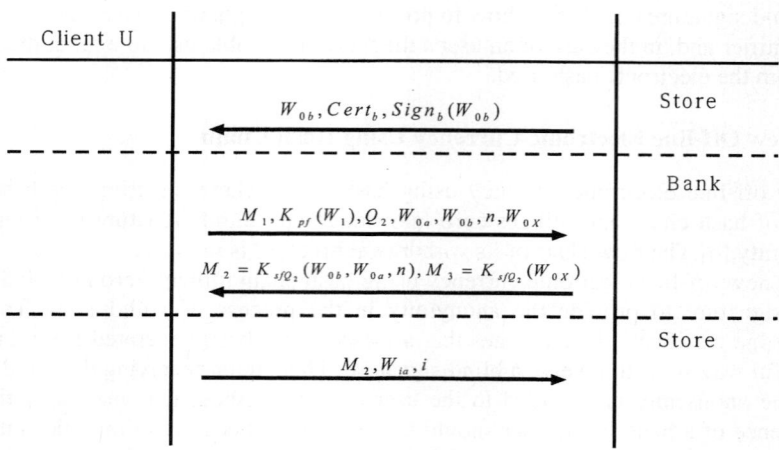

Fig. 2. Safe micropaymetn system providing effective anonymity during protocol flowchart

function[3]. However, the weaknesses of this method include that it cannot provide off-line transactions and that electronic cash can be utilized only at specific stores. In addition, a third party can commit an illegal act during a flow of the next protocol. While a user is dealing at the same time with the currency to be used at a bank and the currency to be used on the Internet, a third party also can use the electronic currency after manipulating as he/she pleases the information on the currency. It is because the third party can have access to the value of M3. Then, the third party can freely use the money of a rightful user to pay the price of a commodity.

4 Methods Suggested in This Thesis

This method comprises user certification, issue protocol, payment protocol, and transfer protocol and is described in detail in the following.

4.1 System Coefficients

- M : store
- R_* : random number generated by the objects of * (* : objects of User, Bank, M)
- h(*) : hash value of *; The hash function is a one-direction one and will not be collided with others.
- sig_* : signature of *
- X_* : personal key of *
- Y_* : Public-key of * (g^{X_*} mod n)
- N : commodity name
- NP : commodity price
- ID_u : user's ID
- PW : user's password
- W_u : principal money the user has deposited in a bank
- P : amount of money to issue
- Q : remainder of the principal minus the issued money
- R_u : random number selected by the user.

4.2 User Certification

A user shall be certified using his/her ID and password (PW1) registered by himself/herself off-line in a safe way, after having access to the website of his/her bank.

   1st Step : A user has access to the website of his/her bank.
        U -> B : Request
   2nd Step : The bank generates a random number(R_{B1}) and transmits it to the user.
        B -> U : R_{B1}
   3rd Step : The user hashes and transmits his/her ID and password and the random number forwarded by his/her bank.
        U -> B : IDu, h(h(PW_1)|R_{B1})

4th Step : The bank certifies a user after verifying his/her ID and the value of the hashed password forwarded by the user.

B : $IDu \stackrel{?}{=} ID'u$, $h(h(PW_1)|R_{B1}) \stackrel{?}{=} h(h(PW_1)|R_{B1})$.

4.3 Protocol for Electronic Cash Issuance

A user will receive electronic cash in the amount he/she requests after user certification has been completed.

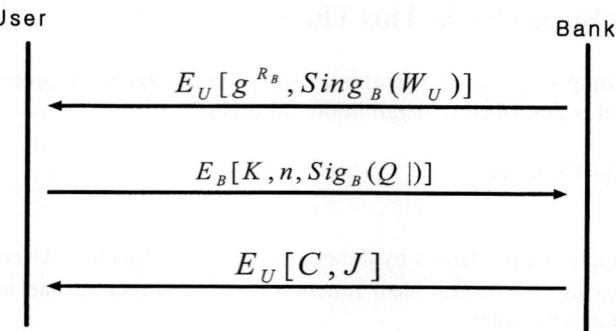

Fig. 3. Protocol for electronic cash issuance flowchart

1st Step : A user receives his/her principal money from his/her bank.
$$B \rightarrow U ; E_u[g^R, W_u, sing_B(W_u)]$$

2nd Step : The user calculates how much he/she will request of the principal transmitted by his/her bank, and then forwards the amount to be withdrawn from the bank (user's operation process). The bank receives data($E_B[K, N, Sig_U(Q)]$) to the user.

$$Q = W_U - P$$
$$K = g^{RP} \bmod n$$
$$S_1 = h(p)$$
$$S_2 = h(p+1)$$
$$\vdots$$
$$S_n = h(p+(n-1))$$
$$S_{root} = S_1 ... S_n$$

3rd Step : In accordance with the formula below, the bank verifies the transmitted data and issues its signature and electronic cash (bank's operation process).

$$P' : W_u - Q$$
$$k' = g^{RP} \bmod n$$

$$C = a^{-1}(aX_B + S_{root})$$
$$J = g^{a^{-1}} \mod n$$

4th Step : The bank verifies the transaction according to the following formula.
$$h(Q|h(R_u)) \stackrel{?}{=} h(Q'|h(R_u))$$
$$B \to U : C, J$$

4.4 Payment Protocol

Payment protocol is composed of the following steps.

1st Step : A user requests a store to certify him.
$$U \to M : \text{Request}$$
2nd Step : The store generates a random number and transmits it to the user.
$$M \to U : R_{M1}$$
3rd Step : The user hashes his ID, password, and the random number given from the store and sends them.
$$U \to M : \text{IDu}, h(h(PW_2)|R_{M1})$$
4th Step : The store verifies the ID and password sent by the user and certifies them.
$$M : ID_u \stackrel{?}{=} ID'_u, h(h(PW_2)|R_{M1}) \stackrel{?}{=} h(h(PW_2)|R_{M1})'$$
5th Step : Then, the store forwards the information (including the price) on the item the user selects.
$$M \to U : N, NP$$
6th Step : In accordance with the formula below, the user pays C' and J' to
$$U \to M : E_M[C', J']$$
$$C' = NPa^{-1}(aX_B + S_{root})$$
$$J' = g^{a^{-1}SNP} \mod n$$
the store.

7th Step : The store sends a commodity or provides a service after verifying the payment using the Public-key of the bank. The verification is made according to the formula below.
$$g^{C'}?g^{X_B NP} g^{j'} = g^{X_B NP + a^{-1}S_{root} NP} \mod n$$
$$C' = NPa^{-1}(aX_B + S_{root})$$
$$J' = g^{a^{-1}S_3 ... S_n NP} \mod n$$
$$g^{C'}?g^{X_B NP} g^{j'S_1 S_2} = g^{X_B NP + a^{-1}S_1 S_2 S_3 ... S_n NP} \mod n$$

The store can verify the signed value and the value of C' and J' through calculating the commodity price using the Public-key of the bank. If user wants division, transmit as following. In accordance with the formula below, the user pays C', J', S_1 and S_2 to the store. The verificantion is made according to the formula below.

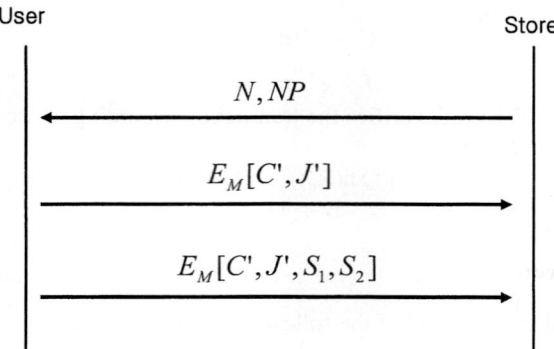

Fig. 4. Payment protocol flowchart

4.5 Deposit Protocol

The store transmits and deposits in the bank the money sent from the user.

> 1st Step : The store sends to the bank the following data.
> $$M \to B : E_B[C', J', NP, NP^{-1}]$$
> 2nd Step : The bank saves C' after verifying the bank's signature using the value sent by the store.
> 3rd Step : After saving C', the bank transmits the money into the store's account.

5 Analysis of Suggested Methods

The safety of the methods suggested in this thesis is analyzed in the following.

5.1 Analysis of User Certification

The qualification of a user is to be certified. A qualified user can generate $h(h(PW_1)|R_{B1})$ using the random number sent from the bank. If a third party it to generate this value, he has to know the hash value of the user's password. But, the third party cannot get the hash value of the user's password, by using only the data sent including the random number. Then, the third party may try again to send the data. However, the second try can be successful only in case the third party uses the same data or the value of the same data is used to make a request. In the method suggested by this thesis, a new random number, which is equivalent to a password, is issued by the bank whenever the user gains access to the bank. Therefore, it almost impossible that the second trial will results in a success. Sometimes, the third party tries to be certified with a random number generated by himself/herself. But, only one value can be generated using the random number and the hash value of password sent by the bank. So, it is also impossible that the third party is successful for this time .

5.2 Analysis of Issuance Protocol

This protocol is needed to protect against illegal generation of electronic currency. The illegal generation is classified into two cases: illegal generation by a rightful user and one by a third party.

The rightful signature of a user shall not be provided for the illegal generation by a third party. So, the third party cannot generate electronic currency in an illegal way.

In case a user generates different cash in an illegal way, there is a change in the value of K generated by the operations of the bank and the user, because the value of K is generated by balancing the principal. In this case, it is impossible to verify through S the value signed by the bank. Moreover, if the value of a random number is substituted by another, the nature of hash function will lead to the original random number.

5.3 Analysis of Payment Protocol

First, a third party may try to pay to other store using C' and J' generated halfway. In this case, the store has to verify the payment by operating the Public-key of the bank using the value from the previous store. Subsequently, C' and J', once used, cannot be used for a second time. Second, a user may try to use electronic cash for a second time. It cannot be prevented at a time when the user uses it at the store, but it can be detected when the money is deposited with a bank. Third, a random electronic currency will be generated in order to provide the anonymity to a store.

5.4 Analysis of Deposit Protocol

It is a step for the bank verifies the used electronic currency and pays the value to the store. In this process, double use should be detected. In the case of double use, the same value as the saved C' will be generated. Then, no deposit will be allowed since all the transactions will be regarded as illegal. When the double use is reported to the store, the information of the illegal user will be revealed. The illegal use can be verified through the value of generated by the bank in the initial stage. The value of K will be generated from the principal of the user according to the formula: K' =h(h(Ru)|Q').

6 Conclusion

This thesis has described general characteristics and security requirements of electronic currency including how to provide the anonymity using a random number for the payment of small sum of money. In the past, the user used a blind signature for the purpose of securing the anonymity; in contrast, in this new system, the anonymity is provided by generating a random number. The electronic cash transacted between the user and the store is to be verified by the signature of the bank. It is also verifiable through the data sent by the issuer for the purpose of confirming the message. The issuer provides its signature for a fixed hash value, in order to prevent excessive use, or forgery or fabrication of the electronic currency. The anonymity of the user will be removed in the case of double use. In the future, electronic currency will be developed to provide wireless services, services for micropayment to a great sum, along with changes in its security services. It calls for research into transference services as well as research into the anonymity and divisiveness applicable to micropayment, not to mention security services.

References

1. Stefan Brands, "Untraceable Off-line Cash in Wallets with Observers", Advances in Cryptology, CRYPTO'93, pp.302-318, 1993
2. Yaug Bo, Liu Dongsu, Wang Yumin, "An anonymity revoking e-payment system with a smart card", International Journal on Digital Libraries, pp.291-296, 2002
3. Magdalena Payeras-Capella, Josep Lluis Ferrer-Gomila, Ll. Huguet-Rotger, "An Efficient Anonymous Scheme For Secure Micropayments", ICWE2003, LNCS 2722, pp80-83, 2003
4. Toru Nakanishi, Yuji Sugiyama "Unlinkable Divisible Electronic Cash", ISW, pp.121-134, 2000
5. Weidong Qiu, Kefei Chen and Dawu Gu, "A New Off line Privacy Protecting E-cash System with Revokable Anonymity", ISC2002, LNCS 2433, pp.177-190, 2002
6. Sangjin Kim, Heekuck Oh, "New off-line electronic currency using hash chain", Korea Information Science Society, Vol 30, NO 02, pp 207-221

An Authenticated Key Exchange Mechanism Using One-Time Shared Key

Yonghwan Lee[1], Eunmi Choi [2,**], and Dugki Min [1,*]

[1] School of Computer Science and Engineering, Konkuk University,
Hwayang-dong, Kwangjin-gu, Seoul, 133-701, Korea
`(yhlee,dkmin)@konkuk.ac.kr`
[2] School of Business IT, Kookmin University,
Chongung-dong, Songbuk-gu, Seoul, 136-702, Korea
`emchoi@kookmin.ac.kr`

Abstract. In order to exchange secure information over the Internet, it is necessary to provide a shared encryption key after dual authentication between the communication parties for data confidentiality. To find an effective authenticated key exchange scheme, many researchers have studied improvement of the Diffie-Hellman key exchange scheme to overcome the weakness of computation complexity and man-in-the-middle attacks. This paper proposes an efficient authentication and key exchange scheme that does not use certificates and public key cryptography, while protecting against man-in-the-middle attacks, replay attacks, DOS attacks and privacy intrusion. This scheme performs a dual authentication using one-time shared authentication key and generates an encryption key which is used in a symmetric block cipher. Our mechanism also includes a secure method that generates an initial seed for creating a one-time shared secret key. In addition, it solves the problem of identity privacy as well as perfect forward secrecy for future data confidentiality.

Index Terms: One-Time Shared key, Authentication, Key Exchange, Diffie-Hellman.

1 Introduction

Secure communications over the Internet need encryption of communication channels. In general, symmetric block cipher methods like the DES are preferred for data confidentiality than asymmetric public key cryptography methods such as the RSA since the symmetric ones require a smaller amount of computing power and memory space than the asymmetric ones. However, symmetric methods should solve a problem of exchanging a shared encryption key after a dual authentication between any two communication parties. The Diffie-Hellman key exchange scheme [1] provides such a key exchange mechanism in symmetric cryptography.

---

* Corresponding author.
** This work was supported by the Korea Science and Engineering Foundation (KOSEF) under Grant No. R04-2003-000-10213-0. This work was also supported by research program 2005 of Kookmin University and Kookmin research center UICRC in Korea.

The Diffie-Hellman key exchange scheme makes use of difficulty in computing discrete logarithms over a finite field. Since this scheme does not authenticate the participants while exchanging messages, it is vulnerable to man-in-the-middle attacks. For this reason, various authenticated key exchange schemes based on the Diffie-Hellman have been studied by many researchers [10, 11, 12]. These schemes can be categorized into two kinds of classes. The first class employs 'certificates' in its key exchange protocol, which foil man-in-the-middle attacks. Certificate-based schemes require additional cost and complexity in key exchange that they are not widely accepted in the market. The other class proposes its authenticated key exchange protocol with an assumption that a pre-shared secret password or a secret key exists between two communication parties. Most of these authenticated key exchange schemes are not efficient because they use a public key cryptography mechanism which requires high computing power. Recently proposed ones like the IKE [2, 8] consider privacy of personal identity and DOS attacks, which require much more computing power.

Recently, mobile computing environment requires low computing power and small memory space even for security service. That is, authenticated key exchange schemes that do not use certificates and public key cryptography are preferable to the mobile environment. This paper proposes an efficient authentication and key exchange scheme that does not use certificates and public key cryptography, while protecting against man-in-the-middle attacks, replay attacks, DOS attacks and privacy intrusion. Characteristics of our scheme are as follows. First, it uses a symmetric block cipher with using a one-way hash function, but without using certificates for dual authentication and key exchange. Since symmetric block cipher requires smaller computing amount and memory space, our scheme is more adaptable to modern distribution environment, such as in ubiquitous and mobile computing. Next, due to the authentication key's one-time property used at each session, our scheme can detect various attacks, such as DOS attacks and man-in-the-middle attacks, without severe computing and memory overhead which overcomes the weakness of Diffie-Hellman. In addition, it solves the problem of identity privacy as well as perfect forward secrecy for future data confidentiality.

The remainder of this paper is organized as follows. In section 2, we present two related schemes: P-SIGMA and Signal-based key exchange schemes. Section 4 proposes our mechanism, which contains seed generation and dual authenticated key exchange phases. Section 5 compares our scheme to the P-SIGMA and Signal methods. Finally, Section 5 draws conclusions.

2 Related Work

In this section, we introduce two existing methods which provide schemes of authentication and key exchange using pre-shared secret information. They are P-SIGMA [3, 7] and a deviation of P-SIGMA [9].

2.1 P-SIGMA

The P-SIGMA solves the problems of personal privacy exposure and DOS attacks by using One-Time-ID or simply OID. The OID is an identity that can be used only once for identifying a user. In P-SIGMA, all OID values are unique by means of sequence numbers and one-way hash functions with collision resistance. Thus, an adversary who does not know a secret key cannot predict the OID that will be used in the future, while both users can calculate any OID values.

However, an adversary can guess all OIDs that have been used previously and will be used in the future if he can obtain a secret key, say K, because a fixed secret key is used for calculating all OID values. It implies the impossibility of perfect forward secrecy for OID. Moreover, an adversary who obtains K can impersonate a user in any future session. If a user shares the same OID with multiple communication partners, he cannot decide whom the connection is from, even when he checks the One-Time-ID.

2.2 P-SIGMA Based Key Exchange Method Using Signal

As a derivation of P-SIGMA, this method uses signals for authenticating users and exchanging keys that are used in symmetric cipher methods. The perfect forward secrecy of One-time-ID can be realized by using the shared secret information which is generated through the Diffie-Hellman key exchange mechanism instead of using a secret key used in P-SIGMA. That is, the seed of OID assures the perfect forward secrecy of One-Time-ID, solving the duplication problem of OID. It also provides a simple key exchange protocol that requires only two rounds, while P-SIGMA requires three rounds. It also extends the application of using the OID method to encrypted communication.

However, this method still has some problems. Since it uses the Diffie-Hellman key exchange mechanism for generating the private information of the next session, it utilizes computing resources a lot. Responders are charged with some computing overhead in calculating OID. This method also cannot generate a dynamic seed for dual-authenticated key exchange that could be changed according to the given security level and the client environment. That is, it uses a fixed initial seed.

3 Proposed Key Exchange Mechanism

The key exchange mechanism proposed in this paper has three phases. The first is a preliminary setup phase. In this phase, public information and a random number matrix of an initiator are delivered to its responder. In the second phase, the initial seed is generated which is used to create a shared secret key for the following data communication session. The third phase performs dual authentication between communication parties and creates a data encryption key that are shared. This section describes our key exchange mechanism in detail.

3.1 Key Exchange Model and Notation

In our key exchange model, an entity who initiates a key exchange mechanism is called an *initiator*, and an entity who responds to the initiator's request is called *an responder*. Both of initiators and responders are kinds of users. There is another type of entity who is not a user, but an attacker, called *an adversary*. An adversary is a

Polynomial-Time Machine that attacks the secrecy of key exchange mechanism. Table 1 summarizes the notations used in our proposed mechanism.

Table 1. Notation

$g_I(k)$	A generator created by the initiator at the k-th time
$h(\ )$	A cryptographic hash function
MX[m,n]	A random number at the [m,n] cell of the random number matrix MX[,]
$Y_I(1), Y_I(2)$	Initiator's first and second public information
$X_I(1), X_I(2)$	Initiator's first and second secret information generated at random
$E_X(Y)$	Encryption Y using X
$OID_I(k)$	One-Time-ID generated by Initiator at the k-th step
C_I	Challenge generated by initiator
$R_I^{C_R}$	Response to the responder's challenge, generated by initiator
CM_I	Encrypted challenge message generated by Initiator
$AK_I^{C_I}$	First authentication key generated by initiator for initiator's challenge
WK_s	Data encryption key (working key) in the s-th session
$AK_I^{'C_R}$	Second authentication key generated by initiator for responder's challenge
$EWK_I(s)$	Encrypted working key generate by initiator in the s-th session

3.2 One-Time-ID (OID)

The OID is an identity that can be used only once for identifying a user. One-time-ID can be used to protect DOS attacks and man-in-the-middle attacks. To prevent DOS and man-in-the-middle attacks, $OID_I(i)$ is attached at every i-th message transmission

$OID_I(i) = h([m,n], MX[m,n], j)$

$OID_I(i)$ is a hash function of [m,n], MX[m,n], and j, where [m,n] is a random position among the random numbers assigned to the initiator within the random number matrix, MX[M,N], and j is just a random number in i-th message transaction. By using [m,n], we can detect DOS attacks and also decide who the initiator is. Man-in-the-middle attacks are detected by checking whether the transferred hash value is correct or not. For message integrity, we attach the hash value of all of the parameters transferred together at each message transmission.

3.3 The Preliminary Setup Phase

In the preliminary setup phase, preliminary numbers are generated and delivered to the opposite side for the next phase. For the Diffie-Hellman key exchange, the initiator side generates $X_I(1), g_I(1), p_I(1)$ and deliver its public values $Y_I(1) = g_I(1)^{X_I(1)} \mod p_I(1)$ and $g_I(1), p_I(1)$ to the responder. For the purpose of protecting DOS attacks and Man-in-the-middle attacks using One-time-ID, the responder generates a MxN random number matrix, MX (M,N) and randomly selects a number of cells in the random number matrix and assigns them to the initiator. Be careful that the same cell should not be assigned to different initiators. For example, we can assign all the columns of the first row to an initiator. These selected random numbers are delivered to the initiator in a safe way. The responder should save the cell assignment information. This information is used for protecting against DOS attacks. Figure 1 describes the steps of the preliminary phase in detail.

Step	Initiator Action/ID_I	Message	Responder Action
Step 1	Select $g_I(1), p_I(1), X_I(1)$ $Y_I(1) = g_I(1)^{x_I(1)} \mod p_I(1)$	$Y_I(1), g_I(1), p_I(1)$ →	Save $Y_I(1), g_I(1), p_I(1)$
Step 2	Save (MX(i,j), ...,MX(i+x, J+y))	← $(MX (i, j)...$ $MX (i + x, j + y))$ (In a secure way)	Assign to the initiator a sequence of random numbers in MX[M,N], i.e. (MX(i,j), ...,MX(i+x, J+y))

Fig. 1. The Preliminary Setup Phase

3.4 The Initial Seed Generation Phase

This phase generates an initial seed that is used to create a shared secret key for the first session. Figure 2 describes the steps of the phase.

In the first step, the initiator sends a challenge message to the responder in order to authenticate the responder according to the Hughes method. In contrast, in the second step the responder sends a challenge message to the initiator for authentication according to the Diffie-Hellman method. In the third step, the initiator and the responder are authenticated according to the corresponding methods, respectively.

3.5 The Authentication and Key Exchange Phase in s-th Session

In the previous phase, each party has the same seed that can be used to create the shared secret key. In the first session, $AK_I(1) = h(AK_I^{C_I})$ and $AK_I R(1) = h(AK_R^{C_I})$. Figure 3 describes the steps of the authentication and key exchange phase in the s-th session. $AK_I(s) = h(AK_I(s-1))$ and $AK_R(s) = h(AK_R(s-1))$. At the first step, the initiator sends $OID_I(1)$ that includes $AK_I(s)$ as a member of hash input. Using this value of $OID_I(1)$, the responder can know whether the sender has the correct shared secret key or not, authenticating the initiator. The working key WK_s is used as the shared data encryption key during the s-th session. Also, the WK_s is used to created $AK_I(s+1)$ and $AK_R(s+1)$ for the $(s+1)$-th session.

Step	Initiator Action	Message	Responder
Step1	Select Randomly $j, X_I(2), g_I(2), p_I(2), C_I$ $AK_I^{C_I} = Y_I(2) = g_I(2)^{X_I(2)} \bmod p_I(2)$ $CM_I = E_{AK_I^{C_I}}(C_I)$	$OID_I(1), [m,n], j, CM_I,$ $g_I(2), p_I(2)$	By using $[m,n]$, detect Dos attack and decide who he/she is. Detect man in the middle attacks by checking whether $OID_I(1) = h([m,n], MX[m,n], 1+j)$
Step2	$AK_I^{C_R} = (Y_R(1))^{x_I(1)} = (g_I(1)^{x_R(1)x_I(1)}) \bmod p_I(1)$ $R_I^{C_R} = D_{AK_I^{C_R}}(CM_R)$ $AK_I^{\prime C_R} = (Y_R(2))^{x_I(2)} = g_I(2)^{X_R(1)X_I(2)} \bmod p_I(2)$	$OID_R(2), [m',n'], CM_R,$ $Y_R(1), Y_R(2)$	Select Randomly $X_R(1), C_R$ $Y_R(1) = g_I(1)^{X_R(1)} \bmod p_I(1)$ $Y_R(2) = g_I(2)^{X_R(1)} \bmod p_I(2)$ $AK_R^{C_R} = (Y_I(1))^{x_R(1)} = (g_I(1)^{x_I(1)x_R(1)}) \bmod p_I(1)$ $CM_R = E_{AK_R^{C_R}}(C_R)$
		$OID_I(3), [m'',n''],$ $R_I^{C_R}, AK_I^{\prime C_R}$	If $C_R = R_I^{C_R}$ then the initiator is authenticated $Z = X_R(1)^{-1}$ $AK_R^{C_I} = (AK_I^{\prime C_R})^Z$
Step3	If $C_I = R_R^{C_I}$ the responder is authenticated	$OID_R(4), [m''',n'''],$ $R_R^{C_I}$	$R_R^{C_I} = D_{AK_R^{C_I}}(CM_I)$

Fig. 2. The Initial Seed Generation Phase

Step	Initiator Action	Message	Responder Action
Step1	$OID_I(1) = h(AK_I(s), [m,n], MX[m,n], 1+j)$ $EWK_I(s) = E_{AK_I(s)}(WK_s)$	$OID_I(1), [m,n],$ $j, EWK_I(s)$	By using $[m,n]$, detect Dos attack and decide who he/she is. If $OID_I(1) = h(AK_R(s), [m,n], MX[m,n], 1+j)$ Then the Initiator is authenticated and we detect man-in-the-middle attacks. $WK_s = D_{AK_R(s)}(EWK_I(s))$
Step2	By using $[m,n]$, detect Dos attack and decide who he/she is. If $OID_R(2) = h(WK_s, [m',n'], MX[m,n], 2+j)$ Then the responder is authenticated and we detect man-in-the-middle attacks.	$OID_R(2), [m',n']$	$OID_R(2) = h(WK_s, [m',n'], MX[m',n'], 2+j)$

Fig. 3. The Authentication and Key Exchange Phase in the s-th Session

4 Analysis of Comparison

We compare our proposed scheme with previous key exchange methods. First of all, our proposed scheme provides dual-authenticated key exchange mechanism as well as data integrity and data confidentiality. Next, the existing methods like IKE and P-

SIGMA are based on the fixed seed of shared key like One-Time-ID and an authentication key. Accordingly, if the adversary knows the fixed secret information, he can impersonate the initiator in the future session. In contrast, our proposed scheme can regenerate the initial seed dynamically depending on the current client environment or adapting to the change in the security level. Third, because our proposed system does not use public key cryptograph like Diffie-Hellman and RSA, which requires much computing power, it can be used for thin clients like mobile or ubiquitous computing devices. Moreover, our proposed system is so efficient as to finish within two messages round for authenticated key exchange. Besides, our scheme provides more concrete protection against DOS attacks, man-in-the-middle attacks, and replay attacks than previously proposed methods. Table 2 summarizes the analysis of comparison.

Table 2. Analysis of Comparison

	Perfect Forward Secrecy	One-Time-ID's Input	Seed	Overlap Of One-Time-ID	Rounds	Algorithms
IKE	unachieved	None	static	None	3 or 6	Digital Signature
P-SIGMA	unachieved	K,RN	Static	Unsolved	3	Diffie-Hellman
SIGNAL	achieved (partially)	DK,RN	Static	Solved	2	Diffie-Hellman
Proposed Key Exchange	achieved (partially)	DK, RN,MX	renewable	Solved	2	Asymmetric Block Cipher

K: Shared Key
DK: Dynamic Shared Key
RN: Random Number
MX: Random Number Matrix

5 Conclusions

We proposed an efficient dual authentication and key exchange mechanisms that can finish within two rounds. Because our proposed system does not require the public key cryptograph like Diffie-Hellman and RSA and certificates, it can be used in a thin client environment like in the mobile and ubiquitous computing. Because the IDs of the initiator and responder are not revealed on the Internet, our scheme protects personal privacy of identity information. It also provides an effective method to protect against DOS attacks with the scope information of initiator's random number table sent by the responder.

References

1. W. Diffie and M.E. Hellman. New directions in cryptography. IEEE Trans. Inform. Theory, IT-22; 644-654, 1976.
2. D. Harkins and D. Carrel. The Internet key Exchange (IKE), RFC2409, 1998.
3. H. Krawczyk, The IKE-SIGMA Protocol, Internet Draft, 2001.
4. Radia Perlman, Charlie Kaufman, Analysis of IPSec Key Exchange Standard, WETICE2001, 2001.

5. Radia Perlman, Charlie Kaufman, Key Exchange in IPSec: Analysis of IKE, 2001.
6. V.Shop, On formal models for secure key exchange, IBM Research Report RZ3120, 1999
7. Hugo Krawczyk, The SIGMA Family of Key-Exchange protocols.
8. Charlie Kaufman, Internet Key Exchange (IKEv2) Protocol, IPSEC Working Group INTERNET-DRAFT, 2003.
9. Kenji IMAMOTO, Kouichi SAKURAI, A Design of Diffie-Hellman Based Key Exchange Using One-Time ID in Pre-shared Key Model, AINA'04. IEEE, 2004.
10. Liqun Chen, Caroline Kudla, Identity Based Authenticated Key Agreement Protocols from Pairings, CSFW'03, IEEE, 2003
11. Dong Hwi Seo, and P. Sweeney, Simple authenticated key agreement algorithm, Electronic Letters 24th, Vol. 35, IEEE, 1999.
12. EIJI OKAMOTO, KAZUE TANAKA, Key Distribution System Based on Identification Information, IEEE Journal On SELECTED AREAS In Communications. Vol. 7. No. 4, IEEE, 1999.

Creation of Soccer Video Highlight Using the Caption Information

Oh-Hyung Kang and Seong-Yoon Shin

Department of Computer Science, Kunsan National University,
68, Miryong-dong, Kunsan, Chonbuk 573-701, South Korea
{ohkang, s3397220}@kunsan.ac.kr

Abstract. A digital video is usually very long and would require large storage capacity. Users may want to watch pre-summarized video before they watch a large long video. Especially in the field of sports video, they want to watch a highlight video. Consequently, highlight video is used that the viewers decide whether it is valuable for them to watch the video or not. This paper proposes how to create soccer video highlight using the structural features of the caption such as temporal and spatial features. Caption frame intervals and caption key frames are extracted by using those structural features. And then, highlight video is created by using scene relocation, logical indexing and highlight creation rule. Finally, retrieval and browsing of highlight and video segment is performed by selecting item on browser.

1 Introduction

In video database, video contents are described by the structure of shots and scenes [1]. Shot is a valid unit for constructing video information and is a set of one or more continuous frames, showing continuous motions in the fixed time and space. Currently, many researchers have interest in automatic and semiautomatic methods for video shot detection and characterization.

Video indexing can be performed efficiently using extraction and recognition of caption text. For example, there has been tremendous success in the automatic conversion of hard-copy documents via optical character recognition (OCR) technology [2,3] and the transcription of speech via voice recognition (VR) technology [4]. In both case, although typically less than perfect, the output is an ASCII text representation that can be indexed with traditional information retrieval techniques. At this point, we can find that some information-rich video sources such as newscasts, commercials, movies and sports events that containing meaningful content in the form of voice, caption text and/or text in the image.

Video summaries are useful to decide whether it is valuable for video viewers to watch this video or not. Video summary is classified in two types; these are video summary sequence and video highlight. Video summary sequence is suitable for documentaries because it provides the significant abstract of whole video. Video highlight is suitable for video trailers or sports highlights because it contains only the interesting segments of video [5].

This paper proposes a creation method of soccer video highlight using structural features of caption. The structural features are temporal appearing features and spatial position, size and color features. Caption frame intervals and caption key frames are extracted from input frames using those structural features of caption. And then caption key frames are indexed physically and logically. Soccer video highlight is created by pre-defined highlight creation rule. And also, easy retrieval and browsing method is provided for selection and watching of video fast and efficiently.

This paper is organized into five additional sections. Section 2 states the related works. Section 3 presents structural features of caption. Section 4 presents extraction of the caption frame interval and key frame. Section 5 presents creation of soccer video highlight. Section 6 presents experimental result, and finally, Section 7 concludes the work.

2 Related Work

Video Summarization is a field of study that many researchers have work continuously with a great interest. It was classified in two types as summary sequence and highlight. There are video skimming [6], scene transition graph [7], cluster validity analysis [8] and video Manga [9] in creation of summary sequence, and movie trailer [10, 11] and event-based sports summarization [12] in creation of highlight.

Christal et. al. [6] proposed video skimming for summarizing documentaries or news videos. Video and transcription of video are aligned by the order of word, and the significant words are identified by linguistic analysis of transcription. Therefore, video clips are selected according to the priority of these words. Yeung et al. [7] proposed scene transition graph that was shot-based structure using the flow of story. Hanjalic et al. [8] extract the key frame and set up the video shots. Then cluster validity analysis is used for creating of video summary sequence including those key frames. Uchihashi et al. [9] proposed a video summarization method of comic or cartoon-like, called Video Manga. They used the measure of importance degree based on the scarcity and persistence of video segments.

Another type of video summarization is the extraction of highlight. Lienhart et al. [10] and Liang-Hua C. [11] proposed the automatic creation method of movie trailer using the tracking of low-level visual and audio features, motion information and color information. They are used the heuristic features of the basic physical parameters for selecting of video clips with the important object, man, action, dialog, title text and title music. Babaguchi [12] proposed the summarization of sports video using the event-based video indexing. It is useful for video summarization but many important features for describing semantics are lost in the process.

3 Structural Features of Caption

3.1 Analysis of Caption Region

The caption appeared in soccer video is important element that gives the core contents of a soccer game. The caption region of soccer video is different from the region of drama and documentary, and it takes the following clear features:

- The position of each caption region is fixed individually according to its type.
- Each caption region has its own size.
- Every caption region appeared immediately (after the event). It remains in some period of time and then disappeared.
- The position of text that can be changed is fixed in the caption region.
- Each caption region has its own uniform colors.
- The event caption region appeared right after event occurrence.
- The appearing sequence of the caption can be different on each and every game.

Based on the above features of caption, the caption key frame is extracted and the video segment is indexed. It plays an important role for the creation of the soccer video highlight.

3.2 Classification of Caption Scene

The caption of soccer video appeared as a limited number of types. In this paper, caption scenes are classified in 13 limited types according to the meaning of contents, as follows:
- teams(*Ctem*), grounds(*Cgnd*), broadcasting(*Cbct*), referee(*Cref*), player list(*Clst*), game beginning(*Cbgn*), score board(*Csco*), bench(*Cbch*), player(*Cplr*), a change of players(*Cchg*), goal(*Cgol*), foul(*Cfol*), game ending(*Cend*)

Beside the above caption scene, the captions that were not very related to current game, such as the caption of sports, game class, league, team rank, weather, other grounds and record, appeared in the video. In this paper, those additional captions were removed from the classification and extraction of the caption.

3.3 Temporal Features of Caption

The captions are classified as starting caption, event caption and ending caption in accordance with its appearing sequence as shows in Table 1. The captions have the temporal structure generally as shown in Figure 1.

Table 1. Temporal Classification of Caption

Temporal class	The component captions
Starting Caption	*Ctem, Cgnd, Cbct, Cref, Clst, Cbgn*
Event Caption	*Csco, Cbch, Cplr, Cchg, Cgol, Cfol*
Ending caption	*Cend*

Before the beginning of the substantial game, the most captions appeared in order of *Ctem, Cgnd, Cbct, Cref, Clst* and *Cbgn*. This appearing sequence of the captions can be varied from each and every game. When the substantial game is in the beginning, the event caption appeared. And when it is in the ending, *Cend* caption appeared.

As shown in Figure 1, event caption shows the event such as *Csco, Cbch, Cplr, Cchg, Cgol* and *Cfol*. It can appear in irregular sequence and must appear right after the event. At the last time, the result of the game was shown by the appearance of the ending caption *Cend*.

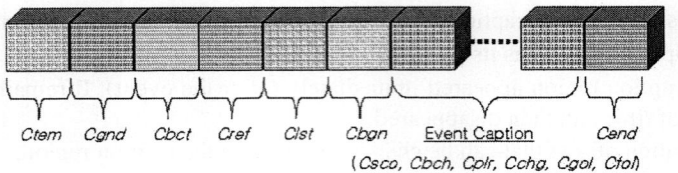

Fig. 1. One Temporal Flow of Caption

3.4 Spatial Features of Caption

The spatial feature of the caption is the information of the region that is the appearing position of the caption in the soccer video. The caption appeared in 6 limited regions as shown in Figure 2.

In Figure 2, the regions of *Csco, Clst, Cbgn, Cend* and *Cref* are overlapped, but have it's own size that was different with one another. The captions of *Ctem, Cgnd, Cbct, Cbch, Cplr, Cgol, Cfol* and *Cchg* are hardly different from its size, and appeared at the center bottom of the frame.

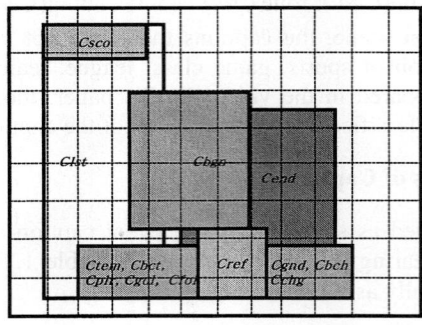

Fig. 2. Spatial Structure of Caption

The caption key frames are extracted from 13 caption scenes based on the features of the caption region. And then the key frames are indexed for the retrieval of the video segment according to the subject. Therefore, the caption region extraction and indexing play an important role in the construction of the soccer video database.

4   Extraction of Caption Frame Interval and Key Frame

4.1   Extraction of Caption Frame Interval

Every caption of soccer video has its own features such as position, size and color. Caption frame intervals are extracted by similarity measure based on its features.

ending caption *Cend*.

To extract caption frame intervals, the predefined structural features such as position, size, and color of captions are compared with those of input frames. If those are similar, an input frame becomes a candidate of caption frame. This similarity measure is performed continuously through whole frames. The consecutive frames that similarity threshold is satisfied becomes caption frame intervals. Similarity measures are performed as follow:

(1) **Similarity Measure of Caption Position**

$$x_{f_i} - x_t < T_x \quad and \quad y_{f_i} - y_t < T_y \tag{1}$$

In Equation 1, x_{fi} and y_{fi} are positions of the input frame, x_t and y_t are the reference values obtained from the structural features. T_x and T_y are the positions threshold.

(2) **Similarity Measure of Caption Size**

$$\frac{A_{m_i}}{A_{f_i}} > T_A \quad and \quad \frac{A_{m_i}}{A_t} > T_A \tag{2}$$

where $A_{m_i} = area(\max(x_{f_i}, x_t), \max(y_{f_i}, y_t), \min(x_{f_i}, x_t), \min(y_{f_i}, y_t))$.

In Equation 2, A_{mi} is the area of the overlapping caption region between input and reference frame. A_{fi} and A_t each is the area of the caption region of input and reference frame. T_A is the size threshold.

(3) **Similarity Measure of Caption Color**

$$C_{f_i}^R - C_t^R < T_C \quad AND \quad C_{f_i}^G - C_t^G < T_C \quad AND \quad C_{f_i}^B - C_t^B < T_C \tag{3}$$

In Equation 3, each C_{fi}^{RGB} and C_t^{RGB} is the maximum number of RGB pixel that is of the caption region of input and reference frame. T_c is the color threshold.

4.2 Extraction of Caption Key Frame

The extracted caption frame intervals indicate significant events, but all that frames can not be key frames. Therefore, one representative frame of caption frame interval is a caption key frame.

In Figure 3, caption key frame is the first frame of caption frame interval. Caption key frames play a significant role for video retrieval, browsing and highlight creation.

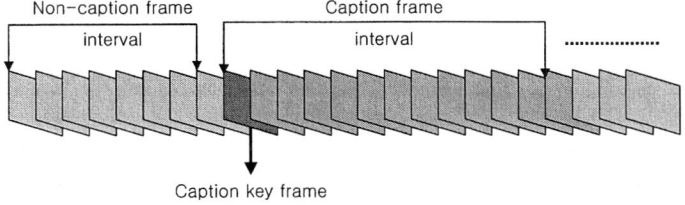

Fig. 3. Caption Key Frame Extraction

5 Creation of Highlight

5.1 Scene Relocation

5.1.1 The Need of Scene Relocation

Event captions such as *Cplr*, *Cchg*, *Cgol*, and *Cfol* appeared right after event occurrence. Therefore, real events exist in the preceding segments of caption key frame. Scene relocation reassigns event scenes for that they have the real events.

In soccer video, except for caption of starting, ending and intermittent score board, event captions appeared right after event occurrence. Therefore, as shown in Table 2, scene relocation was performed to relocation-needed scene.

Table 2. Classification of Scene Relocation

Classification	Scene Details
Relocation-no-need scene	*Ctem, Cgnd, Cbct, Cref, Clst, Cbgn, Csco, Cbch, Cend*
Relocation-need scene	*Cplr, Cchg, Cgol, Cfol*

In Table 2, relocation-no-needed scenes are not performed scene relocation and leave as it is.

5.1.2 The Technique of Scene Relocation

Relocation-need scenes must contain its caption frame intervals and preceding events. Figure 4 shows the technique of scene relocation of relocation-need scenes. Caption key frame interval contains all frames that including its own captions. Preceding events must be sufficient for presenting significant events.

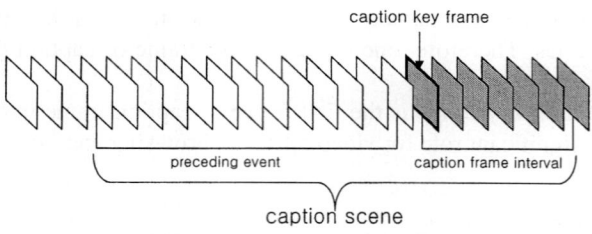

Fig. 4. Scene Relocation

In general, a sufficient presenting time of preceding event is from 10 to 20 seconds. Therefore, in this paper, preceding event time has 20 second. It shows that preceding event has 600 frames in accordance to standard of 30 frames per second. Therefore, caption scene consists of preceding event (20sec, 600frame) and caption frame interval.

5.2 Creation of Highlight

First of all, the reference number of highlight scenes must be determined for creating soccer video highlight. In soccer video, the most significant event is the goal. Therefore, the reference number of highlight scenes is determined flexibly after due consideration of the number of goals. Figure 5 shows the flow of highlight creation.

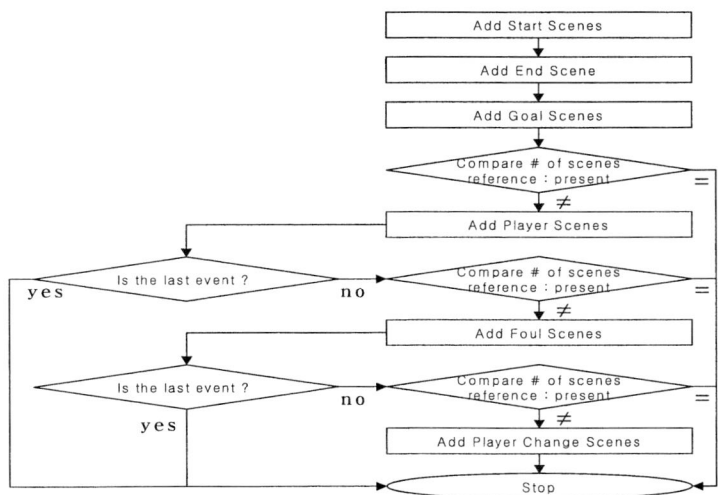

Fig. 5. Shows the flow of highlight creation

In 2004 pro soccer game, the number of goals was made from 0 to 6 goals for each game. Therefore, there can have no goals or maximum 6 goals in highlight. Other scenes are included in highlight according to highlight creation rule.

5.3 Retrieval and Browsing

Indexed video and highlight video must be easily browsed by user who wants to retrieve the video segment. Therefore, video browser has the function of retrieval and VCR together.

It is desirable that retrieval function is performed by the selection of user. Users can select date, play ground and event for retrieved video segment. This is done on video that they want watch, and the result is displayed in the window. While the VCR function can play, stop and move the selected video segments.

6 Experimental Results

In this paper, the experiment was implemented on Pentium IV 2.8GHz PC using Windows XP OS and Visual C++ 6.0 programming language. Experimental video data was the first half of four games from K-League 2004 Pro Soccer. It was used as normalized form of AVI file format and 320 X 240.

Figure 6 shows the extraction of the caption key frames through the extraction of the caption regions and caption frame intervals. In Figure 6, the left part plays the selected video, and right part displays the extracted caption key frames. And the lower part is simple operator for select, play and stop video segments.

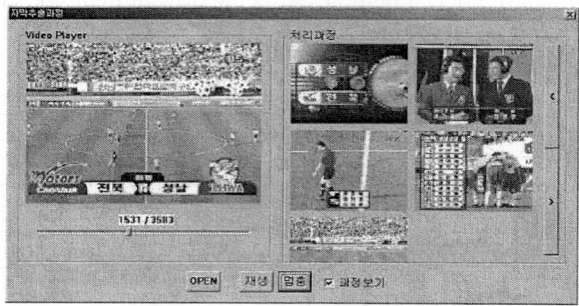

Fig. 6. Window of Caption Key Frame Extraction

The extracted caption key frames are significant elements for the creation of soccer video highlight. Table 3 shows the result of caption key frame extraction.

Table 3. Result of Caption Key Frame Extraction

Games	The number of caption key frame
Video A	62
Video B	54
Video C	67
Video D	58

Fig. 7. Window of Highlight Creation

Figure 7 shows the creation of the highlight according to the highlight creation rule in soccer video. The upper part shows video player, simple operator and frame indicator. And the lower part shows caption key frames of the created highlight.

In highlight creation, we set the reference number of highlight scenes as 20. Therefore, a highlight video was composed of 20 caption scenes. The average playing time of input video was 58.3 minutes, and the average playing time of highlight video was 6.7 minutes. Video viewers can watch and understand whole video contents in 6.7 minute without viewing the whole video.

Figure 8 shows the video browser in which user can easily select and retrieve the video segment that the user wants to watch.

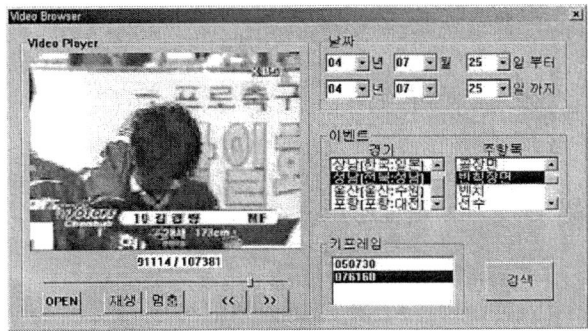

Fig. 8. Browsing Window

Users can select a date or the period that they want to see, and then the events that are satisfied it is listed below. At this time, users can select the events that they want, and can make narrow the scope of retrieval. Also key frames of the selected event are listed below, then, the users can select the key frames and watch the selected shot.

7 Conclusion

This paper proposed the efficient creation method of soccer video highlight using the structural features of caption such as temporal and spatial features. The structural features of captions are extracted by analysis of caption information, and caption frame intervals and caption key frames are extracted using those features. Also, an efficient highlight creation rule is used for highlight creation. Users can watch and understand, fast and easily, the videos that they want to watch. In the experiment, the average playing time of input video was 58.3 minutes, and the average playing time of highlight video was 6.7 minutes. Therefore, video viewers can watch and understand whole video contents in 6.7 minute without viewing the whole video. And it gives a wide range of video selection and reduce the time and cost. Video browser was designed for efficient and easy retrieval too. This paper provides the foundation for the implementation of enhanced sports video management system.

In future, we recommend that a continuous study on automatic and active caption detection and semantic event extraction should be performed by other researchers.

References

1. Davenport G., Smith T., Pincever N.: Cinematic Primitives for Multimedia, Computers and Graphics, Vol. 15, (1991) 67-74.
2. Guyon I.: Applications of Neural Networks to Character Recognition, International Journal of Pattern Recognition and Artificial Intelligence, Vol. 5, (1991) 353-382.
3. Li F., Yu S. S.: Handprinted Chinese Character recognition Using Probability Distribution Feature, International Journal of Pattern Recognition and Artificial Intelligence, Vol. 8, (1994) 1241-1258.
4. Pan Y., Wu J., Tamura S., Okazaki K.: Neural Network Vowel-Recognition jointly Using Voice Features and Mouth Shape Image, Pattern Recognition, Vol. 24, (1991) 921-927.
5. Hang-Bong K.: Generation of Video Highlights Using Video Context and Perception, Proc. of SPIE, Storage and Retrieval for Media Databases 2001, Vol. 4315, (2001) 320-399.
6. Christal M., Smith M., Taylor C., Winkler D.: Evolving Video Skims into Useful Multimedia Abstractions, Proc. CHI'98, (1998) 171-178.
7. Yeung M., Yeo B., Liu B., Segmentation of Video by Clustering and Graph Analysis, Computer Vision and Image Understanding, Vol. 71, No. 1, (1998) 94-109.
8. Hanjalic A., Zhang H.: An Integrated Scheme for Automated Video Abstraction Based on Unsupervised Cluster-Validity Analysis, IEEE Taans. Cir. & Sys. for Video Tech., Vol. 9, No. 8, (1999) 1280-1289.
9. Uchihashi S., Foote J., Girgenshon A., Boreczky J.: Video Manga: Generating Semantically Meaningful Video Summaries, Proc. ACM MM'99, (1999).
10. Lienhart R., Pfeiffer S., Effelsberg W.: Video Abstracting, Communications of the ACM, Vol. 40, No. 12, (1997) 54-62.
11. Liang-Hua C., Chih-Wen S., Hong-Yuan M. L., Chun-Chieh S.: On the preview of digital movies , Visual Communication and Image Representation, Vol. 14, (2003) 358-368.
12. Babaguchi N.: Towards Abstracting Sports Video by Highlights, Proc. IEEE Int. Conf. Multimedia and Expo (III), (2000) 1519-1522.

The Information Search System Using Neural Network and Fuzzy Clustering Based on Mobile Agent

Jaeseon Ko, Bobby D. Gerardo, Jaewan Lee, and Jae-Jeong Hwang

School of Electronic and Information Engineering, Kunsan National University,
68 Miryong-dong, Kunsan, Chonbuk 573-701, South Korea
{kojsno1, bgerardo, jwlee, hwang}@kunsan.ac.kr

Abstract. Nowadays, Internet is generally useful for intelligent information search system that would satisfy user's need. But most of the current agents are each independent and depend on a platform in the information retrieval system. In addition, it is possible to get unreliable information in a distributed environment because of inefficient communication and cooperation among agents. To solve these problems, we propose the searching mechanism based on ontology using mobile agent which is platform independent and uses intelligent classifier among agents. This mechanism improves the efficiency of cooperation and information processing so that the user who requests for the location of information can find it accurately and rapidly. Our experiment showed that the proposed system provided higher rate of accuracy for searching information.

1 Introduction

The advancement of Internet technology leads people to exchange information through the Internet. There are efforts that have been made currently for the development of techniques for storing and searching various information on the Internet. Most of these searching techniques are used only for storing data in the web, and usually using simple keyword search methods. The keyword searching method could identify keywords that had been previously identified and provides both exact and extra information from such search technique. These result to problems of obtaining meaningless or unrelated information.

A typical distributed system uses fixed agents to gather information which use fixed communication, message passing and Remote Procedure Call (RPC). The typical mobile agents have the characteristics of finding nodes to work and executing its functions [2]. They have some features like autonomy, intelligence, mobility, social ability, reactivity, and veracity [5].

Ontology is a concept of knowledge using the XML language which the computer processes understands. Web ontology is the one which complies with semantic web and can be processed by computer and uses several intelligent agents for information gathering [6]. Other ontology is preconditioned of computer processing, but web ontology used several intelligent agents for important communication methods. Ontology can be described based on XML. As used in knowledge modeling, ontology automates the consistent processing of knowledge, can infer it and help us search easily. In addition, ontology can be used for the analysis of unspecific data, accommodating many users query and investigation of easy search process. Ontology based database can provide the characteristics of reusability and can be used in various application programs [7].

We propose in this paper a mechanism for solving the above problems using ontology, neural network, and fuzzy clustering techniques which classifies the ontology document and gather the information in data mining through mobile agents.

2 Intelligent Information Searching System

We propose an Intelligent Information Searching (IIS) system which user can rapidly get the desired information from vast information resources in a distributed system. The architecture of the proposed system consists of 3-tier as shown in Figure 1.

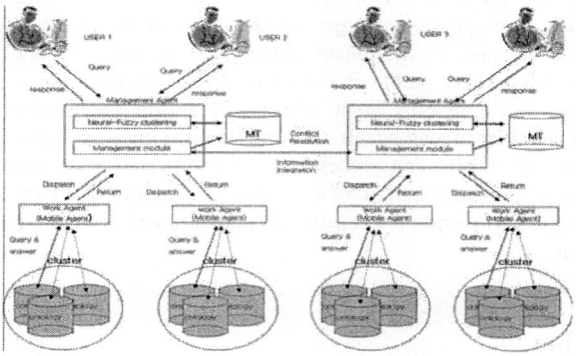

Fig. 1. Architecture of Intelligent Information Search System

The user agent performs a query from user while the work agent manages the information resources. The management agents as shown in the middle-tier process user's query through the cooperation among other agents.

We use the Neural-Fuzzy clustering to cluster similar information in the network and search for ontology document using mobile agent in that clustered information. As a result, this technique can provide fast and exact searching on user's queried information. The knowledge database based on ontology enable us to search knowledge for user's query.

2.1 User Agent

The 3 User Agent has the responsibility of information exchange between the system and client. It receives the user's query which includes the interest topics and related constraints. It transfers the user's query to management agent after preprocessing dealt with section.1. It present user with the results in the form of natural language or graph or picture.

2.2 Management Agent

The Management Agent consist of Neural-Fuzzy Clustering (NFC) for grouping of ontology documents and management module for taking charge of joining and leaving work agents and ontology documents. Information is stored in management table for fast access to specific data or information.

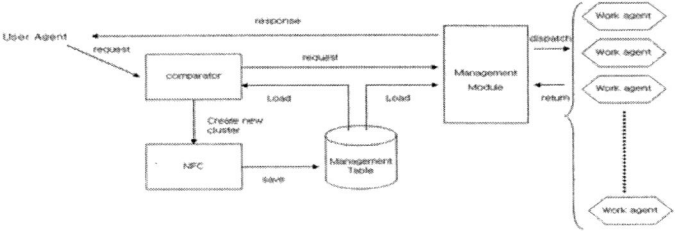

Fig. 2. Architecture of Management Agent

2.2.1 Management Module

2.2.1.1 Comparator. The Comparator refers to management table on a new request from the user. If the query has the same or high similarity rates with table item is high, then Management module will dispatch the work agent which is mobile agent according to the weight through the management module.

Otherwise, after reconstructing management table using NFC, management module dispatch mobile agent to the cluster including the corresponding ontology.

2.2.1.2 Management Table. Initially, system clusters ontology by training the NFC and stores the results to the management table. The Management Table trains the NFC initial clustering ontology and stores it in the training result. It caches the information, such as location of the information, frequency, and keywords. It is used to determine whether the ontology join the clustering or not.

2.2.2 Neural-Fuzzy Cluster (NFC)

The NFC clusters the ontology document combining 3-Layer Feed-Forward neural network and Fuzzy c-means in distributed systems for efficient information clustering.

It approximates short distance data through competitive neural network for input data, transfers to the input node of the fuzzy clustering, and clusters the documents as output.

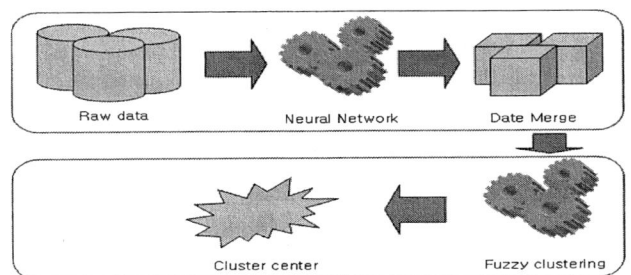

Fig. 3. Architecture of the Neural-Fuzzy Clustering

By adding the user's preference to input node, documents are clustered according to it.

2.3 Work Agent

The work agent acts as information provider that process stored information using ontology. The work agents use the ontology search engine to retrieve the related information.

The work agent searches the ontology moving in the specified cluster by management agent. By finding the information needed by user, the work agents store the result and search the ontology. After gathering all the information that the user needs, work agents remove duplicate documents and transfer the results to the user.

3 The NFC Algorithm

3.1 Clustering Ontology Document

The Work agent extracts Subject, Object, and Predicate from each ontology and removes meaningless words to reduce the element by using the stemming and stop list algorithms. After preprocessing like above, the next example illustrates the value of the Normalized Term Frequency (TF), Inverse Document Frequency (IDF) as a table on the ontology documents including the word of "multimedia computer." Normalized TF represents term frequency which is the number of word (W) of the document (d) as the value of 0.5~1.0.

The max TF means the maximum frequency of the word in the document and TF is the frequency in the corresponding document. The IDF is the number of documents which include the word (w) at least one. Equation (1) calculates for the Normalized TF and IDF.

$$\text{Normalized TF} = 0.5 + 0.5(\text{tf} / \text{Max Tf}) \quad (1)$$

$$\text{IDF} = \log(|D| / DF(w))$$

Normalized TF = 0.5+0.5(5/10) = 0.75
IDF = log(20/5) = 0.60205

Equation (2) considers the user's preference. Ui is the user's preference, Fi is the Normalized TF, Ti is the temporary value, and Xi is the NFC input value.

$$Ti = Fi + Fi*Ui$$

$$Xi = \frac{Ti}{MaxTi} \quad (2)$$

Table 2 shows the normalized TF and IDF using equation (1) and (2) on the ontology documents including the word of "multimedia computer."

Table 2. Normalized TF, IDF value for "multimedia computer" after preprocessing

	Term frequency	Normalized TF	IDF
computer	10	1.0	0.30102
VGA card	5	0.75	0.60205
Sound card	4	0.7	0.69897
speaker	4	0.7	0.69897
O/S	3	0.65	0.82390
:	:	:	:

We train multi-layer perceptron on the ontology documents using the result of preprocessing. For example, for the term of "VGA card" and "Sound card" next illustrates the course of training of multi-layer perceptron given the initial values of w31=0.4, w32=2, w41=0.1, w42=1, w51=0.6, w52=0.5, w61=0.45, w62=0.5 and Θ=0. As the vector values of each item input into multi-layer perceptron, the node of hidden layers are calculated using Equation (3).

$$I_j = \sum_i w_{ij} O_i + \theta_j ; \qquad (3)$$

Equation (4) calculates the output of the previous layer.

$$O_j = \frac{1}{1+e^{-I_j}} \qquad (4)$$

α3 = ω31χ1+ω32χ2=(0.4)(0.75)+(2)(0.7)=1.7, O3= 0.85
α4 = ω41χ1+ω42χ2=(0.1)(0.75)+(1)(0.7)=0.78, O4= 0.69
α5 = ω51O3+ω52O4=(0.85)(0.6)+(0.5)(0.69)=0.855, O5= 0.7016
α6 = ω61O3+ω62O4=(0.45)(0.85)+(0.5)(0.69)=0.7275, O6= 0.674

Equation (5) calculates errors from output value of each node.

$$Err_j = O_j(1-O_j)(T_j-O_j) \qquad (5)$$

Err5=O5(1-O5)(T-O5) = 0.70(1-0.70)(1-0.70) = 0.063
Err6=O6(1-O6)(T-O6) = 0.67(1-0.67)(1-0.67) = 0.073
....

Equations (1) to (5) calculate the weight of node, and the equation (6) updates the weight of node.

$$\Delta w_{ij} = (l) Err_j O_i;$$
$$w_{ij} = w_{ij} + \Delta w_{ij} \qquad (6)$$

Δw_ij=(l)Err_jO_i; = (1)Err5O3 = (1)(0.063)(0.85) = 0.05355
Δw_ij=(l)Err_jO_i; = (1)Err6O3 = (1)(0.073)(0.85) = 0.062

Table 3 shows the result of multi-layer perceptron training using equation (1) to (6) referring to normalized TF and Ui on the selected words from table 2.

Table 3. Result after MLP processing

Doc \ Term	multimedia computer	Servers	Software	industrial computer
VGA card	0.76	0.90	0.0	0.87
Sound card	0.74	0.0	0.0	0.0
speaker	0.72	0.0	0.0	0.0
O/s	0.78	0.8	0.86	0.82
scsi device	0.25	0.82	0.0	0.0
printer	0.52	0.0	0.0	0.0
scanner	0.21	0.0	0.0	0.0

The data sizes are reduced by eliminating the items with less than 0.5 among the weight of keyword from multi-layer perceptron.

The result of the multi-layer perceptron is the input to Fuzzy cluster. Equations (7) to (11) calculate similarity among the set of cluster, and equation (7) illustrates similarity y for x in the domain.

$$\mu_{A,B}(x,y) = \text{if } y \leq (\text{Min}(\mu_A(x), \mu_B(x))) \qquad (7)$$

Similarity of multimedia computer for computer class is as follows:

$$\mu_{A,B}(x,y) = \text{MIN}(0.71, 1.0) = 0.71$$

Equation (8) reflects the domain preference and equation (7) becomes equation (8) of integral type.

$$f(y) = \int_{\text{Domain}} \mu_{A,B}(x,y) d\text{Domain}$$
$$f(y) = \int_{\text{Domain}} \mu_{A,B}(x,y) f_{\text{Domain}}(x) dx \qquad (8)$$

The similarity from Equation (8) is the results for the specific value that satisfies y. Equation (9) calculates the similarity in all satisfaction area.

$$\text{Value}_{A,B}(x,y) = \int_{MV} f(y) dMV = \int_{y=0}^{0} f(y) f_{MV}(y) dy \qquad (9)$$

Equation (10) searches the nearest set (S) in input fuzzy sets and the set of N in sample space. And, Equation (11) calculates the error.

$$S = \text{Max}(\text{Value}_{A,i}, \text{Value}_{B,i}, \ldots, \text{Value}_{Z,i}) \qquad (10)$$

Equation (10) calculates the maximum values among the Fuzzy set. Below is an example for the computation of the maximum value using quation (10).

Max(0.71, 0.21, 0.0 ....) = 0.71

$$\text{Value}_{c,i} \cdot \frac{A_c \cap A_i}{A_c} \qquad (11)$$

Finally, Table 4 represents the result of clustering using the above equation. This table shows that the items of multimedia computer, servers, software, and industrial computer are included in computer class.

Table 4. Result after Fuzzy Clustering

Doc \ class	computer	business	Health	...
multimedia computer	0.71	0.21	0.0	...
Servers	0.75	0.19	0.0	...
Software	0.77	0.20	0.0	...
industrial Computer	0.75	0.22	0.0	...
. . .	. . .	. . .	. . .	. . .

After clustering all ontology documents, the result are stored in management table. And user's request for "information search," dispatches a mobile agent in cluster for gathering information.

3.2 NFC Clustering Algorithms

NFC algorithm using the equations presented in section 3.1 as follows:

Step 1 : Input user preference and feature values to input nodes.
Step 2 : Calculate the weight from hidden layer and determine the output nodes.
Step 3 : Repeat Step 2 at next hidden layer.
Step 4 : Update weight after calculating classified value from the final output.
Step 5 : Repeat steps 1 to 4 until the convergence criterion is satisfied
Step 6 : Input the result to Fuzzy function
Step 7 : Find similarity among sets.
Step 8 : Calculate the closest distance among sets and calculate an error ratio.
Step 9 : Repeat step 6 to step 8 until the convergence criterion is satisfied.

```
Ti = Fi + Fi*Ui ;              // calculate feature + user value
Xi = Ti / Max Ti;              // normalize input value
Initialize all weights and biases in network;
while terminating condition is not satisfied {
   for each training sample X in samples { //Propagate the inputs forward
      for each hidden or output layer unit j {
         I_j = Σ_i w_{ij} O_i + Θ_j; //compute the net input of unit j with respect to the
                                     //previous layer, I
         O_j = 1 / ( 1+e^{-1}_j ); } //compute the output of each unit j
// for each unit j in the hidden layers, from the last to the first hidden layer
         Err_j = O_j(1-O_j)(T_j-O_j); //compute the error with respect to the next higher layer,
      k for each weight wij in network {
         Δw_{ij} = (l)Err_j O_i; // weight increment
         w_{ij} = w_{ij} + Δw_{ij}; } //weight update
   }
}
While terminating condition is not satisfied {
         similarity set of A and B();
         calculate distance from near set();
         calculate errors();
}
// compute the similarity using prefer to domain
function similarity set of A and B(){
if y ≤ (MIN(U_A(X), U_B(X)))
         sim_i = U_{A,B}(x,y) ;  else ............. ; }
function calculate distance from near set() {
         S = Max(Value_{A,l}, Value_{B,l}, ......, Value_{Z,l})
         }
function calculate errors(){ err = Value_{C,l} X (A_c ∩ A_l / A_c)  }
function calculate distance from near set() {
S = Max(Value_{A,l}, Value_{B,l}, ......, Value_{Z,l})
}
function calculate errors(){
err = Value_{C,l} X (A_c ∩ A_l / A_c)
}
```

Fig. 4. Algorithms for Neural-Fuzzy Clustering

4 Experimental Evaluations

In our experiments for performance evaluation, we compare the performance between proposed NFC algorithm and existing clustering algorithm for clustering of information in distributed environment. For evaluation, the 7 class are selected and each class includes 20 ontology documents like in Table 5.

Table 5. Date set

class	# of document
Computer	20
Business	20
Arts	20
Games	20
Health	20
Science	20
Sports	20

Figure 5 shows the performance evaluation for document clustering among proposed NFC and Back Propagation (BP) and the Single layer perceptron. For the efficient comparison, we select 100 ontology documents and evaluate the correctness of classification according to feature words. This figure shows that if the feature word increases, then the proposed NFC accuracy is improving.

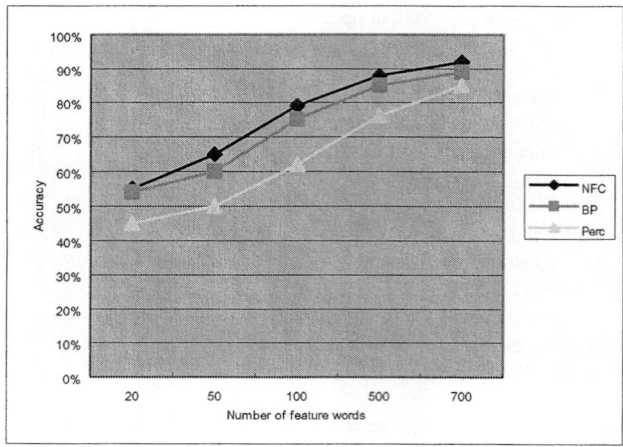

Fig. 5. Result of classified by feature word

Figure 6 shows the result of comparison on whether the user preference is included or not. This evaluation is focused on the user searching word of "computer."

The correctness of classification with considering the user's preference is better than without considering it.

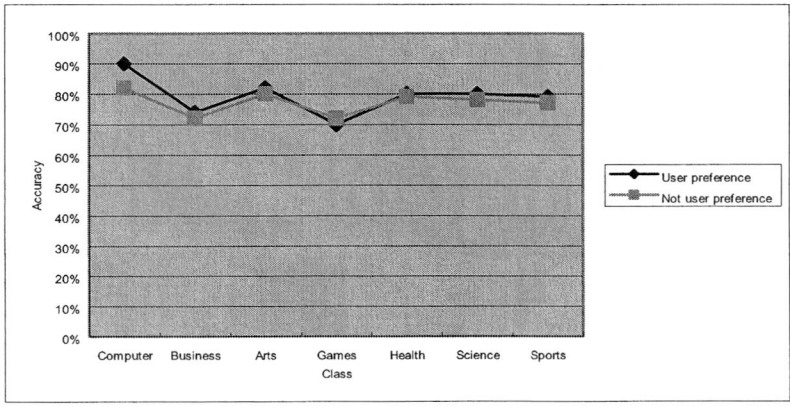

Fig. 6. Result of comparing the user's preference

6 Conclusions

This paper proposed the intelligent information search mechanism using NFC and mobile agent based on ontology document in distributed environment.

Since the proposed mechanism NFC clusters information resource using neural network and fuzzy, it reduces the data duplication and provides correct information to users.

We evaluated the performance of the proposed mechanism through simulation. The simulation results show that if the feature word increases, then the proposed NFC accuracy is improving and the correctness of classification considering the user's preference is better than without considering it.

The result of this paper can be applied in intelligent information search system and data mining fields. Further researches are needed for implementing large database mining in a ubiquitous environment and speeding up the searching of information.

References

1. Pilato, Giovanni and Vitabile, Salvatore.: A Neural Multi-Agent Based System for Smart Html Pages Retrieval, IEEE (2003).
2. Marques, Paulo, Simos, Paulo, Silvam, Luis Fernando Boavida.: Providing Applications with Mobile Agent Technology, IEEE (2001).
3. Pacheco, Peter S.: Parallel Programming with MPI
4. Fausett, Laurene: Fundamentals of neural networks, Premtice Hall (2000) 59-79.
5. Cockayne, William T., Zyda, Michael: Mobile Agents, Prentice-Hall (1997).
6. Gennari, J,. Musen, M.A., Fergerson, R.W., Grosso, W.E., Cruby M., Eriksson, Noy, N.F.: The Evolution of protege an Evironment for Knowledge-Based systems Development, Technical Report (2002).
7. Noy, N.F, Fergerson, R.W.. : The knowledge model of protege-2000: Combining interoperability and flexibility, EKAW 2000 Conference.

8. White, J.: Telescript Technology : Mobile Agents, MIT Press (1997)
9. Mobile Agent, OMG. : Mobile Agent Facility Specification, available at http://www.omg.org/. (2000)
10. Mobile Agent.: available at http://www.gurugail.com/Agent/page.html? subject=mobileAgent.html.
11. Joshi, Anupam, Krishnapuram, Raghu.: Robust Fuzzy Clustering Methods to Support Web Mining, Tech. Rep., Department of Computer Engineering and Computer Science, University of Missouri Columbia (1998).

A Security Evaluation and Testing Methodology for Open Source Software Embedded Information Security System

Sung-ja Choi, Yeon-hee Kang, and Gang-soo Lee

Hannam University, Dept. of Computer Science,
Daejon, 306-791, Korea
irecomm@dreamwiz.com
{dusi82@se, gslee@eve}.hannam.ac.kr

Abstract. Many of Information Security Systems (ISS) have been developed by using and embedding Open Source Software(OSS) such as OpenSSL. The "OSS-embedded ISS" should be tested and evaluated when it will be used as a security product or system for an organization. In this paper,we present a test and evaluation procedure for an OSS-embedded ISS, and ROSEM(real-time OpenSSL execution monitoring system) that is a testing tool in according to presented methodology. The main function of ROSEM such as an execution path generator for OpenSSL is useful for test case generation in the CC evaluation scheme.

1 Introduction

Recently, it has been being increased to introduce an Open Source Software (OSS) such as Apache, Linux, BSD, Mozilla, MySQL, OpenSSL, Crypto++ and so on, that contains security functions and cryptography modules, for the purpose of shorten the development duration of Information Security System (ISS)[1]. OpenSSL which is a well known OSS of cryptography component is mostly used for IDS or VPN development[11]. OSS based components (e.g., cryptography component, communication functions) embodied in various forms, and they are offered as a form of OSS. It is possible that most of ISS developers use components, which are in the form of OSS, without a formal analysis to shorten the period of development.Thus, they could be loaded and embedded to source code of ISS without the assurance of security. Therefore, the safety and security of OSS-embedded ISS is not guaranteed. Also, the most developers and security evaluators in Common Criteria(CC, ISO/IEC 15408) evaluation scheme should know the details about inner structure and source code as well as development information of OSS, because they use and evaluate some cryptography components in OSS. However, it is very hard, because most of OSS do not have any documentation and development information. Thus, we should obtain deliverables for evaluation by means of reverse-engineering from OSS.

From those backgrounds, we have researched and developed the following topics for developers and evaluators in CC evaluation scheme:

- Research of a new test method and paradigm for an OSS embedded ISS.
- Development of a ROSEM as a test case generation for cryptography components in OpenSSL.
- As a case study, generation of test case for testing cryptography function such as rc4 in OpenSSL by using ROSEM.

Next section presents a test method for OSS embedded ISS. Section 3 presents development of ROSEM. Section 4 presents the result of test case generation for OpenSSL by using ROSEM, as a case study. In Section 5, we summarize and conclude.

2 Evaluation and Testing Method for an OSS Embedded ISS

2.1 Procedure of Development and Evaluation

Fig. 1 presents the paradigm for development and evaluation of TOE that is consisted of three process such as OSS development process, TOE development process and TOE evaluation process.

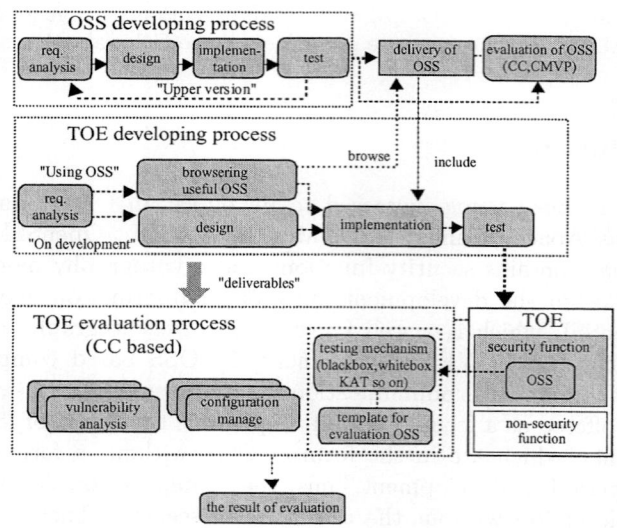

Fig. 1. The paradigm of development and evaluation of TOE

- OSS development process: The most of OSS does not have the development and evaluation process. Because it has been modified by developers without uniform development process and evaluation process whenever new function is required. Then, new version of OSS has been distributed through network. Note that, the integrity of configuration of the OSS should be preserved and validated in case of insecure distribution environment such as internet. So, some cryptography component in OSS such as OpenSSL and Crypto++ have been evaluated under CMVP (Cryptography Module Validation Program) scheme according to FIPS 140-2 of NIST[3]. Linux which is a famous OSS has been evaluated and certified as EAL2+ by CC scheme[12].

- TOE development process: OSS have been greatly used to reduce the development time and cost. A development process of OSS-embedded TOE is consisted of two part, self development and development by using OSS.
- TOE evaluation process: An evaluator need not consider whether a TOE is built by using some OSS or not, because he evaluate all part of TOE. Therefore, if he has evaluated OSS-embedded TOE, the OSS part in TOE would be not evaluated. Our paradigm will provides evaluation methodology of OSS-embedded TOE to be consider evaluation of OSS part.

2.2 Test Process

A test process for evaluation of OSS-embedded ISS has below test process phase. It is test planning, test specification, test execution, test result recording and test result validation [4].

[Step 1] Test planning
Test planning phase has be consist of the establishment test plan and the building test bed for evaluation / authentication.

- [Step 1.1] *Establishment of test plan* : In the establishment of test plan, it should be described to test the OSS-embedded ISS such as the name of vendor, test date and version information of OSS which has been used, and so on.
- [Step 1.2] *Build of test bed* : It could be specified details information to test such as environment construction, the kind of OS and forms of compile so on. It should be built of the test bed according to specified environments.

[Step 2] Test specification
The tester should specify the test mechanism which is software test techniques such as component test methodology [4], security test, and test validation tool [5] in test specification phase. Black-box[2], white-box and KAT testing are well known software test techniques. Then, it could be applied to test, also the tester should derived the test item from the function classes in order to test OSS part in a TOE as like Fig.2. The slicing of OSS-embedded ISS is very important and hard but we have applied the slicing method to test each component. It is possible to test independently.

- [Step 2.1] *Derivation of test item*: The phase of function specification has the test items. It is regarded a test item as a class have been divided according to function approximately.
- [Step 2.2] *Family and component (function A1, A2..) specific* : All of the function could be divided sub-function in details. The several sub-functions

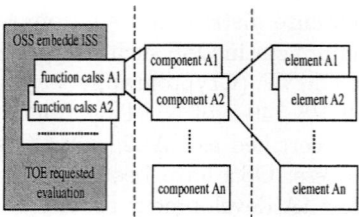

Fig. 2. A Test of OSS and An Evaluation Procedure

could be makes a component to separate OSS-embedded ISS, also it could be classified family unit to slice according to function requirements. It could be details module in order to test modules after components had been gotten [9]. It may be the structure design in reverse engineering.

- **[Step 2.3]** *Detail module (security module)* : Each component is divided element units and these units present each API modules as like OpenSSL. Note that, it may be called non-divided function unit. It may be the detail design in reverse engineering.
- **[Step 2.4]** *Derivation of interface* : Each component has inputs and outputs and it could be drawn a state transition because of interaction among them. Interface among components would be operated independently because a module is possible to extract from each component. It have been presented mapping table or call graph.
- **[Step 2.5]** *Test cases generation* : Test cases are generated by each elements has been independent function unit. A test case, non-divided function, could be operated independently. The test case should be tested by real-time testing tool(as like ROSEM), the result of testing could be used for evaluation OSS-embedded ISS.

[Step 3] Test execution

In test execution, it could be applied according to black-box technique and KAT[7][8][9] so on, which is based on test specification. It could be classified the evaluation/authentication level by analyzing the test execution resulting. We offer the real-time monitoring tool, ROSEM, which had been applied the harness mechanism to test and show the table of execution resulting and the monitor log screen. It will be generate a call graph in the elements testing and it could be validated OSS-embedded ISS.

- **[Step 3.1]** *Test execution* : The generated test case should be execute according to specified test mechanism at test specific step[10]. We apply a state transition and show a table from a state transition. Each test case generates a table for the test specification. State transition table includes state number, API number, Function name, Function family(module) and Function form(it states related calls which are divided into security modules, SSL connections, and applications).

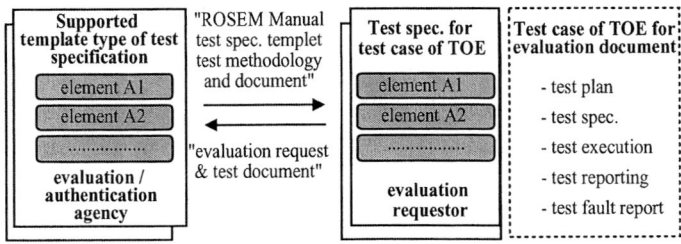

Fig. 3. The evaluation method of TOE test case

- [**Step 3.2**] *Test coverage calculation* : Test coverage could be calculated by the result of test execution of test cases. Test coverage could be applied effectively by applying test measurement technique.

  Test coverage = number of test cases completed / number of total test cases x 100

  This numerical formula is calculated in order to analyze test range, and the aim of the test execution is to test all of the specified test cases.

[Step 4] Test result recording

The recording of the test result, whenever every single test is operated, should be recorded about information such as component version of test, component test specific, test date, number of test cases, number of test discordant, range measurement, fault report, and so on. Using testing in order to compare between actual test result and expected test result should check a test log file. And a fault report should be presented about the differences.

[Step 5] Test result validation

In test result validation, the result of testing execution is used to evaluate and classified evaluation level for OSS-embedded ISS and then should be report evaluation of TOE to ISP.

3 ROSEM – A Real Time OpenSSL Evaluation Monitoring System

We have developed a real-time monitoring tool for the purpose of evaluating and testing "OpenSSL embedded ISS" to which "harness code for the monitor" is included, in order to monitor the calling sequence of set of functions(or modules) during the operation of OpenSSL.

This is a technique of so called the "test harness". The harness code plays a role as a proxy and it has report all sorts of information to be collected to the analyzer in ROSEM through window messaging mechanism. The analyzer has enabled to analyze whatever the tester wants to test. We suggest "OpenSSL +" by inserting harness code into OpenSSL in order to monitor as like Fig 4. ROSEM has a function which receives and analyzes information from harness

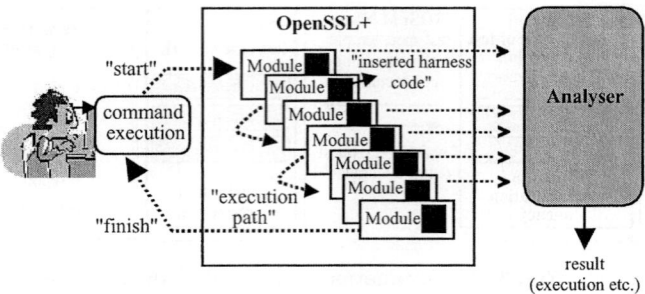

Fig. 4. An operation theory of ROSEM

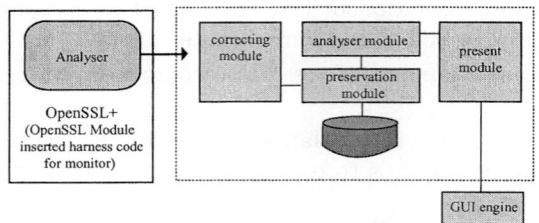

Fig. 5. The structure design of ROSEM

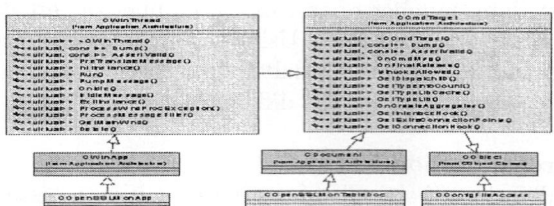

Fig. 6. Class design

code of OpenSSL+, and the result of analyzed information have been output which certain orders are being operated. The ROSEM shows the monitoring procedure of the operation for OpenSSL embedded ISS.

This method have been perfectly executed the every test case on real time. ISS which used certain function of OpenSSL will be call a set of function library in OpenSSL and the called function library sends its own operation information to monitoring tools and returns the operating result to the next code. After it chould be analyzed the received logging information through log analyzer, also the monitoring tool shows that on the screen by using GUI engine.

ROSEM have been made up correcting module, analyzing module, preservation module and presentation module of OpenSSL-embedded ISS as like Fig. 5. We have designed classes for implementation of ROSEM. In Fig 6. is a sample of generated class design.

Below clause shows a generated test case and an evaluation method by applying monitor tools had been developed.

4 Case Study – Test Case for rc4 in OpenSSL

By applying test method and ROSEM for evaluation OpenSSL-embedded ISS, we show the example of testing for OpenSSL. RC2 function testing in the OpenSSL is operated through test bed already had been built up. It could be possible to analyze testing module while RC2 application program is being run. RC2 cryptography program has been run on the Console window. In Fig 7, testfile.txt had been cryptography RC2 by cbc and then it had been created testfile.rc2.

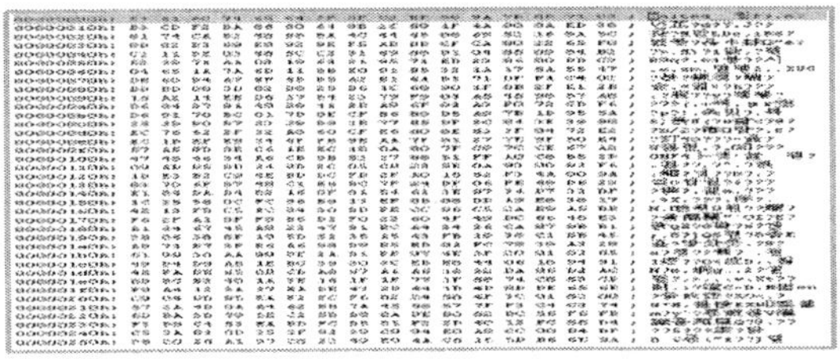

Fig. 7. The captured hex value by RC2 Function Cryptograph

The cryptography resulting of the testfile.txt had been done the gathering of hex value. It is certain that the cryptography function of RC2 had been achieved normally. The RC2 cryptography had been used EVP as well. The testing of Module, which had been used in this cryptography function, it could be analyzed on the monitor tool with the information of logging, too. It had been occurred about 15,600 logs during running this RC2 cryptography function. We present logs screen which is captured the testing result as like Fig. 8.

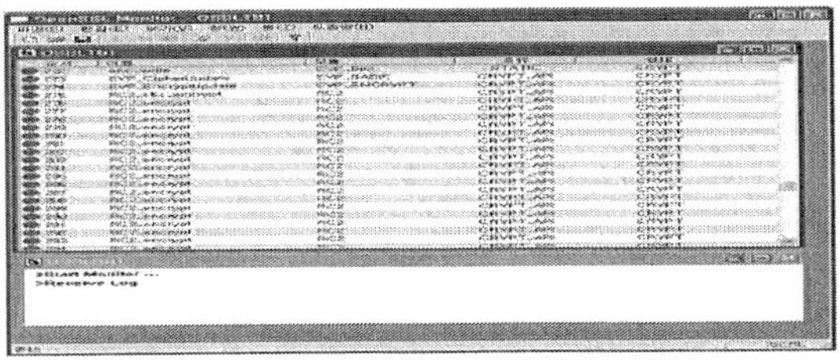

Fig. 8. The captured of RC2 function test result log screen (TS-37)

We can see that mainly five modules are operating apart from basic modules to cryptography RC2. That is, BIO module for input/output and user interface module to ask codes and take them, RAND module which is generating random

RC2 component test specification
• test purpose : RC2 component function test • test case : TS-37 • component test specification : SP-37

Element (API NUMBER Function)		Kind of function type	Function Bum	
16873	2365	RC2_set_key	RC2	CRYPT_API	CRYPT
16874	715	BIO_push	Bio_BASIC	CRYPT_API	CRYPT
16875	711	BIO_ctrl	Bio_BASIC	CRYPT_API	CRYPT
16879	1869	EVP_CipherUpdate	EVP_BASIC	CRYPT_API	CRYPT
16880	1876	EVP_EncryptUpdate	EVP_ENCRYPT	CRYPT_API	CRYPT
16881	2361	RC2_cbc_encrypt	RC2	CRYPT_API	CRYPT
16882	2362	RC2_encrypt	RC2	CRYPT_API	CRYPT

Function call 16873 - 16882 : the text omitted

Fig. 9. RC2 fuction test case - test specification

numbers to encode, MD for message digest, and SHA module, and RC module for cryptography.

An evaluation requestor has to generate the specification of test through test bed as like Fig. 9 and should be evaluated it with other deliverables of TOE when he applies to evaluation. It is similar to procedure of existing certification, the OSS part in TOE should be compared the template of OSS testing with test specification of TOE in deliverable which provided from the requestor of evaluation. It has been assured to be correctly loaded the OSS part in TOE and to be trusted OSS-embedded TOE.

5 Summary and Conclusion

In this paper, we present testing methodology for correctness using of Open source and evaluation of OSS-embedded ISS. It could be achieved reliability and trust in using OSS. Also, we had been developed real-time monitoring too, ROSEM, it had been applied a example in order to test OpenSSL-embedded ISS which is used to develop ISP which has short sort code like smart card or RFID.

Main findings of this study are as follows.

- **Investigation open source related security:** By investigating and analyzing security test tool of open source form and down location and character of cryptography library, it makes easy to use them.
- **Investigation and analysis of software component test method:** We have investigated and analyzed general software test methodology and explicated BS 7925 of English which is the standard of this field. From this finding, the product developer and the evaluator can get general knowledge for the test and have the ability of choosing the most suitable test method.
- **Investigation of test of cryptography module and ISP and evaluation method:** CMVP, the system inspecting cryptography module of open source, has been investigated carefully and evaluation standard of open source-embedded product which shown in CC/CEM has been investigated as well. The result from these investigations can be used as a guide in the CC evaluation system.

- **Analysis of OpenSSL version 0.9.7:** We have analyzed module structure and function of OpenSSL and presented each security module component and interface. It has also presented instruction codes with option and presented 19 weaknesses well known. And it has presented security function classification result inside of OpenSSL and a countermeasure with CC security function.
- **Test of OpenSSL product and presentation evaluation methodology:** We have presented the body of test and evaluation methodology and how to draw report from OpenSSL in each step.

As a conclusion, we provide a draft of "The Guide line for evaluation of OSS-embedded ISS" to the evaluation executor and the vendor for evaluation. It is expected to use as follows. On the developer's side, this study is information needed when the report is drawn for development and evaluation of ISP through OpenSSL and all sorts of open source. The concept of open source, software test technique, structure analysis of OpenSSL, and ROSEM will be very useful information and a tool for developer. On the side of evaluator, it can be used as a guide and a tool for the test and evaluation of ISP that is developed through not only OpenSSL but all sorts of open source. However, the function of ROSEM should be extended to apply to the other security function of open source. Various counterattacking tools those are drawn more various information for evaluation from counterattack against from open source like OpenSSL should be developed. There are a few similar tools to these already. So, methods have to be studied to use them. Just a few functions of OpenSSL have been studied about running condition but more functions should be studied. Also, the environment which is applied evaluation technology of CMVP should be expanded.

Acknowledgements

This work was supported by KISA(Korea Information Security Agency)&RRC(a grand No.R12-2003-004-01001-0 from Ministry of Commerce a Industry and Energy).

References

1. A. Wheeler, *Why Open Source Software / Free Software (OSS/FS)? Look at the Numbers!*. July 23, 2004, http://www.dwheeler.com.
2. B. Beizer, *Black-Box Testing*, John Wiley & Sons, 1995.
3. NIST Special Publication 800-29, *A Comparison of the Security Requirements for Cryptographic Modules*, FIPS 140-1 AND FIPS 140-2, Ray Snouffer, Annabelle Lee and Arch Oldehoeft, NIST, June, 2001.
4. British Computer Society Specialist Interest Group in Software Testing(BCS SIGIST), *Standard for Software Component Testing*, Working Draft 3.4, April 2001. 2001.

5. C. Kaner, *Quality Cost Analysis: Benefits and Risks*, http://www.kaner.com/qualcost.htm.
6. Frank Tip, *A survey of program slicing technique*, Journal of Programming Languages, Vol.3, No.3, pp.121-189, September 1995.
7. W. E. Perry, *Effective Methods for Soft-ware Testing*, 2nd Ed., John Wiley & Sons, Inc., 1999.
8. G. Myers, *The Art of Software Testing*, John Wiley & Sons, Inc., 1979. 1979.
9. K. J. Ross, *Practical Guide to Software System Testing*, K. J. Ross & Associates Pty. Ltd., 1998. 1998.
10. C. Kaner, J. Falk, Q. Nguyen, *Testing Computer Software*, 2nd Ed., Thomson Computer Press, 1993. 1993.
11. OpenSSL, http://www.openssl.org/ .
12. Common Creteria for Information Technology Security Evaluation(CC), CCIMB-2004-01-003, version 2.2, ISO/IEC 15408, Jan. 2004.

An Effective Method for Analyzing Intrusion Situation Through IP-Based Classification

Minsoo Kim[1], Jae-Hyun Seo[1], Seung-Yong Lee[2], Bong-Nam Noh[2],
Jung-Taek Seo[3], Eung-Ki Park[3], and Choon-Sik Park[3]

[1] Dept. of Information Security, Mokpo Nat'l Univ.,
Mokpo, 534-729, Korea
{phoenix, jhseo}@mokpo.ac.kr
[2] Div. of Electr-Comp. & Inform-Engine., Chonnam Nat'l Univ.,
Gwangju, 500-757, Korea
birch@athena.jnu.ac.kr, bbong@jnu.ac.kr
[3] National Security Research Institute,
Daejeon, 305-348, Korea
{seojt, ekpark, csp}@etri.re.kr

Abstract. Due to a false alert or mass alerts by current intrusion detection systems, the system administrators have difficulties in real-time analysis of an intrusion. In order to solve this problem, it has been studied to analyze the intrusion situation or correlation. However, the existing situation analysis method is grouping with the similarity of measures, and it makes hard to respond appropriately to an elaborate attack. Also, the result of the method is so abstract that the raw information before reduction must be analyzed to realize the intrusion. In this paper, we reduce the number of alerts using the aggregation and correlation and classify the alerts by IP addresses and attack types. Through this method, our tool can find a cunningly cloaked attack flow as well as general attack situation, and more, they are visualized. So an administrator can easily understand the correct attack flow.

1 Introduction

Intrusion detection system (IDS) has been studied to protect a system and a network from cyber crimes. The IDS generally uses the predefined rules to detect intrusions and the result can be diversely shown according to the detection measures. The alerts of current IDS are naturally diversified by using different rules and measures. For holding intrusion information, the IDS generate too many alerts, and this is a heavy burden to administrators. So, the interest in reducing alerts and in easy analyzing intrusion situation has been increasing.

The representative method for alert reduction and intrusion situation analysis can be looked for in Aggregation and Correlation Console (ACC) [1]. ACC reduces alerts through comparing and aggregating of three measures (attack type, source IP, destination IP) in intrusion alerts and it helps the administrator easily perceive attack flows that represent the reduced information as a case. This method outdoes in reduction rate and situation grouping owing to the abstract situation class of intrusion alerts, but it does not express a coordinated attack well and is not sufficient for explanation of attack flows. In this respect, administrators have the burden that they should reanalyze the unreduced raw information.

In this paper, we propose classifying method based on IP address, then we show intrusion situation can be effectively analyzed through this method. First, it is executed to aggregate the alerts with similar measures and to integrate the correlated alerts. This helps for easy analyzing situation, as the reduction is more accurate and the data applying is decreased. The attack path and method can be easily shown through classifying the alerts by property and IP address, too. After all, the administrator can understand the attack situation through attack flow, and he doesn't have to analyze the data again.

2 Related Studies

2.1 Intrusion Detection Method and Alert Reduction

For accurate detection of intrusion, the most current IDSs use the misuse detection method which is made of rules of experts' empirical knowledge. The misuse detection method can produce good results against known and analyzed attacks, but it may not show any reactions or produce various alerts against varied attacks. To detect the abnormal attacks, the anomaly detection method by applying datamining or statistics theory is published, but it is shirked because of low accuracy [2].

Various IDSs produce various alerts and the compatibility among the alerts makes it hard for administrators to analyze the results of IDSs. Each IDS produces many alerts against arbitrary attacks, but it provide another stick for IDSs to produce the invalid alerts because the logical integration method of alerts is not presented. The attack which generates the mass packets, such as Distributed Deny of Service (DDoS), also causes IDSs to produce the mass alerts. A security manager by reason of them faces difficulties for analyzing the intrusion situation from the mass alerts [3].

As a result, the reduction of mass intrusion alerts and the integration of alerts of heterogeneous system or varied attack are the challenges facing IDSs. Typical studies on integrating the results of detection systems are ACC of Tivoli Enterprise Console (TEC) and EMERALD [4, 5].

2.2 The Analysis Method of Intrusion Situation

For easy analyzing the intrusion situation in mass alerts, various reduction methods are proposed. ACC defines the 7 situations, reduces by their characteristics, and shows the intrusion situation. In other words, it presents mass alerts as a simple situation through integration based on alert class, destination IP, and source IP like followed figure 1. This method helps to grasp the current alerts' disposition by classifying the mass alerts in real time.

This simple aggregation method has an advantage that it shows an intrusion situation as a simple form, but it needs more effort for analyzing a coordinated attack or grasping the attack flow. The raw information prior to the reduction must be analyzed after all. For example, as the figure 2, DDoS agent is installed on G system by R2L attackabusing samba vulnerability [6]. Then, DDoS attack is delivered to G system,

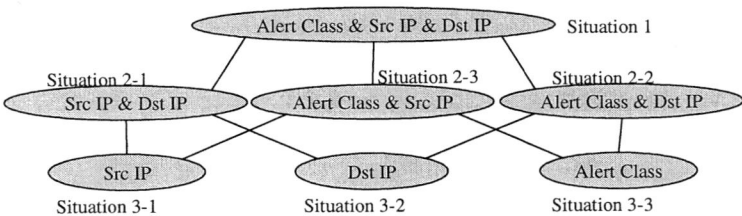

Fig. 1. Aggregation level and situation class in ACC

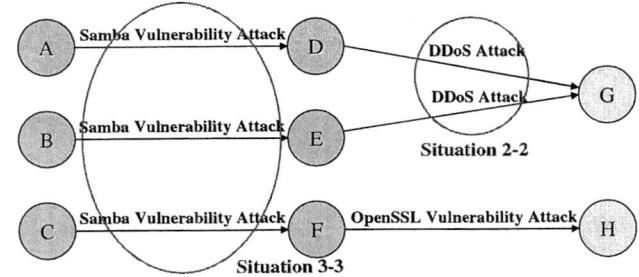

Fig. 2. Stepping stone attack hidden in DDoS attack

and R2L attack abusing OpenSSL vulnerability is given to H system at the same time. This is the situation that the administrator is blinded by DDoS attack, while H system is invaded in essence. The method like ACC shows only the aggregation information of the attacks abusing samba vulnerability and one of DDoS attacks in this example. Therefore, the attack flow of intrusion into H system is consequently included in other situation. Of course the attack information is naturally found in raw data prior to aggregation, but administrator is not willing to analyze the information. An additional cost is also demanded. Therefore, it is needed to take out the attack flow exactly in even this sample.

3 The Proposed Analysis Method for Attack Flow

3.1 IP Address Based Method

In general, the method of tracing the attacker's IP address is used for detecting the attackers. The stepping stone attack is commonly that attacker progresses to occupy the several systems for intruding the final target like stepping stone [7]. To grasp the flow of this attack is achieved by analyzing the intrusion alerts among the systems in the stepping stones. Another merit of grasping the attack flow based on IP address is that the attack such as Probe or DDoS is also easy to classify by ACC's aggregation.

In this paper, we propose the situation analysis method based on destination IP address advancing the past studies. In this method, for guaranteeing the important information of attack flow, the reduction of intrusion alerts is accomplished by adapting the situation classes corresponded with 3 measures among the alerts at least. If three

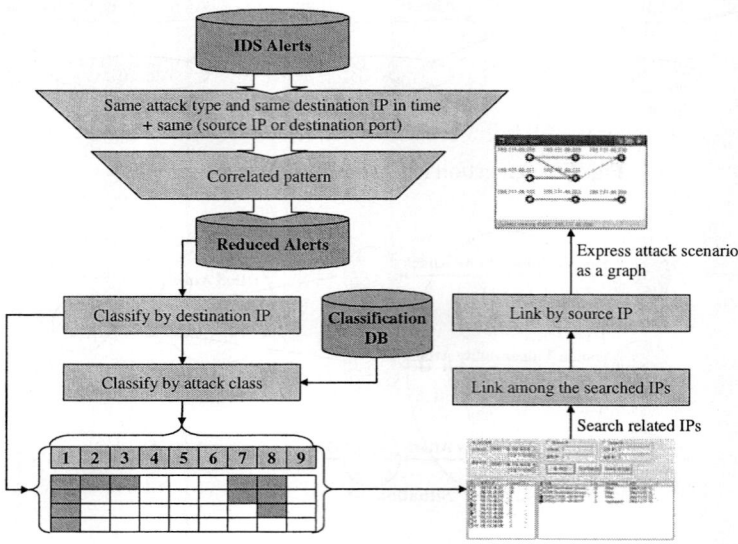

Fig. 3. The process of analyzing attack flow

measures are the same in four measures - in attack type, source IP address, destination IP address, and service type of intrusion alerts -, the alerts is able to be aggregated. The reduction rate of this method is inferior to ACC, but to compensate for decreased reduction rate we use the correlation algorithm which is able to reduce the alerts with different attack type.

3.2 The Analysis of Attack Correlation

The association rule in datamining becomes known as the useful method for finding the correlated pattern from the known facts. Therefore, this has an effect on reducing intrusion alerts of different attack type to one alert, and it can generate high level intrusion information for analyzing the situation. The association rules for reduction is preliminary generated through learning from intrusion data in off-line and is applied to intrusion alerts occurred in real-time. This is the correlation analysis with datamining algorithm and it's easy to analyze the varied attacks.

We process the situation analysis by using the high level intrusion information generated by reduction and correlation analysis for easy grasping the intrusion situation like figure 3. On situation analysis step, the intrusion alerts are grouped by destination address, and the grouped alerts are divided into 9 attack classes. The 9 classes are IP scan, port scan, vulnerability scan, DoS, DDoS, worm virus, Remote-To-Local (R2L), information leakage, and guessing attack. If the source address of classified alert is identical with the destination address of other alert, two alerts are linked and marked as an intrusion flow.

3.3 The Analysis of Attack Flow

The situation analysis gives preferentially notice what are attacker host and victim host. It is also important how the attack get realized. The cases requiring analysis are following;

- when severity level was high in real-time alerts,
- when an abnormal phenomenon presented in statistical analysis,
- and when the alerts with high severity level were found in requested alerts list.

The objects of the situation analysis above are decided by inputting a condition on Viewer. Viewer gets the result of Query and the content is listed according to the destination address.

If the address of important alert is selected in Viewer, then ➲ is marked on the selected address and ⊙ is marked on the addresses correlated with the selected alert. The information of the address has a severity and the number of alerts, and it shows the property of the alerts in detail. The severity of selected alert is represented as the highest severity among the grouped alerts and this helps to grasp the damage state of destination host. This method catches what are attacker host and victim host, and how an intruder breaks in the host.

4 The Development on Situation Analysis Module

4.1 System Architecture

Figure 4 shows the system architecture which is developed by the method proposed in this paper. The IDS used in this architecture is Snort [8]. The flows of attacks are caught in the situating module based on the reduced information through filtering, aggregating and correlating modules. The results are viewed in the GUI viewer. In the intrusion correlation server, the situation module searches the database and analyzes the attack situations at the requests from the GUI viewer. The intrusion correlation server gets the scope of analysis, such as period and IP class, from the GUI viewer and searches the intrusion alerts which correspond to the scope. The situation module sends the scoped alerts as well as all the correlated alerts. It can be known which modules the alerts are reduced in and which classes they are categorized into.

Figure 5 shows the process of identifying an attack and searching the paths by GUI viewer, when the attack executes the probing attack to the R2L attack from any host. First, the lists of addresses can be obtained by inputting the time of attack execution. In the lists, it is necessary to analyze the lists which have high severity level. If the attacks in the lists are marked with 5 or higher level, it is imperative to analyze the lists to identify the attack. Figure 5 indicates that 172.16.112.20 was attacked from 168.131.48.215 with the R2L attack because the severity level of the attack is very high as 9. It can be noticed that 168.131.48.215 was attacked from 168.131.48.219 with the R2L attack by investigating the 168.131.48.215 site again. As the result, it is identified that the attacker was searching hosts to install DDoS Trin00 daemon by analyzing the 168.131.48.219 site. The attack flows can trace back with these methods and be expressed with graphs in the GUI viewer.

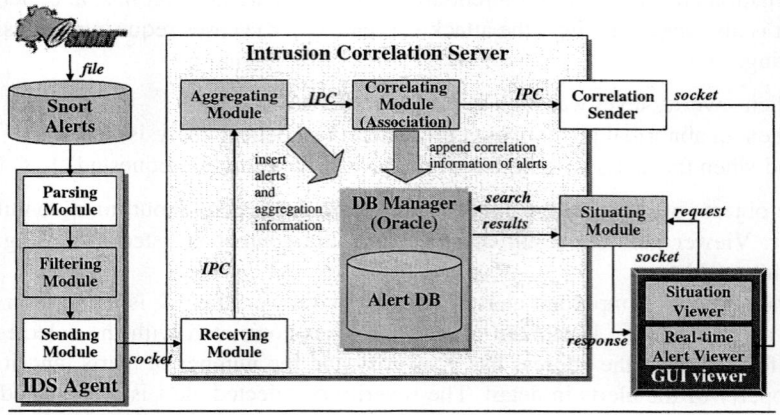

Fig. 4. The system architecture proposed in this paper

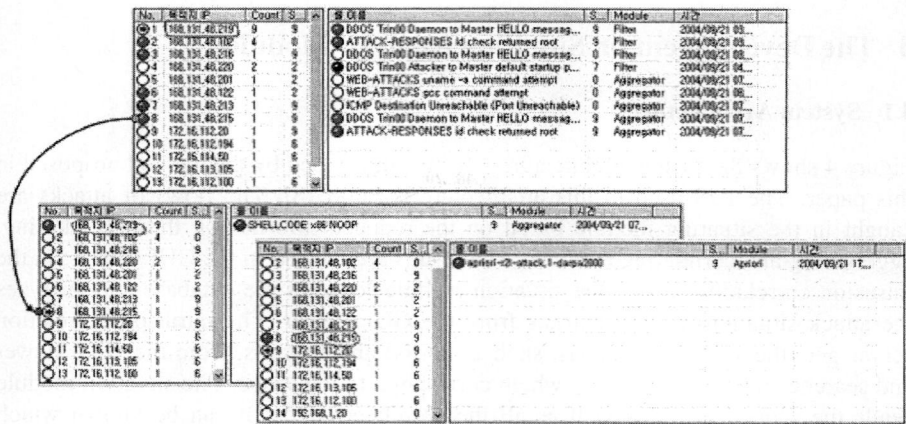

Fig. 5. An example of analyzing attack flows with GUI Viewer

4.2 The Product of IP Based Situation Analysis Method

In this sub-section, we will test the attack scenario by reproducing the situation as illustrated in Figure 2, in order to describe the difference with the proposed method in this paper. Figure 6 describes that DDoS attack and R2L attack are occurred at the same time. An attacker installs the master and agent system for DDoS attack in 168.131.48.210 and 168.131.48.220 respectively, then he is ready for attacking 168.131.48.230. His co-attacker is ready for intruding the 168.131.48.208 system from 168.131.48.102. This is a scenario that the attackers try to intrude the system while executing DDoS attack. We will compare the differences between our method and the existing situation analysis method under this situation.

Figure 7 illustrates analysis of the result of executing the attack following the scenario of figure 6. The DDoS agent was installed in 168.131.48.220 through the R2L attack in figure 7(a). Similarly, figure 7(b) shows that the attacker installs the DDoS agent in 168.131.48.221. Figure 7(c) and 7(d) describe that there was another attack on 168.131. 48.222 and there is an intrusion from 168.131.48.222 to 168.131.48.208 respectively.

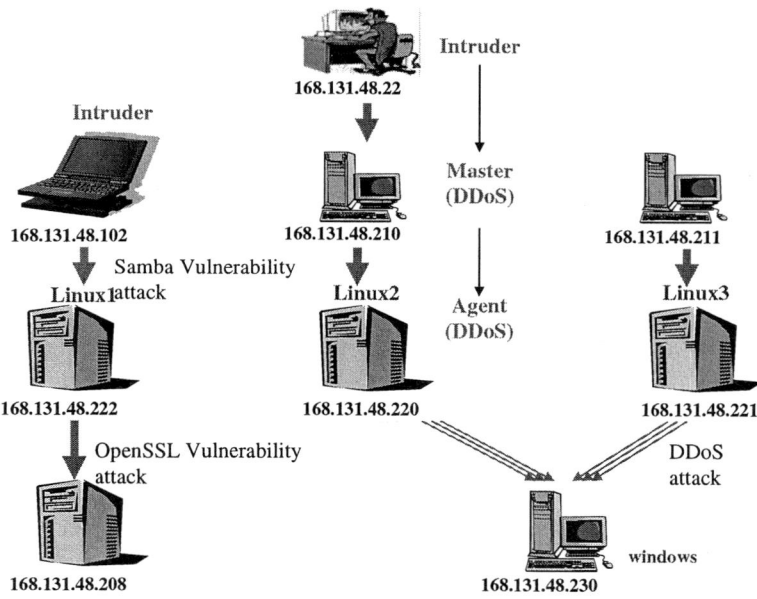

Fig. 6. An attack scenario for testing the situation analysis

The flow of attacks can be shown overall, due to that a graph such as figure 8 is automatically drawn while the procedures of figure 7 are being executed. The arrows have different colors in accordance with the level of danger. The R2L and DDoS attacks are drawn with a red line and a dotted blue line respectively. In the case of the

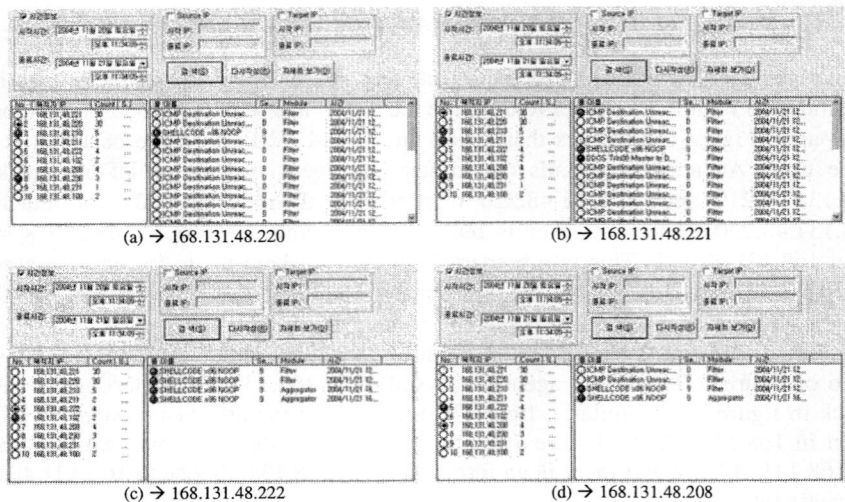

(a) → 168.131.48.220 (b) → 168.131.48.221

(c) → 168.131.48.222 (d) → 168.131.48.208

Fig. 7. Result pictures of analyzing the test scenario

DDos attack, there is a need to change the direction of the arrow when the flow of attacks is drawn, because the alert of the DDoS attack is commonly occurred as the messages to ICMP and the flow of the message is expressed in a reverse direction.

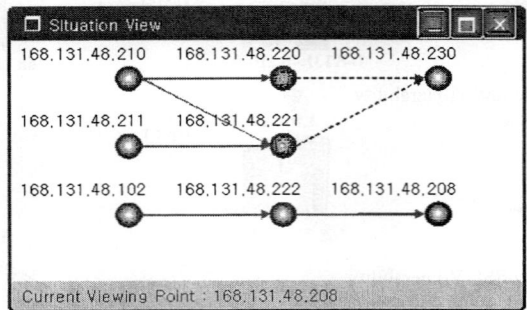

Fig. 8. Graph of attack situation

The result of testing the scenario explains that ACC expresses several attacks as one situation class like table 1. The destination address can not be identified in situation 3-3 and situation 2-2 from the attack information of the situation class. The destination addresses can be searched with the unreduced attack information, but it is overloaded. The proposed method exactly identifies source and destination addresses and shows the connections between the addresses. It shows each situation with an independent path. Our method generates the reduced information and shows the attack flows in graphs in order to analyze the attack among hosts with ease. Therefore, it can help security administrators promptly respond to attack situations.

Table 1. A comparison the our method with ACC situation class

	210→220	211→221	102→222	220→230	221→230	222→208
ACC	Situation 3-3			Situation 2-2		Situation 1
Our method	R2L	R2L	R2L	DDoS		R2L
ACC Situation Class		Situation 3-3		Situation 2-2		Situation 1
Intrusion information	Attack type		ShellCode x86 Noop	ICMP Destination Unreachable		ShellCode x86 Noop
	Source IP		*	*		168.131.48.222
	Destination IP		*	168.131.48.230		168.131.48.202

5 Conclusion

The problem of current IDS is that it generates too many alerts and sometimes the alerts are false messages. This makes it difficult for administrator to cope with an intrusion in real time and to analyze the attack flow. The existing situation analysis method is also difficult to grasp an elaborate attack, even if it gets a simple result by comparing the measures. This method has an effect on reduction of intrusion alerts, but it is not good for analyzing the attack flow.

In this paper, we classified the intrusion alerts by destination IP address and the attack class. When all the intrusion alerts are used to analyze the situation, it takes long and it has too many showing data. For this reason, it must go through the apt reduction procedures. So, we made the aggregation done according to similarity of alerts' measures and the reduction done by correlating the alert types. As the result of this procedure, the effect of reduction was similar to ACC.

We confirmed that the correlation of attacks can be searched based on classification of IP address and attack type and the attack flow can be understood through the graph of attack flow. The ACC is a useful method for grasping an attack situation comprehensively. On the other hand, our method is more useful for reporting the result of intrusion and for inspecting the compromised system, because it can grasp even path of coordinated attacks.

Acknowledgment

This research was supported by the MIC(Ministry of Information and Communication), Korea, under the ITRC(Information Technology Research Center) support program supervised by the IITA(Institute of Information Technology Assessment).

References

1. Debar, H. and Wespi, A., "Aggregation and Correlation of Intrusion-Detection Alerts," RAID 2001, 2001
2. Ning, P., "Techniques and Tools for Analyzing Intrusion Alerts," ACM Transactions on Information and System Security, Vol.7 No.2, pp 274-318, 2004

3. Moh, W., Kim, M., Cheong, I., Noh, B., Seo, J., Park, E. and Park, C., "An Analysis on the Correlation of Network-based Alerts with Association Rule Algorithm," WISA 2004, pp.705-712, 2004
4. Porras, P. and Neumann, P., "EMERALD: Event Monitoring Enabling Responses To Anomalous Live Disturbances," Proc. of the 20th National Information Systems Security Conference, pp 1-13, 1997
5. Valdes, A. and Skinner, K., "Probabilistic Alert Correlation," RAID 2001, 2001
6. Insecure.Org, "Bugtraq: Samba 'smbprint' script tmpfile vulnerability," http://seclist.org/list/bugtraq/2004/Mar/0189.html
7. Yang, Y. and Paxton, V., "Detecting Stepping Stones," USENIX Security Symposium, 2000
8. Beale, J., Foster, J., Posluns, J. and Caswell, B., *Snort 2.0 Intrusion Detection*, SynGress, 2003.

A New Stream Cipher Using Two Nonlinear Functions

Mi-Og Park and Dea-Woo Park

Dept. of Computer Science, Soongsil University,
Sangdo 5-dong Dongjak-gu, Seoul, 156-743, Korea
{mopark777, prof1}@hanmail.net

Abstract. The most widely used digital mobile standards are GSM(Global System for Mobile Communications) and CDMA(Code Division Multiple Access). These systems use data encryption prior to data transference, but these stream ciphers used by data encryption are unsecured. In this paper, in order to protect more securely a data, we propose a new stream cipher based on the summation generator. The proposed algorithm uses four linear feedback shift registers as an input and takes the property of several keystream cycle sequences by usage more than two nonlinear functions, that is, an S-box and a nonlinear combining function. This property makes the proposed algorithm more secure against attack such as correlation attack.

1 Introduction

The demand for wireless communications systems has increased dramatically in the last few years and wireless communications has become more convenient. Since the openness of wireless communications, it is easy to eavesdrop on such systems without detection. So, how to protect the privacy between communicating parties is becoming a very important issue. In order to protect privacy and avoid fraud, cryptographic algorithms have been employed to provide a more secure mobile communications environment. GSM(Global System for Mobile Communications) used mainly in Europe and CDMA(Code Division Multiple Access) used in North America are the most popular mobile phone systems in the world. Both the GSM and the CDMA employ encryption algorithms to protect data. However, recently A5/1 and A5/2 used in GSM were reverse-engineered and published by Briceno, Goldberg and Wager at [1], [2]. Afterwards A5/2 was also cryptanalyzed by Wagner[2]. The structure of A5 is also unsafe and its linear complexity is low. CDMA employ CAVE, ORYX, CMEA algorithm and so on. CAVE algorithm is for challenge-response authentication protocols and key generation, ORYX for wireless data services, and CMEA for encryption message data on the traffic channel. The attack methods on these security algorithms were published at [3], [4]. The E0 stream cipher of Bluetooth based on summation generator was analyzed[5].

As the current being used security algorithms cannot protect the data securely, we propose a new stream cipher whose basic structure is the summation generator. This paper is organized as follows. We describe the summation generator in Section 2 and propose a new stream cipher with enhanced security level by using two nonlinear functions in Section 3. Section 4 shows experimental results and Section 5 concludes this paper.

2   Summation Generator

The summation generator proposed by Rueppel[6] is a nonlinear combiner with memory whose internal state variable, the carry, takes integer values from the set [0, n-1], where n is the number of inputs.

The adder takes the n bits resulting of n different LFSRs and counts how many of these bits are one. The value of the sum is stored in a carry register, whose least significant bit is the output and the rest of bits of the register are feedbacked with those of LFSRs.

Integer addition, when considered as a function of variables in the $GF(2)$, is as highly nonlinear operation and simultaneously provides the property of maximum immunity correlation.

Rueppel' summation generator output z_t and c_t

$$z_t = a_t \oplus b_t \oplus c_{t-1} \tag{1}$$

$$c_t = a_t b_t \oplus (a_t \oplus b_t) c_{t-1} \tag{2}$$

Where a_t is the output sequence of LFSR1, b_t is the output sequence of LFSR2, c_t is the carry sequence, with carry initialization value $c_{t-1} = 0$.

Period. Rueppel proved the period $P = \prod_{i=1}^{n} p_i$ if each period p_i is relatively prime.
Thus, the period of the summation generator is the same as the product of the period of the respective LFSRs if each period is prime relatively.

Linear complexity. It is known that the linear complexity of the summation generator consisted of two m-LFSR is conjectured to be close to the period if L_1 and L_2 are coprime in case that lengths of LFSRs are L_1 and L_2, respectively. Thus, linear complexity is $LC \leq (2^{l_1} - 1)(2^{l_2} - 1)$. However, it was not analyzed in terms of cryptographical security[6].

Correlation property. The summation generator based on a combination of n LFSRs takes maximum order of correlation immunity by the fact that real adder has (n-1) order of correlation immunity. For example, the order of correlation immunity of the summation generator consisted of two LFSRs is 1 and it is maximum order of correlation immunity. However, it is neither secure against and nor immune to correlation attack between its output sequences and carry sequences when it outputs consecutive zeros and ones.

Because Rueppel's summation generator in table 1 has a probability of a input-output correlation probability of 1/2, balanced and a carry-output correlation of 1/4, highly correlated, it can be vulnerable to correlation attack[8][9] when it outputs consecutive zeros or ones.

Table 1. Rueppel's summation generator

a_t	b_t	c_t	c_{t-1}	z_t	Correlation Probability
0	0	0	0	0	
0	0	1	0	1	$P[a_t = z_t] = 1/2$
0	1	0	0	1	
0	1	1	1	0	$P[b_t = z_t] = 1/2$
1	0	0	0	1	
1	0	1	1	0	$P[c_{t-1} = z_t] = 1/2$
1	1	0	1	0	$P[c_t = z_t] = 1/4$
1	1	1	1	1	

It is known that the Rueppel's summation generator produces sequences whose period and correlation immunity are maximum, and whose linear complexity is conjectured to be close to the period[7]. However, it does not serve as a good building block for stream ciphers by carry-output correlation.

3 A New Stream Cipher

3.1 Basic Configuration

The proposed stream cipher consists of four maximal length linear feedback shift registers(m-LFSRs), S-box, nonlinear combining function, and summation generator. For each bit of output, each LFSR is clocked once, and their output bits are inputted into the summation generator through an S-box and any nonlinear combining function. The final output value produced after passing the summation generator is generated through XOR operation with plaintext.

The S-box and the nonlinear combining function used in this algorithm are changeable by the requests of the user and the network to achieve enhanced security level by increasing the number of keystream cycle sequences. Thus, these parameters have to be maintained as secret information.

Besides the used S-box and the nonlinear combining function must be satisfied S-box design condition and nonlinear combining function design condition to maintain the good security level of stream cipher.

The meanings of the respective variables used in a new algorithm are described as follows.

Table 2. The meaning of the variables

a_t, b_t, c_t, d_t : sequences of individual LFSRs at time t
k_{t-1} : carry at time t-1, initial value $k_{t-1} = 0$
k_t : carry at time t, initial value $k_t = 0$
f_i : nonlinear combining function at time t
x_t : variable that saves the output of f_i at time t
s_k : S-box at time t
y_t : variable that saves the output of s_k at time t
z_t : output bits of the summation generator at time t

The operation of the proposed algorithm is as follows:

[Step 1] Four LFSRs are stepped once.
[Step 2] [Step 2-1] and [Step 2-2] are simultaneously operated.
[Step 2-1] It calculates x_t value by nonlinear combining function f_i.
[Step 2-2] It calculates y_t value by S-box s_k.
[Step 3] The output z_t and carry k_t of summation generator are computed by inputs from s_k and f_i.
[Step 4] XOR operation is performed on the value z_t and p_t plaintext.

The notations of the related equations are denoted as follows.

$$x_t = f_i(a_t, b_t, c_t, d_t) \tag{3}$$

$$y_t = s_k[a_t, b_t, c_t, d_t] \tag{4}$$

$$z_t = x_t \oplus y_t \oplus k_{t-1}, \quad t=0,1,2,\cdots \tag{5}$$

$$k_t = y_t \oplus (x_t \oplus y_t)k_{t-1} \tag{6}$$

Rueppel's generator is neither secure against and nor immune to correlation attack between its output sequences and carry sequences in special cases. So, it is not proper stream cipher to protect a data. The proposed algorithm uses Dawson's carry method as a basic model to achieve cryptographically good properties. The carry method in equation (6) is based on Dawson's carry method in equation (7). The meaning of variables in equation (7) is same as in section 2.

$$c_t = b_t \oplus (a_t \oplus b_t)c_{t-1}, \quad t=0,1,2,\ldots \qquad (7)$$

In Dawson's summation generator in table 2, input-output correlation probability is 1/2, and a carry-output correlation probability is 1/2 too. Therefore, it is secure in terms of input-output correlation probability or carry-output correlation probability.

Table 3. Dawson's summation generator

a_t	b_t	c_t	c_{t-1}	z_t	Correlation Probability
0	0	0	0	0	
0	0	1	0	1	$P[a_t = z_t] = 1/2$
0	1	0	1	1	
0	1	1	0	0	$P[b_t = z_t] = 1/2$
1	0	0	0	1	
1	0	1	1	0	$P[c_{t-1} = z_t] = 1/2$
1	1	0	1	0	$P[c_t = z_t] = 1/2$
1	1	1	1	1	

3.2 Transmission of Two Nonlinear Functions

Two nonlinear functions, that is, an S-box s_k and any nonlinear combining function f_i used in the proposed model can be exchanged for new nonlinear functions according to a change request. An S-box and a nonlinear combining function are simultaneously used namely, [Step 2-1] and [Step 2-2] are simultaneously used at the certain time t. But multiple S-boxes themselves are not simultaneously used and only one S-box is used at time t. If the change request for the S-box is happened, then the current S-box is replaced by a new S-box. The multiple nonlinear combining functions are also not used at the same time and only one nonlinear combining function is used at time t.

We supposed that the transmission of S-box and nonlinear combining function is encrypted. It is also assumed that mutual authentication is carried out between the user and the network before S-box and nonlinear combining function are transferred. Whenever the S-box and the nonlinear combining function that are currently being used are changed for new ones, they must be encrypted, because that the proposed algorithm can be attacked by using the S-box and the nonlinear combining function being transferred if they are transmitted without encryption.

In addition, mutual authentication must be provided for a secure transmission of S-box and nonlinear combining function because the secret parameters e.g. S-box and nonlinear combining function are exposed to the attacker who masquerades as a real user. The secure encryption transmission of S-box and nonlinear combining function summation generator and mutual authentication decrease the probability to be attacked by some attack methods that analyze the typical summation generator.

The encryption process and the mutual authentication process are based the each process of the used network. For example, if the proposed algorithm is used in the GSM system, the data encryption method used in the GSM system will be used to encrypt an S-box and a nonlinear combining function. The mutual authentication protocol will also be used mutual authentication method used in GSM system. And if the CDMA system is used, then the encryption and the mutual authentication methods of CDMA system will be used. However, it is assumed that all parameters as *RAND* and *RES* that are transmitted in clear types like the GSM system are transferred after encryption.

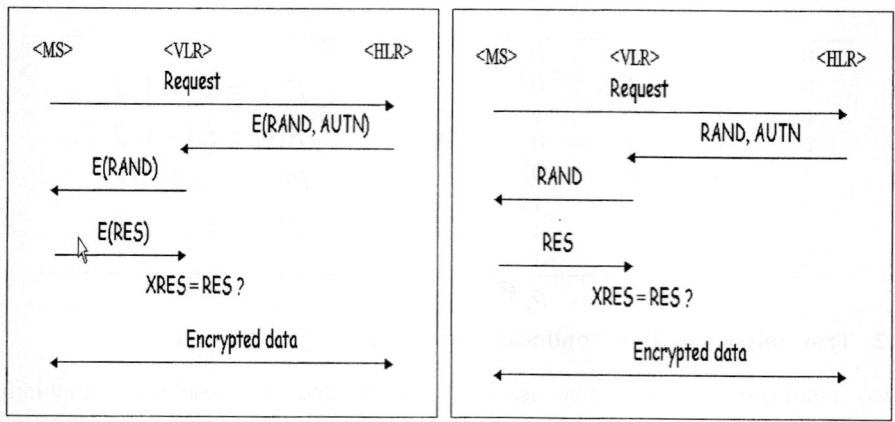

Fig. 1. Procedure of the proposed transmission and the typical transmission

4 Performance Analysis

4.1 Period and Linear Complexity

In order to achieve a maximum period, LFSRi of length l_i has a primitive characteristic polynomial and they are relative prime. Hence, its period is defined as following equation (8).

$$P = (2^{l_1} - 1)(2^{l_2} - 1)(2^{l_3} - 1)(2^{l_4} - 1) \qquad (8)$$

Linear complexity of the proposed algorithm becomes equation (9) according to the characteristic of summation generator because the proposed algorithm uses the summation generator as a basic structure.

$$LC \leq (2^{l_1} - 1)(2^{l_2} - 1)(2^{l_3} - 1)(2^{l_4} - 1) \qquad (9)$$

If a pair of S-box and nonlinear combining function at time *t* uses a good pair that has no correlation, linear complexity is close to the period such as the characteristic of summation generator. Otherwise, linear complexity is less than period.

4.2 Security Analysis on Correlation Attack

A correlation attack exploits the weakness in some combining function, which allows information about individual input sequences to be observed in the output sequence. In such a case, there is a correlation between the output sequence and one of the internal sequences. This particular internal sequence can then be analyzed individually before attention is turned to one of the other internal sequences.

Dawson proved it is possible to analyze the ciphertext by performing XOR operation the previous ciphertext with the current ciphertext on the assumption that the cryptanalyst knows the previous and the current ciphertext on stream cryptosystem uses the same key. It was shown that equation (10) is satisfied from the assumption that K and K' are same when it is the assumption that the previous plaintext $p' = (p_0', p_1', p_2', \cdots, p_n')$, the previous keystream $k' = (k_0', k_1', k_2', \cdots, k_n')$, the previous ciphertext $c' = (c_0', c_1', c_2', \cdots, c_n')$, the current plaintext $p = (p_0, p_1, p_2, \cdots, p_n)$, the current keystream $k = (k_0, k_1, k_2, \cdots, k_n)$, the current plaintext $c = (c_0, c_1, c_2, \cdots, c_n)$ [11].

$$C = (p_0 \oplus k_0, p_1 \oplus k_1, p_2 \oplus k_2, \cdots, p_n \oplus k_n)$$

$$C' = (p_0' \oplus k_0', p_1' \oplus k_1', p_2' \oplus k_2', \cdots, p_n' \oplus k_n') \qquad (10)$$

$$C \oplus C' = (p_0' \oplus p_0, p_1' \oplus p_1, p_2' \oplus p_2, \cdots, p_n' \oplus p_n)$$

The key value can be supposed by using redundancy of plaintext because the type of the result is existed in the type of two plaintexts if the previous ciphertext C and current ciphertext C' perform XOR.operation. It is always initialized with new session key to avoid these attacks whenever synchronization is set up.

We can represent the conventional summation generator and the proposed algorithm by equation (10) which is considered internal states on the assumption that both the conventional stream cipher(summation generator) and the proposed algorithm are initialized with new session key whenever synchronization is set up against Dawson correlation attack.

The conventional summation generator can be represented by equation (11). $k_i[IS_i]$ denotes encryption is processed by key k in the internal state(IS) of each algorithm.

$$C = (p_0 \oplus k_0[IS_0], p_1 \oplus k_1[IS_1], p_2 \oplus k_2[IS_2], \cdots, p_n \oplus k_n[IS_n]) \qquad (11)$$

$$C' = (p_0' \oplus k_0[IS_0], p_1' \oplus k_1[IS_1], p_2' \oplus k_2[IS_2], \cdots, p_n' \oplus k_n[IS_n])$$

All of the internal states IS are same as IS_0 to be same internal states during the n times of encryption process because the conventional summation generator is not changed into internal state. So, if the current plaintext and the previous plaintext perform XOR operation, it is same as the $C \oplus C'$ result of equation (10). Therefore, an attacker can guess the key stream by using redundancy of the current plaintext and the previous plaintext.

The proposed algorithm can be represented by equation (12) because it can change the S-box or nonlinear function.

$$C = (p_0 \oplus k_0[IS_{now0}], p_1 \oplus k_1[IS_{now1}], p_2 \oplus k_2[IS_{now2}], \cdots, p_n \oplus k_n[IS_{nown}] \quad (12)$$

$$C = (p_0 \oplus k_0[IS_{now0}], p_1 \oplus k_1[IS_{now1}], p_2 \oplus k_2[IS_{now2}], \cdots, p_n \oplus k_n[IS_{nown}]$$

Internal states $k_i[IS_i]$ used in the previous ciphertext and the current ciphertext are not same for the internal states which are changed according to the proposed mechanism. The internal state of the previous ciphertext and current one denote $k_i[IS_{pasti}]$ and $k_i[IS_{nowi}]$ respectively in equation (12).

$$C \oplus C' = p_0 \oplus p_0' \oplus k_0[IS_{past0}] \oplus k_0[IS_{now0}], \cdots$$
$$p_1 \oplus p_1' \oplus k_1[IS_{past1}] \oplus k_1[IS_{now1}], \cdots, \quad (13)$$
$$p_n \oplus p_n' \oplus k_n[IS_{pastn}] \oplus k_n[IS_{nown}])$$

The internal state of the previous ciphertext, $k_i[IS_{pasti}]$ and the internal state of the current ciphertext, $k_i[IS_{nowi}]$ is founded in the result expression if the previous ciphertext performs XOR operation with the current ciphertext because internal states are different. To suppose key value, it is need to analyze the previous plaintext, the current plaintext, and correlation of each internal states $k_i[IS_{nowi}]$ and $k_i[IS_{pasti}]$. Therefore, by equation (13) and (12) the proposed algorithm is much more secure as much as redundancy of the previous plaintext, the current plaintext, and correlation of each internal states $k_i[IS_{nowi}]$ and $k_i[IS_{pasti}]$ in terms of correlation attack for deducing the key value. Therefore, the proposed algorithm is more secure than the typical algorithms because it is need analysis of $k_i[IS_i]$ through the equation (10) and (13).

4.3 Randomness Test

The proposed algorithm was simulated by C language on a Sun Ultra SPAC-II 400MHz(2) CPU, 2048M memory. It is tested randomness proposed by FIPS 140-1[11]. The used bits are about 36000 bits.

As the result, the two kinds of sampled data display a good randomness, as shown in Table 4, in terms of the result of frequency test, serial test, generalized serial test, poker test, and autocorrelation test.

Table 4. Test results

Items	p-value(sample 1)	p-value(sample 2)
Frequency	0.484646	0.015065
Block-frequency	0.105618	0.141256
Serial Test	0.078086	0.788728
Run	0.392456	0.105628
Discrete Fourier Transform	0.242986	0.186566
Approximate	0.186566	0.035770
Cumulative Entropy	0.105618	0.141256
Linear Complexity	0.141256	0.392456
Autocorrelation	max ≤ 0.05	max ≤ 0.05

5 Conclusion

In this paper, we have proposed a new stream cipher based on summation generator. In order to protect privacy and prevent fraud, the proposed model was expanded the number of keystream cycle sequences by usage multiple S-boxes and nonlinear combining functions. This property makes the proposed model strong to correlation attack such as Dawson's attack. Because the general stream ciphers use one nonlinear function, the number of keystream cycle sequences is only one. It is vulnerable to general correlation attack.

By the results, the proposed model was proved good randomness by passing the test of FIPS randomness. As the result, the proposed model can protect more securely data than general stream cipher or conventional summation generator because it satisfies cryptographically good properties with maximum period and linear complexity, and so on. So, it could be applicable to the mobile and ubiquitous environments that needed for more secure random number generator.

References

1. Marc Briceno, Ian Goldberg, David Wagner: A Pedagogical Implementation of A5/1, http://jya.com/a51-pi.htm
2. Marc Briceno, Ian GoldBerg, David Wagner: A Pedagogical Implementation of the GSM A5/1 and A5/2 voice privacy encryption algorithms, http://www.scard.org.gsm/a51.html
3. Ravi Chandra N: Cryptanalysis of Security Algorithms in CDMA, http://www.csa.iisc.ernet.in/ academics/projects/
4. D. Wager, L. Simpson, E. Dawson, J. Kelsey, W. Millan, and B.Schneier: Cryptanalysis of ORYX, SAC'98, LNCS 1556, pp.296-305, 1999
5. C. Canniere, T. Johansson, and B. Preneel: Cryptanalysis of the Bluetooth Stream Cipher, http://www.esat.kuleuven.ac.be/~cosicart/ps/CDC-0101.ps.gz

6. R.A. Rueppel: Correlation Immunity and the summation generator, Advances in Cryptology. Crypto'85, Lecture Notes in Computer Science 218, Springer-Verlag (1986) 260-272
7. Dukjae Moon, B. Roy and W. Meier: Algebraic Attacks on Summation Generators, FSE 2004, Lecture Notes in Computer Science 3017 (2004) 34–48
8. J. Golic: Correlation properties of a general combiner with memory, Journal of Cryptology, Sep. (1996) 111-126
9. T. Siegenthaler: Correlation Immunity of Nonlinear Combining Functions for Cryptographic Applications, IEEE Trans. On Infor. Theo., Vol.IT-30, No.5, Sep. (1984) 776-780
10. National Institute of Standards and Technology, FIPS PUB 140-1: Security Requirements for Cryptographic Modules, Jan. 1994.
11. V. Chepyzhov and B. Smeets: On a fast attack on stream ciphers, Advances in Cryptology Eurocrypt'91, Lecture Notes in Computer Science, Vol.547, Springer-Verlag, Berlin, (1991) 176-185

New Key Management Systems for Multilevel Security

Hwankoo Kim[1], Bongjoo Park[1], JaeCheol Ha[2], Byoungcheon Lee[3], and DongGook Park[4]

[1] Information Security Major, Div. of Computer Science and Engineering,
Hoseo University, Asan 336-795, Korea
{hkkim, bjpark}@office.hoseo.ac.kr
[2] Dept. of Information and Communication, Korea Nazarene University,
Cheonan 330-718, Korea
jcha@kornu.ac.kr
[3] Dept. of Information Security, Joongbu University, Kumsan-Gun 312-702, Korea
sultan@joongbu.ac.kr
[4] School of Information Technology, SunChon University, SunChon 540-742, Korea
dgpark6@sunchon.ac.kr

Abstract. In this paper, we review briefly Akl and Taylor's cryptographic solution of multilevel security problem. We propose new key management systems for multilevel security using various one-way functions.

1 Introduction

Secret data should be managed for access to authorized people only. In order to do so, secret keys must be distributed solely to those with access to the pertaining information. However, this is not a simple problem to solve.

First, let us recall notation and terminology from [1]. Assume that the users of a computer system are divided into a number of disjoint sets $S = \{U_1, U_2, \ldots, U_n\}$. The term *security class* (or *class*, for short) will be used to designate each of the U_i. The meaning of $U_i \leq U_j$ in the partially ordered set $(S, \leq)$ is that users in U_i have a *security clearance* lower than or equal to those in U_j. Simply put, this means that users in U_j can have access to information held by (or destined to) users in U_i, while the opposite is prohibited.

Let x_m be a piece of information, or object, that a central authority (CA) desires to store (or broadcast over) the system. The meaning of the subscript m is that object x is accessible to users in class U_m. The partial order on S implies that x_m is also accessible to users in all classes U_i such that $U_m \leq U_i$. It is required to design a system which, in addition to satisfying the above conditions, ensures that access to the information is as decentralized as possible. This means that authorized users should be able to retrieve x_m independently as soon as it is stored or broadcast by CA.

This access control problem arises in organizations where a hierarchical structure exists. Government, the diplomatic corps, and the military are examples of such hierarchies. Applications also exist in business and in other areas of the private sector, for example in the management of databases containing sensitive information, or in the protection of industrial secrets. Finally, the model is employed in the design of computer operating systems to control information flow from one program to another. General references for any undefined terminology and notion are [5, 6, 10, 13, 14].

In sections 2 and 3, we review briefly Akl and Taylor's cryptographic solution of multilevel security problem from [1]. We propose new key management systems for multilevel security using various one-way functions from section 4 to 7.

2 Cryptographic Solution

Let E (resp., D) be an encryption (resp., a decryption) algorithm of a cryptosystem, such as DES (Data Encryption Standard) or AES (Advanced Encryption Standard). Then the simplest cryptographic solution to access control problem may be obtained as follows ([1]). The CA generates n keys $\{K_i\}$ and distributes to U_i its own key K_i and all keys K_j belonging to U_j below U_i in the hierarchy. When an object x_m is to be stored (or broadcast), it is first encrypted with K_m to obtain $x' = E_{K_m}(x_m)$ and then stored (or broadcast) as the pair $[x', m]$. This guarantees that only users in possession of K_m will be able to retrieve x_m from $x_m = D_{K_m}(x')$.

As pointed out in [1], this solution has the advantage that only one copy of x_m is stored or broadcast and the operations of encryption and decryption are performed just once. Its disadvantage is the large number of keys that must be held by each user. To avoid this problem, Akl and Taylor proposed in [1] a new scheme to manage keys such a way that a system is used by which K_i can be feasibly computed from K_j if and only if $U_i \leq U_j$.

3 Overview of Known Schemes

In the case where the structure of classes is totally ordered, we can distribute multilevel security keys using a function $H(M) = M^2 \bmod pq$ as shown in Fig. 1. Instead of this function, we can also use $H(M) = M^3 \bmod pq$. In 1982, applying these functions, Akl and Taylor proposed a cryptographic solution of key management for multilevel security for any poset. The following is a brief review of their method.

For any given poset, we add a top class if there is no any. The CA assigns an integer t_i to each class U_i so that $U_i \leq U_j$ if and only if $t_j | t_i$. The CA chooses a random K, computes $K^{t_i} \bmod pq$, and then distributes it to each class U_i. If $U_i \leq U_j$, then $t_i = d \cdot t_j$ for some integer d. Thus a user in U_j can get the key of U_i by computing $(K^{t_j})^d = K^{t_j d} = K^{t_i} \bmod pq$. This key management system (KMS for short) is not secure. For example, as shown in Fig. 2, U_3 and U_1 can conspire together to find the (master) key K of the top class. For $K^9/(K^4)^2 = K$.

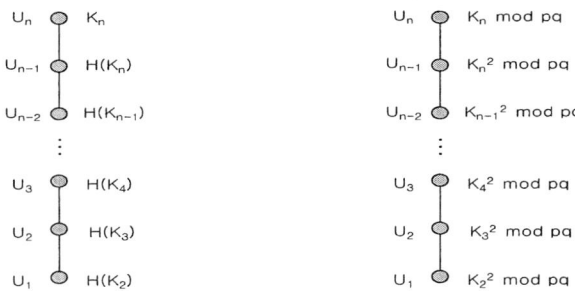

Fig. 1. Key management system for a totally ordered set

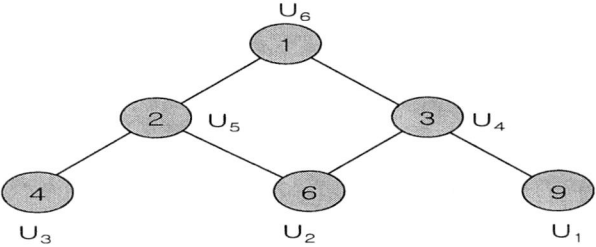

Fig. 2. Key management system using RSA - Conspired version

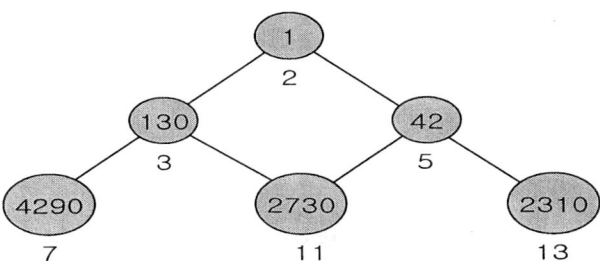

Fig. 3. Key management system using RSA - Improved version

Thus Akl and Taylor proposed a new method to avoid such a problem. They suggested a new algorithm for the assignment of t_i's. Each class U_i is assigned a distinct prime p_i and $t_i = \prod_{U_j \not\leq U_i} p_j$. (See Fig. 3.)

In [2], Akl and Taylor proposed a time-versus-storage trade-off for addressing their key management system. It was shown in [8, 9] that the key generation algorithm of [2] became inefficient when the number of users was large. As a result, an improved algorithm can be described and its optimality is shown.

4 KMS Using a One-Way (Cryptographic) Hash Function

Since Akl and Taylor's KMS for multilevel security uses exponentiation, the overload of computation of keys is high. However, we can compute keys faster than Akl and Taylor's method as shown below if we use a one-way hash function.

Once again, if there is no top class, we add a top class. Then the CA assigns a name to each class as in Fig. 4. Note that, in a lower tree, there is a unique class, denoted by $c(U_k)$, which covers a class U_k of a lower tree with top class.

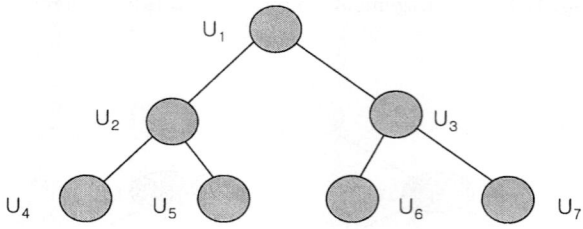

Fig. 4. A lower tree

Now the CA selects a key K belonging to the top class and a one-way hash function H. The CA computes $K_{U_k} = H(U_k, K_{c(U_k)})$ and distributes it together with H to each class U_k. Then the users in an upper class can compute all keys belonging to the classes lower than theirs using their keys, hash function H, and names of lower classes. (See Table 1.) Because of the one-wayness of the hash function, a user in U_k can not compute others' keys belonging to upper classes.

Table 1. Distribution of multilevel security keys for a lower tree using a hash function

Class	Keys
U_1	$K_{U_1} = K$
U_2	$K_{U_2} = H(U_2, K_{U_1})$
U_3	$K_{U_3} = H(U_3, K_{U_1})$
U_4	$K_{U_4} = H(U_4, K_{U_2})$
U_5	$K_{U_5} = H(U_5, K_{U_2})$
U_6	$K_{U_6} = H(U_6, K_{U_3})$
U_7	$K_{U_7} = H(U_7, K_{U_3})$

5 KMS Using RSA Algorithm

While we can manage multilevel security keys for a lower tree using one-way hash functions and names of classes, there is no known algorithm of multilevel security key management for an upper tree. Now we propose a multilevel security key management for an upper tree using the RSA algorithm [12].

Table 2. Distribution of multilevel security keys for an upper tree using the RSA

Class	Keys
U_1	$K_{U_1} = K$
U_2	$K_{U_2} = E(f_{U_2}(K_{U_1}))$
U_3	$K_{U_3} = E(f_{U_3}(K_{U_1}))$
U_4	$K_{U_4} = E(f_{U_4}(K_{U_2}))$
U_5	$K_{U_5} = E(f_{U_5}(K_{U_2}))$
U_6	$K_{U_6} = E(f_{U_6}(K_{U_3}))$
U_7	$K_{U_7} = E(f_{U_7}(K_{U_3}))$

If there is no bottom class, we add a bottom class. Now, we assign a name to each class. Since an upper tree is the dual poset of a lower tree, there is a unique class, denoted by $b(U_k)$, which is covered by a class U_k of an upper tree with bottom class.

Then, we select two large primes p and q and an encryption parameter e for the RSA algorithm and compute a decryption parameter d corresponding to e. Let $n = pq$ and we can define an encryption function E and a decryption function D as follows:

$$E(M) = M^e \bmod n \qquad D(C) = C^d \bmod n.$$

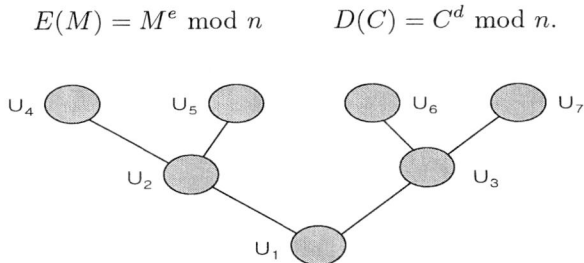

Fig. 5. An upper tree

Thereafter, the CA selects a key K belonging to the bottom class and a function f_{U_i}, the inverse of which is easy to compute. The CA distributes n, d and all f_{U_i}'s to each class and computes $K_{U_k} = E(f_{U_k}(K_{b(U_k)}))$ and distributes it to the class U_k. Here, the parameters p, q and e are secret to all users. Then the users in an upper class can compute all keys belonging to the classes lower than theirs using the decryption function D.

For example, assuming that the keys are distributed (See Table 2) for the poset in Fig. 5, a user belonging to U_7 can get $f_{U_7}(K_{U_3})$ from his/her key K_{U_7} using the RSA decryption function D. Furthermore, he/she can easily compute K_{U_3} by finding the inverse of f_{U_7}. Similarly he/she can get K_{U_1} from $K_{U_3} = E(f_{U_3}(K_{U_1}))$. To conclude, any user in class U_7 can compute the key K_{U_3} belonging to class U_3 and the key K_{U_1} belonging to class U_1.

6 KMS Using Poset Dimension

In this section, we propose a key management system using poset dimension. First we recall the notions of dimension and realizer in a poset from [15]. For any two posets (X, P) and (X, Q), Q is called an *extension* of P if $P \subseteq Q$. In particular, an extension Q of P is called a *linear extension* of P if Q is totally ordered. Let $\mathcal{E}(P)$ be the set of all linear extensions of P. Then it is easy to see that $P = \bigcap \mathcal{E}(P)$. Now, we define the *dimension* of any poset P as follows:

$$\dim(X, P) = \min\{|\Theta| : \Theta \text{ is a family of linear extension of } P, P = \bigcap \Theta\}.$$

A family Θ of linear extensions of P is called a *realizer* if $P = \bigcap \Theta$. It is easy to see that $\dim(X, P) = 1$ if and only if (X, P) is totally ordered.

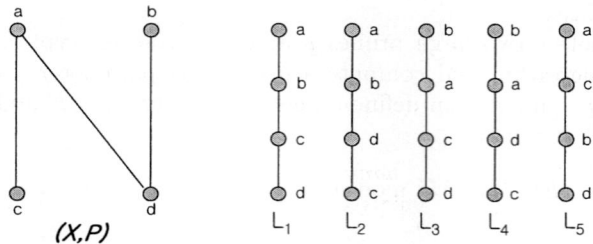

Fig. 6. Poset (X, P) and its linear extensions

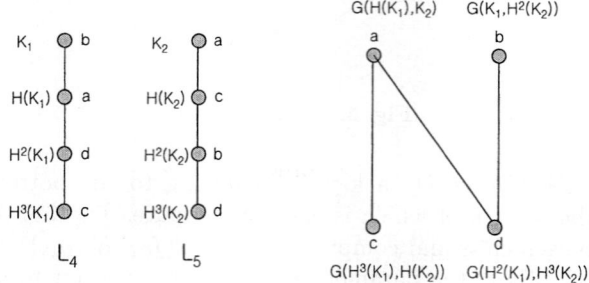

Fig. 7. Key management system using poset dimension

Example 1. Let $X = \{a, b, c, d\}$ be a set and let $P = \{(a, a), (b, b), (c, c), (d, d), (c, a), (d, a), (d, b)\}$ be a partial order on X. Then (X, P) is a poset and the number of extensions of P is 14. In particular, the number of linear extensions of P is 5, as indicated in Fig. 6. Since $L_4 \cap L_5 = P$, $\Theta = \{L_4, L_5\}$ is a realizer of (X, P), and $\dim(X, P) = 2$ since (X, P) is not totally ordered.

Let (X, P) be a poset such that $\dim(X, P) = n$. Then by the definition of dimension, there exist realizers $\{L_1, L_2, \ldots, L_n\}$. The CA selects two one-way hash functions H, G and generates n keys $K_i (1 \leq i \leq n)$ to be assigned to the top class of each $L_i (1 \leq i \leq n)$, and then in each $L_i (1 \leq i \leq n)$ keys for multilevel security are distributed using the hash function H. Then n keys and two hash functions G, H are assigned to each class of the poset (X, P). The users in a class can get the key for multilevel security as the output of the hash function G.

Unfortunately, this key management system is vulnerable to cooperative attacks. However, if a user's key is stored in a smart card or a PCMCIA card, there will be no threat against conspiracy attacks, since it can not be read.

7 KMS Using a Clifford Semigroup

In this section, we propose a key management system using a Clifford semigroup. First we recall the notion of a commutative Clifford semigroup from [16]. A commutative semigroup S is said to be a *Clifford semigroup* if S is a semilattice of groups. For ease of understanding, this condition means: S is a disjoint union of groups $G_\lambda, \lambda \in \Lambda$, where Λ is a semilattice; if $\alpha, \beta \in \Lambda$ and $\beta \leq \alpha$, there exists a homomorphism $\phi_{\alpha,\beta} : G_\alpha \to G_\beta$; if $x, y \in S$, the multiplication $x.y$ in S is defined as follows: If $x \in G_\alpha, y \in G_\beta$ and $\alpha \wedge \beta = \inf\{\alpha, \beta\}$, then $x.y = \phi_{\alpha, \alpha \wedge \beta}(x) \phi_{\beta, \alpha \wedge \beta}(y)$, where the second member is a product in the group $G_{\alpha \wedge \beta}$; $\phi_{\alpha, \beta}$ is called a *bonding homomorphism*.

If S is any commutative semigroup and e is an idempotent of S, the subgroup G_e generated by e is $G_e = \{x \in S : xe = x \text{ and } \exists\, y \in S \text{ such that } xy = e\}$. Let us denote by E the set of idempotents of S; the union of disjoint groups $S_0 = \bigcup \{G_e : e \in E\}$ is the maximal Clifford semigroup contained in S. In fact, the set E is a semilattice with the partial order defined by: $e \leq f$ if and only if $ef = e$; therefore the family $\{G_e : e \in E\}$ inherits the semilattice structure of E; we write $G_e \preceq G_f$ if G_e is less than or equal to G_f in this partial order; in this case the bonding homomorphism $\phi_{f,e} : G_f \to G_e$ is given by $x \mapsto xe$.

To give a concrete example, we refer to [4, 7, 11] for some notions and terminologies of ideal theory in number fields.

Example 2. Let R be a non-maximal order of an imaginary quadratic number field K. We denote by D its integral closure and the index f of R in D as an abelian group, i.e., $f = |D/R|$. Assume that f is sufficiently large. Let $\text{CL}(R)$ be the class semigroup of R. Then by [16–Theorem 11] $\text{CL}(R)$ is a Clifford semigroup and by [16–Proposition 13] the idempotents of $\text{CL}(R)$ are the equivalent classes of ideals of the form $E = (k, \eta)$, where $k \in \mathbb{Z}$ divides f. Note that any idempotent E which is not equivalent to R is not invertible. If E and F are idempotents where $E \leq F$, then the bonding homomorphism $\phi_{F,E} : G_F \to G_E$ is defined by $\phi_{F,E}(K_0) = EK_0$, where K_0 is the key ideal. A representation for an ideal is given in [7], while an efficient algorithm for multiplication of ideals is given in [3–pp. 113]. Note that the computation of K_0 from EK_0 seems to be difficult unless E is equivalent to R. Now, the CA assign an idempotent E_i to each class U_i. For example, we consider the diagram in Fig. 4. The CA selects a random key K_0 and computes E_2K_0, E_3K_0 and distributes each of them to each class U_2, U_3. The CA computes $E_2E_4K_0$ and distributes it to U_4. Similarly the CA computes keys of all classes and distributes each of them respectively. Then the users in an upper class can compute all keys belonging to classes lower than itself.

8 Conclusion and Further Study

In this paper, we proposed new key management systems for multilevel security using various one-way functions and mathematical notions. In particular, we proposed a key management system for multilevel security for an upper tree structure using RSA. In section 7, we also proposed a KMS for multilevel security using poset dimension. However, this system can be vulnerable to cooperative attacks. Thus, it will take further work to solve this problem. In the final section, we proposed a KMS for multilevel security using a Clifford semigroup. Nevertheless, parameter sizes need to be considered for precise and efficient performance of our systems, which aim to provide practical security.

Acknowledgements

This research was supported by the MIC, Korea, under the ITRC support program supervised by the IITA.

References

1. Akl, Selim G., Taylor, Peter D.: Cryptographic Solution to a Multilevel Security Problem. CRYPTO 1982: 237-249.
2. Akl, Selim G., Taylor, Peter D.: Cryptographic Solution to a Problem of Access Control in a Hierarchy, ACM Trans. Comput. Syst. **1**(3): (1983) 239–248
3. Buchmann, J., Willams, H. C.: A key-exchange system based on imaginary quadratic fields. J. Cryptology **1** (1988) 107–118.
4. Cohen, H.: A course in Computational Algebraic Number Theory, Springer, Berlin, 1995.
5. Delfs, H., Knebel, H.: Introduction to Cryptography: Principles and Applications, Springer-Verlag, Berlin, 2002.
6. Denning, D. E.: Cryptography and Data Security, Addison-Wesley, Reading, Massachusetts, 1982.
7. Kim, H., Moon, S.: Public-key cryptosystems based on class semigroups of imaginary quadratic non-maximal orders, in *Information Security and Privacy – ACISP 2003*, LNCS 2727, Springer-Verlag, Berlin, 2003, pp. 488–497.
8. MacKinnon, Stephen J., Akl, Selim G.: New Key Generation Algorithms for Multilevel Security. IEEE Symposium on Security and Privacy (1983) 72–78.
9. MacKinnon, Stephen J., Taylor, Peter D., Meijer, Henk, Akl, Selim G.: An optimal algorithm for assigning cryptographic keys to control access in a hierarchy, IEEE Trans. computers **34** (1985) 797–802.
10. Menezes, A. J., Oorschot, P. C., Vanstone, S. A.: Handbook of Applied Cryptography, CRC Press, Boca Raton, 1997.
11. Mollin, R. A.: Quadratics, CRC Press, Boca Raton, 1996.

12. Rivest, R. L., Shamir, A., Adelman, L.: A method for obtaining digital signatures and public key cryptosystems. Communications of the ACM **21** (1978) 120–126.
13. Schneier, B.: Applied Cryptography, John Wiley & Sons, Inc., 1996.
14. Stinson, D. R.: Cryptography: Theory and Practice, CRC Press, Boca Raton, 2002.
15. Trotter, William T.: Combinatorics and Partially Ordered Sets: Dimension Theory, The Johns Hopkins Univ. Press, Baltimore and London, 1992.
16. Zanardo, P., Zannier, U.: The class semigroup of orders in number fields. Math. Proc. Camb. Phil. Soc. **115** (1994) 379–391.

Neural Network Techniques for Host Anomaly Intrusion Detection Using Fixed Pattern Transformation

ByungRae Cha[1], KyungWoo Park[2], and JaeHyun Seo[3]

[1] Dept. of Computer Eng., Honam Univ., Korea
chabr@honam.ac.kr
[2] Dept. of Computer Eng., Mokpo Univ., Korea
[3] Dept. of Information Security, Mokpo Univ., Korea
{kwpark, jhseo}@mokpo.ac.kr

Abstract. The weak foundation of the computing environment caused information leakage and hacking to be uncontrollable. Therefore, dynamic control of security threats and real-time reaction to identical or similar types of accidents after intrusion are considered to be important. As one of the solutions to solve the problem, studies on intrusion detection systems are actively being conducted. To improve the anomaly intrusion detection system using system calls, this study focuses on techniques of neural networks and fuzzy membership function using the Soundex algorithm which is designed to change feature selection and variable length data into a fixed length learning pattern. That is, by changing variable length sequential system call data into a fixed length behavior pattern using the Soundex algorithm, this study conducted neural networks learning by using a back-propagation algorithm and fuzzy membership function. The proposed method and N-gram technique are applied for anomaly intrusion detection of system calls using Sendmail data of UNM to demonstrate its performance.[†]

1 Introduction

As recent information and communication infrastructure is based on an open structure of the Internet, it is hard to assure service quality and management of the network, and due to weak infrastructure, the network is exposed to such threats as hacking and information leakage. Various methods such as intrusion blocking, and authorization and access control to control problems causing damage to computer systems through illegal or intentional access have been suggested, but are not satisfying. Therefore, dynamic control of security threat and real-time reaction to identical or similar types of accidents after intrusion

---

[†] This research was supported by the Ministry of Information and Communication, Korea, under the Information Technology Research Center support program supervised by the Institute of Information Technology Assessment.

are considered to be important. As one of the solutions to solve the problems, studies on intrusion detection system are actively being conducted.

The host-based anomaly intrusion detection is categorized into enumerative type, frequency-based type, data mining access type, and finite state machine type. The enumerative type detects unknown patterns by tracing normal behaviors based on experiences. The frequency-based type detects intrusion based on frequency distribution of various events. The data-mining type detects intrusion by finding features from common elements that occur in normal behavior data and describing them as a group of rules. And the finite state machine-type, one of machine learning techniques, detects anomaly intrusion by a finite state machine that recognizes through trace of programs[1]. Many detection systems based on the supervisor learning separate learning from detection. Therefore, for intrusion detection, a course of learning should necessarily be given and much expenses are needed to achieve stable performance. The collection and classification of a great amount of learning data are very difficult and the performance of the detection system depends on the quality of learning data. It is very difficult for many algorithms that have been currently used for re-intrusion detection to perform a great amount of data processing and gradual learning at the same time. And it is also difficult for them to detect the intrusion and provide information on intrusion types[2,3,4,5].

This study is to apply a Soundex algorithm to solve the problem in variable length system call data that are used for learning in an intrusion detection system based on supervisor learning neural networks. It is thought that the Soundex algorithm makes simple learning algorithm possible and decrease complexity of learning through transforming variable length data into a fixed length pattern. To detect the host-based intrusion, sessions are identified by process ID, and using a system call, the behavior pattern of host is transformed by the Soundex algorithm as follows: transformation of variable length data into a fixed length pattern. This study is to profile normal behaviors using a normal behavior pattern, and detect abnormal behaviors by the back-propagation learning of neural networks and the Neuro-Fuzzy.

2 Related Works

The soundex is a combination word of sound and index. When customer service is processed by phone calls as in airline companies, some problems such as inarticulate pronunciation or wrong search of customers' names often occur. Though such problems do not happen, in case that there are a great number of customers' names saved in database, a linear search that customers' names are checked one by one needs excessive time and efforts. To solve the problem, Margaret K. Odell and Robert C. Russell developed the Soundex algorithm. It has been used for personal record of U.S army and demographical research. And it has been used for the engines of spell checkers and by Ancestor search website. The Soundex algorithm is composed of the following four rules: Rule 1 saves the first letter of the name and removes a, e, i, o, u, w, y out of the remaining letters except the first letter. Rule 2 gives the following numbers to the letters existing in the name : b, f, p, v : 1, c, g, j, k, q, s, x, z : 2, d, t : 3, l : 4, m, n : 5, r : 6. Rule 3 removes sequentially neighboring letters in the name except the first letter. Rule 4 omits the remaining ones when numbers are above three to arrange in a sequence of letter, number, number and number, and when numbers are below three, 0 is put to the last to complete the form[6].

The assumption of the program behavior-based intrusion detection system is that most of the intrusions may occur due to program defects or bugs, and the behavior is abnormal compared with normal use of the program. Therefore if program's behavior is properly expressed, it can be used as a behavior feature for detection. A representative study for automatic collection and definition of normal behaviors of programs is N-gram technique developed by Forrest research team in the University of New Mexico. This is an example adopting the concept of immunology for detection[7, 8]. The N-gram technique constructs a profile database using system call sequences with a certain length produced by programs, that is, N-gram or a string N-gram. After construction of profile database, if there is no a series of system calls with a specific length within the system calls generated by the programs, it is considered as anomaly behavior to be counted. If the proportion of the numbers of strings that are considered as anomaly is very high, the session is judged as anomaly. The N-gram has a high detection rate using a simple algorithm, but it has a disadvantage that the size of profile data and overhead is very big.

3 Back-Propagation NN and Neuro-Fuzzy Using Fixed Pattern Transformation

To detect anomaly intrusion in a system call base, this study applied techniques of neural networks and fuzzy membership function using a Soundex algorithm. To detect host-based intrusion using system calls, first, we classified sessions by a process ID. The session is a unit of behaviors, and one session is transformed into a fixed length behavior pattern. In use of a normal system call data with variable length, a normal behavior pattern with a fixed length is generated, followed by construction of a normal behavior profile. Through a normal behavior pattern, techniques of the neural network and the fuzzy membership function were performed to detect anomaly intrusion detection. This chapter consists of fixed length pattern transformation section, and fuzzy membership function generation and learning of back-propagation NN and Neuro-Fuzzy section.

3.1 Fixed Length Pattern Transformation

For host-based intrusion detection, the system call information of the host was used. This study used system call information to profile a normal behavior and detect intrusions through neural network techniques. For session division by process ID using system call information, the size of the session is variable, not stable. The kinds of system calls used in sessions ranged from a minimum of 2 to a maximum of 40. And the size of the sessions variably ranges from a minimum of 7 to a maximum of 31927. For the variable length session data, data processing is difficult and application as a learning pattern for neural network learning is also difficult as well. This study intended to construct a fixed pattern profiles while maintaining session information through application of the Soundex algorithm in a variable session made of system call data. To apply variable length data that are components of a session to neural networks learning techniques, first, feature selection to generate fixed length patterns is necessary. The fixed length pattern vectors are generated by selection of three features such as size of session, kinds of system calls, and system calls arrangement. Of the features selected, arrangement field is composed of 40 items in conformity with the Soundex algorithm. As a result of the examination of all the behavior patterns, it was found that system calls are 182 kinds and system calls used in UNM sendmail data are 53 kinds. The total number of system calls in a session used were 40 or less. Therefore, size of the arrangement field is decided to be 40. To apply soundex algorithm for fixed 40 length arrangement pattern transformation modificated alphabet sounds into system calls number. The fixed length patterns to describe session profiles of host, is presented in Figure 1.

The training pattern to be used in neural networks learning is normal and classified as a trace pattern to be used for evaluation after learning. System calls used in a normal pattern were 53 kinds and the ones for a test pattern were 43 kinds. The sessions of the normal pattern were 199 and the ones of the test pattern were 10. For the training pattern, 199 normal behavior patterns were used to perform a neural networks learning techniques.

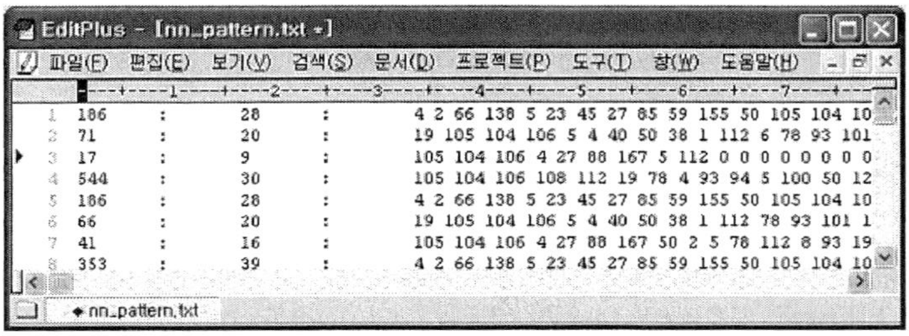

Fig. 1. Fixed length pattern transformation of UNM sendmail data using Soundex Algorithm

3.2 Fuzzy Membership Function Generation and Learning of Back-Propagation NN and Neuro-Fuzzy

In this section, the fixed length patterns of Host based normal behavior apply learning algorithms of Back-propagation NN and Neuro-Fuzzy. Neural network is a field of artificial intelligence by which mechanism of brain activity is mathematically reproduced. The neural network imitates the brain of humans to learn intelligent ability and constructs knowledge base of computers. Using the

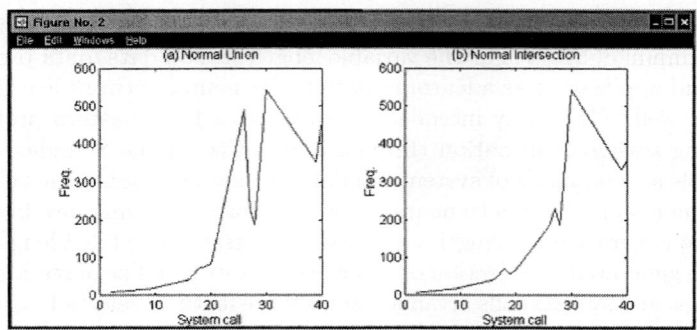

Fig. 2. Union and intersection of fuzzy member function

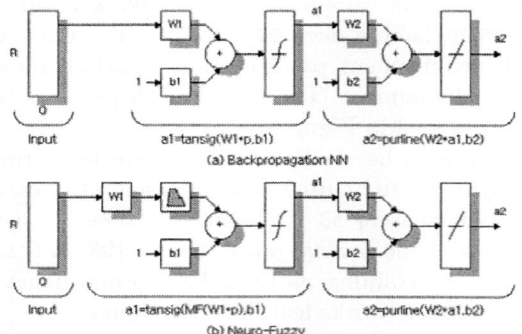

Fig. 3. Software models of BPN and Neuro-Fuzzy

constructed knowledge base, it infers through the given data, and predicts and explains the results. The back-propagation algorithm is the most common supervisor learning technique to be considered as non-linear expansion of the least mean square algorithm. Figure 3 shows a learning model using 2 layers back-propagation neural network software model and Neuro-Fuzzy software model.

In Neuro-Fuzzy software model, The session data of normal behavior make a fuzzy membership function for host based normal behavior. A Fuzzy set is a set without a crisp, clearly defined boundary. It can contain elements with only a partial degree of membership. A Membership Function(MF) is a curve that defines how each point in the input space is mapped to a membership value(or degree of membership) between 0 and 1.

In NF Anomaly Intrusion Detection, the fuzzy MF was generated using normal behavior patterns. To generate fuzzy MF uses the kinds of system calls and their frequency. The fuzzy MF was generated by the kinds of system calls, which were used in sessions ranged from a minimum of 2 to a maximum of 40. The union and intersection operation of system calls membership function was presented in Figure 2. Almost operation of fuzzy sets was performed by Max and Min operations. The MF of union and intersection was presented the max and min value of system calls frequency. If frequency of a session exists between upper and lower boundary, pattern vector was converted scaled value by fuzzy membership function, or else not.

In Table 1, Eq. (1) - (9) are presented in the order in which they would be used during training for fixed length normal patterns. when the error is acceptably small for each of the norml patterns, training can be discontinued.

The Neuro-Fuzzy model of 2 layers represent (b) in Figure 3. Eq. (1) in the backpropagation learning algorithm corrected Eq. (10) in Table 1. the net input was scaled by the fuzzy membership function in Figure 2. The fuzzy membership function was applied to hidden layer of neural networks. That is, the fuzzy membership function supported the many information to neural networks learning.

Table 1. Learning algorithms of BPN and NF

Section	Expression of Learning Algorithms
Back-propagation NN	$net_{pj}^h = \sum_{i=1}^{N} w_{ji}^h x_{pi} + \theta_j^h$ (1) $i_p^h = f_j^h(net_{pj}^h)$ (2) $net_{pk}^o = \sum_{j=1}^{L} w_{kj}^o i_{pj} + \theta_k^o$ (3) $o_{pk} = f_k^o(net_{pk}^o)$ (4) $\delta_{pk}^o = (y_{pk} - o_{pk}) f_k^{o'}(net_{pk}^o)$ (5) $\delta_{pj}^h = f_j^{h'}(net_{pj}^h) \sum_k \delta_{pk}^o w_{kj}^o$ (6) $w_{kj}^o(t+1) = w_{kj}^o(t) + \eta \delta_{pk}^o i_{pj}$ (7) $w_{ji}^h(t+1) = w_{ji}^h(t) + \eta \delta_{pj}^h x_i$ (8) $E_p = \frac{1}{2} \sum_{k=1}^{M} \delta_{pk}^2$ (9)
Neuro-Fuzzy	$net_p^h j = \sum_{i=1}^{N} MF(w_{ji}^h x_{pi}) + \theta_j^h$ (10)

4 Simulation and Analysis

To detect system call anomaly intrusion, this study applied techniques of neural networks and fuzzy membership function using a Soundex algorithm and a N-gram technique. To construct a normal behavior profile, we constructed a profile using N-gram technique and the one using a Soundex algorithm and techniques of neural networks and fuzzy membership function and then compared performances between the three models.

4.1 N-Gram Technique

A normal behavior pattern was generated by application of N-gram technique for normal behavior data. With indicating the size of window of N-gram technique, being changed, intrusion detection rates were compared by trace data. First, through a construction of the profiles of normal behaviors, intrusion behaviors and trace behaviors, the results were obtained as presented in Table 2. If window size N was increased, the number of normal patterns decreased except part of them, and if redundancy of the patterns was excluded, the number of patterns tended to increase. And through comparison of intrusions and trace data based on normal behavior data, the redundancy of the patterns which do not exist in normal behavior was removed.

As window size N increased, the number of the patterns detected in intrusion and trace data increased, but the number of redundancy-removed patterns also increased. In particular, in case that N=4, the number of redundancy-removed patterns in intrusion and trace data suddenly increased, and the largest value was obtained. With window size N being changed from 3 to 10, N with the largest number of undetected anomaly patterns compared with normal data was optimal. That the number of the patterns with redundancy removed is large means that there are more anomaly detection information to be differentiated from normal data. With window size N of normal data, being changed, anomaly detection of the trace data composed of ten sessions was performed. As the size of N in N-gram increased from 3 to 10, the number of the detected patterns increased. When the size of window increased from 3 to 4, the detection rate increased from 80% to 90%. Of ten sessions of trace data with N=3, the two sessions were undetected, but with N=4, only one session was undetected. Intuitively, for the N-gram technique, when window size was more than four, the highest detection rate, 90%, was obtained. Though N was increased more, the detection rate did not increase any longer, and only the number of the detected anomaly patterns increased.

Table 2. Number of the anomaly patterns according to window size of normal, intrusion and trace data

windows size (N)	the number of normal patterns	redundancy removed patterns		
		normal	intrusion	trace
3	809,997	440	21	18
4	227,584	570	176	177
5	227,385	693	55	38
6	227,186	811	66	46
7	226,987	910	72	55
8	226,788	996	78	64
9	226,640	1076	84	73
10	326,179	1153	90	80

4.2 Techniques of BPN and NF

For back-propagation and Neuro-Fuzzy learning, the number of neurons in a hidden layer was changed from 10 to 40 to investigate learning rate and errors. To overcome over-fitting and under-fitting, which are disadvantages of neural network learning, the number of the neurons in the hidden layer should be decided. The over-fitting means learning even noise including learning data, and under-fitting means learning is not perfectly achieved. For back-propagation and Neuro-Fuzzy learning, the number of the neurons in a hidden layer was decided as 12 and learning of 199 normal behavior patterns was progressed. The normal behavior pattern generated system call data as 42 items of the learning pattern by a Soundex algorithm, and learning was performed with 0.01 of error rate, 0.2 of learning rate and 5000 epoch times or fewer. The back-propagation learning is achieved by 428 epoch. And the Nero-Fuzzy learning is achieved by 1776 epoch. Figure 4 and 5 show the results of detection by inputting intrusions and trace data to the learned back-propagation neural networks and Neuro-Fuzzy.

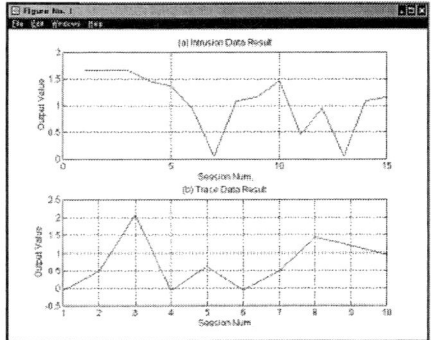
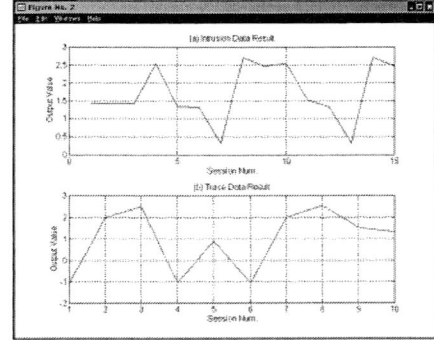

Fig. 4. Output Value of BPN in Intrusion and Trace Data

Fig. 5. Output Value of Neuro-Fuzzy in Intrusion and Trace Data

4.3 Comparison of Anomaly Detections Between Neural Networks Techniques and N-Gram Technique

This study simulated an anomaly detection from system call data sets of Sendmail Deamon by UNM by neural networks techniques using Soundex algorithm and a N-gram technique. The system call data were transformed into 42 items of learning patterns by Soundex algorithm and learning was performed. And for N-gram technique, window size was changed from 3 to 10 to detect anomaly and then the results were compared as in Table 3.

MDL(Minimum Description Length)[9] is composed of the loss of errors and the loss of complexity. MDL is a more effective model as it has the less value. Table 3 compares the N-gram, Back-propagation neural networks and Neuro-Fuzzy technique by MDL. It was demonstrated that with N=4, the N-gram technique was the most effective. However, when the Back-propagation neural networks and Neuro-Fuzzy techniques were compared with N-gram technique with N=4, their detection rates were identical, but in an aspect of model complexity, it was demonstrated that the Back-propagation neural networks and Neuro-Fuzzy techniques was more effective. When window size of N-gram was changed from 3 to 4 and 10, detection rates were 80%, 90% and 90%. And as window size of N-gram increased, space to describe patterns and time to process patterns increased. However, when a Soundex algorithm and neural networks were used, detection rate was 90%. As detection of the suggested neural networks techniques was compared with that of N-gram technique with 3 of window size, the suggested neural networks techniques was absolutely superior in detection rate and complexity aspect. With window size changed from 4 to 10, detection rates of the N-gram, the back-propagation neural networks and Neuro-Fuzzy techniques were same, but the suggested neural networks techniques was superior in time and space complexity aspects. The back-propagation neural networks and Neuro-Fuzzy were same detection rate of 90%, and Neuro-Fuzzy technique needs many learning epoch than the back-propagation neural networks. But Neuro-Fuzzy technique shows that generate scaled output values to discriminate anomaly patterns from normal patterns.

Table 3. Comparative Analysis of the N-gram, BPN and Neuro-Fuzzy

Items		N-gram				BPN	Neuro-Fuzzy
		3	4	6	10		
repetition	# of pattern	809,997	227,584	227,186	326,179	199	
	Data amount	9.6MB	3.47MB	4.97MB	12.08MB	22KB	
repetition remove	# of pattern	440	570	811	1,153	41	
	Data amount	5KB	7KB	14.4KB	34KB	5KB	
MDL	Error	0.2	0.1	0.1	0.1	0.1	
	Complexity	0.997	0.999	1.000	1.000	0.999	
	Total	1.197	1.099	1.100	1.100	1.099	
Epoch #		-	-	-	-	428	1776
Detection Rate		8/10	9/10	9/10	9/10	9/10	9/10

5 Conclusion

This study applied the Soundex algorithm to solve the problems of variable length data to be used for detection system using the neural network techniques of supervisor learning, a machine learning. By transformation of variable length system call data into a fixed length patterns using the Soundex algorithm, learning algorithm of neural networks techniques could be simple, and complexity in space and time required for learning aiming at intrusion detection could be overcome. To detect host-based anomaly intrusion, first, we classified sessions, and generated hosts' behavior patterns by transforming the variable length data into a fixed length pattern. For normal behavior pattern, we detected anomaly behaviors by learning normal behavior patterns using back-propagation neural networks and Neuro-Fuzzy of supervisor learning. By solving difficulties of a variable length data processing, a learning algorithm became simple and complexity in space and time for learning was overcome, which contributed to improvement of anomaly intrusion detection. This study used Sendmail data sets of UNM for simulation. From the Sendmail deamon behaviors, three features such as length of session, kinds of system call and arrangement of system calls used were selected. Compared with the N-gram technique under the condition that neural networks and window size were 3, the suggested back-propagation neural networks and Neuro-Fuzzy techniques showed a higher detection rate, but when window size was changed from 4 to 10, the detection rate of the suggested method was the same as that of the N-gram technique, which was 90%. However, in the complexity of time and space for algorithm performance, intrusion detection of back-propagation neural networks and Neuro-Fuzzy using a Soundex algorithm was superior. And Neuro-Fuzzy technique needs many learning epoch than the back-propagation neural networks. But Neuro-Fuzzy technique shows that generate scaled output values to discriminate anomaly patterns from normal patterns.

References

1. Christina Warrender, Stephanie Forrest, Barak Pearlmutter, "Detecting Intrusion Using System Calls : Alternative Data Models", 1998.
2. Leonid Portnoy, "Intrusion detection with unlabeled data using clustering", Undergraduate Thesis, Columbia University, 2000.
3. Jack Marin, Daniel Ragsdale, and John Shurdu, "A Hybrid Approach to the Profile Creation and Intrusion Detection", Proceedings of DARPA Information Survivability Conference and Exposition, IEEE, 2001.
4. Nong Ye, and Xiangyang Li, "A Scalable Clustering Technique for Intrusion Signature Recognition", Proceedings of 2001 IEEE Workshop on Information Assurance and Security, 2001.
5. Wenke Lee, Salvatore J. Stolfo, Philip K. Chan, Eleazar Eskin, Wei Fan, Matthew Miller, Shlomo Hershkop, and Junxin Zhang, "Real Time Data Mining - based Intrusion Detection", IEEE, 2001.
6. http://www.archives.gov/research_room/genealogy/census/soundex.html
7. S. Forrest, S. Hofmeyr, A. Somayaji and T. Longstaff, "A sense of self for unix processes", In IEEE Symposium on Security and Privacy, pp.120-128, 1996.
8. Steven A. Hofmeyr, Stephanie Forrest, Anil Somayaji, "Intrusion Detection using Sequences of System Calls", Journal of Computer Security, Vol.6, pp.151-180, August 18, 1998.
9. Christopher M. Bishop, "Neural Networks for Pattern Recognition", Oxford University Press, p.429-433, 1995.
10. A. Wespi, M. Dacier and H. Debara, "Intrusion detection using variable-length audit trail patterns", Recent Advances in Intrusion Detection(RAID 2000), pp.110-129, 2000.
11. Matthew V. Mahoney and Philip K. Chan, "Learning Nonstationary Models of Normal Network Traffic for Detecting Novel Attacks", 2002.
12. D. Anderson, T. Lunt, H. Javitz, A. Tamaru, and A. Valdes. "Detecting unusual program behavior using the statistical component of the next-generation intrusion detection expert system (nides)", In Technical Report SRI-CSL-95-06, SRI, 1995.

The Role of Secret Sharing in the Distributed MARE Protocols*

Kyeongmo Park

School of Computer Science and Information Engineering,
The Catholic University of Korea, Yeouido P.O. Box 960,
Yeongdeungpo-Gu, Seoul, 150-010, Korea
kpark@catholic.ac.kr

Abstract. In a mobile agent based distributed system, agents must survive malicious failures of the hosts they visit, and they must be resilient to the potentially hostile actions of other hosts. The replication and voting are necessary to survive malicious behavior by visited hosts. However, faulty hosts that are not visited by agents can confound a naive replica management scheme by spoofing. This problem can be solved by cryptographic protocols. This paper describes the role of cryptographic methods in the protocols for the MARE architecture, which is a fault-tolerant mobile agent replication system. In this system, secret sharing takes on an important role in facilitating mobile processes by distributed authentication.

1 Introduction

Without giving a formal definition, a software agent is a piece of program code that can execute autonomously without the supervision of a central authority. Intelligent agents are also capable of interacting and learning from their environment and can react to change in their environment. Mobility is also attractive feature of software agents as it allows an agent to move a remote location and continue its thread of execution on the remote host machine. Mobile agents are particularly attractive for designing distributed and decentralized applications as they can reduce the processing time and network bandwidth usage by moving the code closer to the data located on a remote host. They are sent by end-users and visit a series of hosts. The mobile agents are executed locally on these hosts to perform their tasks, and will return to the end-users to report their results.

However, the autonomy of a mobile agent on the remote sites can be used maliciously either by creator of the agent or by other entities to subvert and sabotage the entire system. For example, rogue mobile agents can be used encapsulate apparently harmless code that is capable of installing viruses on the sites that the mobile agent visits. Mobile agents have been proposed for a variety of applications in the Internet and other large distributed systems [2], [6], [8], [10].

---

* This work was supported by the Catholic University of Korea Research Fund 2005.

A mobile agent differs from a traditional operating system process. Unlike a process, a mobile agent knows where it is executing in a distributed system at any point in time. A mobile agent is aware of the communication network and makes an informed decision to move asynchronously and independently from one node to another during execution.

In a process-migration system the system decides when and where to move the running to balance workload, whereas the agents move when they choose through a jump or go statement. It has been shown in [9] that agent migration is much faster than migration of traditional processes. Mobile agents are a very attractive paradigm for distributed computing over the Internet, for several reasons [8], including reducing vulnerability to network disconnection and improvements in latency and bandwidth of client-server applications. Mobile agents carry the application code with them from the client to the server, instead of transferring data between a client and a server. Since the size of the code is often less than the amount of data interchanged between the client and the server, mobile agent system provide considerable improvement in performance over client-server computing. Thus, the use of mobile agents is expanding rapidly in many Internet applications [2], [6], [8].

The Internet is unreliable. Hosts connected via the Internet constantly fail and recover. The communication links go down at any time. Due to high communication load, link failures, or software bugs, transient communication and node failures are common in the Internet. Information transferred over the Internet is insecure and the security of an agent is not guaranteed. Fault tolerance guarantees the uninterrupted operation of a distributed software system, despite network node failure. Therefore, reliability is an important issue for Internet applications [2], [10], [13].

In this paper, we address the security and fault tolerance issues for mobile agent systems running across the Internet. We present a cryptographic method in the replication and voting protocols for the agent replication extension system. The system makes mobile agents fault-tolerant and also detects attacks by malicious hosts.

The rest of this paper is organized as follows. Section 2 present a fault-tolerant mobile agent replication system and discusses the replication and voting protocols. Section 3 discusses the role of cryptographic techniques in our protocols. Section 4 describes experiments we ran to explore performance of replication and voting in the presented system setting. Finally, our conclusions are presented in Section 5.

2 Dependable Distributed Computing

2.1 Fault-Tolerant Mobile Agents

Fault tolerance for mobile agent systems is an unsolved topic in dependable distributed computing, to which more importance should be attached. Our approach offers a user-transparent fault tolerance in agent environments. The user can select a single application given to the environment and can decide for every application whether it has to be treated fault-tolerant or not. That is, the user or the application itself can decide individually, if and when fault tolerance is to be activated. The execution of fault-tolerant and non-fault-tolerant applications is possible. Thus, to enable fault-tolerant execution, it is not necessary to change the application code. The separation between application and agent kernel platform facilitates user transparency. Once mobile agents are injected into the network, the users do not have much control over their execution. If the action for fault tolerance was dictated by a monitor instance, the autonomy was limited. All decisions that are made by an autonomous agent would need to be coordinated with the monitor. To enable activation of fault tolerance during runtime, the complete agent is replaced with one that carries those functionalities with it. This would increase demands for memory and computing time. So a modular exchangeable composition of mobile agents is required. The required modularity and separation between the application and agent platform imply that the functional modules should wok independently and in parallel. The application can influence the agent behavior. It is possible to affect the behavior of a mobile agent during runtime.

2.2 Replication and Voting

Replication of agents and data at multiple computers is a key to providing fault tolerance in distributed systems. Replication is a technique used widely for enhancing Internet services. The motivations for replication are to make the distributed system fault-tolerant, to increase its availability, and to improve a service's performance.

One of the goals in this work is to provide fault tolerance to mobile multi-agents through selective agent replication. Multi-agent applications reply on the collaboration among agents. If one of the involved agents fails, the whole computation can get damaged. The solution to this problem is replicating specific agents. One must keep the solution as independent and portable as possible from the underlying agent platform, so as to be still valid even in case of drastic changes of the platform. This offers interoperability between agent systems. The properties of agent systems are dynamic and flexible. This increases the agent's degree of proactivity and reactivity. Note that replication may often be expensive in both computation and communication. A software element of the application may loose at any point in progress. It is important to be able to go back to the previous choices and replicate other elements.

In the passive model of replication, there is a single 'primary' or 'master' replica manager (RM) at any time and one or more secondary RMs - 'backups (slaves)'. Front-end(FE)s communicate only with the primary RM to obtain the service. The primary RM executes the operations and sends copies of the updated data to the backups. If the primary fails, one of the backups is promoted to act to the primary. The passive replication system implements linearizability if the primary is correct, since the primary sequences all the operations upon the shared objects. If the primary fails, then a backup becomes the new primary and the new system configuration takes over: the primary is replaced a unique backup and the RMs that survive agree on which operations had been performed when the replacement primary takes over. The passive model is used in the Sun NIS (Network Information Service), where the replicated data is updated at a master server and propagated from the master to slave servers using one-to-one rather than group communication. In NIS, clients communicate with either a master or slave server but they may not request updates. Updates are made to the master's files.

In the active model, the RMs are state machines that play equivalent roles and are organized as a group. Front-ends multicast their requests to the group of RMs and all the RMs process the request independently but identically and reply. If any RM crashes, then this need have no impact upon the performance of the service, because the remaining RMs continue to respond in the normal way. Schneider [15] proposes active replication with majority voting to obtain a consensus on the computation performed by a set of replicated nodes. This active replication system achieves sequential consistency. All correct RMs process the same sequence of requests. The reliability of multicast ensures that they process them in the same order. Since they are state machine, they all end up with the same state as one another after each request. Front end's requests are served in FIFO order, which is the same as program order. The active system does not achieve linearizability. This is because the total order the RMs process requests is not necessarily the same as the real-time order the clients made their requests. We assume a solution to reliable and totally ordered multicast.

A simple agent computation might visit a succession of hosts, delivering its result messages to an actuator. The difficulties here arise in making such a computation fault-tolerant. The agent computation of interest can be viewed as a pipeline, depicted in the shaded box of Fig. 1. Nodes represent hosts and

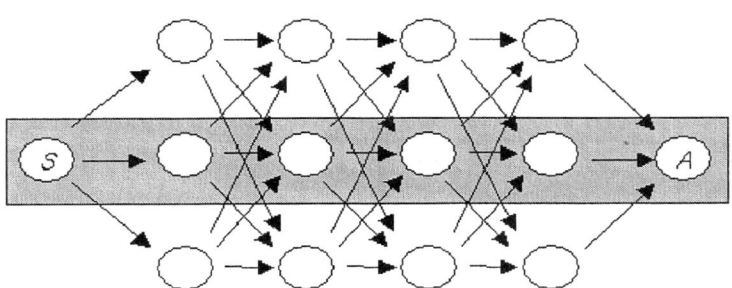

Fig. 1. Fault-tolerant agent computation using replication and voting

edges represent movement of an agent from one host to another. Each node corresponds to a *stage* of the pipeline. *S* is the *source* of the pipeline; *A* is the *actuator*. The computation is not fault-tolerant. The correctness of a stage depends on the correctness of its predecessor, so a single malicious failure can propagate to the actuator. Even if there are no faulty hosts, some other malicious host could disrupt the computation by sending an agent to the actuator first.

One step needed to make fault-tolerant is replication of each stage. We assume that execution of each stage is deterministic, but the components of each stage are not known *a priori* and they depend on results computed at previous stages. A node m in stage k takes as its input the majority of the inputs it receives from the nodes comprising stage $k-1$. And then, m sends its output to all of the nodes that it determines consisting of $k+1$. Fig. 1 illustrates such a fault-tolerant execution. The replicated agent computation with voting tolerates more failures than an architecture where the only voting occurred just before the actuator. The voting at each stage makes it possible for the computation to recover by limiting the impact of a faulty host in one stage on hosts in subsequent stages.

2.3 The MARE System Architecture

The MARE[17] architecture is a Mobile Agent Replication Extension system with voting that makes mobile agents fault-tolerant and reliable. The system (Fig. 2) is similar to several other agent systems including Agent Tcl [5], DaAgent [10], DarX [7], FANTOMAS [11], and FATOMAS [12]. The Replication Group (RG) consists of multiple Agent Replication Tasks (ARTs). The system provides group membership management to add or remove replicas. The number of replicas and the internal details of a specific task are hidden from the other tasks. Each RG has exactly one master communicating with the other ART tasks. The master acts as a fixed sequencer, providing totally ordered multicast within its RG.

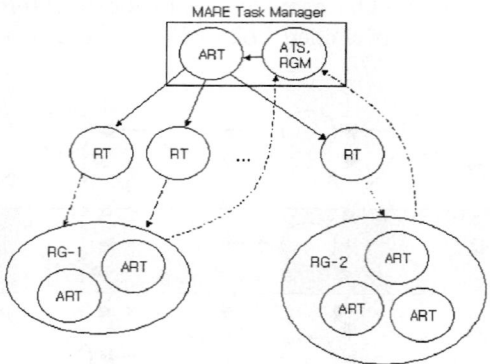

Fig. 2. The fault-tolerant mobile agent replication extension system

Agents are allowed to inherit the functionalities of other ART objects, enabling the underlying system to handle the agent computation and communication. Therefore, it is possible for MARE to act as a middleware for agents. In Fig. 2, each ART is wrapped in an Application Task Shell (ATS) that acts as a Replication Group Manager (RGM), and is responsible for delivering messages to all the members of the RG. RGM is associated each agent (ART). It keeps track of all the replicas in the group, and of the current replication method in use. RGM can change the replication policy and tune its parameters, such as the number of replicas or the periods between backups in case of passive replication. ATS intercepts input messages and enables caching. All messages are processed in the same order within a RG. When an agent is replicated, its RG is suspended and the corresponding ART is copied to a new ATS on the requested host system. A task can communicate with a Remote Task (RT) by using a local proxy with the RT interface. Each RT references a distinct remote entity considered as the master of its RG.

MARE system uses both passive and active replication schemes. It is possible to switch to any user-defined replication method. MARE uses a fault tolerance mechanism to detect attacks by malicious hosts. It is assumed for every stage, i.e., an execution session on one host, a set of independent replicated hosts, i.e., hosts that offer the same set of resources, but do not share the same interest in attacking a host, because they are operated by different organizations. Every execution step is processed in parallel by all replicated hosts. After execution, the hosts vote about the result of the step. At all hosts of the next step, the votes (the resulting agent states) are collected. The execution with the most votes wins, and the next step is executed.

3 Cryptographic Support for MARE

The computation of Fig. 1 should tolerate one malicious host per stage of the pipeline. It does not. Any two faulty hosts could claim to be in the last stage and foist a majority of bogus agents on the actuator. These problems are avoided if the actuator can detect and ignore such bogus agents. This could be accomplished by having agents carry a privilege from the source to the actuator.

The privilege can be encoded as a secret initially known to the source and the actuator. It is necessary to defend against two attacks. First, a malicious host might misuse the secret and launch an arbitrary agent to the actuator. Second, a malicious host could destroy the secret, making it impossible for the remaining correct agents to deliver a result.

To simplify the discussion, we start with a scheme that prevents misuse of the secret by a malicious host. This scheme ensures that if a majority of replicas visit only correct hosts, no hosts except the source and the actuator will learn the secret. Agent replicas cannot carry copies of the secret, since the secret could then be stolen by any faulty host visited by a replica. It is tempting to circumvent this problem by the use of an (n,k) threshold secret sharing scheme[14] to share the secret embodying the privilege. In an (n,k) threshold scheme, a secret is divided into n fragments, where possession of any k fragments will reveal the secret, but possession of fewer fragments reveals nothing. In a system having *2k-1* replicas, the source would create fragments using a *(2k-1, k)* threshold scheme, and send a different fragment to each of the hosts in the first stage. Each of these hosts would then forward its fragment to a different host in the next stage.

However, this protocol fails, due to the voting at intermediate stage of the pipeline. The voting should ensure that the faulty hosts encountered before a vote cannot combine with faulty hosts encountered after the vote to corrupt the computation. However, in the scheme, minorities before and after the vote can steal different subsets of the secret fragments. If they together hold a majority of the secret fragments, then the faulty hosts can collude to reconstruct the secret.

One way to stop collusion between hosts separated by a vote is to redivide the secret fragments at each vote. For a system with *(2k-1)* replicas, the protocol based on this insight outlines, as follows.

- The source divides the secret into *2k-1* fragments and sends each fragment to one of the hosts of the first stage.
- A host in stage i takes the fragments it receives, concatenates them, and divides that into *2k-1* fragments using a *(2k-1, k)* threshold scheme. Each fragment is then sent to a different host in stage *i+1*.
- The actuator uses the threshold scheme backwards and recovers the original secret.

This protocol is inefficient because secret sharing scheme require that each fragment be the same size as the original secret. Thus, messages get longer at every stage of the pipeline. In fact, the message size is multiplied by *2k-1* at every stage.

Two protocols have been developed. They do not suffer from the exponential blowup in message size. In one protocol, message size grows linearly with the number of pipeline stages and the number of replicas. In the other one, message size remains constant but an initialization phase is required. The first scheme uses chains of authentication, instead of a secret privilege, to prevent masquerading. The second scheme renews [16] the secret after each round, making it impossible for fragments from before a vote to be used together with fragments constructed after that vote. To address the second attack - destruction of the secret by a faulty host - it is necessary to replace the secret sharing in the protocols described with verifiable secret sharing. This scheme allows for correct reconstruction of a secret even when hosts including the source are faulty.

4 Simulation Study

To explore voting performance issues, we performed experiments. The system we tested consists of 4 Sun workstations. An agent moving from node to node was simulated by sending a message between these node. In this experiment, we looked at the behavior for 1, 4, and 8 replicas and were interested in how synchronization delay can be amortized by voting less frequently. This experiment examined the cost of voting in the case that host speeds are uniform. In this experiment, agents visited a sequence of N hosts before voting, rather than voting at the end of each stage.

Fig. 3 is the graph of the average time per host visit when N ranges from 1 to 32. The data depicted reports averages from runs of 320 rounds. The time spent per host had a variance .1 percent. We found remarkable improvements as N advanced from 1 to 8. For N greater than 8, the further improvements were not significant. It is interesting to note synchronization delay versus voting tradeoff. When voting is infrequent, replica completion time drift apart, so the synchronization delay increases. A voter need wait for a correct majority, so a vote-delimited stage will complete as soon as the median correct replica votes. Therefore, the completion time for a replicated computation that votes infrequently should approximate the completion time when there is a single replica. The experimental result of Fig. 3 shows this behavior.

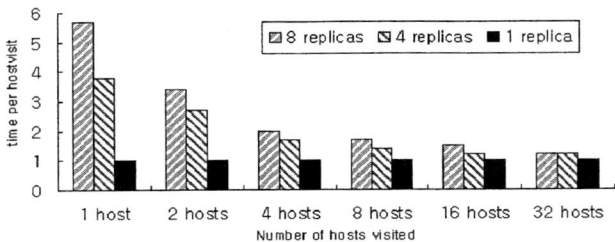

Fig. 3. Voting performance with various voting frequencies; the x-axis is the number of hosts visited between votes and the y-axis is time per host visit, normalized to speed of a single replica

Voting can lead to a replicated computation being faster than the corresponding non-replicated one. Suppose there is a small probability that any given host will be slow. Over a long non-replicated execution, an agent is bound to encounter a slow host. Accordingly, the computation will be slowed. However, with replication and periodic voting, it is likely that a majority of the agents a voter will have encountered no slow hosts. Because the voter waits for this majority, the MARE system's execution time will be independent of the speed of the slow hosts.

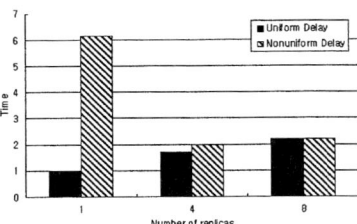

Fig. 4. Voting performance with uniform and nonuniform delays; The probability of encountering a slow host to 1 percent was set. The addition of replicas further reduces the slowdown

Fig. 4 shows the performance of an agent which voted every five moves. The probability of encountering a slow host to 1 percent was set, and such a host was 195 times slower than a normal host. The un-replicated computation experienced the full effect of the impaired hosts, resulting in a sixfold slowdown. The slowdown was reduced in the presence of any replication, dropping to 23 percents for four replicas. The addition of eight replicas reduced the negligible slowdown.

5 Conclusion

This paper describes cryptographic support for MARE, which is a replication extension system with voting. The system makes mobile agents fault-tolerant and also detects attacks by malicious hosts. Replication and voting do not suffice for the security purpose. Cryptographic protocols are required. Cryptographic techniques have been used in distributed protocols. In particular, secret sharing has been employed for asynchronous Byzantine agreement[1] and secure auctioning[3]. However, all of the work depends on computations being immobile. In the MARE system, secret sharing takes on an important role in facilitating mobile processes by providing a form of distributed authentication. As a part of the experimental studies, the effects of the voting frequencies with uniform and nonuniform delays were examined. We found that synchronization delays caused by voting could be made insignificant by making voting less frequent. In some cases, replicated computation with voting improved performance by ensuring that slow hosts do not make progress difficult.

References

1. R. Canetti and T. Rabin: Optimal asynchronous Byzantine agreement. In *25th Symposium on Theory of Computing*, 1993, 42–51.
2. P. Dasgupta: A Fault Tolerance Mechanism for MAgNET: A Mobile Agent based E-Commerce System. In *Proc. Sixth Int. Conf. Internet Computing*, Las Vegas, NV, June 24-27, 2002, 733–739.
3. M.K. Franklin and M.K. Reiter: The design and implementation of a secure auction service. In *IEEE Symposium on Security and Privacy*, 1995.
4. E. Gendelman et al.: An Application-Transparent Platform-Independent Approach to Rollback-Recovery for Mobile Agent Systems. In *Proc. Int. Conf. Distributed Computing Systems*, Taipei, Taiwan, Apr. 2000, 564–571.
5. R.S. Gray: Agent Tcl: A Transportable Agent System. In *Proc. CIKM Workshop Intelligent Information Agents*, Baltimore, MD, Dec. 1995.
6. D. Kotz and R.S. Gray: Mobile Agents and the Future of the Internet. *ACM Operating Systems Review*, 33(3): (Aug. 1999) 7–13.
7. G. Lacote et al.: Towards and Fault-Tolerant Agents. In *Proc. ECOOP'2000 Workshop Distributed Objects Programming Paradigms*, Cannes, Italy.
8. D. B. Lange and M. Oshima: Seven Good Reasons for Mobile Agents. *Communications of the ACM*, 42(3): (Mar. 1999) 88–89.
9. D.S. Milojicic: Trend Wars Mobile Agent Applications. *IEEE Concurrency*, 7(3): (1999) 80–90.

10. S. Mishra et al.: DaAgent: A Dependable Mobile Agent System. In *Proc. of the 29th IEEE Int. Symp. Fault-Tolerant Computing*, Madison, WI, June 1999.
11. H. Pals et al.: FANTOMAS, Fault Tolerance Mobile Agents in Clusters. In *Proc. Int. Par. Distr. Proc. Symp. Workshop*, May 2000, Cancun, Mexico.
12. S. Pleisch and A. Schiper: FATOMAS: A Fault-Tolerant Mobile Agent System Based on the Agent-Dependent Approach. In *Proc. IEEE Int. Conf. Dependable Systems and Networks*, Jul. 2001, Goteborg, Sweden, 215–224.
13. M. Strasser and K. Rothermel: Reliability Concepts for Mobile Agent. *Int. J. Cooperative Information Systems*, 7(4): (Dec. 1998) 355–382.
14. A. Shamir: How to share a secret. *Communications of the ACM* 22: (Nov. 1979), 612–613.
15. F. Schneider: Toward Fault-Tolerant and Secure Agentry. In *Proc. 11th Int. Workshop Distributed Algorithms*, Sep. 1997, Saarbrucken, Germany, 1–14.
16. S. Jarecki: Proactive Secret Sharing and Public Key Cryptosystems. MIT, Master's thesis, Sep. 1995.
17. K. Park and A. Sood: MARE: A Fault-Tolerant Mobile Agent System. *LNCS 3090*, 2004, Springer-Verlag, 1035–1044.

Security Risk Vector for Quantitative Asset Assessment

Yoon Jung Chung, InJung Kim, NamHoon Lee,
Taek Lee*, and Hoh Peter In*

Electronics and Telecommunication Research Institute,
Republic of Korea Daejon, South Korea
{yjjung, cipher, nhlee}@etri.re.kr
* Department of Computer Science and Engineering,
Korea University, Seoul, South Korea
{comtaek, hoh_in}@korea.ac.kr

Abstract. There are standard risk analysis methodologies like GMITS and ISO17799, but new threats and vulnerabilities appear day by day because the IT organizations, its infrastructure, and its environment are changing. Accordingly, the methodologies must evolve in step with the change. Risk analysis methods are generally composed of asset identification, vulnerability analysis, safeguard identification, risk mitigation, and safeguard implementation. As the first process, the asset identification is important because the target scope of risk analysis is defined. This paper proposes a new approach, security risk vector, for evaluating assets quantitatively. A case study is presented.

1 Introduction

Risk management includes many processes such as asset identification, threat and vulnerability analysis, safeguard implementation, and risk mitigation. After the "protective law of information communication infrastructure" was enforced in Korea on July 2001, many organizations have recognized the importance and necessity of security risk analysis. They referred many security risk analysis guidelines and methodologies such as Fips 65 [7], SRAG, and BS7799 [2]. However, they describe *what*, but no *how* in details. Thus, it is difficult for many organizations to apply the risk management techniques. Especially, asset identification and assessment, the first step of risk management, is more important to make risk management successful.

In this paper, a simple, yet effective asset risk assessment method of information systems, called *Security Risk Vector*, is proposed. The information systems composed by data, application, and server are quantitatively evaluated with criteria such as confidentiality, integrity, and availability. These three criteria are represented as vectors. This method assists to understand the criticality of assets.

Section 2 presents the overview of risk analysis methodologies as the context of our research. Section 3 proposes our approach of asset assessment. Section 4 and 5 shows a case study and related work respectively. Section 5 concludes this paper.

2 The Context: Risk Analysis Process

Fig.1 presents a standard risk analysis process based on OCTAVE [1], which is developed by Carnegie Mellon University (CMU) in USA. It is used as a base of our risk analysis. It includes asset identification, threat analysis, vulnerability analysis, damage impact analysis, risk assessment, and safeguards for controlling risk.

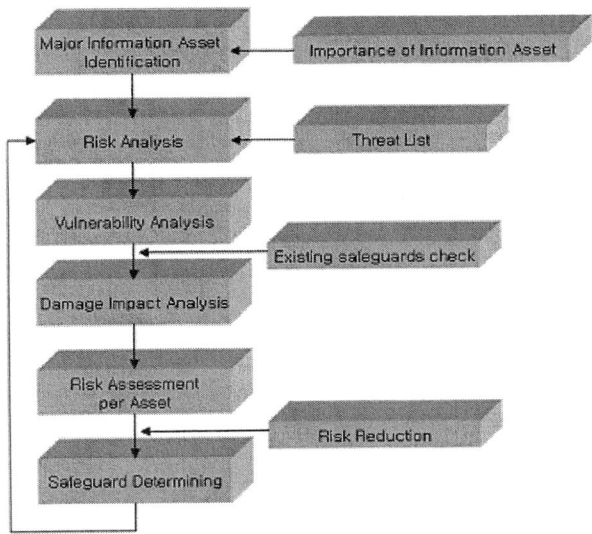

Fig. 1. Risk Analysis Process

Step 1. Major Asset Identification: In order to analyze a target information system, as the first step, core business processes and core assets are investigated. In this paper, we propose a new quantitative approach of asset assessment using *Security Risk Vector*. The detailed of our quantitative asset assessment is explained in Section 3.

Step 2. Threat Analysis: Threat analysis of the identified assets is needed based on the predefined threat list. The threat analysis of data assets investigates not only data, but also the server and the database, which store and distribute the data. Therefore, threat analysis for the server and the database should be also performed simultaneously with one for the data assets.

Step 3. Vulnerability Analysis: The diagnosis tools of network and host vulnerability can be used to perform vulnerability analysis. Vulnerability analysis is performed for the server and database assets, but not for data asset. Furthermore, vulnerabilities of applications used in server should be analyzed as well as the database program.

Step 4. Damage Impact Analysis: Damage will be evaluated based on the results of threat and vulnerability analyses. Safeguards and their appropriateness are investigated and analyzed in this process. The validation of the new safeguard applied in the future need to be also checked.

Step 5. Risk Assessment per each asset: Risk of each asset is calculated and then the assets are prioritized in order.

Step 6. Safeguard Determining: Through threat analysis and vulnerability analysis, safeguards to control identified risk need to be set up, and then after determining the target risk level, the final decision of risk safeguard is made.

Among these above general processes for risk analysis, we will discuss assess assessment process in the rest of this paper.

3 Proposed Method to Evaluate Information System Asset

The asset assessment process is briefly explained based on the international standard ISO 17799, improved by BS7799[2]. The proposed method to assess information system assets is presented in this paper including analyzing information security requirements, understanding criticality of asset, and checking sensitivity for data asset.

3.1 Asset Identification to Be Protected

The British BS7799 suggests the asset classification as follows: [2]

- Information Asset: DB, data file, system document, user manual, study and training materials, regulations for management, plan document, provision for alternative system
- Documents: contracts, guidelines, company documents, important business documents
- Software Asset: applications S/W, system S/W, development tool and utility
- Physical Asset: computer and communication equipment, magnetic tape, magnetic disk, power supply, air conditioner, furniture, facilities
- Personnel Asset: individuals, customer, subscriber
- Image and Reputation of a Company
- Service: computer and communication service, warm, light, air conditioning

This paper doesn't deal with all the listed assets in an organization but assets related only to major information system. Thus, we consider only the following asset domains:

- Server: server, PC, network H/W, OS or essential services
- Application: applications to support the objectives and business processes in an organization
- Data: data to achieve the objectives and business process

3.2 Analysis for Information Security Requirements, Criticality, Sensitivity

The following three factors are common in all the standards of risk analysis of information systems: confidentiality, integrity, and availability. This paper proposes a method for identifying and assessing major assets of the information systems, but not all the assets of an organization. Criticality of the information systems is calculated by the value of confidentiality, integrity, and availability as shown in Table 1.

Table 1. Classification for information security factors in information system

Information System	Confidentiality	Integrity	Availability
Data	√	√	√
Application	×	√	√
Server	×	×	√

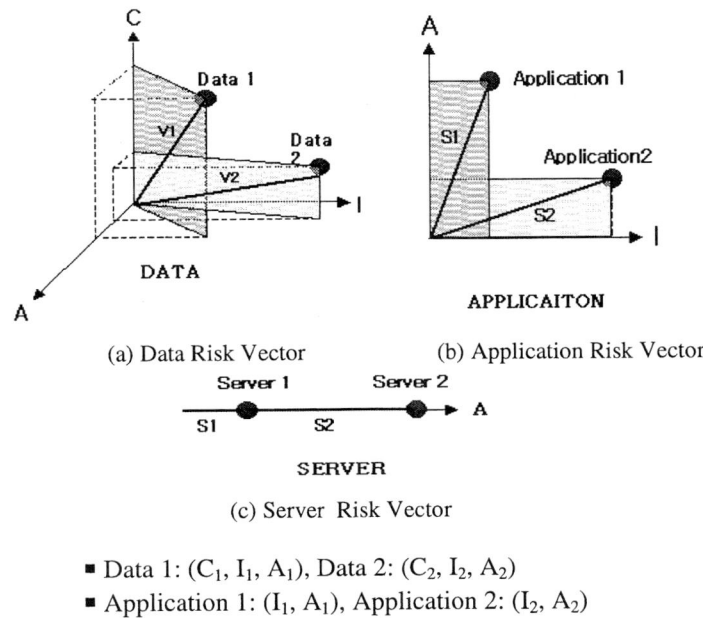

- Data 1: (C_1, I_1, A_1), Data 2: (C_2, I_2, A_2)
- Application 1: (I_1, A_1), Application 2: (I_2, A_2)
- Server 1: A_1, Server 2: A_2

Fig. 2. Criticality analysis for Data, Application, and Server

Data risk vector, V_D, shown in Fig.2 (a) represents the degree of the data risk (the length of the vector) and types of the data risk (the direction of the vector). The data risk vector is determined by the following vectors: V_C *(Confidentiality)*, V_I *(Integrity)*, and V_A *(Availability)*. The degree of data risk vector is calculated by the following equation:

$$V_D = \sqrt{V_C^2 + V_I^2 + V_A^2}$$

We suggest a non-linear scale of the vector values for V_C *(Confidentiality)*, V_I *(Integrity)*, and V_A *(Availability)*, as shown in Table 2. This non-linear scheme is used for the risk management of the NASA space programs. The benefit is to have distinctive risk levels. If all the vectors have the same values (i.e., $V_C = V_I = V_A$), the data risk vector, V_D, is $\sqrt{3V_C^2}$ as shown in the following equation:

$$V_D = \sqrt{V_C^2 + V_I^2 + V_A^2} = \sqrt{3V_C^2}$$

The high-level of V_D has the range from $\sqrt{3V_C^2} = \sqrt{3 \times 64^2} = 110$ to $\sqrt{3V_C^2} = \sqrt{3 \times 256^2} = 433$. Table 3 summarizes the range of each level. Note that V_D has ranges, not distinctive numbers because each V_C, V_I, V_A is able to have the different values due to the mixed levels (e.g., $V_C = 1$, $V_I = 16$, $V_A = 128$).

Table 2. Confidentiality, Integrity, and Availability Level

Level (Color)	C, I, A	Description
High (Red)	64, 128, 256	If the information security requirements of a targeted system are not satisfied, life-threat or critical mission failure is predicted
Medium (Yellow)	8, 16, 32	If the information security requirements are not satisfied, business loss are expected.
Low (Green)	1, 2, 4	If the requirements are not satisfied, little damage is expected, but the critical mission is not affected by the damage or the damage is easily recoverable.

Table 3. Criticality Level in Data Asset

Level (Color)	V_D value	Description
High (Red)	111 ~ 433	The unsatisfied data requirements cause life-threat or critical mission failure.
Medium (Yellow)	14 ~ 55	The unsatisfied data requirements cause business loss.
Low (Green)	2 ~ 7	The unsatisfied data requirements do not affect on critical mission failure and the damage is easily recoverable.

Application risk vector, V_{AP}, shown in Fig.2 (b) represents the degree and types of the application risk. Like data risk vector, the application risk vector is calculated by the following equation using: V_I *(Integrity)*, and V_A *(Availability)*:

$$V_{AP} = \sqrt{V_I^2 + V_A^2}$$

Like Table 3 for data risk vector, Table 4 presents the level of application assets.

Table 4. Criticality Level in Application Asset

Level (Color)	V_{AP} value	Description
High (Red)	91 ~ 362	The unsatisfied application requirements cause life-threat or critical mission failure.
Medium (Yellow)	11 ~ 45	The unsatisfied application requirements cause business loss.
Low (Green)	1 ~ 6	The unsatisfied application requirements do not affect on critical mission failure and the damage is easily recoverable.

Server risk vector, V_{sr}, is expressed by a point in the line unlike data or application risk vector. Server risk vector consider only V_A *(Availability)*. That is, V_{sr} and V_A have the same value.

In order to understand the relationship between server, application, and data, the following mapping approach is proposed as shown in Fig 3.

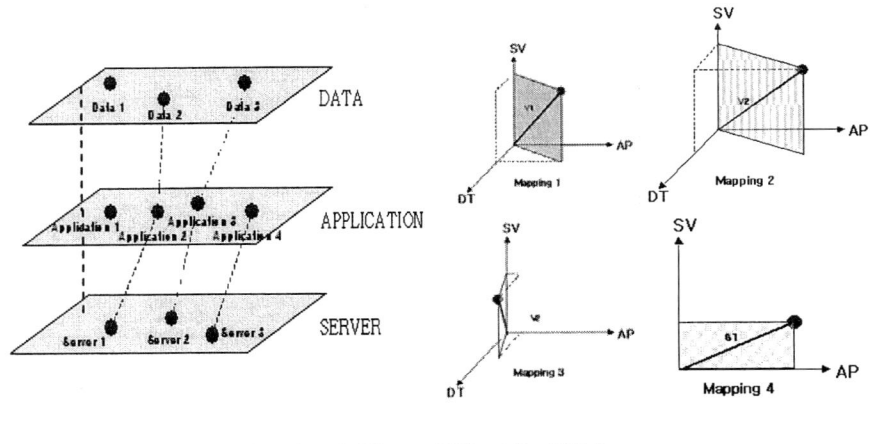

Mapping 1: $V_{D1} = (SV_1, AP_1, DT_1)$
Mapping 2: $V_{D2} = (SV_1, AP_2, DT_2)$
Mapping 3: $V_{D3} = (SV_2, AP_3, DT_3)$
Mapping 4: $V_{AP1} = (SV_3, AP_4)$

Fig. 3. Relationship Analysis between Data, Application, and Server

The 3-dimension diagram shown in Fig. 3 provides the visualized intuition of criticality of risk (i.e., scalar value of the vector) and types of risk (i.e., direction of the vector). For example, mapping 1 and mapping 3 have similar risk criticality value, but they have different sources (types) of the risk. Mapping 3 does not have much application risks.

In order to provide a summary of the analyzed risk, a color-coded table shown in Table 5 is suggested. By combining all the scalar value of each vector, the overall criticality of the mapping is calculated. For example, if the total score is in a range between 266 (=$V_D + V_{AP} + V_{SR}$ = 111 + 91 + 64) and 1151 (= 433 + 362 + 256), the overall risk is determined to High (Red). High (Red) risk assets are classified as major information system assets in a corresponding organization. Likewise, Medium (Yellow, 33 ~ 132) to Low (Green, 4 ~ 17) are determined. If the value is in the range between the High level and the Medium level (i.e., 133 ~ 266) or between the Medium level and the Low level (17 ~ 33), the human experts judge the level or the closed one is selected. Otherwise, the color map can be used as shown in Fig. 4. This summary table assists security risk manager to determine the priority to protect the assets.

Table 5. Criticality Value of Data, Application, and Server

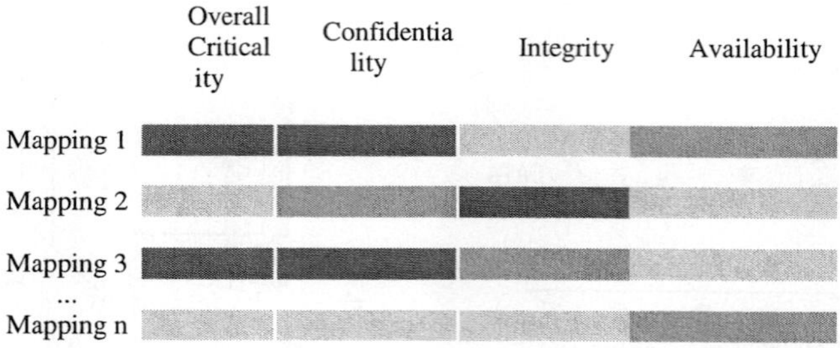

Fig. 4. Continuous Color Code Scheme

4 A Case Study

We first analyzed the organization mission, provided services, the architecture of the information system in Company X as shown in Fig. 5 and Table 6. The critical mission for this organization is public procurement service.

The asset assessment process is as follows: we identified the information assets in terms of servers, applications, and data. The mapping (vector) relationships between the assets are also identified. Then, we analyzed them with the proposed security criteria such as confidentiality, integrity, and availability.

Security Risk Vector for Quantitative Asset Assessment   281

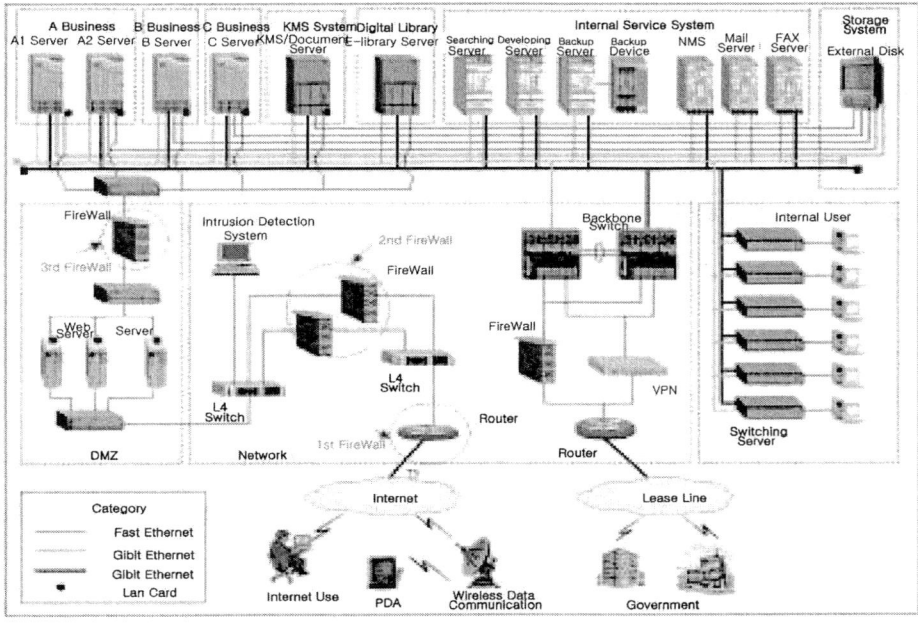

Fig. 5. An Example of the Information System in Company X

Table 6. Analysis Results

Group	Sub system	Categoty	Content	Con-fidentiality	Inte-girity	Avail-ability	Vector value
KMS	2 Server	Server	HP Ultra 6.0			64	64.0
		Application	Handy 3.0		16	128	129.0
		Data	Human Resources	128	128	128	221.7
	Mapping 1						414.7
	KMS Server	Server	CompaQ			64	64.0
		Application	e-payment system		32	128	131.9
		Data	Payment transaction	128	128	8	181.2
	Mapping 2						377.1
	......						
DMZ	Web Server	Server	Samsung v2			32	32.0
		Application	IIS 5.0		4	64	64.1
		Data	Company information	4	128	64	143.2
	Mapping 3						239.3
		Server	LG 3.2			64	64.0
		Application	Apache 3.2		128	128	181.0
		Data	Contract data	64	256	128	293.3
	Mapping 4						538.3
	......						
L4	L4 Switch	Server	3Com			128	128.0
		Application	Routing Algoritm		64	64	90.5
		Data	Routing Table	4	128	64	143.2
	Mapping 5						361.7
	3rd Firewall	Server	HP Ultra 6.0			128	128.0
		Application	ERP 2.0		128	32	131.9
		Data	Rule	16	64	64	91.9
	Mapping 6						351.9
	......						

5 Related Work

SP 800-30, Fips 65 [7] from the National Institute of Standards and Technology (NIST) [6,7], Guidelines for the management of IT security (GMITS) from ISO/IEC JTC1/SC7 [4, 5] and Executive Guide on Information Security Management from the Government Accounting Office(GAO) [3]. British government encourages industry using certification based on BS7799, which is the information security management standard. They regulate that risk analysis should be included in information security management system establishment. Risk analysis is achieved regularly using a risk analysis automation tool of CRAMM [8], RA and etc.

The above risk management has focused on process/method of general risk analysis, but our focus is to assess criticality of assets using *Security Risk Vectors*. It assists many organizations to apply risk analysis in their information assets.

6 Conclusion

This paper proposes a new asset assessment method called *Security Risk Vector*. A case study is investigated for the validation of the usefulness of the Security Risk Vector. It contributes an effective process for assessing the major information system assets such as server, application, and data in terms of confidentiality, integrity, and availability. These three factors are expressed by vectors. This vector representation has the following benefits:

- **Intuitive understanding of the criticality of the assets:** The 3-dimension vector representation provides intuition the degree (scalar value) and types (direction of the vector) of asset criticality.
- **Clearly distinctive importance of assets (sensitiveness):** As we adapt non-linear evaluation values such as 1, 2, 4, 8, 16, etc., the rank of asset criticality is clearly distinguished. In addition, even though two vector lengths are the same, distinctive sensitiveness of an asset is shown by the direction of the vector in axes of confidentiality, integrity, and availability. It helps recognize more important assets easily.

In future, we plan to develop various visualization tools of these vector scalar values and its directions. We are considering that threat and vulnerability analysis can be applied by this vector method.

Acknowledgement. Hoh Peter In is a corresponding author.

References

[1] Carnegie Mellon Software Engineering Institute, *"OCATVE Criteria, Version 2.0*
[2] BSI, BS7799 - Code of Practice for Information Security Management, British Standards Institute, 1999.
[3] General Accounting Office, *Information Security Risk Assessment - Practices of Leading Organizations, Exposure Draft*, U.S. General Accounting Office, August 1999.

[4] ISO/IEC JTC 1/SC27, Information technology - Security technique - Guidelines for the management of IT security (GMITS) - Part 3: Techniques for the management of IT security, ISO/IEC JTC1/SC27 N1845, 1997. 12. 1.
[5] Solm, R., "Information Security Management (2): Guidelines to The Management of Information Technology Security (GMITS)", Information Management & Computer Security, Vol. 6, No. 5, 1998, pp.221-223.. 11-21, 1999.
[6] NIST, " Risk Management Guide for Technology Systems", NIST-SP-800-30, Oct 2001
[7] FIPS-65, "Guidelines for Automatic Data Processing Risk Analysis, NIST, 1975
[8] CRAMM, "A Practitioner's View of CRAMM", http://www.gammassl.co.uk/

A Remote Video Study Evaluation System Using User Profile

Seong-Yoon Shin and Oh-Hyung Kang

Department of Computer Science, Kunsan National University,
68, Miryong-dong, Kunsan, Chonbuk 573-701, South Korea
{s3397220, ohkang}@kunsan.ac.kr

Abstract. We propose an efficient remote video study evaluation system which is suitable to the personalized characteristic of the individual student using an information filtering based on user profile. For the setting questions to use the video, we extract a key frame based on the location, size and color information and extract a setting questions interval using gray-level histogram and time windows. Also, for efficient evaluation, we set questions which compose a category based system and a keyword based system. Consequently, students can enhance their study achievements as supplement to the insufficient knowledge and maintain their interest towards the subject.

1 Introduction

Information filtering is the function to process and to filter the information so that it is suitable to the user requirement, and it is the filtering of dynamic information stream based on the long-term profile according to the user interest are produced and be maintained by the system. Most dedicated filtering systems automatically produce and maintain the user interested profile using a learning technique [1,2]. An information filtering is the important course of an individualism of the information; traditionally, it is classified in three kinds, these are content-based, social and economic filtering, and mixed each other and used [3].

A content-based filtering calls a cognitive filtering. The object is selected by the relationship between the content of objects and priorities of user. Representative example of a content-based filtering is a keyword-based filtering [2,4]. Social filtering, also called collaborative filtering, where objects are filtered for a user upon the preference of other people with similar tastes [5]. Social filtering systems need a critical mass of participants and objects to work efficiently which appear to be their major draw-back. Representative example is as follows. In Tapestry system [6], users give the comment directly and determine the judgment about an interested field. Then, composite filter of a program itself accomplish the filtering about the documents which is saved continuously. Stanford Information Filtering Tool (SIFT) [7] offers a filtering service on the Web; users are provided a filtering service through the profile over one which states the keyword to use a matching strategy. GroupLens [8] is the system for a distributed collaborative filtering of Usenet News; Users could give the weight by themselves about the news to read. Economic filtering is the method to filter the information based on a cost element [8]. Cost elements which are used here are the relationship between cost to be used and profit, or the relationship between network bandwidth and object size, and etc.

The above three kinds of filtering methods mix each other and are used. Representative example is the method that NewsWeeder system [9] mixes content-based filtering and social filtering for Usenet News.

For some years recently, many researches which use the user profile have been performed. XFilter [10] provides highly efficient matching of XML documents to large numbers of user profiles. The XFilter engine uses a sophisticated index structure and a modified Finite State Machine (FSM) approach to quickly locate and examine relevant profiles. Franklin et al. [11] used the user profile for data recharging. They presented an automatic data recharging method based on user profile which is produced in meaningful profile language. Schwab et al. [12], through clear user observation, presented the method that potent an interested user profile study.

A remote study evaluation field on Web can not accomplish the evaluation considering the characteristic and interest of student individual and accomplishes the evaluation. That is, we can solve these problems efficiently if we use the personalized user profile. We will filter the question from the question bank database so that we are suitable to the characteristic and interest of the individual using the individual user profile of students. And, we will provide and solve the problem of "insufficient knowledge through" supplements and "superior area can be developed further" [13].

2 Scene Change Detection

2.1 Key Frame Extraction

If we set questions of the problem to use the video, then we use a JANGHAK quiz program to be broadcasted in EBS. A JANGHAK quiz program accomplishes the evaluation of a total 3 round, but we use only 1 round to divide the video in this paper. A problem area to be divided classifies Language, Mathematics, Society, Science and Foreign Language, and uses at setting questions. We summarize the structural feature about 1 round evaluation of JANGHAK quiz program.

- The question number and content appears always when it sets questions.
- When each question number and content appears in the beginning, the effectiveness of question number blinks.
- The question number and content have each fixed size.
- The question number area has fixed color.
- The question number and content disappear if setting questions is completed.

Based on the above features, key frames of a setting questions scene extract through a similarity measurement using the information of location, size, and color of question number area. A similarity measurement method is equal to Equation 1.

$$\text{Similarity} = CI_i(p, s, c) - TI(p, s, c), \text{ where } i = 1\ldots m \tag{1}$$

In Equation 1, $TI(p, s, c)$ is the template to have a location (p), size (s) and gray color values(c) of question number area producing based on a prior knowledge. And $CI_i(p, s, c)$ has information of each about question number area of input frames.

2.2 Extraction of Setting Questions Interval

The setting questions interval (SQI) is the interval of question number and content which question is from appear to disappear for setting questions. For the extraction of a setting questions interval, we accomplish first the computation of a gray level histogram difference (D_i) about the question number area as shown in Equation 2.

$$D_i = \sum_{j=1}^{Bins} |H_i(j) - H_{i-1}(j)| \qquad (2)$$

In Equation 2, $H_i(j)$ means a j-th bin of a gray level histogram of a frame i. D_i implies histogram difference between currently frame F_i and former frame F_{i-1}. And we set each question, because of a blink phenomenon which appears in the beginning prevent the setting questions interval and extracted wrong, we give a time windows (W_T) of key frame. Detailed setting questions interval extraction algorithm is equal to as follows: where D_T is threshold, W_k is key frames view, W_i is input frame view, W_f is frame difference between W_k and W_i and GFI is the general frame interval.

```
For (i=2;i < n;i++){
    D_i = gray level histogram difference;
    W_f = |W_k - W_i|;
    If (D_i ≤D_T) F_i = SQI;
    Else IF (D_i > D_T AND W_f ≤ W_T) F_i = SQI;
    Else F_i = GFI;}
```

3 A Study Evaluation Which Uses the User Profile

In this paper, first, we construct a question bank DB and can accomplish the evaluation on the web. Secondly, questions are filtered using the user profile. The last, we enhance the efficiency of the evaluation to the setting questions which considers the individual variation, characteristic and interest of student. Overall system structure is presented in Figure 1.

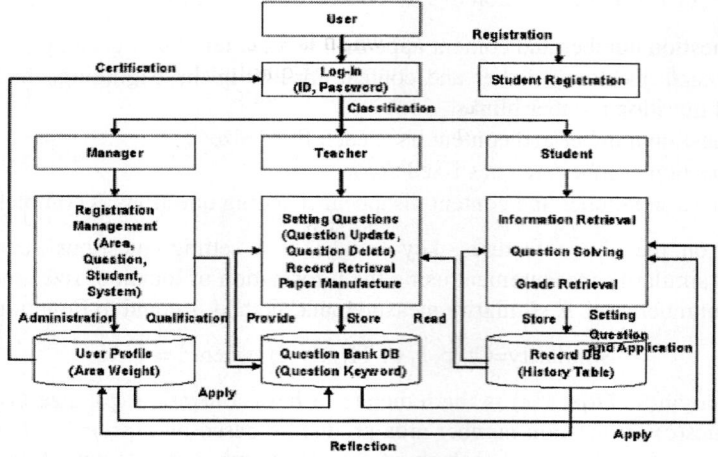

Fig. 1. Overall system structure

3.1 A System Details Structure

(1) User Registration
A user registration does the log in using a user ID and the password through Web page. We distinguish the user by the manager, teacher or student in the user profile and give the authority.

(2) Manager Module
The manager has all authority about the management of system, user, subject, etc. The management of the user accomplishes an information retrieval, query, modification and deletion about the user. The management of the subject accomplishes the addition and deletion of the subject. We manage overall system of the management of whole database and an inadaptable profile threshold ($Cp(V)$) adjustment in the management of the system.

(3) Teacher Module
The teacher has the authority of manufacturing and setting questions, grade retrieval, etc. It is saved at a question bank database if the question is made by the manager. We set questions together with the help when we make the question and can give the aid. The question to apply the multimedia as image and music can set questions. The teacher must select the keyword when it sets questions to support a keyword-based user profile system. We implemented this to select the keyword when making the question and input most important subject (or theme) of that question.

(4) Student Module
Students use the user profile and solve the question to be optimized, and new profile is reflected again in a next examination according to the result. The student to do the log-in is provided an examination schedule such as examination turnaround, question number of area, examination time, etc. and undergo practical examination from the system.

The problem to be set questions must be optimizing the individual characteristic and interest of the student according to the user profile. To produce the user profile of the best suited, we do not reflect the user profile in the examination in all area as well as inadaptable profile threshold ($C_p(V)$) which is set by the manager. At this time, we are given the question as well as the number of question ($_iP_q(N)$) which is an initialized inadaptable profile and accumulate the user profile through the question explanation. If $C_p(V)$ is satisfied through such course, system gets an area weight($weight[i]$) and a user interest word ($I_u(W)$), and produce and maintains the user profile.

3.2 Question Filtering Using the User Profile

(1) Category-Based Method
This evaluation system grasped an individual characteristic and the interest of students through the area relative evaluation. That is, we convert an area marks not as total marks into relative rate and reflect this in the next evaluation. It is the method we put the weight of the plus and give the number of question about a fragility area.

In this paper, we do the supplementation of the defect of a category-based system according to the composing and accomplishing a keyword-based system.

(2) Keyword-Based Method

A keyword-based system find the subject, most important interest in individual question in the detailed element, and does this to the keyword and inserts at the question. Then, we select most the keyword to come out frequently with the question in the result of the marking which the user is correct to make. And we produce new keyword profile by inserting at the queue which existing keyword data have been saved. Afterwards, we set questions above all the questions to have user interest words which reflect a keyword profile again at the number of question of each area which is assigned to a category-based system in the next assessment.

(3) Question Filtering Process

1) We determine the output format using YES or NO and reflect the user profile when we set the questions.

In this paper, we set an inadaptable profile threshold ($C_p(V)$) for the accumulation of the user profile. As shown below, if the total evaluation factor ($T_t(C)$) of the user is smaller than ($C_p(V)$), then the profile is not evaluated but that user is evaluated in general and we accumulate the profile through this manner.

```
IF (T_t(C) < C_p(V))
    THEN FOR(i = 1; i • T_a(N); i++)
        Q_a(N)[i] = _iP_q(N)
```

$C_p(V)$: An inadaptable profile threshold, $T_t(C)$: A total evaluation factor
$T_a(N)$: The number of total areas, $Q_a(N)$: The number of area questions
$_iP_q(N)$: The number of an inadaptable profile questions

2) We calculate the number of area to give to the user.

We bring the user profile which has been saved to a form of the number of question in the database. The number of each area setting questions is decided in this routine.

```
FOR(i = 1; i <= T_a(N); i++)
    Q_a(N)[i] = Query(Select Area[i] From User_Profile_DB where
                ID = userid
```

3) We assign the question of $Q_a(N)$ at that area. This routine implements a keyword-based system. $I_u(W)$ are produced and modified by marking and profile reflection, we choose the question above all about a user interest word.

```
FOR(i = 1; i <= Q_a(N); i++)
    Question_Array += Query(Select Question_No
                From Question_DB
                where Question_Keyword Like %I_u(W)%)
```

$I_u(W)$: A user interest word

If $I_u(W)$ is Null then we added all problem record set which correspond to $I_u(W)$, we use Random() function and choose it randomly. For both cases, when we insert the selected record set to Question_Array to be provided to the user, and we all go via a duplicate inspection process and choose problems to be given to the user by applying a search algorithm. This paper used the selection search.

4) The system shows the user the direction and the problem and receives the examination paper which the user inputs to an array form.

5) Marking a routine which compares right answer brings from the question bank and the examination paper which the user inputs, we calculate an area total of one's marks and whole total of one's marks.

6) Calculate an area weight to apply at a category-based system.

An area weight is important data which decides characteristic and interest of a student. The weight is based on the marks which the student gets in an examination area, it uses minus value for superior subject areas and plus value for the insufficient subject areas.

```
For(i=1; i <= T (N); i++)
            a
```

$$weight[i] = \left\{ \left(\frac{T_t(S)}{T_t(N)} - \frac{T_a(S)[i]}{Q_a(S)[i]} \right) \bullet T_a(N) \right\}$$

$T_t(N)$: The total of one's marks become a total full marks criteria.
$T_t(S)$: The total examination marks which gets $T_t(N)$ to the full marks criteria.
$Q_a(S)$: The total of one's marks to become the area full marks criteria.
$T_a(S)$: The area examination marks which gets $Q_a(S)$ to the full marks criteria.

This weight comes to multiply with a question rate (decimal point conversion) and the number of question of a preceding examination. This result to multiply becomes the round up and produces the number of question of the next examination which come to plus or minus with the number of question of preceding examination.

7) We get $I_u(W)$ and apply at a keyword-based system.

```
For(i = 1; i <= T (N); i++)
                     a
     I (W) = Query( Select Question_Keyword From Area[i]
      u            Where Question_No = Area[i][right_j])
```

right_j : the number of question of right answer set of belonging area

8) We update weight which gets in a category-based system and $I_u(W)$ which gets in a keyword-based system in the user profile database. Through Plus weight at existing profile data and $I_u(W)$ updates data to get newly through Push operation after inserting existing profile data at Queue.

9) We present an area total and a whole total of one's marks through the marking to get to the user and store at a user history database.

4 The Implementation of a Video Study Evaluation System

4.1 The Implementation Environment

In this paper, a video study evaluation system which uses the user profile was implemented on Apache Web server through TCP/IP environment, and on a Pentium 4-2.0 GHz PC. Scene change detection was implemented using Visual C++6.0 with EBS JANGHAK quiz video. Also, to construct a medium size database system in Windows 2000 Servers, we used MySQL and PHP3 for a database access.

4.2 Implementation of Scene Change Detection

A setting questions interval extracted by using a gray-level histogram difference and time windows. Scene change detection is performed as shown in Figure 2.

Table 1 expresses the criteria value and the threshold about the location, size, average gray color, time windows (frame) and gray-level histogram difference of a question number area for scene change detection.

Fig. 2. Scene change detection

Table 1. The threshold for scene change detection

Features	Criteria value (CV)	Threshold
Location (x, y)	25, 171	CV±1
Size (area)	1739	CV±83
Average gray color	170	CV±10
W_T	60	CV±5
D_i	100	CV±10

4.3 Implementation of User Module

The user can connect to the system using an individual ID and Password to get the certification from the manager.

4.4 Implementation of Manager Module

The manager controls and manages general tasks related setting questions and areas. We set $C_p(V)$ and $_iP_q(N)$ to need for information management of student and teacher and profile accumulation. Figure 3 is showing the screen of a manager module.

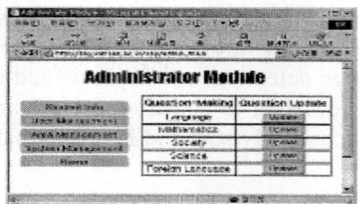

Fig. 3. Screen of manager module

4.5 Implementation of Teacher Module

The teacher can accomplish the management of the student information, setting questions and modification task except the management of operator, system and subject among the function of the manager. To implement a keyword-based method when we set questions, we can insert the question (direction) and the keyword of video form as Figure 4.

Fig. 4. Question (direction) of teacher module

4.6 Implementation of Student Module

If the student connects to a video study evaluation system, the initial screen shows the number of question and rate of the area for a study evaluation by the user profile analysis as shown in Figure 5. After the students read an examination schedules as presented in Figure 4, they solve an examination question continuously as shown in Figure 6.

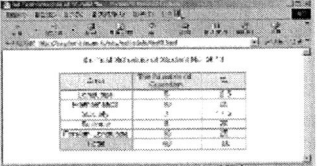 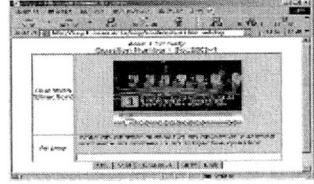

Fig. 5. Examination schedules   **Fig. 6.** Solve an examination question

In a question explanation of examination, we present an area and question number about each question to the user, and input the answer which the user chooses. If we completed a problem explanation which is given, we move automatically to a result authentication page of the examination as shown in Figure 6 and confirm own examination result.

In this way, the result of the examination updates the user profile of a category-based system by area relative marks. We look into a mathematics area in Figure 10, $T_t(N)$ has 100 as the total 100 point full marks and $T_t(S)$ has a total of one's marks of 62.5 of the examination which gets $T_t(N)$ to full marks criteria. $Q_a(S)$ has 100 as an area total of one's marks to a 100 point full marks and $T_a(S)$ has a total of one's marks of 50 of the area which gets $Q_a(S)$ to a full marks criteria. And $T_a(N)$ has 5 of a total area number. Figure 8 is the result that gives the weight by the area based on examination result as shown in Figure 7.

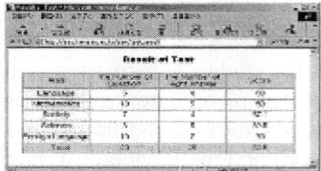
Fig. 7. Examination result

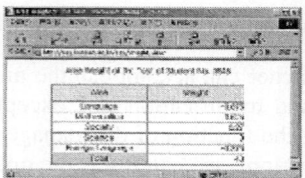
Fig. 8. Weight by the area

Table 2 compared and expressed the change of the number of question between the third and the forth examination which the user profile of student 3545 is applied.

Table 2. Change of the number of question between 3rd and 4th examination

Area	3rd examination		4th examination	
	The number of question	area/total	The number of question	area/total
Language	5	12.5%	5	12.5%
Mathematics	10	25%	11	27.5%
Society	7	17.5%	7	17.5%
Science	8	20%	8	20%
Foreign Language	10	25%	9	22.5%

In Table 2, based on the personalized characteristic about an each area, we accomplished more efficient study evaluation to students by applying the user profile and modifying the number of question.

Figure 9 is the screen to read examination information of student 3546 in a manager module. $Iu(W)$ is applied to the keyword-based system and is reflected at a next examination evaluation.

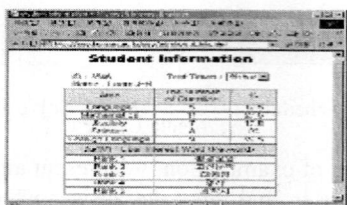
Fig. 9. Examination information of student in a manager module

4.7 Comparison and Estimation

Characteristics between simple uniform traditional study evaluation system and this new study evaluation system which is presented in this paper are compared and estimated as presented in Table 3.

Table 3. compare and estimate of system

Item	Traditional system	Proposed system.
Forms of setting questions and evaluation	Simple, overall and uniform	Different form by the individual
The number of question per area	Total identity	Different by the individual
Criteria of setting questions and evaluation	Examiner and appraiser	Characteristic and interest of the individual
Priority rank of setting questions and evaluation	Identity without the priority rank	Priority to lack area
Media of setting questions	Text	Video
Evaluation purpose	Simple study achievement evaluation	Study achievement enhance

5 Conclusion

In this paper, we presented efficient video study evaluation system which is well suitable with the characteristic and interest of a student individual using an information filtering based on the user profile. To apply the user profile at the evaluation, we used the setting questions method which composes a category-based system and keyword-based system. Also, for the setting questions of a video form in this paper, we extract the key frame of the setting questions scene using the structural features based on the location, size and color information. And the setting questions interval extracted by using a gray-level histogram difference and a time windows. Students can enhance a study achievement as supplement to insufficient subject areas and maintain an interest area using this system. We observed the efficiency of the study evaluation very much using the user profile and induced the interest of the study using user interest word. In addition, this would facilitate teachers in helping many students in the education process.

References

1. Feynman, C.: Nearest neighbor and maximum likelihood methods for social information filtering, International Document, MIT Media Lab, Fall (1993).
2. Sheth, B. D.: A Learning Approach to Personalized Information Filtering, SM Thesis, Department of EEVS, MID, Feb. (1994).
3. Malone, T. W., et al.: Intelligent Information Sharing System," Communications of the ACM, Vol. 30, No. 5, (1987) 390-402.
4. Salton, G. and McGill, M. J.: Introduction to Modern Information Retrieval, McGraw-Hill, (1993).
5. Thomas Kahabka, Mari Korkea-aho, Günther Specht: GRAS : An Adaptive Personalization Scheme for Hypermedia Databases, Proc. of the 2nd Conf. on Hypertext-Information Retrieval-Multimedia(HIM '97), (1997) 279-292.
6. Goldberg D., Nicholas D., Oki B., Terry D.: Using Collaborative Filtering to Weave an Information Tapestry, CACM, Vol. 35, No. 12, Dec. (1992) 61-70.
7. Tak Y. Yan and Hector Garcia-Molina: SIFT-A tool for wide-area information dissemination, In Proc. of the 1995 USENIX Technical Conf., (1995) 177-186.

8. Paul Resnick, Neophytos Iacovou, Mitesh Suchak, Peter Bergstrom and John Riedl: GroupLens : An open architecture for collaborative filtering of netnews, In Proc. of ACM 1994 Conf. on Computer Supported Cooperative Work, (1994) 175-186.
9. Lang K.: NewsWeeder : An Adaptive Multi-User Text Filter, Research Summary, Aug. (1994).
10. Altinel M., Franklin M. J.: Efficient Filtering of XML documents for Selective Dissemination of Information, Proc. VLDB Conf., Sep. (2000).
11. Cherniack M., Franklin M. J., Zdonik S.: Expressing User Profiles for Data Recharging, IEEE Personal Communications, (2001) 6-13.
12. Schwab I.., Kobsa A.: Adaptivity through Unobstrusive Learning, KI 3(2003), Special Issue on Adaptivity and User Modeling, (2002) 5-9.
13. Lee E. B, Kwak D. H, Ryu K. H.: Understanding of Computer, KNOU press center, (1999).

Performance Enhancement of Wireless LAN Based on Infrared Communications Using Multiple-Subcarrier Modulation

Hae Geun Kim

School of Computer and Information Communication,
Catholic University of Daegu,
330 Kumrak-ri, Hayang-up, Kyungsan-si, 712-702, Korea
kimhg@cu.ac.kr

Abstract. Infrared communications employing the 4-dimensional Multiple-Subcarrier Modulation with fixed bias for wireless LAN is introduced. In the proposed system, computer simulated 4-dimensional vectors having the largest Euclidean distances are used. After the vectors are applied to the 4-dimensional modulation, the power and the bandwidth efficiencies of the infrared communication system are improved. From the performance evaluation results, the normalized power and the bandwidth requirements of the proposed system are improved up to 1.8 dB and 1.4 dB compared to those of the conventional schemes using QPSK, respectively. The proposed system needs less power and bandwidth for optical wireless connection to fulfill the requirements of the standard.

1 Introduction

Wireless communications are an emerging technology and become an essential feature of computer networks. The IEEE 802.11 study group proposed a standard for WLAN (Wireless Local Area Network) which transmits and receives data over the air, minimizing the need for wired connections. One of the main advantages of WLAN is scalability. That means we can scale it very small room or up to very large building. This standard provides two different technologies for the physical layer of WLAN. First one is based on infrared and the other one is based on Spread spectrum using radio frequency (RF) [1]. Infrared wireless local access links provide an attractive alternative because infrared radiation offers higher speed, wider band, and less interference between adjacent channels than RF. In infrared communication for indoor WLAN, we encounter two major issues considering the communication quality of the system that are the scattering of the light by the interior of a room and power efficiency. Besides, concerns of eye safety limit the average transmission power, further restricting the operating range and deteriorating the Quality of Service (QoS) of the whole network.

To overcome those impairments, Multiple-Subcarrier Modulation (MSM) systems with Intensity Modulation / Direct Detection (IM/DD) in infrared communications are popularly researched. IM/DD MSM systems are attractive not only for minimizing inter-symbol interference (ISI) on multi-path channels between narrowband subscribers, but also for providing immunity to ambient light inducing in an infrared receiver [2][3][4][5][6]. Yet, the efficiency of average optical power intensity is decreased as the larger number of subcarriers is used in an MSM system. Since the electrical MSM signal is a sum of sinusoids containing positive and negative values, a dc bias must be added to an electrical MSM signal to modulate the intensity of optical carrier. As the number of the subcarriers increases, the required dc bias of an MSM system increases.

In this paper, the MSM system employing the computer-simulated 4-dimensional (4-D) vectors is introduced where the optimization technique of signal waveforms [7] are used to derive the vectors. The 4-D vectors have the largest Euclidean distances to generate the output amplitude of a block coder, so that the power and the bandwidth efficiency in the Multiple-Subcarrier Modulator are improved. The 4-D vectors in the proposed system maps the information bits to be generated to the symbol amplitudes modulated on to the subcarriers. The 4-D vectors for 16 symbols are symmetrically constructed on the surface of a 4-D sphere. Also the fixed bias is used for all symbols so that the power used for each symbol is constant and equals the average transmitted power. In the proposed system, one symbol is transmitted with 4 orthogonal subcarrier signals, while one symbol is transmitted with one subcarrier in On-OFF Keying (OOK) and with two subcarriers in Quadrature Phase Shift Keying (QPSK).

2   4-D IM/DD MSM System in Optical Channel

In the IM/DD channel with impulse response, $h(t)$, for optical wireless communication, the received photocurrent, $y(t)$, is defined as

$$y(t) = \int_{-\infty}^{\infty} rx(\tau)h(t-\tau)d\tau + n(t) \quad (1)$$

where r is the responsivity of photodetector, $n(t)$ is the channel noise, and $x(t)$ is the instantaneous optical intensity. If the channel is modeled by a linear system, the noise can be models as a white Gaussian with two-sided spectrum. $x(t)$ is nonnegative and an average optical power is the mean of $x(t)$ while an average power electrical signal is the mean of $x^2(t)$.

Fig. 1 depicts the transmitter and receiver design used in the proposed MSM transmission scheme with 4-D orthogonal modulation. The transmitter in Fig. 1 (a) consists of N 4-D MSMs where 2 subcarriers are used in each 4-D MSM. During each symbol interval of duration T, it transmits Nk information bits. In the block coder of each 4-D MSM, k-input bit is transformed into one of M symbols A_i where $M = 2^k$ and, $i = 1, ..., M$, then, each symbol A_i is mapped to a corresponding vector of 4-D symbol amplitudes, $a_i = (a_{i1}, a_{i2}, a_{i3}, a_{i4})$.

When a_i pass through the rectangular transfer pulse shape $g(t)$ followed by the 4-D orthogonal modulator, the electrical MSM signal $s(t)$ in Fig. 1 (a) can be

$$s(t) = \sum_{j=-\infty}^{\infty} \sum_{n=1}^{N} [a_{i1} \cos \omega_{1n} t + a_{i2} \sin \omega_{1n} t + a_{i3} \cos \omega_{2n} t + a_{i4} \sin \omega_{2n} t] g(t - jT) \quad (2)$$

where $g(t)$ is 1 for $0 \leq t < T$, and is 0 for $t < 0$ or $t \geq T$. If A_i is a set of equal energy symbols and each symbol A_i has a corresponding 4-D vector $\boldsymbol{a}_i$, we can have

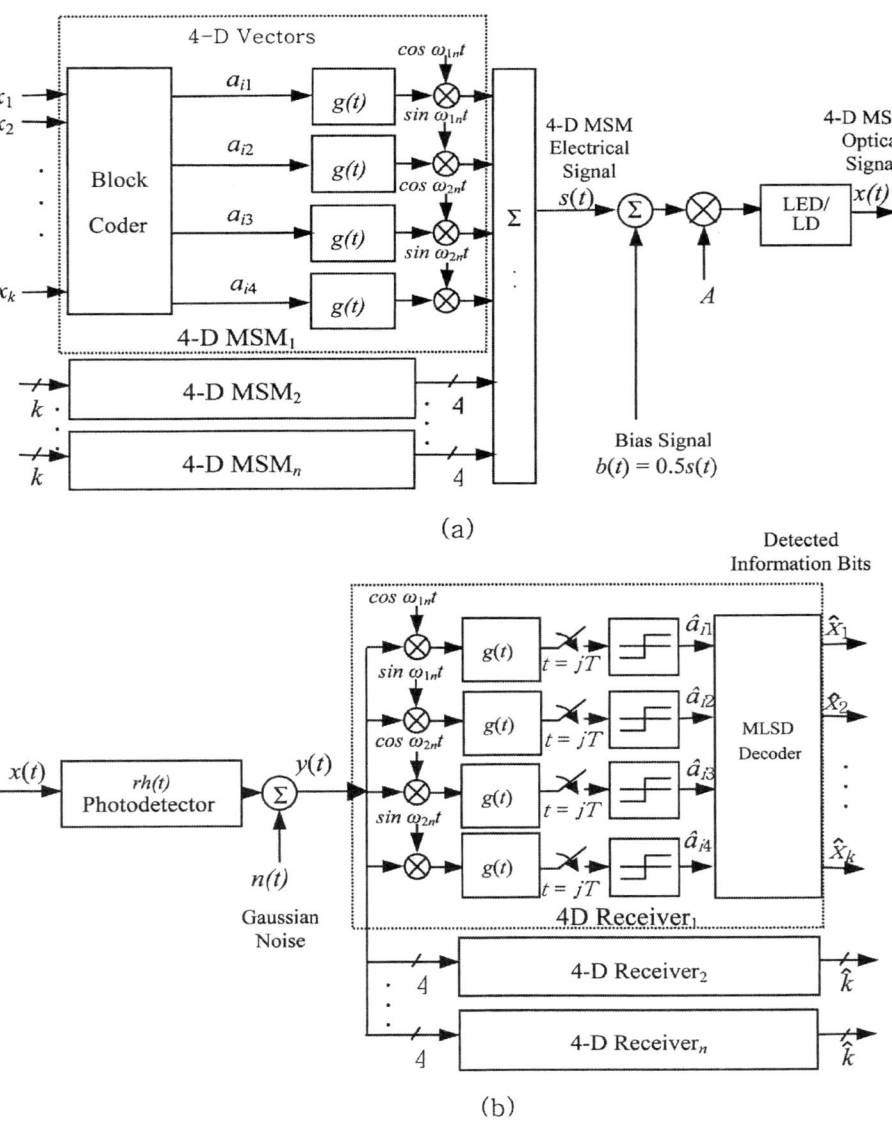

Fig. 1. MSM (Multiple–Subcarrier Modulation) system: (a) transmitter and (b) receiver with 4-D orthogonal modulation scheme where $n = 1, 2, \ldots, N$

$$\int_0^T A_i^2(t)dt = T\sum_{j=1}^{4}(a_{ij})^2 = E(\text{independent of } i) \qquad (3)$$

where 4-D vectors a_i are derived by the optimization technique of signal waveforms described in details in Chapter 3. In the proposed system, one symbol is transmitted with 2 subcarrier frequencies composed of 2 cosine pulses and 2 sine pulses compared with QPSK transmitting one symbol per one subcarrier frequency.

To modulate an optical signal, the electrical MSM signal should be nonnegative. Since the electrical MSM signal $s(t)$ can be negative or positive, a baseband dc bias $b(t)$ must be added. After optical modulation using a LED or a LD, the MSM optical output can be expressed as $x(t) = A[s(t) + b(t)] \geq 0$ where A is a nonnegative scale factor. The average optical power is $P = AE[s(t)] + AE[b(t)]$ where $E[x]$ represents an expected value of x.

If the subcarrier frequency $\omega_n = n(2\pi/T)$ and $g(t)$ is used in the MSM, $E[s(t)]$ is always 0 and the optical power P only depend on $b(t)$. Hence the average optical power can be given by $P = AE[b(t)]$. When the bias signal is properly chosen, the average power requirement of the MSM system can be decreased.

There are two biasing schemes: fixed bias and time-varying bias. For fixed bias, the bias has the same magnitude with the smallest allowable value of electrical MSM signal $s(t)$ given by $b_0 = -\min s(t)$. For the time-varying bias, the bias has the smallest allowable symbol-by-symbol value. In general, the average optical power with time-varying bias is smaller than that with the fixed bias. On the other hand, the system using the fixed bias is simple and easy to implementation. In proposed system, the fixed bias is used and has the bias value of $0.5s(t)$ that ensures nonnegative MSM signal to modulate an optical signal.

The receiver shown in Fig. 1(b) uses 4 hard decision devices to obtain a 4-D vector of detected symbol amplitudes $\hat{\boldsymbol{a}}_i = (\hat{a}_{i1}, \hat{a}_{i2}, \hat{a}_{i3}, \hat{a}_{i4})$.

For MSM systems, several block codes employing QPSK such as normal block code, reserved-subcarrier block code, and minimum-power block code to improve the power efficiency of an MSM optical communication system have been introduced [1]. Under the normal block coder, all N subcarriers are used for transmission of information bit where the number of input bits $k = 2N$, and the number of symbol $M = 2^{2N}$ for QPSK. Each information bit can be mapped independently to the corresponding symbol amplitudes. At the receiver, the detected symbol amplitudes can be mapped independently to information bits.

Under the reserved-subcarrier block code, L subcarriers are reserved for minimizing the average optical power P. Hence, the number of input bits $k = 2(N-L)$, and the number of symbol $M = 2^{2(N-L)}$ for QPSK. An information bit vector is encoded by freely choosing the symbol amplitudes on the reserved subcarriers.

Under the minimum-power block code, no fixed set of subcarrier is reserved, but $L > 0$ subcarriers are reserved for the minimum value of the average optical power P. Also, the number of input bits $k = 2(N-L)$, and the number of symbol $M = 2^{2(N-L)}$ for QPSK and the average optical power always lower bounds the average optical power requirement.

3 Optimization of Signal Waveforms

Let the signal points be considered as particles constrained to the surface of an n-D sphere in a conservative force field. The particles will be activated and then move in a manner that will cause the total system potential approach to a local minimum. Correspondingly, it is possible to have the error rate approach a local minimum by choosing the potential to be equal to the system error rate. By the time the system comes near stable state, the final position of particles will become the a_i coefficients in (2) that yields a local minimum in the error rate. This technique is applied to equal energy signals whose source and channel statistics are equally likely [7].

The error probability is almost entirely contributed by the nearest pair of signal points in an M-ary PSK scheme. For the case of high signal-to-noise ratio, we can choose a law of force expression, which can be given by

$$F_{ik} = C(k_0) e^{-|d_{ik}|^2 / 4k_0 / T} d_{ik} / |d_{ik}| \qquad (3)$$

where F_{ik} is the force between particle i and k, k_0 is the bandwidth of noise, and d_{ik} is the vector from particle i to particle k.

We have made the program for the problem based on {3}. For example, if the force, F_{ik}, between two particles, A and B, in a conservative force field is repulsive, the positions of two signal points become A''' and B''' after executing one iteration of the program as illustrated in Fig. 2.

The program has run for 16 points on the surface of a 4-D sphere as shown in Table 1 where the symbols A_i in (2) are corresponding to 4-input bits ($k = 4$) and the symmetric 4-D vectors are listed in an ascending order of the squared distances. The derived minimum distance is 0 for A_0 itself, 1.223 for A_0 to A_1, A_0 to A_2 and so forth, and the magnitude of each vector $|A_i| = \sqrt{a_{i1}^2 + a_{i2}^2 + a_{i3}^2 + a_{i4}^2} = 1$ so that this block code is symmetry and each 4-D symbol has the same energy as any other character symbol with the minimum Euclidean distance.

In the proposed system, 4-D vector, $a_j = (a_{i1}, a_{i2}, a_{i3}, a_{i4})$, in Table 1 is used in transmitting 4 information bits for the block coder in Fig. 1(a).

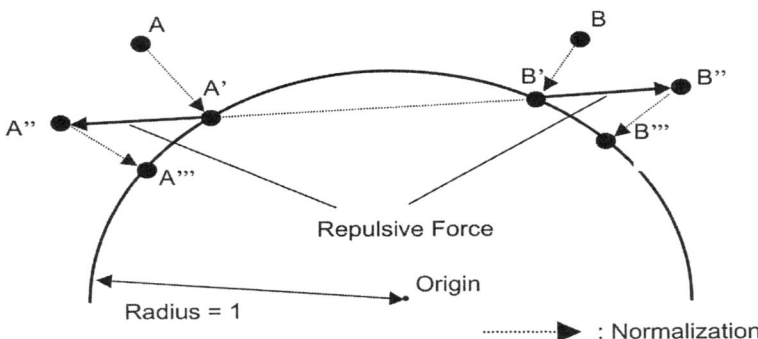

Fig. 2. Action of two signal points on the surface of a sphere in a conservative force field

Table 1. 16-point vectors with maximized minimum squared distance on the surface of 4-D sphere

A_i	Input Bits	a_{i1}	a_{i2}	a_{i3}	a_{i4}	d^2 to A_0
A_0	0 0 0 0	-0.658161	-0.727785	0.184561	0.055596	0.000
A_1	0 0 0 1	-0.611962	0.142302	0.263295	0.732071	1.223
A_2	0 0 1 0	-0.926621	0.252992	-0.127305	-0.247309	1.223
A_3	0 0 1 1	-0.290729	-0.388937	-0.643680	0.591507	1.223
A_4	0 1 0 0	-0.305343	-0.454497	-0.600646	-0.582600	1.223
A_5	0 1 0 1	-0.322096	-0.182096	0.478185	-0.796514	1.223
A_6	0 1 1 0	0.209338	-0.508136	0.834258	0.044604	1.223
A_7	0 1 1 1	0.368454	-0.910504	-0.183474	0.039525	1.223
A_8	1 0 0 0	0.405778	-0.257479	0.160944	0.862059	2.004
A_9	1 0 0 1	-0.191508	0.519871	0.831972	-0.029688	2.201
A_{10}	1 0 1 0	0.624431	-0.188116	-0.092000	-0.752485	2.666
A_{11}	1 0 1 1	-0.157128	0.468879	-0.867366	-0.056034	2.802
A_{12}	1 1 0 0	0.029424	0.737302	-0.011063	-0.674831	3.191
A_{13}	1 1 0 1	0.780091	0.017110	-0.592842	0.199259	3.248
A_{14}	1 1 1 0	0.124137	0.869804	-0.030985	0.476519	3.388
A_{15}	1 1 1 1	0.827956	0.326276	0.454868	0.033600	3.393

4 Numerical Analysis

Generally, in N-independent communication system ($N \times$ system) as shown in Fig. 3, let p_{e1}, ..., p_{eN} and p_{c1}, ..., p_{cN} be the probability of error and the probability of a correct decision for System 1, ..., System N, respectively. If the input statistics for all systems are same, the combined probability of a correct decision p_c and the combined probability of error p_e at the summed output can be $p_c = p_{c1}^N = (1 - p_{e1})^N$ and $p_e = 1 - (1 - p_{e1})^N \cong N p_{e1}$.

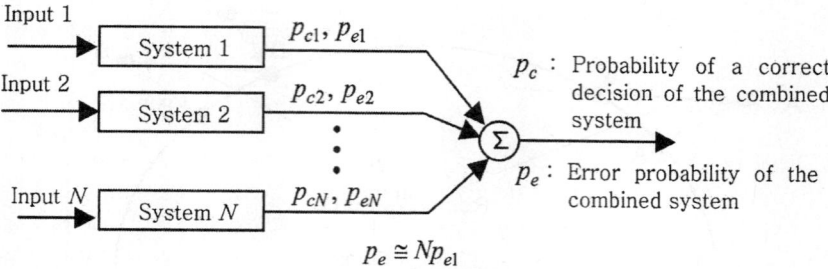

Fig. 3. N-independent communication system

The power requirement of the 4-D MSM systems is compared with that of QPSK as a reference system where the fixed bias scheme is used for all the modulation schemes. For the M-ary PSK including the proposed system and QPSK, the signal is composed of a sum of modulated sinusoids so that the bit error probability can be

$$P_b = Q\left(\sqrt{r^2 A^2 T / 2N_0}\right) \qquad (4)$$

where T represents the rectangular pulse duration, r is the responsivity of photodetector in (1) and $Q(x)$ is the Gaussian error integral, and A is a nonnegative scale factor. In the proposed system, one symbol is transmitted with 4 orthogonal subcarrier signals, while one symbol is transmitted with two subcarriers in Quadrature Phase Shift Keying (QPSK). When we consider the number of input bits and subcarriers, N 4-D MSM system is equivalent to $2N$ x QPSK as described in Fig. 5. The error probability of each scheme is easily calculated with the minimum Euclidean distance. We set the required bit error probability to $P_b=10^{-6}$ [4].

The power requirement in terms of the number of subcarriers is compared, because it affects the required speed of the demodulation electronics and influences the multipath immunity of the signal [2]. Fig. 3 represents the numerical results of the normalized power requirement in optical dB versus the number of subcarriers with fixed bias for three block codes employing QPSK and the proposed scheme. Here, the normalized power requirement for three block codes employing QPSK is the result of [1] for fixed bias. The normalized power requirement of the proposed 4-D MSM is up to 1.8 dB smaller than those of above three QPSK schemes when the number of subcarriers is 2. For the number of subcarriers is 4, 6, and 8, the power requirement is reduced to 1.4 dB.

The power requirement in terms of the bandwidth requirement is compared to measure the electrical bandwidth efficiency of the optical signal. Fig. 4 represents the normalized power requirement in optical dB versus the normalized bandwidth requirement with fixed bias for above three block codes employing QPSK and the proposed scheme. In the range of 1.125 ~ 1.25 of the normalized bandwidth requirement, the proposed system reduces up to 1.4 dB in bandwidth requirement compare to normal QPSK, Res. Subcarrier, Min. Power schemes, respectively.

The proposed system has a larger minimum value than that of QPSK scheme and large Euclidean distances for 16 signal points, so that the required dc bias is minimized and the error rate performance is improved. Hence, for wireless LAN using infrared communication, the 4-D MSM system can be more efficient than a conventional QPSK for transmission of high-speed data via the narrowband channel.

5 Conclusions

This paper has described the basic principles and characteristics of multiple subcarrier modulation techniques in infrared communication for wireless LAN. The optimization of signal waveform technique is used in deriving 4-D vectors for 16 points on the surface of Euclidean sphere having minimum distances between signal points. The 4-D MSM with fixed bias for optical wireless system using 4-D block coder improves the power efficiency. Therefore, the proposed system can operate in wider range with broader band for an optical wireless connection.

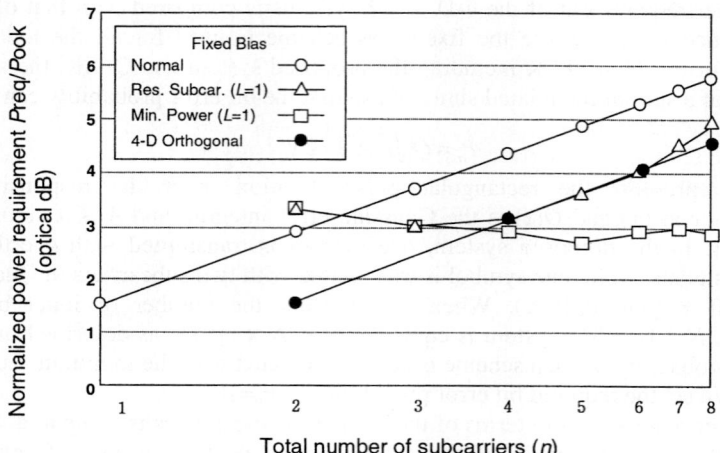

Fig. 4. Normalized power requirement versus total number of subcarriers for Normal QPSK, Res. Subcarrier, Min. Power, and the proposed 4-D MSM system

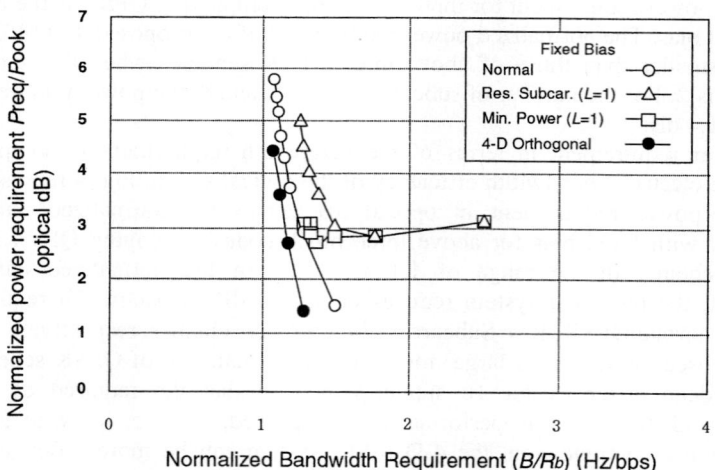

Fig. 5. Normalized power requirement versus normalized bandwidth requirement for Normal QPSK, Res. Subcarrier, Min. Power, and the proposed 4-D MSM system. Each number denotes total number of subcarriers

References

1. IEEE standard: http://standards.ieee.org/getieee802/802.11.html
2. You, R., Kahn, J.: Average Power Reduction Techniques for Multiple-Subcarrier Intensity-Modulated Optical Signals, IEEE Trans. Commun., Vol. 12 (2001) 2164-2171
3. Ohtsuki, T.: Multiple-Subcarrier Modulation in Optical Wireless Communications, IEEE Commun. Mag., Vol. 3 (2003) 74-79

4. Kahn, J., Berry, J.: Wireless Infrared Communications, Proc. IEEE, Vol. 2 (1997) 265-298
5. Teramoto, S., Ohtsuki, T.: Multiple-subcarrier Optical Communication System with Subcarrier Signal Point Sequence, IEEE GLOBECOM 2002, Taipei, Taiwan, Nov. (2002)
6. Yamaguchi, H., Ohtsuki, T., Sasase, I.: Multiple Subcarrier Modulation for Infrared Wireless Systems using Punctured Convolutional Codes and Variable Amplitude Block Codes, IEEE GLOBECOM 2002, Taipei, Taiwan, Nov. (2002)
7. Lachs, G.: Optimization of Signal Waveforms, IEEE Trans. Info. Theory, Vol. 4 (1963) 95-97

Modeling Virtual Network Collaboration in Supply Chain Management

Ha Jin Hwang[*]

Department of Management, Catholic University of Daegu,
Kyungsan-si, Kyungbuk, Korea
hjhwang@cu.ac.kr

Abstract. This study is designed to suggest a collaboration network model and apply it to establish a desirable framework for the textile supply chain management. The challenge in textile supply chain management is the development of collaboration network which accommodates diverse concerns of various participants while explicitly recognizing interdependencies and promoting effective relationship management. Major contents of the study are as follows. First, ideal collaboration network model which can draw a positive collaboration from the supply chain of the textile industry is suggested. Second, utilizing the collaboration model, e-Textile Supply Chain Management (e-TSCM) is designed to improve customer services and delivery time, to promote information sharing, and shorten product life cycle time. e-TSCM is expected to promote corporate innovation and information sharing, generate infrastructure which reduces the gap of the competitiveness across the textile supply chain and enhance the collaboration, which in turn improve the competitiveness of the textile industry.

Keywords: supply chain, competitiveness of textile industry, collaboration network.

1 Introduction

Recently, the local textile industry has experienced a rapid down of competitive edge in global market. The local textile companies, mainly small-medium sized. Companies, have focused on exporting fabric cloths based on mass production. The industry structure of heavy dependence on small-medium sized companies because an obstacle for restructuring and the mass production of simple fabrics oriented products prevented the industry from differentiating products and introducing high value-added products. Furthermore, China and South-east Asian countries which benefited from low wages made inroads into existing overseas markets. In an effort to overcome this kind of problems, various attempts such as the development of new products and production techniques, shifting to industrial materials, restructuring of the industry, and enhancing the overseas marketing campaigns have been tried. However, the

---

[*] This Research is Supported by 2005 Research Fund of Catholic university of Daegu.

significant outcome has not yet been realized except in some large-size companies. This was due to the fact that the small-medium sized companies had to deal with lack of money and enterprise capabilities. With the rapid expansion of internet, e-business has come up as a candidate to solve the down-sloping of competitive edge of the local textile industry, especially, the supply chain management of the textile industry which constitutes very complex supply-demand structure and value chain. The supply chain management has been accepted as an alternative to improve the competitive power. SCM became a general and strategic concept of dealing with efficient logistics and network collaboration within a same value chain. Attempts to apply Quick Response system to some Korean textile companies has not resulted in favorable outcome. Rather, misunderstanding of Quick Response as the introduction of new information technology, short delivery time improvement, and small-lot production has emerged with unexpected failures. The emphasis was given to increase productivity and improve efficiency level of logistics without concentrating on the entire supply chain and the collaboration of business partners. Research findings confirm that SCM has contributed to reduction of inventory and purchasing cost, shortening the business process, lead-time and sales promotion planning time, and enhancing delivery time, increasing sales revenue and decreasing defective rates. This study is designed to explore SCM as an innovative alternative to improve competitive power of the local textile industry. In order to fulfill research objective, the desirable SCM model is suggested with ideal collaboration network so that the local textile companies can find a solution to handle their structural drawbacks and strategic problem in value chain management.

2 Literature Review

2.1 Supply Chain and SCM of Textile Industry

In recent years, companies are in the race for improving their corporate competitiveness in order to compete in the 21st century global market. This market is electronically connected and dynamic in nature. Therefore, companies are trying to improve their strategic response level with the objective of being flexible and responsive to meet the changing market requirements. In an effort to achieve this, many companies have decentralized their value-adding activities by outsourcing and developing virtual enterprise (VE). All these highlight the importance of information technology (IT) in integrating suppliers/partnering firms in virtual enterprise and supply chain. Supply chain management (SCM) is an approach that has evolved out of the integration of these considerations. SCM is defined as the integration of key business processes from end user through original suppliers that provides products, services, and information and hence add value for customers and other stakeholders (Lambert et al., 1998). The concepts of supply chain design and management have become a popular business paradigm in these days. This has intensified with the development of information and communication technologies that include electronic data interchange (EDI), the Internet and World Wide Web(WWW) to overcome the ever-increasing complexity of the systems driving buyer-supplier relationships. The complexity of SCM has also forced companies to improve online network communication systems. Supply chain management emphasizes the overall and long-term benefit of all parties on the chain through co-operation and information sharing. This signifies the importance of communication and the application of IT in SCM. Information sharing between members of a supply chain using EDI technology should be increased to reduce uncertainty and enhance shipment performance of suppliers and greatly improve the performance of the supply chain system (Srinivasan et al., 1994). Companies need a large investment for redesigning internal organizational and technical processes, changing traditional and fundamental product distribution channels and customer service procedure and training staff to achieve IT-enabled supply chain (Motwani et al., 2000). The followings are some of the problems often cited in the literature both by the researchers and practitioners when developing an IT-integrated SCM: lack of integration between IT and business model, lack of proper strategic planning, poor IT infrastructure, insufficient application of IT in virtual enterprise, and inadequate implementation knowledge of IT in SCM. There is no comprehensive framework available on the application of IT for achieving and effective SCM. Considering the importance of such a framework, an attempt has been made in this paper to develop such a framework to provide more effective management of whole supply chain. In a supply chain world, suppliers, finished goods producers, service providers, and retailers are required to create and deliver the best products and services possible. Collaboration enables a company to do exceptionally well a few things for which it has unique advantages. Other activities are shifted to channel members that possess superior capabilities. However, there are several underlying themes. Outstanding supply chain companies stay customer-centric, focus on process management, invest in IT as a capability enabler, and are obsessed with performance measurement. Supply chain management is the collaborative design and management of seamless value-added processes to meet the real needs of the end customer. The development and integration of people and technological resources as well as the coordinated management of materials, information and financial flows are critical to successful supply chain integration. SCM's goal is to establish unique value-added processes that satisfy customers better and more efficiently than the competition. Managing outstanding processes across functional and organizational boundaries require dramatic and often painful changes in both thinking and behavior.

The supply chain of the textile industry consists of the distribution structure which includes manufacturer, wholesaler, retailer, and consumer of raw silk, 원단 , dyeing, and apparel. The textile industry has a very complex value chain structure and requires a complex processes to supply products to consumers. Even though the structure of value chain is very complicated, the links of the value chain is independently separated and the communication among the companies are carried out as needed without any particular methods.

Looking into the level of information systems based on the streams, most companies are very weak in sharing information and information technology capabilities. In terms of the IT infrastructure, broadband internet, point of sale(POS) and EDI(electronic data interchange) are common while ERP, SCM, CRM, and KM are at its stage of beginning. It was found that the textile companies are not actively utilizing the information sharing and therefore management of the generation, storage and distribution of information, is not systematically done. Furthermore, due to the short life cycle of the textile products, it is very difficult to standardize the products except some categories such as raw materials, yarn, gray fabrics, and the level of innovation is also extremely low.

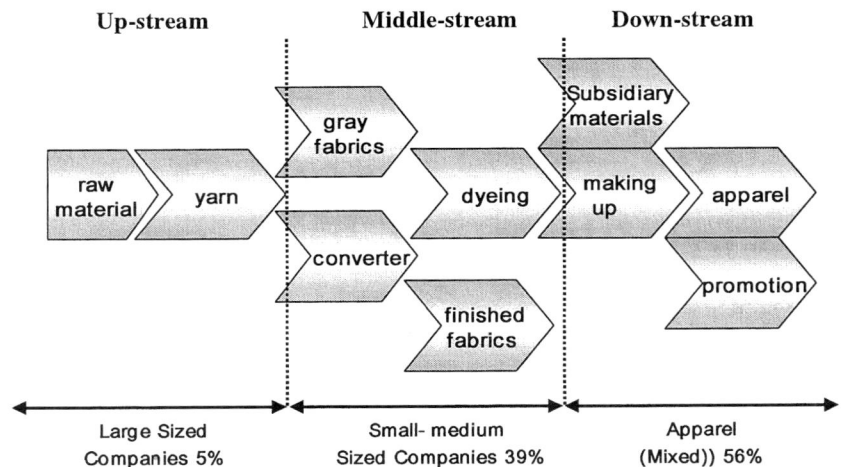

Fig. 1. Supply chain of textile industry

2.2 Network Collaboration

A firm's position in the network is dependent upon the nature of the direct and indirect relationships it has with other actors in the network (Johanson & Mattsson, 1992). As firms are as much the product of their relationships and network position as they are the result of the firm's own strategic actions and intentions (hakansson & Ford, 2002), attention is gradually shifting from the control of business networks to one of greater participation and adaptation in which the participating firms must be more flexible and adaptable (Wilkinson & Young, 2002). As a network is a set of connected relationships between firms (hakansson & Johanson, 1993), effects will flow through the various relationships that the focal firm has established with other connected actors. Connectedness is the extent to which exchange in one relation is contingent upon exchange in another (Cook & Emerson, 1978). Moreover, two connected relationships can be directly or indirectly connected to many other relationships that may have some bearing on each firm as part of a larger business network (Anderson, Hakansson, & Johanson. 1994). Thus, collaboration within one relationship will affect relationships with other closely connected actors, making the collaboration process and its outcomes contingent upon the goals of the network rather than the dyad. Hakansson and Ford(2002) describe how firms embedded in business networks are interdependent on other firms in the network. This interdependence implies that firms have limited discretion to act or to build independent strategy (Gadde et al., 2003). As a result, the outcomes of the firm's actions are strongly influenced by the attitudes and actions of those firms with whom the focal firm has relationships. But who are the relevant others and how can they be determined? Network structure and network position affect how network collaboration will occur and between which network actors' collaboration will take place. Developing an understanding of network structure will enable firms to consider with whom they may be directly or indirectly affected and affected by. However, a business network does not have a natural centre or clean borders making network structure a fluid concept that invariably changes over time (Hakansson & Snehota, 1995). Networks organizations can be described by the density, multiplexity, and reciprocity of ties, and a shared value system that defines membership roles and responsibilities(Achrol, 1997). If the overall collaborative efforts of the network are well directed, the network may become more of a network organization than a network of linkages. Examples can be derived from technology networks where R+D organizations, products, and distributors closely coordinate their activities to provide new products to the market in a timely manner. Networks have both economic and social dimensions that are important for the optimal operation of the network. This implies that many aspects of business relationships cannot be formalized or based on legal criteria (contracts) (Gadde et al., 2003). Collaboration involves both aligning the economic goals and aims of the network and the development of the social dimensions - in particular, mutual trust and commitment. Trust is the critical determinant of a good relationship (Dwyer, Schurr, & Oh, 1987). Anderson and Narus(1990) view trust as the belief that the partner will perform actions that will result in positive outcomes for the firm and not to take unexpected actions that may result in negative outcomes. Moorman, Deshpande, and Zaltman(1993) define trust as the willingness to rely upon an exchange partner in whom one has confidence. They describe trust as a belief, a sentiment, or an expectation abort an exchange partner that results from the partner's expertise, reliability, and intentionality. Power is an essential characteristic of social organization and an inevitable instrument for interorganizational coordination. While the power to coordinate is the prerogative of the dominant firm, the use of reward power, coercive power, and legitimate authority is seldom conducive to the evolution of network organizations (Achrol, 1997). Furthermore, the more a single firm seeks to control the network, the less effective and innovative the network will become (Hakansson & Ford, 2002). Where development processes are directed by just one firm, there is a greater risk that the network will become a hierarchy with the reduced potential for innovation(Gadde et al., 2003). Communication has been described as the glue that holds together a channel of distribution (Mohr & Nevin, 1990). Communication enables information to be exchanged that may reduce certain types of risk perceived by either one of the parties to the transaction (McQuiston, 1989). Any uncertainty about a customer's or supplier's organizational structure, viability, methods of operation, technical expertise, or competence can be resolved by communication between the parties. Communication not only improves a firm's credibility but may also provide a convenient and simple means of gaining knowledge about the market (Cunningham & Tumbull, 1982). Communication may also facilitate other elements of the interaction, such as adaptations by suppliers and customers to the design or application of a product, or the modification of production, distribution, and administrative systems by either party. While effective communication may enable the firm to differentiate its product from the competitor's offerings (McQuiston, 1989), meaningful communication and cooperation between firms is a necessary antecedent of trust (Anderson & Narus, 1990).

3 e-Textile Supply Chain Management (e-TSCM)

3.1 Concept of e-TSCM

The e-TSCM is the communications and operations framework of a collaboration network that links textile suppliers, business partners and customers together as one cohesive, collaborating entity. A collaboration network is a series of value added-processes/stages owned by one or more enterprises, starting with material/information suppliers and ending with consumers (Papazoglou et al., 2000; Gek Woo et al., 2000). Each intermediate stage is a supplier to its adjacent downstream stage and a customer to its upstream stage. That means that participants may assume many different roles in a supply chain network, but all relationships come down to a supplier and a customer role. e-TSCM efficiently utilizes information and knowledge, competes on agility and speed, and views collaboration as a competitive strategic weapon. A supply chain must coordinate with each other in order to optimize the process within a supply chain (Cooper et al., 1997). Collaboration between suppliers, manufacturers and retailers can improve the number of satisfied customers by reducing lead times, improving service levels and decreasing costs. In this paper a e-TSCM model is developed in order to integrate the technical and organizational infrastructure, to facilitate business communication between the participant members, to identify and synchronize the specific roles and responsibilities of the partners, to organize the relationship interface between the partners, and to enable intelligent decisions based on knowledge acquisition. The model provides a useful framework for the planning, implementation and evaluation of supply chain collaboration in practice. The primary objectives of the model are as follows: First, it is developed to coordinate the activities of each partner and the transition between partner exchanges. Second, it is designed to facilitate the efficient flow of products, services down the supply chain minimizing the cost and the time while maximizing the quality, service and credibility. Third, it seeks to match the supply with the market demand, based on partners' relationship management and knowledge.

3.2 e-TSCM Architecture

e-TSCM enables members of the textile supply chain to be equipped with the collaborative management and monitoring of disparate companies-members of the supply chain. It captures the required information and sets the procedures and accountability, performance measurement criteria, and capabilities to resolve exceptional cases. This provides companies with flexibility and control for effective business models, and generates a mechanism to analyze and understand the impact of collaborative business processes on its own operations (Mamoukaris, et al., 2000). Business partners in a collaboration networks make cooperative efforts in the forecasting, purchasing, production and inventory management and synchronize delivery and distribution schedules. e-TSCM standardizes best practices through out supply chain, using appropriate technology to reinforce relationships between business partners. An integrated e-supply chain e-TSCM captures and stores partners' transactions and supply chain activities through various touch points, and data from transactional systems and external sources. A centralized partner data warehouse with a reliable, scalable and highly available storage infrastructure solves the problem of data integration of diverse data assets. Figure 2 shows the basic components and architecture of an integrated e-TSCM solution. The output of the intelligence module should be delivered as an extensible application that uses a set of partners' profile and profitability models and reports. Partners' analysis results should integrate with supply chain management decisions in order to transform partners' information into building better relationships. A data mart, in the context of a PRM system, is a decision support system incorporating a subset of the partners' data focused on specific supply chain applications or activities. Data marts allow for greater flexibility or increased performance. However, the data mart must be incorporated into the overall partners' data warehouse and managed and populated from this central data warehouse.

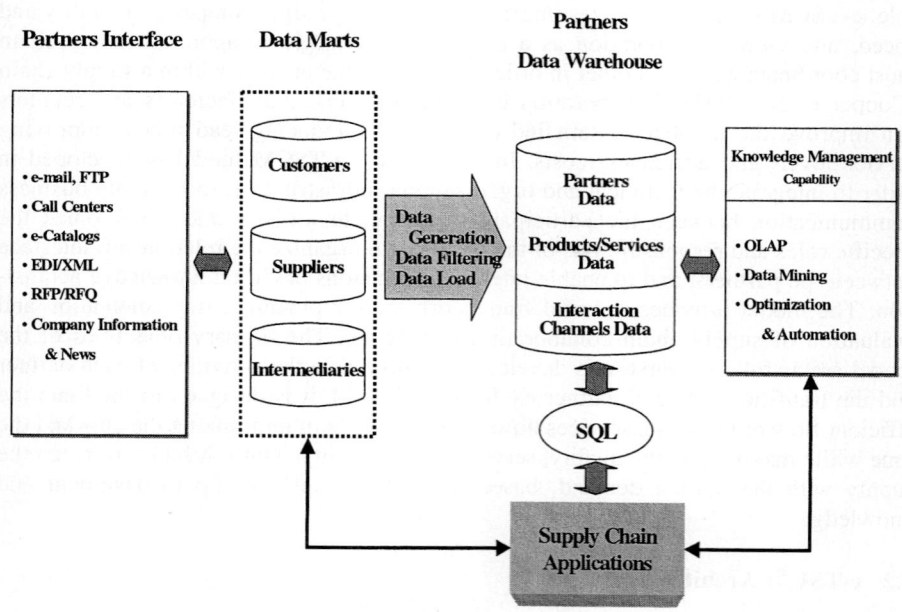

Fig. 2. e-TSCM architecture

Supply chain partners in a virtual network need a comprehensive view of their business, and greater insights into supply chain channels and process to improve decision making and business operations, as well as to adapt systematically and rapidly to market fluctuations. The e-supply chain intelligence module tracks collaborative channel events and processes, and extracts and presents decision oriented information. Partners' data analysis processing allows e-supply chain network members to derive information and partners' intelligence from data warehouse systems by providing tools for querying and analyzing data, leading to multidimensional view of the specific partners. Knowledge management capabilities such as analysis software, data mining software, optimization and automation software, web-enabled technology, and campaign management software can be used in order to transform the data from the partners' data warehouse and the supply chain applications int useful partners' knowledge. (Warkentin et al., 2001). The output of partners' data analysis are useful to evaluate partners' readiness to collaborate and to compare and analyze real-time business performance and customer satisfaction. A service level agreement(SLA) is a contracting tool keyed to a client's service performance expectations identifying the respomsibilities of both the service supplier and customer (Folinas et al., 2001). The e-supply chain value proposition for customers consists of the following service level criteria: service price, credit policy, quality, fulfillment time, creditability and legal issues. The establishment of the customer requirements, the determination of service supplier capacity to meet them are performed in the e-supply chain PRM module, whereas, the development of SLA system, the negotiation, the evaluation and maintenance are accomplished in the intelligence module.

4 Conclusions

As the global competition becomes more fierce and the customer expectation gets higher, many companies have turned to supply chain management to leverage the resources and build more collaborative business relationships. It is evident that textile industry facing revere competition and losing competitive power needs to develop an effective strategy which delvers innovative, high quality, low-cost products on time with shorter product life cycle time and better customer services. e-supply chain has become recognized as a core competitive strategy. As organizations continuously seek to provide their products and services to customers faster, cheaper and better than the competition, managers have come to realize that they cannot do it alone; rather, they must work on a cooperative basis with the best organizations in their supply chains in order to succeed. Members of a supply chain network in a virtual environment use technology and management collaboratively to improve business operations in terms of speed, agility, real-time control, and customer response. Moving from traditional supply chains to virtual chain networks requires that partners focus on communications, relationships, and knowledge. Business is about an integrated set of relationships. Technology, quality, cost availability and collaborative business practices are important to each business partner in the supply chain network. Once partners enter into a business relationship mutual success will depend on trust, information and knowledge sharing, communication, and co-owned product service design and performance measures. The proposed model introduces an e-supply chain collaboration framework where the necessary modules are designed in order to guide partners of a virtual network to achieve strategic and tactical capabilities. Information technologies, integrated telecommunications networks, multimodel transportation systems, knowledge centers, commercial and service support, technical and organizational infrastructure, are the main elements supporting TSCM networking.

The TSCM allows companies to build or dissolve relationships quickly and efficiently as appropriate, and measure channel performances to boost and improve profitability and deliver transactions and customer satisfaction. As companies in a supply chain shift their business models to work in virtual networks, the form of relationships becomes critical to success. A network's success will strongly depend on its relationships with its business partners and on its customer/partner knowledge assets. The success of an e-supply chain will depend upon the choice of the specific partners in the supply chain and on the way in which they co-operate efficiently and effectively with each other.

References

[1] Anderson, J. C., Hakansson, H., & Johanson, J. (1994), Dyadic Business Relationships Within a Business Network Context, Journal of Marketing, 58, pp. 1-15.
[2] Batt, P. J., & Purchase, S. (2004), Managing Collaboration Within Networks and Relationships, Industrial Marketing Management, 33, pp. 169-174.
[3] Boubekri, N. (2001), Technology Enablers for Supply Chain Management, Integrated Manufacturing Systems, 12(6), pp. 394-399.
[4] Croom, S. (2001), Restructuring Supply Chains Through Information Channel Innovation, International Journal of Operations & Production Management, 21(4), pp. 504-515.
[5] Eng T. Y. (2004), The Role of e-Marketplaces in Supply Chain Management, Industrial Marketing Management, 33, pp. 97-105.
[6] Fawcett, S. E., & Magnan, G. M. (2004), Ten Guiding Principles for High-Impact SCM, Business Horizons, 47(5), pp. 67-74.
[7] Gadde, L. E., Hoemer, L., & Hakansson, H. (2003), Strategizing in Industrial Networks. Industrial Marketing Management, 32, pp. 357-364.
[8] Gulati, R., Nohria, N., & Zaheer, A. (2000), Strategic Networks, Strategic Management Journal, 21, pp. 203-215.
[9] Gunasekaran, A., & Ngai, E. W. T. (2003), Information Systems in Supply Chain Integration and Management, European Journal of Operational Research, 159, pp. 269-295.
[10] Hakansson, H., & Ford, D. (2002), How Should Companies Interact in Business Network, Journal of Business Research, 55, pp. 133-139.
[11] Humphreys, P. K., Lai, M. K., & Sculli, D. (2001), An Inter-organizational Information System for Supply Chain Management, International Journal of production Economics, 70, pp. 245-255.
[12] Lambert, D. M., & Cooper, M. C. (2000), Issues in Supply Chain Management, Industrial Marketing Management, 29, pp. 65-83.
[13] Lau, H. C. W., & Lee, W. B. (2000), On a responsive Supply Chain Information System, International Journal of Physical Distribution & Logistics, 30(7/8), pp. 598-610.
[14] Manthou, V., Vlachopoulou, M., & Folinas, D. (2004), Virtual e-Chain(VeC) Model for Supply Chain Collaboration, International Journal of Production Economics, 87, pp. 241-250.
[15] Motwani, J., Madan, M., & Gunasekaran, A. (2000), Information Technology in Managing Supply Chains, Logistics Information Management, 13(5), pp. 320-327.
[16] Murillo, L. (2001), Supply Chain Management and The International Dissemination of e-Commerce, Industrial Management & Data Systems, 101(7), pp. 370-377.
[17] Overby, J. W., & Min, S. (2001), International Supply Chain Management in an Internet Environment, International Marketing Review, 18(4), pp. 392-420.
[18] Salcedo, A., & Grackin, A. (2000), The e-Value Chain, Supply Chain Management Review, 3(4), pp. 63-70.
[19] Spekman, R. E., Spear, J., & Kamauff, J. (2002), Supply Chain Competency: Learning As a Key Component, Supply Chain Management: An International Journal, 7(1), pp. 41-55.
[20] Srinivasan, K., Kekre, S., & Mukhopadhyay, T. (1994), Impact of Electronic Data Interchange Technology on JIT Shipments, Management Science, 40, pp. 1291-1304.

[21] Tan, K. C. (2001), A Framework of Supply Chain Management Literature, European Journal of Purchasing & Supply Management, 7, pp. 39-48.
[22] Tan, K. C., Handfield, R. B., & Krause, D. R. (1998), Enhancing Firm's Performance Through Quality and Supply Base Management: An Empirical Study. International Journal of Production Research, 36(10), pp. 2813-2837.
[23] Teo, T. S. H., Ang, J. S. K. (1999), Critical Success Factors in The Alignment of IS Plans With Business Plans, International Journal of Information Management, 19, pp. 173-185.
[24] Tracey, M., Smith-Doerflein, K. A. (2001), Supply Chain Management: What Training Professionals Need to Know. Industrial and Commercial Training, 33(3), pp. 99-104.
[25] Wilkinson, I., & Young, L. (2002), On Cooperating: Firms, Relationships, Networks, Journal of Business Research, 55. pp. 123-132.
[26] Yu, Z., Yan, H., & Cheng, T. C. E. (2001), Benefits of Information Sharing With Supply Chain Partnerships, Industrial Management & Data Systems, 101(3), pp. 114-119.

SPA-Resistant Simultaneous Scalar Multiplication

Mun-Kyu Lee

School of Computer Science and Engineering,
Inha University, Incheon 402-751, Korea
mklee@inha.ac.kr

Abstract. The Simple Power Analysis (SPA) attack against an elliptic curve cryptosystem is to distinguish between point doubling and point addition in a single execution of scalar multiplication. Although there have been many SPA-resistant scalar multiplication algorithms, there are no known countermeasures for simultaneous scalar multiplication. In this paper, we propose an SPA-resistant simultaneous scalar multiplication algorithm using scalar recoding. The computational and memory overheads of our scheme are almost negligible.

1 Introduction

Since Koblitz [1] and Miller [2] proposed to use elliptic curves in cryptography, a lot of attention has been paid to efficient and secure implementation of elliptic curve cryptosystems (ECCs). For an ECC, the most dominant operation is a scalar multiplication kP, where k is an integer and P is a point on an elliptic curve. Note that k should be kept secret in many cases.

Power analysis attacks, which are introduced by Kocher et al. [3], are techniques that enable an adversary to learn secret information contained inside a device by monitoring its power consumption. There are two kinds of power analysis attacks, i.e., Simple Power Analysis (SPA) and Differential Power Analysis (DPA). An SPA is to analyze a single execution of a cryptographic algorithm, while a DPA is to observe many executions and to analyze them using a statistical method.

As for power analysis attacks on ECCs, Coron [4] first showed that the naive implementations of scalar multiplication algorithms may leak information on secret k if SPA or DPA is applied. Since that time, various countermeasures using randomization and scalar recoding have been proposed [4, 5, 6, 7, 8], and also various countermeasures using special forms of curves have been proposed [9, 10, 11].

Although there have been many SPA- or DPA-resistant scalar multiplication algorithms, there are no known countermeasures for simultaneous scalar multiplication $kP + lQ + mR + \cdots$. In this paper, we propose an SPA-resistant simultaneous scalar multiplication algorithm with very small overheads.

2 Preliminaries

2.1 Elliptic Curves and Scalar Multiplication

An *elliptic curve* is the set of points (x, y) that satisfy an equation of the form [12]:

$$y^2 + a_1 xy + a_3 y = x^3 + a_2 x^2 + a_4 x + a_6.$$

It is well known that the set of points on an elliptic curve, together with a special point O called the *point at infinity*, forms an Abelian group under *point addition* operation. A point addition is to compute $P + Q$ according to a geometric rule called the *chord-and-tangent rule* when two points P and Q on the curve are given. A special case of a point addition is a *point doubling*, which is to compute $2P$ when a point P is given. Using these operations, we can construct a basic algorithm for scalar multiplication as follows.

Algorithm 1 Double-and-add scalar multiplication

Input: point P; integer k in the binary representation $(k_{l-1}, k_{l-2}, \ldots, k_0)$, where $k_{l-1} = 1$ is the MSB.
Output: point $R = kP$.
  1. $R \leftarrow P$.
  2. for i from $l - 2$ to 0 do
  3. $\quad R \leftarrow 2R$.
  4. $\quad$ if $k_i = 1$ then $R \leftarrow R + P$.
  5. od.

If we want to compute the sum of several scalar multiples $k^{(1)}P_1 + k^{(2)}P_2 + \cdots + k^{(n)}P_n$ when n integers $k^{(1)}, k^{(2)}, \ldots, k^{(n)}$ and n points $P_1, P_2, \ldots, P_n$ are given, then we can use simultaneous scalar multiplication ([13],p.618). For simplicity, we give an algorithm for $n = 2$. See Algorithm 2.

Algorithm 2 Simultaneous scalar multiplication for $n = 2$

Input: points P and Q; integers $k = (k_{l-1}, k_{l-2}, \ldots, k_0)$ and $m = (m_{l-1}, m_{l-2}, \ldots, m_0)$
Output: point $R = kP + mQ$.
  1. Construct a pre-computation table T s.t.
     $T[0, 0] \leftarrow 0P + 0Q = O$; $T[0, 1] \leftarrow 0P + 1Q = Q$;
     $T[1, 0] \leftarrow 1P + 0Q = P$; $T[1, 1] \leftarrow 1P + 1Q = P + Q$.
  2. $R \leftarrow T[k_{l-1}, m_{l-1}]$.
  3. for i from $l - 2$ to 0 do
  4. $\quad R \leftarrow 2R$.
  5. $\quad R \leftarrow R + T[k_i, m_i]$.
  6. od.

2.2 Simple Power Analysis and Coron's Countermeasure [4]

Power analysis attacks against scalar multiplication are based on the observation that it is possible for an attacker to distinguish between point doubling and point addition through power consumption analysis, although he has no direct access to the internal private information k. These attacks are particularly effective for customized small devices such as smart cards, where most of the power is consumed only for cryptographic operations. For example, assume that a smart card executes a scalar multiplication using Algorithm 1, and that its power consumption trace is as Fig.1(c).[1] At the beginning, an attacker doesn't know the exact behavior of an addition or a doubling, i.e., (a) and (b) of Fig.1. However, if he obtains the complete power trace (c) by monitoring the smart card, then he can recognize doublings, since he knows the fact that generally doublings are much more frequently performed than an addition. Hence, he can find out that the power trace (c) represents $DADADDAD\ldots$, where A and D mean an addition and a doubling, respectively. Then he decomposes this operation sequence into blocks $DA, DA, D, DA, D, \ldots$, where each block corresponds to an iteration of the for loop in Algorithm 1. Finally, he recovers the key $k = 11010\ldots$.

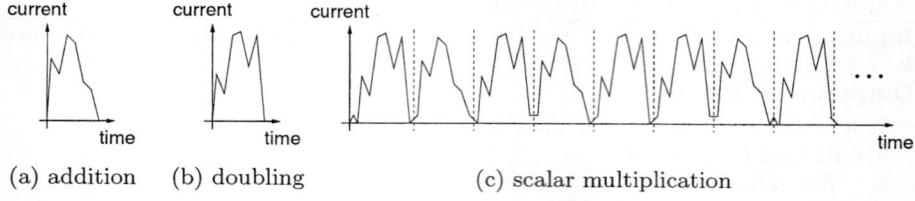

Fig. 1. Examples of power consumption traces

Coron [4] proposed to modify Algorithm 1 so that the sequence of doublings and additions may not depend on the data being processed and power consumption behavior is independent of individual bits of k. In Algorithm 3, every iteration of for loop contains one doubling and one addition, and the operation sequence is $DADADADA\ldots$, irrespective of the value of k.

Algorithm 3 Double-and-add scalar multiplication resistant against SPA

Input: points P and integer $k = (k_{l-1}, k_{l-2}, \ldots, k_0)$.
Output: point kP.
1. $R[0] \leftarrow P$.
2. for i from $l - 2$ to 0 do
3. $R[0] \leftarrow 2R[0]$.
4. $R[1] \leftarrow R[0] + P$.
5. $R[0] \leftarrow R[k_i]$.   // no branch instruction
6. od.
7. Output $R[0]$.

Note that, however, since a dummy operation is inserted in line 4, Algorithm 3 becomes slower than Algorithm 1. If we assume that the half of the bits of k are

---

[1] Note that this is not a measurement from a real-world implementation, but it is a simple example fabricated for explanation purpose.

ones on the average, then Algorithm 1 requires $(l-1)$ doublings and $(l-1)/2$ additions, while Algorithm 3 requires $(l-1)$ doublings and $(l-1)$ additions. Hence the computational overhead of Algorithm 3 over Algorithm 1 is about 33.3%, if we set the computational costs of an addition and a doubling as the same according to the literature.

3 SPA Countermeasure Using a Scalar Recoding Method

In this section, we consider SPA against Algorithm 2, and propose a countermeasure to prevent it. We begin by seeing the possibility of SPA. At a glance, it seems that lines 4 and 5 in Algorithm 2 correspond to a doubling and an addition, respectively. Note that, however, line 5 becomes just an assignment if $k_i = 0$ and $m_i = 0$, since $R \leftarrow R+O$ is equivalent to $R \leftarrow R$. Hence, Algorithm 2 reveals the information on integers k and m, enabling an attacker to distinguish the case $(k_i, m_i) = (0,0)$ from the other cases.

To prevent an SPA, we should convert $R+T[0,0]$ to $R+Z$ with a certain point $Z \neq O$ so that a 'real addition' occurs. In advance of a scalar multiplication, we scan the two scalars k and m in parallel from the least significant bits, and recode them so that an addition with O may never appear. That is, we transform $(k_i, m_i) = (0,0)$ to another adequate digit pair with at least one non-zero digit. For this, we should consider the adjacent pair (k_{i+1}, m_{i+1}) together. Fixing $(k_i, m_i) = (0,0)$, there are four possible cases according to (k_{i+1}, m_{i+1}) as follows.

1. If $(k_{i+1}, m_{i+1}) = (0,0)$, transform $[(k_{i+1}, m_{i+1}), (k_i, m_i)] = [(0,0), (0,0)]$ into $[(0,1), (0,-2)]$. Note that this transformation does not affect the result of scalar multiplication.
2. If $(k_{i+1}, m_{i+1}) = (0,1)$, transform $[(k_{i+1}, m_{i+1}), (k_i, m_i)] = [(0,1), (0,0)]$ into $[(0,2), (0,-2)]$.
3. If $(k_{i+1}, m_{i+1}) = (1,0)$, transform $[(k_{i+1}, m_{i+1}), (k_i, m_i)] = [(1,0), (0,0)]$ into $[(1,1), (0,-2)]$.
4. If $(k_{i+1}, m_{i+1}) = (1,1)$, transform $[(k_{i+1}, m_{i+1}), (k_i, m_i)] = [(1,1), (0,0)]$ into $[(1,0), (0,2)]$.

Now we modify Algorithm 2 so that the sequence of doublings and additions may not depend on the data being processed. See Algorithm 4.

Note that the table can be implemented using a one-dimensional array by introducing another alphabet set, e.g. $\{[0,1], [1,0], [1,1], [0,2], [0,-2]\} \rightarrow \{1,2,3,4,5\}$, so that there may not be any conditional branch to fetch a table element in lines 13 and 16.

Now we consider the overhead of new algorithm. Note that Algorithm 2 requires $(l-1)$ doublings and $1+3(l-1)/4$ additions, if we assume that about a quarter of the bit pairs (k_i, m_i) are $(0,0)$ on the average. Also, it requires memory space for four pre-computed points. On the other hand, Algorithm 4 requires $1+(l-1)$ doublings and $1+(l-1)$ additions. The number of points to be pre-computed and stored is five, i.e., a zero pair $(0,0)$ is replaced with two non-zero pairs $(0,2)$ and $(0,-2)$.

Algorithm 4 Simultaneous scalar multiplication resistant against SPA

Input: points P, Q; integers k, m.
Output: point $kP + mQ$.
// Recoding of scalars
1. $i \leftarrow 0$;
2. while $(i < l - 1)$ do
3.   if $(k_i, m_i) = (0, 0)$ then do
4.   if $(k_{i+1}, m_{i+1}) = (0, 0)$ then set $(k_{i+1}, m_{i+1}) \leftarrow (0, 1)$ and $(k_i, m_i) \leftarrow (0, -2)$;
5.   else if $(k_{i+1}, m_{i+1}) = (0, 1)$
         then set $(k_{i+1}, m_{i+1}) \leftarrow (0, 2)$ and $(k_i, m_i) \leftarrow (0, -2)$;
6.   else if $(k_{i+1}, m_{i+1}) = (1, 0)$
         then set $(k_{i+1}, m_{i+1}) \leftarrow (1, 1)$ and $(k_i, m_i) \leftarrow (0, -2)$;
7.   else set $(k_{i+1}, m_{i+1}) \leftarrow (1, 0)$ and $(k_i, m_i) \leftarrow (0, 2)$.
8.   set $i \leftarrow i + 2$ and goto step 2.
9.   od.
10.   else set $i \leftarrow i + 1$.
11. od.
// Scalar multiplication
12. Construct a pre-computation table T s.t.
    $T[0, 1] \leftarrow 0P + 1Q = Q$; $T[1, 0] \leftarrow 1P + 0Q = P$; $T[1, 1] \leftarrow 1P + 1Q = P + Q$;
    $T[0, 2] \leftarrow 0P + 2Q = 2Q$; $T[0, -2] \leftarrow 0P - 2Q = -T[0, 2]$.
13. $R \leftarrow T[k_{l-1}, m_{l-1}]$.
14. for i from $l - 2$ to 0 do
15. $R \leftarrow 2R$.
16. $R \leftarrow R + T[k_i, m_i]$.
17. od.

According to the assumption used in the literature, set the computational costs for a doubling and an addition are equivalent. Then the computational overhead of Algorithm 4 over Algorithm 2 is only about

$$\frac{2l}{l + 3(l-1)/4} - 1 = \frac{l+3}{7l-3}$$

and the memory overhead is just one additional point.

4 Generalization to $n \geq 3$

In this section, we generalize our result to the simultaneous scalar multiplication $k^{(1)}P_1 + k^{(2)}P_2 + \cdots + k^{(n)}P_n$ with $n \geq 3$. First, we see that Algorithm 2 can be generalized to the simultaneous scalar multiplication with $n \geq 3$ as follows.

- In line 1, 2^n points $T[0, \ldots, 0] = 0P_1 + \cdots + 0P_n$, ..., $T[1, \ldots, 1] = 1P_1 + \cdots + 1P_n$ are pre-computed.
- Lines 2 and 5 are changed into $R \leftarrow T[k_{l-1}^{(1)}, \ldots, k_{l-1}^{(n)}]$ and $R \leftarrow R + T[k_i^{(1)}, \ldots, k_i^{(n)}]$, respectively.

Algorithm 5 Scalar recoding for resistance against SPA ($n \geq 2$)

Input/Output: n scalars $k^{(1)}, \ldots, k^{(n-1)}, k^{(n)}$.
1. $i \leftarrow 0$;
2. while $(i < l - 1)$ do
3.   if $(k_i^{(1)}, \ldots, k_i^{(n-1)}, k_i^{(n)}) = (0, \ldots, 0, 0)$ then do
4.   if $k_{i+1}^{(n)} = 0$, then set $k_{i+1}^{(n)} \leftarrow 1$, $k_i^{(n)} \leftarrow -2$;
5.   else if $(k_{i+1}^{(1)}, \ldots, k_{i+1}^{(n-1)}, k_{i+1}^{(n)}) = (0, \ldots, 0, 1)$ then set $k_{i+1}^{(n)} \leftarrow 2$, $k_i^{(n)} \leftarrow -2$;
6.   else set $k_{i+1}^{(n)} \leftarrow 0$, $k_i^{(n)} \leftarrow 2$.
7.   set $i \leftarrow i + 2$ and goto step 2.
8.   od.
9.   else set $i \leftarrow i + 1$.
10. od.

Note that this algorithm is vulnerable to an SPA since an attacker can distinguish an addition with $T[0, \ldots, 0]$ from an addition with any other table element. Hence, we should recode the n-tuple $(k_i^{(1)}, \ldots, k_i^{(n)}) = (0, \ldots, 0)$ to an n-tuple with at least one non-zero element. Now we give a recoding algorithm which produces the n-tuple set

$$\{(0, \ldots, 0, 1), (0, \ldots, 1, 0), \ldots, (1, \ldots, 1, 1), (0, \ldots, 0, 2), (0, \ldots, 0, -2)\}$$

with $(2^n + 1)$ elements. Algorithm 5 scans n scalars $(k_i^{(1)}, \ldots, k_i^{(n)})$ in parallel from the LSBs, and recodes tuples $(k_i^{(1)}, \ldots, k_i^{(n-1)}, k_i^{(n)}) = (0, \ldots, 0, 0)$ to non-zero tuples. As in the example of $n = 2$ in the previous section, we consider $(k_i^{(1)}, \ldots, k_i^{(n-1)}, k_i^{(n)})$ and its adjacent tuple $(k_{i+1}^{(1)}, \ldots, k_{i+1}^{(n-1)}, k_{i+1}^{(n)})$ together.

Table 1. Number of point additions or doublings for a simultaneous scalar multiplication

		number of point operations		computational
		original algorithm*	SPA-resistant algorithm**	overhead
$l = 160$	$n = 2$	279	320	14.7%
	$n = 3$	302	323	7.0%
	$n = 5$	339	345	1.8%
$l = 192$	$n = 2$	335	384	14.6%
	$n = 3$	362	387	6.9%
	$n = 5$	402	409	1.7%
$l = 256$	$n = 2$	447	512	14.5%
	$n = 3$	482	515	6.8%
	$n = 5$	528	537	1.7%

* $(2^n - n - 1) + (l - 1) + (l - 1)(2^n - 1)/2^n$
** $(2^n - n) + 2(l - 1)$

Now we consider the computational and memory overheads of SPA-resistant scalar multiplication algorithm which uses Algorithm 5. First, we consider a simultaneous scalar multiplication without SPA-resistance. It requires $(2^n - n - 1)$ additions to construct T, and $(l-1)$ doublings and $(l-1)(2^n-1)/2^n$ additions on the average to compute the scalar multiple. On the other hand, a simultaneous scalar multiplication with recoded scalars requires only one additional doubling to construct T, i.e., computation of $T[0, 0, \ldots, 2]$. (The cost to compute $T[0, 0, \ldots, -2]$ and the cost for scalar recoding are negligible.) It also requires $(l-1)$ doublings and $(l-1)$ additions to compute the scalar multiple. The following table 1 shows the computational overheads for SPA-resistance in various practical settings.

As shown in Table 1, the computational overhead of the new algorithm is very small, and it becomes almost zero for relatively larger n. Also, note that the memory overhead is only one point, regardless of n.

5 Discussion

In this paper, we have proposed an SPA-resistant scalar multiplication algorithm with very small computational and memory overheads. Although our algorithm is aimed at securing a simultaneous scalar multiplication, it is also applicable to single scalar multiplications on curves with efficient endomorphism. For example, in Kobayashi's algorithm [14], a scalar k is decomposed into $k^{(s)}\phi^s + k^{(s-1)}\phi^{s-1} + \cdots + k^{(1)}\phi + k^{(0)}$ using the Frobenius map ϕ, and a single scalar multiplication kP is equivalent to a simultaneous scalar multiplication $k^{(s)}(\phi^s P) + k^{(s-1)}(\phi^{s-1}P) + \cdots + k^{(1)}(\phi P) + k^{(0)}P$. Hence it is possible to use our SPA-countermeasure. GLV method [15] and Park el al.'s method [16] are other examples where our algorithm can be used for a single scalar multiplication.

Finally, we remark that in this work, we have not considered the problem that the use of a fixed table may allow adversaries to find out which of the digits have the same values [8]. Although projective randomization can solve this problem by changing the table values every time the table is accessed, this solution requires some overhead for randomization. Hence, it could be a further research direction to develop an SPA-resistant simultaneous scalar multiplication algorithm which does not use any fixed table.

References

1. Koblitz, N.: Elliptic curve cryptosystems. Mathematics of Computation **48** (1987) 203–209
2. Miller, V.: Use of elliptic curves in cryptography. In: Advances in Cryptology-CRYPTO '85. Volume 218 of LNCS., Springer (1986) 417–428
3. Kocher, P., Jaffe, J., Jun, B.: Differential power analysis. In: Advances in Cryptology-CRYPTO '99. Volume 1666 of LNCS., Springer (1999) 388–397
4. Coron, J.: Resistance against differential power analysis for elliptic curve cryptosystems. In: Cryptographic Hardware and Embedded Systems-CHES '99. Volume 1717 of LNCS., Springer (1999) 292–302

5. Hasan, M.: Power analysis attacks and algorithmic approaches to their countermeasures for Koblitz curve cryptosystems. IEEE Transactions on Computers **50** (2001) 1071–1083
6. Oswald, E., Aigner, M.: Randomized addition-subtraction chains as a countermeasure against power attacks. In: Cryptographic Hardware and Embedded Systems-CHES 2001. Volume 2001 of LNCS., Springer (2001) 39–50
7. Möller, B.: Securing elliptic curve point multiplication against side-channel attacks. In: Information Security-ISC 2001. Volume 2200 of LNCS., Springer (2001) 324–334
8. Möller, B.: Parallelizable elliptic curve point multiplication method with resistance against side-channel attacks. In: Information Security-ISC 2002. Volume 2433 of LNCS., Springer (2002) 402–413
9. Okeya, K., Kurumatani, H., Sakurai, K.: Elliptic curves with the montgomery-form and their cryptographic applications. In: Public Key Cryptography-PKC 2000. Volume 1751 of LNCS., Springer (2000) 238–257
10. Liardet, P., Smart, N.: Preventing SPA/DPA in ECC systems using the Jacobi form. In: Cryptographic Hardware and Embedded Systems-CHES 2001. Volume 2162 of LNCS., Springer (2001) 391–401
11. Joye, M., Quisquater, J.: Hessian elliptic curves and side-channel attacks. In: Cryptographic Hardware and Embedded Systems-CHES 2001. Volume 2162 of LNCS., Springer (2001) 402–410
12. Menezes, A.: Elliptic Curve Public Key Cryptosystems. Kluwer Academic Publishers (1993)
13. Menezes, A., van Oorschot, P., Vanstone, S.: Handbook of Applied Cryptography. CRC Press (1997)
14. Kobayashi, T.: Base-ϕ method for elliptic curves over OEF. IEICE Trans. Fundamentals **E83-A** (2000) 679–686
15. Gallant, R., Lambert, R., Vanstone, S.: Faster point multiplication on elliptic curves with efficient endomorphisms. In: Advances in Cryptology-CRYPTO 2001. Volume 2139 of LNCS., Springer (2001) 190–200
16. Park, T., Lee, M., Kim, E., Park, K.: A general expansion method using efficient endomorphisms. In: Information Security and Cryptology-ICISC 2003. Volume 2971 of LNCS., Springer (2004) 112–126

HSEP Design Using F2m HECC and ThreeB Symmetric Key Under e-Commerce Environment

Byung-kwan Lee[1], Am-Sok Oh[2], and Eun-Hee Jeong[3]

[1] Dept. of Computer Engineering, Kwandong Univ., Korea
  bklee@kd.ac.kr
[2] Dept. of Multimedia Engineering, Tongmyong Univ., Korea
  asoh@tit.ac.kr
[3] Dept. of Economics, Samcheok National Univ., Korea
  jeh@samcheok.ac.kr

Abstract. SSL(Secure Socket Layer) is currently the most widely deployed security protocol, and consists of many security algorithms, but it has some problems on processing time and security. This paper proposes an HSEP(Highly Secure Electronic Payment) Protocol that provides better security and processing time than an existing SSL protocol. As HSEP consists of just F2mHECC, ThreeB(Block Byte Bit Cipher), SHA algorithm, and Multiple Signature, this protocol reduces handshaking process by concatenating two proposed F2mHECC public key and ThreeB symmetric key algorithm and improves processing time and security. In particular, Multiple signature and ThreeB algorithm provides better confidentiality than those used by SSL through three process of random block exchange, byte-exchange key and bit-xor key.

1 Introduction

This paper proposes an HSEP(Highly Secure Electronic Payment) protocol whose characteristic are the followings.

First, The HSEP uses HECC instead of RSA to improve the strength of encryption and the speed of processing. The resulting value which is computed with the public key and the private key of HECC becomes a shared secret key, that is, the values is become a master key. Second, The shared secret key is used as input the proposed ThreeB(Block Byte Bit Cipher) algorithm which generates session key for the data encryption. Finally, HSEP protocol uses multiple signatures instead of MAC(message authentication code) to improve the reliability of EC.

Therefore, HSEP protocol reduces handshaking process by concatenating a shared private key of HECC. Also, Multiple signature and ThreeB algorithm provides better confidentiality than those by SSL through three process of random block exchange, byte-exchange key and bit-xor key.

This paper is structured as follows. Section 2 provides some basic concepts of encryption and decryption, HECC, and SSL. Section 3 describes the structure of HSEP protocol and ThreeB algorithm. An performance comparison of HSEP protocol with SSL protocol are presented in Section 4. Finally, our conclusions are summarized in Section 5.

2 Basic Concepts

2.1 Encryption and Decryption Algorithm

As shown in Fig. 1, the user A computes a new key $k_A(k_B P)$ by multiplying the user B's public key by the user A's private key k_A. The user A encodes the message by using this key and then transmits this cipher text to user B. After receiving this cipher text, The user B decodes with the key $k_B(k_A P)$, which is obtained by multiplying the user A's public key, $k_A P$ by the user B's private key, k_B. Therefore, as $k_A(k_B P) = k_B(k_A P)$, we may use these keys for the encryption and the decryption.

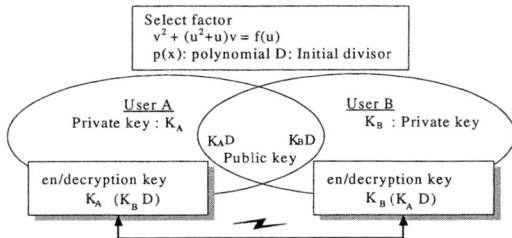

Fig. 1. Concept of en/decryption of HECC

2.2 SSL (Secure Socket Layer) Protocol

SSL is a commonly-used protocol for managing the security of a message transmission on the internet. SSL uses a program layer located between the internet's Hypertext Transfer Protocol and Transport Control Protocol layers.

The SSL protocol includes two sub-protocols : the SSL record protocol and the SSL handshake protocol. The SSL record protocol defines the format used to transmit data. The SSL handshake protocol involves using the SSL record protocol to exchange a series of messages between an SSL-enabled server and an SSL-enabled client when they first establish an SSL connection. This exchanges of messages is designed to facilitate the following actions:

- Authenticate the server to the client.
- Allow the client and server to select the cryptographic algorithms, or ciphers, that they both support.
- Optionally authenticate the client to the server.
- Use public-key encryption techniques to generate shared secrets.
- Establish an encrypted SSL connection.

3 Proposed HSEP (Highly Secure Electronic Payment) Protocol

3.1 HSEP Protocol

The existing SSL uses RSA in key exchange and DES in message encryption. Our proposed HSEP protocol uses HECC instead of RSA, Because of this, the strength of encryption and the speed of processing are improved. Besides, in message encryption, HSEP utilizes ThreeB algorithm to generate session keys and cipher text. The encryption and decryption processes are shown in Fig. 2 respectively.

First, Select a private key x and an initial point D of HECC, and then computes xD. Using the result of addition, generates a session key of ThreeB.

Second, the ThreeB algorithm encodes the message applying these keys. Since the receiver has his own private key, HSEP can reduce handshake procedure without pre-master key exchange, which enhances the speed for processing a message, and strengthens the security for information. Therefore, HSEP simplifies a handshake and decreases a communicative traffic over a network as compared with the existing SSL.

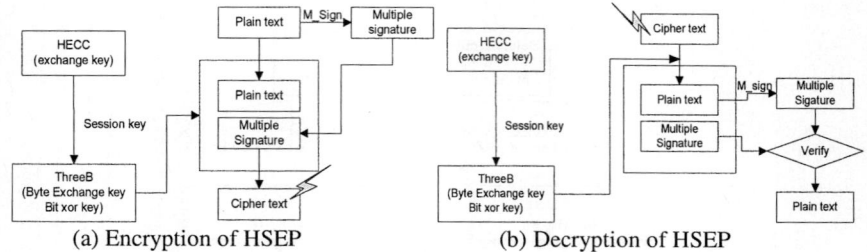

(a) Encryption of HSEP (b) Decryption of HSEP

Fig. 2. The flow of HSEP

3.1.1 The Basic Algorithm for HSEP Protocol

HSEP protocol proposed in this paper uses the same hash function, SHA as SSL protocol, HECC and ThreeB algorithms are no used in SSL. So this section shows that the process of the keys, sk1 and sk2 are generated by using the shared private key of HECC, and data is encrypted by using sk1 and sk2.

1. Key generation

The process of generation of sk1 and sk2 is shown in Fig. 3.

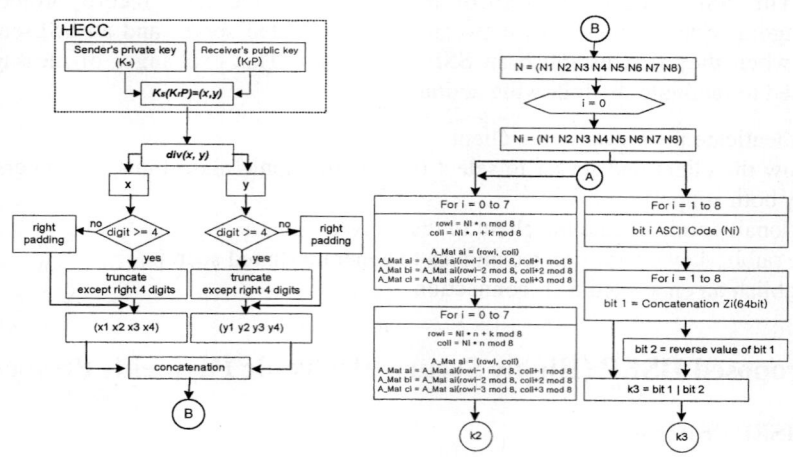

Fig. 3. Key generation

2. Data Encryption

Fig. 4 explains the process of data encryption and is treated in section 3.3.2 in detail.

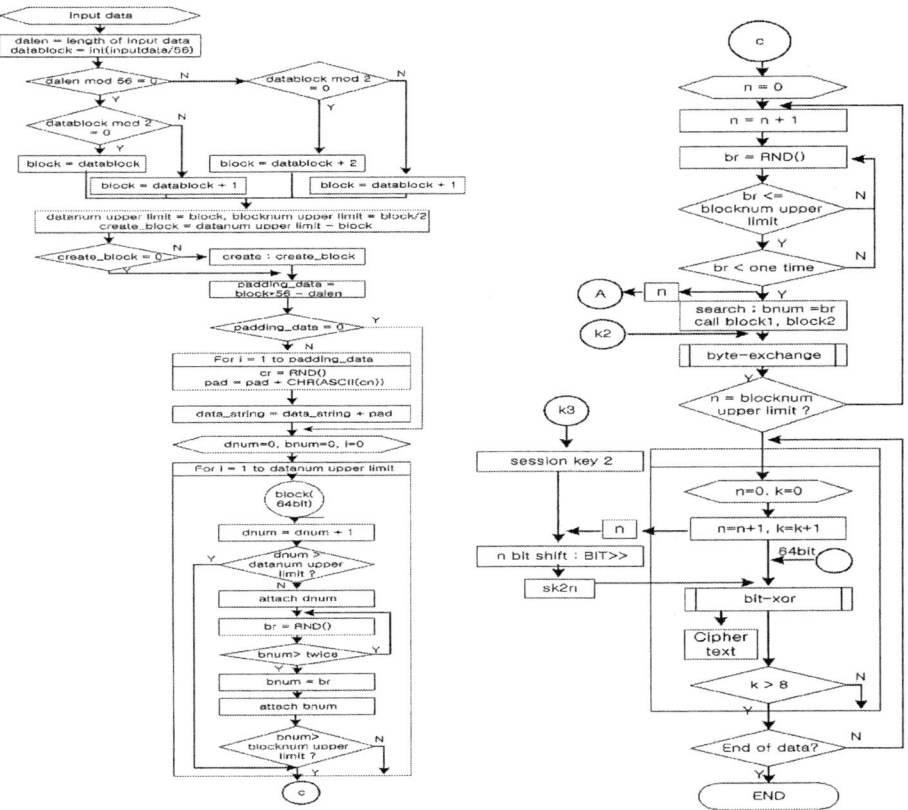

Fig. 4. Data Encryption

3.2 F2m HECC (Hyper Elliptic Curve Cryptosystem)

Nowadays, in the area of cryptology, using hyperelliptic curves is eagerly studied, because it gives the same security level with a smaller key length as compared to cryptosystems using elliptic curves. From the fact it is expected to be possible to use hyperelliptic curves to factor integers, since elliptic curve method exploits the property of the Abelian groups in the same way as the cryptosystems.

A hyperelliptic curve H of genus $g(g \geq 1)$ over a field F is a nonsingular curve that is given by an equation of the following form:

$$H : v^2 + h(u)v = f(u) \quad \text{(in F[u, v])}$$

where $h(u) \in F[u]$ is a polynomial of degree $\leq g$, and $f(u) \in F[u]$ is a monic polynomial of degree $2g+1$.

3.2.1 Divisors

Divisors of a hyperelliptic curve are pairs denoted div(a(u), b(u)), where a(u) and b(u) are polynomials in $GF(2^n)$ [u] that satisfy the congruence $b(u)^2 + h(u)b(u) \equiv f(u) \pmod{a(u)}$.

They can also be defined as the formal sum of a finite number of points on the hyperelliptic curve. Since these polynomials could have arbitrarily large degree and still satisfy the equation, the notion of a reduced divisor is needed. In a reduced divisor, the degree of a(u) is no greater than g, and the degree of b(u) is less than the degree of a(u).

3.2.2 Reduced Divisors

Let H be a hyperelliptic curve of genus g over a field F. A reduced divisor(defined over F) of H is defined as a form div(a, b), where a, b∈ F[u] are polynomial such that
  (1) a is monic, and deg b < deg a $\leq$ g,
  (2) a divides $(b^2 - bh - f)$.

In particular div(1,0) is called zero divisor.

Input : A semi-reduced divisor, D=div(a, b)
Output : The equivalent reduced divisor, $D' = div(a', b') \sim D$
1. Set $a' = (f - bh - b^2)/a$ and $b' = (-h-b) \pmod{a'}$
2. If $\deg_u a' > g$ then set $a = a'$, $b = b'$ and go to step 1.
3. Let c be the leading coefficient of a'. Set $a' = c^{-1}a'$.
4. Output $D' = div(a', b')$

Fig. 5. Reduction of a divisor to a reduced divisor

3.2.3 Adding Divisors

If $D_1 = div(a_1, b_1)$ and $D_2 = div(a_2, b_2)$ are two reduced divisors defined over F, then Fig. 6 finds a semi-reduced divisor or reduced divisor $D_3 = div(a, b)$. To find the unique divisor, $D_3 = div(a, b)$, Fig. 5 should be used just after the addition of two divisors.

Input : Two reduced divisors, $D_1 = div(a_1, b_1)$ and $D_2 = div(a_2, b_2)$
Output : A reduced divisor or semi-reduction divisor, $D_3 = div(a, b)$
1. Compute d_1, e_1 and e_2 which satisfy $d_1 = GCD(a_1, a_2)$ and $d_1 = e_1 a_1 + e_2 a_2$
2. If $d_1 = 1$, then $a := a_1 a_2$, $b := (e_1 a_1 b_2 + e_2 a_2 b_1) \bmod a$
    otherwise do the following:
    (1) Compute d, c_1 and s_3 which satisfy
        $d = GCD(d_1, b_1 + b_2 + h)$ and $d = c_1 d_1 + s_3 (b_1 + b_2 + h)$.
    (2) Let $s_1 := c_1 e_1$ and $s_2 := c_1 e_2$, so that $d = s_1 a_1 + s_2 a_2 + s_3 (b_1 + b_2 + h)$.
    (3) Let $a := a_1 a_2 / d^2$, $b := (s_1 a_1 b_2 + s_2 a_2 b_1 + s_3 (b_1 b_2 + f))/d \bmod a$
3. output $D_3 = div(a, b)$

Fig. 6. Addition defined over the group of divisors

3.3 ThreeB (Block Byte Bit Cipher) Algorithm

In this paper, the proposed ThreeB algorithm consists of two parts, which are session key generation and data encryption. And the data encryption is divided into three phases, which are inputting plaintext into data blocks, byte-exchange between blocks, and bit-wise XOR operation between data and session key.

3.3.1 Session Key Generation
As we know that the value which is obtained by multiplying one's private key by the other's public key is the same as what is computed by multiplying one's public key to the other's private key. The feature of EC is known to be almost impossible to estimate a private and a public key. With this advantage and the homogeneity of the result of operations, the proposed ThreeB algorithm uses a 64-bit session key to perform the encryption and decryption. Given the sender's private key $X = X_1 X_2 ... X_m$ and the receiver's public key, $Y = Y_1 Y_2 ... Y_n$, we concatenate X and Y to form a key N (i.e., $N = X_1 X_2 ... X_m Y_1 Y_2 ... Y_n$), and then compute the session keys as follows:

i) If the length (number of digits) of X or Y exceeds four, then the extra digits on the left are truncated. And if the length of X or Y is less than four, then they are padded with 0's on the right. This creates a number $N' = X_1' X_2' X_3' X_4' Y_1' Y_2' Y_3' Y_4'$. Then a new number N'' is generated by taking the modulus of each digit in N' with 8.

ii) The first session key sk1 is computed by taking bit-wise OR operation on N'' with the reverse string of N''.

iii) The second session key sk2 is generated by taking a circular right shift of sk1 by one bit. And repeat this operation to generate all the subsequent session keys needed until the encryption is completed

3.3.2 Encryption
The procedure of data encryption is divided into three parts, inputting plaintext into data block, byte-exchange between blocks, and bit-wise XOR operation between data and session key.

1. Input plaintext into data block

The block size is defined as 64 bytes. A block consists of 56 bytes for input data, 4 bytes for the data block number, and 4 bytes for the byte-exchange block number (1 or 2, see Fig. 7). During the encryption, input data stream are blocked by 56 bytes. If the entire input data is less than 56 bytes, the remaining data area in the block is padded with each byte by a random character. Also, in the case where the total number of data blocks filled is odd, then additional block(s) will be added to make it even, and each of those will be filled with each byte by a random character as well. Also, a data block number in sequence) is assigned and followed by a byte-exchange block number, which is either 1 or 2.

Data	Area
Data Block Number	Byte-exchange block no

Fig. 7. Structure of block

2. Byte-exchange between blocks

After inputting the data into the blocks, we begin the encryption by starting with the first data block and select a block, which has the same byte-exchange block number for the byte exchange. In order to determine which byte in a block should be exchanged, we compute its row-column position as follows:

For the two blocks whose block exchange number, n = 1, we compute the following:

   byte-exchange row = $(N_i * n)$ mod 8 (i = 1,2 ...,8)
   byte-exchange col = $N_i * n) + 3$) mod 8 (i = 1,2 ...,8),

where N_i is a digit in N" These generate 8 byte-exchange positions. Then for n = 1, we only select the non-repeating byte position (row, col) for the byte-exchange between two blocks whose block exchange numbers are equal to 1. Similarly, we repeat the procedure for n = 2.

3. Bit-wise XOR between data and session key

After the byte-exchange is done, the encryption proceeds with a bit-wise XOR operation on the first 13 byte data with the session sk1 and repeats the operation on every 8 bytes of the remaining data with the subsequent session keys until the data block is finished. Note that the process of byte-exchange hides the meaning of 56 byte data, and the exchange of the data block number hides the order of data block, which needs to be assembled later on. In addition, the bit-wise XOR operation transforms a character into a meaningless one, which adds another level of confusion to the attackers.

3.3.3 Decryption

Decryption procedure is given as follows. First, a receiver generates a byte exchange block key sk1 and a bit-wise XOR key sk2 by using the sender's public key and the receiver's private key. Second, the receiver decrypts it in the reverse of encryption process with a block in the input data receiving sequence. The receiver does bit-wise XOR operation bit by bit, and then, a receiver decodes cipher text by using a byte-exchange block key sk1 and moves the exchanged bytes back to their original positions. We reconstruct data blocks in sequence by using the decoded data block number.

3.4 Multiple Signature

In the proposed HSEP protocol, the multiple signature is used instead of MAC.

(1) User A generates message digests of OI(order information) and PI(payment information) separately by using hash algorithm, concatenates these two message digests; produces $MD_B MD_C$; and hash it to generates MD(message digest). Then the user A encrypts this MD by using an encryption key, which is obtained by multiplying the private key of user A to the public key of the receiver. The PI to be transmitted to user C is encrypted by using ThreeB algorithm. The encrypted PI is named CPI.

(2) User B generates message digest MD_B' with the transmitted OI from user A. After having substituted MD_B' for the MD_B of $MD_B MD_C$, the message digest MD is generated by using hash algorithm. User B decrypts a transmitted DS_B, and extracts MD from it. User B compares this with MD generated by user B, certificates user A

and confirms the integrity of message. Finally, user B transmits the rest of data, $MD_B MD_C$, CPI, DS_C to user C.

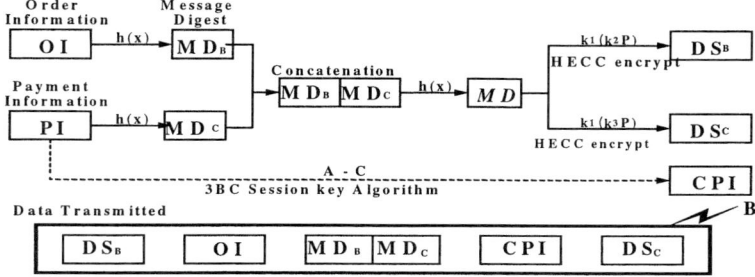

Fig. 8. Encryption of user A

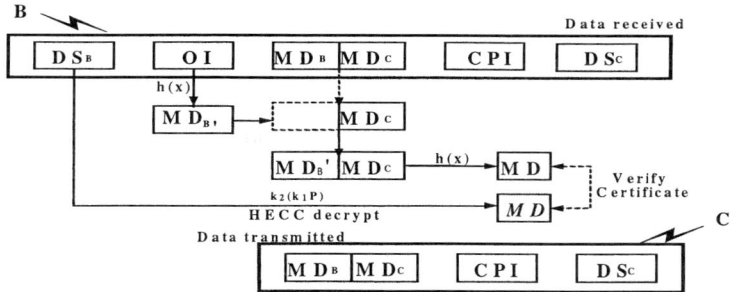

Fig. 9. Decryption of user B and data transmitted to C

(3) User C decrypts the CPI transmitted from user B, extracts PI, and generates message digest (MD_C) from this by using hash algorithm; substitutes this for MD_B of $MD_B MD_C$ transmitted from user B, and produces message digest (MD) by using Hash algorithm. Then the user C decrypts the DS_C transmitted from user B and extracts message digest (MD). Again, the user C compares this with the MD extracted by user C, verifies the certificate from the user A, and confirms the integrity of the message. Finally, the user C returns an authentication to the user B.

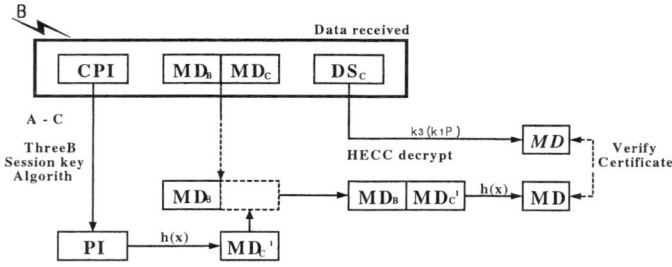

Fig. 10. Decryption of an user C

4 Performance Evaluation

4.1 HECC and RSA

In this paper, the proposed HSEP protocol uses HECC instead of RSA. In comparison with RSA, the results of the encryption and decryption times are shown in Fig. 11 respectively, which indicate that encryption and decryption time of HECC are much less than those of RSA.

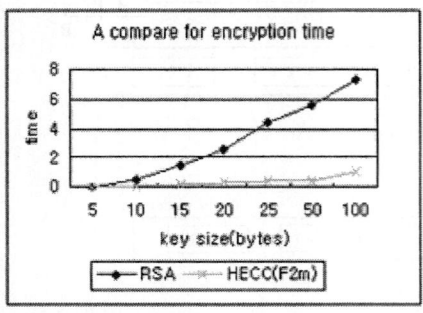 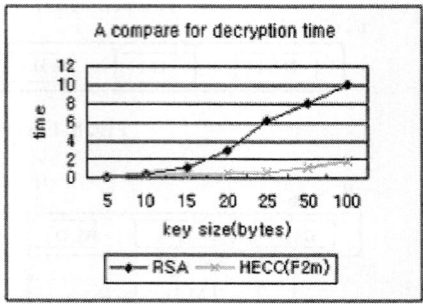

(a) A comparison for encryption time   (b) A comparison for decryption time

Fig. 11. The comparison of HECC and RSA(unit : Φs)

4.2 ThreeB and DES

Fig. 12 show the mean value of encryption time of ThreeB and DES by executing every number of block about message twenty times. According to Fig. 12, we can conclude that ThreeB is faster than the existing DES in encryption time. In addition, the security of ThreeB is enhanced by using Byte-exchange and Bit-wise XOR Therefore, the strength of the encryption is improved and more time is saved for encryption and decryption than DES.

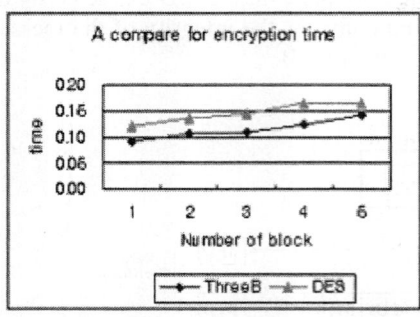 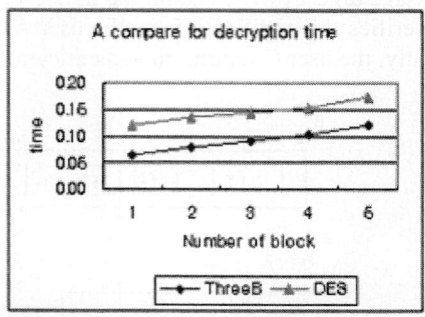

(a) A comparison for encryption time   (b) A comparison for decryption time

Fig. 12. A comparison of ThreeB and DES(unit : Φs)

5 Conclusion

The proposed HSEP protocol employs HECC, Multiple Signature and ThreeB algorithm other than the existing SSL. HSEP protocol removes a pre-master key exchange, and replaces the master key exchange by the shared secret key. As a result, it speeds up the handshaking process by reducing communication traffic for transmission. The proposed ThreeB, which uses byte-exchange and the bit operation increases data encryption speed. Even though cipher text is intercepted during transmission over the network. Because during the encryption process, the ThreeB algorithm performs byte exchange between blocks, and then the plaintext is encoded through bit-wise XOR operation, it rarely has a possibility for cipher text to be decoded and has no problem to preserve a private key.

Moreover, the proposed HSEP protocol has a simple structure, which can improve the performance with the length of session key, byte-exchange algorithm, bit operation algorithm, and so on. From the standpoint of the supply for key, the CA (Certificate authority) has only to certify any elliptic curve and any prime number for modulo operation, the anonymity and security for information can be guaranteed over communication network.(See Table 1.)

Table 1. Decryption of HSEP protocol

protocol	Digital signature	Encryption for message	Digital envelope
SET	RSA	DES	Use
ECSET	ECC	DES	Use
HSEP	HECC	ThreeB	Unnecessary

References

1. CHO, I.S., and B.K. Lee, ASEP (Advanced Secure Electronic Payment) Protocol Design, in Proceedings of International Conference of Information System, pp. 366-372, Aug. (2002).
2. CHO, I.S., D.W. Shin, T.C. Lee, and B.K. Lee, SSEP (Simple Secure Electronic Payment) Protocol Design, in Journal of Electronic and Computer Science, pp. 81-88, Vol. 4, No. 1, Fall (2002).
3. ECOMMERCENET, <http://www.ezyhealthmie.com/Service/Editorial/set.htm>.
4. HARPER, G., A. Menezes, and S. Vanstone, Public-key Cryptosystem with Very Small Key Lengths, in Advances in Cryptology-Proceedings of Eurocrypt '92, Lecture Notes in Computer Science 658, pp. 163-173, Springer-Verlag, (1993).
5. IEEE P1363 Working Draft, Appendices, pp. 8, February 6, 1997.
6. KOBLITZ, N., Elliptic Curve Cryptosystems, in Math. Comp. 48 203-209 (1987).
7. MILLER, V.S., Use of Elliptic Curve in Cryptography, in Advances in Cryptology-Proceedings of Crypto '85, Lecture Notes in Computer Science 218, pp. 417-426, Springer-Verlag, (1986).

A Fault Distance Estimation Method Based on an Adaptive Data Window for Power Network Security

Chang-Dae Yoon[1], Seung-Yeon Lee[2], Myong-Chul Shin[3], Ho-Sung Jung[4],
and Jae-Sang Cha[5]

[1] SungKyunKwan Univ., Department of Information & Communication Eng.,
Kyonggi, Korea
phasors@hanmail.net
[2] SungKyunKwan Univ., Department of Information & Communication Eng.,
Kyonggi, Korea
syoun@ece.skku.ac.kr
[3] SungKyunKwan Univ., Department of Information & Communication Eng.,
Kyonggi, Korea
mcshin@skku.edu
[4] Korea Railroad Research Institute, Uiwang-City, Kyonggi, Korea
hsjung@krri.re.kr
[5] SeoKyeong Univ., Dept. of Information & Communication Eng., Seoul, 136-704, Korea
chajs@skuniv.ac.kr

Abstract. This paper presents a rapid and accurate algorithm for fault location estimation in a power transmission line. This algorithm uses the least square error (LSE) method for fault impedance estimation. After interrupting a fault, an adaptive data window technique using LSE estimates the fault impedance. Since it changes its data length according to the convergence degree of fault impedance, it can find an optimal data window length and estimate fault impedance rapidly. To prove the performance of the algorithm, the authors have tested relaying signals obtained from EMTP simulation. Test results show that the proposed algorithm estimates fault location by calculating fault impedance within a half cycle of the fault, regardless of fault type and various fault conditions. Compared to traditional techniques, it can protect parallel transmission lines more quickly and reliably.

1 Introduction

As the power system grows bigger in capacity and higher in voltage, and since faults in transmission lines may exert far-reaching effects on the whole power system, it is essential to supply power with fault elimination and security through quicker fault detection.

Traditional transmission line protection methods can distinguish convergence availability by calculating line impedance between relay and fault locations using a Fourier Transform [1].

Although the results of the Fourier Transform are relatively accurate, the data may burden the relay's processor because high calculation time and quantity of the one cycle or half cycle data window of fixed size required calculating impedance.

To mitigate calculation burden of a relay, several methods have been suggested. Of those with flexible data windows, one estimates impedance by calculating the coefficients of each component after approximating the DC component, a fundamental wave, and sum of low order harmonic components from voltage and current signal using the least square error method (LSE); another is to estimate impedance directly by expressing the fault line with differential equations [2]. These methods can choose the size of the data window flexibly and have higher sampling rates per cycle than those using Fourier Transforms, but there is difficulty in deciding optimum data window depending on various fault conditions.

In this paper, a method is suggested that can estimate fault impedance by making the size of the data window flexible depending on various conditions of fault voltage and current that are measured in a relay. That is, it estimates variables with using a small data window before the fault occurs, however, it will change the size of the data window gradually until variables converge after the fault is detected.

To verify the method, a power system was modeled using EMTP (Electromagnetic Transient Program), and fault data with various fault distances, fault impedances, fault occurrence angles, source impedances, source phase angles, etc., were generated. The major purpose of the present study is to evaluate the general performance of the method, and to test the comparative validity, including working speed and accuracy, between this method and others, such as LSE with fixed size or Discrete Fourier Transform (DFT).

2 Fault Distance Estimation Method with Adaptive Data Window

2.1 Fault Distance Estimation Using LSE

The least square error method (LSE) is a technique to estimate the best approximate value by minimizing the error between estimated value and actual value in the case of not obtaining the correct value when the number of given equations is more than that of unknown quantities. For applying LSE, a line with a single line to ground fault can be modeled in the form of resistance and reactance as shown in Figure 1.

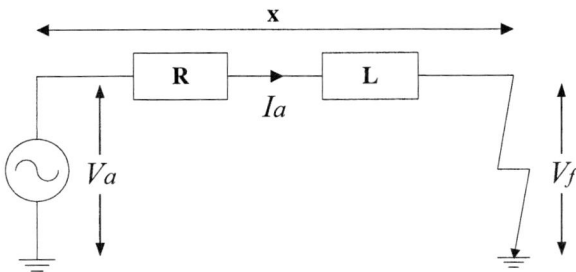

Fig. 1. Single line to ground fault

In the case of the single line to ground fault (in phase A) of Figure 1 above, the voltage equation that applies to the distance relay side becomes a first-order differential equation of the form of Equation (1) below.

$$V_a = R_1(I_a + K_r I_0) + L_1\left(\frac{dI_a}{dt} + K_l \frac{dI_0}{dt}\right) + V_f \quad (1)$$

where V_a and I_a are voltage and current values measured in the relay, and R_0, R_1, L_0 and L_1 are zero and positive sequence line resistances and reactances, respectively. K_r and K_l are given by

$$K_r = (R_0 - R_1)/R_1, \quad K_l = (L_0 - L_1)/L_1$$

In the case of the single line to ground fault, if it is known that the voltage at fault location V_f is '0', then the two unknown terms are R_1 and L_1.

If Equation (1) is divided into a matrix of known and unknown terms at the kth time point, the result is given by Equation (2).

$$y(k) = A(k) \cdot x(k) = [a_1(k) a_2(k)] \cdot x(k) \quad (2)$$

$$\text{where, } y(k) = V_a(k)$$

$$x(k) = [R_1(k) L_1(k)]$$

$$a_1(k) = I_a(k) + K_r I_0(k)$$

$$a_2(k) = \frac{dI_a(k)}{dt} + K_l \frac{dI_0(k)}{dt}$$

The differential terms of Equation (2) are calculated using the difference between two signals as given by Equation (3) below.

$$\frac{dI_a(k)}{dt} = \frac{I_a(k+1) - I_a(k-1)}{2 \cdot \Delta T} \quad (3)$$

where, ΔT is the interval of sampling time. The unknown matrix $x(k)$ in (2) can be calculated using Equation (4), which uses a pseudo-inverse matrix.

$$x(k) = (A(k)^T \cdot A(k))^{-1} \cdot A(k)^T \cdot y(k) \quad (4)$$

Calculating Equation (4) allows us to estimate values of resistance and reactance and to calculate the distance to the fault location. LSE enables us not only to calculate unknown values directly in the time domain without transformation to the frequency domain, but also, since the size of the data window for estimation is flexible, it gives faster fault location estimation. That is, impedance can be estimated by successive data points because there are only two unknown quantities. But, because of stability, it is necessary to calculate values with a data window greater than a certain size because transitory noise or oscillation of signals in the early transient state can cause serious errors when the calculation is done using a small data window. On the other hand, it is necessary to limit the data window size because improvement of convergence speed can not occur when the size of data window is too big due to the increase in calculation quantity [4, 5].

2.2 Adaptive Data Window Using LSE

Before a fault occurs on a line, impedance can be estimated exactly with a relatively small data window because the variations of voltage and current signals are so light. When a fault that causes great variations in voltage and current signals occurs, however, a bigger data window is required because it is difficult to estimate the exact fault location with a small data window. And since transient state characteristics, which depend on the fault situations, are so various, fault location estimation with a fixed data window would either lower the stability or delay the speed of solution convergence. Therefore, fault impedance calculations can converge more quickly and stably by increasing the size of the data window gradually according to the fault situations.

In case the difference between the two impedance values calculated at the *(k-1)*th and *k*th time points by Equation (5) is higher than the set value, the fault location could be estimated by increasing the size of the data window and then the optimum size of the data window can be chosen at the time point that has a smaller value than that already set up.

$$Z_{set} \rangle Z_{diff}(k) = \left| Z_{(k)} - Z_{(k-1)} \right| \tag{5}$$

So, the size of the data window will finally be determined when the difference between successive impedance estimation values falls below the set value within a cycle window. In other words, it would be fixed to the cycle window size in case the difference doesn't fall below the set value even if a cycle passes by. As a result, the relay doesn't need to process much data for impedance estimation. Figure 2 shows a flowchart for our distance relaying technique that uses an adaptive data window algorithm based on LSE.

In the below flowchart, while impedance is estimated by a relatively small data window, 8 samples (1/8 cycle), before a fault occurs, it could be estimated until convergence by increasing the size of the data window if a fault is detected. This algorithm not only enables a relay to estimate fault location quickly and stably, but also gives fewer quantities to calculate with a flexible data window depending on the situation.

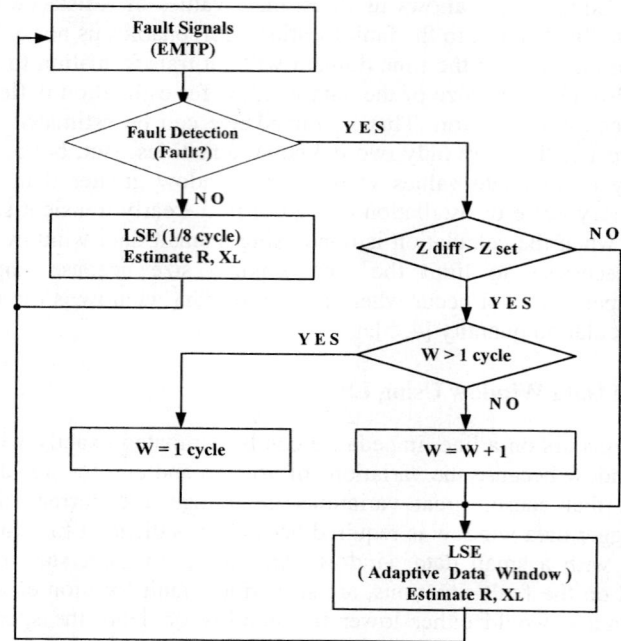

Fig. 2. Flowchart of the adaptive data window algorithm

In the above flowchart, while impedance is estimated by a relatively small data window, 8 samples (1/8 cycle), before a fault occurs, it could be estimated until convergence by increasing the size of the data window if a fault is detected. This algorithm not only enables a relay to estimate fault location quickly and stably, but also gives fewer quantities to calculate with a flexible data window depending on the situation.

3 Case Studies

3.1 Power System Model

Figure 3 shows a single line diagram and the line constants of a single 100 km transmission line with two 345kV sources. A single line to ground fault has been simulated for phase A of this model line.

The sampling frequency is 3840 [□]; that is, 64 samples per cycle are adopted and various fault data have been generated for various fault distances, fault occurrence angles, fault impedances, source impedances, and phase angles between the two sources. It has been assumed that faults occur at the 176th and 192nd sample points, corresponding to 0° and 180° faults, respectively, considering the source G1 as a reference.

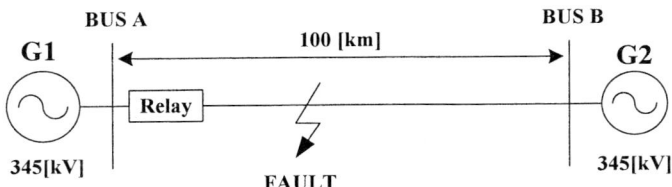

	R [Ω/km]	X_L [Ω/km]	C [μF/km]
Zero Sequence	0.3434	1.3158	0.0052
Positive or Negative Sequence	0.1342	0.4765	0.0090

Fig. 3. Model System

3.2 Result of Fault Distance Estimation

It is assumed that the fault occurs at the middle point of the line, at an angle of $0°$, with $0\,\Omega$ of fault impedance. In addition, the preset value of impedance (Z_{set}), which is used for deciding the size of the adaptive data window, has been set to 0.2475 to have it converge to not more than ±0.5% of the whole line impedance.

The result of simulation, on the assumption that the fault is detected at the 182nd sample (after the lapse of 7 samples), is that resistance and reactance values converged to within 1% of actual values after 30 samples and 25 samples, respectively. The convergence ratio mentioned above has been calculated based on Equation (6) below.

$$\text{convergence ratio} = \frac{Z_{real} - Z_{estimate}}{Z_{total}} \times 100 \quad [\%] \tag{6}$$

Where, Z_{total}, Z_{real}, $Z_{estimate}$ are the magnitudes of the whole line impedance, the actual value, and the estimation value using the least square error method, respectively.

Figure 4 shows changes of the adaptive data window accompanied by changes of estimated impedance. This shows that the algorithm estimates impedance with a window size of 8 data samples (1/8 cycle) before a fault occurs, but it is on the increase gradually in its size of data window after a fault occurs until impedance is lower than 0.2475, which is 1% of whole line impedance.

If the differences between the last and present values of impedance estimated become lower than the set value, impedance would be estimated using the last data window size. In this case, we see that the size of the data window has been optimized to a value of 21.

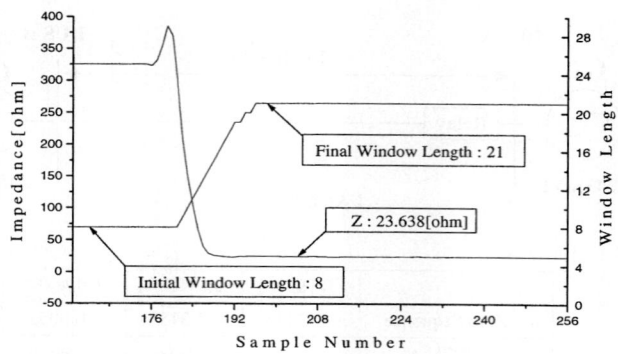

Fig. 4. Change of the adaptive window length according to impedance

Figure 5 and 6 compare the convergence characteristics of resistance and reactance by the DFT method (with a one cycle data window), the least square error method

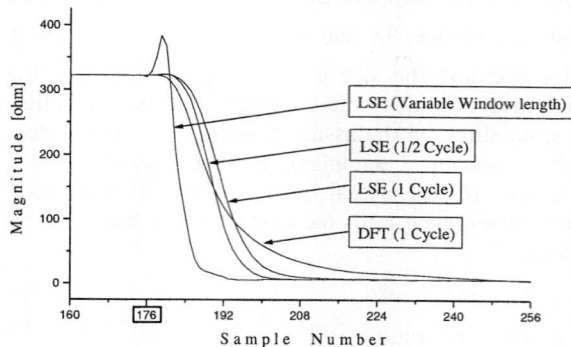

Fig. 5. Resistance Estimation

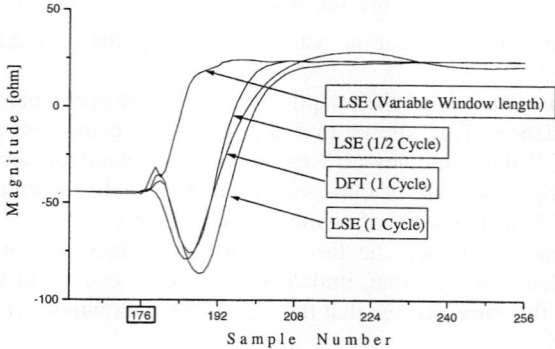

Fig. 6. Reactance Estimation

(with half and one cycle data windows) using the same fault situations used previously to evaluate performance of adaptive data window method. Table 1 shows the comparisons for each convergence speed by showing the number of samples required after fault occurrence. Hence, estimation values of resistance and reactance converge within 1% and 5% of the actual values.

Table 1. Comparisons of convergence speed

Method [samples] Ratio [%]		1 Cycle DFT	1 Cycle LSE	1/2 Cycle LSE	Adaptive LSE
5 %	R	76	54	32	25
	L	34	50	32	18
1 %	R	88	63	31	30
	L	35	60	31	25

As we can see from the convergence characteristics and comparisons of the speed above, it is not until a cycle later that the resistance value estimated by the one-cycle window DFT method converges within 1% to 5% of actual value. On the other hand, the reactance value converges within about half a cycle, but oscillations persist even after convergence. The one-cycle window LSE method couldn't convergence for impedance quickly enough (i.e., within the allowable error) until after one cycle passed, but the value was estimated within only about half a cycle by the LSE method (with a half cycle data window). On the other hand, although LSE with an adaptive data window may have large errors in convergence initially because the size of the data window is small, it can increase the convergence speed remarkably by increasing the size of the data window properly according to fault situations.

Figure 7 shows the impedance loci described by LSE with an adaptive data window when a ground fault occurs. The ground fault established for this figure occurred at 30% and 70% of the length of the whole line, and at the fault occurrence angles of 0° and 90°.

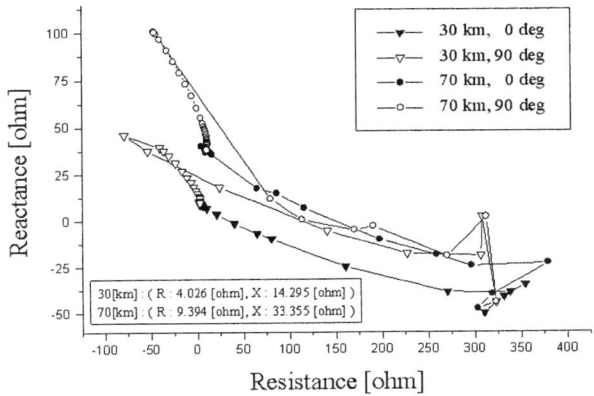

Fig. 7. Impedance loci for ground fault

4 Conclusion

In this paper, a faster and more accurate fault location estimation method for transmission line protection has been suggested. By applying a flexible data window size (adaptive data window) to the least square error method, data window size can optimally adjusted according to fault situations for faster fault location estimation. In other words, an algorithm for transmission line protection is suggested that uses the least square error method (LSE), which can estimate the location of faults to ground, so that it always calculates an optimum data window size by applying an adaptive data window technique that can be varied according to transient phenomena. Simulation results for various fault situations show that the fault location can be estimated exactly within a half cycle after a fault occurs.

References

[1] P. G. McLaren, "Fourier-Series Techniques Applied to Distance Protection", Proceedings of the IEE, Vol. 122, No. 11, pp. 1301-1305, 1975.
[2] M. B. Djuric, "Time Domain Solution of Fault Distance Estimation and Arcing Fault Detection on Overhead Lines", IEEE Transactions on Power Delivery, Vol. 14, No. 1, pp. 60-67, 1999.
[3] Ho-Sung Jung, "A Neural Networks Fault Patterns Estimator for the Digital Distance Relaying Technique", Journal of KIEE, Vol. 47, pp. 1804-1811, 1998.
[4] K. K. Li "An adaptive window length algorithm for accurate high speed digital distance protection", Electrical Power & Energy Systems, Vol. 19, No. 6, pp. 375-383, 1997.
[5] M. S. Sachdev, "Design of a Distance Relay Using Adaptive Data Window Filters", Electrical and Computer Engineering, 2000 Canadian Conference, Vol. 2, pp. 610-614, 2000.

Distribution Data Security System Based on Web Based Active Database

Sang-Yule Choi*, Myong-Chul Shin**, Nam-Young Hur**, Jong-Boo Kim*, Tai-hoon Kim***, and Jae-Sang Cha****

*Dept.of Electronic Engineering , Induk Institute of Technology,
San 76 Wolgye-dong, Seoul, Korea
**School of Electrical and Computer Engineering, Sungkyunkwan University,
Suwon 440-746, Korea
***3San 7, Geoyeou-dong, Songpa-gu, Seoul, Korea
****Dept.of Information and Communication Engineering,
SeoKyeong University, Seoul, Korea

Abstract. The electric utility has the responsibility to provide a good quality of electricity to their customers. Therefore, they have introduced DDSS(Distribution Data Security System) to automate the power distribution data security. DDSS engineers need a set of state-of-the-art applications, eg. managing distribution system in active manner and gaining economic benefits from a flexible DDSS architectural design. The existing DDSS functionally could not handle these needs. It has to be managed by operators whenever feeder faults data are detected. Therefore, it may be possible for propagating the feeder overloading area, if operator makes a mistake. And it utilizes closed architecture, therefore it is hard to meet the system migration and future enhancement requirements. This paper represents web based, platform-independent, flexible DDSS architectural design and active database application. The recently advanced internet technologies are fully utilized in the new DDSS architecture. Therefore, it can meet the system migration and future enhancement requirements. And, by using active database, DDSS can minimize feeder overloadings area in distribution system without intervening of operator, therefore, minimizing feeder overloadings area can be free from the mistake of operator.

1 Introduction

The electrical utility has the responsibility to provide a good quality of electricity to their customers. Therefore, the DDSS (Distribution Data Security System) is introduced to control and operate complex power distribution system in an economical and reliable fashion. It includes important functions such as data security, feeder automation, load control and telemetering. In korea, the electrical utility company is now facing deregulation and privatization. Several privatized distribution companies will be appear in few years. And they will require new way of thinking and new solutions for open-access competitive electricity market. But the existing DDSS could not meet these new requirements, because it has to be managed by passive manner and it utilizes closed system architecture.

In closed system architecture, DDSS has utilized proprietary software which is tightly coupled to a particular operating environment. Therefore, system integration and data migration in a heterogeneous environment give pressures to DDSS developers and operators. And, due to the passive DDSS management, DDSS is always exposed to unintentional mistake from operators. Therefore, whenever feeder overloadings and faults data are detected, inexperienced operator may get worse feeder overloadings situation.

Despite of the above defects, a closed architecture and a passive management have been successfully applied to DDSS until the present time. However, under open-access competitive market environment, data migration in a heterogeneous environment and reliable system management by active manor become critical issues to distribution utility companies. This environment gives DDSS engineers require a set of state-of-the-art applications, eg using internet application for convenient data exchange and system openness and adopting active database application for reliable system management by active manor. Based on these requirements, the authors present new DDSS architecture based on open system and an active rule application. An open DDSS system architecture utilizes recently matured internet technology and relational active database.

2 An Active Database for DDSS

An active database system is the result of combination of active techniques and DBMS, Production rules in active database system allow specification of data manipulation operations that are executed automatically whenever certain events occur or conditions are satisfied. Active database rules provide a general and powerful mechanism for maintaining many database features, including integrity constraint, data consistency, and so on, in addition, active database system provide a convenient platform for large and efficient knowledge bases and expert system [1]

2.1 Active Database Rules

The heart of active database systems is active database rule. In general form, active database rule consist of three parts.

- Event : External_operation
- Condition : Condition
- Action : Action .

Each part can be expressed as follows:

- Event

It specifies the rule to be triggered. Events are data modification operations ("Update""Insert", "Delete") data retrieval, and general application procedures.

- Condition:

It is checked when the rule is triggered. If condition is true then the rule's action is executed. Conditions are database predicates, restricted predicates and database queries.

- Action:

It is executed when the rule is triggered and its condition is true. Actions are data modification operations, data retrieval operations and general application procedures.

Once a set of rules is defined, the active database system monitors the relevant events. For each rules

On signaling of an event, a condition is evaluated and if this evaluation is true, an action is executed. The relation between DDSS and active database system is that active database technology can ideally used to solve typical tasks within DDSS database system : Using appropriate rules, we can satisfy the radial distribution systems topology constraint and integrity constraints during data modification operations, and minimize line losses when distribution feeders are overloaded. The example of appropriate rule for minimizing line losses is as follows.

Assumption. Feeder reconfiguration program for minimizing line losses is executed when feeder loadings exceed 80 [%] of line capacity.

This assumption can define the following rule

<Rule for line>
Event : Update to loadings
Condition : Updated (line (L)),
NEW L.loadings> L.capacity * 0.8
Action : Reconfiguration ()

In the above rule, if the attribute loadings value is updated, and updated value is 80 [%] over than the attribute capacity value then feeder reconfiguration is executed to minimize line losses. If updated loadings value is lower than 80 [%] of feeder capacity then action can not be executed because condition is false.

Active database system architecture be usually specified as three ways : Layered architecture, built-in architecture, compiled architecture.

In this paper, layered architecture and built-in architecture are used.

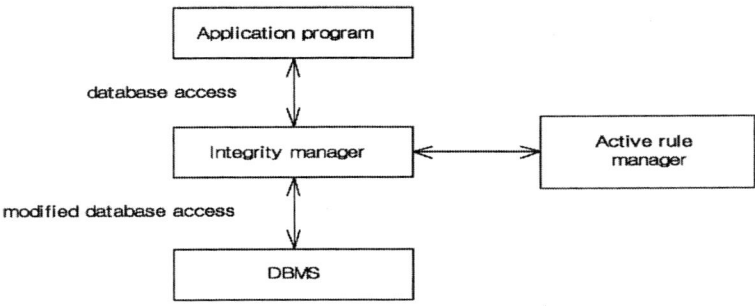

Fig. 1. Layered and built-in architecture

In fig.1, integrity manager is in the between application program and DBMS, and it detects data modification signals from application program. If data modification signals violate integrity constraints then integrity manager sends signals to active rule manager. Active rule manager determines which rules to be fired, and it sends back another signals to integrity manager. Integrity manager checks if signals sending from active rule manager satisfy the integrity constraints or not. If data operation signals are satisfied then it goes to DBMS .

2.2 Active Rule Manager

In ECA rules, one event may triggers several defined rules, and some actions of triggered rules may trigger another rules. Therefore, forward-chaining rules can be fired by one event.

Active rule manager controls the firing of rules. It interfaces with integrity manager, which is used to check if integrity constraints are satisfied or not. It also responsible for aggregating composite primitive events that are signaled from integrity manager, and determine which rules are to be fired. When integrity manager sends signals, rules are fired.

2.3 Data Requirements for DDSS Database

Data requirements for DDSS database are substation, transformer, feeder, feeder section, switch, and information. And attributes of these requirements have to be easily applied to feeder reconfiguration program.

After feeder reconfiguration is executed, resulted data are restored to information table.

- SUBSTATION

Substation number(psn), transformer number(tsn).

- TRANSFORMER

Transformer number(tsn), substation number (psn), transformer capacity(tac), feeder number1 (fsn1), Feeder number2(fsn2), feeder number3 (fsn3), transformer loadings(premva), open/close status(close : 1, open : 0).

Transformer is connected to three feeders, and attribute transformer loadings value is sum of all feeder loadings of feeder number1, 2, 3 .

- FEEDER

Feeder number(fsn), connected transformer number (tsn), feeder loadings (fsnmva),Start feeder section number (fsnsec), close/open status (close:1, open:0).

- FEEDER SECTION

Feeder section number (fsnsec), switch number (ssn), feeder capacity 1(fnc), feeder capacity 2 (fanc), Resistance(rr), reactance(xx), feeder number(fsn), fault

flag(normal :0 ,fault :1), start point(fsnfbsn), end point(fsntbsn), section loadings(ssnmva), voltage(vv), minimum voltage(min_vv).
Where, voltage (vv) = VL—Vdrop
VL : Voltage of appropriate switch
VL : Voltage drop of feeder section
Fnc is 80 [%] of feeder capacity and fanc is 100 [%] of feeder capacity.
Feeder section loadings and close/open status are the same value of switch because, by using active rule, they are updated by the same quality of the appropriate switch which close(or open) feeder section.

- SWITCH

Switch number(ssn), feeder section number (fsnsec), loadings(ssnmva), current(aa), voltage (vv), close/open status (close:1, open:1).
It is assumed that detected loadings, current, voltage values are updated by data acquisition system that is located in real distribution systems.

- INFORMATION

Identification (rsn), total losses (ploss), the amount of loss change (dploss).

2.4 Conceptual Design for DDSS Databases

In this paper, conceptual design is represented by entity-relationship diagram, and relationship between entities is as follows:

- SUBSTATION :TRANSFORMER

One substation can have several transformers, therefore, a relationship type is "have" and 1 : N relationship between the two entity types SUBSTATION and TRANSFORMER.

- TRANSFORMER : FEEDER

One transformer can distribute power among several feeders, therefore, a relationship type is "distribute " and 1 : N relationship between the two entity types TRANSFORMER : FEEDER.

- FEEDER : FEEDER SECTION

One feeder is consisted of feeder sections, therefore, a relationship type is "consists-of" and 1 : N relationship between the two entity types FEEDER : FEEDER SECTION.

- FEEDER : INFORMATION

Information contains resulted data obtained from feeder reconfiguration program. And feeder reconfiguration program can be executed several times, therefore, a relationship type is "execute" and 1 : N relationship between the two entity types FEEDER : INFORMATION.

● FEEDER SECTION : SWITCH

one switch can close(or open) a appropriate feeder section, therefore, a relationship type is " close_open" and 1 : 1 relationship between the two entity types FEEDER SECTION : SWITCH.

The above entity-relationship is displayed by means of the graphical notation known as ER diagram in fig.2.

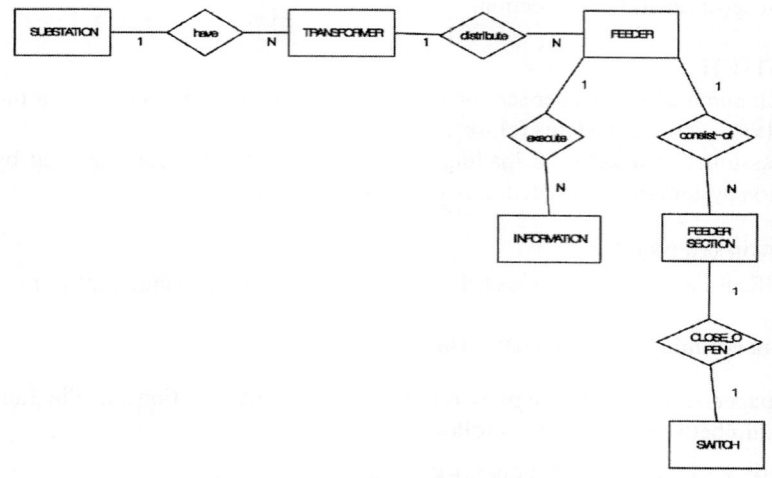

Fig. 2. ERD of DDSS Database

2.5 Rules Definition for DDSS Database

Data acquisition system which is located in distribution networks detects currents, section loadings, voltage and the status of close/open in real time, and the present attribute values of switch is updated by detected ones. DDSS active database monitors the present state of distribution networks by using updated values: If updated values trigger some events and condition are true, then they imply that distribution networks is in overloaded state. Therefore action for feeder reconfiguration is executed to relive feeder overloadings and minimize line losses. Switches to be open(or closed) are selected after feeder reconfiguration is executed.

Feeder reconfiguration can fire other rules which update attribute close/open status value of selected switchs and send close/open signal to appropriate intelligent switches installed in distribution networks.

Due to maintaining radial distribution networks structure, updated value of switch can trigger other rules that update close/open status value of another switch(not a selected one from feeder reconfiguration). And updated attribute value of switch can trigger rules which updates the close/open status value of appropriate feeder section that is connected to updated switch, because close/open status of feeder section is dependent on attribute values of appropriate switch object. Distribution networks can be operated with minimum losses, and radial structure can be maintained without intervening of operator with the definition of active rules which have update propagation characteristics.

The definition of active rules for DDSS are as follows:

Rule 1) If initially attribute close/open status value of switch is changed from open to close then attribute close/open status value of appropriate feeder section which is connected to the switch is changed from open to close.

Rule R1 for switch
Event : update to st
Condition : updated(switch(S)), NEW S.st=close
Action: update feeder section.st=close where feeder section.ssn = switch.ssn

Rule 2) If initially attribute close/open status value of switch object is changed from close to open then close/open status value of appropriate feeder section object, which is connected to the switch is changed from close to open.

Rule R2 for switch
Event : update to st
Condition : updated(switch(S)), NEW S.st=open
Action: update feeder section.st=open where feeder section.ssn = switch.ssn

Rule 3) If attribute loadings value of switch object is updated then section loadings value of feeder section is updated by the same value of loadings of switch.

Rule R3 for switch
Event : update to ssnmva
Condition : True
Action : update feeder section.ssnmva = switch.
ssnmva where feeder section.ssn = switch.ssn

Rule 4) If updated section loadings value of feeder section is exceeded by 80[%] of feeder capacity then feeder reconfiguration program is executed to minimize line losses and relive overloadings.

Rule R4 for feeder section
Event : update to ssnmva
Condition : updated(feeder section(FD)),
NEW FD.ssnmva > FD.fnc
Action : reconfiguration()

Rule 5) If reconfiguration() is executed and switch to be closed(or opened) is selected, then attribute close/open status of selected switch is updated by close(or open).

Rule R5 for reconfiguration()
Event : reconfiguration()
Conditon : TRUE
Action : (update switch.st=open where switch.ssn
= result of open reconfiguration())
&&(update switch.st=close where switch.ssn=
result of close reconfiguration())
where, result of open reconfiguration() : a selected switch to be opened resulting from feeder reconfiguration.
result of close reconfiguration() : a selected switch to be closed resulting from feeder reconfiguration.

Rule 6) If attribute voltage value of switch is updated then the attribute voltage value of appropriate feeder section is updated
Rule R6 for switch
Event : update vv
Condition : updated(switch(S)), TRUE
Action : update feeder section.vv =
switch.vv—Vdrop
where, feeder section.ssn = switch.ssn
Where, Vdrop : Voltage drop of feeder section

Rule 7) If updated voltage value of feeder section is lower than minimum voltage value of feeder section then feeder reconfiguration program is executed to minimize line losses and relive overloadings.
Rule R7 for feeder section
Event : update to vv
Condition : updated(feeder section(FD)),
NEW FD.vv > FD.min_vv
Action : reconfiguration()

Rule 8) If initially attribute close/open status value of switch is changed from open to close then Active database sends signal to intelligent switch to be closed
Rule R8 for switch
Event : update to st
Condition : updated(switch(S)), NEW S.st=close
Action: signal to intelligent switch to be closed

Rule 9) If initially attribute close/open status value of switch is changed from close to open then active database sends signal to appropriate intelligent switch to be opened
Rule R9 for switch
Event : update to st
Condition : updated(switch(S)), NEW S.st=open
Action: signal to intelligent switch to be opened

Rule 10) If feeder reconfiguration executed then information is updated by the resulted data form feeder reconfiguration.
Rule R10 for reconfiguration
Event : reconfiguration()
Condition : TRUE
Action :(update information.rsn = history) &&
(update information.ploss= result loss of reconfiguraion())&&(update information.dploss = result of change_of_loss of reconfiguration())
Where, history = reconfiguration times
result loss of reconfiguraion(): total loss after feeder reconfiguration
result of change_of_loss of reconfiguration() : the amount of loss change resulting from feeder reconfiguration.

2.6 Active Rule Manager for DDSS Active Database

The execution of a rule's action may trigger another rule, whose action may trigger other rule's event. Active rule manager coordinates active rule interaction and the execution of active rules during transaction execution by interfacing constraint manager. In this paper, rule interaction is represented by means of Triggering Graph[2].

DEFINITION : Let R be an arbitrary active rule set. The triggering Graph is a directed graph $\{V,E\}$, where each node vi $\in$ V corresponds to a rule ri $\in$ R, A directed arc <jrj , rk > $\in$ E means that the action of rule rj generates events which trigger rule rk . Fig 3 represents rule interaction using Triggering Graph.

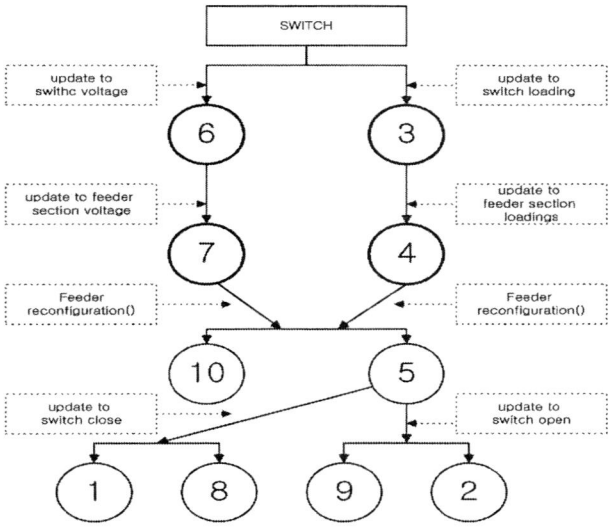

Fig. 3. Triggering graph for Feeder automation

In fig.3, when intelligent switch loadings of distribution networks is changed because of the increasing of customer's consuming, data acquisition system detects the amount of loadings change and sends updating signals switches of DDSs database. Due to the updating signals, rule R3 is triggered to update section loadings of appropriate feeder section that is connected to the switch. If this updated value exceeds 80[%] of feeder capacity then R4 is triggered to minimize line losses and to relive feeder overloadings by using feeder reconfiguration program. R4's action trigger R5 and R10 simultaneously. Due to R10's action, data resulted from feeder reconfiguration program is stored in information object. R5's action updates attribute close/open status value of switch selected from feeder reconfiguration program and triggers R1, R2, R8 and R9 simultaneously. Because of R1, R2, R8 and R9, close/open status of feeder section is updated by the same status of appropriate switch and close/open status of Intelligent switch located in distribution networks is changed by the same status of appropriate switch.

3 Web Based DDSS Active Database System Architecture

A flexible open DDSS system architecture has to satisfy the two-fold requirements[3] It is built to vendor neutral standards(easy in the use of software package).

It will provide the ability to enhance an existing DDSS without relaying one vendor(easy in the continuing development and maintenance of the software package) These requirements can be met by using the internet as the operating environment.

Using internet technology for DDSS architecture will gain following benefits

First, it support cross-platform architecture:

In a standardized internet browser environment with HTML and TCP/IP protocols, users will continue using the platforms with which they are most familiar without conscious of different hardware platform.

Second, it follows open system standard :

By following Structured Query Language(SQL), HTML, HTTP, FTP, TCP/IP, and PPP, data exchange and system expansion are easily done with minimum efforts[4].

The proposed new web based DDSS architecture make uses of the Java 2 Enterprise Edition architecture which is a Web-based multitier architecture, as presented in fig. 4.

The architecture can be expressed by subdividing two parts in fig.4The one is web based architecture which support open system and the other is active database architecture which support DDSS feeder automation

For web based architecture, Its structure can be divided into three tiers[3]

Client tier : It provides user interface.
Middle tier: It is subdivided into the Web server and the application server.
Data tier: It handles the information storage

In the middle tier, application server is transferring request from the web into appropriate functions in the system and also provide for interfacing different kinds of database. Web server acts as the gateway for Web-based clients to access the database.

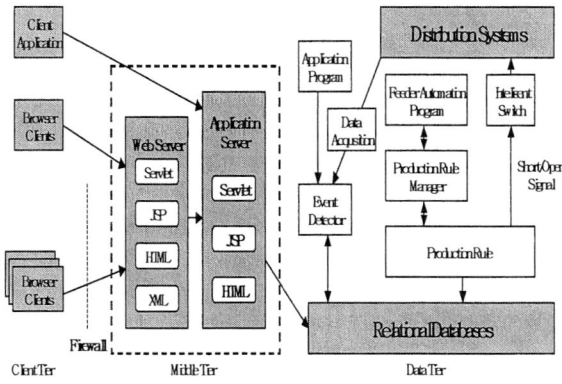

Fig. 4. Web based DDSS active database system architecture

Power distribution engineers and operators sent HTTP requests to Web server through the industrial standard web browser. If requested web page containing database SQL command, the database interprets SQL commands and returns matched data in the database back to Web server. Matched data is formulated as a web page and displayed web page in the client window[4].

For active database architecture shown in the right-side of fig.4.

Its structure can be expressed specifically by dividing into three parts
- Event detector :
  It checks telemetered data and sends signal to production rule manager if integrity constraints are violated.
- Production rule manager
  It accept signal from event detector and determines which rules to be fired.
- Production rule
  It allows specification of data manipulation operations that are executed automatically whenever certain events occur or conditions are satisfied.

Data acquisition system located in distribution network detects switch's close/open status, current, voltage and loadings, and it sends a detected data to central DAS database system.

In DDSS database system, event detector accepts detected data sending from data acquisition system and sends data modification operation to database. If integrity constraints is violated because of data modification then event detector send signal to production rule manager. By using production rule manger, production rules are fired and resulted data is updated in database. After updating values in database, event detector checks if updating values satisfies integrity constraints or not. If integrity constraints are violated then event detector sends signals to production rule manager. And, when Rule R10 and Rule R5 are triggered, close/open signals are send to appropriate intelligent switch located in distribution networks.

4 Conclusion

This paper presents a new Web-based active database architectural design to the Distribution data security system applications. This architecture includes web-based architecture and active database parts. For web based parts, the authors utilize the Java 2 Enterprise Edition architecture for open system, which will easy in the continuing development and maintenance of the software package. Therefore, it is easy to meet open-access competitive market environment. For active database architecture, the author design production rule and production rule manager for DDSS feeder automation By utilizing proposed rules, distribution network can be operated reliably with minimum operators intervening.

Acknowledgment

This work has been supported by 2004-0726-000, which is funded by Gyeonggi Province, Korea.

References

1. J. Widom, S.Ceri, "Active Database Systems : Triggers and rules for Advanced Database Processing", MORGAN KAUFMANN PUBLISHERS"
2. E. Baralis, S. Ceri and S.Paraboshi, "Compile-Time and Runtime Analysis of Active Behavior", IEEE Trans, On Knowledge and Data Engineering, Vol. 10, No.3, pp 353 – 370, 1998
3. S.Chen, F.Y. LU, "Web-Based Simulations of Power Systems", IEEE Computer Application in Power, January, 2002, pp35-40
4. J.T. Ma, T. Liu, L.F. Wu, "New Energy Management System Architectural Design and Intranet/Internet Applications to Power Systems" Conference Proc. Power industry computer application conference, 1995, pp 207- 212
5. E.Baralis, A.Bianco, "Performance Evaluation of Rule Execution Semantics in Active Database" Tech. Rep. DAI.EB. 96.1, Aug. 1996
6. IEEE Task Force on Power System Control Center Database, "Critical Issues Affecting Power System Control Center Database", IEEE Trans. On Power System, vol, 11, no.2 , May, 1996
7. G..S. Martire, D.J.H. Nuttall, "Open Systems and Database", IEEE Trans, On Power System, Vol. 8, NO. 2, May. 1993.

Efficient DoS Resistant Multicast Authentication Schemes

JaeYong Jeong[1], Yongsu Park[2], and Yookun Cho[1]

[1] School of Computer Science and Engineering,
Seoul National University,
San 56-1 Shilim-Dong Gwanak-Ku,
Seoul 151-742, Korea
{jjy, cho}@ssrnet.snu.ac.kr
[2] The College of Information and Communications,
Hanyang University,
Seoul 133-791, Korea
nikolayeva@naver.com

Abstract. To enable widespread commercial stream services, authentication is an important and challenging problem. As for multicast authentication, recently proposed schemes well-operate in adversarial network environment where an enemy can inject a large amount of invalid packets to choke the decoding process in the receivers, at the expense of a large communication overhead. In this paper, we present two efficient DoS resistant multicast authentication algorithms. To detect DoS attack, they require loose time-syncronization or delay of sending the packets, respectively. Compared with the previous schemes, they have much lower communication overhead and smaller computation cost on the receivers.

Keywords: network security, authentication, multicast network.

1 Introduction

To enable widespread commercial stream services, it is crucial to ensure data integrity and source authentication [3, 4, 10], e.g., a listener may feel the need to be assured that news streams have not been altered and were made by the original broadcast station.

There are three issues to consider for authenticating live streams. For a sender, the scheme must have low computation cost to support fast packet rates. For receivers, it must assure a high verification probability, which is defined as a ratio of verifiable packets to received packets, in spite of a large packet loss [10]. Moreover, communication overhead should be small.

Related work [3, 4, 9, 10, 13, 14] deals with two aspects: designing faster signature schemes and amortizing each signing operation by making use of a single signature to authenticate several packets [4]. This paper deals with the latter approach. Most of previous works tried to minimize communication overhead or verification probability in the restricted environment where only the small number of the packets can be modified or forged. Since they are vulnerable to the case where a large amount of the packets are forged or injected, recent schemes deal with these problems.

In this paper, we propose two efficient DoS-resistant multicast authentication schemes. They well-operates even in the adversarial envionments where a large amount of the packets are forged or injected. To check DoS attack, the first one uses MAC (Message Authentication Code) while the second relies on hash chaining. They require loose time-syncronization or delay of sending the packets, respectively. As for authenticating mechanism, both of them uses Pannetrat and Molva's authentication method to minimize the communication overhead. Compared with the previous schemes, they have much lower communication overhead and smaller computation cost on the receivers.

This paper is organized as follows. Section 2 covers related work. In Section 3, we describe the proposed algorithms. In Section 4, we analyze the computation overhead and verification delay of our schemes and we show the comparison results between our schemes and the previous works. Finally, conclusions are made in Section 6.

2 Related Work

In this section, we briefly review the previous works. Researches on stream authentication can be classified into two types. First, researches for designing faster signature schemes and second, researches for amortizing each signing operation by making use of a single signature to authenticate several packets [4].

2.1 Researches Related to the Fast Signature Algorithms

In [14], Wong and Lam proposed methods to speed up FFS (Feige-Fiat-Shamir) signature scheme [5] by using CRT (Chinese Remainder Theorem) [5], reducing the size of verification key, and using precomputation with large memory. They showed that the verification in the scheme was as fast as that of RSA with small exponent and the signing operation was much faster than other schemes (RSA, DSA, ElGamal, Rabin). Moreover, they extended FFS to allow "adjustable and incremental" verification, in which a verifier could verify the signature at different levels, so that he can verify it at a lower security level with a small computation cost, and later increase the security level with larger computation time.

In [3], Gennero and Rohatgi proposed the on-line stream authentication method by using one-time signatures. Compared with the ordinary signature schemes, one-time (or k-time) signature scheme shows much faster sign/verification rates. However, the size of one-time (or k-time) signature is proportional to the size of the input message and is quite large, i.e., the size of Lamport one-time signature [5] of the SHA-1 hashed message is about 1300 bytes. In [13], Rohatgi used k-time signature based on TCR (Target Collision Resistant) function, which reduced the size of the signature to less than 300 bytes.

Recently, Perrig proposed an efficient one-time signature scheme, BiBa, where the size of the signature was much smaller than those of the previous one-time signature schemes [9]. However, in this scheme the computation cost for signature generation is higher and the public key size is larger than those of the previous one-time signature schemes such as Merkle-Winternitz [6] or Bleichenbacher-Maurer [2].

2.2 Researches Related to Amortizing Each Signing Operation

To the best of our knowledge, the first stream authentication scheme was proposed by Gennaro and Rohatgi [3]. In this scheme, the kth packet includes the hash value of the $(k+1)$th packet and only the first packet is signed in each packet group so each signing operation is amortized over several packets. However, it has two weak points: the first is that it does not tolerate packet loss because successive packets in the group cannot be verified when a packet loss occurs. Moreover, the sender must compute authentication information from the last packet in a group to the first packet. Therefore, it is unlikely this scheme will be used to authenticate on-line streams.

In [14], Wong and Lam proposed a method for amortizing signing operation by using Merkle's authentication tree [7]. In this scheme, the sender constructs an authentication tree for each group of stream chunks and signs the root's value. Each packet includes the signature of root's value and the values of siblings of each node in the packet's path to the root. A received packet can be verified by reconstructing the values of nodes in the packet's path to the root and comparing the reconstructed root's value with the value in the signature. This scheme has an advantage in that all received packets are verifiable. But, it requires large communication overhead because each packet involves a lot of hashes and a signature. Another disadvantage is that all the packets in a packet group cannot be computed and sent until the root's value is signed, which in turn incurs bursty traffic patterns. In [14], the authors proposed another scheme that uses authentication stars. The authentication star is a special authentication tree such that its $n+1$ nodes consist of one root and n leaves. This scheme has the shortcomings described above, too.

Perrig, Canetti, Song, and Tygar proposed TESLA (Timed Efficient Stream Loss-tolerant Authentication) and EMSS in [10]. TESLA uses MAC (Message Authentication Code) to authenticate packets. A fixed time interval after sending the packet that contains MAC, the key used for generating the MAC is revealed. This method is very efficient in that it requires low computation overhead and communication overhead. But, it has limitations in that the clocks of the sender and receivers must be synchronized within the allowable margin. Moreover, it does not provide non-repudiation. Recently, Perrig proposed the broadcast authentication protocol using BiBa [9], which has similar constraints of TESLA, such as time-synchronization or not providing non-repudiation.

In EMSS, the hash value of kth (index k) packet is stored in several packets whose indexes are more than k. After sending all the packets for a packet group, the sender transmits the signature packet that has the hashes of last packets in the group and their signature. The algorithm is very simple but is non-deterministic in the sense that the indexes are chosen at random. Moreover, the verification of each packet can be delayed and the delay bound cannot be determined. Particularly, although a packet cannot be verified by using all received packets and signature packet in one group, it can be verified after receiving packets and signature packets in the next groups. To increase the verification probability, the authors take the size of group to be small. So, even though a packet is unverifiable through the use of the packets of one group, it can be verified by receiving the packets of the next groups until the delay deadline. Authors also suggested extended EMSS, which is similar to EMSS but it uses IDA (Information Dispersal Algorithm) [11] to increase the verification probability.

Golle and Modadugu pointed out the fact that packet loss occurs quite bursty on the Internet and proposed GM scheme [4]. They proved that the scheme can resist the longest single burst packet loss assuming that there is a bound on memory size of hash buffer and packet buffer for the sender. Because each packet has only two hashes on the average, this scheme has low communication overhead. Moreover, they showed that it is close to optimally resist burst packet loss under the practical condition such as constraining the average capacity of the buffer on the sender or estimating the endurance on the longest average burst loss. It has deterministic algorithm and its computation cost on the sender is quite low.

Recently, Park, Chong and Siegel devised an efficient stream authentication scheme, SAIDA (Signature Amortization using IDA) [8]. As in extended EMSS, SAIDA is based on Rabin's IDA algorithm. They mathematically analyzed the verificaton probability of SAIDA. Moreover, the simulation result showed that under the same communication overhead, the verification probability of SAIDA is much higher than those of the previous schemes.

3 Proposed Schemes

In prior to the explanation of the proposed schems, we explain broadcast/multicast authentication schemes that rely on the erasure codes. Then, we explain DoS (Denial of Service) attacks by injecting or forging the packets. Finally, we describe our two schemes.

We will use the following notations. $h(X)$ denotes a one-way hash function. $C||D$ is the concatenation of strings C and D. $|E|$ denotes the bit size of E. The sender and the receiver are denoted by S and R, respectively. $SIG_F(G)$ stands for the digital signature of G signed by a signer F.

The erasure code contains the following two modules. $Disperse(F, n, r)$ splits the data F with some amount of redundancy resulting in $n+r$ pieces F_i ($1 \leq i \leq n+r$), where $|F_i|$ is $|F|/n$. Reconstruction of F is possible with any combination of n pieces by calling $Merge(\{F_{i_j}|(1 \leq j \leq n), (1 \leq i_j \leq n)\}, n, r)$.

3.1 Erasure Code Based Multicast Authentication

To the best of our knowledge, erasure code based multicast authentication schemes are the best algorithms in terms of the communication overhead and verification probability. In this subsection, we briefly describe typical erasure code based multicast authentication schemes. We assume that a live stream is divided into fixed-size chunks $M_1, M_2, \ldots$. For the first n chunks (we call this a *group*), S generates n packets which include authentication information (by using the digital signature, hash function, and erasure codes). After sending all these packets, S repeats this procedure for the next group of n chunks. When R receives at least $(1-p)n$ $(0 < p \leq 1)$ packets for a group, R can verify all the received packets.

3.2 DoS (Denial of Service) Attack

The above schemes work well in the environment where only the small number of the packets can be modified or forged. However, they are vulnerable to the case where a large amount of the packets are forged or injected. Let us consider a simple example. Assume that an adversary A forges and injects a lot of different i-th packet P_i. Then, to find a correct P_i, R should perform decodeing operation of the erasure code and signature verification operation for every candidate of P_i. Since their computation costs are very high, A can success the DoS attack. Moreover, if A forges the packets for $P_i, P_j, P_k, \ldots$, R should perform the operations for all the possible combinations, which require tremendous CPU load.

3.3 Proposed Scheme 1

Proposed scheme 1 uses MAC (Message Authentication Code) to detect DoS attack. First, S generates a random number K_{n+1} and then computes $K_i = h(K_{i+1})$ $(1 \leq i \leq n)$. Assume that K_1 is successfully transmitted to R. For the chunks $B_{n(i-1)+1}, B_{n(i-1)+2}, \ldots, B_{n(i-1)+n}$ of group G_i, S computes $M_{n(i-1)+j} = MAC(K_i, B_{n(i-1)+j})$. Then, by the algorithm described in Section 3.3, S computes authentication information $A_{n(i-1)+j}$. S makes a packet $P_{n(i-1)+j} = (B_{n(i-1)+j}, M_{n(i-1)+j}, A_{n(i-1)+j})$ and transmits it to R.

Note that K_i is sent with the packets of G_{i+d}. If R receives at least one packet of G_{i+d}, he can surely receive the K_i. Then, R checks the validity of K_i by checking that $K_{i-1} = h(K_i)$. If succeeds, he verifies the equation $M_{n(i-1)+j} = MAC(K_i, B_{n(i-1)+j})$ to check whether the packet was forged/modified by DoS attack. If there are at least $(1-p)n$ packets that successfully passed the tests, by $A_{n(i-1)+j}$ and the algorithm of Section 3.3, R verifies $B_{n(i-1)+j}$.

Proposed scheme 1 has little delay for sending packets since S can send a packet immediately after a chunk is generated. However, in spite of receiving a packet of group G_i, R should wait to receive the packets of group G_{i+d} to compute the MAC value.

As mentioned, checking the MAC value is for detecting DoS attack, which would not occur frequently. Hence, if the scheme optimistically operates as follows, verification delay of R will be significantly reduced.

STEP 1. For a group G_i, R checks whether there are the packets having the same sequence number. If exist, this means that there was a DoS attack and go to STEP 3.

STEP 2. If there are at most $(1-p)n-1$ packets, then the verification algorithm fails. Otherwise, R randomly selects $(1-p)n$ packets among them and verifies them by the algorithm of Section 3.3. If verification fails, go to STEP 3.

STEP 3. R waits to receive the packets of G_{i+d} to get K_i. Then, he computes MAC to check the forged/modified packets by the DoS attack and removes them. Finally, he verifies the remaining packets by the algorithm of Section 3.3.

In Proposed scheme 1, if an attacker A eavesdrops a packet P of group G_i, he can forges any packet of the group G_{i-d} by using K_i in P. To prevent this attack, S and R should synchronize their clock within an allowable margin and if R receives a packet whose transmission delay is larger than the pre-defined value, he should drop it.

3.4 Proposed Scheme 2

To detect the DoS attack, this scheme uses hash chaining. For the groups $G_2,\ldots,G_k$, S first computes the concatenation of the hash values of the each chunk, as follows: $h(B_{n+1})||h(B_{n+2})||\cdots||h(B_{kn})$. Then, S gets $F_1,\ldots,F_n$ by calling $Disperse$ ($h(B_{n+1})||h(B_{n+2})||\cdots||h(B_{kn}), (1-p)n, pn)$. He makes $B'_j = B_j||F_j$ and computes A_j by the algorithm of Section 3.3. After S transmits a packet $P_j = (B'_j, A_j)$ to R, he performs the above work for the next group.

Assume that R receives the packets for the group G_i. If R has already verified the packets for at least one group among $G_{i-1},\ldots,G_{i-k}$, he can reconstruct $h(B_{n(i-1)+1}),\ldots,h(B_{n(i-1)+n})$ by using $Merge()$. For each packet of G_i, if the hash value of the chunk is equal to one of these values, then it has not been forged/modified by the DoS attack. If the number of such packets are at least $(1-p)n$, R can verify it by using the algorithm of Section 3.3.

3.5 Pannentrat and Molva's Authentication Scheme

In this section, we describe the erasure code based authentication method that our scheme uses. Recall that the erasure code consists of the following two modules. $Disperse(F, n, r)$ splits the data F with some amount of redundancy resulting in $n+r$ pieces F_i $(1 \leq i \leq n+r)$, where $|F_i|$ is $|F|/n$. Reconstruction of F is possible with any combination of n pieces by calling $Merge(\{F_{i_j}|(1 \leq j \leq n), (1 \leq i_j \leq n)\}, n, r)$. The code in which $\{F_1,\ldots,F_n\} = F$ is denoted as the *systematic* code.

To the best of our knowledge, among the erasure code based authentication schemes [8, 15, 16], Pannetrat and Molva's method has the smallest communication overhead and the highest verification probability. This algorithm consists of the following two modules. The first modules generates authentication information for a single group of chunks and the second module tries to verify the received packets only if the number of the received packets is at least $(1-p)n$.

Authentication information generation: For the n chunks $B_1,\ldots,B_n$, S computes the hashed values $h_1,\ldots,h_n$. Then, S generates $G = SIG_S(h_1||\cdots||h_n)$.

Then, he computes $Disperse(\{h_1, \ldots, h_n\}, n, pn) = \{F_1, \ldots, F_{n+pn}\}$ and $Disperse(\{F_{n+1}, \ldots, F_{n+pn}, G\}, (1-p)n, pn) = \{A_1, \ldots, A_n\}$. Finally, the packets are generated as follows: $P_i = \{B_i, A_i\}$.

Verification: Assume that at least $(1-p)n$ packets are received: $P_1, \ldots, P_{(1-p)n}$. Then, R computes $\{F_{n+1}, \ldots, F_{n+pn}, G\}$ by calling $Merge(\{A_1, \ldots, A_{(1-p)n}\}, (1-p)n, pn)$. Then, R computes $F_1, \ldots, F_n$ by hashing the received chunks. He gets $\{h_1, \ldots, h_n\}$ by calling $Merge(\{F_1, \ldots, F_{(1-p)n}, F_{n+1}, \ldots, F_{n+pn}\}, npn)$. R verifies the digital signature G. If successfully verified, all the packets are correctly verified.

4 Analysis

In Section 4.1, we analyze the communication overhead of the proposed schemes. In Section 4.2, we analyze the computation cost on the receiver. In Section 4.3 we analyze the verification delay. Finally, in Section 4.4, we show the comparison result between the previous schemes and our schemes.

4.1 Communication Overhead

In this section, we analyze the communication overhead of the proposed schemes. For each packet, Scheme 1 has one key and a single $MAC()$ and one A_i. We does not need a large size of the key or $MAC()$ since we does use them not for authenticaiton but for checking the DoS attack. According to [17], it suffice the security requirement that the key size and the output size of $MAC()$ are 40 bits. As for the size A_i, by the Theorem 2 of [16], $|A_i| = \frac{(|SIG()|+pn|h()|)}{(1-p)n}$. Hence, the communication overhead of Scheme 1 is $40 + \frac{(|SIG()|+pn|h()|)}{(1-p)n}$ bits.

Scheme 2 has a parameter k and each packet consists of a chunk, A_i, and the output of $Disperse()$ whose input is the concatenation of kn hash values. For DoS detection, it suffices the security requirement that the size of each hash value is 40 bits. Hence, the communication overhead is $40kn/(1-p) + \frac{(|SIG()|+pn|h()|)}{(1-p)n}$.

4.2 Computation Cost

In this section, we analyze the computation cost for generating a single group in the sender or receiver. First, let us consider the Scheme 1. To generate n packets for a group, S computes 2 $Disperse()$, n $MAC()$, and one signing operation. Note that $MAC()$ can be computed by applying hash operations twice. Hence, the computation cost on the sender is $2nC_H + 2C_{Disperse} + C_{Sign}$ where $C_H, C_{Disperse}$, and C_{Sign} are the computation costs of $h()$, $Disperse()$, and the signing operation, respectively. Similarly, the computation cost on the receiver is $2nC_H + 2C_{Merge} + C_{Verify}$ where C_{Merge} and C_{Verify} are the computation costs of $Merge()$ and the verification operation, respectively. Note that the hash operation is very efficient and C_H is about 1000 times smaller than $C_{Disperse}, C_{Merge}, C_{Sign}$, and C_{Verify}.

In Proposed scheme 2, the computation cost on the sender is $3C_{Disperse} + C_{Sign}$ and the computation cost on the receiver is $3C_{Merge} + C_{Verify}$.

4.3 Verification Delay

In this section, we analyze the verification delay of the proposed schemes. In Scheme 1, if there was no DoS attack, R can verify the received packets immediately. However, if R detects the DoS attack, the verification of the packets are delayed until another d groups are received. Hence, the verfication delay is nd.

Scheme 2 has a parameter k such that R can verify the received packet immediately only if he has verified at least one packet for the previous k group. Hence, the verification delay is 0.

5 Comparison

In this section, we analyze the communication cost of our proposed schemes and show the comparison results of the previous schemes. The compared schemes are SAIDA [8], Pannetrat and Molva's scheme [16], PRAB [17], and Lysyanakaya, Tamassia and Triandopolous's scheme [15]. According to the parameters in [17, 8], let the group size $n = 128$, packet loss rate $p = o.5$, $|h()| = 80bits$, $|SIG()| = 1024bits$. Also, let the maximum flood rate $\beta = 10$ as in [15]. Let this condition be called as the case A. Table 1 shows the communication cost of above schemes.

Table 1. Comparison results

Scheme		Communication overheads (Bytes)	
		formula	case A
Schemes vulnerable to DoS	hash chaining scheme	$40|h| + \frac{|SIG|}{n}$ ~ $50|h| + \frac{|SIG|}{n}$	40~50
	SAIDA [8]	$\frac{|SIG|+n|h|}{(1-p)n}$	22
	Pannetrat & Molva [16]	$\frac{|SIG|+pn|h|}{(1-p)n}$	12
Schemes resistant to DoS	PRABS [17]	$2|h|logn + \frac{|SIG|+n|h|}{(1-p)n}$	92
	Lysyanskaya et. al [15]	$\frac{(|h|+|SIG|/n)\beta(1+\epsilon)}{(1-p)^2}$	440
	Proposed scheme 1	$|K| + |MAC()| + \frac{|SIG|+pn|h|}{(1-p)n}$	32
	Proposed scheme 2	$\frac{k|h|}{(1-p)} + \frac{|SIG|+pn|h|}{(1-p)n}$	72 (k = 3)

As can be seen in Table 1, the schemes resistant to DoS have rather higher communication overheads than those vulnerable to DoS. This is due to overheads to detect DoS attacks: PRABS uses authentication information of authentication trees, Lysyanskaya's scheme uses error correction codes, proposed scheme 1 uses MAC(), and proposed scheme 2 uses hash chain. But, we can see that proposed scheme 1 and 2 have lower communication overheads than those of PRABS and Lysyanskaya's scheme.

6 Conclusion

To enable widespread commercial stream services, authentication is an important and challenging problem. As for multicast authentication, recently proposed schemes well-operate in adversarial network environment where an enemy can inject a large amount of invalid packets to choke the decoding process in the receivers, at the expense of a large communication overhead. In this paper, we present two efficient DoS resistant multicast authentication algorithms. To detect DoS attack, they require loose time-syncronization or delay of sending the packets, respectively. Compared with the previous schemes, they have much lower communication overhead and smaller computation cost on the receivers.

References

1. FIPS 180-1. Secure Hash Standard. Federal Information Processing Standard (FIPS), Publication 180-1, National Institute of Standards and Technology, US Department of Commerce, Washington D.C., April 1995.
2. D. Bleichenbacher and U. Maurer. Optimal tree-based one-time digital signature schemes. In *STACS'96*, pages 363–374, 1996.
3. Rosario Gennaro and Pankaj Rohatgi. How to Sign Digital Streams. In *CRYPTO'97*, pages 180–197, 1997.
4. Philippe Golle and Nagendra Modadugu. Authenticating Streamed Data in the Presence of Random Packet Loss. In *NDSS'01*, pages 13–22, 2001.
5. Alfred J. Menezes, Paul C. van Oorschot, and Scott A. Vanstone. *Handbook of Applied Cryptography*. CRC Press, 1997.
6. R. C. Merkle. A digital signature based on a conventional encryption function. In *CRYPTO'87*, pages 369–378, 1987.
7. Ralph C. Merkle. A Certified Digital Signature. In *CRYPTO'89*, 1989.
8. J. M. Park, E. K. P. Chong, and H. J. Siegel. Efficient Multicast Packet Authentication Using Signature Amoritization. *ACM Transactions on Information and System Security*, 6(2):258–285, 2003.
9. Adrian Perrig. The BiBa One-Time Signature and Broadcast Authentication Protocol. In *8th ACM Conference on Computer and Communication Security*, pages 28–37, November 2001.
10. Adrian Perrig, Ran Canetti, Dawn Song, and J. D. Tygar. Efficient Authentication and Signing of Multicast Streams over Lossy Channels. In *Proceedings of IEEE Security and Privacy Symposium*, May, 2000.
11. Michael O. Rabin. Efficient dispersal of information for security, load balancing and fault tolerance. *Journal of the Association for Computing Machinery*, 36(2):335–348, 1989.
12. R. L. Rivest, A. Shamir, and L. M. Adelman. A Method for obtaining digital signatures and public-key cryptosystems. *Communications of the ACM*, 21(2):120–126, 1978.
13. Pankaj Rohatgi. A Compact and Fast Hybrid Signature Scheme for Multicast Packet Authentication. In *6th ACM Conference on Computer and Communication Security*, November 1999.
14. Chung Kei Wong and Simon S. Lam. Digital Signatures for Flows and Multicasts. *IEEE/ACM Transactions on Networking*, 7(4):502–513, 1999.

15. A. Lysyanskaya, R. Tamassia, and N. Triandopoulos. Muticast Authentication in Fully Adversarial Networks. In *IEEE S&P 2004*, 2004.
16. A. Pannetrat and R. Molva. Efficient Multicast Packet Authentication. In *NDSS'2003*, 2003.
17. C. Karlof, Naveen Sastry, Yaping Li, Adrian Perrig, and J. D. Tygar. Distillation Codes and Applications to DoS Resistant Multicast Authentication. In *NDSS'2004*, 2004.

Development System Security Process of ISO/IEC TR 15504 and Security Considerations for Software Process Improvement

Eun-ser Lee[1] and Malrey Lee[2]

[1] TQMS, 370 Dangsan-dong 3ga, Youngdeungpo-gu, Seoul, Korea
eslee@object.cau.ac.kr
http://object.cau.ac.kr/selab/index.html
[2] School of Electronics & Information Engineering, Chonbuk National University, 664-14, DeokJin-dong, JeonJu, ChonBuk, Korea
mrlee@chonbuk.ac.kr

Abstract. This research is intended to develop the system security process. The IT products like as firewall, IDS (Intrusion Detection System) and VPN (Virtual Private Network) are made to perform special functions related to security, so the developers of these products or systems should consider many kinds of things related to security not only design itself but also development environment to protect integrity of products. When we are making these kinds of software products, ISO/IEC TR 15504 may provide a framework for the assessment of software processes, and this framework can be used by organizations involved in planning, monitoring, controlling, and improving the acquisition, supply, development, operation, evolution and support of software. But, in the ISO/IEC TR 15504, considerations for security are relatively poor to other security-related criteria such as ISO/IEC 21827 or ISO/IEC 15408 [10-12]. In this paper we propose some measures related to development process security by analyzing the ISO/IEC 21827, the Systems Security Engineering Capability Maturity Model (SSE-CMM) and ISO/IEC 15408, Common Criteria (CC). And we present a Process of Security for ISO/IEC TR 15504. This enable estimation of development system security process by case study.

1 Introduction

ISO/IEC TR 15504, the Software Process Improvement Capability Determination (SPICE), provides a framework for the assessment of software processes [1-9, 15]. This framework can be used by organizations involved in planning, monitoring, controlling, and improving the acquisition, supply, development, operation, evolution and support of software. But, in the ISO/IEC TR 15504, considerations for security are relatively poor to others. For example, the considerations for security related to software development and developer are lacked.

In this paper, we propose a process related to security by comparing ISO/IEC TR 15504 to ISO/IEC 21827 and ISO/IEC 15408. The proposed scheme may be contributed to the improvement of security for IT product or system. And in this paper, we propose some measures related to development process security by analyzing the ISO/IEC 21827, the Systems Security Engineering Capability Maturity Model (SSE-

CMM) and ISO/IEC 15408, Common Criteria (CC). And we present a Process for Security for ISO/IEC TR 15504.

2 ISO/IEC TR 15504

2.1 Framework of ISO/IEC TR 15504

The SPICE project has developed an assessment model (ISO/IEC 15504: Part 5) for software process capability determination. The assessment model consists of process and capability dimensions. Figure 1 shows the structure of the process and capability dimensions. In the process dimension, the processes associated with software are defined and classified into five categories known as the *Customer-Supplier*, *Engineering*, *Support*, *Management*, and *Organization*. The capability dimension is depicted as a series of process attributes, applicable to any process, which represent measurable characteristics necessary to manage a process and to improve its performance capability. The capability dimension comprises of six capability levels ranging from 0 to 5. The higher the level, the higher the process capability is achieved.

2.2 Process Dimensions

The process dimension is composed of five process categories as follows (See Appendix for detailed processes in each category).

The Customer-Supplier process category (CUS) - processes that have direct impact on the customer, support development and transition of the software to the customer, and provide the correct operation and use software of products and/or services.

The Engineering process category (ENG) - processes that directly specify, implement, or maintain the software product, its relation to the system and its customer documentation.

The Support process category (SUP) - processes that may be employed by any of the other processes (including other supporting processes) at various points in the software life cycle.

The Management process category (MAN) - processes which contain generic practices that may be used by those who manage any type of project or process within a software life cycle.

The Organization process category (ORG) - processes that establish business goals of the organization and develop processes, products, and resource assets which, when used by the projects in the organization, will help the organization achieve its business goals.

3 A New Process for Development System Security

3.1 Work Products of ISO/IEC TR 15504 Related to Development System Security

As mentioned earlier, ISO/IEC TR 15504 provides a framework for the assessment of software processes, and this framework can be used by organizations involved in planning, managing, monitoring, controlling, and improving the acquisition, supply, development, operation, evolution and support of software. ISO/IEC TR 15504 does not define any Process related to security, but the security-related parts are expressed in some Work Products (WP) as like;

Table 1. Security-related Work Products

ID	WP Class	WP Type	WP Characteristics
10	1.3	Coding standard	- Security considerations
51	3.2	Contract	- References to any special customer needs (i.e., confidentiality requirements, security, hardware, etc.)
52	2.2	Requirement specification	- Identify any security considerations/constraints
53	2.3	System design/architecture	- Security/data protection characteristics
54	2.3	High level software design	- Any required security characteristics required
74	1.4/2.1	Installation strategy plan	- Identification of any safety and security requirements
80	2.5	Handling and storage guide	- Addressing appropriate critical safety and security issues
101	2.3	Database design	- Security considerations
104	2.5	Development environment	- Security considerations

ISO/IEC TR 15504 may use these work products as input materials, and these may be the evidence that security-related considerations are being considered. But this implicit method is not good for the 'security' and more complete or concrete countermeasures are needed. Therefore, we propose some new processes which deal with the security.

3.2 A New Process for Development System Security

The processes belonging to the Engineering process category are ENG.1 (Development process), ENG.1.1 (System requirements analysis and design process), ENG.1.2 (Software requirements analysis process), ENG.1.3 (Software design process), ENG.1.4 (Software construction process), ENG.1.5 (Software integration process), ENG.1.6 (Software testing process), ENG.1.7 (System integration and testing process), and ENG.2 (Development process).

These processes commonly contain the 52nd work product (Requirement specification), and some of them have 51st, 53rd, 54th work products separately. Therefore, each process included in the ENG category may contain the condition, 'Identify any security considerations/constraints'. But the phrase 'Identify any security considerations/constraints' may apply to the 'software or hardware (may contain firmware) development process' and not to the 'development site' itself.

In this paper we will present a new process applicable to the software development site. In fact, the process we propose can be included in the MAN or ORG categories, but this is not the major fact in this paper, and that will be a future work. We can find the requirements for Development security in the ISO/IEC 15408 as like;

Development security covers the physical, procedural, personnel, and other security measures used in the development environment. It includes physical security of the development location(s) and controls on the selection and hiring of development staff.

Development system security is concerned with physical, procedural, personnel, and other security measures that may be used in the development environment to protect the integrity of products. It is important that this requirement deals with measures to remove and reduce threats existing in the developing site (not in the operation site). These contents in the phrase above are not the perfect, but will suggest a guide for development system security at least.

The individual processes of ISO/IEC TR 15504 are described in terms of six components such as Process Identifier, Process Name, Process Type, Process Purpose, Process Outcomes and Process Notes. The style guide in annex C of ISO/IEC TR 15504-2 provides guidelines which may be used when extending process definitions or defining new processes.

Next is the Development System Security process we suggest.

(1) Process Identifier: ENG.3
(2) Process Name: Development System Security process
(3) Process Type: New
(4) Process purpose:

The purpose of the Development System Security process is to protect the confidentiality and integrity of the system components (such as hardware, software, firmware, manual, operations and network, etc) design and implementation in its development environment. As a result of successful implementation of the process:

(5) Process Outcomes:

- access control strategy will be developed and released to manage records for entrance and exit to site, logon and logout of system component according to the released strategy
- roles, responsibilities, and accountabilities related to security are defined and released
- training and education programs related to security are defined and followed
- security review strategy will be developed and documented to manage each change steps

(6) Base Practices:

ENG.3.BP.1 Develop project measures. Develop and release the project measures for protecting the access to the development site and product.
ENG.3.BP.2 Develop platform measures. Develop and release the platform measures for protecting execution time, storage and platform volatility.
ENG.3.BP.3 Development personnel measures. Develop and release the personnel measures for selecting and training of staffs.
ENG.3.BP.4 Develop procedural measures. Develop the strategy for processing the change of requirements considering security.

ENG.3.BP.5 Development internal & external environment measures. Develop and release the environment security measures for processing threaten factor of virus and improper administration.

ENG.3.BP.6 Development processing measures. Develop and strategy for processing threaten factor of weak encryption and server spoof.

ENG.3.BP.7 Development application measures. Develop and release the environment security measures for processing threaten factor of poor programming & weak authentication.

ENG.3.BP.8 Develop client measures. Develop and release client measures for processing threaten factor of virus.

ENG.3 Development Security process may have more base practices (BP), but we think these BPs will be the base for future work. For the new process, some work products must be defined as soon as quickly. Next items are the base for the definition of work products.

Table 2. Work products of Development Security process

WP category number	WP category	WP classification number	WP classification	WP type
1	ORGANIZATION	1.1	Policy	Access control to site and so on
		1.2	Procedure	Entrance and so on
		1.3	Standard	Coding and so on
		1.4	Strategy	Site open and so on
2	PROJECT	Future work	Future work	Future work
3	RECORDS	3.1	Report	Site log and so on
		3.2	Record	Entrance record and so on
		3.3	Measure	Future work

4 Security Requirements for Software

4.1 General Software Development Process

There are many methodologies for software development, and security engineering does not mandate any specific development methodology or life cycle model. Fig.1 depicts underlying assumptions about the relationship between the customer's requirements and the implementation. The figure is used to provide a context for discussion and should not be construed as advocating a preference for one methodology (e.g. waterfall) over another (e.g. prototyping).

It is essential that the requirements imposed on the software development be effective in contributing to the objectives of consumers. Unless suitable requirements are established at the start of the development process, the resulting end product, however well engineered, may not meet the objectives of its anticipated consumers. The process is based on the refinement of the customer's requirements into a soft-ware implementation. Each lower level of refinement represents design decomposition with additional design detail. The least abstract representation is the software implementation itself.

In general, customer does not mandate a specific set of design representations. The requirement is that there should be sufficient design representations presented at a sufficient level of granularity to demonstrate where required:

a) that each refinement level is a complete instantiation of the higher levels (i.e. all functions, properties, and behaviors defined at the higher level of abstraction must be demonstrably present in the lower level);

b) that each refinement level is an accurate instantiation of the higher levels (i.e. there should be no functions, proper-ties, and behaviors defined at the lower level of abstraction that are not required by the higher level).

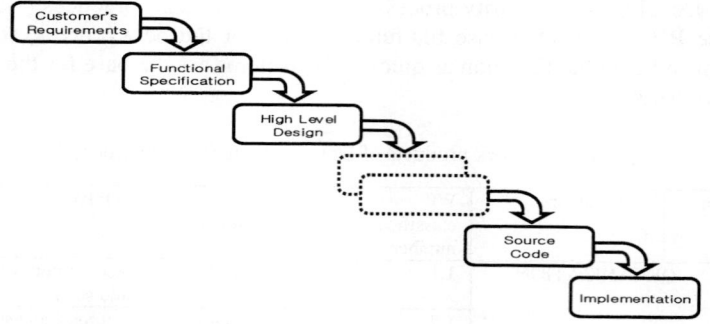

Fig. 1. The relationship between the customer's requirements and the implementation

4.2 Append Security Requirements

For the development of software, the first objective is the perfect implementation of customer's requirements. And this work may be done by very simple processes. However, if the software developed has some critical security holes, the whole network or systems that software installed and generated are very vulnerable.

Therefore, developers or analyzers must consider some security-related factors and append a few security-related requirements to the customer's requirements. Fig.2 depicts the idea about this concept.

The processes based on the refinement of the security-related requirements are considered with the processes of soft-ware implementation.

4.3 Implementation of Security Requirements

Developers can reference the ISO/IEC 15408, Common Criteria (CC), to implement security-related requirements appended.

The multipart standard ISO/IEC 15408 defines criteria, which for historical and continuity purposes are referred to herein as the CC, to be used as the basis for evaluation of security properties of IT products and systems. By establishing such a common criteria base, the results of an IT security evaluation will be meaningful to a wider audience.

The CC will permit comparability between the results of independent security evaluations. It does so by providing a common set of requirements for the security functions of IT products and systems and for assurance measures applied to them during a security evaluation. The evaluation process establishes a level of confidence that the security functions of such products and systems and the assurance measures applied to them meet these requirements. The evaluation results may help consumers to determine whether the IT product or system is secure enough for their intended application and whether the security risks implicit in its use are tolerable.

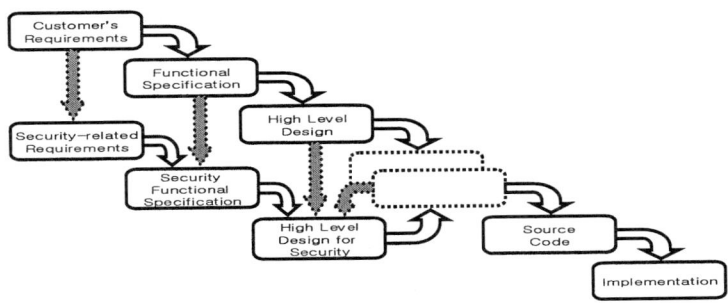

Fig. 2. Append security-related requirements

5 Introduction to Effectiveness Analysis of Development System Security Process (Case Study)

We analyzed aspect that defect removal is achieved efficiently to analyze effectiveness of development system security process[14].

Also, we present effectiveness of introduction through if productivity improves because defect of whole project is reduced through development system security process.

5.1 Defect Removal Efficiency Analysis

Purpose of development system security process design improves quality of product and heighten productivity.

Therefore, when we applied development system security process in actuality project, we wish to apply defect exclusion efficiency (Defect Removal Efficiency). to measure ability of defect control activity.

After apply development system security process, defect exclusion efficiency analysis investigated defect number found at relevant S/W development step and defect number found at next time step in terms of request analysis, design and coding stage. Production of defect exclusion efficiency is as following. DRE = E(E+D)

E= Number of defect found at relevant S/W development step(e.g : Number of defect found at request analysis step)

D= Number of defect found at next S/W development step (e.g : Defect number that defect found at design step is responsible for defect of request analysis step)

Ideal value of DRE is 1, and this displays that any defect does not happen to S/W.

Table 3. Table of defect removal efficiency

	Number(%) of defect found at relevant S/W development step (E)	Number(%) of defect found at next S/W development step (D)
Requirement	10	3
Design	15	3
Coding	5	1

<Table 3> is a table to inspect S/W development step defect number after development system security process application.

0.769 = 10(10+3) (Requirement phase)
0.833 = 15(15+3) (Design phase)
0.833 = 5(5+1) (Coding phase)

If we save DRE at each S/W development step by <Table 3>, it is as following.

Therefore, because DRE is approximated to 1, when we remove defect by development system security process, defect exclusion efficiency was analyzed high.

5.2 Size-Based Software Metrics

After apply Defect Trigger, we investigate by <Table 4> to compare and analyze how productivity improved with last project[20].

Table 4. Size based software metrics

	Last project	The present project
SLOC	40,000	120,620
Project Cost	400,282,000	1,500,000,000
Effort (Man-Month)	55.2	381.6
Defect number	400	810
Project People	11	50

If depend to <Table 4>, last project decreased more remarkably than number of defect found in the present project.

And in case defect happens, it is decreased effort (Man-Month) and human strength to solve this.

Being proportional in scale of project, Contents of each item are increasing. Therefore, based on whole SLOC, project expense and Effort(Man-Month), we compared number of found defect. By the result, scale of project increased by 30% than previous project but number of found defect decreased by 20% than whole scale.

5 Conclusions

In this paper we proposed a new Process applicable to the software development site. In fact, the Process we proposed is not perfect not yet, and the researches for improving going on. Some researches for expression of Base Practice and development of Work Products should be continued. But the work in the paper may be the base of the consideration for security in ISO/IEC TR 15504. And This paper proposes a method appending some security-related requirements to the customer's requirements. For the development of software, the first objective is the perfect implementation of customer's requirements. However, if the software developed has some critical security holes, the whole network or systems that software installed and generated may be very vulnerable. Therefore, developers or analyzers must consider some security-related factors and append a few security-related requirements to the customer's requirements.

ISO/IEC TR 15504 provides a framework for the assessment of software processes, and this framework can be used by organizations involved in planning, monitoring, controlling, and improving the acquisition, supply, development, operation, evolution and support of software. Therefore, it is important to include considerations for security in the Process dimension.

In this paper we did not contain or explain any component for Capability dimension, so the ENG.3 Process we suggest may conform to capability level 2. Therefore, more research efforts will be needed. Because the assessment cases using the ISO/IEC TR 15504 are increased, some processes concerns to security are needed and should be included in the ISO/IEC TR 15504.

For the future work, the processes based on the refinement of the security-related requirements must be considered with the processes of software implementation.

References

1. ISO. ISO/IEC TR 15504-1:1998 Information technology – Software process assessment – Part 1: Concepts and introductory guide
2. ISO. ISO/IEC TR 15504-2:1998 Information technology – Software process assessment – Part 2: A reference model for processes and process capability
3. ISO. ISO/IEC TR 15504-3:1998 Information technology – Software process assessment – Part 3: Performing an assessment
4. ISO. ISO/IEC TR 15504-4:1998 Information technology – Software process assessment – Part 4: Guide to performing assessments
5. ISO. ISO/IEC TR 15504-5:1998 Information technology – Software process assessment – Part 5: An assessment model and indicator guidance
6. ISO. ISO/IEC TR 15504-6:1998 Information technology – Software process assessment – Part 6: Guide to competency of assessors
7. ISO. ISO/IEC TR 15504-7:1998 Information technology – Software process assessment – Part 7: Guide for use in process improvement
8. ISO. ISO/IEC TR 15504-8:1998 Information technology – Software process assessment – Part 8: Guide for use in determining supplier process capability
9. ISO. ISO/IEC TR 15504-9:1998 Information technology – Software process assessment – Part 9: Vocabulary
10. ISO. ISO/IEC 15408-1:1999 Information technology - Security techniques - Evaluation criteria for IT security - Part 1: Introduction and general model
11. ISO. ISO/IEC 15408-2:1999 Information technology - Security techniques - Evaluation criteria for IT security - Part 2: Security functional requirements
12. ISO. ISO/IEC 15408-3:1999 Information technology - Security techniques - Evaluation criteria for IT security - Part 3: Security assurance requirements

13. Tai-Hoon Kim, Byung-Gyu No, Dong-chun Lee: Threat Description for the PP by Using the Concept of the Assets Protected by TOE, ICCS 2003, LNCS 2660, Part 4, pp. 605-613
14. Eun-ser Lee, KyungWhan Lee, KeunLee Design Defect Trigger for Software Process Improvement, Springer-Verlag's LNCS, 2004
15. Hye-young Lee and Dr. Ho-Won Jung, Chang-Shin Chung, Kyung Whan Lee, Hak Jong Jeong, Analysis of Interrater Agreement In ISO/IEC 15504-based Software Process Assessment1, the second Asia Pacific Conference on Quality Software (APAQS2001), Dec. 10-11, 2001, Hong Kong
16. Eun-ser Lee, Kyung Whan Lee, Tai-hoon Kim, Il-Hong Jung Introduction and Evaluation of Development System Security Process of ISO/IEC TR 15504, ICCSA2004 LNCS 3043, 2004, Italy
17. Eun-ser Lee, Tai-hoon Kim, Introduction of Development Site Security Process of ISO/IEC TR 15504, Knowledge-Based Intelligent Information and Engineering Systems, LNAI 3215, 2004, New zealand

Flexible ZCD-UWB with High QoS or High Capacity Using Variable ZCD Factor Code Sets

Jaesang Cha[*], Kyungsup Kwak[**], Changdae Yoon[***], and Chonghyun Lee[****]

[*]Dept. of Information and Communication Eng. Seokyeong Univ. Seoul, Korea
chajs@skuniv.ac.kr
[**]UWB Wireless Communications Research Center(UWB-ITRC),
Inha Univ. Incheon, Korea
[***]Dept. of Elect Eng. Sungkyunkwan Univ. Suwon, Korea
[****]Dept. of Electronic Eng. Seokyeong Univ. Seoul, Korea

Abstract. In this paper, we propose a Variable ZCD(Zero Correlation Duration) Factor Code Sets. Additionaly, Flexible ZCD-UWB(Ultra Wide Band) with High QoS or High Capacity using Variable ZCD Factor Code Sets are presented with their BER(Bit Error Rate) performance in this paper. Flexible ZCD-UWB could be a very useful solution of the wireless home network applications (WHNA) having always High QoS or High Capacity.

1 Introduction

Recently, UWB technique has been paid much attention and been debated by IEEE 802.15.3a[1] or IEEE 802.15.4a[2] for standardization.

Currently, the transmission method of various form have been proposed in DS-CDMA (Direct Sequence-Code Division Multiple Access) based UWB system[1] which is used in WPAN. The performance of DS-CDMA based UWB is influenced by the orthogonal property of the spreading code. And it must consider the MAI (Multiple Access Interference) and the MPI (Multi Path Interference) problem in the multi-path and multiple access environments[3].

In order to solve these MPI and MAI problems without adopting complicated MUD scheme[4] or other interference cancellation schemes, we presented a flexible ZCD-UWB defined as a DS-CDMA based UWB system using ternary ZCD code set, We can solve the interference to the characteristic of ZCD spreading code[5]-[12].

Especially, TZCD (Ternary Zero Correlation Duration) spreading codes[5]-[8] presented in this paper have superior high ZCD property than that of Binary ZCD code (Binary Zero Correlation Duration)[9]-[11] or Walsh code[1][2]. Also, they have many Family sizes as well as wide ZCD and variable ZCD length. Therefore, flexible ZCD-UWB using variable ZCD factor code sets characterized by high Capacity or high QoS is proposed in this paper.

In the first chapter, of this paper we will shortly refer about proposed system. Then, we will introduce the variable ZCD spreading codes for flexible ZCD-UWB.

In the third chapter, we will set up a modeling of flex ZCD-UWB system having flexibility for the variable channel environments. Moreover, we will comment the analysis through variable BER simulation results in the chapter four.

2   Variable ZCD Spreading Code for Flexible ZCD-UWB

2.1   ZCD Characteristics and Variable Ternary ZCD Spreading Codes

For any two spreading codes of period N ; $S_N^{(x)} = (s_0^{(x)}, \cdots, s_{N-1}^{(x)})$ and $S_N^{(y)} = (s_0^{(y)}, \cdots, s_{N-1}^{(y)})$, the periodic correlation and aperiodic correlation for a shift τ is defined as equation (1) and (2), respectively.

$$Periodic \; R_{x,y}(\tau) = \sum_{n=0}^{N-1} s_n^{(x)} \, s_{(n+\tau, \bmod N)}^{(y)} \tag{1}$$

$$Aperiodic \; R_{x,y}(\tau) = \sum_{n=0}^{N-\tau-1} s_n^{(x)} \, s_{(n+\tau)}^{(y)} \tag{2}$$

Here, N is one period of spreading code and $\oplus$ appear modulo N operation. This function becomes the autocorrelation function(ACF) when x=y and the cross-correlation function(CCF) when x≠y. Since the maximum magnitude of periodic ACF sidelobes(θ_{as}) and the maximum magnitude of periodic CCF(θ_c) are bounded by theoretical limits[13], binary codes with both zero θ_{as} and zero θ_c at the local duration around $\tau = 0$.

Therefore, we will present a ZCD-UWB system having ZCD property. Since ZCD means continuous local orthogonal time duration between spreading codes, ZCD-UWB system could overcome the conventional MAI and MPI problems[3] of DS-CDMA based UWB systems. On the other hand, in order to implement flexible ZCD-UWB, we applied TZCD codes[5]-[8] proposed by author Cha. We will abbreviate the explanation of construction method of TZCD codes because their detailed code generation methods and variable ZCD property can be easily found in the references[5]-[8]. TZCD codes applied for ZCD-CDMA schemes could easily control ZCD length by changing the spreading factor or family sizes. Variable Ternary ZCD code sets and their properties are explained in the following subsections.

2.2   Variable Ternary ZCD Code Sets and Their Properties

Ternary ZCD code sets have a family size of M must satisfy ZCD ≤ (0.75N+1). This Ternary ZCD code sets can be constructed by the chip-shift operation using ternary ZCD preferred pair, $\{C_N^{(a)}, C_N^{(b)}\}$.

Let T^l be the chip-shift operator, which shifts a sequence cyclically to the left by l chips, a set of M ternary ZCD sequences of period N can be generated from $\{C_N^{(a)}, C_N^{(b)}\}$ as

$$\{C_N^{(a)}, C_N^{(b)}, T^\Delta[C_N^{(a)}], T^\Delta[C_N^{(b)}], T^{2\Delta}[C_N^{(a)}], T^{2\Delta}[C_N^{(b)}], \cdots \\ , T^{(k-1)\Delta}[C_N^{(a)}], T^{(k-1)\Delta}[C_N^{(b)}], T^{k\Delta}[C_N^{(a)}], T^{k\Delta}[C_N^{(b)}]\} \quad (3)$$

where Δ is a chip-shift increment and k the maximum number of chip-shifts for a sequence.

On the other hand, Δ and k of TZCD codes and BZCD codes[8] satisfy the following equation (4) and equation (5), respectively, where Δ is a positive and k a nonnegative integer. Moreover, M and ZCD of binary and ternary ZCD codes become equation (6).

$$|(k+1)\Delta| \leq \left|\frac{3N}{8}+1\right| \quad (4)$$

$$|(k+1)\Delta| \leq \left|\frac{N}{4}+1\right| \quad (5)$$

$$M = 2(k+1) \text{ or } ZCD = |2\Delta - 1| \quad (6)$$

Using equation (4),(5),(6), we extract a relation equation between TZCD codes and BZCD codes as following.

$$|M_T(ZCD_T+1)| = \left|\frac{3N_T+8}{2N_B+8}\right| \cdot |M_b(ZCD_b+1)| \quad (7)$$

Where, M_T, ZCD_T and N_T is a family size, ZCD length and period of TZCD codes, respectively. And M_b, ZCD_b and N_b a family size, ZCD length and period of BZCD codes, respectively. Assuming that the period of TZCD codes and BZCD codes and length of ZCD are same, maximum capacity (Family size) of TZCD code is 1.5 times larger than that of BZCD code as sown in the equation (7), figure 1. As mentioned in equation (4) to (7), longer period of variable TZCD code offer higher QoS based on the enlarged ZCD length. And shorter period of variable TZCD offer large capacity due to their large family sizes.

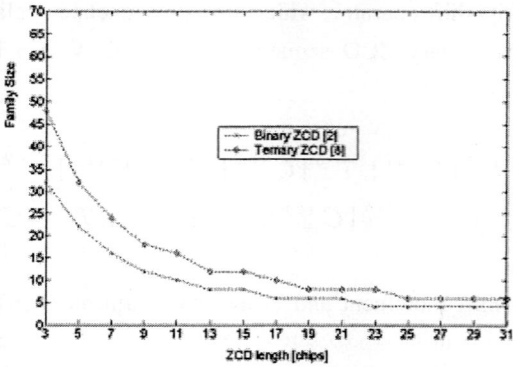

Fig. 1. ZCD vs Family size Comparison of BZCD codes[8]-[10] and proposed Ternary ZCD codes for N=128

3 The System Concept and Analysis of Flexible ZCD-UWB

In this section, we consider ZCD-UWB system by assuming antipodal modulation for transmitted binary symbols. Then UWB transmitted waveform of ZCD-UWB is defined as

$$s^k(t) = \sum_{i=-\infty}^{\infty} \sum_{n=0}^{N_r-1} \sqrt{P_k} b_i^k a_n^k z(t - iT_b - nT_c) \qquad (8)$$

Where, Nr is the period of spreading code, $b_i^k \in \{\pm 1\}$ are the modulated data symbols for the kth user, $a_n^k \in \{\pm 1\}$ are the spreading code for the kth user, $z(t)$ is the sinusoidal waveform or transmitted pulse waveform, T_b is the bit period and, T_c is the chip period. If we assume a UWB system using sinusoidal carrier system, z(t) becomes a sinusoidal signal. Then if we assume the UWB system be a no-carrier system, z(t) be a short pulse.

In this paper, for simplicity, we assume that the multi-path components arrives at the some integer multiple of a minimum path resolution time. By assuming the minimum path resolution time T_m ($T_m \sim 1/Bs$), we can write the received waveform as follows:

$$r(t) = \sum_{l=0}^{L-1} c_l^0 s^0(t - lT_m - \tau^o) + \sum_{k=1}^{K} \sqrt{P_k} \sum_{l=0}^{L-1} c_l^k s^k(t - lT_m - \tau^k) + n(t) \qquad (9)$$

Where L is the number of multi-paths, c_l^k is the amplitude of the lth path, T_m is the pulse period, and $n(t) \sim N(0,1)$ is the AWGN(additive white gaussian noise).

The multi-path delay is described as $\tau^k = q_k T_m$, q_k being an integer uniformly distributed in the interval [0, $N_r N_c$-1] and N_r is the processing gain of spreading code and $N_c = T_c/T_m$ which results in $0 \le \tau^k < T_r$. Here note that T_r is the maximum time delay to be considered. For the detection of received signal, this paper used the pulse matched filter instead of code matched filter[3].

After all, various interference channels that influence to existing DS-UWB system appear according to coefficients of l and k of equation (9) and if ZCD property is kept in such various interference channel, correlation of spread signal amount to 0 in ZCD.

There is a trade-off between interference immunity(QoS) and system capacity. Thus Flexible ZCD-UWB proposed in this paper is adaptively select ZCD length according to the channel condition. For example, flexible ZCD systems select a longer TZCD code to maintain higher QoS In the severe MAI and MPI environment. On the other hand, flexible ZCD systems select a shorter TZCD code to maintain higher code capacity in the static channel environments. Regardless of High QoS or High capacity, ZCD length designed to cover the delay time of various MAI or MPI components. By using reference of [4], BER of the flexible ZCD-UWB with k^{th} user and matched filter in AWGN channel can be written as

$$P^k(\sigma) = \frac{1}{2} \sum_{e l \in \{-1,1\}} \cdots \sum_{\substack{e j \in \{-1,1\} \\ j \neq k}} \cdots \sum_{e k \in \{-1,1\}} Q(\frac{c_k}{\sigma} + \sum_{j \neq k} e_j \frac{c_k}{\sigma} \beta \rho_{jk}) \quad (10)$$

Where Q(x) is the complementary cumulative distribution function of the unit normal variable. When Rayleigh fading channel is considered for the flexible ZCD-UWB system using IF carrier signal, the BER can be written as

$$P^{Fk}(\sigma) = \frac{1}{2}(1 - \frac{c_k}{\sqrt{\sigma^2 + \sum_j c_j (\beta \rho_{jk})^2}}) \quad (11)$$

Since the cross-correlations of flexible ZCD-scheme are zeros, the performance of flexible ZCD-system is as good as no delay environment, thus the BER probabilities both in AWGN and Rayleigh fading can be written as follows:

$$P^k(\sigma) = \frac{1}{2} Q(\frac{c_k}{\sigma}) \quad (12)$$

$$P^{Fk}(\sigma) = \frac{1}{2}(1 - \frac{c_k}{\sqrt{\sigma^2 + c_k}}) \quad (13)$$

Since the Q(x) function is monotonically decreasing function with respect to x, the probabilities of the flexible ZCD-UWB system are always smaller than the conventional system using Walsh spreading codes which have some cross-correlation in the MAI and MPI environments.

4 Case Study for BER Simulation and the Result

In this section, we simulated the BER performance of the flexible ZCD-UWB system via computer simulation to the following case:

Table 1. Parameters of simulation

	Case1.	Case2.
Condition	High QoS with longer ZCD and small family size.	High Capacity with shorter ZCD and large family size
Spreading codes	Walsh code, TZCD code	Walsh code with no ZCD, BZCD code with ZCD of 9 chip, TZCD code with ZCD of 11 chip
Spreading factor	64	64
Required ZCD length for QoS	17 chip	11 chip
User number	6	Variable user number (1 to 6)
Muti-path number	10	3
Channel environment	AWGN channel	AWGN channel
Etc.		Eb/No: 10 dB

The simulation results of case 1 and 2 can be shown in figure 2 and 3, respectively. In figure 2, we present the BER performance under case 1 condition by using matched filter scheme. Here, we can observe that the proposed flexible ZCD-UWB system exhibits better performance than that of Walsh-UWB system. Moreover, we can observe that the flexible ZCD-UWB system able to reject MAI and MPI perfectly if it has a sufficient ZCD length such as 17 chip example. From the results in figure 2, we certified that flexible ZCD-UWB system able to reject MAI and MPI in the condition of high QoS with longer ZCD and small family size. Figure 3 is a simulation result of Flexible TZCD-UWB system for High Capacity with shorter ZCD and large family size under MAI and MPI environments with a 10 dB hold value of 10^{-3} of BER. Even if user's number increase in system, However, UWB system of Walsh code base could not correct value, regardless of User number variation since Walsh codes have no any

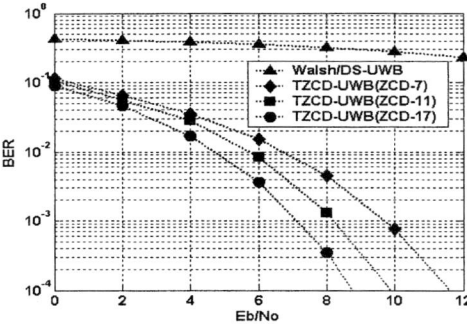

Fig. 2. Simulation of Flexible ZCD-UWB system with High QoS

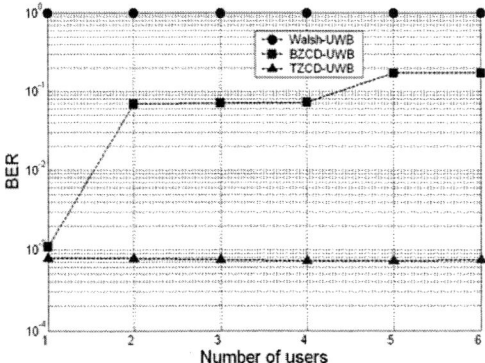

Fig. 3. Simulation of Flexible ZCD-UWB system for High Capacity

ZCD property. On the other hand, the other systems using ZCD codes have superior BER performance according to the user number variation since they have ZCD property. Furthermore, we certified TZCD-UWB have the better BER performance than that of BZCD-UWB since TZCD codes have sufficient ZCD length in the high capacity environments while BZCD codes have insufficient ZCD in the same high capacity. From the simulation results, we can note that proposed flexible ZCD-UWB always could maintain optimal BER performance by changing ZCD length and family size according to the channel environments.

5 Conclusion

In this paper we proposed flexible ZCD-UWB system using ternary ZCD spreading codes. From a few case study and simulation results, we certified that the proposed flexible ZCD-UWB could maintain an optimal BER performance just changing ZCD length and family size according to the channel environments.

Flexible ZCD-UWB system proposed in this paper, not only has novel interference immunity, but also has high system capacity or high QoS. We expect the proposed Flexible ZCD-UWB system could be an efficient solution for interference free wireless home network applications.

Acknowledgment

This work was supported in part by University IT Research Center Project (INHA UWB-ITRC) in Korea.

References

1. http://www.ieee802.org/15/pub/TG3a.html
2. http://www.ieee802.org/15/pub/TG4a.html
3. J. Foerster, "The Performance of a Direct-Sequence Spread ULTRA-Wideband System in the Presence of Multipath, Narrowband Interference, and Multiuser Interference," IEEE UWBST Conference Proceedings, May, 2002.
4. Sergio Verdu, Multiuser Detection, Cambridge university press, 1998
5. Cha,J.S. and Tsubouchi, K, "New ternary spreading codes with with zero-correlation duration for approximately synchronized CDMA", Proc. IEEE ISIE 2001, Pusan, Korea, Vol. 1, pp.312-317, June 12, 2001.
6. Cha,J.S, "Class of ternary spreading sequences with zero correlation duration", IEE Electronics Letters, Vol. 36, no.11, pp. 991-993, 2001.5.10
7. Cha,J.S., Song,S.I.,Lee,S.Y.,Kyeong,M.G. and Tsubouchi, K, "A class of Zero-padded spreading sequences for MAI cancellation in the DS-CDMA systems", Proc. IEEE VTC01 Fall, Atlantic City, USA,October 6, 2001.
8. Jae-sang Cha, Sang-yule Choi, Jong-wan Seo, Seung-youn Lee, and Myung-chul Shin, "Novel Ternary ZCD Codes With Enhanced ZCD Property and Power-efficient MF Implementation", Proc. IEEE ISCE'02, Erfurt, Germany, Vol. 1, pp.F117-122, 2002.
9. Cha, J.S., Kameda, S., Yokoyama, M., Nakase, H., Masu, K., and Tsubouchi, K.: 'New binary sequences with zero-correlation duration for approximately synchronized CDMA '. Electron. Lett., 2000, Vol. 36, no.11, pp.991–993
10. Cha,J.S., Kameda,S., Takahashi,K., Yokoyama,M., Suehiro,N., Masu, K. and Tsubouchi, K, "Proposal and Implementation of Approximately synchronized CDMA system using novel biphase sequences", Proc. ITC-CSCC 99, Vol. 1, pp.56-59, Sado Island, Japan, July13-15, 1999.
11. Cha,J.S. and Tsubouchi, K, "Novel binary ZCD sequences for approximately synchronized CDMA", Proc. IEEE 3G Wireless01, Sanfransisco, USA, Vol. 1, pp.810-813, May 29, 2001.
12. Jaesang Cha, Kyungsup Kwak, Jeongsuk Lee, Chonghyun Lee, "Novel interference-cancelled ZCD-UWB system for WPAN", 2004 IEEE International Conference on, Vol. 1, pp.95-99, June 2004
13. Sarwate, D.V. : 'Bounds on Crosscorrelation and AutoCorrelation of Sequences', IEEE Trans., 1979, IT-25, pp. 720-724

Fine Grained Control of Security Capability and Forward Security in a Pairing Based Signature Scheme

Hak Soo Ju[1], Dae Youb Kim[2], Dong Hoon Lee[2], Jongin Lim[2], and Kilsoo Chun[1]

[1] Korea Information Security Agency(KISA), Korea
{hsju, kschun}@kisa.or.kr
[2] Center for Information and Security Technologies
{david_kdy, donghlee, jilim}@korea.ac.kr

Abstract. Recently, Libert and Quisquater showed that the fast revocation method using a SEcurity Mediator(SEM) in a mRSA can be applied to the Boneh-Franklin identity based encryption and GDH signature scheme. In this paper we propose a mediated identity based signature(mIBS) scheme which applies the SEM architecture to an identity based signature. The use of a SEM offers a number of practical advantages over current revocation techniques. The benefits include simplified validation of digital signatures, efficient and fast revocation of signature capabilities. We further propose a forward mediated signature scheme with an efficient batch verification and analyze their security and efficiency.

1 Introduction

Boneh et al. [4] introduced an efficient method for obtaining instantaneous revocation of a user's public key called the mediated RSA(mRSA). Their method was to use a SEcurity Mediator(SEM) which has a piece of each user's private key. In such a setting, a signer can't decrypt/sign a message without a token information generated by the SEM. Instantaneous revocation is obtained by instructing the mediator to stop helping the user decrypt/sign messages. This approach has several advantages over previous certification revocation techniques such as Certificate Revocation Lists(CRLs) and the Online Certificate Status Protocol(OCSP) : fast revocation and fine-grained control over users' security capabilities [3].

We note that mRSA still relies on conventional public key certificates to store and communicate public keys. To solve this problem, Boneh el al. described how to transform mRSA into an identity based mediated RSA scheme(IB-mRSA) [2]. Their method combine the features of identity based and mediated RSA. Identity-based cryptography greatly reduces the need for and reliance on public key certificates and certification authorities. Recently Libert and Quisquater showed that the SEM architecture in a mRSA can be applied to the Boneh-Franklin identity based encryption and GDH signature schemes [11]. Despite these recent results, this SEM architecture has not been applied to ID-Based Signature schemes and no forward secure mediated pairing based signature scheme has been proposed.

Our Contribution. Several ID-Based Signature (IBS) schemes [7, 8] do not explicitly provide revocation of users' security capabilities. This is natural since it aims to avoid the use of certificates in the course of digital signatures. On the other hand, revocation is often necessary and even imperative. The only way to obtain revocation in IBS is to require time dependent public keys, e.g., public keys derived from identities combined with time or date stamps. This has an unfortunate consequence of having to periodically reissue all private keys in the system. Moreover, these keys must be periodically and securely distributed to individual users. In contrast, our mediated ID-Based Signature (mIBS) scheme inherits its fine-grained revocation as in mRSA. It dose not demand the Private Key Generator (PKG) to periodically reissue new private keys. Additionally a dishonest user who corrupts the SEM can not sign a message instead of other users. This is not the case for the IB-mRSA signature scheme that can be designed similarly with IB-mRSA encryption scheme [9]. It is pointed out in [11] that IB-mRSA scheme is completely broken if a user can corrupt a SEM.

In a security system, the key exposure problem is one of important problems to be solved. This problem is rapidly emerging as cryptographic primitives are used in lightweight, easy-to-lose, portable and mobile devices. To reduce the damage caused by exposure of secret keys stored on such devices, the concept of key updating was introduced by Anderson [1] as a forward security. Since then, a number of schemes were proposed. The basic idea is that the signature scheme guarantees the security of previous time periods even if the adversary compromises the current private key. We concentrate on *weak forward security* in a pairing based mediated signature scheme. Informally, weak forward security means that an adversary is unable to forge past signatures if she compromises only one (of the two) share-holders of the private key [12]. Since the security of the pairing based mediated signature scheme is based on the non-compromise of both key shares, weak forward security is sufficient for our scheme like mRSA. Our scheme inherits a aggregation for k signatures and its efficient batch verification since it is based on IBS proposed by Cheon et. al. [8]. Owing to pairing computation our scheme is inefficient than a forward secure mRSA signature scheme (FS+mRSA). However, we expect that the batch verification property of our scheme has an efficient performance in case of verifying many signatures.

This paper is organized as follows. The next section introduces our mediated ID-Based Signature scheme and its security analysis is given. A forward secure mediated GDH signature scheme and its security analysis is given in section 3. Section 4 describes the efficiency of our schemes. Finally, we conclude in Section 5.

2 The Mediated Identity Based Signature Scheme

Moni Naor has observed that an ID Based Encryption(IBE) scheme can be immediately converted into a signature scheme [10]. This observation can be extended to a mediated infrastructure: an IBE scheme can be immediately converted to a mediated IBE scheme. The decryption key d in the IBE scheme is divided into the user's partial decryption key $d_{user} = s_{user}H(ID)$ and the SEM's partial decryption key $d_{sem} = s_{sem}H(ID)$. The encryption is the same as in the original IBE scheme. To decrypt a ciphertext C, the signer has to obtain a decryption token on the ciphertext C from the SEM. This IBE scheme can be also converted to a mediated public key signature scheme. The master key s is divided into the signer's partial private key s_{user} and the SEM's partial private key s_{sem}. The signature on a message M is the combination of the user's partial decryption key $d_{user} = s_{user}H(M)$ and SEM's partial decryption key $d_{sem} = s_{sem}H(M)$ for $H(ID) = H(M)$. Verification is the same as in the above signature scheme. The exact schemes are recently suggested by Libert and Quisquater [11]. In this section, we introduce a new mediated version of the IBS scheme which extends the mediated signature scheme.

2.1 The Scheme

Our mediated ID-Based Signature(mIBS) scheme uses two private keys. One key issued by PKG is used to prove user's identity by inheriting one to one mapping between public ID and private key. We use the second private key in order to sign a message while maintaining one to one mapping. Two secret keys are split into shares via a one out of two secret sharing scheme, with one share held by the user and the other by the SEM. That is, $d_{ID} = d_{ID,user} + d_{ID,sem}$ and $s_l \equiv s_{l,user} + s_{l,sem}$, where user's secret key is $(d_{ID,user}, s_{l,user})$, and the sem's secret key is $(d_{ID,sem}, s_{l,sem})$. The signing is then performed as a function of ID, K_{pub} as the IBS scheme.

Our construction is based on a hierarchical identity based signature scheme (HIDS) [10] and a mediated GDH signature scheme [11]. We emphasize that there is no security proof for HIDS scheme and the proof in [11] does not fit our scheme directly, because our scheme is an identity-based signature rather than an encryption scheme. The details of our scheme are :

1. **Setup.** Given a security parameter k, the PKG :
   - Generates groups G_1, G_2 of prime order q and an admissible bilinear map $\hat{e} : G_1 \times G_1 \to G_2$.
   - Chooses a generator $P \in G_1$
   - Picks secret keys $s, s_l \in_R Z_q^*$ and sets $P_{pub} = sP$, $Q_l = s_l P$.
   - Chooses cryptographic hash functions $H_1 : \{0,1\}^* \to G_1^*$ and $H_2 : \{0,1\}^* \to G_1^*$.

   The system public parameters are $K_{pub} = (q, G_1, G_2, \hat{e}, P, P_{pub}, Q_l, H_1, H_2)$ while the secret key s and s_l are kept secret by the PKG.

2. **Keygen.** Given a user of identity ID, the PKG computes $Q_{ID} = H_1(ID)$ and $d_{ID} = sQ_{ID}$. Then it chooses random numbers $d_{ID,user} \in_R G_1$, $s_{l,user} \in Z_q$ and computes $d_{ID,sem} = d_{ID} - d_{ID,user}$ and $s_{l,sem} = s_l - s_{l,user}$. The PKG gives the partial private keys $d_{ID,user}, s_{l,user}$ to the user and $d_{ID,sem}, s_{l,sem}$ are given to the SEM.

3. **Sign.** To sign a message M, user U sends to the SEM a hash $P_M = H_2(ID, M) \in G_1$ of a message and a identity. They perform the following protocol in parallel.

- SEM: 1. Check if the user's identity ID is revoked. If it is, return "Error".
  2. Compute $S_{M,sem} = d_{ID,sem} + s_{l,sem}P_M$ and send it to the user U.
- USER: 1. U computes $S_{M,user} = d_{ID,user} + s_{l,user}P_M$.
  2. When receiving $S_{M,sem}$ from the SEM, U computes $S_M = S_{M,user} + S_{M,sem}$.
  3. U verifies that S_M is a valid signature on M. If it holds, U returns the pair message-signature (M, S_M).

4. **Verify.** On inputting a message M and a signature S_M, the verifier confirms that: $\hat{e}(P, S_M) = \hat{e}(P_{pub}, H_1(ID)) \cdot \hat{e}(Q_l, H_2(ID, M))$.

This completes the description of our mIBS scheme. Consistency is easily proved as follows: If S_M is a valid signature of a message M for an identity ID, then $S_M = d_{ID} + s_l P_M$ and $Q_l = s_l P$ for $P_M = H_2(ID, M)$. Thus

$$\hat{e}(P, S_M) = \hat{e}(P, d_{ID} + s_l P_M) = \hat{e}(P, d_{ID}) \cdot \hat{e}(P, s_l P_M)$$
$$= \hat{e}(P_{pub}, H_1(ID)) \cdot \hat{e}(Q_l, H_2(ID, M))$$

As in mIBE [11], the SEM never sees the user's partial key $d_{ID,user}, s_{l,user}$ and can not sign messages instead of him. User never see the SEM's partial key $d_{ID,sem}, s_{l,sem}$ and can not compute it from the token $S_{M,sem}$ he receives. The token $S_{M,sem}$ is useless to any user other than the corresponding one and it does not provide any useful information to any other users about a specific user's full private key since it is a random element of G_1.

We may consider a case that the user can use the same token $S_{M,sem}$ twice at the another message M'. However, this is not possible because the inputs of H_2 are a message M and a user's identity information ID. If we use only $sH_2(ID, M)$ instead of $sH_1(ID) + s_l H_2(ID, M)$, it can cause a problem that PKG must know a message M on which the user sign.

2.2 The Security Analysis

To analyze the security of mIBS, we first define a new identity based signature scheme. This scheme is exactly the same as the HIDS with $t = 1$ where t is the level of the recipient, but we present it again here because our notation is slightly different. First, we will use a Cha and Cheon's lemma [7]-proving that breaking IBS for an adaptive chosen message and ID attack (we denote by $EF\text{-}ID\text{-}CMA$ this security notion) is as hard as breaking IBS for an adaptive chosen message and given ID attack. We then prove that breaking IBS for an adaptive chosen message and given ID attack is as hard as solving an instance of the CDH problem- to show that the security of IBS is based on the difficulty of the CDH problem.

Our IBS scheme is specified by four algorithms : *Setup, Keygen, Sign, Verify.*
1. **Setup.** As in the mIBS scheme.
2. **Keygen.** Given a user of identity ID, the PKG computes $Q_{ID} = H_1(ID)$, $d_{ID} = sQ_{ID}$ and secretly sends d_{ID}, s_l to the user. We remark that $Q_{ID} = H_1(ID)$ and Q_l play the role of the associated public key.
3. **Sign.** Given a secret key d_{ID} and a message M, user U output a signature S_M where $S_M = d_{ID} + s_l H_2(ID, M)$.
4. **Verify.** To verify a signature S_M of a message M for an identity ID, the verifier confirms that :

$$\hat{e}(P, S_M) = \hat{e}(P_{pub}, H_1(ID)) \cdot \hat{e}(Q_l, H_2(ID, M)).$$

At first, we can reduce the adaptively chosen ID attack to the given ID attack as in [7].

Lemma 1 ([7]). *If there exists a forger $\mathcal{F}_0$ for an existential forgery against an adaptively chosen message and ID attack to our IBS scheme with running time t_0 and advantage ϵ_0, then there exists a forger $\mathcal{F}_1$ for an existential forgery against an adaptively chosen message and given ID attack with running time $t_1 \leq t_0$ and advantage $\epsilon_1 \leq \epsilon_0(1 - 1/q)/q_{H_1}$. Additionally, the number of queries to Hash functions, Extract, Signing asked by $\mathcal{F}_1$ are the same as those of $\mathcal{F}_0$.*

The security of our IBS is based on the difficulty of the CDH problem, as stated in the following theorem.

Theorem 1. *If there exists a forger $\mathcal{F}_0$ for an existential forgery against an adaptively chosen message and ID attack to our IBS scheme with running time t_0 and advantage ϵ_0, then CDHP can be solved with at least the same advantage ϵ_1 in a running time $t' = t_1 + t_A(q_E + q_{H_1} + q_{H_2} + q_S) + t_B q_S$ where t_A denotes the time to multiply two elements on Z_q^* and t_B is the time to add two elements of G_1.*

Proof. (Sketch) Using the Lemma 1, we can reduce the forger $\mathcal{F}_0$ to $\mathcal{F}_1$ an adaptively chosen message and given ID attack with running time $t_1 \leq t_0$ and advantage $\epsilon_1 \leq \epsilon_0(1 - 1/q)/q_{H_1}$. We construct an algorithm $\mathcal{A}$ using $\mathcal{F}_1$ to solve the CDHP. To break CDH in the additive group G_1 with the order q, $\mathcal{A}$ is given P, aP and bP, where $a, b \in Z_q$ are randomly chosen and remains unknown to $\mathcal{A}$. The target of $\mathcal{A}$ is to derive $S' = abP$ with the help of the forger $\mathcal{F}_1$. To derive S', $\mathcal{A}$ runs $\mathcal{F}_1$ as the subroutine. $\mathcal{A}$ provides $\mathcal{F}_1$ the public key and answers its hash queries, signing queries and extract queries. $\mathcal{A}$ embeds the CDH problem into the public key and then answers the foregoing queries. After $\mathcal{F}_1$ forges a signature successfully, $\mathcal{A}$ is able to derive the answer to the CDH problem using the forged signature. □

We will now show that the mediated IBS scheme is weakly secure. That means it is secure against inside attackers that do not have the user part of the private key corresponding to the attacked public key. This provides the same level of security as the one achievable by IB-mRSA signature scheme. For the mIBS scheme, we will slightly modify the notion of security for an existential forgery against an adaptively chosen message and ID attacks (*EF-ID-CMA*) given in [7, 11].

Definition 1. *We say that a mediated identity based signature scheme is weakly secure for an existential forgery against an adaptively chosen message and ID attack (we denote by EF-mID-wCMA this security notion) if no polynomial time algorithm $\mathcal{A}$ has a non-negligible advantage against a challenger $\mathcal{C}$ in the following game (EF-mID-wCMA game) :*

1. $\mathcal{C}$ runs the Setup algorithm of the scheme. The resulting system parameters are given to $\mathcal{A}$.
2. $\mathcal{A}$ issues the following queries as he wants :
   − User Key Extraction Query : Given an identity ID_i, $\mathcal{C}$ returns the user part of the extracted private key corresponding to ID_i.
   − SEM Key Extraction Query : Given a target identity ID, $\mathcal{C}$ returns the sem part of the private key d_{ID} corresponding ID.
   − SEM Query : Given an identity ID_i, $\mathcal{C}$ returns the token allowing the user of identity ID_i to sign.
   − Hash Function Query : Given an identity ID_i or (ID_i, M), $\mathcal{C}$ computes the value of the hash function for the requested inputs and sends the values to $\mathcal{A}$.
   − Sign Query : Given an identity ID_i and a message M, $\mathcal{C}$ generates both pieces of the private key corresponding to ID_i and sends the result signature to $\mathcal{A}$.

   $\mathcal{A}$ can present its requests adaptively : every request may depend on the answer to the previous ones.
3. $\mathcal{A}$ outputs (ID, M, σ), where ID is a target identity, M is a message, and σ is a signature such that ID and (ID, M) are not equal to the inputs of any query to User Key Extract and Sign, respectively.

This is a weak notion of security against inside attackers that have access the user part of the private key corresponding to any identity but the one on which they are challenged. The weak security notion implies that no coalition of dishonest users which the SEM can allow them to sign a message instead of a honest user. We just prove the weak security against inside attackers as it is shown in [11] that the mediated IBE scheme is weakly semantically secure against inside attackers.

Theorem 2. *Let H_1, H_2 be random oracles from $\{0, 1\}^*$ to G_1^*. Let $\mathcal{A}$ be an attacker against the mediated IBS scheme. We assume this attacker is able to win the EF-mID-wCMA game with a non-negligible advantage ϵ when running in a time t and asking at most q_E user key extract queries and q_S SEM queries. Then there exits an adversary $\mathcal{B}$ performing EF-ID-CMA against the underlying our IBS scheme with at least the same advantage. Furthermore, the running time of $\mathcal{B}$ is $t' = t + q_E t_A + q_S t_A$ where t_A denotes the time to add two elements of G_1.*

Proof. (Sketch) We will use the attacker $\mathcal{A}$ to build an algorithm $\mathcal{B}$ that is able to perform *EF-ID-CMA* of our IBS scheme. At the beginning of the game,

$\mathcal{B}$ receives the system parameters from its challenger. It is allowed to ask a polynomial number of key extraction, sign and hash queries to its challenger but it first initializes the *EF-mID-wCMA* game it plays with $\mathcal{A}$ by giving him the same system parameters it received from its challenger. $\mathcal{B}$ will act as $\mathcal{A}$'s challenger in the *EF-mID-wCMA* game and control the SEM. It maintains a list L_{sem} to store information about the answers to key generation queries. $\mathcal{A}$ performs a first series of queries and $\mathcal{B}$ answers to these queries. After the query stage, $\mathcal{A}$ produces (ID, M, σ). When $\mathcal{A}$ outputs (ID, M, σ) as a signature, $\mathcal{B}$ outputs (ID, M, σ). Since $\mathcal{B}$ simulates $\mathcal{A}$'s environment in its attack against IBS, $\mathcal{B}$ wins as long as $\mathcal{A}$ does. □

3 The Forward Secure Mediated Gap Diffie Hellman Signature Schemes

3.1 Forward Secure+mGDH Signature Scheme

We now devise a new forward secure mediated GDH signature scheme with efficient batch verification. The main idea in forward secure mGDH signature scheme is for both SEM and user to evolve their private key shares in parallel. The evolution is very simple: each party uses the current period index i in the hashing of the input message. Like most forward secure signature methods, FS+mGDH signature scheme is composed of the following algorithms: *Setup, Keygen, UpdSEM, UpdUSER, Sign, Verify.*

1. **Setup.** Given a security parameter k, the $\mathcal{TA}$:
   - Generates groups G_1, G_2 of prime order q and an admissible bilinear map $\hat{e}: G_1 \times G_1 \to G_2$.
   - Chooses a generator $P \in G_1$.
   - Picks a secret key $x \in_R Z_q^*$ and sets $P_{pub} = xP$.
   - Chooses hash functions $H_1 : \{0,1\}^* \times G_1 \to Z_q$, $H_2 : \{0,1\}^* \to G_1$.
   The system's public parameters are $K_{pub} = (q, G_1, G_2, \hat{e}, P, P_{pub}, H_1, H_2)$.
2. **Keygen.** Let (t, T) be the length of the update interval and the maximum number of update intervals, respectively. To generate user U's key pair, the $\mathcal{TA}$ chooses random numbers $x_{user} \in_R Z_q^*$ and computes $x_{sem} = x - x_{user}$. The $\mathcal{TA}$ gives the partial private key x_{user} to the user and x_{sem} to the SEM.
3. **UpdSEM.** Let $i(0 \leq i \leq T)$ is the current interval index and M is the current input message. At the start of stage i, the SEM uses x_{sem} to compute $x_{i,sem} = x_{sem} H_2(i, M) \in G_1$.
4. **UpdUSER.** At the start of stage i, the USER uses x_{user} to compute $x_{i,user} = x_{user} H_2(i, M) \in G_1$.
5. **Sign.** To sign a message M, a user U generates a random number r and computes $R = rP$. Then U sends to the SEM a hash $h = H_1(M, R) \in Z_q$ of a message and a interval i. They perform the following protocol in parallel.
   - SEM: 1. Check if the user's identity ID is revoked. If it is, return "Error".
     2. Compute $S_{i,M,sem} = h \cdot x_{i,sem}$ and sends it to the user U.

- USER: 1. U computes $S_{i,M,user} \leftarrow h \cdot x_{i,user}$.
  2. When receiving $S_{i,M,sem}$ from the SEM, U computes $S_{i,M} \leftarrow rH_2(i,M) + S_{i,M,sem} + S_{i,M,user}$.
  3. He verifies that $\sigma = (R, <S_{i,M}, i>)$ is a valid signature on M at time i. If it holds, he returns the pair message-signature (M, σ).
6. **Verify.** The verifier return "Error" if $(i < 0)$ or $(i > T)$. If it is verified, then he checks whether $(P, P_{pub}, R + hP_{pub}, S_{i,M})$ is a valid Diffie Hellman tuple by checking if $\hat{e}(P, S_{i,M}) = \hat{e}(R + hP_{pub}, H_2(i,M))$.

3.2 The Security Analysis

All security aspects other than forward security of the proposed scheme is based on the ID-based signature scheme of [8]. The forward security of FS+mGDH signature scheme is based on the difficulty of computing discrete logarithm in a elliptic curve group which is also the foundation of the ECC cryptosystem. Now we provide the following intuitive security validation of our scheme. We consider informal argument as follows:

Assume that the adversary compromises the user at an interval j and, as a result, learns $x_{j,user}$. In order to violate forward security, it suffices for the adversary to generate a new signature for some new message M':

$$S_{i,M'} = r \cdot H_2(i, M') + S_{i,M',sem} + S_{i,M',user} = r \cdot H_2(i, M') + h \cdot x_{i,sem} + h \cdot x_{i,user}$$

where $h = H_1(M', R)$ and $i < j$. Computing $h \cdot x_{i,user}$ is trivial. But computing $h \cdot x_{i,sem}$ requires solving ECDLP (Elliptic Curve Discrete Logarithm Problem).

Forward security offered by our scheme is weak, it means that the adversary is allowed to compromise only one of the parties' secrets, i.e., only x_{user} or x_{sem} but not both. Since the security of mGDH signature scheme is based on the non-compromise of both key shares, weak forward security is sufficient for the mGDH signature scheme like mRSA.

There are two types of attacks considered in FS+mRSA [12] : a future dating attack, a oracle attack. In a future dating attack, an adversary obtains a valid signature from the user $(M, \sigma = (R, S_{i,M}, i))$ under the current public key (P, P_{pub}, i). He then takes advantage of the private key structure to construct a valid signature $\sigma' = (R, S_{j,M}, j)$ in some future interval $j(i < j \leq T)$. This attack is not possible because our signature involves hashing the index of the current time interval together with the message :

$$S_{i,M} = r \cdot H_2(i,M) + H_1(M, rP)(x_{sem} + x_{user})H_2(i,M)$$

In a oracle attack, an adversary impersonating as the user sends signature requests to the SEM during the time interval i. The adversary collects a number of "half-signatures" of the form $(M, S_{i,M,sem} = h \cdot x_{i,sem})$. At a later interval $j(i < j \leq T)$, the adversary compromises the user's secret key $x_{j,user}$. He can use the perviously acquired half signatures to forge signatures from period i. But this attack is not possible in our scheme because our scheme have the homomorphic property :

$$x_{i,user} + x_{i,sem} = (x_{user} + x_{sem}) \cdot H_2(i,M) = s \cdot H_2(i,M).$$

4 Efficiency

From the implementation results in [6], we find that a GDH signature generation is much faster than a RSA signature generation. Therefore, we can be assured that our schemes are able to achieve more efficient signing when compared with RSA based approaches. Signature verification in our scheme is still limited by the efficient computation involved in pairing operations. However, the results shown in [6] have demonstrated that the computation time for a Tate pairing with the prime field size of 512 bits is now comparable to one RSA signing operation with a 1024 bits modular and a 1007 bits exponent. Additionally, our forward secure scheme inherit an aggregation and batch verification for k signatures. With this inherent features, our schemes are applicable in a number of areas. For example, for mobile device authentication, the signature verification is done at the server side, which has more computation power to do the pairing operations. Signing generation can be done efficiently on the mobile device. The comparison the performance of our schemes with the previous mRSA schemes is shown in Tab.1.

Table 1. The previous schemes vs our schemes, SM : Scalar Multiplication, PC: Paring Computation

	mRSA [4]	FS+mRSA [12]	mIBS	FS+mGDHS
Sign(SEM)	d_{sem}	$d_{sem} \times e^i$	1 SM	2 SM
Sign(User)	$d_u + e$	$(d_u \times e^i) + (e \times e^{T-i})$	1 SM	3 SM
Verify	e	$e \times e^{T-i}$	3 PC	2PC + 1SM
Security	-	weak forward security	-	weak forward security

5 Conclusion

We have shown that the method of mRSA to allow fast revocation of RSA keys can be used by IBS scheme. Rather than revoking the user's private key by concatenating valid interval to identities in IBS, our approach revokes the user's ability to perform signature operations as in mRSA. This approach provides instantaneous revocation since the private key privileges of the user are instantaneously removed. It does not demand the PKG to periodically re-issue new private keys.

In this paper, we also described a mediated Gap Diffie Hellman signature scheme with a forward security. The degree of forward security is weak since we assume that only the user or the SEM (but not both) is compromised by the adversary. However, this assumption is appropriate for the mGDH signature scheme whose security is based on the inability to compromise both parties as in mRSA. Our FS+mGDH signature scheme also inherits the efficient batch verification property. The batch verification property of our scheme has an efficient performance in case of verifying many signatures. We leave for future research to find an efficient method with more security such as (not weak) forward security.

References

1. R.Anderson, *Invited lecture at the acm conference on computer and communication security (CCS'97)*, 1997.
2. D.Boneh, X.Ding, and G.Tsudik, *Identity based encryption using mediated RSA*, In Proceedings of the 3rd International Workshop on Information Security Applications WISA 2002.
3. D.Boneh, X.Ding, and G.Tsudik, *Fine-grained control of security capabilities*, ACM Transactions on Internet Technology (TOIT) Volume 4,Issue 1, February 2004.
4. D.Boneh, X.Ding, G.Tsudik, and C.Wong, *A method for fast revocaiton of public key certificates and security capabilities*, In Proceedings of the 10th USENIX Security Symposium, USENIX 2001.
5. D.Boneh and M.Franklin, *Identity Based Encryption From the Weil Pairing*, In Advances in Cryptology-CRYPTO 2001, Lecture Notes in Computer Science, pages 213-229. Springer-Verlag, 2001.
6. P.Barreto, H.Kim, B.Lynn, and M.Scott, *Efficient algorithms for pairing-based cryptosystems*, In Advances in Cryptology-CRYPTO 2002, Lecture Notes in Computer Science, pages 354-368. Springer-Verlag, 2002.
7. J. Cha and J. Cheon, *An Identity-Based Signature from Gap Diffie-Hellman Groups*, In Proceedings of PKC 2003, Lecture Notes in Computer Science. Springer-Verlag, 2003.
8. H. Yoon, J. Cheon and Y. Kim, *Batch Verifications with ID-based Signatures*, In Pre-Proceedings of ICISC 2004, Lecture Notes in Computer Science. Springer-Verlag, 2004.
9. X. Ding and G.Tsudik, *Simple Identity based Cryptography with Mediated RSA*, In Proceedings of CT-RSA 2003, Lecture Notes in Computer Science. Springer-Verlag, 2003.
10. C.Gentry and A.Silverberg, *Hierarchical ID Based Cryptography*, In Advances in Crypotology-ASIACRYPT 2002, Lecture Notes in Computer Scinece, pages 548-566, Springer-Verlag, 2002.
11. B. Libert, J.-J. Quisquater, *Efficient revocation and threshold pairing based cryptosystems*, Symposium on Principles of Distributed Computing-PODC 2003, 2003.
12. G. Tsudik, *Weak Forward Security in Mediated RSA*,In Security in Communication Networks (SCN 2002), LNCS 2576, pp. 45-54. Springer-Verlag, 2002.

The Large Scale Electronic Voting Scheme Based on Undeniable Multi-signature Scheme

Sung-Hyun Yun[1] and Hyung-Woo Lee[2]

[1] Division of Information and Communication Engineering,
Cheonan University, Anseo-dong,
Cheonan, 330-704, Korea
shyoon@cheonan.ac.kr
[2] Dept. of Software, Hanshin University,
Osan, Gyunggi, 447-791, Korea
hwlee@hs.ac.kr

Abstract. In undeniable multi-signature scheme, a multi-signature can not be verified and disavowed without cooperation of all signers. The proposed voting scheme consists of four stages which are preparation, registration, voting and counting stages. Existing voting schemes assume that the voting center is trustful and untraceable channels are exist between voters and the voting center. To minimize the role of the voting center, the proposed scheme let multiple administrators to manage voting protocol. It also provides fair voting and counting stages. In voting and counting stages, a ballot can not be opened without help of all administrators. Before counting the ballot, they must confirm the undeniable multi-signature on it. Due to the undeniable property of the proposed scheme, voters can change their mind to whom they vote in registration stage. They can restart voting process by simply rejecting signature confirmation protocol launched by the voting manager.

1 Introduction

Election is one of the most important social activities in a democratic society. If a voting scheme in real life is replaced with a computerized scheme, voting expenses can be reduced fairly. Many researchers have proposed secure electronic voting schemes suitable for a large scale election[3,4,5,6,7].

Typically, existing voting schemes generally consist of four stages which are preparation, registration, voting and counting stages. To provide privacy of voters, blind signature scheme and pseudonym technique is used during the preparation and registration stages[4,5]. And untraceable channels proposed by D.Chaum[8] are used during the voting and counting stages so that voters can anonymously votes to the voting center without IP traces[3,4,5,7]. During the voting and counting stages, a challenge-response protocol between voters and the voting center is used to ensure the fairness requirement[5,7]. However, disadvantage of existing schemes is that voters and each party must trust voting and counting process managed by the voting center.

In large scale election such as a national referendum where candidates are nominated by several parties, it is in danger of strife between several parties if the election is managed by only one administrator, the voting center. Therefore, to minimize the risk of strife in large scale election, it is preferable to manage the election with several administrators controlled by each party.

In US election 2000 in florida county, opening and counting of ballots are uncertain which makes voters and parties have less confidence on their voting system. We need new type of electronic voting scheme where all candidates' parties must agree and engage in opening and counting of ballots during the voting and counting stages. If the voting system provides these scheme, participating voters and candidates can get more confidence on their voting system.

In this paper, we propose the large scale electronic voting scheme based on undeniable multi-signature scheme. To minimize the role of the voting center, we use administrators managed by each party to control the voting and counting protocol. It also provides the method of fair counting of ballots. Ballots can not be opened without help of all administrators due to the undeniable property of our multi-signature scheme. Before opening the ballot, all administrators must verify the multi-signature on the ballot.

The proposed scheme additionally provides user convenience during the registration stage. Voters can change their mind to whom they vote, by simply rejecting signature confirmation protocol launched by the voting manager. The undeniable signature on the ballot that is not verified through the signature confirmation protocol carries no legal binding force. Therefore, voters can restart registration steps and can make another ballot.

In section 2, we review existing voting protocols and describe motivation of our research. In section 3, the proposed scheme is presented. In section 4, we analyze our voting scheme according to requirements of large scale election. Conclusion and future works are in section 5.

2 Related Works and Motivation of Our Research

With importance of the election in democratic society and especially the development of internet, many researchers have proposed secure electronic voting scheme suitable for large scale election[3,4,5,6,7]. Those schemes assume the existence of trusted voting center to make practical voting scheme. For the privacy of voters, pseudonyms are used instead of personal ID and the tracing of the ballots from the public network is protected by the assumption of the existence of anonymous communication channel[8]. The followings are basic requirements of the large scale electronic voting scheme.

- Unreusability: An eligible voter cannot vote more than once.
- Privacy: No one can determine who voted for whom.
- Fairness: During the voting stage, intermediate voting results that may influence the entire election can not be obtained by the voting center.
- Unforgeability: Only eligible voters can make authorized ballot.
- Eligibility: Only eligible voters can participate in the election.

The security of the boyd's scheme[6] is based on the difficulty of discrete logarithm problem. Multiple key cipher that defined by the group of exponentiation transformations in a prime field is used to ensure voter's privacy. The voting center, however, can know the intermediate voting results during the voting stage. Fairness property is not satisfied in the boyd's scheme. Fujioka, Okamoto and Ohta proposed electronic voting scheme[5] which fully conforms to requirements of the large scale election. Voting fairness is ensured by the bit commitment scheme. However, once the voters get eligible ballots from the center, they must cast their ballots and no voter abstain from voting. Baraani-Dastjerdi et. al. proposed more practical voting scheme[3]. Threshold scheme is applied the election to minimize cheating by voters, candidates and administrators. Pseudonyms are generated by the center and distributed to all registered voters via secure communication channels. This assumption of the existence of secure communication channels between the center and all voters is not suitable for the large scale election. In order to minimize role of the trusted voting center, horster et. al. proposed voting scheme with multiple administrators[4]. Blind multi-signature scheme and threshold encryption technique are used to distribute role of the voting center to several administrators. At least one administrator is honest then the fairness property of the voting scheme is satisfied. However, if all administrators collude then fairness property is violated. Especially, in registration stage, the number of communication steps between voters and administrators is increased in proportion to the number of administrators.

The undeniable multi-signature scheme consists of multi-signature generation, multi-signature confirmation and disavowal stages[10]. In the multi-signature confirmation stage, the challenge-response protocol is used to confirm the undeniable multi-signature. Since the undeniable multi-signature scheme has function of digital multi-signature scheme and that of challenge-response protocol, it is best suited to the large scale electronic voting scheme. It can solve above mentioned basic requirements of large scale election by itself not using separate protocols for separate stages.

3 The Proposed Scheme

In this paper, we propose the practical electronic voting scheme suitable for large scale election. It consists of preparation stage, registration stage, voting stage and counting stage. In preparation stage, the voting manager generates unique integer pseudonyms for each candidate. Administrators and voters generate their own parameters and register their public keys to a proper public key authentication center. In registration and voting stages, each voter generates his/her own pseudonym and makes the ballot for corresponding candidate. The undeniable multi-signature scheme[10] between administrators and each voter is used to sign the ballot for registration. In voting and counting stages, multi-signature confirmation protocol[10] is used to verify registered ballot and to open the ballot for counting.

3.1 Preparation Stage

As like other existing voting schemes, our scheme is based on two assumptions.

Assumption 1. *The voting manager and at least one administrator are trustful. They have responsibility of registration of eligible voters and counting of ballots.*

Assumption 2. *Before the election day, untraceable communication channel is prepared between the voting manager and voters[8].*

The following parameters are used in the proposed scheme.

Administrators: $a_1, a_2, \ldots, a_n$ Ballot : *ballot* Blinded ballot: *ballot'*
Administrator i's private key: $X_i \in Z_{p-1}$, $1 \leq i \leq n$
Administrator i's public key: $Y_i \equiv g^{X_i} \pmod{p}$, $1 \leq i \leq n$

Step 1: The voting manager publishes the following table of candidate lists on the public bulletin board. Pseudonym is a unique random number generated for each candidate.

Table 1. Candidate Name and its Corresponding Pseudonym

Candidate Name	Corresponding Pseudonym
C_1	cps_1
C_2	cps_2
...	...
C_{n-1}	cps_{n-1}
C_n	cps_n

Step 2: Administrators generate cryptographically secure $GF(p)$, generator g and common public key Y. The common public key Y is obtained by launching following protocol.

Step 2.1: The voting manager requests common public key generation to the first administrator a_1.

Step 2.2: The first administrator a_1 sends public key Y_1 to the second administrator a_2.

Step 2.3: The intermediate administrator a_i ($2 \leq i \leq n$) receives $Y_{i-1} \equiv Y_{i-2}^{X_{i-1}} \pmod{p}$ from the administrator a_{i-1}.

Step 2.4: The a_i computes $Y_i \equiv Y_{i-1}^{X_i} \equiv g^{\prod_{j=1}^{i} X_j} \pmod{p}$.

Step 2.5: The a_i sends Y_i to the next administrator a_{i+1}. If the a_i is the last administrator, the common public key Y is computed as follows. The a_n sends it to all administrators and the voting manager.

$$Y \equiv Y_{n-1}^{X_n} \equiv g^{\prod_{j=1}^{n} X_j} \pmod{p}$$

Step 3: A voter generates his/her own public key y and private key x as follows.

$$y \equiv g^x \ (mod \ p)$$

3.2 Registration Stage

Figure 1 shows the proposed registration stage. A voter and the voter's ballot are registered by administrators. In registration stage, each voter can restart registration steps if the voter wants to vote other candidate.

(1) Generates Blinded Ballot and Makes Undeniable Signature
Step 1: A voter generates pseudonym ps and selects candidate i's pseudonym cps_i. Then, the voter makes the ballot and blinds it as follows.

$$ballot \equiv (cps_i \cdot ps)^{ps} \ (mod \ p)$$

$$ballot' \equiv ballot^{bf} \ (mod \ p), \ (bf \cdot bf^{-1} \equiv 1 \ (mod \ p-1))$$

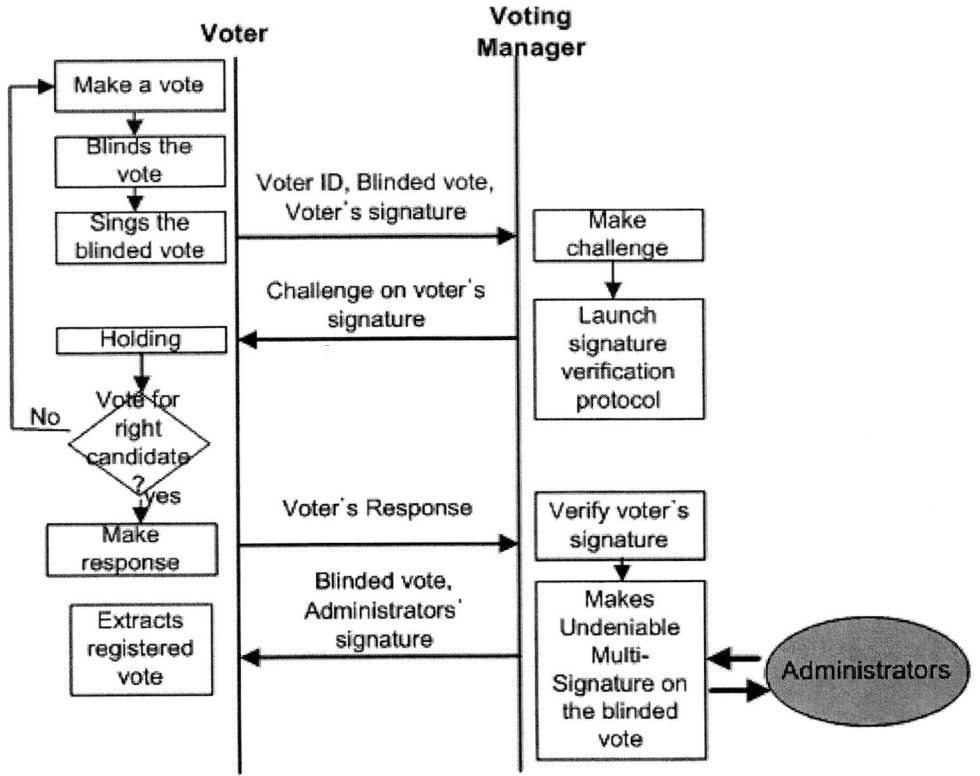

Fig. 1. Registration Procedure of the Proposed Scheme

Step 2: The voter makes undeniable signature on the blinded ballot. In this study, undeniable signature scheme in [7,9] is applied to the blinded ballot as follows. The voter generates the signature (s, r) on the $ballot'$.

$$r \equiv ballot'^k \ (mod \ p), \quad k \in Z_{p-1}, \quad k \cdot (ballot' + s) \equiv x \cdot r \ (mod \ p - 1)$$

Step 3: The voter sends identification information, blinded ballot and undeniable signature to the voting manager.

(2) Registration of the Voter
Step 1: The voting manager verifies the voter's signature by launching signature confirmation protocol[7]. If the signature is not valid then the voting manager rejects the ballot. Otherwise, the voting manager checks the voter's ID to identify whether the voter applied the registration more than once. If the voter has applied the registration more than once, the voting manager rejects the ballot. Otherwise, the following step 2 is proceeded. If the voter wants to vote other candidate, then the voter can restart registration stage by rejecting the signature confirmation protocol. Undeniable signature cannot be verified without signer's cooperation. If the voter doesn't answer the challenge of the voting manager, voter's signature can not be verified.
Step 2: The voting manager keeps the voter's ID to prevent the voter's multiple registrations.
Step 3: The voting manager sends the voter's blinded ballot to all administrators to register the ballot.

(3) Makes Undeniable Multi-signature on the Blinded Ballot
Step 1: The voting manager requests generation of the common random number R on the blinded ballot to the first administrator a_1.
Step 1.1: The first administrator a_1 selects random number k_1 in Z_{p-1}. The a_1 makes $R_1 \equiv ballot'^{k_1} \ (mod \ p)$ and sends it to the second administrator a_2.
Step 1.2: The intermediate administrator a_i ($2 \leq i \leq n$) receives $R_{i-1} \equiv R_{i-2}^{k_{i-1}} \ (mod \ p)$ from the administrator a_{i-1}.
Step 1.3: The a_i computes $R_i \equiv R_{i-1}^{k_i} \equiv g^{\prod_{j=1}^{i} k_j} \ (mod \ p)$.
Step 1.4: The a_i sends R_i to the next administrator a_{i+1}. If the a_i is the last administrator, the common random number R on the blinded ballot is computed as follows. The a_n sends it to all administrators and the voting manager.

$$R \equiv R_{n-1}^{k_n} \equiv g^{\prod_{j=1}^{n} k_j} \ (mod \ p)$$

Step 2: The administrator a_i ($1 \leq i \leq n$) computes the undeniable signature s_i and sends it to the voting manager. Since k_i and $p - 1$ are relatively prime integers, there exists s_i satisfying the following equation.

$$k_i \cdot s_i \equiv x_i \cdot R - k_i \cdot ballot' \ (mod \ p - 1), \quad 1 \leq i \leq n$$

Step 3: The voting manager computes the undeniable multi-signature S as follows.
$$S \equiv \prod_{j=1}^{n}(ballot' + s_j) \ (mod \ p)$$

Step 4: The voting manager sends undeniable multi-signature (S, R) to the voter.

(4) Extract Registered Ballot Signed by Administrators
Step 1: The voter extracts the registered ballot $S_A(ballot)$ from the multi-signature (S, R) as follows.

$$R^{S \cdot bf^{-1} \cdot (R^n)^{-1}} \equiv ballot'^{bf^{-1} \cdot (R^n)^{-1} \cdot \prod_{j=1}^{n} k_j \cdot (ballot' + s_j)} \ (mod \ p)$$
$$\equiv (cps_i \cdot ps)^{ps \cdot \prod_{j=1}^{n} X_j} \ (mod \ p) \equiv ballot^{\prod_{j=1}^{n} X_j} \ (mod \ p) \equiv S_A(ballot)$$

3.3 The Voting Stage

Step 1: The voter generates the challenge value on the registered ballot $S_A(ballot)$ as follows and sends it to the voting manager. (a, b) are random numbers selected in Z_{p-1} and Y is the common public key of administrators.

$$ch \equiv S_A(ballot)^a \cdot Y^b \ (mod \ p)$$

Step 2: The voter sends $(S_A(ballot), ch)$ to the voting manager via untraceable communication channel[8].

Step 3: The voting manager sends $(S_A(ballot), ch)$ to the first administrator a_1. Administrators sequentially make response on the voter's challenge through following steps.

Step 3.1: Administrator i receives authenticated ballot and response from administrator i-1 as follows.

$$S_A(ballot)^{\prod_{j=1}^{i-1} X_j^{-1}} \equiv ballot^{\prod_{j=i}^{n} X_j} \ (mod \ p)$$
$$rsp_{i-1} \equiv ch^{\prod_{j=1}^{i-1} X_j^{-1}} \equiv ballot^{a \cdot \prod_{j=i}^{n} X_j} \cdot g^{b \cdot \prod_{j=i}^{n} X_j} \ (mod \ p)$$

Step 3.2: Administrator i extracts his/her own private key X_i as follows.

$$(S_A(ballot)^{\prod_{j=1}^{i-1} X_j^{-1}})^{X_i^{-1}} \equiv ballot^{\prod_{j=i+1}^{n} X_j} \ (mod \ p)$$
$$rsp_i \equiv (ch^{\prod_{j=1}^{i-1} X_j^{-1}})^{X_i^{-1}} \equiv ballot^{a \cdot \prod_{j=i+1}^{n} X_j} \cdot g^{b \cdot \prod_{j=i+1}^{n} X_j} \ (mod \ p)$$

Step 3.3: Administrator i sends results from step 3.2 to the next administrator i+1. The last administrator extracts the ballot of the voter and makes response of all administrators as follows.

$$S_A(ballot)^{\prod_{j=1}^{n} X_j^{-1}} \equiv ballot^{\prod_{j=1}^{n} X_j \cdot X_j^{-1}} \equiv ballot \ (mod \ p)$$

$$rsp_n \equiv ch^{\prod_{j=1}^n X_j^{-1}} \equiv ballot^a \cdot g^b \pmod{p}$$

Step 4: The voting manager publishes $(S_A(ballot), ballot, rsp_n)$ on the public bulletin board.

3.4 The Counting Stage

Step 1: The voter checks whether response rsp_n is correct. If the response is incorrect disavowal protocol[10] is launched to discriminate whether multi-signature is invalid or some administrators have cheated. If the response is correct, the voter sends pseudonym ps to the voting manager to open the ballot.

Step 2: The voting manager counts the ballot using pseudonym ps as follows.

$$cps_i \equiv \frac{ballot^{ps^{-1}}}{ps} \pmod{p} \equiv \frac{(cps_i \cdot ps)}{ps} \equiv cps_i \pmod{p}$$

If cps_i is in table 1, the candidate lists table, the voting manager add counts of the corresponding candidate. Otherwise, the voting manager rejects the ballot.

Step 3: The voting manager publishes following results on the public bulletin board.

$$(ps, cps_i, S_A(ballot), ballot)$$

4 Security Analysis

In this section, we analyze the security of the proposed scheme according to requirements of large scale electronic election.

(1) Unreusability

In registration stage, the voting manager compares each voter's ID to prevent multiple registrations. Therefore, a dishonest voter who wants to vote more than once, should solve the following equations to make the authorized ballot.

$$S_A(ballot) \equiv ballot^{\prod_{j=1}^n X_j} \pmod{p}, \quad \prod_{j=1}^n X_j \equiv \log_{ballot} S_A(ballot) \pmod{p} \quad (4.1)$$

To solve the equation 4.1, the dishonest voter must solve the discrete logarithm problem of large prime number p. It's proven that solving discrete logarithm of large prime number p is computationally infeasible[1,2]. Therefore, the authorized voter can not vote more than once.

(2) Privacy

In registration stage, a voter generates the ballot using pseudonyms and the voter blinds the ballot by applying blinding factor bf. The voter sends the blinded ballot $ballot'$ to the voting manager via untraceable communication channel[8]. The voting manager makes undeniable multi-signature on the voter's blinded ballot by using each administrator's undeniable signature on it. The voting manager sends undeniable multi-signature to the voter. The voter extracts the registered ballot from the blinded ballot signed by all administrators. The dishonest participants who want to know who voted for whom must find the blinding factor bf as follows.

$$ballot' \equiv ballot^{bf} \pmod{p}, \quad bf \equiv log_{ballot} ballot' \pmod{p} \quad (4.2)$$

To find the blinding factor bf, the dishonest ones must solve equation 4.2. It's a discrete logarithm problem of large prime number p. It is computationally infeasible to find bf from equation 4.2. We also assume the existence of untraceable channel in section 3. Therefore, dishonest ones can not trace the ballot in order to know who voted for whom.

(3) Fairness
In voting stage, a voter sends the registered ballot $S_A(ballot)$ to the voting manager. After the voting stage, the voter sends pseudonym ps to the voting manager to open the ballot. The ballot is generated with pseudonyms of the candidate and the voter in registration stage. Therefore, the voting manager cannot open the voter's ballot during the voting stage. To open the ballot during the voting stage, the voting manager must find the voter's pseudonym ps as follows.

$$ps \equiv log_{(cps_i \cdot ps)} ballot \pmod{p} \quad (4.3)$$

Solving equation 4.3 is a discrete logarithm problem of large prime number p. Therefore, it's computationally infeasible to find the voter's pseudonym ps during the voting stage.

(4) Unforgeability
To forge the ballot, a dishonest voter must find the secret keys of all administrators. Then, they could make following eligible record on their own will.

$$(ps, \, cps_i, \, (cps_i \cdot ps)^{ps} \pmod{p}, \, (cps_i \cdot ps)^{\prod_{j=1}^{n} X_j \cdot ps})$$

However, the problem of finding $\prod_{j=1}^{n} X_j$ is a discrete logarithm problem of large prime number p. Therefore, if at least one administrator is trustful, dishonest voters can not forge eligible ballot.

(5) Eligibility
Undeniable digital signature scheme[7,9] is used in the proposed voting scheme. The security of the signature scheme is proved in [9]. The voter generates the digital signature on the blinded ballot to get the authorized ballot from administrators. The voting manager determines voter's eligibility by verifying the digital signature on the blinded ballot. Therefore, unregistered voter who wants to vote must make verifiable digital signature. This is only possible if the unregistered voter colludes with the voting manager. However, this contradicts assumption 1 in section 3.

5 Conclusion

In this paper, we propose the large scale electronic voting scheme based on undeniable multi-signature scheme. A multi-signature on the voter's ballot can not be verified and disavowed without cooperation of all administrators. The proposed voting scheme consists of four stages as preparation, registration, voting and counting stages. To minimize the role of the voting center, we use administrators controlled by each party to manage voting protocol. Undeniable multi-signature scheme is used to register the voter's ballot and to manage voting and counting stages fairly. To count the ballot, it can not be opened without help of all administrators. Before counting, they must confirm the undeniable multi-signature on the ballot. Therefore, all parties can fairly manage voting and counting stages. The proposed scheme also provides user convenience during the registration stage. Voters can change their mind to whom they vote, by simply rejecting signature confirmation protocol launched by the voting manager.

References

1. W.Diffie, M.E.Hellman, "New Directions in Cryptography," IEEE Transactions on Information Theory, Vol. IT-22, No. 6, pp.644-654, 1976.
2. T.Elgamal, "A Public Key Cryptosystem and a Signature Scheme Based on Discrete Logarithms," IEEE Transactions on Information Theory, Vol. IT-31, No. 4, pp.469-472, 1985.
3. A.Baraani-Dastjerdi, J.Pieprzyk, R.Safavi-Naini, "A Secure Voting Protocol Using Threshold Schemes," Proceedings of COMPSAC'95, pp. 143-148, 1995.
4. P.Horster, M.Michels, H.Petersen, "Blind Multisignature Schemes and Their Relevance for Electronic Voting," Proceedings of COMPSAC'95, pp. 149-155, 1995.
5. A.Fujioka, T.Okamoto, K.Ohta, "A Practical Secret Voting Scheme for Large Scale Elections," In Advances in Cryptology, Proceedings of AUSCRYPT'92, 1992.
6. C.Boyd, "A New Multiple Key Cipher and an Improved Voting Scheme," In Advances in Cryptology, Proceedings of EUROCRYPT'89, LNCS 434, pp. 617-625, 1990.
7. S.H.Yun, S.J.Lee, "An electronic voting scheme based on undeniable signature scheme," Proceedings of IEEE 37th carnahan conference on Security Technology, pp.163-167, 2003.
8. D.Chaum, "Untraceable Electronic Mail, Return Addresses and Digital Pseudonyms ," Communications of the ACM, Vol. 24, No. 2, pp. 84-88, 1981.
9. S.H.Yun, T.Y.Kim, "Convertible Undeniable Signature Scheme," Proceedings of IEEE High Performance Computing ASIA'97, pp. 700-703, 1997.
10. S.H.Yun, H.W.Lee, "The Undeniable Multi-signature Scheme Suitable for Joint Copyright Protection on Digital Contents," Advances in Multimedia Information Processing - PCM2004, LNCS 3333, pp.402-409, 2004.

IPv6/IPsec Conformance Test Management System with Formal Description Technique

Hyung-Woo Lee[1], Sung-Hyun Yun[2],
Jae-Sung Kim[3], Nam-Ho Oh[3], and Do-Hyung Kim[3]

[1] Dept. of Software, Hanshin University,
Osan, Gyunggi, 447-791, Korea
hwlee@hs.ac.kr
[2] Div. of Information and Communication Engineering,
Cheonan University, Anseo-dong,
Cheonan, Chungnam, 330-704, Korea
shyoon@cheonan.ac.kr
[3] Korea Information Security Agency,
Garak, Songpa, Seoul, 138-803, Korea
{jskim, nhooh, kdohyung}@kisa.or.kr

Abstract. To ensure the correctness of IPv6/IPsec implementations, it is very necessary to introduce the technique of "protocol testing". Secure IPsec protocol on IPv6 are possible through proper use of the *Encapsulating Security Payload*(ESP) header and the *Authenticated Header*(AH). IPv6/IPsec test tools must be able to perform a wide variety of functions to adequately test its conformance on standard and validate mechanism on IPv6 devices. In this study, we developed *advanced conformance test management system* on FreeBSD 4.8/5.0 with *KAME kit* based on the compatibility function with TAHI project. Proposed system adopts *Formal Description Technique* for proving its correctness on overall testing sequences and suites.

Keywords: IPSec, IPv6, Formal Description, Conformance Test, Security, Architecture. [1]

1 Introduction

IPv6 was developed specifically to address these deficiencies, enabling further Internet growth and development[1]. The IPv6 header has been streamlined for efficiency. The new format introduces the concept of an extension header, allowing greater flexibility to support optional features[2]. The most important issue addressed by IPv6 is the need for increased IP addresses: IPv4 32-bit address space is nearly exhausted, while the number of Internet users continues to grow exponentially. This need is exacerbated by the continual introduction of address hungry Internet services and applications.

---

[1] This work is supported by the University IT Research Center(ITRC) Project.

Within the IPv6 address space, the implementation of a multi-leveled address hierarchy provides more efficient and scalable routing. This hierarchical addressing structure reduces the size of the routing tables Internet routers must store and maintain. Security on IPv6 are possible through proper use of the *Encapsulating Security Payload*(ESP)[3] header and the *Authenticated Header*(AH)[4]. AH provides mechanisms for applying authentication algorithms to an IP packet, whereas ESP provides mechanisms for applying any kind of cryptographic algorithm to an IP packet including encryption, digital signature, and/or secure hashes.

Network operators and service providers need to understand how well new IPv6 equipment will behave in multi-vendor environments. For conformance testing, the test solution must be able to fully exercise the control plane of the device or system under test[5,6]. The conforming implementations can be directly related to the specification by means of an *implementation relation*: given a set of potential implementations, this relation indicates which ones are conforming to a given specification. More representative methods must be designed to derive tests from *Formal Descriptions*(FDs) of a IPv6/IPsec system based on checking experiments[7,8].

In this study, we proposed a FD based IPv6/IPsec conformance test management system with sequence DB on Web interface. We developed it on FreeBSD 4.8/5.0 with *KAME kit*[9] with modified a kernel module to provide additional encryption and hashed MAC function such as *SEED* and *HAS-160*, etc. And this test system provides compatibility function with TAHI project[10]. Proposed suite provided an outstanding results on the fully IPv6/IPsec conformance tests.

2   Overview of IPv6

2.1   IPv6 Structure: Extended IPv6 Header

IPv6 extension headers The extension header is optional in IPv6. If present, extension headers immediately follow the header field. IPv6 extension headers have the following properties: They are 64-bit aligned, with much lower overhead than IPv4 options. They have no size limit as with IPv4. The only limitation is the size of IPv6 packet. They are processed only by destination node. The only exception is the Hop-by-Hop header option. The Next Header field of the base IPv6 header identifies the extension header.

Optional in IPv4, IPSec is a mandatory part of the IPv6 protocol suite. IPv6 provides security extension headers, making it easier to implement encryption, authentication, and virtual private networks (VPNs). By providing globally unique addresses and embedded security, IPv6 can provide end-to-end security services such as access control, confidentiality, and data integrity with less impact on network performance[1,2].

2.2   IPsec over IPv6

IPsec as defined in RFC2460 provides a security architecture for the Internet Protocol(IP). IPsec defines security services to be used at the IP layer, both for IPv4 and IPv6. The IPsec provides an interoperable and open standard for building security into the network layer rather than at the application or transport layer. The most important application IPsec enable is the creation of *virtual private network*(VPNs) capable of securely carrying data across the open Internet. IPsec allows maintenance of the following[2]:

- **Access Control** allowing authentication of users with secure exchange of keys.
- **Connectionless Integrity** allowing nodes to validate each IP packet independent of any other packet through the using of secure hashing techniques.
- **Data Origin Authentication** Identifying the source of the data contained in an IP packet.
- **Defense Against Replay Attacks** Providing a packet counter mechanism.
- **Encryption** Providing a data confidentiality through the use of encryption function.

3 IPv6 Conformance Test

3.1 Existing Conformance Test Suite

Providing standard and interoperable products is a key element to success with the introduction of any new technology. The fear of incompatibility problems between legacy IPv4 infrastructure and multiple vendors's IPv6 systems can only be dealt with via a thorough test methodology to ensure conformance test based on interoperability primitive.

IPv6 is defined by over several IETF RFCs. The implementation of very large and complex RFCs is prone to misunderstanding and misinterpretation. Conformance testing such as TAHI project[10], with a comprehensive and rigorous test methodology, increases product quality and customer confidence. Conformance testing also saves time and money, by allowing vendors to verify a product's design throughout the entire product life cycle. Problems can be identified earlier in development, reducing costly last-minute rework and post-deployment problems.

3.2 Model of Testing System

For conformance testing, the test solution must be able to fully exercise the control plane of the device or system under test. Based on the TTCN, TTCN-2[11] makes an extension of supporting protocol testing. Abstract testing architecture includes :

1. Test Component(TC): includes one MTC(Main Test Component - Test tool) and zero or more STCs(Sub Test Component). STCs communicate with IUT(Implementation Under Test). MTC is mainly responsible to control and coordinate STCs by control commands and collect the local verdicts of STCs.

2. Point of Control and Observation (PCO) : communicate with IUT so that tester can not only observe the behavior of IUT but also control IUT by sending "packets" to it.
3. Coordination Point (CP) : exchange Coordinate Message (CM) with each other via CP for coordination and synchronization of test components.

Definition 1 (Formal Model of Test Component). *a 'i'th test component can be modeled as a finite state machine with queues.* $TC_i = \{M_i; eq_j; iq_j | i = 1, ..., n, j = 1, ..., n, i \neq j\}$

- n : *total number of test components.*
- eq_j, iq_j : *'j'th external or internal input queues, which buffer external or internal inputs from other test components.*
- $M_i = (S, I, O, \delta, \lambda, s_0)$ *where, S is the set of states of M_i, $I = I_e \cup I_i$ is the set of input of M_i where $I_e \rightarrow eq : a$ is the set of external input a and $I_i \rightarrow iq : b$ is the set of internal input. O is the set of output symbols of M_i. δ is the state transition function as $\delta : S \times I \rightarrow S$. λ is the output function as $\lambda : S \times I \rightarrow O$ with initial state s_0.*

Using proposed test suite, we could find the detailed cause of several errors more easily if conformance test was executed on a product beforehand. This would make interoperability testing more efficient in the near future. To achieve a smooth IPv4 to IPv6 migration it is essential that the IPv6 nodes support those features that enable them to communicate with IPv4 ones. This can be assured with executing conformance tests on the appropriate features.

4 Conformance Testing Theory

4.1 Definition of Conformance Test

A protocol is a collection of rules to establish the communication between the components of these networks. A statement of a collection of such rules is called a specification for the protocol. *Reliable communication therefore depends on the conformance of actual implementations of the protocols to their specifications.* In order to maintain the operation of a system in a heterogeneous environment, standards must be enforced for the interoperability of different components. The area of research and application that deal with such testing methods is called *conformance testing*[5,6,11].

Definition 2 (Conformance Testing). *the 'black box' implementation of a protocol conforms "exactly" to its specification. The specification is a formal entity usually modeled as a finite state automation.*

Whereas testing conformance relates a "black box", namely an implementation into which the tester has access to input and output events to a formal specification of the protocol residing in one definite site in a communication environment, verification involves testing the entire design of the system residing in all the sites involved in the system.

4.2 Assessment Mechanism on Conformance Testing

There are several types of an assessment mechanism on conformance testing, such as *Formal Description Techniques*(FDT) and *Finite State Machines*(FSM) based methods[7,8,12].

Definition 3 (FDT based Conformance Testing). *More precise and unambiguous derivation of tests than the natural language specifications with improvements in the concepts of automatic generation of the tests and reduction in their complexity.*

The conforming implementations can be directly related to the specification by means of an *implementation relation*: given a set of potential implementations (for instance described in the same FDT as the specification), this relation indicates which ones are conforming to a given specification. We call this approach a behavioral mode of conformance. It is more adapted to study conformance to specifications which are expressed in FDTs.

Testing is a way to assess conformance of an implementation to its specification by means of a test experiment. In order to know the result of one test case, several test runs (or executions) may be necessary. An important point in testing is the choice of a particular *test architecture*.

More representative methods designed to derive tests from Formal Descriptions (FDs) of a protocol based on checking experiments. These methods have their origins in the *checking experiment problem* from automata theory whether a given state table describes the behavior of a FSM implementation as *intermediate models* for test.

Definition 4 (FSM based Conformance Testing). *Systematically probe the implementation with input test sequences and observe the outputs to establish whether its internal structure, considered as a finite state machine, conforms to the structure of the specification, by focusing on the control structure of the system.*

5 A Formal Description for IPv6/IPsec

5.1 Formal Description on the IPv6 Packet

An FD describes an abstract machine generating some observable behavior. A basic assumption of the methods described in this section is, that the FD is a finite state machine (FSM). It is also assumed, that the IUT behaves like a FSM, although their states may not be explicitly observable. In first, we can define IP packet as follows.

Definition 5 (IPv6 packet datagram). *An IP packet(datagram) is abstractly defined as a (type, hashed-MAC–data, encrypted-data, payload-data, data-list) tuple.*

- type : Null(*NoProtection*), ESP or AH;
- hashed-MAC-data : a list of indices into *data-list* that indicate the data in the IP packet that is covered by a hashed MAC function;

- `encrypted-data` : a data in the IP packet that is covered by an encryption function;
- `payload-data` : a TCP/UDP data in the IP packet;
- `data-list` : a list of all of the other fields in the IP packet.

A primary use of a cryptographic hash function is to provide an *integrity check* over data. Using the abstract IP packet, the integrity of the hash-data, hash-value, IP source address and SPI[13] can be verified as a unit. If A and B share a secret key, K, for use with a cryptographic hashed MAC function, H, then if A creates a message $z = H_K(x)$, B can verify the integrity of x and z combined by computing $H_k(x) = z$. If either of these are changed then, assuming a cryptographically strong keyed hash function, B will detect the change. Since the source IP address and SPI are used to index the security association information and determine what key is used with the HMAC function, then if any of these fields were modified, it follows that the computation of the hash-value will be incorrect.

```
entity RECEIVED datagram
datagram HMAC_OK SA_table
message INDEXED (hashed-MAC-data datagram)
∼ INTEGRITY message SA_table entity
```

Authentication of data is shown by proving that there is integrity over both the data and the source. This can be done directly, as in the AH header where the source IP address is included in the hash-data or indirectly as in the ESP. Directly integrity is provided over the hashed-MAC-data and indirectly integrity is provided over the source IP address and SPI, therefore, the hashed-MAC-data is authenticated.

```
INTEGRITY message SA_table entity
sender INDEXED (hashed-MAC-data datagram)
∼ AUTHEN sender message SA_table entity
```

Informally, if A and B share a secret key, K, for use with a cipher module, then if A creates a message $E_K(p) = c$ where E is the encryption function such as DES, AES on the IP packet p, B can decrypt c by computing $D_K(c) = D_K(E_K(p))$. Since the decryption key is retrieved from the *security association*(SA)[14] table using the source IP address and SPI as an index, thn integrity over these fields implies that the correct key, K, is gotten. This is the method by which *confidentiality* is supplied using current version of the ESP header.

```
entity RECEIVED datagram
INTEGRITY SPI SA_table entity
INTEGRITY message SA_table entity
ESPkey datagram SA_table
message INDEXED (encrypted-data datagram)
∼ CONFID message SA_table entity
```

5.2 Formal Description on the IPv6/IPsec AH Header

The AH header can be used to do several functions such as strong integrity, authentication and non-repudiation for IP datagram with protecting against replay attaks. In IPv6 packet, the AH header consists of the following fields :

NextHeader | PayloadLen | Reserved | SPI | SeqNo | AuthData

and is proceded by an IP header and followed by the data. The *AuthData* field is the `hashed-MAC-data`. In IPv6/IPsec transport/tunnel mode an AH packet is represented as following fields :

IPHeader | ExtHeader | NextHeader | PayloadLen | Reserved | SPI | SeqNo | AuthData | Options | payload-data

NewIPHeader | ExtHeader | NextHeader | PayloadLen | Reserved | SPI | SeqNo | AuthData | IPHeader | ExtHeader | payload-data

Since all fields in a packet using the AH header are included in the hashed-MAC-data, it is easy to prove that integrity and authentication are provided over the entire packet.

5.3 Formal Description on the IPv6/IPsec ESP Header

The ESP header allows IP nodes to exchange datagrams whose payloads are encrypted with confidentiality, authentidation of data origin, *antireplay services* through the same sequence number machanism, limited traffic flow confidentiality. The ESP header can be used un conjunction with an AH header. The ESP header consists of the following fields :

SPI | SeqNo | InitVector | PayloadData | Pad | PadLen | NextHeader | AuthData

In transport/tunnel mode an ESP packet is :

IPHeader | ExtHeader | SPI | SeqNo | InitVector | encrypted-data | Pad | PadLen | NextHeader | AuthData

NewIPHeader | ExtHeader | SPI | SeqNo | InitVector | IPHeader | encrypted-data | Pad | PadLen | NextHeader | AuthData

5.4 IPv6/IPsec Conformance Test Sequences

Proposed suite sends packets to the router being tested, receives the packets sent in response, and then analyzes the response to determine the next action to take. This allows to test complicated situations or reactions in a much more intelligent and flexible way than can be done by simple packet generation and capture devices. IPv6/IPsec conformance test suite(CTS) must be able to perform a wide variety of functions to adequately test and validate IPv6/IPsec devices and systems. For conformance testing, the test suite must be able to fully exercise the control plane of the device or system under test.

```
ipsecCheckNUT(host | router); initialization;
ipsecSetSAD(src,dst,SPI, mode={transport,tunnel},
    protocol={Null,AH,ESP}, Algo={encrypt(),HMAC()}, Key);
ipsecSetSPD(src,dst,upperspec, direction={in,out},
    protocol={Null,AH,ESP}, mode={transport,tunnel});
vCapture($NIC);
set $ret = ICMPv6_ping(dst_AH);
if ($ret eq 'PASS') $ret = ICMPv6_ping(src_AH);
if ($ret eq 'PASS') ipv6_ipsec_Pass(); else
    ipv6_ipsec_Fail();
```

Test packet is defined by Perl script and test sequences for IPv6/IPsec AH and ESP on host/router mode are also defined. Proposed suite checks its DUT system with initialization module. And both SAD and SPD are set for proper test. And then the test suite sends ICMPv6 packet to the NUT, on which AH or ESP header is attached for verifying its correctness. This CTS runs a number of test cases against the DUT based on the direct interpretation of various IPv6 RFCs.

6 Implementation of Conformance Test Management System

6.1 Integrated Conformance Test Suite

We have taken the advantage of open source operating systems like FreeBSD 4.8/5.0 with *KAME kit* for generating IPv6 packets. Then, we have modified a kernel module to provide additional encryption and hashed MAC function such

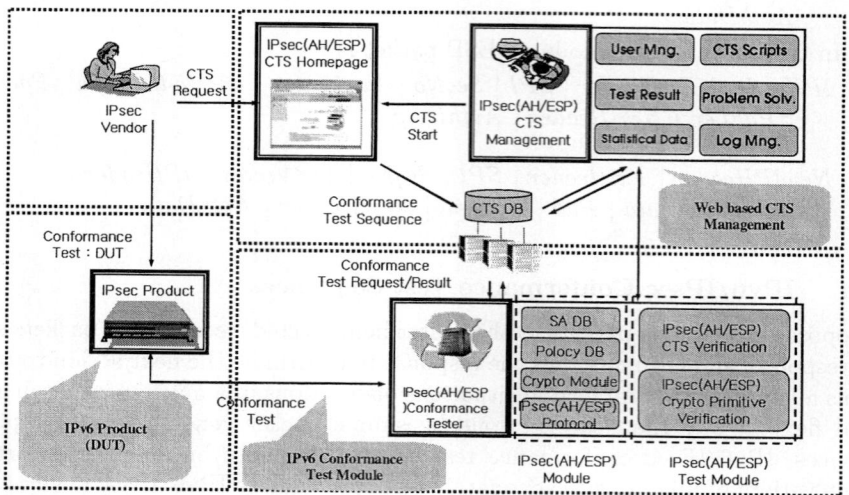

Fig. 1. Detailed IPv6/IPsec Conformance Test Structure

as *SEED* and *HAS-160*, etc. And this test suite provides compatibility function with TAHI project. In management module, we developed a test sequence DB with Web based management system for user interface. Selected sequence from the DB was sent to the tester system by CTS daemon on Windows system. The CTS tester system generated a IPsec packet and sent it to the DUT. Finally, the test results are stored into DB for reporting on our Web system. The overall architecture is shown as follow figure.

6.2 IPv6/IPsec Conformance Test Procedure

The objective of IPv6/IPsec test is to verify the DUT's compliance with the following features defined in various RFCs. An FreeBSD based workstation connects directly to the DUT with one or two test interfaces. The proposed test suite will emulate either hosts or routers in IPv6 mode, depending on the configuration of each test case.

If an IPv6 node processing a packet finds a problem with a field in the IPv6 header or extension headers such that it cannot complete processing the packet, it must discard the packet and should send an ICMPv6[15] *Parameter Problem* message to the packet's source, indicating the type and location of the problem. Every node must implement an ICMPv6 Echo responder function that receives *Echo Requests* and sends corresponding *Echo Replies*. A node should also implement an application-layer interface for sending Echo Requests and receiving Echo Replies, for diagnostic purposes.

1. Configure each network interface with the appropriate network parameters.
2. Specify configuration of the DUT.
3. Select a set of test cases to run from the test sequence DB through Web based management interface.
4. Run script in a batch mode from the CTS daemon by sending test script with the command ID.
5. Return test result to the CTS daemon.
6. Store each test result as passed/failed in DB, including reasons for failed cases.

7 Conclusions

The global need for IP addresses has even added political force to the drive for IPv6 implementation. For latecomers to the Internet explosion, IPv6 is the only solution that will accommodate billions of new users. Optional in IPv4, IPSec is a mandatory part of the IPv6 protocol suite. IPv6 provides security extension headers, making it easier to implement encryption, authentication, and virtual private networks (VPNs). IPv6 test tools must be able to perform a wide variety of functions to adequately test and validate IPv6 devices and systems.

For conformance testing, the test solution must be able to fully exercise the control plane of the device or system under test. Testing is a way to assess conformance of an implementation to its specification by means of a test experiment. More representative methods designed to derive tests from *Formal Descriptions*(FD) of a protocol based on checking experiments. In this study, we proposed a FD based IPv6/IPsec conformance test mechanism and implemented it in sequence DB based Web interface for efficiency. As a result, proposed suite reported about 98% of success from total 116 test cases of IPv6/IPsec with AH/ESP packet on Host/Router mode.

References

1. S. Deering, R. Hinden, "Internet Protocol Version 6 (IPv6) Specification", RFC 2460, IETF Network Working Group, December 1998.
2. P. Loshin, "IPv6 : Theory, Protocol and Practice", 2nd Edition, Morgan Kaufmann, Elsevier, 2003.
3. S. Kent, R. Atkinson, "IP Encapsulating Security Payload(ESP)", Network Working Group, Internet Draft : draft-ietf-ipsec-esp-04.txt, 1997.
4. S. Kent, R. Atkinson, "IP Authentication Header", Network Working Group, Internet Draft : draft-ietf-ipsec-auth-05.txt, 1997.
5. R. Gecse, "Conformance testing methodology of Internet protocols", Testing of Communicating Systems, Tomsk, Russia, September 1998.
6. University of NewHampshire, InterOperability Lab, "IP Consortium Test Suite, Internet Protocol Version 6", Technical Document, January 2000.
7. B. S. Bosik, M. U. Uyar, "Finite State Machine based Formal Methods in Protocol Conformance Testing : From Theory to Implementation", Computer Networks and ISDN Systems, Vol. 22, No. 7, 1991.
8. A. Rezake, "Implementation of a Protocol Conformance Test Sequence Generation Methodology Using Distingushing Sequences", Thesis, Middle East Technical University, Ankara, 1993.
9. TAHI project, http://www.tahi.org/
10. KAME project, http://www.kame.org/
11. ISO/IEC 9646-3, "Open System Interconnection, Conformance Testing Methodology and Framework, IS-9646, Part 3 - The Tree and Tabular Combined Notation", 1996.
12. W. Zhiliang, W. Jianping, L. Zhongjie, Y. Xia, "Communication Mechanism in Distributed Protocol Testing System", Proceeding of ICCT2003, pp.149-152, 2003.
13. D. Maughan, M. Schertler, M. Schneider, J. Turner, "Internet Security Associations and Key Management Protocol (ISAKMP)", IPSEC Working Group, Internet Draft, 1997.
14. S. Kent, R. Atkinson, "Security Architecture for the Internet Protocol", Network Working Group, Internet Draft : draft-ietf-ipsec-.txt, 1997.
15. A. Conta, S. Deering, "Internet Control Message Protocol (ICMPv6) for the Internet Protocol Version 6 (IPv6) Specification", RFC 2463, IETF Network Working Group, December 1998.

Interference Cancellation Algorithm Development and Implementation for Digital Television

Chong Hyun Lee[1] and Jae Sang Cha[2]

[1] Department of Electronics,
Seokyeong University Seoul, Korea
chonglee@skuniv.ac.kr
[2] Department of Information and Communication Engineering,
Seokyeong University Seoul, Korea
chajs@skuniv.ac.kr

Abstract. In this paper, we introduce newly developed DSP HW and SW module which is applicable to DTV-OCR and for cancellation of the interference signal. In general, RF repeater has problems of system oscillation and signal quality degradation due to feedback interference signal coming from transmit antenna. In this paper, we demonstrate newly developed DSP HW and SW module for cancelling the interference signal by investigating the field data measured through a RF repeater. Also, the structure and signal processing method for non-regenerative repeater system based on the newly developed DSP HW and SW module is illustrated as well.

1 Introduction

The HDTV broadcasting has begun since October in 2001 in Korea and the field test had been done since 2000 for secure broadcasting signal transmission. With HDTV signal transmission, the lack of spectrum resource is apparent since analog TV broadcasting is also in service. Thus, in order to use the spectrum resource efficiently, DTV-OCR(Digital TV-On Channel Repeater) is suggested and the filed test has been conducted [2], [3].

DTV-OCR of on-channel frequency can be divided into two categories, one regenerative repeater and non-regenerative repeater. The former re-transmits the received signal after demodulation and error correction and the latter only re-transmit the signal after amplification of the received signal. The former produces different output during the DMV of TCM(Trellis Coded Modulation) and thus results in trellis ambiguity. This ambiguity makes no correlation between transmitted and received signal thus make time-domain processing impossible. However, the latter can keep the correlation between transmitted and received signal, we can use time-domain cancellation algorithm. To this end, we propose signal processing HW module nd SW algorithm for non-regenerative DTV-OCR.

In section 2, we explain a structure of DSP HW and SW applicable to DTV-OCR. In section 3, we demonstrate the performance of the proposed system by generating 6 MHz PN(Pseudo Noise) sequence in DTV field synchronous segment and analyzing the received data with interference cancellation algorithm. Finally, section 4 concludes our paper.

2 Design of Interference Cancellation DSP Module for DTV-OCR

2.1 Overview of the Developed System

In this paper, we have selected non-regenerative RF repeater system to apply ICS (Interference Cancellation System) since we can develop and modify the system with low cost and minimum time. The ICS is composed of DSP hardware module and software module. The overall block diagram of the RF repeater system including ICS is shown in Fig.1

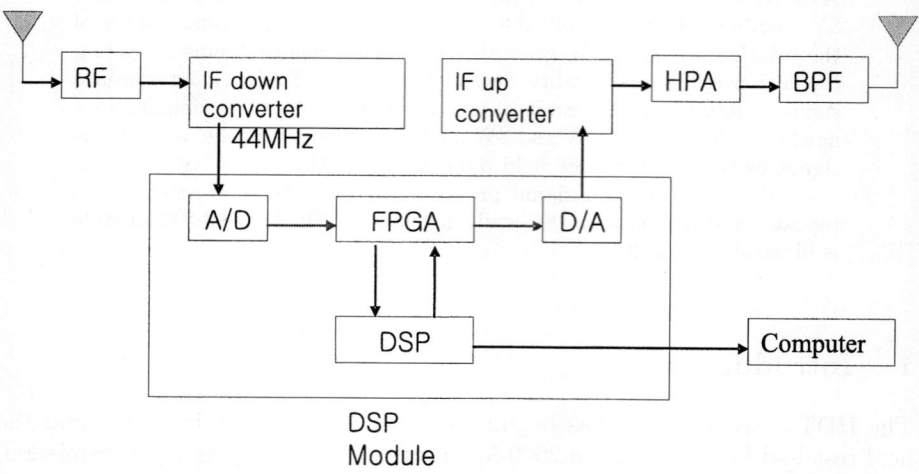

Fig. 1. DSP implemented RF repeater

As shown in the Fig. Fig.1 the RF repeater system is composed of typical RF repeater system and ICS system. The function of each module can be describes ad follows:

1. RF: This module receives the signal and passes the signal with bandpass filter giving input the IF module.
2. IF down converter: This module receives the RF signal and converts the signal into signal of IF frequency of 44MHz.
3. IF up converter: This module receives analog baseband signal, converts the signal into signal of IF frequency of 44MHz and then into signal of RF frequency.

4. HPA(High Power Amplifier): This amplifier amplifies input signal to specified level.
5. BPF(BandPass Filter): This filter passes the signal of interested channel while blocking the unused or out-of-interest channel signal.
6. DSP: This module is one is absent in typical RF repeater system. This module is charge of analysis of input signal and applying the interference cancellation algorithm. This module is composed of AD(Analog to Digital) converter, one FPGA(Field Programmable Gate Array) for real time processing, Digital Signal Processor and DA (Digital to Analog) converter. With DSP, the processed data is transferred from or to host computer with serial port communications.
7. Host computer: This host computer is implemented with software which can verify capability of interference cancellation with off-line operation.

2.2 The Characteristics of DSP Module and SW Algorithm

The Characteristics of DSP Module

The board of DSP module for interference cancellation algorithm is designed for cancelling the feedback interference signal via applying numerous algorithms. In this board, two FPGA are used for FIR (Finite Impulse Response) filter output filtering input In-phase and Quadrature phase signal. The DSP is used for calculating FIR filter coefficients and AD & DA converter are also included in this module. The functional block diagram of the is shown in Fig. 2. By using the SW in DSP module, we also regulate the input gain of IQ modulator and demodulator. The stored data in FPGA memory can be transformed to PC.

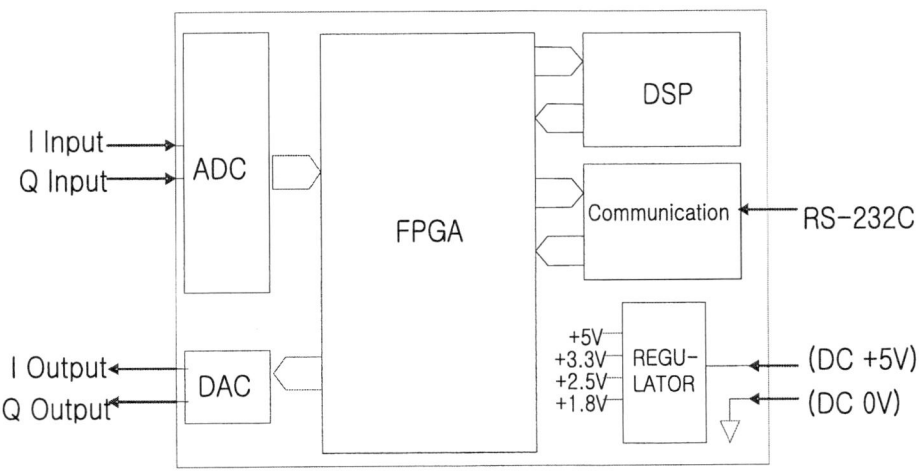

Fig. 2. Functional Block Diagram of DSP module

The Characteristics of DSP SW

In order to obtain channel parameters such as overall multi-pass delay time and tap length of filter via time domain interference cancellation algorithm with DSP and FPGA, we need to collect analyze data via off-line processing. To this end, we develop GUI(Graphic User Interface) which can receive multi-path fading of transmitted training signal and then transfer to host computer via DSP processor. In this paper, by considering communications protocols, the amount and specification of data to be transferred and software tool, we design and develop software. The specific parameters of software are described as follows:

1. Software Platform: Visual Basic and Matlab
2. Serial Port Communication Protocol: Asynchronous Communications
3. Contents of data: 12bit of baseband I and Q data
4. Amount of data: 256 or 512 of I and Q data

The proposed DTV-OCR structure for mobile reception and static reception is shown in the Fig. 3.

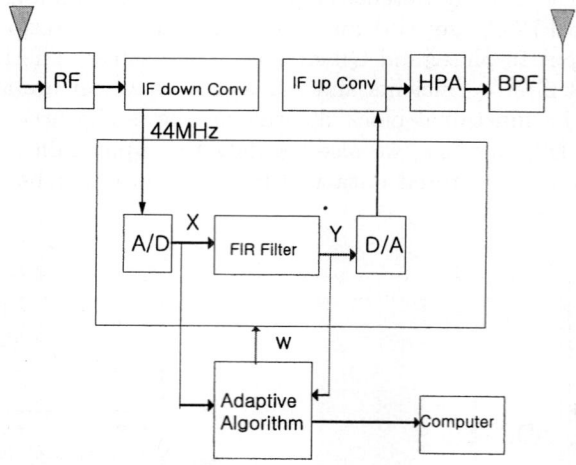

Fig. 3. RF repeater with Interference Cancellation DSP module

The GUI software is developed by using Visual Basic and Matlab. The overall flow of SW is described as follows:

1. Initialize DSP module board.
2. Set the speed of serial port communication.
3. Transfer the transmitted training data and the received data.
4. Execute Matlab and analyze the data by using Wiener algorithm, RLS, LMS, QR-RLS algorithms.
5. Once the data analysis is completed, then collect and process new data.

The one of example of the above process is illustrated in the following Fig. 4. In the figure, FIR filter coefficients, the dB difference of transmitted and received

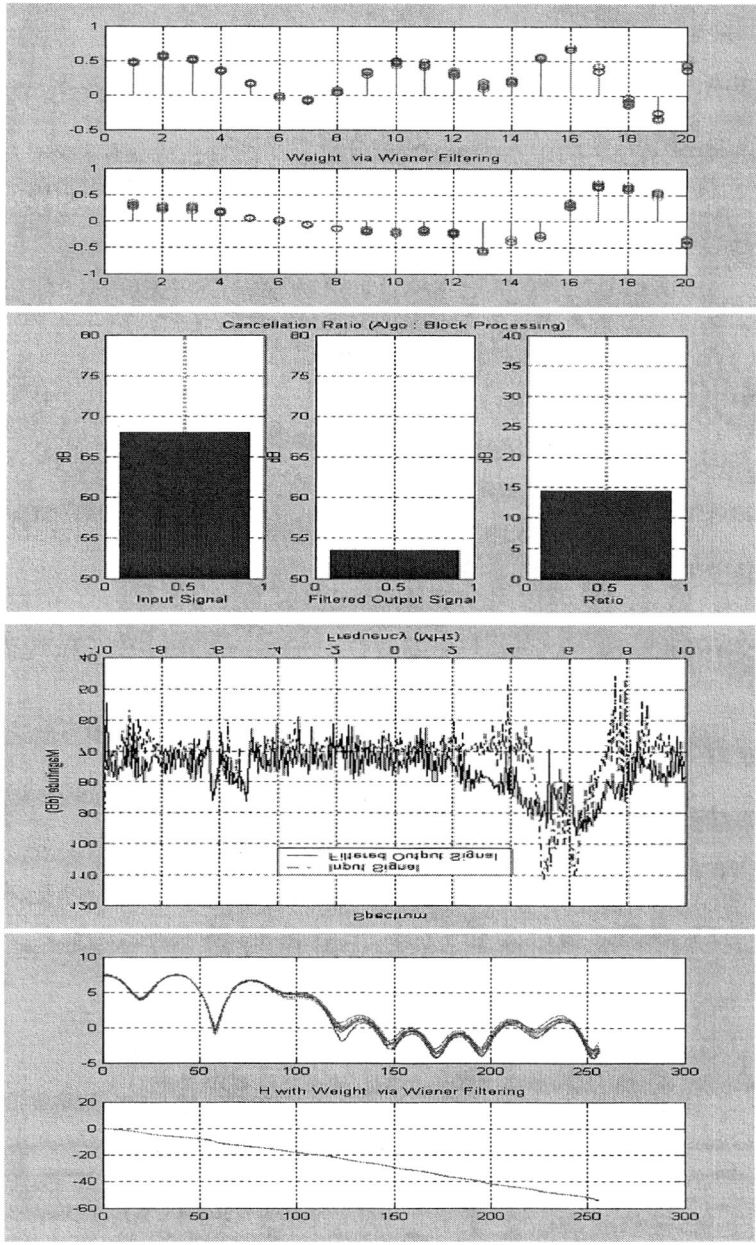

Fig. 4. Output of Interference Cancellation SW module

signal, and the spectrum of input and filtered output and filter magnitude and phase responses are depicted.

3 Experiments

In this section, we verified our DSP module and SW algorithm by using RF repeater for mobile communications. The field measurement data is collected at top of building by placing the transmit antenna and receive antenna at opposite direction. With this configuration, the receive antenna signal are composed of Non LOS(Line Of Sight) only. The collected signal is AD sampled at 20MHz and then FIR filter coefficients are calculated. The system delay time taken passing

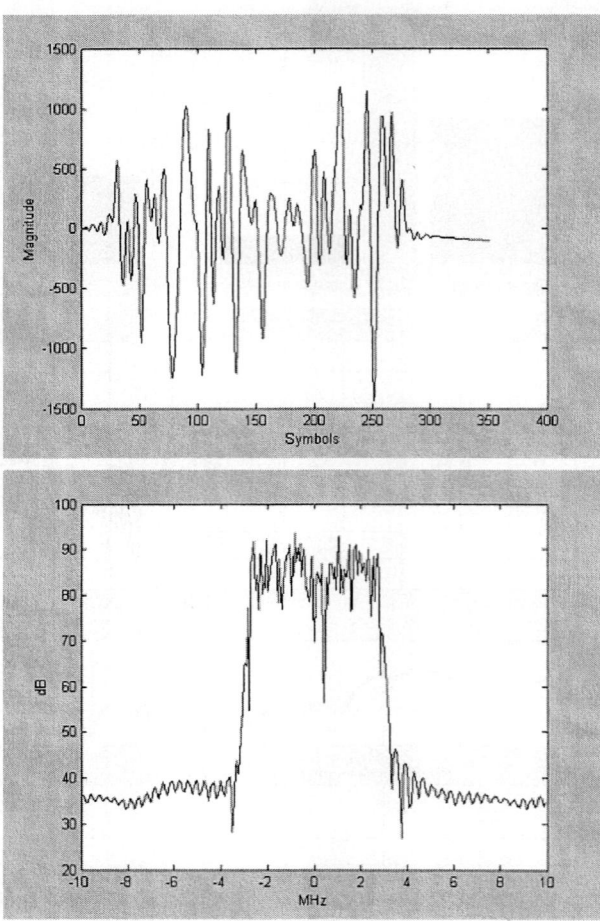

Fig. 5. Transmitted Training Signal and its spectrum

the RF and Digital circuits is assumed 7 μsec and the corresponding 140 digital samples are neglected in computing filter coefficients.

In the experiments, in order to generate transmitted training signal of 6 MHz bandwidth, PN sequence generator is used. The transmitted signal and its power spectrum is shown in Fig. 5.

In order to time varying characteristics of channel, we measure the data for ten consecutive time interval. The FIR filter coefficients of 10 data samles are shown in Figure 6 . In this figure, we can observe that over all delay spread is about 20 samples or 1 μsec. Also in this figure, even transmit and receive antenna are fixed, the channel characteristics are time varying and thus adaptive interference cancellation algorithms should be implemented.

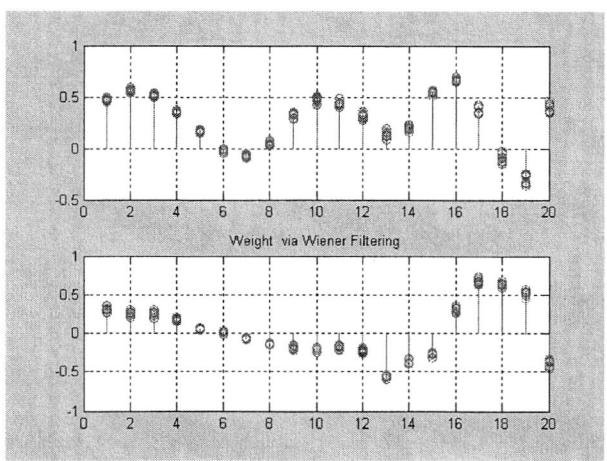

Fig. 6. FIR filter Taps of 10 data set

Next, we simulate the data by using LMS and QR-RLS algorithms. The simulation parameters are listed in the following table 1.

Table 1. Parameters of LMS and RLS algorithms

Algorithm	LMS	RLS
Tap Length	20	20
μ	0.1 & 1.0	X
λ	X	0.999
δ	X	0.01

The results are depicted in Figure 7 when LMS algorithm is appled. As shown in the figure, the speed of adaptation is dependent of adaptation constant.

The RLS algorithm is better in convergence speed at the cost of high implementation cost. However, when the channel is not highly time varying, the

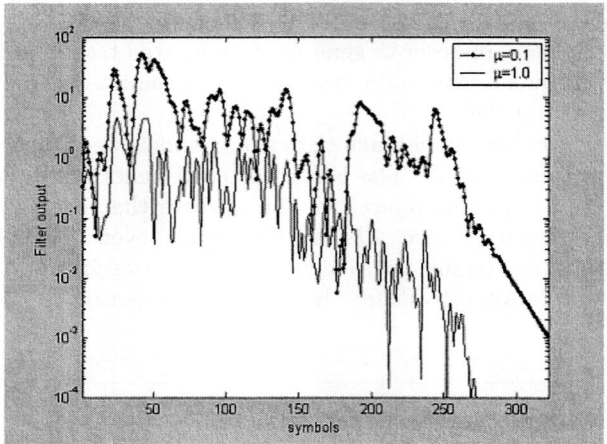

Fig. 7. Filtered output of LMS algorithm

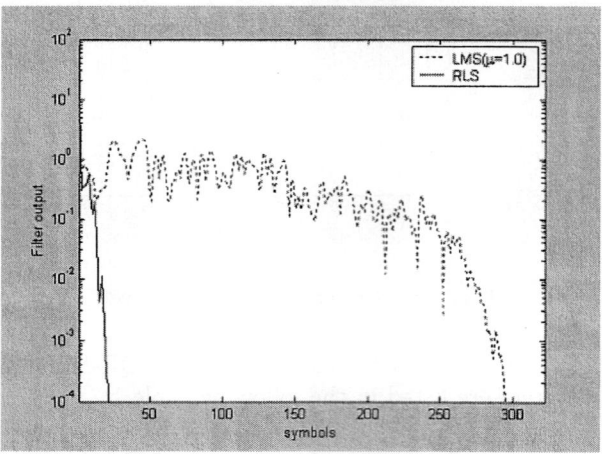

Fig. 8. Filtered output of LMS and RLS algorithm

LMS algorithm of low cost would guarantee cancellation performance. Finally, we verified adaptation algorithm by using all 10 sets of data. In the experiment, RLS algorithm is used by using the parameters in table 1. The comparison of the LMS and RLS algorithms is depicted in Figure 8.

4 Conclusion

In this paper, we design and verify the DSP HW and SW algorithm which can be applied to DTV-OCR for interference cancellation. The proposed system is based on adaptive interference algorithm such as LMS and RLS algorithms and the performance of the system is verified by using field synchronous signal in ATSC DTV data frame. Also, the adaptive interference cancellation algorithms are verified via field measured data. With the experiment result, we conclude that the proposed DSP HW and SW algorithm can be applied DTV-OCR repeater.

References

1. Mok, H, Seo Y, Hwang H: The field measurement results of HDTV in Korea, Jounal of Broadcasting Engineering, Vol 5, No.2, Dec. (2000)142–158 (in Korean)
2. Seo Y, Mok, H,Hwang H: The analysis of field measurement of DTV On channel repeater, Jounal of Broadcasting Engineering, Vol 7, No.1, Mar.(2002)10–20 (in Korean)
3. W Husak, H. Helm: Design and Construction of a Commercial DTV On-Channel Repeater, NAB Broadcasting Engineering Conference Proceedings, Apr. (2000).
4. Simon Haykin: Adaptive Filter Theory, Prentice Hall, (1996)

Algorithm for ABR Traffic Control and Formation Feedback Information

Malrey Lee[1], Dong-Ju Im[2], Young Keun Lee[3], Jae-deuk Lee [4], Suwon Lee[5], Keun Kwang Lee[6], and HeeJo Kang[7]

[1] School of Electronics & Information Engineering , Chonbuk National University, 664-14, DeokJin-dong, JeonJu, ChonBuk,, Korea, 561-756
mrlee@chonbuk.ac.kr
[2] Department of Multimedia, Yosu National University
[3] Department of Orthopedic Surgery, Chonbuk National University Hospital
yklee@chonbuk.ac.kr
[4] Chosun College of Science & Technology, Korea
jdlee@mail.chosun-c.ac.kr
[5] Kunsan National University
lsw91@kunsan.ac.kr
[6] Dept of Skin Beauty Art, Naju Collage
kklee@naju.ac.kr
[7] Division of Computer & Multimedia Contents Engineering, Mokwon University
hjkang@mokwon.ac.kr

Abstract. The feedback control method proposed in this paper predicts the queue length in the switch using the slope of queue length prediction function and queue length changes in time-series. The predicted congestion information is backward to the node. NLMS and neural network are used as the predictive control functions, and they are compared from performance on the queue length prediction. Simulation results show the efficiency of the proposed method compared to the feedback control method without the prediction. Therefore, we conclude that the efficient congestion and stability of the queue length controls are possible using the prediction scheme that can resolve the problems caused from the longer delays of the feedback information.

Keywords: multimedia communication, ABR traffic control, information prediction, neural network.

1 Introduction

ABR service should use an appropriate control for an unpredictable congestion due to a feature of data traffic. A feedback mechanism is used for a dynamic control of the transmission rate of each source to a present network state in order to guarantee the quality of a required service [1][2]. ABR service has been also devised for a fair distribution of an available bandwidth for ABR users. As it were, it should maintain a better packet loss rate and a fair share of given resources by an adaptive adjustment to a network state. In addition to ATM cell, the identity management cell having control information is called RM (Resource Management) cell in ATM network. The feedback mechanism for ABR service uses RM cell in order to provide traffic control information. This RM cell having a detailed description of control information is transmitted to a source, which adjusts a cell transmission rate suitable for a present network by using the information cell [2]. As a standard for a traffic control of ABR service, TM 4.0 has been approved in ATM forum, and the basic regulation has been established for ABR service parameter, RM cell structure, and the operation of a switch and a transmitting/receiving terminal. Many researches have been made for a traffic control of ABR service [2][3]. Especially, most studies of feedback congestion control schemes for ABR traffic control tend to focus on the control algorithms using a threshold of internal queue of ATM switch [2].

However, an effective control of a source traffic already transmitted before controlled would be impossible in the existing algorithms, because the transmission time of a backward RM delayed due to the congestion between a source and a destination [6][7][8]. Congestion at the switch can occur due to a control information delay, and thus a variation of queue length can also occur over time. A variation of queue length impedes an efficient ATM traffic control. A delay of feedback information transmission can be caused not only by a long physical transmission time of a network but also by network congestion.

This paper proposes a predictive control function and feedback algorithm improved for an even more effective traffic control than the algorithms [2] for a long feedback delay within a time-out period after the establishment of a dynamic connection. The algorithm implemented at a switch predicts a future value of queue length, sends a queue length of a switch to a source in advance, and prevents congestion. It also controls a variation of a queue length to the utmost. That is, it uses feedback information, as it increases or decreases a transmission rate of a source beforehand in a computation of a future queue length at a switch.

In order to predict a future queue length, it monitors periodically a cell input rate to a switch and a recent queue length. It adapts periodically a predictive function of a future queue length to an optimized value using NLMS (Normalized Least Mean Square) [4][5] and an optimized adaptation of a neural network [9]. A new transmission rate of a source is computed with a feedback algorithm using an existing threshold value and a predictive function of a future queue length. As a predictive function of a future queue length, NLMS method and an optimized adaptation of a neural network method predict a queue length using a linear function and non-linear function respectively. I studied a predictive control method of ABR traffic that was even more efficient through a simulation using the two methods described above. The section 2 explains NLMS, a neural network adaptive algorithm, and a feedback model which are proposed above. The section 3 presents a simulation environment and result.

2 A Predictive Feedback Control Model

2.1 A Proposed Predictive Feedback Control Model

A proposed predictive feedback control model is presented in Fig. 1 above. N sources transmit packet data cells in a single switch, a cell transmission rate is constant, and a queue state is monitored regularly. It is assumed that a transmission delay time of packet data between a source and a switch is di, and that sources is added or deleted randomly for ABR service. A network state is specified in time n by Q(n) of a queue length at a switch node. For a given ABR traffic processing buffer, TH and TL show high and low thresholds respectively. A predictive control function computes a future queue length in response to time-series by a queue length. When a future predictive queue size exceeds the high threshold TH, the switch is considered to be in congestion and the switch computes the explicit rate (ER) at which sources have to send a backward RM cell to the switch in order to avoid a congestion. If it is less than the high threshold TH, however, a source changes its transmission rate in its computation of ACR (Available Cell Rate) by being informed of non-congestion situation instead of ER. The next section presents predictive control functions. One of them is a method predicting a future queue length using NLMS to adapt a linear function, and the other is the one using back propagation to adapt non-linear function in a neural network structure.

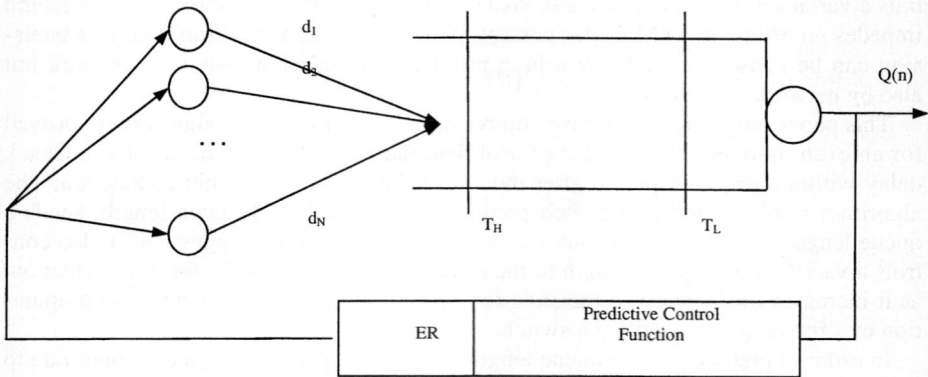

Fig. 1. Feedback predictive control model

2.2 A Predictive Control Function Using NLMS

NLMS control estimates buffer size in the next k steps using a linear function with a current value of the buffer size and weighting factor (slope) at time n. Let Q(n) denote the buffer size at time n. The k-step predictor is formulated such that the buffer size at k steps in the future is estimated from the Q(n), as given by

$$Q(n+k) = a^k(n)Q(n) \qquad (1)$$

where $a(n)$ is an estimated weighting factor at time instant n, and $k = 1,2,3,\ldots,t$ and t is a maximum prediction interval. Error of the prediction at time n is

$$e(n) = Q(n) - \hat{Q}(n) \qquad (2)$$

where

$$\hat{Q}(n) = a(n-1)Q(n-1) \qquad (3)$$

The prediction scheme uses the error to modify the weighting factor whenever the error is available at each time step. Furthermore, the weighting factor $a(n)$ is affected in time as sources are added or removed and as the activity levels of source changes. We thus put the problem into the one of estimating the weighting factor and use the normalized least mean square error (NLMS) linear prediction algorithm. Given an initial value for $a(0) = 0$, the weighting factors are updated by

$$a(n) = a(n-1) + \frac{\mu e(n) Q(n-1)}{|Q(n-1)|^2}$$

(4)

where μ is a constant. If $Q(n)$ is stationary, $a(n)$ is known to converge in the mean squared sense to the optimal solution [1][4][5]. The NLMS is known to be less sensitive to the factor μ. The estimated weighting factor $a(n)$ in each time-step will be used to predict the buffer size $\hat{Q}(n)$. Therefore each time step, the weighting factor indicates the direction of evolution of the functions for buffer size increases/decreases in term of recent residual $e(n)$ computed by the estimated buffer size $\hat{Q}(n)$ and actual buffer sizes $Q(n)$.

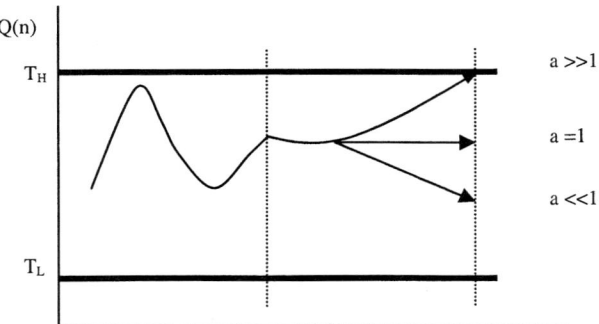

Fig. 2. Changed graph in applying a predictive

Fig. 2 presents a predictive scheme described in this paper [6]. If $a \gg 1$, a predictive queue length increased by the expression $Q(n+k) = a^k(n)Q(n)$. Therefore k, a time to hit the T_H, is predicted at time n using $Q(n)$ and $a(n)$ which are clearly known at time n [6].

2.3 A Predictive Control Function of Neural Network Using BP

A non-linear predictive function using neural network adjusts to predict a optimized value using BP algorithm [9]. It computes optimized variables of a non-linear equation (sigmoid) included in neural network nodes, and adjusts to get minimal errors to be occurred in a predictive value. That is, as in Fig. 3, BP is a kind of delta learning method to adjust adaptively the degree of a connection in order to minimize the differential error between required output and predictive output. Input layer x_i got continuously changing queue length $Q(n), Q(n-1), \cdots, Q(n-m-1)$ in time units, and output layer got a predictive value of queue length $Q(n+k)$ after n+k.

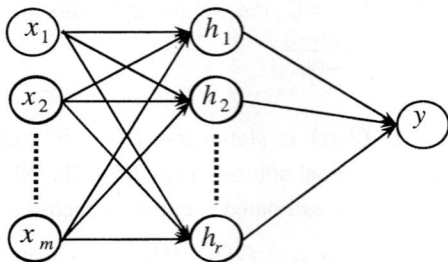

Fig. 3. Multi-layer Neural Network Structure

To explain back propagation in Fig. 3, let input layer pattern vector **x**, hidden layer output vector **z**, output layer output vector **y**, and connection degree **v** and **w**. A learning pattern is repeated as follows; **z** and **y** are computed using sigmoid function as an active one, output layer error signal δ_y and hidden layer error signal δ_z are computed using **z** and **y**, and connection degree w^{l+k} between hidden layer and output one, and connection degree v^{l+k} between input layer and hidden one, at *l+1* step, are computed in following expression

$$w^{l+1} = w^l + \Delta w^l = w^l + \alpha \delta_y z^l$$
$$v^{l+1} = v^l + \Delta v^l = v^l + \alpha \delta_z x^l$$
(5)

A case using neural network as in using NLMS also predicts a future queue length through monitoring queue length at a switch. However, the case is more complicated than the case of NLMS, because a weighted value for each connection link should be computed in advanced for optimal adaptation. The detailed computation processing of BP algorithm is consulted in Reference [9].

2.4 A Feedback Algorithm Using a Predictive Value

A feedback algorithm is explained for an implementation of ABR feedback control using a predictive control function. A predictive queue length is computed using a predictive control function with consulting high and low thresholds and with monitoring present queue length at ATM switch. If a predictive value is over high threshold, it sends minimal cell transmission rate to each source in advance after its computation as congestion is impending. By performing a prediction, it prevents congestion due to cell inflow from sources having long transmission delay of feedback information.

In Fig. 4, at congestion, if all the sources get cell transmission rate computed at ATM switch through RM cell, the cell transmission rate cannot be over ER. ACR is the next cell transmission rate computed at each source, in case congestion does not occur at the switch (it is not specified at RM cell).

[1] : Initialize ABR parameter
   FS = Link_speed_Switch / Number_sources
   Interval ← initial value
[2] : For n do [3-4] until max n
[3] : IF n=1 *Initialize training parameter*
   ELSE *NLMS or BP algorithm*
[4] : IF
   Congestion=1
   $ER \leftarrow FS \times ERF$
   $ACR \leftarrow min[ER, ACR(1-RDF)]$
   $\hat{Q}(n+k) \geq Q_H$
   ELSE IF
   and
   $ACR \leftarrow min[ER, PCR, ACR+RIF \times PCR]$
   goto [2]
[5] : Stop condition

$$\hat{Q}(n+k) < Q_H$$
$$\hat{Q}(n+k) > Q_L$$

Fig. 4. Feedback algorithm using a predictive function

3 Simulation

3.1 Simulation Environment

As in Fig. 5, the simulation model of a control algorithm presented in this paper is that the link speed of the switch is set to 150 Mbps, and the link speed for each source to the switch is set to 150 Mbps/N for N sources. The control algorithm is experimented in Visual C++ for a single bottleneck switch with a buffer of high and low thresholds of 5000 and 1000 cells respectively.

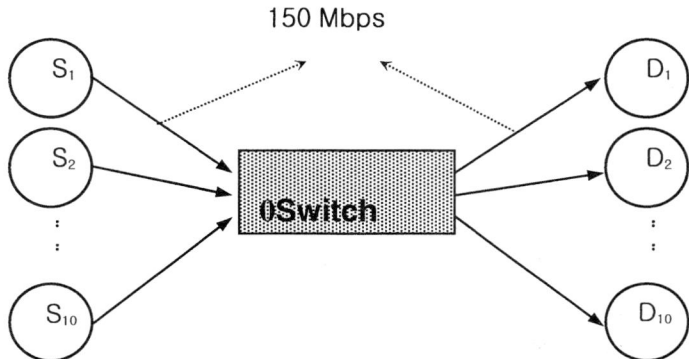

Fig. 5. Simulation Model

Following parameters are used for the simulation: peak cell rate (PCR) is set to 150 Mbps, additive increase rate (AIR) is set to 0.1 Mbps, and explicit reduction factor (ERF) is defined to 4/5. Ten active sources with various packet data cell generation times are used for simulation. In order to examine the transitional behaviors, an abrupt change of active sources is made. Initially, sources with cell generation times at {2, 4, 6, 9, 11, 13, 16, 20, 21, 23} are active, which the numbers represents the time-delay d_i from current time-unit n at the switch to the sources. At time-unit 1000, sources with time-delay d_i {14, 16, 17, 19, 20, 22, 23, 26, 28, 30} are active, which it includes the active sources with long delays. Two cases are compared in terms of stabilization and congestion of queue length at the switch through the change of transmission delay. The first case uses only feedback control method, and the second one does feedback predictive control method.

3.2 Simulation Results

Fig. 6 presents the change of queue length size at each switch, one of which uses a feedback predictive control algorithm using NLMS proposed in this paper, and the other of which uses only a feedback control one. A predictive interval(k) for NLMS was 10. Fig 6 presents that A feedback control algorithm only always brings about a congestion, and that a variation of queue length is considerably severe. It also shows that a variation of the length size Q(n) is severe after time 1000. It means that the sources with much longer delay time than other ones are incoming at the same time.

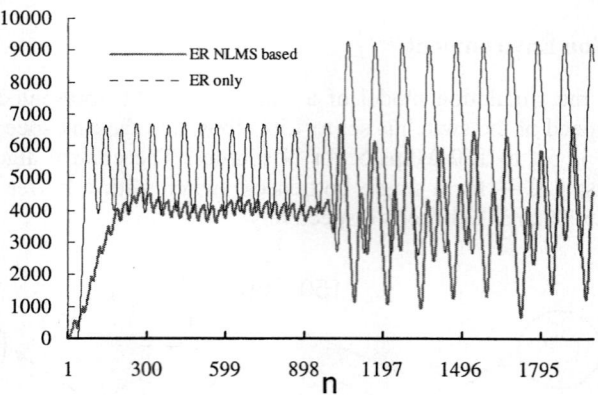

Fig. 6. Comparison of Q(n) for ER only and ER with NLMS

Fig 6 shows that NLMS feedback predictive control algorithm presents no variation of queue length before time unit 1000, close to high and low threshold, compared with a feedback control one only. After time unit 1000, however, as the sources with much longer delay time than other ones are incoming at the same time, the variation occurs more severely than before 1000 even with the predictive control method, and cases exceeding over high and low threshold also occur. The reason is that feedback delay of transmission sources is longer from time unit 1000, and that as a worst condition any traffic does not occur during time delay 1 through 13. That is, a predictive control function responds inappropriately to constant long-term interval, or sudden and random change. However, It is concluded that a predictive control algorithm caused a stability of the change and not severe congestion during simulation, compared with a feedback control one only. The use of neural network structure brought about similar results. It also responds inappropriately after n=1000.

Neural network structure needs a training using long-term BP, compared with NLMS, to adapt non-linear predictive control function. As in Fig. 7, a variation of queue length can be stabilized by a rapid drop of error rate. A systematic establishment for neural network structure should be preceded [9]. The decision on how many nodes of hidden layer are needed is required. In order to solve the problem about an inappropriate response of a predictive control function to occur after n=1000, a following method was tested. In the algorithm proposed in Fig. 4, increase rate control computation was used with constant instead of linear increase in ACR computation when predicting normal queue state. Therefore, traffic occurrence could not be detected, and a constant was increased in case normal queue state predicted. A simulation result for it is presented in Fig. 8. A change of queue length occurred when neural network structure had input node 10, hidden layer node 10, and k=10. NLMS predictive control function also had k=10.

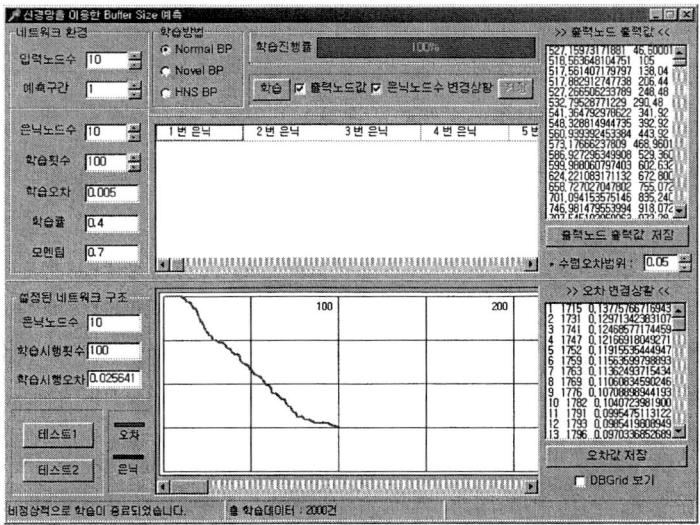

Fig. 7. Buffer size prediction learning result using BP

For comparison in Fig. 8, NLMS presents the most severe change of queue length, and normal predictive case of queue state is the result of using linear increase method. The other two graphs of queue length change represent the use of NLMS and neural network respectively as a predictive control function, and a constant increase method is used in normal predictive state of queue length. The figure shows a stability of queue length and non-congestion. NLMS represents a traffic control result similar to neural network.

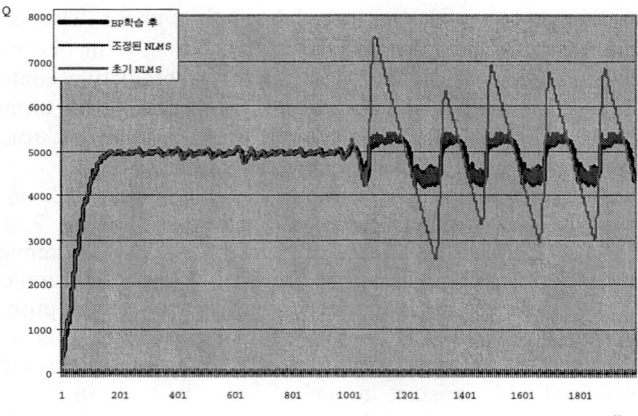

Fig. 8. Simulation result using constant increase

4 Conclusion

This paper studied that congestion at switch was predicted in advance and thus traffic could be controlled. Also, by making an active use of the result of predicted queue length as feedback control information, the sources could be informed of prompt and precise congestion situation. A predictive algorithm based on NLMS prediction scheme estimated the buffer size in the next k steps by using NLMS, and the control algorithm based on ER algorithm was applied. The case of NLMS predictive algorithm used proved to be effective.

Neural network structure also proved to be effective in controlling a congestion and queue length variation as in case NLMS predictive control function. In order to apply neural network, the establishment of an optimal neural network structure should be preceded, and many variables in it requires much more computation time needed for training than in NLMS. Therefore, it is not suitable for ATM switch requiring a real-time processing.

However, the experimental comparison of a control algorithm based on a predictive algorithm with the other ER control algorithm is different in terms of the establishment of the input variables. Thus, it is difficult to assert that the algorithm presented in this paper proves to be effective by a simple comparison. The simulation in a different environment is required for the experimental verification of its effectiveness.

References

[1] Adas, A., "Supporting Real Time VBR Video Using Dynamic Reservation Based on Linear Prediction," Infocom 1999.
[2] ATM Forum, Traffic Management v.4.0 Aug. 1996.
[3] Black, Uyless., ATM Volume I Foundation for Broadband Networks, New Jersey: Prentice Hall PTR, Feb. 1999.

Interference-Free ZCD-UWB for Wireless Home Network Applications

Jaesang Cha[1], Kyungsup Kwak[2], Sangyule Choi[3], Taihoon Kim[4], Changdae Yoon[5], and Chonghyun Lee[6]

[1] Dept. of Information and Communication Eng. Seokyeong Univ.
Seoul, Korea
chajs@skuniv.ac.kr
[2] UWB Wireless Communications Research Center(UWB-ITRC),
Inha Univ. Incheon, Korea
[3] Dept.of Electronic Engineering , Induk Institute of Technology,
Seoul,Korea
[4] San 7, Geoyeou-dong, Songpa-gu, Seoul, Korea
[5] Dept. of Elect Eng. Sungkyunkwan Univ. Suwon, Korea
[6] Dept. of Electronic Eng. Seokyeong Univ. Seoul, Korea

Abstract. Interference-free ZCD-UWB for Wireless Home Network Applications(WHNA) are presented in this paper. Interference-free ZCD-UWB could be a very useful solution for an enhanced wireless home network applications having interference-cancellation property without employing complicated multi user detection (MUD) or other interference cancellation techniques.

1 Introduction

Recently, UWB technique has been paid much attention in the high rate WHN and been debated by IEEE 801.15.3a[1] for standardization. A code domain UWB scheme of DS-UWB[2] has been proposed and studied by many researchers. In the DS-UWB, since the short pulse signal is modulated by phase shift keying (PSK) and spreaded with spreading code, BER performance of DS-UWB can be largely affected by the orthogonal characteristic of the spreading codes. Furthermore, system performance of DS-UWB based multiple access system are determined by multiple access interference (MAI) and multi-path interference (MPI) environment. In order to solve these MPI and MAI problems without adopting complicated MUD scheme or other interference cancellation schemes, we proposed ZCD-UWB defined as the DS-based UWB system using enhanced ZCD spreading codes. The conventional ZCD[3]-[13] codes are not suitable for the high data rate transmission due to their low code capacity and narrowband environments. Thus we proposed the construction methods of a class of enhanced ternary ZCD codes and applied them to the DS-UWB system for WHNA. Using system analysis and computer simulation, we also certified the interference cancellation property and high system capacity of the proposed ZCD-UWB system. To verify the system performance, we estimate the BER performances using computer simulation under various conditions. The results show that performance of the proposed systems are superior to the systems of the conventional scheme and have interference-cancellation property without employing complicated MUD or other interference cancellation techniques.

2   System Model of ZCD-UWB

We define ZCD-UWB as the DS-based UWB system using ZCD spreading code. In this section, we consider ZCD-UWB system by assuming antipodal modulation for transmitted binary symbols. Then UWB transmitted waveform of ZCD-UWB is defined at the following:

$$s^k(t) = \sum_{i=-\infty}^{\infty} \sum_{n=0}^{N_r-1} \sqrt{P_k} b_i^k a_n^k z(t - iT_b - nT_c) \quad (1)$$

where, Nr is the period of spreading code, $b_i^k \in \{\pm 1\}$ are the modulated data symbols for the kth user, $a_n^k \in \{\pm 1\}$ are the spreading code for the kth user, $z(t)$ is the sinusoidal waveform or transmitted pulse waveform, T_b is the bit period and, T_c is the chip period. If we assume a UWB system using sinusoidal carrier system, z(t) becomes a sinusoidal signal. Then if we assume the UWB system be a no-carrier system, z(t) be a short pulse. For the no-carrier system, z(t) is transformed in the receiver as a $w(t)$ which include the differential effects in the transmitter and receiver antenna systems. A typical pulse employed in the literature [1],[2] is the second derivative of a Gaussian pulse given by

$$w(t) = \left[1 - 4\pi\left(\frac{t}{T_m}\right)\right] \exp\left[-2\pi\left(\frac{t}{T_m}\right)^2\right] \quad (2)$$

Where the T_m is the pulse period.

In the receiver, for simplicity, we assume that the multi-path components arrives at the some integer multiple of a minimum path resolution time. By assuming the minimum path resolution time Tm (Tm ~ 1/Bs), we can write the received waveform as follows:

$$r(t) = \sum_{l=0}^{L-1} c_l^0 s^0(t - lT_m - \tau^o) + \sum_{k=1}^{K} \sqrt{P_k} \sum_{l=0}^{L-1} c_l^k s^k(t - lT_m - \tau^k) + n(t) \quad (3)$$

Where L is the number of multi-paths, c_l^k is the amplitude of the lth path, and $n(t)$ is the additive white Gaussian noise. The multi-path delay is described as $\tau^k = q_k T_m$, q_k being an integer uniformly distributed in the interval [0, NrNc-1] and Nr corresponds to the process gain of spreading code and $Nc = Tc/Tm$, which results in $0 \le \tau^k < T_r$. Here note that Tr is the maximum time delay to be considered. For the detection of received signal, we use code matched filter rather than pulse matched filter[2]. Then we could form the vector of sufficient statistics **y** obtained by collecting the outputs of the K individual code matched filters over one symbol with the input $r(t)$, written as follows:

$$\mathbf{y} = \mathbf{RWCb} + \mathbf{n} \quad (4)$$

Where the cross-correlation matrix of normalized signature waveform vector is formed as

$$\mathbf{R} = \int_0^{Tr} \mathbf{d}(t)\mathbf{d}^H(t)dt \quad (5)$$

Here, **n** is a Gaussian zero-mean K-vector with covariance matrix equal to **R** and the **C** is a multi-channel matrix of Rayleigh random variables. The matrix of **C**, **W**, the vector **d** are given by

$$\mathbf{C} = \begin{bmatrix} c^0 & 0 & 0 & \cdots \\ 0 & c^1 & 0 & \cdots \\ 0 & 0 & \ddots & \cdots \\ \cdots & 0 & 0 & c^K \end{bmatrix}, \mathbf{W} = \begin{bmatrix} W^0 & 0 & 0 & \cdots \\ 0 & W^1 & 0 & \cdots \\ 0 & 0 & \ddots & \cdots \\ \cdots & 0 & 0 & W^K \end{bmatrix} \quad (6)$$

$$\mathbf{d}(t) = [\mathbf{a}_0^T(t) \quad \mathbf{a}_1^T(t) \quad \mathbf{a}_K^T(t)]^T \quad (7)$$

Where $\mathbf{W}^k = \sqrt{P_k}\mathbf{I}_L$, $\mathbf{c}_k = [c_0^k \quad c_1^k \quad \cdots \quad c_{L-1}^k]^T$ and

$$\mathbf{a}_k(t) = [a_0^k z(t) \quad a_1^k z(t - T_c) \quad \cdots \quad a_{Nr}^k z(t - (Nr-1)T_c)]^T$$

The element of the cross-correlation matrix R defined in equation (5) can be found as follows

$$\mathbf{R}_{i,j} = \beta\rho_{i,j} = E\{\tilde{\mathbf{d}}(t)\tilde{\mathbf{d}}^H(t)\} \tag{8}$$

Where β is defined as a variable coefficient decided by normalized process gain, i.e $\beta = 1$, for binary code, and $\beta \leq 1$, for ternary code, and $\tilde{\mathbf{d}}(t) = [\mathbf{p}_0^T(t) \quad \mathbf{p}_1^T(t) \cdots \mathbf{p}_k^T(t)]^T$

Here the waveform $\mathbf{p}_k(t)$ is given by

$$\mathbf{p}_k^T(t) = [\tilde{s}_k(t) \quad \tilde{s}_k(t-Tc) \quad \tilde{s}_k(t-LTc)]^T \tag{9}$$

Where $\tilde{s}_k(t) = \sum_{n=0}^{N_r-1} a_n^k z(t - nT_c)$

As seen in (5), if the ZCD property is maintained, the cross-correlations of any signals in ZCD are zeros. With the proposed ZCD-UWB system, the effective interference cancellation can be obtained in the MAI and MPI environments.

3 Analysis

In this section, we present the performance analysis of the proposed system. As a performance evaluation, we use theoretical bit-error-rate (BER) of the system by using the results in [14].

3.1 SUD Under Multiple Access Interference

The BER of the single user detection (SUD) is quite straightforward. By referring the theoretical BER of SUD, we rewrite BER of the kth user in AWGN channel as

$$P^k(\sigma) = \frac{1}{2} \sum_{e_1 \in \{-1,1\}} \cdots \sum_{\substack{e_j \in \{-1,1\} \\ j \neq k}} \cdots \sum_{e_k \in \{-1,1\}} Q(\frac{c_k}{\sigma} + \sum_{j \neq k} e_j \frac{c_k}{\sigma} \beta\rho_{jk}) \tag{10}$$

where $Q(x)$ is the complementary cumulative distribution function of the unit normal variable. When Rayleigh fading channel is considered for the UWB system using IF carrier signal, the BER can be written as

$$P^{Fk}(\sigma) = \frac{1}{2}(1 - \frac{c_k}{\sqrt{\sigma^2 + \sum_j c_j (\beta \rho_{jk})^2}}) \qquad (11)$$

Since the cross-correlations of our proposed scheme are zeros, the performance of proposed ZCD-UWB system is as good as single user case, or no MAI and thus the BER probabilities both in AWGN and Rayleigh fading can be written as follows:

$$P^k(\sigma) = \frac{1}{2}Q(\frac{c_k}{\sigma}) \qquad (12)$$

$$P^{Fk}(\sigma) = \frac{1}{2}(1 - \frac{c_k}{\sqrt{\sigma^2 + c_k}}) \qquad (13)$$

Since the Q(x) function is monotonically decreasing function with respect to x, the probabilities of the proposed system are always smaller than the conventional system using Walsh spreading codes which have some cross-correlation in the MAI and MPI environments.

3.2 MUD Under Multiple Access Interference

In this subsection, we consider the BER performance of our proposed ZCD-UWB system when decorrelator–based multi user detection (MUD) is applied to the system. As described in [14], the probability of error in AWGN can be written as

$$P^k_d(\sigma) = \frac{1}{2}Q\left(\frac{c_k}{\sigma}\sqrt{1 - \mathbf{r}_k^t \mathbf{R}_k^{-1} \mathbf{r}_k}\right) \qquad (14)$$

Where $\mathbf{a}_k$ is the kth column of $\mathbf{R}$ without the diagonal element, and $\mathbf{R}_k$ is the (KL-1) x (KL-1) matrix that by striking out the kth row and column from $\mathbf{R}$. With the property of ZCD of the proposed system, $\mathbf{a}_k^T \mathbf{R}^{-1} \mathbf{a}_k$ becomes zero and thus the error probability of MUD applied system becomes as follows:

$$P^k_d(\sigma) = \frac{1}{2}Q\left(\frac{c_k}{\sigma}\right) \qquad (15)$$

Which is the same as that of SUD system.

Similarly, the BER performance of MUD applied system under Rayleigh fading can be written as follows:

$$P^k{}_d(\sigma) = \frac{1}{2}\left(1 - \frac{1}{\left(\sqrt{1 - \frac{\sigma^2}{c_k{}^2} \frac{1}{(1 - \mathbf{r}_k^t \mathbf{R}_k^{-1} \mathbf{r}_k)}}\right)}\right) \tag{16}$$

$$= \frac{1}{2}\left(1 - \frac{c_k}{\sqrt{\sigma^2 + c_k}}\right)$$

which is the same as that of SUD system.

In term of BER performance, we compare the results of our proposed system when SUD and MUD schemes are applied to the system. Consequently, we can conclude that the proposed system eliminates the necessity of the MUD while keeping the same performance obtained with MUD.

When we consider the performance of our system under MPI condition, the analysis is quite straightforward. In short, the cross correlation matrix in the equation (10) ~ (16) becomes autocorrelation matrix in MPI. Thus, we can conclude that the proposed system is not affected by the MPI and shows the perfect interference cancellation performance under MAI and MPI.

4 Results

In this section, we present the BER performance of the proposed system via computer simulation based on the fading channel model with described in [14]. The BER performance is presented in Fig. 1 through Fig 4. . In Fig. 1, we present the BER performance under MPI condition by using SUD scheme. Here, we can observe that the

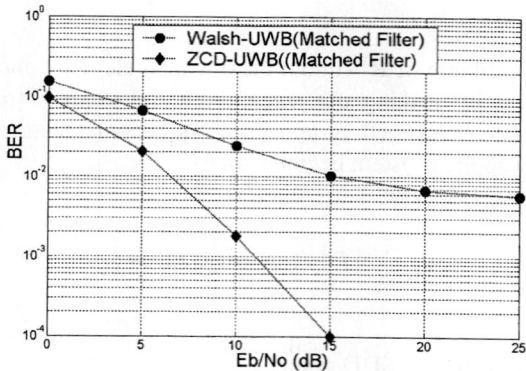

Fig. 1. Comparison of BER performance of ZCD-UWB and Walsh-UWB with SUD only. (Uplink with MAI and MPI, two user multi-path, $N_r = 64$, EGC with four Fingers)

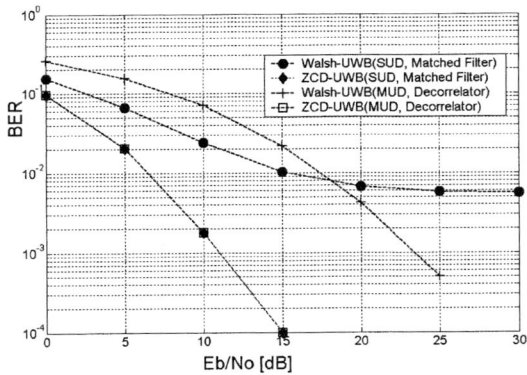

Fig. 2. Comparison of BER performance of ZCD-UWB and Walsh/DS-UWB with SUD and MUD. (MAI and MPI, 2 user multi-path, N_r =64, EGC with 4 Fingers)

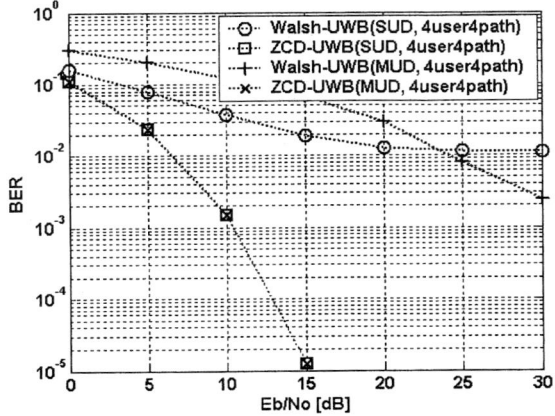

Fig. 3. Comparison of BER performance of ZCD-UWB, Walsh/DS-UWB, in the MAI and MPI environment with SUD and MUD (4user 4path, Uplink, Spreading Factor=64chips, Path delays for Each users=[0 1 2 4], [1 2 3 4], [0 1 3 2], [1 3 2 4])

proposed ZCD-UWB system exhibits better performance than that of Walsh-UWB system. Next, we present the BER performance under MPI and MAI conditions of two-users by using SUD and MUD scheme in Fig. 2. In the presence of MPI, ZCD-UWB shows robust performance compared with that of Walsh-Hadamard code-based UWB (i.e, Walsh-UWB) as the path number is increased. And we can observe that the proposed ZCD-UWB system exhibits better performance than that of Walsh-UWB

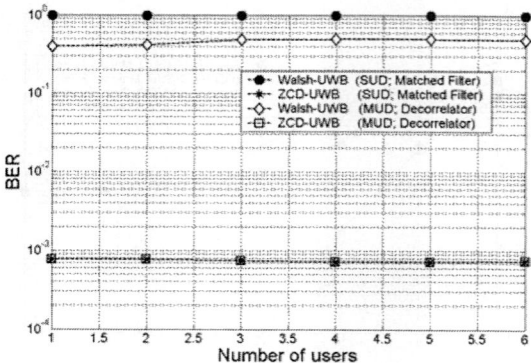

Fig. 4. BER performance of ZCD-UWB and Walsh-UWB. (The number of users varies from 1 to 6, where all users have the same received power.)

system. With the results in Fig. 2, we observed that the ZCD-UWB outperforms Walsh-UWB in the MAI and MPI environments. In Fig. 3, the results obtained using SUD and MUD under MAI of four users and MPI of four multi-paths are presented. We fix the Eb/No to 10 dB, number of delay path to 3, and plot the performance of various receivers versus the number of users in order to certify MAI cancellation property under the MAI conditions with no fading in Fig. 4. From the results in Fig. 4, we certified that ZCD-UWB system able to reject MAI compared to Walsh-UWB which have heavy MAI loads. It is noted that proposed ZCD-UWB exhibits same BER performance both in SUD and MUD conditions, which means MPI-cancellation can be attained without using complicated MUD scheme in ZCD-UWB as discussed in section 3.

5 ZCD-UWB Design Example for WHNA

We consider a ZCD-UWB system for WHNA with a MAI free Pico-cell. For Uplink under MAI environment, the time delay differences of signals arriving at access point (AP) within a Pico-cell are due to the different propagation delay between the mobile stations and AP. The implementation of MAI-cancelled system is corresponds to the construction of intra-cell without MAI. Thus, the MAI-free Pico-cell of WHN can be designed as

$$R = \frac{c \cdot \delta}{2} = \left(\frac{c \cdot (ZCD-1)}{(4 \cdot R_c)} \right) = \frac{c \cdot T_p \cdot PF \cdot (ZCD-1)}{4} \qquad (17)$$

where R is the radius of MAI-free Pico-cell, δ is the maximum propagation delay time in the cell, Rc is chip rate, c is the speed of light, T_p is the pulse width, and PF is defined as Pulse factor, i.e., Pulse number/Chip.

Next, we can implement MPI cancelled system by designing the ZCD-length covering the delay-path-length of MPI. In other word, if the delay-path-length of MPI is included in the 0.5×(ZCD-1) duration, MPI cancelled system can be obtained by using ZCD property. The system capacity proposed system could be represented by the peak bit rate as

$$Rb = \frac{Mr \cdot rT \cdot M \cdot B_N}{N_r \cdot T_p \cdot PF} \qquad (18)$$

Table 1. Specification Example for the ZCD-UWB system

	System A	System B	System B
Access, Duplex	Multi-band DS-UWB, TDD		
Given BW(bandwidth)	Low band(3.15 to 5GHz) + High band(5.825 to 10.6GHz) [2]		
BW/channel	500MHz plus Guard Band/channel		
Data modulation	BPSK		
Error correction coding	No		
Pulse width Tp	2 nsec		
Rake combining	EGC or MRC		
Network, Cell radius R	Pico-net, within 10m		
BW number for Ch. B_N	14 (4Ch./Low band + 10Ch./High band)		
Mono pulse type	Gaussian pulse with 2 nsec		
Pulse Factor	1		
Chip rate R_C	500Mcps		
Spreading sequence	Enhanced ZCD preferred pair		
	$N_r = 12$ ZCD = 11	$N_r = 32$ ZCD = 31	$N_r = 72$ ZCD = 71
Receptible Time to the Delay path / cell radius	10nsec / 1.5m	30nsec/ 4.5m	70nsec/ 10.5m
Peak Bit rate R_b	700Mbps	263 Mbps	117 Mbps

> ZCD : Zero correlation duration, δ : Maximum propagation delay time Gc: Guard chip, M : Family size of sequence, Tp : Pulse width, PF : Pulse factor, i.e. Pulse number/Chip, Nr : Spreading factor = Sequence period N R_c : Chip rate, M_r : Mary-Phase level factor; i.e, M_r of BPSK case =1, rT : Time share ratio of TDD, e.g., 0.6, R: Cell radius, R_b: Peak bit rate, B_N : BW number for channelization.
>
> $Gc = (ZCD-1)/2, \; \delta = Gc/Rc, \; R = (c \cdot \delta)/2, \; R_c = 1/(T_p \cdot PF)$
>
> $R_b = (Mr \cdot rT \cdot M \cdot B_N)/(N_r \cdot T_p \cdot PF)$

And the terminology definition in (18) and the specification examples for ZCD-UWB systems are listed in the Table1 in detail. In the system examples of Table1, TDD-based Multi-band ZCD-UWB systems with ZCD property for MPI cancellation are considered. From the results of estimated system specification, we can verify the usability of the proposed system and thus the proposed system have ZCD capability and Peak bit rate of 117Mbps to 700Mbps using simple BPSK modulation within the WHN Pico-net area.

6 Conclusion

In this paper, we propose a ZCD-UWB system for WHNA with interference cancellation property and high system capacity. After constructing an enhanced PG-corrected ternary ZCD spreading codes for ZCD-UWB, we built a mathematical model for ZCD-UWD wireless system including new spreading codes. To verify the system performance, we estimate the BER performances using computer simulation under various conditions. The results show that performance of the proposed systems are superior to the systems of the conventional scheme and have interference-cancellation property without employing complicated MUD or other interference cancellation techniques.

Acknowledgment

This work was supported in part by University IT Research Center Project (INHA UWB-ITRC) in Korea.

References

1. http://www.ieee802.org/15/pub/TG3.html
2. J. Foerster, "The Performance of a Direct-Sequence Spread ULTRA- Ultra-Wideband System in the Presence of Multipath, Narrowband Interference, and Multiuser Interference," IEEE UWBST Conference Proceedings, May, 2002

3. Cha, J.S., Kameda, S., Yokoyama, M., Nakase, H., Masu, K., and Tsubouchi, K.: 'New binary sequences with zero-correlation duration for approximately synchronized CDMA '. Electron. Lett., 2000, Vol. 36, no.11, pp.991–993
4. Cha,J.S.,Kameda,S.,Takahashi,K.,Yokoyama,M., Suehiro,N.,Masu,K.and Tsubouchi, K, "Proposal and Implementation of Approximately synchronized CDMA system using novel biphase sequences", Proc. IEICE ITC-CSCC 99, Vol. 1, pp.56-59, Sado Island, Japan, July13-15, 1999.
5. Fan, P, Suehiro, N., Kuroyanagi, N and Deng, X.M.: "Class of binary sequences with zero correlation zone," Electron. Lett., 1999, Vol. 35, no.10, pp. 777–779
6. Deng, X., and Fan, P.: "Spreading sequence sets with zero correlation zone," Electron. Lett., 2000, Vol. 36, no.11, pp. 993–994
7. Cha,J.S. and Tsubouchi, K, "Novel binary ZCD sequences for approximately synchronized CDMA", Proc. IEEE 3G Wireless01, Sanfransisco, USA, Vol. 1, pp.810-813, May 29, 2001.
8. Cha,J.S, "Class of ternary spreading sequences with zero correlation duration", IEE Electronics Letters , Vol. 36, no.11, pp. 991-993, 2001.5.10
9. Cha,J.S. and Tsubouchi, K, "New ternary spreading codes with with zero-correlation duration for approximately synchronized CDMA", Proc. IEEE ISIE 2001, Pusan, Korea, Vol. 1, pp.312-317, June 12, 2001.
10. Cha,J.S., Song,S.I.,Lee,S.Y.,Kyeong,M.G. and Tsubouchi, K, "A class of Zero-padded spreading sequences for MAI cancellation in the DS-CDMA systems", Proc. IEEE VTC01 Fall, Atlantic City, USA,October 6, 2001.
11. Jae-sang Cha, Sang-yule Choi, Jong-wan Seo, Seung-youn Lee, and Myung-chul Shin, "Novel Ternary ZCD Codes With Enhanced ZCD Property and Power-efficient MF Implementation", Proc. IEEE ISCE'02, Erfurt, Germany, Vol. 1, pp.F117-122, 2002.
12. H .Donelan and T.O"Farrell, "Families of ternary sequences with aperiodic zero correlation zones for MC-DS-CDMA", IEE Electronics Letters , Vol. 38, no.25, pp. 1660-1661, 2002.12.05
13. Mun Geon Kyeong, Suwon park, Jae Kyun Kwon, Dan Keun Sung, and Jae-sang Cha, "3G Enhancements with a view towards 4G", Tutorial. CIC 2002, Seoul, Korea, Tutorial-3, pp.91~245,2002.
14. Sergio Verdu, Multiuser Detection, Cambridge university press, 1998

Appendix

New Ternary ZCD Codes Enlarged ZCD and Various Sequence Period

1) Type1: Ternary ZCD Code Having (N-1) Chip-ZCD and Perfect Periodic CCF Property

A TPP $\{A_{Nr}^{(a)}, A_{Nr}^{(b)}\}$ is generated for the period of N_r chip as

$$\{A_N^{(a)}, A_N^{(b)}\} = \{\pm Pa \times (+Z_i - Z_i), \pm Pa \times (+Z_i + Z_i)\}.$$

where $\pm Pa$ means $+Pa$ or $-Pa$, +, -, Z_i and pa denote 1, -1, inserted i zeros, and PG correction factor of 1 to $\sqrt{i+1}$.

By using accurate PG correction factor, $pa = \sqrt{i+1}$, PG of the proposed sequence could be 1. The sequence period, i.e., spreading factor N_r, positive integer i and maximum ZCD denoted as ZCD_{max} have the relations as

$$N_r = 2 \cdot (i+1) \quad (i=1,2,3\cdots)$$

$$ZCD_{max} = N_r - 1 = 1 + 2 \cdot i$$

Aperiodic ZCD = Periodic ZCD = (N_r-1) chips

$$Periodic\ R_{x,y}(\tau) = 0,\ \forall \tau$$

By using the chip shift operation [1] to $\{A_{Nr}^{(a)}, A_{Nr}^{(b)}\}$, Enhanced ternary ZCD set with various ZCD and family size can be constructed.

2) Type 2: Ternary ZCD Code Having (N-1) or (N-3) Chip-ZCD and Perfect Periodic ACF Property

A TPP $\{C_N^{(a)}, C_N^{(b)}\}$ with a period of $N_r = N = 2 \cdot (i+2)$ is constructed as

$$\{\pm Pc \times (++Z_i +-Z_i),\ \pm Pc \times (+-Z_i ++Z_i)\} \quad \text{or}$$

$$\{\pm Pc \times (++Z_i -+Z_i),\ \pm Pc \times (+-Z_i --Z_i)\} \quad \text{or}$$

$$\{\pm Pc \times (--Z_i -+Z_i),\ \pm Pc \times (-+Z_i --Z_i)\} \quad \text{or}$$

$$\{\pm Pc \times (--Z_i +-Z_i),\ \pm Pc \times (-+Z_i ++Z_i)\}$$

where $\pm Pc$ means $+Pc$ or $-Pc$, Pc is a PG correction factor which have the values of 1 to $\sqrt{0.5i+1}$. By applying $Pc = \sqrt{0.5i+1}$ to $\{C_N^{(a)}, C_N^{(b)}\}$, PG of proposed TPP could be 1. Then, using the chip shift operation to $\{C_N^{(a)}, C_N^{(b)}\}$, ternary ZCD set with various ZCD chips and a various family sizes can be constructed. A class of new Type-2 codes takes the following properties

$$\text{Aperiodic ZCD} = (N_r - 3) \text{ chips}$$

$$\text{Peirodic ZCD} = (N_r - 1) \text{ chips}$$

$$Periodic\ R_{x,x}(\tau) = 0, \qquad \tau \neq 0$$

Safe Authentication Method for Security Communication in Ubiquitous

Hoon Ko[1], Bangyong Sohn[1], Hayoung Park[2], and Yongtae Shin[2]

[1] Department of Computer Engineering Daejin University,
Sundan-dong, Pocheon-Si, 487-711, Korea
{skoh21, bysohn}@daejin.ac.kr
[2] Department of Computer Science Soongsil University,
Dongjak-gu, Seoul, 156-743, Korea
hayoung13@hotmail.com, shin@comp.ssu.ac.kr

Abstract. Ubiquitous computing environment has a lot of different things as for applying existing security technical. It needs authentication method which is different kinks of confidence level or which satisfies for privacy of user's position. Using range localizes appoint workstation or it uses assumption which is satify environment of client in Kerberos authentication method which is representation of existing authentication method but it needs new security mechanism because it is difficult to offer the condition in ubiquitous computing environment. This paper want to prove the result which is authentication method for user authentication and offering security which are using wireless certificate from experiment in ubiquitous environment. Then I propose method which is offering security and authentication in ubiquitous environment.

1 Introduction

Ubiquitous period is coming from developing network and wireless communication. Security problem of Ubiquitous-Computing-Environment is more complicate than Security problem of internet-period and easier attracting but counter-plan- establishing is rudimentary level in present.

Ubiquitous-Computing-Environment is that computer or different kind of devices comes into daily life so user has help from those between themselves can't recognize about it. The application should be efficiently used available resource and service for supporting user activity without intervention of user.

Essential elementary of Ubiquitous-Computing-Realizing with context- recognition is security. Resource and service in Ubiquitous-Computing-Environment can be in area where performs application but most of time it exists in physically distributed environment. So ubiquitous-computing-environment has to necessarily consider about resource in distributed environment and security about service besides security in area where performs application. After all, Security-Technology of Ubiquitous-Environment has many kind of different things to apply security-technology of existence. It needs authentication method which is different kinks of confidence level or which satisfies for privacy of user's position. Security- service of ubiquitous is confidentiality, integrity, availability, anonymity, non-repudiation and so on.

In this research, it should be transmitted by using specification value for resource in ubiquitous computing and for giving user-certification. And we should decide whether using- permission of service for judging integrity of right-value.

We explain about treat-method and Requestment of Security in section , and explain existence method in section 3. And in section 4, we explain about safe authentication method which proposed in ubiquitous environment in this paper. In section 5, We analyse safety which wrote about Treat-Method in section 2.

2 Treat Method and Demanding Items of Security

In this chapter, we write about treat-method and demanding item of security in ubiquitous computing environment[3].

Main mean of attack which happens in ubiquitous computing, follows below.

- Eavesdropping: Attacker(invader) can hear without big effort because communication method which is between tag and leader, is wireless.
- Traffic Analysis: Attacker(invader) analyse contents which get from eavesdropping and then can predicts tag's answer which is about leader's inquiry.
- Location Tracking: Attacker(invader) or leader who is sinister, perceives position of tag . So it is a type which disturbs user's privacy by method which grips moving path of tag- owner.
- Spoofing: It is a method which passes the authentication-process that individual which is unfair, deceives like to fair.
- Message loss: It can lose a part of communication method which reciprocates between tag and leader, cause by intention of attacker or error of system .
- Denial of Service: It is a mean of attack which emits obstructive wave having special frequency for normally not to operate.
- Physical Attack: It gets rid of temper-defense-package in chip and then explains main information to put on prove on IC(Integrated Circuit) chip. We analyze electron-wave which emits from attacking prove, communication devices and computer .

But it is weak about TEM PEST attacking which can eavesdrop contents, which is transmitting between those. Treat method which explained above, is safely available by authentication processing but case of physical attack method is unavaliable cause by character of system. Contents which explains belows, is consideration-item for designing authentication system in ubiquitous environment.

- User doesn't have to think about security system how to operate and is not disturbed by security system in ubiquitous computing environment.
- Security structure should be able to offer difference kinds of security level accoding to system police , context information, environment condition and time condition.
- It should offers dynamic security with combining context.

- Security system should be flexible and be able to adapt and fit by special demand.
- User who hasn't received authentication about contents in communication, should be able to look at contents.
- Database, leader and individual should not offer any information which can grips moving path of tag, for preventing position cracking.
- It should process authentication about mate's inquiry for safe by spoofing attack
- It should minimize calculation-quantity and storage-space in needed authentication cause by character of ubiquitous-computing.

When this kinds of demanding item of security design, it reflect that it can come true safely ubiquitous-computing from treat-method which explained above.

3 Previous Authentication Method

There is an existing hash based authentication method[Fig 1] which is proposed by Henrici and Muller[4]. This method is a protocol which prevents location tracking by updating ID based on hash. Manufacturer constructs database which can save h(ID), ID, TID, LST, AE and save ID, TID and LST in TAG. The TAG which received query increases 1 of TID and calculates h(ID), T=h(TID xor ID), △TID and transmits to READER. The database searches ID with h(ID) and calculates T' which is added pertinent TID to △TID.

In [Fig 1], If T and T' are same in Behaving of (3), Database calculates and transmits Q and xor calculates randomly generated R for updating ID. The tag which received (5) also calculates Q' and compares with Q. If both Q' and Q are same, the tag updates its ID. AE is designed safe from errors in system or losing messages by attacker. Because AE has previous ID information.

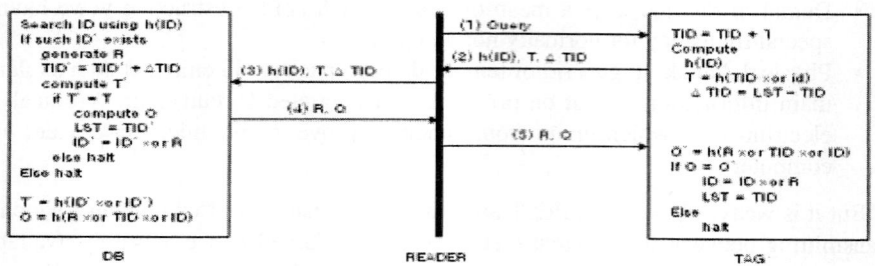

Fig. 1. Authentication Protocol based on Hash

This method looks safe from location tracking attack because of ID is updated when authentication process is over. In case of abnormal authentication processing between TAG and database that is attacker send query to TAG for attack, The attacker can do location tracking attack to TAG because The TAG always replies corresponding h(ID).

Also, It is not safe from spoofing attack. Attacker can gain (2) through query. If the TAG transmit (2) in session with database before opening normal authentication session, database is authenticated to attacker as normal TAG.

Attacker give R that continuous character string which consists of 0 in the value of (5) that READER transmits to TAG in the middle of session. And then, attacker transmits T instead of Q. Therefore, the TAG can't notice error. When next authentication processing, server can find existed ID with h(ID). However, there is an disadvantage that TAG can't receive authentication, existing ID about LST is not corresponded with saved TAG and database.

There is another method that READER generates random value S with Pseudo random number generator and query to TAG previously[5]. However, this method have an disadvantage. If the 3rd person send spoofing query to TAG as READER, the TAG can't notice normal user or not.

Of course, several advanced methods are proposed to solve the disadvantage . But they can't solve original problem.

4 Safe Authentication Method

We proposed new authentication method which is safe about spoofing attack and reducing hash time that 2 of hash function time reduce one in tag calculation time[Fig 2].

[Notation]

ID : Tag identification
h : Hash function
∥ : String of character connection calculation
xor : Exclusive or calculation
R : Random value

This method is similar to ID transformation protocol based on advanced hash[5].

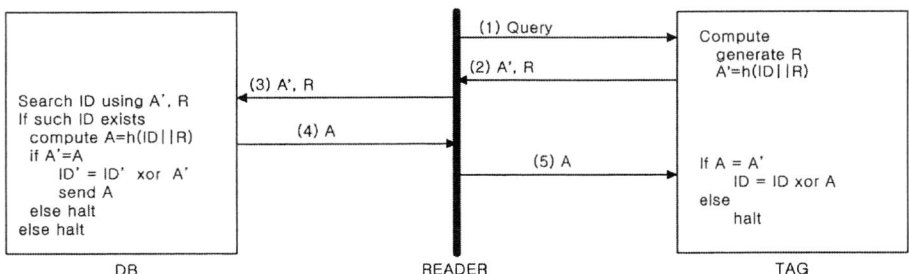

Fig. 2. Authentication Method

However, READER who is proposed [5], transmits for authentication by creating specific value is weak about man in middle attack above figure.

So, main that creating random value R, changed to TAG which is not to READER.

First, TAG makes random value R from pseudo random number generator and then, creation A'=h(ID‖R) and then, A' and R transmit to DB through READER. DB is searching ID through A' and R and creating A=h(ID‖R). DB compares A with A'. If both are same, DB authenticate right TAG and updating with XOR calculating ID to A'. And then transmit it to TAG. TAG compares A with A'. If both are same, TAG calculates ID=ID xor A.

The authentication process is over as like above order.

[STEP 1] *query to TAG;*   /* Reader */
[STEP 2] *generate R;*   /* Tag */
         compute A'=h(ID ‖ R);
         send (A', R) to READER;
[STEP 3] *bypass to DB;*   /* Reader */
[STEP 4] *search ID using (A', R);* /* DB */
         if such ID exist
            compute A=h(ID ‖ R);
            if A'=A
               ID'=ID' xor A';
               send A to READER;
[STEP 5] *bypass to TAG;*   /* Reader */
[STEP 6] *if A=A' then ID=ID xor A;* /* Tag */

5 Safety Analysis

Proposed authentication method is safe about man in the middle attack problem that READER creating random value R. The 3rd person can not creat R. Because it use R that created by TAG. So, It is safe about spoofing attack which can occur in [4]. And It can efficient. Because, Reducing hash time that 2 of hash function time reduce one.

In this section, we explain safety of proposed method of explained risk in section.

Eavesdropping is an attack method which analyze acknowledgement which query to TAG. It compares creation value in TAG with transmitted hash value of ID and random value to DB. If both values are same, it process authentication. However, attacker can't know ID of TAG because attacker don't know this value. Location Tracking is knowing method that moving path of TAG owner. The random value is continuously change with moving TAG so attacker can't do location tracking.

Spoofing is an attack of authentication processing that invalid object spoof as valid object. However, It is impossible because don't know ID.

Table 1. Analysis of Authentication Method

Classification	Authentication method based on hash[4]	Advanced authentication method based on hash[5]	Expanded authentication method based on hash[3]	Proposed method
Safety of location tracking attack	×	×	○	△
Safety of spoofing attack	×	×	○	○
Possibility of message recovery	○	○	○	○
Using quantity of TAG memory(L)	3L	1L	2L	1L
Calculation time of TAG	3 times of hash	2 times of hash	3 times of hash	1 time of hash
Calculation quantity of Database	×	×	△	△
Possibility of processing of READER	×	○	○	×

6 Conclusion

It needs a lots of element for constructing ubiquitous computing. In this paper, we analyzed and explained about authentication method of RFID which is one of this kind of element. There were proposed a lot of existing authentication method and analyzed those shortcoming and then there were proposed method which solved the problem. Specially, in this paper, we designed position-privacy for recognizing weakness in Location Tracking and Spoofing of existing problem, and for and satisfying it. Also, we reduced Hash which is orderly processed, from twice or third to once for authentication.

Therefore, if uses proposed authentication method in this paper, we reduce each element's load. we are looking forward to realizing safely and effective ubiquitous authentication. But, if creates continuously random-value for receiving authentication , if authentication frequently happens, and then creating-random-value increases to compare with authentication-frequency-number. It can be increasing load of Tag. Therefore, we should research mechanism for solving this kinds of problem in the future.

References

1. Mark Weiser, "Some Computer Science Problems in Ubiquitous Computing," *Communications of the ACM*, July, 1993.
2. Mark Weiser, "Ubiquitous Computing," *Nikkei Electronics*, pp. 137-143, December, 1993.
3. Sungho Yoo, Kihyun Kim, Yongho Hwang and Piljoong Lee, H. "Satus-Based RFID Authentication Protocol," *Journal of The Korean Institute of Information Security and Cryptology*, Volume 14, Number 6, pp. 57-67, December 2004.

4. Dirk Henrici and Paul Muller, "Hash based enhancement of location privacy for radio frequency identification devices using varying identifiers," *PerSec'04*, pp. 149-153, March 2004.
5. Youngjoo Hwang, Misoo Lee, Donghoon Lee and Jongin Lim, "Low-Cost RFID Authentication Protocol on Ubiquatous." *CISC'S04*, pp. 120-122, June 2004.
6. Stephen Weis, Sanjay Sarma, Ronald Rivest and Daniel Engels, "Security and Privacy aspects of low-cost radio frequency identification system," *SPC'03*, pp.457-469, March 2003.
7. Alastair Beresford and Frank Stajano, "Location Privacy in Pervasive Computing," *IEEE Pervasive Computing 2003,* pp.46-55, 2003.
8. F. Stajano, "Security fot Ubiquitous Computing," *Halsted Press*, 2002.
9. M. Langheinrich, "Privacy by Design Principles of Privacy-Aware Ubiquitous System," *presented at ACM UbiComp 2001,* Atlanta, GA, 2001

Pre/Post Rake Receiver Design for Maximum SINR in MIMO Communication System

Chong Hyun Lee[1] and Jae Sang Cha[2]

[1] Department of Electronics,
Seokyeong University Seoul, Korea
`chonglee@skuniv.ac.kr`
[2] Department of Information and Communication Engineering,
Seokyeong University Seoul, Korea
`chajs@skuniv.ac.kr`

Abstract. In this paper, we present a receiver for maximum signal-to-interference plus noise ratio (SINR) for MIMO channel in DS/CDMA communications systems in frequency-selective fading environments. The proposed system utilizes full degree of freedom in FIR filters at both the transmitter and receiver antenna array. We develop at our system by attempting to find the optimal solution to a general MIMO antenna system. Also we derive a single user joint optimum scenario and a multiuser SINR enhancement scenario and prove that the existing algorithm is a special case of our method. In addition, we propose a system which uses zero correlation duration spreading sequence, with which superior performance is guaranteed over the existing algorithm. Extensive numerical results reveal that significant system performance and capacity improvement over conventional approaches are possible.

1 Introduction

In wireless systems, there are three major impairments caused by the radio channel and these are fading, delay spread, and co-channel interference. In order to achieve high-speed communications, high-quality communications or high-capacity communications, countermeasures should be employed to combat these impairments. One solution is the use of adaptive antenna array systems. In this paper, we present a multiple-input.multiple-output (MIMO) system with pre-Rake and Rake adaptive filters (FIR filters) to enhance signal-to-interference plus noise ratio (SINR) for DS/CDMA communications in the downlink for frequency-selective fading environments. Improved performance is demonstrated when compared to a conventional system with a Rake receiver at the mobile station (MS) and to a pre-Rake MRC transmit diversity system. Previous studies of adaptive transmit antenna arrays and adaptive receive antenna arrays for DS/CDMA systems show that they can mitigate multipath fading effects and suppress co-channel interference, thereby, increasing the capacity or improving system performance. However, transmit antenna arrays and receive antenna arrays have been mainly used at the base station (BS) only, and a single weight is used at each antenna.

In this paper, we consider a MIMO CDMA system similar to [1], which utilizes a transmit antenna array at the BS and a receive antenna array at the MS. However, our work is different in such way that our MIMO CDMA system operates the transmit array and receiver antenna array jointly for SINR enhancement by using full degree of freedom in channel information so that the algorithm in [1] is a special case of our algorithm. In the proposed system, each antenna adopts an adaptive finite-impulse response (FIR) filter, whose complexity is fairly low. In this paper, a single user joint optimum scenario and a multiuser SINR enhancement scenario are derived by trying to find the optimal solution to a MIMO antenna system. This paper is organized as follows. In Section II, the system model and preliminaries are introduced. Section III provides the derivation of the algorithm for maximum SINR based on SVD(Singular Value Decomposition). Comparative simulation results are then presented in Section IV. Finally, Section V concludes our work.

2 System Model

The configuration of our CDMA antenna system is shown in Fig. 1, where M_t transmit antennas are located at the BS and M_r receive antennas are located at the MS. At the BS, the signal transmitted by the mth antenna for user k passes through a L_t tap transmit transversal filter whose characteristics are described by the complex response vector $\mathbf{V}_m^k$ as shown in Fig. 2(a). Likewise, at the MS, the signal received by the it mth antenna of user it k passes through a tap receiver transversal filter whose characteristics are described by the complex response vector $\mathbf{U}_n^k$ as shown in Fig. 2(b). The weighed signals from all antennas are summed to form a scalar output. The tap weights $u_{n,l}^k$ and $v_{n,l}^k$ are complex quantities and the tapped delay time is T_c.

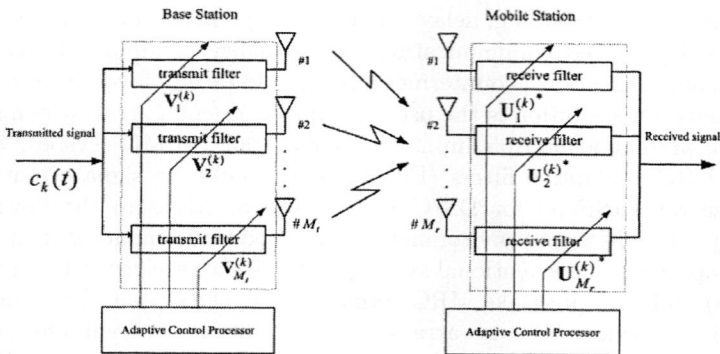

Fig. 1. General CDMA System Configuration

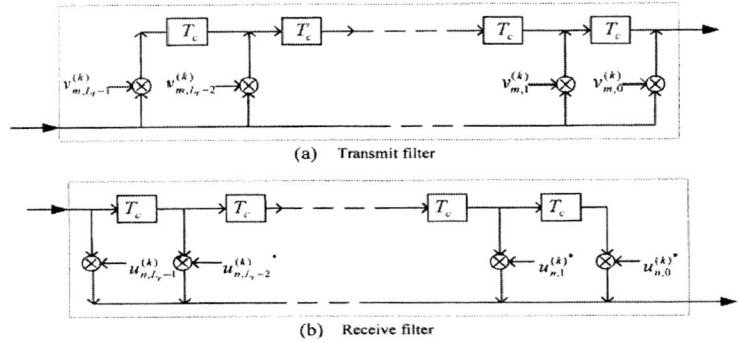

Fig. 2. Transmit and Receive Filter Structure

Then the transmitted baseband signal for user k can be written as

$$c(t) = \sqrt{P_k} b_k(t) a_k(t),$$

where P_k is the transmitted signal power, $b_k(t)$ is the binary data signal for user k and $a_k(t)$ is the PN sequence for user k with chip duration T_c and code length $N = T/Tc$.

The channel impulse response of user k is

$$h_{n,m}^k(t) = \sum_{l=0}^{L-1} h_{n,m,l}^k(t) \delta(t - lT_c),$$

where $L = \lfloor T_{max}/T_c \rfloor$ and it m and n refer to channel between nth receiver antenna at MS and the mth antenna at the BS. The overall channel matrix $\mathbf{H}^k$ size $M_r(L - L_t - 1) \times (L_t \times M_t)$ of user k can be expressed as

$$\mathbf{H}^k = \begin{bmatrix} \mathbf{H}_{1,1}^k & \mathbf{H}_{1,2}^k & \cdots & \mathbf{H}_{1,M_t}^k \\ \mathbf{H}_{2,1}^k & \mathbf{H}_{2,2}^k & \cdots & \mathbf{H}_{2,M_t}^k \\ \vdots & & \ddots & \vdots \\ \mathbf{H}_{M_r,1}^k & \mathbf{H}_{M-r,2}^k & \cdots & \mathbf{H}_{M_r,M_t}^k \end{bmatrix}$$

where

$$\mathbf{H}_{n,m}^k = \begin{bmatrix} h_{n,m,0}^k & 0 & \cdots & 0 \\ h_{n,m,1}^k & h_{n,m,0}^k & \cdots & \vdots \\ \vdots & \vdots & \ddots & \vdots \\ h_{n,m,L-1}^k & h_{n,m,L-2}^k & \ddots & \\ 0 & h_{n,m,L-1}^k & \ddots & \\ 0 & 0 & \cdots & h_{n,m,L-1}^k \end{bmatrix}$$

With the channel and signal model, we can write the received signal at nth antenna for user $\acute{k}$ after despreading as

$$x_n(t) = \eta_n^{\acute{k}} + \mathbf{u}_n^{\acute{k}H}\{\sum_{k=1}^{K}\mathbf{P}_{n,n}^{(\acute{k},k)}\mathbf{H}_{n,m}^{(\acute{k})}\mathbf{v}_m^{(k)}\},$$

where $\eta_n^{\acute{k}}$ is AWGN and $\mathbf{P}_{n,n}^{(\acute{k},k)}$ represents a correlation matrix of $L_r \times (L+L_t-1))$, which is given as

$$\mathbf{P}_{n,n}^{(\acute{k},k)} = \begin{bmatrix} \rho^{(\acute{k},k)}(0) & \rho^{(\acute{k},k)}(1) & \cdots & \rho^{(\acute{k},k)}(L+L_t-2) \\ \rho^{(\acute{k},k)}(-1) & \rho^{(\acute{k},k)}(0) & \cdots & \rho^{(\acute{k},k)}(L+L_t-3) \\ \vdots & & \ddots & \vdots \\ \rho^{(\acute{k},k)}(-L_r+1) & \rho^{(\acute{k},k)}(-L_r+2) & \cdots & \end{bmatrix}$$

The overall received signal using M_t Transmit antennas for user $\acute{k}$ then expressed as

$$x(t) = \eta_n^{\acute{k}} + \mathbf{u}^{\acute{k}H}\{\sum_{k=1}^{K}\mathbf{P}^{(\acute{k},k)}\mathbf{H}^{(\acute{k})}\mathbf{v}^{(k)}\},$$

where $\eta_n^{\acute{k}}$ is AWGN and

$$\mathbf{u}^k = \begin{bmatrix} \mathbf{u}_1^{k^T} & \mathbf{u}_2^{k^T} & \cdots & \mathbf{u}_{M_r}^{k^T} \end{bmatrix}^T$$

$$\mathbf{v}^k = \begin{bmatrix} \mathbf{v}_1^{k^T} & \mathbf{v}_2^{k^T} & \cdots & \mathbf{v}_{M_t}^{k^T} \end{bmatrix}^T.$$

Here, the cross correlation matrix between k and $\acute{k}$ user, $\mathbf{P}(\acute{k},k)$ represents a correlation matrix of $(M_r L_r \times M_r(L+L_t-1))$, which is given as

$$\mathbf{P}_{n,n}^{(\acute{k},k)} = diag\left(\begin{bmatrix} \mathbf{P}_{1,1}^{(\acute{k},k)} & \mathbf{P}_{2,2}^{(\acute{k},k)} & \cdots & \mathbf{P}_{M_r,M_r}^{(\acute{k},k)} \end{bmatrix}\right)$$

3 Algorithm

In this section, with the configuration of CDMA antenna system in [1], the SINR of the user $\acute{k}$ is given by

$$SINR_{\acute{k}} = \frac{|\mathbf{u}^{(\acute{k}H)}\mathbf{P}_I\mathbf{H}^{(\acute{k})}\mathbf{v}^{(\acute{k})}|^2}{\mathbf{u}^{(\acute{k}H)}\mathbf{Q}\mathbf{u}^{(\acute{k})}},$$

where $\mathbf{P}_I$ is matrix depending on the receiver structure and $\mathbf{Q}$ represents covariance matrix of MPI and MAI. For example, $\mathbf{P}_I$ becomes $\mathbf{I}$ for a receiver that is matched all paths of channel output. Neglecting MPI, the $\mathbf{Q}$ becomes as follow:

$$\mathbf{Q} = (\mathbf{P}^{(\acute{k},k)} - \mathbf{T}\sqrt{P_{\acute{k}}}\mathbf{P}_I)\mathbf{H}^{(\acute{k})}\mathbf{v}^{(\acute{k})}\mathbf{v}^{(\acute{k}H)}\mathbf{H}^{(\acute{k}H)}(\mathbf{P}^{(\acute{k},k)} - \mathbf{T}\sqrt{P_{\acute{k}}}\mathbf{P}_I)^H$$

Note that MAI in the **Q** can be written as

$$MAI_Q = \sum_{k=1, k \neq \acute{k}}^{K} \mathbf{P}^{(\acute{k},k)} \mathbf{H}^{(\acute{k})} \mathbf{v}^{(k)} \mathbf{v}^{(k)H} \mathbf{H}^{(k)H} \mathbf{P}^{(\acute{k},k)H}$$

With the formulation described above, the SINR can be written as follows:

$$SINR_{\acute{k}} = \frac{|\tilde{\mathbf{u}}^H \mathbf{P}_I \mathbf{H}^{(\acute{k})} \mathbf{v}^{(\acute{k})}|^2}{\tilde{\mathbf{u}}^H \tilde{\mathbf{u}}},$$

where $\tilde{\mathbf{u}} = \mathbf{Q}^{1/2} \mathbf{u}^{(\acute{k})}$. Given $\mathbf{v}^{(\acute{k})}$, the SINR can be written as follows:

$$SINR_{\acute{k}} = \frac{|\tilde{\mathbf{u}}^H \mathbf{P}_I \mathbf{H}^{(\acute{k})} \mathbf{v}^{(\acute{k})}|^2}{\tilde{\mathbf{u}}^H \tilde{\mathbf{u}}} \leq \mathbf{v}^{(\acute{k}^H)} \mathbf{H}^{(\acute{k}^H)} \mathbf{P}_I^H \mathbf{P}_I \mathbf{H}^{(\acute{k})} \mathbf{v}^{(\acute{k})},$$

where the upper bound achieved by $\tilde{\mathbf{u}} = \mathbf{P}_I \mathbf{H}^{(\acute{k})} \mathbf{v}^{(\acute{k})}$ or $\mathbf{u}(\acute{k}) = \mathbf{Q}^{-1/2} \tilde{\mathbf{u}} = \mathbf{Q}^{-1/2} \mathbf{P}_I \mathbf{H}^{(\acute{k})} \mathbf{v}^{(\acute{k})}$.

The Optimum $\mathbf{u}^{(\acute{k})}$ for max SINR can be obtained by SVD of the following matrix:

$$\mathbf{B}(\acute{k}) = \mathbf{Q}^{-1/2} \mathbf{P}_I \mathbf{H}^{(\acute{k})} = \mathbf{SDW}.$$

The optimum $\mathbf{u}^{(\acute{k})}$ and $\mathbf{u}^{(\acute{k})}$, can be obtained as follows:

$$\mathbf{u}^{(\acute{k})} = left\ singular\ vectors\ of\ \mathbf{B}(\acute{k}) = vectors\ of\ \mathbf{S}$$

$$\mathbf{v}^{(\acute{k})} = right\ singular\ vectors\ of\ \mathbf{B}(\acute{k}) = vectors\ of\ \mathbf{W}$$

The number of sub-channels can be selected by choosing the left and right singular vectors. We may expect that as the number of sub-channels increases, the higher performance can get at the cost of implementation cost. The transmit and receive antenna structure which can adopt the proposed algorithm are shown in the as seen the Fig. 3(a) and (b).

The proposed algorithm can be implemented iteratively. The iterative procedure can be summarized as follows:

1. Loop for $\hat{k}$=1:K
2. Initialize weight vectors $\mathbf{v}^{\hat{k}}$
3. Loop, for k=1 to $\hat{k}$
   (a) Find Rx weight vectors $\mathbf{u}^k$ using $\mathbf{v}^k$
   (b) Find Tx weight vectors $\mathbf{v}^k$ using $\mathbf{u}^k$
   (c) If improvement of SINR is obtained, goto step (a)
   (d) When no improvement of SINR, goto step 1.

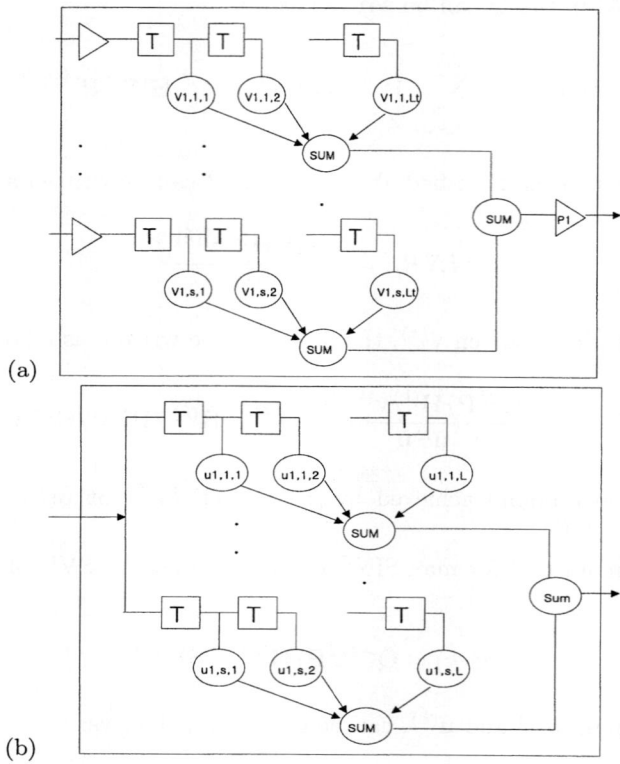

Fig. 3. (a) The proposed Transmit FIR filter Structure (b) The proposed Receive FIR filter Structure

4 Simulation

The performance of the SINR enhancement system is investigated in this section. For each bit-error rate (BER) simulation, more than 100 data packets are transmitted with independent channels. One data packet consists of 100 data symbols and BPSK / QPSK are used. In the simulation, we assume that the transmit power of all users is the same, with a spreading factor of 32, are used. Furthermore, a 3-ray equal gain model channel is utilized where the path delay is taken as T_c. Comparisons are made with the conventional SINR maximization method in [1]. The simulation parameters are described as follows:

- Number of Users = 3
- Number of Tx antenna = 2
- Number of Rx antenna = 2
- Number of Tx FIR Tap = 2
- Number of Rx FIR Tap = 3

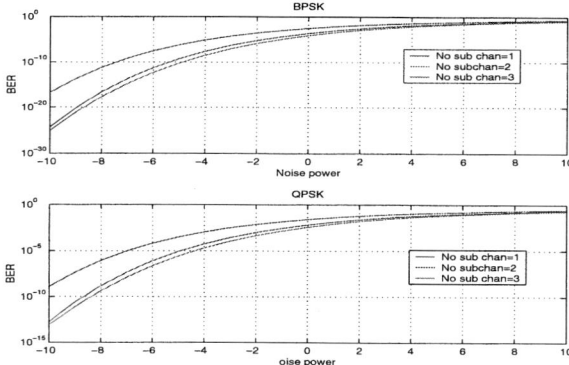

Fig. 4. BER vs Noise power

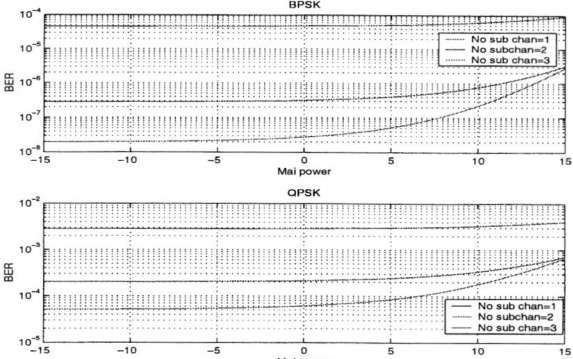

Fig. 5. BER vs MAI power

- Number of chips = 32
- Multi-path Delay = 1:3

When we used Gold code sequence, the performance of the proposed algorithm with varying noise power is illustrated in the the Fig. 4. As seen the figure, we can observe that when number of sub-channel is 2, the performance is much greater than the previous method and not much improvement can be achieved compared with 3 sub-channels.

With Gold code sequence used, the performance of the proposed algorithm with varying MAI power is illustrated in the the Fig. 5. As seen the figure, we can observe that the performance of using 2 sub-channels is much greater than the one using 1 sub-channel. In this figure, we can observe that much improvement can be achieved when using 3 sub-channels.

We conclude this section by presenting Fig. 6, where we provide results for our BER enhancement using different codes such as Gold code, Walsh code and Zcd code [3]. In this figure, we showed the results of 1 and 2 sub-channels. As

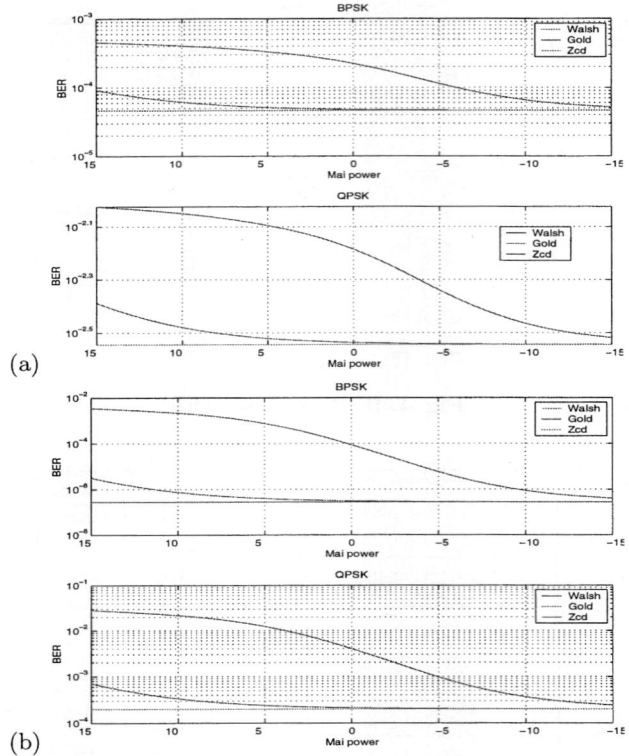

Fig. 6. (a) BER vs. Walsh, Gold, ZCD codes using one sub-channel (b) BER vs. Walsh, Gold, ZCD codes using two sub-channels

seen the figure, we can observe that the performance of using ZCD code exhibits superior performance over other code regardless of the number of sub-channels.

5 Conclusion

In this paper, we have presented a MIMO system to enhance SINR and corresponding BER for DS/CDMA communications, in which improved performance is demonstrated when compared to a conventional MIMO system in [1]. We present a SVD based pre/post rake receiver which maximizes SINR and finds the optimal solution to a general MIMO antenna system. Simulation results have demonstrated that significant system performance and capacity improvements are possible. Also, we propose a system which uses zero correlation duration spreading sequence in [3]. Computer simulation shows that the performance of the proposed system continues to improve if we adopt the zero correlation code.

References

1. R. L. Choi, R. D. Murch, and K. B. Letaief, MIMO CDMA Antenna System for SINR Enhancement , *IEEE Trans. Wireless Commun.*, vol. 2, NO. 2, p. 240-249, March 2003.
2. R. L. Choi, K. B. Letaief, and R. D. Murch, MISO CDMA transmission with simplified receiver for wireless communication handsets, *IEEE Trans. Commun.*, vol. 49, pp. 888.898, May 2001.
3. Cha, J.S., Kameda, S., Yokoyama, M., Nakase, H., Masu, K., and Tsubouchi, K., "New binary sequences with zero-correlation duration for approximately synchronized CDMA". it Electron. Lett., Vol. 36, no.11, pp.991-993, 2000.

SRS-Tool: A Security Functional Requirement Specification Development Tool for Application Information System of Organization

Sang-soo Choi[1], Soo-young Chae[2], and Gang-soo Lee[1]

[1] Hannam University,
Dept. of Computer Science,
Daejon, 306-791, Korea
gcss09@se.hannam.ac.kr, gslee@eve.hannam.ac.kr

[2] National Security Research Institute,
Daejon, 305-718, Korea
sychae@etri.re.kr

Abstract. An application information system (IS) of public or private organization should be developed securely and cost-effectively by using security engineering and software engineering technologies, as well as a security requirement specification (SRS). We present a SRS-Process that is a development process for SRS of IS, and a SRS-Tool that is a development tool for SRS in accordance with the SRS-Process. Our approach is based on the paradigm of Common Criteria (ISO/IEC 15408), that is an international evaluation criteria for information security products, and PP which is a common security functional requirement specification for specific types of information security product.

1 Introduction

Construction and operation of evaluation scheme have been actively operated because importance of information security and demand for evaluation and certification increases rapidly. Especially, public or private organizations construct and integrate their IS by using evaluated security products (e.g., Smart card and Firewall) and validated cryptographic algorithms or modules (e.g., OpenSSL). Therefore, various IS have been developing and operating by using security products and cryptographic modules as well as risk management concept.

CMVP (cryptographic module validation program) [1] and CC (common criteria) [2,3,4] are used as a base of evaluation scheme for security product of every country. Also, ISO/IEC 17799 [5] and ISO/IEC 13335 [6] that are international standards of risk management and risk evaluation, are used as a base of evaluation scheme for risk management. But, evaluation scheme for more sophisticate system rather than individual security product is required, because factors that threaten information security become more complicated. Especially, IS of public or private organization should developed securely and cost-effectively by using security engineering and software engineering technology as well as a SRS which is a blueprint for organization's IS.

To cope with those problems, we propose a SRS-Process that is a development process of SRS for organization's IS, and a SRS-Tool that is a support tool for develops in accordance with the SRS-Process. SRS Process is based on the paradigm of CC, that is an international evaluation criteria for information security products, and Protection Profile (PP) which is a common security functional requirement specification for specific types of information security product. CC-ToolBox developed by NIST is a tool for developing PP for specific information security product, as SRS-Tool is a tool for developing SRS for specific IS. SRS-Process and SRS-Tool have some contributive idea and methods, such as object-oriented generation of policy and assumption statement, well-formed threat statement, a new asset classification schema, these are useful for development of SRS.

2 SRS-Process: A Development Process for SRS

In a CC environment, PP is a common security functional requirement specification for a specific type of security product. Therefore, we define SRS (security functional requirement specification) as follow: "SRS is a security functional requirement specification for IS of a specific organization". We develop the SRS-Process which is a process to develop SRS. It consists of four steps as shown in Fig.1.

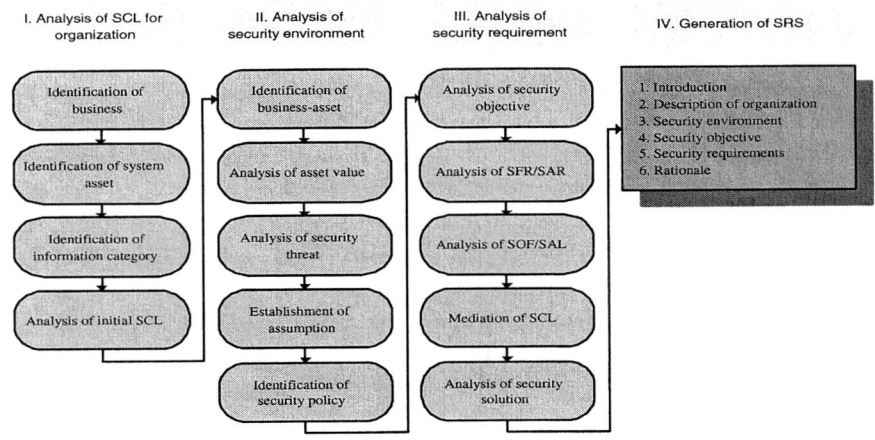

Fig. 1. An architecture of SRS-Process

2.1 Step 1: Analysis of SCL for Organization

In the first step, a SRS developer identifies business and critical system asset that is required to achieve business. Then, he identifies main information category that is performed by system asset, to decide initial security classification level (SCL) of entire organization (or IS). Note that, initial SCL effect to selection of statement in security environment and security requirement analysis step.

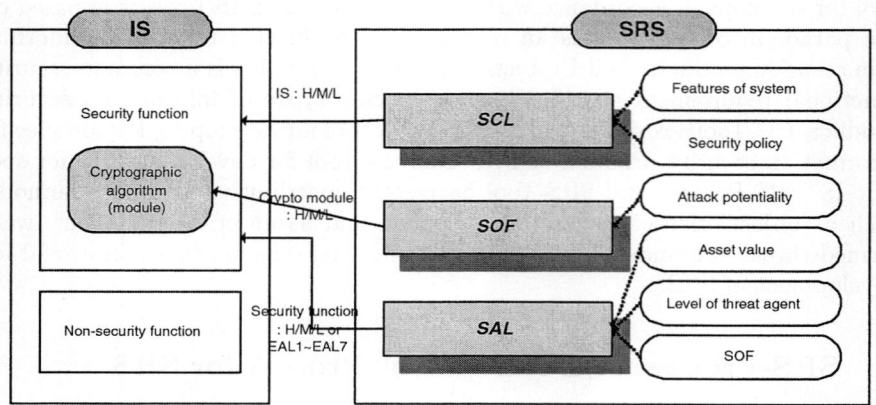

Fig. 2. Relation and difference of SCL, SAL and SOF

Analysis of security level: We define three level of security for SRS development as shown in Fig.2. Strength of function (SOF) is a security level for security function which related to cryptographic function. We apply SOF concept of CC [7], and SOF is represented by three levels (High/Medium/Low). And, Security assurance level (SAL) is an assurance level for entire organization or specific business. We apply concept of CC to SAL, and SAL is represented by seven levels (EAL1 ~ EAL7) or three levels (High/Medium/Low). SCL is a security level for entire organization or IS. We extend security category concept of FISMA project [8,9,10] and the robustness concept in CC and PP, and SCL has many features as follows:

- SCL is based on security policy and security treatment level that are operating currently in organization (e.g., security treatment of national defense business is 'High').
- Whole security level of system decided by the highest value of security level of subsystem (e.g., if system is consisted of three subsystems, that have 'High', 'Medium' and 'Low' security level respectively, then whole security level of system is 'High').
- SCL is decided by referencing FISMA project (NIST SP-800-60): Let C, I, A and S are confidentiality, integrity, availability and whole security level, respectively. And $*@$ and $*\#$ are security level and value of security level (High = 3, Medium = 2 and Low = 1), respectively.

If $3 \leq (C\#+I\#+A\#) \leq 4$, then $S@ = $ Low
If $5 \leq (C\#+I\#+A\#) \leq 7$, then $S@ = $ Medium
If $8 \leq (C\#+I\#+A\#) \leq 9$, then $S@ = $ High

For example, if $C@$, $I@$ and $A@$ of Ministry of Construction Transportation are Low, Medium and High, respectively, then whole security level $S@$ is Medium ($C\#+I\#+A\#=5$).

2.2 Step 2: Analysis of Security Environment

A developer identifies critical asset that is important to business process. Then, he identifies security threat that is related to the critical asset. Also, he identifies security policy related to security that is operated currently in organization, and assumption related to security.

Analysis of critical asset: Assets of organization should be identified and classified for the purpose of analysis of security threat. Especially, security function should be implemented in accordance with value and class of asset, because an asset is a target which must be prevented from threat by means of security function. But, ISO/IEC PDTR 15446 [7] does not provide detail method for analysis and evaluate of asset. Also, risk analysis and security management field does not provide general asset classification system. Therefore, we develop ABCS (asset and business classification schema) that integrates ACS (asset classification schema) and BCS (business classification schema). Also, we develop difference method to calculate asset value in accordance with acquisition method and class of asset. Especially, deflection of result is very huge when many developers evaluate value of asset. Thus, we use "beta-distributed delphi evaluation method based on web" as follow to improve accuracy of evaluation result:

- Developer evaluate three values such as a(best value), m(suitable value) and b(worst value) for specific asset. Then, developer select final value by using mean value ($mean\ value = (a + 4m + b) / 6$) of beta-distribution.
- Many developers can evaluate asset value mutual independently.
- Evaluation result coverages in fixed value according to many evaluation rounds.
- Evaluation activity and result are managed by using internet.

Analysis of security threat: Threat statement should be mapping to predefined threat statement in PKB [11,12]. Also, generated statement must have equal sentence level, and well-defined as well as generated by semi-automatically. Therefore, we develop generation model for threat statement based on security attribute as shown in Fig.3. Note that, Korean ($S + O + V$) have different structure with English ($S + V + O$), thus, developed model is based on Korean. Developed model generate threat statement as below:

- Threat statement: "Authorized user modified hostile asset maliciously, so availability is damaged seriously."
- Attributes: agent type=*human*, authentication=*authenticated*, attitude =*hostile*, motive=*N/A*, sophistication level=*N/A*, locality=*local*, force=*N/A*, lifecycle phase=*operation*, human role=*N/A*, action=*modification*, loss category = *availability*, IT capabilities=*user data*, location=*N/A*

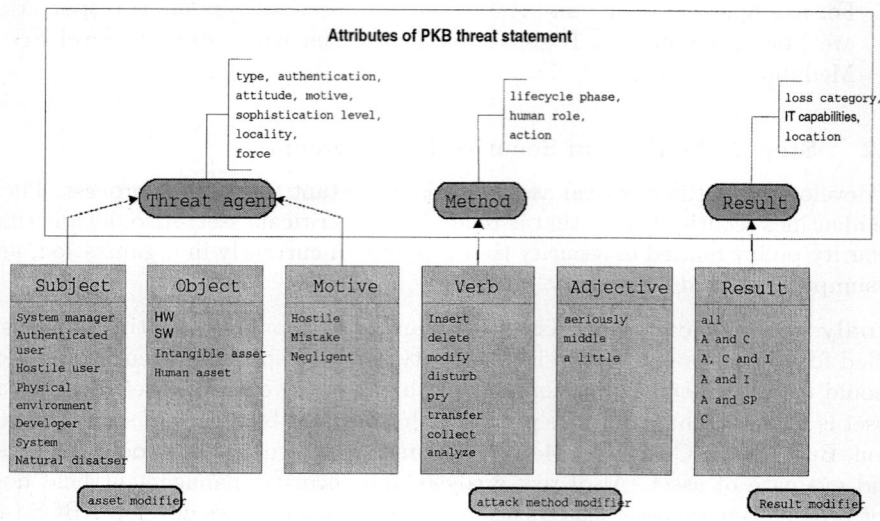

Fig. 3. Generation model for threat statement based on attribute

Especially, attributes that are managed with generated statement, are used by reference data for mapping between generated threat statement and PKB threat statement. We define the "mapping similarity function" and "mapping similarity" for this purpose as follow:

- Mapping similarity function: MS(i,j)= $\frac{2A_{ij}}{T_i+T_j}$, i=attributes of generated threat statement, j=attributes of PKB threat statement, A_{ij}=mapping count for ij, T_i=total count of attribute i, and T_j=total count of attribute j.
- Mapping similarity: μ_A=i, j, $MSA(i,j)$, $i \in I$, $j \in J$, $0 \leq \mu_A \leq 1$ I=total attribute count of generated threat statement, J=total attribute count of PKB threat statement, $MSA(i,j)$=degree that i, j belonging to μ_A. Also, μ_A=degree that similarity i, j.

If μ_A value is nearer to 1, then similarity value is being correct more. (If μ_A=0, then similarity is perfectly discordant. If μ_A=1, then similarity is perfectly agreement.)

Analysis of security policy and assumption: We have been developed CSSL-DB (common security statement lists database) through analyze 37 class of PP and PKB DB. Especially, we consider consistency among security statements, and develop security statement list that have fixed level. Therefore, PP/ST or SRS developer can use easily to write security-related statement by selecting common statement. Then, developer can append more detail information to selected statement. Note that, a common statement means a general statement. We named this method as OOSSM (Object-oriented security statement method). Fig.4 presents method to create security assumption statement using OOSSM.

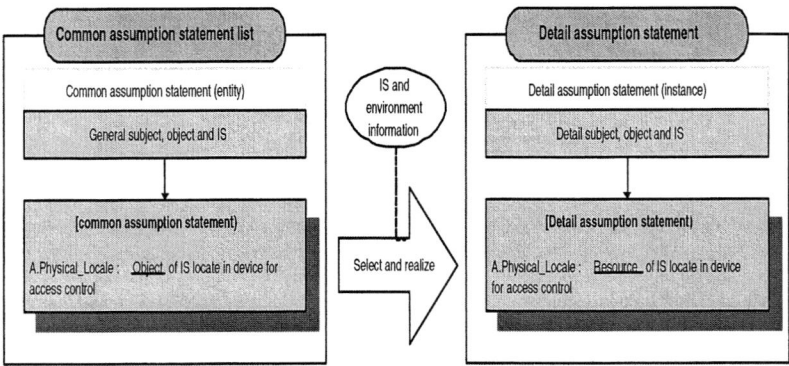

Fig. 4. Method to create security assumption statement by using OOSSM

2.3 Step 3: Analysis of Security Requirement

A developer selects security objective to implement security environment in accordance with identified security threat and policy statement. Then, he selects security requirement such as security functional and assurance requirement (SFR/SAR) in CC according to selected security objective statement. Then, he analyzes SOF and SAL that are illustrated in Fig.2. And, he modifies SCL according to analyzed security environment, SFR/SAR, SOF and SAL. Finally, he selects security solution to implement IS in accordance with developed SRS.

Analysis of relation between security objective and security requirement: PKB presents security objective statement list which based on CC. Also, PKB presents relation among security threat and security objective, security policy and security objective as well as security objective and security requirement. But, PKB does not consider some security functional requirements such as FDP_UCT, FIA_AFL, FIA_SOS, FPT_ITA, FPT_RPL, FPT_STM, FTA_LSA. Therefore, we extends security objective statement list to cope with those problems. Especially, we improves relationships between security objective and security requirement as well as added some requirement objective statements by analyzing 37 kinds of PP.

Analysis of security solution: SRS developer wants to know what kind of security solution (security products) is required to implement IS in accordance with SRS. Therefore, we develop SSDM (security solution derivation model) to recommend cost-effectively security solution which has minimum function repetition for developer. Fig.5 presents an architecture of SSDM.

2.4 Step 4: Generation of SRS

A developer writes SRS for IS in accordance with analyzed security environment, security requirement and determined SCL. Especially, we apply and extend format of PP/ST, that has well-formed form and generally used in CC evaluation

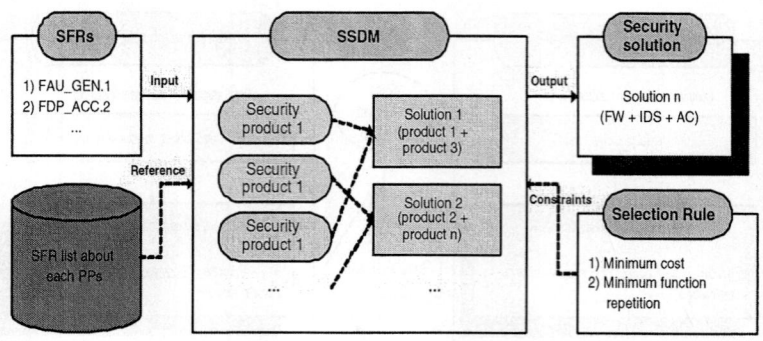

Fig. 5. An architecture of SSDM

environment, for SRS. Therefore, index of SRS is similar to PP/ST. Also, it allows SRS could be used system PP in CC.

3 SRS-Tool: A Development Tool for SRS

SRS-Tool is a support tool for developing SRS in accordance with SRS-Process. Especially, we apply CC-ToolBox and PKB database to develop SRS-Tool. SRS-Tool has been implemented by using Windows, PowerBuilder 9.0 and MySQL 3.23.38. SRS-Tool is consisted of DB server module client module. Fig.6 presents DB-schema of SRS-Tool. Especially, DB-schema of SRS-Tool is based on PKB DB of CC-ToolBox. Fig.7 presents Korean version of SRS-Tool. Especially, SRS-Tool provides the following features to system manager of some organization,

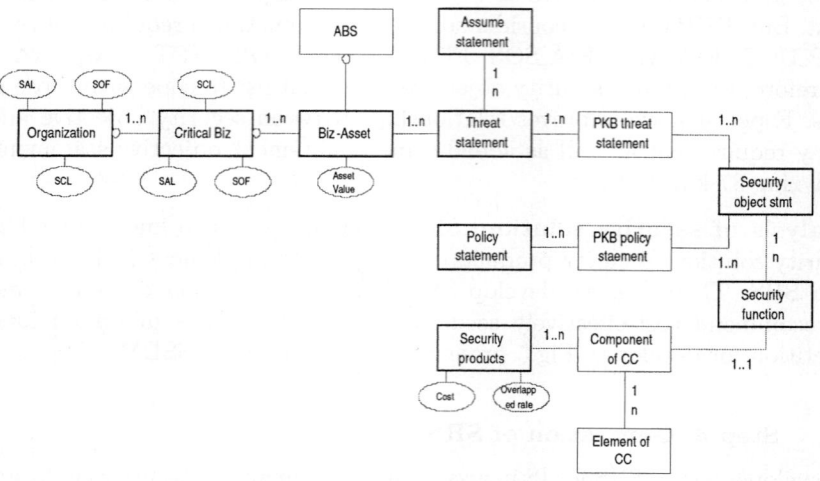

Fig. 6. DB-schema of SRS-Tool

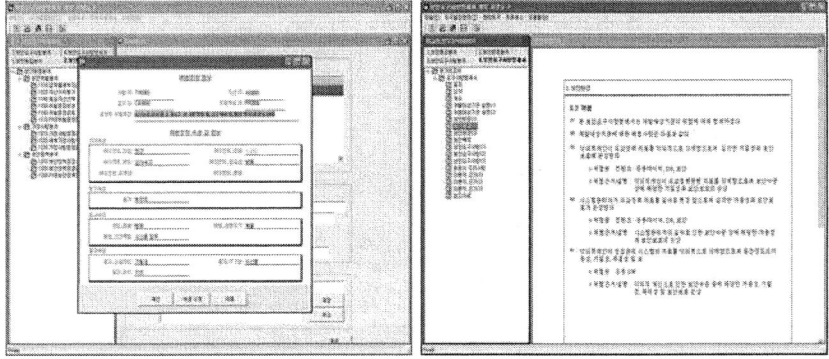

Fig. 7. Some screen shots of SRS-Tool (Korean version)

who wants to develop SRS but does not have expert knowledge of CC, security technology and requirement specification, software-engineering, and so on:

- Convenience of use: SRS-Tool has a tree-based menu. Thus, general user who is security novice can use SRS-Tool more easily because users used to Explorer of Windows.
- Intuition of GUI: User can confirm some result immediately at screen. Also, user can confirm and display the created SRS in real-time.

4 Analysis and Conclusion

4.1 Analysis with Related Work

Most of organization is developing and operating various IS to process quickly and efficiently various business process. Especially, most of organization develop IS by using evaluated and certificated security products. And, security products are developed by using validated cryptographic modules and algorithms. Especially, security-level in all kinds of system must be evaluated, because security-level in lower-level system depends on security-level in upper-level system.

Therefore, various evaluation schemes are operating for cryptographic modules and security products, as well as evaluation scheme for security management, that include risk management and development, operation of whole system. Table 1 presents example of evaluation scheme by Target of Evaluation (TOE)'s level.

Comparison of security level: FISMA project defines security category of application system by each three levels about confidentiality, integrity and availability. But, it is very difficult that decide each level independently, because confidentiality, integrity and availability are interconnected concept. Therefore,

Table 1. Example of evaluation scheme by TOE's level

Level of TOE	Evaluation Scheme	Evaluation Criteria	Development Tool
Crypto Algorithm and Module	- CMVP of NIST(USA)	- FIPS 197, 46-3, 186-2, 180-1, 185 - FIPS 140	N/A
Security Product	- CCEVS of NIST(USA) - CC Evaluation Scheme(KOREA) [13]	- CC, ITSEC, TCSEC, STCPEC etc. - Evaluation Criteria for IDS/FW(KOREA)	CCToolBox
Application System	- SPP Project of FAA(USA) [14] - SYS Assurance Package of CESG(UK) [15]	- Common Criteria	SPPT (template) N/A
Security Management	- FISMA Project of NIST(USA) - ISMS of KISA(KOREA) [13]	- ISO/IEC 1799, 13335 - FIPS 1999, NIST SP 800-53(USA) - Criteria for ISMS (KOREA)	N/A

we integrated confidentiality, integrity and availability, and classified SCL by three levels.

Comparison of support tool: FISMA project does not provide supporting tool like a SRS-Tool. Also, FAA provide only document-level template to develop PP easily. SRS-Tool is based on CC-ToolBox and PKB DB. But, Table 2 presents main difference between SRS-Tool and CC-ToolBox.

Table 2. SRS-Tool vs. CC-ToolBox

	CCToolBox/PKB	SRS-Tool
Target	PP/ST	SRS, PP/ST
Process	ISO/IEC PDTR 15446	SRS-Process
Asset analysis	X	O
Classification of Security level	X	O
DB	PKB DB Threat : 109 Policy : 35 Assumption : 38 Security Objective : 157	SRS DB Common Threat : 6720 Common Policy : 150 Common Assumption : 213 Security Objective : 489
Generation method of statement	X	O
Security function for tool	X	O
GUI	GUI based on text	GUI based on form
Report form	Self definition	PDF

Comparison with security management: Evaluation schemes for security management such as ISMS and FISMA project are used in high-level security control (e.g., risk management, implementation of countermeasure and post management, etc.). But, SRS focuses in implementation of countermeasure. Therefore, SRS can be used to develop request for proposal as well as security requirement for IS.

4.2 Conclusion

In this paper, we present SRS-Process and SRS-Tool. SRS which is developed through proposed SRS-Tool and SRS-Process can be used RFP and requirement specification for development of organization's IS. Also, we expect that IS of organization will be developed and constructed securely and cost-effectively by using a SRS-Tool. SRS-Tool has the following features:

- Organization can develop easily SRS by itself. Also, Reusability of SRS between similar organizations is promoted.
- Organization develops IS by suitable assurance level (main concept of security engineering).
- SRS-Tool promotes systematic development of IS by using standard concept of CC and PP that have history more than 10 years.

Acknowledgement

This work was supported by a grant No. R12-2003-004-01001-0 from KOSEF and NSRI.

References

1. Web site of CMVP, htp://csrc.nist.gov/cryptval/.
2. Web site of CCEVS, http://niap.nist.go/cc-scheme/index.html.
3. CC, Common Criteria for Information Technology Security Evaluation, CCIMB-2004-03, Version 2.2, Jan. 2004.
4. CEM, Common Methodology for Information Technology Security Evaluation (CEM), CCIMB-2004-01-04, Version 2.2, Jan. 2004.
5. ISO/IEC 17799, ISO/IEC 17799: 2000 - Code of Practice for Information Security Management, ISO17799/BS7799, Dec. 2000.
6. ISO/IEC TR 13335, Guidelines for the Management of IT Security, 1998.
7. ISO/IEC PDTR 15446, "Information technology - Security techniques - Guide for the production of protection profiles and security targets," Draft, Apr. 2000.
8. Web site of FISMA project, http://csrc.nist.gov/sec-cert/.
9. William C. Barker, NIST SP-800-60, Guide for Mapping Types of Information and information Systems to Security Categories, March. 2004.
10. Ron Ross, et al., NIST SP-800-53, Recommended Security Controls for Federal Information Systems, Oct. 2003.
11. NIAP, CC Toolbox Reference Manual, Version 6.0f, http://niap.nist.gov/tools/cctool.html, 2000.
12. NIAP List of Threat, Attack, Policy, Assumption, and Environment Statement Attribute, CC Profiling Knowledge base Report, 2002.
13. Web site of KISA, http://www.kisa.or.kr/.
14. Web site of FAA, http://www.faa.gov/SciefSci/.
15. UK IT Security Evaluation and Certification Scheme, SYSn Assurance Packages Framework, Issue 1.0, Sep. 2002.

Design Procedure of IT Systems Security Countermeasures

Tai-hoon Kim[1] and Seung-youn Lee[2]

[1] San-7, Geoyeo-Dong, Songpa-Gu, Seoul, Korea
taihoon@empal.com
http://home.paran.com/taihoonn
[2] SKK Univ., Dept. of Information & Communication Eng. Kyonggi, 440-746, Korea
syoun@ece.skku.ac.kr

Abstract. The developers of the security policy should recognize the importance of building security countermeasures by using both technical and non-technical methods, such as personnel and operational facts. Security countermeasures may be made for formulating an effective overall security solution to address threats at all layers of the information infrastructure. This paper uses the security engineering principles for determining appropriate technical security countermeasures. It includes information on threats, security services, robustness strategy, and security mechanism. This paper proposes a countermeasure design flow that may reduce the threats to the information systems.

1 Introduction

When we are making some kinds of software products, by considering some approaches and methods we may provide a framework for the assessment of software quality or security characteristics. And this framework can be used by organizations involved in planning, monitoring, controlling, and improving the acquisition, supply, development, operation, evolution and support of software.

But in the general cases, security countermeasures may be regarded as buying and installing some security products or system such as Firewall, IDS and Antivirus system.

As you know, most of the threat agents' primary goals may fall into three categories: unauthorized access, unauthorized modification or destruction of important information, and denial of authorized access. Though any cases, if the compromises are occurred, your money and your job may be missed. Therefore, Security countermeasures must be implemented to prevent threat agents from successfully achieving these goals [1-4].

This paper proposes a countermeasure design flow considering attacks (including threat), security services, and appropriate security technologies.

Security countermeasures should be considered with consideration of applicable threats and security solutions deployed to support appropriate security services and objectives. Subsequently, proposed security solutions may be evaluated to determine if residual vulnerabilities exist, and a managed approach to mitigating risks may be proposed.

2 Measuring the Level of Security Countermeasures

Implementation of any security countermeasure may require economic support. If your security countermeasures are not sufficient to prevent the threats, the existence of the countermeasures is not a real countermeasure and just considered as like waste. If your security countermeasures are built over the real risks you have, maybe you are wasting your economic resources.

For considering what the security level is needed, we are now researching next model. We call this model "Block Model for Security Countermeasure." Next time we will report about the model more precisely.

First step is dividing IT systems we will protect and making block (See the Fig.1):

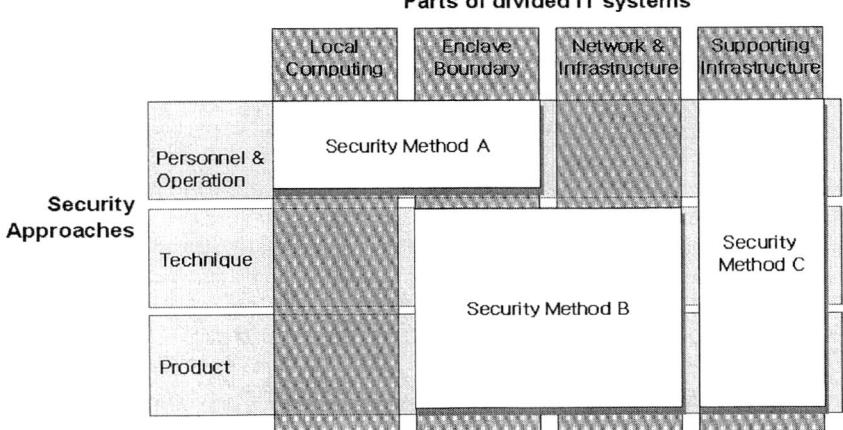

Fig. 1. Security Approaches and Methods for divided IT systems

Second step is building security countermeasures by using Block Region (See the Fig.2).

3 Threat Identification

A 'threat' is an undesirable event, which may be characterized in terms of a threat agent (or attacker), a presumed attack method, a motivation of attack, an identification of the information or systems under attack, and so on. In order to identify what the threats are, we need to answer the following questions [5-6]:

- What are the assets that require protection? (e.g., sensitive data or secret),
- Who or what are the threat agents? (e.g., an authorized user who wants to access to the system),
- What are the motivations of attack? (e.g., delete some information),
- What attack methods or undesirable events do the assets need to be protected from? (e.g., impersonation of an authorized user of the TOE).

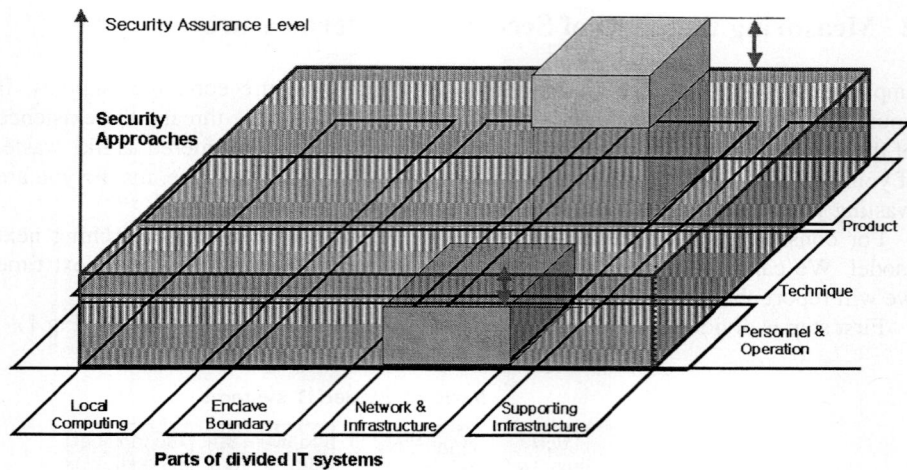

Fig. 2. Building security countermeasures by using Block Region

Threat agents come from various backgrounds and have a wide range of financial resources at their disposal. Typically Threat agents are thought of as having malicious intent. However, in the context of system and information security and protection, it is also important to consider the threat posed by those without malicious intent. Threat agents may be Nation States, Hackers, Terrorists or Cyber terrorists, Organized Crime, Other Criminal Elements, International Press, Industrial Competitors, Disgruntled Employees, and so on.

Most attacks maybe aim at getting inside of information system, and individual motivations of attacks to "get inside" are many and varied. Persons who have malicious intent and wish to achieve commercial, military, or personal gain are known as hackers (or cracker). At the opposite end of the spectrum are persons who compromise the network accidentally. Hackers range from the inexperienced Script Kiddy to the highly technical expert.

4 Determination of Robustness Strategy

The robustness strategy is intended for application in the development of a security solution. An integral part of the process is determining the recommended strength and degree of assurance for proposed security services and mechanisms that become part of the solution set. The strength and assurance features provide the basis for the selection of the proposed mechanisms and a means of evaluating the products that implement those mechanisms.

Robustness strategy should be applied to all components of a solution, both products and systems, to determine the robustness of configured systems and their component parts. It applies to commercial off-the-shelf (COTS), government off-the-shelf (GOTS), and hybrid solutions. The process is to be used by security requirements developers, decision makers, information systems security engineers, customers, and others involved in the solution life cycle. Clearly, if a solution component is modified, or threat levels or the value of information changes, risk must be reassessed with respect to the new configuration.

Various risk factors, such as the degree of damage that would be suffered if the security policy were violated, threat environment, and so on, will be used to guide determination of an appropriate strength and an associated level of assurance for each mechanism. Specifically, the value of the information to be protected and the perceived threat environment are used to obtain guidance on the recommended strength of mechanism level (SML) and evaluation assurance level (EAL) [8].

5 Consideration of Strength of Mechanisms

SML (Strength of Mechanism Levels) are focusing on specific security services. There are a number of security mechanisms that may be appropriate for providing some security services. To provide adequate information security countermeasures, selection of the desired (or sufficient) mechanisms by considering particular situation is needed. An effective security solution will result only from the proper application of security engineering skills to specific operational and threat situations. The strategy does offer a methodology for structuring a more detailed analysis. The security services itemized in these tables have several supporting services that may result in recommendations for inclusion of additional security mechanisms and techniques.

6 Selection of Security Services

In general, primary security services are divided five areas: access control, confidentiality, integrity, availability, and non-repudiation. But in practice, none of these security services is isolated from or independent of the other services. Each service interacts with and depends on the others. For example, access control is of limited value unless preceded by some type of authorization process. One cannot protect information and information systems from unauthorized entities if one cannot determine whether that entity one is communicating with is authorized. In actual implementations, lines between the security services also are blurred by the use of mechanisms that support more than one service.

7 Application of Security Technologies

An overview of technical security countermeasures would not be complete without at least a high-level description of the widely used technologies underlying those countermeasures. Next items are some examples of security technologies.
- Application Layer Guard, Application Program Interface (API).
- Common Data Security Architecture (CDSA).
- Circuit Proxy, Packet Filter, Stateful Packet Filter.
- CryptoAPI, File Encryptors, Media Encryptors.
- Cryptographic Service Providers (CSP), Certificate Management Protocol (CMP).
- Internet Protocol Security (IPSec), Internet Key Exchange (IKE) Protocol.

- Hardware Tokens, PKI, SSL, S/MIME, SOCKS.
- Intrusion and Penetration Detection.
- Virus Detectors.

8 Determination of Assurance Level

The discussion of the need to view strength of mechanisms from an overall system security solution perspective is also relevant to level of assurance. While an underlying methodology is offered by a number of ways, a real solution (or security product) can only be deemed effective after a detailed review and analysis that consider the specific operational conditions and threat situations and the system context for the solution.

Assurance is the measure of confidence in the ability of the security features and architecture of an automated information system to appropriately mediate access and enforce the security policy. Evaluation is the traditional method that ensures the confidence. Therefore, there are many evaluation methods and criteria exist. In these days, the ISO/IEC 15408, Common Criteria, replaces many evaluation criteria such as ITSEC and TCSEC.

The Common Criteria provide assurance through active investigation. Such investigation is an evaluation of the actual product or system to determine its actual security properties. The Common Criteria philosophy assumes that greater assurance results come from greater evaluation efforts in terms of scope, depth, and rigor.

9 Countermeasure Design Flow

Until now we identified some components we should consider for building security countermeasures. In fact, the Procedure we proposed in this paper is not perfect one yet, and the researches for improving are going on.

Next figure is the summarized concepts we proposed in this paper.

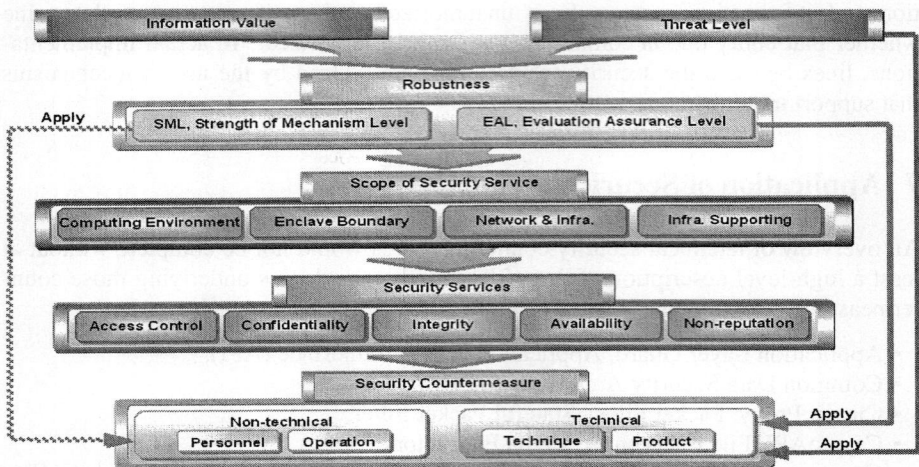

Fig. 3. Example of Design Procedure for Security Countermeasures

But the Fig.3 is an example of design procedures for security countermeasures, but as you know, this sample is not a perfect one. Now we are researching about this model for more perfect one.

10 Conclusions

This paper proposes a countermeasure design flow considering threats, robustness, and strength of mechanism, security service, technology, and appropriate assurance level. But this is not a framework yet. More efforts are needed for refining the flow and making framework.

For more detail work, we are now researching about the Block Model for Building Security Countermeasures.

References

1. Tai-Hoon Kim, Byung-Gyu No, Dong-chun Lee: Threat Description for the PP by Using the Concept of the Assets Protected by TOE, ICCS 2003, LNCS 2660, Part 4, pp. 605-613
2. Tai-hoon Kim and Haeng-kon Kim: The Reduction Method of Threat Phrases by Classifying Assets, ICCSA 2004, LNCS 3043, Part 1, 2004.
3. Tai-hoon Kim and Haeng-kon Kim: A Relationship between Security Engineering and Security Evaluation, ICCSA 2004, LNCS 3046, Part 4, 2004.
4. Eun-ser Lee, Kyung-whan Lee, Tai-hoon Kim and Il-hong Jung: Introduction and Evaluation of Development System Security Process of ISO/IEC TR 15504, ICCSA 2004, LNCS 3043, Part 1, 2004
5. Tai-hoon Kim, Tae-seung Lee, Kyu-min Cho, Koung-goo Lee: The Comparison Between The Level of Process Model and The Evaluation Assurance Level. The Journal of The Information Assurance, Vol.2, No.2, KIAS (2002).
6. Tai-hoon Kim, Yune-gie Sung, Kyu-min Cho, Sang-ho Kim, Byung-gyu No: A Study on The Efficiency Elevation Method of IT Security System Evaluation via Process Improvement, The Journal of The Information Assurance, Vol.3, No.1, KIAS (2003).
7. Tai-hoon Kim, Tae-seung Lee, Min-chul Kim, Sun-mi Kim: Relationship Between Assurance Class of CC and Product Development Process, The 6th Conference on Software Engineering Technology, SETC (2003).
8. Ho-Jun Shin, Haeng-Kon Kim, Tai-Hoon Kim, Sang-Ho Kim: A study on the Requirement Analysis for Lifecycle based on Common Criteria, Proceedings of The 30th KISS Spring Conference, KISS (2003)
9. Haeng-Kon Kim, Tai-Hoon Kim, Jae-sung Kim: Reliability Assurance in Development Process for TOE on the Common Criteria, 1st ACIS International Conference on SERA

Similarity Retrieval Based on Self-organizing Maps

Dong-Ju Im[1], Malrey Lee[2], Young Keun Lee[3], Tae-Eun Kim[4],
SuWon Lee[5], JaewanLee[5], Keun Kwang Lee[6], and Kyung Dal Cho[7]

[1] Dept. of Multimedia, Yosu National University, Korea
[2] School of Electronics & Information Engineering and , Chonbuk National University, 664-14,
1Ga, DeokJin-Dong, JeonJu, CnonBuk, 561-756, Korea
mrlee@chonbuk.ac.kr
[3] Dept. of Orthopedic Surgery , Chonbuk National University Hospital
yklee@chonbuk.ac.kr
[4] Dept. of Multimedia, Namseoul University, Korea
tekim@nsu.ac.kr
[5] Dept. of Information Communication, KunSan National University, Korea
{lsw91, jwlee@kunsan.ac.kr}
[6] Dept of Skin Beauty Art, Naju Collage
kklee@naju.ac.kr
[7] Dept of Computer Science, Chung-Aang Unicersity

Abstract. The features of image data are useful to discrimination of images. In this paper, we propose the high speed k-Nearest Neighbor search algorithm based on Self-Organizing Maps. Self-Organizing Maps provides a mapping from high dimensional feature vectors onto a two-dimensional space. The mapping preserves the topology of the feature vectors. The map is called topological feature map. A topological feature map preserves the mutual relations in feature spaces of input data. and clusters mutually similar feature vectors in a neighboring nodes. Each node of the topological feature map holds a node vector and similar images that is closest to each node vector. In topological feature map, there are empty nodes in which no image is classified. We experiment on the performance of our algorithm using color feature vectors extracted from images.

Keywords: Self-organizing maps, Image databases, Similarity retrieval, Content-based image retrieval.

1 Introduction

First of all, CBIR approach extracts the unique visual feature of image such as color, texture, and shape out of image registered in database. The feature helps determine image discrimination, i.e., which category a given image belongs to. And then, feature vector is generated for each image, which is mapped to a point on feature space. An essential processing of a similar retrieval by preprocessing in image database is a search for the nearest neighbor ranking image data nearest to a query using distance measure provided by a system for a query of feature space.

An essential processing of a similar retrieval by preprocessing in image database is a search for the nearest neighbor ranking image data nearest to a query using distance measure provided by a system for a query of feature space(10). In general, an information quantity of image data is tremendous. Therefore, a similar image retrieval for a query image given in large scale database should be finished in a practical time. CBIR approach using the above technique has been verified in many systems (6,14,15), and the system such as QBIC(7,6) has been commercialized. In this paper, Kohonen Self-Organizing Maps(SOM)(2,4) is used for realization of automatic classification by image similarity. SOM generates two dimensional feature map with topological relationships with feature vector mapping a set of data with high dimensional feature vector on two dimension. The vector generated in each node of topological feature map is generally called codebook vector. It is also called node vector in a sense of vector each node has. Neighboring nodes on a map with similarity between feature vectors can be clustered using SOM. With this clustering, two or more images are mapped in one node, because similar images are included in database. As a result, empty node in which image is not classified in a node of map layer is generated. The empty node deteriorates retrieval performance of a whole system, because it performs unnecessary access to a disk.

In this paper, we propose k-NN search algorithm which can perform high speed similar image search in SOM considering it. k-NN search, which searches as retrieval candidates for k images most similar to a given query image, has been applied for GIS(Geo grapic Information System)(10).

A search method for realization of k-NN search is defined for each node of map layer. A search algorithm is implemented for the definition, and its effectiveness is verified through simulation.

2 Related Work

SOM has been applied for a visual information retrieval(1). Especially, the application to similar image retrieval has two types. One is the retrieval by visualization of map(18) as a method to perform information retrieval navigating map, for which relevance feedback is suitable to perform interactive information retrieval for user's information requirement. The above system, however, has the weakness that user cannot retrieve by inputting an arbitrary image as a query image.

The other type is a query by example(QBE)(19,17). Netra(17) expresses visually similar pattern by dividing feature space into different parts combining neural network to perform unsupervised learning and supervised learning. We proposed feature extraction and classification technique valid in similar retrieval by combining wavelet transformation and SOM(19). In retrieval technique, we propose a retrieval by example image or a retrieval by partial space, and we evaluate its performance. In this retrieval technique, however, we can point out the problem that retrieval results do not correspond to each other compared with other retrieval techniques(e.g. linear retrieval) depending on image, because image obtained from winner node on map is regarded as retrieval result. Learning of SOM can cluster similar feature vectors into neighboring node of map by discovering a similarity between feature vectors(3). In spite of the property, the problem occurs that retrieval results do not correspond with ones of other retrieval techniques. In order to get rid of the above problem, all the nodes of map should be searched for. It is not, however, practical that it takes so much time for them to be retrieved. In this paper, we propose k-NN search algorithm to perform high speed image retrieval based on SOM in order to solve the above problem.

3 Classification of Similar Images by SOM

In this paper, we extracted features for color of each image using Haar wavelet transformation in image data, and we used them for similar image retrieval(19). We obtained wavelet coefficient by performing 5-leveled Haar wavelet transformation for each channel of color space using YIQ space(16) as a color expressing space. As a result, we extracted 4×4 of the lowest banded element representing average color element of whole image, which is regarded as 48 dimensional color feature vector.

FV_s	ELEMENTS OF VECTORS
FV_1	$fv_{11}, fv_{12}, ..., fv_{1j}, ..., fv_{1m}$
FV_2	$fv_{21}, fv_{22}, ..., fv_{2j}, ..., fv_{2m}$
...	......
FV_i	$fv_{i1}, fv_{i2}, ..., fv_{ij}, ..., fv_{im}$
...	......
FV_n	$fv_{n1}, fv_{n2}, ..., fv_{nj}, ..., fv_{nm}$

many-to-one SOM

CBV_s	ELEMENTS OF VECTORS
CBV_1	$cv_{11}, cv_{12}, ..., cv_{1j}, ..., cv_{1m}$
CBV_2	$cv_{21}, cv_{22}, ..., cv_{2j}, ..., cv_{2m}$
...	......
CBV_i	$cv_{i1}, cv_{i2}, ..., cv_{ij}, ..., cv_{im}$
...	......
CBV_n	$cv_{n1}, cv_{n2}, ..., cv_{nj}, ..., cv_{nm}$

Fig. 1. The relationship between feature vector and node vector. (a) Feature vector generated by wavelet transform, (b) node vector generated by SOM

For classification of similar image, learning of SOM is performed using feature vector F_{col} of color obtained from wavelet transformation. SOM is a neural network with 2-layered structure. 1 layer is an input one of n dimension, and its node number n is composed of n=48 which is the same as dimensional number of feature vector. 2 layer comprises of map layer in which multiple nodes are in 2 dimensional array, and given random weighting value is initialized before learning is performed.

Classification of similar image consists of generation process of topological feature map and of optimally matched image list.

(1) Generation process of topological feature map

For generation of topological feature map, parameters needed for learning (learning rate, neighborhood radius, size of map layer, neighborhood function, learning iteration) are given, and learning by SOM is performed.

Generation process of topological feature map is as follows. First, a node with weighting vector nearest to given feature vector as an input is selected from map layer. Next, a node and the node within its neighboring area are updated to access feature vector. It is repeated as times as learning iteration designated for this process. As a result of learning, vector generated at each node of map layer is called node vector, and it is represented as

$$CBV_i = [cv_{i1}, cv_{i2}, ..., cv_{ij}, ..., cv_{im}]^T$$

Here, $i(1 \leq i \leq k)$ represents node number of map layer. m means node number of input layer (dimensional number of feature vector of image), and k represents node number of map layer. The obtained topological feature map means that similar relationship between feature vectors is mapped on two dimensions.

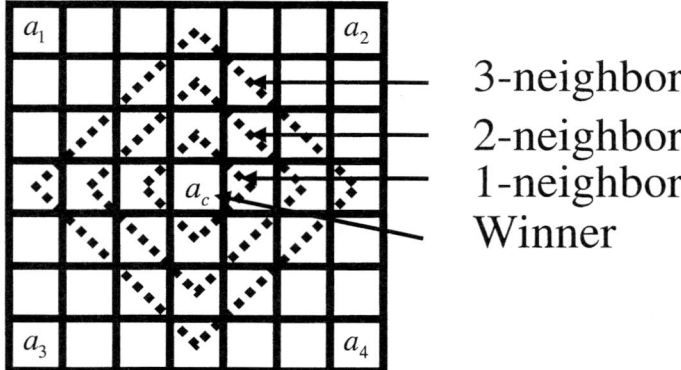

Fig. 2. City block distance between a_c and a_i

(2) Generation process of best matching image list

In generation process of best matching image list, the distance is computed between feature vector and node vector using topological feature map, and similar image is classified with the nearest node (winner node) obtained. The similarity between feature vector and node vector is computed by Euclidean distance. Winner node BMN_i is computed by

$$BMN_i = \min_{1 \leq i \leq k} \{\|F_{col} - CBV_i\|\}$$

The relationship between feature vector and node vector is shown as in Fig. 1. These two vectors have many-to-one relationship by distance measure.

For the above processing, topological feature map and best matching image list for color are generated using color feature vector for all the images in database.

4 High Speed Nearest Neighbor Search by SOM

Nearest neighbor search is the method to search for a point nearest to a given query point. It is often used for multimedia application by performing similar retrieval using feature vector obtained from image. A basic idea of k-NN search is that search processing is finished at a point of time completing all the nearest searching area among collected k candidates, which result from sequential retrieval from the area nearest to a query point. At the point of time, a set of retrieved candidates becomes a result of retrieval.

Based on the above processing, efficient k-NN retrieval is realized for topological feature map and optimally matched image list obtained from a learning result of SOM. In order to do so, minimum bounding rectangle (MBR)(10) widely used for construction of conventional spatial index is applied to each node of topological feature map.

(1) Search in a Neighboring Area of Winner Node

Initially visited node needs to be determined for k-NN search. Here, let the node be a winner node. SOM also has a similarity between winner node and neighboring node. Therefore, it is an effective method to retrieve image closer to query image by proceeding to search for neighboring nodes of winner node.

A neighboring area of map layer is defined as in Fig. 2. The distance between nodes within an radius of winner node and neighboring area is computed by City Block Distance of L_1 space (2). In searching, images similar to a given query image is retrieved sequentially extending a radius of a neighboring node. A maximum radius of a neighboring area to be searched for on map is defined as

$$\Gamma = \max_{1 \leq i \leq 4}(\|a_c - a_i\|)$$

Figure 1 Phase-1:Find Winner node and initialize retrieval candidates.
Figure 2 Phase

Here, a_c and a_i represent location vector of winner node and location vector of 4 points on a lattice respectively.

(2) MBR Definition of Each Node

In order to define *MBR* R for each node of map, an object (image) classified in each node, $O = \{o_{ij} \mid 1 \leq i \leq N, 1 \leq j \leq n\}$, is used. Here, N and n represent object number classified in each node and the elements of feature vector each node has respectively. *MBR* R is defined as

$$\begin{cases} R = (S,T) \\ s_j = \min_{1 \leq i \leq N}(O_{ij}), \quad j = 1,2,3,\ldots,n \\ t_j = \max_{1 \leq i \leq N}(O_{ij}), \quad j = 1,2,3,\ldots,n \quad (s_j \leq t_j) \end{cases}$$

Here, $S = \{s_j \mid 1 \leq j \leq n\}$, $T = \{t_j \mid 1 \leq j \leq n\}$ represent minimum and maximum points in *MBR R* respectively. They are computed by Computation of *MBR R* is performed for all the nodes of map layer except for an empty node.

Next, minimum distance *MINDIST* between given query point *P* and *MBR R* is defined as

$$MINDIST(P, R) = (\sum_{i=1}^{n} (|p_i - r_i|)^{1/2}$$

$$r_i = \begin{cases} s_i & \text{if } p_i < s_i \\ t_i & \text{if } p_i > t_i \\ p_i & \text{otherwise} \end{cases}$$

That is, *MINDIST* is a point of minimum distance between all the points on boundary of rectangle and *P*. Distance computation of *MINDIST*, which is distance measure between feature vectors, uses Euclidean distance of L_2 space. This minimum distance *MINDIST(P,R)* is used for decision of visiting unnecessary nodes

(3) Pruning Strategy

High speed of the nearest search is inevitable in order to retrieve tremendous data within practical time. Here, in order to judge visiting unnecessary nodes, (i) empty node with no object (image) among nodes of map, (ii) minimum distance *MINDIST(P,R)* which is a node larger than maximum distance in retrieval candidate list, are processed for pruning. For this processing, high speed of searching time can be tried.

(4) *k*-NN Search Algorithm in SOM

k-NN search algorithm proposed in this paper consists of two phases.

First, in phase-1, winner node nearest to a given query image is obtained from topological feature map (Step4-11). *k*(*k*>0) of images classified in this winner node most similar to a query image are sorted and stored in retrieved list as candidates to be retrieved(step 14). *k*-th candidate in retrieved list is the farthest from a query image in

Table 1. Map size vs. empty nodes for each data set

Data set	Map size	Empty nodes	(%)
1000	32×32	434	42
5000	70×70	2634	53
10000	100×100	5395	54
20000	140×140	10819	55
30000	175×175	18377	60
40000	200×200	23706	59

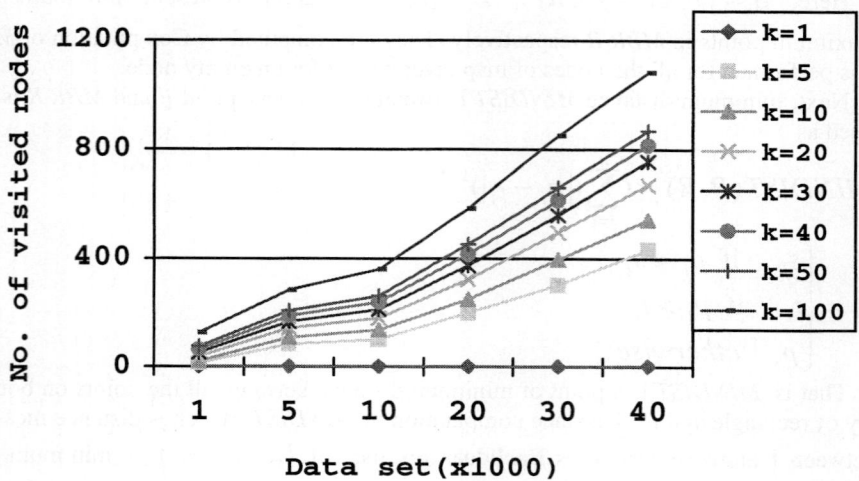

Fig. 3. Total number of visiting nodes for increasing k

the result of retrieval, because retrieved list is stored with it sorted from candidate nearest to a query image. In this processing, maximum neighbor area Γ to be searched for on map layer is obtained.

In Phase-2 the nearest search is performed with Γ radius of neighbor area of winner node extended. In this processing, two kinds of pruning strategy are applied in searching as discussed before (Step11-19). **RetrievedList** are updated with images classified in the node to be searched for(Step20). This processing is executed recursively for $\gamma \leq \Gamma$(Step24).

5 Experimental Results and Study

k-NN algorithm was implemented in order to verify an effectiveness of NN search algorithm in SOM, and similar retrieval of image was performed with examples for its performance evaluation. Image data used in the experiment was composed of 40000 images added with image collection being open in Stanford Univ. and with image data of H^2soft Company. For the size of image database, 6 kinds of data set different from each other in its size was used as represented in Table-1 (Data set). The size of each image data was fixed by 128×128 pixel. We used feature vector of each image for color of 48 dimension described in section 3.

The performance of image retrieval was evaluated with 20 query images selected randomly for each database, and result of the measure was its average.

Table 2. Retrieval time cost of each data set for increasing k (sec)

k	Data set(×1000)					
	1	5	10	20	30	40
1	0.07	0.25	0.62	1.57	2.54	4.26
10	0.08	0.26	0.62	1.57	2.56	4.30
50	0.08	0.26	0.63	1.58	2.57	4.30
100	0.08	0..2	0.64	1.60	2.58	4.30

Ultra5(CPU:UltraSPARC-□i, OS:Solaris8, Memory:512MB) of SUN Microsystems was used along with C++ of program language. Learning by SOM was performed with feature vector for color of image data in input data. The size of map layer in SOM was established almost equal to image number of each data set as represented in Table-1 (Map Size). Weighting vector each node of map layer has is randomly initialized 48 dimensional vector which has the same dimension as feature vector. Learning iteration T, initial value of neighbor set in learning γ_0, and initial value of learning rate α_0 were 10000, 30, and 0.9 respectively.

Neighbor set and learning rate is reduction function updated continuously while learning is performed. As a result of learning of SOM, generally, many images are mapped to one node in each node of map layer, because similar images are included in database. Therefore, empty node in which image is not classified at all occurs, if node number of map layer is established almost equally to data set. In this experiment, Table-1 (Empty nodes,(%)) represents rate of empty nodes of map layer obtained from 6 kinds of data sets performing learning. MBR is obtained for each node of map layer except for empty node.

k-NN retrieval is performed for a given query image. In this experiment, values of retrieval candidate k were experimented for 8 kinds of k=1,5,10,20,30,40,50,100. Here, k=1 means a query image itself, because image data in database are used in query image in this experiment.

Distance measure between a query image and an image in database uses Euclidean distance of L_2 space. Through the experiment, operation of k-NN search algorithm was investigated in an increase of the size between k value and data set. We also reviewed the effect of *MINIDIST*.

6 Conclusion

In this paper, k-NN search algorithm is proposed to perform highspeed similar retrieval of image in a result of learning of SOM. MBR for each node of map layer was applied for an implementation of highspeed search algorithm, and minimum distance function between a given query point and MBR, *MINDIST* was defined. We implemented a proposed algorithm and evaluated its effectiveness through the experiment of a similar image retrieval using actual image data. The realization of highspeed similar image retrieval by *MINDIST* was verified through this experiment. The effect of k-NN search by SOM was confirmed in that the retrieval time was almost not changed regardless of an increase of retrieval candidate k.

It became possible similar image to be retrieved in practical time, even though retrieval technique by example image was used. The problem is that the accessed image number was too many for the number of retrieval candidate k. It makes an effect on retrieval time. Considering this, we are going to improve the algorithm, and to evaluate its performance with an implementation of actual application program.

Acknowledgment

This paper is supported by Nam Seoul University.

References

1. Guido Deboeck, and Teuvo Kohonen: Visual Explorations in Finance with Self-organizing Maps, Springer-Verlag, London(1998);Japanese ed., Springer-Verlag, Tokyo (1999).
2. Kohonen, T.: Self-Organizing Maps, Series in Information Sciences, vol.30, Springer-Verlag, second edition, Berlin (1997).
3. Baba, N., Kojima F., and Ozawa M.: Basis and Application of Neural Network, Kyoritu Shuppan, Tokyo, Japan (1994).
4. Kohonen, T., Hynninen, J., and Laaksonen, J.: SOM_PAK:The Self-Organizing Map Program Package, In Technical Report A31,Helsinki University of Technology, Laboratory of Computer and Information science (1996).
5. Gudivada, V. N.and Raghavan, V. V. eds.: Content-based Image Retrieval System, IEEE Computer, Vol.28, No.9, pp. 18--22 (1995).
6. Flickner, M. et al.: Query by Image and Video Content: The QBIC System, IEEE Computer, Vol.28, No.9, pp. 23--32 (1995).
7. Faloutsos, C. et al.: Efficient and Effective Query by Image Content, J. Intell. Inform. Syst.}, Vol.~3, pp.\ 231--262 (1994).
8. Eakins, J. P. and Graham, M. E.: Content-based image retrieval. Report to the JISC Technology Applications Programme (1999).
9. Rui, Y., Huang, T. and Chang S-F.:Image Retrieval: Current Techniques, Promising Directions, and Open Issues, J. Visual Communication and Image Representation(JVCIR), Vol.10, No.1, pp. 39--62 (1999).
10. N. Roussopoulos, S. Kelley, and F. Vincent. "Nearest neighbor queries," In Proceedings of the ACM SIGMOD Conference, pp.71--79, San Jose, CA, May (1995).
11. Yong Rui, Thomas S. Huang, Michael Ortega, and Sharad Mehrotra, ``Relevance Feedback: A Power Tool in Interactive Content-Based Image Retrieval," IEEE Trans. on Circuits and Systems for Video Technology, Special Issue on Segmentation, Description, and Retrieval of Video Content, pp.644--655, Vol.8, No.5(1998).
12. Mallat,S. G.: Multifrequency channel decompositions of images and wavelet models, IEEE. Trans., Acoust.,Speech and Signal Proc., Vol.37, No.12, pp. 2091--2110 (1989).
13. Niblack, W. et al.: The QBIC project:Query Image by content using color, texture and shape, SPIE Storage and Retrieval for Image and Video Databases, San Joes, pp. 173--187 (1993).

An Expert System Development for Operating Procedure Monitoring of PWR Plants

Malrey Lee[1], Eun-ser Lee[2], HeeJo Kang[3], and HeeSook Kim[4]

[1] School of Electronics & Information Engineering, Chonbuk National University, 664-14, DeokJin-dong, JeonJu, ChonBuk, Korea
`mrlee@chonbuk.ac.kr`
[2] TQMS, 370 Dangsan-dong 3ga, Youngdeungpo-gu, Seoul, Korea
`eslee@object.cau.ac.kr`
`http://object.cau.ac.kr/selab/index.html`
[3] Division of Computer & Multimedia Contents Engineering, Mokwon University
`hjkang@mokwon.ac.kr`
[4] Dep.of Multimedia Asan Information Polytechnic College
`prima@jopo.or.kr`

Abstract. An expert system for operating-procedure monitoring during plant startup, from cold shutdown to power operation, has been chosen as the object of this study. The knowledge base for the system is principally derived from the written form of the standard procedures. The programmed procedures of the system are displayed on a CRT screen to assist the operators' understanding. The main features of the system include the following: (1) computerization of the procedures for checking the initial conditions, the precautions, and the operational procedures; (2) plant status monitoring to detect any plant malfunctions or to help with the operators' decision-making; (3) the provision of access to technical specifications of the CRT screen at any time during plant operation; and (4) easy prediction of the estimated critical point prior to reactor criticality, without complicated hand calculation. The results of system verification indicate that the system performs the intended functions, and can be used as an effective tool to minimize the operators' burden during plant startup.

Keywords: Expert systems, general operating procedures, PWR, knowledge bases.

1 Introduction

The operation of complex nuclear power plants is potentially a very promising area for the application of computerized systems technology. Computerized operator-aid systems(expert system) can be used effectively to eliminate much of the uncertainty in operator decisions, by providing expert advice and rapid access to a large information base. Much development has taken place, and several prototypes have been built to support the operation of nuclear power plants in abnormal of accident conditions. For the operation of a nuclear power plant, a great amount of both numerical and symbolic information is handled by the operators, even during normal operation, and the operators are some- times affected by fatigue, stress, and emotional and environmental factors that may have varying degrees of influence on their performance[1,2]. Nevertheless, computerized aid systems to guide normal operational procedures have not attracted much attention, due to the belief that experienced of well-trained operators can do their task well.

The need for computerized operator aid systems to reduce the operators' burden or to assist in their decision making during plant startup operation has been recognized because of information overload, the complexities of the procedures, and the tedious calculations because of which operators may lose concentration on fine detail, especially under stress of fatigue conditions. Also, because many of the processes are complex and are not well understood, the probability always exists that operators may misunderstand or omit some important items that need to be checked during the process. A computerized operator aid system for monitoring the operation procedures has been developed here for a pressurized water reactor. The system performs the following functions to assist the operators during heatup and startup, from cold shutdown to power generation.

The main program has been written using the Systems Designers Prolog. The programs for data aquisition or graphical display were programmed using C. Prolog is a high-level language that was specifically designed for expert-system development. System Designers Prolog is a powerful optimized implementation of the Prolog programming language, suitable for applications on personal computers, that provides a complete programming environment for developing and maintaining Prolog programs on personal computers[3,4].

2 Structure of the General Operating Procedure

The general operating procedure (GOP) describes the plant's operational methods or procedures necessary for plant heatup, startup, power operation, shutdown, cool down, and so forth[5]. The computerized aid system developed for this study treats procedures describing the plant operation from cold shutdown to full power operation[6,7]. The overall architecture of the system is shown in Fig. 1. In the GOP module, the procedures necessary for heating up the plant to power operation are displayed on a CRT, and proceed step by step according to the operators' response. It consists of four components, GOP1-GOP4 and each component is separated from the others. That is, after one component completes execution, the next component is selected from the menu. In the ECP module, the boron concentration necessary to make the reactor critical is calculated. In the shutdown margin module, the shutdown margin is evaluated with a minimum input. In the last module, the real-time axial flux difference (AFD) is displayed on the CRT, to monitor the current status of the axial power distribution. When starting up the system, the main title displaying the program name appears on the screen. On striking any Key, a menu for selecting the desired modules appears on the screen and by moving the cursor up or down the module under selection is high lighted to help the user make a choice.

The user interface must also provide a level of ease of use that ensures a smooth interaction between the user and the system. The system mainly employs a menu driven approach as a user interface in each module, thus allowing the user to choose

An Expert System Development for Operating Procedure Monitoring of PWR Plants

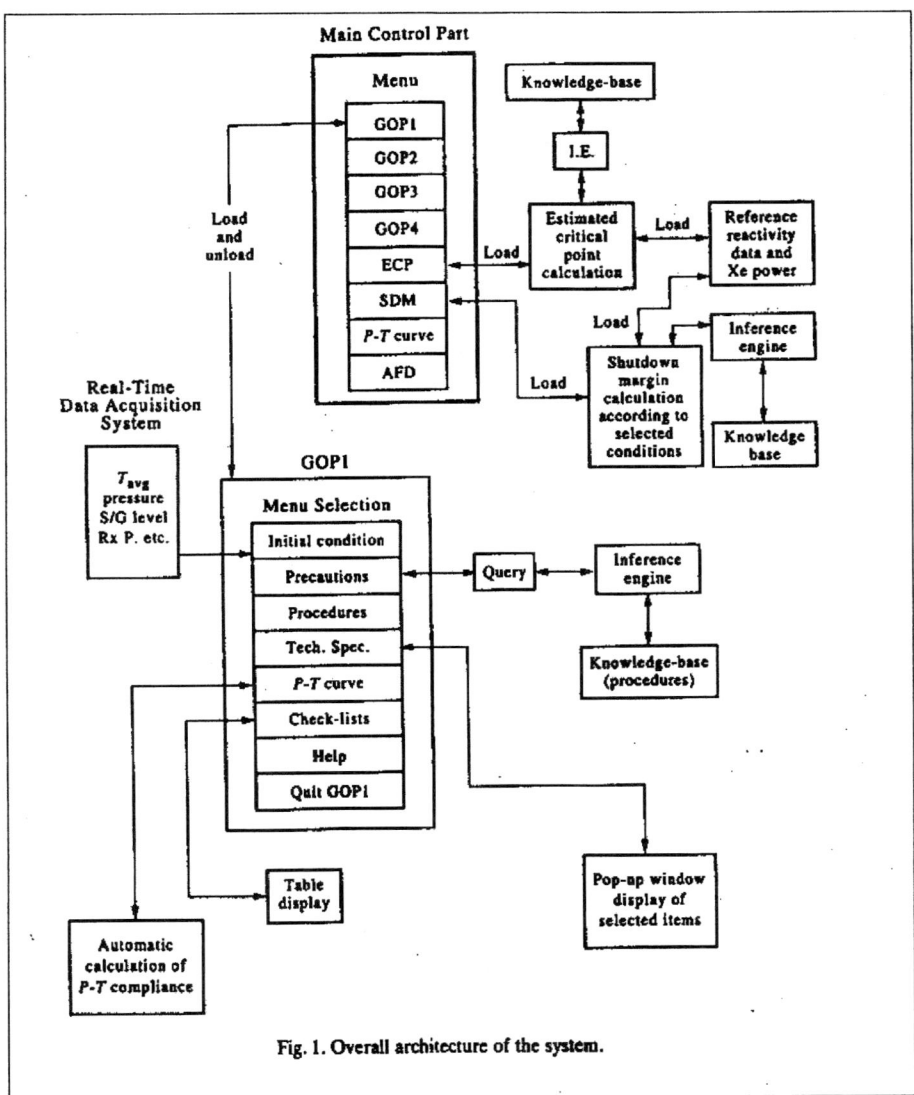

Fig. 1. Overall architecture of the system

what he wants the system to perform. After the selection of any item from the menu, a window appears on the screen and displays messages, queries, or results of a calculation. The messages provide instructions on how to operate the system, or descriptions of some pertinent commands, and how to use them. If the user does not know what command to use, or wants a description of a command, the help function can be used to provide a list of commands and descriptions.

3 An Expert System for the GOP

3.1 GOP Module

In conventional plant operation, the operator does not have much computerized assistance, and relies chiefly on written forms of procedures, long lists, and his own knowledge of the plant, to understand the functional relationships of the equipment in the plant. In such systems the database is widely distributed. Use of this distributed information requires the operator to gather all the data, analyze it, and arrive at appropriate decisions. This type of stressful situation can increase the possibility of human errors. A well-optimized man-machine interface during plant operation may be a good way of solving the problems mentioned above. So, in the GOP module, all the materials necessary for plant startup, for instance procedures, technical specifications, checklists, etc., are computerized, and the required procedures are performed in a sequence by handling them simultaneously on the same CRT screen with windows or menus. This reflects the general trend in the nuclear industry, toward the increased use of CRT displays.

- Knowledge base

The knowledge base for processing the procedures is derived from the written form of the standard procedures. Each GOP component consists of four parts, namely the initial conditions, the precautions and limitations, the procedures, and the appendices (including many checklists or curves). The startup procedures from cold shutdown to power operation, used as the knowledge base, are constructed as follows[7].

These procedures are displayed on the CRT step by step, and any procedure resulting from the operators' answer is collected into the database. During the process, in order to select any other item in the menu, the operator may leave the current task and select the desired item, for example technical specifications. When he returns to the previous task, the procedures he has already answered are skipped over.

Procedure processing strategy

In this system, procedure numbers are used as patterns. Procedure numbers are processed one after another, and then the resulting number searches for its matching pattern, After matching of the relevant procedure are displayed on the user-defined window, and the operator's response, yes or no, is added to the database. In the processing procedure, cautions are displayed, when necessary, to inform operators of warnings needed for plant operation.

- Plant status monitoring

The accident at Three Mile Island (TMI) and the subsequent investigations have demonstrated the need for improving the presentation to operators of the plant status, and of information – processing methods. During plant operation, a reactor operator is required to monitor and process large amounts of data to verify the operating status of the safety of the plant. The primary function of plant status monitoring is to help operators make a quick assessment of a plant's safety. In terms of this primary function, the monitoring system developed for this study can provide operators with plant information by means of an integrated display. The operator is a key subsystem in a plant, who can synthesize the plant process and assess the important plant functions from the data provided on the display, A computer driven CRT display allows more flexibility in the data display format and data display enhancement than to analog meters and analog chart recorders.

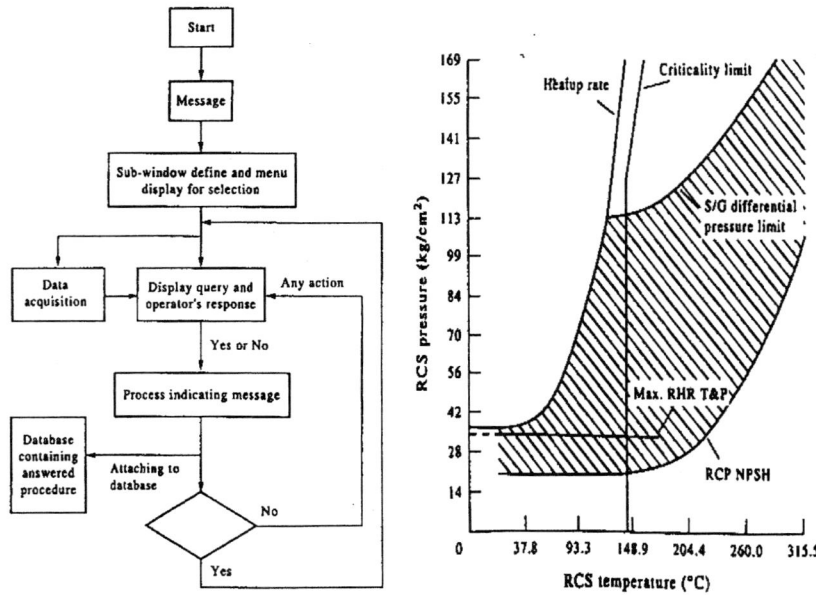

Fig. 2. Schematic diagram of the processing procedure

Fig. 3. Pressure-temperature limit in RCS

The system includes two verification methods for compliance with the RCS P-T limits. One is to calculate P-T limit with T_{avg}, the pressurizer pressure, obtained through the data-acquisition system. The curves shown is Fig. 3 are divided into several regions that can be approximately represented by straight lines. Then, an equation is derived to represent each of these lines. The T_{avg} obtained from the plant is entered in the<if> part, and the <then> part generates the limiting value of the PZR pressure, within which the RCS pressure should remain. Then the actual PZR pressure from the data-acquisition system is compared with the limiting value, If the RCS P-T is satisfactory, then compliance has been achieved. If it does not fall within the limits, the system displays the action to be taken. An example of an <if-then> rule search for a solution is as follows:

Rule : IF T_{avg} belongs to any region from among the identified regions
  THEN the upper limit of the pressure (Pr1) is determined by the equation
      on the upper side
  AND the lower limit of the pressure (Pr2) is determined by the equation on
      the lower side.

where Pr_1 is the upper – limit pressure and Pr_2 is the lower – limit pressure, and each equation represents the respective straight line dividing the regions.

3.2 Estimated Critical-Point Calculation Module

When operators want to manipulate the reactor under hot standby conditions, they should determine the control rod position and the boron concentration needed to take the reactor to "critical". The concept of the estimated critical-point calculation is as follows:

The core is critical ($k_{eff} = 1$) prior to shutdown, at a given percentage of power, xenon concentration, samarium concentration, boron concentration, and rod position. The algebraic sum of the changes in reactivity, after shutdown, will determine the amount of positive or negative reactivity that has been added to the core due to shutdown. A positive of negative algebraic sum will correspond to the respective boration or dilution required to compensate for the reactivity change of the core. In general, operators will calculate the necessary boron concentration at a pre-determined rod position. To determine the required boron concentration, operators usually refer to the curve book, and then read tables or curves describing nuclear parameters such as moderator temperature coefficient, power defect, rod worth, etc. The operator can lose concentration during he process of hand calculation, because it needs many input values and is a long process. This new system therefore automatically calculates the ECP with minimum inputs, and will be useful to operators in determining the necessary boron concentration to make the reactor critical, by removing the tedious hand-calculation process. A curve representing any nuclear parameter is divided into several regions that can each be approximately described as straight line. After reception of an input, for example T_{avg} by pattern matching, the output (for example, the moderator temperature coefficient) is generated from the equation describing the identified region. The overall operating/calculating structure of the ECP module is shown in Fig. 4. First, the system subtracts the total reactivity at startup from the total reactivity at shutdown, calculated from the reference reactivity data. The user inputs a differential boron worth (pcm/ppm) for an estimated boron concentration at the anticipated temperature. The required boron change is calculated by dividing the estimated difference in reactivity (pcm) between the shutdown and the startup by the differential boron worth (p cm/ p pm). Finally, the boron concentration required to make the reactor critical is obtained by algebraically adding the last critical boron concentration prior to shutdown, obtained from the reference reactivity data, to the required boron concentration change, calculated from the above step.

3.3 Shutdown Margin Evaluation Module

The objective of the shutdown margin calculation is to confirm whether the current shutdown margin (after a reactor trip, at startup, or at power operation) is over the limit value as specified in the technical specifications. The shutdown margin is defined in the technical specifications. The shutdown margin is defined as the amount of negative reactivity by which a reactor is maintained in a subcritical state at hot zero power (HZP) conditions after a control rod trip with no change in pre-trip boron or xenon parameters. In computing the total rod worth at HZP, it is assumed that the most reactive rod remains stuck out of the core. In addition, an uncertainty of 10% is considered. Like the ECP calculation, the shutdown margin calculation procedure requires many input values which can be obtained from the curve book, and is a time-consuming process. The probability that human error will occur exists as mentioned in the ECP module. The inference strategy is the same method as that used in the ECP module.

4 System Verification and Validation

Generally, as a matter of policy, verification and validation (V&V) should always be carried out by a group completely independent of the developer of a system. However, at the present stage, the system developer has performed tests to verify all the required functions. System verification and validation have been performed in the following fields.

- Straight-line approximation

The programs such as the P-T curve display, the ECP module, and the SDM module principally use straight lines to infer any result. The results calculated using these straight – line equations have been compared with the exact values which are used to make the real curves. The compared results show a good representation of the actual values, with negligible errors. Consequently, straight-line approximation is considered as a good method of modeling complex curves.

- Monitoring of plant parameters

In each GOP component, the essential plant parameters to be monitored are slightly different. The data acquisition programs are coded differently in each GOP program, and these parameters have been mentioned in the previous section. First, a data-acquisition program (coded in C) is executed after receiving a command from the main program (coded in Prolog) and it makes up the data files containing the parameters. Then the main program gets these values from the files, and displays them on the screen. That is, the monitoring of plant parameters needs two steps. These programs are correctly executed, without any problems.

- Procedure processing

Each GOP component consists of several parts, such as the initial conditions, precautions and procedures. In addition to the above parts, there are other parts, such as technical specification displays, checklist displays, help functions, etc. In the process of performing procedures, technical specifications or checklists can be referred to whenever necessary. The procedure number appears at the bottom of the screen in green if all requirements are satisfied, otherwise in red. All of the intended functions are correctly performed.

5 Conclusions and Recommendations

An expert system for monitoring operating procedures during plant star up, from cold shutdown to power operation has been developed for PWR plants. The knowledge base for the system is principally derived from the written form of the standard procedures. A computerized display of this knowledge base has proved to be useful for pointing out ambiguities and inconsistencies in the current operating procedures. Often, the operators are put under stress by complexities or ambiguous clauses in the written procedures. However, using the developed system, the procedure can be followed step by step according to the operators' responses, and thus the probability that any important steps are skipped will be eliminated. Therefore, the use of this system during plant startup will be valuable in minimizing the operators' burden and the errors made.

The system principally offers programmed operational guidelines to operators as well as programmed displays of technical specifications and checklists, by means of pop-up windows. In order to help the operator's decision making, or to perform automatic calculation of the pressure-temperature limits for the reactor's coolant system, plant parameters are acquired through the data-acquisition system and are displayed on the screen in the form of sub-windows, using various colors. The plant parameters used for system verification are arbitrarily generated from the DC power source because plant data acquisition is very limited. The real plant data could be used in a future application stage without much difficulty.

The system is capable of some calculations, which are performed efficiently by the estimated critical-point calculation module and the shutdown margin evaluation module. By removing the tedious hand calculation process, these modules are useful for operators in determining the boron concentration required to make the reactor critical, of in confirming whether the required shutdown margin has been secured.

Finally, the system developed here can aid operators during the complex startup operation. Using the system will improve the operability of nuclear power plants and reduce the operators' burden and errors and therefore, will contribute to enhancing the reliability, safety, and efficiency of nuclear power plants.

References

1. Erdmann R. C. and Sum B. K.-H. *An Expert System Approach for Safety Diagnosis*, pp. 162 – 172. Nuclear Technology (1988).
2. Cheon S. W. and Chang, S. H. Neural network application to fault diagnosis by pattern recognition of multiple alarms in nuclear power plant. Proc. Conf. on Expert Systems, pp.160-176 (1993)
3. Townsend, C. *Mastering Expert Systems with Turbo Prolog*, pp.21. Howard W. SAMS & Co.(1987)
4. General Operation Procedure, Procedure No. : GOP-1, Procedure Title : Cold Shutdown to Hot Shutdown(1984)
5. Ohga Y. A Computer Program for Assessment of Emergency Operation Procedures under Non-LOCA Transient Conditions in BWRs, pp. 465-474. Nuclear Technology (1983)
6. Ogha Y. and Utena S. An event-Oriented Method for Determining Operation Guides under Emergency Condition in BWRS, pp. 229-236. Nuclear Technology(1984)
7. Lee J. K. and Kwon S. B. An expert system development planner using a constraint and rule-based approach. Proc. Conf. On Expert Systems, PP. 333-351(1999)

Security Evaluation Targets for Enhancement of IT Systems Assurance

Tai-hoon Kim[1] and Seung-youn Lee[2]

[1] San-7, Geoyeo-Dong, Songpa-Gu, Seoul, Korea
`taihoonn@empal.com`
[2] SKK Univ., Dept. of Information & Communication Eng. Kyonggi, 440-746, Korea
`syoun@ece.skku.ac.kr`

Abstract. The general systems of today are composed of a number of components such as servers and clients, protocols, services, and so on. Systems connected to network have become more complex and wide, but the researches for the systems are focused on the 'performance' or 'efficiency'. While most of the attention in system security has been focused on encryption technology and protocols for securing the data transaction, it is critical to note that a weakness (or security hole) in any one of the components may comprise whole system. Security engineering is needed for reducing security holes may be included in the IT systems. This paper proposes a method for securing the IT systems. This paper proposes IT system security evaluation and certification for achieving some level of assurance each owners of their IT systems want to get.

1 Introduction

In general, threat agents' primary goals may fall into three categories: unauthorized access, unauthorized modification or destruction of important information assets, and denial of authorized access. Security countermeasures are implemented to prevent threat agents from successfully achieving these goals.

Information assets consist of many components. The physical systems and the information stored or processed in the systems are examples of the information assets. Therefore, the strategy protecting only information is not good one. If someone wants to protect or secure his valuable information, he must build some countermeasures for IT systems itself.

Security countermeasures should be considered with consideration of applicable threats and security solutions deployed to support appropriate security services and objectives. Subsequently, proposed security solutions may be evaluated to determine if residual vulnerabilities exist, and a managed approach to mitigating risks may be proposed.

But there is a problem about the security countermeasures. How can we believe that the countermeasures implemented may protect our IT systems? About this question, some answers may exit. And the answers may assure that the countermeasures can protect IT systems from the threat. Evaluation is a method of them and has been the traditional means of providing assurance.

So the researches about the evaluation have proceeded and many evaluation criteria were developed. The Trusted Computer System Evaluation Criteria (TCSEC), the European Information Technology Security Evaluation Criteria (ITSEC), and the Canadian Trusted Computer Product Evaluation Criteria (CTCPEC) existed, and they have evolved into a single evaluation entity, the CC. The CC is a standard for specifying and evaluating the security features of IT products, and is intended to replace previous security criteria such as the TCSEC.

The criteria listed above mainly concerned with IT products. But we may not obtain perfect security of IT systems only by products. Therefore, in these days, operational and personnel security-related considerations are considered as important components for secure IT systems. So the ISO/IEC 17799 and 13335 are used to complement these things.

This paper identifies some components should be evaluated and certified to assure that IT systems are secure. Security objective of IT systems will be obtained by protecting all areas of IT systems, so not only visible parts but also non-visible parts must be protected. And for verifying all the parts of IT systems are protected, we should check the scope of evaluation and certification covers all necessary parts.

2 Overview of the CC

The multipart standard ISO/IEC 15408 defines criteria, which for historical and continuity purposes are referred to herein as the Common Criteria (CC), to be used as the basis for evaluation of security properties of IT products and systems. By establishing such a common criteria base, the results of an IT security evaluation will be meaningful to a wider audience.

The CC will permit comparability between the results of independent security evaluations. It does so by providing a common set of requirements for the security functions of IT products and systems and for assurance measures applied to them during a security evaluation. The evaluation process establishes a level of confidence that the security functions of such products and systems and the assurance measures applied to them meet these requirements. The evaluation results may help consumers to determine whether the IT product or system is secure enough for their intended application and whether the security risks implicit in its use are tolerable.

The CC is presented as a set of distinct but related parts as identified below.

Part 1. Introduction and General Model, is the introduction to the CC. It defines general concepts and principles of IT security evaluation and presents a general model of evaluation. Part 1 also presents constructs for expressing IT security objectives, for selecting and defining IT security requirements, and for writing high-level specifications for products and systems. In addition, the usefulness of each part of the CC is described in terms of each of the target audiences.

Part 2. Security Functional Requirements, establishes a set of functional components as a standard way of expressing the functional requirements for TOEs (Target of Evaluations). Part 2 catalogues the set of functional components, families, and classes.

Part 3. Security Assurance Requirements, establishes a set of assurance components as a standard way of expressing the assurance requirements for TOEs. Part 3 catalogues the set of assurance components, families and classes. Part 3 also defines evaluation criteria for PPs (Protection Profiles) and STs (Security Targets) and presents evaluation assurance levels that define the predefined CC scale for rating assurance for TOEs, which is called the Evaluation Assurance Levels (EALs).

In support of the three parts of the CC listed above, it is anticipated that other types of documents will be published, including technical rationale material and guidance documents [1].

3 Overview of the ISO/IEC 17799

The ISO/IEC 17799 standard gives recommendations for information security management for use by those who are responsible for initiating, implementing or maintaining security in their organization. ISO 17799 is "a comprehensive set of controls comprising best practices in information security," and is essentially an internationally recognized generic information security standard. It has ten sections of security controls with various perspectives. Each section has two levels of sub-items. The key objectives of these sections are described in below.

Sec.4.1 - Security policy: to provide management direction and support for information security.

Sec.4.2 - Security organization: to manage information security within the company; to maintain the security of organizational information processing facilities and information assets accessed by third parties; to maintain the security of information when the responsibility for information processing has been out-sourced to another organization.

Sec.4.3 - Asset classification & control: to maintain appropriate protection of corporate assets and to ensure that information assets receive an appropriate level of protection.

Sec.4.4 - Personnel security: to reduce risks of human error, theft, fraud or misuse of facilities; to ensure that users are aware of information security threats and concerns, and are equipped to support the corporate security policy in the course of their normal work; to minimize the damage from security incidents and malfunctions and learn from such incidents.

Sec.4.5 - Physical & environmental security: to prevent unauthorized access, damage and interference to business premises and information; to prevent loss, damage or compromise of assets and interruption to business activities; to prevent compromise or theft of information and IPFs.

Sec.4.6 -Communication & operation management: to ensure the correct and secure operation of information processing facilities; to minimize the risk of systems failures; to protect the integrity of software and information; to maintain the integrity and availability of information processing and communication; to ensure the safeguarding of information in networks and the protection of the supporting infrastructure; to prevent damage to assets and interruptions to business activities; to prevent loss, modification or misuse of information exchanged between organizations.

Sec.4.7 - Access control: to control access to information; to prevent unauthorized access to information systems; to ensure the protection of networked services; to prevent unauthorized computer access; to detect unauthorized activities; to ensure information security when using mobile computing and tele-networking facilities.

Sec.4.8 - System development & maintenance: to ensure security is built into operational systems; to prevent loss, modification or misuse of user data in application systems; to protect the confidentiality, authenticity and integrity of information; to ensure IT projects and support activities are conducted in a secure manner; to maintain the security of application system software and data.

Sec.4.9 - Business continuity planning: to counteract interruptions to business activities and to critical business processes from the effects of major failures or disasters.

Sec.4.10 - Compliance: to avoid breaches of any criminal or civil law, statutory, regulatory or contractual obligations and of any security requirements; to ensure compliance of systems with organizational security policies and standards; to maximize the effectiveness of and to minimize interference to/from the system audit process [2].

4 IT Systems and Threat to Them

In these days, most companies and governments have their own IT systems that store, process and transmit data. Each IT systems have their own unique characteristics and common requirements.

As IT systems deal with important data, threats to the data itself and the systems contain that data are increasing. The Fig. 1 is an example of IT systems and an objective of that system is supporting decision-making.

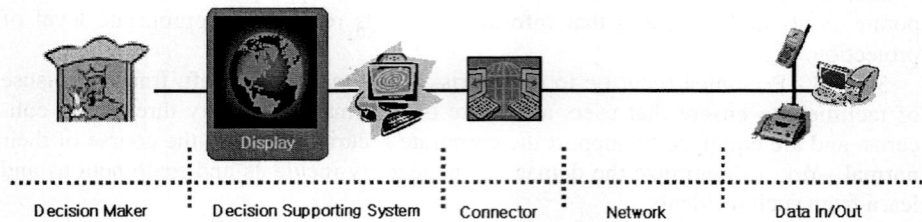

Fig. 1. The example of general IT systems

The main problem is occurred when the decision makers of that system can't trust the display screen or back data reported. In fact, threat agent may compromise one or more components of the IT systems and replace some data of that system.

Therefore, the owners of IT systems make some security countermeasures to protect their systems from threat agents. Next Fig. 2 is the concept of this idea. This figure is expressed in CC and we modified some parts of them.

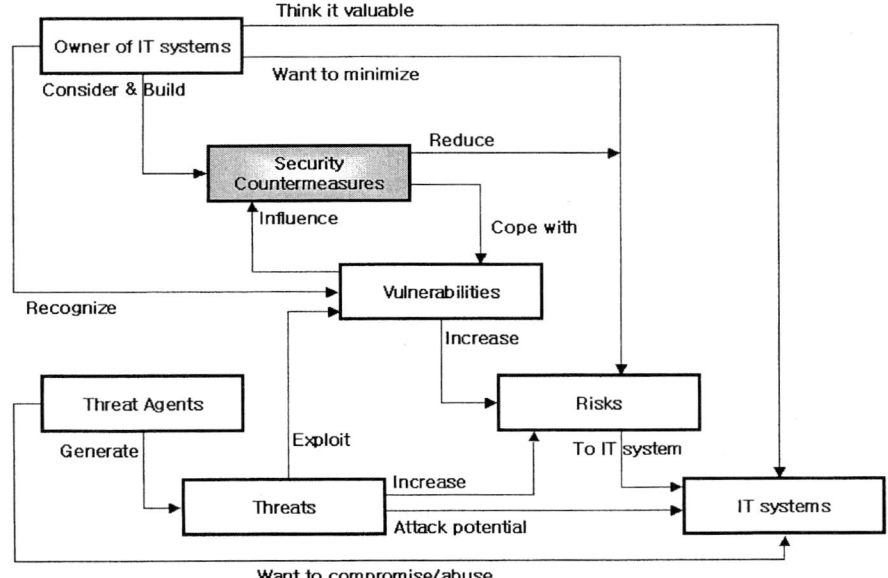

Fig. 2. The concept of security countermeasures and risks

5 IT Systems Security Countermeasures and Threat to Them

The security countermeasures are built, and though they working well, some problems remain as usually. The representative problem is that how we can trust that the security countermeasures may protect IT systems from compromising.

In general, after the security countermeasures are built and start working, owners and administrators may think that IT systems are safe and secure. This is a very big security hole not solved by physical system.

For example, although we bought and installed the IDS (Intrusion Detection System), it is impossible that all of the electronic intrusions are detected. If any developer who has malicious mind made that IDS? And if he or she inserted backdoor codes in that IDS? In any case, who can estimate the results of the IDS installation?

Therefore, we need some assurance for security countermeasures. Next Fig. 3 depicts this concept. This figure is expressed in CC and we modified some parts of them.

6 IT Systems Security Countermeasures and Related Components

IT systems consist of very many components, and have all kinds of security countermeasures. Now all we want to get is assurance for the security countermeasures of IT systems. Fig. 4 depicts the relationship between Security and Assurance.

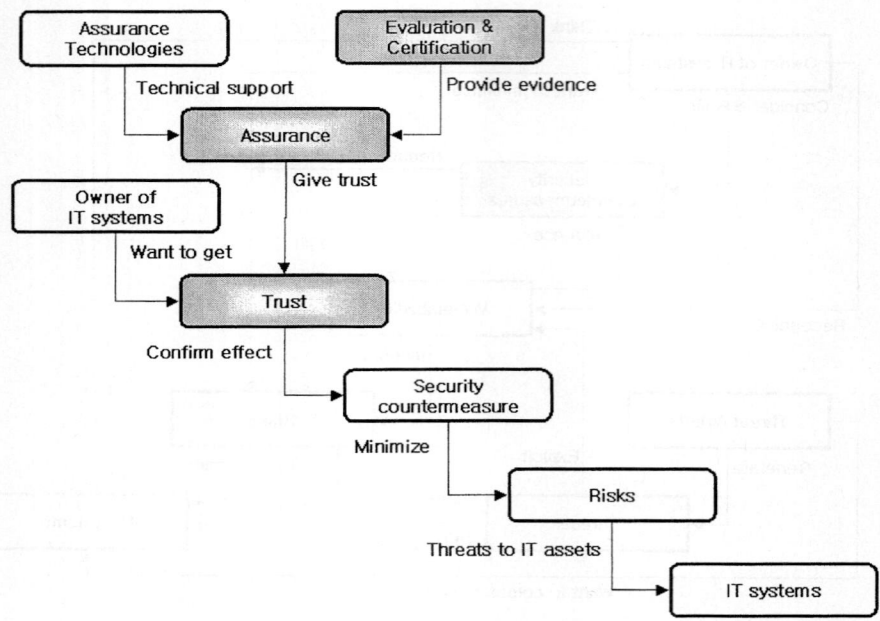

Fig. 3. The concept and necessity of assurance

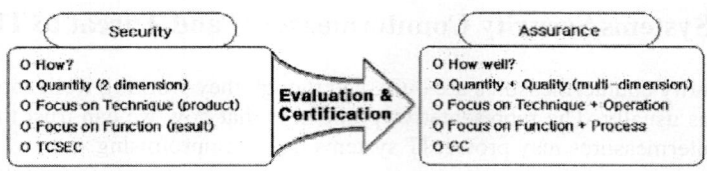

Fig. 4. The relationship between Security and Assurance

To get assurance, we must consider more precisely about the security countermeasures. Next items should be considered to assure that IT systems are secure.

- Are the organizational <u>security policies</u> reflected properly?
- Are the security objectives identified properly according to the security policies?
- The value of assets, threat/vulnerability, and impact of compromise are reflected properly?
- Security objectives are defined according to the result of <u>risk assessment</u>?
- Some places need <u>security countermeasures</u> are identified?
- Security countermeasures are connected to each other to obtain security objectives?
- Security countermeasures are enough to achieve security objectives?
- Security countermeasures reduce the vulnerabilities or threats really?
- The <u>residual risk</u> is under security baseline?
- <u>IT products used</u> in the IT systems are made via secure process?
- Can you trust that the <u>development site and developers</u> are believable?

7 Security Evaluation Targets of IT Systems

To assure that IT systems are secure, you should prove that all the parts of those IT systems are secure. If a part or a component of IT systems contains a security hole, the whole IT systems are not secure.

Next Fig. 5 depicts the parts should be evaluated to assure that IT systems are secure. All blocks of Fig. 5 may be the security evaluation targets for enhancement of IT systems assurance.

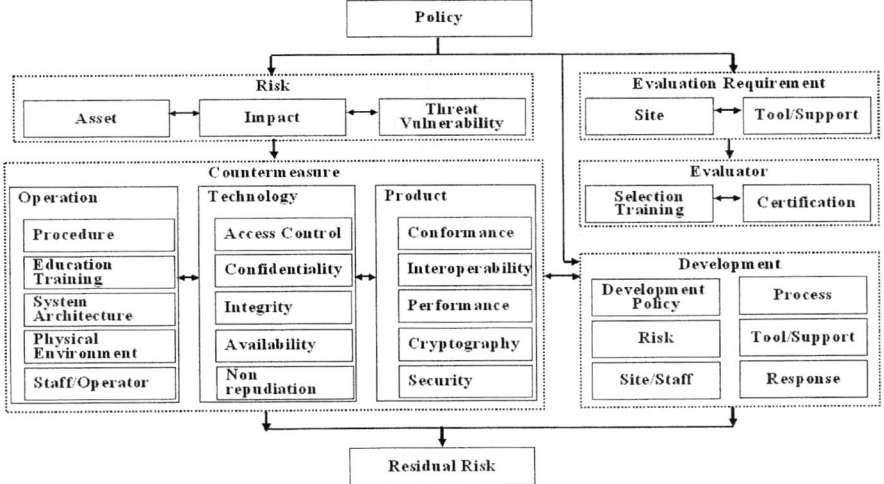

Fig. 5. Security evaluation targets for enhancement of IT systems assurance

8 Conclusions

The base of this paper is a research that makes the security countermeasures used in IT systems more believable. In other words, how can we believe that the countermeasures implemented may protect our IT systems?

We find the answer in the security engineering and the assurance methodology using evaluation and certification. This paper contains only the basic concept of our research and the start point of future work.

This paper identifies some components should be evaluated and certified to assure that IT systems are secure. Although the model proposed in this paper may be not a perfect one, the concept and ideal object of our research is contained in this paper.

References

1. Tai-hoon Kim, Tae-seung Lee, Kyu-min Cho, Koung-goo Lee: The Comparison Between The Level of Process Model and The Evaluation Assurance Level. The Journal of The Information Assurance, Vol.2, No.2, KIAS (2002).

2. Sangkyun Kim, Hong Joo Lee, Choon Seong Leem: Applying the ISO17799 Baseline Controls as a Security Engineering Principle under the Sarbanes-Oxley Act, ICCMSE 2004
3. Tai-hoon Kim, Yune-gie Sung, Kyu-min Cho, Sang-ho Kim, Byung-gyu No: A Study on The Efficiency Elevation Method of IT Security System Evaluation via Process Improvement, The Journal of The Information Assurance, Vol.3, No.1, KIAS (2003).
4. Tai-hoon Kim, Tae-seung Lee, Min-chul Kim, Sun-mi Kim: Relationship Between Assurance Class of CC and Product Development Process, The 6th Conference on Software Engineering Technology, SETC (2003).
5. Ho-Jun Shin, Haeng-Kon Kim, Tai-Hoon Kim, Sang-Ho Kim: A study on the Requirement Analysis for Lifecycle based on Common Criteria, Proceedings of The 30th KISS Spring Conference, KISS (2003)
6. Tai-Hoon Kim, Byung-Gyu No, Dong-chun Lee: Threat Description for the PP by Using the Concept of the Assets Protected by TOE, ICCS 2003, LNCS 2660, Part 4, pp. 605-613
7. Haeng-Kon Kim, Tai-Hoon Kim, Jae-sung Kim: Reliability Assurance in Development Process for TOE on the Common Criteria, 1st ACIS International Conference on SERA

Protection Profile for Software Development Site

Seung-youn Lee and Myong-chul Shin

SungKyunKwan Univ., Department of Information & Communication Eng.,
Kyonggi-do, Korea
{syoun, mcshin}@ece.skku.ac.kr

Abstract. A PP defines an implementation-independent set of IT security requirements for a category of TOEs. Consumers can therefore construct or cite a PP to express their IT security needs without reference to any specific TOE. Generally, PPs contain security assurance requirements about the security of development environment for IT product or system and they are described in ALC_DVS (Development Security) family in the part 3 of the Common Criteria (CC). This paper proposes some security environments for development site by analyzing the compliance between ALC_DVS.1 of the CC and Base Practices (BPs) of the Systems Security Engineering Capability Maturity.

1 Introduction

The Common Criteria (CC) philosophy is to provide assurance based upon an evaluation of the IT product or system that is to be trusted [1]. Evaluation has been the traditional means of providing assurance. In fact, there are many evaluation criteria. The Trusted Computer System Evaluation Criteria (TCSEC), the European Information Technology Security Evaluation Criteria (ITSEC), and the Canadian Trusted Computer Product Evaluation Criteria (CTCPEC) existed, and they have evolved into a single evaluation entity, the CC. The CC is a standard for specifying and evaluating the security features of IT products and systems, and is intended to replace previous security criteria such as the TCSEC.

The CC is presented as a set of distinct but related 3 parts. Part 3, Security assurance components, establishes a set of assurance components as a standard way of expressing the assurance requirements for Target of Evaluation (TOEs), catalogues the set of assurance components, families, and classes, and presents evaluation assurance levels that define the predefined CC scale for rating assurance for TOEs, which is called the Evaluation Assurance Levels (EALs). But about the evaluation with the CC, for example, the evaluation for EAL 3, whenever each product or system is evaluated, same evaluation process is required for all component included in EAL 3. Especially, ALC_DVS.1 component which means development security and included in EAL 3, may be evaluated each time evaluation proceeded. Therefore, many kinds of methods are researched to solve this problem [2].

In this paper we propose some assumptions about the development site to describe security environments of PP for software development site.

2 Overview of Related Works

2.1 Common Criteria

The multipart standard ISO/IEC 15408 defines criteria, which for historical and continuity purposes are referred to herein as the Common Criteria (CC), to be used as the basis for evaluation of security properties of IT products and systems. By establishing such a common criteria base, the results of an IT security evaluation will be meaningful to a wider audience.

The CC will permit comparability between the results of independent security evaluations. It does so by providing a common set of requirements for the security functions of IT products and systems and for assurance measures applied to them during a security evaluation. The evaluation process establishes a level of confidence that the security functions of such products and systems and the assurance measures applied to them meet these requirements. The evaluation results may help consumers to determine whether the IT product or system is secure enough for their intended application and whether the security risks implicit in its use are tolerable.

The CC is presented as a set of distinct but related parts as identified below.

Part 1. Introduction and General Model, is the introduction to the CC. It defines general concepts and principles of IT security evaluation and presents a general model of evaluation. Part 1 also presents constructs for expressing IT security objectives, for selecting and defining IT security requirements, and for writing high-level specifications for products and systems. In addition, the usefulness of each part of the CC is described in terms of each of the target audiences.

Part 2. Security Functional Requirements, establishes a set of functional components as a standard way of expressing the functional requirements for TOEs (Target of Evaluations). Part 2 catalogues the set of functional components, families, and classes.

Part 3. Security Assurance Requirements, establishes a set of assurance components as a standard way of expressing the assurance requirements for TOEs. Part 3 catalogues the set of assurance components, families and classes. Part 3 also defines evaluation criteria for PPs (Protection Profiles) and STs (Security Targets) and presents evaluation assurance levels that define the predefined CC scale for rating assurance for TOEs, which is called the Evaluation Assurance Levels (EALs).

In support of the three parts of the CC listed above, it is anticipated that other types of documents will be published, including technical rationale material and guidance documents.

2.2 Protection Profile

A PP defines an implementation-independent set of IT security requirements for a category of TOEs. Such TOEs are intended to meet common consumer needs for IT security. Consumers can therefore construct or cite a PP to express their IT security needs without reference to any specific TOE.

The purpose of a PP is to state a security problem rigorously for a given collection of systems or products (known as the TOE) and to specify security requirements to address that problem without dictating how these requirements will be implemented. For this reason, a PP is said to provide an implementation-independent security description. A PP thus includes several related kinds of security information (See the Fig. 1).

A description of the TOE security environment which refines the statement of need with respect to the intended environment of use, producing the threats to be countered and the organizational security policies to be met in light of specific assumptions.

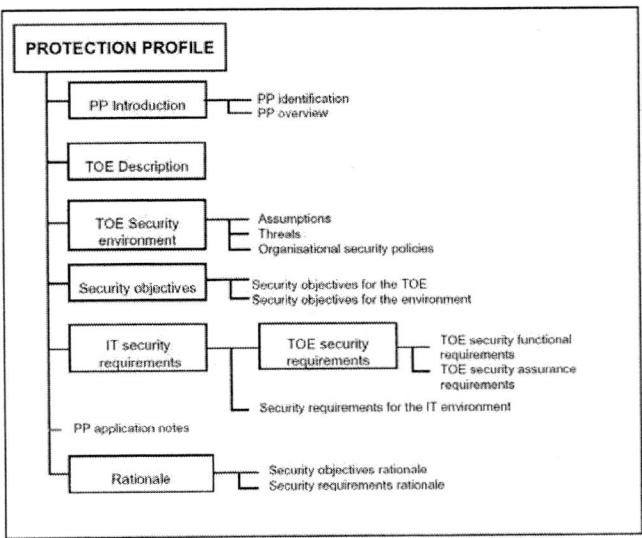

Fig. 1. Protection Profile content

2.3 ALC_DVS

ALC_DVS.1 component consists of one developer action element, two evidence elements, and two evaluator action elements. Evidence elements contains following contents:

- The evidence required,
- What the evidence shall demonstrate,
- What information the evidence shall convey

Contents and presentation of evidence elements of ALC_DVS.1 component are described as like following (Requirements for content and presentation of evidence are identified by appending the letter 'C' to the element number):

ALC_DVS.1.1C. The development security documentation shall describe all the physical, procedural, personnel, and other security measures that are necessary to protect the confidentiality and integrity of the TOE design and implementation in its development environment.

ALC_DVS.1.2C. The development security documentation shall provide evidence that these security measures are followed during the development and maintenance of the TOE.

2.4 SSE-CMM

Modern statistical process control suggests that higher quality products can be produced more cost-effectively by emphasizing the quality of the processes that produce them, and the maturity of the organizational practices inherent in those processes.

More efficient processes are warranted, given the increasing cost and time required for the development of secure systems and trusted products. The operation and maintenance of secure systems relies on the processes that link the people and technologies. These interdependencies can be managed more cost effectively by emphasizing the quality of the processes being used, and the maturity of the organizational practices inherent in the processes.

The SSE-CMM model is a standard metric for security engineering practices covering:

- The entire life cycle, including development, operation, maintenance, and decommissioning activities
- The whole organization, including management, organizational, and engineering activities
- Concurrent interactions with other disciplines, such as system, software, hardware, human factors, and test engineering; system management, operation, and maintenance
- Interactions with other organizations, including acquisition, system management, certification, accreditation, and evaluation

3 Performance Analyses

3.1 Comparison in Process Area

The SSE-CMM has two dimensions, "domain" and "capability." The domain dimension is perhaps the easier of the two dimensions to understand. This dimension simply consists of all the practices that collectively define security engineering. These practices are called Base Practices (BPs).

The base practices have been organized into Process Areas (PAs) in a way that meets a broad spectrum of security engineering organizations. There are many ways to divide the security engineering domain into PAs. One might try to model the real world, creating process areas that match security engineering services. Other strategies attempt to identify conceptual areas that form fundamental security engineering building blocks. The SSE-CMM compromises between these competing goals in the current set of process areas.

Each process area has a set of goals that represent the expected state of an organization that is successfully performing the PA. An organization that performs the BPs of the PA should also achieve its goals.

There are eleven PAs related to security in the SSE-CMM, and we found next three PAs which have compliance with ALC_DVS.1 component:

- PA01 Administer Security Controls
- PA08 Monitor Security Posture
- PA09 Provide Security Input

3.2 Comparison in Base Practice

All of the BPs in each PA mentioned earlier need not have compliance with the evidence elements of ALC_DVS.1. But if any BP included in the PA is excluded or failed when the evaluation is preceded, the PA itself is concluded as fail.

Evidence element ALC_DVS.1.1C requires that the development security documentation shall describe all the physical, procedural, personnel, and other security measures that are necessary to protect the confidentiality and integrity of the TOE design and implementation in its development environment. But ALC_DVS.1.1C dose not describe what are the physical, procedural, personnel, and other security measures. Evidence element ALC_DVS.1.2C requires that the development security documentation shall provide evidence that the security measures described in ALC_DVS.1.1C are followed during the development and maintenance of the TOE.

Some BPs contains example work products, and work products are all the documents, reports, files, data, etc., generated in the course of performing any process. Rather than list individual work products for each process area, the SSE-CMM lists Example Work Products (EWPs) of a particular base practice, to elaborate further the intended scope of a BP. These lists are illustrative only and reflect a range of organizational and product contexts. As though they are not to be construed as mandatory work products, we can analysis the compliance between ALC_DVS.1 component and BPs by comparing evidence elements with these work products. We categorized these example work products as eight parts:

1. Physical measures related to the security of development site and system.
2. Procedural measures related to the access to development site and system.
3. Procedural measures related to the configuration management and maintenance of development site and system.
4. Procedural measures (contain personnel measures) related to the selection, control, assignment and replacement of developers.
5. Procedural measures (contain personnel measures) related to the qualification, consciousness, training of developers.
6. Procedural measures related to the configuration management of the development work products.
7. Procedural measures related to the product development and incident response in the development environment.
8. Other security measures considered as need for security of development environment.

Categorized eight parts above we suggested are based on the contents of evidence requirement ALC_DVS.1.1C, and contains all types' measures mentioned in ALC_DVS.1.1C. But the eight parts we suggested may contain the possibility to be divided to more parts.

We can classify work products included in BPs according to eight parts category mentioned above. Next table 1 describes the result.

Table 1. Categorization of work products

Number of category	Work Products	Related BP
1	control implementation	BP.01.02
	sensitive media lists	BP.01.04
2	control implementation	BP.01.02
	control disposal	BP.01.02
	sensitive media lists	BP.01.04
	sanitization, downgrading, & disposal	BP.01.04
	architecture recommendation	BP.09.05
	implementation recommendation	BP.09.05
	security architecture recommendation	BP.09.05
	users manual	BP.09.06
3	records of all software updates	BP.01.02
	system security configuration	BP.01.02
	system security configuration changes	BP.01.02
	records of all confirmed software updates	BP.01.02
	security changes to requirements	BP.01.02
	security changes to design documentation	BP.01.02
	control implementation	BP.01.02
	security reviews	BP.01.02
	control disposal	BP.01.02
	maintenance and administrative logs	BP.01.04
	periodic maintenance and administrative reviews	BP.01.04
	administration and maintenance failure	BP.01.04
	administration and maintenance exception	BP.01.04
	sensitive media lists	BP.01.04
	sanitization, downgrading, and disposal	BP.01.04
	architecture recommendations	BP.09.05
	implementation recommendations	BP.09.05
	security architecture recommendations	BP.09.05
	administrators manual	BP.09.06
4	an organizational security structure chart	BP.01.01
	documented security roles	BP.01.01
	documented security accountabilities	BP.01.01
	documented security authorizations	BP.01.01
	sanitization, downgrading, and disposal	BP.01.04

5		user review of security training material	BP.01.03
		logs of all awareness, training and education undertaken, and the results of that training	BP.01.03
		periodic reassessments of the user community level of knowledge, awareness and training with regard to security	BP.01.03
		records of training, awareness and educational material	BP.01.03
6		documented security responsibilities	BP.01.01
		records of all distribution problems	BP.01.02
		periodic summaries of trusted software distribution	BP.01.02
		sensitive information lists	BP.01.04
		sanitization, downgrading, and disposal	BP.01.04
7		periodic reassessments of the user community level of knowledge, awareness and training with regard to security	BP.01.03
		design recommendations	BP.09.05
		design standards, philosophies, principles	BP.09.05
		coding standards	BP.09.05
8		philosophy of protection	BP.09.05
		security profile	BP.09.06
		system configuration instructions	BP.09.06

From the table above, we can verify that some BPs of SSE-CMM may meet the requirements of ALC_DVS.1.1C by comparing the contents of evidence element with work products.

The requirements described in ALC_DVS.1.2C can be satisfied by records express the development processes, and some BPs can meet the requirements of evidence element ALC_DVS.1.2C. We researched all BPs and selected the BPs which can satisfy the requirements of evidence element ALC_DVS.1.2C. We list BPs related to ALC_DVS.1.2C as like:

- BP.08.01: Analyze event records to determine the cause of an event, how it proceeded, and likely future events.
- BP.08.02 Monitor changes in threats, vulnerabilities, impacts, risks, and the environment.
- BP.08.03 Identify security relevant incidents.
- BP.08.04 Monitor the performance and functional effectiveness of security safeguards.
- BP.08.05 Review the security posture of the system to identify necessary changes.
- BP.08.06 Manage the response to security relevant incidents.
- And all BPs included in PA01.

Therefore, if the PA01, PA08 and PA09 are performed exactly, it is possible ALC_DVS.1 component is satisfied. But one more consideration is needed to meet the requirements completely.

4 Assumptions

Now, we can describe some assumptions like as:

- Adequate communications exist between the component developers and between the component developers and the IT system developers.
- The development site will be managed in a manner that allows it to appropriately address changes in the IT System.
- The security auditor has access to all the IT System data it needs to perform its functions.
- The threat of malicious attacks aimed at entering to site is considered low.
- There will be one or more competent individuals assigned to manage the environments and the security of the site.
- Administrators are non-hostile, appropriately trained and follow all administrator guidance.
- There will be no general-purpose computing or storage repository capabilities (e.g., compilers, editors, or user applications) not used for developing in the site.
- Anybody cannot gain access to recourses protected by the security countermeasures without passing through the access control mechanisms.
- Physical security will be provided within the domain for the value of the IT assets.
- The security environment is appropriately scalable to provide support to the site.

5 Conclusions

In general, threat agents' primary goals may fall into three categories: unauthorized access, unauthorized modification or destruction of important information, and denial of authorized access. Security countermeasures are implemented to prevent threat agents from successfully achieving these goals.

This paper proposes some assumptions about the development site to describe security environments of PP for software development site.

In these days, some security countermeasures are used to protect development site. But the security countermeasures should be considered with consideration of applicable threats and security solutions deployed to support appropriate security services and objectives. Maybe this is one of our future works.

References

1. ISO. ISO/IEC 15408-1:1999 Information technology - Security techniques - Evaluation criteria for IT security - Part 1: Introduction and general model
2. Tai-hoon Kim, Tae-seung Lee, Kyu-min Cho, Koung-goo Lee: The Comparison Between The Level of Process Model and The Evaluation Assurance Level. The Journal of The Information Assurance, Vol.2, No.2, KIAS (2002).

3. Tai-hoon Kim, Yune-gie Sung, Kyu-min Cho, Sang-ho Kim, Byung-gyu No: A Study on The Efficiency Elevation Method of IT Security System Evaluation via Process Improvement, The Journal of The Information Assurance, Vol.3, No.1, KIAS (2003).
4. Tai-hoon Kim, Tae-seung Lee, Min-chul Kim, Sun-mi Kim: Relationship Between Assurance Class of CC and Product Development Process, The 6th Conference on Software Engineering Technology, SETC (2003).
5. Ho-Jun Shin, Haeng-Kon Kim, Tai-Hoon Kim, Sang-Ho Kim: A study on the Requirement Analysis for Lifecycle based on Common Criteria, Proceedings of The 30th KISS Spring Conference, KISS (2003)
6. Tai-Hoon Kim, Byung-Gyu No, Dong-chun Lee: Threat Description for the PP by Using the Concept of the Assets Protected by TOE, ICCS 2003, LNCS 2660, Part 4, pp. 605-613
7. Haeng-Kon Kim, Tai-Hoon Kim, Jae-sung Kim: Reliability Assurance in Development Process for TOE on the Common Criteria, 1st ACIS International Conference on SERA, 2004

Improved RS Method for Detection of LSB Steganography

Xiangyang Luo, Bin Liu, and Fenlin Liu

Information Engineering Institute,
The Information Engineering University,
Zhengzhou 450002, China
{xiangyangluo, liubin2110}@126.com

Abstract. This paper presents a new dynamical RS steganalysis algorithm[1] to detect the least significant bit (LSB) steganography. The novel algorithm dynamically selects an appropriate mask for each image to eliminate the initial error. Experimental results show that our algorithm is more accurate than the conventional RS method. Meanwhile, the theoretical deduction of the length-estimate equation of RS method is given.

1 Introduction

Steganography is a new research hotspot in the field of information security these years. It makes secrete communication available by embedding messages in the texts, images, audio, video files or other digit carriers. Compared with the Cryptography, modern steganography not only encrypts messages but also masks the very presence of the communication. Among all the image information hiding methods, LSB embedding is widely used for its high hiding quality and quantity, and simpleness to realize. So it's with great significance to detect the images with hidden messages produced by LSB embedding effectively, accurately and reliably. And many research have been done by experts these years.

Westfeld et al. [2] performed the blind steganalysis on the basis of statistical analysis of PoVs(pairs of values). This method, so-called χ^2-statistical analysis, gave a successful result to a sequential LSB(least signi.cant bit) embedding steganography. Provos[3] extended this method by re-sampling test interval and re-pairing values. Fridrich et al.[4] developed a steganographic method for detection of LSB embedding in 24-bit color images (the Raw Quick Pairs-RQP method). The RQP method is based on analyzing close pairs of colors created by LSB embedding. It works reasonably well as long as the number of unique colors in the cover image is less than 30% of the number of pixels. Stefan Katzenbeisser[5] proposed a steganalysis method based on Laplace transform. However, this method needs training and its decision precision is low. Fridrich et al.[1] also presented a powerful RS method (regular and singular groups method) for detection of LSB embedding which utilizes sensitive dual statistics derived from RS correlations in images. This method counts the numbers of the regular group and the singular one respectively, describes the RS chart, and constructs a quadratic equation. The length of message embedded in image is then estimated by solving the equation. This approach is suitable for color and gray-scale images.The literature [6] introduced a steganalytic method for detection of LSB embedding via different histograms of image. If the embedding ratio is higher (more than 40%), the result is more accurate than conventional RS method. The speed of this method is faster and the detection result is better than RS method for uncompressed images. However, if the embedding ratio is lower than 40%, the performance is not as good as conventional RS method. Sorina Dumitrescu et al. [7] proposed SPA, a method to detect LSB steganography via sample pair analysis. When the embedding ratio is more than 3%, this method can estimate it with relatively high precision. Being enlightened by RS method, we improved it by changing the mask actively. The novel algorithm dynamically selects an appropriate mask for each image to eliminate the initial error. Experimental results show that our algorithm is more accurate than conventional RS method. Meanwhile, some theoretical deductions of RS method will be given.

This paper is structured as follows. In Section 2, we introduce the principle of RS method as the foundation of our new method. In Section 3, we describe the principle of DRS method. Section 4 presents the detailed detection steps of the new approach. Then, in Section 5, we present our experimental results.

2   Principle of RS Method

Fridrich et al. [1] introduced the RS steganalysis which is based on the partition of an image's pixels as three disjoint groups; Regular, Singular and Unusable groups. Fridrich found that the RS ratio of a typical image should satisfy a certain rule through large amount of experiments. To explain the details of RS steganalysis, we need to define some notations. Let $\mathcal{C}$ be the test image, which has $M \times N$ pixels with pixel values from the set $\mathcal{P}$. As an example,for an 8-bit grayscale image,$\mathcal{P}= \{0, 1, \cdots, 255\}$.Then divide $\mathcal{C}$ into disjoint groups G of n adjacent pixels $G = \{0, 1, \cdots, 255\} \in \mathcal{C}$ (in RS method G is built by 4 pixels in line). The discrimination function is defined as follows,

$$f(x_1, x_2, \cdots, x_n) = \sum_{i=1}^{n-1} |x_{i+1} - x_i|. \tag{1}$$

Generally, the noisier the group of pixels $G= \{x_1, x_2, \cdots, x_n\}$, the larger the value of the discrimination function becomes.The invertible operation $\mathcal{F}$ on x called flipping is also defined like that

$$F_1 : 0 \leftrightarrow 1, 2 \leftrightarrow 3, \cdots, 254 \leftrightarrow 255, F_{-1} : -1 \leftrightarrow 0, 1 \leftrightarrow 2, \cdots, 255 \leftrightarrow 256. \tag{2}$$

In RS method, a mask M, which is an n-tuple with values -1, 0, and 1 can decide the operation of flipping to pixels. So the flipped group F(G) can be defined as $F_M(F_{M_{(1)}}(x_1), F_{M_{(2)}}(x_2), \cdots, F_{M_{(n)}}(x_n))$.

Then the group G is determined on one of three types of pixel groups,
Regular groups: $G \in R$ $f(F(G)) > f(G)$
Singular groups: $G \in S$ $f(F(G)) < f(G)$
Unusable groups: $G \in U$ $f(F(G)) = f(G)$

Middle points $R_M(1/2)$ and $R_{-M}(1/2)$ can be obtained by randomizing the LSB plane of the test image. Because these two points depend on the particular randomization of the LSBs, $R_M(1/2)$ and $R_{-M}(1/2)$ can be estimated from the statistical samples. $F_M(F_{M_{(1)}}(x_1), F_{M_{(2)}}(x_2), \cdots, F_{M_{(n)}}(x_n))$ is also determined on one of types in the R, S and U, for every Mask M_i. Fridrich experimentally verified the following two statistical assumptions for a large database of images with unprocessed raw BMPs, JPEGs, and processed BMP images.

$$R_M \cong R_{-M} \text{ and } S_M \cong S_{-M} \qquad (3)$$

$$R_M(1/2) = S_M(1/2) \qquad (4)$$

where the mask M denotes $M = [F_0, F_1; F_1, F_0]$ and -M denotes $-M = [F_0, F_{-1}; F_{-1}, F_0]$ respectively. By extensive experiments, Fridrich got the estimation of RS-diagram in Fig 1.

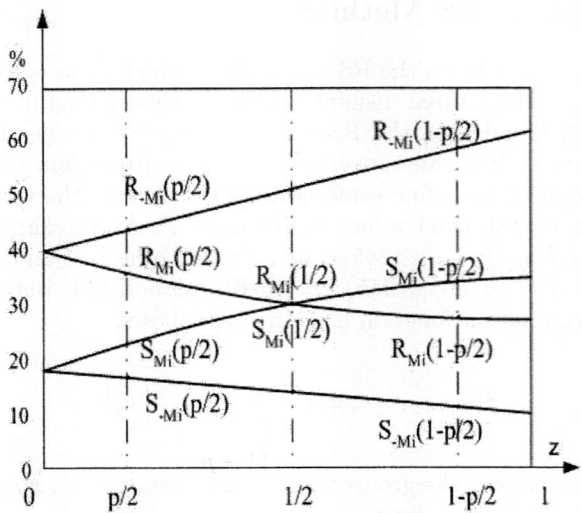

Fig. 1. RS-diagram of an image

When a message with length p (in percent of pixels) is embedded in a test image, generally, $p/2$-assuming the message is a random bit stream-of a test image pixels would be flipped with their corresponding values. Then the four points are acquired $R_M(p/2)$, $R_{-M}(p/2)$, $S_M(p/2)$, $S_{-M}(p/2)$. By applying the flipping F_1 and the shift flipping F_{-1} to all pixels, the next following four points $R_M(1-p/2)$, $R_{-M}(1-p/2)$, $S_M(1-p/2)$, $S_{-M}(1-p/2)$ are also calculated.

The assumptions (3) and (4) make it possible to derive a length-estimate equation (5) for calculating the embedding ratio p

$$2(d_1+d_0)x^2+(d_{-0}-d_{-1}-d_1-3d_0)x+d_0-d_{-0}=0 \qquad (5)$$

$$d_0=R_M(p/2)-S_M(p/2),\ d_1=R_M(1-p/2)-S_M(1-p/2)$$

$$d_{-0}=R_{-M}(p/2)-S_{-M}(p/2),\ d_{-1}=R_{-M}(1-p/2)-S_{-M}(1-p/2)$$

The embedding ratio is calculated from the root whose absolute value is smaller by

$$p=x/(x-1/2) \qquad (6)$$

3 Deduction of RS Length-Estimate Equation

According to Fig.1 R_{-M},S_{-M} is linear while R_M and S_M can be simulated by two quadratic equations. We will deduce the equation (5), which is not proved in [1].

Firstly, we use a transformation of coordinates $T_1 : [p/2, 1-p/2] \to [0,1]$, and the coordinates 0, $p/2$, $1/2$, $1-p/2$ in original coordinates system will be turned into $p/(p-2)$, 0, $1/2$,1 in new coordinates system.

In new coordinates system, points $(0, R_{-M}(p/2))$ and $1, R_{-M}(1-p/2)$ decide the equation:

$$R_{-M}(x)=[R_{-M}(1-p/2)-R_{-M}(p/2)]x+R_{-M}(p/2) \qquad (7)$$

Similarly, $S_{-M}(x)$ can be decided by points $(0, S_{-M}(p/2))$ and $(1, S_{-M}(1-p/2))$.

$$S_{-M}(x)=[S_{-M}(1-p/2)-S_{-M}(p/2)]x+S_{-M}(p/2) \qquad (8)$$

$S_M(x)$ can be decided by points $(0, S_M(p/2))$, $(1/2, S_M(p/2))$ and $(1, S_M(1-p/2))$

$$S_M(x)=2[S_M(p/2)+S_M(1-p/2)-2S_M(1/2)]x^2$$
$$+[-S_M(1-p/2)-3S_M(p/2+4S_M(1/2))]+S_M(p/2) \qquad (9)$$

$R_M(x)$ can be decided by points $(0, R_M(p/2))$, $(1/2, R_M(p/2))$ and $(1, R_M(1-p/2))$

$$R_M(x)=2[R_M(p/2)+R_M(1-p/2)-2R_M(1/2)]x^2$$
$$+[-R_M(1-p/2)-3R_M(p/2+4R_M(1/2))]+R_M(p/2) \qquad (10)$$

According to hypothesis (3), $R_M(x)$ and R_{-M} intersect at $x = 0$ in origin coordinates system, so $x = p/(p-2)$ is in new coordinates system, and we can obtain,

$$2[R_M(p/2) + R_M(1-p/2) - 2R_M(1/2)]x^2$$
$$+[-R_M(1-p/2) - 3R_M(p/2 + 4R_M(1/2))] + R_M(p/2)$$
$$= [R_{-M}(1-p/2) - R_{-M}(p/2)]x + R_{-M}(p/2) \qquad (11)$$

Similarly, S_{-M} and S_M also intersect at $p/(p-2)$, then,

$$2[S_M(p/2) + S_M(1-p/2) - 2S_M(1/2)]x^2$$
$$+[-S_M(1-p/2) - 3S_M(p/2 + 4S_M(1/2))] + S_M(p/2)$$
$$= [S_{-M}(1-p/2) - S_{-M}(p/2)]x + S_{-M}(p/2) \qquad (12)$$

Subtracting (12) from (11) yields

$$2[R_M(p/2) - S_M(p/2) + R_M(1-p/2) - S_M(1-p/2)]x^2$$
$$+[R_{-M}(p/2) - S_{-M}(p/2) - R_{-M}(1-p/2) - S_{-M}(1-p/2)$$
$$-R_M(1-p/2) - S_M(1-p/2) - 3R_M(p/2) - S_M(p/2)]x$$
$$+R_M(p/2) - S_M(p/2) - R_{-M}(p/2) - S_{-M}(p/2) \qquad (13)$$

Let

$$d_0 = R_M(p/2) - S_M(p/2), \quad d_1 = R_M(1-p/2) - S_M(1-p/2)$$

$$d_{-0} = R_{-M}(p/2) - S_{-M}(p/2), \quad d_{-1} = R_{-M}(1-p/2) - S_{-M}(1-p/2)$$

$$2(d_1 + d_0)x^2 + (d_{-0} - d_{-1} - d_1 - 3d_0)x + d_0 - d_{-0} = 0$$

4 The Dynamical RS Steganalysis Algorithm

The precision of RS is based on the hypotheses (3) or (4). Once the hypotheses do not hold, the quadratic equations (5) above will not hold. Hence, when the embedding ratio is low, the errors of those hypotheses will make decision inaccurate. And when there is no embedded message in images, the false alarm rate is high. In this section, we will give further analysis on hypothesis (3), and present an improved algorithm to estimate the embedding ratio precisely.

Fixed mask $M : [0110]$ is used in conventional RS method. However, $R_M(0)$ is not absolutely equal to for this mask, and neither is equal to $R_{-M}(0)$ actually. This initial deviation may lead a serious estimate error. In our new algorithm, we establish a mask set by selecting N kinds of different masks. For example, one 4-tuple mask in the mask set can be expressed by M_i $i = 1, 2, \cdots, N$.

Let $\epsilon_{R_{M_i}} = R_{-M_i}(0) - R_{M_i}(0)$ and $\epsilon_{S_{M_i}} = S_{-M_i}(0) - S_{M_i}(0)$, so equation (5) will be turned into

$$2(d_{1_{M_i}} + d_{0_{M_i}})x^2 + (d_{-0_{M_i}} - d_{-1_{M_i}} - d_{1_{M_i}} - 3d_{0_{M_i}})x + d_{0_{M_i}} - d_{-0_{M_i}}$$
$$= \epsilon_{R_{M_i}} + \epsilon_{S_{M_i}} \qquad (14)$$

According to RS method, we can select a series of different masks, for each M_i we will get a set of values of $R_{M_i}(1-p/2)$, $S_{M_i}(1-p/2)$, $R_{-M_i}(1-p/2)$, $S_{-M_i}(1-p/2)$, $R_{M_i}(p/2)$, $S_{M_i}(p/2)$, $R_{-M_i}(p/2)$, $S_{-M_i}(p/2)$. Then we can calculate the parameters in (5) and obtain an equation corresponding M_i sequentially.

Only when the initial deviation $|\epsilon_{R_{M_i}} + \epsilon_{S_{M_i}}|$ is very close to 0, the detection result can be accurate.

Let
$$a = 2(d_{1_{M_i}} + d_{0_{M_i}}), b = d_{-0_{M_i}} - d_{-1_{M_i}} - d_{1_{M_i}} - 3d_{0_{M_i}}, c = d_{0_{M_i}} - d_{-0_{M_i}}$$
then the left side of equation (14)

$$2(d_{1_{M_i}} + d_{0_{M_i}})x^2 + (d_{-0_{M_i}} - d_{-1_{M_i}} - d_{1_{M_i}} - 3d_{0_{M_i}})x + d_{0_{M_i}} - d_{-0_{M_i}} \qquad (15)$$

is changed into $ax^2 + bx + c$, After being squared, it becomes

$$a^2x^4 + 2abx^3 + (2ac + b^2)x^2 + 2bcx + c^2 \qquad (16)$$

then calculate the differentiate of (16), we can get

$$4a^2x^3 + 6abx^2 + (4ac + 2b^2)x + 2bc \qquad (17)$$

To get the minimum value of $|\epsilon_{R_{M_i}} + \epsilon_{S_{M_i}}|$, we should let the value of (16) is equal to 0 as follows,

$$4a^2x^3 + 6abx^2 + (4ac + 2b^2)x + 2bc = 0 \qquad (18)$$

Selecting an appropriate one from the three roots of (18) as p, and substituting it into (15), we can obtain the value of (15) Q_{M_j} under current mask M_j. Similarly, we also can obtain others Q_{M_j} under corresponding masks M_i, $i = 1, 2, \cdots, N, i \neq j$.

Choose the minimum value from all the ones of Q, and record the corresponding p and mask. Then this mask is optimal for the current image and p is the most accurate estimate value.

5 Experimental Results

5.1 Absolute Average Error Analysis

Firstly, we selected 50 standard test images (such as Lena, peppers and so on, see Fig.2) with 512 × 512 pixels and 200 images come from Sony digital camera with 800 × 600 pixels. We created a series of test images by embedding secret messages into the four images using random LSB replacement method with embedding ratios $0, 3\%, 5\%, 10\%, 20\%, \cdots, 100\%$. Then we estimated the embedding ratio from those test images using RS method and our DRS method, respectively. Table 1 lists the estimated results which indicate that our new algorithm is more effective and reliable than RS method.

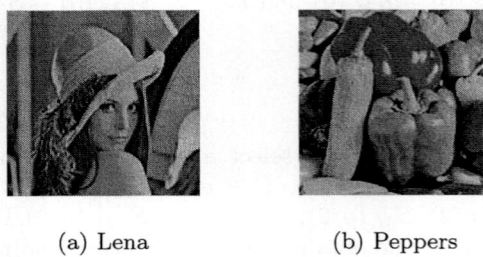

(a) Lena   (b) Peppers

Fig. 2. Sample of standard test images

Table 1. Test results of 50 standard images (in percent)

	standard images(50)		images of ours(200)	
	RS	DRS	RS	DRS
0	1.48	0.26	1.89	0.76
3	4.40	3.34	1.55	2.19
5	6.55	5.09	3.40	4.17
10	11.66	10.56	8.71	9.45
20	22.90	20.83	18.91	19.54
30	32.70	30.07	29.42	29.51
40	41.90	40.20	39.37	39.52
50	52.98	50.50	50.56	50.20
60	59.81	60.00	58.55	58.01
70	72.07	70.33	72.44	70.87
80	79.67	79.04	80.61	78.88
90	91.08	90.19	90.85	89.69
100	96.95	99.16	97.37	99.78

5.2 Correct Rate Test

To compare the correct rate of DRS method with conventional RS method, we did the same experiments for above 250 images. For minimize the false alarm rate and missing report rate, we selected 0.03 as the threshold based on experiments. The correct rates of experimental results are shown in the Table 2. Compared with conventional RS method, our method can greatly decrease the false alarm rate, which is about 8% now. Meanwhile, the missing rate is decreased. From Table 2, we can also find that the estimate accuracy is higher than conventional RS method when the embedding ratio is 5%. If the embedding ratio is higher than 5%, the DRS method's missing ratios are all about 0.

5.3 Standard Deviation Analysis

Fig.3 describes the absolute average error of the three types of test image sets respectively.

Table 2. Correct rates of judgments (in percent)

	standard images (50)		others images (200)	
	RS	DRS	RS	DRS
0	90	96	76	96
3	68	94	78	98
5	84	100	96	100
10	96	100	98	100
15	98	100	100	100
20	98	100	100	100
30	100	100	100	100
40	100	100	100	100
50	100	100	100	100
60	100	100	100	100
70	100	100	100	100
80	100	100	100	100
90	100	100	100	100
100	100	100	100	100

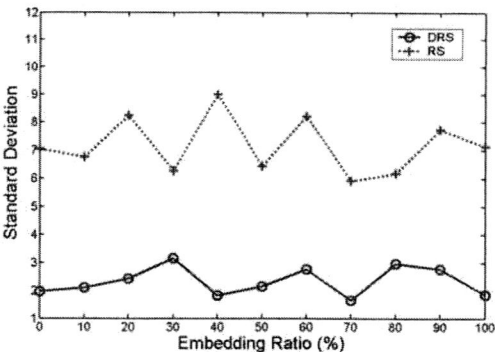

Fig. 3. Standard deviation comparison between DRS and RS

From Fig.3, we can obtain that the absolute average error and the standard deviation of DRS analysis method are smaller than these of conventional RS method.

6 Conclusions

In this paper, we improve the conventional RS steganalysis method by selecting mask dynamically. The novel method has following advantages: the lower false alarm rate and missing report rate, the more accurate estimating embedding ratio and the faster detection speed. The theoretical deduction of the length-estimate equation of RS method is also given.

Acknowledgements. We thank the anonymous reviewers for their valuable comments. The work is supported partially by the Nation Natural Science Foundation of China (Grant No. 10171017, 90204013 and 60374004), Henan Science Fund for Distinguished Young Scholar (Grant No. 0412000200) and HAIPURT (Grant No. 2001KYCX008).

References

1. J. Fridrich, M. Goljan.: Practical Steganalysis of Digital Images - State of the Art. http://www.ssie.Binghamton.edu/fridrich
2. A. Westfeld, A. P.tzmann.: Attacks on Steganographic Systems. Information Hiding 1999 : pp.61-76.
3. N.Provos.: Defending Against Statistical Steganalysis. 10th USENIX Security Symposium, Washington, DC, 2001
4. J. Fridrich, R.Du, and L. Meng.: Steganalysis of LSB Encoding in Color Images. Proceedings IEEE International Conference on Multimedia and Expo, July 30-Auguest 2,2000, New York City, NY.
5. Stefan Katzenbeisser.: Breaking PGMStealth Using Laplace Filters. access from http://stud3.tuwien. ac.at/ e9625414/
6. Tao Zhang, Xijian Ping.: Reliable Detection of LSB Steganography Based on the Difference Image Histogram. ICASSP 2003, Vol. I, pp.545-548.
7. Sorina Dumitrescu, Xiaolin Wu, and Zhe Wang,.: Detection of LSB Steganography via Sample Pair Analysis. IEEE Transactions on Signal Processing, VOL.51, NO.7, July 2003, pp.1995-2007.

Robust Undetectable Interference Watermarks*

Ryszard Grząślewicz[1], Jarosław Kutyłowski[2], Mirosław Kutyłowski[1], and Wojciech Pietkiewicz[1]

[1] Institute of Mathematics, Wrocław University of Technology
{Ryszard.Grzaslewicz, Miroslaw.Kutylowski}@pwr.wroc.pl
pietkiew@im.pwr.wroc.pl
[2] International Graduate School of Dynamic Intelligent Systems, University of Paderborn
jarekk@upb.de

Abstract. We propose a digital watermarking method for gray-tone images in which each watermark consists of a collection of single points and each point is encoded in the spatial domain of the whole image. The method is somewhat related to physical digital holograms and interference images. Reconstruction of such watermarks is based on a similar principle as the reconstruction of physical holograms.

While encoding a watermark in the spatial domain one of the major problems is to avoid a textured appearance due to the encoding scheme. We avoid a recognizable pattern by creating pseudorandom keyed watermarks, which resemble random noise.

The method proposed yields robust watermarks that are resistant against many attacks which preserve the distance between points (filtering, rotation, JPEG compressing). The watermarking scheme provides means for detection and reversal of scaling transformations, thus making the watermark resistant to this attack.

The original picture is not required for reconstruction. The watermark is quite hard to detect, which prohibits easy violation of watermark protection.

Our method guarantees exact reconstruction provided that the watermark image consists of a limited number of white pixels on a black background.

1 Introduction

Digital watermarks are used for protecting digital images and contain information necessary for exercising the rights of the owner. They need to be robust in the sense that digital image operations occurring during normal use do not destroy the watermark.

Two types of watermarks can be distinguished – watermarks that can be easily detected by everyone and the information they carry can be extracted (e.g.

---

* The third author was partially supported in years 2003-2005 by KBN, project 0 T00A 003 23.

the name of the copyright owner) and such that are undetectable. Our scheme constitutes an undetectable watermark, which cannot be easily recognized in an image without the proper key. In such a scenario it is not crucial to encode a lot of information in the watermark. The most important issue is whether an image carries a watermark created with a given key.

There has been a lot of research on watermark schemes in the last decade (see for instance [1]). Despite enormous efforts there is no ultimate solution of this problem. To the best of our knowledge, every scheme proposed so far has some weakness that can be exploited by an adversary attacking the watermark.

Geometric distortions in the image are particularly dangerous. A standard set of operations, called RST, consist of rotations, scaling and translations (shifting the pixels some number of positions). However, there are further geometric attacks such as cropping the image or nonlinear transformations. Each of these techniques destroys watermarks for which the pixel positions cannot be changed.

A general idea to resist RST attacks is to find image characteristics invariant to the RST operations. Then we may encode a watermark into these characteristics – obviously this requires some freedom to manipulate the image without influencing it so that the changes become detectable by a human eye. A solution of this kind based on Fourier-Mellin transform is proposed in [3]. The method requires the original image and implementation problems have been claimed [2]. Fourier-Mellin transform is used again in [4], a characteristic vector of a digital image is defined so that it remains unchanged during RST operations. Again, implementation problems due to inaccuracies during embedding and detection computations have been reported. A similar approach [2] based on Radon transform provides a scheme that is claimed to be practical. Nevertheless, it has a weak point: the characteristic vector changes substantially if the watermarked image is not "homogeneous" and we crop it.

Digital watermarks proposed in [5] are constructed in a way that mimics physical holograms. They have a remarkable property that the watermark image can be reconstructed from each reasonably large block of pixels. However, the scheme is based on Fourier transform, with the watermark reconstructed in the frequency domain. Our experiments have shown that the watermark can be easily removed by blurring the watermarked image slightly.

Features of the New Scheme. We propose a scheme that is based on a very simple idea, but surprisingly yields very good results. Let us point to its major features:

- Watermark retrieval does not require the original image.
- Watermarks are resistant to translations, cropping and symmetries.
- Watermarks are resistant to scaling after such an operation has been detected. The watermarking scheme provides means for detecting scaling operations.
- Watermarks are resistant to standard image processing operations, such as changing the contrast, changing the brightness, blurring, adding random noise and JPEG compression.
- Watermarks can be retrieved from image parts, without even knowing the original location of the part within the image.

- The robustness of the scheme has been tested against the Stir Mark benchmark suite and satisfactory results have been obtained.
- A watermark is composed of single white pixels on a black background with white point positions encoding the watermark information.
- Once a watermark is known, it can be easily removed from the image by superposing the same watermark but with the negative sign.
- Watermarks are created with a secret key. The key is necessary for detecting, restoring and removing the watermark image.
- Retrieving a watermark is computationally intensive. In most practical cases this prohibits examining the images for watermarks without knowledge of the key and without access to enormous computing resources.

The paper is organized as follows: in Section 2 we present the physical motivation for our scheme and give a theoretical background for the watermarking encoding and decoding algorithms, which are described in Section 3. The results of our experimental evaluation are presented in Section 4.

2 Interference Images

Let us describe the basic idea of our approach. Assume that we have to encode a watermark image which can be represented as a matrix $(a_{i,j})_{i,j \leq n}$, where $a_{i,j} = 1$ if pixel (i,j) is white, and $a_{i,j} = 0$ otherwise. We assume that the white pixels are used to encode the information, so the main issue is how to represent a single white pixel. For this purpose let us recall the phenomenon of interference images.

Physical Motivation. Let us recall the Young experiment: coherent monochromatic light is passing through two small slits H_1, H_2 lying close to each other on plane P. The light is diffracted when passing through the holes and it goes into all possible directions. Consider a single point A on plane Q parallel to P. The distances between A and H_1 and between A and H_2 are slightly different. Let λ be the length of the light wave. The waves passing through H_1 and H_2 are in the same phase, but they have different distances to reach A. So when they reach A, they are shifted in phase – the shift corresponds to the additional distance one of these waves has to go. If the difference equals $\lambda \cdot i$, for $i \in \mathbb{N}$, then the waves sum up. But if the difference is $\lambda \cdot i + \lambda/2$, for $i \in \mathbb{N}$, then the waves cancel themselves out. It follows that on screen Q we get an interference pattern consisting of dark and bright lines. This effect is called two-source light interference. If there is more than one pair of slits in P, the values corresponding to different pairs sum up.

General Framework. First we pick a certain function F (later we discuss the necessary properties of F). Let $F_{i,j}(x,y) = F(x-i, y-j)$. Then we represent a watermark image $(w_{i,j})$ by the sum

$$\sum_{i,j} w_{i,j} \cdot F_{i,j} .$$

That is, in the resulting image the pixel with coordinates (a, b) has the value

$$h_{a,b} = \sum_{i,j} w_{i,j} \cdot F(a-i, b-j) \ . \tag{1}$$

Reconstruction of the watermark image from $H = (h_{i,j})$ will be performed by computing for every point of the watermark

$$w_{a,b} := \sum_{i,j} |h_{i,j} - F(a-i, b-j)| \ . \tag{2}$$

Let us explain informally the motivation for such a reconstruction rule. Let us assume that the watermark image consists of points (e_1, f_1), (e_2, f_2), (e_3, f_3). Then

$$h_{i,j} = F(i - e_1, j - f_1) + F(i - e_2, j - f_2) + F(i - e_3, j - f_3) \ . \tag{3}$$

So for the point (e_1, f_1) the reconstruction rule yields

$$\sum_{i,j} |F(i - e_2, j - f_2) + F(i - e_3, j - f_3)| \ . \tag{4}$$

On the other hand, if we perform reconstruction at a point (u, v) that is different from (e_1, f_1), (e_2, f_2), (e_3, f_3), then the negative term in Eq. 2 does not cancel any of $F(i - e_k, j - f_k)$ and we get the expression:

$$\sum_{i,j} |F(i - e_1, j - f_1) + F(i - e_2, j - f_2) + F(i - e_3, j - f_3) - F(i - u, j - v)| . \tag{5}$$

For summation over all integer i, j and after removing the absolute values we would get exactly the same result for Eq. (4) and Eq. (5). However, it may happen that $F(i-u, j-v) > F(i-e_1, j-f_1) + F(i-e_2, j-f_2) + F(i-e_3, j-f_3)$. Then in Eq. (5) we get an extra positive value that does not cancel out due to the use of absolute values. If we assume that $F(i - e_l, j - f_l)$, $l = 1, 2, \ldots$ and $F(i - u, j - v)$ are independent random variables uniformly distributed over $[0, M]$, then one can prove that the difference between expected values of the corresponding additive factors in by Eq. (5) and Eq. (4) is of order $M/k!$, where k is the number of white points in the watermark image. Certainly, the assumption about stochastic independence is not valid mathematically, but for our choice of function F similar phenomena can be observed. An important point is that for large k we cannot hope for a good reconstruction due to the factor $M/k!$.

An alternative way of computing reconstruction values based on orthogonal functions could be

$$w_{a,b} := \sum_{i,j} h_{i,j} \cdot F(a-i, b-j) \ .$$

However, our experiments have shown that for the employed functions F and images occurring in practice, reconstruction with equality (2) yields significantly better results.

In order to get reasonable resistance against attacks on watermarks we need some properties of F. The first point is that an adversary could crop the image, for instance take only one quarter of it, that is, put value 0 at all points except the chosen quarter. The crucial issue then is how much of the value $w_{a,b}$ is contained in the chosen quarter. We need a property that each rectangular block of points B contributes a value into $w_{a,b}$ which is roughly proportional to the area of B, provided that B is not too small. Informally speaking, the "energy" of a white watermark pixel should be dispersed quite evenly on the whole transformed image. The last property would automatically yield resistance against local editions in image H.

In order to get resistance against operations like replacing a pixel value through the average in its closest neighborhood, we need the property that the image H does not consist of waves of high frequency only. This objective contradicts the previous objective, where high frequency waves are preferred. So we need to find a proper compromise.

Another point is that the objects that are likely to appear on the images to be watermarked, should be almost orthogonal to functions $F_{i,j}$ (in the sense that such objects do not contribute many small values in the sum from Eq. (2)). For instance lines, treated as functions with value 1 on the line points and zero elsewhere, should be orthogonal to functions $F_{i,j}$. This excludes the functions such as $F(x,y) = \cos(\max(x,y))$, since vertical and horizontal lines of the image would coincide with constant values of F. Ideally, if we take a curve at which the value of F is constant, then the pixel values on the image to be watermarked should form a quasi-random multiset of values. Of course, this should be true for images that can occur in practice.

Artificial Pseudorandom Interference Images. We consider "interference images" created by changing the functions – from those describing physical reality to more handy ones suitable for watermarking. First we consider an interference image for a pair of holes located at point $(0,0)$ with interference value at point (x,y) described by a function

$$F(x,y) = \cos\left(\sqrt{x^2 + y^2 + z^2}\right)$$

where z is a parameter which can be thought of as the distance between the planes P and Q introduced in the description of the physical motivation. This image is not suitable for watermarking, since it contains visible circles. Thus, we introduce pseudorandomness to the interference images. Let H be a secure hash function and K be a secret key used for watermarking. We define

$$F_K(x,y) = P_K\left(\angle(x,y) \bmod r, \sqrt{x^2 + y^2 + z^2}\right) \cdot \left(\frac{\sqrt{x^2+y^2}}{f} + 1\right)^{-1}$$

where $\angle(x,y) = 90\frac{2}{\pi}\arctan\left(\frac{x}{y}\right)$, and $P_K(a,b) = H(K,a,b)$.

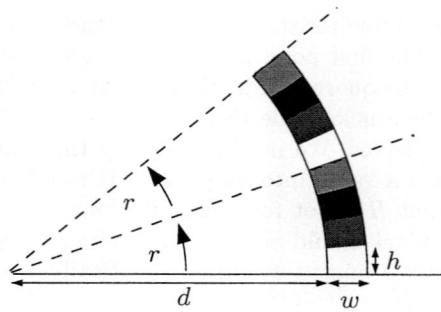

Fig. 1. Interference image values with function F_K

Thus the value of $F_K(x, y)$ depends both on the distance of point (x, y) from $(0, 0)$ and on the angle $\angle(x, y)$ of the vector (x, y). Basing on the security of the hash function, it is impossible to predict the value of $F_K(x, y)$ without knowledge of key K.

Fig. 1 depicts a circle of points in distance d from point $(0, 0)$. As the values of F_K depend on the residue of the angle $\angle(x, y)$ modulo r, the same pattern is repeated every r degrees. Thus when the interference image is rotated by a multiple of r degrees, reconstructing the watermark is possible. In order to deal with rotations by non-multiples of r, we have to try to reconstruct the watermark for r different rotations. (Note that it is not necessary to know around which point the image has been rotated.)

The "width" w of the circle (see Fig. 1) is determined by arithmetic precision of the computation of the distance from point $(0, 0)$. The height h of each sector of the circle is determined by the arithmetic precision of the angle computation. This can be assumed to be one degree.

The term $\left(\frac{\sqrt{x^2+y^2}}{f} + 1\right)^{-1}$ influences intensity of $F_K(x, y)$ based on the distance between $(0, 0)$ and (x, y) and a constant parameter f.

In order to obtain integer values as arguments of P_K (and therefore of H), both input parameters are divided by, respectively, w and h and truncated to integer values.

3 Watermarking Algorithm

Creating a Watermark. The watermark is created by inserting the vertices of $z/3$ equilateral triangles with edge length T equally distributed on the watermark image. Thus the watermark image consists of z white pixels.

Inserting a Watermark into an Image. Let $h_{i,j}$ denote the value of pixel (i, j) of the interference image. In order to construct this image and embed it into the cover image we execute the following steps:

1. We start with a black interference image of size $n \times n$, which is equal to the size of the watermark.
2. For each white point (a, b) of the watermark and every point (i, j) within the interference image the value of $F(a-i, b-j)$ is computed and added to the current value of $h_{i,j}$.
3. The intensity of the interference image is normalized, so that the minimum intensity of a pixel equals $-\frac{\alpha}{2}\mathcal{M}$ and the maximum equals $+\frac{\alpha}{2}\mathcal{M}$, where $\mathcal{M}$ is the maximum intensity of the cover image. Values in between are scaled linearly. Parameter α determines the strength of the embedding of the interference image within the cover image. A bigger value of α results in a stronger watermark but lowers image quality.
4. The interference image and the cover image are superposed by adding the intensities of the pixels. The result represents the watermarked image.
5. The normalization procedure is applied to the watermarked image with the bounds 0 and $\mathcal{M}$.

In Point 2, we compute interference values as given by Eq. (1). Time complexity of inserting a watermark is proportional to n^2 times the number of white pixels.

Watermark Reconstruction. Our goal is to determine whether an image has been watermarked with a given key K. The reconstruction process consists of two phases. First, the scale factor of the image is detected and the image is re-scaled properly. In the second phase, the actual reconstruction takes place.

1. The watermarked image is rotated by $0, 1, 2, \ldots, r$ degrees (recall that r is the angular periodicity of F_K). The next algorithm steps are executed for each of these rotations until a watermark is found.
2. If the watermarked image has been scaled so that the distance between points is changed, this scaling must be detected and reversed. So, prior to the actual reconstruction the detection of the performed scaling is done in the following way:
   (a) An image part is chosen, such that there has been at least one watermark point encoded in this part. If the z points have been distributed uniformly over the whole watermark, then every image part of appropriate size should contain such a point.
   (b) A reconstruction for this part is computed for every reasonable scale factor. This reconstruction is performed as in step 3.
   (c) The value of each point is considered as a function of the scale factor. For each point the maximum peak is detected. The peaks among all points are ordered according to their size. The scale factor of the largest peak is chosen as the proper scaling factor and used for the reconstruction.
   (d) The image is scaled by the discovered scale factor.
3. The actual reconstruction process is as follows:
   (a) For each position (a, b) of the watermark we compute
   $$w_{a,b} := \sum_{i,j} |o_{i,j} - F_K(a-i, b-j)|, \qquad (6)$$

where $o_{i,j}$ is the pixel value of the watermarked image at point (i,j).
(b) The difference between $o_{i,j}$ and $F_K(a-i, b-j)$ is near to zero for pixels which have been white in the watermark (since $o_{i,j}$ is closer to $F_K(a-i, b-j)$ in this case). Thus we invert the reconstructed watermark, so that the brightest pixels in the reconstructed watermark stand for those which have been white in the watermark.
(c) We find the δz brightest pixels. The δ factor allows for some error-correction, if due to noise there are invalid bright pixels. Additionally, if more than one bright pixel is found in a small area, only one representative of them is chosen to be used in the computation. Such a situation can occur if due to scaling the intensity of a watermark pixel goes over to its neighboring pixels.
(d) We look for equilateral triangles with vertices among the brightest pixels. If enough triangles with an edge length of T are found, the watermark is detected.

The presented algorithms work with low-precision arithmetic implemented using integer numbers, with a resolution of 10^{-2}.

4 Experimental Results

We have implemented our scheme and checked the results to examine the practical relevance of the algorithm. We have looked for appropriate values of parameters giving good reconstruction results. The values has been chosen to be $d = 2000$, $k = 6$, $r = 10$, $\delta = 1.3$, $T = 100$, $z = 8$ and $f = 25\sqrt{2n^2}$. Embedding of watermarks into the cover image is performed with the factor $\alpha = 0.2$, which leads to a PSNR value larger than 32dB for all tested images. All images used in the test process had an original size of 1024×1024 pixels and were encoded using 8-bit grayscale values.

The watermarking process takes about one second on a modern-class PC whereas the reconstruction can last up to a few hours on a comparable computer.

Fig. 2(a) presents one of the cover images from the StirMark suite, whereas Fig. 2(b) is the cover image with the watermark embedded. The interference image is shown in Fig. 3.

Watermark Robustness. We have examined the resistance of the watermark against several common attacks by invoking the Stir Mark Benchmark 3.1 [6] suite. We summarize the results

- Cropping – All tests (from 1% to 75%) passed. (9 of 9 successful)
- Remove rows/columns – Tests 1/1, 1/5 and 5/1 passed. (3 of 5 successful)
- Flip – Test passed. (1 of 1 successful)
- Scaling – All tests (from 0.5 to 2.0) passed. (6 of 6 successful)
- Change aspect ratio – All tests passed. (8 of 8 successful)
- Rotation with cropping – Tests with rotation smaller than 30% passed. (13 of 16 successful)

(a) original (b) watermarked

Fig. 2. Cover image prior to watermarking and afterwards

Fig. 3. Interference image for watermark

- Rotation with cropping and scaling – Tests with rotation smaller than 30% passed. (13 of 16 successful)
- Shearing – Tests with shearing of 1% on one axis passed. (2 of 6 successful)
- General linear transformation – No tests passed. (0 of 3 successful)
- StirMark – Test not passed. (0 of 1 successful)
- Gaussian filtering – Test passed. (1 of 1 successful)
- Sharpening – Test passed. (1 of 1 successful)
- Median filtering – All tests passed. (3 of 3 successful)
- LRAttack – Test passed. (1 of 1 successful)
- JPEG compression – All tests passed. (12 of 12 successful)

In overall 77 of 89 tests have been successful. In the present implementation there are no countermeasures against linear transformations applied, however a similar approach as for scaling may be used.

The StirMark test consists among others of distortions like shearing, stretching and rotating and nonlinear transformations like bending and random displacement. Its success against our scheme is mainly due to nonlinear transformations. However, there are techniques that help to trace which nonlinear transformations have been used. They might enable recovery of the image before transformations and in this way – recovery of the watermark. These countermeasures have not been included yet in the implementation tested.

5 Conclusion

The presented watermarking method is robust against several common attacks and provides an innovative technology for embedding undetectable watermarks in the spatial domain of images. A nice feature of the scheme is that the watermarks are reconstructed exactly – so it enables direct encoding of digital information.

Its asymmetric behavior – easy insertion and time-costly reconstruction even with a known key – might be a useful tool for copyright protection in the Web. Massive and automatic coping of digital images would require removing the watermarks – which is computationally intensive.

The main concern of the scheme remain nonlinear transformations. Future work should encompass the application of detection schemes for these transformations.

References

1. Hartung, F., Kutter, M.: Multimedia Watermarking Technique. Proc. IEEE 87, 1999, 1079-1107
2. Hyung-Shin Kim, Yunju Baek, Heung-Kyu Lee, Young-Ho Suh: Robust Image Watermark Using Radon Transform and Bispectrum Invariants. Information Hiding '2002, Lecture Notes in Computer Science 2578, Springer-Verlag, 145-159
3. O'Ruanaidh, J.J.K., Pun, K.: Rotation, Scale and Translation Invariant Spread Spectrum Digital Image Watermarking. Signal Processing 66, 1998, 303-317
4. Lin, C.Y., Wu, M., Bloom, J.A., Cox, I.J., Miller, M.L., Lui, Y.M.: Rotation, Scale, and Translation Resilient Watermarking for Images. IEEE Trans. Image Processing 10, 2001, 767-782
5. Takai, N., Mifune, Y.: Digital Watermarking by a Holographic Technique. Applied Optics 41(5), 2002, 865-873
6. Petitcolas, F., Anderson, R., Kuhn, M.: Attacks on Copyright Marking Systems. Information Hiding '1998. Lecture Notes in Computer Science 1525, Springer-Verlag, 219-239

Equidistant Binary Fingerprinting Codes. Existence and Identification Algorithms

Marcel Fernandez, Miguel Soriano, and Josep Cotrina*

Department of Telematics Engineering,
Universitat Politècnica de Catalunya,
C/ Jordi Girona 1 i 3, Campus Nord,
Mod C3, UPC, 08034 Barcelona, Spain
{marcelf, soriano, jcotrina}@entel.upc.es

Abstract. The fingerprinting technique consists in making the copies of a digital object unique by embedding a different set of marks in each copy. However, a coalition of dishonest users can create pirate copies that try to disguise their identities. We show how equidistant, binary codes that can be used as fingerprinting codes.

1 Introduction

The concept of fingerprinting was introduced by Wagner in [4] as a method to protect intellectual property in multimedia contents. The fingerprinting technique consists in the embedding of marks into an object in order to be able to distinguish it from other objects of the same kind. If the owner of a fingerprinted object misbehaves and illegally redistributes his object then the embedded fingerprint will allow to trace him back.

Since fingerprinted objects are all different from each other, they can be compared one another and some of the embedded marks detected. So, to attack a fingerprinting scheme, a group of users collude [1], compare their copies and produce another copy that hides their identities. Therefore a fingerprinting scheme must take this situation into account, and place marks in a way that allows to trace back the members of such treacherously behaving collusions. Error correcting codes can be used in fingerprinting schemes, since one can take advantage of their structure.

In this paper we first discuss the use of equidistant codes as collusion secure fingerprinting codes against collusions of size 2. Secondly, we show how by giving structure to a code, tracing dishonest users can be accomplished by going through the trellis of the code. Moreover, we also show that this process can be improved if the code is systematic.

---

* This work has been supported in part by the Spanish Research Council (CICYT) Projects TIC2002-00818 (DISQET) and TIC2003-01748 (RUBI) and by the IST-FP6 Project 506926 (UBISEC).

The paper is organized as follows. In Section 2 we provide an overview of the coding theory concepts used throughout the paper. The use of equidistant codes as fingerprinting codes is discussed in Section 3. Section 4 deals with the tracing process and how it can be efficiently accomplished. In Section 5 we summarize our work.

2 Previous Results

2.1 Coding Theory Overview

Let $\mathbb{F}_q^n$ be the vector space over $\mathbb{F}_q$, then $C \subseteq \mathbb{F}_q^n$ is called a *code*. The field, $\mathbb{F}_q$ is called the *code alphabet*. A code C is called a *linear code* if it forms a subspace of $\mathbb{F}_q^n$. If $q = 2$ then the code is called a binary code. An element $\mathbf{x} = \{x_1, \ldots, x_n\}$ of $\mathbb{F}_q^n$ is called a *word*. The number of nonzero coordinates in $\mathbf{x}$ is called the *weight* of $\mathbf{x}$ and is commonly denoted by $w(\mathbf{x})$. The *Hamming distance* $\mathbf{d}(\mathbf{a}, \mathbf{b})$ between two words $\mathbf{a}, \mathbf{b} \in \mathbb{F}_q^n$ is the number of positions where $\mathbf{a}$ and $\mathbf{b}$ differ. The *minimum distance* d of C, is defined as the smallest distance between two different codewords. If the dimension of the subspace is k, and its minimum Hamming distance is d, then we call C an $[n,k,d]$-code. A code whose codewords are all the same distance apart is called an *equidistant code*.

A $(n-k) \times n$ matrix $\mathbf{H}$, is a *parity check matrix* for the code C, if C is the set of codewords $\mathbf{c}$ for which $\mathbf{Hc} = \mathbf{0}$, where $\mathbf{0}$ is the all-zero $(n-k)$ tuple. Each row of the matrix is called a *parity check equation*. A code whose codewords satisfy all the parity check equations of a parity check matrix is called a *parity check code*.

For any two words $\mathbf{a}, \mathbf{b}$ in $\mathbb{F}_q^n$ we define the *set* of *descendants* $D(\mathbf{a}, \mathbf{b})$ as $D(\mathbf{a}, \mathbf{b}) := \{x \in \mathbb{F}_q^n : x_i \in \{a_i, b_i\}, 1 \leq i \leq n\}$. For a code C, the *descendant code* C^* is defined as: $C^* := \bigcup_{\mathbf{a} \in C, \mathbf{b} \in C} D(\mathbf{a}, \mathbf{b})$.

If $\mathbf{c} \in C^*$ is a descendant of $\mathbf{a}$ and $\mathbf{b}$, then we call $\mathbf{a}$ and $\mathbf{b}$ *parents* of $\mathbf{c}$.

Note that the concepts of descendant and parents model the situation of a collusion attack, the descendant being the word in the pirate copy, and the parents being the participants in a collusion.

Let C be an equidistant binary code, and let $\mathbf{z} \in C^*$. Then there are three possible configurations for the parents of $\mathbf{z}$.

1. *Star* : there is a single codeword, say $\mathbf{u}$, such that $\mathbf{d}(\mathbf{u}, \mathbf{z}) \leq (d/2) - 1$.
2. *"Degenerated" star*: there is a single pair of codewords, say $\{\mathbf{u}, \mathbf{v}\}$, such that $\mathbf{d}(\mathbf{u}, \mathbf{z}) = \mathbf{d}(\mathbf{v}, \mathbf{z}) = d/2$.
3. *Triangle*: there three possible pairs of codewords, say $\{\mathbf{u}, \mathbf{v}\}$, $\{\mathbf{u}, \mathbf{w}\}$ and $\{\mathbf{v}, \mathbf{w}\}$, such that $\mathbf{d}(\mathbf{u}, \mathbf{z}) = \mathbf{d}(\mathbf{v}, \mathbf{z}) = \mathbf{d}(\mathbf{w}, \mathbf{z}) = d/2$.

2.2 Trellis Representation of Block Codes

The contents of this section are based on [5].

For a binary linear block code, a *trellis* is defined as a graph in which the nodes represent states, and the edges represent transitions between these states.

The nodes are grouped into sets S_t, indexed by a "time" parameter t, $0 \leq t \leq n$. The parameter t indicates the *depth* of the node. The edges are unidirectional, with the direction of the edge going from the node at depth t, to the node at depth $t+1$. Each edge is labeled using an element of $\mathbb{F}_2$.

In any depth t, the number of states in the set S_t is at most $2^{(n-k)}$. The states at depth t are denoted by $\mathbf{s}_t^i$, for certain values of i, $i \in \{0, 1, \ldots, 2^{(n-k)} - 1\}$. The states will be identified by binary $(n-k)$-tuples. In other words, if we order all the binary $(n-k)$-tuples from 0 to $2^{(n-k)} - 1$, then $\mathbf{s}_t^i$ corresponds to the ith tuple in the list. Using this order, for each set of nodes S_t, we can associate the set I_t that consists of all the integers i, such that $\mathbf{s}_t^i \in S_t$. The set of edges incident to node $\mathbf{s}_t^i$ is denoted by $\mathcal{I}(\mathbf{s}_t^i)$.

In the trellis representation of a code C, each distinct path corresponds to a different codeword, in which the labels of the edges in the path are precisely the codeword symbols. The correspondence between paths and codewords is one to one, and it is readily seen from the construction process of the trellis, that we now present.

The construction algorithm of the trellis of a linear block code, uses the fact that every code word of C must satisfy all the parity check equations imposed by the parity check matrix $\mathbf{H}$. In this case, the codewords are precisely the coefficients $c_1, c_2, \ldots, c_n$ of the linear combinations of the columns $\mathbf{h}_i$ of $\mathbf{H}$, that satisfy

$$c_1 \mathbf{h}_1 + c_2 \mathbf{h}_2 + \cdots + c_n \mathbf{h}_n = \mathbf{0}, \qquad (1)$$

where $\mathbf{0}$ is the all zero $(n-k)$-tuple.

Intuitively, the algorithm first constructs a graph, in which all linear combinations of the columns of $\mathbf{H}$ are represented by a distinct path. Then removes all paths corresponding to the linear combinations that do not satisfy (1).

1. Initialization (depth $t = 0$):
   $S_0 = \{\mathbf{s}_0^0\}$, where $\mathbf{s}_0^0 = (0, \ldots, 0)$.
2. Iterate for each depth $t = 0, 1, \ldots, (n-1)$.
   (a) Construct $S_{t+1} = \{\mathbf{s}_{t+1}^0, \ldots, \mathbf{s}_{t+1}^{|I_{t+1}|}\}$, using
   $$\mathbf{s}_{t+1}^j = \mathbf{s}_t^i + c_l \mathbf{h}_{t+1}$$
   $\forall i \in I_t$ and $l = 0, 1$.
   (b) For every $i \in I_t$, according to 2a:
   – Draw a connecting edge between the node $\mathbf{s}_t^i$ and the 2 nodes it generates at depth $(t+1)$, according to 2a.
   – Label each edge $\theta_t^{i,j}$, with the value of $c_j \in \mathbb{F}_2$ that generated $\mathbf{s}_{t+1}^j$ from $\mathbf{s}_t^i$.
3. Remove all nodes that do not have a path to the all-zero state at depth n, and also remove all edges incident to these nodes.

According to the convention in 2b, for every edge $\theta_t^{i,j}$, we can define the function **label_of**$(\theta_t^{i,j})$ that, given a codeword $\mathbf{c} = (c_1, c_2, \ldots, c_n)$, returns the c_j that generated $\mathbf{s}_{t+1}^j$ from $\mathbf{s}_t^i$.

There are 2^k different paths in the trellis starting at depth 0 and ending at depth n, each path corresponding to a codeword. Since the nodes (states) are generated by adding linear combinations of $(n-k)$-tuples of elements of $\mathbb{F}_2$, the number of nodes (states) at each depth is at most $2^{(n-k)}$.

2.3 The Viterbi Algorithm

The Viterbi algorithm is an efficient maximum-likelihood decoding method [2], when applied to the trellis of a code. Each path of the trellis has an associated "length". The Viterbi Algorithm (VA) identifies the state sequence corresponding to the minimum "length" path from time 0 to time n. The incremental "length" metric associated with moving from state s_t^i to state s_{t+1}^j, is given by $l[\theta_t^{i,j}]$, where $\theta_t^{i,j}$ denotes the edge that goes from s_t^i to s_{t+1}^j.

We consider time to be discrete. Using the notation of Section 2.2, and since the process runs from time 0 to time n, the state sequence can be represented by a vector $\mathbf{s} = \langle s_0^0, \ldots, s_n^0 \rangle$.

Among all paths starting at node s_0^0 and terminating at the node s_t^j, we denote by ψ_t^j the path segment with the shortest length. For a given node s_t^j, the path ψ_t^j, is called the *survivor* path, and its length is denoted by $L[\psi_t^j]$.

Due to the structure of the trellis, at any time $t = t_1$ there are at most $|S_{t_1}|$ survivors, one for each $s_{t_1}^i$. The key observation is the following one [2]: the shortest complete path ψ_n^0 must begin with one of these survivors, if it did not, but passed through state $s_{t_1}^l$ at time t_1, then we could replace its initial segment by $\psi_{t_1}^l$ to get a shorter path, which is a contradiction.

With the previous observation in mind, we see that for any time $(t-1)$, we only need to mantain m survivors ψ_{t-1}^m ($1 \leq m \leq |I_{t-1}|$, one survivor for each node), and their lengths $L[\psi_{t-1}^m]$. In order to move from time $t-1$ to time t:

- we extend the time $(t-1)$ survivors, one time unit along their edges in the trellis, this is denoted by $\psi_t^j = (\psi_{t-1}^i || \theta_{t-1}^{i,j})$.
- compute the new length $L[\psi_t^i, \theta_t^{i,j}]$, of the new extended paths, and for each node (state) we select as the time t survivor the extended path with the shortest length.

The algorithm proceeds by extending paths and selecting survivors until time n is reached, where there is only one survivor left.

3 Equidistant Codes as Fingerprinting Codes

In this section we discuss the use of equidistant codes as fingerprinting codes. Recall from Section 2 that given a descendant, there are three possible configurations for the parents of a descendant. Note that among these configurations, the only one that defeats the fingerprinting scheme is the triangle one, and therefore it should be difficult for the colluders to achieve it. Below, we show that the probability, that a collusion generates a descendant that "decodes" in a triangle configuration, can be made exponentially small by increasing the length (and reducing the rate) of the code.

The following notation will be useful. Given a codeword $\mathbf{c} = (c_1, \ldots, c_n)$ we define the *support* of $\mathbf{c}$ as $S(\mathbf{c}) = \{i | c_i \neq 0, \text{ for } i = 1, \ldots, n\}$.

Proposition 1. *Let C be an $[n, k, d]$ equidistant binary linear code. The minimum distance of C is an even number. Moreover, $\forall \mathbf{c}, \mathbf{c'} \in C - \{\mathbf{0}\}$ with $\mathbf{c} \neq \mathbf{c'}$ we have that $|S(\mathbf{c}) \cap S(\mathbf{c'})| = d/2$.*

Proof. Let $\mathbf{c} \neq \mathbf{c'} \in C - \{\mathbf{0}\}$. Since C is linear and equidistant $S(\mathbf{c}) = S(\mathbf{c'}) = S(\mathbf{c+c'}) = d$. Also $|S(\mathbf{c+c'})| = |S(\mathbf{c})| + |S(\mathbf{c'})| - 2|S(\mathbf{c} \cap \mathbf{c'})|$, therefore $|S(\mathbf{c} \cap \mathbf{c'})| = d/2$. ∎

Proposition 2. *In a binary, linear and equidistant code we always have that the minimum distance increases exponentially with respect the dimension of the code.*

Proof. In fact we will prove that $d \geq 2^{k-1}$. We start by constructing a base of the code. Let $B = \{\mathbf{b}_1, \ldots, \mathbf{b}_m\}$ be such a base. Since the code is linear and equidistant, $|S(\mathbf{b}_i)| = d$ and $|S(\mathbf{b}_i) \cap S(\mathbf{b}_j)| = d/2$, for all $\mathbf{b}_i \neq \mathbf{b}_j$.

We now define the set $I_0^1 = \{i : i \in S(\mathbf{b}_1)\}$, and recursively given I_j^i we also define $I_{2j}^{i+1} = I_j^i \cap \{s : s \in S(\mathbf{b}_{i+1})\}$ and $I_{2j+1}^{i+1} = I_j^i \cap \{s : s \notin S(\mathbf{b}_{i+1})\}$.

We stop the recursivity when $|I_j^i| = 0$. Note that $I_{2j}^{i+1} \cup I_{2j+1}^{i+1} = I_j^i$. We will see that B will determine a base of a binary, linear and equidistant code if and only if

$$|I_{2j}^i| = \frac{|I_j^{i-1}|}{2}, \text{ for } i = 2, \ldots, m - 1. \quad (2)$$

Note that (2) implies that $2^{i-1} \leq d = |I_0^1|$ and therefore the proposition follows. It is immediate to see that (2) is a sufficient condition. We now show that it is also a necessary condition.

Suppose that it is not necessary to satisfy (2) in order to obtain a binary, linear and equidistant code. This implies that there exists a base that does not satisfy (2). Taking the elements of the base as an ordered set, that is, bases with the same vectors but considered in a different order are different bases, we choose one of the bases $B' = \{\mathbf{b'}_1, \ldots, \mathbf{b'}_m\}$ that does not satisfy (2) for a smallest value i.

Now that i is fixed, we consider the smallest $j = 2k$ such that $|I_{2j}^i| \neq \frac{|I_j^{i-1}|}{2}$. Without loss of generality, we can assume that

$$|I_{2j}^i| < \frac{|I_j^{i-1}|}{2} = \alpha_1 \quad (3)$$

then (3) implies that

$$|I_{2(j+1)}^i| < \frac{|I_{j+1}^{i-1}|}{2} = \alpha_2 \quad (4)$$

because otherwise, if we consider the vector $\mathbf{v} = \mathbf{b}_i + \mathbf{b}_{i+1}$ and redefining I_j^{i-1} and I_{j+1}^{i-1} as a function $\mathbf{v}$, we observe that $|I_j^{i-1}| \neq \frac{|I_k^{i-2}|}{2}$ which is a contradiction of our initial assumption, because changing $\mathbf{b}_{i-1}$ by $\mathbf{v}$ in B', we have found a base that does not satisfy (2) for a value less than i.

Therefore we can assume that (3) and (4) are satisfied. But defining I_j^{i-1} and I_{j+1}^{i-1} as a function of $\mathbf{b}_i$, we see that necessarily $|I_j^{i-1}| < \alpha_1 + \alpha_2 = \frac{|I_k^{i-2}|}{2}$ which is another contradiction. ∎

We will use the following proposition.

Proposition 3. *Let C be a binary, linear and equidistant code with parameters $[n, k, d]$. Let $\mathbf{u}, \mathbf{v}, \mathbf{w} \in C - \{\mathbf{0}\}$, then $|(S(\mathbf{u}) \cap S(\mathbf{w})) - (S(\mathbf{u}) \cap S(\mathbf{v}))| = |(S(\mathbf{w}) \cap S(\mathbf{v})) - (S(\mathbf{u}) \cap S(\mathbf{v}))| \leq d/2$.*

Proof. Suppose that $|S(\mathbf{w}) \cap S(\mathbf{u}) \cap S(\mathbf{v})| = d/2 - r$ with $0 \leq r \leq d/2$. Then, since $|S(\mathbf{w}) \cap S(\mathbf{u})| = |S(\mathbf{w}) \cap S(\mathbf{v})| = d/2$ we have that $|(S(\mathbf{u}) \cap S(\mathbf{w})) - (S(\mathbf{u}) \cap S(\mathbf{v}))| = |(S(\mathbf{u}) \cap S(\mathbf{w}))| - |S(\mathbf{w}) \cap S(\mathbf{u}) \cap S(\mathbf{v})| = r$. ∎

Theorem 1. *Let C be a binary, linear and equidistant code with parameters $[n, k, d]$. Let $\mathbf{u} \neq \mathbf{v} \in C$ and let $\mathbf{z}$ be a descendant of $\mathbf{u}$ and $\mathbf{v}$, with $d(\mathbf{u}, \mathbf{z}) = d(\mathbf{v}, \mathbf{z}) = d/2$. Then the probability p that a codeword $\mathbf{w} \in C - \{\mathbf{u}, \mathbf{v}\}$ satisfies $d(\mathbf{w}, \mathbf{z}) = d/2$ is*

$$p \leq 2^{k - 2^{k-2}}$$

Proof. If $\mathbf{u} = \mathbf{0}$ then to have $d(\mathbf{w}, \mathbf{z}) = d/2$ we need that $|S(\mathbf{z})| = d/2$ and since $|S(\mathbf{v})| = d$ there are $N_{tri} = \binom{d}{d/2}$ of such $\mathbf{z}$. In the case that $\mathbf{u} \neq \mathbf{v} \neq \mathbf{w} \neq \mathbf{0}$, by Proposition 3 we know that $|(S(\mathbf{u}) \cap S(\mathbf{w})) - (S(\mathbf{u}) \cap S(\mathbf{v}))| = |(S(\mathbf{w}) \cap S(\mathbf{v})) - (S(\mathbf{u}) \cap S(\mathbf{v}))|$, and since $|S(\mathbf{w})) - (S(\mathbf{u}) \cup S(\mathbf{v}))| = d/2 - |(S(\mathbf{u}) \cap S(\mathbf{w})) - (S(\mathbf{u}) \cap S(\mathbf{v}))|$, if $d(\mathbf{w}, \mathbf{z}) = d/2$ we have that $d/2 = S(\mathbf{w} + \mathbf{z}) \geq d/2 - |S(\mathbf{w}) \cap S(\mathbf{u}) \cap S(\mathbf{v})| + |S(\mathbf{w})) - (S(\mathbf{u}) \cup S(\mathbf{v}))| = d/2$, that is, the descendant $\mathbf{z}$ satisfies $|(S(\mathbf{u}) \cap S(\mathbf{z})) - (S(\mathbf{u}) \cap S(\mathbf{v}))| = |(S(\mathbf{v}) \cap S(\mathbf{z})) - (S(\mathbf{u}) \cap S(\mathbf{v}))|$.

If $\mathbf{w} = \mathbf{0}$, it is clear that $|(S(\mathbf{u}) \cap S(\mathbf{z})) - (S(\mathbf{u}) \cap S(\mathbf{v}))| = |(S(\mathbf{v}) \cap S(\mathbf{z})) - (S(\mathbf{u}) \cap S(\mathbf{v}))| = 0$.

With this, the number of descendants that we can construct is

$$N_{tri} = \sum_{i=0}^{d/2} \binom{d/2}{i}^2 = \binom{d}{d/2}.$$

It is clear that $N_{tri} \geq 2^{d/2}$ and therefore, the probability p that a codeword $\mathbf{w}$ satisfies $d(\mathbf{w}, \mathbf{z}) = d/2$ can be bounded by $p \leq \frac{2^k}{2^{d/2}} = 2^{k-d/2}$, but since by the proof of Proposition 2 we have that $d \geq 2^{k-1}$ it follows that

$$p \leq 2^{k-d/2} < 2^{k - 2^{k-2}}.$$

∎

Intuitively Theorem 1 has the following explanation. The quantity $\binom{d}{d/2}$ is the number of descendants that can be generated by choosing $d/2$ positions from each parent. The greater the quantity, the smaller the probability of a collusion of coming up with a descendant that forms a triangle configuration. The probability of a triangle configuration is thus lowered by increasing the distance and, of course, the length of the code.

4 Tracing – Identifying the Guilty

We now tackle the problem of how to recover the guilty in case of a collusion attack. As stated before, this is the same as searching for the parents of a descendant $\mathbf{z}$.

In order to do the search efficiently, we will add structure to an equidistant code, and work with an equidistant parity check matrix $[n, k, d]$ code C. Such a code can be represented by a trellis using the results in Section 2.2.

Note that, from Proposition 1 it follows that d is an even number. And since we don't know in advance if we have to deal with a star, degenerated star or triangle configuration, we have to design an algorithm that outputs all codewords of a (2,2)-separating code within distance $d/2$ of $\mathbf{z}$. Since the error correcting bound of the code is $\lfloor \frac{d-1}{2} \rfloor$ we have that in the cases, "degenerated" star and triangle, we need to correct one more than the error correcting bound of the code. As it is shown below, this can be done by modifying the Viterbi algorithm.

4.1 A Tracing Viterbi Algorithm

In [5] it is shown that maximum likelihood decoding of any $[n, k, d]$ block code can be accomplished by applying the VA to a trellis representing the code. However, the algorithm discussed in [5] falls into the category of unique decoding algorithms since it outputs a single codeword, and is therefore not fully adequate for our purposes. In this section we present a modified version of the Viterbi algorithm that when applied to a descendant, outputs a list that contains all codewords within distance $d/2$ of the descendant. If the list is of size 3, then there are three possible pairs of parents, whose intersection is disjoint. In a fingerprinting scheme, this basically means that the colluders cannot be traced. The algorithm we present falls into the category of list Viterbi decoding algorithms [3].

We first give an intuitive description of the algorithm.

Recall that, in order to search for the parents of a given descendant $\mathbf{z}$, we find, either the unique codeword at a distance less or equal than $\frac{d}{2} - 1$ of $\mathbf{z}$, or the two or three codewords at a distance $\frac{d}{2}$ of $\mathbf{z}$. Let $\mathbf{z} = (z_1, z_2, \ldots, z_n)$ be a descendant. Let $\boldsymbol{\theta}_c = \{\theta_0^{0,l}, \ldots, \theta_{t-1}^{i,j}, \ldots \theta_{n-1}^{k,0}\}$ the sequence of edges in the path associated with codeword $\mathbf{c} = (c_1, \ldots, c_t, \ldots, c_n)$. As defined in Section 2.2, we have that $\textbf{label\_of}(\theta_{t-1}^{i,j}) = c_t$. Each distinct path of the trellis corresponds to a distinct codeword, and since we need to search for codewords within a given distance of $\mathbf{z}$, it seems natural to define the "length" of the edge $\theta_{t-1}^{i,j}$, $l[\theta_{t-1}^{i,j}]$, as
$$l[\theta_{t-1}^{i,j}] := \mathbf{d}(z_t, c_t) = \mathbf{d}(z_t, \textbf{label\_of}(\theta_{t-1}^{i,j})).$$

Since we expect the algorithm to return all codewords within distance $d/2$ of $\mathbf{z}$, we can have more than one "survivor" for each node. For node $\mathbf{s}_t^j$, we denote the lth "survivor" as $\psi_t^{j,l}$.

Using $l[\theta_{t-1}^{i,j}]$, we define the length of the path $\psi_t^{j,c}$ associated with codeword $\mathbf{c}$, as the Hamming distance between $\mathbf{z}$ and $\mathbf{c}$, both truncated in the first t symbols, $L[\psi_t^{j,c}] := \mathbf{d}(\mathbf{z}, \mathbf{c}) = \sum_{m=1}^{t} \mathbf{d}(z_m, \textbf{label\_of}(\theta_{m-1}^{i,j}))$.

Then, whenever $L[\psi_t^{j,c}] > d/2$ we can remove the path $\psi_t^{j,c}$ from consideration. Note that, for a given node the different "survivors" do not necessarily need to have the same length. For each node (state) s_t^j, in the trellis, we maintain a list Ψ_t^j of tuples $(\psi_t^{j,k}, L[\psi_t^{j,k}])$, $k \in \{1, \ldots, |\Psi_t^j|\}$, where $\psi_t^{j,k}$ is a path passing through s_t^j and $L[\psi_t^{j,k}]$ is its corresponding length.

Tracing Viterbi Algorithm. (TVA)

Variables:

t	time index.
$\psi_t^{j,m}, \forall j \in I_t$	mth survivor terminating at s_t^j.
$L[\psi_t^{j,m}], \forall j \in I_t$	mth survivor length.
$L[\psi_t^{j,m}, \theta_{t-1}^{i,j}]$	Length of the path $(\psi_{t-1}^{i,k} \| \theta_{t-1}^{i,j})$.
$\Psi_t^j, \forall j \in I_t$	List of "survivors" terminating at s_t^j.

Initialization:

$t = 0$;
$\psi_0^{0,1} = s_0^0$; $L[\psi_0^{0,1}] = 0$; $\Psi_0^0 = \{(\psi_0^{0,1}, L[\psi_0^{0,1}])\}$;
$\Psi_t^j = \{\emptyset\}\ \forall t \neq 0$

Recursion: $(1 \leq t \leq n)$

    for every $s_t^j \in S_t$ do
        $m := 0$
        for every s_{t-1}^i such that $\theta_{t-1}^{i,j}$ is defined do
            for every $\psi_{t-1}^{i,k} \in \Psi_{t-i}^i$
                Compute $L[\psi_t^{j,m}, \theta_{t-1}^{i,j}] = L[\psi_{t-1}^{i,k}] + l[\theta_{t-1}^{i,j}]$
                if $L[\psi_t^{j,m}] <= d/2$
                    add $(\psi_t^{j,m}, L[\psi_t^{j,m}])$ to Ψ_t^j
                $m := m + 1$

Termination:
The codewords associated with each path $\psi_n^{0,m} \in \Psi_n^0$ are all within distance $d/2$ of **z**.

4.2 Improvement

If we allow a simple substraction to be performed at each node, then for *systematic* codes, the number of maintained paths in the recovery Viterbi algorithm can be reduced. In a *systematic* code, the message is found unchanged in the codeword.

The TVA discards a path, whenever its length exceeds $d/2$. We will now show, that for systematic codes, at time $t \geq k$ we are in a position to give a lower bound on the total length of a path. Therefore, if this lower bound is greater than $d/2$, the path can be immediately discarded, without the need for more computations.

In the construction of the trellis, we saw that every node s_t^i can be represented by an $(n-k)$-tuple, that is a linear combination of the first t columns of H.

More precisely, for the path corresponding to codeword $\mathbf{c} = \{c_1, \ldots, c_n\}$, we have that $\mathbf{s}_t^i = c_1 \mathbf{h}_1 + \cdots + c_k \mathbf{h}_t$. Moreover, according to (1) we have that $\mathbf{s}_t^i + c_{n-t} \mathbf{h}_{n-t} + \cdots + c_n \mathbf{h}_n = \mathbf{0}$. If the code is systematic, then it can seen that the columns $\mathbf{h}_{n-k}, \ldots, \mathbf{h}_n$ conform an identity matrix, that is $(n-k)$-tuples of weight 1, and therefore, at time $t = k$, we have that the tuples $\mathbf{s}_k^i$ and $(c_{n-k}, \ldots, c_n)$ must be identical.

As before let $\psi_n^{1,m}$ be a complete path. Suppose that at time k the path $\psi_n^{1,m}$ passes through node $\mathbf{s}_k^i$ and that at that time the length of the path is $L(\psi_k^{i,m})$. We now give a lower bound on the total length of the path. We have seen that at a given time t, the length of the path is increased whenever the label of the edge $\theta_t^{i,j}$ differs from the symbol in the tth position of the descendant. Therefore, from time $t = k$ to time $t = n$, the length of the path will be increased by the distance $\mathrm{dist}(\mathbf{s}_k^i, \mathbf{z}^{(n-k,n)})$ between the $(n-k)$-tuple representing state $\mathbf{s}_k^i$ and the tuple $\mathbf{z}^{(k+1,n)} = (z_{k+1}, \ldots, z_n)$ containing the last $n - k$ bits of $\mathbf{z}$. Since $\mathrm{dist}(\mathbf{s}_k^i, \mathbf{z}^{(k+1,n)}) \leq |w(\mathbf{s}_k^i) - w(\mathbf{z}^{(k+1,n)})|$, the total length of the path $L(\psi_n^{1,m})$ is lower bounded by $L(\psi_k^{i,m}) + |w(\mathbf{s}_k^i) - w(\mathbf{z}^{(k-1,n)})|$. Note that this reasoning is also valid for $t > k$ by considering the distance between the last $n - t$ symbols of $\mathbf{s}_k^i$, that is $\mathbf{s}_k^{i(t-k+1,n-k)}$, and $\mathbf{z}^{(t+1,n)}$.

The algorithm below, for time $t \geq k$, computes for each path $\psi_t^{i,m}$ at each node $\mathbf{s}_t^i$, the weight difference $|w(\mathbf{s}_t^i) - w(\mathbf{z}^{(t+1,n)})|$ and adds it to the length of the path $L(\psi_t^{i,m})$. If the sum is greater than $d/2$, the path is discarded, otherwise the path and its new length are added to the list of paths at that node Ψ_t^j.

Improved Tracing Viterbi Algorithm.

Variables:

r_w_t weight of the positions $t+1, \ldots, n$ of $\mathbf{z}$; $r\_w_t = w[(z_{t+1}, \ldots, z_n)]$

Initialization:

$t = 0$;
$\psi_0^{0,1} = \mathbf{s}_0^0$; $L[\psi_0^{0,1}] = 0$; $\Psi_0^0 = \{(\psi_0^{0,1}, L[\psi_0^{0,1}])\}$;
$\Psi_t^j = \{\emptyset\} \; \forall t \neq 0$
$r\_w_t = w[(z_{t+1}, \ldots, z_n)]$; $0 \leq t \leq (n-1)$

Recursion: $(1 \leq t \leq (k-1))$

     for every $\mathbf{s}_t^j \in S_t$ do
         for every $\mathbf{s}_{t-1}^i$, such that $\theta_t^{i,j}$ is defined do
           $m := 0$
           for every $\psi_{t-1}^{i,k} \in \Psi_{t-i}^i$
             compute $L[\psi_t^{j,m}] = L[\psi_{t-1}^{i,k}] + l(\theta_t^{i,j})$
             if $L[\psi_t^{j,m}] <= d/2$
               add $(\psi_t^{j,m}, L[\psi_t^{j,m}])$ into Ψ_t^j
             $m := m + 1$

Recursion: $(k \leq t \leq (n-1))$

    for every $\mathbf{s}_t^j \in S_t$ do
      m:=0
      for every $\mathbf{s}_{t-1}^i$, such that $\theta_t^{i,j}$ is defined do
        for every $\psi_{t-1}^{i,k} \in \Psi_{t-i}^i$
          compute $L[\psi_t^{j,m}] = L[\psi_{t-1}^{i,k}] + l(\theta_t^{i,j})$
          if $L[\psi_t^{j,m}] > d/2$
            discard $\psi_t^{j,m}$
          else
            set $\Delta w = |w(\mathbf{s}_k^{i(t-k+1,n-k)}) - w(\mathbf{z}^{(t+1,n)})|$
            if $L[\psi_t^{j,m}] + \Delta w <= d/2$
              add $(\psi_t^{j,m}, L[\psi_t^{j,m}])$ into Ψ_t^j
              m:=m+1

Termination:

The codeword associated with each path $\psi_n^{0,m} \in \Psi_n^0$ is within distance $d/2$ of $\mathbf{z}$.

5 Conclusions

This paper discusses the use of equidistant codes as fingerprinting codes. It is shown that equidistant codes are robust against size 2 collusion attacks. We also discuss a tracing algorithm for such codes. The algorithm traces the colluders by going through the trellis of the code. Moreover, we show how the algorithm can be improved when the underlying code is systematic.

References

1. D. Boneh and J. Shaw. Collusion-secure fingerprinting for digital data. *Advances in Cryptology-Crypto'95, LNCS*, 963:452–465, 1995.
2. G. D. Forney. The Viterbi algorithm. *Proc. IEEE*, 61:268–278, 1973.
3. Nambirajan Seshadri and Carl-Erik W. Sundberg. List Viterbi decoding algorithms with applications. *IEEE Trans. Comm.*, 42:313–323, 1994.
4. N. Wagner. Fingerprinting. *Proceedings of the 1983 IEEE Symposium on Security and Privacy*, pages 18–22, April 1983.
5. Jack K. Wolf. Efficient maximum likelihood decoding of linear block codes using a trellis. *IEEE Trans. Inform. Theory*, 24:76–80, 1978.

Color Cube Analysis for Detection of LSB Steganography in RGB Color Images

Kwangsoo Lee, Changho Jung, Sangjin Lee, and Jongin Lim

Center for Information Security Technologies, Korea University, Korea
kslee@cist.korea.ac.kr

Abstract. This paper introduces a new distribution model of RGB colors in RGB color images that can be easily breakable by LSB embedding. Regarding the RGB colors as the points of a 3-dimensional lattice space, we consider the noisiness of some special sets called the δ-cubes. The model is based on the symmetrical patterns of noise vectors in some of the comparable δ-cubes. Our new steganalytic technique based on the model, called by the color cube analysis, was capable to detect reliably the low-rate LSB embedding, even when it did not have any false detection, in our experiment.

1 Introduction

Steganography is the art of hiding information by embedding them into innocuous looking cover signals. Digital images are frequently used as the carrier media of steganographic systems because they contain a lot of redundancies that could be modulated without having any significant impacts on the visual properties of the images. Early Steganographers observed that the least significant bits (LSBs) of image data are extremely random, and considered that the LSBs can be substituted by random message bits unsuspiciously. The substituting method is called LSB embedding or LSB steganography.

Steganalysis is the art and science of detecting the steganographic use in signals. It takes some advantages of statistical or perceptual distinction of stego signals from cover signals. Several steganalytic algorithms[4,7,9] for detecting LSB steganography are modelled in gray-scale images and can be extended to RGB color images simply by gathering featured information from each one of the three projected images into R, G and B planes. These simple extension techniques, however, can use only either the marginal distributions of RGB colors or the spacial correlations in each one of the three projected images.

Other types of steganalyses[5,8] are modelled in RGB color space. The methods are commonly based on the fact that a steganographic embedding can increase the number of the colors presented in RGB color images. They well discriminate between the stego images and the cover images, when the cover images are obtained by the conversion from lossy compressed format such as JPEG and have the small number of colors. However, since the images obtained by the scanning process contain fully abundant colors involving complex noise ingredients and have similar properties like the stego images, the steganalytic methods cannot discriminate between the scanned cover images and the stego images.

In this paper, we introduce a new distribution model of the RGB colors. Regarding the RGB colors as the points (or vectors) of a lattice space Z^3, the new concept of the noisiness of a special set called the δ-cube is defined by the complexity measure in the space. The proposed distribution model is based on the symmetrical patterns of noise vectors in some of the comparable δ-cubes called the left and right δ-cubes. Briefly, the left δ-cubes and the right δ-cubes have the same noisiness levels in high probability. LSB flipping causes the point shifting of RGB vectors and consequently violates the symmetrical property of the noisiness levels. Our new steganalytic technique, called by the color cube analysis, is based on the assumption of the distribution model of the RGB colors.

We notice that the experimental results have shown the remarkably high detecting rate for the stego images, even when the analysis did not have any false detection. In the experiment, without false detection, the color cube analysis detected not only 85% of stego images generated by 10% LSB embedding but also 24% of stego images generated by 5% LSB embedding. When embedded message lengths are larger than 20, all of the stego images were detected by our method. For demonstrating the detecting performance of the color cube analysis, we will compare the experimental results of our method with the results of Regular-Singular analysis[7] and Sample Pair analysis[9]. Although our results are very remarkable, the theoretical reason of the performance is not described. Alternately, we give some consideration of the impact of LSB embedding on the distribution model.

This paper is organized as follows. In Section 2, we describe some basic concepts required to our model in the lattice space Z^3, that include the complexity measure and the δ-cubes. In Section 3, we introduce the new distribution model of RGB colors and give some consideration for the impact of LSB steganography on the model. Section 4 presents the new steganalytic algorithm based on the model and the experimental results are displayed in section 5. Finally, we finish with the conclusion in section 6.

2 Basic Concepts

In this section, we give the basic concepts to be necessary for describing the proposed approach.

2.1 RGB Color Space

A digitized RGB color image S can be represented by the succession of 3-tuples $s = (\sigma_1, \sigma_2, \sigma_3)$ whose coordinates σ_i are integral values which represent the relative intensities of the components. We treat the RGB colors s as the points (or vectors) of the 3-dimensional lattice space Z^3 and then S is a subset of Z^3. We define a function $f_S : Z^3 \to \{0,1\}$ as follows: for every $p \in Z^3$,

$$f_S(p) = \begin{cases} 1 & \text{if } p \in S \\ 0 & \text{otherwise} \end{cases} \qquad (1)$$

A point $p \in Z^3$ is said to be filled with S if $f_S(p) = 1$, and the point p is said to be empty with S otherwise. The function $f_S(\cdot)$ will be called by the state function with S.

Let $\mathcal{F}$ be the collection of subsets of Z^3. If a image S is given, we can define a new measure $\gamma_S : \mathcal{F} \to [0, \infty]$, which plays an important role in our approach, as follows: for every $A \in \mathcal{F}$,

$$\gamma_S(A) = \sum_{s \in A} f_S(s) \qquad (2)$$

The measure γ_S will be called by the complexity measure with S. A set $A \in \mathcal{F}$ is said to be m-complex with S if $\gamma_S(A) = m$. From the definition, it follows that $0 \leq \gamma_S(A) \leq |A|$ for every $A \in \mathcal{F}$. The ratio $\gamma(A)/|A|$ is used to measure the noisiness level of S in the set A.

2.2 Color Cube Approach

For a positive integer δ, we shall call the set of the form

$$Q(a; \delta) = \left\{ s \in Z^3 : \sigma_i = \alpha_i \text{ or } \sigma_i = \alpha_i + \delta, \ 1 \leq i \leq 3 \right\} \qquad (3)$$

by the δ-cube (or specially the color δ-cube) with corner at a. Here, $a = (\alpha_1, \alpha_2, \alpha_3)$. If an image S is given, the δ-cubes can be classified by the complexity measure γ_S. Since a δ-cube has $8(= 2^3)$ points, it can have 8-complexity and 2^8 patterns to the maximum (See Fig. 1). Let Ω_δ be the collection of δ-cubes, and let $\Omega_\delta(m)$ be the sub-collection of Ω_δ that consists of m-complex δ-cubes with S. Then, the collection Ω_δ is partitioned into the sub-collections $\Omega_\delta(m)$:

$$\Omega_\delta = \Omega_\delta(0) \cup \Omega_\delta(1) \cup \cdots \cup \Omega_\delta(8) \qquad (4)$$

The inner points of $Q(a; \delta)$ mean the points s of Z^3 such that $\alpha_i \leq \sigma_i \leq \alpha_i + \delta$ for all $i = 1, 2, 3$. A cube is called by the inner cube of $Q(a; \delta)$ if all the points of the cube are inner points of $Q(a; \delta)$. For example, $Q(a; 1)$ has a unique inner 1-cube of itself and $Q(a; \delta)$ has 8 adjacent inner 1-cubes when $\delta \geq 2$. On the other hand, the outer points of $Q(a; \delta)$ mean the points s of Z^3 that are not inner points of $Q(a; \delta)$. A cube is called by the outer cube of $Q(a; \delta)$ if all the points of the cube are outer points of $Q(a; \delta)$ with the exception of one point correspondence. For example, $Q(a; \delta)$ has 8 adjacent outer 1-cubes.

3 Distribution Model of RGB Colors

In this section, we present the proposed distribution model of RGB colors and give some consideration of the impact of LSB embedding on the model.

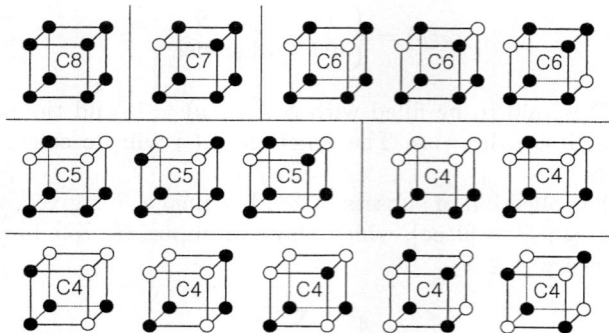

Fig. 1. Pattern inventory of δ-cubes with different complexities in unoriented figure (floating in three dimension). The black and white colors of points represent the filled and empty states of the points respectively. If the colors are changed to each other, C8, C7, C6 and C5 become C0, C1, C2, and C3 respectively

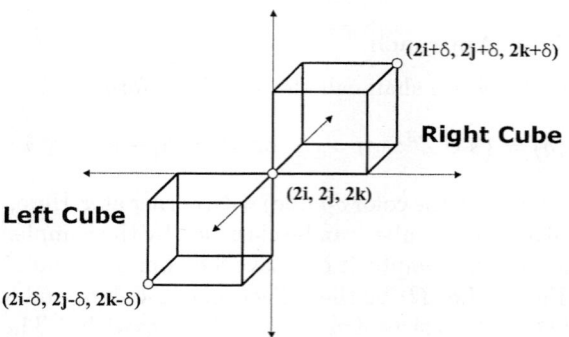

Fig. 2. Left δ-cube and right δ-cube

3.1 Color Cube Distribution Model

From now, we assume that the δ is a positive odd integer. Let P be the set of the points a of Z^3 whose coordinates α_i are even integers. For a point $a \in P$, we shall call the set

$$Q_L(a;\delta) = \{s \in Z^3 : \sigma_i = \alpha_i \text{ or } \sigma_i = \alpha_i - \delta, \ 1 \leq i \leq 3\} \tag{5}$$

by the left δ-cube with corner at a. Also, we shall call the set

$$Q_R(a;\delta) = \{s \in Z^3 : \sigma_i = \alpha_i \text{ or } \sigma_i = \alpha_i + \delta, \ 1 \leq i \leq 3\} \tag{6}$$

by the right δ-cube with corner at a. Here, $a = (\alpha_1, \alpha_2, \alpha_3)$.

It is clear that every point of Z^3 is contained in exactly one of the left δ-cubes. So the left δ-cubes are pairwise disjoint and cover the lattice space Z^3.

The same is true for right δ-cubes.

$$Z^3 = \bigcup_{a \in P} Q_L(a; \delta) \text{ and } Z^3 = \bigcup_{a \in P} Q_R(a; \delta) . \tag{7}$$

Let $\mathcal{L}_\delta$ be the collection of left δ-cubes with corners at the points of P, and let $\mathcal{R}_\delta$ be the collection of right δ-cubes with corners at the points of P.

$$\mathcal{L}_\delta = \{Q_L(a; \delta) : a \in P\} \text{ and } \mathcal{R}_\delta = \{Q_R(a; \delta) : a \in P\} . \tag{8}$$

For a given image S, let $\mathcal{L}_\delta(m)$ be the sub-collection of $\mathcal{L}_\delta$ that consists of left m-complex δ-cubes with S, and let $\mathcal{R}_\delta(m)$ be the sub-collection of $\mathcal{R}_\delta$ that consists of right m-complex δ-cubes with S. Then $\mathcal{L}_\delta$ and $\mathcal{R}_\delta$ are partitioned as follows:

$$\mathcal{L}_\delta = \mathcal{L}_\delta(0) \cup \cdots \cup \mathcal{L}_\delta(8) \text{ and } \mathcal{R}_\delta = \mathcal{R}_\delta(0) \cup \cdots \cup \mathcal{R}_\delta(8) . \tag{9}$$

The noise vectors of RGB colors are usually assumed to be symmetrically distributed with **0** mean vector. Our view point is that the noise vectors are symmetrically distributed in the left and right δ-cubes and have the symmetrical noisiness or patterns in the δ-cubes. So the left and right δ-cubes have the same complexity in high probability. For the practical use for detection of LSB steganography, we assume that the following statement is hold:

Assumption of Color Cube Distribution Model: Let δ be a positive odd integer, and let m be an integer in the interval $[1, 8]$. Then every cover image S satisfies the following equation:

$$E\Big[\big|\mathcal{L}_\delta[m]\big|\Big] = E\Big[\big|\mathcal{R}_\delta[m]\big|\Big] . \tag{10}$$

The model is a key observation in developing our steganalysis techniques. The model indicates that the left and right δ-cubes have the similar complex levels with S. The assumption is reasonable in our view point. LSB embedding, however, results in the significant modification of the color cube distribution (See Fig. 3).

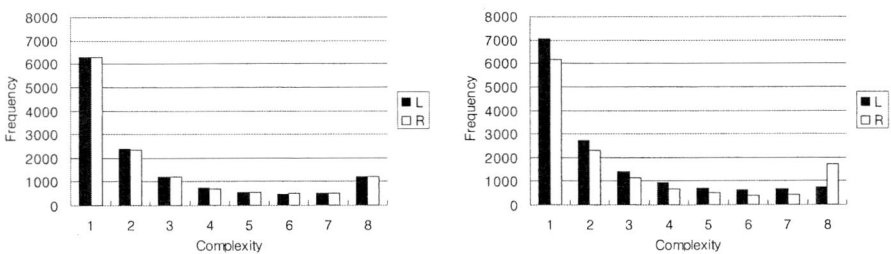

Fig. 3. The frequency comparison between left and right m-complex 1-cubes in racing.bmp before(left) and after(right) LSB embedding of a random message with 100% capacity. One can observe that the cover image has the approximate statistic of the color cube distribution model. However, the stego image shows the significant violation for the symmetrical properties

3.2 Impact of LSB Embedding on the Model

LSB embedding in a digital image changes the RGB colors drawn from the image. In a lattice space Z^3, it can make a change of the point state; an empty point becomes a filled point or a filled point becomes an empty point. In fact, it occurs when the point has a small frequency value such as $0, 1, 2, \cdots$. The points with large frequency values, however, remain to be filled. The more a δ-cube has the newly generated filled points, the more complex the δ-cube becomes. On the contrary, the more a δ-cube has the newly generated empty points, the more simple the δ-cube becomes.

It is clear that the right 1-cube is closed under the LSB modulation, *i.e.*, the RGB colors on the right 1-cube are changed into the points of the right 1-cube of itself. If a heavy-weighted right 1-cube contains an empty point, the point tends to be filled in high probability from the modification of other points having large frequency values. Furthermore, since random bits are uniformly distributed, the frequencies of the points of the right 1-cube will be equalized after the LSB embedding to the RGB colors in the right 1-cube. As the result, the right 1-cube tends to be more complex after the LSB embedding.

Assuming that the distribution of RGB colors on a lattice space Z^3 has a 3-dimensional bell-shaped curve, we discuss the change of the complexities of δ-cubes by LSB embedding according to their locations in the distribution as follows:

Center: If a δ-cube is located in the center of the distribution, the points of the δ-cube have large frequency values. The points of centered δ-cube remain to be filled in high probability after the embedding. So, the complexities of the δ-cubes will not be changed by LSB embedding. This case does not make a significant impact on the color cube distribution.

Slope: If a δ-cube is located in a slope of the distribution, some of the points of the δ-cube have small frequency values such as $0, 1, 2, \cdots$, that are frequently observed at lower ends of the δ-cube slope. **In case of the right δ-cubes located in slopes**, a relatively small frequency valued point s of a right δ-cube Q_R is contained in an inner right 1-cube of Q_R in which most of other points have larger or equal frequency values than the point s in high probability. By the histogram equalization of right 1-cubes, the frequency value of the point s tends to be increased after the embedding. In particular, the empty points of the right δ-cubes become filled points in high probability. Hence the right δ-cubes will be more complex. On the contrary, **in case of the left δ-cubes located in slopes**, a relatively small frequency valued point s of a left δ-cube Q_L is contained in an outer right 1-cube of Q_L in which most of other points have smaller or equal frequency values than the point s in high probability. In this case, the histogram equalization of right 1-cubes results in the decrement of the frequency value of the point s. In particular, it is possible that some filled points of the left δ-cubes, having small frequency values, become empty points after the embedding. Hence, the left δ-cubes will be more simple. As the result, the color cube distribution model will not be satisfied for the stego images generated by the LSB embedding.

Tail: If a δ-cube is located in the tail of the distribution, lots of the points of the δ-cube have 0 frequency values and few points of the δ-cube have small frequency values. The complexity changes both of the left and right δ-cubes are negligible. This case does not make a significant impact on the color cube distribution.

4 Detection Algorithm

4.1 Color Cube Analysis

For a given RGB color image S, one can use some statistical measures for checking the similarity of complex levels of the left and right δ-cubes with S. We use simply the χ^2-test[1] as the statistical measure. The following is the formal procedure for calculating the statistic.

1. Make the state function $f_S(\cdot)$ for S from Eqn. (1).
2. Accumulate the distribution of left and right δ-cubes as the cube complexities with S: For every $a \in P$, calculate the complexity c_l of the left δ-cube $Q_L(a;\delta)$ and increase $|\mathcal{L}_\delta(c_l)|$ by 1. In the same manner, for every $a \in P$, calculate the complexity c_r of the right δ-cube $Q_R(a;\delta)$ and increase $|\mathcal{R}_\delta(c_r)|$ by 1.
3. Calculate the p-value for the similarities between $|\mathcal{L}_\delta(m)|$ and $|\mathcal{R}_\delta(m)|$ for $m = 1, \cdots, 8$: We use the χ^2-test to determine whether the image S satisfies Eqn. (10). The expected distribution $Y_\delta^*(m)$ for the χ^2-test is computed by the arithmetic mean:

$$Y_\delta^*(m) = \frac{|\mathcal{L}_\delta(m)| + |\mathcal{R}_\delta(m)|}{2} \qquad (11)$$

The χ^2 value for the differences between the distributions is given as

$$\chi^2 = \sum_{m=1}^{\nu+1} \frac{\left(|\mathcal{L}_\delta(m)| - Y_\delta^*(m)\right)^2}{Y_\delta^*(m)}, \qquad (12)$$

where ν are the degrees of freedom. Here, $\nu = 7$. Then the p-value p is then given by the cumulative distribution function,

$$p = \int_0^{\chi^2} \frac{t^{(\nu-2)/2} e^{-t/2}}{2^{\nu/2} \Gamma(\nu/2)} dt, \qquad (13)$$

where Γ is the Euler Gamma function. return the p-value p.

4.2 Detection Strategy on Multiple δ Values

There are a lot of the positive odd integers. The selection of the δ value affects the performance of the detection. The well-detecting δ values depend on the noisiness of a stego image. It happens that a stego image is detected for some

δ values, although it is not detected for other δ values. Some large δ values, however, are frequently unavailable for detection of LSB embedding. For the efficiency of the analysis, it is necessary that some available δ values with good performances should be obtained.

We assume that the available δ values $(\delta_1, \cdots, \delta_{\max})$ and the decision thresholds $(T_1, \cdots, T_{\max})$ are obtained. If a suspicious image S is given, for every $i = 1, \cdots, \max$, run the color cube analysis on the input (S, δ_i) and store the output p-values p_i. If there exists a δ_i such that $p_i > T_i$, then return "the image is a stego data"

5 Experimental Results

We display the experimental results of our detection method by ROC (Receiver Operating Characteristic) curves, and compare them with those of Regular-Singular analysis[7] and Sample Pair analysis[9].

We used a database consisting of 100 RGB color images of size 960 × 1296 obtained by scanner. The stego images were generated by embedding messages in randomly selected LSBs of the scanned images, where the message lengths are in 5%, 10%, 15% and 20% of the capacity of the images.

Fig. 4 shows the experimental results of Regular-Singular analysis[7] and Sample Pair analysis[9] for our tested imges. One can see that the two methods have extremely low detecting rates when any false detections are not existed.

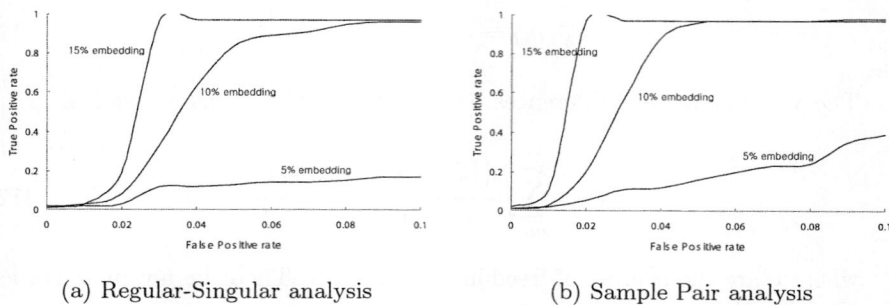

(a) Regular-Singular analysis (b) Sample Pair analysis

Fig. 4. ROC curves for our test images by Regular-Singular analysis and Sample Pair analysis, where the stego images are generated by LSB steganography in scattering mode with hidden message length 5%, 10% and 15%

For the δ values over 21, the color cube analysis did not well discriminate between the cover images and the stego images by low-rate embedding such as 5% and 10%. So we simulated the color cube analysis with the $\delta = 1, 3, 5, \cdots, 21$. Fig. 5 shows the experimental results of the color cube analysis. One can see that the color cube analysis achieved high detecting rate for 10% and 15% embedding, even when it had no false detection. Although the detecting performance for

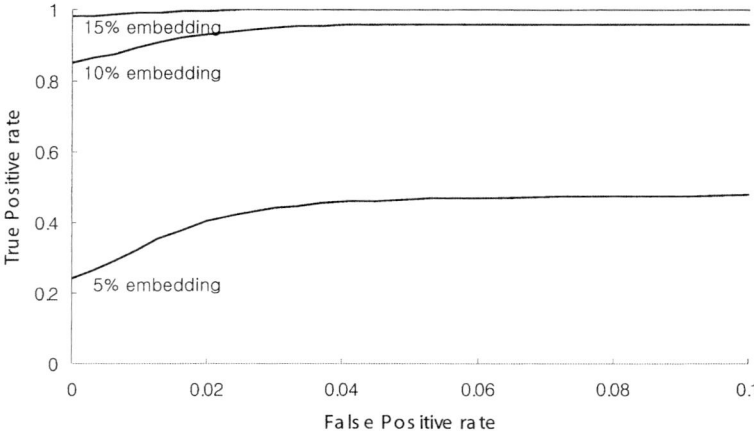

Fig. 5. ROC curves showing the results of the detection of LSB steganography by the color cube analysis, where the hidden message lengths are 5%, 10%, and 15%

Table 1. Comparison of experimental detecting performances

Message length	0%	5%	10%	15%	20%
Regular-Singular analysis	0%	1%	2%	2%	100%
Sample Pair analysis	0%	1%	1%	2%	99%
Color Cube analysis	0%	24%	85%	98%	100%

5% embedding are relatively poor, it is better than the above two methods when any false detections are not existed. Table 1 is to compare the detecting performance of Color Cube analysis with those of Regular-Singular analysis and Sample Pair analysis. The results indicate that the proposed detection method is highly effective.

6 Conclusion

In this paper, we introduced a new distribution model in the RGB color images that are easily breakable by LSB embedding. The color cube analysis based on the model can have reliably detected LSB steganography, even when it did not have any false detection, in our experiment. Although the experimental results show the high reliability of the color cube analysis without having any false detection, these do not mean the reliability for all RGB color images via the internet. The performance of the color cube analysis needs to be more investigated for numerous images. The improvement and theorization of the color cube analysis will be our further research.

Acknowledgements

This research was supported by the MIC(Ministry of Information and Communication), Korea, under the ITRC(Information Technology Research Center) support program supervised by the IITA(Institute of Information Technology Assessment).

References

1. Ueli M. Maurer.: A Universal Statistical Test for Random Bit Generators. *Journal of Cryptology,* 5(2):89–105, (1992)
2. Anderson, R. J. and Petitcolas, F.A.P.: On The Limits of Steganography. IEEE Journal of Selected Areas in Communications. Special Issue on Copyright and Privacy Protection, vol.16(4)(1998) pp.474–481
3. Cachin. C: An Informaition-Theoretic Model for Steganography. In: Aucsmith, D. (Ed.): Information Hiding. 2nd International Workshop. Lecture Notes in Computer Science, vol.1525, Springer-Verlag, Berlin Heidelberg New York(1998) pp.306–318
4. Westfeld, A., Pfitzmann, A.: Attacks on steganographic systems. Information Hiding. 3rd International Workshop. Lecture Notes in Computer Science, vol.1768. Springer-Verlag, Berlin Heidelberg New York(1999) pp.61–76,
5. Fridrich, J., Du, R., and Meng, L.: Steganalysis of LSB Encoding in Color Images. Proceedings IEEE International Conference on Multimedia and Expo ICME 2000. New York(2000)
6. Provos N. and Peter Honeyman.: Detecting Stegnaographic Content on the Internet. CITI Technical Report 03–11. (2001)
7. Fridrich, J., Goljan, M., and Du, R.: Detecting LSB stegangoraphy in color and grayscale images. Magazine of IEEE Multimedia. Special Issue on Security, October-November issue. (2001) pp.22–28
8. Westfeld, A.: Detecting Low Embedding Rates. Information Hiding. 5th International Workshop. Lecture Notes in Computer Science, Vol.2578. Springer-Verlag, Berlin Heidelberg New York(2002) pp.324–339.
9. Dumitrescu, S., Wu, X., Wang, Z.: Detection of LSB Steganography via Sample Pair Analysis. In: F.A.P. Peticolas (Ed.): Information Hiding. 5th International Workshop. Lecture Notes in Computer Science, vol.2578. Springer-Verlag, Berlin Heidelberg(2003). pp.355–372

Compact and Robust Image Hashing*

Sheng Tang[1,2], Jin-Tao Li[1], and Yong-Dong Zhang[1]

[1] Institute of Computing Technology,
Chinese Academy of Sciences,
100080, Beijing, China
[2] Graduate School of the Chinese Academy of Sciences,
100039, Beijing, China
{ts, jtli, zhyd}@ict.ac.cn

Abstract. Image hashing is an alternative approach to many applications accomplished with watermarking. In this paper, we propose a novel image hashing method in the DCT Domain which can be directly extended to MPEG video without DCT transforms. A key goal of the method is to produce randomized hash signatures which are unpredictable for unauthorized users, thereby yielding properties akin to cryptographic MACs. This is achieved by encryption of the block DCT coefficients with chaotic sequences. After applying Principal Components Analysis (PCA) to the encrypted DCT coefficients, we take the quantized eigenvector matrix (8 × 8) and 8 eigenvalues together as the hash signature, the length of which is only 72 bytes for any image of arbitrary size. For image authentication, we also present an algorithm for locating tampering based on the hashing method. Experiments on large-scale database show that the proposed method is efficient, key dependent, pairwise independence, robust against common content-preserving manipulations.

1 Introduction

With the rapid growth of multimedia applications, protection of intellectual property is becoming more prominent. Image hashing (also known as fingerprinting, digital signature, and passive or noninvasive watermarking) is useful in protection of intellectual property. It can be used for multimedia authentication, indexation of content, and management of large database [2, 15]. It is an emerging research area that is receiving increased attention [2]. An image hash function maps an image to a short binary signature based on the image's appearance to the human eye [20]. In general, an image hash function requires the following desirable properties [2,4]:

1. Robustness (Invariance under perceptual similarity): Images can be represented equivalently in different forms, and undergo various manipulations during

---

* This work was supported by National Nature Science Foundation of China under grant number 60302028.

distribution that may carry the same or similar perceptual information. Therefore, the signatures resulting from degraded versions of an image should result in the same or at least similar signatures with respect to that of the original image. This renders traditional cryptographic schemes using bit-sensitive hash algorithms, such as MD5 and SHA-1 not applicable [1, 2, 18], since even one bit change of the input will alter the output signature dramatically.

2. Pairwise independence (Discriminability or collision free): If two images are perceptually different, the signatures from the two images should be considerably different.

3. Key dependence: In some applications such as image authentication, it is required that the hash function $H()$ depends on a key K, i.e., for two different keys K_1 and K_2, $H_{K_1}(C) \neq H_{K_2}(C)$ for any image C.

4. Short bit length: The hash function should map an input image of arbitrary size to an output signature of short bit length. In some cases such as the method in [4] and our proposed method, the fixed bit length of the signature is preferable for its convenience in signature matching.

Significant attention has been given to robust hashing techniques. Up to now, many image hashing methods have been proposed [2, 3, 4, 5, 6, 7, 8, 9, 10, 11, 12, 13, 14, 15, 16, 17, 18, 19, 20]. These methods can be roughly classified into statistics-based [9, 10, 11, 12], relation-based [6, 7, 8], edge or feature point based [12, 13, 14, 15], coarse representation based [16, 17, 18], radon-based [2, 3, 4, 5], mesh-based [19], and clustering-based [20]. As to statistics-based methods, some are not secure because their signatures can be easily forged due to the easiness of modifying images maliciously without changing the signatures such as block-histogram-based method in [9], and block-mean-based method in [10], and moment-based in [12]. Most current feature-point-based approaches have limited utility as they have poor robustness properties [15]. Additionally, the signature lengthes of many existing methods such as those in [6,7,9,10,11,12] etc., are not short and depend on image sizes. Although the signature length of the method proposed by [4] is only 180 real numbers regardless of image size, it is not very short compared with our method (only 72 bytes). Recently, several radon-based signatures have been proposed [2, 3, 4, 5], which take the advantage of invariant features of the transform to provide robustness, but few address the problem of how to locate tampered regions, and can not directly extended and applied to MPEG video for real-time processing, which is the key motivation of this work. The method in [19] is somewhat complex in that most of the time is consumed in mesh normalization, and can not directly extended to MPEG video either. Additionally, most existing methods have focussed extensively on the problem of capturing image characteristics but randomization of the hash are not explicitly analyzed [15].

In this paper, we present a novel image hashing scheme in the DCT domain. To increase the signature's discriminability, we use the DC and 7 low-frequency terms of block DCT coefficients as the distinguishing features of an image called DCT data matrix. On the other hand, for the purpose of making the signature robust to minor pixel modifications that arise from blurring and compression operations, we apply PCA to the matrix, and quantize the eigenvector matrix and eigenvalues to get compact signature. Before PCA, we encrypt the DCT coefficients with chaotic sequences to achieve the randomization or key dependence of the hash. Based on the new scheme, we also present an algorithm for locating tampering. Experimental results show that our proposed method is effective. The paper is organized as follows. Section 2 and 3 describes signature generation and matching respectively. Section 4 addresses how to locate tampering. Section 5 reports experimental results, and conclusions are drawn in the last section.

2 Hash Algorithms

We propose two algorithms, Algorithm A and Algorithm B. We present Algorithm A first as it is simpler and deterministic, and forms the backbone of the main, and it is aimed for image indexing which requires no motivation to randomization [15]. The second algorithm uses randomization to increase the output entropy and achieve key dependence of the hash function for authentication.

2.1 Algorithm A – Deterministic

The procedure for generating hash signature is described as follows. First, we transform the image C into 8×8 block-DCT domain. Then, we prepare the DCT data matrix A for PCA: divide the DC and 7 low-frequency AC terms (as shown in Fig.10.10 in [21]) by the corresponding values of the quantization matrix used in JPEG, and place the 8 quantized coefficients of each block into the N×8 matrix A in row or column order, where N is the total number of blocks. This can be represented as (1), where D_{ij} denotes the jth quantized DCT coefficients of the ith block of the image C.

$$A = \begin{pmatrix} D_{11}, D_{12}, \cdots, D_{18} \\ D_{21}, D_{22}, \cdots, D_{28} \\ \cdots\cdots\cdots\cdots\cdots\cdots \\ D_{N1}, D_{N2}, \cdots, D_{N8} \end{pmatrix} \quad (1)$$

Before PCA, for the purpose of achieving high speed, instead of using the covariance matrix of A adopted by conventional PCA algorithm, we adopt the centered and scaled matrix B of A [22], i.e., standardizing A by removing the mean of each column and dividing each column by its standard deviation.

Finally, we apply PCA to the standardized matrix B [22], and use the resultant 8×8 eigenvector matrix V and 8 eigenvalues $\lambda_i (i = 1, \ldots, 8)$ as signature. To get more compact signature, we quantize each element $a \in [-1, 1]$ of V and λ_i to an one-byte integer a_q and λ_{qi} according to (2) and (3) respectively.

$$a_q = \lfloor 127(1 + a) \rfloor \quad (2)$$

$$\lambda_{qi} = \lfloor \frac{255\lambda_i}{\sum_{j=1}^{8} \lambda_j} \rfloor \quad (3)$$

2.2 Algorithm B – Randomized

For image authentication, the security of the hash algorithm is an issue. More precisely, it is required that the hash function depends on the private key K [2]. We propose a novel method by using chaotic sequences to achieve this as follows.

Chaotic systems are very sensitive to initial conditions, have noise-like behaviors and compact description [23]. So we use chaotic sequence to randomize (encrypt) the DCT data matrix A before PCA. The logistic map for generating the chaotic sequence is:

$$x_{n+1} = 1 - 2x_n^2. \tag{4}$$

where $x_n \in (-1, 1)$ is a real number, $n \in [0, 8N - 1]$, and the initial value $x_0 \in (0, 1)$ is returned by a random function using the key K as its seed. Thus, we can easily convert the chaotic sequence (column vector) $\{x_n\}$ to an $N \times 8$ encryption matrix G in row major order. Therefore we can calculate the encrypted matrix E from the DCT data matrix A by:

$$E = A .* G \tag{5}$$

where the operator ".*" means the scalar multiplication of two matrices. Finally, we substitute E for A in the deterministic algorithm. The remainder of the algorithm is the same as the deterministic algorithm, that is, applying PCA to the standardized matrix of E and subsequent quantization of eigenvector matrix and eigenvalues.

Because it is impossible to deduce E from the signature, it is very hard to find the private key to forge the signature after encryption, even if the original image is available. The adoption of encryption is to ensure that only the right source can generate the authentication signature, i.e., the hash function depends on the private key. Therefore, different signatures generated with different keys do not match due to their different eigenvectors and eigenvalues. In the extreme hypothetical case, the private key used by the original source may be known to the attacker. This is is a general problem for any secure communication and is out the scope of this paper.

3 Signature Matching

Two images are declared similar (for indexation) or authentic (for authentication) if the similarity S between their signatures is above a certain threshold T, which can be determined by experiments or by user's demands according to various applications. The main idea of signature matching is that if two images are considered similar or authentic, corresponding eigenvectors from the two signatures should be high correlative. Thus, S can be calculated by computing correlation between corresponding pairs of eigenvectors, that is, the cosine of the angle between them since the two eigenvector matrices are orthogonal matrices.

After dequantizing each element a_q of eigenvector matrices by:

$$a = \frac{a_q}{127} - 1 \tag{6}$$

we let $V_o=(\alpha_{o1},\alpha_{o2},\ldots,\alpha_{o8})$, $\lambda_o=(\lambda_{o1},\lambda_{o2},\ldots,\lambda_{o8})$ and $V_t=(\alpha_{t1},\alpha_{t2},\ldots,\alpha_{t8})$, $\lambda_t=(\lambda_{t1},\lambda_{t2},\ldots,\lambda_{t8})$ be the dequantized eigenvector matrices and quantized eigenvalue vector of the original image C_o and the image C_t to be tested respectively. S can be calculated by computing the eigenvalue-weighted summation of the correlations of all the pairs as:

$$S = \sum_{i=1}^{8} \omega_i |\alpha'_{oi}\alpha_{ti}| \tag{7}$$

where α'_{oi} denotes the transpose of column vector α_{oi}, and ω_i is the eigenvalue factor defined as (8). The factor is used for considering different contribution of each compared pair of eigenvectors.

$$\omega_i = \frac{\lambda_{oi} + \lambda_{ti}}{2 \times 255} \tag{8}$$

4 Locating Tampered Blocks

For authentication, locating of tampering, such as detecting modification of licence plate is useful [1,21]. Based on the randomized hash algorithm, we present an algorithm for locating tampering if the calculated S between the images C_o and C_t is below the authentic threshold T. If a malicious tampering is occurred, the DCT coefficients of tampered blocks changes significantly, hence remarkable altering of corresponding HTS values. Therefore, by comparing the HTS values of the corresponding blocks of the image C_o and C_t, we can easily determine which block is most possibly tampered. The algorithm is described as follows.

First, after PCA, according to [22], HTS$_t$ vector of the image C_t can be calculated from the $N \times 8$ standardized matrix B_t, 8×8 eigenvector matrix V_t, and 8×8 diagonal eigenvalue matrix λ_t as:

$$\text{HTS}_t = |\tfrac{1}{\sqrt{\lambda_t}}(B_t V_t)'|' \tag{9}$$

where the operator "||" returns a row vector of the Euclidian length of each column.

For authentication without original images, since the B_o can not be accessed, we use B_t to estimate HTS$_o$ of the original image C_o. So we substitute λ_o and V_o for λ_t and V_t in (9) to estimate HTS$_o$:

$$\text{HTS}_o = |\tfrac{1}{\sqrt{\lambda_o}}(B_t V_o)'|'. \tag{10}$$

Finally, determine which block is the region most possibly tampered by computing the difference vector δ between the HTS$_t$ with HTS$_o$ according to:

$$\begin{cases} \delta = (\text{HTS}_t - \text{HTS}_o)^2 \\ (i,j) = argmax\{\delta\} \end{cases} \tag{11}$$

where the returned (i,j) denotes the indices (or location) of the required block.

5 Experimental Results

In evaluating our proposed method, we tested it on the well-known image "Lena" (512×512) and "bmw" (800 × 600) downloaded from www.bmw.com, and 10000 test images randomly selected from the Corel Gallery database (www. corel.com) including many kinds of images (256 × 384 or 384 × 256). All the colour images are transformed into 8 bits/pixel gray level images. To do experiments, we first extracted signatures from all the 10000 images. Although we implemented the method with Matlab C++ Math Library, it took only about 350 seconds for Algorithm B to extract the 10000 signatures on the PC of Pentium IV 2.4G, which shows the method is efficient.

5.1 Robustness Test

To test robustness of the method, the original images were subjected to various image processing steps which are detailed in [24]. We first compressed the 10000 images to various JPEG images with different quality levels Q ranging from 20% to 90%, and calculated S between images and their corresponding JPEG images. The means and standard deviations (Std) of the measured S were shown in Table.1. Compared with the mean and Std of S in the following pairwise dependence test, this table shows that two proposed algorithms are fairly robust against compression.

Table 1. Means and Std of the measured S between 10000 images and corresponding JPEG images

JPEG Compression	Algorithm A		Algorithm B	
	Mean	Std	Mean	Std
JPEG(Q=20%)	0.9620	0.0555	0.9158	0.0879
JPEG(Q=30%)	0.9738	0.0453	0.9325	0.0750
JPEG(Q=40%)	0.9753	0.0469	0.9404	0.0732
JPEG(Q=50%)	0.9670	0.0542	0.9387	0.0758
JPEG(Q=60%)	0.9867	0.0304	0.9647	0.0515
JPEG(Q=70%)	0.9890	0.0290	0.9713	0.0473
JPEG(Q=80%)	0.9949	0.0179	0.9849	0.0340
JPEG(Q=90%)	0.9950	0.0160	0.9915	0.0197

For image authentication, we set the mean S (0.9158) of JPEG(Q=20%) as the threshold T for authentication.

As to Algorithm A, we added noise to the image "Lena" in various noise levels ranging from 1 to 5, and the calculated S between the original image and noised ones are 0.9980, 0.9514, 0.9848, 0.8939, 0.8370 respectively. We rotated the image "Lena" with small angles varying from 1 to 6 degree. The calculated S between the original image and rotated ones are 0.7792, 0.7459, 0.8031, 0.7950,

0.8378, 0.6670 respectively. We also scaled the image "Lena" with scaling factors ranging from 20% to 200%. The mean and Std of the calculated S between the original image and scaled ones are 0.7337 and 0.1360 respectively. The above results show that the Algorithm A is fairly robust against noising when noise levels is less than 4, while not robust against geometric manipulations such as scaling and rotation.

As to Algorithm B and the same noised image "Lena", the calculated S between the original image and noised ones are 0.9456, 0.9203, 0.8946, 0.8735, 0.7512, which shows that Algorithm B is robust against noising when noise levels is less than 3.

The Algorithm B is sensitive to geometric manipulations which can be clearly specified by users [6]. The reason is that the encryption matrix G is sensitive to the altering of image sizes.

5.2 Pairwise Independence Test

As to Algorithm B, we randomly selected 10×2^{20} pairs of signatures from the 10000 images, and calculated S between each pair. As shown in Fig.1, all the measured S were within the range between 0.0305 and 0.7491. The mean μ and Std σ were 0.2995 and 0.0801. As the histogram closely approaches the ideal random i.i.d. case $N(\mu, \sigma)$, we can conclude that the proposed Algorithm B is pairwise independent, and can calculate the false alarm rate P_{FA} (the probability that declare different images as authentic) according to:

$$P_{FA} = \int_T^\infty \frac{1}{\sqrt{2\pi}\sigma} e^{\frac{-(x-\mu)^2}{2\sigma^2}} = \frac{1}{2} erfc(\frac{T-\mu}{\sqrt{2}\sigma}). \qquad (12)$$

Substituting μ=0.2995, σ=0.0801, T=0.9158, we got very low false alarm rate: $P_{FA} = erfc(5.4406)/2 = 7.1229 \times 10^{-15}$. It shows that our method is fairly discriminative, i.e., collusion-free.

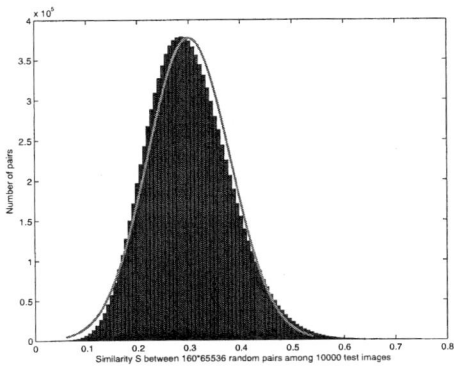

Fig. 1. Pairwise independence test of Algorithm B: Histogram of the measured S between 10×2^{20} pairs of signatures randomly selected from the 10000 images. The red line represents the ideal random i.i.d. case N(0.2995, 0.0801)

As to Algorithm A, we got the similar result. The mean μ and Std σ were 0.3648 and 0.1004. So we can conclude that the two proposed Algorithms are pairwise independent.

5.3 Key Dependence Test

To test key dependence of the proposed Algorithm B, we used the image "Lena" to generate 65536 different signatures by different keys ranging from 0 to 65535, and randomly selected 25×65536 pairs of signatures to calculate S between each pair. As shown in Fig.2, all the measured S were within the range between 0.0349 and 0.7542. The mean and Std were 0.3067 and 0.0807. According to (12), we got the probability P_F that declare different signatures generated by different keys as same: $P_F = erfc[(0.9158 - 0.3067)/(\sqrt{2} \times 0.0807)]/2 = erfc(5.3362)/2 = 2.2347 \times 10^{-14}$, which shows the method is key dependent.

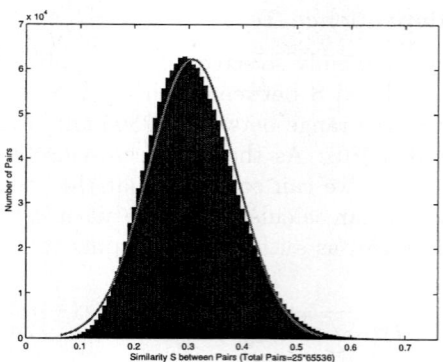

Fig. 2. Key dependence test of Algorithm B: Histogram of the measured S between 25×65536 pairs of signatures randomly selected from 65536 signatures generated by different keys from the image "Lena". The red line represents the ideal random i.i.d. case N(0.3067, 0.0807)

5.4 Authentication Test

We made modifications within the region of the license plate in "bmw" as shown in Fig.3(b). The measured S between the original and tampered images was 0.5614 ($< T$), so we successfully detected that the image was tampered, and located the tampered regions as shown in Fig.3(d). The row and column indices of the most possibly tampered block returned by the propose method are 59 and 47 respectively. It shows that Algorithm B can detect malicious modifications and locate tampering.

6 Conclusion

In this paper, we present a compact image hashing method based on PCA of block DCT coefficients. Experiments show that the proposed method is efficient,

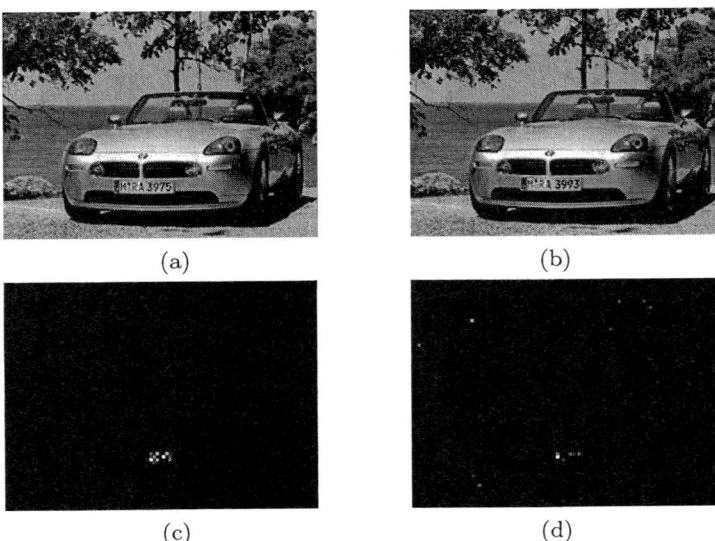

Fig. 3. Authentication test: (a)original image (800 × 600); (b) tampered image (800 × 600) with changing licence "3975" to "3993"; (c) the highlights indicate the real changes of block DCT coefficients between (a) and (b); (d) block-based HTS difference vector δ map (100 × 75) between (a) and (b), the highlight intensity is proportional to the possibility of being tampered

pairwise independent, and robust against common content-preserving manipulations. The randomized algorithm is key dependent, and can detect malicious modifications and locate tampering. It is convenient to extend our method to verify MPEG video streams without DCT transforms. Additionally, since the signature length is only 72 bytes long regardless of image size, it is of great importance to embed the signature into the image itself (such as into the middle-frequency terms of block DCT coefficients) for providing solutions to self-authentication watermarking system [7] in that watermarking capacity is greatly limited [12].

References

1. B.B.Zhu, M.D.Swanson, A.H.Tewfik, "When seeing isn't believing [multimedia authentication technologies]", *Signal Processing Magazine, IEEE,* 21(2):40-49, 2004.
2. J.S.Seoa, J.Haitsmab, T.Kalkerb, C.D.Yoo, "A robust image fingerprinting system using the Radon transform," *Signal Processing: Image Communication,* 19(4):325-339, April 2004.
3. Z.Yao, N.Rajpoot, "Radon/Ridgelet Signature for Image Authentication", *Proc. IEEE ICIP 2004, Singapore,* October 2004
4. F. Lefebvre, J. Czyz, and B. Macq, "A robust soft hash algorithm for digital image signature", *In: Proceedings of IEEE ICIP 2003, Barcelona,* II:495-498, Sept. 2003.
5. F. Lefebvre, B. Macq, and JD Legat, "RASH: RAdon Soft Hash Algorithm", *In 11th European Signal Processing Conference,* Toulouse, France, Sept. 2002.

6. C.-Y. Lin and S.-F. Chang, "A robust image authentication method distinguishing JPEG compression from malicious manipulation", *IEEE Trans. Circuits Syst. Video Technol.*, 11(2): 153-168, 2001.
7. C.-Y. Lin and S.-F. Chang, "Robust digital signature for multimedia authentication", *Circuits and Systems Magazine, IEEE,* 3(4):23-26, 2003.
8. C.-S.Lu, H.-Y. M.Liao, "Structural digital signature for image authentication:an incidental distortion resistant scheme", IEEE Trans. Multimedia,pp.161-173,June 2003.
9. M. Schneider, S.-F. Chang, "A robust content based digital signature for image authentication", *In: Proceedings of IEEE ICIP 96, Lausanne, Switzerland,* Vol.3:227-230, October 1996.
10. Lou DC, Liu JL. "Fault resilient and compression tolerant digital signature for image authentication", *IEEE Trans. Consumer Electronics,* 46(1):31-39, 2000.
11. C. Kailasanathan and R. Safavi Naini, "Image authentication surviving acceptable modifications using statistical measures and k-mean segmentation", IEEE-EURASIP Work. Nonlinear Sig. and Image Proc., June 2001.
12. M.P. Queluz, "Authentication of digital images and video: generic models and a new contribution", *Signal Processing: Image Communication,* 16(5):461-475, 2001.
13. S. Bhattacharjee, M. Kutter, "Compression tolerant image authentication", *In: Proceedings of the IEEE ICIP 1998, Chicago, IL,* October 1998.
14. J. Dittman, A. Steinmetz, and R. Steinmetz, "Content based digital signature for motion picture authentication and content-fragile watermarking, Proc. IEEE Int. Conf. Multimedia Comp. and Sys., pp. 209-213, 1999.
15. Vishal Monga and Brian L. Evans, "Robust Perceptual Image Hashing Using Feature Points", *Proc. IEEE ICIP 2004, Singapore,* 3:677-680, Oct. 2004
16. J. Fridrich and M. Goljan, "Robust hash functions for digital watermarking", *Proc. IEEE Int. Conf. Info. Tech.: Coding and Comp.,* Mar. 2000.
17. R. Venkatesan, S. M. Koon, M. H. Jakubowski, and P. Moulin, "Robust image hashing", *Proc. IEEE Conf. Image Proc.,* Sept. 2000.
18. K. Mihcak and R. Venkatesan, "New iterative geometric techniques for robust image hashing, *Proc. ACM Work. Security and Privacy in Dig. Rights Man.,* Nov. 2001.
19. Chao-Yong Hsu and Chun-Shien Lu, "Geometric Distortion-Resilient Image Hashing System and Its Application Scalability", *In Proceedings of the ACM 2004 multimedia and security workshop on Multimedia and security, Magdeburg, Germany,* pages:81-92, Sept., 2004.
20. Vishal Monga, Arindam Banerjee, and Brian L. Evans, "Clustering Algorithms for Perceptual Image Hashing", *Proc. IEEE Work. on Digital Signal Processing,* Aug. 1-4, 2004, pp. 283-287, Taos, NM.
21. I. J. Cox, M. L. Miller, and J. A. Bloom, *Digital Watermarking,* NewYork: Morgan Kaufmann, 2001.
22. J. Edward Jackson, *A User's Guide to Principal Components,* John Wiley & Sons, Inc., pp. 1-25, 1991.
23. Hui Xiang, Lindong, Wang Hai Lin, Jiaoying Shi, "Digital Watermarking Systems with Chaotic Sequences", *SPIE Conference on Security and Watermarking of Multimedia Contents,* SPIE Vol.3657, pp.449-457, San Jose, California,January,1999.
24. Fabien A. P. Petitcolas, "Watermarking schemes evaluation", *IEEE Signal Processing,* 17(5):58-64, September 2000.

Watermarking for 3D Mesh Model Using Patch CEGIs

Suk-Hwan Lee[1] and Ki-Ryong Kwon[2]

[1] Tongmyong University of Information Technology,
Department of Information Security, 535 Yongdang-dong, Nam-gu,
Pusan, 608-711, Republic of Korea
skylee@tit.ac.kr
[2] Pusan Univ. of Foreign Studies, Division of Digital and Information Engineering,
55-1 Uam-dong, Nam-gu, Pusan, 608-738, Republic of Korea
krkwon@pufs.ac.kr

Abstract. This paper proposes a watermarking for 3D mesh using the CEGI distribution that is robust against mesh simplification, cropping, vertex randomization, and rotation. In the proposed algorithm, a 3D mesh model is divided into patches using a distance measure, then the watermark bits are embedded into the normal vector direction of the meshes that are mapped into cells with large complex weights in the patch CEGIs. The watermark can be extracted based on two watermark keys, the known center point of each patch and a rank table of the cells in each patch. Experimental results verified the robustness of the proposed algorithm based on watermark extraction after various types of attack.

1 Introduction

Digital media, such as images, audio, and video, can be readily manipulated, reproduced, and distributed over information networks. Therefore, a lot of research has been carried out to protect the copyright of digital media, and digital watermarking is one such copyright protection technique. Recently, 3D graphic models, such as 3D geometric CAD data, MPEG-4, and VRML, have become very popular, leading to the development of various 3D watermarking algorithms to protect the copyright of 3D graphic models [1]-[7].

Ohbuchi et al. proposed an algorithm that embeds a watermark in the mesh spectral domain based on the connectivity of the vertices [2], while Praun et al. proposed an algorithm that provides a scheme for constructing a set of scalar basis functions over the mesh vertices on the basis of the spread-spectrum principle [3]. Both of these algorithms are robust against the various geometrical attacks, yet if the mesh connectivity is altered by remeshing or mesh simplification, these algorithms require the suspect mesh model to be resampled for the watermark extraction to obtain the geometry of the original mesh model with a given connectivity information. Benedens proposed an algorithm that em-beds a watermark by modifying the mesh normal distribution that is included in randomly selected bins of the EGI [4]. However, in the case of partial geometric deformation, such as cropping, the meshes included in these directional bins disappear along with the embedded watermark. Therefore, mesh watermarking is required that is not only robust against remeshing, mesh simplification, and cropping, but also allows watermark extraction without the vertices and connectivity of the original mesh model. On the other hand, Kang et al. extended the EGI [8] to a Complex EGI (CEGI) by adding the surface distance as a phase component in the complex function, making the weight associated with a particular surface normal a complex value Ae^{jd}, where the magnitude component is the area A of the corresponding surface and the phase component is the distance d of the surface from the designated origin [9].

The current paper proposes a watermarking for 3D mesh models using patch CEGIs. First, the meshes of a 3D mesh model are clustered into certain patches using a distance measure. The patch number clustered in a model is determined by the magnitude distribution of complex weight in CEGI considering the robustness of watermark. The patch CEGIs are obtained by mapping the normal vectors of meshes in each of patches into the cells with the same direction. The cells with the high rank of the magnitude in patch CEGIs are selected as the embedding target and permuted. Each of the watermark bit is embedded respectively into the cells with the same permuted order in patch CEGIs by using the step searching based on SAD. The watermark is extracted by using the patch center points and the rank table. The results of experiment verify that the proposed algorithm is imperceptible and robust against geometrical attacks of cropping, rotation, and vertex randomization as well as topological attacks of remeshing and mesh simplification.

2 Proposed Mesh Watermarking

This paper considers a 3D mesh model $\mathbf{M}$ that has triangular meshes $\mathbf{m} = \{m_i | 0 < i \leq N_m\}$ of N_m number and vertices $\mathbf{v} = \{v_i | 0 < i \leq N_v\}$ of N_v number, plus $\boldsymbol{n}_{m_i}$ is the unit normal vector of a mesh m_i in a visible direction. All the vectors in this paper are unit vectors.

2.1 Embedding Target

Clustering meshes of 3D mesh model into patches. The meshes in a 3D mesh model are clustered into various patches using a distance measure. The current study selected 6 as the initial number of patches number, then changed the patch number according to the magnitude of the cell distribution in the patch CEGIs. Let the 6 initial center points $I_{i \in [1,6]}$ be vertices randomly selected from among all the vertices of a 3D mesh model. Then, all the vertices v are clustered into a patch $P_{i \in [1,6]} = \{v | d(v, I_i) < d(v, I_j), \text{ all } i \neq j, 1 \leq i, j \leq 6\}$ that has the minimum distance among the initial center points $I_{i \in [1,6]}$, where $d(v, I_i)$ is the distance between v and $I_{i \in [1,6]}$. $I_{i \in [1,6]}$ are updated to the center points of the clustered vertices in each patch $P_{i \in [1,6]}$. The clustering and updating process are iterated until $\sum_{i=1}^{N_p} \| I'_i - I_i \| < 10^{-5}$, where I' is the patch center

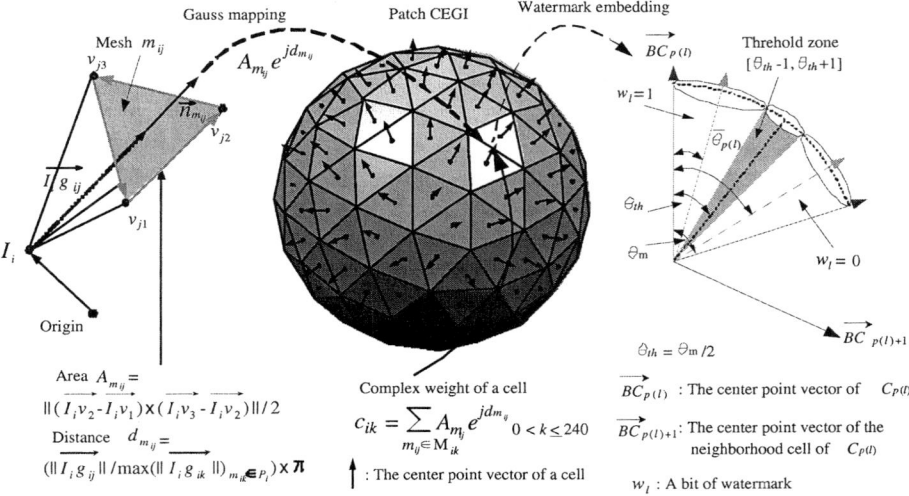

Fig. 1. Normal vector $\vec{n}_{m_{ij}}$ and complex weight $A_{m_{ij}}e^{jd_{m_{ij}}}$ of jth mesh m_{ij} in P_i and mapping into the cell with closest direction to normal direction of mesh in left part and Embedding a watermark bit w_l into average angle $\bar{\theta}_l$ of cell $C_{p(l)}$ in right part

point in the previous iteration and I is the patch center point in the current iteration. If the ratio of the average magnitude $\|\bar{c}_{rank^{-1}(N)}\|$ of Nth ranked cells $\|\bar{c}_{i,rank^{-1}(N)}\|_{i\in[1,N_p]}$ in the patch CEGIs to the total magnitude is not within $[0.03\%, 0.05\%]$, the patch number N_p is changed until the above condition is satisfied. The meshes are then clustered into patches that include their three vertices. The patch center points also need to match the patch CEGIs of the original model for the watermark extracting, so that in the case the models are deformed or cropped in any direction, the watermark can still be extracted from the remaining parts.

Cell selection for embedding target in Patch CEGIs. In the present study, a *pentakis dodecahedron* divided into 240 cells was used as the unit sphere for the CEGI and the mesh normal vectors were mapped into 240 cells. The patch center points are used as the predefined origin in each patch, so that the patch CEGIs $PE_{i\in[1,N_p]}$ have different distributions according to the patch center points. Thus, the normal vector $\vec{n}_{m_{ij}}$, distance $d_{m_{ij}}$, and area $A_{m_{ij}}$ of the jth mesh m_{ij} consisting of (v_{j1}, v_{j2}, v_{j3}) in the ith patch P_i are calculated as shown in Fig. 1. $\overrightarrow{I_i g_{ij}}$ is the vector from to the mass center point of mesh m_{ij}. The complex weight of mesh m_{ij} is $A_{m_{ij}}e^{jd_{m_{ij}}}$. m_{ij} is mapped into the cell that has the closest direction to $\vec{n}_{m_{ij}}$ among the 240 cells. A particular cell C_{ik} in P_i has set M_{ik}

$$\mathrm{M}_{ik} = \{m_{ij}|\cos^{-1}(\vec{n}_{m_{ij}} \cdot \overrightarrow{BC}_{ik}) < \cos^{-1}(\overrightarrow{BC}_{ik} \cdot \overrightarrow{BC}_{i,k+1})/2, 0 < i \leq n_{im}\} \quad (1)$$

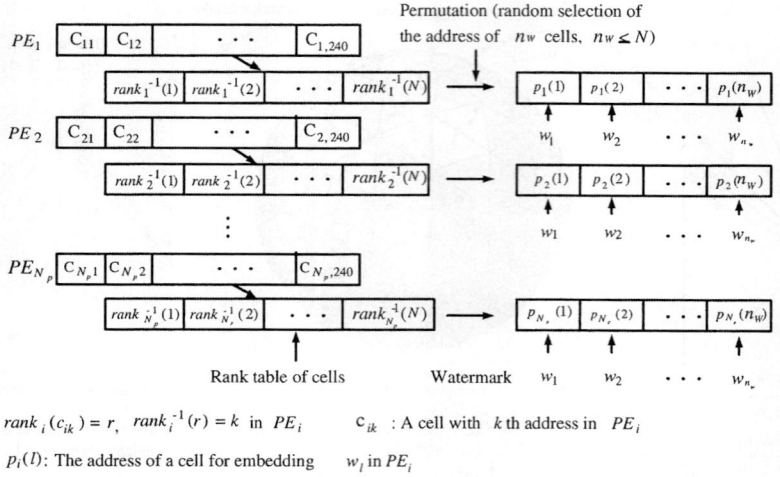

Fig. 2. Select permutated cell address in rank table for embedding target

of the meshes mapped into this cell, where n_{im} is the number of meshes in a patch P_i, and $\overrightarrow{BC}_{ik}$ and $\overrightarrow{BC}_{ik+1}$ are the center point vectors for cell C_{ik} and the neighborhood cell C_{ik+1}. The complex weight magnitude $\| c_{ik} \|$ of C_{ik}

$$\| c_{ik} \| = \sqrt{\sum_{m_{ij} \in M_{ik}} A^2_{m_{ij}} + 2 \sum_{m_{ij} \neq m_{il}} A_{m_{ij}} A_{m_{il}} \cos(d_{m_{ij}} - d_{m_{il}})} \quad (2)$$

is the magnitude component in the sum of the complex weights of the meshes included in set $\mathbf{M}_{ik}$. When $\| c_{ik} \|$ is high, the mesh areas mapped into cell C_{ik} are large and the meshes very close to each other. These cells can be targets for the watermark embedding.

The cells in the patch CEGIs are ranked in descending order according to the magnitude of the complex weight. A rank table of cells below the Nth rank in the patch CEGIs $PE_{i \in [1,N_p]}$ is shown in Fig. 2, where N is the 120th rank that is the maximum rank for embedding a watermark in this paper. In this table, the rank $rank_i(c_{ik})$ of C_{ik} in PE_i is r and the inverse $rank_i^{-1}(r)$ is the address k of cell C_{ik} with rth rank in PE_i. The address of the cell in the table is permuted and the address is selected as the target for watermark embedding. A bit w_l of the watermark is embedded into $C_{1,p_1(l)}, C_{2,p_2(l)}, \cdots, C_{N_p,p_{N_p}(l)}$ with the permuted address $p_1(l), p_2(l), \cdots, p_{N_p}(l)$ in each of the patch CEGIs. The rank table with the information of the permuted cell address is used as the key for the watermark extracting and realignment process.

2.2 Watermark Embedding

The watermark bit w_l is embedded into the average angle $\bar{\theta}_{il}$ of the lth permuted cell $C_{i,p_i(l)}$ in PE_i. $\bar{\theta}_{il}$ is defined as $1/N_{i,p_i(l)} \sum_{m_{ij} \in M_{i,p_i(l)}} \cos^{-1}(\vec{n}_{m_{ij}} \cdot$

$\overrightarrow{BC}_{i,p_i(l)}$), the average angle between the normal vectors of the meshes in set $M_{i,p_i(l)}$ and the center point vector $\overrightarrow{BC}_{i,p_i(l)}$ of $C_{i,p_i(l)}$. $\bar{\theta}_{il}$ moves to the reference angle θ_{w_l} that is $0°$ if w_l is 1 or θ_m otherwise, as shown in Fig. 1. $N_{p_i(l)}$ is the number of meshes in set $M_{i,p_i(l)}$. To change $\bar{\theta}_{il}$ according to the watermark bit, the direction of the mesh normal vectors $\overrightarrow{n}_m$ in $M_{i,p_i(l)}$ must be moved. If the average angle $\bar{\theta}$ is within the threshold range $[\theta_{th} - \triangle\theta_{max}, \theta_{th} + \triangle\theta_{max}]$ in Fig. 1, a bit error will occur in the case of attacks. Thus, since $\bar{\theta}$ should not be within the threshold range, $\triangle\theta_{max}$ was experimentally determined to be $1°$. The threshold angle θ_{th} determining whether the watermark bit is 0 or 1 is $\sum_{i=1}^{N_p} \sum_{r=1}^{N} \bar{\theta}_{ir}/(N_p \times N)$, based on using the average angle of the $N(=120)$th ranked cells in the patch CEGI.

All vertices move to the position of the minimum SAD within the search range, which must be below the coordinate values of the valence vertices when considering the invisibility of the watermark. SAD is the sum of the absolute difference between the reference angle of the watermark bit and the angle of the mesh normal vector and the center point vector of the cell. The search range of a vertex $v_{x,y,z}$ is $[x-\triangle x, x+\triangle x]$, $[y-\triangle y, y+\triangle y]$, and $[z-\triangle z, z+\triangle z]$. $\triangle x$, $\triangle y$, and $\triangle z$ are respectively $0.5 \times min|x - v_k(x)|_{v_k \in valv(v)}$, $0.5 \times min|y - v_k(y)|_{v_k \in valv(v)}$, and $0.5 \times min|z - v_k(z)|_{v_k \in valv(v)}$. $valv(v)$ represents the valence vertices that are connected to v and $v_k(x)$, $v_k(y)$, and $v_k(z)$ are the coordinate values at the axis x, y, z of the valence vertex $v_k \in valv(v)$, respectively.

Let the current vertex be the initial center point $v_o = (x_0, y_0, z_0)$. In the ith step, the search points $\mathbf{v}_i$ are set to 27 points within the search range with the center point $v_{i-1} = (x_{i-1}, y_{i-1}, z_{i-1})$ that is a point with the minimum SAD in the previous step; $\mathbf{v}_i = \{(x_i, y_i, z_i) | x_i \in \{x_{i-1} - \triangle x/2, x_{i-1}, x_{i-1} + \triangle x/2\}, y_i \in \{y_{i-1} - \triangle y/2, y_{i-1}, y_{i-1} + \triangle y/2\}, z_i \in \{z_{i-1} - \triangle z/2, z_{i-1}, z_{i-1} + \triangle z/2\}\}$. The vertex v_{i-1} moves to point v_i with the minimum SAD,

$$v_i = \arg[min_{v_i}\{SAD(\mathbf{v}_i)\}] \qquad (3)$$

among these points. In the ith step, let v_i be the center point of the search range, which is decreased by half. v_i then moves to v_{i+1} with the minimum SAD among the 27 points within its range. This process is performed iteratively until $\|v_i - v_{i+1}\| < 10^{-5}$. The SAD of a particular point is defined as $SAD(v) = \sum_{m_i \in valm(v)} a_{m_i} |\theta_{w_i} - \theta_{m_i}|$, where $valm(v)$ represents the valence meshes connected to v and θ_{m_i} is the angle between the normal vector $\overrightarrow{n}_i$ of the ith mesh m_i in $valm(v)$ and the center point vector of a cell that mesh m_i is mapped into inside a patch CEGI. a_{m_i} is 1 if the cell that m_i is mapped into is the cell selected for embedding the watermark in the rank table, otherwise a_{m_i} is 0. θ_{w_i} is $0°$ if this cell embeds a watermark bit, otherwise it is θ_m. θ_{w_i} is the reference angle of SAD according to a watermark bit.

2.3 Watermark Extracting

The watermark is extracted from the watermarked model using the patch center points $I_{i \in [1, N_p]}$ and the rank table with the permuted cell address without the original model. Two keys are also needed to realign the rotated model to the orientation of the original model. The vertices of the watermarked model $\mathbf{M}'$ are clustered into the patches using the patch center points $I_{i \in [1,N_p]}$, then the patch CEGIs $PE'_{i \in [1,N_p]}$ are calculated. The watermark can be extracted using the rank table obtained from the patch CEGIs $PE_{i \in [1,N_p]}$ in the watermark embedding process. The lth watermark bit w_l is 1 if the average $\bar{\theta}_l = 1/N_p \sum_{i=1}^{N_p} \bar{\theta}_{il}$ of $\bar{\theta}_{il}$ of the lth permuted cell $C_{i,p_i(l)}$ in each of the patch CEGIs $PE'_{i \in [1,N_p]}$ is below θ_{th}. Otherwise, w_l is 0. $\bar{\theta}_{il}$ is $\sum_{m'_{ij} \in M_{i,p_i(l)}} cos^{-1}(\vec{n}_{m'_{ij}} \cdot \vec{BC}_{i,p_i(l)})/N'_{ij}$. N'_{ij} is the number of elements in set $M_{i,p_i(l)}$. Since there are no vertices in the parts where the model has been cropped, if the complex weight of the cell with the embedded watermark is below 0.03% of the total complex weight, this cell is not considered.

If the watermarked model is rotated, the watermark cannot be extracted, as the model has different CEGIs compared to the original model. Thus, in this case, the model must be realigned before extracting the watermark. The realignment process searches for the Euler angle (α, β, γ) that determines the degree of rotation between the attacked model and the original model. All the vertices in the rotated model are clustered into new patch center points $\hat{\mathbf{I}}$ that vary from the patch center points $\mathbf{I} = \{I_1, I_2, \cdots, I_{N_P}\}$ of the watermark key in the α, β, γ direction of ZYZ. The rank table for the new patch CEGIs $PE'_{i \in [1,N_p]}$ clustered to $\hat{\mathbf{I}}$ is then compared with the rank table of the watermark key. The Euler angle $(\alpha^*, \beta^*, \gamma^*)$ that minimizes the difference between the two rank tables is then the rotation angle of the attacked model.

3 Experimental Results

The 3D VRML [10] data for the Stanford bunny, Knots, Venus, and Agrippa bust model was used as the test models to evaluate the performance of the proposed algorithm. The watermark was a 50 bit stream generated by a Gaussian random sequence. Therefore, the bits of the watermark were embedded into 50 cells among the permuted cells below the N=120th rank in each of the patch CEGIs. The Stanford bunny, Knots, and Venus model with meshes in all directions of the 3D space were divided into 6 patches and the watermark embedded into a total of 300 cells in the models. In contrast, the Agrippa bust model was divided into 4 patches, as it had no meshes in the direction of the -z axis, and the watermark embedded into a total of 200 cells in this model. The SNR of the vertex coordinates was used for the invisibility evaluation, similar to the SNR of the pixel intensity in the image data. The SNR is defined as $10\log_{10}(var(\| v - M \|)/var(\| v - v' \|))$. $var(x)$ is the variance of the random variable x, M is the coordinates of the mass center point in the original model, and v and v' are the vertex coordinates in the original model and watermarked model, respectively. However, this SNR can only be adopted in models that have the same number and order of vertices. Table 1 shows the SNRs for the watermarked models. Since the proposed algorithm embeds the watermark into the

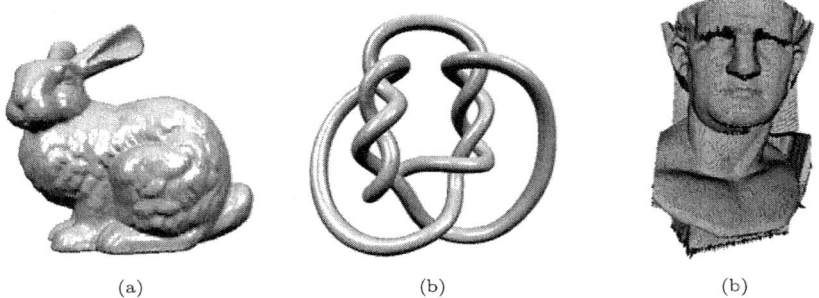

Fig. 3. (a) Stanford bunny, (b) Knots, and (b) Agrippa bust model

Table I. SNR of the models watermarked by the proposed and Benedens' algorithm.

Model	Number of vertices	SNR [dB]		Percentage of the embedding mesh number [%]	
		Proposed	Benedens	Proposed	Benedens
Stanford bunny	35,947	42.69	42.77	62	22
Knots	23,232	39.87	40.77	40	19
Venus	33,591	39.12	39.12	60	20
Agrippa bust	38,000	41.88	-	58	-

patches to improve the robustness, the number of meshes that are changed for the watermark embedding in the proposed algorithm is 2-3 times of the number of meshes in Benedens' algorithm. However, we can obtain a similar SNR of Benedens' algorithm because using the step searching within the search range of each vertex. Beneden's algorithm could not be used with the Agrippa bust model, as it is 40% open 3D space and the algorithm randomly selects the normal distribution in the original model for the watermark embedding.

Table II. Results of the experiment for mesh simplification.

Model	Percentage of vertex number [%]	Bit loss [%]		Percentage of vertex number [%]	Bit loss [%]	
		Proposed	Benedens		Proposed	Benedens
Stanford bunny	77.5	0	4	24.9	2	22
	51.2	0	6	14.8	8	28
Knots	61.0	0	0	29.6	0	4
	39.4	0	2	24.3	2	8
Venus	61.0	0	2	33.0	0	16
	42.5	0	14	27.0	8	18
Agrippa Bust	65.0	0	-	33.2	4	-
	45.5	0	-	20.4	8	-

To evaluate the robustness of the proposed algorithm, the watermarked models were intentionally attacked by mesh simplification, cropping, the addition of random noise, rotation, and multiple attacks. The robustness measure was the bit loss that represented the percentage of bit error in the bit stream of the extracted watermark.

In the mesh simplification experiment, the vertices or meshes of the watermarked model were simplified to vary the vertex number and L-2 norm distance using a MeshToSS tool [10]. Mesh simplification changes the connectivity or topology based on removing vertices and meshes from the model and creating new meshes. Fig. 4(a)-(c) shows that the mesh-simplified models were not smooth as regards the mesh connectivity, yet the shape of the original model was preserved. The results after mesh simplification are shown in Table II. This table shows that the bit loss of the proposed algorithm was 2-20% lower than that of Beneden's algorithm. Furthermore, the proposed algorithm extracted all the bits of the watermark without any bit error until a 30% vertex number percentage, while 92% of the watermark bits remained without any bit error until a 14% vertex number percentage.

Table III. Results of the experiment for random noise addition.

Model	Percentage of number of sampled vertices [%]	Bit loss [%]		Percentage of number of sampled vertices [%]	Bit loss [%]	
		Proposed	Benedens		Proposed	Benedens
Stanford bunny	50	0	6	100	4	16
Knots	50	0	0	100	2	2
Venus	50	0	4	100	6	16
Agrippa Bust	50	0	-	100	2	-

Table IV. The experimental results for the cropping and the multiple attacks.

Model	Attack	Bit loss [%]		Attacks with rotation	Bit loss [%]	
		Proposed	Benedens		Proposed	Benedens
Stanford bunny	Crop	0	18	Crop+Simplify	2	26
Knots	Crop	0	18	Crop+Simplify	6	32
Venus	Crop	0	20	Crop+Simplify	6	32
Agrippa Bust	Crop	0	-	Crop+Simplify	4	-

In the experiment for random noise addition, uniform random noise was added to the x, y, z coordinates of 50% and 100% of the vertices selected randomly from among the vertices in the mesh model; $v' = v \times (1 + \alpha \times uniform())$. The modulation factor α was 0.01 and $uniform()$ was the uniform random function with [-0.5 0.5]. It can change up to 3 decimal points of the coordinate value. As the addition of random noise preserves the connectivity of mesh yet changes the vertex coordinates, the mesh normal vector is slightly changed. However, the proposed algorithm had a 0-6% bit loss, as the watermark was embedded into the patches based on a combination of cells with high complex weights. The experimental results for random noise addition are shown in Table III, which verifies that the bit loss when using the proposed algorithm was 4-12% lower than that when using Beneden's algorithm.

In the cropping experiment, the vertices were removed on the left or right side of the watermarked model. Fig. 4-(e),(f) shows the cropped model with only 60% of the vertices, except for those in the +x axis direction, in the watermarked model. The results for the cropping experiment are shown in table IV, where Beneden's algorithm had a 20% bit loss, because 20% of the watermark

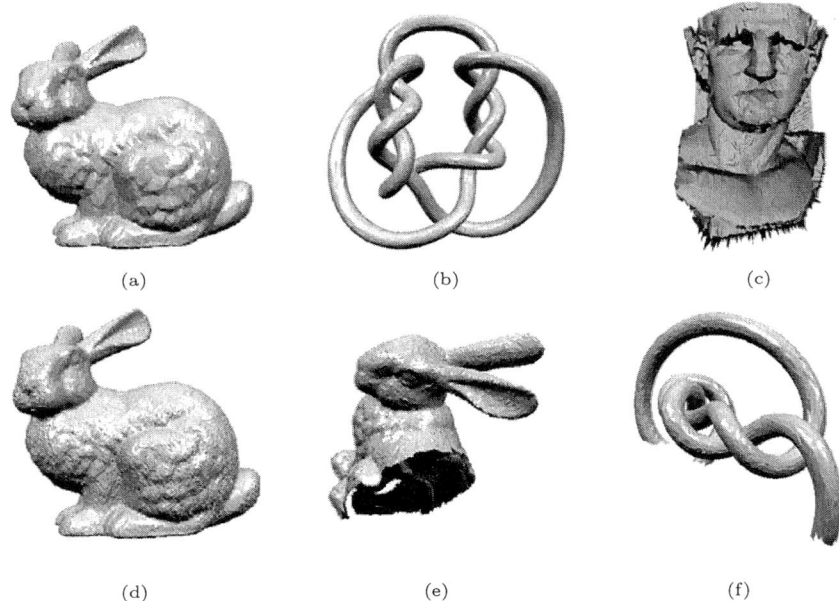

Fig. 4. (a) Stanford bunny simplified to 51.2%, (b) Knots simplified to 39.4%, and (c) Agrippa bust simplified to 33.2% of vertices using mesh simplification, (d) Stanford bunny simplified to 51.2%, (e) Knots simplified to 39.4%, and (f) Agrippa bust simplified to 33.2% of vertices using mesh simplification

had disappeared in the cropping direction. However, when using the proposed algorithm, the watermark could still be extracted from the other patches even though some patches had no vertices and meshes.

To evaluate the robustness to various attacks, multiple attacks including mesh simplification, cropping, and rotation performed. The results for the multiple attacks experiment are shown in Table VI, where the proposed algorithm had a 24% lower bit loss compared to Beneden's algorithm for the mesh-simplified and cropped models. When using Beneden's algorithm, the bit loss was 20% after cropping and 10% after mesh simplification, yet, when using the proposed algorithm the bit loss was just 2-6% only after mesh simplification. Furthermore, the mesh-simplified and cropped models were also rotated in a particular direction. Before extracting the watermark, the models are realigned to the direction of the original model. Plus, a detailed realignment process was performed by slightly varying the Euler angle for exact extraction of the watermark. Although calculating the patch CEGIs and then comparing them to the rank table of the watermark key whenever the Euler angle was slightly varied took about 10-15 minutes using an Intel Celeron 800MHz, the watermark in the rotated model was extracted as if it had not been rotated.

4 Conclusions

This paper presented a method of a watermarking for 3D mesh models using patch CEGIs. The 3D mesh model is divided into a certain number of patches based on the shape of the model, then the CEGI is obtained for each patch. Next, the address of the cells with a high ranking magnitude component in the patch CEGIs is permuted. The water-mark bit is then embedded into the cells that are permuted to the same order in the patch CEGIs using SAD based on step searching. Experimental results verified that the pro-posed algorithm has a similar invisibility to Beneden's algorithm, yet a 4-20% lower bit loss than Beneden's algorithm. The proposed watermarking was also shown to be robust against remeshing, mesh simplification, cropping, and additive random noise, plus the watermark can be extracted using two keys instead of the original model without any resampling or registration process.

Acknowledgements

This work was supported by grant No. (R01-2002-000-00589-0) from the Basic Receach Program of the Korea Science & Engineering Foundation.

References

1. R. Ohbuchi, H. Masuda, M. Aono (1998) Watermarking three-dimensional polygonal models through geometric and topological modification. IEEE JSAC: 551-560
2. R. Ohbuchi, A. Mukaiyama, S. Takahashi (2002) A frequency-domain approach to watermarking 3D shapes. EUROGRAPHICS: 373-382
3. E. Praun, H. Hoppe, A. Finkelstein (1999) Robust mesh watermarking. ACM SIGGRAPH: 49-56
4. O. Benedens (1999) Geometry-based watermarking of 3D models. IEEE CG&A: 46-55
5. F. Cayre, B. Macq (2003) Data hiding on 3D triangle meshes. IEEE Trans. on Signal Processing 51: 939-949
6. B-L, Yeo, Minerva. M (1999) Watermarking 3D Objects for Verification. IEEE Computer Graphics, Special Issue on Image Security: 36-45
7. K.-R. Kwon, S.-G. Kwon, S.-H. Lee, T.-S. Kim, K.-I. Lee (2003) Watermarking for 3D polygonal meshes using normal vector distributions of patch. IEEE International conference on Image Processing 2: 499-502
8. B. K. P. Horn (1986) Robot Vision: The MIT Electrical Engineering and Computer Series, MIT Press, Cambridge, Mass.
9. S.B. Kang, K. Ikeuchi (1993) The Complex EGI: A new representation for 3-D pose determination. IEEE Trans. on Pattern Analysis and Machine Intelligence 15: 707-721
10. T. Kanai, MeshToSS Version 1.0.1, http:// graphics.sfc.keio.ac.jp/MeshToSS/ indexE.html.

Related-Key and Meet-in-the-Middle Attacks on Triple-DES and DES-EXE

Jaemin Choi[1], Jongsung Kim[1], Jaechul Sung[2], Sangjin Lee[1], and Jongin Lim[1]

[1] Center for Information Security Technologies(CIST),
Korea University, Seoul, Korea
{koreamath, joshep, sangjin, jilim}@cist.korea.ac.kr
[2] Department of Mathematics, University of Seoul, Seoul, Korea
jcsung@uos.ac.kr

Abstract. Recently, at CT-RSA 2004, Phan [14] suggested the related-key attack on three-key triple-DES under some chosen related-key condition. The attacks on three-key triple-DES require known plaintext and ciphertext queries under a chosen related-key condition. He also presented related-key attacks on two-key triple-DES and DES-EXE, which require known plaintext and adaptively chosen ciphertext queries under some related-key conditions. In this paper, we extended the previous attacks on the triple-DES and DES-EXE with various related-key conditions. Also we suggest a meet-in-the-middle attack on DES-EXE.

1 Introduction

Data Encryption Standard [13] was developed at IBM and had been adopted by the U.S. National Bureau of Standards as the standard cryptosystem over 20 years. Because of the DES' small key size of 56 bits, some variants of the DES under multiple encryption have been considered, including triple-DES under two or three 56-bit keys and a DESX variant named DES-EXE [5] which switches outer XOR operations and inner DES encryptions. In this paper we focus on the security of triple-DES and DES-EXE against related-key attacks. Also we consider a meet-in-the-middle attack on DES-EXE without related-key conditions.

Related-key attacks are well-known to be very powerful and useful tools for evaluating the security of cryptographic primitives [1, 2, 3, 4, 6, 7, 8, 9, 10, 11, 12, 15, 16]. These kinds of attacks seem to be hard to mount in real protocols, however it has been studied that they can be applied in real protocols [6, 15].

In order to compare our cryptanalytic results with previous ones on triple-DES, let us describe the triple-DES. Let K_i be denoted 56-bit key and E_K be denoted the DES encryption with the key K. Then, two-key triple-DES and three-key triple-DES (which use two 56-bit keys (K_1, K_2) and three 56-bit keys (K_1, K_2, K_3), respectively) can be described as follows.

$$\text{Two-key triple DES} = E_{K_1}(E_{K_2}^{-1}(E_{K_1}(P)))$$

$$\text{Three-key triple DES} = E_{K_3}(E_{K_2}^{-1}(E_{K_1}(P)))$$

where P is a 64-bit plaintext.

Table 1. Summary of attacks on Triple-DES and DES-EXE

Block Cipher	Data / Memory / Time	Type of attack		RKC
Two-Key Triple-DES	2^{56}CP/2^{56}/2^{56}	MITM	[12]	-
	2^{32}KP/2^{32}/2^{88}	KPA	[16]	-
	2^{32}KP,2^{32}RK-KC/$2^{89.5}$/2^{88}	RKA	[14]	(6)
	2^{32}KP,2^{32}RK-KP/2^{33}/2^{89}	RKA	This paper	(7)
	2^{32}KC,2^{32}RK-KC/2^{33}/2^{89}	RKA	This paper	(7)
Three-Key Triple-DES	3CP/2^{56}/2^{112}	MITM	[12]	-
	1KP,1RK-ACC/2^{56}/2^{56}	RKA	[6]	(3)
	2^{32}KP,2^{32}RK-KC/2^{33}/2^{88}	RKA	[14]	(1)
	2^2 RK-CP/ - /$2^{58.5}$	RKA	[15]	(3)
	2^{32}KP,2^{32}RK-KP/2^{56}/2^{57}	RKA	This paper	(2)
	2^{32}KC,2^{32}RK-KC/2^{56}/2^{57}	RKA	This paper	(3)
	2^{32}KP,2^{32}RK-KP/2^{33}/2^{113}	RKA	This paper	(4)
	1KP,1RK-ACP/ - /2^{113}	RKA	This paper	(5)
DES-EXE	2^{32}KP,2^{32}RK-KP/$2^{90.5}$/2^{89}	RKA	[14]	(8)
	1KP,1RK-ACC/2^{56}/2^{56}	RKA	[14]	(9)
	2^{32}KP,2^{32}RK-KP/2^{56}/2^{56}	RKA	This paper	(9)
	1KP,1RK-ACP/2^{56}/2^{56}	RKA	This paper	(11)
	2^{32}KP,2^{32}RK-KP/2^{56}/2^{57}	RKA	This paper	(12)
	1KP,1RK-ACP/2^{56}/2^{57}	RKA	This paper	(13)
	1KP,1RK-ACP/2^{56}/2^{57}	RKA	This paper	(10)
	3KP/$3 \cdot 2^{57}$/2^{59}	MITM	This paper	-

One of the main cryptanalytic results obtained on triple-DES so far is the related-key attack. The related-key attack allows the cryptanalyst to choose a specific relation of related keys whose goal is to recover the related keys by using plaintext or ciphertext queries.

In [6], Kelsey et al. showed that three-key triple-DES is vulnerable to a related-key attack which requires known plaintext and adaptively chosen ciphertext queries under a chosen related-key condition. Recently, Phan [14] presented an extended version of Kelsey et al.'s attack which requires known plaintext and ciphertext queries under another related-key condition. He also presented a related-key attack on two-key triple-DES, which requires known plaintext and adaptively chosen ciphertext queries under a chosen related-key condition. In this paper, we present related-key attacks on two-key and three-key triple-DES, which require adaptively chosen plaintext queries or adaptively chosen ciphertext queries under various related-key conditions. The followings are the related-key conditions used in the previous related-key attacks and our attacks. Here, $K = (K_1, K_2, K_3)$ and $K' = (K'_1, K'_2, K'_3)$ (resp., $K = (K_1, K_2)$ and $K' = (K'_1, K'_2)$) represent chosen related keys of three-key triple-DES (resp., two-key triple-DES), where Δ and ∇ are nonzero arbitrary fixed differences.

$$\text{Three-key}: K_1 \oplus K'_1 = 0, \ K_2 \oplus K'_3 = 0, \ K_3 \oplus K'_2 = 0 \quad (1)$$

$$K_1 \oplus K'_1 = \Delta, \ K_2 \oplus K'_2 = 0, \ K_3 \oplus K'_3 = 0 \quad (2)$$

$$K_1 \oplus K_1' = 0, \ K_2 \oplus K_2' = 0, \ K_3 \oplus K_3' = \Delta \quad (3)$$
$$K_1 \oplus K_1' = \Delta, \ K_2 \oplus K_2' = \nabla, \ K_3 \oplus K_3' = 0 \quad (4)$$
$$K_1 \oplus K_1' = 0, \ K_2 \oplus K_2' = \Delta, \ K_3 \oplus K_3' = \nabla \quad (5)$$
$$\text{Two-key}: \ K_1 \oplus K_2' = 0, \ K_2 \oplus K_1' = 0 \quad (6)$$
$$K_1 \oplus K_1' = \Delta, \ K_2 \oplus K_2' = 0 \quad (7)$$

See Table 1 for a summary of our results and comparison with the previous attacks. We use the following notations in Table 1.

- RKC: Related-Key Condition
- MITM: Meet-In-The-Middle Attack
- SLA: Slide Attack
- KPA: Known-Plaintext Attack
- RKA: Related-Key Attack
- KP: Known Plaintext
- CP: Chosen Plaintext
- RK-CP: Related-Key Chosen Plaintext
- RK-KP: Related-Key Known Plaintext
- RK-KC: Related-Key Known Ciphertext
- RK-ACP: Related-Key Adaptive Chosen Plaintext
- RK-ACC: Related-Key Adaptive Chosen Ciphertext

Another variant of DES, DES-EXE, uses two 56-bit keys and one 64-bit key. This is performed as follows.

$$\text{DES-EXE} = E_{K_c}(K_b \oplus (E_{K_a}(P))), (K_a, K_c : 56 - bit \ keys \text{ and } K_b : 64 - bit \ key)$$

As like the extensions of related-key attacks on triple-DES, we can extend related-key attacks on DES-EXE under various related-key conditions. The followings are the related-key conditions used in the previous related-key attacks and our attacks.

$$\text{DES-EXE}: \ K_a \oplus K_c' = 0, \ K_b \oplus K_b' = 0, \ K_c \oplus K_a' = 0 \quad (8)$$
$$K_a \oplus K_a' = \Delta, \ K_b \oplus K_b' = 0, \ K_c \oplus K_c' = 0 \quad (9)$$
$$K_a \oplus K_a' = 0, \ K_b \oplus K_b' = \Delta, \ K_c \oplus K_c' = 0 \quad (10)$$
$$K_a \oplus K_a' = 0, \ K_b \oplus K_b' = 0, \ K_c \oplus K_c' = \Delta \quad (11)$$
$$K_a \oplus K_a' = \Delta, \ K_b \oplus K_b' = \nabla, \ K_c \oplus K_c' = 0 \quad (12)$$
$$K_a \oplus K_a' = 0, \ K_b \oplus K_b' = \Delta, \ K_c \oplus K_c' = \nabla \quad (13)$$

Furthermore, we will show that our related-key attack on DES-EXE can be converted into a meet-in-the-middle attack which does not need related-key conditions. See Table 1 for the comparison of our results and the previous ones.

2 Related-Key Attacks on Triple-DES

2.1 Related-Key Attack on Two-Key Triple-DES

In this subsection, we request plaintext/ciphertext pairs under two related keys, and look for a pair of these plaintext/ciphertext pairs which has the same input to the middle DES decryption. Since the two related keys use the same key for the middle DES decryption, a pair that has the same input for the middle DES decryption also has the same output from that decryption; this allows a way to test whether a guessed first DES key is correct. Let us assume that we have two-key triple-DES algorithm with two keys of which one uses a key $K = (K_1, K_2)$ and the other uses a key $K' = (K_1 \oplus \Delta, K_2)$ where Δ is a nonzero arbitrary fixed known value. Then we can break two-key triple-DES by using the related-key attack. The attack procedure is as follows. Refer to the left-side of Fig. 1.

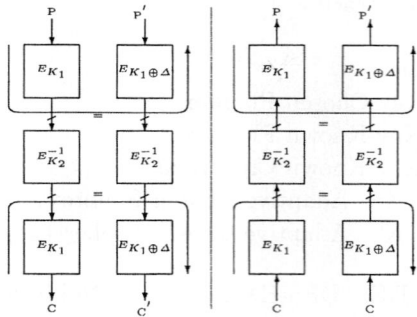

Fig. 1. Related key attack on Two-Key Triple-DES

1. To get a single collision pair, we first collect 2^{32} known-plaintexts P and encrypt P under the key $K = (K_1, K_2)$ to get 2^{32} ciphertexts C.
2. We also collect other 2^{32} related-key known-plaintexts P' and encrypt P' under the key $K' = (K_1 \oplus \Delta, K_2)$ to get 2^{32} ciphertexts C'. We keep the obtained ciphertexts C' together with the corresponding plaintexts P' in a hash table.
3. We guess all values of K_1 for all values of (P,C) and check equations on (P',C'), i.e. $E^{-1}_{K_1 \oplus \Delta}(E_{K_1}(P))=P'$ and $E_{K_1 \oplus \Delta}(E^{-1}_{K_1}(C))=C'$, values in hash table. If there exist $((P,C),(P',C'))$ satisfying these equations, keep the guessed key K_1. Otherwise, restart this step.
4. For the suggested key K_1, we do an exhaustive search for the remaining key K_2 using trial encryption.

This attack requires 2^{32} known plaintexts in step 1 and 2^{32} related-key known plaintexts in step 2. Since the memory requirements of this attack are dominated by step 2, this attack requires 2^{33} 64-bit memories. Since the probability which satisfies the collision test of step 3 for a wrong key is about 2^{-128}, the expectation of wrong keys which pass the collision test is about $2^{-128} \times 2^{32} \times 2^{32} \times 2^{56} = 2^{-8}$ where the $(2^{32} \times 2^{32} \times 2^{56})$ value represents the total number of the collision tests performed in step 3. On the other hand, the right key K_1 passes the collision test with a high probability by the argument of birthday paradox. It follows that if a key K_1 passes the collision test, it will be the right key with a high probability. The time complexity of this attack is dominated by step 3, and thus this attack requires about $2^{32} \times 2^{56} \times 2^{-1} \times 4 = 2^{89}$ DES encryptions on average.

This attack can be also applied to the decryption procedure (refer to the right-side of Figure 1). As like the above analysis, we can compute the complexity of this attack. The memory and time complexities of this attack are same as those of the above attack, but this attack requires 2^{32} known ciphertext and 2^{32} related-key known ciphertexts.

2.2 Related-Key Attack on Three-Key Triple-DES

Let us assume that we have three-key triple-DES algorithm with two keys of which one uses a key $K = (K_1, K_2, K_3)$ and the other uses a key $K' = (K_1 \oplus \Delta, K_2, K_3)$. Then we can break three-key triple-DES by using the related-key attack. The attack procedure is as follows (refer to the left-side of Fig. 2).

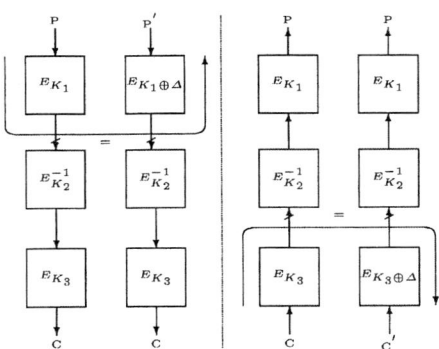

Fig. 2. Related-Key Attack I on Three-Key Triple-DES

1. To get a single collision pair, we first collect 2^{32} known-plaintexts P and encrypt P under the key $K = (K_1, K_2, K_3)$ to get 2^{32} ciphertexts C. We keep the obtained (P, C)-pairs in a hash table.
2. We also collect other 2^{32} related-key known-plaintexts P' and encrypt P' under the key $K' = (K_1 \oplus \Delta, K_2, K_3)$ to get 2^{32} ciphertexts C'. During this procedure, we check whether $C = C'$ for each ciphertext C' and C.
3. Using the collision (P, C) and (P', C) obtained from step 2, we get equation $P' = E^{-1}_{K_1 \oplus \Delta}(E_{K_1}(P))$, and thus we recover the 56-bit key K_1 by exhaustive search for the key K_1.

4. Using the key K_1 obtained from step 3, we compute $E_{K_1}(P)$ and $E_{K_1 \oplus \Delta}(P')$ for any P and P'. We recover the K_2 and K_3 by using the meet-in-the-middle attack on $(E_{K_1}(P), C)$ and $(E_{K_1 \oplus \Delta}(P'), C')$.

This attack requires 2^{32} known plaintexts in step 1 and 2^{32} related-key known plaintexts in step 2. The memory requirements of this attack are dominated by step 4, so this attack requires about 2^{56} 64-bit memories. The time complexity of this attack is also dominated by step 3 and 4 which require about 2^{56} DES encryptions on average, respectively. Thus, it requires $2^{57} (= 2^{56} \cdot 2)$ DES encryptions.

We can also exploit the above method for the case of related keys such that $K = (K_1, K_2, K_3)$ and $K' = (K_1, K_2, K_3 \oplus \Delta)$. See the right-side of Fig. 2. Since these attacks are similar to the above attack, we omit the details.

Furthermore, in the similar way, we can take into account other cases of related-keys whose two subkeys are different. See the appendix A.

3 Attacks on DES-EXE

3.1 Related-Key Attack on DES-EXE

Now we begin our related-key attack on DES-EXE using the meet-in-the-middle technique. Assume that we have DES-EXE algorithms with two keys of which one uses a key $K = (K_a, K_b, K_c)$ and the other uses a key $K' = (K_a, K_b \oplus \Delta, K_c)$. Then, we can break DES-EXE by using the related-key attack which is performed as follows (refer to Fig. 3).

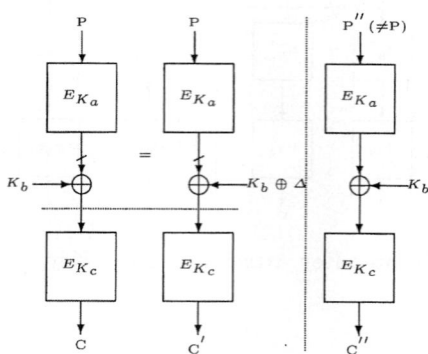

Fig. 3. Related-Key Attack I on DES-EXE

1. First we collect a single known plaintext P and encrypt P to get the corresponding ciphertexts C and C' under keys $K = (K_a, K_b, K_c)$ and $K' = (K_a, K_b \oplus \Delta, K_c)$, respectively. Additionally, by collecting another plaintext $P'' (\neq P)$, we get the corresponding ciphertext C'' under key K.

2. We guess a key K_c and check whether $E_{K_c}^{-1}(C) \oplus E_{K_c}^{-1}(C') = \Delta$. If so, we compute $E_{K_c}^{-1}(C) \oplus K_b$ for each key K_b and keep $(K_b, E_{K_c}^{-1}(C) \oplus K_b)$ in a hash table. Otherwise, we restart this step.
3. We guess a key K_a and check that $E_{K_a}(P) = E_{K_c}^{-1}(C) \oplus K_b$ for each stored value $E_{K_c}^{-1}(C) \oplus K_b$, and if so, we also check that $E_{K_c}(E_{K_a}(P'') \oplus K_b) = C''$ for the guessed key (K_a, K_b, K_c). If the last test is satisfied, we consider the guessed key (K_a, K_b, K_c) as the right key. Otherwise, we restart this step.

This attack requires two known plaintexts and one related-key adaptive chosen plaintext. The memory requirements of this attack are dominated by step 2, so it requires 2^{56} 64-bit memories. Step 2 requires about 2^{55} DES encryptions in average for checking the Δ test. Since the expectation of keys (K_a, K_b, K_c) satisfying the first test of Step 3 is about $2^{56} \times 2^{64} \times 2^{-64} = 2^{56}$, we should perform the last test of step 3 by 2^{56} times. So, step 3 requires about $2^{56} \cdot 2 = 2^{57}$ DES encryptions. Therefore, the total time complexity of this attack is about 2^{57} DES encryptions.

Furthermore, we can use various related keys to mount related-key attacks on DES-EXE which require chosen plaintext queries or ciphertext queries. See the appendix B for the other related-key attacks on DES-EXE.

3.2 Meet-in-the-Middle Attack on DES-EXE

Using the similar technique, we can also mount a meet-in-the-middle attack on DES-EXE without related-key conditions. The attack procedure is as follows.

1. We collect three distinct plaintexts P, P' and P'', and encrypt P, P' and P'' to get the corresponding ciphertexts C, C' and C'', respectively
2. For each key K_a, we compute $S_1 = E_{K_a}(P) \oplus E_{K_a}(P')$, $T_1 = E_{K_a}(P') \oplus E_{K_a}(P'')$ and keep the triple (K_a, S_1, T_1) in a hash table. Similarly, for each key K_c, we compute $S_2 = E_{K_c}^{-1}(C) \oplus E_{K_c}^{-1}(C')$, $T_2 = E_{K_c}^{-1}(C') \oplus E_{K_c}^{-1}(C'')$ and keep the triple (K_c, S_2, T_2) in a hash table.

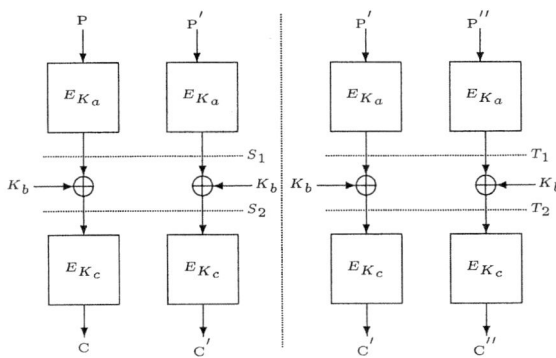

Fig. 4. Meet-in-the-Middle Attack on DES-EXE

3. We check that $S_1 = S_2$ and $T_1 = T_2$ for all 2^{112} quartets $((S_1, T_1), (S_2, T_2))$. If there exist such quartets, keep the keys (K_a, K_c) related to the quartets.
4. For the suggested keys K_a and K_c, we do an exhaustive search for the remaining key K_b using trial encryption.

This attack requires 3 known plaintexts and the memory requirements of this attack are dominated by step 2, so it requires about $3 \cdot 2^{57}$ 64-bit memories. Since the probability which satisfies the collision test step 3 for a wrong key is about 2^{-128}, the expectation of wrong keys which pass the collision test is about $2^{-128} \times 2^{112} = 2^{-16}$ where the $2^{112}(=2^{56} \times 2^{56})$ value represents the total number of the collision tests performed in step 3. On the other hand, the probability which satisfies the collision test for the right keys (K_a, K_c) is one. Thus, if K_a and K_c pass the two collision test, it will be the right keys with a high probability. Moreover, the time complexity of this attack is dominated by step 2, and thus this attack requires $2^{59}(= 2^{56} \times 8)$ DES encryptions.

4 Conclusion

In this paper, we have discussed security of triple-DES and DES-EXE against related-key attacks. We have found various related-key conditions, which allow the cryptanalyst to mount related-key attacks on those ciphers. Furthermore, we have shown that DES-EXE is vulnerable to the meet-in-the-middle attack.

Acknowledgements. This research was supported by the MIC(Ministry of Information and Communication), Korea, under the ITRC(Information Technology Research Center) support program supervised by the IITA(Institute of Information Technology Assessment). We also thank anonymous referees for the valuable comments of our research.

References

1. E. Biham, *New Types of Cryptanalytic Attacks Using Related Keys*, Journal of Cryptology, Vol. 7, No. 4, pp.229–246, 1994.
2. A. Biryukov and D. Wagner, *Advanced Slide Attacks*, Advances in Cryptology - EUROCRYPT 2000, LNCS 1807, pp. 589–606, Springer-Verlag, 2000.
3. M. Blunden and A. Escott, *Related-Key Attacks on Reduced Round KASUMI*, FSE 2001, LNCS 2355, pp. 277–285, Springer-Verlag, 2001.
4. G. Jakimoski and Y. Desmedt, *Related-Key Differential Cryptanalysis of 192-bit Key AES Variants*, SAC 2003, LNCS 3006, pp. 208–221, Springer-Verlag, 2004.
5. B.S. Kaliski and M.J.B. Robshaw, *Multiple Encryption: Weighting Security and Performance*, Dr. Dobb's Journal, 1996.
6. J. Kelsey, B. Schneier and D. Wagner, *Key-Schedule Cryptanalysis of IDEA, G-DES, GOST, SAFER, and Triple-DES*, Advances in Cryptology - CRYPTO'96, LNCS 1109, pp. 237–251, Springer-Verlag, 1996.
7. J. Kelsey, B. Schneier and D. Wagner, *Related-Key Cryptanalysis of 3-WAY, Biham-DES,CAST, DES-X, NewDES, RC2, and TEA*, ICICS'97, LNCS 1334, pp. 233–246, Springer-Verlag, 1997.

8. J. Kilian and P. Rogaway, *How to Protect DES Against Exhaustive Key Search (an Analysis of DESX)*, Journal of Cryptology, Vol. 14, No. 1, pp.27–35, 2001.
9. J. Kim, G. Kim, S. Hong, S. Lee and D. Hong, *The Related-Key Rectangle Attack - Application to SHACAL-1*, ACISP 2004, LNCS 3108, pp. 123–136, Springer-Verlag, 2004.
10. Y. Ko, S. Hong, W. Lee, and J. Kang, *Related key Differential Attacks on 26 Rounds of XTEA and full Rounds of GOST*, FSE 2004, LNCS 3017, pp. 299–316, Springer-Verlag, 2004.
11. Y. Ko, C. Lee, S. Hong and S. Lee, *Related Key Differential Cryptanalysis of Full-Round SPECTR-H64 and CIKS-1*, ACISP 2004, LNCS 3108, pp. 137–148, Springer-Verlag, 2004.
12. R.C. Merkle and M.E. Hellman, *On the Security of Multiple Encryption*, Communications of the ACM, Vol. 24, No.7, 1981.
13. National Bureau of Standard, *Data Encryption Standard*, National Bureau of Standard, FIPS Pub. 46, 1977.
14. R.C.-W. Phan, *Related-Key Attacks on Triple-DES and DESX Variants*, CT-RSA 2004, LNCS 2964, pp.15–24, Springer-Verlag, 2004.
15. R.C.-W. Phan and H. Handschuh *On Related-Key and Collision Attacks: The Case for the IBM 4758 Cryptoprocessor*, ISC 2004, LNCS 3225, pp.111–122, Springer-Verlag, 2004.
16. P.C Van Oorschot, M.J Wiener, *A Known-Plaintext Attack on Two-Key Triple-Encryption*, Advances in Cryptology - EUROCRYPT'90, LNCS 473, pp. 318–325, Springer-Verlag, 1990.

A Related-Key Attack II on Three-Key Triple-DES

We take into account other related keys whose two subkeys are different, i.e., we exploit related keys $(K = (K_1, K_2, K_3), K' = (K_1 \oplus \Delta, K_2 \oplus \nabla, K_3))$ or $(K = (K_1, K_2, K_3), K' = (K_1, K_2 \oplus \Delta, K_3 \oplus \nabla))$ to mount our related-key attacks on three-key triple-DES. We first consider the attack which uses the first case of related-key conditions. Fig. 5 represents the attack procedure. Detailed attack procedure and calculating attack complexities are almost similar to the attack I on three-key triple-DES.

B Related-Key Attack II and III on DES-EXE

As like the extensions of related-key attacks on triple-DES, we can extend related-key attacks on DES-EXE under various related-key conditions. Since these attacks are similar to related-key attacks on three-key triple-DES, we omit the details. Figure 6 and 7 show the outline of all these attacks.

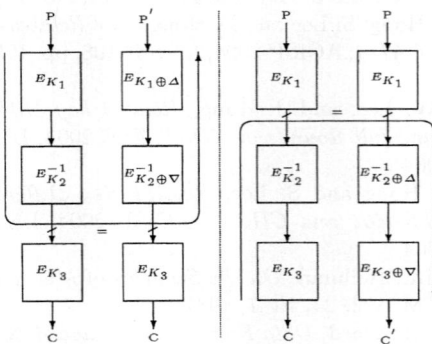

Fig. 5. Related-Key Attack II on Three-Key Triple-DES

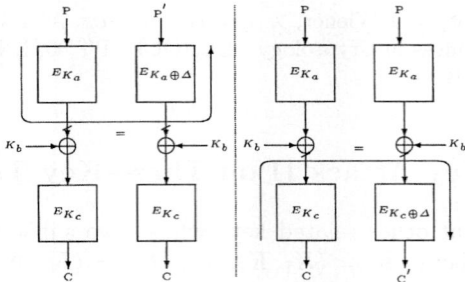

Fig. 6. Related-Key Attack II on DES-EXE

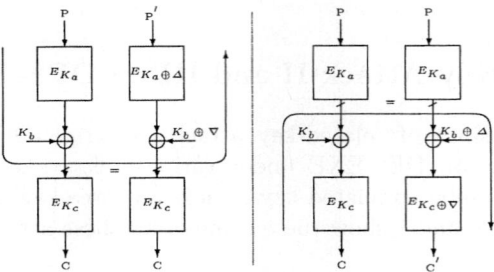

Fig. 7. Related-Key Attack III on DES-EXE

Fault Attack on the DVB Common Scrambling Algorithm

Kai Wirt

Technical University Darmstadt
Department of Computer Science,
Darmstadt, Germany
wirt@informatik.tu-darmstadt.de

Abstract. The Common Scrambling Algorithm (CSA) is used to encrypt streams of video data in the Digital Video Broadcasting (DVB) system. The algorithm uses a combination of a stream and a block cipher, apparently for a larger security margin. However these two algorithms share a common key.

In this paper we present a fault attack on the block cipher which can be launched without regarding the stream cipher part. This attack allows us to reconstruct the common key and thus breaks the complete Algorithm.

Keywords: block cipher, cryptanalysis, fault attack, dvb, pay tv.

1 Introduction

The DVB Common Scrambling Algorithm is used to secure MPEG-2 transport streams. These are used for example for digitally transmitted pay tv in Europe. The algorithm was specified by ETSI and adopted by the DVB consortium in May 1994. However the exact origin and date of the design is unclear. Interestingly, licensees were not allowed to implement the algorithm in software and it was only available under a Non-Disclosure Agreement from an ETSI custodian. As was pointed out, this was due to "security reasons". Only very little information like an ETSI Technical Report [Eur96] and patent applications [Bew98], [WAJ98] were available to the public until 2002. In the fall of 2002 a Windows program called FreeDec which implemented the CSA in software was released and quickly reverse–engineered. The results were published on a web site [Pse03] and details on the algorithm became available to the public.

For keying the CSA, so called *control words* are used. These control words are generated from encrypted control messages contained in the DVB transport stream by a *conditional access mechanism*. Examples for these mechanisms are Irdeto, Betacrypt, Nagravision, Cryptoworks and many others. They vary between broadcasters and are usually implemented on a smart card which is required to view encrypted pay tv transmissions.

The actual key for the CSA is called *common key* and is usually changed every 10–120 seconds. The great relevance of CSA lies in the fact, that every encrypted digital pay tv transmission in Europe is secured using CSA. A practical break of CSA would thus affect all broadcasters which would have to exchange the hardware used to decrypt the transport streams.

The scrambling algorithm is a combination of two cryptographic primitives: a 64-bit block cipher and a stream cipher which both are keyed with the same common key. Thus a key recovery attack on one of the two primitives would break the complete algorithm.

In this paper we present a fault attack on the block cipher part which allows the recovery of the key.

The rest of this paper is organized as follows. In Section 2 we present the notation used in this paper, section 3 gives a short overview over side-channel attacks and sections 4 and 5 describe CSA resp. the block cipher part. Our attack is presented in section 6 and final remarks are given in section 7. Tables and figures are combined in an appendix.

2 Definitions

In the rest of this paper we use the following notation:

K	the common key. A 64 bit key used for both the stream and the block cipher
$[k]_i$	denotes the i-th bit of k
$[k]_{i...j}$	denotes bits i through j of k
K^E	denotes the running key which is derived through the key schedule of the block cipher
$P = (p_0, \ldots, p_7)$	is the plain text
$C = (c_0, \ldots, c_7)$	is the cipher text
$S = (s_0, \ldots, s_7)$	is the state of the block cipher
$S^r = (s_0^r \ldots, s_7^r)$	is the state in round r
$\bar{x}$	is the faulted value x

We number the rounds from $0 \ldots 56$ for encryption resp. $56 \ldots 0$ for decryption i.e. $P = S^0$, $C = S^{56}$.

3 Side-Channel Attacks

Conventional attacks try to find weaknesses in a cipher construction itself. There are various methods to do so, like observing the distribution of cipher texts or attacking the structure of a cipher with algebraic methods. In contrast, side channel attacks are used to find weaknesses in an actual implementation of a cipher system. They are more powerful than conventional attacks, because of the fact, that the attacker can get additional information by observing side channels like the time required to encrypt certain plain texts or the power usage of the encryption device.

One certain type of side channel attacks are so called fault attacks, where the attacker introduces errors in the encryption or decryption process. The attacker then gains informations on the key by observing the difference between the actual and the faulty result. There are various results showing, that these attacks are very powerful and feasible like [BDJ97] and [ABF+02].

Fault attacks are often combined with observations on other side channels like for instance time measurements etc., because the faults have to be introduced at specific points in the encryption/decryption process or at a specific register value. To simplify things one specifies what values can be affected when by the attacker.

This type of side channel attacks was first applied to symmetric cryptosystems by Eli Biham and Adi Shamir in [BS97]. Our attack is a variant using a slightly different setting than in [BS97] and [BDJ97]. In the setting we investigate in this paper, the attacker is capable of changing the value of a specific register to a random value in one specific round. However, we will show, that if the attacker is only able to introduce a random error where the exact error location is evenly distributed over the whole decryption process (i.e. the setting used by Boneh et. al), our attack still works.

There are various possibilities to inject errors like applying voltage peaks or other glitches and/or modifying the clock of the encryption device. Another possibility is to do it by laser. Using this method, it is possible to target at a specific part of the device performing the cryptographic operations and thus affect for example a certain register. We believe, that changing the value stored in such a register to a random value is possible. Moreover we believe, that using short flashes of the laser and precise equipment, it is possible to do so at a specific encryption/decryption operation resp. at a certain round in iterated ciphers. We therefore believe, that the presented attack is an actual threat to the common scrambling algorithm. An overview and further references on how to realize fault attacks can be found in [BS03], where the first fault attack on the Advanced Encryption Standard is given.

4 Overview over CSA

The common scrambling algorithm can be seen as a cascade of two different cryptographic primitives, namely a block cipher and a stream cipher. Both ciphers use the same 64-bit key K, which is called the *common key*. In this section, we will describe how the block and the stream cipher are combined, whereas the next section is focusing on the block cipher.

In the encryption process a m-byte packet is first divided into blocks (DB_i) of 8 bytes each. It is possible, that the length of the packet is not a multiple of 8 bytes. If so, the last block is called *residue*.

The sequence of 8-byte blocks is encrypted in reverse order with the block cipher in CBC mode. The initialization vector is always equal to zero. Note, that the residue is left untouched in this encryption step.

The last output of the chain IB_0 is then used as a nonce for the stream cipher. The first $m - 8$ bytes of key stream generated by the stream cipher are XORed to the encrypted blocks $(IB_i)_{i \geq 1}$ followed by the residue to produce the scrambled blocks SB_i. Figure 1 depicts the descrambling process.

Note, that since we are interested in introducing errors in the decryption process of the block cipher and in comparing the actual decrypted output with the faulty output, we can completely ignore the chaining mode and the stream cipher part, taking only the decryption process of the last block cipher application into account. For more details on the overall design and an analysis of the stream cipher as well as an overview of properties of the block cipher we refer to [WW04].

5 The DVB CSA Block Cipher

CSA uses an iterated block cipher that operates byte-wise on 64-bit blocks of data. In each round of the cipher the same round transformation is applied to the internal state. We will denote this transformation by ϕ. ϕ takes the 8-byte vector representing the current internal state, along with a single byte of the running key, to produce the next internal state. This round transformation is applied 56 times.

The Key Schedule. Let ρ be the bit permutation on 64-bit strings as defined in table 2. The 448-bit running key $K^E = ([k^E]_0, \ldots, [k^E]_{447})$ is recursively computed as follows:

$$[k^E]_{0,\ldots,63} = [k]_{0,\ldots,63}$$
$$[k^E]_{64i,\ldots,64i+63} = \rho([k^E]_{64(i-1),\ldots,64i-1}) \oplus \texttt{0x0i0i0i0i0i0i0i0i} \quad \text{for all } 1 \leq i \leq 6$$

where the expression $\texttt{0x0i0i0i0i0i0i0i0i}$ is to be interpreted as a hexadecimal constant.

Encryption/Decryption. A plain text $P = (p_0, \ldots, p_7)$ is encrypted according to

$$S^0 = P$$
$$S^r = \phi(S^{r-1}, ([k^E]_{8(r-1)}, \ldots, [k^E]_{8(r-1)+7})) \quad \text{for all } 1 \leq r \leq 56$$
$$C = S^{56}$$

which yields the cipher text $C = (c_0, \ldots, c_7)$. For decrypting this cipher text the inverse round transformation is used and therefore the following operations have to be carried out:

$$S^{56} = C$$
$$S^r = \phi^{-1}(S^{r+1}, ([k^E]_{8r}, \ldots, [k^E]_{8r+7})) \quad \text{for all } 55 \geq r \geq 0$$
$$P = S^0$$

where ϕ is the round function described below.

The Round Function. The round transformation uses two non-linear permutations on the set of all byte values π and π'. These permutations are related by another permutation σ, i.e. $\pi' = \sigma \circ \pi$. The bit permutation σ maps bit 0 to 1, bit 1 to 7, bit 2 to 5, bit 3 to 4, bit 4 to 2, bit 5 to 6, bit 6 to 0 and bit 7 to 3. See table 3 for the actual values described by π.

Let $S = (s_0, \ldots, s_7)$ be the vector of bytes representing the internal state of the block cipher in an arbitrary round and k the next 8-bit round key. The function ϕ taking the internal state S from round i to round $i+1$ is given by

$$\phi(s_0, \ldots, s_7, k) = (s_1, s_2 \oplus s_0, s_3 \oplus s_0, s_4 \oplus s_0,$$
$$s_5, s_6 \oplus \pi'(k \oplus s_7), s_7, s_0 \oplus \pi(k \oplus s_7))$$

The inverse round transformation for the decryption of a message block is then

$$\phi^{-1}(s_0, \ldots, s_7, k) = (s_7 \oplus \pi(s_6 \oplus k), s_0,$$
$$s_7 \oplus s_1 \oplus \pi(s_6 \oplus k), s_7 \oplus s_2 \oplus \pi(s_6 \oplus k),$$
$$s_7 \oplus s_3 \oplus \pi(s_6 \oplus k), s_4, s_5 \oplus \pi'(s_6 \oplus k), s_6)$$

6 Fault Attack on the Block Cipher

Our attack is a fault attack on the decryption of the last block from the block cipher part of CSA, which yields the first eight round keys i.e. the bits $[k^E]_0 \ldots [k^E]_7$. These round key bits are equal to the common key.

Note, that since we are only interested in the decryption of the last block from the block cipher part, the stream cipher and the chaining mode used with the block cipher are irrelevant as pointed out before.

The attacker starts by introducing a random error in the last round of the decryption process in s_6^1 which changes this value to $\overline{s}_6^1$. Since these two values appear unchanged in the decrypted plain text, the attacker can calculate

$$s_6^1 = s_7^0 = p_7$$
$$\overline{s}_6^1 = \overline{s}_7^0 = \overline{p}_7$$
$$g([k^E]_{0\ldots 7}) := \pi(s_6^1 \oplus [k^E]_{0\ldots 7}) \oplus \pi(\overline{s}_6^1 \oplus [k^E]_{0\ldots 7}) = s_0^0 \oplus \overline{s}_0^0 = p_0 \oplus \overline{p}_0$$

from the faulted and the actual output.

Now we verify for every possible round key k' if $g(k') = g([k^E]_{0\ldots 7})$. Table 1 shows, how many possible round keys are expected to fulfill this equation. As we can see, we can expect that the number of possible round keys is approximately two for every introduced error. Therefore, if we repeat the attack for two or three different errors, the round key can be uniquely determined.

After recovery of the round keys for the rounds $0 \ldots i$, the attacker introduces an error at round $i+1$ of the decryption process and uses the known round keys to perform $i+1$ rounds of the encryption process with the plain text and the faulted plain text. Doing so, the attacker gets the values

$$s_6^{i+1} = s_7^i$$
$$\overline{s}_6^{i+1} = \overline{s}_7^i$$
$$g([k^E]_{8i} \ldots [k^E]_{8i+7}) := \pi(s_6^{i+1} \oplus [k^E]_{8i} \ldots [k^E]_{8i+7})$$
$$\oplus \pi(\overline{s}_6^{i+1} \oplus [k^E]_{8i} \ldots [k^E]_{8i+7}) = s_0^i \oplus \overline{s}_0^i$$

He can thus retrieve the key bits $[k^E]_{8i} \ldots [k^E]_{8i+7}$ as pointed out above and therefore iteratively recover the required 8 round keys.

The common key is then given by $K = [k^E]_0 \ldots [k^E]_{63}$. Since this key is shared among the stream and the block cipher parts of CSA this attack breaks the complete CSA - Cipher.

To perform this basic version of our attack, the attacker has to introduce approximately two errors per round key, that sums up to a total of 16 errors. Additionally to uniquely determine one round key the attacker has to evaluate $g(k')$ for all 256 different values of k' for every introduced error. Therefore the overall complexity is 16 error introductions and $8 \cdot 2 \cdot 256 = 4096$ evaluations of g.

Another possibility to recover the round keys is, that the attacker takes every possible key retrieved through the equation $g(k') = g([k^E]_{8i} \ldots [k^E]_{8i+7})$ into account. With this method, the attacker does not have to repeat the error introduction. From table 1 we conclude, that the attack is only little more expensive. The wrong round keys can then be discovered by testing all calculated common keys for the correct one.

In this version the number of required error introductions decreases to 8. However the attacker now has to evaluate the g-function approximately $\sum_{i=0}^{7} 2^i \cdot 256 = 65280$ times which leaves him with 256 possible keys.

Table 1. Probability for the Number of Round Keys for the Attack

Possible number of keys	0	1	2	3	4	5	6	7	8	> 8
Probability	0.61	0.00	0.31	0.00	0.07	0.00	0.01	0.00	0.00	0.00

6.1 Improvements

The presented attack allows a time-memory trade-off. It is possible to calculate a table which contains all the possible round keys for every combination of $s_6, \overline{s}_6$ and $g(k)$. This table uses approximately $2^8 \cdot 2^8 \cdot 2^8 \cdot 2 = 2^{25}$ bytes.

Using this improvement the attack requires only one table lookup per introduced error, resp. per possible round key in the above scenarios.

One additional possibility is, that the adversary does not calculate all 8 round keys. He can also retrieve only some of the first round keys and then perform an exhaustive search on the missing bits. Since the common key is 64 bits, the costs of an exhaustive search can be reduced to $2^{64-8 \cdot j}$, where j is the number of round keys calculated. Clearly, this variant requires $2 \cdot j$ introduced errors and $j \cdot 2 \cdot 256$ evaluations of the g-function in the basic setting.

6.2 Evenly Distributed Errors

In the case that the attacker is not able to introduce errors at a specific register at a certain round, but only an error evenly distributed over the whole decryption process, i.e. the error can affect either register at either round, the presented attack still works. This is due to the fact, that the attacker can determine if the error has affected the desired value by comparing the values s_7^i and $\overline{s}_7^i$. If these values are not equal and $s_1^i = \overline{s}_1^i$, $s_5^i = \overline{s}_5^i$ and $s_2^i \oplus \overline{s}_2^i = s_3^i \oplus \overline{s}_3^i = s_4^i \oplus \overline{s}_4^i = s_0^i \oplus \overline{s}_0^i$ the correct register and the correct round have been modified.

Assuming that the errors are evenly distributed, this should occur every $56 \cdot 8$ tries. Therefore the costs for the attack in terms of the number of introduced errors only increase by a constant factor.

One further improvement would be, that the attacker records all the faulted outputs, even if the wrong round and/or register have been altered. Before introducing an error targeting the next round key, the attacker then checks if one of the recorded values is a modification of the correct register in this round. With this method the number of required faults can be decreased.

7   Conclusion

In this paper we presented a fault attack on the DVB common scrambling algorithm. Although the overall design, especially the combination of the stream and the block cipher, makes simple attacks difficult [WW04], it is possible to easily break the cipher using a fault attack. This again proves, that it is important to include countermeasures against fault attacks like the verification of the result of the encryption respectively decryption process in an implementation of a cryptographic system. Additional countermeasures against the presented attack should include different keys for the stream and the block cipher part, a non-linear key schedule and modifications on the round function of the block cipher to make the recovery of the round keys more difficult.

References

[ABF+02]   Christian Aumueller, Peter Bier, Wieland Fischer, Peter Hofreiter, and Jean-Pierre Seifert. Fault attacks on rsa with crt: Concrete results and practical countermeasures. In B. Kaliski, editor, *Cryptographic Hardware and Embedded Systems – CHES 2002*, volume 2523 of *Lecture Notes in Computer Science*, pages 260–275. Springer-Verlag, 2002.

[BDJ97]   Dan Boneh, Richard A. DeMillo, and Richard J.Lipton. On the importance of checking cryptographic protocols for faults. In W. Furny, editor, *Advances in Cryptology – Eurocrypt 1997*, volume 1233 of *Lecture Notes in Computer Science*, pages 37–51. Springer-Verlag, 1997.

[Bew98]   Simon Bewick. Descrambling DVB data according to ETSI common scrambling specification. UK Patent Applications GB2322994A / GB2322995A, 1998.

[BS97]   Eli Biham and Adi Shamir. Differential fault analysis of secret key cryptosystems. In B. Kaliski, editor, *Advances in Cryptology – Crypto 1997*, volume 1294 of *Lecture Notes in Computer Science*, pages 513–525. Springer-Verlag, 1997.

[BS03]   Johannes Bloemer and Jean-Pierre Seifert. Fault based cryptanalysis of the advanced encryption standard (aes). In R. Wright, editor, *Financial Cryptography*, volume 2742 of *Lecture Notes in Computer Science*, pages 162–181. Springer-Verlag, 2003.

[Eur96]   European Telecommunications Standards Institute. ETSI Technical Report 289: Support for use of scrambling and Conditional Access (CA) within digital broadcasting systems, 1996.

[Pse03]   Pseudononymous authors. CSA – known facts and speculations, 2003. http://csa.irde.to.

[WAJ98] Davies Donald Watts, Rix Simon Paul Ashley, and Kuehn Gideon Jacobus. System and apparatus for blockwise encryption and decryption of data. US Patent Application US5799089 , 1998.

[WW04] Ralf-Phillip Weinmann and Kai Wirt. Analysis of the dvb common scrambling algorithm. In *Eighth IFIP TC-6 TC-11 Conference on Communications and Multimedia Security, CMS 2004. Proceedings.* Kluwer Academic Publishers, 2004.

A Appendix

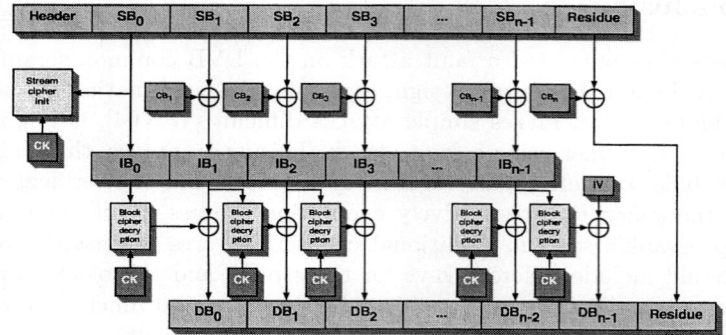

Fig. 1. Combination of Block- and Stream Cipher

Table 2. Key Bit Permutation

i	0	1	2	3	4	5	6	7	8	9	10	11	12	13	14	15
$\rho(i)$	17	35	8	6	41	48	28	20	27	53	61	49	18	32	58	63
i	16	17	18	19	20	21	22	23	24	25	26	27	28	29	30	31
$\rho(i)$	23	19	36	38	1	52	26	0	33	3	12	13	56	39	25	40
i	32	33	34	35	36	37	38	39	40	41	42	43	44	45	46	47
$\rho(i)$	50	34	51	11	21	47	29	57	44	30	7	24	22	46	60	16
i	48	49	50	51	52	53	54	55	56	57	58	59	60	61	62	63
$\rho(i)$	59	4	55	42	10	5	9	43	31	62	45	14	2	37	15	54

Table 3. S-Box of the Block Cipher. Output Arranged Row-wise; Lower Nibble on Horizontal, Upper on Vertical

	0x00	0x01	0x02	0x03	0x04	0x05	0x06	0x07	0x08	0x09	0x0A	0x0B	0x0C	0x0D	0x0E	0x0F
0x00	0x3A	0xEA	0x68	0xFE	0x33	0xE9	0x88	0x1A	0x83	0xCF	0xE1	0x7F	0xBA	0xE2	0x38	0x12
0x01	0xE8	0x27	0x61	0x95	0x0C	0x36	0xE5	0x70	0xA2	0x06	0x82	0x7C	0x17	0xA3	0x26	0x49
0x02	0xBE	0x7A	0x6D	0x47	0xC1	0x51	0x8F	0xF3	0xCC	0x5B	0x67	0xBD	0xCD	0x18	0x08	0xC9
0x03	0xFF	0x69	0xEF	0x03	0x4E	0x48	0x4A	0x84	0x3F	0xB4	0x10	0x04	0xDC	0xF5	0x5C	0xC6
0x04	0x16	0xAB	0xAC	0x4C	0xF1	0x6A	0x2F	0x3C	0x3B	0xD4	0xD5	0x94	0xD0	0xC4	0x63	0x62
0x05	0x71	0xA1	0xF9	0x4F	0x2E	0xAA	0xC5	0x56	0xE3	0x39	0x93	0xCE	0x65	0x64	0xE4	0x58
0x06	0x6C	0x19	0x42	0x79	0xDD	0xEE	0x96	0xF6	0x8A	0xEC	0x1E	0x85	0x53	0x45	0xDE	0xBB
0x07	0x7E	0x0A	0x9A	0x13	0x2A	0x9D	0xC2	0x5E	0x5A	0x1F	0x32	0x35	0x9C	0xA8	0x73	0x30
0x08	0x29	0x3D	0xE7	0x92	0x87	0x1B	0x2B	0x4B	0xA5	0x57	0x97	0x40	0x15	0xE6	0xBC	0x0E
0x09	0xEB	0xC3	0x34	0x2D	0xB8	0x44	0x25	0xA4	0x1C	0xC7	0x23	0xED	0x90	0x6E	0x50	0x00
0x0A	0x99	0x9E	0x4D	0xD9	0xDA	0x8D	0x6F	0x5F	0x3E	0xD7	0x21	0x74	0x86	0xDF	0x6B	0x05
0x0B	0x8E	0x5D	0x37	0x11	0xD2	0x28	0x75	0xD6	0xA7	0x77	0x24	0xBF	0xF0	0xB0	0x02	0xB7
0x0C	0xF8	0xFC	0x81	0x09	0xB1	0x01	0x76	0x91	0x7D	0x0F	0xC8	0xA0	0xF2	0xCB	0x78	0x60
0x0D	0xD1	0xF7	0xE0	0xB5	0x98	0x22	0xB3	0x20	0x1D	0xA6	0xDB	0x7B	0x59	0x9F	0xAE	0x31
0x0E	0xFB	0xD3	0xB6	0xCA	0x43	0x72	0x07	0xF4	0xD8	0x41	0x14	0x55	0x0D	0x54	0x8B	0xB9
0x0F	0xAD	0x46	0x0B	0xAF	0x80	0x52	0x2C	0xFA	0x8C	0x89	0x66	0xFD	0xB2	0xA9	0x9B	0xC0

HSEP Design Using F2m HECC and ThreeB Symmetric Key Under e-Commerce Environment

Byung-kwan Lee[1], Am-Sok Oh[2], and Eun-Hee Jeong[3]

[1] Dept. of Computer Engineering, Kwandong Univ., Korea
bklee@kd.ac.kr
[2] Dept. of Multimedia Engineering, Tongmyong Univ., Korea
asoh@tit.ac.kr
[3] Dept. of Economics, Samcheok National Univ., Korea
jeh@samcheok.ac.kr

Abstract. SSL(Secure Socket Layer) is currently the most widely deployed security protocol, and consists of many security algorithms, but it has some problems on processing time and security. This paper proposes an HSEP(Highly Secure Electronic Payment) Protocol that provides better security and processing time than an existing SSL protocol. As HSEP consists of just F2mHECC, ThreeB(Block Byte Bit Cipher), SHA algorithm, and Multiple Signature, this protocol reduces handshaking process by concatenating two proposed F2mHECC public key and ThreeB symmetric key algorithm and improves processing time and security. In particular, Multiple signature and ThreeB algorithm provides better confidentiality than those used by SSL through three process of random block exchange, byte-exchange key and bit-xor key.

1 Introduction

This paper proposes an HSEP(Highly Secure Electronic Payment) protocol whose characteristic are the followings.

First, The HSEP uses HECC instead of RSA to improve the strength of encryption and the speed of processing. The resulting value which is computed with the public key and the private key of HECC becomes a shared secret key, that is, the values is become a master key. Second, The shared secret key is used as input the proposed ThreeB(Block Byte Bit Cipher) algorithm which generates session key for the data encryption. Finally, HSEP protocol uses multiple signatures instead of MAC(message authentication code) to improve the reliability of EC.

Therefore, HSEP protocol reduces handshaking process by concatenating a shared private key of HECC. Also, Multiple signature and ThreeB algorithm provides better confidentiality than those by SSL through three process of random block exchange, byte-exchange key and bit-xor key.

This paper is structured as follows. Section 2 provides some basic concepts of encryption and decryption, HECC, and SSL. Section 3 describes the structure of HSEP protocol and ThreeB algorithm. An performance comparison of HSEP protocol with SSL protocol are presented in Section 4. Finally, our conclusions are summarized in Section 5.

2 Basic Concepts

2.1 Encryption and Decryption Algorithm

As shown in Fig. 1, the user A computes a new key $k_A(k_BP)$ by multiplying the user B's public key by the user A's private key k_A. The user A encodes the message by using this key and then transmits this cipher text to user B. After receiving this cipher text, The user B decodes with the key $k_B(k_AP)$, which is obtained by multiplying the user A's public key, k_AP by the user B's private key, k_B. Therefore, as $k_A(k_BP) = k_B(k_AP)$, we may use these keys for the encryption and the decryption.

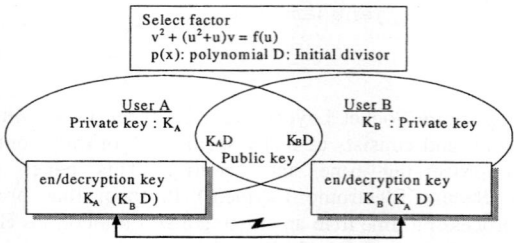

Fig. 1. Concept of en/decryption of HECC

2.2 SSL (Secure Socket Layer) Protocol

SSL is a commonly-used protocol for managing the security of a message transmission on the internet. SSL uses a program layer located between the internet's Hypertext Transfer Protocol and Transport Control Protocol layers.

The SSL protocol includes two sub-protocols : the SSL record protocol and the SSL handshake protocol. The SSL record protocol defines the format used to transmit data. The SSL handshake protocol involves using the SSL record protocol to exchange a series of messages between an SSL-enabled server and an SSL-enabled client when they first establish an SSL connection. This exchanges of messages is designed to facilitate the following actions:

- Authenticate the server to the client.
- Allow the client and server to select the cryptographic algorithms, or ciphers, that they both support.
- Optionally authenticate the client to the server.
- Use public-key encryption techniques to generate shared secrets.
- Establish an encrypted SSL connection.

3 Proposed HSEP (Highly Secure Electronic Payment) Protocol

3.1 HSEP Protocol

The existing SSL uses RSA in key exchange and DES in message encryption. Our proposed HSEP protocol uses HECC instead of RSA, Because of this, the strength of encryption and the speed of processing are improved. Besides, in message encryption, HSEP utilizes ThreeB algorithm to generate session keys and cipher text. The encryption and decryption processes are shown in Fig. 2 respectively.

First, Select a private key x and an initial point D of HECC, and then computes xD. Using the result of addition, generates a session key of ThreeB.

Second, the ThreeB algorithm encodes the message applying these keys. Since the receiver has his own private key, HSEP can reduce handshake procedure without pre-master key exchange, which enhances the speed for processing a message, and strengthens the security for information. Therefore, HSEP simplifies a handshake and decreases a communicative traffic over a network as compared with the existing SSL.

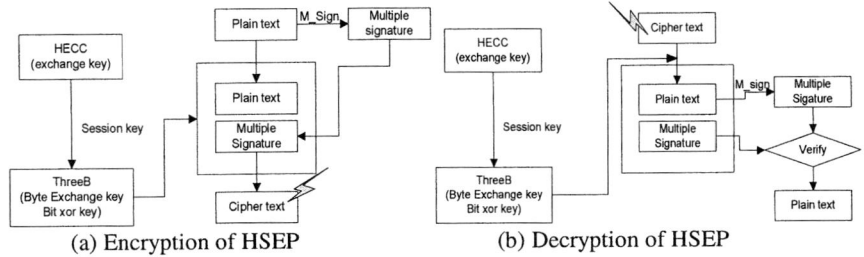

(a) Encryption of HSEP (b) Decryption of HSEP

Fig. 2. The flow of HSEP

3.1.1 The Basic Algorithm for HSEP Protocol

HSEP protocol proposed in this paper uses the same hash function, SHA as SSL protocol, HECC and ThreeB algorithms are no used in SSL. So this section shows that the process of the keys, sk1 and sk2 are generated by using the shared private key of HECC, and data is encrypted by using sk1 and sk2.

1. Key generation

The process of generation of sk1 and sk2 is shown in Fig. 3.

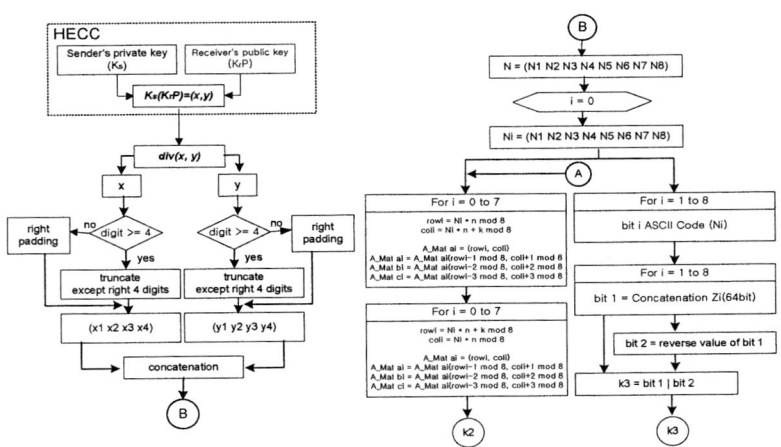

Fig. 3. Key generation

2. Data Encryption

Fig. 4 explains the process of data encryption and is treated in section 3.3.2 in detail.

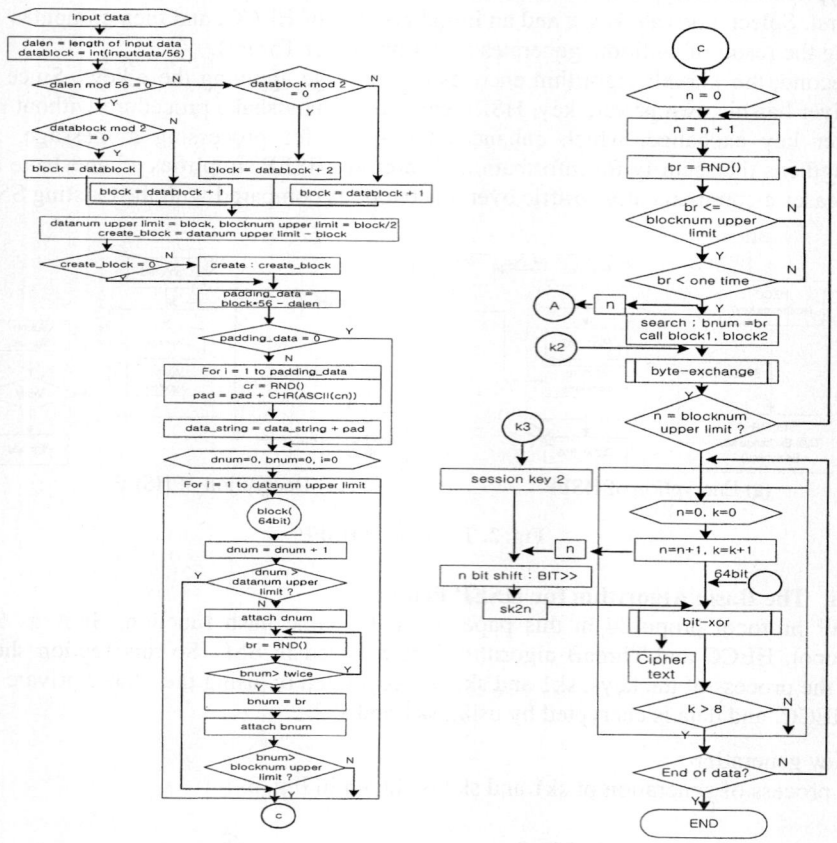

Fig. 4. Data Encryption

3.2 F2m HECC (Hyper Elliptic Curve Cryptosystem)

Nowadays, in the area of cryptology, using hyperelliptic curves is eagerly studied, because it gives the same security level with a smaller key length as compared to cryptosystems using elliptic curves. From the fact it is expected to be possible to use hyperelliptic curves to factor integers, since elliptic curve method exploits the property of the Abelian groups in the same way as the cryptosystems.

A hyperelliptic curve H of genus $g(g \geq 1)$ over a field F is a nonsingular curve that is given by an equation of the following form:

$$H : v^2 + h(u)v = f(u) \quad (\text{in } F[u, v])$$

where $h(u) \in F[u]$ is a polynomial of degree $\leq g$, and $f(u) \in F[u]$ is a monic polynomial of degree $2g+1$.

3.2.1 Divisors

Divisors of a hyperelliptic curve are pairs denoted div(a(u), b(u)), where a(u) and b(u) are polynomials in $GF(2^n)$ [u] that satisfy the congruence $b(u)^2 + h(u)b(u) \equiv f(u) \pmod{a(u)}$.

They can also be defined as the formal sum of a finite number of points on the hyperelliptic curve. Since these polynomials could have arbitrarily large degree and still satisfy the equation, the notion of a reduced divisor is needed. In a reduced divisor, the degree of a(u) is no greater than g, and the degree of b(u) is less than the degree of a(u).

3.2.2 Reduced Divisors

Let H be a hyperelliptic curve of genus g over a field F. A reduced divisor(defined over F) of H is defined as a form div(a, b), where a, b∈ F[u] are polynomial such that
 (1) a is monic, and deg b < deg a $\leq$ g,
 (2) a divides $(b^2 - bh - f)$.
In particular div(1,0) is called zero divisor.

Input : A semi-reduced divisor, D=div(a, b)
Output : The equivalent reduced divisor, $D' = div(a', b') \sim D$
 1. Set $a' = (f - bh - b^2)/a$ and $b' = (-h-b) \pmod{a'}$
 2. If $\deg_u a' > g$ then set $a = a'$, $b = b'$ and go to step 1.
 3. Let c be the leading coefficient of a'. Set $a' = c^{-1}a'$.
 4. Output $D' = div(a', b')$

Fig. 5. Reduction of a divisor to a reduced divisor

3.2.3 Adding Divisors

If $D_1 = div(a_1, b_1)$ and $D_2 = div(a_2, b_2)$ are two reduced divisors defined over F, then Fig. 6 finds a semi-reduced divisor or reduced divisor $D_3 = div(a, b)$. To find the unique divisor, $D_3 = div(a, b)$, Fig. 5 should be used just after the addition of two divisors.

Input : Two reduced divisors, $D_1 = div(a_1, b_1)$ and $D_2 = div(a_2, b_2)$
Output : A reduced divisor or semi-reduction divisor, $D_3 = div(a, b)$
 1. Compute d_1, e_1 and e_2 which satisfy $d_1 = GCD(a_1, a_2)$ and $d_1 = e_1 a_1 + e_2 a_2$
 2. If $d_1 = 1$, then $a := a_1 a_2$, $b := (e_1 a_1 b_2 + e_2 a_2 b_1) \bmod a$
    otherwise do the following:
    (1) Compute d, c_1 and s_3 which satisfy
        $d = GCD(d_1, b_1 + b_2 + h)$ and $d = c_1 d_1 + s_3 (b_1 + b_2 + h)$.
    (2) Let $s_1 := c_1 e_1$ and $s_2 := c_1 e_2$, so that $d = s_1 a_1 + s_2 a_2 + s_3 (b_1 + b_2 + h)$.
    (3) Let $a := a_1 a_2 / d^2$, $b := (s_1 a_1 b_2 + s_2 a_2 b_1 + s_3 (b_1 b_2 + f))/d \bmod a$
 3. output $D_3 = div(a, b)$

Fig. 6. Addition defined over the group of divisors

3.3 ThreeB (Block Byte Bit Cipher) Algorithm

In this paper, the proposed ThreeB algorithm consists of two parts, which are session key generation and data encryption. And the data encryption is divided into three phases, which are inputting plaintext into data blocks, byte-exchange between blocks, and bit-wise XOR operation between data and session key.

3.3.1 Session Key Generation

As we know that the value which is obtained by multiplying one's private key by the other's public key is the same as what is computed by multiplying one's public key to the other's private key. The feature of EC is known to be almost impossible to estimate a private and a public key. With this advantage and the homogeneity of the result of operations, the proposed ThreeB algorithm uses a 64-bit session key to perform the encryption and decryption. Given the sender's private key $X = X_1 X_2 ... X_m$ and the receiver's public key, $Y = Y_1 Y_2 ... Y_n$, we concatenate X and Y to form a key N (i.e., $N = X_1 X_2 ... X_m Y_1 Y_2 ... Y_n$), and then compute the session keys as follows:

i) If the length (number of digits) of X or Y exceeds four, then the extra digits on the left are truncated. And if the length of X or Y is less than four, then they are padded with 0's on the right. This creates a number $N' = X_1' X_2' X_3' X_4' Y_1' Y_2' Y_3' Y_4'$. Then a new number N" is generated by taking the modulus of each digit in N' with 8.
ii) The first session key sk1 is computed by taking bit-wise OR operation on N" with the reverse string of N".
iii) The second session key sk2 is generated by taking a circular right shift of sk1 by one bit. And repeat this operation to generate all the subsequent session keys needed until the encryption is completed.

3.3.2 Encryption

The procedure of data encryption is divided into three parts, inputting plaintext into data block, byte-exchange between blocks, and bit-wise XOR operation between data and session key.

1. Input plaintext into data block

The block size is defined as 64 bytes. A block consists of 56 bytes for input data, 4 bytes for the data block number, and 4 bytes for the byte-exchange block number (1 or 2, see Fig. 7). During the encryption, input data stream are blocked by 56 bytes. If the entire input data is less than 56 bytes, the remaining data area in the block is padded with each byte by a random character. Also, in the case where the total number of data blocks filled is odd, then additional block(s) will be added to make it even, and each of those will be filled with each byte by a random character as well. Also, a data block number in sequence) is assigned and followed by a byte-exchange block number, which is either 1 or 2.

Data	Area
Data Block Number	Byte-exchange block no

Fig. 7. Structure of block

2. Byte-exchange between blocks
After inputting the data into the blocks, we begin the encryption by starting with the first data block and select a block, which has the same byte-exchange block number for the byte exchange. In order to determine which byte in a block should be exchanged, we compute its row-column position as follows:
  For the two blocks whose block exchange number, n = 1, we compute the following:
    byte-exchange row = $(N_i *n)$ mod 8 (i = 1,2 …,8)
    byte-exchange col = $N_i *n) + 3)$ mod 8 (i = 1,2 …,8),
where N_i is a digit in N″ These generate 8 byte-exchange positions. Then for n = 1, we only select the non-repeating byte position (row, col) for the byte-exchange between two blocks whose block exchange numbers are equal to 1. Similarly, we repeat the procedure for n = 2.

3. Bit-wise XOR between data and session key
After the byte-exchange is done, the encryption proceeds with a bit-wise XOR operation on the first 13 byte data with the session sk1 and repeats the operation on every 8 bytes of the remaining data with the subsequent session keys until the data block is finished. Note that the process of byte-exchange hides the meaning of 56 byte data, and the exchange of the data block number hides the order of data block, which needs to be assembled later on. In addition, the bit-wise XOR operation transforms a character into a meaningless one, which adds another level of confusion to the attackers.

3.3.3 Decryption
Decryption procedure is given as follows. First, a receiver generates a byte exchange block key sk1 and a bit-wise XOR key sk2 by using the sender's public key and the receiver's private key. Second, the receiver decrypts it in the reverse of encryption process with a block in the input data receiving sequence. The receiver does bit-wise XOR operation bit by bit, and then, a receiver decodes cipher text by using a byte-exchange block key sk1 and moves the exchanged bytes back to their original positions. We reconstruct data blocks in sequence by using the decoded data block number.

3.4 Multiple Signature

In the proposed HSEP protocol, the multiple signature is used instead of MAC.

(1) User A generates message digests of OI(order information) and PI(payment information) separately by using hash algorithm, concatenates these two message digests; produces $MD_B MD_C$; and hash it to generates MD(message digest). Then the user A encrypts this MD by using an encryption key, which is obtained by multiplying the private key of user A to the public key of the receiver. The PI to be transmitted to user C is encrypted by using ThreeB algorithm. The encrypted PI is named CPI.

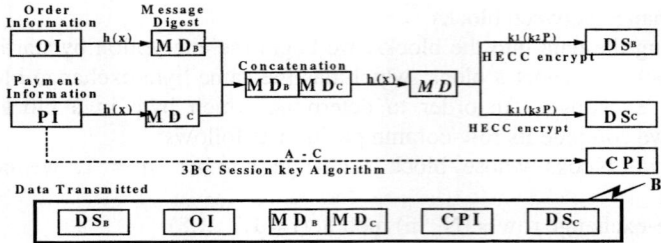

Fig. 8. Encryption of user A

(2) User B generates message digest MD_B' with the transmitted OI from user A. After having substituted MD_B' for the MD_B of $MD_B MD_C$, the message digest MD is generated by using hash algorithm. User B decrypts a transmitted DS_B, and extracts MD from it. User B compares this with MD generated by user B, certificates user A and confirms the integrity of message. Finally, user B transmits the rest of data, $MD_B MD_C$, CPI, DS_C to user C.

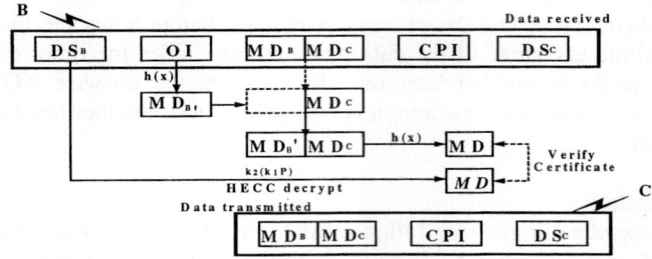

Fig. 9. Decryption of user B and data transmitted to C

(3) User C decrypts the CPI transmitted from user B, extracts PI, and generates message digest (MD_C) from this by using hash algorithm; substitutes this for MD_B of $MD_B MD_C$ transmitted from user B, and produces message digest (MD) by using Hash algorithm. Then the user C decrypts the DS_C transmitted from user B and extracts message digest (MD). Again, the user C compares this with the MD extracted by user C, verifies the certificate from the user A, and confirms the integrity of the message. Finally, the user C returns an authentication to the user B.

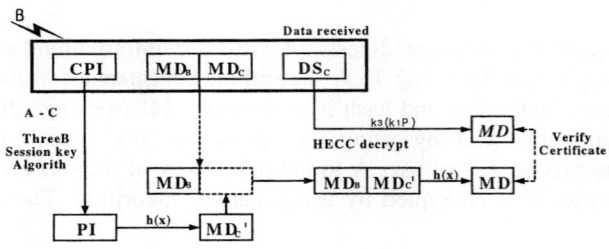

Fig. 10. Decryption of an user C

4 Performance Evaluation

4.1 HECC and RSA

In this paper, the proposed HSEP protocol uses HECC instead of RSA. In comparison with RSA, the results of the encryption and decryption times are shown in Fig. 11 respectively, which indicate that encryption and decryption time of HECC are much less than those of RSA.

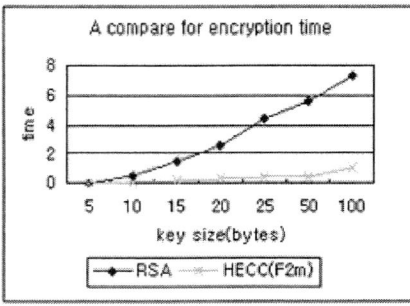
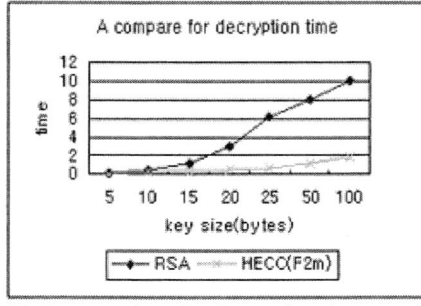

(a) A comparison for encryption time   (b) A comparison for decryption time

Fig. 11. The comparison of HECC and RSA(unit : Φs)

4.2 ThreeB and DES

Fig. 12 show the mean value of encryption time of ThreeB and DES by executing every number of block about message twenty times. According to Fig. 12, we can conclude that ThreeB is faster than the existing DES in encryption time. In addition, the security of ThreeB is enhanced by using Byte-exchange and Bit-wise XOR Therefore, the strength of the encryption is improved and more time is saved for encryption and decryption than DES.

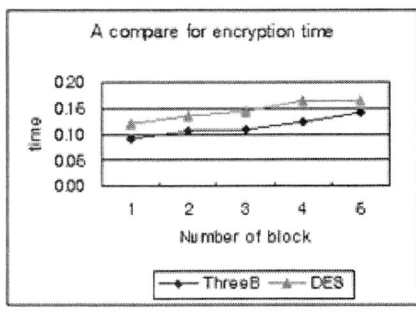
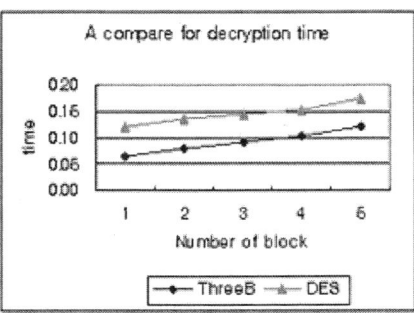

(a)   A comparison for encryption time   (b) A comparison for decryption time

Fig. 12. A comparison of ThreeB and DES(unit : Φs)

5 Conclusion

The proposed HSEP protocol employs HECC, Multiple Signature and ThreeB algorithm other than the existing SSL. HSEP protocol removes a pre-master key exchange, and replaces the master key exchange by the shared secret key. As a result, it speeds up the handshaking process by reducing communication traffic for transmission. The proposed ThreeB, which uses byte-exchange and the bit operation increases data encryption speed. Even though cipher text is intercepted during transmission over the network. Because during the encryption process, the ThreeB algorithm performs byte exchange between blocks, and then the plaintext is encoded through bit-wise XOR operation, it rarely has a possibility for cipher text to be decoded and has no problem to preserve a private key.

Moreover, the proposed HSEP protocol has a simple structure, which can improve the performance with the length of session key, byte-exchange algorithm, bit operation algorithm, and so on. From the standpoint of the supply for key, the CA (Certificate authority) has only to certify any elliptic curve and any prime number for modulo operation, the anonymity and security for information can be guaranteed over communication network.(See Table 1.)

Table 1. Decryption of HSEP protocol

protocol	Digital signature	Encryption for message	Digital envelope
SET	RSA	DES	Use
ECSET	ECC	DES	Use
HSEP	HECC	ThreeB	Unnecessary

References

1. CHO, I.S., and B.K. Lee, ASEP (Advanced Secure Electronic Payment) Protocol Design, in Proceedings of International Conference of Information System, pp. 366-372, Aug. (2002).
2. CHO, I.S., D.W. Shin, T.C. Lee, and B.K. Lee, SSEP (Simple Secure Electronic Payment) Protocol Design, in Journal of Electronic and Computer Science, pp. 81-88, Vol. 4, No. 1, Fall (2002).
3. ECOMMERCENET, <http://www.ezyhealthmie.com/Service/Editorial/set.htm>.
4. HARPER, G., A. Menezes, and S. Vanstone, Public-key Cryptosystem with Very Small Key Lengths, in Advances in Cryptology-Proceedings of Eurocrypt '92, Lecture Notes in Computer Science 658, pp. 163-173, Springer-Verlag, (1993).
5. IEEE P1363 Working Draft, Appendices, pp. 8, February 6, 1997.
6. KOBLITZ, N., Elliptic Curve Cryptosystems, in Math. Comp. 48 203-209 (1987).
7. MILLER, V.S., Use of Elliptic Curve in Cryptography, in Advances in Cryptology-Proceedings of Crypto '85, Lecture Notes in Computer Science 218, pp. 417-426, Springer-Verlag, (1986).

Perturbed Hidden Matrix Cryptosystems

Zhiping Wu[1,*], Jintai Ding[2], Jason E. Gower[2], and Dingfeng Ye[1,*]

[1] State Key Laboratory of Information Security,
Graduate School of Chinese Academy of Sciences,
100039-08, Beijing, China
zpwu@mails.gscas.ac.cn, ydf@is.ac.cn
[2] Department of Mathematical Sciences,
University of Cincinnati,
Cincinnati, OH 45221-0025 USA
{ding, gowerj}@math.uc.edu

Abstract. We apply internal perturbation [3] to the matrix-type cryptosystems $[C_n]$ and HM constructed in [9]. Using small instances of these variants, we investigate the existence of linearization equations and degree 2 equations that could be used in a XL attack. Our results indicate that these new variants may be suitable for use in practical implementations. We propose a specific instance for practical implementation, and estimate its performance and security.

Keywords: public key, multivariate, perturbation, hidden matrix, XL attack.

1 Introduction

Public key cryptography plays an important role in many modern communication systems. In the last few years, great effort has been made to develop cryptosystems based on systems of multivariate polynomials over a finite field. The results of these efforts include C^*, HFE, $[C]$, $[C_n]$ and HM [7, 8, 6, 9]. Recently, the idea of "perturbation" was proposed to improve the security of C^* and HFE [3, 4] without much loss of efficiency. In this paper we study the effect of perturbation on the matrix-type schemes $[C_n]$ and HM.

To construct $[C_n]$ or HM, we begin by choosing secret invertible affine transformations $s : K^{n^2} \longrightarrow \mathcal{M}_n(K)$ and $t : \mathcal{M}_n(K) \longrightarrow K^{n^2}$, where K is a finite field and $\mathcal{M}_n(K)$ is the set of $n \times n$ matrices with entries in K. If we have an "invertible" quadratic map $g : \mathcal{M}_n(K) \longrightarrow \mathcal{M}_n(K)$, we can build a cipher for encryption as follows: $x \xmapsto{s} A \xmapsto{g} g(A) \xmapsto{t} y$, where $x, y \in K^{n^2}$, $A \in \mathcal{M}_n(K)$, and K^{n^2} is the plaintext/ciphertext space. If the inverse of the mapping g can be computed in polynomial time then the decryption can be performed efficiently. However, $[C_n]$ is vulnerable to the linearization attack, and HM may be vulner-

---

* Supported by Natural Science Foundation of China No. 60473026.

able to XL-type attacks [1] due to that fact that such systems may produce a large number of new quadratic equations.

To create perturbation we choose a set of linear polynomials $z_j = \sum \alpha_{ij} x_i + \beta_j$ (for $j = 1, \ldots, r$) in the variables x_i of the original system such that the $z_j - \beta_j$ are linearly independent. A set of randomly chosen secret quadratic polynomials in the z_j are added to $[C_n]$ to produce the first PHM system. Similarly, randomly chosen secret linear and quadratic polynomials in the z_j are applied to HM to produce the second PHM system. For several small instances of both variants, we made a direct search for potentially fatal linearization and quadratic equations. Our results indicate that for proper choices of parameters these variants are very likely to be resistant to linearization and XL-type attacks.

In the first section we introduce $[C_n]$ and HM, along with the known attacks on these systems. We then describe in Section 3 a method for constructing two new variants using perturbation. In Section 4 we analyze the security of these new variants against the known attacks, and then in Section 5 we use this analysis to suggest some choices of parameters for use in practical implementations. We summarize our work in Section 6.

2 Hidden Matrix Cryptosystems

The first multivariate cryptosystem based on matrices, $[C]$, was proposed by Imai and Matsumoto [6]. This system and its generalization, $[C_n]$, were defeated by Patarin, Goubin and Courtois using linearization equations [9]. In this same paper, they suggested an improved scheme that they named the Hidden Matrix (HM) cryptosystem. Though HM is resistant to the linearization attack, it is sometimes possible to generate several new quadratic equations that can be used in a XL attack.

2.1 Description of $[C_n]$ and HM

Let K be a finite field of cardinality $q = 2^m$ and let $\mathcal{M}_n(K)$ denote the set of $n \times n$ matrices with entries in K. Recall that $\mathcal{M}_n(K)$ can be considered as a vector space of dimension n^2 over K. Plaintext and ciphertext are elements in K^{n^2}.

Public/Private Keys: The private key consists of the invertible affine transformations $s : K^{n^2} \longrightarrow \mathcal{M}_n(K)$ and $t : \mathcal{M}_n(K) \longrightarrow K^{n^2}$. The public key includes the field structure of K, and $f : K^{n^2} \longrightarrow K^{n^2}$, where $f(x) = (t \circ g \circ s)(x) = (f_1, \ldots, f_{n^2})$ and $g : \mathcal{M}_n(K) \longrightarrow \mathcal{M}_n(K)$ is quadratic (hence the f_i are quadratic as well). If $g(x) = x^2$, then we have $[C_n]$; if $g(x) = x^2 + Mx$, for some nonzero secret matrix M, then we have HM.

Encryption/Decryption: For any plaintext $(x'_1, \ldots, x'_{n^2})$, the corresponding ciphertext $(y'_1, \ldots, y'_{n^2})$ can be computed by $y'_i = f_i(x'_1, \ldots, x'_{n^2})$. To decrypt a given ciphertext $y' \in K^{n^2}$, we solve the equation $g(A) = B$, where $B = t^{-1}(y')$, and then compute the plaintext as $x' = s^{-1}(A)$. For more about the decryption of $C[n]$ and HM, see [9].

2.2 Attacks on $[C_n]$ and HM

Patarin, Goubin and Courtois used the linearization attack to defeat $[C_n]$, and then suggested HM as a possible improvement. However, the improved scheme also has a potential defect that frequently allows attackers to produce many new quadratic equations satisfied by the plaintext/ciphertext pairs.

Linearization Attack: In $[C_n]$, if $A = s(x_1, \ldots, x_{n^2})$, and $B = g(A) = A^2$, then we have $AB = BA$. This equation can be used to generate linearization equations

$$\sum a_{ij} x_i y_j + \sum b_i x_i + \sum c_j y_j + d = 0 \tag{1}$$

satisfied by any plaintext/ciphertext pair. If enough of these equations can be found then we can find the plaintext for a given ciphertext.

Degree 2 Equation Attack: The authors in [9] noticed that for HM we have the equation $AB - BA = AMA - MA^2$, where $g(A) = A^2 + MA$. This yields n^2 quadratic equations of the form:

$$\sum \alpha_{ij} x_i x_j + \sum \beta_{ij} x_i y_j + \sum \gamma_i x_i + \sum \delta_i y_i + \mu = 0 , \tag{2}$$

and so for a given ciphertext we can generate n^2 quadratic equations satisfied by the plaintext. These "new" quadratic equations can be combined with the public key equations and used in a XL-type attack. We refer to this generation of new quadratic equations as a *degree 2 equation attack*.

3 Perturbed Hidden Matrix Cryptosystems

$[C_n]$ and HM may not be suitable for practical use due to the attacks outlined in the previous section. In this section we show how to apply the idea of perturbation to these two schemes as a way to create resistance to these attacks.

3.1 Perturbation of $[C_n]$

Let r be a small positive integer and $z_j = \sum \alpha_{ij} x_i + \beta_j$ (for $j = 1, \ldots, r$) be randomly chosen degree 1 polynomials in the x_i over K such that the $z_j - \beta_j$ are linearly independent. Let $Z : K^{n^2} \longrightarrow K^r$ be the map defined by $Z(x_1, \ldots, x_{n^2}) = (z_1(x_1, \ldots, x_{n^2}), \ldots, z_r(x_1, \ldots, x_{n^2}))$. Randomly choose n^2 quadratic polynomials $f_1, \ldots, f_{n^2}$ in the variables $z_1, \ldots, z_r$, and define the map $f : K^r \longrightarrow K^{n^2}$ by $f(z_1, \ldots, z_r) = (f_1(z_1, \ldots, z_r), \ldots, f_{n^2}(z_1, \ldots, z_r))$. Let $u : K^{n^2} \longrightarrow \mathcal{M}_n(K)$ be another secret invertible affine transformation and compute $B' = u \circ f$. Let $P = \{(\lambda, \mu) : \lambda \in (u \circ f)(K^r), \mu = (u \circ f)^{-1}(\lambda)\}$. The set P is called the *perturbation set*. We construct the first perturbed hidden matrix (PHM) scheme as illustrated in Figure 1, where we define $\bar{B} = B + B'$. We say that that $\bar{B}$ is the *perturbation* of B by B', and that the number r is the *perturbation dimension*.

Public/Private Keys: The private key includes the three affine transformations s, u, and t; the set of degree 1 polynomials $z_1, \ldots, z_r$; and the set P, or

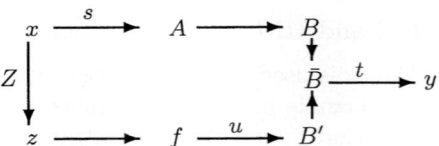

Fig. 1. Construction of the first PHM

equivalently, the set of the polynomials $f_i(z_1,\ldots,z_r)$. The public key includes the field structure of K and the n^2 quadratic polynomials $y_1,\ldots,y_{n^2}$.

Encryption/Decryption: Given a plaintext message $x' = (x'_1,\ldots,x'_{n^2})$, the ciphertext is $y' = (y'_1,\ldots,y'_{n^2})$, where $y'_i = y_i(x'_1,\ldots,x'_{n^2})$. To decrypt a given ciphertext $(y'_1,\ldots,y'_{n^2})$, we first compute $\bar{B} = t^{-1}(y'_1,\ldots,y'_{n^2})$. For each $(\lambda,\mu) \in P$ we compute $(x'_{\lambda 1},\ldots,x'_{\lambda n^2}) = (g \circ s)^{-1}(\bar{B} - \lambda)$, and then check if $Z(x'_{\lambda 1},\ldots,x'_{\lambda n^2})$ is the same as the corresponding μ. If it is not then we discard it; otherwise $(x'_{\lambda 1},\ldots,x'_{\lambda n^2})$ may be the plaintext. It is possible that there may be more than one candidate for the plaintext. However, we can use the same technique suggested in [8] to find the true plaintext.

3.2 Perturbation of HM

First let $h_1,\ldots,h_{n^2}$ be randomly chosen degree 1 polynomials in the variables $z_1,\ldots,z_r$, where the z_i are as above, which defines a map $h: K^r \longrightarrow K^{n^2}$. Let $v: K^{n^2} \longrightarrow \mathcal{M}_n(K)$ be an invertible affine transformation and define $A' = v \circ h$. Let $f_1,\ldots,f_{n^2}$ be randomly chosen quadratic polynomials in the variables $z_1,\ldots,z_r$, which defines a map $f: K^r \longrightarrow K^{n^2}$. We choose $u: K^{n^2} \longrightarrow \mathcal{M}_n(K)$ to be another secret invertible affine transformation and define $A'' = u \circ f$. Let $P = \{(\lambda_h,\lambda_f,\mu) : \lambda_h \in (v \circ h)(K^r),\ \lambda_f \in (u \circ f)(K^r),\ \mu = (v \circ h)^{-1}(\lambda_h) \cap (u \circ f)^{-1}(\lambda_f)\}$. We construct the second PHM scheme as defined in Figure 2, where

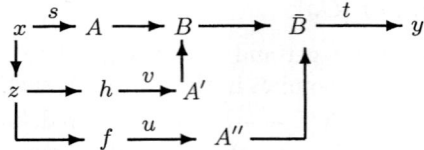

Fig. 2. Construction of the second PHM

$\bar{B} = g(A) + A'' = A^2 + A'A + A''$ is the perturbation by A' and A'', and $g(x) = x^2 + A'x$. Here we must include u,v,h and f in the private key, and modify the decryption process as follows. For each $(\lambda_h,\lambda_f,\mu) \in P$, compute $(x_{\lambda 1},\ldots,x_{\lambda n^2}) = (g \circ s)^{-1}(\bar{B} - \lambda_f)$; in other words, solve the equation $\bar{B} - \lambda_f = g(A) = A^2 + \lambda_h A$ for A, where λ_h and λ_f are known, and then find $s^{-1}(A)$. Check if $Z(x_{\lambda 1},\ldots,x_{\lambda n^2})$ is the same as the corresponding μ. If it is not then discard it; otherwise $(x_{\lambda 1},\ldots,x_{\lambda n^2})$ may be the plaintext.

4 Security of PHMs

In this section, we investigate the security of the two PHM schemes. The existing attacks on hidden matrix cryptosystems mainly use either the linearization attack or the degree 2 equation attack using the XL method (see [9]). We now consider the application of these attacks to the PHM schemes.

4.1 Linearization Attacks on PHM

We can obtain linearization equations to attack $[C_n]$ from $BA = AB$. The analogous equation that we should consider in either of the PHM system is $\bar{B}A - A\bar{B} = 0$. Of course, this equation need not be true; however, we may be able to find non-trivial linear relations among the n^2 entries of the left-hand side of this equation, which could potentially yield many linearization equations. Even if this is impossible, we are still not guaranteed that linearization equations do not exist. Therefore, in order to test the two PHM schemes for the existence of linearization equations, we search directly for all equations of the form of Equation (1), in the variables a_{ij}, b_i, c_i, d, which hold for all plaintext/ciphertext pairs $x = (x_1, \ldots, x_{n^2}), y = (y_1, \ldots, y_{n^2})$, for small values of n and r. Table 1 summarizes our findings for 100 randomly chosen instances for each choice of parameters (n, r) with $3 \leq n \leq 6$ and $3 \leq r \leq 9$. The entry in the nth row and r^{th} column is the probability that a particular instance with parameters (n, r) had no linearization equations.

Table 1. Linearization Attack Failure Probabilities

First PHM							
$n \backslash r$	3	4	5	6	7	8	9
3	0.39	0.49	0.73	0.80	0.90	0.96	0.98
4	0.30	0.50	0.77	0.91	0.95	0.97	0.99
5	0.98	1	1	1	1	1	1
6	1	1	1	1	1	1	1
Second PHM							
$n \backslash r$	3	4	5	6	7	8	9
3	0.84	0.88	0.95	0.97	0.97	0.99	0.99
4	0.88	0.98	0.97	1	1	1	1
5	1	1	1	1	1	1	1
6	1	1	1	1	1	1	1

4.2 Degree 2 Equation Attack on PHM

In order to resist the degree 2 equation attack, we need to show that it is not easy to generate new quadratic equations. First notice that if the linear space spanned by the degree 3 terms in the entries of $\bar{B}A - A\bar{B}$ has maximum dimension of $n^2 - 1$, then no new degree 2 equations can be found from linear combinations of the degree 3 entries. (Note that the -1 comes from the trivial relations derived from the trace of $\bar{B}A - A\bar{B}$.) Table 2 shows the probability that the linear space

Table 2. First Degree 2 Equation Attack Failure Probabilities

First PHM							
$n\backslash r$	3	4	5	6	7	8	9
3	0.18	0.48	0.73	0.79	0.88	0.98	0.98
4	0.17	0.53	0.69	0.83	0.92	0.97	1
5	0.15	0.54	0.65	0.89	0.92	0.96	1
6	0.08	0.49	0.66	0.81	0.95	0.97	0.98
Second PHM							
$n\backslash r$	3	4	5	6	7	8	9
3	0.95	0.97	0.99	0.99	1	1	1
4	0.94	0.99	0.99	0.99	1	1	1
5	0.93	0.95	0.99	1	1	1	1
6	0.93	0.96	1	1	1	1	1

spanned by the degree 3 terms in $\bar{B}A - A\bar{B}$ is of maximum dimension for the instances considered in the linearization attack of the previous section.

Once again we note that even if we cannot use $\bar{B}A - A\bar{B}$ to find new quadratic equations this does not imply that there are no new quadratic equations. Therefore we performed experiments to directly check whether or not there are new nontrivial solutions to Equation (2) in the variables $\alpha_{ij}, \beta_{ij}, \gamma_i, \delta_i, \mu$. Table 3 shows the probability that the above equation has no new nontrivial solutions for the instances considered in the linearization attack of the previous section.

Table 3. Second Degree 2 Equation Attack Failure Probabilities

First PHM							
$n\backslash r$	3	4	5	6	7	8	9
3	0	0.03	0.03	0.16	0.18	0.40	0.63
4	0	0.05	0.18	0.37	0.53	0.72	0.84
5	0.69	0.87	0.95	1	1	1	1
6	1	1	1	1	1	1	1
Second PHM							
$n\backslash r$	3	4	5	6	7	8	9
3	0.19	0.35	0.54	0.66	0.63	0.77	0.82
4	0.59	0.86	0.93	0.99	1	1	1
5	1	1	1	1	1	1	1
6	1	1	1	1	1	1	1

5 A Practical Implementation

We note that when $r = 0$, the first and second PHM reduce to $[C_n]$ and HM, respectively. On the other hand, any system with $r = n^2$ would simply be a system of n^2 randomly chosen quadratic polynomials. Since the decryption is slower by a multiple of q^r, we must not choose r too large. For any given choice of n and r, we suggest using the second PHM rather than the first due to the former's superior resistance to both linearization and degree 2 attacks.

Parameters and Security: With $q = 2$ (i.e., $m = 1$), our experiments suggest that if we take $n = 11$, then $r = 5$ should be large enough so that the probability that either a linearization or degree 2 attack will be successful is extremely small. Based on preliminary experiments using F_4 [5], we project that the time and memory requirements of a Gröbner basis attack will be prohibitively large, and that implementations of the second PHM with $n = 11$ and $r = 5$ will enjoy a security level of 2^{121} 3-DES.

Public/Private Key Size: An implementation of the second PHM with parameters $q = 2, n = 11$ and $r = 5$ will have a public key which consists of 121 quadratic polynomials. Each polynomial has $\binom{121}{2} = 7,260$ quadratic terms, 121 linear terms, and one constant term, so the public key size is roughly 109 KB. The private key includes the four affine transformations s, t, u, and v, the perturbation vector z, and the perturbation set P. The four affine maps and their inverses together require $121 \cdot 121 \cdot 2 \cdot 4 = 117, 128$ bits of storage, the five linear polynomial components of z require $(121 + 1) \cdot 5 = 610$ bits of storage, and P requires $32 \cdot (5 + 2 \cdot 121) = 7904$ bits of storage. Therefore the private key requires roughly 15.3 KB of storage.

Encryption/Decryption Computational Complexity: For encryption, we need to compute the value of 121 quadratic polynomials for a given plaintext $x' = (x'_1, \ldots, x'_{121})$. Calculating the value of each polynomial needs 14,641 multiplications and 122 additions when we rewrite each quadratic polynomial as $\sum x_i (b_i + \sum a_{ij} x_j) + c$. The decryption will be slower than it is for HM due to the perturbation set P. If ν is the time required to compare the value of $Z(x'_1, \ldots, x'_{n^2})$ with μ in the decryption step, then the extra time spent will be at most 32ν, though we expect it to be much smaller on average.

6 Conclusion

In this paper, we illustrate how to perturb the matrix-type cryptosystems $[C_n]$ and HM. Computer experiments with small parameter choices indicate that the resulting two variants seem to be very resistant to both linearization attacks and degree 2 attacks. We propose a practical implementation scheme for the second PHM system with an estimated security of 2^{121} 3-DES. We note in passing that these new variants can be easily modified for use as signature schemes. We believe that our results, though experimental in nature, are promising and warrant further investigation.

Acknowledgements

We would like to thank the anonymous referees for their valuable comments and suggestions.

References

1. N. Courtois, A. Klimov, J. Patarin and A. Shamir. Efficient Algorithms for Solving Overdefined Systems of Multivariate Polynomial Equations. In *EUROCRYPT 2000*, LNCS 1807:392–407.
2. N. Courtois and J. Patarin. About the XL Algorithm over $GF(2)$. In *CT-RSA 2003*, LNCS 2612:141–157.
3. J. Ding. A New Variant of the Matsumoto-Imai Cryptosystem Through Perturbation. In *PKC 2004*, LNCS 2947:305–318.
4. J. Ding. Cryptanalysis of HFEv and Internal Perturbation of HFE. In *PKC 2005*, LNCS 3386:288–301.
5. J.-C. Faugère. A New Efficient Algorithm for Computing Gröbner Bases (F_4). In *Journal of Applied and Pure Algebra*, 139:61–88, June 1999.
6. H. Imai and T. Matsumoto. Algebraic Methods for Constructing Asymmetric Cryptosystems. In *AAECC-3*, LNCS 229:108–119, 1985.
7. T. Matsumoto and H. Imai. Public Quadratic Polynomial-Tuples for Efficient Signature-Verification and Message-Encryption. In *EUROCRYPT 1988*, LNCS 330:419–453.
8. J. Patarin. Hidden Fields Equations (HFE) and Isomorphisms of Polynomials (IP): Two New Families of Asymmetric Algorithms. In *EUROCRYPT 1996*, LNCS 1070:33–48.
9. J. Patarin, L. Goubin and N. Courtois. C_{-+}^{*} and HM: Variations Around Two Schemes of T. Matsumoto and H. Imai. In *ASIACRYPT 1998*, LNCS 1514:35–50.

Identity-Based Identification Without Random Oracles

Kaoru Kurosawa[1] and Swee-Huay Heng[2]

[1] Department of Computer and Information Sciences,
Ibaraki University,
4-12-1 Nakanarusawa, Hitachi,
Ibaraki 316-8511, Japan
kurosawa@cis.ibaraki.ac.jp
[2] Multimedia University,
Jalan Ayer Keroh Lama,
75450 Melaka, Malaysia
shheng@mmu.edu.my

Abstract. This paper shows identity-based (ID-based) identification schemes which are provably secure in the standard model. The schemes are derived from Boneh-Boyen signature scheme, a signature scheme which is provably secure in the standard model based on the strong Diffie-Hellman assumption. More precisely, we present two canonical schemes, namely, a scheme which is secure against impersonation under passive attack, and a scheme which is secure against impersonation under active and concurrent attacks.

Keywords: ID-based cryptography, identification scheme.

1 Introduction

An *identification* scheme assures one party (through acquisition of corroborative evidence) of both the identity of a second party involved, and that the second party was active at the time the evidence was created or acquired. Informally speaking, an identification protocol is an interactive process that enables a prover holding a secret key to identify himself to a verifier holding the corresponding public key. One of the primary purposes of identification is to facilitate access control to a resource, when an access privilege is linked to a particular identity.

The fundamental paper of identification was due to Fiat and Shamir [8]. Some other famous identification schemes follow such as [6, 10, 13]. However, there is no rigorous definition as well as security proof for "ID-based" identification (IBI) schemes until the work in [11] and [2]. In [11], the authors proposed a transformation from any standard digital signature (DS) scheme having 3-move honest verifier zero-knowledge proof of knowledge protocol to an IBI scheme. They further proved that if the underlying DS is existentially unforgeable under adaptive chosen message attack then the newly derived IBI scheme is secure against impersonation under passive attack. However, their transformation method cannot give rise to IBI schemes which are secure against impersonation under active and concurrent attacks. Moreover, the two IBI schemes proposed by them are provably secure in the *random oracle model* only.

[2] provided security proofs or attacks for a large number of IBI and ID-based signature (IBS) schemes defined either explicitly or implicitly in existing literature. The approach in [2] is different from that in [11] in that a framework that reduces proving security of IBI or IBS schemes to proving security of an underlying standard identification scheme is first given. Moreover, all the IBI schemes proposed or surfaced in [2] are provably secure in the random oracle model only. Therefore, we can say that in the existing literature, no IBI scheme which is provably secure in the *standard model* is known.

Kurosawa, Heng and Furukawa showed another (unpublished) IBI scheme in the standard model [12]. This scheme makes use of Cramer-Shoup signature scheme and hence is secure under the strong RSA assumption. In [12], it is also shown that one-way functions are equivalent to IBI schemes.

Our Contribution. In this paper, we propose the first (published) IBI schemes which are provably secure in the *standard model*. Our schemes are derived from Boneh-Boyen signature scheme [4], a scheme which is existentially unforgeable under adaptive chosen message attack in the standard model based on the strong Diffie-Hellman assumption.

Firstly, we construct an efficient canonical (3-move) IBI scheme. We prove that this scheme is provably secure against impersonation under passive attack if Boneh-Boyen signature scheme is existentially unforgeable under adaptive chosen message attack.

Secondly, we construct another canonical IBI scheme wherein each user must possess two independent public private key pairs. The core of this new scheme lies on its "witness indistinguishable" property. The concept of witness indistinguishable and witness hiding protocols was introduced by Feige and Shamir [7]. More precisely, in a witness indistinguishable protocol, the prover demonstrates the knowledge of 1-out-of-2 witnesses corresponding to two problem instances without revealing which is known. We further prove that this scheme is secure against impersonation under active and concurrent attacks if Boneh-Boyen signature scheme is existentially unforgeable under adaptive chosen message attack.

2 Preliminaries

2.1 The Strong Diffie-Hellman Assumption

Let G_1, G_2 be two cyclic groups of prime order p, where possibly $G_1 = G_2$. Let g_1 be a generator of G_1 and g_2 a generator of G_2.

q-Strong Diffie-Hellman Problem. The q-strong Diffie-Hellman problem (SDH problem) in (G_1, G_2) is defined as follows:

given a $(q+2)$-tuple $(g_1, g_2, g_2^w, g_2^{(w^2)}, \ldots, g_2^{(w^q)})$ as input, output a pair $(n, g_1^{1/(w+n)})$ where $n \in Z_p^*$. An algorithm A has advantage ϵ in solving q-SDH in (G_1, G_2) if

$$\Pr[A(g_1, g_2, g_2^w, \ldots, g_2^{(w^q)}) = (n, g_1^{1/(w+n)})] \geq \epsilon$$

where the probability is over the random choice of w in Z_p^* and the random bits consumed by A.

Definition 1. *We say that the (q, t, ϵ)-SDH assumption holds in (G_1, G_2) if no t-time algorithm has advantage at least ϵ in solving the q-SDH problem in (G_1, G_2).*

2.2 Digital Signatures

Definition 2 (Digital signature). *A digital signature scheme $\mathcal{DS}$ is denoted by a triple* (Gen, Sign, Verify) *of polynomial time algorithms, called key generation algorithm, signing algorithm and verification algorithm, respectively. The first two algorithms are probabilistic.*

- **Key Generation.** *On input 1^k, the algorithm produces a pair of matching public and secret keys (pk, sk).*
- **Signing.** *On input (sk, m), the algorithm returns a signature $\sigma = \text{Sign}(sk, m)$, where m is a message.*
- **Verification.** *On input (pk, m, σ), the algorithm returns 1 (accept) or 0 (reject). We require that $\text{Verify}(pk, m, \sigma) = 1$ for all $\sigma \leftarrow \text{Sign}(sk, m)$.*

We consider two security notions under adaptive chosen message attack, namely, the standard notion called existential unforgeability [9] and a slightly stronger notion called strong existential unforgeability [1].

In the security analysis for the standard notion, a forger F takes as input a public key pk, where $(pk, sk) \leftarrow \text{Gen}(1^k)$, and tries to forge signatures with respect to pk. F is allowed to query messages m_i adaptively to the signing oracle to obtain the corresponding signatures σ_i. A valid forgery is a message-signature pair (m, σ) such that $\text{Verify}(pk, m, \sigma) = 1$ but m has never been queried by F.

Definition 3. *We say that a digital signature scheme $\mathcal{DS}$ is (t, q_S, ϵ)-secure against existential forgery under adaptive chosen message attack if for any forger F who runs in time t,*

$$\Pr[F \text{ can output a valid forgery}] < \epsilon,$$

where F can make at most q_S signing queries.

The strong existential unforgeability notion captures a stronger version than the standard one, that is, we require that the forger F cannot even generate a new signature on a previously signed message. Informally, this means that it is infeasible for the forger F to produce a valid forgery (m, σ) such that $\text{Verify}(pk, m, \sigma) = 1$ but (m, σ) is not any of the message-signature pair (m_i, σ_i) queried by F earlier.

3 ID-Based Identification Scheme

Definition 4 (ID-based identification). *An IBI scheme $\mathcal{IBI} = (\mathcal{S}, \mathcal{E}, \mathcal{P}, \mathcal{V})$ is specified by four probabilistic polynomial time (PPT) algorithms, called setup algorithm, extract algorithm, proving algorithm and verification algorithm, respectively.*

- **Setup.** $\mathcal{S}$ *takes as input the security parameter 1^k and generates the global system parameters* params *and the* master-key. *The system parameters will be publicly known while the* master-key *will be known to the PKG only.*
- **Extract.** *An algorithm used by the PKG to extract a private key corresponding to a given public identity. $\mathcal{E}$ receives as input the* master-key, params *and a public identity* ID, *it returns the corresponding private key d.*
- **Identification Protocol.** $\mathcal{P}$ *receives as input* (params, ID, d) *and $\mathcal{V}$ receives as input* (params, ID), *where d is the private key corresponding to the public identity* ID. *After an interactive execution of $(\mathcal{P}, \mathcal{V})$, $\mathcal{V}$ outputs a boolean decision 1 (accept) or 0 (reject). A legitimate $\mathcal{P}$ should always be accepted.*

Specifically, we consider the following IBI scheme having 3-move protocol which is commonly called *canonical*. That is, $\mathcal{P}$ first sends a *commitment* CMT to $\mathcal{V}$. $\mathcal{V}$ returns a *challenge* CH which is randomly chosen from some challenge set and $\mathcal{P}$ provides a *response* RSP. Finally, $\mathcal{V}$ either accepts or rejects the proof.

3.1 Notions of Security

We consider the security notion as those defined in [11, 2], which was the adaptation of the notion first proposed by Feige, Fiat and Shamir [6] to the ID-based setting. That is, We consider three types of attacks on the honest, private key equipped prover, namely, passive attack, active attack and concurrent attack.

Normally, a two-phase game is considered between a challenger and an adversary. In the standard identification model, the above attacks should take place and complete before the impersonation attempt, i.e. the attacks should be completed in Phase 1. However, in the ID-based setting, it is natural to allow the adversary to interact with the real provers with identities other than the challenge identity ID even in Phase 2.

In the passive attacks, the adversary can eavesdrop and she is in possession of transcripts of conversations between the provers and the verifiers. In the active and concurrent attacks, the adversary first plays the role of a cheating verifier, interacting with the provers several times before the impersonation attempt and even in Phase 2 (for provers with identities $\neq$ ID). The difference between active and concurrent attacks is that in an active attack, the cheating verifier interacts serially with prover "clones"; while in a concurrent attack, the cheating verifier is allowed to interact with many different prover "clones" concurrently. The clones all have the same secret key but are initialized with independent coins and maintain their own state. Clearly, security against impersonation under concurrent attack implies security against impersonation under active attack.

The two-phase attack game between a passive or active/concurrent impersonator I and the challenger is described below.

- **Setup.** The challenger takes as input 1^k and runs the setup algorithm $\mathcal{S}$. It gives I the resulting system parameters params and keeps the master-key to itself.
- **Phase 1.**
  1. I issues some key extraction queries $\mathsf{ID}_1, \mathsf{ID}_2, \ldots$. The challenger responds by running the extract algorithm $\mathcal{E}$ to generate the private key d_i corresponding to the public identity ID_i. It returns d_i to I.
  2. I issues some transcript queries (in passive attack) or some identification queries on ID_j (in for active/concurrent attack).
  3. The queries in step 1 and step 2 above can be interleaved and asked adaptively. Without loss of generality, we may assume that I will not query the same ID_i that has been issued in the key extraction queries in the transcript queries or identification queries again.
- **Phase 2.**
  1. I outputs a challenge identity $\mathsf{ID} \neq \{\mathsf{ID}_i$ in the extraction queries$\}$ on which it wishes to impersonate. Next, I plays the role as a cheating prover (impersonation attempt on the prover holding the public identity ID), trying to convince the verifier.
  2. I can still issue some key extraction queries as well as transcript queries or identification queries in Phase 2, with the restriction that no queries on the challenged identity ID are allowed.

We say that I succeeds in impersonating if it can make the verifier accepts.

Definition 5. *We say that an ID-based identification scheme $\mathcal{IBI}$ is (t, q_I, ϵ)-secure under passive (active and concurrent) attacks if for any passive (active and concurrent) impersonator I who runs in time t,*

$$\Pr[I \text{ can impersonate}] < \epsilon,$$

where I can make at most q_I key extraction queries.

4 Proposed Scheme Secure Against Passive Attack

In this section, we propose the first canonical IBI scheme which is provably secure against impersonation under passive attack in the standard model. This scheme builds on Boneh-Boyen signature scheme [4], a scheme which is known to be secure in the standard model. We show that our proposed scheme is secure against impersonation under passive attack if Boneh-Boyen signature scheme is existential unforgeable under adaptive chosen message attack. The latter is indeed the case follows from Proposition 1 which says that Boneh-Boyen signature scheme is strongly existential unforgeable. We remark that the standard existential unforgeability is enough for our purpose, which is naturally satisfied by Boneh-Boyen signature scheme.

Basically, our proposed IBI scheme employs the key generation algorithm of Boneh-Boyen signature scheme as the setup algorithm and its signing algorithm as the extract algorithm.

4.1 Boneh-Boyen Signature Scheme [4]

Let (G_1, G_2) be bilinear groups where $|G_1| = |G_2| = p$ for some prime p. As usual, g_1 is a generator of G_1 and g_2 is a generator of G_2. If we assume that the messages m to be signed are elements in Z_p^*, then the collision-resistant hash function H in the scheme can be omitted. However, we adopt the scheme with collision-resistant hash function H here.

- **Key Generation.** Pick random $x, y \in Z_p^*$ and compute $u = g_2^x \in G_2$ and $v = g_2^y \in G_2$. Choose a collision-resistant hash function $H : \{0,1\}^* \to Z_p^*$. The public key is (g_1, g_2, u, v, H) and the secret key is (x, y).
- **Signing.** Given a secret key $x, y \in Z_p^*$ and a message $m \in \{0,1\}^*$, pick a random $r \in Z_p^*$ and compute $\sigma = g_1^{1/(x+H(m)+yr)} \in G_1$. Here $1/(x+H(m)+yr)$ is computed modulo p. In the unlikely event that $x + H(m) + yr = 0 \bmod p$, we try again with a different random r. The signature is (σ, r).
- **Verification.** Given a public key (g_1, g_2, u, v, H), a message $m \in \{0,1\}^*$, and a signature (σ, r), verify that

$$e(\sigma, u \cdot g_2^{H(m)} \cdot v^r) = e(g_1, g_2).$$

The security of Boneh-Boyen signature scheme is given as follows.

Proposition 1. *[4-Theorem 1] Suppose the (q, t', ϵ')-SDH assumption holds in (G_1, G_2) and that H is a collision-resistant hash function. Then the above signature scheme is (t, q_S, ϵ)-secure against strong existential forgery under adaptive chosen message attack provided that*

$$t \leq t' - o(t), \quad q_S < q \quad \text{and} \quad \epsilon \geq 2(\epsilon' + q_S/p) \approx 2\epsilon'.$$

4.2 Proposed Scheme

Let $\mathcal{IBI} = (\mathcal{S}, \mathcal{E}, \mathcal{P}, \mathcal{V})$ be four PPT algorithms as follows.

Let (G_1, G_2) be bilinear groups where $|G_1| = |G_2| = p$ for some prime p. As usual, g_1 is a generator of G_1 and g_2 is a generator of G_2.

- **Setup.** Given a security parameter 1^k, pick random $x, y \in Z_p^*$ and compute $u = g_2^x \in G_2$ and $v = g_2^y \in G_2$. Choose a collision-resistant hash function $H : \{0,1\}^* \to Z_p^*$. The system parameters params is (g_1, g_2, u, v, H) and the master-key is (x, y).
- **Extract.** Given a master-key (x, y) and an identity $\mathsf{ID} \in \{0,1\}^*$, pick a random $r \in Z_p^*$ and compute $\sigma = g_1^{1/(x+H(\mathsf{ID})+yr)} \in G_1$. Here $1/(x+H(\mathsf{ID})+yr)$ is computed modulo p. In the unlikely event that $x + H(\mathsf{ID}) + yr = 0 \bmod p$, we try again with a different random r. The user private key is (σ, r).

- **Identification Protocol.**
  1. $\mathcal{P}$ chooses $R \in G_1$ randomly and computes $X = e(R, u \cdot g_2^{H(\mathsf{ID})} \cdot v^r)$. It then sends (r, X) to $\mathcal{V}$.
  2. $\mathcal{V}$ chooses $c \in Z_p$ randomly and sends c to $\mathcal{P}$.
  3. $\mathcal{P}$ sends $S = R + c\sigma$ to $\mathcal{V}$.
  4. $\mathcal{V}$ accepts if and only if $e(S, u \cdot g_2^{H(\mathsf{ID})} \cdot v^r) = X \cdot e(g_1, g_2)^c$.

4.3 Security Analysis

Theorem 1. *The above IBI scheme is (t, q_I, ϵ)-secure against impersonation under passive attack in the standard model if H is a collision-resistant hash function and Boneh-Boyen signature scheme is (t', q_S, ϵ')-secure against existential forgery under adaptive chosen message attack where*

$$t \geq (t'/2) - \mathrm{poly}(k), \quad q_I = q_S, \quad \epsilon \leq \sqrt{\epsilon'} + (1/p).$$

Proof. Let I be an impersonator who (t, q_I, ϵ)-breaks the IBI scheme. Then we show that Boneh-Boyen signature scheme is not (t', q_S, ϵ')-secure. That is, we present a forger F who (t', q_S, ϵ')-breaks the Boneh-Boyen signature scheme by running I as a subroutine.

The forger F receives the public key (g_1, g_2, u, v, H) as its input. F then gives params $= (g_1, g_2, u, v, H)$ as the system parameters to the impersonator I. In Phase 1, the impersonator I starts the key extraction queries. If I issues a key extraction query ID_i, then the forger F queries ID_i to its signing oracle. F returns the answer (σ_i, r_i) obtained from the signing oracle as the private key to I.

If I issues a transcript query on ID_j, then F chooses $r_j \in Z_p^*, c_j \in Z_p, S_j \in G_1$ randomly and computes X_j such that $e(S_j, u \cdot g_2^{H(\mathsf{ID}_j)} \cdot v^{r_j}) = X_j \cdot e(g_1, g_2)^{c_j}$. F then gives (r_j, X_j, c_j, S_j) to I as the transcript.

Eventually, I decides that Phase 1 is over and it outputs a public identity $\mathsf{ID} \neq \{\mathsf{ID}_i$ in the key extraction queries$\}$ on which it wishes to be challenged. I can still issue some key extraction queries as well as transcript queries in Phase 2, with the restriction that no queries on the challenged identity ID are allowed. I plays the role as the cheating prover now, trying to convince the verifier $\mathcal{V}$ that it is the holder of public identity ID. F plays the role of $\mathcal{V}$ now. Immediately after the first run, F resets the prover I to the step whereby I has sent the message (r, X). F then runs the protocol again. Let the conversation transcripts for the first run and second run be (r, X, c, S) and (r, X, c', S') respectively. Based on the Reset Lemma proposed by Bellare and Palacio [3], the forger F can extract σ from the two conversation transcripts with probability more than $(\epsilon - 1/p)^2$. σ can be extracted from the above transcripts easily as follows. Since we have $S = R + c\sigma$ and $S' = R + c'\sigma$ and thus $\sigma = (c - c')^{-1}(S - S')$.

Finally, the forger F returns the message-signature pair $(\mathsf{ID}, (\sigma, r))$ as its forgery. Thus it is clear that

$$t' \leq 2t + \mathrm{poly}(k), \quad q_S = q_I, \quad \epsilon' \geq (\epsilon - 1/p)^2.$$

□

5 Proposed Scheme Secure Against Active and Concurrent Attacks

In this section, we proposed the first canonical IBI scheme which is provably secure against impersonation under active and concurrent attacks in the standard model. This scheme is also derived from Boneh-Boyen signature scheme [4]. We show that our proposed scheme is secure against impersonation under active and concurrent attacks if Boneh-Boyen signature scheme is existentially unforgeable under adaptive chosen message attack.

This new scheme possesses the "witness indistinguishable" property. The concept of witness indistinguishable and witness hiding protocols was introduced by Feige and Shamir [7]. More precisely, the prover demonstrates the knowledge of 1-out-of-2 witnesses corresponding to two problem instances without revealing which is known.

The core of this transformation lies in that in our proposed IBI scheme, the system parameters params and the corresponding master-key consist of two components, (pk_1, pk_2) and (sk_1, sk_2) respectively, where they can be seen as obtaining from the two independent executions of the key generation algorithm of Boneh-Boyen signature scheme. We treat the user private key as the corresponding signatures on a message m signed using sk_1 and sk_2 respectively, where $m = \mathsf{ID}$.

This is important because in the security analysis, as in the previous scheme, reduction needs to be done from Boneh-Boyen signature scheme to the proposed IBI scheme. Now, the forger in Boneh-Boyen signature scheme has the knowledge of a public secret key pair, say (pk_1, sk_1), thus it can choose another public secret key pair (pk_2, sk_2) randomly. Therefore, it can answer the identification queries in the active/concurrent attack and thus responds interactively to the cheating verifier (the impersonator) with the knowledge of sk_2, a component of the master-key.

5.1 Proposed Scheme

Let $\mathcal{IBI} = (\mathcal{S}, \mathcal{E}, \mathcal{P}, \mathcal{V})$ be four PPT algorithms as follows. Let (G_1, G_2) be bilinear groups where $|G_1| = |G_2| = p$ for some prime p. As usual, g_1 is a generator of G_1 and g_2 is a generator of G_2.

- **Setup.** Given a security parameter 1^k, pick random $x_1, y_1, x_2, y_2 \in Z_p^*$ and compute $u_1 = g_2^{x_1}, v_1 = g_2^{y_1}, u_2 = g_2^{x_2}$ and $v_2 = g_2^{y_2}$ such that $u_1, v_1, u_2, v_2 \in G_2$. Choose a collision-resistant hash function $H : \{0,1\}^* \to Z_p^*$. The system parameters params is $(g_1, g_2, u_1, v_1, u_2, v_2, H)$ and the master-key is (x_1, y_1, x_2, y_2).
- **Extract.** Given the master-key (x_1, y_1, x_2, y_2) and an identity $\mathsf{ID} \in \{0,1\}^*$, pick random $r_1, r_2 \in Z_p^*$ and compute $\sigma_1 = g_1^{1/(x_1+H(\mathsf{ID})+y_1 r_1)} \in G_1$ and $\sigma_2 = g_1^{1/(x_2+H(\mathsf{ID})+y_2 r_2)} \in G_1$. Here $1/(x_1+H(\mathsf{ID})+y_1 r_1)$ and $1/(x_2+H(\mathsf{ID})+y_2 r_2)$

are computed modulo p. In the unlikely event that $x_1 + H(\mathsf{ID}) + y_1 r_1 = 0 \bmod p$ and $x_2 + H(\mathsf{ID}) + y_2 r_2 = 0 \bmod p$, we try again with different random r_1 and r_2 respectively. The user private key is $(\sigma_1, r_1, \sigma_2, r_2)$.

- **Identification Protocol.** Assume that the prover $\mathcal{P}$ proves that it knows (σ_1, r_1).
  1. Since $\mathcal{P}$ knows the value of r_2, it needs to choose only $c_2 \in Z_p$ and $S_2 \in G_1$ randomly. It next computes X_2 such that $e(S_2, u \cdot g_2^{H(\mathsf{ID})} \cdot v^{r_2}) = X_2 \cdot e(g_1, g_2)^{c_2}$. It also chooses $R \in G_1$ randomly and computes $X_1 = e(R, u \cdot g_2^{H(\mathsf{ID})} \cdot v^{r_1})$. It then sends (r_1, X_1, r_2, X_2) to $\mathcal{V}$.
  2. $\mathcal{V}$ chooses $c \in Z_p$ randomly and sends c to $\mathcal{P}$.
  3. $\mathcal{P}$ first finds $c_1 = c - c_2 \bmod p$ and computes $S_1 = R + c_1 \sigma_1$. It then sends (c_1, c_2, S_1, S_2) to $\mathcal{V}$.
  4. $\mathcal{V}$ accepts if and only if $c = c_1 + c_2 \bmod p$, $e(S_1, u_1 \cdot g_2^{H(\mathsf{ID})} \cdot v_1^{r_1}) = X_1 \cdot e(g_1, g_2)^{c_1}$ and $e(S_2, u_2 \cdot g_2^{H(\mathsf{ID})} \cdot v_2^{r_2}) = X_2 \cdot e(g_1, g_2)^{c_2}$.

5.2 Security Analysis

Theorem 2. *The above IBI scheme is (t, q_I, ϵ)-secure against impersonation under active and concurrent attacks in the standard model if H is a collision-resistant hash function and Boneh-Boyen signature scheme is (t', q_S, ϵ')-secure against existential forgery under adaptive chosen message attack, where*

$$t \geq (t'/2) - poly(k), \quad q_I = q_S, \quad \epsilon \leq \sqrt{2\epsilon'} + (1/p).$$

Proof. Let I be an impersonator who (t, q_I, ϵ)-breaks the IBI scheme. Then we show that Boneh-Boyen signature scheme is not (t', q_S, ϵ')-secure. That is, we present a forger F who (t', q_S, ϵ')-breaks the Boneh-Boyen signature scheme by running I as a subroutine.

The forger F receives the public key (g_1, g_2, u, v, H) as its input. F next chooses $x_2, y_2 \in Z_p^*$ randomly and computes $u_2 = g_2^{x_2} \in G_2$ and $v_2 = g_2^{y_2} \in G_2$. F then sets $u_1 = u$ and $v_1 = v$ and gives params $= (g_1, g_2, u_1, v_1, u_2, v_2, H)$ as the system parameters to the impersonator I.

In Phase 1, the impersonator I starts the key extraction queries. If I issues a key extraction query on ID_i, then the forger F queries ID_i to its signing oracle. F returns the private key as $(\sigma_{1,i}, r_{1,i}, \sigma_{2,i}, r_{2,i})$ to I, where $(\sigma_{1,i}, r_{1,i})$ is the output of the signing oracle of F and $(\sigma_{2,i}, r_{2,i})$ is generated by F since it knows (x_2, y_2).

If I issues an identification queries on ID_j, since F knows (x_2, y_2), it can respond to I by executing the identification protocol with I interactively for any ID_j. This is because F can always produce one component of the private key $(\sigma_{2,j}, r_{2,j})$ with the knowledge of (x_2, y_2) and further convincing the cheating verifier I. Therefore, F manages to simulate the environment for I perfectly. (Remember that our protocol is witness-indistinguishability.)

Eventually, I decides that Phase 1 is over and it outputs a public identity $\mathsf{ID} \neq \{\mathsf{ID}_i$ in the key extraction queries$\}$ on which it wishes to be challenged. I can still issue some key extraction queries as well as identification queries in Phase 2, with the restriction that no queries on the challenged identity ID are allowed. I

plays the role as the cheating prover now, trying to convince the verifier $\mathcal{V}$ that it is the holder of public identity ID. F plays the role of $\mathcal{V}$ now. Immediately after the first run, F resets the prover I to the step whereby I has sent the message (r_1, X_1, r_2, X_2). F then runs the protocol again. Let the conversation transcripts for the first run and second run be $[(r_1, X_1, r_2, X_2), c, (c_1, c_2, S_1, S_2)]$ and $[(r_1, X_1, r_2, X_2), c', (c'_1, c'_2, S'_1, S'_2)]$ respectively.

Based on the Reset Lemma proposed by Bellare and Palacio [3], one can extract σ_1 or σ_2 from the two conversation transcripts with probability more than $(\epsilon - 1/p)^2$. It is not difficult to extract σ_1 or σ_2 from the transcript. Remember that $S_1 = R + c_1\sigma_1$ and $S'_1 = R + c'_1\sigma_1$, and therefore $\sigma_1 = (c_1 - c'_1)^{-1}(S_1 - S'_1)$. Also, since $S_2 = R' + c_2\sigma_2$ and $S'_2 = R' + c'_2\sigma_2$ for some R', $\sigma_2 = (c_2 - c'_2)^{-1}(S_2 - S'_2)$ can be extracted easily.

Finally, the forger F returns the message-signature pair $(\mathsf{ID}, (\sigma_1, r_1))$ as its forgery with probability more than $\frac{1}{2}(\epsilon - 1/p)^2$. Thus it is clear that

$$t' \leq 2t + poly(k),\ q_S = q_I,\ \epsilon' \geq 1/2(\epsilon - 1/p)^2.$$

□

References

1. J. H. An, Y. Dodis and T. Rabin. On the security of joint signature and encryption. *Advances in Cryptology — EUROCRYPT '02*, LNCS 2332, pp. 83–107, Springer-Verlag, 2002.
2. M. Bellare, C. Namprempre and G. Neven. Security proofs for identity-based identification and signature schemes. *Advances in Cryptology — EUROCRYPT '04*, LNCS 3027, pp. 268–286, Springer-Verlag, 2004.
3. M. Bellare and A. Palacio. GQ and Schnorr identification schemes: proofs of security against impersonation under active and concurrent attacks. *Advances in Cryptology — CRYPTO '02*, LNCS 2442, pp. 162–177, Springer-Verlag, 2002.
4. D. Boneh and X. Boyen. Short signatures without random oracles. *Advances in Cryptology — EUROCRYPT '04*, LNCS 3027, pp. 56–73, Springer-Verlag, 2004.
5. D. Boneh, B. Lynn and H. Shacham. Short signatures from the Weil pairing. *Advances in Cryptology — ASIACRYPT '01*, LNCS 2248, pp. 514–532, Springer-Verlag, 2001.
6. U. Feige, A. Fiat and A. Shamir. Zero-knowledge proofs of identity. *Journal of Cryptology*, vol. 1, pp. 77–94, Springer-Verlag, 1988.
7. U Feige and A. Shamir. Witness indistinguishable and witness hiding protocols. *ACM Symposium on Theory of Computing — STOC '90*, pp. 416–426, 1990.
8. A. Fiat and A. Shamir. How to prove yourself: practical solutions to identification and signature problems. *Advances in Cryptology — CRYPTO '86*, LNCS 263, pp. 186–194, Springer-Verlag, 1987.
9. S. Goldwasser, S. Micali and R. Rivest. A digital signature scheme secure against adaptive chosen-message attacks. *SIAM Journal of Computing*, vol. 17, no. 2, pp. 281–308, 1988.
10. L. Guillou and J. Quisquater. A practical zero-knowledge protocol fitted to security microprocessors minimizing both transmission and memory. *Advances in Cryptology — EUROCRYPT '88*, LNCS 330, pp. 123–128, Springer-Verlag, 1989.

11. K. Kurosawa and S.-H. Heng. From digital signature to ID-based identification/signature. *Public Key Cryptography — PKC '04*, LNCS 2947, pp. 248–261, Springer-Verlag, 2004.
12. K. Kurosawa, S.-H. Heng and J. Furukawa. Digital signatures are equivalent ID-based identifications/signatures. Manuscript.
13. C. Schnorr. Efficient signature generation by smart cards. *Journal of Cryptology*, vol. 4, pp. 161–174, Springer-Verlag, 1991.

Linkable Ring Signatures: Security Models and New Schemes

(Extended Abstract)

Joseph K. Liu[1] and Duncan S. Wong[2],*

[1] Department of Information Engineering,
The Chinese University of Hong Kong, Shatin, Hong Kong
ksliu@ie.cuhk.edu.hk
[2] Department of Computer Science,
City University of Hong Kong, Kowloon, Hong Kong
duncan@cityu.edu.hk

Abstract. A ring signature scheme is a group signature scheme but with no group manager to setup a group or revoke a signer's identity. The formation of a group is spontaneous in the way that diversion group members can be totally unaware of being conscripted to the group. It allows members of a group to sign messages on the group's behalf such that the resulting signature does not reveal their identity (anonymity). The notion of linkable ring signature, introduced by Liu, et al. [10], also provides signer anonymity, but at the same time, allows anyone to determine whether two signatures have been issued by the same group member (linkability). In this paper, we enhance the security model of [10] for capturing new and practical attacking scenarios. We also propose two polynomial-structured linkable ring signature schemes. Both schemes are given strong security evidence by providing proofs under the random oracle model.

1 Introduction

A group signature scheme [7, 6, 2] allows members of a group to sign messages on behalf of the group without revealing the identity of the signer (anonymity). It is also not possible to decide whether two signatures have been issued by the same group member (unlinkability). Only a designated group manager which also maintains the group membership can revoke the authorship of signatures. For anyone else, signatures are anonymous and unlinkable.

A ring signature scheme [13, 5, 1, 4, 8] can be considered as a group signature scheme without a group manager. The formation of a group is spontaneous. That is, under the assumption that each user is already associated with the public key

---

* The work described in this paper was supported by a grant from the Research Grants Council of the Hong Kong Special Administrative Region, China (Project No. 9040904 (RGC Ref. No. CityU 1161/04E)).

of some standard signature scheme, a user (group creator) can spontaneously create a group by collecting the public keys of some other users and his own public key. The diversion group members (i.e. those users whose public keys are included in the group by the group creator) can be totally unaware of being conscripted into the group. Similar to group signature schemes, ring signature schemes are also anonymous and unlinkable. But also unlike group signature schemes, ring signature schemes do not have any group manager to revoke the anonymity of a signature or maintain the group membership.

A linkable ring signature scheme, introduced by Liu, et al. [10], is a ring signature scheme that provides signer anonymity but at the same time allows one to determine whether two signatures have been issued by the same group member. Linkable ring signature schemes can be used for constructing efficient e-voting systems [10]. To cast a vote, a voter generates a linkable ring signature for his vote. Anonymity is maintained and linkability helps detect double voting if a voter casts two votes. In addition, linkable ring signatures eliminate the involvement of voters in the registration phase of each voting event and at the same time prevent information leak on which voters have cast votes and which voters have not.

We enhance the security model of [10] by providing a stronger notion of signer anonymity and redefining linkability. The new security model captures new and practical attacking scenarios, and properties more thoroughly.

In this paper, we use an approach based on the proof of knowledge system called Witness Indistinguishable Proof of Equality of a Discrete Logarithm to construct two linkable ring signature schemes. We also discuss a subtlely on linkability between these two schemes. For both of the schemes proposed in this paper, we give strong evidence of their security by providing security proofs in our enhanced security model under the random oracle model [3].

The paper is organized as follows. In Sec. 2, a linkable ring signature scheme and a security model are defined. In Sec. 3, the basic techniques of our constructions are described and two polynomial-structured linkable ring signature schemes are proposed. In Sec. 4, we conclude the paper. There are several appendices at the end of the paper containing the proofs of the theorems stated in the paper body.

2 Linkable Ring Signature Schemes

A linkable ring signature scheme is a quadruple $(Gen, Sig, Ver, Link)$.

- $(x, y) \leftarrow Gen(1^k)$ is a probabilistic algorithm which takes security parameter k and outputs private key x and public key y.
- $\sigma \leftarrow Sig(1^k, 1^n, x, L, m)$ is a probabilistic algorithm which takes security parameter k, group size n, private key x, a list L of n public keys which includes the one corresponding to x and message m, produces a signature σ.
- $1/0 \leftarrow Ver(1^k, 1^n, L, m, \sigma)$ is a boolean algorithm which accepts as inputs security parameter k, group size n, a list L of n public keys, message m and

signature σ, returns 1 or 0 for accept or reject, respectively. We require that for any message m, any $(x,y) \leftarrow Gen(1^k)$ and any L that includes y,

$$Ver(1^k, 1^n, L, m, Sig(1^k, 1^n, x, L, m)) = 1.$$

- $1/0 \leftarrow Link(1^k, 1^n, L, m_1, m_2, \sigma_1, \sigma_2)$ is a boolean algorithm which takes security parameter k, group size n, a list L of n distinct public keys, messages m_1, m_2, and signatures σ_1, σ_2, such that $Ver(1^k, 1^n, L, m_1, \sigma_1) = 1$ and $Ver(1^k, 1^n, L, m_2, \sigma_2) = 1$, returns 1 or 0 for linked or unlinked, respectively. We require that for any messages m_1, m_2, any $L = (y_1, \cdots, y_n)$, each of them is generated by $Gen(1^k)$ with fresh coin flips, any integers (indices) π_1, π_2 over the range $[1, n]$, and any $\sigma_1 \leftarrow Sig(1^k, 1^n, x_{\pi_1}, L, m_1)$, $\sigma_2 \leftarrow Sig(1^k, 1^n, x_{\pi_2}, L, m_2)$,

$$Link(1^k, 1^n, L, m_1, m_2, \sigma_1, \sigma_2) = \begin{cases} 1 & \text{if } \pi_1 = \pi_2 \\ 0 & \text{otherwise} \end{cases}$$

x_π represents the private key corresponding to y_π in L.

2.1 Security Model

The security of a linkable ring signature has three aspects: unforgeability, signer anonymity and linkability.

Unforgeability. Let $\mathcal{U} = \{y_1, \cdots, y_N\}$ be a set of public keys, each is generated by $Gen(1^k)$ with fresh coin flips. To support adaptive chosen message attack, we provide the adversary a signing oracle $\mathcal{SO}$. $\mathcal{SO}(L', m')$ takes as inputs any $L' \subseteq \mathcal{U}$, $|L'| = n'$, and any message m', produces a signature σ' such that $Ver(1^k, 1^{n'}, L', m', \sigma') = 1$. The following definition, currently the strongest one for ring signature schemes, is due to Abe, et al. [1].

Definition 1 (Existential Unforgeability against Adaptive Chosen Message and Chosen Public-key Attacks). *Let $\mathcal{U} = \{y_1, \cdots, y_N\}$ be a set of public keys, each is generated by $Gen(1^k)$ with fresh coin flips. A linkable ring signature scheme is unforgeable if, for any probabilistic polynomial-time algorithm $\mathcal{A}$ with signing oracle $\mathcal{SO}$ such that $(L, m, \sigma) \leftarrow \mathcal{A}^{\mathcal{SO}}(1^k, \mathcal{U})$, its output satisfies $Ver(1^k, 1^n, L, m, \sigma) = 1$ only with negligible probability in k, where $L \subseteq \mathcal{U}$ and $|L| = n$. Restriction is that (L, m, σ) should not be in the set of oracle queries and replies between $\mathcal{A}$ and $\mathcal{SO}$.*

A real-valued function ϵ is negligible if for every $c > 0$ there exists a $k_c > 0$ such that $\epsilon(k) < k^{-c}$ for all $k > k_c$.

Signer Anonymity. Given a signature with respect to a group of n members and supposing that the actual signer is chosen at random over these n group members, by signer anonymity, an adversary should not be able to identify the identity of the actual signer with probability significantly greater than $1/(n-t)$ when t private keys of the group are known. The security model of Liu, et al. [10]

captures the adaptive chosen message attack using the signing oracle $\mathcal{SO}$ defined above. It allows the adversary to corrupt up to t group members by revealing their private keys. In the following, we enhance their model by also allowing the adversary to adaptively choose a group to attack. The need of this enhancement will be explained shortly after giving the definition of signer anonymity.

Consider an experiment of two stages: choose and guess. In the choose stage, the adversary $\mathcal{A}$ with signing oracle $\mathcal{SO}$ chooses a subgroup of $\mathcal{U}$ and a message $m \in \{0,1\}^*$. This is denoted by $(L, n, m, State) \leftarrow \mathcal{A}^{\mathcal{SO}}(1^k, \mathcal{U}, \text{choose})$ where $State$ is some state information which can be passed to the guess stage and $L \subseteq \mathcal{U}$, $|L| = n$. Let $L = \{y_{i_1}, \cdots, y_{i_n}\}$. In the guess stage, $\mathcal{A}$ is given access to not only the signing oracle $\mathcal{SO}$, but also a corruption oracle $\mathcal{CO}$. $\mathcal{CO}(\pi')$ takes as input any $\pi' \in \{i_1, \cdots, i_n\}$ and returns the private key $x_{\pi'}$ corresponding to the public key $y_{\pi'} \in L$. The objective of $\mathcal{A}$ in the guess stage is to determine the public key in L whose private key is used to generate a given signature σ with respect to L. This is denoted by $\xi \leftarrow \mathcal{A}^{\mathcal{SO},\mathcal{CO}}(1^k, 1^n, L, m, \sigma, State, \mathcal{U}, Priv_{N-n}, \text{guess})$ where $\xi \in \{i_1, \cdots, i_n\}$ and $Priv_{N-n}$ is the set of private keys corresponding to the public keys in $\mathcal{U} \setminus L$. Below is the complete description of the experiment.

Experiment $\mathbf{Exp}_{\mathcal{A}}^{\text{anon}}(k, N)$
  For $i = 1, \cdots, N$, $(x_i, y_i) \leftarrow Gen(1^k)$ with fresh coin flips
  Set $\mathcal{U} = \{y_1, \cdots, y_N\}$
  $(L, n, m, State) \leftarrow \mathcal{A}^{\mathcal{SO}}(1^k, \mathcal{U}, \text{choose})$
  $\pi \xleftarrow{R} \{i_1, \cdots, i_n\}$, $\sigma \leftarrow Sig(1^k, 1^n, x_\pi, L, m)$.
  $\xi \leftarrow \mathcal{A}^{\mathcal{SO},\mathcal{CO}}(1^k, 1^n, L, m, \sigma, State, \mathcal{U}, Priv_{N-n}, \text{guess})$
  If $\mathcal{A}$ failed, the experiment halts with failure
  If $\mathcal{A}$ did not query $\mathcal{CO}$ with π then
      return 1 if $\xi = \pi$, otherwise return 0
  Else the experiment halts with failure

An experiment succeeds if it halts with no failure. We denote by

$$\mathbf{Adv}_{\mathcal{A}}^{\text{anon}}(k, N) = \Pr[\mathbf{Exp}_{\mathcal{A}}^{\text{anon}}(k, N) = 1 \mid \text{Experiment succeeds}] - \frac{1}{n-t}$$

the advantage of the adversary $\mathcal{A}$ in breaking the anonymity of a linkable ring signature scheme where t is the number of $\mathcal{CO}$ queries made by $\mathcal{A}$ in the experiment. We say that a linkable ring signature is *signer anonymous* if for a constant N, for any probabilistic polynomial-time adversary $\mathcal{A}$, the function $\mathbf{Adv}_{\mathcal{A}}^{\text{anon}}(\cdot, N)$ is negligible.

$\mathcal{A}$ can obtain signatures for any messages and subgroups of $\mathcal{U}$ by querying $\mathcal{SO}$. Note that this also captures a new attacking scenario where an adversary can try to find out the authorship of a signature instance through collecting signatures generated on behalf of different subgroups of $\mathcal{U}$. For example, if it is possible for $\mathcal{A}$ to determine whether two ring signatures are generated by the same ring creator even the diversion members of the two corresponding rings are different (i.e. the two signatures are corresponding to two different subgroups of $\mathcal{U}$), then $\mathcal{A}$ can determine the identity of any signature σ with respect to a subgroup $L = \{y_1, \cdots, y_n\}$ of $\mathcal{U}$ easily by collecting n signature-subgroup pairs

(σ_i, L_i), $1 \leq i \leq n$, such that $L_i = \{y_i\}$ and then check on which σ_i has the same ring creator as that of σ. If σ_j and σ share the same ring creator, then the actual signer is indexed by j in L since the ring creator is the actual signer.

Linkability. We specify two experiments for capturing the following attacks:

1. A group member generates two signatures such that $Link$ returns 0.
2. (Framing) After learning a signature and the identity of the group member who generates the signature, a different group member generates a signature such that $Link$ returns 1 on these two signatures.

Experiment $\mathbf{Exp}_{\mathcal{A}}^{\text{link1}}(k, N)$
 For $i = 1, \cdots, N$, $(x_i, y_i) \leftarrow Gen(1^k)$ with fresh coin flips
 Set $\mathcal{U} = \{y_1, \cdots, y_N\}$
 $\pi \leftarrow \mathcal{A}^{\mathcal{SO}}(1^k, \mathcal{U}, \text{choose})$ where $\pi \in \{1, \cdots, N\}$
 $(L, n, m_1, m_2, \sigma_1, \sigma_2) \leftarrow \mathcal{A}^{\mathcal{SO}}(1^k, x_\pi, \mathcal{U}, \text{sign})$ where $L \subseteq \mathcal{U}$, $|L| = n$,
  $y_\pi \in L$, $Ver(1^k, 1^n, L, m_1, \sigma_1) = 1$ and $Ver(1^k, 1^n, L, m_2, \sigma_2) = 1$.
 If (L, m_i, σ_i), $1 \leq i \leq 2$, were not in the set of oracle queries and replies
  between $\mathcal{A}$ and $\mathcal{SO}$ then
  return $1 - Link(1^k, 1^n, L, m_1, m_2, \sigma_1, \sigma_2)$
 Else the experiment halts with failure

Experiment $\mathbf{Exp}_{\mathcal{A}}^{\text{link2}}(k, N)$
 For $i = 1, \cdots, N$, $(x_i, y_i) \leftarrow Gen(1^k)$ with fresh coin flips
 Set $\mathcal{U} = \{y_1, \cdots, y_N\}$
 $(L, n, \pi_1, m_1, State_1) \leftarrow \mathcal{A}^{\mathcal{SO}}(1^k, \mathcal{U}, \text{choose1})$ where $L \subseteq \mathcal{U}$, $|L| = n$, $y_{\pi_1} \in L$,
  m_1 is some message and $State_1$ is some state information.
 Let $L = \{y_{i_1}, \cdots, y_{i_n}\}$
 $\sigma_1 \leftarrow Sig(1^k, 1^n, x_{\pi_1}, L, m_1)$
 $(\pi_2, State_2) \leftarrow \mathcal{A}^{\mathcal{SO}}(1^k, \mathcal{U}, L, \pi_1, m_1, \sigma_1, State_1, \text{choose2})$
  where $\pi_2 \in \{i_1, \cdots, i_n\} \setminus \{\pi_1\}$ and $State_2$ is some state information.
 $(m_2, \sigma_2) \leftarrow \mathcal{A}^{\mathcal{SO}}(1^k, \mathcal{U}, x_{\pi_2}, L, \pi_1, m_1, \sigma_1, State_2, \text{sign})$
  such that $Ver(1^k, 1^n, L, m_2, \sigma_2) = 1$
 If (L, m_i, σ_i), $1 \leq i \leq 2$, were not in the set of oracle queries and replies
  between $\mathcal{A}$ and $\mathcal{SO}$ then
  return $Link(1^k, 1^n, L, m_1, m_2, \sigma_1, \sigma_2)$
 Else the experiment halts with failure

We denote by

$$\mathbf{Adv}_{\mathcal{A}}^{\text{link}}(k, N) = \Pr[\mathbf{Exp}_{\mathcal{A}}^{\text{link1}}(k, N) = 1 \mid \text{Experiment succeeds}] + \Pr[\mathbf{Exp}_{\mathcal{A}}^{\text{link2}}(k, N) = 1 \mid \text{Experiment succeeds}]$$

the advantage of the adversary $\mathcal{A}$ in breaking the linkability of a linkable ring signature scheme. We say that a linkable ring signature scheme is *linkable* if for a constant N, for any probabilistic polynomial-time adversary $\mathcal{A}$, the function $\mathbf{Adv}_{\mathcal{A}}^{\text{link}}(\cdot, N)$ is negligible.

Note that the definition above does not consider the scenario when two group members are *working jointly* (or someone knows two private keys) to generate three signatures such that *Link* returns 0 for all pairs of the three signatures (i.e. pairwise unlinkable). In fact, this extends the first attack above by considering an adversary who knows two private keys instead of one. The attack can be generalized to a situation where an adversary, who knows k private keys, produces $k+1$ signatures such that they are pairwise unlinkable. We leave further discussions to the end of Sec. 3.2.

3 Polynomial-Structured Schemes

Our construction approach is based on Honest-Verifier Zero-Knowledge (HVZK) proof of knowledge protocols and in particular their signature variants, here we follow Camenisch and Stadler [6] to call these signature schemes *"signatures of knowledge"*. A signature of knowledge is a signature scheme transformed from a HVZK proof by setting the challenge to the hash value of the commitment together with the message to be signed [9]. Here is an example from [6]: Let $G = \langle g \rangle$ be a group of prime order q. Let $\mathcal{H} : \{0,1\}^* \to \mathbb{Z}_q$ be a hash function viewed as a random oracle. The symbol $||$ denotes the concatenation of two binary strings (or of binary representations of integers and group elements). A pair $(s,c) \in \mathbb{Z}_q \times \mathbb{Z}_q$ satisfying $c = \mathcal{H}(m||g||y||g^s y^c)$ is a signature of knowledge of the discrete logarithm x of the element $y \in G$ to the base g on message m. We denote it by

$$SPK[x : y = g^x](m) \qquad (1)$$

Now consider the following signature of knowledge.

Definition 2 (Witness Indistinguishable Proof of Equality of a Discrete Logarithm).

$$SPK[x : y_0 = h^x \wedge (y_1 = g_1^x \vee y_2 = g_2^x)](m) \qquad (2)$$

Given $h, g_1, g_2 \in G$ such that $h \neq g_1$, $h \neq g_2$, and $\log_h g_i$ is unknown for $i = 1, 2$. A signer shows his knowledge of x such that $y_0 = h^x$ and $y_i = g_i^x$ for at least one i, $i = 1, 2$, without revealing the value of i in the case of only one of y_1 and y_2 shares the discrete logarithm to the base g_1 and g_2, respectively, with y_0 to the base h. This can be done by releasing a quadruple (s_1, s_2, c_1, c_2) such that $c_0 = c_1 + c_2 \mod q$ and $c_0 = \mathcal{H}(m||h||g_1||g_2||y_0||y_1||y_2||g_1^{s_1} y_1^{c_1}||h^{s_1} y_0^{c_1}||g_2^{s_2} y_2^{c_2}||h^{s_2} y_0^{c_2})$.

In the following, we describe a generic DL-type linkable ring signature scheme which is based on the generalization of the SPK in (2).

Our Generic Linkable Ring Signature Scheme. Let $(g_i, y_i) \in G \times G$ be the public key of group member i, $1 \leq i \leq n$, and the corresponding private key be $x_i \in_R \mathbb{Z}_q$ such that $y_i = g_i^{x_i}$. Let $L = \{(g_1, y_1), \cdots, (g_n, y_n)\}$. L defines the group. Let $\mathcal{H}' : \{0,1\}^* \to G$ be a hash function viewed as a random oracle. A group member π, knowing x_π, can generate a signature on message m by producing $y_0 = \mathcal{H}'(L)^{x_\pi}$ and a proof of the following.

$$SPK[x_\pi : y_0 = \mathcal{H}'(L)^{x_\pi} \wedge (y_1 = g_1^{x_\pi} \vee \cdots \vee y_n = g_n^{x_\pi})](m) \qquad (3)$$

We now present two realizations of this generic linkable ring signature scheme and show security proofs under the model given in Sec. 2.

3.1 Realization 1: A Practical Scheme

We start with proposing a polynomial-structured realization of (3) for the case when all g_i's are the same. Let $G = \langle g \rangle$ be a cyclic group of prime order q such that the underlying discrete logarithm problem (DLP) is hard. Let $H_1 : \{0,1\}^* \to \mathbb{Z}_q$ and $H_2 : \{0,1\}^* \to G$ be distinct hash functions viewed as random oracles. Assume that for any $\alpha \in \{0,1\}^*$, the discrete-log of $H_2(\alpha)$ to the base g is intractable. For $i = 1, \cdots, n$, user i randomly picks a private key $x_i \in_R \mathbb{Z}_q$ and sets the public key to $y_i = g^{x_i}$. Let $L = \{y_1, \cdots, y_n\}$. Sometimes, we may pass in the set L for hashing and we implicitly assume that certain appropriate encoding method is applied.

Signature Generation. For message $m \in \{0,1\}^*$ and the group defined by L, a signer π, $1 \leq \pi \leq n$, generates a signature $\sigma = (y_0, s_1, \cdots, s_n, c_1, \cdots, c_n)$ as follows.

1. Compute $h \leftarrow H_2(L)$ and set $y_0 \leftarrow h^{x_\pi}$.
2. Find c_π such that
$$c_1 + \cdots + c_\pi + \cdots + c_n \bmod q = H_1(L||y_0||m||z_1'||\cdots||z_n'||z_1''||\cdots||z_n'')$$
where for $i = 1, \cdots, n$, $i \neq \pi$, $s_i, c_i \in_R \mathbb{Z}_q$, $z_i' = g^{s_i} y_i^{c_i}$, $z_i'' = h^{s_i} y_0^{c_i}$, and $r \in_R \mathbb{Z}_q$, $z_\pi' = g^r$, $z_\pi'' = h^r$.
3. Compute $s_\pi = r - c_\pi \cdot x_\pi \bmod q$.

Signature Verification. A signature $\sigma = (y_0, s_1, \cdots, s_n, c_1, \cdots, c_n)$ on message m and public-key set L is valid if
$$\sum_{i=1}^{n} c_i \bmod q = H_1(L||y_0||m||g^{s_1}y_1^{c_1}||\cdots||g^{s_n}y_n^{c_n}||h^{s_1}y_0^{c_1}||\cdots||h^{s_n}y_0^{c_n})$$
where $h \leftarrow H_2(L)$.

Linkability Verification. For any two valid signatures $\sigma' = (y_0', \cdots)$ and $\sigma'' = (y_0'', \cdots)$ with respect to the same L, if $y_0' = y_0''$, then return 1 for concluding that they are generated by the same signer; otherwise, return 0.

We leave the security analysis to the next section after describing a variant.

3.2 Realization 2: A Variant

The scheme above can be optimized by reducing the number of s_i's in the signature σ to one and modifying the checking equation to the following.
$$\sum_{i=1}^{n} c_i \bmod q = H_1(L||y_0||m|| \, g^s \prod_{i=1}^{n} y_i^{c_i} || \, h^s y_0^{\sum_{i=1}^{n} c_i})$$

This can be done during signature generation as follows: First randomly pick $r \in_R \mathbb{Z}_q$ and $c_i \in_R \mathbb{Z}_q$ for $i = 1, \cdots, n$, $i \neq \pi$. Then find c_π such that

$$c_1 + \cdots + c_\pi + \cdots + c_n \bmod q = H_1(L\|y_0\|m\| \; g^r \prod_{i=1, i\neq\pi}^{n} y_i^{c_i} \; \| \; h^r y_0^{\sum_{i=1, i\neq\pi}^{n} c_i})$$

Finally, compute $s = r - c_\pi x_\pi \bmod q$ and the signature is $\sigma = (y_0, s, c_1, \cdots, c_n)$.

The variant can be shown to be existentially unforgeable against adaptive chosen message and chosen public-key attacks by reducing to DLP using rewind simulation [9, 12, 11]. The reduction involves at most N (total number of public keys) successful rewinds to extract one of the N secret keys but in the worst case, all the N secret keys can be extracted at once.

Theorem 1 (Existential Unforgeability). *Given N public keys $y_i \leftarrow g^{x_i}$, $1 \leq i \leq N$, where $x_i \in_R \mathbb{Z}_q$, and supposing $\mathcal{A}$ is a (t, ϵ)-algorithm which takes the public-key set $\mathcal{U} = \{y_i\}_{1 \leq i \leq N}$ and forges a linkable ring signature on any $L \subseteq \mathcal{U}$ in time at most t with success probability at least ϵ, there exists an algorithm $\mathcal{M}$ which solves the DLP for at least one out of N random instances in time at most Nt with success probability at least $(\epsilon/4)^N$.*

The proof is given in the full paper. The proof technique can also be used to show the existential unforgeability of the original, unoptimized scheme described in Sec. 3 under the same model of adaptive chosen message and chosen public-key attacks. We omit the details in this paper.

To argue the signer anonymity of the variant, we consider the following setup for the Decisional Diffie-Hellman Problem (DDHP) first.

Randomly generate $\ell_0, \ell_1, \ell_2, \ell'_0, \ell'_1 \in_R \mathbb{Z}_q$. Randomly pick $b \in_R \{0, 1\}$. Set $\alpha_0 = g^{\ell_0}$, $\beta_0 = g^{\ell_1}$, $\gamma_0 = g^{\ell_2}$, $\alpha_1 = g^{\ell'_0}$, $\beta_1 = g^{\ell'_1}$ and $\gamma_1 = g^{\ell'_0 \ell'_1}$. Given $(\alpha_b, \beta_b, \gamma_b)$, find b.

Theorem 2 (Signer Anonymity). *Given N public keys $y_i \leftarrow g^{x_i}$, $1 \leq i \leq N$, where $x_i \in_R \mathbb{Z}_q$, and suppose there exists a (T, ϵ, q_{H_2})-algorithm $\mathcal{A}$ which runs in time at most T and makes at most q_{H_2} queries to H_2 random oracle in the experiment $\mathbf{Exp}_\mathcal{A}^{\text{anon}}(k, N)$ for some $k \in \mathbb{N}$ such that T is a polynomial of k and $\mathbf{Adv}_\mathcal{A}^{\text{anon}}(k, N)$ is at least ϵ, there exists an algorithm $\mathcal{M}$ which solves the Decisional Diffie-Hellman Problem (DDHP) in the expected time of at most $Nq_{H_2}T$ with success probability at least $1/2 + \epsilon/4$.*

The proof is given in the full paper. Our original, unoptimized scheme above can also be shown, in the same way, to be signer anonymous in $\mathbf{Exp}_\mathcal{A}^{\text{anon}}$, provided that the DDHP is intractable. We skip the details.

Theorem 3 (Linkability). *On input $L = \{y_1, \cdots, y_n\}$, $h \in G$, if a probabilistic polynomial-time algorithm (PPT) $\mathcal{A}$ who knows one private key x_π, $\pi \in \{1, \cdots, n\}$, produces a signature $\sigma = (y_0, s, c_1, \cdots, c_n)$ with success probability $> 1/q$, then $y_0 = h^{x_\pi}$ provided that the DLP is hard.*

The proof is given in the full paper. This theorem implies that the variant is secure against the two linkability attacks described in $\mathbf{Exp}_\mathcal{A}^{link1}$ and $\mathbf{Exp}_\mathcal{A}^{link2}$ in Sec. 2.1. However, it does not consider the attack of having two group members *work jointly* to generate three signatures such that they are pairwise unlinkable. In fact, it is easy to find that if two private keys corresponding to public keys, say y_1 and y_2 in L are given, one can generate a signature $\sigma = (y_0, s, c_1, \cdots, c_n)$ such that $y_0 \neq h^{x_1} \neq h^{x_2}$. Hence it is possible for an adversary who knows two private keys to generate three signatures such that they are pairwise unlinkable. This is the third type of linkability attacks described in Sec. 2.1. If this attack is included in the security model, the variant is no longer secure. For the original, unoptimized scheme, it can be shown that if a PPT algorithm who knows all the private keys corresponding to L, produces a signature $\sigma = (y_0, s_1, \cdots, s_n, c_1, \cdots, c_n)$ with success probability $> 1/q$, then $y_0 = h^{x_\pi}$ for some $\pi \in \{1, \cdots, n\}$ provided that the DLP is hard. Hence the original scheme is secure against this attack. We omit the details as it can be shown easily by following the proof for Theorem 3.

4 Conclusions and Open Problems

We provide a stronger notion of signer anonymity and a new formalization of linkability in the security model when comparing to that of [10]. We propose two polynomial-structured linkable ring signature schemes. A subtlety on the linkability between these two schemes is also discussed. We give strong evidence of the security of our schemes by providing security proofs in our enhanced security model under the random oracle model.

Besides linkability, we can see that all linkable ring signature schemes, including ours proposed in this paper and the one in [10], allow diversion group members (i.e. group members who are not the actual signer) to come forward and repudiate signatures not signed by them. This can be done by revealing their private keys (this is referred as a property called *culpability* in [10]), or using standard proof-of-knowledge techniques. It would be interesting to have a linkable ring signature scheme where diversion group members cannot repudiate the signature as this could give a better protection on the privacy of the whistleblower.

References

1. M. Abe, M. Ohkubo, and K. Suzuki. 1-out-of-n signatures from a variety of keys. In *Proc. ASIACRYPT 2002*, pages 415–432. Springer-Verlag, 2002. Lecture Notes in Computer Science No. 2501.
2. M. Bellare, D. Micciancio, and B. Warinschi. Foundations of group signatures: Formal definitions, simplified requirements, and a construction based on general assumptions. In *Proc. EUROCRYPT 2003*, pages 614–629. Springer-Verlag, 2003. Lecture Notes in Computer Science No. 2656.
3. M. Bellare and P. Rogaway. Random oracles are practical: A paradigm for designing efficient protocols. In *Proc. 1st ACM Conference on Computer and Communications Security*, pages 62–73. ACM Press, 1993.

4. D. Boneh, C. Gentry, B. Lynn, and H. Shacham. Aggregate and verifiably encrypted signatures from bilinear maps. In *Proc. EUROCRYPT 2003*, pages 416–432. Springer-Verlag, 2003. Lecture Notes in Computer Science No. 2656.
5. E. Bresson, J. Stern, and M. Szydlo. Threshold ring signatures and applications to ad-hoc groups. In *Proc. CRYPTO 2002*, pages 465–480. Springer-Verlag, 2002. Lecture Notes in Computer Science No. 2442.
6. J. Camenisch and M. Stadler. Efficient group signature schemes for large groups. In *Proc. CRYPTO 97*, pages 410–424. Springer-Verlag, 1997. Lecture Notes in Computer Science No. 1294.
7. D. Chaum and E. Van Heyst. Group signatures. In *Proc. EUROCRYPT 91*, pages 257–265. Springer-Verlag, 1991. Lecture Notes in Computer Science No. 547.
8. Y. Dodis, A. Kiayias, A. Nicolosi, and V. Shoup. Anonymous identification in ad-hoc groups. In *Proc. EUROCRYPT 2004*. Springer-Verlag, 2004.
9. A. Fiat and A. Shamir. How to prove yourself: Practical solutions to identification and signature problems. In *Proc. CRYPTO 86*, pages 186–199. Springer-Verlag, 1987. Lecture Notes in Computer Science No. 263.
10. J. K. Liu, V. K. Wei, and D. S. Wong. Linkable and anonymous signature for ad hoc groups. In *The 9th Australasian Conference on Information Security and Privacy (ACISP 2004)*, pages 325–335. Springer-Verlag, 2004. Lecture Notes in Computer Science No. 3108.
11. K. Ohta and T. Okamoto. On concrete security treatment of signatures derived from identification. In *Proc. CRYPTO 98*, pages 354–369. Springer-Verlag, 1998. Lecture Notes in Computer Science No. 1462.
12. D. Pointcheval and J. Stern. Security proofs for signature schemes. In *Proc. EUROCRYPT 96*, pages 387–398. Springer-Verlag, 1996. Lecture Notes in Computer Science No. 1070.
13. R. Rivest, A. Shamir, and Y. Tauman. How to leak a secret. In *Proc. ASIACRYPT 2001*, pages 552–565. Springer-Verlag, 2001. Lecture Notes in Computer Science No. 2248.

Practical Scenarios for the Van Trung-Martirosyan Codes

Marcel Fernandez, Miguel Soriano, and Josep Cotrina*

Department of Telematics Engineering, Universitat Politècnica de Catalunya,
C/ Jordi Girona 1 i 3, Campus Nord, Mod C3, UPC, 08034 Barcelona, Spain
{marcelf, soriano, jcotrina}@mat.upc.es

Abstract. Traitor tracing schemes are used in to detect piracy in broadcast encryption systems, in case that a bounded number of authorized users are dishonest. In this paper we present, by solving a variant of the guessing secrets problem defined by Chung, Graham and Leighton [3], a traitor tracing scheme based in the recently discovered van Trung-Martirosyan traceability codes.

1 Introduction

In the original "I've got a secret" TV game show [6], a contestant with a secret was questioned by four panelists. The questions were directed towards guessing the secret. A prize money was given to the contestant if the secret could not be guessed by the panel.

A variant of the game called "guessing secrets" was defined by Chung, Graham and Leighton in [3]. In their variant , there are two players **A** and **B**. Player **A** draws a subset of $c \geq 2$ secrets from a set S of N objects. Player **B** asks a series of boolean (binary) questions. For each question asked **A** can adversarially choose a secret among the c secrets, but once the choice is made he must answer truthfully. The goal of player **B** is to come up with an strategy, that using as few questions as possible allows him to unveil the secrets.

The problem of guessing secrets is related to several topics in computer science such as efficient delivery of Internet content [3] and the construction of schemes for the copyright protection of digital data [1]. In [1] Alon, Guruswami, Kaufman and Sudan realized there was a connection between the guessing secrets problem and error correcting codes. Using this connection they provided a solution to the guessing secrets problem equipped with an efficient algorithm to recover the secrets.

In this paper we modify the condition established by Chung, Graham and Leighton, and allow questions over a larger alphabet. We call this modified version the q-ary guessing secrets problem. We show below that this version of the

---

* This work has been supported in part by the Spanish Research Council (CICYT) Project TIC2002-00818 (DISQET) and TIC2003-01748 (RUBI) and by the IST-FP6 Project 506926 (UBISEC).

problem will lead us to a class of codes called *traceability* (TA) codes, in particular to the van Trung-Martirosyan code [11], that in turn leads to the *traitor tracing* problem [2].

The paper is organized as follows. Section 2 introduces the coding theory concepts that will be used. In Section 3, a description of the game of guessing secrets is presented. In Section 4, we present a new tracing algorithm. Finally, in Section 5 we show the equivalence between the guessing secrets problem and the tracing traitors problem.

2 Overview of Coding Theory Concepts

If C is a set of vectors of a vector space, $\mathbb{F}_q^n$, then C is called a *code*. The field, $\mathbb{F}_q$ is called the *code alphabet*. A code C with length n, size $|C| = M$, and alphabet $\mathbb{F}_q$ is denoted as a (n, M) q-ary code. We will also denote code C as a (n, M, q) code. The *Hamming distance* $\mathbf{d}(\mathbf{a}, \mathbf{b})$ between two words $\mathbf{a}, \mathbf{b} \in \mathbb{F}_q^n$ is the number of positions where $\mathbf{a}$ and $\mathbf{b}$ differ. The *minimum distance* of C, denoted by d, is defined as the smallest distance between two different codewords. A code C is a *linear code* if it forms a subspace of $\mathbb{F}_q^n$. A code with length n, dimension k and minimum distance d is denoted as a $[n, k, d]$-code.

We use the terminology in [10] to describe *identifiable parent property* (IPP) codes and *traceability* (TA) codes. Let $U \subseteq C$ be any subset of codewords such that $|U| = c$. The set of *descendants* of U, denoted $\mathbf{desc}(U)$, is defined as

$$\mathbf{desc}(U) = \{\mathbf{v} \in \mathbb{F}_q^n : v_i \in \{a_i : \mathbf{a} \in U\}, 1 \leq i \leq n\}.$$

The codewords in U are called *parents* of the words in $\mathbf{desc}(U)$.

For a code C and an integer $c \geq 2$, let $U_i \subseteq C$, $i = 1, 2, \ldots, t$ be all the subsets of C such that $|U_i| \leq c$. A code C is a c-IPP (identifiable parent property) code, if for every $\mathbf{z} \in \mathrm{desc}(C)$, we have that $\bigcap_{\{i:\mathbf{z}\in\mathrm{desc}(U_i)\}} U_i \neq \emptyset$, in other words, C is a c-IPP code if the intersection of all possible sets of parents U_i of $\mathbf{z}$ is non-empty.

An important subclass of IPP codes are *traceability* (TA) codes. For $\mathbf{x}, \mathbf{y} \in \mathbb{F}_q^n$ we can define the set of *matching positions* between $\mathbf{x}$ and $\mathbf{y}$ as $M(\mathbf{x}, \mathbf{y}) = \{i : x_i = y_i\}$. Let C be a code, then C is a c-TA code if for all i and for all $\mathbf{z} \in \mathrm{desc}_c(U_i)$ $|U_i| \leq c$, there is at least one codeword $\mathbf{u} \in U_i$ such that $|M(\mathbf{z}, \mathbf{u})| > |M(\mathbf{z}, \mathbf{v})|$ for any $\mathbf{v} \in C \backslash U_i$.

Theorem 1 ([10]). *Let C be a Reed-Solomon $[n,k,d]$-code, if $d > n(1 - 1/c^2)$ then C is a c-traceability code(c-TA).*

2.1 Asymptotically Good TA Codes

We now present an "asymptotically good" family of TA codes due to van Trung and Martirosyan [11]. Their construction is based on IPP code concatenation.

A concatenated code is the combination of an *inner* $[n_i, k_i, d_i]$ q_i-ary code, C_{inn}, ($q_i \geq 2$) with an *outter* $[n_o, k_o, d_o]$ code, C_{out} over the field $\mathbf{F}_{q_i^{k_i}}$. The combination consists in a mapping ϕ, from the elements of $\mathbf{F}_{q_i^{k_i}}$ to the codewords of the inner code C_{inn}, $\phi : \mathbf{F}_{q_i^{k_i}} \to C_{inn}$ that results in a q_i-ary code of length $n_i n_o$ and dimension $k_i k_o$.

Theorem 2. *[11] Let $c \geq 2$ be an integer. Let $n_0 > c^2$ be an integer and let s_0 be an integer with the prime factorization $s_0 = p_1^{e_1} \cdots p_k^{e_k}$ such that $n_0 \leq p_i^{e_i}$ for all $i = 1, \ldots, k$. Then, for all $h \geq 0$ there exists an (n_h, M_h, s_0) c-IPP code, where*

$$n_h = n_{h-1} n_{h-1}^*, M_h = M_{h-1}^{\lceil \frac{n_h^* - 1}{c^2} \rceil}, n_{h-1}^* = n_{h-2}^* \lceil \frac{n_h^* - 2}{c^2} \rceil, M_0 = s_0^{\lceil \frac{n_0}{c^2} \rceil}, n_0^* = n_0^{\lceil \frac{n_0}{c^2} \rceil}.$$

As stated in [11], the codes in Theorem 2 have the best known asymptotic behavior in the literature. Also the results of Theorem 2 can be applied to TA-codes, if the IPP codes used in the recursion are replaced by TA-codes.

2.2 The Guruswami-Sudan Soft-Decision List Decoding Algorithm

The concept of *list decoding* offers a potential way to recover from errors beyond the error correction bound of the code, by allowing the decoder to output, instead of a single codewords, a list of candidate codewords of being the sent codeword.

In *soft-decision* decoding, the decoding process takes advantage of "side information" generated by the receiver by using probabilistic reliability information about the received symbols. The simplest form of soft-decision decoding is called *errors-and-erasures* decoding. An erasure is an indication that the value of a received symbol is in doubt. In this case, when dealing with a q-ary transmission, the decoder has $(q+1)$ output alternatives: the q symbols from $\mathbb{F}_q$, $\gamma_1, \gamma_2, \ldots, \gamma_q$ and $\{*\}$, where the symbol $\{*\}$ denotes an erasure.

Next theorem gives the condition that a codeword must satisfy in order to appear in the output list of the Guruswami-Sudan (GS) algorithm.

Theorem 3. *[7] Consider an $[n, k, n-k+1]$ Reed-Solomon code with messages being polynomials f over $\mathbb{F}_q$ of degree at most $k-1$. Let the encoding function be $f \mapsto \langle f(x_1), f(x_2), \ldots, f(x_n) \rangle$ where $x_1, \ldots, x_n$ are distinct elements of $\mathbb{F}_q$. Let $\epsilon > 0$ be an arbitrary constant. For $1 \leq i \leq n$ and $\alpha \in \mathbb{F}_q$, let $r_{i,\alpha}$ be a non-negative rational number. Then, there exists a deterministic algorithm with runtime polynomial in n, q and $1/\epsilon$ that, when given as input the weights $r_{i,\alpha}$ for $1 \leq i \leq n$ and $\alpha \in \mathbb{F}_q$, finds a list of all polynomials $p(x) \in \mathbb{F}_q[x]$ of degree at most $k-1$ that satisfy*

$$\sum_{i=1}^n r_{i,p(x_i)} \geq \sqrt{(k-1) \sum_{i=1}^n \sum_{\alpha \in \mathbb{F}_q} r_{i,\alpha}^2 + \epsilon \max_{i,\alpha} r_{i,\alpha}} \qquad (1)$$

3 The q-Ary Guessing Secrets Problem

In this section, we discuss our version of the guessing secrets problem. The best asymptotic solution to the problem will interestingly lead us to the van Trung-Martirosyan codes.

In the q-ary guessing secrets game, there are two players **A** and **B** and a set S of N objects known to both players. Player **A** draws a subset of c objects, $s_1, \ldots, s_c$ $c \geq 2$ from S. Since $s_1, \ldots, s_c$ are not known by **B** we will call them secrets. Now player **B** has to guess the secrets by asking questions whose answers are from a q-ary alphabet that, without loss of generality, we take to be the finite field $\mathbb{F}_q$. For each question asked **A** can adversarially choose a secret among the c secrets, but once the choice is made he must answer truthfully.

The series of questions that **B** asks will be called a *strategy*. We will impose the requirement that any strategy that **B** uses must be *invertible*, that is, given the sequence of answers given by **A** there exists an efficient algorithm to recover the secrets.

First of all, we observe that guessing all the secrets is a very strong requirement for player **B**. In fact, in the worst case situation, the best that **B** can hope for is to guess at most one of the c secrets. This is because, without breaking any rule, **A** can always answer according to the same secret. So the requirement we impose on **B** is that he should be able to recover *at least one* of the secrets chosen by **A**.

We model each question **B** asks as a function p_i from the pool of objects to the set of potential answers, $p_i : S \to \mathbb{F}_q$. We denote the number of questions that **B** asks by n. For every question p_i, $1 \leq i \leq n$ each object **u** in the pool has an associated sequence of answers $\mathbf{s} = (s_1, \ldots, s_n)$, where $s_i := p_i(\mathbf{u})$, $s_i \in \mathbb{F}_q$. We call the sequence **p** of functions p_i, $1 \leq i \leq n$ **B**'s strategy. A strategy will solve the q-ary guessing secrets problem, if from the answers to the p_i questions we are able to "reduce" all possible sets of c secrets down to one of the c secrets chosen by **A**.

Since every object in S can be represented unambiguously by a q-ary vector of length $\lceil \log_q |S| \rceil$, we can define a mapping from the universe of objects to the sequences of answers, $\mathcal{C} : \mathbb{F}_q^{\lceil \log_q |S| \rceil} \to \mathbb{F}_q^n$. The mapping $\mathcal{C}$ illustrates the connection between strategies and codes, since $\mathcal{C}$ maps q-ary sequences of length $\lceil \log_q |S| \rceil$ into sequences of length n. Since in general $n > \lceil \log_q |S| \rceil$, then the mapping $\mathcal{C}$ adds redundancy or encodes. Note that this reasoning allows us to refer to an strategy using its associated code. In the rest of the paper, we will refer to a given strategy by its associated code $\mathcal{C}$.

We will now infer the parameters of such codes. The length of the code will be equal to the number of questions n. Since **A** keeps c secrets $U = \{\mathbf{s}^1, \ldots, \mathbf{s}^c\}$, then no matter how **A** plays, at least $\lceil n/c \rceil$ of the answers will correspond to one of the secrets, say $\mathbf{s}^j$. To guarantee the identification of secret $\mathbf{s}^j$, we have to ensure that any sequence of answers given by **A** about the secrets in U only agrees with at most $\lceil n/c \rceil - 1$ of the corresponding answers associated with any of the secrets not in U.

More precisely, let us denote by $\mathbf{d}(\mathbf{u}, \mathbf{v})$ the number of answers that are different between objects **u** and **v**. Let us also define $d = \min_{(\mathbf{u},\mathbf{v}) \in S} \mathbf{d}(\mathbf{u}, \mathbf{v})$.

Using this notation, if we want to be able to recover $\mathbf{s}^j \in U$ (that as we said above is a secret from which at least $\lceil n/c \rceil$ of the answers are given) we need to ensure that for all $\mathbf{x} \notin U$ $\sum_{\mathbf{s}_i \in U}(n - \mathbf{d}(\mathbf{s}_i, \mathbf{x})) \leq c(n - d) < \frac{n}{c} \leq \lceil \frac{n}{c} \rceil$ so $d > n - \frac{n}{c^2}$.

We observe that the parameter d, indicates the minimum distance of the code associated with the strategy, then according to Theorem 1 any code associated with a strategy must be a TA code. Using the results in Section 2.1, we can assert that the van Trung-Martirosyan family of codes, provide the best known solution for the questions to be asked in the q-ary guessing secrets problem.

4 Recovering Secrets in the Van Trung-Martirosyan Code

We have seen in the previous section, that van Trung-Martirosyan codes provide the best known asymptotic solution to the q-ary guessing secrets problem. But in the guessing secrets problem, besides providing a "good" strategy for the questions, there is also the requirement of being able to recover the secrets given the sequence of answers in an efficient manner, in other words, the strategy must be invertible.

As we pointed out in Section 3 for the game of guessing secrets, we cannot expect to find all secrets, since some of them may correspond to a not sufficiently large number of answers and therefore cannot be recovered. So given a sequence of answers, we call any secret kept by **A** in an unambiguous way a *positive secret*. The sequence of answers associated to a given secret corresponds to a codeword of a c-TA code, using this correspondence the condition for a codeword to be a positive secret is given in Theorem 4 below.

Theorem 4. *Let C be a c-TA (n, M) q-ary Reed-Solomon code with minimum distance d, if a codeword agrees in at least $c(k-1)+1$ positions with a given sequence of answers then this codeword must be a positive secret.*

Proof. If the code has minimum distance d then two codewords can agree in at most $n-d$ positions, therefore a given sequence of answers can agree in at most $c(n-d)$ positions with a codeword that is not associated with any of the secrets kept by **A**. Then any codeword that agrees with the sequence of answers in at least $c(n-d)+1$ positions is a positive secret. The theorem follows from the fact that for Reed-Solomon codes $d = n - k + 1$. ∎

Corollary 1. *Let C be a c-TA (n, M) q-ary Reed-Solomon code with minimum distance d. Let $\mathbf{z}$ be a sequence of answers. Suppose that j already identified positive secrets ($j < c$) jointly match less than $n - (c-j)(k-1)$ positions of $\mathbf{z}$, then any codeword that agrees with $\mathbf{z}$ in at least $(c-j)(k-1)+1$ of the unmatched positions is also a positive secret.*

Using the above results we are in the position to present a secret recovering algorithm when the questions are asked according to a van Trung-Martirosyan code.

If we recall the code construction from Theorem 2, we started with codes:

$C_0 : (n_0, M_0, s_0)$ c-TA code with $M_0 = s_0^{\lceil \frac{n_0}{c^2} \rceil}$, and

$C_1^* : (n_0^*, M_1, M_0)$ c-TA code with $n_0^* = n_0^{\lceil \frac{n_0}{c^2} \rceil}$ and $M_1 = M_0^{\lceil \frac{n_0^2}{c^2} \rceil}$.

Denoting code concatenation with the symbol $||$, we have the following sequence of codes: $C_1 = C_0 || C_1^*$; $C_2 = C_1 || C_2^*$; ... ; $C_h = C_{h-1} || C_h^*$, where the C_j are the inner codes, the C_k^* are the outter codes, and each C_k^* is an $(n_{k-1}^*, M_k, M_{k-1})$ c-TA code.

Due to the recursive nature of the code, the decoding will be done in two stages. In the first stage we will need to decode the code C_0, this will be accomplished by an algorithm that we call **Decoding_C0**.

Since at the start there is no side information at all, we set up the weights $r_{s,\alpha}$ of (1) supposing that all symbols in the descendant are "correct". We apply these weights to the GS algorithm. Note that at the output of the GS algorithm we can identify at least one positive secret.

Once some positive secrets are identified, the algorithm computes the number of remaining secrets to be found. Also all symbol positions where these already identified secrets match the sequence of answers are erased. Then we set up the weights again and make another run of the GS algorithm to see if any other positive secrets can be identified. This step is repeated until it becomes clear that there are no more positive secrets.

In the following algorithm we consider the ordering $\{\alpha_1, \alpha_2, \ldots, \alpha_q\}$ of the elements of the field $\mathbb{F}_q$.

Decoding_C0(C,z):
Input: C: Reed-Solomon c-TA code of length n; Word $\mathbf{z} \in \mathbf{desc}_c(U)$, with $U \subset C$ and $|U| \leq c$.
Output: A list L_l of all positive parents of $\mathbf{z}$.

1. Set $i := 1$, $c_i := c$ and $E_i := \{\emptyset\}$.
2. $j := 0$.
3. Using the descendant $\mathbf{z}$, compute the $n \times q$ weights $r_{s,\alpha}$, $1 \leq s \leq n$, $\alpha \in \mathbb{F}_q$ as follows:
$$r_{s,\alpha} := \begin{cases} 1/q & \text{if } s \in E_i \\ 1 & \text{if } z_s = \alpha \text{ and } s \notin E_i \\ 0 & \text{otherwise} \end{cases} \quad (2)$$
4. Apply the $n \times q$ values $r_{s,\alpha}$ to the GS soft-decision algorithm. From the output list take all codewords $\mathbf{u}_{i_1}, \ldots, \mathbf{u}_{i_{j_w}}$, that agree with $\mathbf{z}$ in at least $(c_i(k-1)+1)$ of the positions not in E_i, and add them to L_l.
Set $j := j + j_w$.
5. If $j_w \neq 0$ then
   (a) $E_i := \{m : (z_m = u_m) \; \forall \; \mathbf{u} \in L_l\}$.
   (b) Go to Step 3
6. Set $i := i+1$, $c_i := c_{i-1} - j$ and
$E_i = \{m : (z_m = u_m) \; \forall \; \mathbf{u} \in L_l\}$.
7. If $j = 0$ or $c_i = 0$ or if $|E_i| \geq (n - c_i(k-1))$ output L_l and quit, else go to Step 2.

In the second stage of the recovering algorithm we will decode the code C_h by first decoding codes $C_1^*, \ldots, C_{h-1}^*$. To decode code C_i^*, we will use the function **Decoding_Ci**. This function has the particularity that instead of accepting a codeword at its input, it accepts a set of lists of symbols from the alphabet code. These lists can be processed by using soft-decision decoding techniques, and this is were our advantage comes from, being able to deal with more that one symbol for each position extends the tracing capabilities of previous hard-decision decoding algorithms.

In the following function we consider the ordering $\{\alpha_1, \alpha_2, \ldots, \alpha_{|M_{i-1}|}\}$ of the elements of the alphabet M_{i-1}.

Decoding_Ci(C_i^*, $S\_list_1, \ldots, S\_list_{n_{i-1}^*}$):
Input: C_i^*: Reed-Solomon $(n_{i-1}^*, M_i\ M_{i-1})$-ary c-traceability code; n_{i-1}^* lists, $S\_list_l = \{m_1^l, \ldots, m_{|S\_list_l|}^l\}$ $1 \leq l \leq n_{i-1}^*$ where $m_i^l \in M_{i-1}$.
Output: A list O_list of codewords of C_i^*.

1. Set $j := 0$, $i := 0$ and $c_i := c$.
2. Using the lists $S\_list_1, \ldots, S\_list_{n_{h-1}^*}$ set up the $r_{s,\alpha}$ $1 \leq s \leq n_{h-1}^*$, $\alpha \in M_{i-1}$) weights as follows:

$$r_{s,\alpha} := \begin{cases} \frac{1}{|S\_list_l|} & \text{if } \exists\ m_t^l = \alpha \\ 1/q & \text{if } |S\_list_l| = 0 \\ 0 & \text{otherwise} \end{cases} \qquad (3)$$

3. Apply the $r_{s,\alpha}$ weights to the GS soft-decision algorithm. From the output list take all codewords $\mathbf{u}^j$, such that $u_l^j \in S\_list_l$ for at least $(c_i(k-1)+1)$ values of l, and add them to O_list.
4. Set $i := i+1$, $c_i := c_{i-1} - j$ and
   $S\_list_l := S\_list_l - \{m_i^l : (m_i^l = u_l^j)$ for some $\mathbf{u}^j \in O\_list\}$.
5. If $j = 0$ or $c_i = 0$ output O_list and quit, else go to step 2.

The overall tracing algorithm uses the algorithms **Decoding_C0** and **Decoding_Ci** to identify the secrets $U \subset C_h$ associated with a sequence of answers $\mathbf{z}$.

Tracing Algorithm:
Input: c: positive integer; C_h: Reed-Solomon c-TA van Trung-Martirosyan code; Sequence of answers $\mathbf{z} \in \operatorname{desc}_c(U)$, with $U \subset C_h$ and $|U| \leq c$.
Output: A list L_l of all positive secrets of $\mathbf{z}$.

1. For $cont = 1$ to $\prod_{k=1}^{h} n_{h-k}^*$.
   - Take symbols $\mathbf{z}_{cont} = (z_{(cont-1)n_0+1}, \ldots, z_{(cont)n_0})$
   - $Out_PList_{cont} := $ **Decoding_C0**$(C_0, \mathbf{z}_{cont})$
2. Set $j := h - 1$.
3. For $cont' := 1$ to $\prod_{k=1}^{j} n_{h-k}^*$. ($\prod_{k=1}^{0} n_{h-k}^* := 1$)
   - Set $PList_{(cont'-1)n_{h-1-j}^*+1} := Out\_PList_{(cont'-1)n_{h-1-j}^*+1}, \ldots,$
     $PList_{(cont')n_{h-1-j}^*} := Out\_PList_{(cont')n_{h-1-j}^*}$
   - With the lists $PList_{(cont'-1)n_{h-1-j}^*+1}, \ldots, PList_{(cont')n_{h-1-j}^*}$,
     use the mapping $\phi_{h-j} : M_{h-1-j} \to C_{h-1-j}$
     to obtain the lists of symbols $SL_{(cont'-1)n_{h-1-j}^*+1}, \ldots, SL_{(cont')n_{h-1-j}^*}$,
     where $SL_l = \{h_1, \ldots, h_{|SL_l|}\}$, $h_i \in M_{j-1}$.
   - $Out_PList_{cont'} := $ **Decoding_Ci**$(C_{h-j}^*,$
     $SL_{(cont'-1)n_{h-1-j}^*+1}, \ldots, SL_{(cont')n_{h-1-j}^*})$

4. Set $j := j - 1$.
5. If $j < 0$ output Out_PList_1 (there is only one "surviving" list) else go to Step 3.

Note that since for the code concatenation $C_h = C_{h-1} || C_h^*$, the size of codes C_h and C_h^* is the same, we output the secrets as codewords of the code C_h^*.

4.1 Analysis and Correctness of the Algorithm

For c-IPP codes the runtime complexity of the tracing algorithm is in general $O(\binom{M}{c})$, whereas for c-traceability codes this complexity is in general $O(M)$, where M is the size of the code. This is where the advantage of c-traceability codes over c-IPP codes comes from. In the van Trung-Martirosyan construction the code C_0 and all C_i^* codes are c-traceability codes, this implies that there exists a tracing algorithm with running time complexity $O(M)$. We achieve the running time $poly(\log M)$ promised in [11], by using the GS algorithm that runs in time polynomial.

To prove the correctness of the algorithm we need to show that given a sequence of answers corresponding to questions asked according to the recursive van Trung-Martirosyan code, there is at least one of the secrets present in the output list Out_PList_1. To do this, it suffices to show that both of the algorithms **Decoding_C0** and **Decoding_Ci** identify at least one secret.

To show that the algorithm **Decoding_C0** outputs at least one of the secrets, we consider the worst case situation, which is when there are still $c - j$ secrets unidentified, and each secret contributes equally to the sequence of answers and with the minimum amount of information that allows their identification, so there are $M_i = n - (c - j)[(c - j)(k - 1) + 1]$ erased positions. Then it is clear that at least a positive secret **u** agrees with the sequence of answers in at least $(c - j)(k - 1) + 1$ positions. Setting up the weights $r_{s,\alpha}$ $1 \leq s \leq n$, $\alpha \in \mathbb{F}_q$ as in (2), we have for secret **u** that

$$\sum_{s=1}^{n} r_{s,u_s} = (c-j)(k-1) + 1 + \frac{n - (c-j)[(c-j)(k-1)+1]}{q}$$

$$\sum_{s=1}^{n} \sum_{\alpha \in \mathbb{F}_q} r_{s,\alpha}^2 = (c-j)[(c-j)(k-1)+1] + \frac{n - (c-j)[(c-j)(k-1)+1]}{q}$$

Therefore $\left(\dfrac{\sum_{s=1}^{n} r_{s,u_s}}{\sqrt{\sum_{s=1}^{n} \sum_{\alpha \in \mathbb{F}_q} r_{s,\alpha}^2}} \right)^2 \geq \dfrac{((c-j)(k-1)+1)^2}{(c-j)((c-j)(k-1)+1)} > k - 1$

It follows that (1) is satisfied and so the algorithm outputs all positive secrets.

For the **Decoding_Ci** algorithm, we have that the worst case situation is when in the input lists there is the minimum information required to trace the remaining unidentified positive secrets. If there are $c - j$ unidentified secrets, this worst case situation is clearly the one in which there are $(c - j)(k - 1) + 1$

lists of size $c-j$ and $n-[(c-j)(k-1)+1]$ empty lists. Since we set the entries of the weights $r_{s,\alpha}$ $1 \le s \le n^*_{h-1}$, $\alpha \in M_{i-1}$ according to (3), and we have $\sum_{s=1}^{n} \sum_{\alpha \in M_{i-1}} r^2_{s,\alpha} = \frac{(c-j)(k-1)+1}{c-j} + \frac{n-[(c-j)(k-1)+1]}{M_{i-1}}$.

Since a positive secret, say **u** must contribute in at least $(c-j)(k-1)+1$ of the positions in the sequence of answers, we have that $\sum_{s=1}^{n} r_{s,u_s} = \frac{(c-j)(k-1)+1}{c-j} + \frac{n-[(c-j)(k-1)+1]}{M_{i-1}}$. It follows that

$$\frac{\sum_{s=1}^{n} r_{s,u_s}}{\sqrt{\sum_{s=1}^{n} \sum_{\alpha \in \mathbb{F}_q} r^2_{s,\alpha}}} = \sqrt{(k-1) + \frac{1}{c-j} + \frac{n-[(c-j)(k-1)+1]}{M_{i-1}}}$$

so again (1) is satisfied and therefore **Decoding_Ci** identifies all positive secrets for all codes C_i^*.

5 Tracing Traitors by Guessing Secrets

In this section we show how our solution to the guessing secrets problem fits into the problem of "tracing traitors" [2, 8, 5].

Traitor tracing schemes are generally used in the context of broadcast encryption systems [4]. Broadcast encryption systems allow the delivery of encrypted messages to a selected group of registered users. Each registered user in the selected group owns a decoder, equipped with a set of keys, that allows him to recover the encrypted messages.

A traitor tracing scheme consists of an initialization scheme, an encryption-decryption scheme and an identification, or tracing algorithm. In the initialization scheme each one of the M registered users is assigned a personal key u. This personal key is a unique ordered set of n symbols over an alphabet of size q. The sensitive data is encrypted and transmitted to the users over a channel. The encrypted message consists of two parts, the enabling block and the sensitive data. The enabling block contains a secret key $s = \sum s_i, 1 \le i \le n$ encrypted in such a form that every registered user is able recover it using his personal key.

The enabling block is an $n \times q$ matrix such that the part s_i of the secret key, is encrypted under every element of the ith row. A given user is allowed to access one element in each row. This element is the one in the column corresponding to the ith entry in the user's personal key.

In order to use the system to their own benefit, a coalition of c malicious users might try to build and resell pirate decoders that are able to decode the secret key. The uniqueness of each personal key clearly rules out plain redistribution. The real threat comes from the fact that the coalition can put together parts of their personal keys to create a pirate decoder. In this case the goal of the distributor is to find at least one of the participants of the coalition.

5.1 Tracing Traitors by Guessing Secrets

We now show the relationship between the guessing secrets and traitor tracing problems. We first need to establish a relationship between the set of registered users and the set of secrets. In fact, the relationship should be established between the set of personal keys P and the set of answers to the secrets S. Without loss of generality, we assume that both sets have the same size, $|S| = |P| = M$, and that the length of the keys is equal to the length of the sequences of answers associated with each secret.

So given a set of c secrets and a set of c personal keys, the process of choosing one of the c participants in the collusion and placing one of its symbols in the pirate decoder, is analogous to the process of choosing a secret among the set of c secrets and answering the corresponding question. Therefore, any code that solves the q-ary guessing secrets problem also solves the traitor tracing problem.

6 Conclusions

The importance of the q-ary guessing secrets problem lies in its connection to the traitor tracing problem. As pointed out in [9] traceability schemes are a worth addition to a system provided its associated algorithms provide sufficiently little cost. The focus of this paper is on the efficient decoding traceability codes. In particular we apply soft-decision decoding techniques to decode the van Trung-Martirosyan code construction.

References

1. N. Alon, V. Guruswami, T. Kaufman, and M. Sudan. Guessing secrets efficiently via list decoding. *ACM, SODA 02*, pages 254–262, 2002.
2. B. Chor, A. Fiat, and M. Naor. Tracing traitors. *Advances in Cryptology-Crypto'94*, LNCS, 839:480–491, 1994.
3. F. Chung, R. Graham, and T. Leighton. Guessing secrets. *The Electronic Journal of Combinatorics*, 8(1), 2001.
4. A. Fiat and M. Naor. Broadcast encryption. *Advances in Cryptology - CRYPTO '93*, LNCS, 773:175 ff., 1994.
5. A. Fiat and T. Tassa. Dynamic traitor tracing. *Advances in Cryptology-Crypto 1999*, LNCS, 1666:354–371, 1999.
6. I've got a secret. A classic '50's and '60's television gameshow. See http://www.timvp.com/ivegotse.html.
7. V. Guruswami and M. Sudan. Improved decoding of Reed-Solomon and algebraic-geometry codes. *IEEE Trans. Inform. Theory*, 45(6):1757–1767, 1999.
8. R. Safavi-Naini and Y. Wang. Sequential traitor tracing. *IEEE Trans. Inform. Theory*, 49(5):1319–1326, May 2003.
9. A. Silverberg, J. Staddon, and J. Walker. Efficient traitor tracing algorithms using list decoding. *Advances in Cryptology - ASIACRYPT 2001*, 2248:175 ff., 2001.
10. J. N. Staddon, D. R. Stinson, and R. Wei. Combinatorial properties of frameproof and traceability codes. *IEEE Trans. Inform. Theory*, 47(3):1042–1049, 2001.
11. T. V. Trung and S. Martirosyan. New constructions for ipp codes. *Proc. IEEE International Symposium on Information Theory, ISIT '03*, page 255, 2003.

Obtaining True-Random Binary Numbers from a Weak Radioactive Source*

Ammar Alkassar[1], Thomas Nicolay[2], and Markus Rohe[3]

[1] Sirrix AG security technologies,
Homburg, Germany
`a.alkassar@sirrix.com`
[2] Saarland University, Radio-Frequency Research Group,
Saarbrücken, Germany
`th.nicolay@mx.uni-saarland.de`
[3] Saarland University, Cryptography Research Group,
Saarbrücken, Germany
`mail@markus-rohe.de`

Abstract. In this paper, we present a physical random number generator (RNG) for cryptographic applications. The generator is based on alpha decay of Americium 241 that is often found in common household smoke detectors. A simple and low-cost implementation is shown to detect the decay events of a radioactive source. Furthermore, a speed-optimized random bit extraction method was chosen to gain a reasonable high data rate from a moderate radiation source (0.1 μCi). A first evaluation by applying common suits for analysis of statistical properties indicates a high quality of the data delivered by the device.

1 Introduction

Today, random numbers are well employed in numerical simulations and computations (e.g. Monte-Carlo simulations) as well as in cryptographic applications.

Many essential cryptographic primitives are are considered as probabilistic functions, or at least need a random input, e.g., the generation of challenges and session keys. Making random numbers available in deterministic environments as computers is a crucial task, hence, the security of the cryptographic systems highly depends on the quality of the employed random numbers. In this context, the main property is that the random sequence is *unpredictable*. Based on the source of randomness, we can distinguish between three classes of random generators:

---

* This work was partially funded by by the European Union in the Network of Excellence FIDIS and the German Federal Ministry of Education, Science, Research and Technology (BMBF) in the framework of the Verisoft project under grant 01 IS C38. The responsibility for this article lies with the authors.

Pseudo Random Numbers: A pseudo random number generator (PRNG) is a hard- or software instantiation of a deterministic algorithm that generates a long-periodic sequence of numbers from an initial value, called seed. Roughly speaking, a pseudorandom generator expands a short random seed into much longer random bit sequences that *appear* "random" (although they are not). In other words, the pseudorandom bit sequences have to be unpredictable, hence they are indistinguishable from true random sequences of the same length.

The notion of indistinguishability is strongly related to computational difficulty and the properties of these generators do not apply unconditionally, rather than for computationally restricted attackers. PRNG could be constructed from various intractability assumptions and good generators are proven under such assumptions.

Random Numbers Based on Complex Processes: Another possibility to obtain random numbers is given by relying on complex processes which are in principle physically deterministic but cannot be computed efficiently. For example, the random fluctuations caused by air turbulence within a disk drive or deriving randomness from a microphone/video camera signals [1]. Even "software-based" random generators rely on complex processes. Examples for random bases for such generators are the elapsed time between keystrokes or mouse movement, content of input/output buffers and operating system values such as system load and network statistics.

True Random Numbers: A true physical random number generator (TRNG) generates random numbers by observing a stationary physical phenomenon, like the elapsed time between the emission of particles during radioactive decay or the thermal noise from a semiconductor diode or resistor. The underlying phenomena are characterized by the fact that the basic quantity only could be described in a statistical manner. That is not because of inaccuracies in the used physical measurement methods, rather than because of the physical model of our world. A classic example is quantum theory, as it is intrinsically random. Hence, a quantum process like thermal noise in a semiconductor or the radioactive decay of an atomic nucleus provides an ideal base for a TRNG.

Devices based on those principles meet the definition of information-theoretic secrecy in cryptography: an attacker is unable to predetermine the bit sequence even with unbounded memory and time resources. Thus, an unconditionally secure system is only information-theoretic secure if its non-deterministic functions are founded on unpredictable random data.

Many TRNG devices for research and commercial purposes, based on quantum processes, have been constructed so far. Clipped white noise gained by thermal noise of resistors or semiconductors is used in the design of [2]. Another approach is to use noise from neon tubes [3]. The amplified noise is evaluated by a comparator, sampled and digitally de-skewed, i.e., processed to suppress correlations and statistical errors. Nevertheless these devices are very sensitive to high frequency electromagnetic disturbances. In an alternative approach [4] noise voltage modulates a voltage controlled oscillator (VCO). The output voltage is compared with a stabilized oscillator that runs at a higher frequency. Each time a zero-crossing occurs in the VCO's amplitude, the current value of the fast running oscillator is interpreted as the next random number.

In [5] a physical random number generator with fairly high data rates based on optical quantum processes is presented. Radioactive decay can be evaluated by using a Geiger-Müller tube [6, 7]. For proper operation of the tube, high voltages are necessary and recovery time after a detected pulse is very long which leads to low data rates.

The true random number generator presented in this paper, is based on radioactive decay and utilizes radioactive Americium 241 from a common household smoke detector. According to [8] the total amount of radiation is not critical for humans. For quantitative radiation monitoring a standard optical PIN diode without glass covering is used as sensor and mounted right above the radioactive material of the ion chamber. The blank diode generates very accurate pulses and is sensitive to any radiation, e.g. infra red, visible light, X-ray, $\alpha-$, $\beta-$ and γ-particles. Since both devices are placed in a shielded metal box, only the alpha particles can cause an ionization event. The result is a ready-to-use random data generator with a solid consumer market design, RS232 interface, delivering high quality random data for cryptographic applications and fast enough for single user applications. The data rate is approx. 1600 Bit/s which ties up with the registered decay events.

The outlook of the paper is as follows. The statistical background of the underlying physical process is described in Section 2. In Section 2.2 the binary number extraction method is explained. The circuit description is found in Section 3 followed by the statistical evaluation of the derived binary random numbers in Section 4. We conclude with an outline of our further work.

2 Theoretical Background of Random Number Generation

The TRNG presented in this paper mainly consists of a radioactive source and a standard PIN photo diode as sensor. The decay pulses are detected, filtered and amplified for further digital processing. The random data is obtained by deciding whether the time interval between two consecutive pulses has the length of an even or odd number of timing units. The signal processing is done by a microcontroller that sends the random data via RS232 to a host computer where it is captured by a standard terminal program. The following sections give a more detailed description of the statistic modelling of the radioactive decay and the speed enhanced method to extract the random data. The aim is to obtain a correlation free and uniformly distributed binary bit stream, i.e., the probability that an arbitrary chosen bit from the stream is either zero or one has the value 0.5.

2.1 Properties of Radioactive Decay

The radioactive decay is mathematically described by the Poisson distribution [9]. The distribution of distances between two consecutive decay events is modelled through the negative exponential function $p(t) = \lambda e^{-\lambda t}$ $(t \geq 0)$.

λ represents the intensity parameter of the process and the mean value μ is defined as $\mu = \frac{1}{\lambda}$. This formula holds if the half-life of the radioactive material is very large compared to the measuring time, i.e., the intensity parameter λ remains constant. Furthermore, the amount of atomic nuclei in the radioactive substrate has to be large enough to be considered as constant during the measurement time.

The probability (cumulative distribution) that a decay event X occurs within the time interval [0,t] is described by

$$P(0 \leq X \leq t) = \int_0^t \lambda \cdot e^{-\lambda x} dx = 1 - e^{-\lambda t},$$

while the probability that during the interval $[0, t]$ no impulse occurs is

$$P(X > t) = 1 - P(0 \leq X \leq t) = 1 - (1 - e^{-\lambda t}) = e^{-\lambda t}.$$

Another important aspect of the exponential distribution is its *memoryless property* or *Markov property*. It states that the distribution of the distance between two consecutive events is the same as the distribution of the distance between an arbitrary chosen point and the next event point.

$$\begin{aligned} P(X > T + \Delta t \mid X > T) &= \frac{P(X > T + \Delta t) \cap P(X > T)}{P(X > T)} \\ &= \frac{P(X > T + \Delta t)}{P(X > T)} \quad (since\ \Delta t > 0) \\ &= \frac{e^{-\lambda(T+\Delta t)}}{e^{-\lambda T}} = e^{-\lambda \Delta t} \\ &= P(X > \Delta t) \end{aligned}$$

After the registration of a decay impulse the sensor is insensitive for a short time. During this dead time a particle hitting its surface will not cause an impulse. Due to the Markov property the probability of detecting the next event in a given time interval Δt does not change after the sensor's dead time. In fact it is extended to a constant length by a monostable flip-flop (Fig. 5) to obtain a constant signal quality for the processing in the microcontroller.

Not every decaying atomic nucleus reaches the sensor and causes an evaluable impulse. The dispersion of the α-particles is spherical such that at least half of the particles vanish into the ground where the substrate is fixed. With an increasing distance between radioactive material and sensor surface the α-particles lose kinetic energy when interacting with the air molecules. Additionally, the radioactive disintegration generates new radioactive daughter nuclides that cause peaks of different height in the signal processing part (Fig. 6). But all these factors do not affect the statistical behaviour of the observed process. In fact, it can be subdivided into several statistical independent Poisson processes with different intensity parameters λ_i. According to the *additivity property* we summarise the parameters of all processes that provide a sufficient peak to be detected [10]. Let I be the index set of all relevant intensity parameters. Then the intensity parameter of the relevant sub-processes λ_{rel} is defined as $\lambda_{rel} = \sum_{i \in I} \lambda_i$. Consequently, for the cumulative probability holds:

$$P(0 \leq X \leq t) = 1 - e^{-(\sum_{i \in I} \lambda_i)t} = 1 - e^{-\lambda_{rel} t}$$

Note that the shift in λ_{rel} is assumed to be long-term and does not change within seconds but over the years since Americium 241 has a half-life of more than 400 years. Throughout the following parts of this article, every intensity parameter λ is implicitly referred as λ_{rel}.

2.2 Binary Number Extraction

Method 1: The generators by Gude [7] and Vincent [11] are based on the fact that in a Poisson process the amount of impulses or elementary events within a fixed time interval cannot be predetermined. The probability that k impulses are registered within the interval Δt is

$$P(k) = \frac{(\lambda \Delta t)^k}{k!} \cdot e^{-\lambda \Delta t}$$

Both generators trigger a toggle flip flop with the decay pulses and evaluate its state after a constant amount of time (Fig. 1). Afterwards the flip flop is set back at the end of the time window to guarantee a new memoryless measurement in the following interval.

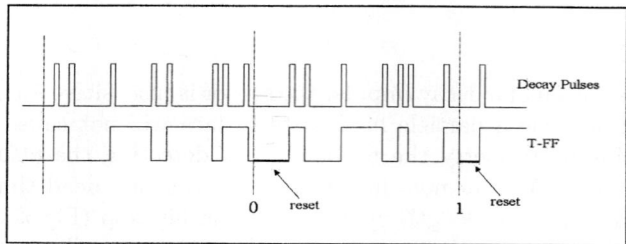

Fig. 1. Counting the pulses within a constant time window

According to Gude [12] the time window Δt should be chosen at least 10 times larger than the mean decay rate μ of the radioactive substance to minimize the influence of autocorrelation and hence, minimizes the error in the equal distribution. The error decreases exponentially by the additive factor: $P(0) - P(1) = e^{-2\lambda \cdot \Delta t}$. Vincent [11, 13] and Kraus [14] obtain the same results.

PURAN1 by Gude [7] produced considerable true random values implementing this algorithm but a lot of decay impulses are wasted which is ineligible if a low-radiation source is used.

Method 2: Another approach is used by Walker's "hotbits" [15] and also suggested by [6]. In both generators the length of two consecutive decay intervals is compared such that the bits are obtained as shown in Figure 2. The exploit of this method is higher than in the first one.

- $t_1 > t_2$: interpret this result as a **1**
- $t_1 < t_2$: interpret this result as a **0**
- $t_1 = t_2$: discard this event, continue with the next pair of intervals

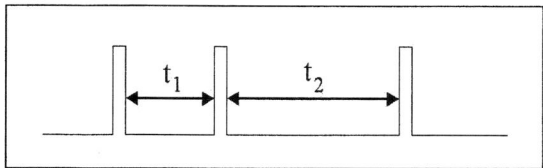

Fig. 2. Time measurement between two consecutive decays

Two decay events are necessary to extract one bit. But here, two registers of appropriate size are necessary depending on the clock speed and the mean decay rate μ. This causes a problem if a long-term shift in intensity occurs and the timers produce an overflow as a consequence. As far as we know, there exist no statistical evaluations of an implementation of this extraction algorithm.

Method 3: Our approach doubles the rate of yield of Method 2. The output of a binary counter is captured each time a new decay impulse occurs. This is equivalent to a time measurement between two impulses taken with a high resolution clock as described in Figure 3: This is the highest gain of information

- the length t of the interval consists of an odd amount of timing units Δt:
  $t \equiv 1\ units\ mod\ 2$
- the length t of the interval consists of an even amount of timing units Δt:
  $t \equiv 0\ units\ mod\ 2$

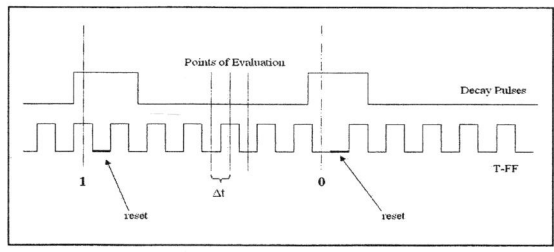

Fig. 3. Time measurement between two consecutive decays

so far compared to the other two methods: *one decay delivers one random bit*, and therefore well-suited to generate reasonable high data rates out of a low-radiation source. Even a slight shift in the mean decay rate, e.g., caused by aging of the radioactive source does not affect the evaluation of the time interval and

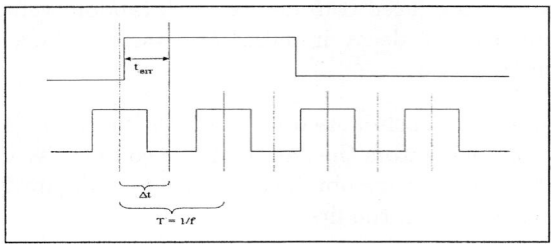

Fig. 4. Error when polling for the next decay impulse

makes this method very robust. The method can be implemented by a toggle flip flop (T-FF) in practice. After a fixed time period Δt it is checked whether a new impulse has occurred. In this case, the current value at the T-FF's output is captured as the new random bit. The T-FF has to be set back to a defined value, e.g. Low. This is analog to a new time measurement where the clock also has to be set back to zero. Hence, each measurement is completely memoryless and therefore uncorrelated [12]. Note that the decay pulses are much longer than Δt such that a random bit followed by a zero generated by the reset of the flip-flop cannot occur. Both the output-ratio of the T-FF has to keep an exact ratio of 1:1 and the time distances Δt between two pollings for a new decay event are obliged to have equal length (Fig. 4). Otherwise the bits would be coloured, i.e., the balance between 0s and 1s would not be equal any more. However, Appendix B cites a simple method by von Neumann [16] to eliminate such a bias in the equal distribution but at the price of a lower data rate; about 75% of the bits would be lost. In our implementation the statistical evaluation indicates that such a step is unnecessary.

2.3 Error Discussion

In order to obtain random bits, the exponential distribution has to be transformed into the binary uniform distribution. The numerical error during this transformation using the mod 2 time measurement decreases exponentially as shown in Appendix A. This is hardware-independent and can be ignored since an error also occurs in time measurements when polling for the next decay event (Fig. 4). If this event occurs shortly after an evaluation point then it is registered to the next sampling point and interpreted some time t_{err} later (quantization noise). In the worst case the maximal error is up to one timing unit Δt and affects the measured result linear reciprocal to the applied sampling frequency f: $t_{err} \leq \Delta t = \frac{T}{2} = \frac{1}{2 \cdot f}$.

Since the occurrence of the decay pulses is unpredictable the error is considered to be uniformly distributed, i.e., the amount of wrong-classified zeros is as large as the amount of wrong-classified ones. This assumption is also justified by the statistical evaluation.

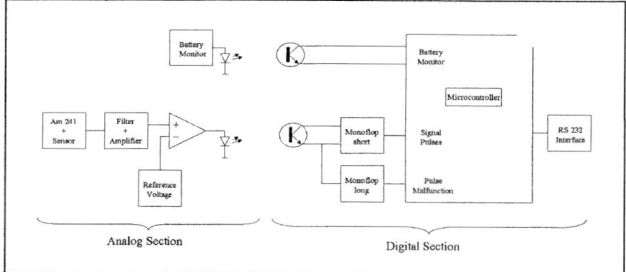

Fig. 5. Block diagram of the true random binary number generator

3 Circuit Description

Figure 5 shows the analog and digital parts of the circuit design. The low-radiation[1] source consists of Americium 241 and is dismantled from a common ionisation smoke detector. The sensor, a standard PIN photo diode BPX61 with removed glass cover is placed directly above the radiation source. In the analog section the decay pulses are detected and amplified. RF-shielding and battery powered operation reduces the influence of any noise and guaranties the functionality of this section even in such a hazardous environment as a PC. Spikes from digital ICs via supply or ground lines are excluded by using optical transmission for pulses and battery monitor. Battery monitoring prevents a failure of the TRNG in consequence of a loss in power supply. In the digital section pulses, now on TTL level, are analyzed, the binary numbers are extracted and transferred by RS232 interface to a host computer. A supplementary monostable flip-flop with a timing constant of about ten times larger that the mean decay rate triggers if the pulse stream breaks down.

The microcontroller (PIC16F628) runs at a clock frequency of 18.432 MHz, a well-suited speed to generate both the clock for serial transmission and the timing units[2] Δt. The diode registers with its $7mm^2 (2.65 \times 2.65mm^2)$ radiant sensitive area about 1700 decays per second. This number can be increased by using a sensor with a larger surface. We omit further technical details due to space restrictions, however, the schematics are available at [17].

Figure 6 shows the analog signal of encountered decay events at the input of the comparator. It can be seen that the alpha particles do not have the same energy level if they are receipted at the PIN diode and time intervals between two consecutive pulses are strongly varying. Behind the mono-flop all pulses have the same amplitude, same time length and are ready for microcontroller processing.

---

[1] $0.1\mu Ci \,\widehat{=}\, 33000 Bq$, i.e. 33000 decays per second.
[2] $\Delta t = 0.217 \mu s \,\widehat{=}\, f = 4.608$ MHz.

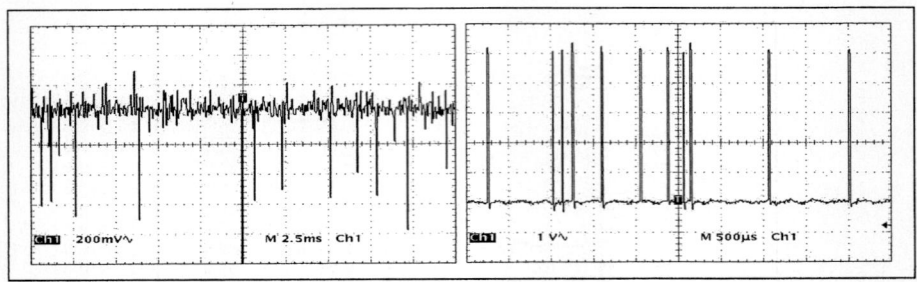

Fig. 6. Signals at the comparator's input (left) and at the μC's input (right)

4 Experimental Results and Statistical Evaluation

Each of the following tests have been performed on several data sets. Depending on the requirements of each test, the length of the data stream varies between 1,000,000 and almost 100,000,000 bit.

Simple counting indicates whether the amount of '0' and '1' in the bit stream is uniformly distributed over the data set [18]. The concatenation of all five data sets (each consists of 10 Million bit) leads to a distribution rate of 50.00126% for the event P(1).

Table 1. Uniform Distribution Test

	Set 1	Set 2	Set 3	Set 4	Set 5
amount of 0s	5,000,050	4,999,910	5,000,280	5,000,654	4,999,846
amount of 1s	4,999,950	5,000,090	4,999,720	4,999,346	5,000,154
P(1)	0.499995	0.5000009	0.500028	0.4999346	0.5000154

Diehard Suite of Tests: The DIEHARD [19] test suite was applied to three sets of data to get a first statement about the quality of the random data produced by the TRNG. The obtained P-values have to be uniformly distributed on (0,1). The description of the test states that it fails big, if 6 or more values are very close to 0 or 1. Such cumulations did not occur. According to these results of Table 2, the extracted random data pass the DIEHARD tests.

NIST Statistical Tests: In order to ensure the quality of our obtained random data, parts of the NIST [20] suite of tests were performed on a continuous data stream, recorded from our TRNG. Our test and result interpretations were performed on five different data sets with the instructions given in [20]. For every test a new data set was recorded. The obtained P-values were within the expected set of confidence for all applied test and were uniform distributed in (0,1). Some typical P-values are shown in Table 3. We say, the test has been bypassed positively if the received P-values are > 0.01 or < 0.99 [20].

Table 2. Diehard Test

p	File1	File2	File3
0.0 - 0.1	24	20	25
0.1 - 0.2	25	22	24
0.2 - 0.3	24	19	22
0.3 - 0.4	18	29	37
0.4 - 0.5	28	20	16
0.5 - 0.6	17	25	24
0.6 - 0.7	23	19	28
0.7 - 0.8	27	28	17
0.8 - 0.9	21	28	15
0.9 - 1.0	27	24	26

Table 3. Nist Test Suite

Statistical Test	Set 1	Set 2	Set 3	Set 4	Set 5
Monobit	0.638423	0.316748	0.255474	0.401319	0.715166
Block Freq.	0.270766	0.591946	0.775216	0.186047	0.352641
Runs	0.018076	0.695193	0.163838	0.515480	0.309741
Long-Runs	0.859042	0.661134	0.633951	0.352641	0.053309
Serial	0.018073	0.325066	0.815960	0.465480	0.309761
CumSum	0.520113	0.495486	0.306135	0.372302	0.921552
Rank	0.852830	0.561925	0.679633	0.434292	0.631485
FFT	0.830444	0.682688	0.363646	0.024256	0.917311
Maurer	0.658820	0.787850	0.302802	0.352204	0.897875

5 Conclusion

In this paper, we introduce a true-random binary number generator based on radioactive decay of Americium 241. Moreover, we propose a method for improved bit extraction to achieve high data rates from a low-radiation source, meeting the demands for cryptographic applications. Several test have been applied on different sets of random binary numbers. The experimental results show a good quality in uniform distribution and randomness of the binary data.

Future work is marked by integrating online tests [21] monitoring the extracted random numbers in order to preclude a failure of the random data source and to guarantee the randomness of the data. Additionally, we will consider the use of modulo 2^n-counters instead of of a toggle flip-flop (modulo 2) to obtain more than one random bit per decay. This can be done up to some theoretic boundary which also should be analyzed. Each radioactive particle causes some damage to the crystal structure of a semiconductor that, as a consequence, change the electrical characteristic of the diode. Hence, a long-term analysis of that effects have to be performed.

References

1. Monrose, F., Reiter, M.K., Li, Q., Wetzel, S.: Cryptographic key generation from voice. In: Proceedings of the IEEE Symposium on Research in Security and Privacy. (2001) 202–212
2. Murry, H.F.: A general approach for generating natural random variables. IEEE Transactions on Computers (1970) 1210–1214
3. Thomson, W.E.: Ernie - a mathematical and statistical analysis. Journal of the Royal Statistical Society, Series B **101** (1959) 301–333
4. Jun, B., Kocher, P.: The intel number generator. White paper prepared for intel corporation, Cryptography Research, Inc. (1999)
5. Stefanov, A., Gisin, N., Guinnard, O., Guinnard, L., Zbinden, H.: Optical quantum random number generator. online (1999)
6. Figotin, A., Vitebskiy, I., Popovich, V., Stetsenko, G., Molchanov, S., Gordon, A., Quinn, J., Stavrakas, N.: Random number generator based on the spotaneous alpha-decay. U.S. patent Appl. No.: 10/127,221 (2003)
7. Gude, M.: Ein quasi-idealer Gleichverteilungsgenerator basierend auf physikalischen Zufallsphänomenen. PhD thesis, RWTH Aachen (1987)

8. Lide, D., ed.: CRC Handbook of Chemistry and Physicas, 84th Edt. CRC Press (2003) ISBN 0849304849.
9. Haight, F.A.: Handbook of the Poisson Distribution. John Wiley & Sons, Inc. (1967)
10. Koutsky, Z.: Theorie der Impulszähler und ihre Anwendung. Aplikace Matematiky **7** (1962) 116–138
11. Vincent, C.H.: The generation of truely random binary numbers. Journal of Physics E: Scientific Instruments **3** (1970) 594–598
12. Gude, M.: Concept for a high performance random number generator based on physical random phenomena. Frequenz **39** (1985) 187–190
13. Vincent, C.H.: Precautions for accuracy in the generation of truely random binary numbers. Journal of Physics E: Scientific Instruments **4** (1971) 825–828
14. Kraus, G.: Stochastische Abhängigkeit von in schneller Folge erzeugten gleichverteilten diskreten Zufallsereignissen. Frequenz **35** (1981) 274–277
15. Walker, J.: Hotbits. online : http://www.fourmilab.ch/hotbits/ (2003)
16. von Neumann, J.: Various techniques used in connection with random digits. Applied Mathematics Series (1951) 36–38
17. Rohe, M.: RANDy - A True-Random Generator Based on Radioactive Decay. Technical report, Saarland University http://www-krypt.cs.uni-saarland.de/projects/randy/ (2003)
18. Pitman, J.: Probability. Springer Verlag, New York (1993)
19. Marsaglia, G.: Diehard battery of tests of randomness. (The Marsaglia random number CDROM)
20. Rukhin, A., Sato, J., Nechvatal, J., Smid, M., Barker, M., Leigh, S., Levenson, M., Vangel, M., Banks, D., Heckert, A., Dray, J., Vo, S.: A statistical test suite for random and pseudorandom number generators for cryptographic applications. NIST Special Publication (2001)
21. Schindler, W.: Efficient online tests for true random number generators. In: CHES: International Workshop on Cryptographic Hardware and Embedded Systems, CHES, LNCS. (2001)

A Appendix

In this section we prove how to transform the exponential distribution into a binary uniform distribution. Let P(0) denote the probability that the next bit extracted from the bitstream is 0 and P(1) is defined analogous. According to the additivity of the Riemann-Integral we divide the cumulative distribution $P(0 \leq X \leq \infty)$ into two sums of integrals representing each summand either an even or an odd timing unit Δt. Note that the first sum of integrals represents $P(0)$ and the second sum $P(1)$.

$$\int_0^\infty \lambda \cdot e^{-\lambda x} dx = \sum_{N=0}^\infty \int_{(2N)\cdot \Delta t}^{(2N+1)\cdot \Delta t} \lambda \cdot e^{-\lambda \cdot s} \, ds + \sum_{N=0}^\infty \int_{(2N+1)\cdot \Delta t}^{(2N+2)\cdot \Delta t} \lambda \cdot e^{-\lambda \cdot s} \, ds$$

$$= \underbrace{\sum_{N=0}^\infty \lambda \cdot \left[-\frac{1}{\lambda} \cdot e^{-\lambda \cdot s}\right]_{(2N)\cdot \Delta t}^{(2N+1)\cdot \Delta t}}_{P(0)} + \underbrace{\sum_{N=0}^\infty \lambda \cdot \left[-\frac{1}{\lambda} \cdot e^{-\lambda \cdot s}\right]_{(2N+1)\cdot \Delta t}^{(2N+2)\cdot \Delta t}}_{P(1)}$$

Recall that by definition $\int_0^\infty \lambda \cdot e^{-\lambda x} dx = P(0) + P(1) = 1$. The following two lemmas provide a discrete formula for both $P(0)$ and $P(1)$.

Lemma 1: $P(0) = \sum_{N=0}^{\infty} (-1)^N \cdot e^{-\lambda N \cdot \Delta t}$
proof: The sum representation of $P(0)$ delivers

$$P(0) = \sum_{N=0}^{\infty} \lambda \cdot \left[-\frac{1}{\lambda} \cdot e^{-\lambda s} \right]_{2N \cdot \Delta t}^{(2N+1) \cdot \Delta t}$$

$$= \sum_{N=0}^{\infty} e^{-\lambda(2N) \cdot \Delta t} - e^{-\lambda(2N+1) \cdot \Delta t}$$

Expanding this sum produces: $e^{-0} - e^{-\lambda \cdot \Delta t} + e^{-\lambda 2 \cdot \Delta t} - e^{-\lambda 3 \cdot \Delta t} + \ldots$ which leads to the following series: $\sum_{N=0}^{\infty} (-1)^N \cdot e^{-\lambda N \cdot \Delta t}$. This series converges because $(-1)^N$ is alternating and $e^{-\lambda \cdot N \cdot \Delta t}$ is a monotone sequence leading to 0 (Leibnitz). □

Lemma 2: $P(1) = P(0) \cdot e^{-\lambda \cdot \Delta t}$
proof: Analogous to the previous case $P(0)$, we start with its sum representation of P(1).

$$P(1) = \sum_{N=0}^{\infty} \lambda \cdot \left[-\frac{1}{\lambda} \cdot e^{-\lambda s} \right]_{(2N+1) \cdot \Delta t}^{(2N+2) \cdot \Delta t}$$

$$= \sum_{N=0}^{\infty} e^{-\lambda(2N+1) \cdot \Delta t} - e^{-\lambda(2N+2) \cdot \Delta t}$$

An expansion of the sum leads to: $e^{-\lambda \cdot \Delta t} - e^{-\lambda 2 \cdot \Delta t} + e^{-\lambda 3 \cdot \Delta t} - e^{-\lambda 4 \cdot \Delta t} + \ldots$ summing up to this series: $\sum_{N=0}^{\infty} (-1)^N \cdot e^{-\lambda(N+1) \cdot \Delta t}$. The exponent $\lambda(N+1) \cdot \Delta t$ can be expanded to $-\lambda N \Delta t - \lambda \Delta t$ which is equivalent to

$$\sum_{N=0}^{\infty} (-1)^N \cdot e^{-\lambda N \cdot \Delta t} \cdot e^{-\lambda \cdot \Delta t} = P(0) \cdot e^{-\lambda \cdot \Delta t}$$ □

Lemma 3: $P(1) - P(0) \longrightarrow 0$ for $\Delta t \longrightarrow 0$, *exponentially.*
proof:

$$P(1) - P(0) = \left(\sum_{N=0}^{\infty} (-1)^N \cdot e^{-\lambda N \cdot \Delta t} \right) \cdot e^{-\lambda \cdot \Delta t}$$

$$- \left(\sum_{N=0}^{\infty} (-1)^N \cdot e^{-\lambda N \cdot \Delta t} \right)$$

$$= \sum_{N=0}^{\infty} (-1)^N \cdot e^{-\lambda N \cdot \Delta t} \cdot (e^{-\lambda \cdot \Delta t} - 1)$$

Since $e^{-\lambda \cdot \Delta t} - 1$ converges to 0 and the series remains bounded, $P(1) - P(0)$ becomes negligible small if Δt approaches 0. □

This leads to : $P(0) \approx P(1) \approx 0,5$ if Δt is suitable small. *q.e.d.*

B Appendix

The biasing of the equal distribution of an uncorrelated bit sequence can be achieved by the following algorithm, proposed by von Neumann [16]: The bitstream is divided into groups of two consecutive bits as shown in the example. These two bits can adopt 4 different pairs of values. Each pair delivers either an output bit (0 or 1) or no bit at all, according to the following function:

$$f : \{0,1\}^2 \to \{0,1,-\}$$
$$f(x_1, x_2) := \begin{cases} 0 & : (x_1, x_2) = (0, 1) \\ 1 & : (x_1, x_2) = (1, 0) \\ - & : \text{otherwise} \end{cases}$$

Due to the uncorrelation property, the following probabilities are independent from the position i in the bitstream: $P(x_i = 0) := p$ and $P(x_i = 1) := q$. In this binary case it holds $q = 1 - p$ such that $P(0,1) = p \cdot (1-p)$ and $P(1,0) = (1-p) \cdot p$. Hence, the combination $(0,1)$ has the same probability to be generated than $(1,0)$. The resulting bitstream is uniformly distributed under the assumption that the pairs $(0,0)$ and $(1,1)$ deliver no output bit. The expected input is 8 bits to gain 2 bits which decreases the former generation rate at least by factor 4.

biased stream	0 1	1 1	0 1	1 0	0 1	1 0	0 0	1 0
unbiased stream	0	-	0	1	0	1	-	1

Modified Sequential Normal Basis Multipliers for Type II Optimal Normal Bases*

Dong Jin Yang[1], Chang Han Kim[2], Youngho Park[3], Yongtae Kim[4], and Jongin Lim[1]

[1] Center for Information Security Technologies(CIST), Korea Univ., Seoul, Korea
{djyang76, jilim}@cist.korea.ac.kr
[2] Dept. of Information and Security, Semyung Univ., Jecheon, Korea
chkim@semyung.ac.kr
[3] Dept. of Information Security and System, Sejong Cyber Univ., Seoul, Korea
youngho@cybersejong.ac.kr
[4] Dept. of Mathematics Education, Gwangju National Univ. of Education, Gwangju, Korea
ytkim@gnue.ac.kr

Abstract. The arithmetic in finite field $GF(2^m)$ is important in cryptographic application and coding theory. Especially, the area and time efficient multiplier in $GF(2^m)$ has many applications in cryptographic fields, for example, ECC. In that point optimal normal basis give attractiveness in area efficient implementation. In [2], Reyhani-Masoleh and Hasan suggested an area efficient linear array for multiplication in $GF(2^m)$ with slightly increased critical path delay from Agnew et al's structure. But in [3], S.Kwon et al. suggested an area efficient linear array for multiplication in $GF(2^m)$ without losing time efficiency from Agnew et al's structure. We propose a modification of Reyhani-Masoleh and Hasan's structure with restriction to optimal normal basis type-II. The time and area efficiency of our multiplier is exactly same as that of S.Kwon et al's structure.

Keywords: Finite fields, Massey-Omura multiplier, Gaussian Normal Basis, ECC.

1 Introduction

Finite field arithmetic is very important in the area of cryptographic applications and coding theory. Especially, the multiplication in $GF(2^m)$ has many applications in cryptographic areas such as ECC, XTR and AES. In these days, the fast and small implementation of finite field multiplication is a major concern. To get the area and time efficiency, many authors use normal basis representation

---

* This research was supported by the MIC(Ministry of Information and Communication), Korea, under the ITRC(Information Technology Research Center) support program supervised by the IITA(Institute of Information Technology Assessment).

[1, 2, 3, 4, 5, 6, 7, 9, 10]. The normal basis representation is suitable for hardware implementation and squaring can be done by simple cyclic shift which is free in hardware.

The Massey-Omura multiplier has a parallel-in , serial-out structure and has a very long critical path delay. So Agnew et al. [1] significantly reduced the complexity of Massey-Omura multiplier [11]. Agnew et al's structure is Sequential Multiplier with Parallel Output (SMPO). Recently, Reyhani-Masoleh and Hasan [2], S.Kwon et al [3] proposed SMPOs which significantly reduced the area complexity of Agnew et al [1]. Reyhani-Masoleh and Hasan significantly reduced the area complexity of [1] with slightly increased critical path delay. For example, in the case of type-II optimal normal basis (type-II ONB) the time complexity of the multiplier of Reyhani-Masoleh and Hasan [2] is $m(T_A + 3T_X)$ while that of Agnew et al. [1] is $m(T_A + 2T_X)$, where m is extension degree of $GF(2^m)$ over $GF(2)$ and T_A, T_X are the delay time of a two input AND gate and a two input XOR gate. S.Kwon et al. reduced the area complexity of Agnew et al's SMPO without losing time efficiency [3]. So it is believed that it has more time efficiency and the same area complexity in comparison with Reyhani-Masoleh and Hasan's SMPO. Therefore, among the known SMPO, S.Kwon et al's SMPO [3] is the best known time and area efficient SMPO.

In this paper we want to present a sequential multiplier using type-II ONB in $GF(2^m)$. The critical path delay of our proposed sequential multiplier is reduced from that of the multiplier of Reyhani-Masoleh and Hasan [2], and thus equally comparable to that of the multiplier of S.Kwon et al [3]. And the area complexity of our proposed multiplier equals to that of the multiplier of Reyhani-Masoleh and Hasan. Therefore our proposed multiplier has exactly the same complexity as S.Kwon et al's multiplier [3] when $k = 2$.

2 Type-II Optimal Normal Basis

A type-II ONB in $GF(2^m)$ is constructed using the normal element $\alpha = \gamma + \gamma^{-1}$, where γ is a primitive $(2m+1)th$ root of unity, i.e. $\gamma^{2m+1} = 1$ and $\gamma^i \neq 1$ for any $1 \leq i < 2m + 1$.

A type-II ONB can be constructed if $p = 2m+1$ is prime and if either of the following two conditions also holds[14]:

1. 2 is primitive in $\mathbb{Z}_{2m+1}$, or
2. $2m + 1 \equiv 3 \mod 4$ and 2 generates the quadratic residues in $\mathbb{Z}_{2m+1}$

The second condition means that (-1) generates a quadratic non-residue modulo p and 2 generates the quadratic residues modulo p.

Optimal normal basis representation makes the implementation of ECC efficient in hardware. So ANSI recommended type-II ONB cases ($m = 191$ EX4,5 and $m = 239$ EX4,5) and NIST recommended one type-II ONB case ($m = 233$) (See [12, 13]) Note that all finite fields $GF(2^m)$ has odd m.

3 Reyhani-Masoleh and Hasan's Multiplier

Let α be a normal element as defined above and $\{\alpha_0, \alpha_1, \cdots, \alpha_{m-1}\}$ be a normal basis in $GF(2^m)$ with $\alpha_i = \alpha^{2^i}$ and then

$$\alpha\alpha_i = \sum_{j=0}^{m-1} \lambda_{ij}\alpha_j,$$

where λ_{ij} is in $GF(2)$.

Reyhani-Masoleh and Hasan[2] approached differently in comparison with Agnew et al. They used $\alpha\alpha_i$ instead of $\alpha_i\alpha_j$ and wisely utilized the symmetric property between $\alpha\alpha_i$ and $\alpha\alpha_{m-i}$. They proposed two different architectures i.e. XESMPO and AESMPO [2]. Both XESMPO and AESMPO have the same critical path delay. So we will deal with XESMPO and discuss AESMPO. In [2], the product C of $A = \sum_{i=0}^{m-1} a_i\alpha_i$ and $B = \sum_{j=0}^{m-1} b_j\alpha_j$ is computed as follows.

$$C = \sum_{i,j} a_i b_j \alpha_i \alpha_j = \sum_{i=0}^{m-1} a_i b_i \alpha_{i+1} + \sum_{i=0}^{m-1}\sum_{j\neq i}^{m-1} a_i b_j (\alpha\alpha_{j-i})^{2^i}$$

$$= \sum_{i=0}^{m-1} a_i b_i \alpha_{i+1} + \sum_{i=0}^{m-1}\sum_{j\neq 0}^{m-1} a_i b_{j+i} (\alpha\alpha_j)^{2^i}$$

When m is odd, the second term of the right side of the above equation is written as

$$\sum_{i=0}^{m-1}\sum_{j=1}^{v} a_i b_{j+i}(\alpha\alpha_j)^{2^i} + \sum_{i=0}^{m-1}\sum_{j=m-v}^{m-1} a_i b_{j+i}(\alpha\alpha_j)^{2^i}$$

and when m is even, it is written as

$$\sum_{i=0}^{m-1}\sum_{j=1}^{v} a_i b_{j+i}(\alpha\alpha_j)^{2^i} + \sum_{i=0}^{m-1}\sum_{j=m-v}^{m-1} a_i b_{j+i}(\alpha\alpha_j)^{2^i} + \sum_{i=0}^{m-1} a_i b_{v+1+i}(\alpha\alpha_{v+1})^{2^i},$$

where $v = \lfloor \frac{m-1}{2} \rfloor$, i.e. $m = 2v+1$ or $m = 2v+2$.

We will be restricted to odd m because of the explanation of previous chapter. For odd m, due to the following (Refer to [3])

$$\sum_{i=0}^{m-1}\sum_{j=m-v}^{m-1} a_i b_{j+i}(\alpha\alpha_j)^{2^i} = \sum_{i=0}^{m-1}\sum_{j=1}^{v} a_i b_{m-j+i}(\alpha\alpha_{m-j})^{2^i}$$

$$= \sum_{i=0}^{m-1}\sum_{j=1}^{v} a_{i+j} b_i (\alpha\alpha_{m-j})^{2^{i+j}}$$

$$= \sum_{i=0}^{m-1}\sum_{j=1}^{v} a_{i+j} b_i (\alpha\alpha_j)^{2^i}$$

C can be written as

$$C = \sum_{i=0}^{m-1} a_i b_i \alpha_{i+1} + \sum_{i=0}^{m-1} \sum_{j=1}^{v} (a_i b_{j+i} + a_{j+i} b_i)(\alpha \alpha_j)^{2^i}$$

$$= \sum_{i=0}^{m-1} (a_i b_i \alpha_1 + \sum_{j=1}^{v} (a_i b_{j+i} + a_{j+i} b_i) \alpha \alpha_j)^{2^i}$$

$$= \sum_{i=0}^{m-1} (F_i)^{2^i}$$

In above equation, $F_i(A,B) = a_{i-g} b_{i-g} \alpha + \sum_{j=1}^{v} z_{i,j} \delta_j$, where $\delta_j = \alpha \alpha_j$ and $g \in \{0, 1\}$ which determines $z_{i,j}$ as follows.

For $1 \leq j \leq v$,

$$z_{i,j} = \begin{cases} (a_i + a_{i+j})(b_i + b_{i+j}), & g = 0; \\ a_i b_{j+i} + a_{j+i} b_i, & g = 1. \end{cases}$$

In the case of $g = 1$, it was called XESMPO. The other case was called AESMPO. Since $F_{m-t} = F_{m-1}(A^{2^{t-1}}, B^{2^{t-1}})$, the product $C = AB$ can be implemented by $\sum_{i=0}^{m-1} F_i^{2^i}$. Using this property, Reyhani-Masoleh and Hasan proved theorem 1.

Theorem 1 ([2]). *Let A, B be elements of $GF(2^m)$ and $C = AB$. Then*

$$C = (((F_{m-1}^2 + F_{m-2})^2 + F_{m-3})^2 + \cdots + F_1)^2 + F_0.$$

For example, Reyhani-Masoleh and Hasan's multiplier for $m = 5$ is shown in Fig. 1 where a type-II ONB is used. In Fig. 1, the structure has 2 part, i.e. Z-array and XOR-array. Z-array computes $z_{i,j}$, and XOR-array computes $\sum_{j=1}^{v} z_{i,j} (\alpha \alpha_j)^{2^i}$ and accumulate the result in register D. Depending on the Z-array, there are two architectures, i.e. XESMPO, AESMPO.

4 Proposed Multiplier

Reyhani-Masoleh and Hasan's SMPO uses this equation

$$C = \sum_{i=0}^{m-1} a_i b_i \alpha_{i+1} + \sum_{i=0}^{m-1} \sum_{j=1}^{v} z_{ij} (\delta_j)^{2^i},$$

where z_{ij}, δ_j is defined as previously mentioned.

In the above section, δ_j is determined by the multiplication matrix (λ_{ij}). We will set some notations. First, since we consider type-II ONB, there are exactly two 1s in each row and column except the 1st row and the 1st column of (λ_{ij}). Let $l(i)$ denote the distance of 1s of ith row of (λ_{ij}). The distance means the number of 0s between two 1s. Since m is odd, we can count even $l(i)$. That is,

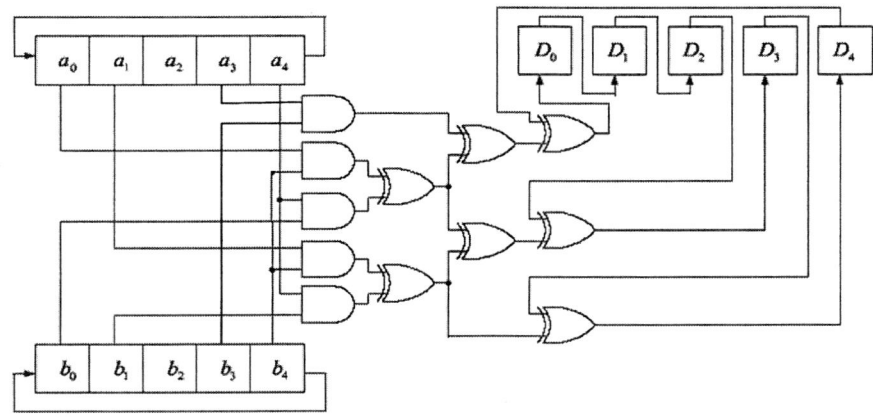

Fig. 1. Reyhani-Masoleh and Hasan's XESMPO in $GF(2^5)$

if $l(i)$ is odd, then $l(i)$ should be a $m - l(i)$. Otherwise, we can set $l(i)$ itself. Before putting second notation, we will give lemma 1. Actually, lemma 1 was mentioned in [2] by Reyhani-Masoleh and Hasan. They suggested experimental results for $m < 5000$ as conjecture.

Lemma 1. In type-II ONB of $GF(2^m)$, for $1 \leq i \neq j \leq v$,

$$l(i) \neq l(j)$$

Proof. [1] The statement is equivalent to say that there is no integer s such that $(\alpha \alpha_i)^{(2^s)} = (\alpha \alpha_j)$ for any different i, j with $1 \leq i \neq j \leq v$. For contradiction, we assume that there is s satisfying above condition for some i and j. Because of section 2 for type-II ONB, we can set $\alpha = \gamma + \gamma^{-1}$.

$$(\alpha \alpha_i)^{(2^s)} = (\alpha \alpha_j)$$
$$((\gamma + \gamma^{-1})(\gamma^{2^i} + \gamma^{-2^i}))^{(2^s)} = (\gamma + \gamma^{-1})(\gamma^{2^j} + \gamma^{-2^j})$$

Then we can get the following equation.

$$(\gamma^{1+2^i} + \gamma^{-(1+2^i)})^{(2^s)} + (\gamma^{-(1-2^i)} + \gamma^{1-2^i})^{(2^s)}$$
$$= (\gamma^{1+2^j} + \gamma^{-(1+2^j)}) + (\gamma^{-(1-2^j)} + \gamma^{1-2^j})$$

By above,

1) $\begin{cases} (1+2^i)2^s \equiv \pm(1+2^j) \\ (1-2^i)2^s \equiv \pm(1-2^j) \end{cases}$ or 2) $\begin{cases} (1+2^i)2^s \equiv \pm(1-2^j) \\ (1-2^i)2^s \equiv \pm(1+2^j) \end{cases}$

---

[1] Proof of Lemma 1 was changed to simple version by anonymous referee. We would like to thank him/her.

Equation 1) is

$$\frac{1+2^i}{1+2^j} \equiv \pm\frac{1-2^i}{1-2^j}$$
$$(1+2^i)(1-2^j) \equiv \pm(1-2^i)(1+2^j)$$
$$2^{i+1} \equiv 2^{j+1} \quad \text{or} \quad 2 \equiv 2^{(i+j+1)}$$

Since $1 \leq i \neq j \leq v$ and $o(2)$ is m in $Z^*_{2m+1} = <-1,2>$ or $2m$ in $Z^*_{2m+1} = <2>$, both of them are impossible. So, s which satisfies 1) doesn't exist. Similarly, equation 2) is

$$2^{i+1} \equiv -2^{j+1} \quad \text{or} \quad 2 \equiv -2^{(i+j+1)}$$

Using similar causes of 1), both of them are impossible. Therefore, there does not exist s that satisfies $(\alpha\alpha_i)^{(2^s)} = (\alpha\alpha_j)$. □

In second lemma, we will put second notation.

Corollary 1. *Let γ_i be the count of right shift of the ith row of (λ_{ij}) for each $1 \leq i \leq v$. If we shift each ith row of (λ_{ij}) γ_i-times, then each column of $(\lambda_{ij})_{1 \leq i \leq v, 0 \leq j \leq m-1}$ has unique 1 except the last column.*

By lemma 1, we know that $l(i)$ are even and distinct. Because 1st to v^{th} row of (λ_{ij}) have exactly two 1s and distinct even $l(i)$, we can shift above half of multiplication matrix in order to have unique 1 in each column. (For understanding, refer example 1) For each $1 \leq i \leq v$, shifting i^{th} row of (λ_{ij}) γ_i-times, you will get the shifted matrix (λ'_{ij}):

$$(\lambda_{ij}) = \begin{pmatrix} \lambda_0 \\ \lambda_1 \\ \vdots \\ \lambda_v \\ \lambda_{v+1} \\ \vdots \\ \lambda_{m-1} \end{pmatrix} \Rightarrow (\lambda'_{ij}) = \begin{pmatrix} \lambda_0 \\ (\lambda_1 \to \gamma_1) \\ \vdots \\ (\lambda_v \to \gamma_v) \\ \lambda_{v+1} \\ \vdots \\ \lambda_{m-1} \end{pmatrix} = \begin{pmatrix} \lambda_0 \\ (\cdots 1 \underbrace{0s}_{l(1)} 1 \cdots 0) \\ \vdots \\ (\cdots 1 \underbrace{0s}_{l(v)} 1 \cdots 0) \\ \lambda_{v+1} \\ \vdots \\ \lambda_{m-1} \end{pmatrix},$$

where $\lambda_i = (\lambda_{i0}, \lambda_{i1}, \cdots, \lambda_{i,m-1})$ and $(\lambda_i \to \gamma_i)$ means γ_i-times right shifted λ_i. Since the entries of the last column of $(\lambda'_{ij})_{1 \leq i \leq v, 0 \leq j \leq m-1}$ are all 0s and the 1st row of (λ'_{ij}) is $(010\cdots 0)$, we will set γ_0 to $m-2$. Then shifted matrix of $(\lambda'_{ij})_{0 \leq i \leq v, 1 \leq j \leq m-1}$ has unique 1 in each column. For understanding γ_i, we give small example for $m = 9$.

Example 1. For $m = 9$, type-II ONB multiplication matrix λ_{ij} and shifted multiplication matrix λ'_{ij} are as below.

$$(\lambda_{ij}) = \begin{pmatrix} 0 & 1 & 0 & 0 & 0 & 0 & 0 & 0 & 0 \\ 1 & 0 & 0 & 0 & 1 & 0 & 0 & 0 & 0 \\ 0 & 0 & 0 & 0 & 1 & 0 & 0 & 1 & 0 \\ 0 & 0 & 0 & 0 & 0 & 0 & 1 & 0 & 1 \\ 0 & 1 & 1 & 0 & 0 & 0 & 0 & 0 & 0 \\ 0 & 0 & 0 & 0 & 0 & 0 & 1 & 1 & 0 \\ 0 & 0 & 0 & 1 & 0 & 1 & 0 & 0 & 0 \\ 0 & 0 & 1 & 0 & 0 & 1 & 0 & 0 & 0 \\ 0 & 0 & 0 & 1 & 0 & 0 & 0 & 0 & 1 \end{pmatrix} \Rightarrow (\lambda'_{ij}) = \begin{pmatrix} 0 & 0 & 0 & 0 & 0 & 0 & 0 & 0 & 1 \\ 0 & 1 & 0 & 0 & 0 & 0 & 1 & 0 & 0 \\ 0 & 0 & 1 & 0 & 0 & 1 & 0 & 0 & 0 \\ 1 & 0 & 0 & 0 & 0 & 0 & 0 & 1 & 0 \\ 0 & 0 & 0 & 1 & 1 & 0 & 0 & 0 & 0 \\ 0 & 0 & 0 & 0 & 0 & 0 & 1 & 1 & 0 \\ 0 & 0 & 0 & 1 & 0 & 1 & 0 & 0 & 0 \\ 0 & 0 & 1 & 0 & 0 & 1 & 0 & 0 & 0 \\ 0 & 0 & 0 & 1 & 0 & 0 & 0 & 0 & 1 \end{pmatrix}$$

Therefore, γ_i, $0 \leq i \leq 4$ as follows.

i	0	1	2	3	4
γ_i	7	6	7	1	2

Theorem 2. *In type-II optimal normal basis, one can find the γ_i for each $0 \leq i \leq v$.*

Proof. By Lemma 1 and Corollary 1, the theorem is clear.

Using above theorem 2 we can modify the equation of Reyhani-Masoleh and Hasan[2].

$$C = \sum_{i=0}^{m-1} a_i b_i \alpha_{i+1} + \sum_{i=0}^{m-1} \sum_{j=1}^{v} z_{ij} \cdot \delta_j^{2^i}$$

$$= \sum_{i=0}^{m-1} a_{i+\gamma_0} b_{i+\gamma_0} \alpha_{i+\gamma_0+1} + \sum_{i=0}^{m-1} \sum_{j=1}^{v} z_{i+\gamma_j,j} \cdot \delta_j^{2^{i+\gamma_j}}$$

$$= \sum_{i=0}^{m-1} (a_{i+\gamma_0} b_{i+\gamma_0} \alpha_{\gamma_0+1} + \sum_{j=1}^{v} z_{i+\gamma_j,j} \cdot \delta_j^{2^{\gamma_j}})^{2^i}$$

$$G_i(A, B) = a_{i+\gamma_0} b_{i+\gamma_0} \alpha_{\gamma_0+1} + \sum_{j=1}^{v} z_{i+\gamma_j,j} \cdot \delta_j^{2^{\gamma_j}}$$

Therefore,
$$C = ((G_{m-1}^2 + G_{m-2})^2 + \cdots + G_1)^2 + G_0$$

By $G_{m-t}(A, B) = G_{m-1}(A^{2^{t-1}}, B^{2^{t-1}})$, we get the multiplication algorithm.

Modified Reyhani-Masoleh and Hasan's Multiplier

INPUT : $A = (a_0, a_1, \cdots, a_{m-1}), B = (b_0, b_1, \cdots, b_{m-1})$
OUTPUT : $C = (c_0, c_1, \cdots, c_{m-1})$
1. Find the proper γ_i for $0 \leq i \leq v$ where each column of $(\lambda'_{ij})_{0 \leq i \leq v, 0 \leq j \leq m-1}$ has exactly one 1s.
2. A, B are loaded in m-bit registers respectively.
   All intermediate values $D_0, D_1, \cdots, D_{m-1}$ are set to zero.
3. For $i = 0$ to $m - 1$, do the following.
   3.1 $D = D^2 + G_{m-1}(A, B)$, where the computation is done in parallel for all α_i's.
   3.2 $A \leftarrow A^2, B \leftarrow B^2$.
4. After m-th iteration, we have $D_i = c_i$ for all $0 \leq i \leq m-1$, where $AB = \sum_{i=0}^{m-1} c_i \alpha_i$

5 Small Examples

5.1 $k = 2$ Example

For example, there is a type-II ONB in $GF(2^5)$. In [3], S.Kwon et al. suggested a method to compute $\alpha \alpha_i$ easily. By the method, we can get the multiplication matrix (λ_{ij}):

$$(\lambda_{ij}) = \begin{pmatrix} 0&1&0&0&0 \\ 1&0&0&1&0 \\ 0&0&0&1&1 \\ 0&1&1&0&0 \\ 0&0&1&0&1 \end{pmatrix} \rightarrow (\lambda'_{ij}) = \begin{pmatrix} 0&0&0&0&1 \\ 1&0&0&1&0 \\ 0&1&1&0&0 \\ 0&1&1&0&0 \\ 0&0&1&0&1 \end{pmatrix}$$

We can set γ_i, $0 \leq i \leq 2$ as follows.

$$\gamma_i = \begin{cases} 3, & \text{if i=0}; \\ 0, & \text{if i=1}; \\ 3, & \text{if i=2}. \end{cases}$$

$$G_i(A, B) = a_{i+\gamma_0} b_{i+\gamma_0} \alpha_{\gamma_0+1} + \sum_{j=1}^{v} z_{i+\gamma_j, j} \cdot \delta_j^{2^{\gamma_j}}$$

Since $m - 1 = 4$ and $v = 2$, we get

$$G_4(A, B) = a_2 b_2 \alpha_4 + z_{4,1} \cdot \delta_1 + z_{2,2} \cdot \delta_2^{2^3}$$

Originally, Reyhani-Masoleh and Hasan [2] computed $z_{4,1}, z_{4,2}$ first. But our structure computes $z_{4,1}, z_{2,2}$ first and then sequentially accumulate the result. Then the target structure is like as Fig.2.

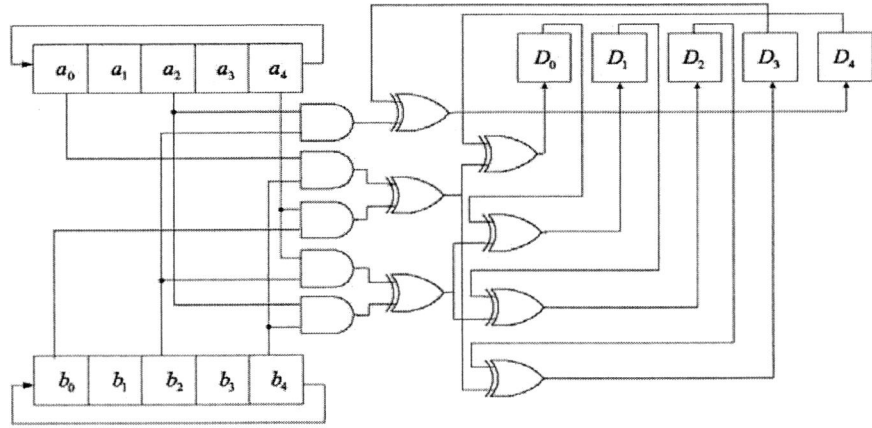

Fig. 2. Proposed modification of Reyhani-Masoleh and Hasan's XESMPO[2]

6 Complexity Analysis

The proposed multiplier has the same area complexity with Reyhani-Masoleh and Hasan[2]. With restriction of type-II ONB, the proposed method improves the time efficiency of Reyhani-Masoleh and Hasan [2]. Therefore, it is more area efficient and has the same time efficient structure as Agnew et al.'s SMPO [1] and it has the same area and time complexity with S.Kwon et al.'s SMPO [3]. Consequently, we can get complexity results of table 1.

Table 1. Complexity Analysis of SMPO Multipliers for type-II ONB case

	Critical path delay	AND	XOR	flip-flop
Agnew et al[1]	$T_A + 2T_X$	m	2m-1	3m
Reyhani-Masoleh and Hasan[2]	$T_A + 3T_X$	m	$\frac{3m-1}{2}$	3m
S.Kwon et al[3]	$T_A + 2T_X$	m	$\frac{3m-1}{2}$	3m
Proposed	$T_A + 2T_X$	m	$\frac{3m-1}{2}$	3m

7 Conclusion

In type-II ONB of $GF(2^m)$, we get the time efficiency of Reyhani-Masoleh and Hasan's multiplier[2]. The area and time efficiency of our proposed multiplier is the same as S.Kwon et al's multiplier[3]. We expect that for general type k it could be possible to expand our method. But to show that, it needs that the each column of $(\lambda'_{ij})_{0 \leq i \leq v, 0 \leq j \leq m-1}$ has the smallest number of 1s possible. Although our proposal applies to XESMPO, it can be also expanded to AESMPO.

References

1. G.B. Agnew, R.C. Mullin, I. Onyszchuk, and S.A. Vanstone,An implementation for a fast public key cryptosystem, J. Cryptology, vol.3, pp. 63-79, 1991.
2. A. Reyhani-Masoleh and M.A. Hasan,Efficient Digit-Serial Normal Basis Multipliers over Binary Extension Fields, ACM Trans. on Embedded Computing Systems (TECS), Special Issue on Embedded Systems and Security, pp.575-592, vol.3, Issue 3, August 2004.
3. S. Kwon, K. Gaj, C.H. Kim, C.P. Hong,Efficient Linear Array for Multiplication in $GF(2^m)$ Using a Normal Basis for Elliptic Curve Cryptography, CHES 2004, LNCS 3156, pp. 76-91, 2004.
4. E.R. Berlekamp,Bit-serial Reed-Solomon encoders, IEEE Trans. Inform. Theory, vol. 28, pp. 869-874, 1982.
5. H. Wu, M.A. Hasan, I.F. Blake, and S. Gao,Finite field multiplier using redundant representation, IEEE. Trans. Computers, vol 51, pp. 1306-1316, 2002.
6. A. Reyhani-Masoleh and M.A. Hasan,A new construction of Massey-Omura parallel multiplier over $GF(2^m)$, IEEE Trans. Computers, vol. 51, pp. 511-520, 2002.
7. A. Reyhani-Masoleh and M.A. Hasan,Efficient multiplication beyond optimal normal bases, IEEE Trans. on Computers, vol.52, pp.428-439, 2003.
8. A. Reyhani-Masoleh and M. A. Hasan, "Low Complexity Word-Level Sequential Normal Basis Multipliers," IEEE Transactions on Computers, pp 98-110, Vol. 54, No. 2, February 2005.
9. C. Paar, P. Fleischmann, and P. Roelse,Efficient multiplier architectures for Galois fields $GF(2^{4n})$, IEEE Trans. Computers, vol.47, pp.162-170, 1998.
10. B. Sunar and C.K. Koc,An efficient optimal normal basis type-II multiplier, IEEE Trans. Computers, vol.50, pp.83-87, 2001.
11. J.L. Massey and J.K. Omura,Computational method and apparatus for finite field arithmetic, US Patent NO. 4587627, 1986.
12. NIST, Digital Signature Standard, FIPS Publication, 186-2, February, 2000.
13. ANSI, Public Key Cryptography for the Financial Services Industry: The Elliptic Curve Digital Signature Algorithm(ECDSA),ANSI x9.62, 1998.
14. S. Gao,Normal Bases over Finite Fields, A thesis for Doctor of Philosophy, 1993.
15. Soonhak Kwon, Chang Hoon Kim and Chun Pyo Hong, Efficient Exponentiation for a Class of Finite Fields Determined by Gauss Periods, CHES 03, LNCS,pp.228-242

A New Method of Building More Non-supersingular Elliptic Curves

Shi Cui, Pu Duan, and Choong Wah Chan

School of Electrical and Electronic Engineering,
Nanyang Technological University,
50 Nanyang Avenue, 639798, Singapore
{PG04063705, PG03460751, ecwchan}@ntu.edu.sg

Abstract. Non-supersingular curves are useful to improve the security of pairing-based cryptosystems. The method proposed by Brezing and Weng is computational inexpensive which can build suitable non-supersingular elliptic curves for pairing-based cryptosystems when the embedding degree is larger than 6. In this paper we propose a new method which extends Brezing and Weng's method to generate more non-supersingular elliptic curves suitable for pairing-based cryptosystems. Furthermore, we show how our proposed method can be used in the method proposed by Scott and Barreto. Some examples are given to show that new non-supersingular curves can be built.

1 Introduction

Since the identity-based encryption scheme [1] and the short signature scheme [5] were proposed, pairing-based cryptosystem has been one of the most active topics in Elliptic Curve Cryptography (ECC). Most of these earlier works were based on supersingular elliptic curves. However, they were less reliable than those of non-supersingular curves [8, 12, 14]. Miyaji, Nakabayashi and Takano [4] proposed the novel idea for building these special curves. Such curves are often cited as the MNT curves. In 2002, Barreto et al. [6] proposed an algorithm, here called BKLS Algorithm, for fast computing pairing. The advantage of the algorithm is that it can use MNT curves to speed up the pairing computing. However, from the work of [18], the authors concluded that the number of such special non-supersingular curves is small.

Miyaji's method [4] for building MNT curves has some strict restrictions as only few curves can be built. In recent years, several new methods [2, 3, 8, 9, 7, 20] have been proposed to find additional non-supersingular curves for pairing-based cryptosystems. However, elliptic curves built by these methods [2, 8] result in longer signatures or larger cipher text in pairing-based cryptosystems [2, 3]. As such, they are rarely used in real world applications. How to find an algorithm to construct non-supersingular elliptic curves with suitable parameters was an open problem for the researchers in the area of Elliptic Curve Cryptography. Scott and

Barreto [3] proposed a method that generates very suitable non-supersingular curves for pairing-based cryptosystems, but the embedding degree must be less than or equal to 6. In 2003, Brezing and Weng [9] proposed a simple and computational inexpensive method to build non-supersingular curves for pairing-based cryptosystems. Their method builds non-supersingular curves over $\mathbb{F}_p$ with embedding degree larger than 6, and the order of elliptic curves has a large prime factor r such that p is much smaller than r^2. Those pairing-based cryptosystems based on such curves have better performance than those stated in [2, 8]. In addition, [7, 20] also gave two methods to generate non-supersingular elliptic curves with embedding degree larger than 6. However, the method is more complicated than that proposed by [9].

In this paper, we extend the work of Brezing and Weng to build more non-supersingular curves suitable for pairing-based cryptosystems. Our method modifies the condition of Brezing and Weng's method such that additional non-supersingular curves suitable for pairing-based cryptosystems are built. The proposed method also can be adopted in other methods [2, 3, 8] to construct more non-supersingular curves for pairing-based cryptosystems.

This paper is organized as follows. In Section 2, the method proposed by Brezing and Weng is reviewed. Section 3 describes our proposed new method, which is an extension of Brezing and Weng's method to obtain new non-supersingular curves. We provide an analysis of the method and give examples. We discuss the effect of the new method when applied to Scott and Barreto's method in Section 4. Finally, the conclusion is drawn in Section 5.

2 Review of Brezing and Weng's Method

Suppose that an elliptic curve E over a finite field $\mathbb{F}_q$ has order n which is the product of a small integer h and a large prime r. If k is the smallest integer such that r divides $q^k - 1$, then E is thought as having embedding degree k. Let t denote the trace of Frobenius endomorphism on E, it is well known that $n = hr = q+1-t$. Since r divides $q^k - 1$ but does not divide $q^i - 1$ for $0 < i < k$, thus $(t-1)^k \equiv 1 \pmod{r}$ can be deduced by $(hr + t - 1)^k \equiv 1 \pmod{r}$. $(t - 1)$ is thought as a primitive kth root ζ_k of unity modulo r. The authors of [9] attributed this contribution to Cock and Pinch.

Definition 1. *An algebraic integer of the form $a + b\sqrt{D}$ forms an imaginary quadratic field, where D is a negative squarefree integer. It is usually denoted as $\mathbb{Q}(\sqrt{D})$.*

An elliptic curve E over $\mathbb{F}_p$ is called ordinary if $E[p] \simeq Z_p$. It is called supersingular if $E[p] \simeq 0$. Let $End(E)$ denote the ring of endomorphism of E. When E is ordinary, $End(E)$ is an order in $\mathbb{Q}(\sqrt{D})$. Then, an ordinary elliptic curve E has complex multiplication in $\mathbb{Q}(\sqrt{D})$. Suppose that $K = \mathbb{Q}(\sqrt{D})$ is an imaginary quadratic field, and $\mathcal{O}_K$ represents the largest subring of K. Brezing and Weng used the following strategy to build non-supersingular curves for pairing-based cryptosystems:

First, randomly choose a negative squarefree integer D, set an imaginary quadratic field $K = \mathbb{Q}(\sqrt{D})$. Second, find a prime r which must satisfy the conditions of r splitting in $\mathcal{O}_K$ and $r \equiv 1 \pmod{k}$, then find a primitive kth root ζ_k of unity modulo r in $\mathbb{Q}(\sqrt{D})$. Third, select an element from some order of discriminant $-D$ in $\mathbb{Q}(\sqrt{D})$, $\omega = \frac{a+b\sqrt{D}}{2}$, where $a = \zeta_k + 1 \pmod{r}$ and $b = \pm\frac{a-2}{\delta} \pmod{r}$ (δ is square root of d modulo r. $d = \frac{D}{4}$ if $D \equiv 0 \pmod{4}$, otherwise $d = D$). Actually, a is equal to the trace t. Finally, if $p = Norm_{K/\mathbb{Q}}(\omega)$ is a prime (or a prime power q), by the complex multiplication method [10], construct elliptic curves over $\mathbb{F}_p$ with embedding degree k. Its discriminant is $-D$, the order of $E(\mathbb{F}_p)$ is:

$$\#E(\mathbb{F}_p) = Norm_{K/\mathbb{Q}}(\omega - 1) = \frac{(a-2)^2 + Db^2}{4}. \tag{1}$$

It is easy to verify $\#E(\mathbb{F}_p) \equiv 0 \pmod{r}$. From [6], it is known that such curves can be used to speed up the Tate Pairing computation.

3 Extension of Brezing and Weng's Method

Brezing and Weng used the relationship between the trace and the primitive kth roots of unity to successfully find a desirable prime p, and then built non-supersingular curves over $\mathbb{F}_p$ by the complex multiplication method. The relationship, $t = \zeta_k + 1 \pmod{r}$, is essential to this method. Here a new relationship between t and ζ_k is proposed, which allows Brezing and Weng's method to build more non-supersingular curves. It is noticed that the first part of the following lemma is summarized from [9].

Algorithm 1 provides the detailed description of the method of Brezing and Weng and its modification under Lemma 1. It is noticed that no modular reduction is needed in Step 5, and in Step 2, $a(x) \leftarrow 1 - g(x) \pmod{f(x)}$ is available only if k is even. Furthermore, r_{min} is at least 160 to resist the Pohlig-Hellman attack [15]. p_{min} must satisfy $k \lg(p) \geq 1024$ to resist the index-calculus attack [15].

Lemma 1. *An elliptic curve E over $\mathbb{F}_q$ with embedding degree k has order n that is the product of a small integer h and a large prime r, t is the trace of Frobenius, then not only $(t-1)$ is the primitive k-th roots of unity modulo r, but also $(1-t)$ is the primitive k-th roots of unity modulo r when k is even.*

Proof: It has been known that $(t-1)^k \equiv \zeta_k \pmod{r}$ for all k in [9]. When k is even, there exists $(1-t)^k = (t-1)^k \equiv 1 \pmod{r}$, thus, $(1-t)$ is the primitive k-th roots of unity modulo r when k is even. □

To illustrate the effect of the new method, we will analyze some examples in the rest of this section. All the examples are obtained under the Magma environment [13] which runs on a Pentium IV PC, 1.7 GHz and 256 Mb RAM. Furthermore, to verify whether the obtained curves are suitable for pairing-based cryptosystems or not, the value of $\lg(p)/\lg(r)$ is required. From [3], the closer the value of $\lg(p)/\lg(r)$ is to 1, the better the performance of pairing-based cryptosystems.

Example 1: $k = 8$, $D = -1$, M is $\mathbb{Q}(\zeta_8, -1)$, $f(x) = x^4 + 1$, the primitive 8th roots of unity are represented by the polynomials:

$$x, -x, x^3, -x^3.$$

The square roots of -1 are denoted by the polynomial: $\pm x^2$. Whatever $t = 1 - \zeta_8$ or $t = 1 + \zeta_8$, there must be $a(x) = (1 \pm x)$ or $(1 \pm x^3)$, all non-supersingular curves based on $t = 1 - \zeta_8$ are identical with those based on $t = 1 + \zeta_8$. Thus, when $k = 8$ and $D = -1$, there are no new non-supersingular curves.

Algorithm 1

| Input: | a positive integer k and a negative squarefree integer D |
| Output: | prime number r, p |

1. Set up a cyclotomic field $M = \mathbb{Q}(\zeta_k)$, if $\sqrt{D} \not\subseteq M$ then $M = \mathbb{Q}(\zeta_{2k})$;
2. Compute the defining polynomial $f(x)$ of M;
3. Compute the primitive k-th roots $g(x)$ of unity;
4. $h(x) \leftarrow \sqrt{D}$, $a(x) \leftarrow 1 \pm g(x) \pmod{f(x)}$, $b(x) \leftarrow \frac{a(x)-2}{h(x)} \pmod{f(x)}$;
5. $p(x) \leftarrow \frac{a(x)^2 - Db^2(x)}{4}$;
6. Find an congruence class $x_0 \bmod (-D)$ such that $b(x_0) \equiv 0 \pmod{-D}$;
7. If $p(x)$ is irreducible and $p(x_0)$ is an integer then
8.   For i from 1 to i_{max} do
9. $x_i \equiv x_0 \pmod{-D}$;
10.   if $b(x_i) \equiv 0 \pmod{-D}$ then
11. $x_1 \leftarrow x_i$
12.   if $f(Dy + x_0)$ is irreducible in $Z[y]$, then
13. $r \leftarrow f(x_1)$ and $p \leftarrow p(x_1)$
14.   if p is prime and r is prime then
15.   if $bits(r) \geq r_{min}$ and $bits(p) \geq p_{min}$ then
16. $\#E \leftarrow \frac{(a(x_1)-2)^2 - Db^2(x_1)}{4}$;
17.   output r, p, $\#E$;
18.   end if;
19.   end if;
20.   end if;
21.   end if;
22.   end for;
23. end if.

Example 2: $k = 10$, $D = -1$, M is $\mathbb{Q}(\zeta_{20}, -1)$, $f(x) = x^8 - x^6 + x^4 - x^2 + 1$, the primitive 10th roots of unity are represented by the polynomials:

$$x^2, -x^4, x^6, -x^6 + x^4 - x^2 + 1.$$

The square roots of -1 are denoted by the polynomial: $\pm x^5$. When $\zeta_{10} = -x^6 + x^4 - x^2 + 1$, $a(x) = 1 - \zeta_{10} = 1 - (-x^6 + x^4 - x^2 + 1) = x^6 - x^4 + x^2$, $h(x) = \pm x^5$, then

$b(x) = \pm\frac{a(x)-2}{h(x)} = \pm(x^5 + x^3) \pmod{f(x)}$. Compare it with $b(x) = \pm(x^5 - x^3)$ $\pmod{f(x)}$ when $a(x) = 1 + \zeta_{10} = -x^6 + x^4 - x^2 + 2$ in [9]. It is easy to find that these two $b(x)$ are different. Search suitable integer x_1 such that $r = f(x_1) = x_1^8 - x_1^6 + x_1^4 - x_1^2 + 1$ and $p = Norm_{K/\mathbb{Q}}(\frac{a(x_1)+b(x_1)\sqrt{-1}}{2})$ are prime. Since $b(x)$ is different, p must be different. Thus we can build new non-supersingular curves over $\mathbb{F}_p$ for pairing-based cryptosystems, where $\lg(p)/\lg(r) \approx \lg(x_1^{12})/\lg(x_1^8) = \frac{3}{2}$.

Example 3: $k = 10$, $D = -5$, M is $\mathbb{Q}(\zeta_{20}, -5)$, $f(x) = x^8 - x^6 + x^4 - x^2 + 1$, the primitive 10th roots of unity are represented by the polynomials:

$$x^2, -x^4, x^6, -x^6 + x^4 - x^2 + 1.$$

The square roots of -5 are denoted by the polynomial: $\pm(2x^7 - x^5 + 2x^3)$. When $\zeta_{10} = x^2$, $a(x) = 1 - \zeta_{10} = 1 - x^2$, $h(x) = \pm x^7$, then $b(x) = \pm\frac{a(x)-2}{h(x)} = \pm\frac{(3x^7 - x^5 + 4x^3 - 2x)}{5} \pmod{f(x)}$. In [3], $a(x) = 1 + \zeta_{10} = 1 + x^2$, $b(x) = \pm\frac{x^7 - x^5 + 2x}{5}$ $\pmod{f(x)}$ which is different from the above $b(x)$. Search suitable integer $x_1 \equiv 2 \pmod{5}$ such that $r = f(x_1) = x_1^{12} - x_1^{10} + x_1^8 - x_1^6 + x_1^4 - x_1^2 + 1$ and $p = Norm_{K/\mathbb{Q}}(\frac{a(x_1)+b(x_1)\sqrt{-5}}{2})$ are prime. We can build new non-supersingular curves over $\mathbb{F}_p$ for pairing-based cryptosystems, where $\lg(p)/\lg(r) \approx \lg(x_1^{14})/\lg(x_1^8) = \frac{7}{4}$.

Example 4: $k = 14$, $D = -1$, M is $\mathbb{Q}(\zeta_{28}, -1)$, $f(x) = x^{12} - x^{10} + x^8 - x^6 + x^4 - x^2 + 1$, the primitive 14th roots of unity are represented by the polynomials:

$$x^2, x^6, -x^4, -x^8, x^{10}, -x^{10} + x^8 - x^6 + x^4 - x^2 + 1.$$

The square roots of -1 are given by the polynomial: $\pm x^7$. When $\zeta_{14} = x^8$, $a(x) = 1 - \zeta_{14} = 1 - (-x^8) = x^8 + 1$, $h(x) = \pm x^7$, then $b(x) = \pm\frac{a(x)-2}{h(x)} = \pm(x^7 + x)$ $\pmod{f(x)}$. If $a(x) = 1 + \zeta_{14} = -x^8 + 1$, then $b(x) = \pm(x^5 - x) \pmod{f(x)}$. It is easy to find that the second $b(x)$ is different from the first one. Search some integer x_1 such that $r = f(x_1) = x_1^{12} - x_1^{10} + x_1^8 - x_1^6 + x_1^4 - x_1^2 + 1$ and $p = Norm_{K/\mathbb{Q}}(\frac{a(x_1)+b(x_1)\sqrt{-5}}{2})$ are prime, then we can build new non-supersingular curves over finite field $\mathbb{F}_p$ for pairing-based cryptosystems, where $\lg(p)/\lg(r) \approx \lg(x_1^{16})/\lg(x_1^{12}) = \frac{4}{3}$.

Example 5: This example is different from the above. Let M be $\mathbb{Q}(\zeta_{60}, -3)$, the defining polynomial of M is $f(x) = x^{16} + x^{14} - x^{10} - x^8 + x^6 + x^2 + 1$. We consider the case of $k = 10$, 20 and 30, respectively.

(1) $k = 10$, the primitive 10th roots of unity in M are represented by the polynomials:

$$x^6, -x^{12}, x^{14} + x^{12} - x^6 - x^4 + 1, -x^{14} + x^4.$$

The square roots of -1 are denoted by the polynomial: $\pm x^{15}$. When $\zeta_{10} = -x^{12}$, $a(x) = 1 - \zeta_{10} = 1 + x^{12}$, the $b(x) = \frac{(-x^{12} + 2x^{10} + 2x^2 - 1)}{3}$. There is no x_0 satisfying $b(x_0) \equiv 0 \pmod{3}$.

(2) $k = 20$, the primitive 20th roots of unity are represented by the polynomials:

$$x^3, -x^3, x^9, -x^9, x^{11} - x, -x^{11} + x, x^{15} - x^{11} - x^9 + x^3 + x.$$

The square roots of -3 are denoted by the polynomial: $\pm(2x^{10} - 1)$. When $\zeta_{20} = x^{15} - x^{11} - x^9 + x^3 + x$, $a(x) = 1 - (x^{15} - x^{11} - x^9 + x^3 + x) = -x^{15} + x^{11} + x^9 - x^3 - x + 1$, then $b(x) = \frac{(-x^{15}+x^{11}+2x^{10}+x^9+2x^7-x^3-x-1)}{3}$, search $x_1 \equiv 2 \pmod{3}$ such that $r = f(x_1)$ and $p = \text{Norm}_{K/\mathbb{Q}}(\frac{a(x_1)+b(x_1)\sqrt{-3}}{2})$ are prime, we can construct non-supersingular curves over $\mathbb{F}_p$, where $\lg(p)/\lg(r) \approx \lg(x_1^{30})/\lg(x_1^{16}) = \frac{15}{8}$.

(3) $k = 30$, the primitive 30th roots of unity in M are represented by the polynomials:

$$x^2, -x^4, -x^8, x^{14}, x^{12} - x^2, -x^{14} + x^{10} + x^8 - x^2 - 1,$$

$$x^{14} - x^{10} - x^8 - x^6 + x^2 + 1, -x^{14} - x^{12} + x^8 + x^6 + x^4 - 1.$$

The square roots of -3 are denoted by the polynomial: $\pm(2x^{10} - 1)$. When $\zeta_{30} = -x^{14} - x^{12} + x^8 + x^6 + x^4 - 1$, $a(x) = 1 - (-x^{14} - x^{12} + x^8 + x^6 + x^4 - 1) = x^{14} + x^{12} - x^8 - x^6 - x^4 + 2$, then $b(x) = \frac{(x^{14}+x^{12}+2x^{10}+x^8-x^6-x^4)}{3}$, search $x_1 \equiv 1 \pmod{3}$ such that $r = f(x_1)$ and $p = \text{Norm}_{K/\mathbb{Q}}(\frac{a(x_1)+b(x_1)\sqrt{-3}}{2})$ are prime, we can construct non-supersingular curves over $\mathbb{F}_p$, where $\lg(p)/\lg(r) \approx \lg(x_1^{30})/\lg(x_1^{16}) = \frac{15}{8}$.

Cryptosystems based on such non-supersingular curves with the value of $\lg(p)/\lg(r)$ close to 1, will generate shorter signatures and smaller ciphertext. It is obvious from the observation that Example 4 is the most desirable case.

4 For the Other Method

Lemma 1 not only allows the method of Brezing and Weng to build more non-supersingular curves, but also extends other methods to build more non-supersingular curves. The methods included are proposed by Dupont et al. [8] and by Scott and Barreto [3]. The authors of [3, 8] thought that r divides $\Phi_k(t-1)$, where $\Phi_k(x)$ is the kth cyclotomic polynomial [11]. So when k is even, by Lemma 1, r divides $\Phi_k(1-t)$, too. Hence, the proposed method permits [3, 8] to find more non-supersingular curves. Especially, [3] can build non-supersingular curves with $\lg(p)/\lg(r) \approx 1$ when k is 6. Example 6 is the result of the extension of Scott and Barreto method [3].

Example 6: Let $y^2 = x^3 + Ax + B$ denote the corresponding elliptic curve equation, a suitable curves with the following parameter has been constructed by [19]:

$D = 190587$,
$p = 208820892459110891021844986733867354845534222206640829$, (a 174 bits prime)
$r = 248596300546560584549815459537304207955629003396551$, (a 168 bits prime);

$A = -3$;
$B = 19130657142612667912176666428455428219574104052452909$,

where $\lg(p)/\lg(r) \approx \frac{174}{168} \approx 1.04$. Apart from the results presented in Example 6, we are still in the process of looking for new curves and have found some new non-supersingular curves suitable for pairing-based cryptosystems. We will include these curves in our future report.

5 Conclusion

In this paper, we presented a method that extends the Brezing and Weng's method to build more non-supersingular curves. We described the proposed new method and how they are used to build more non-supersingular curves which are suitable for pairing-based cryptosystems. Analysis of some examples were provided. We also applied the new method in Scott and Barreto method. The result obtained is favorable.

Acknowledgements

The authors wish to thank the anonymous reviewers for their helpful comments and the authors of Magma [13] which helps us complete the necessary computation.

References

1. D. Boneh and M. Franklin, "Identity based encryption from the Weil pairing", SIAM J. of Computing, Vol. 32, no. 3, pp. 586-615, 2003.
2. P. S. L. M. Barreto, B. Lynn, and M. Scott, "Constructing Elliptic Curves With Prescribed Embedding Degree", In Security in Communication Networks - SCN'2002, Vol. 2576 of Lecture Notes in Computer Science, pp. 263-273, Springer-Verlag, 2002.
3. M. Scott and P. S. L. M. Barreto, "Generating more MNT elliptic curves", Cryptology ePrint Archive, Report 2004/058, 2004. Available from http://eprint.iacr.org/2004/058.
4. A. Miyaji, M. Nakabayashi and S. Takano, "New explicit conditions of elliptic curve traces for FR-reduction", IEICE Trans. Fundamentals, Vol. E84 A, no.5, pp. 1234-1243, May 2001.
5. D. Boneh, B. Lynn and H. Shacham, "Short signatures from the Weil pairing", J. of Cryptology, Vol. 17, no. 4, pp. 297-319, 2004.
6. P.S.L.M. Barreto, H.Y. Kim, B. Lynn, and M. Scott, "Efficient algorithms for pairing-Based Cryptosystems", CRYPTO 2002, Volume 2442 of Lecture Notes in Computer Science, pp.354-369, Springer-Verlag, 2002.
7. P. S. L. M. Barreto, B. Lynn, and M. Scott, "Efficient algorithms for pairing-based cryptosystems", Journal of Cryptology, Vol. 17, no. 4, pp. 321-334, Springer-Verlag, 2004.

8. R. Dupont, A. Enge and F. Morain, "Building curves with arbitrary small MOV degree over finite prime fields", Cryptology ePrint Archive, Report 2002/094. Available from http://eprint.iacr.org/2002/094.
9. F. Brezing and A. Weng, "Elliptic curves suitable for pairing based cryptography", Cryptology ePrint Archive, Report 2003/143. Available from http://eprint.iacr.org/2003/143.
10. I. F. Blake, G. Seroussi and N.P. Smart, "Elliptic Curves in Cryptography", Vol. 265 of London Mathematical Society Lecture Note Series. Cambridge University Press, 1999.
11. Cyclotomic Polynomial, Available from http://mathworld.wolfram.com/Cyclotomic Polynomial.html.
12. D. Page, N. P. Smart and F. Vercauteren, "A comparison of MNT curves and supersingular curves", Cryptology ePrint Archive, Report 2004/165, 2004. Available from http://eprint.iacr.org/2004/165.
13. Computational Algebra Group of the University of Sydney, Magma version 2.11-10, October 2004. Available from http://magma.maths.usyd.edu.au/.
14. D. Coppersmith, "Fast evaluation of logarithms in fields of characteristic two", IEEE Transactions on Information Theory, Vol. 30, no. 4, pp. 587-594, July 1984.
15. A. M. Odlyzko, "Discrete logarithms: the past and the future", Design, Code and Cryptography, Vol. 19, no.2, pp. 129-145, March 2002.
16. F. Morain, "Building cyclic elliptic curves modulo large primes", Advances in Cryptology-EUROCRYPT'91, Vol. 547 of Lecture Notes in Computer Science, pp. 328-336, Springer-Verlag, 1991.
17. A. O. L. Atkin and F. Morain, "Elliptic curves and primality proving", Math. Comp., 61(203):29-68, July 1993.
18. R. Balaubramanian and N. Koblitz, "The improbability that an elliptic curve has subexponential discrete log problem under the Menezes-Okamoto-Vanstone algorithm", J. of Cryptology, Vol. 11, no.2, pp. 141-145, 1998.
19. Complex multiplication, http://ftp.compapp.dcu.is/pub/crypto/cm.exe, 2002.
20. S. D. Galbraith, J. McKee and P. Valenca, "Ordinary abelian varieties having small embedding degree", Cryptology ePrint Archie, Report 2004/365, 2004. Available from http://eprint.iacr.org/2004/365.

Accelerating AES Using Instruction Set Extensions for Elliptic Curve Cryptography

Stefan Tillich and Johann Großschädl

Graz University of Technology,
Institute for Applied Information Processing and Communications,
Inffeldgasse 16a, A–8010 Graz, Austria
{Stefan.Tillich, Johann.Groszschaedl}@iaik.at

Abstract. The Advanced Encryption Standard (AES) specifies an algorithm for a symmetric-key cryptosystem that has already found wide adoption in security applications. A substantial part of the AES algorithm are the MixColumns and InvMixColumns operations, which involve multiplications in the binary extension field $GF(2^8)$. Recently proposed instruction set extensions for elliptic curve cryptography (ECC) include custom instructions for the multiplication of binary polynomials. In the present paper we analyze how well these custom instructions are suited to accelerate a software implementation of the AES. We used the SPARC V8-compatible LEON-2 processor with ECC extensions for verification and to obtain realistic timing results. Taking the fastest implementation for 32-bit processors as reference, we were able to achieve speedups of up to 25% for encryption and nearly 20% for decryption.

Keywords: Advanced Encryption Standard, Rijndael, 32-bit implementation, software acceleration, instruction set extensions.

1 Introduction

A lot of research has been conducted towards the efficient implementation of AES in both hardware and software. There exists a considerable literature about hardware architectures for AES, which target a wide spectrum of platforms ranging from high-end servers [11] to smart cards [10] and RFID tags [4].

Even highly optimized software implementations of the AES are, in general, orders of magnitude slower than dedicated hardware solutions. This is partly due to the fact that secret-key cryptosystems (and also some public-key systems) have to carry out operations which are not very well supported by general-purpose processors. An example for such operations are multiplications in the binary extension field $GF(2^8)$, which are used in some modern block ciphers like the AES or Twofish. General-purpose processors do not provide an instruction for the multiplication of binary polynomials, and therefore this operation must be "emulated" using Shift and XOR instructions.

A recent trend in embedded processor design is to extend a general-purpose instruction set architecture (ISA) by special instructions for performance-critical operations. The concept of *instruction set extensions*, which may be considered as a hardware/software co-design approach, can significantly improve the performance of certain applications, e.g. secret-key or public-key cryptosystems. In general, such extensions allow to achieve much better performance compared to a "conventional" software implementation, but require less silicon area than a dedicated hardware solution like a cryptographic co-processor [7, 9]. One possible application that could make use of ECC instruction set extensions are secure sensor nodes which authenticate connections and exchange session keys with ECC mechanisms and perform secure data transfer using AES encryption.

In this paper we analyze a typical AES software implementation to show that the MixColumns and InvMixColumns transformations can be optimized with instruction set extensions. We try to estimate the achievable performance gain for two implementation strategies of AES and present timing figures for an extended version of the SPARC V8 LEON-2 embedded processor [5, 8].

The rest of the paper is organized as follows. In Section 2 we consider implementation options of AES for 32-bit processors. Section 3 shows how ECC instruction set extensions can be employed to accelerate AES. We also try to estimate the number of instructions for important parts of MixColumns and InvMixColumns in this section. In Section 4 we present the timing measurements on the LEON-2 processor implemented on an FPGA prototyping board. In Section 5 conclusions are drawn and a short outlook on future work is given.

2 AES Implementations on 32-Bit Processors

The AES transforms the input data in a number of rounds, where each round consists of four individual transformations. For encryption these transformations are SubBytes, ShiftRows, MixColumns and AddRoundKey. Decryption uses the inverse of these functions, where AddRoundKey is its own inverse. The order of functions for a decryption round is InvShiftRows, InvSubBytes, AddRoundKey and InvMixColumns. The final round of both encryption and decryption differs in that it has no MixColumns and InvMixColumns, respectively. An initial AddRoundKey is done at the start of both encryption and decryption.

The 128 bits of the input are grouped into 16 bytes which are logically arranged in a 4×4 matrix. SubBytes and InvSubBytes substitute each byte of the state individually using a non-linear function. ShiftRows and its inverse perform a bytewise rotation of the rows of the state matrix. MixColumns and InvMixColumns calculate new state columns, where each byte completely depends on the respective old column. Finally, the AddRoundKey operation is a bitwise XOR of the current state and the current round key, which is derived from the cipher key in an operation called the key expansion. For more details on the AES transformations refer to the NIST specification [12].

The first four bytes of the input to AES constitute the first column of the state matrix, the next four bytes the second column, and so on. On 32-bit platforms it is therefore a common choice to hold the four columns of the state in four 32-bit words. The AES implementation of Brian Gladman [6], which we have used as reference, is a good example of such a column-oriented implementation.

On 32-bit platforms most of the AES operations can be implemented with table lookups [3]. One or four tables with 256 32-bit entries, i.e. 1 kB per table, can be used. Different tables are required for encryption and decryption so that the total size of the lookup tables can be up to 8 kB. Depending on the acceptable code size, these tables can be static or generated at runtime. However, our experience shows that table lookup does not necessarily deliver the best performance. Especially on systems with slow memory and no or minimal cache, it can be faster to calculate the AES round transformations directly. Cryptographic instruction set extensions are mainly designed for use in embedded processors. Embedded systems are limited in size of working memory, memory access latency and also maximal power consumption. A table lookup implementation of AES for such a system therefore puts a strain on working memory and its performance will depend primarily on the memory access time.

A software AES implementation requires at least a lookup table of 256 bytes for SubBytes and InvSubBytes, respectively. The other AES round transformations can be done without lookups. SubBytes and InvSubBytes can be combined with ShiftRows and InvShiftRows, respectively, if the bytes are arranged accordingly after substitution. This is possible because SubBytes and ShiftRows and their inverses are consecutive operations and their order of execution can be switched arbitrarily. The combination delivers the shifting of the rows at no additional cost. As almost all microprocessors offer bitwise XOR instructions, AddRoundKey can be implemented very efficiently. The MixColumns and InvMixColumns operations interpret the state bytes and state columns as elements of finite fields and require operations which are normally not supported by common microprocessors. These finite field operations must instead be done with logical and integer instructions. Therefore, a considerable part of AES is spent on calculating the MixColumns and InvMixColumns operations.

Bertoni et al. [1] have presented a more effective method of calculating MixColumns and InvMixColumns by operating on the rows of the state matrix instead of the columns. Although this strategy requires a transposition of the state matrix at the beginning and end of AES, a transposition of the cipher key and a more complex key expansion, the whole AES operation is commonly faster than a column-oriented implementation. The performance gains are especially significant for decryption, because InvMixColumns is much easier to calculate with the rows of the state than with the columns. The algorithms for calculating MixColumns and InvMixColumns using the state columns and state rows, as well as possible optimizations using ECC instruction set extensions, will be discussed in the next section.

3 Optimizing AES Using Instruction Set Extensions

MixColumns and InvMixColumns require addition and multiplication of elements of the binary extension field $GF(2^8)$ and of polynomials over $GF(2^8)$. Addition in $GF(2^8)$ is defined as a bitwise XOR. Multiplication in $GF(2^8)$ can be seen as multiplication of binary polynomials (i.e. coefficients mod 2), followed by a reduction with an irreducible polynomial. Arithmetic with polynomials over $GF(2^8)$ follows the conventional rules for polynomials, using addition and multiplication in $GF(2^8)$ for the coefficients. More details on the mathematical background can be found in the original Rijndael specification [3].

For our work we have used three of the instructions described in [7] for the MIPS32 architecture. Table 1 lists the instruction names for SPARC and MIPS32 (as given in [7]) with a short description. In this paper we will use the SPARC names. All three instructions work on a dedicated accumulator whose size must be at least twice the word size, i.e. at least 64 bits in our case.

Table 1. The employed ECC instruction set extension

SPARC	MIPS32	Description
gf2mul	mulgf2	Multiply two binary polynomials
gf2mac	maddgf2	Same as gf2mul with addition to accumulator
shacr	sha	Shift lowest word out of accumulator

The instructions gf2mul and gf2mac interpret the two operands as binary polynomials, multiply them and put the result in the accumulator. They differ in that gf2mul overwrites the previous accumulator value while gf2mac adds the polynomial product to it. The shacr instruction writes the lowest word of the accumulator to a given destination register and shifts the accumulator value 32 bits to the right. All timing estimations presented in this paper are based on the following properties of the SPARC V8 architecture:

- No rotate instruction (rotate must be done with two shifts and an OR/XOR).
- Setting a constant value (> 13 bits) in a register takes 2 instructions.
- There are enough free registers to hold up to three constants throughout calculation of MixColumns or InvMixColumns.

3.1 Column-Oriented Implementation

For MixColumns and InvMixColumns, each new column can be calculated separately from the old column. This property is used if the state is held in four 32-bit words corresponding to its columns. The following code performs MixColumns for a state column contained in the variable column without using extensions.

```
1  byte double, triple;
2  double = GFDOUBLE(column);
3  triple = double ^ column;
4  column = double ^ ROTL(triple,8) ^ ROTL(column,16) ^ ROTL(column,24);
```

MixColumns for a single state column (conventional)

The function GFDOUBLE interprets the four bytes of column as four elements of $GF(2^8)$ and doubles them individually. The function ROTL rotates the word to the left by the given number of bits. The basic idea behind the code is that each byte of the resulting column consists of a weighed summation of the four bytes of the old column. The multiplication with the constant factors is done for four bytes each in line 2 and 3 and the bytes are rotated into the correct positions and summed up in line 5. GFDOUBLE takes about 10 instructions, which will be shown in Section 3.2 in more detail. ROTL takes between one and three instructions depending on whether the processor offers a dedicated rotate instruction. We will consider ROTL to take three instructions in the following. The calculation of a single column in MixColumns therefore takes about 23 instructions for one GFDOUBLE, three ROTL and four XOR.

Using the ECC instruction set extensions it is possible to calculate a column much faster. This can be done using the three instructions described in Table 1. We use the original definition of MixColumns, which is a multiplication of two polynomials of degree 3 with coefficients in $GF(2^8)$ [3]. Hereby the column makes up the first polynomial, while the second polynomial is constant. The following code calculates MixColumns for a single column.

```
1  word mask, low_word, high_word;
2  mask = column & 0x80808080;    // MSBs (s0 s1 s2 s3)
3  mask = mask >> 7;              // logical shift right
4  GF2MUL(column, 0x01010302);    // Polynomial multiplicaton
5  GF2MAC(mask, 0x00011a1b);      // GF(2^8) coefficient reduction
6  SHACR(low_word);               // Degrees 0-3 of polynomial
7  SHACR(high_word);              // Degrees 4-6 (== 0-2) of polynomial
8  column = low_word ^ high_word; // Polynomial reduction mod (x^4 + 1)
```

MixColumns for a single state column (using extensions)

Using the instruction set extensions the functions GF2MUL, GF2MAC and SHACR can be done with the corresponding processor instructions. The main idea behind this code is illustrated in Figure 1. There are three phases in the whole calculation: Polynomial multiplication, coefficient reduction and polynomial reduction. Line 4 performs the multiplication of the column with the constant polynomial $x^3 + x^2 + 03x^1 + 02$. Note that as the bytes of the column represent the polynomial coefficients with ascending degree (i.e. the byte at the word's most significant position is the coefficient of x^0) [3], the coefficients of the constant polynomial have been rearranged accordingly to arrive at a product with ascending coefficient degree. The polynomial multiplication puts the first block of coefficients in Figure 1 in the accumulator.

The coefficients si, which have been multiplied with the value 03 or 02, may no longer be elements of $GF(2^8)$. More specifically, $3si$ or $2si$ are no elements of $GF(2^8)$ if the most significant bit (MSB) of si is 1. Such values are called residue values and they can be reduced to an element of $GF(2^8)$ by adding the reduction polynomial of the finite field, which is $x^8 + x^4 + x^3 + x + 1$ (0x11b) in the case of AES. To get an element of $GF(2^8)$ we need to add a reduction value ri for each

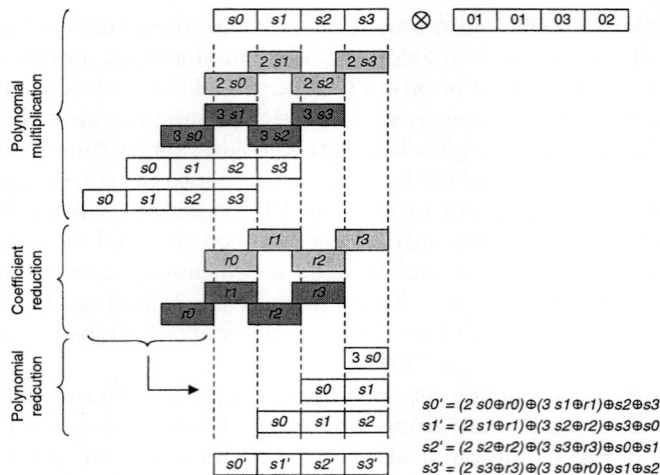

Fig. 1. Polynomial multiplication and reduction to yield a column in MixColumns

coefficient $3si$ and $2si$, which has value 0x11b if MSB(si) is 1 and 0 if MSB(si) is 0. In Figure 1 the reduction values are shown in the middle with corresponding coefficients and reduction values marked in the same grey tone. After adding the reduction values to the result of the polynomial multiplication, all coefficients are reduced to elements of GF(2^8). In order to calculate the reduction values we extract the corresponding MSBs (line 2 and 3) of the coefficients si and multiply them with the value 0x00011a1b (line 5), which is the sum of 0x11b aligned to the two lower bytes of the word. In line 5, the reduction values are also added to the previous multiplication result in the accumulator.

In line 6 and 7, the polynomial is read into two variables and in line 8 the reduction of the polynomial is performed. Due to the special nature of the reduction polynomial $x^4 + 1$, the coefficients for degrees 4 to 6 must be added to the coefficients of degree 0 to 2, respectively. This is easily done by an XOR of the low and the high word of the accumulator. The calculation of a single column for MixColumns requires 13 instructions. This includes the generation of the three constant values which are used in the process (0x80808080, 0x01010302 and 0x00011a1b). But these values need only be generated once per MixColumns, so that in average it takes about 9 instructions to calculate a single column.

The optimizations for InvMixColumns work in a similar fashion, with the exception that the reduction values are generated with an additional GF2MUL operation by performing the polynomial multiplication with the highest three bits of each coefficient alone. It takes approximately 16 instructions to calculate one column, which is much faster than the conventional approach.

3.2 Row-Oriented Implementation

In a row-oriented AES implementation [1] the MixColumns and InvMixColumns operations are calculated for the whole state altogether. The strength of this method lies in the possibility to reuse intermediate results for all four columns of the state. This advantage is especially significant in the relatively complex InvMixColumns operation. The conventional row-oriented MixColumns uses 4 GFDOUBLE operations, while InvMixColumns requires 7. This code shows a conventional implementation of GFDOUBLE for the value in `poly`.

```
1  word mask;
2  mask = poly & 0x80808080;
3  mask = mask >> 7;
4  mask = mask * 0x1b               // reduction mask
5  poly = (poly & 0x7f7f7f7f) << 1;
6  poly = poly ^ mask;
```

GFDOUBLE for row-wise MixColumns/InvMixColumns (conventional)

Here the reduction information is extracted from `poly` in lines 2 to 4. After the actual doubling in line 5, the reduction is performed (line 6). This version of GFDOUBLE takes 10 instructions, but the reuse of the bitmasks in consecutive GFDOUBLE leads to a lower instruction count. There are four consecutive doublings in both MixColumns and InvMixColumns, which can be done in an average of 7 instructions each. Using the GF2MUL, GF2MAC and SHACR instructions, GFDOUBLE can be done slightly faster than with conventional instructions. An optimized GFDOUBLE is shown in the following.

```
1  word mask;
2  mask = poly & 0x80808080;
3  mask = mask >> 7;             // reduction mask
4  GF2MUL(poly, 0x2);
5  GF2MAC(mask, 0x11b);          // GF(2^8) coefficient reduction
6  SHACR(poly);
```

GFDOUBLE for row-wise MixColumns/InvMixColumns (using extensions)

The reduction information is extracted in a similar manner, but doubling and reduction are done with GF2MUL and GF2MAC, respectively. The optimized version takes about 7 instructions. When reusing the bitmask in four consecutive doublings, the average instruction count goes down to approximately 6.

4 Practical Results

We have used a version of the SPARC V8-compatible LEON-2 processor [5] which includes the instructions described in Table 1 developed in course of the ISEC project [8]. The processor has been implemented on the GR-PCI-XC2V FPGA board from Gaisler Research and features a (32×16)-bit integer/polynomial multiplier with a 72-bit accumulator (including 8 guard bits for integer multiply-accumulate). A `gf2mul` instruction executes in three cycles, while a

Table 2. Execution times of AES-128 encryption, decryption and key expansion in clock cycles

	Key expansion	Encryption	Decryption
Gladman NOTABLES	522	1,860	3,125
Gladman NOTABLES optimized	522	1,755	1,906
Column-oriented	497	1,672	2,962
Column-oriented optimized	497	1,257	1,576
Row-oriented	738	1,636	1,954
Row-oriented optimized	738	1,502	1,567
Speedup		23.1%	19.8%

gf2mac instruction takes one cycle[1], provided that the next instruction does not depend on its result. The shacr instruction always finishes in one cycle. To estimate the hardware cost of the extensions, we have compared the synthesis results of a "conventional" LEON-2 with a (32×16)-bit integer multiplier and one with the ECC extensions as described in [7]. The extended version requires about 3.5% more gates than the reference version, whereby the added functionality encompasses not only all the extensions from [7], but also a signed multiply-accumulate instruction. Unfortunately, the LEON-2 does not offer a configuration with a (32×16)-bit multiply-accumulate unit, so the sole cost of the instructions for binary polynomials cannot be determined easily.

We have made tests with Gladman's AES code using it both as reference as well as an instance of a column-oriented implementation. However, Gladman's code only allows to optimize the calculation of a single column and not the whole MixColumns operation. Therefore, we have implemented our own version of a column-oriented AES, which can be better optimized. Furthermore, we have implemented a row-oriented AES. Our column- and row-oriented versions are written in C and support both encryption and decryption for a precomputed key schedule and also for on-the-fly key expansion. Moreover, all versions feature a conventional and an optimized implementation of MixColumns and InvMixColumns using the ECC instruction set extensions.

Timing measurements have been done using the integrated cycle counter of the extended LEON-2. The code which performs the measurements has been derived from Gladman's code. In order to get a fair comparison of the different implementation options, we have used a LEON-2 processor with a very large instruction and data cache (4 sets with 16 kB each, organized in lines of 8 words). This was done to get rid of cache effects with one-way associative caches, where the performance of different implementations depended on the actual memory addresses of their stack variables. The results therefore reflect performance in a environment with fast memory access or with "perfect" cache.

---

[1] This is possible because the LEON-2 processor allows to execute other instructions in parallel to a gf2mac instruction (similar to MIPS32 processors [7]).

4.1 Precomputed Key Schedule

Table 2 lists the timing results for AES-128 encryption and decryption when using a precomputed key schedule. The time for doing the key expansion is also stated. The speedup is calculated between the best conventional implementation and the best optimized implementation. The row-oriented AES is best for both conventional encryption and decryption. For the optimized variants, the column-oriented implementation is best for encryption, while the performance for decryption is nearly identical for the column- and row-oriented versions.

4.2 On-the-Fly Key Expansion

The timing results in Table 3 refer to AES-128 encryption and decryption with on-the-fly key expansion. As Gladman's code does not support this mode, only the results for our column- and row-oriented versions are stated. Note that the last roundkey is supplied to the decryption routine, so that it does not have to do the whole key expansion at the beginning.

Table 3. Execution times of AES-128 encryption and decryption with on-the-fly key expansion in clock cycles

	Encryption	Decryption
Column-oriented	2,254	3,357
Column-oriented optimized	1,674	2,018
Row-oriented	2,328	2,433
Row-oriented optimized	2,230	2,176
Speedup	25.7%	17.0%

For conventional encryption, the column-oriented AES is slightly better, while for decryption, the row-oriented version is fastest. For the versions which use the ECC extensions, the column-oriented AES is better for both encryption and decryption. The speedup is again calculated considering the best conventional and optimized version.

4.3 Code Size and Side-Channel Attacks

The code size for the implementations ranges between 2.5 and 3.5 kB, where the optimized variants are always smaller than the non-optimized ones. Note, however, that the implementations have been optimized for speed and not for code size. Savings through optimizations go up to 15% (column-wise decryption with precomputed key schedule).

The susceptibility to side-channel attacks is not changed through the use of the instruction set extensions. It is therefore necessary to integrate countermeasures into a system which calculates AES using the presented methods, if resistance against side-channel attacks is required. However, a detailed treatment of countermeasures against side-channel attacks is outside the scope of this paper. We refer the interested reader to the relevant literature on this topic, ranging from secure logic styles [14] to masking schemes [2, 13].

5 Conclusions and Future Work

In this work we have demonstrated how to use instruction set extensions originally designed for elliptic curve cryptography to accelerate software implementations of the AES. Although not specifically designed for that purpose, the use of the three instructions gf2mul, gf2mac and shacr allows performance gains of up to 25%. This speedup can be considered as "free" on processors which already feature these instructions. Generally, the column-oriented AES implementations can be optimized very well with the instruction set extensions.

As future work we will investigate the potential of dedicated instructions for AES. A possible instruction could interpret the four bytes in a 32-bit word as elements of $GF(2^8)$ and perform multiplication and reduction in one step. Such an instruction would be very useful for InvMixColumns.

Acknowledgements. The research described in this paper was supported by the Austrian Science Fund (FWF) under grant number P16952-N04 "Instruction Set Extensions for Public-Key Cryptography".

References

1. G. Bertoni, L. Breveglieri, P. Fragneto, M. Macchetti, and S. Marchesin. Efficient Software Implementation of AES on 32-Bit Platforms. In *Cryptographic Hardware and Embedded Systems — CHES 2002*, LNCS 2523, pp. 159–171. Springer Verlag.
2. J. Blömer, J. G. Merchan, and V. Krummel. Provably Secure Masking of AES. Cryptology ePrint Archive (http://eprint.iacr.org/), Report 2004/101.
3. J. Daemen and V. Rijmen. AES Proposal: Rijndael. Available for download at http://csrc.nist.gov/CryptoToolkit/aes/rijndael/Rijndael.pdf, 1999.
4. M. Feldhofer, S. Dominikus, and J. Wolkerstorfer. Strong authentication for RFID systems using the AES algorithm. In *Cryptographic Hardware and Embedded Systems — CHES 2004*, LNCS 3156, pp. 357–370. Springer Verlag, 2004.
5. J. Gaisler. The LEON-2 Processor User's Manual (Version 1.0.24). Available for download at http://www.gaisler.com/doc/leon2-1.0.24-xst.pdf, Sept. 2004.
6. B. Gladman. Implementations of AES (Rijndael) in C/C++. Available online at http://fp.gladman.plus.com/cryptography_technology/rijndael/index.htm.
7. J. Großschädl and E. Savas. Instruction Set Extensions for Fast Arithmetic in Finite Fields $GF(p)$ and $GF(2^m)$. In *Cryptographic Hardware and Embedded Systems — CHES 2004*, LNCS 3156, pp. 133–147. Springer Verlag, 2004.
8. J. Großschädl and S. Tillich. Instruction Set Extensions for Cryptography (ISEC). Available online at http://www.iaik.at/research/vlsi/01_projects/01_isec/.
9. R. B. Lee, Z. Shi, and X. Yang. Efficient permutation instructions for fast software cryptography. *IEEE Micro*, 21(6):56–69, Nov./Dec. 2001.
10. S. Mangard, M. Aigner, and S. Dominikus. A highly regular and scalable AES hardware architecture. *IEEE Transactions on Computers*, 52(4):483–491, Apr. 2003.
11. S. Morioka and A. Satoh. A 10-Gbps full-AES crypto design with a twisted BDD S-box architecture. *IEEE Transactions on Very Large Scale Integration (VLSI) Systems*, 12(7):686–691, July 2004.

12. National Institute of Standards and Technology (NIST). FIPS-197: Advanced Encryption Standard, Nov. 2001.
13. E. Oswald, S. Mangard, N. Pramstaller, and V. Rijmen. A Side-Channel Analysis Resistant Description of the AES S-box. To appear in the Proceedings of the 12th Annual Workshop on Fast Software Encryption (FSE 2005), Feb. 2005.
14. K. Tiri and I. Verbauwhede. Securing Encryption Algorithms against DPA at the Logic Level: Next Generation Smart Card Technology. In *Cryptographic Hardware and Embedded Systems — CHES 2003*, LNCS 2779, pp. 137–151. Springer Verlag.

Access Control Capable Integrated Network Management System for TCP/IP Networks

Hyuncheol Kim[1,*], Seongjin Ahn[2,**], Younghwan Lim[3], and Youngsong Mun[3]

[1] Dept. of Electrical and Computer Engineering, Sungkyunkwan University,
300 Chunchun-Dong, Jangan-Gu,Suwon, Korea 440-746
hckim@songgang.skku.ac.kr
[2] Dept. of Computer Education, Sungkyunkwan University,
53 Myungryun-Dong, Jongro-Gu, Seoul, Korea, 110-745
sjahn@comedu.skku.ac.kr
[3] School of Media, Sungsil University,
Sangdo-Dong, Dongjak-Gu, Seoul, Korea
yhlim@computing.ssu.ac.kr,mun@comp.ssu.ac.kr

Abstract. This paper is on the functional architecture for the access control capable policy based network management system for TCP/IP enterprise networks. The network management system structures proposed in this paper works in parallel with the policy based real-time access control function to make the utmost use of the network resources and to provide a high-quality service to the user which is distinguishable to the network management policy for simple network infrastructure. This paper illustrates an effective interface and interworking functions among the policy server, the access control server, and the network management server. With the proposed policy based network management system, the network operator can recognize network problems in real-time and can effectively figure out how the network infrastructure should be reconfigured in order to resolve the problems.

1 Introduction

With a tremendous increase of Internet users, the Internet backbone, which is a core of the transport network, is getting more complicated along with the lower network structure in order to support diverse traffic characteristics. The network traffic has also continuously increased with remarkable growth. With the advent of various multimedia applications that needs real-time processing and high capacity processing power, the development of the SOC (System On Chip) technology have surpassed the terminal and network equipment processing power which exceeded the Moore's Law level.

---

* This work was supported by grant No. R01-2004-000-10618-0(2004) from the Basic Research Program of the Korea Science & Engineering Foundation.
** Corresponding Author.

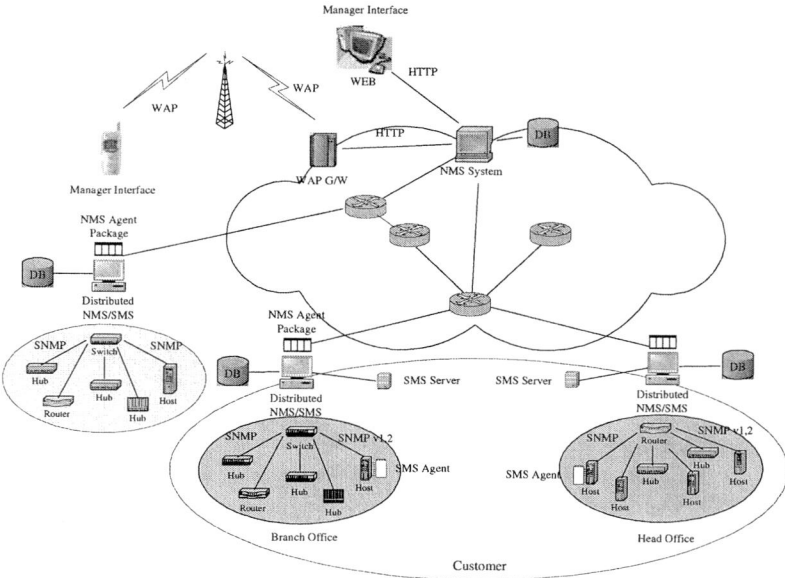

Fig. 1. Traditional Two-tier Network Management

The appearance of the new types of network applications such as diverse intranet, ERP (Enterprise Resource Planning), VPN (Virtual Private Network), and VoIP (Voice over IP), demands effective and rapid network management, access control, and continuous network survivability in order to satisfy quality of services required by the applications.

The main objective of the conventional network management systems is to rapidly process network disorder through network system configuration and through network surveillance. Therefore, the conventional network management function is not capable of reflecting these requirements [1][2]. Most schools, businesses, and research institutes, as shown in Fig. 1, operate its own network that is connected to the Internet and manage the network using the operators' own know-how and experiences. Also, since the conventional network management mostly focuses on network configuration and performances, there exists no effort on managing the network access or the network terminals [3].

It also only provide techniques to recognize and control the problem of the network through real-time/non-real time network monitoring. These solutions do not provide answers in how the network infrastructure should be suitably modified in order to solve the problems in the network. Moreover, the real-time monitoring for network management is one of the factors that can cause network congestion or system performance deterioration.

In these circumstances, the major concerns for the network managers and service provider is to combine network access control technique and network management of all systems that constitutes the network including network terminals to control network disorder or malicious attack to provide high quality service needed by the users and to maximally use the network resources [4].

If it is possible to construct business policy to provide network management policies and services to apply the changes in the network and management environment, converged network management solution is absolutely needed to automatically and effectively reflect the network management policy. Also, these solutions must have the infrastructure to be use effectively in the optical based NGcN (Next Generation converged Network), which will appear in the near future [5][6][7][8].

The rest of this paper is organized as follows. The proposed access control capable network management architecture and functional components are described in section 2 and 3, respectively. Finally, the paper concludes in Section 4.

2 AC-PBNMS

The PBNMS (Policy-Based Network Management System) is one of the most remarkable issues of these days, which is a policy-based network management solution that can be easily applied to both the conventional network infrastructure and the next generation Internet [9][10][11][12].

The PBNM is a network management method that increases processing capability by interoperating the network management, the service management, and the business process with the network infrastructure in quick and efficient manner. The policy mentioned in this paper refers to rules, conditions, and operation method. Each policy consists of more than one rule and also includes conditions. Only when the conditions are met, the operation is executed. The PBNM solution recognizes these policy rules for the entire conditions that can occur in the network and performs automatic management by executing policy defined operation method if the conditions are met.

The AC-PBNMS (Access Control capable PBNMS) that provides access control function is a policy-based network management system. The AC-PBNMS processes Web-based system management function and network management function, thus it can be efficiently used in different network configuration. AC-PBNMS is a network policy and service policy based converged network management system that monitors and analyzes the complex and diverse user system and network and automatically performs configuration, performance, and fault management of the networks and systems.

Especially, AC-PBNMS will provide the MPLS/GMPLS (MultiProtocol Label Switching/Generalized MPLS) based traffic-engineering function in the near future. The traffic-engineering function can be efficiently provided using OTN (Optical Transport Network) or ATM (Asynchronous Transfer Mode) in the Internet backbone and MAN (Metropolitan Area Network) as a transmission system. The conventional networking environment, which does not use the ATM, will optionally use source routing technique to provide the traffic-engineering function.

AC-PBNMS has a web-based system management function and perform distributed possessing in managing the network. AC-PBNMS is a policy-based net-

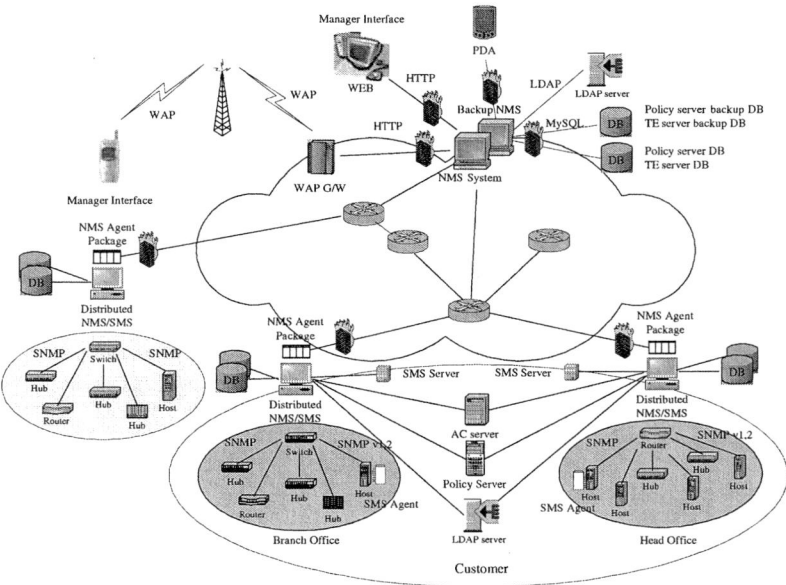

Fig. 2. Configuration of the AC-PBNMS for NGcN

work management system that can be efficiently applied in various network configurations. AC-PBNMS includes the management function of both the conventional method and the method provided by the system management solutions. The network operator can use the AC-PBNMS to efficiently provide a policy based network management in the managed domain. AC-PBNMS has a access control technique that protect network against unauthorized network access.

The Fig. 2 shows the configuration of the Next Generation Internet(NGcN) using AC-PBNMS. Fig. 2 also illustrates the network structure performing converged network management in single a domain or in MSP (Management Service Provider) environment with AC-PBNMS. As shown in Fig. 2, the hierarchical and distributed network management function providing AC-PBNMS will not only relieve the obligation of the network operator but also reduce network management traffic to maximally increase the network performance.

Also, the AC-PBNMS has a network management information processing and reporting function. These information are created when the network operator perform management operation based on network management policy. More efficient network management can be provided through these management knowledge and policies.

AC-PBNMS automatically performs network management by observing the occurring situation of the network and the system in real-time and applies the predefined policy. In these circumstances, the AC-PBNMS report the status to the network operator through user interface and provide function to correct the faults that occurs in the managed domain. Also, the network operator can

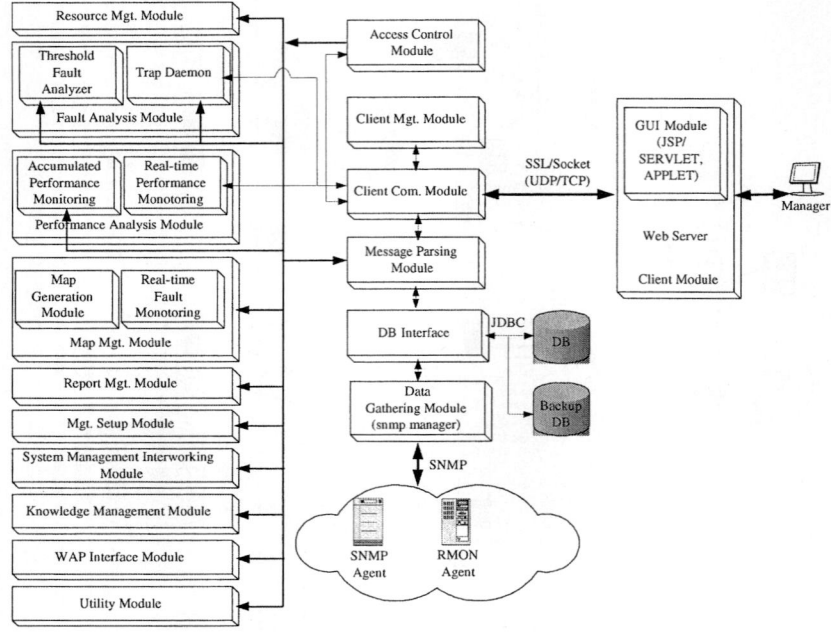

Fig. 3. AC-PBNMS Components

quickly comprehend the network situation single handedly though the visible network topology map provided by the AC-PBNMS.

The AC-PBNMS also supports WAP (Wireless Application Protocol) to provide interface for the wireless and mobile terminal so that the network operator does not have to be restricted in management location. Therefore, the network operator can always perform network management regardless of time and location.

As shown in Fig. 3 and Fig. 4, the proposed AC-PBNMS is consists of many independent software packages and provides function to reconfigure the network suitable for the network situation. Therefore, AC-PBNMS can be easily adapted to the MSP environment such as Trunkey MSP, Internet MSP, and Infrastructure MSP, with out change. Therefore, AC-PBNMS is an efficient solution for the network environment, which consists of independent network configurations.

In the future, the Internet backbone and MAN will evolve to WDM/DWDM (Wavelength Division Multiplexing/ Dense Wavelength Division Multiplexing) based optical Internet. Therefore, the AC-PBNMS will provide MPLS/GMPLS based traffic-engineering function using source routing.

The AC-PBNMS functions can be summarized as follows.

- Hierarchical and distributed management function by distinguishing network management domain,
- Wireless terminal interface function using WAP protocol,

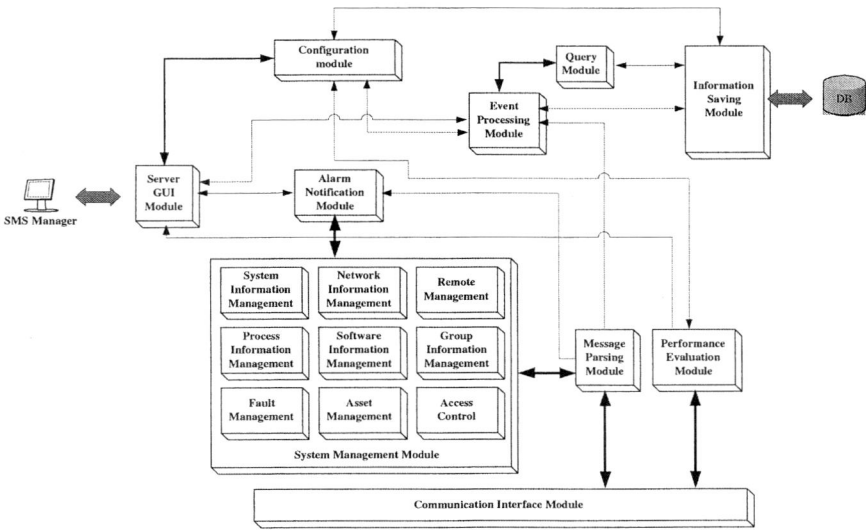

Fig. 4. SMS Architecture

- Real-time fault surveillance and report function,
- Network performance analysis and fault surveillance function,
- Network traffic analysis function,
- Network consultant function based on knowledge based fault analysis,
- System management function,
- Traffic engineering function,
- Routing protocol interworking function,
- Policy based server function.

3 AC-PBNMS Configuration and Functions

NMS is one of the important parts of AC-PBNMS along with the policy server and the access control server and its function is to manage the real network based on the decisions made by the policy server. But, most of the new network management function is correlated with the access control function and the policy based server.

In order to perform distribute network management, NMS can be configured into different level. The conventional system management function is provided through the interoperation with the system management. Fig. 3 and Fig. 4 illustrate the configuration of the NMS used in the AC-PBNMS and the function of each modules that configures in the NMS is as follows.

- Resource Management Module : This module performs registration and cancellation of the network equipment such as switch and router in the managed

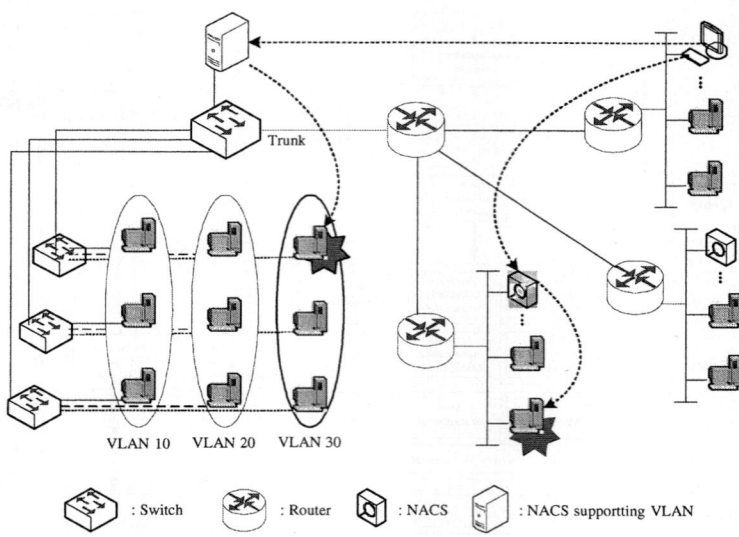

Fig. 5. AC-PBNMS Access Control Function

domain network. It also performs retrieval and modification function of the registered network equipments. The modification history management function is also included in this module.
- Fault Analysis Module : The fault analysis module analyzes failures that may occur in the network equipment or network link. The performance and fault (error rate, packet delivery failure rate, etc.) thresholds are set for those failures. If the threshold exceeds, the threshold fault analysis module determines the fault and the trap analysis module collects the trap information that occurs in the network in real-time.
- Performance Analysis Module and Utility Module : The performance analysis module analyzes network performance parameters, such as usage rate, packet delivery rate, accumulated in the database. There are the accumulated performance analysis module, which examines the accumulated data in the specified amount of time, and the real-time analysis module, which analyze data in real-time. The utility module has lots of tools that are needed by the operator in order to manage the registered network equipments. The tools includes ping function that is used to check the operational state of the equipment, traceroute function to figure out the path of the equipments, telnet-type program that can access the server systems to perform other tasks, and MIB (Management Information Base) browser to retrieve MIB value of the equipment that support MIB-II. The self-implemented tools or other network management tools are executed through single interface.
- Map Management Module : The map management module manages network topology of the managed domain network through network map. There is map creation module to manage basic network topology and monitoring

module that observes the network up/down state in real-time though the use of network tools such as ping.
- Report Management Module : The report management module generates network management report to the operator. The reports include network resource, performance, and fault information for the network equipments and lines in the managed domain. Basically, there are report generation function that generates theses report and report management function that manages the generated reports.
- Management Setup Module and Knowledge Management Module : The management setup module setups environmental parameters of the network management systems. There are user management setup for managing login user and user authority, polling cycle setup in collecting network information, and other setup functions. The knowledge management module records management actions, opinions, and measures made by the operators as a result of the performance analysis and the fault analysis. There is also consultative information input function and retrieval function. The consultative information are stored in the database, and there are many retrieval function.
- Client Management and Communication Module : Client Management Module is a part that manages connections between web applications such as server and applet. This module is needed when the server and the applet needs to send/receive real-time data. The client communication module manages socket communications between server and web applications. The retrieved data are analyzed and stored in the database. If there is a request for the information, the requested information is retrieved from the database and is transmitted to the requester.
- Database Interface Module and Data Gathering Module : Database interface module manages the database. The database can be access using the JDBC. The information of the database is used form of Javabeans. Data Collection Module uses SNMP protocol for network equipment that has SNMP agent or RMON agent and collects information using polling technique. The collected information are stored in the database through database interface module.
- GUI Module : GUI Module is a user interface for the operator to perform operation. This network management interface is created using JSP/servlet and applet. Web server is needed in order to access the web environment, and a driver module such as Tomcat is needed in using JSP/Servlet technique to implement the GUI.
- WAP Interface Module : WAP interface module makes management possible through WAP protocol communication mode by interoperating with the TCP/IP in a wireless Internet environment. This function has to overcome limited bandwidth and limited screen size in the wireless terminal. Through the wireless terminal, network operator can always perform network management function without any restriction to time and places, thus, network operator can have larger activity area and management domain. In order to implement WAP protocol, WML language is used which is an extension of XHTML to configure web page and also need to interoperate with the JSP and database. Therefore, in order to develop WAP interface module,

knowledge in WML Tag is also needed. Dynamical and graphical management information will be also shown in the wireless terminal, in the near future. The wireless application programming technique such as J2ME or Qualcomm's BREW will be used to provide that kind of screen interface.
- Access Control Module : Fig. 5 shows the model of proposed access control function in real LAN environment. There are two types of access control system. The first type is the access control system, which is not support VLAN (Virtual Local Area Network). Another type is the access control system, which is support VLAN. In Fig. 5, the manager executes the policy to block the unauthorized users by AC-PBNMS agent in the network. It updates generic hosts with the incorrect MAC (Media Access Control) address of the IP to be blocked in the LAN. AC-PBNMS agent can be installed in each broadcast domain to monitor packets generated within the domain. The AC-PBNMS can create the ARP (Address Resolution Protocol) packet under the order from the management system to confirm the up/down status of the network nodes, and to obtain the MAC address, additionally shutting down the network against an unauthorized IP. Fig. 5 also shows AC-PBNMS that supports VLAN environment. When all broadcast, multicast, unknown unicast packets destined to the router are forwarded to the default VLAN port, the final trunk link to the router will receive all these traffic. Switch trunk links that are nearer to the router will receive more broadcast traffic than trunk links that are further away from the router. When an AC-PBNMS is run in the network, blocking unauthorized and administration user can be applied same as generic LAN Environment.

4 Conclusions

The main purpose of the conventional network management system is to setup and configure the network equipment and process network faults through continuous surveillance. With the influence of the ATM based MPLS, the emergence of the high-speed network made it possible to actively and effectively control the Internet compared to the conventional method. Also, explosive increase of usage bandwidth request does not only require a simple network management system with configurations and surveillance functions, but also a policy based network management system which operates with the limited network resources and provides high service quality and efficient network resource usage.

The main concern of network management system is not providing individualized network management but establishing network infrastructure that supports high quality service through interoperation between network services providers based on a standardized policy. This paper presents a converged network management system structure that manages network through network management policy. The proposed policy based network management system structure does not use the network management policy for applying it to the simple network infrastructure.

The proposed system provides high quality service to the operators by performing policy-based real-time traffic engineering function and by maximally using the network resources. This paper also describes an efficient interface and interoperating methods between policies based server, traffic engineering server, and network management server. With the use of the policy based network management system proposed in this paper, the network operator can understand the network problem in real-time and acquire answer on how to reconstruct the network infrastructure to resolve the problem.

References

1. Geert Jan Hoekstra, Willem A. Romijn, Harold C. H. Balemans, Abdelkader Hajjaoui, and Gijs G. van Ooijen: An Integrated Network Management Solution for Multi-Technology Domain Networks, Bell Labs Technical Journal, Vol. 7, No. 1, (2002) 115–120
2. John Y. Wei: Advances in the Management and Control of Optical Internet, IEEE J. on Selected Areas in Communicatoins, Vol. 20, No. 4, May, (2002) 768–785
3. Hyuncheol Kim, Seongjin Ahn, Sunghae kim, and Jinwook Chung: A Host Protection Framework Against Unauthorized Access for Ensuring Network Survivability, Advances in Cryptology – NPC'04, Network and Parallel Computing, Springer-Verlag, Lecture Notes in Computer Science 3222, (2004) 635–643
4. Xipeng Xiao, Lionel M. Ni: Internet Qos: A Big Picture, IEEE Network, Mar./Apr., (1999) 8–18.
5. Mark L. Steven, Walter J. Weiss: Policy Based Management for IP Networks, Bell Labs Technical Journal, Oct.-Dec., (1999) 75–04
6. Dinesh C. Verma: Simplifying Network Administration Using Policy Based Management, IEEE Network, Mar., (2002) 20–26
7. Raju Rajan, Dinesh Verma: A Policy Framework for Integrated and Differentiated Services in the Internet, IEEE Network, Sep., (1999) 36–41
8. Christoph Rensing, Hasan, Martin Karsten, and Burkhard Stiller: AAA: A Survey and a Policy-Based Architecture and Framework, IEEE Network, Nov., (2002) 22–27
9. Paris Flegkas, Panos Trimintzios: A Policy Based Quality of Service Management System for IP DiffServ Networks, IEEE Network, Mar., (2002) 50–56
10. Lundy Lewis: Implementing Policy in Enterprise Networks", IEEE Com. Mag., Jan., (1996) 50–55
11. Ashfaq Hossain, Houshing F. Shu, Charles R. Gasman, and Randolph A. Royer, "Policy Based Network Load Management" Bell Labs Technical Journal, Oct.-Dec., (1999) 95–108
12. K. Nahrstedt, A. Hossain, and S. M. Kang: A Probe-Based algorithm for QoS Specification and Adaptation, Proc. of 4th Intl. IFIP workshop on Quality of Service, IFIP, Mar. (1996)

A Directional-Antenna Based MAC Protocol for Wireless Sensor Networks

Shen Zhang and Amitava Datta

School of Computer Science and Software Engineering,
University of Western Australia,
35 Stirling Highway,
Perth, WA 6009, Australia

Abstract. Directional antennas have been extensively used in designing MAC protocols for wireless ad hoc networks in recent years. Directional antennas provide many advantages over the classical antennas which are omnidirectional. These advantages include spatial reuse and increase in coverage range. One of the main considerations in designing MAC protocols for static wireless networks is to reduce power consumption at the sensor nodes. This is usually done by imposing transmission and receiving schedules on the sensor nodes. Since it is desirable for a sensor network to be self managed, these schedules need to be worked out by individual nodes in a distributed fashion. In this paper, we show that directional antennas can be used effectively to design an energy efficient MAC protocol for wireless sensor networks. Our MAC protocol conserves energy at the nodes by calculating a scheduling strategy at individual nodes and by avoiding packet collisions almost completely.

1 Introduction

Traditional wireless devices use omnidirectional antennas. Such an antenna propagate the electromagnetic energy of transmission all around it, i.e., in all directions. Since the transmission range of a signal depends on the power level of transmission, it is usually inefficient to propagate a signal in omnimode when there may be only a few intended recipients for a signal. In other words, energy is wasted in propagating a signal in all directions instead of directing the signal towards the intended recipients.

Directional antennas solve this problem by directing the signal in a narrow angular band. There are two benefits of this type of transmission. First, power can be conserved by not transmitting in an omnimode; and second, the range of the signal can be extended by concentrating power in a narrow region. Moreover, multiple senders and receivers can use the same region of space for communicating without interfering with each other's communication. However, it is necessary to design efficient Medium Access Control (MAC) protocols to take advantage of directional antennas. Inspite of several advantages, directional antennas introduce several difficulties in designing MAC protocols. In particular, directional antennas increase the chance of packet collisions due to enhancing the hidden terminal problem. Moreover, packets can be lost due to the problem of deafness. Korakis et al. [6] discuss these problems in details.

Most of the proposed MAC protocols that take advantage of directional antennas are based on the IEEE 802.11 standard for wireless transmission. The basic mechanism is to use CSMA/CA with Request to Send/Clear to Send (RTS/CTS) handshaking mechanism. The focus in most of these protocols is to design MAC protocols for mobile ad hoc networks.

Several different MAC protocols [11, 10] have been designed specifically for wireless sensor networks. These networks are usually static and they have different requirements for MAC protocols compared to mobile ad hoc networks. Wireless sensor networks are usually deployed in difficult terrains without any infrastructure support. Hence, it is important that they operate as long as possible without the need of changing batteries. Any MAC protocol designed for a sensor network should help the individual nodes to conserve battery power. Moreover, it is desirable to reduce control packet overhead such as ACK, RTS and CTS.

In this paper, we present a MAC protocol using directional antennas for static wireless networks. This protocol is suitable for wireless sensor networks. Since nodes in a sensor network usually participate in regular data exchange, most MAC protocols work out a scheduling strategy in a distributed manner. The aim of such a scheduling strategy is to determine clear time slots when a node needs to either send or receive messages from its neighbors. Our protocol is also based on a similar idea. However, this scheduling is more difficult in our case as each node has several directional sectors and it either listens to or sends message in one of these sectors. Each node needs to synchronize its schedule with its neighbors so that the sending and receiving slots are synchronized among the neighbors.

There are two phases in our protocol. The nodes start working using the normal CSMA/CA MAC protocol in IEEE 802.11 standard in the first phase. The aim of the first phase is to exchange schedules between the neighbors so that we can establish a network-wide schedule. This exchange of schedules is done in a completely distributed fashion. A node enters the second phase of operation once it has stabilized and synchronized its schedule with all its neighbors. Each node sends and receives messages according to its schedule in the second phase.

The rest of the paper is organized as follows. We discuss some related work on MAC protocols using directional antennas and for sensor networks in Section 2. We present our protocol in Section 3. We discuss our experimental results in Section 4. Finally, we make some concluding remarks in Section 5.

2 Related Work

2.1 MAC Protocols Based on Directional Antennas

Most MAC protocols try to take advantage of the enhanced range of directional transmission. Some protocols mix both omnidirectional and directional transmission to transmit packets in order to solve the disadvantages of IEEE 802.11 CSMA/CA. Nasipuri et al. [9] propose a basic directional MAC (DMAC) protocol. In this protocol, the transmission of RTS and CTS are omnidirectional, DATA and ACK are directional. The main purpose of this protocol is to inform all neighbors of the sender and the receiver of a potential transmission.The neighbors that successfully receive the RTS/CTS can determine which node it cannot transmit to at that period of time. Compared with the DMAC protocol, DRTS MAC protocol by Ko et al. [5] uses directional RTS and omnidirectional CTS transmissions before starting transmitting of data. Korakis et al. [6] propose a circular directional RTS MAC protocol. In this protocol, a source node sends directional RTS in consecutive circular directions until it scans all the area around the node. Each node that receives the RTS will defer transmission for a certain period of time. After this, the destination sends a CTS directionally. And then the source transmits data and receives its ACK in a directional manner from the destination.

In the MAC protocol by Roy Choudhury et al. [1], a source uses multihop RTSs to establish links with distant nodes. It then transmits CTS, DATA and ACK over a single hop. All the protocols in [5, 1, 6, 9] are designed by keeping in mind a particular type of directional antenna. All of these protocols have better performance compared to the basic CSMA/CA protocol and this can be seen from the simulation results in [5, 9, 6]. However, none of them can guarantee the delivery of data. They cannot eliminate the hidden terminal problem completely.

2.2 MAC Protocols for Sensor Networks

Since the nodes in a sensor network are static and they usually have periodic exchange of packets with their neighbors, it is possible to design MAC protocols that essentially find efficient scheduling strategies. Most of the MAC protocols designed for sensor networks take this approach. Ye et al. [11] developed the S-MAC protocol for scheduling communication in sensor networks. S-MAC organizes the sensor nodes into slots. Each slot in S-MAC is a fixed duty cycle consisting of a short listen period of 300ms and sleep period. All nodes in the network are free to choose their slots. S-MAC puts a node into sleep mode for a period of time and then wakes it to allow the node to listen to its neighbors. This scheme requires synchronization among neighboring nodes to minimize clock drift. Thus at the beginning of each cycle a node broadcasts a SYNC packet and the neighboring nodes update each other's schedules. To avoid collisions, S-MAC uses RTS/CTS mechanism. However, the fixed listening period of S-MAC may not be sufficient in high traffic loads. Also, it allows a node that has more data to send to monopolize the wireless channel.

Dam and Langendoen [2] proposed the Timeout MAC (T-MAC) protocol to alleviate S-MAC's fixed duty cycle problem. T-MAC implements an adaptive duty cycle that consists of a variable length active period and a sleep period. Lu *et al.* developed a data gathering MAC (D-MAC) protocol for sensor networks. Each slot in this protocol is divided into a receiving, a sending and a sleeping period. D-MAC is limited to data gathering applications and not suitable for general data exchange applications.

There are also several proposed MAC protocols using Time Division Multiple Access (TDMA) scheme. TDMA-based MAC protocols separate nodes in time, i.e., nodes do not interfere with each other's transmissions and hence collision is avoided. The Lightweight Medium Access (LMAC) protocol is developed by van Hoesel and Havinga [10]. In this protocol, a node can own a time slot for transmitting and receiving messages without contending for medium access. When an active node controls a time slot, it can broadcast a control message section of its frame and transmit the data afterwards. Neighboring nodes listen and check the control message. If they are the intended receiver of the packet, they receive the data packet afterwards. Otherwise, they go to the sleep mode and wake up at the next time slot. In contrast to the other MAC protocols for sensor networks discussed above, LMAC does not use RTS/CTS/ACK scheme and puts the issue of transmission reliability at the physical layer. Nodes maintain a scheduling table and select a slot number that is not in use within a two-hop neighborhood. A time slot is only reused at least three hops apart. Kulkarni and Arumugam [7] proposed the Self-Stabilizing TDMA (SS-TDMA) protocol that uses a fixed schedule for the lifetime of a network. SS-TDMA is designed to operate on a regular grid topology such as rectangular, hexagonal and triangular.

In all the MAC protocols designed for sensor networks, the central theme is to save energy and avoid collisions. Hence, all these protocols work by working out a scheduling policy in the network by which each node knows when to send or receive messages. Since each node negotiates a schedule with its neighbors, it is ensured that collisions do not occur while neighbors are communicating. However, collisions still may occur due to transmissions by nodes that are one-hop away. In our protocol, we try to avoid collisions by working out a global schedule by which nodes send and receive messages. This global schedule is worked out in a completely distributed manner.

3 Our Protocol

We divide the space surrounding each node into a number of sectors. In general we can use any number of sectors depending on the capability and number of elements in the directional antenna. However, all our simulations are done assuming six sectors, each with an angular range of $60°$. The protocol has two phases. In the first phase each node tries to work out a schedule according to which it communicates with its neighbors in the second phase. The main aim of our protocol is to reduce energy consumption in each node of the network. This is achieved in two different ways.

Our protocol reduces collisions significantly by scheduling the communication among neighboring nodes (i.e., nodes that are within the transmission range of each other). The nodes start working in the first phase by transmitting in the *omnidirectional mode* according to the normal CSMA/CA protocol possibly enhanced by the RTS/CTS mechanism.

However, this phase is used for working out a schedule and hence much shorter in duration compared to the overall lifetime of a sensor network. The nodes enter the second phase once they have worked out schedules which are consistent with their neighbors' schedules.

The most significant change in terms of wireless transmission between the first and second stages is the direction of transmission. The nodes communicate in the omnidirectional mode in the first stage. As a result, each node knows about its neighbors that are within the omnidirectional transmission range. However, the nodes communicate in directional mode in the second phase, i.e., during the remaining part of the lifetime of the network. The energy spent in the directional mode of transmission is significantly smaller compared to the omnidirectional mode. Hence the nodes can reach the same neighbors that they have discovered in the first phase by spending less energy in the second phase. This prolongs the lifetime of the network.

3.1 Physical Layer Facilities

Each node is assumed to have only one radio transceiver, which can transmit and receive only one packet at any given time. The transceiver is assumed to be equipped with six directional antennas, each antenna responsible for receiving or transmitting in a $60°$ angular sector. It is assumed that the transmissions by adjacent antennas never overlap, that is, complete attenuation of the transmitted signal occurs outside the angular sector of the directional antenna. The MAC protocol is assumed to be able to switch every antenna individually or all the antennas together to the active or passive modes. The radio transceiver uses only the antennas that are in active mode. The receiver node uses receiver diversity while receiving on all antennas. This means that the receiver node uses the signal from the antenna which receives the incoming signal at maximum power. In the normal case, this selected antenna would be the one whose sector pattern is directed towards the source node whose signal it is receiving. It is assumed that the radio range is the same for all directional antennas of the nodes. In order to detect the presence of a signal, a threshold signal power value is used. A node concludes that the channel is active only if the received signal strength is higher than this threshold value. It is assumed that the orientation of sectors of each antenna element remains fixed and same. Moreover, the distance covered by the directional and omnidirectional transmission are same.

3.2 Phase 1 of the Protocol

There are two main tasks in Phase 1 of the protocol. First, each node discovers its neighbors and their locations with respect to its own sectors. Next, each node determines its own receiving and sending schedule by which it either receives from or sends messages to its neighbors. These schedules are exchanged among the neighbors until the point when each node arrives at a satisfactory schedule. All the communication in the first phase is done by using the CSMA/CA protocol.

Location of Neighbors. We assume that each node is equipped with a global position system (GPS) device. Each node knows only its own location. Initially each node informs its neighbors about its location by exchanging *location packets*. When a node receives a location packet, it calculates the receiving sector and the sender's sending sector by using the locations of the sender and its own location. The sectors are calculated with respect to a fixed global coordinate system, so that the sector calculations are uniform across all nodes.

Computing and Exchanging Schedules. The exchange of schedules start once each node has discovered its neighbors. In its pure form, each node sends its schedule to each of its neighbors and each node tries to revise its schedule according to its neighbors' schedules. This process continues until each node arrives at a stable schedule. However, this process is usually very long for a large network. In other words, nodes need to go through many rounds before the schedules are stabilized. Hence we have introduced several measures to arrive at fast schedules for the nodes in this phase.

Nodes in the network are given different priorities for arranging their schedules. Higher priority nodes arrange transmission time first. The nodes are assigned priorities depending on the number of neighbors a node has. The node or nodes with highest number of neighbors are given the highest priority. The other nodes are given priorities according to decreasing number of neighbors. The first phase starts with each node announcing its presence to all its neighbors through a *hello* message. Each node counts the number of its neighbors and then sends another message to its neighbors with this number. As all of these communications are in CSMA/CA mode, each node waits until it has received the neighbor counts from all of its neighbors. Moreover, each node knows its priority in its neighborhood and waits until all the other nodes with higher priority have adjusted their slots.

Each node maintains a location table. The location table contains all the information about a node's neighbors and two-hops-away nodes. The first column and first row of the table represent transmissions about the node itself. Each row number is mapped to a sender identity, and a column number is mapped to a receiver's identity.

A node arranges a suitable communication slot based on its location table. Each node is able to decide at what slot it transmits/receives and in which sector so that the transmissions do not conflict with its neighbors' transmission schedules in the same sector. In the location table of a node, the entries in the first row represent the information of the node transmitting to its neighbors and the entries in the first column represent the information of the node receiving from its neighbors. Table 1 shows the location table of node A. The first row indicates the information of node A transmitting to its neighbors, node B and C. The first column indicates the information of node A receiving from node B and C. The protocol needs to arrange time for those entries for avoidance of collision and packet loss.

Table 1. The location table of node A which is holding the transmission. For example, entry $B \rightarrow A$ means that the transmission from node B to A

	A	B	C
A		$A \rightarrow B$	$A \rightarrow C$
B	$B \rightarrow A$		$B \rightarrow C$
C	$C \rightarrow A$	$C \rightarrow B$	

Every time the node receives a new packet it updates its location table. It checks whether there are more than one entries having same slots in the table. If there is, it means that the node needs to do more than one job in the same slot. For example, if there are two entries in the same slot in a row, the node needs to send two messages at the same slot to two of its neighbors in the same sector. Similarly, if there are two entries in the same slot of a column, the node needs to receive two messages from two of its neighbors at the same slot and in different sectors. These are the conflict situations that a node should avoid while updating its location table. It is necessary for a node to ensure that there is only one entry in each slot. For example, both entry $A \rightarrow B$ and entry $C \rightarrow A$ of node A contains *slot 1* in table 2. Only one of them can be used in one time slot.

The order of arranging time slots depends on the priorities of nodes in their local area. The local area refers to the node and its neighbors. This order can significantly reduce the time of determining time slots among neighbors. However, this does not mean that only one node determines its slots in its local area at a time. It is possible to have more than one node to arrange time slots simultaneously. In this case, two or more nodes with the same priority choose the same slot for transmission in the same sector. This causes collisions or packet loss. When potential collision or packet loss occurs at a node, we need to clarify who is responsible for changing the slots.

First, we clarify that the modifications of communication time for a node may be done by another node. This is because a higher priority neighbor may arrange both transmitting and receiving communication for the lower priority nodes. In a location table, each entry has an ID of a node associated with it that is responsible for changing the entry. The responsibility of changing an entry in a slot is on a node with a smaller number of neighbors. If the transmissions of two nodes occupy the same slot, the node that has lower number of neighbors will change the time. This is a reasonable assumption since the node with a lower number of neighbors has greater flexibility in arranging its slots.

Table 2. Transmission time slots gained in node A's location table. Two entries $A \rightarrow B$ and $C \rightarrow A$ uses time slot 1 to transmit packets. Node A cannot do two actions at one time slot. Therefore, node A must rearrange the conflict

	A	B	C
A		Slot 1	Slot 2
B	Slot 3		
C	*Slot 1*		

3.3 Time Conversion

The nodes communicate with each other in the second phase according to a cyclic or round-robin schedule. Each node uses its sectors in a cyclic fashion Each node in the second phase uses only directional transmission. This directional transmission follows the cycle of tasks that are assigned in phase one. However, each node has different cyclical periods. For example, node A's cycle may include 4 time slots, but node B's cycle may include 8 slots depending on the number of their neighbors. Suppose node A receives node B's schedule with the information that node B will transmit packets to it at B's time slot 2. A does not know which of its own slots corresponds to B's slot 2. Furthermore, although all the nodes in the network have same length cycle, at any time they can be in different position of their own cycles. Hence, it is necessary for a node to understand a neighbor's time slots in terms of its own time slots.

The first task is to normalize the cycles for each node depending on the cycles of its neighbors. The idea here is to choose the maximum cycle in a neighborhood as the normalized cycle for each node in the neighborhood. In other words, if a node has less number of neighbors compared to its neighbors, it still accepts a cycle that is the largest among its neighbors. In this case, a node may use one or more idle slots in its cycle so that it can synchronize its cycle with a neighbor that needs to accommodate more neighbors.

Since the transmission in the first phase is through the CSMA/CA protocol, there is considerable chance of collision while the nodes exchange their schedules. Hence it is very important that the nodes correctly exchange their schedules before they enter the second phase. Since higher priority nodes get the opportunity to fix their schedules earlier than the lower priority nodes, it is possible that higher priority nodes in a neighborhood will be ready to enter the second phase earlier.

Instead, when a higher priority node is ready to enter phase 2, it still continuously broadcasts its schedule omnidirectionally in CSMA mode. This gives more chances to the nodes that did not receive the arranged slots before because of collisions. When the least priority nodes finish their slot arrangements, they broadcast a prepare-to-enter message including a waiting time to their neighbors, and then enter the second phase when the waiting time is reached. The other nodes receive the message with the time and prepare to enter the second phase.

4 Simulation Results

To evaluate the performance of our MAC protocol compared with normal CSMA, we developed a network simulator in C. In our simulator, time is divided into slots. Transmission channel is implemented as a buffer. A packet in the channel is held in one buffer element. A node in a time slot can only take one action, transmit or receive. Furthermore, propagation delay of packet transmission is considered to be zero. That means a node in a time slot transmits a packet to another node. The receiver will receive it at the same time slot. This is a realistic assumption considering the speed of propagation of wireless signals.

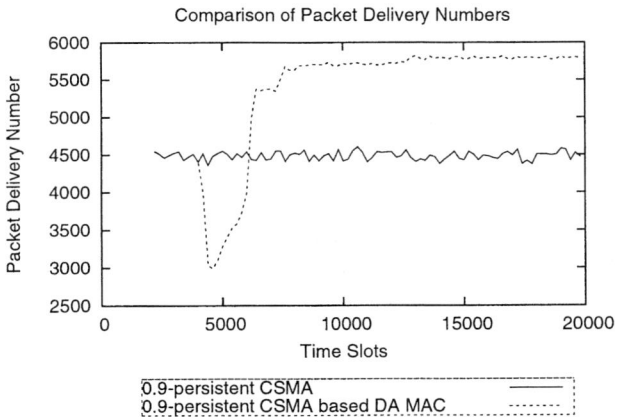

Fig. 1. Comparison of total number of packets delivered when each node has a maximum of eight neighbors

In order to evaluate the performance of our MAC protocol, we compare the packet delivery ratio of our protocol with that of the CSMA/CA protocol. Packet delivery ratio is the number of delivered data packets divided by the total number of transmitted data

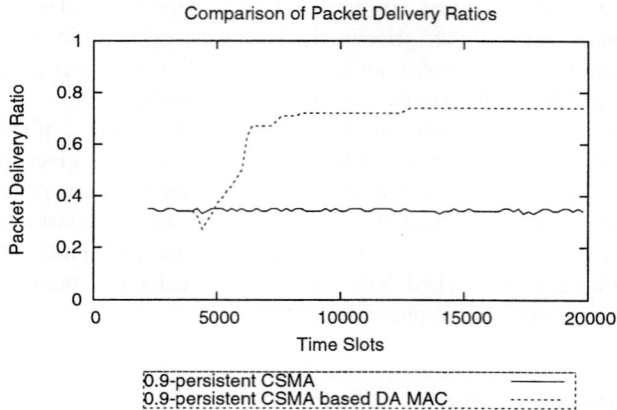

Fig. 2. Comparison of packet delivery ratio when each node has a maximum of eight neighbors

packets. We refer to our protocol as directional MAC or DA-MAC and the normal CSMA/CA protocol as CSMA.

We generated a topology by randomly distributing 200 nodes in a fixed 400x400 square meters area. We ensure that the network is connected. In two sets of experiments, each node was restricted to have a maximum of three and eight neighbors respectively. The traffic in the network is generated in the following way. For our protocol DA-MAC, each node sends a message to its neighbors in each time slot. In other words, a node A sends a message to a node B in the directional mode whenever the sending sector of A matches with the receiving sector of B. For CSMA/CA, each node sends a message to all its neighbors in each time slot in omnidirectional mode. We evaluate the performance on the new protocol compared with 0.9-persistent CSMA. Each node, when it finds the channel unused transmits with a probability of 0.9 in the 0.9 persistent CSMA protocol. We use the 0.9 persistent CSMA protocol in the first phase of our protocol as well. Our simulation is run for 20000 time slots.

Some results from our simulations are shown in Fig. 1 and 2. We compare the packet delivery numbers of our protocol with that of CSMA in Fig. 1. Each point in this graph represents the total number of packets transmitted in 200 time slots. The performance of CSMA is consistent throughout the simulation as the number of collisions in CSMA is almost the same. Our protocol behaves like the CSMA protocol in the first phase as it uses the same 0.9 persistent CSMA. Then it behaves worse than CSMA briefly as the nodes prepare to enter the second phase and broadcast their schedules continuously for a brief period. However, the performance of our protocol improves drastically once the nodes enter the second phase. Fig. 2 shows the corresponding packet delivery ratios. The packet delivery ratio of our protocol is much better in the second phase.

References

1. R. CHOUDHURY, X. YANG, R. RAMANATHAN AND N. VAIDYA Using directional antennas for medium access control in ad hoc networks. In *Proceedings of the 8th annual international conference on Mobile computing and networking* (Atlanta, Georgia, USA, 2002), ACM Press New York, NY, USA, pp. 59–70.
2. T. VAN DAM AND K. LANGENDOEN An adaptive energy-efficient MAC protocol for wireless sensor networks *Proc. 1st Intl. Conference on Embedded Networked Sensor System*, pp. 171-180, ACM Press, 2003.
3. Z. HUANG AND C. SHEN A comparison study of omnidirectional and directional MAC protocols for ad hoc networks. In *Proceedings of IEEE Globecom* (Taipei, Taiwan, Nov. 17-21 2002).
4. C. JAIKAEO AND C. SHEN Multicast communication in ad hoc networks with directional antennas. In *Proceedings of the 12th International conference on Computer Communications and Networks, 2003. ICCCN 2003* (Oct. 20-22 2003), IEEE Xplore, pp. 385–390.
5. Y. KO, V. SHANKARKUMAR, AND N. VAIDYA Medium access control protocols using directional antennas in ad hoc networks. In *Proceedings of IEEE Conference on Computer Communications (INFOCOM)* (Tel Aviv, Israel, 2000), pp. 13–21.
6. T. KORAKIS, G. JAKLLARI AND L. TASSIULAS A MAC protocol for full exploitation of directional antennas in ad hoc wireless networks. In *Proceedings of the 4th ACM international symposium on Mobile ad hoc networking and computing* (Annapolis, Maryland, USA, 2003), ACM Press New York, NY, USA, pp. 98–107.
7. S. KULKARNI AND M. ARUMUGAM TDMA service for sensor networks. *Proc. 24th International Conference on Distributed Computing Systems*, pp. 604-609, IEEE, March 2004.
8. G. LU, B. KRISHNAMACHARI AND C. S. RAGHAVENDRA An adaptive energy-efficient and low-latency MAC for data gathering in wireless sensor networks. *Proc. IPDPS 2004*, Santa Fe, April 2004.
9. A. NASIPURI, S. YE, J. YOU, AND R. HIROMOTO A MAC protocol for mobile ad hoc networks using directional antennas. In *Proceedings of IEEE Wireless Communications and Networking Conference (WCNC)* (Chicago, IL, Sep. 23–28 2000).
10. L. VAN HOESEL AND P. HAVINGA A lightweight medium access protocol (LMAC) for wireless sensor networks. *Proc. 1st Intl. Workshop on Networked Sensing Systems (INSS 2004)*, Tokyo, June 2004.
11. W. YE, J. HEIDEMANN AND D. ESTRIN An energy-efficient MAC protocol for wireless sensor networks. *Proc. IEEE INFOCOM*, pp. 1567-1576, 2002.

An Extended Framework for Proportional Differentiation: Performance Metrics and Evaluation Considerations

Jahwan Koo[1] and Seongjin Ahn[2,*]

[1] School of Information and Communications Engineering, Sungkyunkwan Univ.,
Chunchun-dong 300, Jangan-gu, Suwon, Kyounggi-do, Korea
jhkoo@songgang.skku.ac.kr
[2] Department of Computer Education, Sungkyunkwan Univ.,
Myeongnyun-dong 3-ga 53, Jongno-gu, Seoul, Korea
sjahn@comedu.skku.ac.kr

Abstract. The proportional differentiation model is receiving a lot of attention recently as an effective solution for quantitative service differentiation in IP networks. we introduce the proportional differentiated service and its general framework. This general framework for proportional differentiated service in an IP network consists of three major modules at routers: the packet dropper, the packet scheduler, and the proportional policy unit. In this paper, we propose an extended framework for the proportional differentiated service based on the combination of TCP congestion control (TCC), active queue management (AQM), and packet scheduling to provide a unified, simple, robust, and scalable framework. In our opinion, an appropriate combination of the packet dropper and packet scheduler at intermediate router and the TCC reaction at source/sink system would provide an effective solution for the proportional differentiated service.

1 Introduction

The Internet which originally started as a means of moving files from one computer to another has now evolved into a global communications network. The number and variety of applications on the Internet has increased considerably. The IP based Internet is now serving users ranging from ordinary home users to huge corporations. In spite of this tremendous growth in usage, the Internet architecture has under gone very little change.

The Internet is still based on the best-effort service model. Although the best-effort service, which was used in the early days of the Internet, was adequate as long as the applications using the network were not sensitive to variations in losses and delays, it is no longer adequate, due to the explosion in the number of different applications.

---

* Dr. S. Ahn. is the Corresponding Author.

To solve this problem, in the last few years, there has been a wave of interest in providing network services with performance guarantees and in developing algorithms supporting different levels of services. The various solutions that have been proposed to solve these problems can be summarized under the general heading of Quality-of-Service (QoS). The need for such quality assurance, service guarantee, and reliability drove the Internet Engineering Task Force (IETF) to define two QoS solutions - Differentiated Services (DiffServ) and Integrated Services (IntServ). Of these two methods, the DiffServ gained popularity due to its inherent scalability, ease of implementation, and reduced operation complexity at the core of the Internet.

However, the DiffServ can only guarantee that traffic of a high priority class will receive no worse service than that of a low priority class. Recently, it has been further refined into a quantitative scheme: the proportional differentiated service. For this proportional differentiated service is controllable, consistent, and scalable, it provides network operators convenient management of services and resources even in a large-scale network.

Proportional differentiated service, initially proposed by Dovrolis et al. [1], has been an effort to quantify the differentiation between classes of traffic and to enforce that the ratios of delays or loss rates of successive priority classes be roughly constant.

Most mechanisms for proportional differentiated service have been used independent algorithms for delay and loss differentiation. For example, proportional differentiation of delays has been implemented with appropriate scheduling algorithms [2] [4] [5] and proportional differentiation of loss has been implemented by queue management algorithms [3] [6]. Recently, a joint queue management and scheduling algorithm on proportional differentiation was reported in [9].

Various QoS metrics have been used for proportional differentiation. The first QoS metric discussed in proportional differentiation is average packet delay with the Wait Time Priority (WTP) scheduler [2]. The work in [3] extended proportional differentiation to packet loss rate using a proportional loss rate (PLR) dropper. The performance of real-time applications does not depend on average packet delay. As a result, the proportional differentiation has been applied to deadline violation probability in [7]. Recently, jitter has also been included in this proportional differentiation [8]. However, all these algorithms perform only in terms of not end-to-end but per-hop or edge-to-edge.

The rest of the paper is organized as follows. In section 2 and 3, we introduce the proportional differentiated service and its general framework. This general framework for proportional differentiated service in an IP network consists of three major modules at routers: the packet dropper, the packet scheduler, and the proportional policy unit. In section 4, we provide details of proportional differentiation on various QoS metrics. In section 5, we propose an extended framework for the proportional differentiated service based on the combination of TCP congestion control (TCC), active queue management (AQM), and packet scheduling. This goal is to provide a unified, simple, robust, and scalable framework.

2 Proportional Differentiation Model

In the proportional differentiation model, the service differentiation can be quantitatively adjusted to be proportional to the differentiation factors that a network service provider sets beforehand. If q_i is the QoS metric of interest and s_i is the differentiation factors for class i, $i = 1, \cdots, n$, in the proportional differentiation model, we should have:

$$\frac{q_i}{q_j} = \frac{s_i}{s_j} \tag{1}$$

If the desired differentiation among n classes are quantified as the ratios, δ_i, the proportional differentiation of delay(d_i) and loss(l_i) can be rewritten as

$$\frac{d_i}{d_j} = \frac{\delta_i}{\delta_j}, \frac{l_m}{l_n} = \frac{\delta_m}{\delta_n}. \tag{2}$$

The proportional differentiation model has gained attention recently as an effective solution for quantitative service differentiation in IP networks.

3 General Framework for Proportional Differentiation

In this section, we describe a general framework for the proportional differentiation.

A general framework for proportional differentiated service in an IP network consists of four major modules at routers: 1) the packet classifier that can distinguish packets and group them according to their different requirements, 2) the packet dropper that determines both of the following: how much queue space should be given to certain kinds of network traffic and which packets should

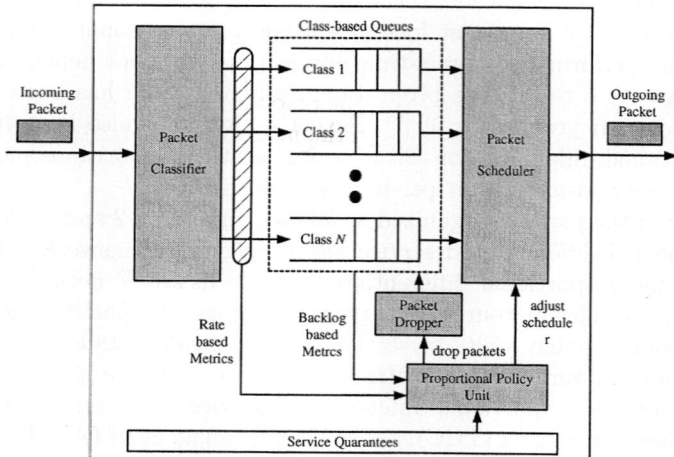

Fig. 1. General framework for proportional differentiated service

be discarded during congestion, 3) the packet scheduler that decides the packet service order so as to meet the bandwidth and delay requirements of different types of traffic, and 4) the proportional policy unit that takes the real-time measurement of the QoS metrics and manages the packet dropper and scheduler.

4 QoS Metrics for Proportional Differentiation

In this section, we provide details of proportional differentiation on various QoS metrics. Broadly, there are two metrics for proportional differentiated service, characterized by the way they observe. One is rate based metrics such as the packet arrival rate and the other is backlog based metrics such as the number of packets in the queue, the average queue size, the average queue delay, and the queue delay jitter. These various metrics directly affect either the packet dropper or the packet scheduler. Recent efforts are reported on proportional differentiation using multiple QoS metrics simultaneously.

Next, we give a formal description of the basic metrics used in the proportional differentiation. We assume that the link has a capacity C and a total queue space Q. We use $a(t)$, $l(t)$, and $r(t)$, respectively, to denote the arrivals, the amount of traffic dropped, and the service rate at time t.

We now introduce the notions of arrival curve, input curve, and output curve for a traffic in the time interval $[t_1,t_2]$. The arrival curve A and the input curve R^{in} are defined as

$$A(t_1, t_2) = \int_{t_1}^{t_2} \lambda(x)dx, \tag{3}$$

where $\lambda(t)$ is the instantaneous arrival rate at time t, and

$$R^{in}(t_1, t_2) = A(t_1, t_2) - \int_{t_1}^{t_2} \xi(x)dx, \tag{4}$$

where $\xi(t)$ is the instantaneous drop rate at time t. From Eqn.(2), the difference between the arrival and input curve is the amount of dropped traffic. The output curve R^{out} is the transmitted traffic in the interval $[t_1,t_2]$, given by

$$R^{out}(t_1, t_2) = \int_{t_1}^{t_2} r(x)dx, \tag{5}$$

where $r(t)$ is the instantaneous service rate at time t. From now on, we will use the following shorthand notations to denote the arrival, input and output curves at a given time t, respectively:

$$A(t) = A(0,t),$$
$$R^{in}(t) = R^{in}(0,t),$$
$$R^{out}(t) = R^{out}(0,t).$$

In figure 2, the vertical and the horizontal distance between the input and output curves, respectively, are the queue length Q and the delay D. The delay

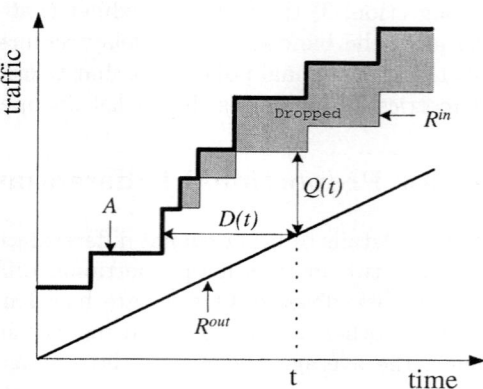

Fig. 2. Queue length, delay and loss

D at time t is the delay of an arrival which is transmitted at time t. Queue length and delay at time t are defined as

$$Q(t) = R^{in}(t) - R^{out}(t), \tag{6}$$

and

$$D(t) = \max_{x<t}\{x | R^{out}(t) \geq R^{in}(t-x)\}. \tag{7}$$

We also obtain the average delay by averaging the instantaneous delay $D(t)$ over a time window of length τ.

$$D_t^{avg}(\tau) = \frac{1}{\tau} \int_{t-\tau}^{t} D(x)dx. \tag{8}$$

We denote the loss rate as $P(t)$, which expresses the fraction of lost traffic since the beginning of the current busy period at time t_0. A busy period is a time interval with a positive queue length of traffic. So, $P(t)$ expresses the fraction of traffic that has been dropped in the time interval $[t_0, t]$, that is,

$$P(t) = \frac{\int_{t0}^{t} l(x)dx}{\int_{t0}^{t} a(x)dx} = 1 - \frac{R_{in}(t_0, t^-) + a(t) - l(t)}{A(t_0, t)}. \tag{9}$$

with

$$t^- = \sup\{x | x < t\}.$$

With the metrics just defined, therefore, we showed that we can now formally describe the QoS metrics for the proportional differentiation. The above three performance metrics do not fully expressed all network characteristics for proportional differentiation, but they will be added if needed.

Table 1. Analytical Models of Packet Scheduler

Packet Scheduler	Analytical Model
WTP	$Max\left\{\frac{t_{current} - t_{i.arrival}}{\delta_i}\right\}$
EDD	$Min\{t_{i.arrival} + d_i\}$
FCFS	$Max\{t_{current} - t_{i.arrival}\}$
Uniform	$Max\left\{\frac{t_{current} - t_{i.arrival} - d_i}{\delta_i}\right\}$
Parameters	$t_{current}$: local time when a packet need to be scheduled $t_{i.arrival}$: class i packet's arrival time d_i: deadline for class i

The analytical models for each packet scheduler are derived in [5], as summarized in Table 1.

The analytical models for each packet dropper are derived in [5], as summarized in Table 2.

Table 2. Analytical Models of Packet Dropper

Packet Dropper	Analytical Model
DropTail	N/A
RED	$P_a = \frac{P_b}{(1 - count \cdot P_b)}, P_b = \frac{max_p}{max_{th} - min_{th}} \times (Q_{avg} - min_{th})$
DRED	$p(t) = p(t-1) + \alpha \widehat{e}(t),$ $\widehat{e}(t) = (1 - \beta)\widehat{e}(t-1) + \beta e(t),$ $e(t) = q(t) - T(n)$
PI	$p(t) = p(t-1) + a(q(t) - q_{ref}) - b(q(t-1) - q_{ref})$
Parameters	P_a: drop probability P_b: intermediate drop probability max_p: maximum drop probability max_{th}: higher threshold, min_{th}: lower threshold $p(t)$: drop probability at time t $\widehat{e}(t)$: filtered error signal at time t $e(t)$: error signals at time t α: control gain, β: filter gain $q(t)$: queue size at time t $T(t)$: control target at time t, i.e., queue threshold

5 Extended Framework: Integration with TCC Mechanism

In this section, we introduce the requirement of the extended framework and propose an extended framework for the proportional differentiated service based on the combination of TCC, AQM, and packet scheduling.

The extended framework for the proportional differentiation is required as the following evaluation considerations: 1) The extended framework should be capable of controlling all two (delay and loss) proportional differentiation simultaneously. 2) Such a framework must be very simple to implement. 3) Although simple, it must be robust to handle the highly varying traffic loads. 4) For a network that is as wide as the Internet, it must be highly scalable.

Therefore, we suggest to combine the TCC mechanism with the packet dropper and packet scheduler described the previous section. This goal is to provide a unified, simple, robust, and scalable framework. The analytical models for each TCC mechanism are derived in [11], as summarized in Table 3.

Table 3. Analytical Models of TCC Mechanisms

TCC Mechanism	Analytical Model
Reno	$\bar{x} = \frac{1-q_i(t)}{\tau_i^2} - \frac{1}{2}q_i(t)x_i^2(t)$
Vegas	$\bar{x} = \frac{1}{(d_i+q_i(t))^2} \operatorname{sgn}\left(1 - \frac{x_i(t)q_i(t)}{\alpha_i d_i}\right)$
Parameters	τ_i: equilibrium round trip time for source i $x_i(t)$: source rate for source i at time t $q_i(t)$: end-to-end marking probability for source i at time t d_i: round trip propagation delay for source i α: control gain $\bar{x}$: average source rate for source i at time t

Finally, we refer to a article [12] on an analysis of the reciprocal relationship between TCC mechanisms and AQM schemes, by considering the average packet loss rate and average delay. It has demonstrated that an appropriate combination of the packet dropper and packet scheduler at intermediate router and TCC source reaction would provide an effective solution for the proportional differentiated service.

6 Conclusion and Future Work

In this paper, we introduced the proportional differentiated service and its general framework. This general framework for proportional differentiated service in an IP network consists of three major modules at routers: the packet dropper, the packet scheduler, and the proportional policy unit, while our proposed framework for the proportional differentiated service was based on the combination of the TCC, AQM, and packet scheduling. The TCC mechanism plays a significant role in the proportional differentiation that it should be included in QoS framework as one of the major components for the proportional differentiation. In future, we will demonstrate that our extended framework is more robust, stable, and scalable than the general framework in the proportional differentiation via simulation.

References

1. C. Dovrolis and P. Ramanthan, "A Case for Relative Differentiated Services and the Proportional Differentiation Model," IEEE Network, 13(5):26-34, 1999.
2. C. Dovrolis, D. Stiliadis, and P. Ramanathan, "Proportional Differentiated Services: Delay Differentiation and Packet Scheduling," Proc. ACM SIGCOMM, pp. 109-120, 1999.
3. C. Dovrolis and P. Ramanathan, "Proportional Differentiated Services, Part II: Loss Rate Differentiation and Packet Dropping," Proc. IWQoS, pp. 52-61, 2000.
4. Chin-Chang Li et al., "Proportional Delay Differentiation Service Based on Weighted Fair Queueing," Proc. IEEE ICCCN, pp. 418-423, 2000
5. Y. Chen, M. Hamdi, D. Tsang, and C. Qiao, "Proportional QoS provision: a uniform and practical solution," Proc. ICC 2002, pp. 2363-2366, 2002
6. J. Aweya, M. Ouellette, and D. Y. Montuno, "Weighted Proportional Loss Rate Differentiation of TCP Traffic," Int. J. Network Mgmt, pp. 257-272, 2004
7. S. Bodamer, "A New Scheduling Mechanism to Provide Relative Differentiation for Real-Time IP Traffic," Proc. GLOBECOM, vol. 1, pp. 646-650, 2000.
8. T. Quynh et al, "Relative Jitter Packet Scheduling for Differentiated Services," Proc. 9th IFIP Conf. Perf. Modeling and Eval. of ATM and IP Networks, 2001.
9. J. Liebeherr and N. Christin, "JoBS: Joint buffer management and scheduling for differentiated services," Proc. IWQoS 2001, pp. 404-418, Karlsruhe, Germany, 2001.
10. B. Lowekamp, B. Tierney, L. Cottrell, R. Hughes-Jones, T. Kielmann, and M. Swany, "A hierarchy of network performance characteristics for Grid applications ans services, GFD-R-P.023," May 2004.
11. S.H. Low, "A duality model of TCP and queue management algorithms," Proc. ITC Specialist Seminar on IP Traffic Measurement, Modeling and Management, 2000.
12. J. Koo, J. Shin, S. Ahn, and J. Chung, "Models and Analysis of TCC/AQM Schemes over DiffServ Networks," Proc. IEEE ICN, 2005

QoS Provisioning in an Enhanced FMIPv6 Architecture

Zheng Wan, Xuezeng Pan, and Lingdi Ping

College of Computer Science, Zhejiang University, Hangzhou, P.R.C, 310027
{chen-hl, xzpan, ldping}@zju.edu.cn

Abstract. Mobility management and QoS provisioning are both key techniques in the future wireless mobile networks. In this paper we propose a framework for supporting QoS under an enhanced "Fast Handovers for Mobile IPv6" (FMIPv6) architecture. By introducing the key entity called "Crossover Router" (CR), we shorten the length of packet forwarding path before the MN completes binding update. For QoS guarantee, we extend the FBU and HI messages to inform the NAR of the MN's QoS requirement and make advance resource reservation along the possible future-forwarding path before the MN attaches to the NAR's link. We keep RSVP states in the intermediate routers along overlapped path unchanged to reduce reservation hops and signaling delays. The Performance analysis shows that the proposed scheme for QoS guarantee has lower signaling cost and latency of reservation re-establishment, as well as less bandwidth requirements in comparison with MRSVP.

1 Introduction

Wireless devices are expected to increase in number and capabilities in the following years. Mobile and wireless access will become more and more popular. Thus Mobile IPv6 (MIPv6) protocol [1] is proposed to manage mobility and maintain network connectivity in the next generation Internet.

However, there are still two problems to be resolved in MIPv6 environment. Firstly, the handover latency and packet loss in basic MIPv6 protocol are not ideal, which raises the need for a fast and smooth handover mechanism. A number of ways of introducing hierarchy into IPv4 as well as IPv6 networks, and realizing the advanced configuration have been proposed in the last few years [2-4]. Secondly, as real-time services grow, the desire for high quality guarantee of these services becomes eager in MIPv6 networks. As we know, two different models are proposed to guarantee QoS in the Internet by IETF: the integrated services (IntServ) [5] and differentiated services (DiffServ) [6] models. However, only IntServ model which uses RSVP protocol [7] to reserve resources can provide end-to-end QoS.

In this paper we propose a scheme for QoS provisioning in an enhanced "*Fast Handovers for Mobile IPv6*" (FMIPv6) architecture [2]. Two enhancements are introduced to improve the performance of basic FMIPv6. To reduce tunnel distance between the *Previous Access Router* (PAR) and the *New Access Router* (NAR), we propose that the *Crossover Router* (CR) intercept packets destined to MN and forward them to the NAR. CR is the first common router of the old path and the new forwarding path. We also use an efficient mechanism to eliminate the long *Duplicate*

Address Detection (DAD) latency. As for QoS guarantee, we use extended FBU and HI messages to inform the NAR of the MN's QoS requirement. Upon receiving the information, the NAR initiates an advance reservation process along the possible future-forwarding path before the MN arrives the NAR's link. Again the CR is used to reduce the length of reservation path.

The rest of the paper is organized as follows. Section 2 presents some related work. Section 3 presents the overview of proposed scheme. Section 4 describes the detailed handover and resource reservation process. Section 5 gives the performance measurement, and Section 6 concludes the paper and presents some areas for future work.

2 Related Work

2.1 Fast Handover for MIPv6

FMIPv6 aims to decrease packet loss by reducing IP connectivity latency and binding update latency. The MN uses L2 triggers to discover available *access points* (APs) and obtain further information of corresponding *access routers* (ARs) when it is still connected to its current subnet. After that, the MN may pre-configure the *New CoA* (NCoA) and register it to the PAR to bind *previous CoA* (PCoA) and NCoA. Through these operations the movement detection latency and the new CoA configuration latency are reduced. To reduce the binding update latency, a bi-directional tunnel between the PAR and the NAR is used to forward packets until the MN completes binding update. When the MN moves to the new subnet link, it will announce its attachment to launch forwarding of buffered packets from the NAR.

However, there are two disadvantages in basic FMIPv6 protocol. One is that the tunnel between the PAR and the NAR is fairly long. The other is that the DAD procedure for NCoA validation causes large handover latency. We'll discuss the solutions later.

2.2 Techniques of QoS Provisioning

Due to host mobility and characteristics of wireless networks, there are several problems in applying RSVP to mobile wireless networks. In the past several years many RSVP extensions were proposed to solve the problems. Talukdar et al. [9] proposed the MRSVP protocol in which resource reservations are pre-established in the neighboring ARs to reduce the timing delay for QoS re-establishment. However, too many advance reservations may use up network resources.

Chaskar et al. [10] proposed a solution to perform QoS signaling during the binding registration process. This mechanism defines the structure of "*QoS OBJECT*" which contains the QoS requirement of MN's packet stream. One or more QoS OBJECTs are carried in a new IPv6 option called "*QoS OBJECT OPTION*" (QoS-OP), which may be included in the hop-by-hop extension header of binding update and acknowledgement messages. Fu et al. [11] applied QoS-OP in the *Hierarchical Mobile IPv6* (HMIPv6) [3] architecture. Both schemes make use of intrinsic mobility signaling and achieve faster response time for effecting QoS along the new path.

Moon et al. [12] explained the concept of CR, which is the beginning router of the common path. And the common path is the overlapped part of the new path and previous path. Fig. 1 presents an example of the common path and the CR. Shen et al. [13] presented an interoperation framework for RSVP and MIPv6 based on the "Flow Transparency" concept, which made use of common path by determining the "Nearest Common Router" (just like CR) too. In both schemes the CR ensures that reservation will not be re-established in the routers along common path. Thus the QoS signaling overheads and delays as well as data packet delays and losses during handover can be greatly reduced.

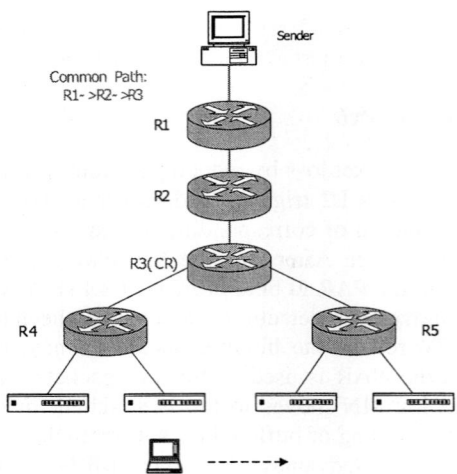

Fig. 1. Common Path and Crossover Router (CR)

3 Overview of Proposed Solution

The proposed solution includes two parts: some improvements to basic FMIPv6 and an efficient framework for end-to-end QoS guarantee in the enhanced FMIPv6 architecture.

Assuming that we have determined the location of CR, data forwarding path using the bi-directional tunnel of FMIPv6 would be CN-CR-PAR-CR-NAR. Obviously we can shorten the path to CN-CR-NAR. The method for CR determination will be introduced later. Though the bi-directional tunnel is eliminated in our scheme, a unidirectional tunnel from PAR to NAR is still included because the CR does not know when to intercept packets that destined to the MN's PCoA. When tunneling process begins, the PAR sends a TUN_BEGIN message which enables the CR to intercept the packets destined to the MN's PCoA and forward them to the NAR. In the opposite direction, the NAR directly sends packets with the CN's address filled in the destination address field. The CR intercepts these packets, sets the source address field to the MN's PCoA and forwards them to the CN.

A further modification to the basic FMIPv6 is the elimination of DAD procedure. We adopt the method of "Address Pool based Stateful NCoA Configuration" [8]. The NCoA pools are established at NAR or PAR. Each NCoA pool maintains a list of NCoAs already confirmed by the corresponding NAR. Thus the NCoA assigned to the MN at each handover event is already confirmed so that the DAD procedure can be ignored.

Now come to the part of QoS guarantee. As we know in FMIPv6 architecture, the NCoA is pre-established. Thus we can set up reservation along several possible future-forwarding paths (one or more NARs may be detected in FMIPv6) in advance when the MN still locates in the PAR's link. Just like MRSVP, *active* and *passive* Path/Resv messages and reservations are defined in our proposal. The NAR, which makes advance reservation and maintains soft state on behalf of the MN, acts as *remote mobile proxy*. To inform the NAR of the MN's QoS requirements, we extend the FBU and HI messages with QoS-OP in the hop-by-hop extension header.

Then we can initiate advance reservation along possible future path. Since there may be more than one NARs detected by the MN, all the possible future-forwarding paths must perform advance reservation. If the MN is a receiver, the CR issues the *passive Path* message to the NAR on behalf of the CN and the NAR in turn sends the *passive Resv* message to the CR. If the MN is a sender, the NAR issues the passive Path message. Upon receiving Path message, the CR immediately replies with a passive Resv message to the NAR. By performing these operations, the passive RSVP messages are restricted within the truly new part of the possible future path, which results in decreased RSVP signaling overheads and delays.

When the MN attaches to certain NAR's link, the packets sent from or destined to it can acquire QoS guarantee without any delay. At the same time advance reservations in other NARs' link must be released immediately. The modified FMIPv6 handover and resource reservation procedures when the MN acts as a receiver are depicted in Fig. 2.

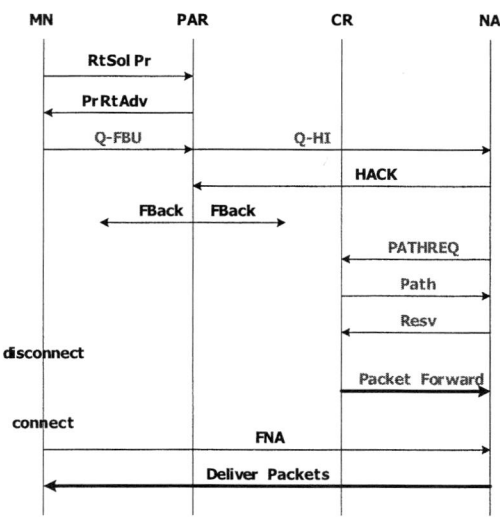

Fig. 2. Handover and Reservation Procedures of a Mobile Receiver

In conclusion, our proposed QoS provisioning scheme has the following advantages:

1. The transmission of QoS requirement makes use of intrinsic mobility signaling of FMIPv6, which results in faster response time for effecting QoS along the new path.
2. The advance reservation along the possible future path decreases the delay of reservation re-establishment and provides QoS guarantee for the MN until it completes binding update.
3. The CR keeps reservation along common path unchanged. Thus the reservation delay and signaling cost can be minimized, which in turn minimizes the handover service degradation.
4. The duration of advance reservations in our proposal is much shorter than MRSVP.

4 Detailed Operations

First of all, we assume that the MN moves into the boundary of the PAR so that the fast handover procedure launches. The procedures of proposed fast handover and resource reservation are as follows:

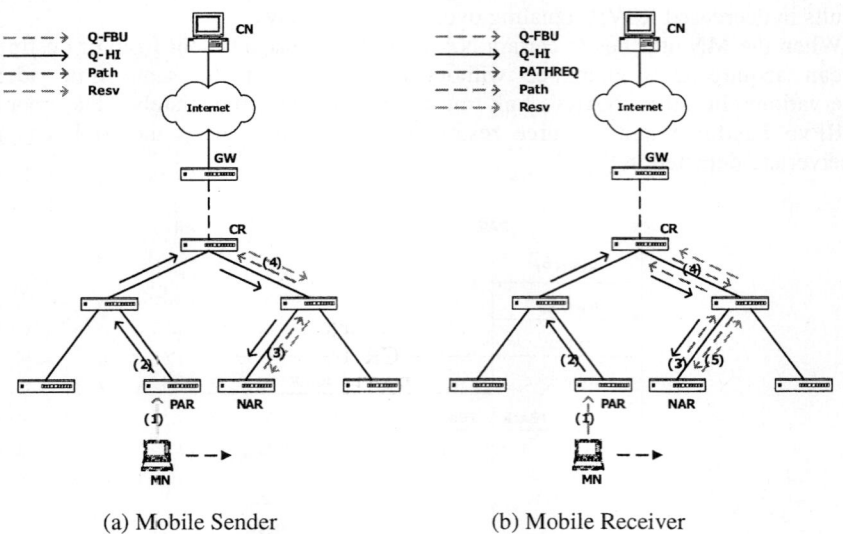

Fig. 3. Procedures of Advance Resource Reservation

1) The MN discovers available APs using link-layer specific mechanisms and then sends a *Router Solicitation for Proxy* (RtSolPr) message including the identifiers of the APs to the PAR.
2) After the reception of the RtSolPr message, the PAR resolves the access point identifiers to subnet router(s) (i.e. the [AP-ID, AR-Info] tuples). Though several

NARs may be discovered, the following description will just focus on the operations of certain NAR. Using the *"PAR-based stateful NCoA configuration"* proposed in [8], the PAR obtains a confirmed NCoA and responds the NCoA as well as the [AP-ID, AR-Info] tuple (via PrRtAdv) to MN.

3) In response to the PrRtAdv message, the MN sends a *Q-FBU* message to the PAR before its disconnection from the PAR's link. The Q-FBU message includes a QoS-OP (contains one or more QoS OBJECTs) in the hop-by-hop extension header. The QoS OBJECT may contain RSVP objects such as FLOW_SPEC, SENDER_TSPEC and FILTER_SPEC.

4) On reception of the Q-FBU message, the PAR again includes the MN's QoS requirement in the *Q-HI* message and sends it to the NAR. The Q-HI message should also contain the CN address corresponding to each QoS OBJECT, which will be used as the destination address of the *PATHREQ* message when the MN acts as a receiver.

Case 1. When the MN acts as a sender,

5a) The NAR directly issues the passive Path message. A RSVP router decides if it is the CR just by comparing the home address, the CoA and the previous RSVP hop carried in the passive Path message against the same information stored in the Path State. If there is a Path state related to the home address of passive Path message, and for the same home address both the CoA and the previous RSVP hop have been changed, then the router decides it is the CR. The binding of PCoA and NCoA is also included in a hop-by-hop extension header of the passive Path message. The CR will use the binding to prevent packet forwarding between the PAR and NAR.

6a) The CR does not forward the Path message further to the CN, but immediately replies with a passive Resv message to the NAR. By performing these operations, the RSVP states in the routers along the common path will not change. Fig. 3a describes the advance reservation process when the MN acts as the sender.

Case 2. Otherwise, the MN acts as a receiver,

5b) The NAR sends a PATHREQ message which has the CN's address as destination address (thus the CR can intercept this message) to request passive Path message. A RSVP router decides if it is the CR by searching the home address in PATHREQ against the same field in PATH state on the downlink direction. If there is a match of the home address in the PATH state in the downlink direction, then the router decides it is the CR. The PATHREQ message, which contains MN's home address and new CoA as introduced in [13], is extended to include the binding of PCoA and NCoA.

6b) The CR then issues the passive Path message to the NAR on behalf of the CN because the path between the CR and the CN is the common path and needn't any change. Finally the NAR will issue the passive Resv message towards the CR. Fig. 3b depicts the advance reservation process when the MN acts as the receiver.

7) At the same time as advance reservation process initiates, the NAR replies with a HACK message to the PAR, which may in turn issue the FBack message. The PAR may ignore sending this message because the NCoA is already confirmed.

8) When packet tunneling launches, the PAR will send a *TUN_BEGIN* message which has the CN's address as destination address. Upon receiving this message the CR begins to intercept packets destined to the PCoA and forward them to the NAR. Reversely, the NAR directly sends packets with the CN's address filled in the destination address field. The CR intercepts these packets, sets the source address field to the MN's PCoA and forwards them to the CN.
9) As soon as the MN attaches to the NAR, it sends the FNA message to the NAR. As a response, the NAR forwards buffered packets to the MN.

Finally, the MN can send a binding update to the HA and the CN. After it completes binding update, the CR stops intercepting packets sent from or destined to the MN. The packets will be forwarded with QoS guarantee along the new RSVP path.

5 Performance Analysis

In this section we study the performance of handover and resource reservation. We consider a network environment with a single domain made up of 16x16 square-shaped subnets and model the MN's mobility as a two-dimensional (2-D) random walk, which is similar to reference [14]. In a 2-D random walk, an MN may move to one of four neighboring subnets with equal probability. Under FMIPv6 architecture, only when the MN moves into the overlapped area of two or more APs, it may achieve information of the possible future NARs. Thus the number of the NARs is less than two. Under other simulated or real environments, the number of possible NARs is always less than the number of neighboring ARs in MRSVP.

Parameters:

N_p average number of possible NARs in FMIPv6;
N_n average number of neighboring ARs of current AR in MRSVP;
d_{x_y} average number of hops between x and y;
B_w bandwidth of the wired link;
B_{wl} bandwidth of the wireless link;
L_w latency of the wired link (propagation delay and link layer delay);
L_{wl} latency of the wireless link (propagation delay and link layer delay);
P_t routing table lookup and processing delay;
s_a average size of a signaling message for resource reservation;
B_r amount of the actual resource requirement of the handover MN;
t_r average time the MN will resident in certain AR's link;
t_{pl} time from completion of reservation to the beginning of L2 switch;
t_{l2} time to complete L2 switch.

With the above parameters, we define $t(s, d_{x\_y})$ as the transmission delay of a message of size s sent from x (an MN always) to y via the wireless and wired links.

$$t(s, d_{x\_y}) = (\frac{s}{B_{wl}} + L_{wl}) + d_{x\_y} \times (\frac{s}{B_w} + L_w) + (d_{x\_y} + 1) \times P_t \qquad (1)$$

5.1 Handover

Our proposed handover scheme affects the handover performance of FMIPv6 in three aspects. Firstly, the elimination of DAD procedure can reduce significant delays in FMIPv6. Secondly, decreased length of packet forwarding path during handover saves packet delivery time. When the MN attaches to the NAR's link, it can receive these packets from the NAR more quickly. This is necessary for real-time applications for that more packets' latency will be less than the threshold so that the application can use them for real-time audio and video playback. However, we should also consider the signaling cost of TUN_BEGIN message and additional overheads of PCoA and NCoA binding notification to the CR.

Finally, our proposed QoS guarantee mechanism also influences the handover performance. The Q-FBU and Q-HI messages size is enlarged to hold QoS requirement. So the signaling cost is larger than the basic FMIPv6 protocol. Since the size of QoS requirement is small in proportion to the total signaling cost, the additional latency introduced by the Q-FBU and Q-HI messages can be ignored.

Further analysis is not presented and we focus the discussion on the performance analysis of resource reservation.

5.2 Resource Reservation

1) *Total signaling cost of resource reservation*: In our proposed scheme, signaling messages for resource reservation include Q-FBU, Q-HI, PATHREQ, Path and Resv messages. s_a is the average size of these messages. Q-FBU message travels from the MN to the PAR; Q-HI message from the PAR to the NAR; PATHREQ, Path and Resv messages from NAR to CR. The total signaling cost of resource reservation is denoted by C and is computed as the following.

$$C_{FMIPv6-R} = s_a \times (d_{MN\_AR} + d_{AR\_AR} + 3 \times d_{AR\_CR}) \times N_p \qquad (2)$$

If MN acts as a sender, the PATHREQ message is not used.

$$C_{FMIPv6-S} = s_a \times (d_{MN\_AR} + d_{AR\_AR} + 2 \times d_{AR\_CR}) \times N_p \qquad (3)$$

In MRSVP, Spec, MSpec, Path, active Resv and passive Resv message are the signaling messages for resource reservation. We consider the scenario that the sender acts as the *receiver_anchor* node [9].

$$C_{MRSVP} = s_a \times (d_{AR\_AR} \times N_n + d_{AR\_CN}) + 2 s_a \times d_{AR\_CN} \times (N_n + 1) \qquad (4)$$

2) *Reservation establishment delay*: We compute the total delay since the MN issues Q-FBU to PAR. As the signaling cost, total delay of QoS establishment delay is affected by the same messages. The total delay of reservation establishment is denoted by D.

$$D_{FMIPv6-R} = t(s_a, d_{MN\_AR}) + t(s_a, d_{AR\_AR}) + 3 \times t(s_a, d_{AR\_CR}) \qquad (5)$$

$$D_{FMIPv6-S} = t(s_a, d_{MN\_AR}) + t(s_a, d_{AR\_AR}) + 2 \times t(s_a, d_{AR\_CR}) \quad (6)$$

$$D_{MRSVP} = t(s_a, d_{AR\_AR}) + 3 \times t(s_a, d_{AR\_CN}) \quad (7)$$

Note that the delays we compute here are reservation establishment delays. Actually, except for switch operation between active and passive reservation, the resource reservation can be used immediately when the MN attaches to the new subnet both in our FMIPv6 based advance reservation mechanism and in MRSVP.

3) *Bandwidth requirements*: The duration of advance reservation along the paths between CR and possible NARs is t_{pl} plus t_{l2}. When the MN arrives certain NAR's link, reservation status on this link changes to active while other passive reservations are released. We use B to denote the total bandwidth requirements including active and passive reservation during the period a MN residents in certain AR's link.

$$B_{FMIPv6} = B_r \times (d_{AR\_CR} + d_{CR\_CN}) \times t_r$$
$$+ B_r \times d_{AR\_CR} \times (t_{pl} + t_{l2}) \times N_p \quad (8)$$

$$B_{MRSVP} = B_r \times d_{AR\_CN} \times t_r \times (N_n + 1) \quad (9)$$

Now we can compare the performance of our proposal and the MRSVP. Let's focus on three pairs of parameters: d_{AR_CR} against d_{AR_CN}, N_p against N_n, and $t_{pl}+t_{l2}$ against t_r. The comparison results are identical: the former is much less than the latter. Thus we can draw the conclusion that the total signaling cost of resource reservation and the reservation establishment delay, as well as bandwidth requirements in our scheme are much less than those in MRSVP.

6 Conclusion

This paper proposes a framework for QoS guarantee based on an enhanced FMIPv6 architecture. We introduce a key entity which called "Crossover Router" (CR) to reduce the length of packet forwarding path before the MN completes binding update. Furthermore we use "Address Pool based Stateful NCoA Configuration" mechanism to eliminate the long DAD latency. The proposed QoS guarantee scheme achieves low signaling cost and reservation re-establishment latency by making use of the FBU and HI signaling messages of FMIPv6 to transmit QoS requirements and adopting the idea of advance reservation and common path. Performance analysis shows that our proposal outperforms MRSVP in terms of signaling cost, reservation re-establishment delay, and bandwidth requirements.

The simulation based on NS2 [15] platform for our scheme will be done soon to achieve the further performance analysis under various environments. When and how to release passive reservations on other NARs' link after the MN attaches to certain NAR's link, should be considered. Furthermore, we are also making efforts to apply the idea of our QoS provisioning scheme to F-HMIPv6 architecture [4].

References

1. D. Johnson, C. Perkins, J. Arkko, "Mobility support in IPv6", IETF RFC 3775, June 2004.
2. R. Koodli (Ed.), "Fast Handovers for Mobile IPv6", Internet Draft, IETF, draft-ietf-mipshop-fast-mipv6-03.txt, October 2004.
3. H. Soliman, C. Castelluccia, K. El-Malki, L. Bellier, "Hierarchical Mobile IPv6 mobility management", Internet Draft, IETF, draft-ietf-mipshop-hmipv6-03.txt, October, 2004.
4. H.Y Jung, S.J. Koh, H. Soliman, K. El-Malki, "Fast Handover for Hierarchical MIPv6 (F-HMIPv6)", Internet Draft, draft-jungmobileip-fastho-hmipv6-04.txt, June 2004.
5. R. Braden, D. Clark and S. Shenker, "Integrated services in the Internet architecture: An overview", IETF RFC 1633, June 1994.
6. S. Blake et al., "An architecture for differentiated services," IETF RFC 2475, December 1998.
7. R. Braden, Ed., L. Zhang, S. Berson, S. Herzog, S. Jamin. "Resource reserVation protocol (RSVP) -- Version 1 Functional Specification", IETF RFC 2205, September 1997.
8. Hee Young Jung, Seok Joo Koh, Dae Young Kim, "Address Pool based Stateful NCoA Configuration for FMIPv6", Internet Draft, draft-jung-mipshop-stateful-fmipv6-00.txt, August 2003.
9. A.K. Talukdar, B.R. Badrinath and A. Acharya, "MRSVP: A resource reservation protocol for an integrated services network with mobile hosts", Journal of Wireless Networks, vol.7, iss.1, pp.5-19 (2001).
10. H. Chaskar, and R. Koodli, "QoS support in mobile IP version 6", IEEE Broadband Wireless Summit (Networld+Interop), May 2001.
11. X. Fu et al., "QoS-Conditionalized binding update in mobile IPv6", Internet Draft, IETF, draft-tkn-nsis-qosbinding-mipv6-00.txt, January 2002.
12. B. Moon and A.H. Aghvami, "Quality of service mechanisms in all-IP wireless access networks", IEEE Journal on Selected Areas in Communications, June 2004.
13. Q. Shen, W. Seah, A. Lo, H. Zheng, M. Greis, "An interoperation framework for using RSVP in mobile IPv6 networks", Internet Draft, draft-shen-rsvp-mobileipv6-interop-00.txt, July 2001.
14. Shou-Chih Lo, Guanling Lee, Wen-Tsuen Chen, Jen-Chi Liu, "Architecture for mobility and QoS support in all-IP wireless networks", IEEE Journal on Selected Areas in Communications, May 2004.
15. The Network Simulator - NS (version 2), http://www.isi.edu/nsnam/ns/

Delay of the Slotted ALOHA Protocol with Binary Exponential Backoff Algorithm

Sun Hur[1], Jeong Kee Kim[1], and Dong Chun Lee[2]

[1] Dept. of Industrial Eng., Hanyang Univ., Korea
hursun@hanyang.ac.kr
[2] Dept. of Computer Science, Howon Univ., Korea

Abstract. We propose a method to compute the delay of the slotted ALOHA protocol with Binary Exponential Backoff (BEB) as a collision resolution algorithm. When a message which tries to reserve a channel collides n times, it chooses one of the next 2^n frames with equal probabilities and attempts the reservation again. We derive the expected access delay until an arbitrary message reserves a channel. Then the expected transmission delays for real time messages and non-real time messages are calculated analytically. The accuracy of our analytic model is checked against simulation.

1 Introduction

Slotted ALOHA(S-ALOHA) protocol has been widely adopted in local wireless communication systems as a random multiple access protocol. In these systems, each frame is divided into small slots and each mobile terminal contends for the slot to transmit its packets at the beginning of each frame. If two or more mobile terminals contend for the same slot, then a collision occurs and none of them can transmit their packets. The colliding packets are queued and retry after a random delay. The way to resolve the collision is called the collision resolution protocol. One of the widely used collision resolution protocol is the BEB algorithm, forms of which are included in Ethernet and wireless LAN standards: Whenever a node's message is involved in a collision, it selects one of the next 2^n frames with equal probabilities, where n is the number of collisions that the message has ever experienced, and attempts the retransmission.

Soni and Chockalingam [1] analyzed three backoff schemes, namely, linear backoff, exponential backoff, and geometric backoff. They calculated the throughput and energy efficiency as the reward rates in a renewal process and illustrated that the truncated BEB, which is considered in this paper, performs better since the idle length should grow only until a maximum value by numerical result. More recently, Chen and Li [2] proposed the quasi-FIFO algorithm, which is another novel collision resolution scheme. They showed, by simulation, that the proposed scheme shares the bandwidth more equally and maximizes the throughput, but no analytic model was given in [2] as well.

Delay distributions of slotted ALOHA and CSMA are derived in Yang and Yum [3] under three retransmission policies. They found the conditions for achieving finite delay mean and variance under the BEB. Their assumption, however, that the combination of new and retransmitted packet arrivals is a Poisson process is not valid because the stream of the retransmitted packets depends on the arrivals of new packets. This dependency makes the Poisson assumption invalid. Chatzimisios and Boucouvalas [4] presented an analytic model to compute the throughput of the IEEE 802.11 protocol for wireless LAN and examined the behavior of the Exponential Backoff (EB) algorithm used in 802.11. They assumed that the collision probability of a transmitted frame is independent of the number of retransmissions. As we will show later in this paper, however, this probability is a function of the number of competing stations and also depends upon the number of retransmissions that this station has ever experienced. Kwak et al. [5] gave new analytical results for the performance of the EB algorithm. Especially, they derived the analytical expression for the saturation throughput and expected access delay of a packet for a given number of nodes. Their EB model, however, is assumed that the packet can retransmit infinitely many times.

Stability is another issue on BEB algorithm and there are many methods dealing with this. As pointed in Kwak et al. [5], however, these studies show contradictory results because some of them do not represent the real system and they adopt different definitions of stability used in the analyses. The dispute is still going on so we do not focus on this topic but on the analytic model to analyze the performance of the BEB algorithm.

In this paper, we propose a new analytical model to find the performance measures to evaluate the system which adopts BEB algorithm, including the throughput, expected medium access delay and transmission delay. A simulation model is performed to verify our analytical model.

2 System Description

The following Fig. 1 illustrates the procedure considered in this paper and the access delay and transmission delay.

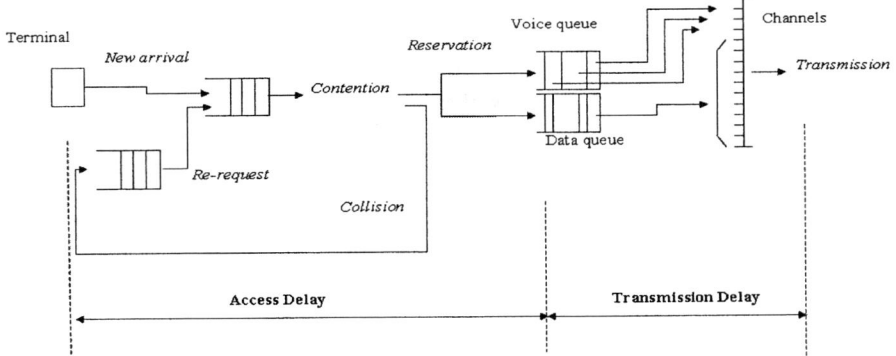

Fig. 1. Transmission Procedure

New messages arrive from infinite number of Mobile Terminals (MTs) forming a Poisson process with rate λ to the system. The time is divided into slots which are grouped into frames of fixed size. A frame is divided into two groups of multiple slots, request slots for reservation of channel and transmission slots for transmission of the actual information. The numbers of request slots and transmission slots in a frame are V and T, respectively.

The types of messages are divided into real time traffic such as voices and multi-media, and non-real time traffic like data. Newly arrived message is assumed to be real time messages and non-real time messages with probabilities α and β, respectively ($\alpha+\beta=1$). For notational simplicity, we use the terms *voice* message or call for real-time message and *data* message or call for non-real time message. Let the numbers of packets in a voice message and a data message are geometrically distributed with means $1/\varepsilon$ and $1/\delta$, respectively. It is essential that a priority is given to the voice traffic, but an effort is made to accommodate data traffic, whenever possible. When a message (whether it is voice or data) arrives at the system, it waits until the beginning of the next frame and randomly accesses one of the request slots to reserve a channel for transmission. If the message succeeds in the reservation, then a channel is allocated in any cell. If, however, two or more messages contend for the same request slot, then a collision occurs and none of the messages can reserve the request slot.

The message which fails to get a request slot retries under the BEB algorithm: whenever a message is involved in a collision and if it was the bth ($b=0,1,\cdots,15$) collision, then it selects one of the next 2^i frames with probability $1/2^i$ and attempts the reservation again, where $i=\min(b, 10)$. If a message collides 16 times, then it fails to transmit and is dropped. Those messages reserved slots then enter the queues and transmit their packets according to the proper scheduling method.

3 Steady-State Analysis of the Contention Phase

We obtain the SSD of the number of messages at the beginning of the frame. Let A_n be the number of new messages arrived during the n th frame and N_n be the total number of messages waiting in the system at the beginning of the n th frame. Also, denote J_n by the number of messages which successfully reserve a request slot at the n th frame. Then it can be shown that

$$N_{n+1} = \begin{cases} N_n - J_n + A_n, & N_n \geq 1, \\ A_n, & N_n = 0, \end{cases} \quad (3.1)$$

and $\{N_n, n \geq 1\}$ is a Markov chain. Let us denote $a_j, j = 0, 1, 2, \cdots$ by the SSD of A_n, where $a_j = \Pr(A_n = j) = e^{-\lambda d}(\lambda d)^j / j!$, $j=0,1,2,\cdots$. Let us introduce a random variable Y_n that is the number of messages which *actually* participate in the contention at the n th frame. Then for $J(y,k) = \lim_{n \to \infty} \Pr(J_n = k | Y_n = y)$ and $Y(i,y) = \lim_{n \to \infty} \Pr(Y_n = y | N_n = i)$, from [6],

$$J(y,k) = \frac{(-1)^k V! y!}{V^y k!} \sum_{m=k}^{\min(V,y)} (-1)^m \frac{(V-m)^{y-m}}{(m-k)!(V-m)!(y-m)!} \qquad (3.2)$$

for $0 \le k \le \min(V,y)$ and

$$Y(i,y) = \binom{i}{y} r^y (1-r)^{i-y}, \qquad y=0,1,\cdots,i, \qquad (3.3)$$

where r is the probability that an arbitrary message participates in the contention. We derive this probability by conditioning the number of collisions that an arbitrary message waiting in the system has experienced as following:

$$r = \left(\frac{1-(\gamma_c/2)^{11}}{1-(\gamma_c/2)}\right) \frac{1-\gamma_c}{1-\gamma_c^{16}} + (\gamma_c/2)^{10} \frac{\gamma_c(1-\gamma_c^5)}{1-\gamma_c^{16}}, \qquad (3.4)$$

where γ_c is the probability that an arbitrarily chosen (tagged) message experiences a collision when it contends for a request slot, which will be derived in the next subsection. Now we can calculate the one-step transition probability $p_{ij} = \Pr(N_{n+1}=j|N_n=i)$ as given below:

$$p_{ij} = \sum_{k=\max(0,i-j)}^{\min(i,V)} a_{j-i+k} \sum_{y=0}^{i} J(y,k) Y(i,y), \qquad (3.5)$$

for $i \ge 1$ and $p_{0j} = a_j$. The Steady State Probability Distribution (SSPD) $\pi_j \equiv \Pr(N=j) = \lim_{n \to \infty} \Pr(N_n=j)$ of the number of messages in system at the beginning of the frame can be obtained by solving the steady state equations $\pi_j = \sum_{i=0}^{\infty} \pi_i p_{ij}$ and $\sum_{i=0}^{\infty} \pi_i = 1$.

Now, we derive the collision probability, γ_c, that a tagged message experiences a collision given that it actually participates in the contention for a request slot in this subsection. This probability has not been found in an analytic form in the previous studies and we calculate it for the first time in this paper. Let M be the number of messages in the system at the beginning of the frame in which the tagged message is included. It is known that M is differently distributed from N because it contains the tagged message [7]. The Probability Distribution (PD) of M is given by

$$\Pr(M=m) = \frac{m\pi_m}{E(N)}, \quad \text{where} \quad E(N) = \sum_{j=0}^{\infty} j\pi_j. \qquad (3.6)$$

When y messages including the tagged message participate in the contention, the probability that the tagged message collides is $\sum_{i=1}^{y-1}\binom{y-1}{i}\left(\frac{1}{V}\right)^i\left(1-\frac{1}{V}\right)^{y-1-i}$. Therefore, we have the following:

$$\gamma_c = \sum_{m=2}^{\infty}\sum_{y=2}^{m}\sum_{i=1}^{y-1}\binom{y-1}{i}\left(\frac{1}{V}\right)^i\left(1-\frac{1}{V}\right)^{y-1-i} \cdot Y(m,y) \cdot \Pr(M=m)$$

$$= \sum_{m=2}^{\infty}\left\{1-\frac{V}{V-1}\left(1-\frac{r}{V}\right)^{m-1}+\frac{1}{V-1}(1-r)^m\right\} \cdot m\pi_m \Big/ \sum_{j=0}^{\infty} j\pi_j .$$

(3.7)

Note that the probability that a message is eventually blocked is γ_c^{16}. In order to obtain γ_c in Equ. (3.7), we need π_j but in turn γ_c should be given to obtain π_j. So we perform a recursive computation, i.e., we initially set γ_c to be an arbitrary value between 0 and 1 and compute π_j, $j = 0,1,2,\cdots$. Then with this π_j, we update γ_c using the Equ. (3.8) and this updated γ_c is utilized to update π_j again. This recursive computation continues until both values converge.

4 Expected Access Delay and Throughput

Now we derive the expected medium access delay of a message which is defined as the time from the moment that a message arrives at the system to the moment that it successfully reserves a request slot. It can be obtained by counting the number of frames from which a newly arrived message contends for a slot for the first time until it successfully reserves a slot. If a message reserves in the first trial with no collision (i.e., $b=0$), then it experiences, on average, 3/2 frame length's delay, which is the sum of 1/2 frame length (average length from the message's arrival epoch to the beginning of the next frame) and 1 frame length. Suppose a message collides exactly $b(1 \leq b \leq 10)$ times then it selects one of 2^b states with equal probability and thus the average number of frames it has spent in the system is $\frac{1}{2}+\sum_{i=0}^{b-1}2^i+\frac{1}{2^b}\sum_{j=0}^{2^b} j$. In the same manner, if $11 \leq b \leq 15$, the average delay is $\frac{1}{2} + \sum_{i=0}^{10} 2^i + (b-11)\cdot 2^{10} + \frac{1}{2^{10}}\sum_{j=0}^{2^{10}} j$. We obtain the expected access delay, $E(D_{\text{Access}})$, in frames, as the following:

$$E(D_{\text{Access}}) = \sum_{b=0}^{15} E(D_{\text{Access}}| b \text{ collisions})\Pr(b \text{ collisions}) = \frac{1}{2}+\frac{1-\gamma_c}{1-\gamma_c^{16}} \times$$

$$\begin{pmatrix} 1+2.5\gamma_c+5.5\gamma_c^2+11.5\gamma_c^3+23.5\gamma_c^4+47.5\gamma_c^5+95.5\gamma_c^6+191.5\gamma_c^7+383.5\gamma_c^8+767.5\gamma_c^9 \\ +1535.5\gamma_c^{10}+2559.5\gamma_c^{11}+3583.5\gamma_c^{12}+4607.5\gamma_c^{13}+5631.5\gamma_c^{14}+6655.5\gamma_c^{15} \end{pmatrix}$$

(4.1)

The PD of the number, Z, of messages which reserve request slots successfully in a frame can be obtained by conditioning Y and N as following:

$$\Pr(Z=x) = \sum_{n=0}^{\infty} \sum_{y=0}^{n} J(y,x) Y(n,y) \pi_n. \tag{4.2}$$

The expected number $E(Z)$ is given by

$$E(Z) = \sum_{n=0}^{\infty} \sum_{y=0}^{n} E(Z|Y=y, N=n) \Pr(Y=y|N=n) \pi_n = \sum_{n=0}^{\infty} nr \left(1 - \frac{r}{V}\right)^{n-1} \pi_n, \tag{4.3}$$

where the second equality comes from the Equ. [6].

5 Transmission Delay

Now we consider the expected transmission delay of a message, which is the time duration elapsed between a message succeeds in the contention and it is successfully transmitted. A fundamental requirement in the voice communication is prompt delivery of information. In our study, we put a buffer of size B for voice messages, with which one can adjust the allowable delay of voice message until its successful transmission. For example, if a longer delay for voice is allowed with low packet dropping probability then we make B bigger. This is in contrast to the data messages which respond to congestion and transmission impairments by delaying packets in queue. So we assume the buffer size for data messages is unlimited.

Each voice message uses one slot in a frame and transmits one packet per frame, which accommodates the real time transmission requirement on the voice messages. All T transmission slots are available for voice transmission, so maximum T voice messages can transmit simultaneously, while the data messages can transmit their packets only when there are less than T slots occupied by the voice messages. Even during a data message is sending its packets, if an arriving voice message finds no idle slots, then the data message interrupts its transmission and hand over one slot to the voice message. That is, the voice messages are preemptive. The interrupted data message resumes its transmission whenever there are any slots available. We analyze transmission delays for each type of messages in the sequel.

5.1 Transmission Delay of Voice Messages

Notice that the transmission of voice messages is independent of the transmission of data messages. Denote K_n by the number of voice messages which succeed in the contention during the n th frame and newly join at the voice transmission queue. Since a message is voice message with probability α, the PD of K_n can be obtained from the Equ. (4.2) as following:

$$\Pr(K_n = l) \equiv k_l = \sum_{x=l}^{\infty} \Pr(Z=x) \binom{x}{l} \alpha^l (1-\alpha)^{x-l}, \quad l = 0,1,2,\cdots. \tag{5.1}$$

Then X_n, the number of voice messages in the system at the beginning of the n th frame has the relationship $X_{n+1} = X_n - \xi(X_n) + K_n$, where $\xi_n(X_n)$ is the number of voice messages which complete the transmission at the n th frame given X_n. The PD of $\xi_n(X_n)$ when $X_n = i$, assuming the number of packets in a voice message is geometric with mean $1/\varepsilon$, is given by

$$\Pr(\xi_n(i) = k) = \binom{\min(i,T)}{k}(1-\varepsilon)^{\min(i,T)-k}\varepsilon^k \equiv \xi_n(\min(i,T), k), \qquad (5.2)$$

for $k = 0, 1, \cdots, \min(i,T)$. Let $\lim_{n \to \infty} X_n = X$, $\lim_{n \to \infty} K_n = K$, $\lim_{n \to \infty} \xi_n = \xi$. It can be shown that $\{X_n, n \geq 1\}$ is a Markov chain. Then we can obtain the one-step transition probabilities q_{ij} of this chain as following:

$$q_{ij} = \begin{cases} \sum_{l=0}^{j} k_l\, \xi(\min(i,T), \min(i,T)-j+l), & j = 0, 1, \cdots, T+B-1, \\ \sum_{l=0}^{\infty} k_l\, \xi(\min(i,T), \min(i,T)-j+l), & j = T+B, \end{cases} \qquad (5.3)$$

for all $i = 0, 1, \cdots, T+B$ and $\xi(i,k) = \xi(i,i)$ if $k \geq i$ and $\xi(i,k) = 0$ if $k < 0$. Now the SSPD, $\eta = (\eta_0, \eta_1, \cdots, \eta_{T+B})$, where $\eta_j = \Pr(X = j)$, of the number of voice messages to be transmitted in the system at the beginning of the frame is given by solving the equations:

$$\eta_j = \sum_{i=0}^{T+B} \eta_i q_{ij}, \quad \sum_{j=0}^{T+B} \eta_j = 1. \qquad (5.4)$$

The expected transmission delay of voice message in frames is

$$E(D_{\text{voice}}) = E(X) / E(K) \qquad (5.5)$$

by applying the well-known Little's result [7].

5.2 Transmission Delay of Data Messages

As explained before, the data message can transmit its packets whenever there are slots available. Therefore, the PD of the number of packets waiting to be transmitted in the queue depends on how many slots are currently being held by voice transmission. Denote U_n by the number of data messages which succeed in the contention during the n th frame and newly join at the data transmission queue. A message is data message with probability β, and thus

$$\Pr(U_n = l) \equiv u_l = \sum_{x=l}^{\infty} \Pr(Z = x) \binom{x}{l}(1-\beta)^{x-l}\beta^l, \qquad (5.6)$$

and the PD, w_j, $j \geq 0$, of the number of data packets arrived during a frame is:

$$w_j = \sum_{l=1}^{\infty} u_l \binom{j-1}{l-1} \delta^l (1-\delta)^{k-l}, \quad j \geq 1, \tag{5.7}$$

and $w_0 = u_0$. Let k be the number of slots which are occupied for voice transmission.

By means of the probabilities w_j, we can find the steady state conditional PD, $v^{(k)} = (v_0^{(k)}, v_1^{(k)}, v_2^{(k)}, \cdots)$, of the number of data packets in the system at the beginning of the frame, when k voice messages are transmitting, by solving the following steady-state equations:

$$\begin{aligned} v_0^{(k)} &= w_0 \left(\sum_{i=0}^{T-k} v_i^{(k)} \right), \\ v_j^{(k)} &= w_j \left(\sum_{i=0}^{T-k} v_i^{(k)} \right) + \left(\sum_{m=1}^{j} v_{T-k+m}^{(k)} w_{j-m} \right), \quad j \geq 1, \text{ and } \sum_{j=0}^{\infty} v_j^{(k)} = 1. \end{aligned} \tag{5.8}$$

Finally, the unconditional PD $v = (v_0, v_1, v_2, \cdots)$ is:

$$v_j = \sum_{k=0}^{\infty} v_j^{(k)} \eta_k, \quad j = 0, 1, 2, \cdots. \tag{5.9}$$

Now we can calculate the expected number of data packets at the beginning of the frame, that is $L_D = \sum_{j=0}^{\infty} j v_j$. Then $(\delta \cdot L_D)/(T - E(X))$ is the expected number of frames to transmit one data message, which gives the transmission delay in frames of a data message.

6 Performance Analysis

In numerical study to verify our approach proposed, we compute the expected (access and transmission) delays for voice and data messages using the parameters $V = 30$, $T = 95$, $B = 10$, $\lambda = 5$, $\alpha = 0.3$ and the expected number of packets of a data message is set to be 1000 (i.e., $\delta = 0.001$). Expected delays are calculated as the expected number of packets in a voice message (i.e., $1/\varepsilon$) varies. The same was found using simulations as well. The plots in Fig. 2 show close agreement between analytic results and the simulation, especially in the voice message, thus validating the analysis. As for the data message, however, analytical results tend to overestimate than the simulation (but are still within the confidence intervals). This is because we first derived the probability distribution of the number of data packets in the system under the condition that there are k voice messages transmitting, and then unconditioned it. This is based on the assumption that the number of data packets reaches steady state before k changes.

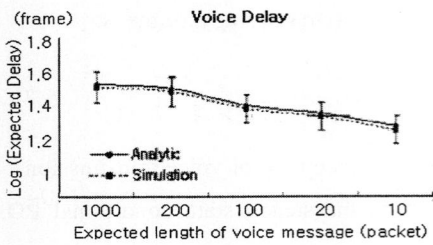 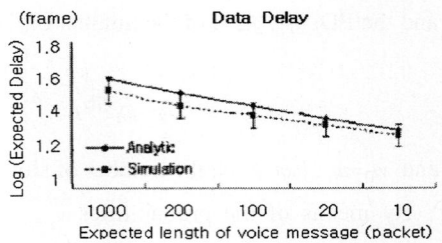

Fig. 2. Expected delays

7 Conclusion

In this paper, we considered the performance evaluation of the BEB policy, which is a collision resolution algorithm often adopted in the random access packet networks. We obtain the SSD of the number of messages waiting in the system, which is utilized to get the probability that a tagged message experiences a collision given that it actually participates in the contention for a request slot in a frame, which has never been investigated in the literature. With these, the expected access and transmission delay of frames that a message experiences from its arrival to the system until the successful transmission are found analytically. In addition, a numerical study to compare the expected delays computed from the analytic model with simulation results is provided. It shows that our analytic method gives an excellent agreement with the simulation.

References

1. P. M. Soni and A. Chockalingam.: Analysis of Link-Layer Backoff Scheme on Point-to-Point Markov Fading Links. IEEE Trans. on Comm., Vol. 51. (2003) 29-32
2. Yung-Fang Chen, Chih-Peng Li.: Performance Evaluation of the Quasi-FIFO Back-off Scheme Wireless Access Networks. IEEE VTC. Vol. 2. (2003) 1344 - 1348
3. Yang Yang, and Tak-Shing Peter Yum.: Delay Distributions of Slotted ALOHA and CSMA. IEEE Trans. on Comm., Vol.51. (2003) 1846-1857
4. P. Chatzimisios, V. Vitsas and A. C. Boucouvalas.: Throughput and Delay Analysis of IEEE 802.11 Protocol. Networked Appliances, Liverpool. Proceedings. (2002) 168 - 174
5. B. J. Kwak, N. O. Song, and L. E. Miller.: Analysis of the Stability and Performance of Exponential Backoff. IEEE WCNC, Vol. 3. (2003) 1754-1759
6. W. Szpankowski.: Analysis and stability considerations in a reservation multiaccess system. IEEE Trans. on Comm., Vol. com-31. (1983) 684-692
7. L. Kleinrock.: Queueing system, Vol. 1. Theory. John Wiley & Sons, (1975)

Design and Implementation of Frequency Offset Estimation, Symbol Timing and Sampling Clock Offset Control for an IEEE 802.11a Physical Layer

Kwang-ho Chun[1], Seung-hyun Min[2], Myoung-ho Seong[1], and Myoung-seob Lim[1]

[1] The Faculty of Electronic & Information Eng., Chonbuk National Univ., Korea
lighttiger@chonubk.ac.kr
[2] The Faculty of Electrical & Computer Eng., Chungbuk National Univ., Korea
imturtle@chungbuk.ac.kr
{smh, mslim}@hslab.chonbuk.ac.kr

Abstract. In this paper, the simulation and the design results about the algorithm of symbol timing recovery and frequency offset using the PLCP preamble, and sampling clock offset using the cyclic prefix in time domain for an IEEE 802.11a high-speed wireless LAN modem are presented. The algorithm of frequency offset estimation and compensation for making the frequency offset converge fast below the allowable limit is proposed. For the efficient implementation of the algorithm, the method that H/W size can be reduced up to 80% in the cross correlation block is designed and the method for the high speed processing of the divider block in the phase estimation is designed. And the newly proposed sampling clock offset estimation method makes it possible to adjust the optimum sampling point.

1 Introduction

In recent years, there has been a lot of interest in applying Orthogonal Frequency Division Multiplexing (OFDM) in wireless systems because of its various advantages in lessening the severe effects of frequency selective fading. However, OFDM systems are vulnerable to synchronization error [1]. Because of the increase of the wireless multimedia service request, an IEEE 802.11a high-speed wireless LAN standard draft in OFDM systems of a 6~54 Mbps transmission speed in 5GHz was achieved [3]. This paper is organized as follows. In Section 2, the advantage of OFDM and the simple overview of signal format in IEEE802.11a physical layer are given. In Section 3, frequency offset synchronization, symbol timing synchronization and sampling clock offset synchronization about synchronization problems as a transmission technique in OFDM systems are observed. Section 4, simulation and design results of proposed methods about the algorithm of symbol timing recovery, frequency offset using the PLCP preamble and sampling clock offset using the cyclic prefix is observed.

2 Signal Format in IEEE802.11a Physical Layer

It is required for the physical layer of wireless LAN to be robust, maintaining the QoS at even high-speed data rate, under the indoor multi-path fading environments. When high rate symbols are transmitted in OFDM system, the symbol duration is increased in IFFT block. As a result, the OFDM signal is affected less by delay spreads and ISI under the frequency selective fading channel, because of the longer symbol duration. The data rate can be changed into 6, 9, 12, 18, 24, 36, 48, 54Mbps and the modulation method is used as BPSK, QPSK, 16-QAM, 64-QAM in each sub-carrier and the FEC is employed as 1/2, 2/3, 3/4 convolutional coding. 64-point IFFT, FFT per one frame in signal part are carried out and the cyclic prefix is added as a guard interval, which can be used in the frequency domain equalizer for compensating the degrading due to fading channel. The short training symbol and long training symbol for signal detection, automatic gain control, frequency offset estimation, and channel estimation are ahead of data frames.

3 The Receiving Environment of OFDM

Problems of synchronization that can be considered in the OFDM systems are divided into timing synchronization and frequency synchronization.

(A) Frequency synchronization
Frequency synchronization is to maintain the coherency of RF sub-carrier frequency between the transmitter and the receiver, and if there is the frequency offset, the shift of frequency happens in the frequency spectrum of a received signal, and it let orthogonality of sub-carrier disappear. Thus two serious impacts on FFT output is presented. That is,

$$Y_k = (X_k H_k) \left\{ \frac{\sin \pi \varepsilon}{N \sin (\pi \varepsilon / N)} \right\} \cdot e^{j\pi \varepsilon (N-1)/N} + I_k + W_K \qquad (1)$$

$$I_k = \sum_{\substack{l=-K \\ l \neq k}}^{K} \frac{(X_l H_l) \cdot \sin \pi \varepsilon}{N \sin(\pi (l-k+\varepsilon)/N)} e^{j\pi \varepsilon (N-1)/N} e^{-j\pi (l-k)/N} \qquad (2)$$

In this part, X_k, Y_k, H_k is the transmitted signal, the received signal and the channel transfer function for k-th sub-carrier respectively. The ε as the normalized frequency offset is defined as the rate of actual frequency offset per the distance of frequency between sub-carriers. The frequency offset can be cause of reducing the size of signal, phase rotation, and interfering with inter sub-carriers. The third item presents the effect by AWGN.

(B) Timing Synchronization
The timing synchronization comes from uncertainty in the two views. First, it consists in restoration of location in symbol window for parallel processing of the accurate signal, as we cannot know the arrival time of OFDM symbol, and the restoration of sampling clock that controls sampling clock of analog to digital converter (ADC), in order to sample the location of the most SNR.

The inaccurate symbol timing can cause inter-symbol interference (ISI) and degrade the performance of OFDM systems [3]. The sampling clock frequency error can cause inter-carrier interference (ICI) [4]. Furthermore, the sampling clock frequency error can cause a drift in the symbol timing and further worsen ISI. For instance, if the sampling clock specification is 10 parts per million (ppm) and the sampling frequency is 5 MHz, then the clock has a drift of about 50 samples per second. Thus, sampling clock synchronization is also an important issue that needs to be addressed in OFDM systems [1].

4 Acquisition Methods of Synchronization

OFDM systems have to carry out the synchronization in the first stage of system. For the synchronization of the receiver, PLCP preamble is transmitted before transmitting data. PLCP preambles consist of ten short training sequences and two long training sequences.

(A) Acquisition technique of frequency offset
In this paper, the method that carries out simultaneously the estimation and compensation of the frequency offset, after a quick signal detection is proposed. The signal detection is designed by cross correlation using short training sequence. A large H/W size is necessary in the cross correlation. Total thirty two multipliers must be used. But, because of the symmetry of the short training sequence, the broadcast structure instead of direct form and also the shift and adder scheme make the H/W size reduced 80% less than the direct form.

Figure 1 shows the simulation of the cross correlation in case of data of 8 bit and 12 bit quantization of the multiplier output under the SNR=7dB, $\varepsilon = 1$ and multi-path fading of indoor. Figure 2 shows case of $\varepsilon = 2$. And Figure 3 shows the design of signal detection.

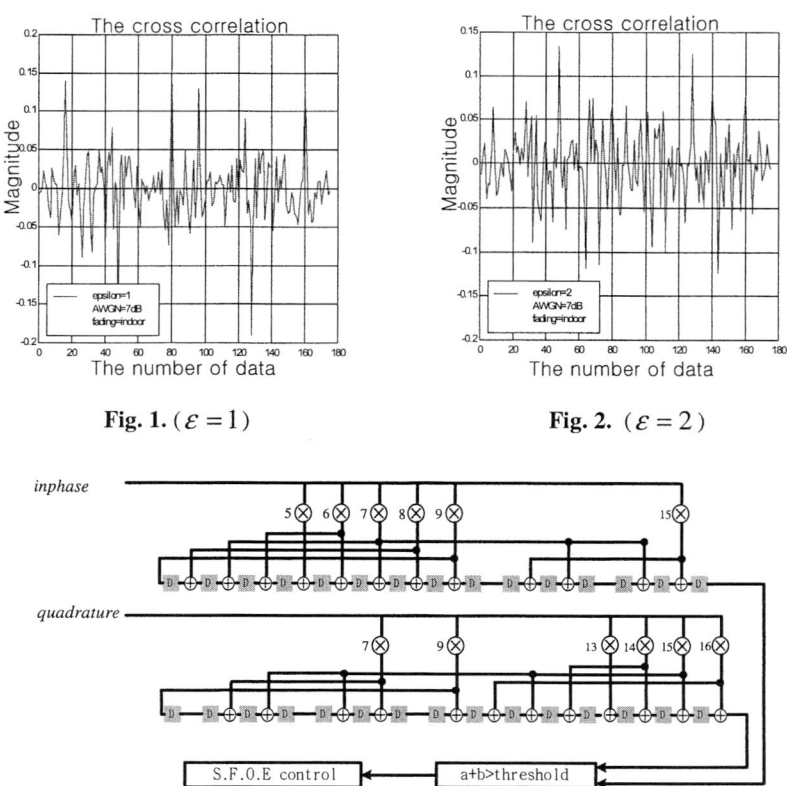

Fig. 1. ($\varepsilon = 1$)   **Fig. 2.** ($\varepsilon = 2$)

Fig. 3. Signal detection

In general, the coarse frequency offset is estimated using short training sequence in the first stage of system and the fine frequency offset is estimated using long training sequence. The general algorithm of frequency-offset estimation is as follows.

$$\varepsilon = \frac{4}{2\pi}\tan^{-1}\left\{\left(\sum_{n=0}^{15}\text{Im}[y_{1n}^* y_{2n}]\right) \Big/ \left(\sum_{n=0}^{15}\text{Re}[y_{1n}^* y_{2n}]\right)\right\} \quad -2<\varepsilon<2 \tag{3}$$

$$\varepsilon = \frac{1}{2\pi}\tan^{-1}\left\{\left(\sum_{n=0}^{63}\text{Im}[y_{1n}^* y_{2n}]\right) \Big/ \left(\sum_{n=0}^{63}\text{Re}[y_{1n}^* y_{2n}]\right)\right\} \quad -0.5<\varepsilon<0.5 \tag{4}$$

In the case of the coarse frequency offset estimation, after the 32 clocks, the first estimation value can be obtained. If the estimation and the compensation are carried out simultaneously by the lookup table that the compensation value corresponding to the output of the divider is stored, after the 16 clocks, the second estimation value can be obtained. Therefore, if the signal detection is carried out quickly, the frequency offset can be estimated many times. Figure 4 shows the design of frequency offset estimator and compensator.

Figure 5 shows the simulation of the frequency-offset estimation and compensation after the four times iteration. In the case of the coarse frequency offset estimation and compensation, the processing speed of the divider is important. In order to reduce the processing speed of the divider, only the integer part of 6 is considered. After quantizing $\tan 6$ corresponding to 6, in the case of the coarse frequency offset estimation, if three times shift and subtraction using the four bit from MSB are carried out, the 6 can be estimated quickly.

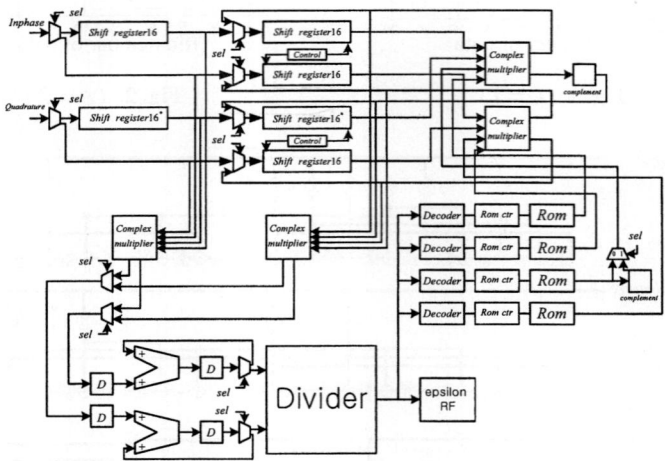

Fig. 4. Frequency offset estimator

Also, In the case of the fine frequency offset estimation, using the six bit from MSB, the more accurate frequency offset can be estimated. Figure 6 shows the algorithm of the divider used in the fine frequency offset.

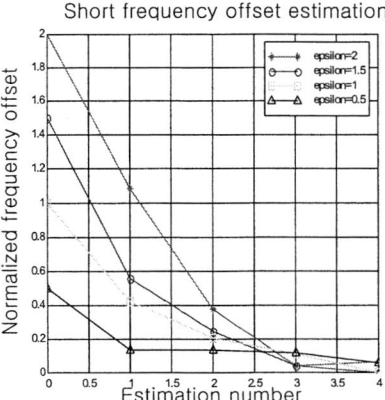

Fig. 5. Compensation values

(B) Acquisition technique of symbol timing

The symbol timing synchronization using the PLCP preamble must be estimated before the long training sequence, because of the channel estimation and the frequency offset estimation using the long training sequence.

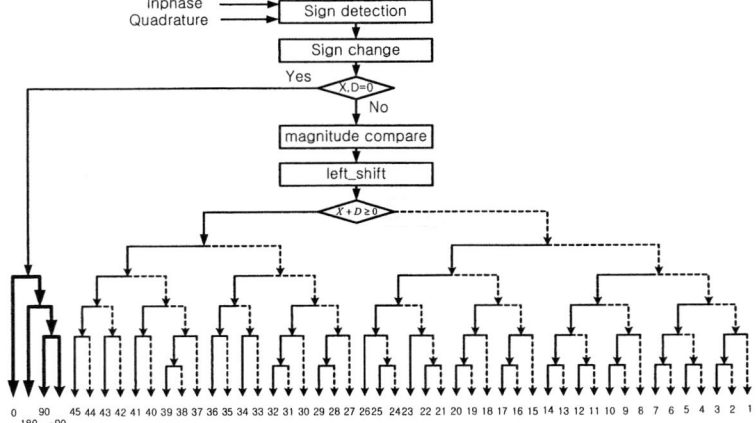

Fig. 6. The algorithm of the divider

In this paper, the efficient design of the cross correlation using the GI (guard interval) is proposed, in order to get the correct symbol timing. If the symbol timing synchronizer is designed as same as the method used in the signal detection, H/W size can also be reduced efficiently. Figure 7 shows the simulation of the cross correlation in case of data of eight bit and twelve bit quantization of the multiplier output under the SNR=7dB, ε =0.5 and no fading. Figure 8 shows the case of the multi-path fading of indoor. Figure 9 shows the algorithm of the symbol timing synchronizer.

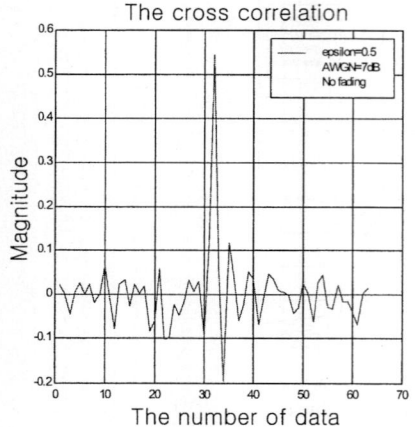

Fig. 7. No fading

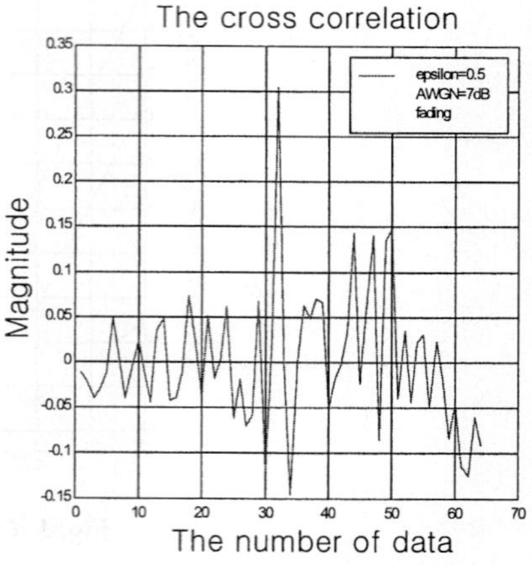

Fig. 8. Fading

(C) Acquisition technique of sampling clock offset

Sampling clock adjustment is the function of maintaining the optimal sampling timing. If the sampling clock offset happens, the skip and duplicate of the symbol occur, so that it results in additional symbol timing offset.

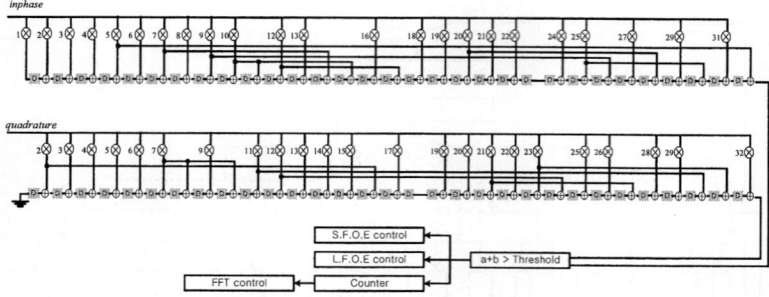

Fig. 9. The symbol timing synchronizer

In order to estimate the skip and duplicate, in this paper, the algorithm using the correlation of the first symbol of the cyclic prefix and the correlation of the last symbol of the cyclic prefix is proposed. If the skip happens, the correlation of the last symbol is less than the correlation of the first symbol.

$$P_{skip} = (C, X) = P\left(\left|\overline{C_k 1 \cdot C_k 2} - \overline{C_k 1 \cdot C_k 2}\right| < \left|\overline{C_k 16 \cdot C_k 1} - \overline{C_k 16 \cdot C_k 1}\right|\right) \tag{5}$$

$$P_{du} = (C, X) = P\left(\left|\overline{C_{k-1} 16 \cdot C_k 1} - \overline{X_k 48 \cdot C_k 1}\right| > \left|\overline{C_k 15 \cdot C_k 16} - \overline{C_k 15 \cdot C_k 16}\right|\right) \tag{6}$$

Where, $P(C, X)$: the probability, k: the frame of k-th, C: the cyclic prefix, X: the valid symbol, $\overline{C \cdot X}$: the sampling symbol between the symbol C and the symbol X. In case of the skip, $P_{skip}(C, X)$ is above $P_{duplicate}(C, X)$, and in case of the duplicate, $P_{skip}(C, X)$ is below $P_{duplicate}(C, X)$. For the large ppm, after the over sampling, using the sum of the on time clock sample and late clock sample, and the sum of the on time clock sample and early clock sample, the number of sample can be increased. First of all, after the absolute value of the subtraction is obtained. Secondly, after comparing of the size, as estimating sum of counting value, the skip and duplicate phenomenon of sampling clock offset can be estimated.

Figure 10 and Figure 11 shows the simulation of the skip in the case of 20 ppm. Figure 12 and Figure 13 shows the simulation of the skip in the case of 100 ppm. The threshold of the sampling clock offset is determined by dint of the worst case. Figure 14 shows the synchronization block model. After signal detection is carried out using short training sequence, firstly, short frequency offset, secondly, long frequency offset and lastly sampling clock offset is estimated by the control signal of the synchronizer.

(D) Channel environment and design
In this paper, the Eb/N0 of 7dB, $-2 < \varepsilon < 2$ and the multi-path fading channel are assumed. Furthermore, the multi-path fading channel of indoor environment with 20 path and Jakes model that maximum Doppler frequency is 24Hz are used. For performance analysis, two EPF10K100ARC240-3 devices of Altera Corporation are used. The total usage of gates is 90%.

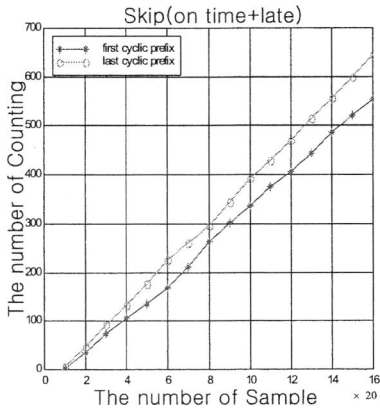

Fig. 10. On time + late

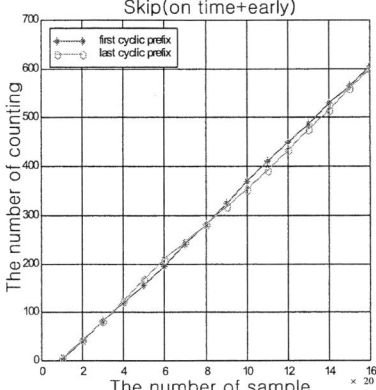

Fig. 11. On time + early

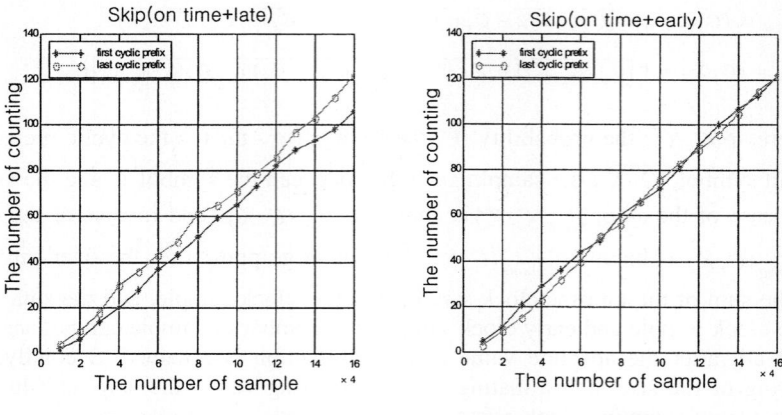

Fig. 12. On time + late   **Fig. 13.** On time + early

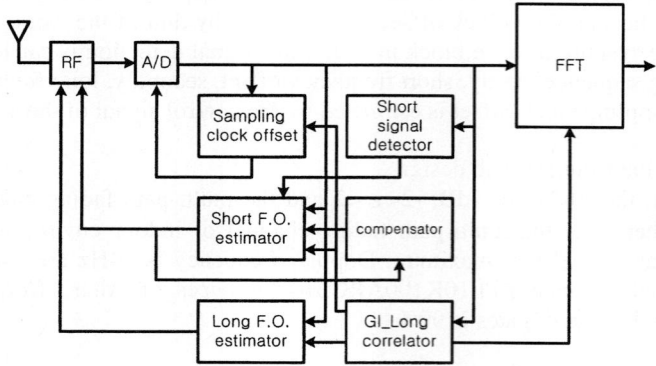

Fig. 14. The Synchronization block model

5 Conclusions

In this paper, the efficient design technique of symbol timing and frequency offset for an IEEE 802.11a high speed wireless LAN modem and the sampling clock offset acquisition algorithm for an OFDM systems with the cyclic prefix is proposed. It is demonstrated that the proposed method improves the efficiency of the H/W design. Furthermore, in the case of the algorithm of sampling clock offset, this method makes it possible to adjust the optimum sample point and can be used in every communication system with the copied cyclic prefix.

Acknowledgements

This work was supported (in part) by the Ministry of Information & Communications, Korea, under the Information Technology Research Center (ITRC) Support Program.

References

1. Baoguo YANG, "An Improved Combined Symbol and Sampling Clock Synchronization Method for OFDM Systems" IEEE Trans. on Comm., 1999
2. 2. IEEE 802.11a/D7.0 "High Speed Physical Layer in the 5GHz Band" , 1999
3. P. H. Moose, "A Technique for Orthogonal Frequency Division Frequency Offset Correction," IEEE Trans. on Comm., Vol.42, No.10, pp2908-2914, 1994
4. Magnus Sandell, Jan-Jaap van de Beek, and Per Ola Borjesson, "Timming and Frequency Synchronization in OFDM Systems Using the Cyclic Prefix," Proc. of International Symposium on Synchronization, pp.16-19, 1995.
5. Richard Van Nee, OFDM for Wireless Multimedea Communications, Artech., 2003

Automatic Subtraction Radiography Algorithm for Detection of Periodontal Disease in Internet Environment

Yonghak Ahn[1] and Oksam Chae[2,*]

[1] Dept. of Mobile Internet, Dongyang Technical College,
62-160, Kochuk-dong, Kuro-ku, Seoul, 152-714, Republic of Korea
yohan@dongyang.ac.kr
[2] Dept. of Computer Engineering, Kyunghee University,
1, Sochen-ri, Giheung-eup, Yongin-si, Gyeonggi-do 449-701, Republic of Korea
oschae@khu.ac.kr

Abstract. In this paper, we propose an automatic subtraction radiography algorithm for detection of periodontal disease in dental radiography. For these goals, this paper proposes the method of an automatic image alignment and detection of minute changes, to overcome defects in the conventional subtraction radiography by image processing technique, that is necessary for getting subtraction image and ROI(Region Of Interest) focused on a selection method using the structure features in target images. Therefore, we use these methods because they give accuracy, consistency and objective information or data to results. In result, easily and visually we can identify minute differences in the affected parts whether they have problems or not, and use application system in a real-time internet environment.

1 Introduction

Recently it is possible to a digital x-ray image because of a development of x-ray image sensor, so a study of analysis a digital x-ray image and a diagnostic method is progressing[1]. The one of theses study is a subtraction to detect minute changes in a sequential radiography. The subtraction radiography to detect minute changes in a series of dental radiography refers to a method where two dental x-ray images take at some intervals and overlapped with each other shows a difference between the two images, serving as a standard of judgment for some cures and diagnoses of almost all dental disease[2-4].

The dental x-ray subtraction radiography contributes to quickly detecting and coping with changes in surrounding bony tissues affected by teeth. But, so far, almost all such works have been manually handled with a use of reference film, which would result in a lack of accuracy, consistency and objectivity in their results. And recently digital x-ray image is generally used, so studies of digital x-ray subtraction radiography needs[5-7].

---

* Corresponding author ; Oksam Chae.

There are three types of methods concerning the works; one is a use of reference film with cross stripes, which is selected reference points by user and then it is followed by the manual subtraction using the selected points[8-9]. Another is the subtraction after image alignment using edge detection by edge operator[10-11], and the other is to measure a bone loss of teeth by a probe and digitalizes it[12].

In the case of the former method using reference film, however, its simple subtraction of the whole x-ray image could lead to difference information due to some movements during acquisition the image or other status of the x-ray equipments. And in the case of the method using edge operator, as had appeared to solve such problems, processes transformation on the whole image to minimize a difference of edge information after edge detection, but it has a lot of processing time and perform image alignment covering the orientation of the image alone so that its position and deformation, and the subtraction of the whole image only after the alignment could result in its detection of difference information more from errors than from real minute changes. Specially, it has high error information because the ROIs segment in manually. In the final case of the method using a probe, it is a usually method but result in a lack of accuracy, consistency and objectivity due to each tooth measures a bone loss using a probe by user.

In this paper to overcome those problems, a structural analysis of tooth is used to segment the region of interests(ROI) on which edge-based matching in the image alignment is worked upon for its automatic subtraction and the resulting accurate segmentation.

2 Automatic Subtraction Radiography Algorithm

Proposed in this study is an algorithm to quickly and automatically detect minute changes in surrounding bony tissues of teeth and to recognize pieces of structural information on the tooth. But a matching on the target of the whole image has a lack of accuracy due to change input environment and it is so difficult that determines an optimal threshold value.

To solve these problem, this study propose that the derived ROI is used to more quickly and precisely process matching and generate relating difference images. Also, considering characteristics that concern minute changes, it predicts their location and marginal region to recognize the ROI and then to display only difference information about the relating region.

After two x-ray images are input, the following procedure is gone through as shown in Fig.1.

2.1 Teeth Contours and Lines Detection

The important region in the digital images from the x-ray dental radiography is the surroundings of the gingiva. So, information about its starting point can be search for based on the lines that include information about the range of either side of the giniva between teeth.

Teeth contours detection is not difficult because it has high intensity in dental radiography, relatively. The lines can be detected as follows; First, it is to detect accuracy teeth edges by canny edge detection algorithm[14] in dental radiography. Detected edges structure the unit of segments and the independent object having relation information between each segments as shown Fig.2. So, using structured edge information removes a noise edge and a small edge segments, and it can improve the performance[15].

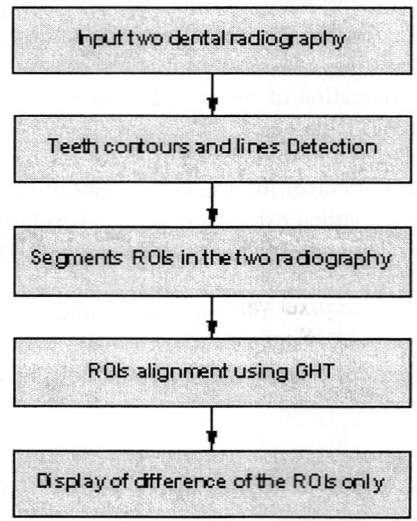

Fig. 1. The proposed algorithm

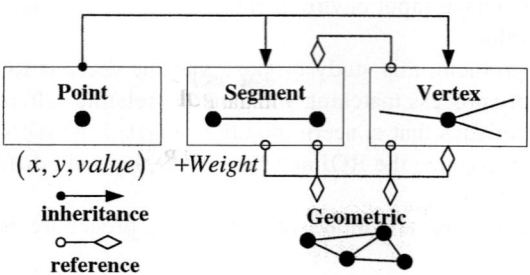

Fig. 2. Structure of edge class

Then, the edge information detected on the x-ray images, it is to detect the contours of teeth left-right only as shown in Fig.3 (b), using gradient direction and edge segments information, which include edge information as edge segments.

And it can detect teeth lines using the line fitting algorithm[16] at the contours as shown in Fig.3 (c).

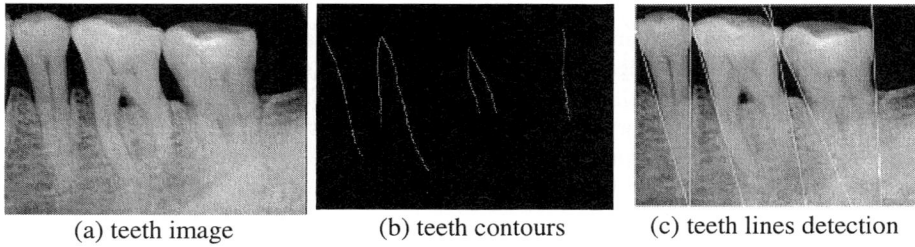

(a) teeth image　　　　　　(b) teeth contours　　　　(c) teeth lines detection

Fig. 3. Teeth contours and lines detection

2.2 Segments ROI

It is so difficult that the conventional methods using a global threshold value are detecting change regions because pixel values surrounding staring the gingiva in the x-ray images have minute changes. So, in this paper proposed summation method to get region information increasing minute differences. The main purpose of summation method is to get ROI having a starting point of the gingiva and to get more reliable results, it is performed the summation method according to rotation angle if images are rotated.

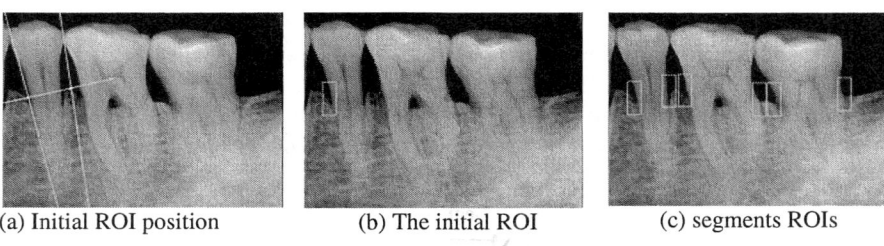

(a) Initial ROI position　　　　(b) The initial ROI　　　　(c) segments ROIs

Fig. 4. Segments initial ROI

In this study, after the middle line between the teeth, the right line, as shown in Fig.4 (a) has been searched for, the values of the pixels of the position corresponding to the line intersecting itself with the middle line at right angles are accumulated. And then, there comes to smooth what is less than their accumulative average, showing that the accumulative value increases at the starting point of the gingiva. An ROI can be obtained with the point set as a starting point of the gingival as shown in Fig.4 (b). If those processes are applied to each teeth of the two x-ray images, all the ROI can be obtained as shown in Fig.4 (c).

2.3 ROIs Alignment

The purpose of ROIs alignment is to match with accuracy between two x-ray images to get with some intervals. It is the important process to get the reliable difference information.

In this paper uses the computed contours and lines information of the teeth images. As the information provides orientation and position information of the teeth, the processing time of the alignment can decreased a lot. To reduce processing time, in this paper, it is that used the GHT(General Hough Transform) for an arbitrary object detection[5] using the computed orientation and position information of teeth. Edges are to get in the ROI of the first image to make a reference pattern while matching is processing on the second image using the GHT, when limiting an extent of the accumulator to the computation location and direction, and using the error range of the computed value enables its processing time to be lessened, elevating its accuracy.

The method has great advantage of search for the position, even if it was some difference between the model and the matching model or lost some information of the matching model.

2.4 Display of Difference Information of the ROIs Only

After the alignment of the ROI in the two images segmented by a structural analysis, a difference image is generated to show the difference between the ROIs only in two images. The difference image is as follows equation (1);

$$sub(x, y)_{ROI} = | ROI(x, y)_{INPUT1} - ROI(x, y)_{INPUT2} | \qquad (1)$$

Where (x, y) refers to each coordinate in a x-ray image, and $ROI(x, y)$ means the ROI segmented from the input images, the first and the second image, respectively.

Because the region infected by bone loss around tooth projects itself as dark on the x-ray image, a region with some change displays its minute changes on the difference image. So, a local threshold value should be computed to segment the gingiva from the background based on the information about approximate gingival boundary obtained through analysis of the structure of teeth. The value is applied to the difference image of ROIs as shown in equation (2) for the thresholding.

$$g(x, y) = \begin{cases} 1 & \text{if } sub_{ROI_i}(x, y) > T_i \\ 0 & \text{if } sub_{ROI_i}(x, y) \leq T_i \end{cases} \qquad (2)$$

Where $sub_{ROI}(x, y)$ refers to a difference image from the i th ROI while T_i and $g(x, y)$ means a computed i th local threshold value and a threshold image, respectively.

The segmented region of bone loss is labeled to compute its area and perimeter which are numerically displayed.

3 Performance Analysis

The proposed algorithm has been implemented under such the environment as is supported by Pentium IV 2.0GHz, RAM 512M, Visual C++ 6.0™ and Microsoft Windows 2000™, in usually. The algorithm has been developed with the internal function of MTES[17], an image processing algorithm development tool.

The algorithm is such method as is detected and recognized quickly minute changes of bone loss surrounding teeth based on subtraction radiography. But the conventional methods, the focus of the whole image, are reduced an accuracy because of a change of the input environments and so difficult that determine an optimal threshold value.

The test results of the algorithm to the x-ray images from six dental patients showed that the proposed algorithm was efficient enough to serve as real-time system as shown in Table.1.

Table 1. The result of performance

Dental patients	P1	P2	P3	P4	P5	P6	P7	P8
Image size (8b)	450×670	450×670	450×670	450×670	450×670	450×670	450×670	450×670
Processing time (second)	1.612	1.804	2.015	2.317	1.914	1.962	2.212	2.003
general average (second)	**1.980**							

Table.2 is a test result of image alignment using RMS(Root Mean Square Difference) as shown in equation (3) compares the conventional methods, a subtraction method using reference film[9] and an image alignment method using edge operator[11], with the proposed algorithm.

$$RMS = \frac{\sqrt{\sum_{j=1}^{M}(I_j^1 - I_j^2)^2}}{M} \quad (3)$$

Where I refers to the j th ROI image and M means total pixels in image. Then, perfect alignment equals 0.

The proposed algorithm could align of teeth accurately compared with the existing methods as shown in Table.2.

The proposed algorithm could detect difference of teeth bony tissue accurately compared with the existing methods as shown in Table.3.

The reason why the algorithm as proposed in Table.2 and Table.3 showed a relatively high level of success rate is that it focuses only the ROI to use a local threshold value and to exactly segment changes in surrounding dental bony tissues alone, compared with the existing methods which detect a change in around noises more greatly than as they are, using a general matching and global threshold value.

Table 2. Comparisons alignment of the existing methods with the proposed algorithm

Patients	The proposed method	A subtraction method[9]	Image alignment method[11]
P1	3.17	16.28	17.78
P2	5.32	15.69	18.49
P3	3.06	15.46	18.74
P4	4.14	15.81	18.34
P5	4.62	17.24	17.96
P6	3.84	16.62	19.28
P7	2.94	16.51	19.72
P8	3.26	17.43	18.46
average	**3.79**	**16.38**	**18.60**

Table 3. Comparisons detection of the existing methods with the proposed algorithm

algorithms rate	The proposed algorithm		A subtraction method[9]		Image alignment method[11]	
bone loss	exist	not exist	exist	not exist	exist	not exist
Successful detection rate(%)	99.7	98.9	23.1	11.6	72.4	63.8
Error detection rate(%)	0.3	1.1	76.9	88.4	27.6	36.2
General success rate(%)	99.3		17.35		68.1	

Successful detection rate = (the number of detected regional pixels of the region with bone loss / the number of the whole detected regional pixels) × 100

Error detection rate = (the number of detected regional pixels of the region without bone loss / the number of the whole detected regional pixels) × 100

General success rate = (Successful detection rate + Error detection rate) / 2

4 Conclusion

This study proposed the automatic subtraction radiography algorithm of a medical image processing to detect minute changes of bony tissue surrounding teeth. The proposed algorithm segments the ROI by structural analysis of teeth, and it solves the problem of the conventional methods, such as include difference information resulting from the erroneously directed projection of x-ray equipments and others.

The result showed that can segment the minute changes of bony tissue surrounding teeth and the affected region only, and solved the problem in the whole matching. And it showed the result of objectivity and accuracy, also.

For its more reliable results, the future works should do upon some experiments and corrections based upon a variety of data so that it could consider changeable variables depending upon thins and make a structure of the whole system to serve internet services.

References

1. Compend Contin Educ Dent, "Computerized Image Analysis in Densitry: Present Status and Future Application", Vol.XIII, No.11.
2. Glenn F. Knoll, "Radiation Detection and Measurement", Wiley, 1988.
3. Grondahl K, Grondahl H-G, Wennstorm J. and Heijl L., "Examiner agreement in estimation changes in periodontal bone from conventional and subtraction readiographs", Journal of Clinicical Periodontal, 1987.
4. McDonnel, D.Price, "An evaluation of the Sens-A-Ray digital dental image system", Dentomaxillofac, Radiol.22, pp.21-26, 1993.
5. Francesco Bassi, Cristiano Marchisella, Gianmario Schierano, Egon Gasser, "Detection of Platelet-Activating Factor in Gingival Tissue Surrounding Failed Dental Imaplants", Journal of Periodontal, Vol.72, No.1, January, 2001.
6. Steenberghe D., Quirynen M., Naert I., Maffei G., Jacobs R., "Marginal bone loss around implants retaining hinging mandibular overdentures, at 4-, 8- and 12-years follow-up", Journal of Clinical Periodontology 2001, 28, pp.628-633, 2001.
7. T.M. Lehmann, H.G.Grondahl, D.K.Benn, "Computer-based for digital subtraction in dental radiography", Dentomaxillofacial Radiography, 29, pp.323-346, 2000.
8. E.H.Verdonschot, A.J.Sanders, A.J.Plasschaert, "A computer-aided image analysis system for area measurement of tooth root surfaces", Journal of Periodontal, Vol.61, No.5, pp.275-280, May, 1990.
9. H.G.Grondahl, K.Grondahl, R.L.Webber, "A digital subtraction technique for dental radiography", Oral Surg., Vol.55, No.1, pp.96-102, January, 1993.
10. Paul F. van der Stelt, Wil G.M.Geraets, "Computer-Aided Interpretation and Quantification of Angular Periodontal Bone Defects on Dental Radiographs", IEEE Transactions on Biomedical, Engineering, Vol.38, No.4, pp.334-338, April, 1991.
11. DC Yoon, "A new method for the automated alignment of dental radiographs for digital subtraction radiography", Dentomaxillofacial Radiography 29, pp.11-19, 2001.
12. M.C.Juan, M.Alcaniz, C.Monserrat, V.Grau, C.Knoll, "Computer-aided periodontal disease diagnosis using computer vision", Computerized Medical Imaging and Graphics 23, pp.209-217, 1999.

13. OK SAM CHAE, "Specialized Parallel Structure For VLSI Implementation of the Hough Transform for Arbitary Shape Detection", Oklahoma State University, Ph.D., a doctoral dissertation, 1982.
14. J.Canny, "A Computational Approach to Edge Detection", IEEE Transactions on PAMI, 8-6, pp.679-698, 1986.
15. Yonghak Ahn, Giok Ahn, Oksam Chae, "Detection of Moving Objects Edges to Implement Home Security System in a Wireless Environment", Computational Science and Its Application-ICCSA2004, International Conference Part.I, LNCS, pp.1044-1051, May, 2004.
16. William H.Press, Saul A.Teukolsky, William T. Vetterling, Brian P.Flannery, "Numerical Recipes in C", Second Edition, pp.659-699, 1992.
17. Ok-sam Chae, Jung-hun Lee, Young-hyun Ha, "Integrated Image Processing Environment for Teaching and Research", Proceedings of IWIE2002, International Workshop on Informations & Electrical Engineering, 2002.

Improved Authentication Scheme in W-CDMA Networks

Dong Chun Lee[1], Hyo Young Shin[2], Joung Chul Ahn[3], and Jae Young Koh[3]

[1] Dept. of Computer Science Howon Univ., Korea
ldch@sunny.howon.ac.kr
[2] Dept. of Internet Information Kyungbok College, Korea
[3] Principal Member of Eng. Staff, National Security Research Institute, Korea

Abstract. In the W-CDMA network authentication, three nodes, the Mobile Station (MS), the VLR/SGSN, and the HE/HLR are involved. We propose another protocol which use public key in VLR/SGSN-HE/HLR link except for MS-VLR/SGSN link. W-CDMA network authentication procedure consists of two-stage. One is MS- Visitor Location Register (VLR)/SGSN link and the other is VLR/SGSN-HE/HLR link. Since VLR/SGSN link is based on implicit trust model, there exist security threats. To complement the threat, we propose two protocols using asymmetric key. First protocol uses asymmetric key in MS, VLR/SGSN, and MS. Second protocol uses secret key in MS-VLR/SGSN link, and public key in VLR/SGSN-HE/HLR link.

1 Introduction

The first generation mobile communication system had little security methods to protect users. The second generation mobile communication system such as GSM, IS-95 CDMA made a partial improvement including confidentiality. In spite of these improvements, there left many weak points in 2G system. With the advent of 3G systems, an effort to make consistent security architecture is being progressed based on several threats.

The technical specification of W-CDMA networks has been developed by 3GPP. The W-CDMA security architecture which is based on 2G security architecture has various new security features. In W-CDMA network security architecture, User Equipment (UE) can authenticate Serving Network (SN) and integrity function of signaling data is added. Using 128 bit key, the confidentiality of mobile data is strengthened [1, 2, 3].

The Authentication and Key Agreement (AKA) consists of two stages in W-CDMA networks. Authentication procedure involves three elements: MS, VLR/SGSN, Home Location Register (HLR)/AuC. In first stage, Authentication Vector (AV) is transferred from the Home Environment (HE) to the Serving Network (SN). The second stage is where the SGSN/VLR performs the challenge-response procedure between the USIM and the SGSN/VLR. The authentication procedure between USIM and the SGSN/VLR is based on the long term preshared secret key K (e.g., 128 bit). The master key K is stored in the AuC/HLR as well as UICC/USIM. To maintain security, it is important that the master key must not be exposed.

But the two stages AKA procedure assumes implicit trust model. The two stages AKA require a level of trust between the roaming partners with respect to handling foreign subscribers. This is persuasive in an environment where a few telephone operators had a monopoly on all telecommunication operators. But this is not effective when there is various telephone operators exist.

The description of this paper is as follows. In Section 2, we discuss the W-CDMA networks authentication scheme. In Section 3, we make some proposals to enhance authentication scheme using asymmetric key. Also the analysis of proposed protocol is given in Section 4. In the final section, the conclusion and some direction for further research are mentioned.

2 Related Work

Fig. 1 gives an overview of the complete 3G security architecture developed by 3GPP. Five security feature groups are defined. Each of these feature groups meets certain threats and accomplishes certain security objectives [3].

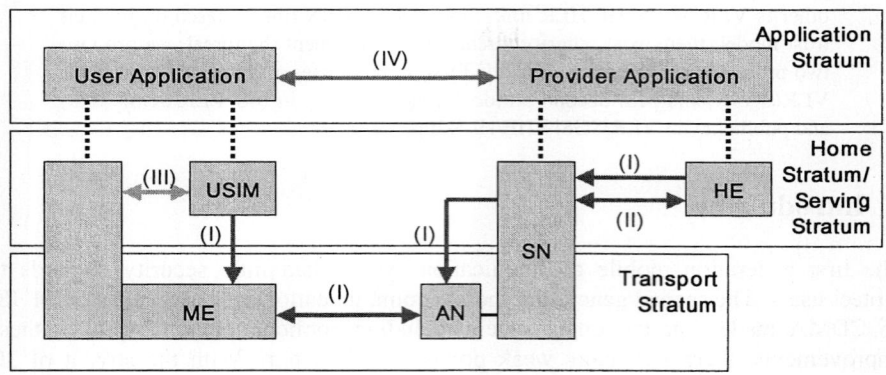

Fig. 1. Overview of the security architecture

- Network access security: the set of security features that provide users with secure access to 3G services, and which in particular protect against attacks on the (radio) access link.
- Network domain security: the set of security features that enable nodes in the provider domain to securely exchange signaling data, and protect against attacks on the networks.
- User domain security: the set of security features that secure access to mobile stations.
- Application domain security: the set of security features that enable applications in the user and in the provider domain.
- Visibility and configurability of security: the set of features that enables the user to inform himself whether a security feature is in operation or not and whether the use and provision of services should depend on the security feature.

2.1 Identification of User

The principal user identity is the International Mobile Subscriber (IMSI) number in Fig. 2. The IMSI number is not the subscriber number. IMSI number is used for system internal identification and routing. If the permanent identity IMSI is visible over-the-air interface, this is undesirable since it can be used for tracking location of user. To solve the problem, a Temporary Mobile Subscriber Identity (TMSI) is used on the radio access link. The TMSI, when available, is normally used to identify the user on the radio access path, for instance in paging requests, location update requests, attach requests, service requests, connection re-establishment requests and detach requests. Since there is no apparent relationship between IMSI and TMSI, the use of TMSI provides identity and location confidentiality.

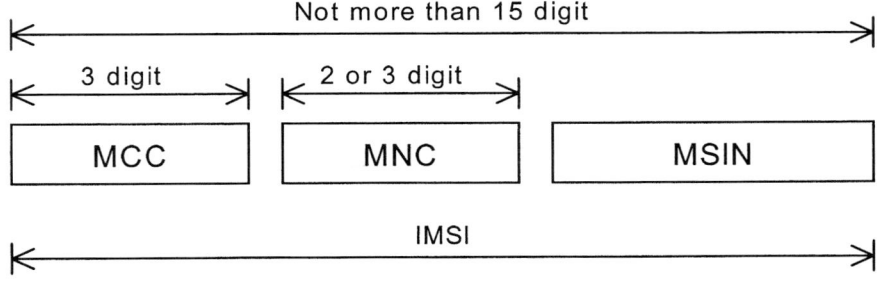

Fig. 2. The structure of IMSI

2.2 Authentication and Key Management

The authentication mechanism achieves mutual authentication by the user and the network showing knowledge of a secret key K which is shared between and available only to the USIM and the AuC in the user's HE. In addition the USIM and the HE keep track of counters SQNMS and SQNHE respectively to support network authentication. The sequence number SQN_{HE} is an individual counter for each user and the sequence number SQN_{MS} denotes the highest sequence number the USIM has accepted.

This mechanism has maximum compatibility with the current GSM security architecture and facilitates migration from GSM to W-CDMA. The method is composed of a challenge/response protocol identical to the GSM subscriber authentication and key establishment protocol.

An overview of the mechanism is shown in Figure 3. Upon receipt of a request from the VLR/SGSN, the HE/AuC sends an ordered array of n authentication vectors (i.e., the equivalent of a GSM "triplet") to the VLR/SGSN. The authentication vectors are ordered based on sequence number. Each authentication vector consists of the following components: a random number RAND, an expected response XRES, a cipher key CK, an integrity key IK and an authentication token AUTN. Each authentication vector is good for one authentication and key agreement between the VLR/SGSN and the USIM.

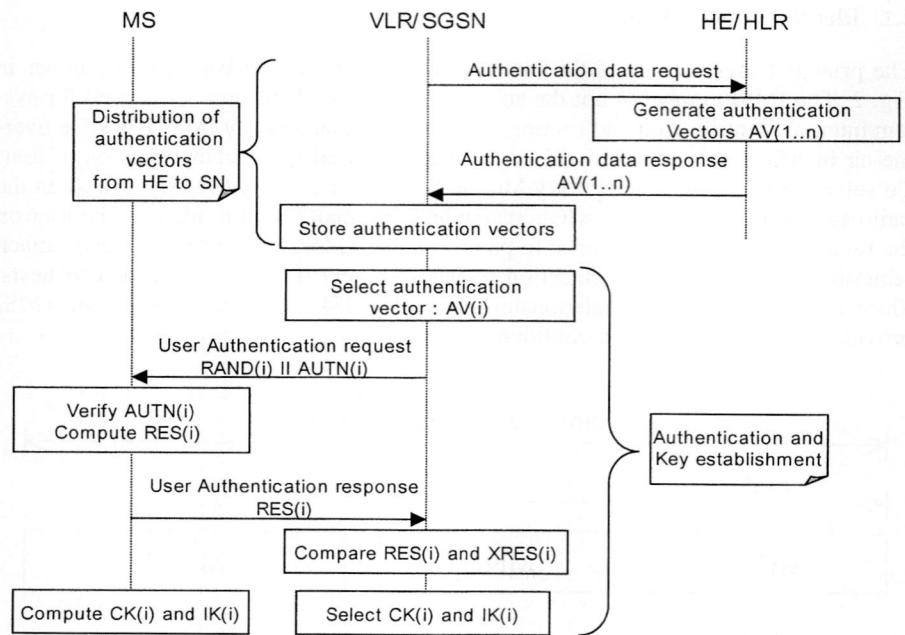

Fig. 3. Authentication and key agreement

When the VLR/SGSN initiates an authentication and key agreement, it selects the next authentication vector from the ordered array and sends the parameters RAND and AUTN to the user. Authentication vectors in a particular node are used on a first-in / first-out basis. The USIM checks whether AUTN can be accepted and, if so, produces a response RES which is sent back to the VLR/SGSN. The USIM also computes CK and IK. The VLR/SGSN compares the received RES with XRES. If they match, the VLR/SGSN considers the authentication and key agreement exchange to be successfully completed. The established keys CK and IK will then be transferred by the USIM and the VLR/SGSN to the entities which perform ciphering and integrity function.

VLR/SGSN can offer secure service even when HE/AuC links are unavailable by allowing them to use previously derived cipher and integrity keys for a user so that a secure connection can still be set up without the need for an authentication and key agreement. Authentication is in that case based on a shared integrity key, by means of data integrity protection of signaling messages.

3 Proposed Protocol

The two-staged AKA approach implicitly assumes a trust model in which the roaming partners must trust each other. This can be effective for distributed processing and load sharing. But the absence of global AKA procedure can be drawback of this system. Although the processing time of global AKA procedure affect call set-up time, an improved mechanism is necessary for security.

3.1 Authentication Mechanism Using Asymmetric Key

The MS, VLR, and HLR use the pair of public and private key for authentication and key agreement. The notation used here are as follows:

- PU_{MS}: public key of MS   - PR_{MS}: private key of MS

- PU_{VLR}: public key of VLR   - PR_{VLR}: private key of VLR

- PU_{HLR}: public key of HLR   - PR_{HLR}: private key of HLR.

VLR/SGSN transmits authentication data request to HE/HLR for acquiring the information of authentication and key agreement. This message is encrypted by HLR's public key. After HLR decrypts this message by its private key, HLR generates key information such as CK, IK, and RAND which is used by MS. HLR transmits the generated key information and MS's private key to the MS after encrypting with HLR's private key.

VLR encrypts the received key information (i.e., CK, IK, and RAND) with public key of MS and transfers to MS. The MS decrypts the received message with its private key. The CK is used for data encryption and the IK is used for integrity of signaling information. The MS encrypts RAND with its private key and sends to the VLR an acknowledgement message. This procedure is shown in Figure 4.

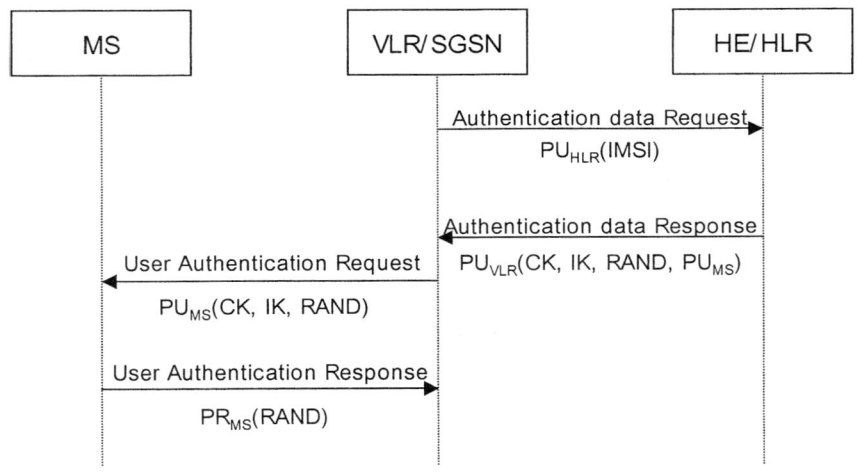

Fig. 4. Authentication using asymmetric key

3.2 Authentication Using Asymmetric Key and Symmetric Key

In authentication process node, the MS-VLR/SGSN link uses symmetric secret key and VLR/SGSN-HE/HLR link uses symmetric key.

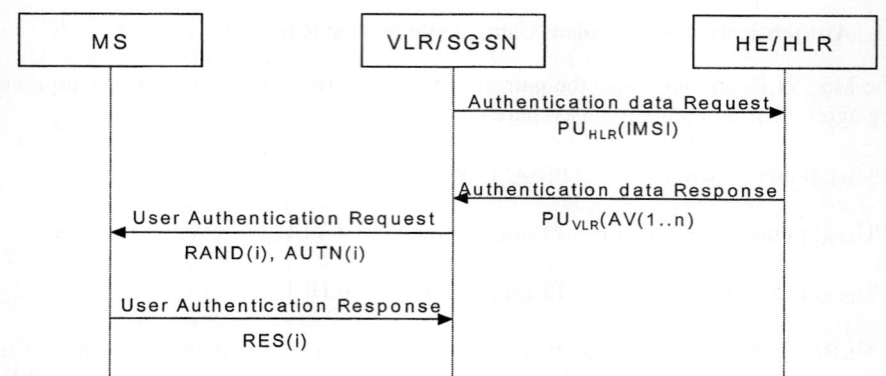

Fig. 5. Authentication using asymmetric key and symmetric key

VLR sends authentication data request to the HLR for acquiring the information of authentication and key agreement. This message is encrypted with HLR's public key. After the HLR decrypts this message with HLR's private key, HLR generates authentication vector (AV) which will be used by MS. The HLR sends the generated key information and AV to the VLR encrypting with HLR's private key.

The VLR retrieves RAND and AUTN from the received key information and sends RAND and AUTN to the MS. The USIM checks AUTN and generates RES if it can be accepted. Also the USIM generates CK and IK. The USIM sends RES to VLR/SGSN. The VLR/SGSN compares the received RES with previously stored XRES. If two values are consistent, VLR/SGSN decides that the authentication and key agreements are successfully completed. The established CK and IK are used for confidentiality and integrity. This proposed procedure is shown in Figure 5.

3.3 Identifying the Attack Methods

Having identified the assets to be protected and the threat agents which may be the subject of attack, the next step is to identify the possible attack methods which could lead to a compromise of the assets.

This will be based on what is known regarding the TOE security environment. There is a very important thing must be considered. If you want to consider all attack methods already known, maybe it is impossible because the attack methods are too various and created or found everyday. Therefore, if you want to describe the threats about the TCP/IP, should use the phrase not flooding, spoofing or DoS but 'TCP/IP vulnerabilities already known'.

4 Performance Analysis

4.1 Analysis of Security

(1) Authentication

The proposed two protocols authenticate user and network using RAND. In the first protocol which uses asymmetric key, HE/HLR sends RAND to VLR/SGSN. Only the

VLR/SGSN can decrypt the received message which is encrypted with PUVLR. VLR/SGSN sends RAND to MS. Since user authentication request message is encrypted with PUMS, only intended MS can decrypt the received message. VLR/SGSN sends RAND to MS. After the VLR/SGSN receives RES through user authentication message, VLR/SGSN compares RES with previously stored XRES. If two values are matched, the VLR/SGSN can authenticate the MS.

In second protocol, MS receives RAND and AUTN through user authentication request. If the MS computes XMAC, VLR/SGSN compares XMAC and MAC in AUTN. If two values are match, the MS can authenticate network. VLR/SGSN receives RES using user authentication response message. VLR/SGSN compares RES with expected response XRES. If they are equal, VLR/SGSN can authenticate MS.

(2) Confidentiality
In first proposed protocol, we use asymmetric protocol. Due to the slow speed of public key, it is inefficient to use public key in data encryption. We use public key for authentication and key agreement. The distributed session key is used for data encryption.

In second protocol, session key CK is used for data encryption. Session key CK is generated in HE/HLR and transferred to user. The f8 function is used for encryption. Both protocols guarantee confidentiality with session key.

(3) Integrity
Integrity Key (IK) is distributed during authentication and key agreement procedure. The f9 function is used for data integrity. Comparing MAC of sender and receiver, integrity can be guaranteed.

The first protocol can implement authentication procedure simply applying asymmetric key. But public key must be distributed to VLR/SGSN, HE/HLR and MS. Also use of public key in MS can be a load due to the slow data rates it can offer.

In second protocol, since MS-VLR/SGSN link uses symmetric key, processing overhead is not loaded to MS. Because only VLR/SGSN-HLR/HE link uses symmetric key, management of public key is relatively easy. Both protocol uses session key when encryption and decryption are executed. The processing time for confidentiality and integrity is equal to both protocols.

5 Conclusions

This paper applied asymmetric key to the W-CDMA authentication and key agreement procedure. This improves the problem of existing authentication scheme which is based on trust of VLR/SGSN and HE/HLR. But the use of public key in the MS has some problems. Because the processing speed of public key is lower than secret key, the use of public key may burden to MS. Although modern smart cards are capable of executing public key algorithm, user faces the extra cost of more expensive smart cards and slightly higher computational delay.

But this problem will be solved near future by the enhancement of smart card, the continual decline of smart card price, and the development of fast public key algorithm, etc. And the use of public key in MS would be very useful for e-commerce purpose.

Since the processing of public key in MS has some burden today, we propose another protocol which use public key in VLR/SGSN-HE/HLR link except for MS-VLR/SGSN link. This protocol can be used until user's terminal has enough processing capability.

Acknowledgement

This work was supported by grand from National Security Research Institute (NSRI) and the Information Technology Research Center (ITRC) Support Program.

References

1. 3GPP TS 21.133 "Security threats and requirement"
2. 3GPP TS 33.120 "Security Objectives and Principles"
3. 3GPP TS 42.009 "Security Architecture"
4. Nawal El Fishway, "An Effective Approach for Authentication of Mobile Users"
5. Geir M. Kohen, "An Introduction to Access Security in UMTS," IEEE Wireless Communications, Feb. 2004, p.8 - 18.
6. Chang-Seop Park, "On Certificate based Security Protocol for Mobile Comm.", IEEE Network, 1977
7. Nawal El Fishwa, "On the Design of Auth. Protocol for 3G Mobile Comm. Systems", Twentieth National Science Conference, 2003.
8. Y-Bing Lin, "Reducing Auth. Signaling Traffic in 3G Mobile Network", IEEE Trans. on Wireless Comm. Vol. 2, No. 3, May 2003, p.493 - 501
9. Zang Bin, "Authentication and Key Distribution in Mobile Comput. Env."
10. Constantinos F. Grecas, "Towards the Intro. of the Asymm. Crypto. in GSM, GPRS, UMTS", p.15 - 21, IEEE 2001
11. Min Lei, "Security Archi. and Mechanism of 3G Mobile Comm.", Proc. of IEEE TENCON '02, p.813 -816
12. Erik Dahlman, "The Evolution of 3G WCDMA", Telecommunications Review, Vol. 13, No. 6, Dec. 2003, p.824 - 833.

Memory Reused Multiplication Implementation for Cryptography System

Gi Yean Hwang, Jia Hou, Kwang Ho Chun, and Moon Ho Lee

Institute of Information & Communication, Chonbuk National Univ., Korea
{infoman,lighttiger}@.chonbuk.ac.kr
houjiastock@hotmail.com
moonho@chonbuk.ac.kr

Abstract. In this paper, we simply present a memory reused multiplication implementation. This scheme could efficiently reduce the number of computations for cryptography system or other computation algorithms.

1 Introduction

To protect the information, Shannon suggested two encryption concepts for frustrating the statistical endeavors of the cryptanalyst. He termed these encryption transformations confusion and diffusion. Confusion involves substitutions and computations that render the final relationship between the key and cipher-text as complex as possible [1, 2]. This makes it difficult to utilize a statistical analysis to narrow the search to a particular subset of the key variable space. Confusion ensures that the majority of the key is needed to decrypt even very short sequences of cipher-text. Diffusion involves transformations that smooth out the statistical differences between characters and between character combinations. For example, in the typical cryptography systems (e.g. DES and RSA algorithm), the multiplications over field are widely used in puzzles [3, 4]. In those cryptography algorithms, the implementation of the multiplications is very complex following the increase of the number and security. Therefore, in this paper, we present a simple modified multiplications implementation, and introduce a memory reuse scheme to build the cryptography system.

2 Modified Multiplications Implementation

In the field $F_{(q^m)}$, the multiplications of two elements could be represented by

$$Z(\alpha) = A(\alpha) \times B(\alpha) = \sum_{i=0}^{m-1} b_i(\alpha^i A(\alpha))$$

$$= b_0 A(\alpha) + b_1 A(\alpha)\alpha + \ldots + b_{m-1} A(\alpha)\alpha^{m-1}, \quad a_i, b_i \in \{0,1,2,\ldots,q-1\}, \quad (1)$$

where,
$$A(\alpha) = a_0 + a_1\alpha + a_2\alpha^2 + ... + a_{m-1}\alpha^{m-1}, \quad B(\alpha) = b_0 + b_1\alpha + b_2\alpha^2 + ... + b_{m-1}\alpha^{m-1},$$
and the primitive element α is a root of the prime polynomial of the field $F(q^m)$, q is prime. The prime polynomial of the field $F(q^m)$ could be given as

$$\alpha^m = p_0 + p_1\alpha + ... + p_{m-1}\alpha^{m-1}, \quad p_i \in \{0,1,2,..,q-1\}. \tag{2}$$

To reduce the number of computation, we can easily modify the (1) by using the formula as below.

$$Z(\alpha) = A(\alpha) \times B(\alpha) = \sum_{i=0}^{m-1} b_i(\alpha^i A(\alpha))$$
$$= (...(b_{m-1}\alpha A(\alpha) + b_{m-2}A(\alpha))\alpha + b_{m-3}A(\alpha))\alpha...)\alpha + b_1 A(\alpha))\alpha + b_0 A(\alpha)). \tag{3}$$

In this formula, we found that there is only one computation pattern, which is given as

$$K(\alpha) + b_h A(\alpha), \quad h \in \{0,1,...,m-2\}, \tag{4}$$

and we can write a recursive function for $K(\alpha)$ as

$$K_n(\alpha) = (K_{n-1}(\alpha) + b_{m-n-1}A(\alpha))\alpha, \quad n \in \{1,2,...,m-1\}, \tag{5}$$

where $K_0(\alpha) = b_{m-1}A(\alpha)\alpha$. Thus the multiplication of two elements could be formed as

$$Z(\alpha) = A(\alpha) \times B(\alpha)$$
$$= K_{m-1}(\alpha) + b_0 A(\alpha). \tag{6}$$

Therefore, the multiplications of two elements will be translated to a recursive circuit on the function $A(\alpha)\alpha$. By using (2), we can simply write that

$$A(\alpha)\alpha = (a_0 + a_1\alpha + ... + a_{m-1}\alpha^{m-1})\alpha = a_0\alpha + a_1\alpha^2 + ... + a_{m-1}\alpha^m$$
$$= a_0\alpha + a_1\alpha^2 + ... + a_{m-2}\alpha^{m-1} + a_{m-1}(p_0 + p_1\alpha + ... + p_{m-1}\alpha^{m-1})$$
$$= a_{m-1}p_0 + (a_0 + a_{m-1}p_1)\alpha + (a_1 + a_{m-1}p_2)\alpha^2 + ... + (a_{m-2} + p_{m-1})\alpha^{m-1}$$
$$= a_{m-1}p_0 + \sum_{l=0}^{m-2}(a_l + a_{m-1}p_{l+1})\alpha^{l+1}. \tag{7}$$

Based on above mathematical model, the efficient implementation diagrams can be drawn as Fig.1 and Fig.2. The implementation of the basic unit of $A(\alpha)\alpha$ only need m bit-wise multiplications and $(m-1)$ bit-wise additions, which saved $(m-1)$ bit-wise additions from the directly form. Otherwise, as illustrated in the Fig.2, we implement $Z(\alpha)$ by using only one $A(\alpha)\alpha$ unit and the recursive functions, which could remarkably reduced $m/(m-1)$ complexity of the directly calculation form.

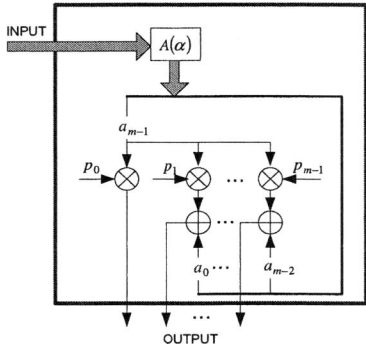

Fig. 1. Unit of implementation of $A(\alpha)\alpha$

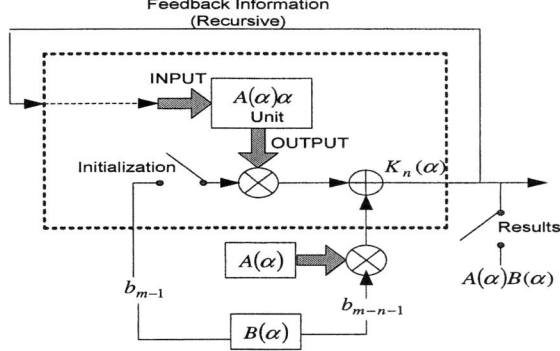

Fig. 2. Multiplication of two elements by using recursive function and the unit of $A(\alpha)\alpha$

3 Memory Reused Multiplications Implementation

Based on the modified unit of multiplications, we now present a novel memory reused multiplications implementation for the RSA algorithm over the Galois Field $GF(2^m)$. The scheme developed by RSA makes use of an expression with exponentials [3, 4]. Plaintext is encrypted in blocks, with each block having a binary value less than some number n. That is, the block size must be less than or equal to $\log_2(n)$; in practice, the block size is k bits, where $2^k < n \leq 2^{k+1}$. Encryption and decryption are of the following form, for some plaintext block M and cipher-text block C:

$$C = M^e \bmod n$$

$$M = C^d \bmod n = (M^e)^d \bmod n . \tag{8}$$

Both sender and receiver must know the value of n. The sender knows the value of e, and only the receiver knows the value of d. And we should set

$$d \equiv e^{-1} \bmod \phi(n), \qquad (9)$$

where $\phi(n)$ is the Euler totient function. The implementation of the RSA algorithm is the mathematical model to calculate the multiplications of M. Now we define that

$$M^e \bmod n = \left(\underbrace{M \cdot M \cdots M}_{e}\right) \bmod n \overset{\Delta}{=} \underbrace{M \circ M \cdots \circ M}_{e} \overset{\Delta}{=} M * e, \qquad (10)$$

where $\circ$ is the multiplications mod n. If given $e = (e_k e_{k-1} e_{k-2} \ldots e_1 e_0)$, a binary representation, we have

$$M * e = (\ldots(M * e_k) * 2 \circ (M * e_{k-1})) * 2 \circ (M * e_{k-2}) \ldots) * 2 \circ (M * e_1)) * 2 \circ (M * e_0), \qquad (11)$$

where we force that $e_k = 1$. Additionally, we set two memory units as

$A = (M * 2) \circ M$ Number of Multiplications = 2

$$B = (A * 2^2) \circ A \quad \text{Number of Multiplications} = 3, \qquad (12)$$

where $M \circ 0 \overset{\Delta}{=} M$, $M * 2 \overset{\Delta}{=} M \circ M$, and this multiplications of two elements could be realized by using the (6), and the modified diagram as shown in section 2

Example: If $e = (1011100111101111)$
The conventional algorithm needs
$M * e = (\ldots(M * 2) \circ 0) * 2 \circ M) * 2 \circ M) * 2 \circ M) * 2 \circ 0) * 2 \circ 0) * 2 \circ M) * 2 \circ M) * 2 \circ M) * 2 \circ M) * 2 \circ 0)$
$* 2 \circ M) * 2 \circ M) * 2 \circ M) * 2) \circ M$

where the number of "$*2$" is 15, and number of "$\circ M$" is 11. Then total number of operation is $15 + 11 = 26$.

Proposal: The binary representation could be modified as $e = (1011100111101111) = (10A100B0B)$, thus we can write

$$M * e = ((M * 2^3) \circ A) * 2 \circ M) * 2^6 \circ B) * 2^5 \circ B)$$

thus the operation of this function is

$$3 + 1 + 1 + 1 + 6 + 1 + 5 + 1 = 19.$$

By adding the operations of the reused memories A, and B, we get

$$19 + A + B = 19 + 2 + 3 = 24 \quad \text{(Total number of operations)}$$

As result, we save 2 operations from the conventional calculation. In general, if the length of e is much longer, then the operations will be saved more. The implementation of the proposed computation can be drawn as Fig.3.

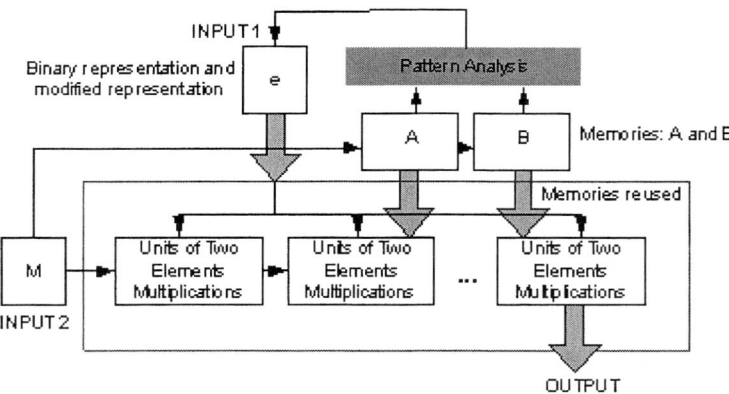

Fig. 3. The implementation of $M^e \bmod n$ by using memories reused

4 Conclusions

This paper proposed a simple implementation, and the memories reused algorithm for multiplications in the cryptography systems. The contributions of this work will efficiently improve the complexity of the encryption and decryption. Moreover, this algorithm can be widely applied for communications and the number theory over finite fields to reduce the operations and computation time.

Acknowledgements

This work was supported (in part) by the Ministry of Information & Communications, Korea, under the Information Technology Research Center (ITRC) Support Program.

References

1. Man Young Rhee *Cryptography and Secure Communications,* published by McGraw-hill book Co. 1994;
2. H.Feistel, "Cryptography and Computer Privacy", *scientific American*, 228(1973), pp. 15-23;
3. H.Feistel, "Some Cryptographic Techniques for Machine-to-Machine Data Communications", proceeding of the *IEEE*, 63(1975), pp.1545-1554;
4. William Stallings: *Cryptography and Network Security.* Pearson Education, Inc. U.S. 2003.

Scheme for the Information Sharing Between IDSs Using JXTA*

Jin Soh, Sung Man Jang, and Geuk Lee

Dept. of Computer Engineering, Hannam Univ., DaeJeon, South Korea
{jsoh01,smjang,leegeuk}@hannam.ac.kr

Abstract. This paper proposes information sharing scheme which is able to change information between Information Security Systems (ISS) with/without changing network topology of ISS. Because it is impossible to communicate directly between adjacent ISS in current hierarchical topology network, we suggest a direct information sharing method using P2P technique based on JXTA. The advantage of this system can make direct communication among ISS without change of hierarchical network topology.

1 Introduction

In the progress of information society, the whole world is connected by internet and the dependency of information system has been deepening. Therefore, intrusion incident frequency is getting higher and the main attack object of intrusion is unspecified information system rather than specified information system [1, 2].

A technique is needed that detects intrusions on the important system and responds with it quickly. However, it is not easy to correspond appropriately under the communication method of current intrusion detection systems [3, 4]. So, new intrusion corresponding technique is needed, and for that reason, this study would like to show a communication method of intrusion detection system which is to inform the whole network when attacks occur. And the damages of information assets will be prevented more effectively through this method.

2 Related Work

2.1 Requirement for Network Construction Using P2P

P2P (Peer-to-Peer) is a solution of communication schemes which are to share information and resource services among multi devices. And it should enable to solve problems as follows [5].
- A method that one device knows the existence of the other one
- A method that devices unites for the common interests

---

* This work was supported by a grant No.R12-2003-004-02003-0 from Korea Ministry of Commerce, Industry and Energy.

- A method that one device shows its own functions
- A Information to identify the devices
- A data exchanging method among devices

The features of P2P are as follows.

- It is a system based on interactions among peers.
- It is a system which does not need services or resources from the central server.
- It is a very flexible system on changes of network topology.
- It is a system that can stand well on network environment which has non-deterministic topology.
- It is easy to expand into a large scale of user.

2.2 Information Sharing System Using JXTA

Both current server/client architecture and hierarchical server architecture require a strong and stable server group for the user management. In this system environment, when there are more users, storage capacity and bandwidth must be increased with computing ability of central server. However, P2P system only depends on interactions among peers because peers communicate with clients (other peers) without passing through central server. And it means that peers have to have both client function and server function for bidirectional communication.

2.2.1 JXTA Framework

JXTA is an abbreviation of juxtapose and it is a library of common functions that are needed to implement P2P system. It reduces ineffectiveness that each P2P system programmer has to develop his own networks and protocols. On the P2P network, there must be available some service while peers are connected [6].

JXTA is open protocol for P2P network. This protocol defines complex operations such as peer discovery, end point routing, connection binding, basic queries/replies message exchange and propagating network through rendezvous peer. JXTA network is consisted of components as shown in Fig.1 [7, 8].

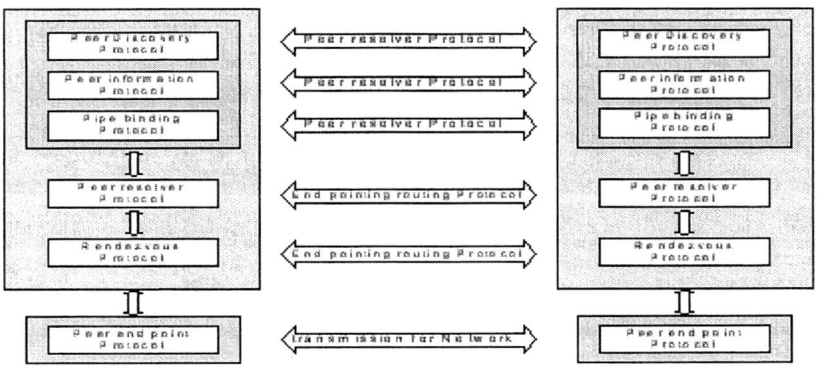

Fig. 1. JXTA Framework

- **Peer**

A peer is any networked device that implements one or more of the JXTA protocols. Each peer operates independently and asynchronously from all other peers, and is uniquely identified by a Peer ID. Peers publish one or more network interfaces for use with the JXTA protocols. Each published interface is advertised as a peer endpoint, which uniquely identifies the network interface. Peer endpoints are used by peers to establish direct point-to-point connections between two peers.

- **Peer group**

A peer group is a collection of peers that have agreed upon a common set of services. Peers self-organize into peer groups, each identified by a unique peer group ID. Peers may belong to more than one peer group simultaneously. By default, the first group that is instantiated is the Net Peer Group. All peers belong to the Net Peer Group. Peers may elect to join additional peer groups. The JXTA protocols describe how peers may publish, discover, join, and monitor peer groups; they do not dictate when or why peer groups are created.

- **Pipes**

JXTA peers use pipes to send messages to one another. Pipes are an asynchronous and unidirectional message transfer mechanism used for service communication. JXTA pipes can have endpoints that are connected to different peers at different times, or may not be connected at all. Pipes are virtual communication channels and may connect peers that do not have a direct physical link. In this case, one or more intermediary peer endpoints are used to relay messages between the two pipe endpoints. Pipes offer two modes of communication, point-to-point and propagate.

A point-to-point pipe connects exactly two pipe endpoints together: an input pipe on one peer receives messages sent from the output pipe of another peer. A propagate pipe connects one output pipe to multiple input pipes. Messages flow from the output pipe (the propagation source) into the input pipes. All propagation is done within the scope of a peer group. That is, the output pipe and all input pipes must belong to the same peer group.

- **Messages**

A message is an object that is sent between JXTA peers; it is the basic unit of data exchange between peers. The JXTA protocols are specified as a set of messages exchanged between peers. There are two representations for messages: XML and binary. The JXTA J2SE platform binding uses a binary format envelop to encapsulate the message payload. Services can use the most appropriate format for that transport . Binary data may be encoded using a Base64 encoding scheme in the body of an XML message.

The use of XML messages to define protocols allows many different kinds of peers to participate in a protocol. Because the data is tagged, each peer is free to implement the protocol in a manner best-suited to its abilities and role. If a peer only needs some subset of the message, the XML data tags enable that peer to identify the parts of the message that are of interest.

- **Advertisements**

All JXTA network resources (such as peers, peer groups, pipes, and services) are represented by an advertisement. Advertisements are language-neutral metadata structures represented as XML documents. The JXTA protocols use advertisements to describe and publish the existence of peer resources. Peers discover resources by searching for their corresponding advertisements, and may cache any discovered advertisements locally.

2.2.2 JXTA Software Architecture

JXTA software architecture is shown in Fig.2. Main component of JXTA software is depicted as follows.

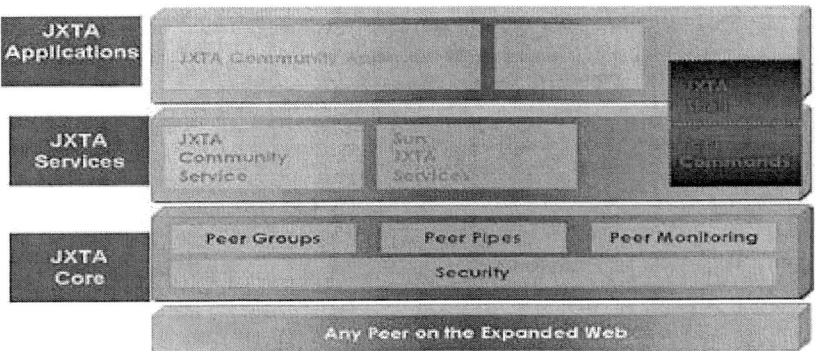

Fig. 2. JXTA Software Architecture

- **Platform Layer (JXTA Core)**

The platform layer, also known as the JXTA core, encapsulates minimal and essential primitives that are common to P2P networking. It includes building blocks to enable key mechanisms for P2P applications, including discovery, transport (including firewall handling), the creation of peers and peer groups, and associated security primitives.

- **Services Layer**

The services layer includes network services that may not be absolutely necessary for a P2P network to operate, but are common or desirable in the P2P environment. Examples of network services include searching and indexing, directory, storage systems, file sharing, distributed file systems, resource aggregation and renting, protocol translation, authentication, and PKI (Public Key Infrastructure) services.

- **Applications Layer**

The applications layer includes implementation of integrated applications, such as P2P instant messaging, document and resource sharing, entertainment content management and delivery, P2P Email systems, distributed auction systems, and many others. The boundary between services and applications is not rigid.

3 System Design

3.1 Prototype Design

The present information protection system network has typical server/client architecture. In this architecture, information sharing is impossible among intrusion detection system nodes. All information should transmit to the highest central server on the network. And also it is a structure that information needed every node starts from central server to terminal node as like an intrusion detection pattern. In hierarchical structure like this, each node can only have hierarchically dependent information in Fig. 3.

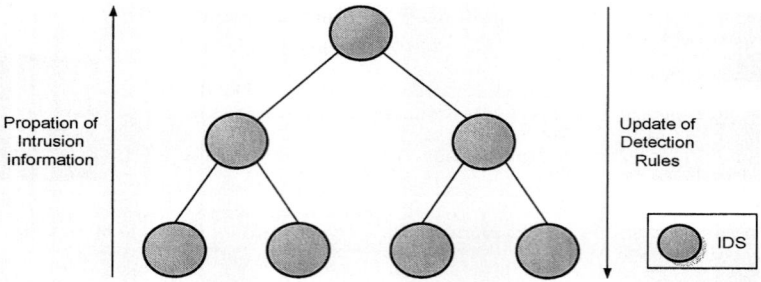

Fig. 3. Hierarchical Network of IDS

This paper proposes a structure that communicates information among the inside nodes horizontally. This causes each node to interoperate organically like a system. This horizontal and organic relationship propagates intrusion detection results which are not offered by current independent systems to adjacent nodes.

The advantage of this system can share attack-signature mutually regardless of existence of central server and it can be applied without adjustment on current network. Unlike Fig 3, ISS shares information among adjoining nodes directly in Fig 4.

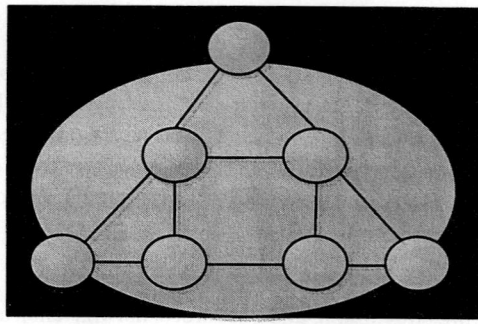

Fig. 4. Information Sharing System

3.2 Characteristics of Prototype

The suggested system uses JXTA which is open P2P protocol and Java language, which makes it possible to port all platforms.

The system designed for information sharing and automatic propagation among the nodes can apply to situations as follows.
- Propagate intrusion information detected to whole network
- Update and propagate the new detection rules to whole network
- Presentation of information propagation method among intrusion detection systems

A prototype for information sharing has functions as follows.
- Management of sharing files in each peer
- Client component that has a function which can query what is in the service.
- Client GUI that shows file lists which are available to be used in each service instance of group.
- Client component that requests and receives files form other peers

A prototype network has a characteristic above and consists of 3 nodes and its operating is as shown in Fig 5.

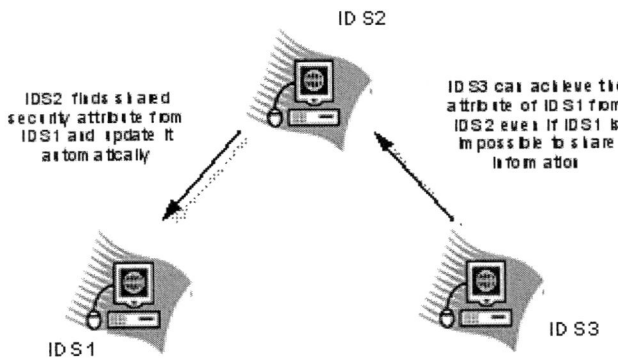

Fig. 5. Operating Process of the System

The characteristics of designed system are as follows.
- It is possible to share information only if there is a peer in the network.
- It can be operated through non-deterministic network topology that can be disconnected and does not reconnect peers anytime.

4 Implementation

In Fig.1, a part of snapshot of this system (agent), contents of each box are information which is obtained from each IDS. Agents installed in each IDS show system information where agents are installed. Also, it shows that they achieve and bring information of other IDS. In Fig.6, inside contents of dashed boxes are internal information of current IDS and outside contents of dashed boxes are external information achieved form other IDS.

As we mentioned above, it is impossible to achieve information of adjacent IDS or internal information of other IDS of the same network system which had a current hierarchical structure. As a result of adding P2P form by using JXTA suggested in this study, it is clear that nodes from the same network share all security attributes. The advantage of this system is that it does not change current network topology and makes each IDS communicate horizontally at the same time. This information sharing structure can apply current implemented IDS without structural changes.

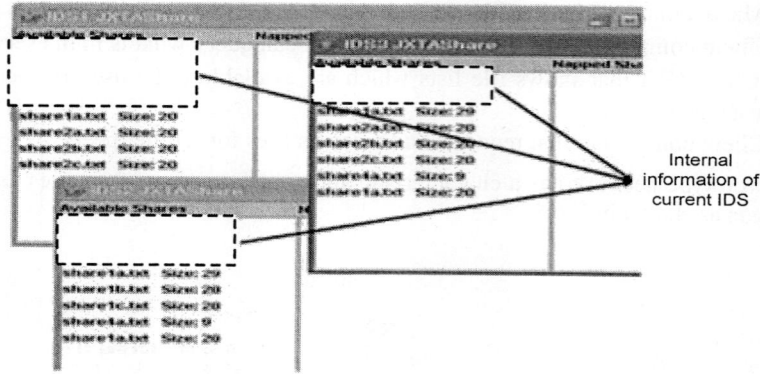

Fig. 6. Partial Snapshot of the System

5 Conclusions

The attacking trends on network have been intellectualized, decentralized and automatized gradually. It makes the exact detection difficult and lowers the reliability on intrusion detection system. However, there are a lot of difficulties for intrusion detection systems based on current communication structure to deal actively with every attack. Therefore, it needs to study about the concept of effective and active communication structure for intrusion detection system.

For the effective intrusion protection network construction, we have studied about an effective communication scheme among systems which are formed of existing intrusion detection system with hierarchical network structure. Current intrusion detection systems with hierarchical network structures depend on central management method in order to propagate attack patterns detected by individual intrusion detection system. To improve this, this paper suggests the communication method of effective intrusion detection system network that keeps pure P2P characters and is linked with resources and file sharing method in the internet environment.

When you apply this method to the network, it is possible to search file effectively using P2P and you can relocate node and sharing resources to adapt to distributed environment using this resource sharing function. Appling this relocation function, it is possible to change current IDS and network to interactive systematic environment. Intrusion detection system network that applied this method is maintained by communicating pre-defined messages (intrusion detection pattern, rules, blocking addresses and so on) and offers expansion abilities and conveniences that P2P network has. And it is clear for network environment organized by this technique is more flexible and simple compared to current hierarchical structure.

This P2P application could be used on the internet environment in any area or on the intranet environment applied to special isolated circumstance. As the result, intrusion detection systems which operate on the base of current network topology would communicate directly and security information among systems would be propagated and offered with ease regardless of topology change. It would be expected that those would improve current hierarchical IDS network environment effectively and operate more effectively and efficiently against attacks.

References

1. Fred Cohen, "50 ways to Defeat Your Intrusion Detection System" http://all.net/
2. White House, "The National Strategy to Secure Cyber Space", http://www.whitehouse.gov/pcipb, 2003.
3. Steve Schupp "Limitation of Network Intrusion Detection", December 1, 2000.,http://www.sans.org
4. T.F. Lunt, "A Survery of Intrusion Detection Techniques", Computer & Security", Vol. 12, No. 4, Jun., 1993.
5. Project JXTA : A Technology Overview", http://www.jxta.org
6. Project JXTA : JavaTM Programmer's Guide", http://www.jxta.org
7. Project JXTA : An Open, Innovative Collaboration", http://www.jxta.org
8. Project JXTA : JXTA v1.0 Protocols Specification", http://www.jxta.org

Workflow System Modeling in the Mobile Healthcare B2B Using Semantic Information

Sang-Young Lee[1], Yung-Hyeon Lee[1],
Jeom-Goo Kim[2], and Dong Chun Lee[3]

[1] Dept. of Health Administration, Namseoul Univ., South Korea
{sylee,skylee}@nsu.ac.kr
[2] Dept. of Computer Science, Namseoul Univ., South Korea
jgoo@nsu.ac.kr
[3] Dept. of Computer Science, Howon Univ., Korea
ldch@sunny.howan.ac.kr

Abstract. The UML activity diagram is useful to model business process and workflow for its suitability to present dynamic aspect of system. However, it is difficult to present precise semantics which is important in mobile workflow system with the guide provided by OMG to the UML activity diagram. This paper suggests mobile workflow system modeling methodology by applying ASM semantics to the healthcare B2B after extending semantics to corresponding workflow system characteristics. To extend the action node timing, the timing element and the state element of the action node are added. Through the exact definition of formal semantics based on ASM the efficient workflow modeling can surely be expected.

1 Introduction

Recently, In the field of information technology, remarkable change has been observed in switch over from the data-driven information technology to the process-driven information technology.

The common core concept based on the process-driven information technology is the business process. The workflow technology enables automation of process management [1, 2]. Unified Modeling Language (UML) is the only OMG standard notation for modeling software, among these notations, the UML Activity Diagram is well-known for describing dynamic behaviors of systems [3]. Activity diagram is applied efficiently for workflow modeling included in business process. Workflow modeling should be exploited understandable, general and easy to work even for unprofessional person. The method of modeling must contain the formal meaning and analyzability.

In this paper we provide the ASM semantics expression for mobile workflow modeling based on UML activity diagram.

2 Related Work

In general, the workflow specification describes how the workflow system to do. Therefore, it is necessary to define the semantics of activity diagram and give the meaning of related workflow system.

The research community tries to formalize UML activity diagram in various ways. The methods of expression for the semantics of activity diagram are OCL (Object Constraint Language) [8], pi-calculus [9], FSP(Finite State Processes) [4] and ASM(Abstract State Machine) [6].

ASM adopted in this paper was presented ten years ago. Since then, ASM has been successfully used in specifying and verifying many software systems [10]. The ASM methodology has the following desirable characteristics of classical mathematical structure to describe states of a computation, thus, distinguishing ASM from informal methodologies. ASM uses extremely simple syntax to read and write for improved understanding. Although existing methods are useful in a specific domain, ASM maintains useful popularity in a wide variety of domains.

Holding positive merits as mentioned in this paper, semantics for mobile workflow modeling activity diagram using ASM provides the definition of semantics with basic formalization.

3 ASM Semantics Expression for Workflow Modeling

3.1 Information Needed for Workflow Modeling

The existing workflow systems are not suitable for modeling the timely category of systems and are not suitable for modeling modern business processes either. In particular, if these are to model interconnected real time asynchronous systems like mobile environment one's modeling ability is required as followings. To exploit entire workflow models anticipated changes from the environment, exceptions, dynamic change, a quantum of workflow that can be chained to develop the full workflow model and timing factors should be modeled. A workflow model also can contain conditions or constraints because activities cannot be executed in an arbitrary way.

Two types of conditions can be applied in a workflow model, these are pre-conditions and post-conditions. Moreover, in a workflow modeling information should be associated with each activity.

The information are required on who control over the activity through the assignment of activities for qualified users or application function, on other activities required to complete the activity, on the input or output of the activity and on the data and control information required for task accomplishment. The necessary information for workflow modeling is presented in the table 1 according to workflow characteristics.

Table 1. Workflow Elements and Definition

Elements	Definition
Data	Represents the data that will be used or produced in an action
Conditions	Represents the conditions of an action
Resources	Represents the resources required for an action done
Participants	Represents employees who will do an action
Action	Represents piece of work, it enables manual action or automatic action
Timing	Represents time elements which involves time or deadline while accomplish the workflow action

3.2 Expression of the ASM Semantics

On the basis of the workflow elements and the definition as mentioned above, we tried to express the extended ASM semantics to extend the existing UML activity diagram.

The existing action node is short of representation on the state of the action node, and could not express the timing element that is important in workflow system. Therefore, to extend the action node timing, the timing element and the state element of the action node are added. The expression of ASM semantics for extended action node is presented in the table 2.

In the Table 2, action node are of form *node (in, in_data, A, out, out_data, isDynamic, dynArgs, dynMult, action_type, time)* where the parameter *in* and *out* denotes the incoming and outcoming arc, *A* an atomic action, *isDynamic* if A may be executed *dynMult* times in parallel, each time with an argument L_i came from a set *dynArgs* of sequences of objects $\{L_1,...,L_n\}$.

Where the parameter *out_data* should be created after action node A accomplished. Where the parameter *action_type* presents either manual action or automatic action, if *action_type* is manual, the participant must be accompanied. Finally, the parameter *time* will be exited if time is exceeds the deadline.

ASM semantics will be extended to meet the information required for the conditioned_element. The pre-conditions need to be satisfied so that an action can be enacted, and pre-conditions must be satisfied prior to action node. If the pre-conditions are existent, then pre-conditions are to be accomplished, and if the pre-conditions are true, the out semantics will be activated. The expression of ASM semantics for extended pre-conditions node is presented in the table 3.

Table 2. The Expression of ASM Semantics for Extended Action Node

if $currTarget$ is $node(in, in\_data, A, out, out\_data,$
$\qquad isDynamic, dynArgs, dynMult, action\_type, time)$
then if $\neg isDynamic$ then A
$\qquad$ elseif $[dynArgs] = dynMult$ then
$\qquad\qquad$ for all $L_i \in dynArgs$
$\qquad\qquad\qquad A(L_i)$
$\qquad$ elseif $time \geq deadline$
$\qquad\qquad$ then $active := exit()$
$\qquad$ elseif $action\_type$ is $manual$ then
$\qquad\qquad active := participant$
$\qquad\qquad active :=\rightarrow out$

Table 3. The Expression of ASM Semantics for Extended Pre-condition Node

if $currTarget$ is $node(PRE, out, isExistent)$
then if $isExistent$ then PRE
$\qquad$ elseif $PRE = true$ then
$\qquad\qquad active :=\rightarrow out$

Post-conditions are needed to be satisfied when its action finished. After the action was accomplished, the post-conditions are to be accomplished contingent to post-conditions are existent, If post-conditions are true, they will provide the next action node accomplished. The expression of ASM semantics for extended post-conditions node is presented in the Table 4.

Table 4. The Expression of ASM Semantics for Extended Post-condition Node

if $currTarget$ is $node(in, POST, isExistent)$
then if $isExistent$ then $POST$
$\qquad$ elseif $POST = true$ then
$\qquad\qquad active := nextnode(in, in\_data, A, out, out\_data,$
$\qquad\qquad\qquad isDynamic, dynArgs, dynMult, action\_type, time)$

Finally, resources node is required to do an action, that describes required machine, tool and device to accomplish the work. Resources node is considered as special action node without parameters of initial node and final node.

4 Extension of the Activity Diagram in Mobile Workflow Modeling

In the present step, activity diagram will be extended by applying the concepts of extended ASM semantics. The inscribed method for activity diagram notation of extended action node is presented in the Fig. 1.

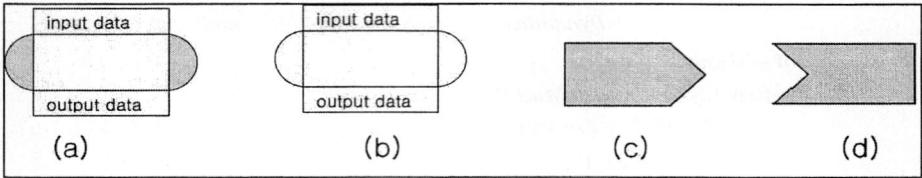

Fig. 1. Extended Nodes

From the figure 1, to distinguish between the manual action type and the automatic action type, if the action type is manual, action node is filled with color as shown in (a), and otherwise is not colored as in (b). Upper and down the rectangle parts present the input data needed by the action node and output data from the action node. Also the item of the timing element can be assigned for action attributes, and, the notation can be extended remarkably. The pre-condition and post-condition notation nodes are presented in the Fig. 1, (c) and (d). (a) represents manual action, (b) represent automatic action, (c) represent pre-conditions and (d) represent post-conditions.

5 Performance Evaluation

In this paper, we defined the workflow as a part of the healthcare B2B which is ordering and purchasing the healthcare items (e.g., goods). As In this business process

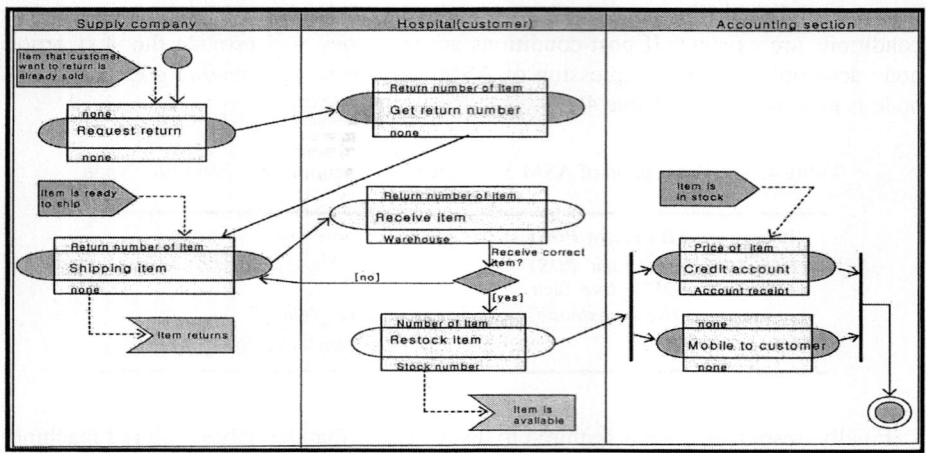

Fig. 2. The Workflow Modeling for Mobile Health B2B

formed by cooperation between the various automatic system and human system is modeled by means of activity diagram. In a case study, we apply the workflow case that supply company rebates the item ordered by mobile tools. The workflow modeling for treating rebated items is presented in the figure 2.

The existing workflow systems are not suitable for modeling the timely category of systems and are not suitable for modeling modern business processes either. In particular, if these are to model interconnected real time asynchronous systems like mobile environment one's modeling ability is required as followings. The steps can be expressed the ASM semantics in table 5.

Table 5. The Steps Expressing the ASM Semantics

step	semantics
(diagram: Item is ready to ship → Return number of item → Shipping item → Item returns)	$makeGraph(PREPPOST, in, out)$ $makeGraph(PRE, out)$ $makeGraph(P, in, out)$ $makeGraph(POST, in)$
(diagram: in → [yes]/[no] branch → Number of item / Restock item / Stock number / Return number of / Shipping item)	$makeGraph\ (IF\ a\ THEN\ P\ ELSE\ Q, in, out)$ $let\ \beta = new(BRANCHNODE)$ $yes = makeLink(\beta, a)$ $no = makeLink(\neg\beta, a)$ $connect(in, \beta)$ $makeGraph\ (P, yes, out)$ $makeGraph\ (O, yes, out)$
(diagram: in → Price of item / Credit account / Account receipt → out; none / Mobile to customer / none)	$makeGraph(P_1, P_2, in, out)$ $let\ \alpha = new(FORKNODE)$ $\alpha_1 = makeLink(\alpha, transition)$ $\alpha_2 = makeLink(\alpha, transition)$ $\beta = new(JOINNODE)$ $\beta_1 = new(transition)$ $\beta_2 = new(transition)$ $connect(in, \alpha)$ $makeGraph\ (P_1, \alpha_1, \beta_1)$ $makeGraph\ (P_2, \alpha_2, \beta_2)$ $connect(\beta_1, \beta)$ $connect(\beta_2, \beta)$ $connect(\beta, out)$

The workflow modeling technology presented in this paper is the ASM semantics adopted from the activity diagram. The existing methods applying Pi-calculus and FSP semantics express only the semantics of the UML activity diagram itself, and is not extended modeling techniques in accordance with the workflow characteristics.

No study on the activity diagram has been extended and customized based on semantics. Up until now, UML activity diagram itself has been expressed by semantics only.

The studies on the part of the UML activity diagram has been expressed partly by the semantics. Jablonske and Bussler discussed the main components of a workflow model [12]. They discussed them as perspectives of the workflow model. There are five perspectives which are so important that every workflow model should obtain.

The function perspective describes the functional units of a work process that will be executed, and the operation perspective describes how a workflow operation is implemented. The Behavior perspective describes the controlled flow of a workflow, and the information perspective describes the data flow of workflow. Finally, the organization perspective describes who has to execute a workflow or a workflow application.

Table 6. Comparison Our Method with The Other Methods

Methods Perspective		Lee et. al. method	Eshuis's study(Petri-Net method)	Yang's study(applying Pi-calculus semantics)
function	dynamic	support	supprot	support
	Goal- driven	the exit of activity daigram is not goal-approach	-	the exit of activity diagram is not goal approach
behavior	conditions and constraints	express the pre-condition and post-condition	the lack of expression	express the constraints using transition
	timing	define the semantics but node notation is not supported	support the perpect timing elements	the lack of the timing elements
information	data flow	data flow using input and output data	data flow is not supported	data flow is not supported
	event- driven	itself is event-driven	event-driven because event modeling	itself is event-driven
organization	participants	expressed by swimlane	participants are not supported	express by swimlane
	resources	use the resource node	resources are not supported	don't use the resources notation

In this paper, we evaluate those of four perspectives mentioned above except the operation perspective. The perspectives include function, behavior, information and organization perspective. Because operation perspective describes how a workflow operation is implemented and bring to focus on the implementation.

The evaluation is presented in the table 6. We compare our method with Petri-net method proposed by Eshuis's study [13] and with the activity diagram method applying Pi-calculus semantics proposed by Yang [9].

6 Conclusion

UML activity diagram has been well-known for describing systems of dynamic behaviors. It has been usefully applicated to model business process and workflow modeling. In this paper, the alternative approach of using Abstract State Machine(ASM) to formalize UML activity diagrams is presented. Therefore, activity models can be of rich process semantics. It is suggested that ASM semantics is extended in correspond with workflow system characteristics in the mobile healthcare B2B.

Through the exact definition on the formal semantics based on ASM, it is possible to model the workflow effectively. If we use our method, we will use the more usable and treat the exception of various variations occurred during workflow accomplishment procedure.

References

1. Fischer, L.: 2003 workflow handbook. Workflow Management Coalition(2003)
2. Kumar, A., Zhao, L.: Workflow Support for Electronic Commerce Applications. Decision Support System, Vol. 32(4). (2002)265-278
3. Chang, E., Gautama, E., Dillon, T.S.: Extended Activity Diagrams for Adaptive Workflow Modeling. IEEE 2001(2001) 413-418
4. Roberto W. S., Rodrigues.: Formalising UML Activity Diagrams using Finite State Processes. UML2000 Workshop(2000)
5. Han, Y., Sheth, A., Blussler, C.: A Taxonomy of Adaptive Workflow Management. Proceedings of the CSCW-98 Workshop Towards Adaptive Workflow Systems(1998)
6. Borger, E., Cavarra, A., Riccobene, E.: An ASM Semantics for UML Activity Diagram. Lecture Notes in Computer Science, Vol. 1816. Springer-Verlag, AMAST 2000(2000) 292-308
7. Hausmann, J. H., Heckel, R., Sauer, S.: Toward Dynamic Meta Modeling of UML Extensions: An Extensible Semantics for UML Sequence Diagrams. IEEE 2001(2001)80-87
8. Gogolla, M., Richters, M.: Expressing UML Class Diagrams Properties with OCL. Lecture Notes in Computer Science, Vol. 2262. Springer-Verlag (2002)85-96
9. Dong, Y., ShenSheg, Z.: Using pi-calculus to Formalize UML Activity Diagram for Business Process Modeling. IEEE ECBS(2003)
10. Shen, W., Compton, K., Huggins, J. K.: A Toolset for Supporting UML Static and Dynamic Model Checking. 26th International Computer Software and Applications Conference(COMPSAC 2002). IEEE Computer Society(2002)147-152

11. Hammer, D., Hanish, K., Dillon, T. S.: Modeling Behavior and Dependability of Object-Oriented Real-Time Systems. Journal of Computer Systems Science and Engineering, Vol. 13(3). (1998)139-150
12. Jablonski, S., Bussler, C.: Workflow Management- Modeling Concepts, Architecture and Implementation. Int. Thomson Publishing. London(1996)
13. Eshuis, R., Wieringa, R.: A Comparison of PetriNet and ActivityDiagram Variants. In Proceedings 2nd International Colloquium on PetriNet Technologies for Modeling Communication Based Systems, Berlin, Germany(2001)

Detecting Water Area During Flood Event from SAR Image

Hong-Gyoo Sohn[1], Yeong-Sun Song[1], and Gi-Hong Kim[2]

[1] School of Civil and Environmental Eng., Yonsei Univ., Seoul, Korea
point196@yonsei.ac.kr
[2] Department of Civil Engineering, Kangnung National Univ., Korea

Abstract. In this paper, efficient and economical methods for water area detection during flood event in mountainous area is proposed. To accomplish this, various case studies were preformed based on SAR image processing methods with the support of additional information such as Gray Level Co-occurrence Matrix (GLCM), Digital Elevation Model (DEM), and Digital Slope Model (DSM). As a result of various test2, the case when Synthetic Aperture Radar (SAR) image was classified with DSM applied by MIN filter gave the best performance, even in small streams of different elevation categories in mountainous terrain.

1 Introduction

Numerous investigations have been carried out to examine the capabilities of microwave sensors for water area mapping and monitoring flooding area. In the previous works, since target areas were performed in relatively flat terrain that has not serious radiometric distortions, it was relatively easy to identify water extent (Giacomelli, 1995; Birkett, 2000; Liu et al., 2002; Costa, 2004). However, the radiometric distortions in SAR imagery increase significantly in mountainous areas, which should be corrected by a backscattering model or other additional information for better classification results.

When the accurate DEM is available, it may be possible to correct topographic effects by backscattering model, but it requires huge amount of time and complicated procedures. Currently most DEM data, however, contain some uncertainties in height information and it is conceivable that more noise could be introduced when local slopes are calculated. In rugged terrain areas such as Korean peninsula radiometric slope correction technique may not completely remove those errors due to very steep slopes (Goering et al., 1995; Goyal et al., 1998; Sun et al., 2002). In the case of radar shadow, there is no signal to normalize, and no improvement is to be expected. Thus SAR images are of limited use for flood monitoring in high terrain relief regions.

SAR image itself is rather hard to use for the analysis of floods because of speckles noise, poor visual interpretation and single radar tonal channel, so additional data such as optical satellite images, texture information referring to the spatial distribution of tonal variation and terrain shape information can improve the classification results. Previous studies have shown that tonal classification of single-date SAR image produced poor classification results. Water area detection of SAR imagery could be significantly improved by using multi-date and multi-sensor images (Shang, 1996; Wang et al., 1998; Milne et al., 2000; Töyrä, 2002). However, multi-date and multi-sensor images are not always available. The texture information of a SAR image is a valuable characteristic for discriminating among different land cover types. Texture measures computed from GLCM have been widely used for land cover classification with optical and radar.

While numerous researches using various images or texture information have been carried out, there are few existing studies that apply SAR images and terrain information for land cover mapping. Peng and others (2003) mapped land covers in mountainous areas of southern Argentina using the texture analysis of radar imagery and a DEM generated from the same data source of a radar stereo pair. The study showed that DEM was useful for land cover mapping in mountainous areas.

In short, SAR image is very applicable in water related research, but the topographic effects by terrain relief may cause a large effect on image quality and mislead the classification results in the rugged areas. For analyzing flood event, it is critical to accurately and efficiently identify water area extent. Therefore, this research investigates an efficient water area detection method based on SAR image during flood event without clumsy and tedious procedures applying in high-relief mountainous area.

2 Area and Data Description

Every year, flood-related disasters have been the most serious and highly frequent events in South Korea and caused substantial suffering, severe losses of life, and economic damage. In 1998, a couple of typhoons, Yanni and Penny, gave tragic effects on the middle of Korean peninsula, and one of affected areas was Ok-Chun and Bo-Eun residential areas and agricultural lands. Penny had visited this area between August 11 and 12, 1998 and had caused heavy rainfalls during those days.

The characteristics of topography in these areas are narrow and high relief region, so the flash flood has been recorded frequently. The elevations over the areas are ranged from 28 to 830m, meaning high relief areas, and the maximum slope is up to 64°. A single-date RADARSAT path image(SGF), acquired on 12 August with HH polarization and standard 6 beam mode, was used for this study. Fig. 1 shows RADARSAT-1 amplitude image and the location of the study area.

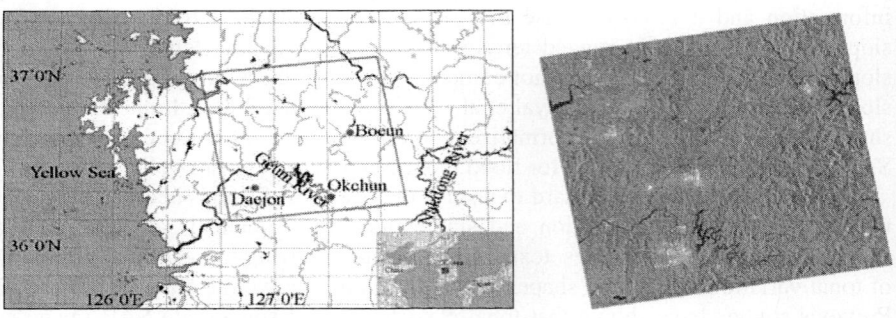

Fig. 1. Study area RADARSAT-1 image on the study area

3 Method for Water Area Detection

As shown in Fig. 2, we tested five cases for the most efficient water area detection applying in mountainous areas; (i) preprocessed SAR image, (ii) GLCM texture measures from preprocessed SAR image (iii) preprocessed SAR image performed radiometric slope correction, (iv) preprocessed SAR image with DEM and (v) preprocessed SAR image with DSM. Preprocess encloses general SAR image processing methods to correct radiometric and geometric distortions of SAR imagery.

Generally water areas are highly possible to have relatively low slope, elevation and radar backscattering coefficient, so water flow direction during a flood event is also greatly affected by the shape of terrain. Therefore, we expect that terrain information is a useful data source for the improvement of water area detection during a flood in mountainous area. The results from SAR image with both DEM and DSM together were similar to the using DEM only as additional information, so the results are excluded here. The maximum likelihood method was applied for the classification of water area. The classification accuracy was analyzed by a visual interpretation, the ratio graph and an error matrix constructed using land use maps and Annual Disaster Report (1998) provided by National Disaster and Prevention and Countermeasure Headquarters.

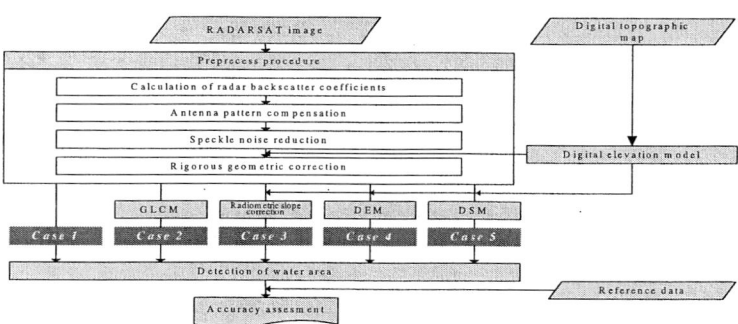

Fig. 2. Schematic diagram for water area detection

Preprocess Procedure

The task of calculating σ^0 (dB) from digital numbers (*DN*) is necessary for quantitative analysis of radar imagery and can be performed by the Equation (1).

$$\sigma_j^0 = 10 log(DN_j^2 + A_0)/ A_j)+10 log(sin(I_j)) \tag{1}$$

A_0 and A_j are the automatic gain control factors, and I_j is the incidence angle of each pixel across the range direction. The calculated backscattering coefficients ranged from −24.2 ~ 6.7 dB and water areas usually have low dB values around −20dB. Most of the dark areas in Figure 1 correspond to water or shadow areas.

The antenna pattern causes different pixel values for same or similar objects in SAR image, and the radiation distortion occurs along the range, that is to say, the central part of the radar image is brightest and its brightness gradually decreases from the central line to its two sides. The second polynomial method was performed for antenna pattern compensation.

The most distinctive characteristic of SAR is speckles. The presence of speckle in an image reduces the detectability of ground targets, obscures the spatial patterns of surface features, and decreases the accuracy of automated image classification. There are a number of filters to reduce speckle noise in SAR image, and Rio and others (2000) analyzed the behavior of various speckle filters according to window size and iteration for RADARSAT images. The study showed that two iterations with a square window size of three were not enough to reduce speckle noise in RADARSAT data. Three iterations for Lee-Sigma filters presented a significant improvement against the unfiltered image. But four iterations improved classification accuracy, however, this was not statistically better and there was greater loss of resolution and edges. Therefore Lee-sigma filter with 3×3 window size was applied three times for the radiometrically correct the image in this study.

Interpretations of SAR images in area with high relief require rigorous geometric correction if one is to be able to perform meaningful multi-source analysis using images acquired with different geometries and geographical data. Accurate geometric correction is immediately necessary to flood monitoring, but no less important is the rapid geometric correction for near real time data process. For this, we applied a single control point geometric correction method using systematic shift error calculated from ephemeris data. The systematic error in satellite orbit can be determined by extrapolation of Kepler elements provided by header information. The RADARSAT image was transformed into map geometry, Transverse Mercator (TM) coordinate system on Bessel 1841 ellipsoid. The geometric corrected image has 20m spatial resolution, which is the same as DEM resolution. The geometric accuracies for 16 check points are 1.5 pixels for range direction and 1.8 pixels for azimuth direction, respectively.

Radiometric Slope Correction

The gray value of the SAR imagery is the backscatter of the ground object recorded in the image. The radar equation describes the relationship among the energy by radar, systematic parameter and ground object parameter as follows:

$$P_r = \frac{P_t G^2 \lambda^2}{(4\pi)^3 R^4} \sigma^0 \Delta A \qquad (2)$$

P_r denotes the energy received, P_t is the energy emitted, λ is the radar wavelength, R is the distance between antenna to the scattering area, G is antenna amplifier, and σ^0 is backscattering coefficient for the resolution unit ΔA. In Equ. (2), ΔA is defined as the Equ. (3), when the ground is flat without terrain relief or spherical earth.

$$\Delta A = \frac{\delta_r \delta_a}{\sin(\eta)} \qquad (3)$$

η is the incidence angle, and δ_r and δ_a are the slant range and azimuth pixel spacing, respectively. In reality, the resolution unit varies resulted from terrain relief according to the ground slope and the azimuth direction when it acquired. The changed resolution unit ($\Delta A'$) is calculated by Equ. (4).

$$\Delta A' = \frac{\delta_r}{\sin(\eta - \theta_r)} \frac{\delta_a}{\cos(\theta_a)} \qquad (4)$$

θ_r is the tilt of the surface in the range direction and θ_a is the tilt of the surface in the azimuth direction. Therefore, the error introduced by using Equ. (3) instead of Equ. (4) is the value of ratio $\Delta A / \Delta A'$. Fig. 3 illustrates the geometry of resolution unit by local terrain relief. The backscattering coefficient was corrected up to 6 dB in high relief areas by Equ. (2), (3), (4) and 20m spatial resolution DEM generated from 1/5,000 scale digital topographic maps.

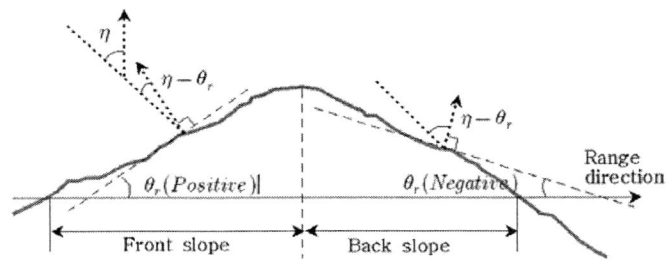

Fig. 3. Geometry of resolution unit

GLCM Texture Measures

Water areas with larger sizes and smoother textures tend to show lower accuracy when a window size for generation of GLCM increases. If the window size is too large, the misclassification seemed to be occurred at the borders of the adjacent. Also, excessively large windows tend to interfere with determination of small streams that have linear shapes. The window size and the combination of texture measures have been considered as two important factors that affect water area detection results using texture channels. To determine which combinations of texture measures produce better results at what window size, we performed many experiments on various combinations of texture measures at different window sizes. The final three texture information selected are the mean, the contrast, and the variance at window size 3×3 for water area classification.

Terrain Information

DEM was generated from 1:5,000 scale digital topographic maps supported by Korea NGII(National Geographic Information Institute). The overall accuracy of the 1/5,000 scale digital topographic map compiled photogrammetically from 1:20,000 scale aerial photos are about 2m horizontally and vertically. The spatial resolution of the final resampled DEM is 20m. The DSM was calculated from DEM and has the same pixel spacing as DEM. If DSM is used for water area detection, normal water extent can be detected effectively but inundated areas by a flood may be lost. In order to resolve the problem, MIN filter that selects the smallest value within the filter window was applied to the DSM. Because the window size and the number of iteration are very critical factors for the best result, we performed many experiments on various the number of iteration at different window sizes. Consequently, the MIN filter with two iterations and widow size 3×3 applied to the original DSM to regard for water boundary expansion factor by a flood.

Water Area Detection

The supervised classification of maximum likelihood method was applied for the detection of water area. Out of many classification methods, the prevailing method in SAR is a neural network algorithm, because it is not affected by the statistical distribution of image data. SAR normally shows the Rayleigh distribution. There were,

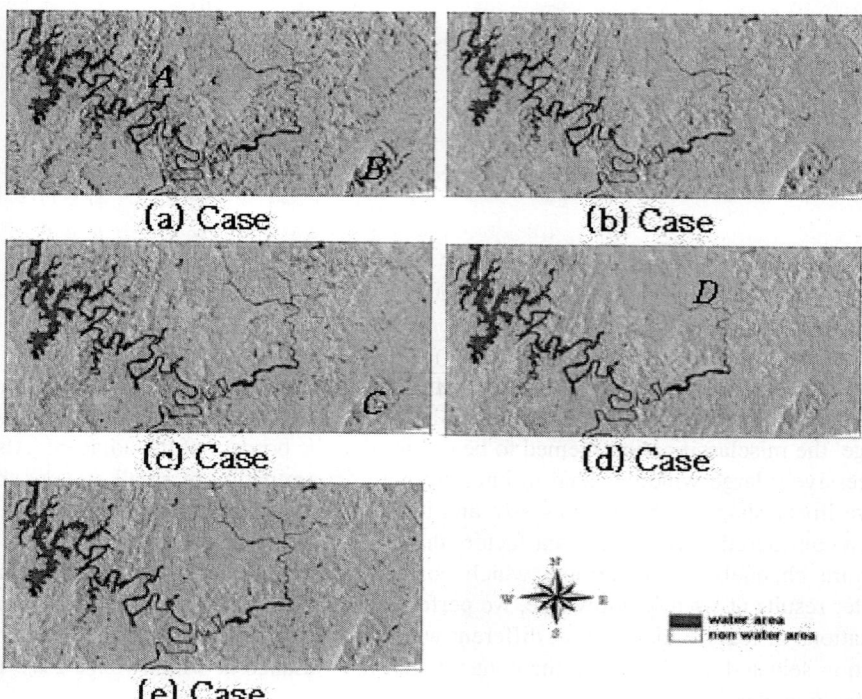

Fig. 4. Results of water area detection

however, frequently met some technical obstacles to apply due to long process time and the divergence in training processes. The method used here is the maximum likelihood method, one of the most commonly used in classification technique in satellite imagery. The rationale of this method here is that the preprocessing through spatial filtering not only reduced the variance but also the skewness of the distribution to an extent that a Gaussian distribution applies (Lee 1981; Giesen, 2000).

For input images of classification process, false color composite images were generated by layer stacking operation in case 2, 4, 5. Case 2 was combined three textures that are mean as red, contrast as blue and variance as green band. In case 4 and 5, preprocessed RADARSAT image was assigned to a green band, DEM or DSM to a red band, a blank layer that has constant value to a blue band, respectively.

The classification for water area detection was performed with 22 training sites. The training sites for classification of water and non-water areas were 12 and 10 sites over an image, respectively. Since this study was merely focused on identifying water or non-water areas, water areas accumulated in various terrain areas were carefully selected over the region. The final results of water area detection are shown in Fig. 4.

4 Numerical Results

The reference data for accuracy estimation were land use map and Annual Disaster Report of NDPCH. We generated manually ground truth ROI(Region of Interest) from the reference data to analyze detection result and constructed an error matrix as shown in Table 1.

Table 1. Error matrix of each case

	Prod. accuracy (%)		User accuracy (%)		Overall accuracy (%)	Kappa coefficients
	Water	Non-water	Water	Non-water		
Case 1	99.25	96.02	86.54	99.80	96.67	0.90
Case 2	88.43	98.90	95.40	97.07	96.75	0.89
Case 3	97.73	97.02	89.42	99.40	97.16	0.91
Case 4	96.20	99.88	99.50	99.03	99.12	0.97
Case 5	98.20	99.66	99.54	98.68	99.36	0.98

The overall accuracy of all cases exceeds 95% and the case 5 records the highest accuracy 99.36%. The kappa coefficients results range from 0.89 to 0.98, and case 5 also records the highest kappa coefficient, 0.98. Although accuracy of case 4 is similar to that of case 5, the result images are drastically different. The detail assessments of visual interpretation on each case are as follows.

The case 1 clearly shows the errors from topographic effect such as shadow areas where the high relief areas lie. The region *A* and *B* in Fig.4 was mainly misclassified into water areas because of topographic effect. This misclassification occurred all over the study area. In case 2, although the classification errors come from topographic effect were somewhat corrected, the misclassification was occurred at the borders of water areas. Regions at the border of water areas were not classified into water area. The case 3 seems to correct most misclassification in high relief areas located in the northwestern part of the test site. The classification errors, however, are not completely removed where severe steep areas cause shadow effects, for example region C in Fig.4. Case 4 greatly improved on steep areas in the northwestern section. The result, however, did not detect water areas in high elevation and steep areas such as the upper middle section, for example region D in Fig.4. As seen on Fig.4 (d), the result significantly is corrected misclassified areas in areas caused by topographic effects such as shadow areas, but lost water area on high elevation. The case 5 gave the best classification results of all cases. This correctly classified the areas caused by topographic effects and also greatly improved on high elevation and steep areas. For better performance assessment, we plotted a chart that showed calculated water area detection ratio values according to slopes. The ratio value calculates the number of pixels with a certain slope classified into water areas to the number of pixels with a certain slope.

$$ratio\ value = \frac{the\ number\ of\ pixels\ with\ slope\ i\ classified\ into\ water\ area}{the\ number\ of\ pixels\ with\ slope\ i}$$

Fig.5 explains for all five cases and the ratio value substantially decreases in slope $0°$ to $24°$ for all cases, but the patterns of graph seemed to be different over slope $30°$. Case 1 contains all candidate pixels of water area, although it has many misclassified pixels in high slope areas. It is obvious that pixels classified as water in low slope region must be actual water areas and pixels classified as water in higher slope area are almost misclassified areas due to topographic effect. Therefore, it would be the optimal classification case that is similar to the pattern of case 1 in low slope area and has low ratio value in high slope area. In lower slope areas, case 3 and case 5 have similar patterns to case 1, but case 3 includes higher classification errors than case 5 in higher slope areas. As a result we concluded that case 5 is most possible case to classify water areas in mountainous regions based on the error matrix, visual interpretations, and the ratio analysis.

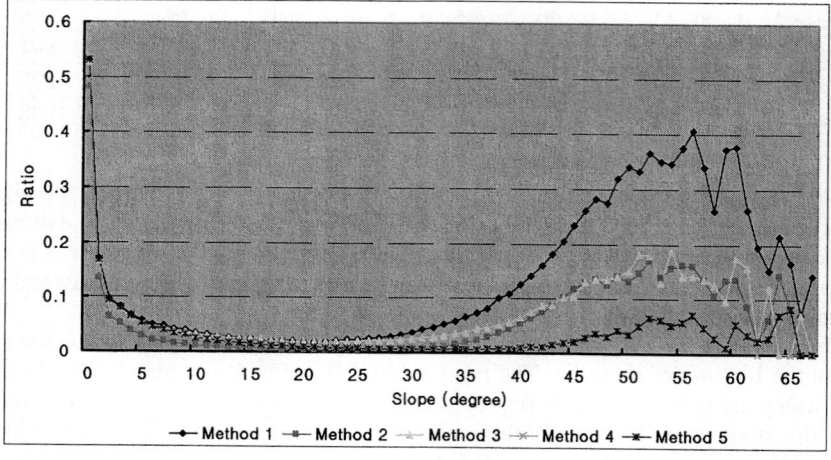

Fig. 5. Water area detection ratio according to slope

5 Conclusions

SAR images have been frequently used in flood analysis since SAR image, an active sensor image, is advantageous on bad weather conditions, such as a thick-clouded condition. This study performed the detection of water area in high relief regions during a flood event based on RADARSAT-1 SAR image.

The radiometric slope correction technique seems to correct most misclassification in high relief areas. The classification errors, however, are not completely removed where severe steep areas cause shadow effects in spite of complicated process. DEM and GLCM texture measures eliminate some steep slope areas from classified water areas, but the former mistakenly eliminates the water areas lying in high elevation areas and the latter bring about underestimation at the border of water areas. These are not appropriate for water area detection under mountainous environment. The water area detection using a SAR image is greatly improved by DSM. Accordingly, the combination of a SAR image and DSM applied by MIN filter is the most time and effort efficient method for water area detection in mountainous area during flood event.

Acknowledgements

The work was supported by grant No. NIDP-2004-14 from practical use of National Institute for Disaster Prevention.

References

1. Birkett, C. M., Synergistic Remote Sensing of Lake Chad: Variability of Basin Inundation, *Remote Sensing of Environment*, 72(2), 218-236(2000).
2. Costa, M. P. F., Use of SAR Satellites for Mapping Zonation of Vegetation Communities in the Amazon Floodplain, *International Journal of Remote Sensing*, 25(10), 1817-1835(2004).
3. Giacomelli, A., Mancini M. M., and Rosso, R., Assessment of Flooded Areas from ERS-1 PRI Data: An Applicaion to the 1994 Flood in Northern Italy, *Phys. Chem. Earth*, 20(5-6), 469-474(1995).
4. Giesen, N. V. D., Characterization of west african shallow flood plains with L- and C-band radar, *Remote sensing and Hydrology 2000 Proceedings of a symposium*, 267, 365-367(2000).
5. Goering, D. J., Chen, H., Hinzman, L. D., and Kane, D. L., Removal of terrain effects from SAR satellite imagery of Arctic tundra, *IEEE Transaction on Geoscience and Remote Sensing*, 33(1), 185-194(1995).
6. Goyal, S. K., Seyfreid, M. S., and O'Neill, P. E., Effect of digital elevation model resolution on topographic correction of airborne SAR, *International Journal of Remote Sensing*, 19(3), 3076-3096(1998).
7. Lee, J. S., Speckle analysis and smoothing of synthetic aperture radar images, *Computer Graphics and Image Processing 17*, 24-32(1981).
8. Liu, Z, F. Huang, Li, L., and Wan, E., Dynamic monitoring and damage evaluation of flood in north-west Jilin with remote sensing, *International Journal of Remote Sensing*, 23(18), 3669-3679(2002).

9. Mline, A. K. Horn, G. and Finlayson, M., Monitoring wetlands inundation patterns using RADARSAT multi-temporal data, *Canadian Journal of Remote Sensing*, 26(2), 133-141(2000).
10. Peng, X., Wang, J., Raed, M., and Gari, J., Land cover mapping from RADARSAT stereo images in a mountainous area of southern Argentina, *Canadian Journal of Remote Sensing*, 29(1), 75-87(2003).
11. Rio, J. N. R., and Lozano-Carcia, D. F., Spatial Filtering of Radar Data (RADARSAT) for Wetlands (Brackish Marshes) Classification, *Remote Sensing of Environment*, 73(2), 143-151(2000).
12. Shang, j., Evaluation of multi-spectral scanner and radar satellite data for wetland detection and classification in the Great Lakes Basin, Masters Thesis, Department of Geography, University of Windsor, Windsor, Ont. (1996).
13. Sun, G., Ranson, K. J., and Kharuk, V. I., Radiometric slope correction for forest biomass estimation from SAR data in the Western Sayani Mountains, Siberia, *Remote Sensing of Environment*, 79(2-3), 279-287(2002).
14. Töyrä, J., Pietroniro, A., Maritz, L. W., and Prowse, T. D., A multi-sensor approach to wetland flood monitoring, *Hydrological Process*, 16(8), 1569-1581(2002).

Position Based Handover Control Method

Jong chan Lee[1], Sok-Pal Cho[2], and Hong-jin Kim[3]

[1] Dept. of Computer Information Science Kunsan National Univ., Korea
chan2000@etri.re.kr
[2] Dept. of C&C Eng. Sungkyul Univ., Korea
[3] Dept. of Computer Information, KyungWon College, Korea

Abstract. It is widely accepted that the coverage with high user densities can only be achieved with small cell such as micro- and pico-cell. The smaller cell size causes frequent handovers between cells and a decrease in the permissible handover processing delay. This may result in the handover failure, in addition to the loss of some packets during the handover. In these cases, re-transmission is needed in order to compensate errors, which triggers a rapid degradation of throughput. In this paper, we propose a new handover scheme in the next generation mobile communication systems, in which the handover setup process is done in advance before a handover request by predicting the handover cell based on mobile terminal's current position and moving direction. Simulation is focused on the handover failure rate and packet loss rate. The simulation results show that our proposed method provides a better performance than the conventional method.

1 Introduction

Next generation wireless communication systems are considered to support various types of high-speed multimedia traffic with packet switching at the same time. To do that, more upgraded quality of service and system capacity are needed. Due to the limitations of the radio spectrum, the next generation wireless networks will adopt micro/pico-cellular architectures for various advantages including higher data throughput, greater frequency reuse, and location information with finer granularity. In this environment, because of small coverage area of micro/pico-cells, the handoff rate grows rapidly and fast handoff support is essential [1]. The handover algorithm for the current 3G system is based on the received signal strength. As a mobile terminal (MT) moves further away, towards the edge of a cell the received signal strength decreases. The MT continuously measures this signal strength and that of the neighboring cells at the allotted broadcast channels. The MT passes this information to the BS. If the signal strength in the current cell is lower than a certain threshold, it fines out which neighboring cell has the highest signal strength and request it to setup a session for the MT. However, in small cell areas, handovers occur more frequently and the permissible handover processing delay is smaller than in large cell areas. These would incur situations where the network cannot complete the handover before the deadline [2], which is when a MT cannot continue communications via the old (i.e., before the handover) base station (BS) because of radio signal degradation (i.e., handover failure).

A major problem arises in providing real-time services due to frequent handoffs resulting from mobility [3]. When a MT moves from one BS to another, packets arrived in the previous one are either dropped or forwarded to the new one without QoS support. The most common way to make this handoff adjustment period faster is to send all packets to the potential future BS in addition to the current BS [4-6]. When the current BS knows the position of the MT, it does not need to send packets to all of its neighboring BSs. Instead, it can specify the BSs that are in the direction of the movement and only send to them. This eliminates the packet loss in handoff and time interval between packets is reduced. This is important for satisfying quality of service for real time data.

We propose a new method that makes it possible to avoid a handover failure by performing the handover setup process in advance before a handover request in order to shorten the handover delay, in which the handover cell is selected based on the direction information from a block information database and the current position information from Global Positioning System (GPS). For predicting the MT's movement to the handover cell, we also propose the use of a block information database composed of block objects for mapping into the positional information provided by GPS.

2 Defining Location

The process of making a block object, which is the constituent of the block information database and comprise of handover cell information and so on, is described in this section. The position of a MT within a cell can be defined by dividing each cell into tracks and blocks, and relating these to the signal level received by it at that point. It is done automatically in two phases of track definition and block definition. Then the block information database is constructed with these results. The system scheme estimates in stepwise the optimal block at which the MT locates with the help of the block information database and the position information for GPS.

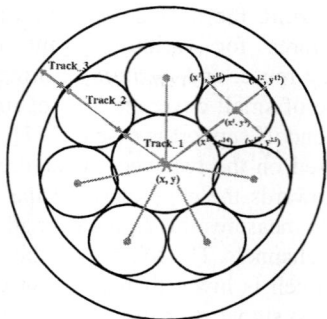

Fig. 1. Dividing a cell into tracks and Identifying the block using the vector

Three classified tracks are used to predict the mobility of the MT as shown in Figure 1. Each cell consists of track_1 as a serving cell area, track_2 as a handover cell selection area, and track_3 as a handover area, where the handover area is defined to be the area where the received signal strength from the BS is between the handover threshold and the acceptable received signal threshold. Within this area, a handover is performed to the BS with the highest signal strength.

The track_1 does not need pre-established sessions since the handover probability of the MT is very low. In track_2, the number of pre-established sessions is dynamically changed according to the MT's moving direction and neighboring cells. The track_3 needs handover. Track_2 is divided into n blocks by the location information, and block objects are created for each block. The MT's position information from GPS is valid only in track_2, and is ignored in other tracks.

The collection of block information is called the *block object* as shown in Figure 2. The *block object* contains the following information: *BlockId*, *BlockLocationInfo* indicating the information on the block's location within a cell comprised of one center point and four of area point; *HandoverCellId* indicating the adjacent cells to which a MT may hand over in this block; *NextBlockId* indicating another block within track_2 which may be traveled by a MT; *VerificationRate* indicating verification rate for the selected handover cells.

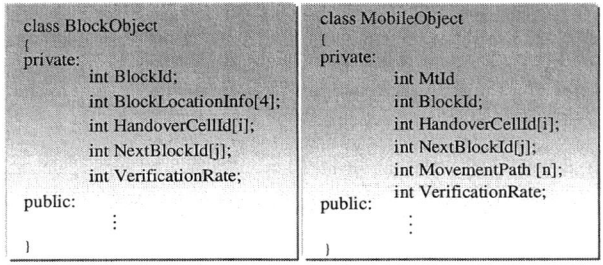

Fig. 2. Object information for a MT and a block

Each MT updates periodically his mobile object which represents his current state for handover as shown in Figure 2. The *Mobileobject* contains the following information: *MtId*, *BlockId*, *HandoverCellId*, *NextBlockId*, *MovementPath*, *VerificationRate*. *BlockId* is ID of the block in which a MT is located, *HandoverCellId* is IDs of the cells which a MT may hand over, and *MovementPath* is the moving course of a MT.

3 Direction Based Handover Method

In this section a new handover method, which is based on the direction information from a block information database described in Section 2 and the current position information from GPS, is presented. The basic principle of the prediction based handover is list as follow:

1. The position of each active MT is detected by active BS based on the use of the position information provided by GPS receiver for predicting the MT's position within a cell.
2. Handover system knows and determines the target cell for handover based on the selection of handover cells from a handover cell selection algorithm.

3. Handover system will inform each MT the information about the BS in the selected handover cells.
4. MT will search the neighboring BSs based on the handover cell information.
5. MT will be synchronized with the target BS based on the execution of the handover pre-processing procedure before handover.

3.1 Selection of a Handover Cell

The basic principle of the handover cell selection is list as follow:

1. *Position information creation and the measurement of the received signal strengths from the active cell and Surrounding cells*, the GPS engine of each MT determines his position from the triangulation based on the distance between the satellites and the GPS receiver. First candidate cell set is obtained from the measurement of the downlink channel quality of the active cell and the surrounding cells.

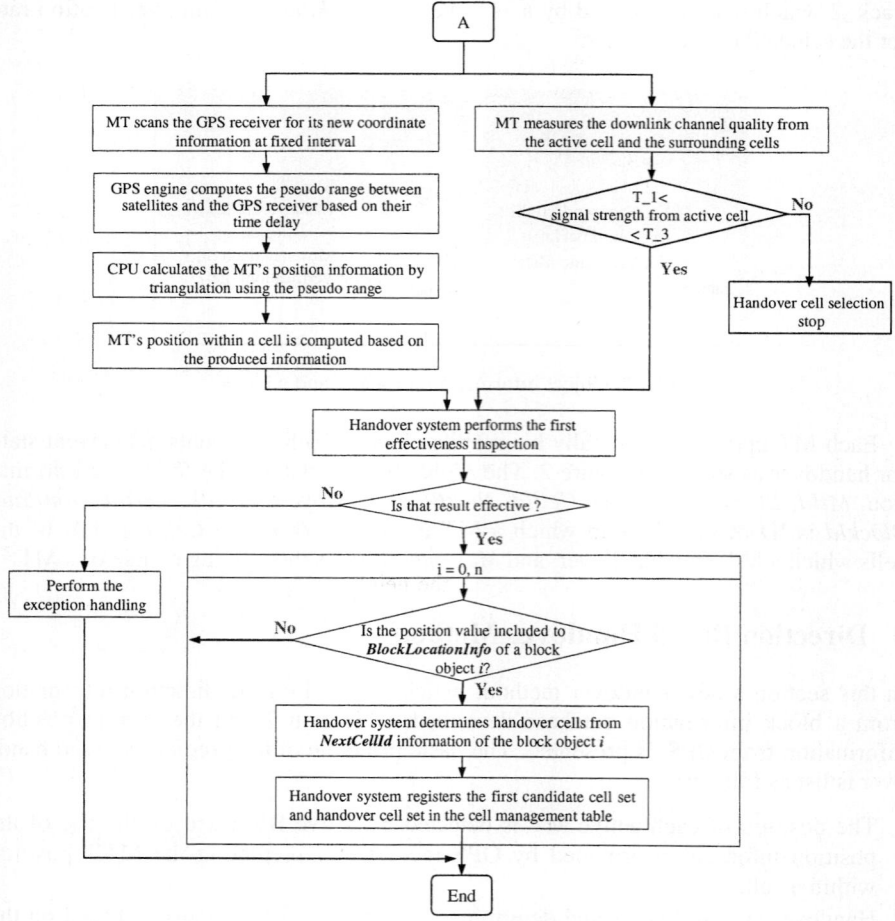

Fig. 3. Flowchart of a handover cell selection algorithm

2. *Block selection*, system selects the corresponding block object by comparing the computed position by GPS with the *BlockLocationInfo* of each block object. Handover cell set is obtained from the selected block object.
3. *The first effectiveness inspection*, the first effectiveness inspection between the first candidate cell set and the handover cell set is done. The exception handling is done if there is no correspondence between the two.
4. *Handover cell selection*, if an effectiveness inspection for the selected block is completed, handover cells are determined from *HandoverCellId* information of the block object.
5. *The registration of the handover cell information*, the handover cell information is registered to a cell management table.

Figure 3 shows a handover cell selection procedure. Each MT's position information from GPS is valid only in track_2. Therefore, the handover cell selection process is terminated if a MT is located at other tracks. The first effectiveness inspection between the first candidate cell set and the handover cell set is done, and if there is no correspondence between the two, the exception handling is performed. If one more cells are same, Handover system selects an optimum handover cell based on the resource availability.

3.2 Handover Pre-processing

Using information on the MT's handover cells determined from the above handover cell selection algorithm, two level handover process, radio level and network level, is performed.

- The radio level handover process is performed for the conversion of radio link – modem reconfiguration, synchronization setting and so forth - from previous access point to new access point.
- The network level handover process is performed for packet buffering and re-routing, for the purpose of supporting the radio level handover.

3.3 Handover Decision

For a handover decision, a MT will search the neighboring BSs using the information on the handover cells selected from the above handover cell selection algorithm. Three types of handover can be provided, namely forward handover, backward handover and reconfiguration as shown in Figure 4. A forward handover is done if the handover cell set is identical with the set of the second candidate cells, and backward handover with MAHO procedure is done if the handover cell set does not correspond with the second candidate cell set, and reconfiguration is done if the position traveled by the MT is another block within track_2. The MT reports his handover completion to a handover system through the target cell, and the handover system requests the release of the connections related with the MT. Old cell releases all the resource allocated for the MT, reports the result to the handover system.

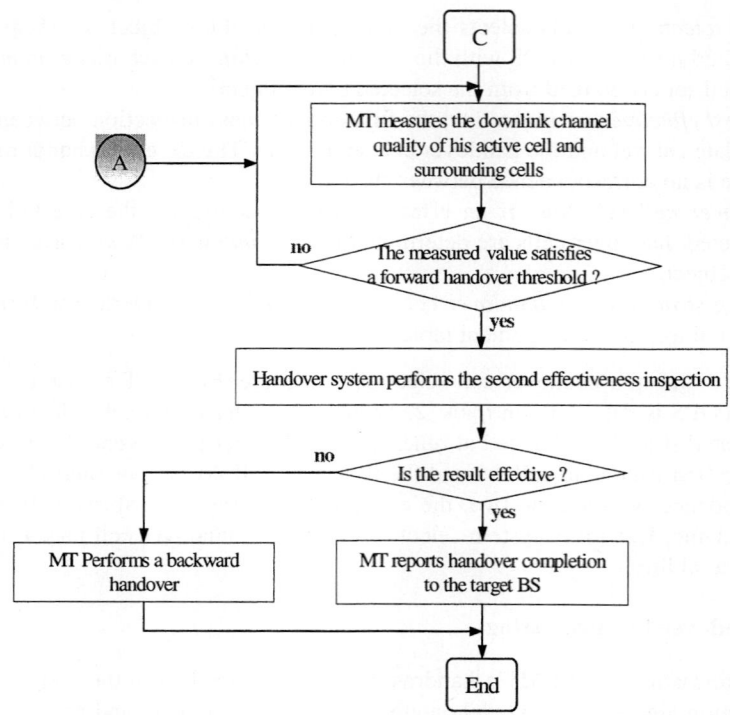

Fig. 4. Flowchart of a handover decision method

4 Simulation Result

Figure 5 shows the handover failure rate versus the session arrival rate. The solid curve represents the prediction based handover method applied and the dashed curve

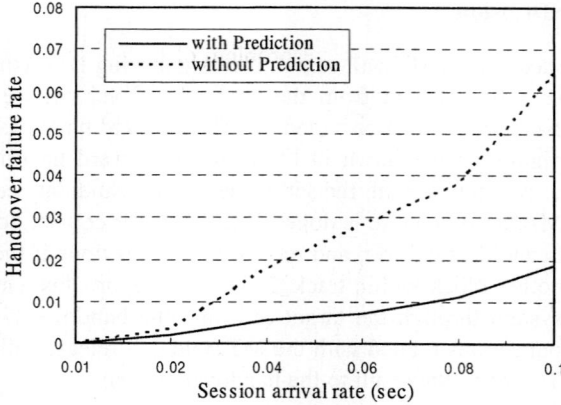

Fig. 5. The comparison of handover failure rate

represents a previous handover method. The major cause of the handover failure is because of the prediction error caused by GPS resolution error. It can be seen that the prediction error does very largely with increase in the GPS resolution error. The proposed method performs the handover setup process in advance before a handover request by predicting the handover cell based on each MT's current position so that the handover failure can be reduced.

Figure 6 shows the effect of the proposed method on packet loss rate. It is observed that the proposed method provides a noticeable improvement over the conventional scheme, because the MT has already established synchronization to the BS in target cell and switches its Tx to target BS while stop communicating with the original BS at the same time after the cell search procedure so that there will be data lost for uplink Of 2~4 frames due to the uplink synchronization and there's no data lost for downlink.

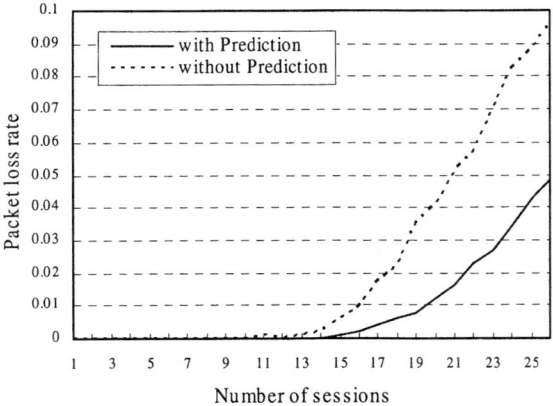

Fig. 6. The comparison of packet loss rate

5 Conclusion

This paper main goal is to address the problem of handover failure for MTs as they move from one position to another at high speeds in small cell environment. This is achieved through mobility information such as the current position and the moving direction that is presented with a set of attributes that describes the user mobility. In this scheme, the handover connection setup process is established prior to the handover request. The handover cell is predicted by the MT's position and direction and a database that includes the MT's position information. We have focused in improving the overall system performance. The proposed scheme shows a great improvement of the handover failure probability and packet loss rate. It is because our handover method is more adaptive than previous handover methods. The determination of the optimal direction should be studied consecutively. Also further researches are required on their implementation and applications to the handover.

References

1. Yu Cheng and Weihua Zhuang, "Diffserv Resource Allocation for Fast Handover in Wireless Mobile Internet," IEEE Communication Magazine, pp. 130-136, 2002.
2. GPS Based Predictive Resource Allocation in Cellular Networks
3. M. Ergen, S. Coleri, B. Dundar, A. Puri, J. Walrand, and P. Varaiya, "Position Leverage Smooth Handover Algorithm", IEEE ICN 2002, Atlanta, 2002.
4. C.L.Tan,S.PinkandK.M.Lye."A Fast Handoff Scheme for Wireless Networks", ACM/IEEE WoW-MoM ,1999.
5. B.Liang and Z. J. Hass, "Predictive Distance-Based Mobility Management for PCS Network", Proc. of IEEE INFORCOM'99, pp. 1377-1384, 1999.
6. T. Liu, P. Bahl, and I. Chlamtac, "Mobility Modeling Location Tracking, and Trajectory Prediction in Wireless ATM Networks," *IEEE J. Select. Areas Commun.,* Vol. 16, No. 6, pp. 922-936, 1998.

Improving Yellow Time Method of Left-Turning Traffic Flow at Signalized Intersection Networks by ITS

Hyung Jin Kim[1], Bongsoo Son[1], Soobeom Lee[2], and Joowon Park[3]

[1] Dept. of Urban Planning and Engineering, Yonsei Univ., Korea
{hyungkim,sbs}@yonsei.ac.kr
[2] Dept. of Transportation Engineering, Univ. of Seoul, Korea
mendota@uos.ac.kr
[3] Researcher, Korea Transport Institute, Korea
sonagi95@nate.com

Abstract. The main purpose of this paper is to present a method for determining yellow time of left-turning traffic flow based on dilemma zone concept. The results of comparative analysis show that estimated yellow times obtained from the model proposed in this paper are much different from actual yellow times obtained from the field. Actual yellow times are considerably shorter than the estimated yellow time. These results would be very significant for signalized intersection networks in terms of improving traffic safety and traffic signal operation. A method for improving yellow time of left-turning movement by Intelligent Transportation Systems (ITS) is presented.

1 Introduction

The initiation of yellow time requires all vehicles approaching the signalized intersection to be clear of the intersection conflicts area by the end of the yellow time and the, each vehicle must either stop prior to entering the intersection or pass through without stopping. If the vehicle cannot stop and cannot pass through the intersection within the yellow time scheduled, it will be caught in a "dilemma zone." Thus, yellow time is very significant in traffic signal operation in terms of preventing collision at the signalized intersection networks. While yellow time of straight moving traffic flow has been determined by using GHM (Gazis-Herman-Maradudin) formulation [1] based on dilemma zone concept, yellow time of left-turning traffic flow has been using the same value as straight moving traffic flow. However, clearance time of left-turning movement is always longer than that of straight moving traffic flow, so yellow time of left-turning traffic flow is not enough for completing safe movements at the signalized intersection. This fact mainly yields vehicles' collision between left-turning and straight moving traffic flows at the signalized intersection.

The main purpose of this paper is to improve the yellow time of left-turning traffic flow. To do this, a model for estimating the yellow time of left-turning traffic flow reasonable well has developed based on the dilemma zone concept. In this study, the model's performance has been assessed by comparing actual yellow times collected from several signalized intersections with estimated yellow times of the model.

2 Related Work

In GHM formulation, minimum yellow time, τ, of vehicles passing straight the signalized intersection at the constant travel speed has been estimated as

$$\tau = \delta + \frac{v_0}{2a} + \frac{W}{v_0} \qquad (1)$$

where δ is the driver perception-reaction time, v_0 is initial approach speed of vehicle, a is constant deceleration rate and W is width of the signalized intersection ,w, including average length of the vehicles, L. LHG (Liu-Herman-Gazis) model [2] has simply substituted v_l that is the speed limit at the signalized intersections for v_0. In the model, yellow time is consisted of entering time into the intersection, $\tau_y = \delta + v_l/2a$, and clearance time from the intersection, $\tau_r = W/v_l$, as

$$\tau = \tau_y + \tau_r = \delta + \frac{v_l}{2a} + \frac{W}{v_l} \qquad (3)$$

This model has tried to solve the problem of vehicles approaching at the speed of $v_0 < v_l$, where the vehicles are supposed to exist within the dilemma zone by the end of the yellow time.

William [3] has attempted to reduce the yellow time estimated by GHM Model, since longer yellow time is not good for improving traffic signal operation. His model is:

$$\tau = \delta + \frac{v_{0.85}}{2a_{0.85}} + \frac{W}{v_{0.85}} - \left(\delta_{cross} + \sqrt{\frac{2d}{acc}} \right) \qquad (2)$$

where $v_{0.85}$ and $a_{0.85}$ are the 85th percentile of approach speed and deceleration rate, respectively, δ_{cross} is the driver perception-reaction time of cross-flow traffic, d is the distance between vehicles and cross-flow traffic, and acc is maximum acceleration rate of cross-flow traffic which was assumed as 4.9 m/sec^2.

LOY (Liu-Oey-Yu) model [4] has applied the trajectory of left-turning vehicles for reflecting various geometric conditions of signalized intersection as (refer to Fig. 1)

$$\tau_L = \tau_{yL} + \tau_{rL} = \frac{\tau v_0}{(v_0 + v_l)/2} + \frac{S}{\bar{v}}$$

$$= \frac{2(\delta + v_0/2a)}{1 + v_0/v_l} + \frac{\beta S_{max} + (1-\beta) S_{min}}{\min[\sqrt{\gamma g S/\theta}, \Phi v_l + (1-\Phi) v_{l-turn})]} \qquad (4)$$

where τ_L is yellow time of left-turning traffic flow, S is the length of the curve section measured from the stop line to L feet ahead of the clearance line, $\bar{v}$ is the average speed of the turning vehicles, $v_{l\text{-}turn}$ is the speed limit of turning movements, g is the acceleration due to gravity, θ is turning angle and β, γ, Φ are coefficients. In the equation, S_{min} and S_{max} are:

$$S_{min} = \sqrt{(w_t + L)^2 + w_l^2 + 2(w_t + L) w_l \cos \theta} \qquad (5)$$

and

$$S_{max} = (w_t + L) + w_l \qquad (6)$$

This model employs many parameters which are not readily available. Moreover, τ_{yL} is derived based on unreasonable assumption that the stopping distance of straight moving vehicles and that of left-turning vehicles are the same. In fact, the stopping distance of left-turning vehicles is relatively shorter than that of straight moving vehicles.

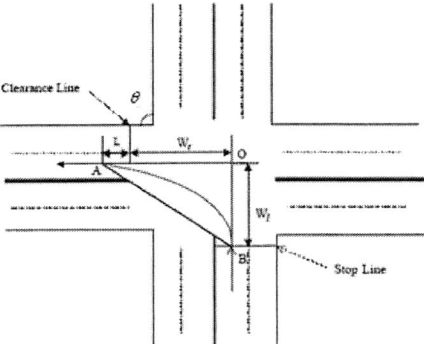

Fig. 1. Typical geometry of signalized intersection for LOY model

3 Proposed Model

3.1 Basic Assumptions

In this study, a model for determining yellow time of left-turning traffic flow based on dilemma zone concept was developed to improve the limitations of the existing models. To do this, several assumptions have been employed for the model formulation. The assumptions are as follows (refer to Figure 2); (1) there is an exclusive lane for left-turning traffic flow, (2) left-turning traffic flow decelerates constantly for approaching the stop line associated with Section B-D and accelerates constantly for traveling Section A-B, and (3) left-turning traffic flow approaches the intersection with constant deceleration rate within Section C-D and with critical deceleration rate within Section B-C for making complete stop at the stop line.

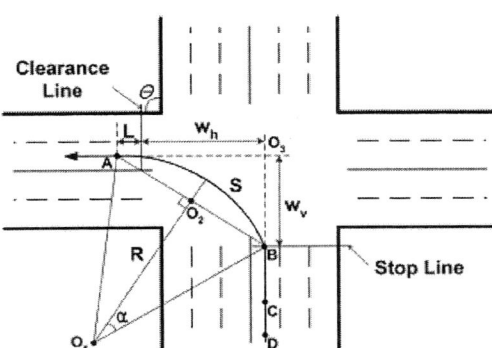

Fig. 2. Typical geometry of signalized intersection for the model proposed

The yellow time for the left-turning traffic flow of the model is consisted of τ_{yL} and τ_{rL} same as LOY model:

$$\tau_L = \tau_{yL} + \tau_{rL} \qquad (8)$$

3.2 Determination of τyL

τyL is the sum of driver perception-reaction time and stopping time. Left-turning traffic flow approaching the intersection at speed v0 starts to decelerate for making turning movement at Point D and reaches speed vB at Point B. Deceleration rate, aB-D, in Section B-D is maintained constant through the section. The length of Section B-D can be calculated as

$$x_{B-D} = \frac{v_0^2 - v_B^2}{2a_{B-D}} \qquad (9)$$

If left-turning traffic flow meets yellow phase before passing stop line, it decelerates and stops at stop line. The left-turning traffic flow approaching intersection at speed v_0 decelerates at Point D and meets yellow phase at speed v_C at Point C and then stops at Point B. The lengths of Section B-C and Section C-D can be estimated as

$$x_{B-C} = \frac{v_C^2}{2a} \qquad (10)$$

and

$$x_{C-D} = \frac{v_0^2 - v_C^2}{2a_{B-D}} \qquad (11)$$

Therefore, the length of Section B-D can be express as

$$x_{B-D} = \frac{v_0^2 - v_B^2}{2a_{B-D}} = \frac{v_C^2}{2a} + \frac{v_0^2 - v_C^2}{2a_{B-D}} \qquad (12)$$

From Equation (12), the vehicles' speed at Point C, v_C, can be estimated as

$$v_C = \sqrt{\frac{a}{a - a_{B-D}}} v_B \qquad (13)$$

When the left-turning traffic flow meets yellow time, the time passing critical distance x_{B-C} without stopping, τ_{yL}, can be expressed as

$$\tau_{yL} = \delta + \frac{x_{B-C}}{(v_B + v_C)/2} = \delta + \frac{v_B}{\sqrt{a(a - a_{B-D})} + a - a_{B-D}} \qquad (14)$$

3.3 Determination of τ_{rL} and τ_L

Left-turning vehicles approaching the stop line at speed vB start to accelerate at Point B and reach the speed vA at Point A. Section A-B is associated with the trajectory of turning vehicles traveling the curve section with the spinning radius, R [5]. Assuming the super-elevation, e, of the curve section is zero, the spinning radius, R, can be calculated as

$$R_{e=0} = \frac{\bar{v}_L^2}{g\left(\frac{0.01e+f}{1-0.01ef}\right)} = \frac{\bar{v}_L^2}{gf} \qquad (15)$$

where $\bar{v}_L$ is average value of the two speeds v_A and v_B. With R, the length of Section A-B, S, can be expressed as

$$S = \frac{\pi R \alpha}{90} = \frac{\pi \bar{v}_L^2 \alpha}{90 \, gf} \qquad (16)$$

Side friction factor, f, is selected using the relation graph between turning speed and side friction factor [5]. Then, α can be explained as

$$\alpha = \sin^{-1}\left(\frac{\sqrt{W_h^2 + w_v^2 + 2W_h w_v \cos\theta}}{2R}\right) \qquad (17)$$

where $W_h = w_h + L$ (refer to Figure 2).

The clearance time, τ_{rL}, that the left-turning vehicles completely pass Section A-B is

$$\tau_{rL} = \frac{S}{\bar{v}_L} = \frac{2S}{v_A + v_B} \qquad (18)$$

Therefore, total yellow time of left-turning traffic flow, τ_L, is as follows;

$$\tau_L = \delta + \frac{v_B}{\sqrt{a(a - a_{B-D}) + a - a_{B-D}}} + \frac{2S}{v_A + v_B} \qquad (19)$$

3.4 Determination Dilemma Zone of Left-Turning Traffic Flow

When the left-turning vehicle that was decelerating and traveling at the speed v_C can stop at the stop line without entering intersection, the distance from the stop line, x_{cL}, is then,

$$x_{cL} = \frac{v_B + v_C}{2}\delta + \frac{v_C^2}{2a} = \frac{\delta(\sqrt{a(a - a_{B-D})} + a - a_{B-D})v_B + v_B^2}{2(a - a_{B-D})} \qquad (20)$$

When the left-turning vehicle can clear intersection, the distance from the stop line, x_{0L}, is

$$x_{0L} = \frac{\sqrt{a(a - a_{B-D})} + 1}{2} v_B \tau_{yL} + \frac{v_A + v_B}{2}\tau_{rL} - S \qquad (21)$$

Therefore, if x_{0L} is longer than x_{cL}, the vehicle can stop safely. However if x_{0L} is shorter than x_{cL}, the vehicle cannot stop at stop line before the end of yellow phase and also cannot clear intersection. Thus, the vehicle cannot help violating a traffic signal. In this case (i.e., $x_{0L} < x < x_{cL}$), we say that the vehicle exists within the dilemma zone. The length of dilemma zone can be estimated as

$$D_L = \frac{v_B + v_C}{2}\left[\delta + \frac{v_C^2}{a(v_C + v_B)} - \tau_{yL}\right] + \frac{v_A + v_B}{2}\left[\frac{2S}{v_A + v_B} - \tau_{rL}\right] \qquad (22)$$

$$= \frac{\sqrt{a(a - a_{B-D})} + 1}{2}(\delta - \tau_{yL}) + \frac{v_B^2}{2(a - a_{B-D})} + S - \frac{v_A + v_B}{2}\tau_{rL}$$

4 Performance Evaluation

The model's performance has been assessed for three cases. To do this, three signalized intersections located in urban area in Seoul, Korea were selected. The layouts of the three study sites and some key input values of the model have presented in Figure 3 and Table 1. In the table, the data for v_A, v_B, τ_{rL}, v_0 and τ_r were collected directly from the filed. The average length of the vehicles, L, was assumed as 5 meters, driver perception-reaction time, δ, as 1.0 second, the deceleration rate of straight moving vehicles for stopping, a, as 5 m/sec^2 [6] and the deceleration rate of left-turning vehicles, a_{B-D}, as 3 m/sec^2 [7]. Side friction factor, f, was selected from Figure 4 [7].

Table 1. Some key input values for the analysis of the model's performance

Sites	geometric conditions (units: meter and degree)				speed (m/sec)			some coefficients			
	w_h	w_v	w	θ	v_B	v_A	v_0	f	α	R (m)	S (m)
#1	21.0	15.0	27.0	90.0	6.6	6.9	9.5	0.25	0.95	18.50	35.02
#2	20.5	31.0	38.8	96.0	6.4	7.8	9.1	0.25	1.17	20.64	48.29
#2	25.7	11.0	38.8	97.0	6.9	8.0	10.6	0.25	0.76	22.73	34.52

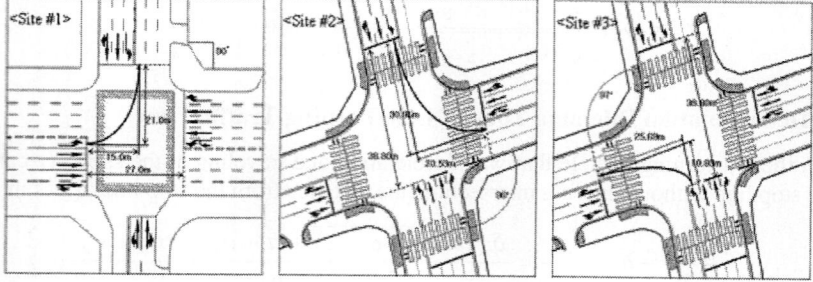

Fig. 3. Layouts of three study sites

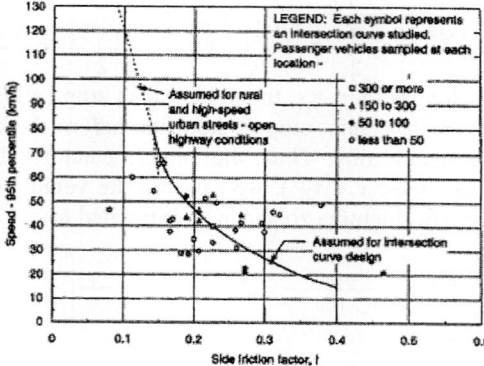

Fig. 4. Relation between side friction factor and the 95th percentile speed

In order to assess the model's performance, the estimated yellow times of left-turning vehicles and straight moving vehicles were compared with the times measured from the filed as well as the times scheduled for the signal traffic. The analysis results were presented in Table 2. In the tables, the values within ( ) were related to the estimated yellow times when the turning vehicles are approaching the intersection at the speed limit. As can be seen from the three tables, the estimated values of the straight moving vehicles, τ_r, are close to the observed reasonably well. This may be a strong evidence to support the model's soundness. Although our target is the yellow time of the left-turning traffic flow, the good agreement of the yellow times for the straight moving traffic flow indicates that the basic assumptions and parameter values employed in the model were reasonably established. With respect to the left-turning traffic flow, the model overestimated the clearance time, τ_{rL}, for the sites #1 and 2, but underestimated for the site #3. However, the difference between the estimated and the observed values ranged from 0.4 to 0.7 second. It should be noted that the difference tends to be increased as the vehicle's speed increases. This result indicates that the slow moving vehicles require longer time for safely passing the intersection than the vehicles traveling at the speed limit. One important thing to be noted here is that the scheduled yellow time which has been used for the signal operation, 3 seconds, are obviously not appropriate for both straight moving and left-turning vehicles for all cases. It seems that the yellow time currently used in the filed should be increased for traffic safety, even though the increase of yellow time may decrease the capacity of signalized intersection networks in urban area.

Table 2. Comparisons of the yellow times for field tests (unit: second)

Site #1		observed	scheduled	estimated
Left-Turning	τ_{yL}	-	-	2.3 (3.0)
	τ_{rL}	4.5	-	5.2 (2.9)
	τ_L	-	3.0	7.5 (5.9)
Straight moving	τ_y	-	-	2.0 (2.7)
	τ_r	3.3	-	3.4 (2.0)
	T	-	3.0	5.4 (4.6)
Site #2		observed	scheduled	estimated
Left-Turn	τ_{yL}	-	-	2.3 (3.0)
	τ_{rL}	6.3	-	6.8 (3.7)
	τ_L	-	3.0	9.0 (6.7)
Straight	τ_y	-	-	1.9 (2.7)
	τ_r	5.1	-	4.8 (2.6)
	T	-	3.0	6.7 (5.3)
Site #3		observed	scheduled	estimated
Left-turn	τ_{yL}	-	-	2.3 (3.0)
	τ_{rL}	5.0	-	4.6 (3.0)
	τ_L	-	3.0	7.0 (6.1)
Straight	τ_y	-	-	2.1 (2.7)
	τ_r	4.8	-	4.2 (2.6)
	T	-	3.0	6.2 (5.3)

This paper has presented a model which is able to estimate the yellow time of left-turning vehicles at the signalized intersection. The model has required some parameters which are not readily available, so it may expected that the model will not be good for traffic engineers to estimate the yellow time reasonably well for the practical purposes. However, it does not necessarily mean that the model is not sound. The model has showed quite promising performances, so the model is good enough for the analysis purpose. Nevertheless, this study has presented another method for improving the yellow time for the practical purpose. That is how to adjust the yellow time considering the time-varying traffic conditions. Recently, signalized intersections in major cities have been implemented by many kinds of vehicle detectors which were developed based on the various detection technologies included video image detection, radar, Doppler microwave, passive acoustic, and a system based on inductive loops. Thus, we can obtain traffic data for the intersection from the vehicle detection facilities. By using the traffic data, it is possible to estimate the actual times of left-turning as well as straight moving vehicles required for passing the intersection under various traffic situations. Previously, traffic data collected from the vehicle detectors have been used for only the purpose of traffic signal operation. However, the data can be very usefully used for improving the yellow time of signalized intersection. The locations of detector installation will be determined based on traffic signal phases. The yellow time can be estimated by simply measured the time difference, called as *trigger time-lag*, between the pair of vehicle detectors when the two detectors were first trigged by the vehicles moving during the same signal phase.

5 Conclusions

A model for estimating the yellow time of left-turning traffic flow was presented in this study. The model has showed quite promising performances, so the model is good enough for the analysis purpose. By using the model, the appropriateness of yellow time traditional used for traffic signal operation have been were assessed based on the results obtained from both the model and field observations. It was confirmed that the scheduled yellow time, typically used 3 seconds, were obviously not appropriate for both straight moving and left-turning vehicles under all traffic conditions; and the slow moving vehicles require longer time for safely passing the intersection than the vehicles traveling at the speed limit. To solve this problem, this study has presented a method for improving the yellow time for the practical purpose considering time-varying traffic conditions. That is to estimate the yellow time by simply measured the time difference between the pair of vehicle detectors when the two detectors were first trigged by the vehicles moving during the same signal phase.

References

1. Gazis, D. C., R. Herman, and A. Maradudin, The problem of amber signal in traffic flow. Operation Research, Vol. 8, 1960, p.112-132.
2. Liu, C., R. Herman, and D. C. Gazis, A review of the yellow interval dilemma. Transportation Research A, Vol. 30, No. 5, 1996, p.333-348.

3. Williams, W. W., Driver behavior during yellow interval. Transportation Research Record No. 644, TRB, National Research council, Washington, D.C., 1977, p.75-78.
4. Liu, C., K. Saksit, H. Oey, and L. Yu, Determination of left-turn yellow change and red clearance intervals. 80th Annual Meeting of the Transportation Research Board, January 2001, #01-0591
5. AASHTO, A Policy on Geometric Design of Highways and Streets, 2001
6. C. Doh, Traffic Engineering Manual - I, Korea, 2001
7. Ministry of Construction, Manual of regulation for standards of structure and facilities of Road, 1990

A Multimedia Database System Using Dependence Weight Values for a Mobile Environment

Kwang Hyoung Lee[1], Hee Sook Kim[2], and Keun Wang Lee[3]

[1] School of Computing, Soongsil University, Seoul, Korea
kwlee@chungwoon.ac.kr
[2] Asan Information Polytechnic College, Chungnam, Korea
[3] Dept. of Multimedia Science, Chungwoon Univ., Chungnam, Korea

Abstract. This paper proposes a semantic-based video retrieval system that supports semantic-based retrieval of large-capacity video data in a mobile environment. The proposed system automatically extracts content information from video data and retrieves video data using an indexing agent. The indexing agent analyzes a user's basic queries while extracting actual keywords from the queries. It then makes the meaning of key frame annotations more concrete through the use of dependence weight values. In addition, the indexing agent compares query images and database key frames through the use of the proposed binary-image histogram technique and retrieves the most similar key frame image that is displayed to the user. In the evaluation of a performance experiment, the proposed semantic-based video retrieval system shows higher retrieval performance than existing techniques in terms of scene retrieval from video data.

1 Introduction

Amid recent Internet developments, rapid progress is being made in image compression technologies as well as multimedia content service technologies. As a result, the demand for information on video data is increasing, and a wide variety of video data need to be effectively managed in order to meet the vast and varying needs of different users [1]. To ensure effective management of video data, there is a need for technologies that allow for systematically classifying and integrating information on large-capacity video data. In addition, video data need to be retrieved and stored in an efficient manner to provide different users with services that meet their varying needs [2].

Studies currently being conducted on content-based video data retrieval are largely classified into the following: 1) a feature-based retrieval scheme that uses similarity by extracting features from key frames; and 2) an annotation-based retrieval scheme in which a comparative retrieval is performed on a user's annotation that has been inputted and saved for a key frame. However, both content-based video data retrieval methods have some disadvantages.

In the feature-based retrieval scheme, retrieval is performed in such as way as to extract low-level feature information (i.e., color, texture, region, and spatial color distribution) from video data [3]. In the case of the feature-based retrieval scheme that focuses on comparative retrieval through extraction of visual features from the videos themselves to calculate similarity, extracting visual features is very important. However, correctly extracting feature information from a lot of videos is quite difficult. In addition, matching extracted feature information to a vast amount of video data is not easy in the performance of retrieval.

In the case of the annotation-based retrieval scheme, retrieval is performed in a way that by using characters, the user attaches annotations to the semantic information of individual video data whose automatic recognition is difficult [4]. This scheme offers the advantage of correctly expressing and retrieving video contents because the user can deal with video contents with annotations while watching a video. However, the user has to attach annotations to each individual video one by one through the use of characters. This not only involves a lot of time and effort, but also causes a vast amount of unnecessary annotations. In addition, it cannot achieve retrieval accuracy since many different annotators attach their own meanings to videos.

2 Video Retrieval Scheme

2.1 Feature-Based Retrieval Scheme

QBIC (Query by Image and Video Content) [5] developed by the IBM Almaden Laboratory is a system that supports image-example-based similarity queries in addition to usersketch-based queries, and color and texture queries. Since it supports both images and video data, QBIC performs data retrieval using feature information such as shot detection, shot-to-key frame creation, and object movement.

VisualSEEK [6] developed by Columbia University is an image database system supporting color and spatial queries in which images are distinguished by features (i.e., color and histogram etc.), and additional features (i.e., image area/color, size, spatial location etc.) are used for image comparison.

Venus [7] developed by Taiwan's National Chinghwa University is used for image retrieval through establishing time and spatial relationships among objects that are displayed in each video frame through the use of metadata.

In the case where feature-based retrieval has few camera techniques or scene conversions, detecting the boundary between two shots is a very challenging task. In particular, it is very difficult to detect the boundary of scenes that make up a single story. Many existing studies have been conducted on detecting the boundary between scenes and shots using color histograms and typical colors. However, they have yet to fully support semantic-based retrieval.

2.2 Annotation-Based Retrieval Scheme

The AVIS (Advanced Video Information System) [8] developed by the University of Maryland in the United States defines metadata such as objects, events, and behaviors that are displayed in a video, and proposes an effective retrieval scheme by linking the metadata with video segments.

VideoSTAR (Video Storage and Retrieval) [9] developed by the Norwegian University of Science and Technology is a database system based on the relational database model in which various attributes such as characters, locations, and events make up a metadata, a structured video data. Those attributes are classified back into basic, primary, and secondary contexts that allow for the easy reusing and sharing of metadata, and enable the user to easily configure queries through the use of fixed attributes.

In the case of the annotation-based retrieval scheme, it is difficult to derive accurate retrieval results because annotators define attributes(i.e., objects, locations, events, etc.) in a simple manner when constructing a system, and attach their own meanings to videos. In addition, inputting detailed annotations into every scene that consist of a vast amount of video data requires a lot of effort and time. This leads to many difficulties in building a system that enables retrieval of a vast amount of video data.

With hindsight, this paper presents a semantic-based multimedia database system in which an annotation-based retrieval scheme and a feature-based retrieval scheme are combined in a mobile environment.

3 System Architecture

The propose system consists of a video server, an agent middleware and a mobile client. The video server stores all information on video data and its metadata.

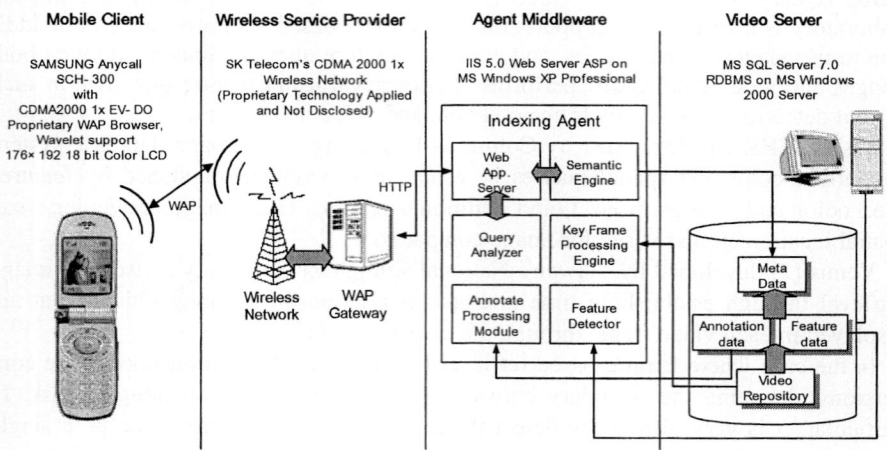

Fig. 1. The System Architecture

If a video data is inputted into the system, the system extracts annotation and feature information while updating its metadata in a continuous manner. The agent middleware processes users' queries coming in through the WAP gateway through the use of a query processor, and creates queries accessing the video server. It then receives responses for the queries to deal with key frame annotations and images, and sends the processed key frame images and WAP responses to the client. The mobile client accesses the agent middleware via wireless networks where mobile service-enabled terminals, its infrastructure, and a WAP gateway are in use. Figure 1 shows the overall architecture of the proposed system.

4 Indexing Technique

4.1 Dependence Weight Value

A dependence weight value indicates the dependence between annotations. If the user inputs a keyword, the inputted keyword and the existing dependence weight value are calculated in the order of dependence to assist the user in inputting the next keyword. The dependence weight value is added and/or subtracted by the formulas (1) and (2).

(1) Addition of Dependence Weight Values
As shown in Formula (1), the dependence is increased due to the effect of learning if there are more than two keywords inputted by the user, and if there is an existing dependence between two annotations. This is useful when a keyword inputted by another user is inputted improperly, or when a wrong keyword is selected.

$$W_{Dep\_ab} = W_{Dep\_ab} + \frac{N_{Kframe\_a \cap b}}{N_{Kframe\_a} + N_{Kframe\_b}} \quad (1)$$

- W_{Dep_ab} : Dependence weight values of the annotations a and b;
- W_{Kframe_a} : Number of key frames that have the annotation a in the whole annotation DB;
- W_{Kframe_b} : Number of key frames that have the annotation b in the whole annotation DB;
- $W_{Kframe\_a \cap b}$: Number of key frames that have annotations a and b in the whole annotation.

(2) Subtraction of Dependence Weight Values
In the case where the keyword inputted by the user contains the annotation a without any annotation b, and there are dependence weight values (W_{Dep_ab}) of the annotations a and b, these dependence weight values should be decreased as shown in Formula (2):

$$W_{Dep\_ab} = W_{Dep\_ab} - \frac{1}{N_{Kframe\_a} + N_{Kframe\_b}} \quad (2)$$

($Kframe\_b$ is a DB annotation, not a keyword inputted by the user.)

- W_{Dep_ab} : Dependence weight values of the annotations a and b;
- W_{Kframe_a} : Number of key frames that have the annotation a in the whole annotation DB;
- W_{Kframe_b} : Number of key frames that have the annotation b in the whole annotation DB;
- $W_{Kframe\_a \cap b}$: Number of key frames that have annotations a and b in the whole annotation.

If an inputted keyword combination isn't included in the DB, the addition of a dependence weight value allows for decreasing the dependence weight value in such a way as to search dependence weight values corresponding to each keyword. In this case, subtracted dependence weight values should be lower than the added ones.

4.2 Image Information Using Binary Histograms

Compared with users' image queries, this similar image retrieval uses color distributions, allowing for extracting similar images in a fast and efficient manner. However, different images with similar color distributions might be mistaken as similar images. This retrieval method can increase similar image detection rates by analyzing shape information through the use of binary image conversions and width-length histograms showing color distribution-based result frames.

To convert images extracted by key frames into binary images, the average of the entire image is calculated. Each pixel value is then converted into either 0 (black) if it is lower than the average, and 1 (white) if it is equal to or above the average. In the case of existing binary image conversions, pixel values are converted into black or white based on the value 128, which is half of the color values between 0 and 255. The existing methods cannot ensure accuracy in extracting shape information in the case of images where darker or brighter colors are widely distributed. In this case, binary images must be searched by means of the average. To draw a width-height histogram for binary images, a histogram should be drawn such that white pixels have a pixel value of 1. Skewness and kurtosis obtained in each width-height histogram can be used as feature information for image shapes. Figure 2 shows image conversion width-height histograms that are used to extract the shape information of binary images.

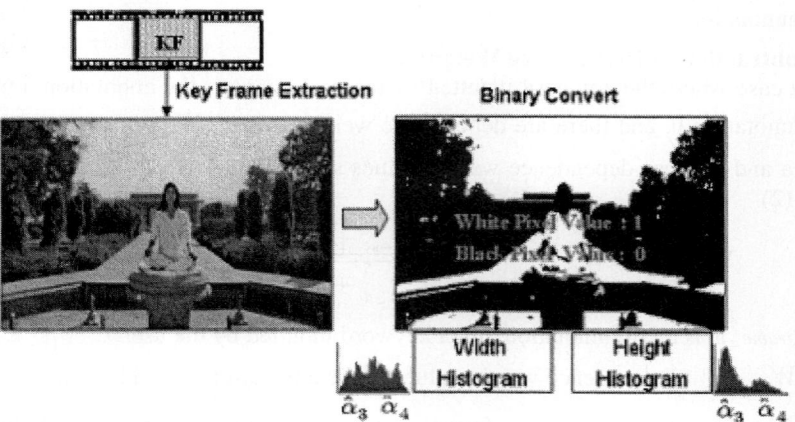

Fig. 2. Binary images conversion and width-height histograms

(1) Conversion of Query Images Into Binary Images

As shown in Formula (3), the conversion of query images into binary images is performed in a manner that finds the average of the entire query image pixel and converts pixels whose values are below the average into black, while converting pixels whose values are above the average into white.

$$Binary_{-img}(x, y) = \begin{cases} 1, & if \ (Frame \ (x, y) \geq \overline{X}) \\ 0, & if \ otherwise \end{cases} \quad (3)$$

- $Binary_{-img}(x, y)$: Value of (x, y) pixels for binary images;
- $Frame \ (x, y)$: Value of (x, y) pixels for inputted frames;
- $\overline{X}$: Average of the entire pixel for inputted frames.

(2) Calculation of skewness and kurtosis using width-height histograms for query images

A binary image displays white and black where white has a value of 1. Skewness and kurtosis can be calculated by drawing a histogram that has a frequency of the number 1 in width. The skewness of a width histogram can be calculated as in Formula (4):

$$BF_{w\hat{a}_3} = \frac{F_{-w}}{(F_{-w}-1)(F_{-w}-2)} \sum_{i=1}^{F_{-w}} \frac{(H_{-wi} - \overline{X})^3}{S^3} \quad (4)$$

- $BF_{w\hat{a}_3}$: Skewness of a width histogram for binary images;
- F_{-w} : Length of the width of a binary image;
- H_{-wi} : Frequency of the ith column in the width histogram;
- S : Standard deviation of a binary image.

Skewness is a relative measure of the asymmetry of a frequency distribution as compared to the normal distribution. If $BF_{w\hat{a}_3} > 0$, the left side is asymmetric, and if $BF_{w\hat{a}_3} < 0$, the right side is asymmetric. If $BF_{w\hat{a}_3} = 0$, both sides are entirely symmetric. A binary image histogram has white pixel values and features image asymmetry. In addition, kurtosis, which describes the peakedness of a frequency distribution, is deemed sharp in the case of higher pixel values. In the case of normal distribution, a pixel value has a value of 0. Formula (5) is used to calculate the skewness of a width histogram.

$$BF_{w\hat{a}_4} = \frac{F_{-w}(F_{-w}-1)}{(F_{-w}-1)(F_{-w}-2)(F_{-w}-3)} \sum_{i=1}^{F_{-w}} \frac{(H_{-wi} - \overline{X})^4}{S^4} - \frac{3(F_{-w}-1)^2}{(F_{-w}-2)(F_{-w}-3)} \quad (5)$$

- $BF_{w\hat{a}_4}$: Kurtosis of a width histogram for binary images;
- F_{-w} : Length of the width of a binary image;
- H_{-wi} : Frequency of the ith column in the width histogram;
- S : Standard deviation of a binary image.

5 Experiments and Evaluations

To assess the retrieval accuracy of the proposed system, user queries were performed over five hundred times. Figure 3 shows the reproduction rate and retrieval accuracy of the proposed system.

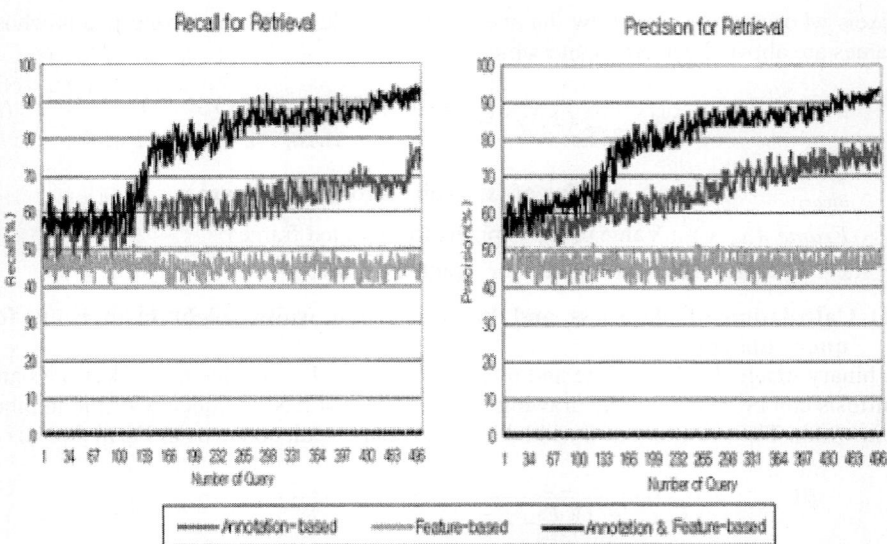

Fig. 3. The Reproduction Rate and Retrieval Accuracy of the Proposed System

As illustrated in Figure 3, the proposed system shows high reproduction rates and retrieval accuracy. As the number of query languages increases, the retrieval accuracy of the proposed system increases because the agent learns query languages as well as it updates annotation information.

6 Conclusion

In this paper, a multimedia database system that allows the user to search various meanings through the use of the annotation- and feature-based retrieval of large-capacity video data has been proposed. In the case of the performance of annotation-based retrieval, the proposed system reduces keyword errors by the user using dependence weight values, allowing the user to make queries using more accurate and concrete keywords. In addition, in the case of feature-based retrieval, this system uses the skewness and kurtosis of a histogram as key frame image features through the use of binary image conversions. It is believed that future research should be focused on the method for extracting voice from video data and performing similarity retrieval, allowing for searching video scenes using both keywords and the user's voice through application of speech recognition technologies.

References

1. N. Dimitrova, A. Zakhor, and T. Huang, "Applications of Video-content Analysis and Retrieval," IEEE Multimedia, Vol. 9 No. 3 pp. 42-55, 2002.
2. N. Dimitrova, A. Zakhor and T. Huang, "Applications of video-content analysis and retrieval," IEEE Multimedia, Vol. 9, No. 3, pp. 42-55, 2002.

3. C. W. Ngo, T. C. Pong and H. J. Zhang, "Clustering and retrieval of video shots through temporal slices analysis," IEEE Trans. on Multimedia, Vol.04, No.04, pp.446-458, 2002.
4. M. S. Kankanhalli and T. S. Chua, "Video Modeling using Strata-based Annotation," IEEE Multimedia, Vol. 7, No. 1, pp. 68-74, 2000.
5. M. Flickner, "Query by Image and Video Content : The QBIC system," IEEE Computer, Vol. 28, No. 9, pp. 23-32, 1995.
6. J. R. Smith and S. F. Chang, "VisualSEEK : A Fully Automated Content-based Image Query System," ACM Multimedia Conference, Boston, 1996.
7. T. Kuo and L. P. Chen, "A Content Based Query Language for Video Database," IEEE Multimedia, pp. 209-214, 1996.
8. S. Adali, "The Advanced Video Information System : Data Structure and Query Processing," ACM Multimedia System, Vol. 4, No. 4, pp. 172-186, 1996.
9. R. Hjelsvold, "VideoSTAR-A Database for Video Information Sharing," Ph.D. Thesis, Norwegian Institute of Technology, 1995.

A General Framework for Analyzing the Optimal Call Admission Control in DS-CDMA Cellular Network

Wen Chen, Feiyu Lei, and Weinong Wang

Network Information Center, Computer Science and Engineering Dept.,
Shanghai Jiao Tong University,
1954 Huashan Road, Shanghai 200030, China
chenwen@cs.sjtu.edu.cn
{fylei, wnwang}@sjtu.edu.cn

Abstract. This paper deals with the optimal Call Admission Control (CAC) problem in DS-CDMA cellular network supporting multiple traffic types with different Quality of Service (QoS) requirements. We present a general analysis framework to solve the problem, that is, Generalized semi-Markov Decision Process (GSMDP). It discards any restrictive unrealistic assumptions and therefore can be applied to any complex cases including non-Markovian environment. Besides, incorporating a weighted linear function of new call and handoff call blocking probabilities for each service type, we attain the goal of maximizing network revenue while minimizing the blocking probabilities. Finally, through a form of reinforcement learning algorithm known as Q-learning, the optimal policy is worked out with requiring neither explicit state transition probabilities nor any assumptions behind the network model.

1 Introduction

With the growing interest in the integration of voice, data, and video traffic in telecommunication networks, direct sequence code-division multiple access (DS-CDMA) appears increasingly attractive as the wireless access method of choice. To efficiently utilize of the scarce radio resources and at the same time provide necessary quality of service (QoS) guarantees for multiple services, it is of critical importance to design efficient call admission control (CAC) policies. Whereas the CAC mechanisms in DS-CDMA systems [1, 2, 3] rely on the unique "soft" capacity nature determined by the level of multi-access interference ratio, often characterized by signal-to-interference ratio (SIR).

Currently, many studies have been engaged in the optimal CAC problem, and Markov Decision Process (MDP) and Semi-Markov Decision Process (SMDP) as popular techniques are applied to this field [1, 4]. However, to hold Markov property, these schemes all have to make the common assumption that new/handoff users arrive according to Poisson processes and have exponential call holding time. Unluckily, recent field investigations [5] are shown that, except for the case of exponentially distributed cell residence time, the call holding time is not exponentially distributed, the handoff traffic is not Poisson, and the merged call traffic to a cell is not Poisson. In fact, due to users mobility and the irregular geographical cell shapes, the cell residence time also typically has a general distribution and therefore does not hold Markov property at all. So the analytical framework based on an Markov/semi-Markov formulation does not hold true for many systems with realistic traffic patterns, especially for the next generation wireless network. Nevertheless, suppose that we drop the classical Markov assumption, the CAC problem can be formulated as a generalized semi-Markov process (GSMP) [1]. And yet it is well known that there are not many quantitative results available [6] for GSMP. The difficulty significantly prevents GSMP from modeling the realistic stochastic discrete systems.

In this paper, we propose a novel approach to formulate and solve the optimal SIR-based CAC problem for the uplink of a DS-CDMA cellular network. The main contributions include:

1. We propose a general solution to the optimal CAC problem with any traffic models via constructing a GSMDP framework dropping any restrictive assumptions.
2. This paper transform GSMDP to a solvable problem in terms of a General State-Space Markov Chain (GSSMC). Since GSSMC satisfies the Markov property, we settle the problem through a real time Reinforcement Learning (RL) technique known as Q-learning not to care about the network dimensionality and accurate state transition probabilities.
3. At the time of maximizing the network revenue, balancing new call blocking and handoff dropping probabilities are considered by joining a weighted linear cost function. Therefore, we can find an optimal CAC policy that gives attention to both network revenue and the blocking costs.

The remaining of this paper is organized as follows. Section 2 describes a SIR-based CAC scheme in DS-CDMA cellular network. In Sect.3 we formulate the optimal CAC problem as a GSMDP framework while satisfying QoS requirements, and the Q-learning solution and its implementation issues are discussed respectively in Sect.4 and 5. Finally, conclusions and future work are given.

2 SIR-Based CAC Scheme

In DS-CDMA multicellular network, all users share the same total frequency band W Hz. The capacity of a cell varies over time with the loading of the considered cell (intra-cell) and the surrounding cells (inter-cell). Then the decision of the admission controller like [2, 3] is based on the measured bit-energy-to-noise density ratio, E_b/I_o denoted by Δ, at receiver (base station, BS).

2.1 Model of Multiple-Class Calls

Suppose the system supports data and k_V classes of voice services, where each class presents different QoS requirements and calls from the same class have the same QoS requirements. Namely, a class-l call requires a bit rate and a minimum bit-energy-to-noise density ratio, denoted by R_l and $\Delta_l^{\min}$, respectively.

Call requests are classified into handoff call and new call requests. We give higher priority to handoff calls than new calls within the same class, and a new call request of high-priority class may have higher priority as compared with a handoff call of low-priority class. If $m < n$, the admission priority of a class-m call is higher than that of a class-n call.

Hereafter, the suffix notation used in our paper are listed below:

d = data users;
v,i = voice users of the class i;
h = handoff;
k_V = the total number of voice classes, $i \in \{1, 2, \cdots, k_V\}$.

Note that if the suffix h combines with the suffix d or v,i, it represents handoff users of relevant traffic classes. And let $C = \{d, (v,1), (v,2), \cdots, (v, k_V)\}$ denote all user classes the network supports.

2.2 Network State Admissibility

We take network states in a cell to consist of the number of users of each class in the cell with s_l mobiles of class l. Let P_l denote the power of a class-l call received at the base station, and $P_{\text{intra},l}$, $P_{\text{inter},l}$ the power of the interference within the cell and from other cells. For a class-l user request, the BS will first compute the relative $P_{\text{intra},l}$ and $P_{\text{inter},l}$:

$$P_{\text{intra},l} = (s_l - 1)P_l + \sum_{j \neq l, j \in C} s_j P_j, \quad P_{\text{inter},l} = \beta \left(\sum_{j \in C} \bar{s}_j P_j \right).$$

Here $\bar{s}_j$ is the average number of class-j calls per cell surrounding the considered cell. And β is a relative other cell interference factor according to [2], which is considered to compute the received interference power from other cells. Then in order to check the network state feasible if accepting the class-l call request, BS will estimate Δ for each class since their values vary by the change of s_l:

$$\Delta_j = \frac{W}{R_j} \frac{P_j}{P_{\text{intra},j} + P_{\text{inter},j}}, \quad \forall j \in C. \tag{1}$$

Assumed that background noise is negligible. Only if the following conditions for each class j are satisfied, the network state is called admissibly:

$$\Delta_j \geq \Delta_j^{\min}, \quad \forall j \in C.$$

Note that the process above as the elementary restriction is used to construct the state space of the optimal CAC problem in the next section.

In brief, the CAC problem is described by the call arrival, traffic, and departure processes; the revenue payments; QoS metrics; and network model [8]. In this paper, the interarrival and the service time distribution of a new/handoff call are all general distributions independently different from the common Markov models. Ultimately, our goal is to find a policy π that for every system state s chooses the correct control actions so that network maximizes revenue.

3 The Proposed GSMDP Framework for the Optimal CAC Problem

In this paper, one of the main contribution is designing a GSMDP framework for the optimal CAC problem. In theory, dropping any restrictive assumptions endows GSMDP sufficient modeling power to capture many other complex discrete-event stochastic systems. The paper just utilizes it to formulate the optimal CAC problem in general traffic environment, which Markov models are not able to handle.

In [9] the definition of GSMDP is adding a decision dimension to the formalism by distinguishing a subset of the events as controllable and adding rewards based on a GSMP model of discrete event systems. Unlike an MDP/SMDP, a GSMDP remembers if an event enabled in the current state has been continuously enabled in previous states without triggering. It holds history dependence and therefore breaks the Markov property.

3.1 Description of a GSMP

First, we introduce the basic concept of a GSMP model. Its basic building components are a countable set of states S and a finite set of events E. At any time, the process occupies some state $s \in S$ in which a feasible subset $E(s)$ of the events is enabled. With each event $e \in E$ is associated a positive distribution G_e governing the time until e triggers if it remains enabled, namely sampling a residual lifetime (or clock) t_e, and a next-state probability distribution $p(s', s, e)$. The enabled events race to trigger, and the event e^* with the smallest clock value that triggers causes a transition to a state $s' \in S$ according to $p(s', s, e^*)$. At each event epoch, in state s, the ensuing interevent time $\gamma(s)$ is given as $\gamma(s) = \min_{e \in E(s)} t_e$. See [9] for details.

3.2 Constructing a GSMDP Framework for the Problem

Later development elaborates on the components of our proposed GSMDP framework one by one so as to analyze the optimal CAC problem in DS-CDMA network. Note that the admission control is performed by a BS of each cell in a distributed way. The following discussion is based on a cell.

State Space. Let S denote the state space and $s(t) \in S \subset Z_+^{1+k_V}$ denote the state of the BS at time t, where $t \in R_+$. The state vector is given by $s(t) = [s_d(t), s_{V,1}(t), s_{V,2}(t), \cdots, s_{V,k_V}(t)]^T$, where it denotes the number of active users of all classes. The other two vectors, $R = [R_d, R_{V,1}, R_{V,2}, \cdots, R_{V,k_V}]^T$

and $\Delta = [\Delta_{\mathrm{d}}^{\min}, \Delta_{\mathrm{v},1}^{\min}, \Delta_{\mathrm{v},2}^{\min}, \cdots, \Delta_{\mathrm{v},k_\mathrm{v}}^{\min}]^T$, respectively characterize the corresponding required bit rate and minimum bit-energy-to-noise density ratio of different classes users. For the current state $s(t) = s$, let $\Delta_j(s)$ denote the estimated bit-energy-to-noise density ratio of all active class-j users, and is computed by (1). The state space S in the considered cell is defined as

$$S = \left\{ s = [s_\mathrm{d}, s_{\mathrm{v},1}, \cdots, s_{\mathrm{v},k_\mathrm{v}}]^T \in Z_+^{1+k_\mathrm{v}} : \Delta_j(s) \geq \Delta_j^{\min}, \forall j \in C \right\}.$$

Obviously the interference restrictive conditions cut down the cardinality of the state space. Besides, bit-energy-to-noise density ratio threshold values may be adjusted in term of [3]; thereby the priorities among call classes are assigned. Note that since the arrival and departure of calls are random, $\{s(t)\}_{t \in R_+}$ is a finite-state stochastic process.

Events, Clocks. The set of all possible event E that can occur in a cell is defined as

$$E = \{a_\mathrm{d}, a_{\mathrm{d},\mathrm{h}}, a_{\mathrm{v},1}, a_{\mathrm{v},1,\mathrm{h}}, \cdots, a_{\mathrm{v},i}, a_{\mathrm{v},i,\mathrm{h}}, \cdots, a_{\mathrm{v},k_\mathrm{v}}, a_{\mathrm{v},k_\mathrm{v},\mathrm{h}}, d_\mathrm{d}, d_{\mathrm{v},1}, \cdots, d_{\mathrm{v},k_\mathrm{v}}\},$$

where $a_\mathrm{d}, a_{\mathrm{v},i}$ denote users' arrival, $a_{\mathrm{d},\mathrm{h}}, a_{\mathrm{v},i,\mathrm{h}}$ denote handoff users' arrival, $d_\mathrm{d}, d_{\mathrm{v},i}$ denote users depart the cell including the service termination or handoff to another cell.

We associate a real-valued clock t_e with each $e \in E(s)$. For $s \in S$, the set $C(s)$ of possible clock-reading vectors in state s is defined as

$$C(s) = \left\{ (t_{e_1}, \cdots, t_{e_j}, \cdots, t_{e_k}) \mid t_{e_j} > 0, \exists j \in [1,k] \right\},$$

assumed that the number of events being enabled in state s is k. In addition, we respectively denote $G_\mathrm{d}^\mathrm{a}, G_{\mathrm{v},i}^\mathrm{a}, G_{\mathrm{d},\mathrm{h}}^\mathrm{a}, G_{\mathrm{v},i,\mathrm{h}}^\mathrm{a}$ the general interarrival time distributions of new and handoff user, and $G_\mathrm{d}^\mathrm{s}, G_{\mathrm{v},i}^\mathrm{s}$ the general service time distributions. In general, each general distribution is represented by two useful concepts consisting of the mean of a distribution μ and the variation σ^2 (their notation is uniform with their corresponding distribution).

Before we discuss actions, policies, and rewards for the GSMDP, we introduce the following concepts [9]:

An extended state-space X: In order to maintain the Markov property, the state space S must be extended with the clock-readings of the enabled events. Thus we obtain an extended state space $X \subset S \times R_{\geq 0}^{|E|}$. We can define a Markov chain $\{(s_n, C(s_n)) : n \geq 0\}$ with state space X (it is called as a GSSMC) that corresponds to a GSMP with state space S.

The next state distribution for the GSSMC $f(x'|x)$: Given that an extended state $x \in X$, and $f_e(t)$ is the probability density function of the distribution G_e associated with event e, $f(x'|x)$ is defined as

$$f(x'|x) = p(s', s, e^*) \prod_{e \in E} \tilde{f}_e(t'_e|s', x),$$

where

$$\tilde{f}_e(t'_e|s',x) = \begin{cases} f_e(t'_e) & , \text{ if } e \in E(s') \cap (\{e^*\} \cup (E \setminus E(s))) \\ \delta(t'_e - (t_e - t_{e^*})) & , \text{ if } e \in E(s') \cap (E(s) \setminus \{e^*\}) \\ \delta(t'_e - \infty) & , \text{ if } e \notin E(s') \end{cases}.$$

Here $\delta(t-t_0)$ is the Dirac delta function with the property that $\int_{-\infty}^{x} \delta(t-t_0)dt$ is 0 for $x < t_0$ and 1 for $x \geq t_0$.

Observation Model: Since the time that an event has been enabled is known to the system, this knowledge is sufficient to provide the system with a probability distribution over extended states. Then we set up an observation model based on GSSMC. Let $O \subset S \times R_{\geq 0}^{|E|}$ be the set of observations. An observation is $o = \langle s, u \rangle \in O$, where a vector u with elements u_e for each $e \in E$ being the time that the event e has been enabled ($u_e = 0$ if $e \notin E(s)$). These two components are both observable.

Function obs: $X \times O \times S \longrightarrow O$: Given an extended state x, an observation of x, and the observable part s' of a successor x' of x, it can obtain the observation of x'. Namely, $obs(x,o,s') = \langle s', u' \rangle$, where u' consists of elements u'_e for each $e \in E$, with

$$u'_e = \begin{cases} u_e + t_{e^*} & , \text{ if } e \in E(s') \cap (E(s) \setminus \{e^*\}) \\ 0 & , \text{ otherwise} \end{cases}.$$

Actions. Given a GSMP with a event set E, we identify a set $A \subset E$ of controllable events, or actions [9]. In our GSMDP model, the set of actions A is the same with E, which can be disabled at will.

Policies. A control policy π determines which actions should be enabled at a given time in a state. We allow the action choice to depend on the entire execution history of the process, which can be captured in an observation $o \in O$. Thus a policy is a mapping from observations to sets of actions:

$$\Pi = \{\pi : O \to 2^A\},$$

where Π denotes the set of admissible CAC policies. Given an observation o, the set of actions being enabled in state s is $\pi(o)$ for the policy π.

A GSMDP controlled by a policy π is a GSSMC with $E(s)$ replaced by $E^\pi(o) = \pi(o)$ in the definition of e^*, $f(x|o)$, and $obs(x,o,s')$. The probability density function over X is defined as

$$f(x|o) = \begin{cases} \prod_{e \in E} \tilde{f}_e(t_e|t_e > u_e, s) & , \text{ if } x = \langle s, t \rangle \\ 0 & , \text{ otherwise} \end{cases},$$

where $\tilde{f}_e(t_e|t_e > u_e, s)$ is $f_e(t_e > u_e)$ if $e \in E(s)$ and $\delta(t_e - \infty)$ otherwise. The next-state distribution is defined as

$$f(x'|x,o) = p(s',s,e^*) \prod_{e \in E} \tilde{f}_e(t'_e|s',x,o),$$

where $\tilde{f}_e(t'_e|s',x,o)$ is defined as $\tilde{f}_e(t'_e|s',x)$ with $E^\pi(o)$ replacing $E(s)$ and $E^\pi(obs(x,o,s'))$ replacing $E(s')$.

Rewards. Based on the actions being enabled in a state, the network earns revenue due to the carried traffic in the cell. So, let the being enabled actions in state S associates a continuous reward rate $r(s,\pi(o))$. Moreover, we assume the transition from state S to S' triggering by the event e associates a lump sum reward $k(s',s,e)$. The expected lump sum reward if e triggers in state s is $\hat{k}(s,e) = \sum_{s'\in S} p(s',s,e)k(s',s,e)$. Since policies can switch actions at arbitrary time points, reward rates can vary over time in a given state.

Balanced Costs Payed by Blocking Operations. From user's opinion, the network needs to pay for the blocking of new and handoff call at the same time. In this paper, we weight these two blocking probabilities appropriately to reflect their relative importance following the idea of [7]. We define a weighted linear function of new call blocking and handoff dropping probability for each service type as a cost during each interevent time $\gamma(s)$,

$$c(\gamma(s)) = \sum_{j\in C} \left(w_j \frac{\mu_j^a}{\mu_j^a + \mu_{j,h}^a} B_j^{\gamma(s)} + w_{j,h} \frac{\mu_{j,h}^a}{\mu_j^a + \mu_{j,h}^a} B_{j,h}^{\gamma(s)} \right),$$

where w_j, $w_{j,h}$ respectively represent the blocking cost rate of new and handoff call request with $w_{j,h} > w_j$ expressing the fact that rejecting a handoff request is more undesirable than blocking a new call attempt, and $B_j^{\gamma(s)}$, $B_{j,h}^{\gamma(s)}$ denote the measured new call blocking and handoff dropping probability respectively during $\gamma(s)$. In particular, we set $w_{j,h} > w_j > w_{j+1,h} > w_{j+1} > \cdots > w_{k_v} > w_{d,h} > w_d$, $j \in \{(v,1),\cdots,(v,k_{v-1})\}$ to show rejecting a higher priority request costs much than rejecting a lower one. Formulating the ranking expenses due to the blocking operations for different priority classes in the reward value function, we can choose actions towards gaining revenue farthest. For convenience, we denote $G(s,\pi(o)) = r(s,\pi(o)) - c(\gamma(s))$.

Value Function. Thereupon, for a fixed policy π, the expected discounted value of an observation o over an infinite-horizon is given by

$$v_\alpha^\pi(o) = \int_X f(x|o) \left(\int_0^{t_{e^*}} e^{-\alpha t} G(s,\pi(o))dt + e^{-\alpha t_{e^*}} \int_X f(x'|x,o)\Big(k(s',s,e^*) \right.$$

$$\left. + v_\alpha^\pi(obs(x,o,s')) \Big) dx' \right) dx$$

$$= \int_X f(x|o) \left(\frac{1}{\alpha}(1 - e^{-\alpha t_{e^*}}) G(s,\pi(o)) \right.$$

$$\left. + e^{-\alpha t_{e^*}} \Big(\hat{k}(s,e^*) + \sum_{s'\in S} p(s',s,e^*) v_\alpha^\pi(obs(x,o,s')) \Big) \right) dx, \quad (2)$$

where the parameter $\alpha (0 \leq \alpha < 1)$ is the discount rate, and indicates the impact or not of future actions and their associates rewards [9]. Via (2), we give attention to both network rewards and the blocking costs; thereby we can find an optimal policy to balance them best.

In this paper, the controller objective is to determine a policy $\pi^* \in \Pi$ that satisfies

$$v_\alpha^{\pi^*}(o) = \max_{\pi \in \Pi} v_\alpha^\pi(o), \quad \forall o \in O. \tag{3}$$

Here π^* is called the optimal policy, and $\pi^*(o)$ is the set of actions chosen to be enabled in state s.

4 Our Solution Using Q-Learning

Although few available quantitative results [6] for GSMP significantly prevent its applications, our GSMDP framework is defined in terms of GSSMC (it satisfies the Markov property). At present, a number of researchers have successfully explored many approaches to solve the optimal control problems in Markov environments, such as dynamic programming with a perfect Markov model [1]. However, they require extremely large state space to model these problems exactly. Consequently, the numerical computation is intractable due to the curse of dimensionality. Also, a priori knowledge of state transition probabilities is required. Alternatively, many researchers turns to use the reinforcement learning (RL) [7, 8] to approach an optimal solution online, which avoided the above disadvantages.

In this paper we introduce a real time RL technique known as Watkin's Q-learning [10] as an efficient method for computing v^{π^*} based on a reformation of a Bellman equation. We denote Q-value, $Q^\pi(s, \pi(o))$, as the expected return starting from s, being enabled the actions $\pi(o)$, and thereafter following policy π:

$$\begin{aligned}
Q^\pi(s, \pi(o)) &= \int_0^{te^*} e^{-\alpha t} G(s, \pi(o)) dt + e^{-\alpha t e^*} \int_X f(x'|x, o) \left(k(s', s, e^*) \right. \\
&\quad \left. + v_\alpha^\pi(obs(x, o, s')) \right) dx' \\
&= \frac{1}{\alpha}(1 - e^{-\alpha t e^*}) G(s, \pi(o)) + e^{-\alpha t e^*} \left(\hat{k}(s, e^*) \right. \\
&\quad \left. + \sum_{s' \in S} p(s', s, e^*) v_\alpha^\pi(obs(x, o, s')) \right).
\end{aligned} \tag{4}$$

The object is to estimate Q-values for an optimal policy. Namely, given optimal Q-values, $Q^{\pi^*}(s, \pi^*(o)) \equiv Q^*(s, \pi^*(o))$, the optimal policy π^* is defined by

$$\pi^*(o) = \arg \max_{\pi^*(o) \in E} Q^*(s, \pi^*(o)), \quad \forall o \in O. \tag{5}$$

To learn optimal Q-values, our Q-value function is updated recursively at each decision epoch as follows. Available information being used consists of current extended state x_n, observation o_n, selecting and being enabled action set $\pi(o_n)$, the subsequent state $s'_n (= s_{n+1})$, etc. The learning rule is:

$$Q_{n+1}(s, \pi(o)) = \begin{cases} Q_n(s, \pi(o)) + \eta_n(s, \pi(o)) \triangle Q_n(s, \pi(o)) & , \text{if } s = s_n \text{ and } \pi(o) = \pi(o_n) \\ Q_n(s, \pi(o)) & , \text{otherwise} \end{cases}$$

$$\triangle Q(s, \pi(o)) = \int_0^{te^*} e^{-\alpha t} G(s_n, \pi(o_n)) dt + e^{-\alpha t e^*} \left(k(s'_n, s_n, e^*) \right.$$
$$\left. + \max_{\pi(obs(x_n, o_n, s'_n)) \in E} Q_n \left(s'_n, \pi(obs(x_n, o_n, s'_n)) \right) \right) - Q_n(s, \pi(o)). \quad (6)$$

$\eta_n(s, \pi(o)) \in (0, 1]$ is the stepsize or learning rate, n is an integer variable to index successive updates.

It has been proved in detail [10] that if the Q-value of each admissible action-state pair is visited infinitely often, and if the learning rate is decayed appropriately, then as $t \to \infty$, the above learning algorithm $Q_n(s, \pi(o))$ converges to $Q^*(s, \pi^*(o))$ with probability 1.

5 Algorithm Implementation Issues

In this paper, we simply mention some key issues in an online implementation of the Q-learning algorithm [12] for solving the optimal CAC problem.

5.1 Q-Values Representation

In practice, an important issue is how to represent and store Q-values. Currently, two different approaches, which are lookup table and neural network, are very popular to be used. Although the lookup table is the most straightforward method, it is prohibitive in the face of very large state space because it requires a huge amount of memory units. Fortunately, neural network, such as the multiplayer perceptron with a single hidden layer architecture [11, 12], can reduce the storage requirement.

5.2 Exploration

Exploration implies that each action should be executed in each state an infinite number of times to guarantee the convergence of a RL algorithm. Therefore, we may take decisions other than that with the highest action value with a small probability [11]. Or we select the being enabled actions that leads to the least visited configuration instead of selecting using (5) in order to overcome slow convergence [12].

Furthermore, to ensure the convergence of the Q-learning algorithm, we may use the Darken-Chang-Moody search-then-convergence procedure [11] to decay the learning rate η_n. Namely, we use the equation: $\eta_n = \theta_0/(1 + \zeta_n)$, where $\zeta_n = n^2/(\theta_r + n)$, and θ_0 and θ_r are constants.

6 Conclusion

This paper shows that the optimal CAC problem for DS-CDMA cellular network can be solved via constructing a general GSMDP framework and learning by a Watkin's Q-learning algorithm. In fact, the formulation is quite general and can be applied to many other cases including realistic non-Markovian cases that cannot be solved by the previous approaches. And we incorporate the QoS guarantee into the measure criteria of the optimal policy. Our future work will look for a better way to enhance the convergence rate of the learning algorithm.

References

1. Sumeetpal Singh, Vikram Krishnamurthy, H. Vincent Poor: Integrated Voice/Data Call Admission Control for Wireless DS-CDMA Systems. IEEE Transactions on Signal Processing. **50** (2002) 1483–1495.
2. Christoph Lindemann, Marco Lohmann, Axel Thmmler: Adaptive Call Admission Control for QoS/Revenue Optimization in CDMA Cellular Networks. Wireless Networks. **10** (2004) 457–472.
3. Wha Sook Jeon, D.G.Jeong: Call admission control for CDMA mobile communications systems supporting multimedia services. IEEE Transactions on Wireless Communications. **1** (2002) 649–659.
4. Bin Li, Lizhong Li, Bo Li, Sivalingam K.M., Xi-Ren Cao: Call Admission Control for Voice Data Integrated Cellular Networks - Performance Analysis and Comparative Study. IEEE Journal on Selected Areas in Communications. **22** (2004) 706–718.
5. Yuguang Fang, Imrich Chlamtac, Yi-Bing Lin: Channel Occupancy Times and Handoff Rate for Mobile Computing and PCS Networks. IEEE transactions on computers. June 1998, 47(6): 679 - 692. **47** (1998) 679–692.
6. Cao Xi-Ren, Ho Yu-Chi: Models of Discrete Event Dynamic Systems. IEEE Control Syst. Mag. **10** (1990) 69–76.
7. El-Sayed El-Alfy, Yu-Dong Yao, Harry Heffes: A learning approach for call admission control with prioritized handoff in mobile multimedia networks. IEEE Vehicle Technology Conference (VTC), Greece. (2001) 972–976.
8. Hui Tong, Timothy X Brown: Adaptive Call Admission Control Under Quality of Service Constraints: A Reinforcement Learning Solution. IEEE Journal on Selected Areas in Communications. **18** (2000) 209–212.
9. Håkan L. S. Younes, Reid G. Simmons: A formalism for stochastic decision processes with asynchronous events. In the AAAI Workshop on Learning and Planning in Markov Processes-Advances and Challenges, San Jose, CA, AAAI Press. (2004) 107–110.
10. Christopher J.C.H. Watkins, Peter Dayan: Q-learning. Machine Learning. **8** (1992) 279–292.
11. Fei Yu, Vincent W.S. Wong, Victor C.M. Leung: A New QoS Provisioning Method for Adaptive Multimedia in Cellular Wireless Networks. In Proc. IEEE INFOCOM'04, HongKong, PR China. (2004).
12. Sidi-Mohammed Senouci, André-Luc Beylot, Guy Pujolle: Call admission control in cellular networks: a reinforcement learning solution. International Journal of Network Management. **14** (2004) 89–103.

Heuristic Algorithm for Traffic Condition Classification with Loop Detector Data

Sangsoo Lee[1], Sei-Chang Oh[1], and Bongsoo Son[2]

[1] Assistant Prof., Division of Environmental, Civil and Transportation Engineering,
Ajou Univ., Suwon, Korea
{sslee,scoh}@ajou.ac.kr
[2] Associate Prof., Dept. of Urban Planning and Engineering, Yonsei Univ., Seoul, Korea
sbs@yonsei.ac.kr

Abstract. This paper presents a heuristic algorithm for detecting traffic conditions from loop detector information at signalized intersections. With use of the characteristic of occupancy data from each cycle, the algorithm determines the level of traffic conditions. Several variables were introduced for this algorithm, and the detailed descriptions of flow chart of the initial algorithm were included in this paper. The proposed algorithm has a simple logic, however, the performance of the algorithm could be improved by performing calibration or by introducing additional variables.

1 Introduction

Traffic congestion is a common problem in urban cities. Although there are some differences in levels of congestion and its duration, most cities have experienced traffic congestion on a recurring basis. Identifying of traffic conditions in real-time is crucial information for effective operation and management of traffic facility. Recently, the concept of intelligent transportation systems (ITS) enables to consider more surveillance systems that provide real-time traffic information. Consequently, a lot of surveillance systems have been installed on highways such as detectors and closed circuit television (CCTV).

Generally, a vision-based detection system provides very reliable and accurate traffic information, but it requires huge investment. Therefore, the detector-based system installed on roadway is a key equipment for traffic information collection in most surveillance systems. However, the traffic information from detector-based system is limited in direct applications due to lack of accuracy and reliability as compared to the vision-based system. The collected detector data should be processed with some types of algorithms to become informative.

At signalized intersections, a presence-type loop detector is commonly installed to collect traffic data. The detector data could be used to generate traffic signal control parameters for either actuated control systems or computer controlled systems. However, the data could be used not only for direct signal control purpose, but also for identifying real-time traffic conditions by adopting an appropriate procedure or algorithm. If detector data can be used for identifying traffic conditions, the effectiveness of management activities and system performance is significantly improved.

This paper investigates a heuristic data processing algorithm for identifying traffic conditions with loop detector data at signalized intersections. The suggested algorithm is based on the occupancy data collected from a single loop detector. By investigating the characteristics of occupancy data and resulting statistics, the status of real-time traffic conditions is determined. The optimal parameter values could be estimated from field data, but the estimation result of the parameters was not included in this paper. The key concepts and working flow of the algorithm are presented as well as the operational characteristics of the algorithm.

2 Literature Review

A study had investigated optimal type and location of loop detector to measure accurate traffic conditions influenced by traffic variation (Lee and Lee, 1996). In this study, the optimal types and location of loop detectors for application of adaptive signal control systems have been proposed for each direction of traffic flows by checking confidences of occupancy time. Estimation of link travel time using loop detector information was investigated (Kim *et al.* 1997). The study proposed three different methodologies; VPLUSKO model, method from fuzzy theory, and method from neural networks theory. The study showed that the fuzzy theory model provided the best performance in term of estimation errors.

A study presented a methodology to estimate link travel times directly from the single-loop loop detector flow and occupancy data without heavy reliance on the flawed speed calculation (Petty *et al.* 1998). It was shown that the single-loop based estimation of travel time could accurately track the true travel time through many degrees of congestion. Recently, an innovative method for speed estimation from traffic count and occupancy data was proposed (Hazelton, 2004). By assuming a simple random walk model for successive vehicle speeds, an MCMC approach to speed estimation can be applied, in which missing vehicle lengths are sampled from an exogenous data set. The test results provided quite reasonable estimation accuracy.

For freeway systems, a vehicle re-identification algorithm for consecutive detector stations on a freeway was proposed to estimate vehicle speed using vehicle length measured at dual loop speed traps (Coifman and Cassidy, 2000). In addition, a new algorithm was proposed for estimating velocity from single loop detector data (Coifman, 2001). The algorithm was simple enough for real-time application, and it could be extended to automatic tests of detector data quality at dual loop speed traps.

An estimation of traffic measurement using a signature from a single loop detector was suggested (Oh *et al.* 2001). It was identified that an individual vehicle speed could be estimated from single loop detectors by integrating existing inductive loop detectors with advanced loop detector cards. A travel time estimation algorithm was developed which used the loop detector data with theoretical considerations (Oh *et al.* 2002). It was also shown that the information from point detectors had a deficiency in estimating travel times under congested traffic condition. Test results showed that the proposed algorithm produced improved travel time estimates as compared to other methods.

3 Algorithm

When the signal turns green, traffic begins to discharge from the stop line, and a loop detector records the on or off status bits. The sum of "on" state bits is referred to as the vehicle occupancy time (May, 1990). Likewise, the duration of absence of vehicles is easily estimated by the sum of "off" status bits. In general, the patterns of these detector outputs are different according to prevailing traffic conditions. For undersaturated conditions, there is enough unused green time during a given phase, but green time is fully utilized at saturated conditions. In oversaturated conditions, occupancy time per vehicle could be increase. Therefore, it is postulated that traffic conditions may be easily identified by analyzing these patterns of the detector outputs.

The proposed algorithm is designed to identify traffic conditions by using variation of occupancy time. From loop detector outputs, two indices could be extracted for control purpose: percent occupancy time and average occupancy time. The percent occupancy time is calculated using Equation (1) (McShane et al. 1990).

$$OCC_{per} = \frac{\sum_{n=1}^{N}(t_{occ})_n}{T} \times 100 \tag{1}$$

Where OCC_{per} = percent occupancy time,

t_{occ} = occupancy time of each vehicle n,

N = number of vehicles detected during time period T, and

T = time period.

The average occupancy time is estimated using Equation (2).

$$OCC_{avg} = \frac{\sum_{n=1}^{N}(t_{occ})_n}{N} \tag{2}$$

Where OCC_{avg} = average occupancy time during time period T.

Generally, percent occupancy time will be increased as the discharge volume increases. When approaching the capacity of an intersection, the increasing pattern of the percent occupancy time versus discharge volume will be scattered. That is, the values of percent occupancy time are distributed within some ranges near capacity because the occupancy time is no longer a function of traffic volumes only. Therefore, the magnitude of percent occupancy time cannot be used as a reliable indicator of traffic conditions.

The traffic condition may be characterized by a mixture use of those two variables: percent occupancy time and average occupancy time. As traffic conditions become worse, occupancy time increases as compared to normal traffic conditions because occupancy time is an indirect measure of traffic density. Applying these characteristics of the detector outputs, the level of traffic conditions could be identified. To make the algorithm working, a set of reasonable threshold values of the two variables should be determined. Considering these characteristics, the algorithm has the following procedures.

Step 1. Check the beginning of green phase i and collect loop detector data.

Step 2. Check the end of the green phase i.

If sum of headway time is equal to green time, then compute the sum of occupancy time. Proceed to Step 3.

Otherwise, perform Step 2.

Step 3. Examine traffic flow characteristics.

Compute the percent occupancy time and average occupancy time.

Step 4. Decide traffic conditions.

Condition 1: If the percent occupancy time is smaller than α, then it is an under-saturated condition. Go to Step 1.

Otherwise check the Condition 2.

Condition 2: If the percent occupancy time is smaller than β, then it is a saturated condition. Go to Step 1.

Otherwise check the Condition 3.

Condition 3: If the percent occupancy time is greater than β and average occupancy time is greater than δ, then it is an oversaturated condition. Go to Step 1.

The flow chart of the proposed algorithm was also illustrated in Figure 1.

4 Characteristics of the Algorithm Validation

To validate the suggested algorithm, extensive data collection should be made. Currently, however, enough data were not collected for the validation of the proposed algorithm. Therefore, important concepts for the validation work are briefly described, and no performance test result is provided in this paper.

First, a set of the threshold values of the parameters should be decided. The selection of reasonable range of values is critical for the good performance of the algorithm. Several methods can be used to estimate the threshold values such as heuristic search techniques, statistical techniques, and neural network models. Each technique tries to capture the certain patterns from the data. It is recommended that the use of a heuristic algorithm unless there is clear performance difference among the techniques.

Second, it is necessary to check the minimum required number of data set. In general, as the number of data points increase, the reliability of the model will be improved. However, there may be practical limitations to obtain enough data set. Therefore, the minimum data set should be defined by using a statistical method before the field study is performed.

Third, the decision interval of the proposed algorithm is dependent on the cycle lengths for a given location. To test the algorithm, loop detector data collected from a given cycle should be sequentially processed by the proposed detection algorithm. Since the test is conducted on a cycle basis, the detection interval of the algorithm is a cycle. This operational characteristic is a good feature for demand responsive traffic signal control systems. To evaluate the performance of the algorithm, both detection rate and false alarm rate should be calculated.

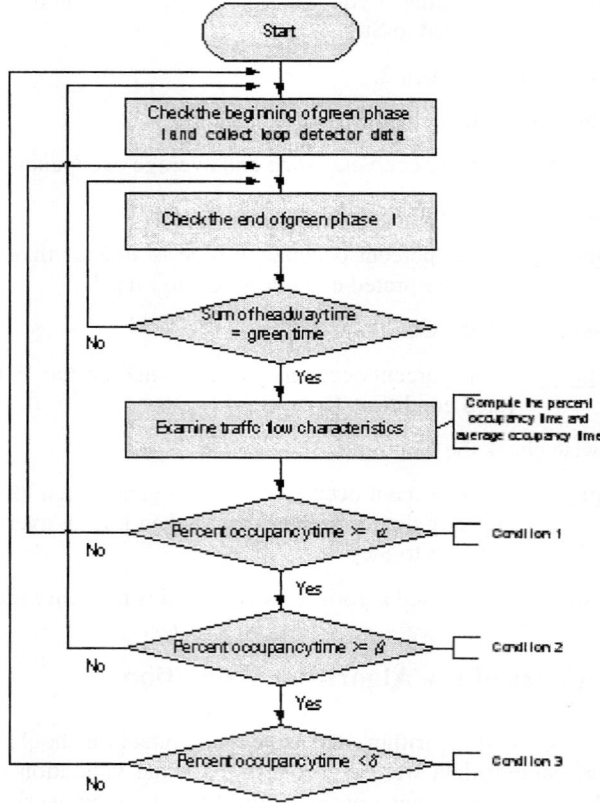

Fig. 1. Working Flow of the proposed algorithm

Last, the proposed algorithm has limited variables and a simple structure, thus, the model may have wide application areas. However, the accuracy of the estimation is also affected from other factors such as overall quality of detection systems, especially for congested traffic conditions. Thus, it is necessary to check the status of the detector system in a regular basis.

5 Conclusions

A heuristic data processing algorithm for identifying traffic conditions at signalized intersections was introduced in this paper. The logic and flow of each element of the algorithm are described. The algorithm mainly uses the characteristics of occupancy data collected from a single loop detector. The estimation of traffic conditions will be obtained through the algorithm introduced by extracting two occupancy-based variables. The model can be used for many real-time application purposes in urban areas because of a simple algorithm structure. However, it is necessary to validate the algorithm with extensive field data set. In addition, the performance of the algorithm could be improved by performing calibration or by introducing additional variables.

References

1. Seung-Hwan Lee and C. Lee, A study on the Loop Detector System for Real-time Traffic Adaptive Signal Control. Journal of Korean Society of Transportation, Korean Society of Transportation, Vol.14. No.2, pp. 59~88, 1996.
2. Yung-Chan Kim, G. Choi, D. Kim, and G. Oh, Estimation of Link Travel Speed Using Single Loop Detector Measurements for Signalized Arterials, Journal of Korean Society of Transportation, Korean Society of Transportation, Vol.15. No.4, pp. 53~71, 1997.
3. Karl Petty, P. Bickel, M. Ostland, J. Rice, F. Schoenberg, J. Jiang, and Y. Ritov, Accurate Estimation of Travel Times From Single-Loop Detectors, Transportation Research Part-A, Vol. 32, No. 1, pp.1~17, 1998.
4. Martin Hazelton, Estimating Vehicle Speed from Traffic Count and Occupancy Data, Journal of Data Science, Vol.2, No. 3, pp. 231~244, 2004.
5. Benjamin Coifman and M. Cassidy, Automated Travel Time Measurement Using Vehicle Lengths From Loop Detector Speed Traps, California PATH Research Report, UCB-ITS-PRR- 2000-12, 2000.
6. Benjamin Coifman, Improved Velocity Estimation Using Single Loop Detectors, Transportation Research Part-A, Vol. 35, No. 10, pp.863~880, 2001.
7. Seri Oh, S. Ritchie, and C. Oh, Real Time Traffic Measurement from Single Loop Inductive Signatures, California PATH Research Report, UCI-ITS-WP-01-15, 2001.
8. Jun-Seok Oh, R. Jayakrishnan, and W. Recker, Section Travel Time Estimation from Point Detection Data, California PATH Research Report, UCI-ITS-WP-02-11, 2002.
9. Adolf May, Traffic Flow Fundamentals, Prentice-Hall Inc., New Jersey, 1990.
10. McShane W. R., Roess R., and E. Prassas, Traffic Engineering, Prentice-Hall, New Jersey, 1990.

Spatial Data Channel in a Mobile Navigation System

Luo Yingwei, Xiong Guomin, Wang Xiaolin, and Xu Zhuoqun

Dept. of Computer Science and Technology, Peking University,
Beijing, P.R.China, 100871
lyw@pku.edu.cn

Abstract. With growing popularity of mobile devices and wireless communications, mobile location-based services that deliver location dependent and context sensitive information to mobile users are emerging as one of the hot topics of mobile. In this paper, we introduce a mobile navigation solution and discuss spatial data management strategies for mobile navigation service to answer the challenge of providing spatial map data efficiently. And a model, Spatial Data Channel, is devised for integrating various strategies including data preparation, transmission and management. Finally, we present some conclusions.

1 Introduction

The convergence of mobile communication and location technology is creating new opportunities for mobile services. Mobile location-based services that deliver location dependent and context sensitive information to mobile users are emerging like one of the topics in the mobile communication area [1][2]. In various services, mobile navigation service is special because location context geographic information, i.e. spatial map data, needs to be provided along with position information.

It is noted that mobile devices use small batteries for their operations, and the bandwidth of wireless communication is in general limited [3]. And moreover, it is difficult for applications on mobile devices to store huge data, i.e. more than 100MB, in local environment because of limited memory. So we cannot put the huge spatial data in local memory. How to provide spatial map data efficiently is one of the key problems in mobile navigation services. Developing proper infrastructure and spatial data management strategies for mobile navigation service has been a major challenge. To answer this challenge, we devise Spatial Data Channel to provide dynamic and efficient access to location context spatial map data.

2 Related Work

Many GIS software vendors provide various Web publishing tools that convert map files into images. Some vendors further give plug-in components to display vector maps in popular Web browsers [4]. Others embed JavaScript code or Java applets in Web pages. Among them, http://map.pku.edu.cn features integration of local and Internet data on the client side. However, they all have a common dilemma between large sizes of GIS datasets and limited bandwidth available for ordinary wireless users. As a result, only simple or small maps are available for general public viewings.

Internet transmission of vector spatial data remains a challenge [5]. Many recent researches have been conducted on how to transmit spatial vector data efficiently. Such methods take advantage of map generalization to realize the efficient transmission of spatial vector data on the Internet. According to the display scale on the client, invisible spatial objects are eliminated to reduce transmission cost and response time [6].

However, our work differs from previous works in which we consider the problem from a global view. We integrate various strategies including data preparation, transmission and management into a data channel. Especially, some efficient caching and pre-fetching strategies are introduced in this paper to enhance spatial data providing.

3 Our Mobile Navigation Solution

JNavigator, our java-based mobile navigation solution, is an integrated system including client applications and remote services. Solution architecture is shown as figure 1.

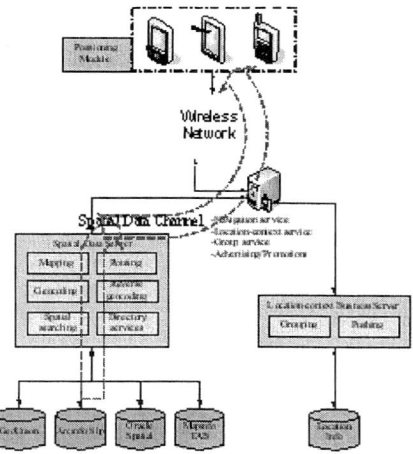

Fig. 1. Architecture of Our Mobile Navigation Solution

Client applications are independent of the target device, yet automatically can exploit specific features of the device and provide customized content depending on the availability of such features. To provide real-time location information for mobile user, we integrate a positioning module with client applications. The GPS satellite-based positioning system requires an unobstructed view of the sky for a positional fix to be made, however, in an urban environment for which a vehicles navigation service is designed for, this may not always be possible. GSM cellular-based positioning systems aren't as accurate as GPS, but have the advantage of being able to work both indoors and outdoors. A hybrid of the two can provide the solution to the system requirements for a location-calculate based tourist application. Currently, an implementation of a vehicles navigation service has been developed that uses GPS positioning technology.

In remote server, JNavigator provides a variety of portal services as listed below.

(1) Navigation. Navigation refers to real-time directions. In practice, this means instead of receiving a simple list of directions all at once, navigation services automatically provide the user with prompts of the correct directions in real-time, as he approaches road junctions and intersections. Navigation service can be combined with vehicles management and tracking services.
(2) Location-context service. This service gives answers to questions such as what's nearby, where am I, how can I get there, etc. This provides the user with access to large datasets that are contextualized by the user's location, and provide maps to a chosen location.
(3) Group service. These services link a community or individual mobile users and allow them to quickly locate each other (subject to individually predefined privacy restrictions) and subsequently communicate, meet and interact.
(4) Advertising/Promotions. This provides alerts to mobile users when they are within certain geographical areas to advertise or give discounts on a product or service available in the vicinity.

Between the portal services described above and various spatial data sources, JNavigator provides a set of geographical enabling functions that enable pre-integrated mobile-enabled services for location related queries. This enables to easily ingest different sources of location services providers worldwide using a single, consistent Java interface. These interfaces allow seamless integration with existing location service vendors. Typical functions include:

- Mapping - Map rendering. Mapping enables users with capable devices to visualize location-related data.
- Geocoding - converting a location/address into a geographic coordinate.
- Reverse geocoding - converting a coordinate into an address.
- Routing - providing a list of directions between two geocoded points. Routers might also provide maps of each turn and of the complete route. The router might also supply a list of point coordinates along the route, to enable the requesting user to perform some spatial analysis (for example, to identify which customers can be visited along the route).
- Spatial searching – searching spatial entities in specified region.
- Directory services - determine a list of businesses matching a specified region.

Location-context business services are also supplied in our solution. Typical services include grouping and pushing service.

- Grouping – different users can be grouped together to provide club-like services. A group can be established or dismissed according to its members' demands. A user can join or quit a group freely.
- Pushing - By keeping track of user's purchasing habits and current location, a very targeted advertising campaign can be performed. Mobile users are informed about various on-going specials. Messages can be sent to all users who are currently in a certain area (identified by advertisers or even by users) or to certain users in all locations.

Spatial data channel can be seen as a pipe. On one side of the pipe, huge spatial data are collected and selected according to specified strategies. After a series of processes, on the other side of the pipe, lightweight spatial data block will be provided for the client applications running on handheld devices.

4 Spatial Data Channel

Spatial data channel is a way to provide data. Through the control of precision and extension of spatial data, huge spatial data that have been arranged by spatial data channel will turn to be lightweight spatial data block. This lightweight spatial data block is preferred by mobile terminal.

The establishment of spatial data channel requires two steps for data:

- Prepare different scales of spatial data.
- Divide the spatial data into virtual partitions to control the extension.

Figure 2 describes the architecture of spatial data channel.

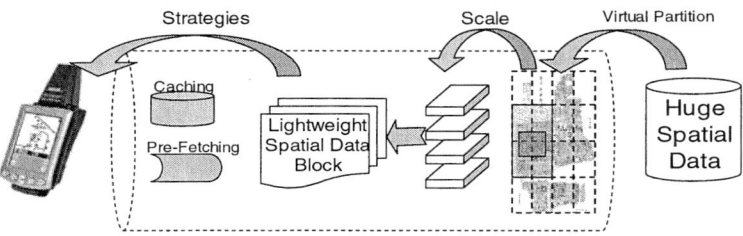

Fig. 2. Spatial Data Channel

4.1 Virtual Partition

Virtual partition means we not only partition the spatial map data into map blocks, but also operate this partition in a virtual way. Map data after the process of virtual partition is the same as before. Through a partition grid, the map data will be viewed as being constituted of many map data blocks, just as virtual partition grid layer being inserted between the map data and data visitor.

Just as the following figure 3 shows, *VP* is a virtual partition of *Real map*. *Real map* will not be divided to many blocks actually.

When receiving request from clients, according to the extension of requested area, the service will calculate which virtual partitions blocks overlapped with the area and return map data located in these blocks as the result. As figure 3 shows, client sends request *Req* and the request's area overlaps virtual partitions 1, 2, 3 and 4, so the map data in these four blocks are sent back to the client as the response.

The size of virtual partition is determined by three factors: the screen size of handheld device, zoom scale of mobile user and map extension.

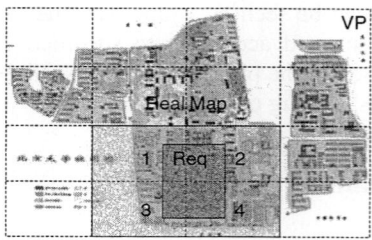

Fig. 3. Virtual Partition

4.2 Precision Strategy

By using the strategy of virtual partition, users only have to request the related data blocks and usually this will reduce the amount of transmit data largely. But when the whole map is requested, all partitions will be transmitted. In that case, only using the strategy of virtual partition is not enough. Different scales of spatial data are prepared to solve this problem. That is the precision strategy of spatial data.

What is the precision strategy of spatial data? The precision of spatial data means the scale that spatial entity shows. We use the term precision to mean both the amount of entities contained in the spatial data and the information stored about such entities, for instance, their geometry. Therefore, for example, changes of detail in a representation can be caused by the addition or elimination of some entities, as well as by the refinement or coarsening of the representation of existing entities (e.g., change of dimension or shape)[7].

Why the precision strategy of spatial data can solve problems? In the perspective of handheld device users, the extension and precision of map can be inverse ratio: When the map extension is large, i.e. users want to view a whole map in the screen, the coarse map is acceptant for users, while when the extension is small, such as in one small section, the highly detailed map is needed.

How to design the precision strategy of spatial data? For each map, we prepare different spatial map data with different scales, i.e. 1:500, 1:10,000 and etc, each scale corresponding to a precision. A solution is to pre-compute a sequence of multiple representations of the data to be stored on the server site. So when receiving requests from mobile client, the remote service will gather spatial data from the specific data source in which the scale matches the requested scale.

4.3 Caching Strategy

The primary goal of caching strategy is to reduce user-perceived response time. Thus, effective use of cached blocks is critical to the performance improvement.

Many contributions in computer science such as processors, databases and web browsers all support caching. The basic unit in our caching strategy is a lightweight spatial data block (LSDB). When a LSDB is gathered from remote service, if not exist in the cache, it can be stored into the cache bundled with a unique key that is identified by block boundary and precision.

Considering the characteristics of area and time localization, the general-purpose caching replacement mechanism such as the LFU (least frequently use) and LRU (least recently use) policy can often work. However, these general-purpose policies sometimes mismatch users' operation interests or navigation trends. Therefore we propose three location-based caching replacement mechanisms: furthest neighbor, least recently navigation and least recently operating.

4.3.1 Furthest Neighbor

LSDBs in current screen are defined as Center LSDBs. In furthest neighbor replacement mechanism, LSDBs that are further from Center LSDBs than others will be replaced earlier. As figure 4 shows, Center LSDBs are provided in current screen. When caching replacement operation is executing under furthest neighbor replacement mechanism, Tier 2 LSDBs will be replaced earlier than Tier 1 because Tier 2 LSDBs are further from Center LSDBs than Tier 1 LSDBs.

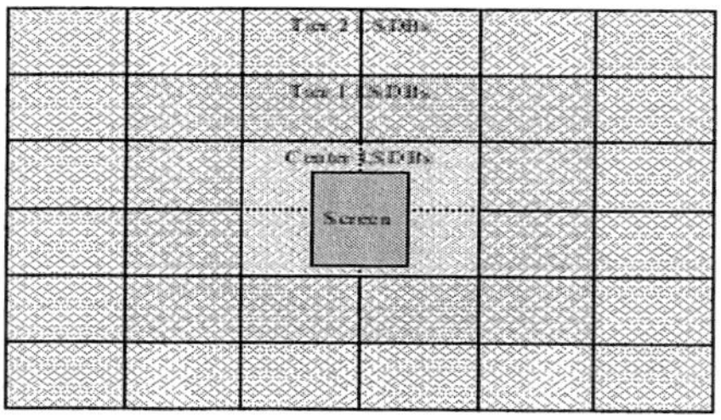

Fig. 4. Furthest Neighbor Replacement

4.3.2 Least Recently Navigation

Vehicles will have definite tracks while being navigated. From navigation tracks we can obtain navigation history, which means we can know which LSDBs are traversed by the navigation tracks. In least recently navigation replacement algorithm, LSDBs are replaced according to their replacement priorities that are determined by navigation history.

LSDB is classified into two types: navigation-sensitive LSDBs (s-LSDBs) and navigation-free LSDBs (f-LSDBs). The replacement priority of any s-LSDB is higher than any f-LSDB. All f-LSDBs have equal replacement priorities while priorities of s-LSDBs are correlative with navigation history. The earlier any s-LSDB is traversed by navigation tracks, the higher replacement priority the s-LSDB has. The rules that determine the replacement priorities of s-LSDBs can be described as follows.

The replacement priority of an s-LSDB can be represented by a value pair (ps, pn). The ps value of each s-LSDB means the replacement priority of the screen that is using the s-LSDB, and the pn value represents replacement priority of the navigation. The smaller the value is, the higher the replacement priority is.

Rules to Determine ps Value. All the LSDBs used in the same screen have equal replacement priorities. Different screens have different replacement priorities that are determined by the order of arrival, that is to say, earlier screen has higher priority. If a LSDB is used in more than one screen, in which case the LSDB has multiple screen priorities, the lower one will be selected as the ps value. As figure 5 demonstrates,

Screen 1 and *Screen 2* both have used LSDB *D*, so the replacement priority of *D* can be 1 or 2. According to the above rules, the ps value of *D* will be 2.

Rules to Determine pn Value. Considering whether the LSDBs are traversed by navigation vehicles, we can classify all the LSDBs into two types: traversed and non-traversed. The non-traversed LSDBs have the same pn value 0 (min value) while the traversed LSDB are ranked in the order of being traversed by navigation vehicles, the earlier being traversed, the higher rank the LSDB has. As figure 5 shows, the navigation vehicle sequentially traverses LSDB *C*, *D*, *E*, *F*, *H*, *I*, so their replacement priorities of the navigation decrease in the traversing order.

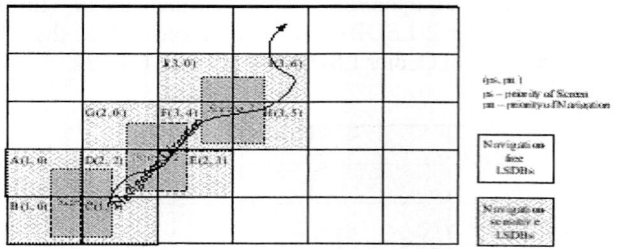

Fig. 5. Least Recently Navigation Replacement

Following above determined rules of replacement priorities, we introduce a replacement algorithm based on priorities, which can be described as following steps:

- Rank all the s-LSDBs according to ps value, those s-LSDBs with smaller ps value will be replaced earlier;
- If ps values of some s-LSDBs are equal, rank these s-LSDBs according to pn value, those s-LSDBs with smaller pn value will be replaced earlier;

So s-LSDBs appeared in figure 5 will be replaced in the order of *A* or *B*, *C*, *G*, *D*, *E*, *J*, *F*, *H*, *I*.

4.3.3 Least Recently Operating

Move, Zoon In and Zoom Out are common GIS operations for which the caching strategies are available. The replacement mechanisms are different for different operations. Replacement mechanisms in different operations are described simply as follows:

- Move - According to users' Move Operations History, the LSDBs used by earlier screens will be replaced earlier. This replacement mechanism is same as Least Recently Navigation.
- Zoon In — According to users' Zoom In Operations History, the LSDBs used by earlier screens will be replaced earlier.
- Zoon Out — The same as Zoom In operation.

4.3.4 Procedure Using Caching Strategy

No matter which replacement mechanism is adopted, the procedure of requesting map data using caching strategy will be the same, which can be described as following steps:

(1) Construct OriginReq [x1, y1, x2, y2][scale]
(2) Search map data in the cache and determine whether the data that not only overlap the request area but also match request scale exist.
  a)   If existed, use and display the searched map data, then calculate the absent data, and form as a new request, turn to b)
  b)   If no, send request to remote service.
(3) Construct RemoteReq [rx1, ry1, ry1, ry2][scale] and send it to remote service.
(4) Match RemoteReq with virtual map block,
  a)   If matched, ReturnResp = spatial entities located in matched map blocks
  b)   If no, ReturnResp = spatial entities located in overlapped map block

(5) Send ReturnResp back to mobile clients, clip out the requisite part of map from ReturnResp and display it.

4.4 Pre-fetching Strategy

A unique strategy in our Spatial Data Channel is the location-based Pre-fetching mechanism. What is Pre-fetching? In the Spatial Data Channel, different screens have different spatial data blocks. That is to say, each spatial data block can only be used for one screen, so when the screen changes, the relevant spatial data block will also change. If the new data block has not been downloaded from remote service, we can only view former old map on the next screen. Therefore to reduce user perceived response time, pre-fetching strategy is adopted. Figure 6 describes the relations among screen, next screen and pre-fetched map. The arithmetic of pre-fetching strategy is same as that described in 4.3.4, just regarding OriginReq there as next screen.

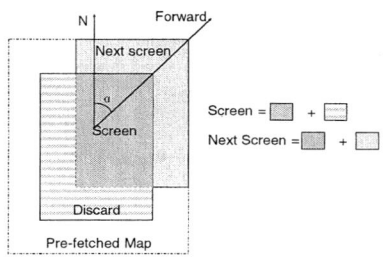

Fig. 6. Relations in Pre-fetching Strategy

Move, Zoon In, Zoom Out and Navigation are common GIS operations for which the pre-fetching strategies are available. The arithmetic of constructing pre-fetching requests are different for different operations. How to construct pre-fetching requests in those operations is described simply as follows:

(1) Move - The move direction and distance of next Move operation are presumed according to user's Move Operations History. Then the extension of data that the Next screen needs will be calculated while the scale equals current value.

(2) Zoon In - Two different situations are concerned:
- For free Zoon In operation, the scaling of next free Zoon In operation is presumed according to user's Zoon in Operation History. Then the extension and scale of data which the Next screen needs can be calculated
- For fixed Zoon In operation, the scaling of next operation is set as a default (such as 2), and then the extension and scale of next screen can be calculated.

(3) Zoon Out - similar to Zoon In.

(4) Navigation - The calculation of pre-fetching extension is somewhat complicated while the pre-fetching request scale is unaltered. When users use navigation, the move track may relate to the counterpoint in map. For example, when I'm walking along the road, the shape of my move track is similar to that of the road entity. Also the move track may be irrelevant with any entity in the map data, i.e. when a navigator is walking through a meadow, no relevant entities matches this track. So two different situations are concerned:

- Entity-sensitive: when move track T is relevant to map entity E, the arithmetic can be described as follows: 1) Determine whether point of intersection between E and the boundary of current map block (MB) exists; 2) If yes, construct the pre-fetching area centered with the point of intersection; 3) If no, it is concluded that the relation between T and E has already ended up before navigating objects reaches the boundary of MB, then E is replaced by next E that is E's next neighboring entity, and return to 1).
- Entity-free: when move track T has no relation to map entity E, the arithmetic can be described as follows: 1) Based on Move Operations History of navigating objects, we conclude the T's simulate curve C; 2) Determine whether point of intersection between C and the boundary of current map block (MB) exists; 3) If yes, construct the pre-fetching area centered with the point of intersection; 4) If no, first calculate the point of intersection between C and Block Median Boundary (BMB) which refers to the boundary in the middle of the center and MB, then calculate the tangent of C, and finally return to 2).

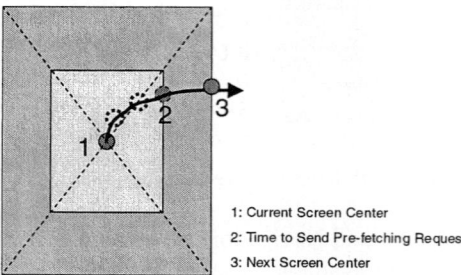

1: Current Screen Center
2: Time to Send Pre-fetching Request
3: Next Screen Center

Fig. 7. Pre-fetching Occasion

Another consideration is pre-fetching occasion. When current operation is in status of Move, Zoom or Zoom Out, pre-fetching request is sent instantly after current operation is executed. While in Navigation status, pre-fetching request is sent when navigating objects move on BMB. Figure 7 describes pre-fetching occasion. Here, when navigating objects move to Location 2 which is on *BMB*, Navigation Prefetching Request which request spatial data block centered by Location 3 will be sent.

5 Conclusion

This paper introduces a mobile navigation solution and discusses spatial data management strategies for mobile navigation service to answer the challenge of providing spatial map data efficiently. JNavigator is an integrated system including client applications and remote services. Spatial Data Channel is devised to provide dynamic and efficient access to location context spatial map data between client and remote sides. We integrate various strategies including data preparation, transmission and management into a data channel. Especially, some efficient caching and pre-fetching strategies are introduced to enhance spatial data providing.

There are two main parts of our future work. One is to define models to compare different caching and pre-fetching strategies in terms of storage and transmission costs. The other is the investigation of how to specify the size of virtual partition dynamically to reduce user perceived response time.

Acknowledgement

This work is supported by the National Research Foundation for the Doctoral Program of Higher Education of China under Grant No. 20020001015; the National Grand Fundamental Research 973 Program of China under Grant No.2002CB312000; the National Science Foundation of China under Grant No.60203002; the National High Technology Development 863 Program under Grant No. 2002AA134030.

References

1. Aloizio P. Silva, Geraldo R. Mateus: Location-Based Taxi Service in Wireless Communication Environment, Proceedings of the 36th Annual Simulation Symposium, PP: 47 (2003).
2. Shiow-yang Wu, Kun-Ta Wu: Dynamic Data Management for Location Based Services in Mobile Environments, Seventh International Database Engineering and Applications Symposium, PP: 180 (2003).
3. Wen-Chih Peng, Jiun-Long Huang, Ming-Syan Chen: Dynamic Leveling: Adaptive Data Broadcasting in a Mobile Computing Environment, Mobile Networks and Applications, 8(4): 355-364 (2003).
4. Shengru Tu, Xiangfeng He, Xuefeng Li, Jay J. Ratcliff: A Systematic Approach to Reduction of User-perceived Response Time for GIS Web Services, Proceedings of ninth ACM international symposium on advances in geographic information system, PP: 47-52 (2003).
5. Butternfield, B. P: Progressive Transmission of Vector Data on the Internet: A Cartographic Solution, http://greenwich.colorado.edu/babs/ottawa/Bfield_Ottawa.htm.
6. Chen Liang, Chung-Ho LEE, Jae-Dong LEE, Hae-Young Bae: Scale-Dependent Transmission of Spatial Vector Data on the Internet, The 3rd International Conference on Information Integration and Web-based Applications & Services (2001).
7. Michela Bertolotto, Max J. Egenhofer: Progressive Vector Transmission, Proceedings of 7th ACM symposium on advances in geographical information system, PP: 152-157 (1999).

A Video Retrieval System for Electrical Safety Education Based on a Mobile Agent

Hyeon Seob Cho[1] and Keun Wang Lee[2]

[1] Dept. of Electronics Engineering, Chungwoon Univ., Chungnam, Korea
[2] Dept. of Multimedia Science, Chungwoon Univ., Chungnam, Korea
{chohs, kwlee}@chungwoon.ac.kr

Abstract. Recently, retrieval of various video data has become an important issue as more and more multimedia content services are being provided. To effectively deal with video data, a semantic-based retrieval scheme that allows for processing diverse user queries and saving them on the database is required. In this regard, this paper proposes a semantic-based video retrieval system that allows the user to search diverse meanings of video data for electrical safety-related educational purposes by means of automatic annotation processing. If the user inputs a keyword to search video data for electrical safety-related educational purposes, the mobile agent of the proposed system extracts the features of the video data that are afterwards learned in a continuous manner, and detailed information on electrical safety education is saved on the database. The proposed system is designed to enhance video data retrieval efficiency for electrical safety-related educational purposes.

1 Introduction

With advancements in network and multimedia data compression technologies, rapid progress has recently been made in technologies that enable multimedia content services. As a result, user demand for large-capacity video data in mobile environments is growing. To meet the diverse needs of different users, a vast amount of video data needs to be effectively managed [1]. The effective and efficient management of video data requires a technology that enables systematic classification and integration of large-capacity video data. In addition, a system allowing for effective retrieval and storage of video data should be in place to provide users with information on video data according to diverse user environments, for example, on mobile terminals [2].

However, compared with desktop computers, mobile terminals have many inherent restraints such as low CPU processing rate/bandwidth/battery capacity, and small screens[3,4]. In particular, low CPU processing rates and bandwidths are key inhibitors to servers that aim to provide seamless multimedia data. To ensure the effective retrieval and playout of video data on mobile terminals, CPU performance of terminals must be improved, and network technologies and systematic video data indexing technologies must be developed. Currently, studies that focus on addressing such restraints on mobile terminals and effectively indexing video data are being actively conducted [5,6]. However, such video indexing methods are based on simply classifying video genres or types, rather than on reflecting user needs.

To attach various types of information to video data is difficult since video data contain no textualized information relative to typical text data. As such, there is a need for semantic-based retrieval using additional information such as frames, key frames, and annotations, in video. It is very important to make information on video data more systematic and concrete so that the user can perform content-based retrieval of such video data [7].

2 Video Retrieval Scheme

Studies currently being conducted on content-based video data retrieval are largely divided into the following: 1) a feature-based retrieval scheme that uses similarity by extracting features from key frames; and 2) an annotation-based retrieval scheme in which a comparative retrieval is performed on a user's annotation that has been inputted and saved for a key frame [8]. However, both content-based video data retrieval methods have some drawbacks.

In the case of the annotation-based retrieval scheme, retrieval is performed such that, if the user attaches annotations using characters to the semantic information of individual video data whose automatic recognition is difficult, the video data are saved on the database and extracted by already attached annotations during retrieval[9]. This scheme offers the advantage of correctly expressing and retrieving video contents because the user can process video contents with annotations while watching a video. However, the user has to attach annotations to each individual video one by one through the use of characters. This not only involves a lot of time and effort, but also causes a vast amount of unnecessary annotations. In addition, it cannot achieve retrieval accuracy since many different annotators attach their own meanings to videos.

In the case of the feature-based retrieval scheme, retrieval is performed in such as way as to extract low-level feature information (i.e., color, texture, regional information, and spatial color distribution) from video data[10]. Since this scheme focuses on performing similarity retrieval through extraction of visual features from the videos themselves, extracting visual features is very important. However, extracting accurate feature information from a lot of videos is a challenging task. In addition, matching extracted feature information to a vast amount of video data is not easy in the performance of retrieval.

This paper proposes a semantic-based video retrieval system that allows an automatic indexing agent to learn users' queries and their results as a means of automatically updating the video server's metadata in a continuous manner.

3 Video Retrieval System Based on Mobile Agent

3.1 Automatical Indexing Processing

The indexing agent extracts keywords from the queries submitted from the user for video retrieval, and matches them with basic annotation information stored in the metadata. Then, it detects the keyframes that has the same keywords as annotation information, and sends them to the user.

Figure 1 presents the architecture of annotation-based retrieval performed by the indexing agent in the Agent Middleware.

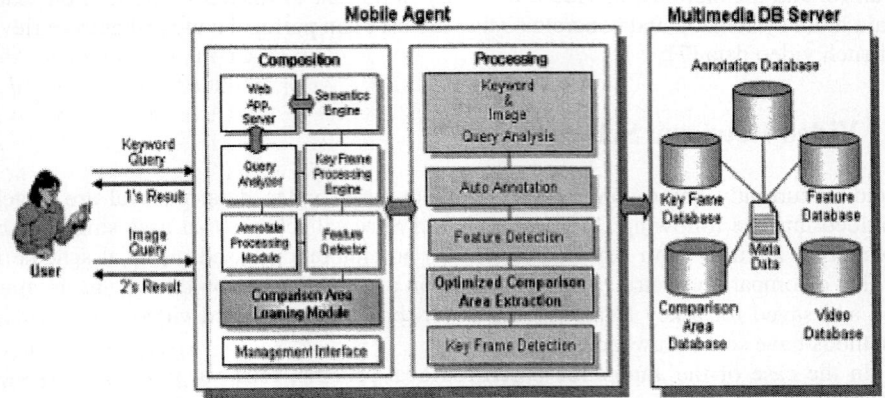

Fig. 1. Architecture for Mobile Indexing Agent & Multimedia DB Server

Once entered, queries submitted from the user are analyzed, and keywords are extracted. The user's keywords extracted are matched with the annotation information of metadata stored in the annotation database. As a result of matching, the keyframes having exactly matched keywords are detected in the database, and then they are sent to the user. Additionally, the keywords that do not exactly match annotation information among queries received from the user are defined as potential keywords.

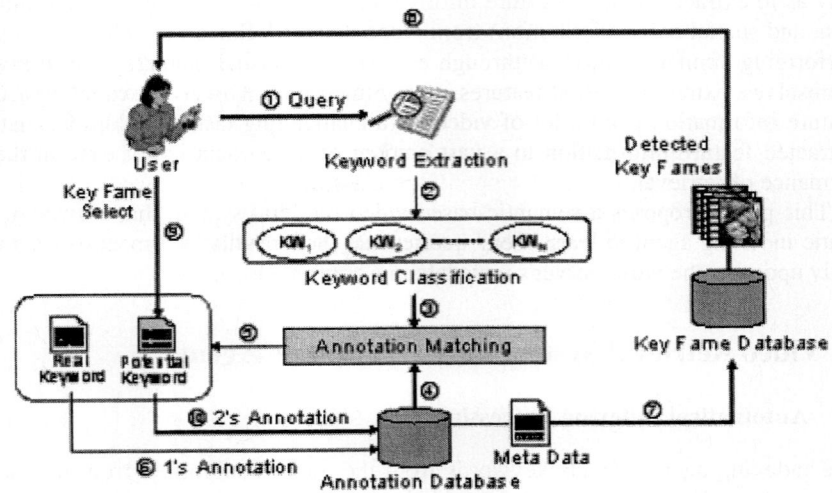

Fig. 2. Automatical Annotation Processing

If the mobile agent receives queries on retrieval of video contents from the user, it extracts keywords among query words and matches them to the basic annotation information of the metadata. It then detects key frames that have the same keywords as the annotation information and sends them to the user. Figure 2 shows an overall schematic of how annotations are automatically processed by the mobile agent.

As illustrated in Figure 2, once a user's query is inputted, a corresponding keyword is extracted following an analysis of the user's query. The extracted user's keyword is matched to the corresponding annotation information of metadata in the annotation database. As a result, the key frames that have correctly matched keywords as annotation information are extracted from the database and forwarded to the user. In addition, the keywords among users' query words that aren't correctly matched to the annotation information are defined as potential keywords.

The candidate key frames that are extracted according to the user's queries sent by the mobile agent are transmitted to the user in the form of exemplified image lists. They are displayed in descending order. If the user selects his/her desired key frame among key frame lists, the selected key frame becomes the secondary query image. Then, potential keywords are included in the annotation information of a corresponding key frame to make the meaning of the key frame more concrete. In addition, every time annotations are automatically processed by users' queries, the keywords extracted from individual users' queries and the selected example images are matched together. The similar weight values of accurately matched keywords are increased while those of unmatched keywords are decreased. This leads to making the meaning of annotations for corresponding key frames more concrete.

Once the annotator has inputted basic annotation information, the annotation information on key frames is automatically updated whenever the user performs a retrieval of video contents. This significantly reduces the problems that may occur when all individual users do not use identical vocabularies to express video scenes.

3.2 Automatic Annotation Processing Mechanism

Once a user's query consisting of one or more words is inputted, a corresponding keyword is extracted. The key frames containing users' key words are searched by the extracted keyword inputted by the user. The users' keywords are then classified into "real keywords" and "potential keywords". The accurately matched keywords among keywords in the annotation information are classified as "same keywords", while the accurately unmatched keywords are classified as "difference keywords". Figure 3 shows the classification of keywords which is a preprocessing step to automatic annotation processing.

The agent extracts key frames containing identical keywords and displays key frame lists to the user. If the user selects a specific key frame among the key frame lists, semantic weight values are calculated for each individual keyword that belongs to the specific key frame.

Where annotation keywords in the keyframe are same keywords, new semantics weight is calculated as Equ (1):

$$W_{Keyword\_new} = W_{Keyword\_old} + \frac{1}{N_{Kframe\_SK}} \quad (1)$$

While $W_{Keyword\_new}$ is the new semantics weight for annotation keywords and $W_{Keyword\_old}$ is the previous semantics weight for annotation keywords. $\frac{1}{N_{Kframe\_SK}}$ is the number of keyframes with same keywords.

In the meantime, where annotation keywords in the keyframe are difference keywords, new semantics weight is calculated as Equ (2):

$$W_{Keyword\_new} = W_{Keyword\_old} - \frac{1}{N_{Kframe\_SK}} \qquad (2)$$

4 Implementation and Experimental Evaluation

4.1 Implementation

Figure 4 shows the interface that allows the user to perform semantic-based retrieval on a mobile terminal. The user can search key frames for his/her desired scenes by inputting several search words.

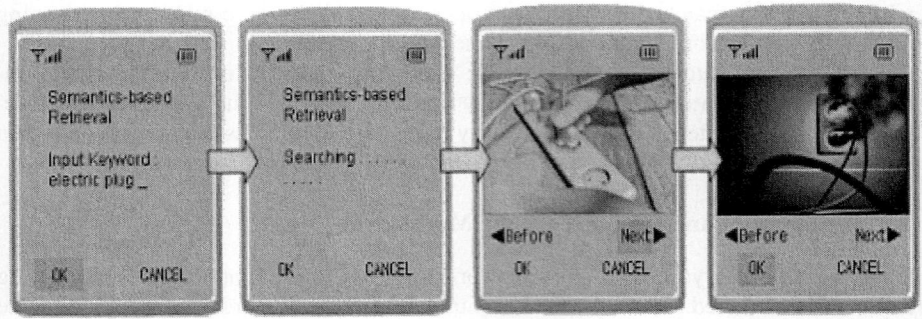

Fig. 4. Mobile Client Interface for Annotation-based Retrieval

4.2 Experimental Evaluation

MPEG formatted movie video files were used as domains for video data in order to evaluate the proposed system. For the purpose of this study, we used some 20 movies for electric safety education that had corresponding videos totaling 38 video clip files, and detected a total of 20,292 keyframes. By default, a single annotation was attached to 3,714 keyframes, except that a keyframe has duplicate portions due to successive cuts or that individual objects of a keyframe are indistinguishable.

In order to evaluate the proposed system for overall retrieval precision and recall, user queries were performed more than 500 times. Figure 5 illustrates the precision and recall of retrieval for this system.

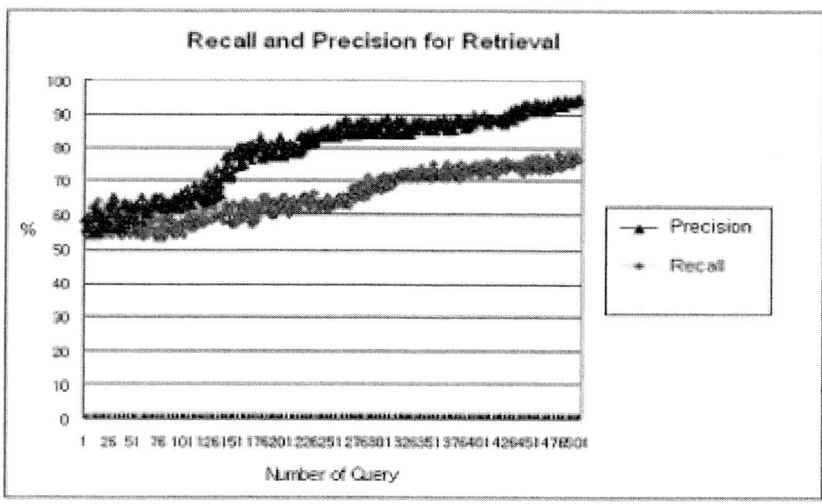

Fig. 5. Retrieval Recall & Precision of the Proposed System

5 Conclusions

This paper presents a video retrieval system that allows the user to perform various semantics retrievals using automatical indexing techniques for large capacity multimedia data in wireless network.

As with experiment results, the proposed system was able to improve retrieval precision for semantics-based video data retrieval as well as produce high precision at an approximate rate of 95.9% as a result of testing and evaluating user queries. Proposed system showed precision that improve more using optimized comparison area detection and reduced overhead of system and retrieval time in mobile phone.

Acknowledgements

This research is supported by the Electric Power Industry R&D Fund 2004 supported by Ministry of Commerce, Industry and Energy in republic of Korea.

References

1. Sibel Adali, et. al., "The Advanced Video Information System : data structure and query processing," *Multimedia System*, pp.172-186, 1996.
2. B. Y. Ricardo and R. N. Berthier, Modern Information Retrieval, *ACM press*, 1999.
3. T. Kamba, S. A. et al., "Using small screen space more efficiently," In P*roceedings of CHI'96*, ACM Press, 1996.
4. O.Buyukkokten, H. et al., "Power Browser: Efficient Web Browsing for PDAs," In *Proceedings of CHI'2000*, ACM Press, 2000.

5. Jiang Li. et al., "Scalable portrait video for mobile video communication," *IEEE Trans on CSVT*, Vol. 13, No. 5 pp.376-384, 2003.
6. C. Dubuc. et al., "The design and simulated performance of a mobile video telephony application for satellite third-generation wireless systems," *IEEE Trans on Multimedia*, Vol. 3, No. 4 pp.424-431, 2001.
7. Rune Hjelsvold, VideoSTAR - A Database for Video Information Sharing, Ph. D. Thesis, Norwegian Institute of Technology, Nov. 1995.
8. H. Rehatsch 다, and H. Muller, "A Generic Annotation Model for Video Database," *The 3rd International Conference on VISUAL '99*, Amsterdam, Netherlands, pp.383-390, 1999.
9. J. R. Smith and S. F. Chang, "VisualSEEK : a fully automated content-based image query system," *ACM Multimedia*, Boston, 1996.
10. Chong-Wah Ngo, Ting-Chuen Pong, Hong-Jiang ZhangOn, "Clustering and retrieval of video shots through temporal slices analysis," *IEEE Trans on Multimedia*, Vol.04 No.04 pp.0446-0458, 2002.

Fuzzy Multi-criteria Decision Making-Based Mobile Tracking

Gi-Sung Lee

Department of Computer Science, Howon University, Korea
ygslee@sunny.howon.ac.kr

Abstract. In this paper, we propose a novel mobile tracking method based on Multi-Criteria Decision Making, in which uncertain parameters the received signal strength, the distance between the mobile and the Base Station (BS), the moving direction, and the previous location are used in the decision process using the aggregation function in the fuzzy set theory. In numerical results, the proposed method provides a better performance than the conventional method.

1 Introduction

To ensure the efficient implementation of future mobile communication networks, it is important to determine the exact location of a Moving Object (MO). One of the major problems of mobile communication networks is the considerable volume of traffic gneated by the continuously changing locations of MO s with high mobility. In particular, severe heavy traffic occurs when configuring a system with small Microcells/Picocells with small diameters as a means of increasing subscriber density since a reduction in cell size results in frequent handoffs. In addition, varying locations of a MO andincorrect speed information cause an increase in incidental forward traffic. Therefore,one of the keys to efficient implementation of future mobile communication networksis in maximizing the efficiency of wireless spectrums.

Location estimation have focused on obtaining information on vehicles for transporting freights or on airline and/or shipping line operations from traffic control systems[1,2]. As shown in [3], a series of BS s determine the signal intensity of the transmitter of a MO through statistical methodologies and uncover the location of a MO through the information on contour lines. As a means of deriving retrieval procedures suitable for practical circumstances, a scheme for estimating the location of a MO is proposed which uses signal intensity and a signal's angle of arrival that are both transmitted to multi-beam antennas placed in multiple stations. In [5], a scheme for estimating the location of a MO through the use of adjoining BS s and each wave's arrive time is proposed. In [6], a scheme for estimating the location of a MO through the use of time difference of arrival of two BS s is offered. Given that IS-95B can estimate distances through the use of CDMA-based PN codes, a scheme for estimating the location of a MO using [5] and [6] is under consideration. However, said schemes provide less estimation accuracy because reflections are generated by buildings in multi-path environments. Existing schemes rely too much on radiowave-related information such as the signal intensity of neighboring BS s, the direction of their signals, and/or the time of arrival measured from a MO, in order to obtain information on the distance and speed of a MO. They are affected by short-term fading, shadowing, and/or diffraction generated by the buildings surrounding a MO, or by other obstacles. As a result, the scheme for estimating the location of a MO using signal information alone that is transmitted from the corresponding BS suffers from reduced accuracy.

2 System Model

Studies on conventional schemes have so far been carried out according to the simple law of propagation of a radiowave and under the assumption that signal intensity is accurately estimated. However, there is a wide gap between virtual models and actual environments. In actual environments, shadowing areas occur due to mountains and buildings. In addition, due to reflections, dots with the same average signal intensity are expressed in the form of distorted contour lines, not circles. As a result, the law of propagation of a radiowave varies as a function of various parameters, not as a function of simple distances. In other words, such models feature radiowave environments that are affected by reflections and diffractions inherent in Microcell environments where no Line of sight exists. Location estimation methodologies currently under study cause greater errors in the estimation of a MO in Microcell environments where little line-of-sight exists, as in the Manhattan model. In this study, which is based on the receiving signal intensity used as an existing estimated parameter, the distance between a MO and a BS, the previous location of a MO, and the direction of travel of a MO are added to the evaluation parameter in order to ensure increased estimation accuracy.

2.1 Multi-criteria Decision-Making Parameters

This paper deals with decision-making items such as the receiving signal intensity, the distance between a BS and a MO, the previous location of a MO, and the direction of travel of a MO. The receiving signal intensity is the most widely used parameter among existing location estimation schemes. It features very irregular profiles due to the scattering or reflection of radiowaves caused by buildings or other obstacles around a MO, weather variations in propagation channels, and/or multi-paths. Therefore, determining the location of a moving target through the use of the signal intensity alone transmitted by a MO from the BS may cause an irregular result. In addition to the receiving signal intensity, the distance between a BS and a MO is considered as a parameter. In general, the reason behind the distance between a BS and a MO being considered as a parameter lies in its relation to block placement schemes. However, determining distance information can also cause inaccuracies due to a radiowave's multi-paths. As such, if the distance itself is considered an evaluation index, inaccurate estimations may inevitably be generated. The previous location of a MO can be considered as a parameter. According to the velocity of a MO, the multitude of the radius of travel of the MO is considered at the previous location of the MO. The location of a MO is updated on a regular basis. Therefore, it is generally estimated that the MO is positioned near its previous location. Using such conditions, a correlation between the previous location of the MO and its estimated location can be determined.

2.2 Definition of a Membership Function

To determine the membership degree of a MO, a trapezoid-shaped membership function is used. The trapezoid-shaped membership function provides more diverse membership degrees for upper and lower limits than the stepwise function. A fuzzy number has many maximum points (e.g., $\alpha=1$) of the membership degree of a MO, becoming trapezoid-shaped.

1) Receiving Signal Intensity and Membership Function

The membership function is defined by using the receiving signal intensity of adjoining BS s. Here, μ_R refers to the membership function of an I the inequality RSS_i is the receiving signal immntensity that the BS I transmitted to the moving target. s_i is the lower limit (e.g., 5.5) for the left side of the inequality, while s_{i+1} is the upper limit (e.g., 7.5) for the right side of the inequality. Figure 1 shows the membership function $\mu_R(RSS_i)$ of RSS_i.

$$\mu_R(PSS_i) = \begin{bmatrix} 0, & PSS_i < s_1 \\ 1 - \frac{PSS_i - s_1}{|s_2 - s_1|}, & s_1 \leq PSS_i \leq s_2 \\ 1, & PSS_i > s_2 \end{bmatrix}$$

Fig. 1. Membership Function of Receiving Signal Intensity

2) Membership Function of the Distance between a BS and a MO

The membership function is defined by using the distance between a BS and a MO. Here, D_i is the distance between the BS i and a MO. d_i refers to the upper_limit (e.g., 90), and d_{i+1} refers to the lower_limit (e.g., 120). Figure 2 shows the membership function $\mu_R(D_i)$ of the distance membership function.

$$\mu_R(D_i) = \begin{bmatrix} 1, & D_i < d_1 \\ 1 - \frac{|D_i - d_2|}{|d_1 - d_2|}, & d_1 \leq D_i \leq d_2 \\ 0, & D_i > d_2 \end{bmatrix}$$

Fig. 2. Distance Membership Function

3) Membership Function of the Previous Location of a MO

The membership function is defined by using the relationship between the previous location of a MO and its estimated location. Here, L_i refers to the previous location of a MO, and E_i refers to the present location of the MO. d_i refers to the spatial distance between the previous location and present location of a MO. Figure 3 shows the membership function $\mu_R(L_i)$ of the location of a MO.

$$\mu_R(L_i) = \begin{bmatrix} 0\,; & L_i < E_1 \\ 1 - \dfrac{L_i - E_1}{|g_i|}, & E_1 \leq L_i \leq E_2 \\ 1, & E_2 \leq L_i \leq E_3 \\ 1 - \dfrac{L_i - E_3}{|g_i|}, & E_3 \leq L_i \leq E_4 \\ 0, & L_i > E_4 \end{bmatrix}$$

Fig. 3. Membership function of the location of a MO

4) Membership Function of the Direction of Travel of a MO

The membership function is defined by using the direction of travel of a MO. Here, C_i refers to the direction of travel of a MO. P_i refers to the receiving signal intensity. S_i refers to the spatial distance between the previous and present location of a MO. Figure 4 shows the membership function $\mu_R(C_i)$ of the direction of travel of the MO.

$$\mu_R(C_i) = \begin{bmatrix} 0, & C_i < PSS_1 \\ 1 - \dfrac{C_i - PSS_1}{|o_i|}, & PSS_1 \leq C_i \leq PSS_2 \\ 1, & PSS_2 \leq C_i \leq PSS_3 \\ 1 - \dfrac{C_i - PSS_3}{|o_i|}, & PSS_3 \leq C_i \leq PSS_4 \\ 0, & C_i > PSS_4 \end{bmatrix}$$

Fig. 4. Membership function of the direction of travel of a MO

3 Location Estimation Using a Fuzzy Theory

Location estimation is the process of estimating a block where a MO is located, and repeated by an estimator. Such a location estimation process is based on the three-stage location estimation scheme [9] where an optimal block is determined by narrowing the area where a MO is located. The estimator initiates the estimation process by means of its timer, performing repeated cycles of estimation. By using a multi-criteria decision-making scheme, the estimated value estimates the sector where a moving target is located in the sector estimation stage; the zone where the moving target is located in the zone estimation stage; and the block where the moving target is located in the block estimation stage

3.1 Multi-criteria Sector Estimation

Signal intensity, the distance between a BS and a MO, and the previous locations of a MO are considered as multi-criteria decision-making parameters in the sector estimation stage. The sector adjoining the BS with the highest totalized membership degree is estimated as the moving target being located. The estimation process can be expressed in the following sector estimation algorithm:

Stage 1. Membership degrees are obtained by referring to their membership functions for evaluated parameters.

Stage 2. For the present and adjoining BS s, the membership degrees obtained in the Stage 1 are totalized by means of connection operators:

$$\mu_i = \mu_R(RSS_i) \cdot \mu_R(D_i) \cdot \mu_R(L_i). \tag{1}$$

The totalizing operators refer to the fuzzy replacement operator (1) and weighted average operator.

$$\varpi\mu_i = \mu_R(RSS_i) \cdot W_{RSS} + \mu_R(D_i) \cdot W_p + \mu_R(L_i) \cdot W_L \tag{2}$$

Here, $W_{RSS} + W_p + W_L = 1$. As shown in Equation (2), the reason for added adding a weight value is that parameters may differ in importance from each other. In this study, the weighted value W_{RSS} of receiving signal intensity is defined as 0.5; the weighted distance value W_P is defined as 0.3; and the weighted location value W_L is defined as 0.2.

Stage 3. Blocks with estimated sector numbers are selected for next-stage estimations following the examination of all blocks within the cell. Examination of a sector number in the block's object information allows for identifying its corresponding block.

3.2 Multi-criteria Zone Estimation

In this the second stage of the estimation process, the blocks belonging to the zone where a corresponding MO is currently located are estimated among the blocks estimated in the multi-criteria estimation stage. The blocks are determined in the optimal zone by the next multi-criteria zone estimation algorithm. Signal intensity, the distance between a BS and a MO, and the direction of travel of a MO, are considered as multi-criteria decision-making parameters in the zone estimation stage.

Stage 1. Membership degrees are obtained by referring to their membership functions for evaluated parameters.

Stage 2. The membership degrees obtained in the Stage 1 are obtained by means of fuzzy connection operators to determine the following fuzzy replacement operator

$$\mu_i = \mu_R(RSS_i) \cdot \mu_R(D_i) \cdot \mu_R(C_i) \tag{3}$$

and the following weighted average operator

$$\varpi\mu_i = \mu_R(RSS_i) \cdot W_{RSS} + \mu_R(D_i) \cdot W_p + \mu_R(C_i) \cdot W_C. \tag{4}$$

Here, the weighted value W_{RSS} of receiving signal intensity is defined as 0.6; the weighted distance value W_P is defined as 0.2; and the weighted location value W_C is defined as 0.2.

Stage 3. Blocks belonging to the estimated zone are selected for next-stage estimations. Examination of a zone number in the blocks estimated in the sector estimation stage allows for identifying its corresponding block.

3.3 Multi-criteria Block Estimation

In this last estimation stage, a block where a corresponding MO is currently located is estimated among the blocks that have been estimated in the multi-criteria zone estimation stage. Signal intensity, the distance between a BS and a MO, and the direction of travel of a MO, are considered as multi-criteria decision-making parameters in the block estimation stage. The membership degree of the signal intensity among the parameters is determined by the receiving signal intensity of two BS s among the pilot signals of adjoining BS s. An optimized block is estimated by the following block estimation algorithm:

Stage 1. The membership degrees for evaluated parameters are obtained.
Stage 2. The membership degrees obtained in the Stage 1 are obtained by means of fuzzy connection operators to determine the following fuzzy replacement operator,

$$\mu_i = \mu_R(RSS_i) \cdot \mu_R(D_i) \cdot \mu_R(C_i) \tag{5}$$

and the following weighted average operator.,

$$\varpi\mu_i = \mu_R(RSS_i) \cdot W_{RSS} + \mu_R(D_i) \cdot W_P + \mu_R(C_i) \cdot W_C \tag{6}$$

Here, the weighted value W_{RSS} of receiving signal intensity is defined as 0.6; the weighted distance value W_P defined as 0.1; and the weighted location value W_C defined as 0.3.
Stage 3. Blocks belonging to the estimated zone are selected for next-stage estimations. Examination of a zone number in the block's object information allows for identifying its corresponding block.

4 Performance Analysis

4.1 Simulation Parameters

For the purpose of this study, we used simulation environments and considerations as follows consider the following: The size of the areas of concern is 10×10(Km), the number of BS installed is 7. We assumed that a moving target travels at a specific speed, regardless of areas of straight lines and intersections, and that the speed of a slowly slow-MO (for example, a pedestrian or a car running at low speeds) is maintained at 10Km/h, and that of a fast- MO (for example, a car running at high speeds) is maintained at 60Km/h, and that a slowly- MO measures signal intensities sent from seven BS s every minute, while a fast- MO measures the signal intensities every 0.45 seconds. Simulation parameters related to the receiving signal intensity are as follows:

- Average signal attenuation due to pass loss is proportional to 3.5 times the propagation distance.
- Shadowing varies as a function of the log-normal distribution that has a standard deviation of $\sigma = 6dB$.

- If the receiving signal intensity is below −16dB, it is regarded as erroneous. As a result, it is excluded in the calculation of the average receiving signal intensity.
- No execution of power control is made.

4.2 Simulation Results

In Figures 5 to 8, the horizontal and vertical axes in the graphs show the size of the areas of concern. Figure 5 shows the estimated probability measured when a slow-MO travels along a straight-line path. It allows for estimation of near-accurate locations when the slow-MO travels along the straight-line path.

Figure 6 shows the estimated probability measured when a slowly MO moves to along a high-speed path. As shown in Figures 5 and 6, the multi-criteria decision-making scheme proposed in this study allows for estimation of near-accurate locations of a MO traveling even in dynamic multi-path environments with severe fading. In particular, estimated parameters such as directions of travel of a MO and its previous locations serve to compensate for signal intensity errors on straight-line paths.

Figure 7 shows the estimated probability measured when a fast- MO travels along the straight-line path. The result is almost similar to that estimated when a slowly slow-MO travels along a straight-line path. Figure 8 shows the estimated probability measured when a fast- MO travels along a curved-line path. Even the turning of the fast- MO traveling along the straight-line path to the left/right gave had little effect on the result of estimation of the location of the moving target. In actuality, turning to the left/right causes a rapid signal distortion. Nonetheless, the distance between the previous location of a MO and a BS served serves to determine the location of the MO to compensate for average signal intensity errors, thereby compensating for a significant portion of rapid left/right signal distortions.

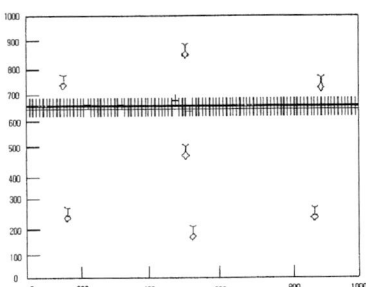 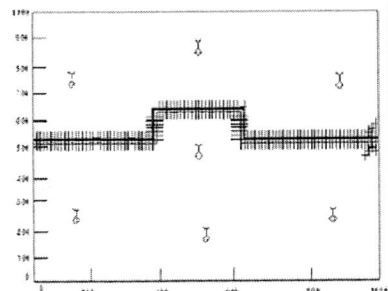

Fig. 5. Estimation of on the Straight-Line Path   **Fig. 6.** Estimation of on the Curved-Line Path

The estimation result described above has been is derived from the assumption that a MO turns to the left/right at the same speed as in the case when it travels along the straight-line path. However, the fast- MO traveling along the straight-line path actually reduces its speed while turning to the left/right. It is expected that, if a MO travels at reduced speeds, estimation accuracy will be increased.

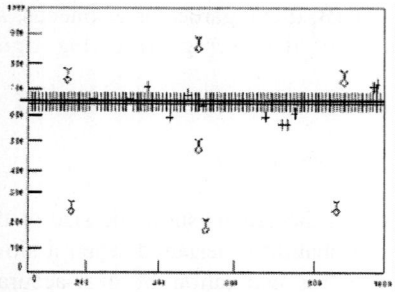

Fig. 7. Estimation of on the Straight-Line Path   **Fig. 8.** Estimation of on the Curved-Line Path

5 Conclusions

In this paper, a fuzzy-based multi-criteria decision-making methodology was used to improve the performance of the Microcell system where estimating the location of a moving target and its velocity is are very difficult. In addition, we analyzed location estimation schemes on in which research has been so far conducted previous researches, thereby presenting pointing out their problems. To increase estimation accuracy, we proposed a multi-criteria decision-making scheme that uses estimated parameters such as distances between MOs and BS s, previous locations of a MO, and directions of travel of a MO, in addition to as well as the existing estimated parameters and, signal intensity.

In Microcell environments, dots with the same average signal intensity are expressed in the form of distorted contour lines, not circles, due to shadowing and reflections. In this paper a fuzzy-based multi-criteria decision-making scheme is used to gradually narrow the area where a MO is located, in order to estimate an optimized block of its location. The proposed scheme allowed for simplifying the existing complex location estimation processes on a step-by-step basis, thereby improving the accuracy of estimation of the location of a MO.

Acknowledgments

This work was supported by Howon University Fund, 2003.

References

1. T. S. Rappaport, J. H. Reed and B. D. Woerner, "Position Location Using Wireless Communications on Highways of the Future," *IEEE Communications Magazine*, pp. 33-41, Oct. 1996.
2. J. C. Lee and Y. S. Mun, "Mobile Location Estimation Scheme," SK Telecommunications Review, Vol. 9, No. 6, pp. 968-983, Dec. 1999.
3. G. N. Senarath and D. Everitt, "Reduction Call Drop-outs During Handoff Using Efficient Signal Strength Prediction Algorithms for Personal Communication Systems," in *Proc. GLOBECOM'95*, Vol. 3, pp. 2308-2312, Nov. 1995.

4. D. J. Cichon, T. C. Becker and M. Dottling, "Ray Optical Prediction of Outdoor and Indoor Coverage in Urban Macro-and Micro-cells," in *Proc. VTC'96*, Vol. 1, pp. 41-45, May 1996.
5. T. S. Rappaport, J. H. Reed and B. D. Woerner, "Position Location Using Wireless Communications on Highways of the Future," IEEE Communications Magazine, pp. 33-41, Oct. 1996.
6. L. J., J. Y. Kuo, and H. W. T., "Fuzzy Decision Making through Relationships Analysis between Criteria, Fuzzy Systems Symposium, Soft Computing in Intelligent Systems and Information Processing", pp. 296 –301, 1996.

Evaluation of Network Blocking Algorithm Based on ARP Spoofing and Its Application

Jahwan Koo[1], Seongjin Ahn[2,*], Younghwan Lim[3], and Youngsong Mun[3]

[1] School of Information and Communications Engineering, Sungkyunkwan Univ., Chunchun-dong 300, Jangan-gu, Suwon, Kyounggi-do, Korea
jhkoo@songgang.skku.ac.kr

[2] Department of Computer Education, Sungkyunkwan Univ., Myeongnyun-dong 3-ga 53, Jongno-gu, Seoul, Korea
sjahn@comedu.skku.ac.kr

[3] School of Media, Sungsil University, Sangdo-Dong, Dongjak-Gu, Seoul, Korea

Abstract. Sometimes network resources including IP address, MAC address, and hostname could be misused for the weakness of TCP/IP protocol suite and the deficiency of network management. Therefore, there is urgent need to solve the problems from the viewpoint of network management and operation. In this paper, we propose a network network blocking algorithm based on ARP spoofing and evaluate the robustness of this algorithm via various experiments. We have performed several experiments on the gratuitous ARP exchange and IP address conflict detection in order to identify the robustness of the network blocking algorithm under both homogeneous and heterogeneous operating system.

1 Introduction

Sometimes we have some experiences on the IP address conflict and network inaccessibility. Without administrator's admission, the unauthorized user abuses the network configuration resources like IP address, MAC address, and hostname on the TCP/IP network environment. It is basically caused by the security weakness of TCP/IP protocols and the deficiency of network administration. Therefore, there is urgent need to solve the problems from the viewpoint of network management and operation.

Ironically, Address Resolution Protocol (ARP) spoofing techniques can be used to prohibit unauthorized network access and resource modifications [1] [2]. The ARP is a dynamic mapping method that finds a physical address given a logical address and a basic protocol in every Internet host and router. Gratuitous ARP is a mechanism used by TCP/IP computers to announce their IP address to the local network and, therefore, avoid duplicate IP addresses on the network. A gratuitous ARP is an ARP request for a node's own IP address. In the gratuitous

---

* Dr. S. Ahn. is the Corresponding Author.

ARP, the sender protocol address (SPA) and the target protocol address (TPA) are set to the same IP address. If a node sends a gratuitous ARP and no ARP reply frames are received, the node determines that other nodes are not using its assigned IP address. If a node sends a gratuitous ARP and an ARP reply frame is received, the node determines that another node is using its assigned IP address. Although ARP and gratuitous ARP are simple protocol and very useful in every Internet host and router, however, it is said that the ARP has many security weaknesses [3] [4].

In this paper, we introduce a network blocking algorithm (NBA) and evaluate this algorithm via various experiments.

The rest of the paper is organized as follows. In section 2, we describe a NBA based on ARP Spoofing technique. In section 3, we evaluate this algorithm via various experiments. The final section offers some concluding remarks.

2 Network Blocking Algorithm and Architecture

The NBA based on ARP spoofing we proposed consists of three major modules as shown Figure 1 and 2: network blocking, preservation, and dissolution modules.

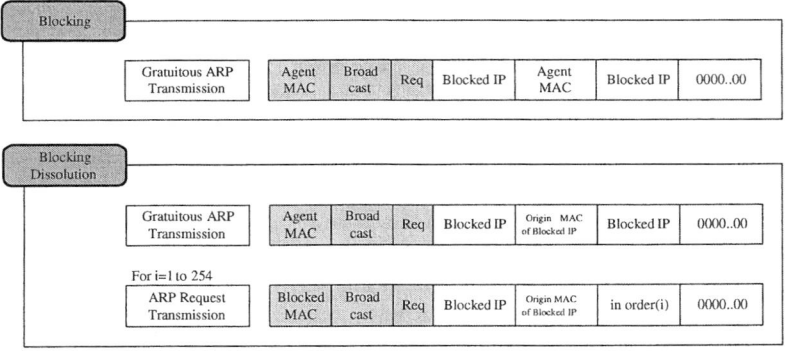

Fig. 1. Network blocking and dissolution of NBA

Typically, the architecture for the network resource and security management is a manager-to-probe or a manager-to-agent model. A node on the entire network is designated as a manager system and at least a probe or an agent system has to be installed on each network management domain. The manager issues a message of the protocol data unit to a probe. The probe interprets the protocol data unit message. If the message is set operation, the probe updates the policy database. After the creation of the policy database is completed, the probe monitors all the packets on its management domain by using the packet capture library, for example, pcap library in Linux and picks up only the ARP request or reply packets. If the ARP packets are in violation of the management policy, the probe issues network block messages by means of NBA.

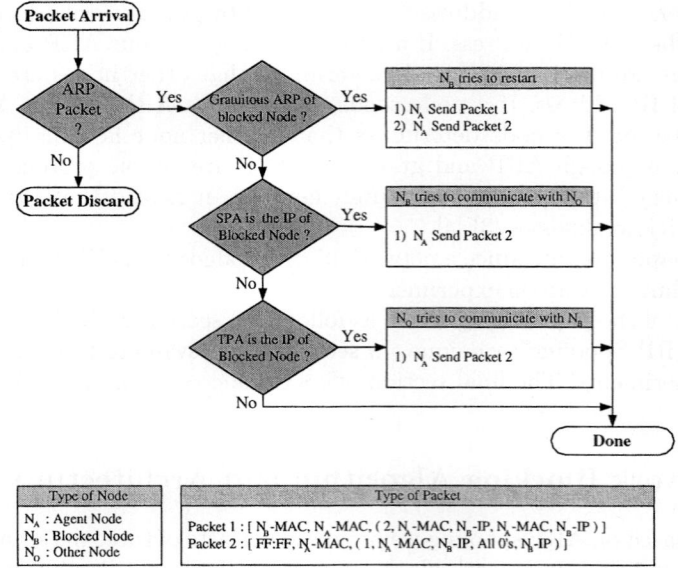

Fig. 2. Blocking preservation module of NBA

The manager system serves as the interface for the human network manager. It has a set of management modules for visualization, management, report, communication, and database. It maintains the network resource management information from the probes via the Inform-PDU message exchange, visualizes current resource status information such as the number of total IP addresses, used IP addresses, and unused IP addresses, and provides the real time information such as the corresponding IP-to-MAC addresses, hostnames, and policies.

The probe system responds to requests for information and actions from the manager system. It has a set of probe modules for communication, control, packet monitoring, policy, and network blocking message module. The manager and probes are communicated with several protocol data unit messages.

3 Evaluation

We have performed several experiments on the gratuitous ARP exchange and IP address conflict detection in order to identify the robustness of the NBA.

3.1 Experiment 1: IP Address Conflict on Homogeneous Environment

The first experiment's objective is to observe the operation process of normal gratuitous ARP in an IP address conflict on homogeneous environment and to examine whether the gratuitous ARP vulnerability exists.

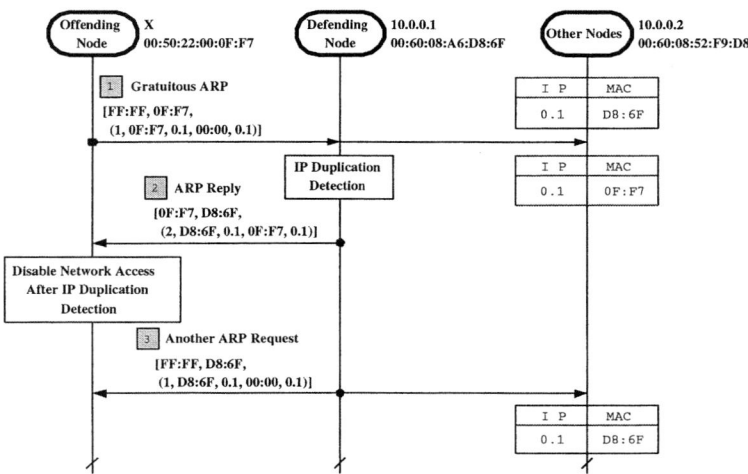

Fig. 3. The gratuitous ARP and address conflict exchange on homogeneous environment

For this experiment, all nodes on the same network segment are the computers running Microsoft Windows operating systems. The defending node means the node that is already successfully configured with the IP address and the offending node means the node that is sending the gratuitous ARP. Let's assume that the other nodes have already an entry for the IP address of the defending node. And we have manually configured the IP address of the offending node to the IP address of the defending node and restarted the offending node in order to invoke IP address conflict.

Figure 3 shows the exchange of the gratuitous ARP in the detection process of IP address conflict. Frame 1 is the offending node's gratuitous ARP request, frame 2 is the defending node's ARP reply, and frame 3 is the defending node's gratuitous ARP request. We have showed the changes of ARP cache table entry in other nodes. At the end of frame 3, all network nodes have been reset to the proper IP-to-MAC address.

We have known that the gratuitous ARP and address conflict detection for the Windows family is an exchange of three frames [5]. The first two frames are the ARP request-reply exchange for the conflicting address. After that, the defending node sends another broadcast ARP request to reset the ARP cache entries that were improperly updated by the offending node's sending of the gratuitous ARP request. However, it seems that the vulnerability of gratuitous ARP does not exist in the first experiment's situation.

3.2 Experiment 2: IP Address Conflict on Heterogeneous Environment

The second experiment's objective is to observe the operation process of normal gratuitous ARP in an IP address conflict on heterogeneous environment and to examine whether the gratuitous ARP vulnerability exists.

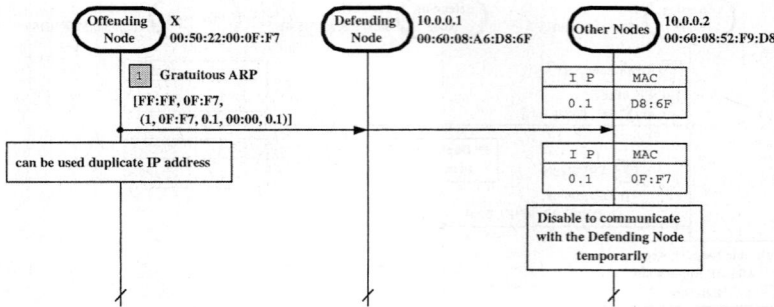

Fig. 4. The gratuitous ARP and address conflict exchange on heterogeneous environment

For this experiment, we have configured the nodes on the same network segment as the following. The offending node and the defending node were installed to Microsoft Windows and Redhat Linux, respectively and other nodes were installed to Microsoft Windows. Let's assume that the other nodes have already an entry for the IP address of the defending node. And we have manually configured the IP address of the offending node to the IP address of the defending node and restarted the offending node in order to invoke IP address conflict.

Figure 4 shows the exchange of the gratuitous ARP in the detection process of IP address conflict when running heterogeneous operating systems on the same network segment. Note that the defending node running Redhat Linux does not send any reply frame even if the offending node sends a gratuitous ARP request frame. To this effect, both the offending node and the defending node have continued to use the conflicted IP address in rotation, and the other nodes having the corresponding ARP cache entry were confused with periodically changing the entry of the ARP cache table. It is certain that the vulnerability of gratuitous ARP exists in the second experiment's situation.

3.3 Experiment 3: Forged Gratuitous ARP

The third experiment's objective is to observe the operation process of forged gratuitous ARP and to examine whether we can arbitrarily modify ARP cache entries and prevent a specific node from accessing to the network segment at the network protocol level.

For this experiment, we have implemented a simple command-line tool in C language which generates forged gratuitous APR packets. The syntax of this tool is fgarp *datafile count interval-sec*, where *datafile* is an ASCII file containing ARP frame information such as destination, source, operation, SHA, SPA, THA, and TPA, *count* is the number of sending ARP packet, and *interval-sec* is the interval time in second between the moment an ARP packet sends and the moment the next ARP packet sends.

One of the main functions is the *sendarp* function which takes a pointer to the data structure of ARP, creates a socket by the socket function setting

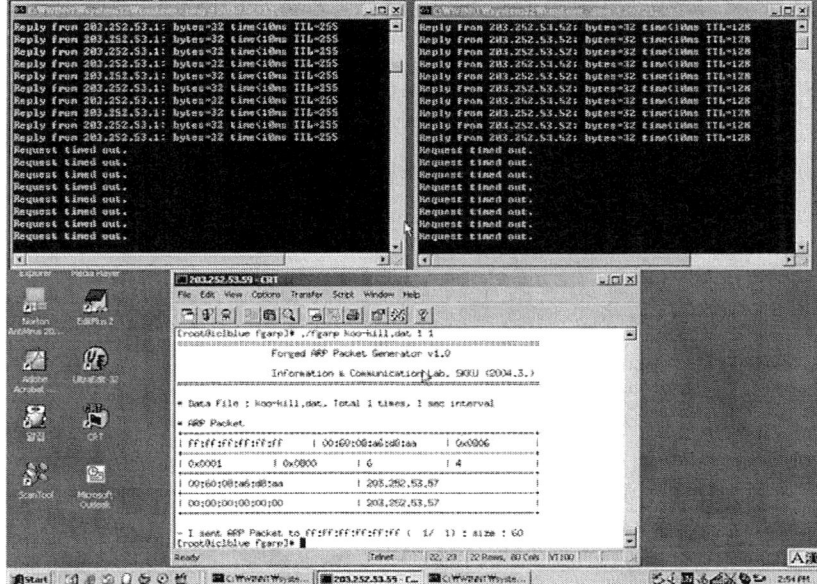

Fig. 5. Experimental result for network blocking

values for three fields (family, type, and protocol) of the socket structure, forms a gratuitous ARP packet for the specific node's IP address, repeatedly transmits it to the target node by variable counter and variable interval and closes the socket.

Figure 5 presents the result for the third experiment. The node on the left window and right window has the IP address of 203.252.53.57. In initial step, we can see that the node was able to connect to the IP address of 203.252.53.1 and .52 using Ping program. After that, the node on the bottom window having the IP address of 203.252.53.59 blocked the network access of node .57 by using the fgarp tool we have implemented. This is to verified that we can arbitrarily modify ARP cache entries and prevent a specific node from accessing to the other node.

4 Integration with Network Blocking Algorithm on Wireless LAN

In case of wireless environment, there are the security considerations that should be applied to different layers of the wireless network- namely, the physical, network, and application layers [6].

In this paper, we focus only on the 802.11-based network security mechanism which is the use of MAC access control lists. A MAC access control list is a list of physical addresses that are allowed to access the wireless network. This security mechanism is found in almost all access points. It enables the network administrator to enter lists of valid MAC addresses into an access control list,

854   J. Koo et al.

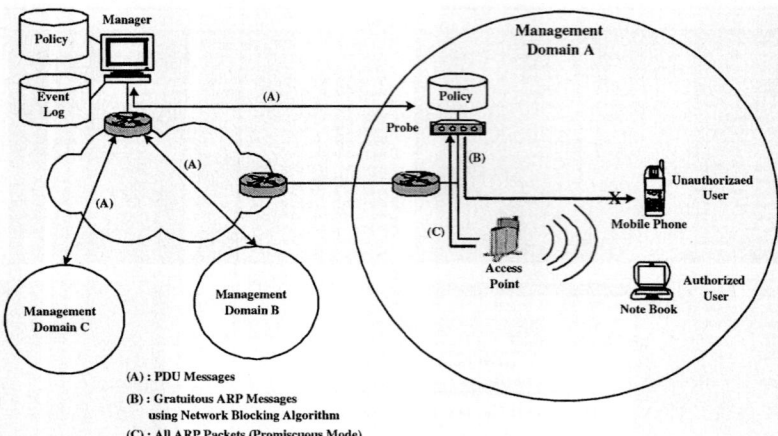

Fig. 6. Security framework with NBA on wireless LAN

limiting network access. However, the administrative overhead needs to be considered because keeping track of valid MACs and updating all access points with the valid address can be a time-consuming task. To make up for the weak point, we propose an application for the network resource and security management with NBA shown in figure 6.

5 Conclusions

Sometimes network resources including IP address, MAC address, and hostname could be misused for the weakness of TCP/IP protocol suite and the deficiency of network management. Therefore, we proposed a NBA and evaluated the robustness of this algorithm via various experiments. The basic concept of the proposed network resource and security management system is that authorized users can access their own network but unauthorized users should not be able to access. The proposed system is an effective tool for managing network resources containing IP address, MAC address and hostname, etc. under diverse and complicated network environment.

References

1. K. Kwon, S. Ahn, and J. Chung, "Network Security Management Using ARP Spoofing," Proc. ICCSA 2004, LNCS 3043, pp. 142-149, 2004.
2. J. Koo, S. Ahn, and J. Chung, "Network Blocking Algorithm and Architecture for Network Resource and Security Management," Proc. ISPC Comm 2004, Bishkek, pp.181-186, 2004.
3. S.M. Bellovin, "Security problems in the TCP/IP protocol suite," Computer Communication Review, Vol. 19, No. 2, pp 32-48, April 1989.

4. N.E. Hastings, P.A. McLean, "TCP/IP spoofing fundamentals, computers and communications," Proc. IEEE Fifteenth Annual International Phoenix Conference, pp. 218-224, 1996.
5. J. Davies and T. Lee. "Microsoft Windows Server 2003 TCP/IP Protocols and Services Technical Reference," Microsoft Press, 2003.
6. M. Maxim and D. Pollino, Wireless security, RSA Press, 2002.

Design and Implementation of Mobile-Learning System for Environment Education

Keun Wang Lee[1] and Jong Hee Lee[2]

[1] Dept. of Multimedia Science, Chungwoon Univ., Chungnam, Korea
Kwlee@chungwoon.ac.kr
[2] Principal Research Engineer, Retail Tech Co., LTD., Seoul, Korea

Abstract. Amid the growing need for efficient and automated educational agents for mobile learning systems, learner demand for customized coursewares is increasing. However, many m-Learning systems that have been studied as of late have yet to support a mobile learning course in a seamless manner, thus failing to meet learner expectations. One of the problems of such m-Learning systems is their weakness in motivating learners, in particular, in the continuous feedback of the course. This paper proposes a mobile learning scheduling system that provides environment-related educational courses for learners. The proposed system monitors and evaluates the learning outcomes of learners on a continual basis, and calculates degrees of accomplishment in learning activities that will be applied to the agent's scheduling in order to provide learners with appropriate courses. Such courses enable learners to experience vigorous learning activities through repetitive programs, according to their ability.

1 Introduction

The fast development of Internet has recently enabled the on-line lecture through the e-learning system, which is now became popular topics in the area of computer education system industry. As this e-learning system is spread widely to the public, the users demand more diverse education service, and that results facilitating study on applied education service being very active [1-3].

Since the agent and broker for the domestic and foreign education software are organized to meet the demands of the average public more rather than customized service for individual learner, it is very difficult to accommodate the various needs for knowledge and evaluated level for each and every individual [4,5].

Although tools to help interaction between learners had been supported in many ways, in instructor's perspective, it is very hard to provide the right course schedule and combinations by analyzing each learner status after facing all registered learners. Hence, agent who can deliver feedback such as effective way of learning, course formation and course schedule to learners is needed in this e-learning system [6-8].

Multi-agent will be proposed in this paper, who can provides the appropriate active course scheduling and feed back to learner after evaluating the learners education level and method.

2 Course Scheduling Multi-agent

2.1 Mobile-Learning System Structure

In this m-learning system, learner and CSMA (Course Scheduling Multi-Agent) are connected via mobile interface, and through Mobile Interface (MI) the request and transfer for course scheduling occurs between learners and CSMA. Learners study the course provided by CSMA in this system.

All the information created by CSMA will be stored into the database and if required, it will be loaded by CSMA and used for reorganizing course. Learner's profile and information obtained by their learning activity as well will be stored into database via MI, then by CSMA it will be regenerated and stored again as necessary information to the learner such as learning achievement level, course, scheduling, evaluation data, feedback and etc. Figure 1 shows the structure for proposed m-learning system.

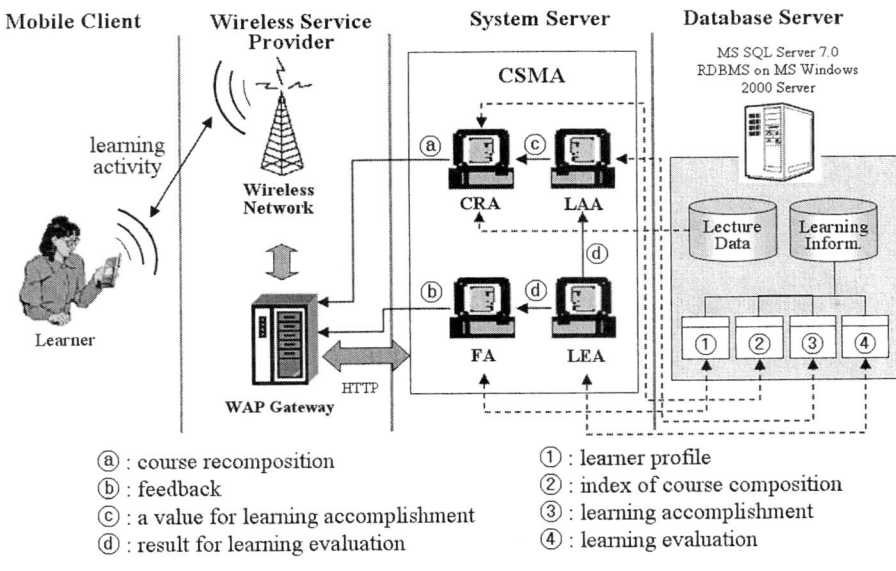

Fig. 1. Mobile-Learning System Structure

The key component of CSMA consists of four agents, CRA (Course Recomposition Agent), LAA(Learning Accomplishment Agent), LEA (Learning Evaluation Agent) and FA (Feedback Agent). The CRA is delivered the information on the degree of accomplishment of learning from the LAA and creates and provides a new and most customized learner-oriented course. The LAA estimates the degree of learning accomplishment based on the test results from the LEA and tracks the effectiveness of learning. The LEA is carrying out learning evaluation at every stage. The FA provides relevant feedback to learners in accordance with the learners profile and calculated degree of accomplishment of learning.

2.2 Course Scheduling Scheme

The degree of learning accomplishment can be calculated by the comparison between the current test result and the previous test result, and the analysis of the learning effectiveness growth. Let the maximum degree of learning accomplishment be 1 and we can give certain amount as a degree of weakness. Therefore, 1 minus the amount is the degree of learning accomplishment. This can be defined by the following equation.

$$A(I, i) = 1 - W(I, i) \tag{1}$$

where $A(I, i)$ is achievement degree for each subsection and $W(I, i)$ is weakness for each subsection.

The reason the degree of learner's weakness is under 1 is to represent it with percentage. It has observation from 0 to 1. Evaluation agent decides the degree of learning accomplishment by test. Evaluation agent also estimates the weighted value of the weak problems by calculating the marking time of individual question. Figure 2 is a diagram in which section test is inserted into the subsection test.

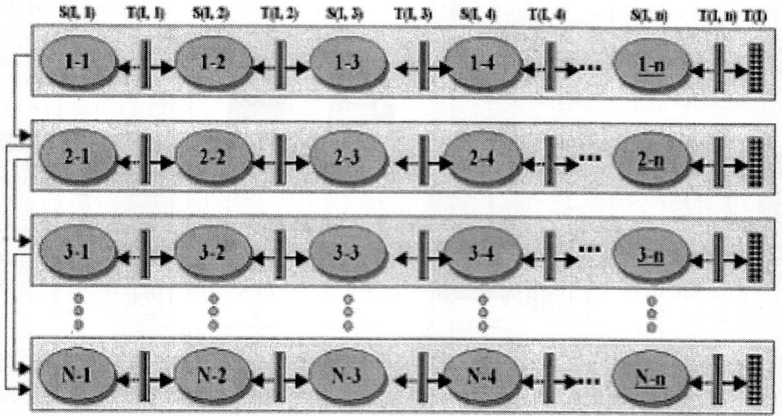

Fig. 2. Diagram of Subsection and Section Test

The degree of weakness of subsections $Wtr(I, i)$ can be represented based on the marking time and ratio of correct answers as following equation.

$$Wt(I,i) = \begin{cases} 0 & : when \quad td(I,i) < ta(I,i) \\ 1 & : when \quad td(I,i) < (4 * ta(I,i)) \\ \dfrac{(td(I,i) - (ta(I,i))}{(3 * ta(I,i))} & : when \quad td(I,i) < (4 * ta(I,i)) \end{cases} \tag{2}$$

$$WtR(I, i) = Wt(I, i) * at + (1 - R(I, i)) * (1 - at) \tag{3}$$

where $td(I, i)$ is the needed time for solving subsection question in section test, $ta(I, i)$ is average required time for solving one subsection question in section test, $R(I, i)$ is average required time for solving one subsection question in section test, $Wt(I, i)$ is weakness of solving time for each subsection, $WtR(I, i)$ is weakness of solving time and correct answer for each subsection and at is Weight value of time weakness applied. The equations to calculate the weakness of subsection analyzing repeated learning is defined by:

$$Wr\,(I, i) = (Lc\,(I, i) - 1) * 0.3 \qquad (4)$$

where $Lc(I, i)$ is count of repeat for subsection learning. Accordingly the degree of learning weakness according to the course test can be calculated as follows equation.

$$W\,(I, i) = WtR\,(I, i) * (1 - ar) + Wr\,(I, i) * ar \qquad (5)$$

where $W(I, I)$ is weakness for each subsection and ar is weight value of repeat learning weakness applied. The degree of learning weakness from the analysis of learning repetition represents the weakness of total subsection along with the marking time. Therefore the degree of learning weakness at each subsection is calculated by the weighted value at between the subsection weakness analyzed by the marking time and the rate of correct answer and the learning weakness analyzing the repetition of subsection study. With this learning weakness, we can estimate the degree of learning accomplishment. And by the degree of learning accomplishment, we track the subsection showing weakness and recompose the course.

3 Experiments and Evaluations

A total of 80 persons were selected among many persons to provide a same courseware in which 40 persons were finally selected to evaluate a CSMA- and PDA-based learning system. Table 1 is a summary of the experimental environments.

Table 1. Comparison of Experimental Evaluation Methods

Learning Method Item	General Learning Method	CSMA Learning Method
Target	Group A: 40 kindergarteners	Group B: 40 kindergarteners
Subject Name	Environmental Education	
Planning Sections of Learning	Number of Major Sections: 2 Number of Subsections per Major Section:4	
Learning Method	General Educational Materials	CSMA Course Learning
Place for Learning	PC Lab	PC Lab
Method of Evaluating Learning	General Evaluation (Objective Test)	General Evaluation (Objective Test)
Learning Time for Subsections	At the Instructor's Discretion	According to CSMA
Evaluation Time	Final Evaluation: 15 minutes	
Evaluation Items	Number of Items: 20	
Relearning Weak Sections	Allow learners to study weak subsections at the instructor's discretion.	Allow learners to study weak subsections according to CSMA-scheduled proposals.

For convenience, let the 40 persons who learn according to general learning methods be referred to as Learner Group A, and the 40 persons who learn according to the CSMA learning method be referred to as Learner Group B. Both the general learning methods implemented according to different items and the elements of the CSMA learning method are identical. For the general learning methods, learners themselves can assign learning times, while the CSMA learning method provides learners with suitable learning times according to the CSMA's course scheduling algorithm.

The subsection-specific weakness is demonstrated with the graph and figures and the final test degree so that learners can compare it with their target score. Learners under the target degree can begin the repetition program by the course schedule provided by CSMA. Figure 3 offers information on learning data and the degree of learning accomplishment in PDA format.

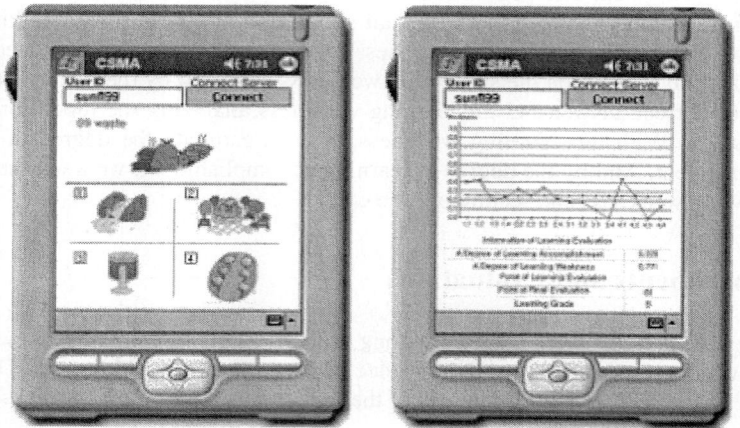

Fig. 3. User Interface in PDA Format

In the experiment, both Learner Groups A and B were allowed to learn under learning environments where, while the learning elements were the same, the learning methods were different. As a method for identifying the difference in evaluation results between the two groups, a chi-square test (a statistical methodology) was used. Formula (6) shows the chi-square distribution.

$$X = \sum_{i=1}^{c} \sum_{j=1}^{2} ((f_{ij} - \frac{f_{is} * f_{cj}}{n})^2 / (\frac{f_{is} * f_{cj}}{n})) \tag{6}$$

Probabilistic verifications are available through comparison between calculated values and the values of statistical distribution tables, as well as through identifying the difference between the two variances. Table 2 shows chi-square values.

As shown in the statistical table above, the probability of there being a difference at a degree of freedom 7 and at a significance level of 95% is $X = 12.59$ due to the existence of seven classes. Therefore, the value (16.20) from the approximate chi-square distribution is deemed a large number. According to the above average (88.3,

81.3), and the value from the chi-square distribution, the CSMA-based learning method can be evaluated as excellent, with above 95% reliability.

Table 2. Chi-square Values

Chi-square values between two experimental groups		
Score Class	Group A	Group B
70	0.73	0.70
75	2.66	2.54
80	0.59	0.56
85	0.06	0.06
90	2.15	2.06
95	1.63	1.56
100	0.47	0.44
Total	16.20	

4 Conclusion

In this paper, a mobile learning scheduling system for environment-related educational purposes has been proposed. The proposed system calculates degrees of learning accomplishment for learners using a multi-agent, and automatically creates course schedules, allowing learners to increase their degree of learning accomplishment.

These agents continuously learn the individual learning course and the feedback on learning, offering customized scheduling course and giving maximum learning effectiveness. Accordingly, the course ordered by the learner is tailored to the learner with the help of the course scheduling agent. Learners can continuously interact with agents until completing the course. If the agents judge the course schedule for learners to be ineffective, they recompose the course schedule and offer a new course to learners. To test and assess the proposed system, environmental education courses were provided to kindergarteners. As a result of the performance of such experiments and through evaluations, the proposed system proved to be excellent.

Acknowledgements

This subject was supported by Ministry of Environment as The Eco-technopia 21 project.

References

1. Moore, M.G and Kearsley, G., "Distance Education," Wadsworth Publishing Company, 1998.
2. Hamalainen, M, Whinston, A, and Vishik, S., "Electronic Markets for Learning: Education Brokerages on the Internet," Communications of the ACM, vol. 39, No. 6, pp. 51-58, 1996.
3. Eyong B. Kim, Marc J. Schniederjans, "The role of personality in Web-based distance education courses," Communications of the ACM, Vol. 47, No. 3, pp. 95-98, 2004.

4. Jason A. Brotherton, Gregory D. Abowd, "Lessons learned from eClass: Assessing automated capture and access in the classroom," ACM Trans on Computer-Human Interaction, Vol. 11, No. 2, pp. 121-155, 2004.
5. Hal Berghel, David L. Sallach, "A paradigm shift in computing and IT education," Communications of the ACM, Vol. 47, No. 6, pp. 83-88, 2004.
6. Whinston, A., "Re-engineering MIS Education.", Journal of Information Science Education, Fall 1994, 126-133, 1994.
7. Badrul H. Khan, Web-Based Instruction (WBI): What Is It and Why Is It?, Education Technology Publications, Inc., 1997.
8. Eyong B. Kim, Marc J. Schniederjans, "The role of personality in Web-based distance education courses," Communications of the ACM, Vol. 47, No. 3, pp. 95-98, 2004.

A Simulation Model of Congested Traffic in the Waiting Line

Bongsoo Son[1], Taewan Kim[2], and Yongjae Lee[2]

[1] Yonsei Univ., 134 Shinchon-Dong, Seodaemun-Gu, Seoul, Korea
sbs@yonsei.ac.kr
[2] Chung-Ang Univ., 72-1 Nae-Ri, Ansung-Si, Kyunggi-Do, Korea
{twkim, yjlee}@cau.ac.kr

Abstract. Congested traffic shows very complicated and stochastic features. One of the most interesting features is the amplification/decay of perturbations, indicating the development of small speed oscillation in the downstream into the large speed oscillation in the upstream. While traffic theories attempt to explain this phenomenon, the mechanism of the speed oscillation is not yet clear in the aspect of drivers' behavior. Similar phenomenon is also found in a long waiting line such as in front of the stadium or theater ticket box. In this paper, the stop-and-go movement in the waiting line, which is relatively easy to understand than the vehicular movement, is modeled and simulated. From the simulation, it is found that the amplification/decay of perturbation exists in the waiting line and shows similar pattern as in the vehicular movement.

1 Introduction

Traffic congestion induces travel time delay, high risk of accident, frequent acceleration/deceleration and increase of emission. To alleviate the traffic congestion, various measures such as ramp metering, incident management system, and VMS (Variable Message Sign) to divert traffic are developed. For these measures to be effective and reliable, it is required to understand the characteristics of congested traffic and model it. Traffic flow can be categorized into free-flow state and congested state. The former is usually found when the traffic demand is less than the capacity and the vehicular speed is constant. However, when the demand exceeds the capacity or an incident occurs, the free-flow traffic switches to the congested state and the vehicular speed changes irregularly and frequently. In congested traffic, a number of vehicular perturbations are generated and then grow or decay. It is required to understand the generation and evolution of these perturbations to describe the features of congested traffic and develop effective measures to control the freeway. One way to explain the evolution of perturbation is through the continuum traffic flow theory (Lighthill and Whitham, 1955; Payne, 1971; Zhang, 1998). However, these models are deterministic and could not well describe the complicated features of amplification/decay of perturbation and non-periodical change of the traffic parameters such as flow, density inside the congested region. In the car-following theory, the evolution of perturbation is explained by drivers' reaction time and sensitivity. However, the speed oscillation modeled in this theory is not only deterministic but also unbounded, which is unrealistic. In this paper, different approach is attempted to explain the evolution of vehicle perturbation in the analogy of waiting line in front of the stadium or theater ticket box, which is relatively easy to understand than the vehicular movement. The movement of people in the waiting line is modeled and through simulation the analogy of perturbation evolution with vehicular traffic is to be investigated.

2 Vehicular Congested Traffic

2.1 Empirical Features of Congested Traffic

It is often observed that in a severely congested traffic, vehicles travel with a periodic speed fluctuation, which is called as a stop-and-go traffic. It occurs when a small perturbation is amplified during its propagation to upstream.

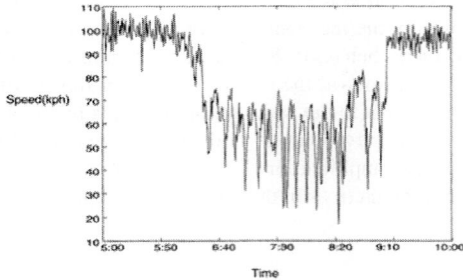

Fig. 1. Speed fluctuation in congestion (Queen Elizabeth Way, Canada)

Figure 1. shows the temporal evolution of speed observed at Queen Elizabeth Way, Toronto, Canada. The congestion has begun at 06:30 and ended at 09:00. During the congestion, the speed fluctuated between 20~80 kph. This plot shows the typical stop-and-go flow pattern. The amplitude and cycle of speed fluctuation in congestion flow is much bigger than that in free flow state. Unfortunately, no theory up-to-date could model this phenomenon in a consistent and quantitative manner.

Edie and Baverez (Edie, 1967) have studied the stop-and-go waves in the region of the bottleneck of the Holland Tunnel. They found that the stoppage wave was present at the density 96 ~139 vpm, and the density after the stoppage wave was 62 ~ 84 vpm. The speed of the wave was around 10 mph, and the number of vehicles in the wave was 5 to 12 vehicles. While this paper showed the general pictures of stop-and-go waves, the result may be site-specific. Mika, Kreer and Yuan (1969) have analyzed the traffic flow data with time series analysis. They found that 'the onset of the oscillatory behavior was abrupt, remarkably regular, and varied by more than a factor of two in amplitude as a function of location'. The frequency of oscillation was 1/4 cycle per minutes and the wave speed was about 16mph, which is higher than the study of Edie.

2.2 Theories of Perturbation Evolution

In conventional car-following model, the stability of the perturbation is explained by the driver's reaction time and sensitivity. When a follower responds slowly to the perturbation and then applies larger acceleration or deceleration rate, the perturbation will be amplified and accordingly the traffic flow will reach to a complete stop or maximum speed. On the other hand, if the driver is very attentive or looks far ahead such that the reaction time is small, the perturbation tends to be smoothed out. In the context of the scope, stability analysis can be divided as local stability and asymptotic stability (May, 1990). The former is concerned with the perturbation evolution for a following vehicle and the latter for a line of vehicles, theoretically for an infinite number of vehicles. Chandler and Herman, *et. al* (1958) and Herman (1959) have investigated the asymptotic stability of a linear car-following model and found that the asymptotic stability is gained when the value of reaction time by sensitivity is smaller than one half. The stability analysis of the car-following model assumes that the driver's are identical. However, in real traffic, drivers apply different reaction time and sensitivity. The randomness of the headway is another decisive factor concerning the evolution of the perturbation as well as reaction time and sensitivity. The stability analysis of car-following theory has limitation on modeling the temporal and spatial speed fluctuation patterns. For example, how can we explain the cycle length of speed fluctuation in stop-and-go traffic is 3-4 minutes?

3 Modeling Waiting Line Movement

3.1 Introduction

In this paper, we investigate people's movement in the waiting line to understand the features of congested traffic for the following reasons. Some features of waiting line problem show much similarity with the traffic flow in congestion. For example, from the real world observation, we can find that that the perturbation occurred at the downstream boundary of the waiting line propagates upstream and sometimes it dissipates and sometimes it is magnified. This observation is similar to the traffic congestion occurred at the upstream of signal.

When people are in the waiting line, sometimes they leave longer gap and sometimes they leave shorter gap. This is also similar to our traffic flow model, in which vehicles have different equilibrium time headways.

Waiting line problem is much simpler than traffic flow. Because we can assume that the speed of people can take only two value, "move" or "stop". The study on people's waiting line problem may provide a clue to our study on the congested traffic.

3.2 Algorithm for the Waiting Line

For the description of waiting line movement, we will consider one line and use cellular automata method. The concept of the algorithm is in Figure 1 (a).

In other words, one people take one cell and they can move one cell at one time step. The initial perturbation will be made by the first people only. To describe the behavior of the follower, an algorithm is developed. Here, we assume that the movement of people in the waiting line follows below cycle consisting of four steps.

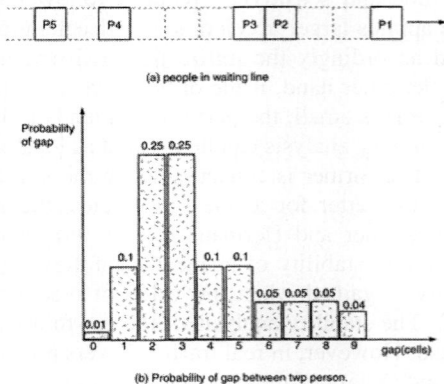

Fig. 2. Basic concept of the algorithm

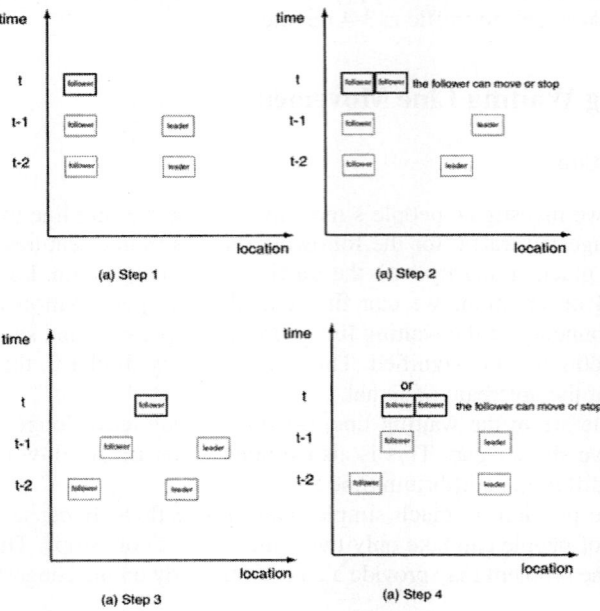

Fig. 3. Four steps of the movements

First step is the case when both the leader and follower do not move. See Figure 3(a). If the leader and follower are in the first step at time $t-1$, then the behavior of the follower at time t would be "stop".

Next step is the case when the leader moves. See Figure 3(b). If the leader and follower are in the second step at time $t-1$, then the behavior of the follower at time t can be "stop" or "move".

The third step is when both the leader and follower move. See Figure 3(c). If the leader and follower are in the third step at time $t-1$, then the follower will move at time t.

The last step is the case that the leader stopped. See Figure 3(d). If the leader and follower are in the fourth step at time $t-1$, then the follower can move or stop at time t.

At every cycle, the follower will choose a preferred gap (the number of cells between the leader and the follower) and this preferred gap does not change throughout the one cycle. At the beginning of each cycle (when the follower does not move and the leader begins to move) the preferred gap is randomly chosen as is in Figure 2(b). In the figure, the preferred gap can take value between 0 to 9 cells and the probability of the gap is left-skewed. For example, the probability that the follower will take 3 cells as the preferred gap is 25%.

4 Simulation Results

In the simulation, 100 people are in the line and the simulation is conducted for 2000 time step. The perturbation is made by the first people. To make the perturbation similar to the traffic signal, the perturbation is repeatedly generated in a same pattern.

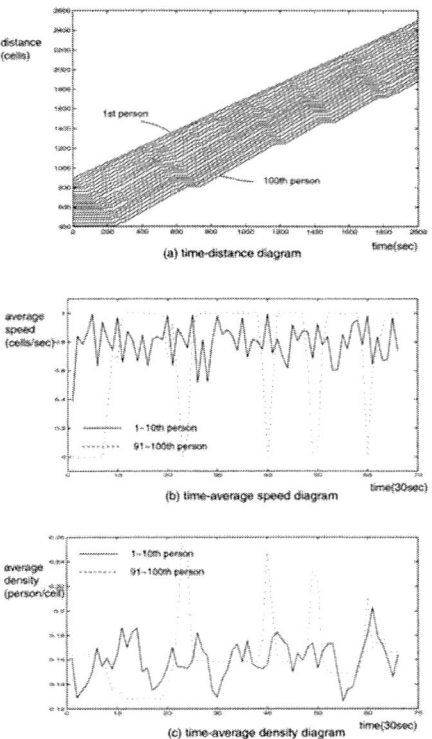

Fig. 4. Simulation results for 40 time step move 10 time step stop

For example, the first people moves during 40 time steps and then stops during 10 time steps and then move again during next 40 time steps. This moving and stop continues through the simulation time.

Figure 4 shows the results. Figure 4(a) is the time-distance diagram for every 5 people. We can see here that most of the perturbation made by the first people is dissipated, but some perturbation become stronger. So, once the upstream people stops, the duration time is much longer. Figures 4(b) and 4(c) are the speed and density plot. Here, the speed is averaged for 10 people for every 30 time steps. The solid line represents the average speed for the first 10 people and the dotted line represents the average speed for the last 10 people. In Figure 4(b), we can see that the frequency of the speed oscillation is smaller at upstream but the amplitude is larger at upstream. In Figure 4(c) the average density is drawn. Here, as a rule of a thumb, the density oscillates along the value 0.16, which corresponds to gap 5.25 cells. Figure 5 shows the results when the first vehicle moves during 25 time steps and then stops during next 25 time steps. So the perturbation is more severe than in Figure 4. Compared with Figure 4, the frequency of speed oscillation become larger and the average speed is lower than in Figure 4. The average density is also higher than in Figure 4.

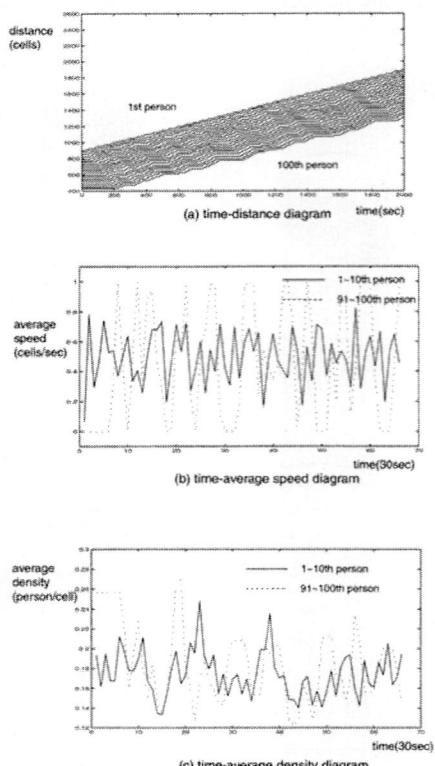

Fig. 5. Simulation results for 25 time step move 25 time step stop

5 Conclusions

The waiting line simulation showed much similarity with the features of traffic in congestion. However, there exist some differences between people's movement and vehicle's movement. For example, in traffic flow the gap can be shorter when the leader decelerates, which does not happen in waiting line case. In addition, since the speed can be in a wide range, the complexity happens when the leader and follower travels with different speed. Despite these differences, the result of the study shows if we can apply the basic ideas of waiting line to the traffic flow, we may get a good model that describes the frequency and amplitude of stop-and-go traffic, which is remained for the future study.

References

1. Chandler, R. E., Herman, R., and Montroll, W., Traffic Dynamics: Studies in Car-Following, Operations Research 6, pp. 165-184, 1958.
2. Edie, L. C. and Baverez, E., Generation and propagation of Stop-Start traffic waves, Vehicular Traffic Science, Proceedings of the third international symposium on the theory of traffic flow, pp.26-37, 1967.
3. Herman, R., Montroll, E. W., Potts, R. B., and Rothery, R. W., Traffic Dynamics: Analysis of Stability in Car Following, Operation Research No. 7, pp. 86-106, 1959.
4. Lighthill, M. J. and Whitham, J. B., On kinematic waves II: A theory of traffic flow on long crowded roads, Proceedings of the Royal; Society A 229, pp.317-345, 1955.
5. May, A. D., Traffic Flow Fundamentals, Prentice-Hall Inc., 1990.
6. Mika, H. S., Kreer, J. B., and Yuan, L. S., Dual mode behavior of freeway traffic, Highway Research Record 279, pp.1-12, 1969.
7. Payne, H. J., Models of freeway traffic and control, Simulation Councils Proc. Series: Mathematical Models of Public Systems, Vol. 1, No. 1, ed., pp.51-61, 1971.
8. Zhang, H. M., A Theory of Non-equilibrium Traffic Flow, Transportation Research Part B. Vol.32, pp.485-498, 1998.

Core Technology Analysis and Development for the Virus and Hacking Prevention

Seung-Jae Yoo

Dept. of Information Security, Joongbu Univ., Korea
sjyoo@joongbu.ac.kr

Abstract. In this paper, we try to construct the searching and treatment modules to count the malignant programs effectively so that the Worm virus detection & blocking program has the functions that it can check the files early to cope the Worm virus attacks by intercepting the value of interrupt vector table of memory structure for checking the action of viruses, and comparing with the character strings of virus. Also check the facts that files are deformed or not by generating the statistics sum of encrypting file and check the facts that files are infected or not by the pattern checking of all executable files.

1 Introduction

In 21st century, most of countries in the world try to invest enormous amount of money to build the complete society of information. As a result of this, IT became the major industry recently. Among this IT area, the business which based on internet produce tremendous amount of economic value and also its infinite potentials can evaluated by many scientists and engineers. Currently, the cyber terrors such as computer virus and harmful hacking etc. are increasing in everywhere in the world and those cyber terror are threatening the main stems of society of information. Especially, current trend of cyber terrors are rely on e-mail transfer method and various types of viruses with hacking techniques are spread out very quickly with the development of open type network environment.

In this paper, the techniques which can effectively cope with the complicate, intelligent worm will be developed. Also the types and characteristics of worm virus will be analyzed and finally worm virus detection & blocking program will be developed.

2 Malignant Code and Worm

Among the various types of system threatening codes such as virus, Worm and Trojan virus, internet worm is the most dominating system damaging factor. In DOS age, Worm was treated as non harmful code even though it copy and reproduced by itself continuously, it never contaminate the existing other files and system. As the development of network and internet, Worm also evolved to produce damages to system but itsoriginal characteristics are never changed at all. The original worm of DOS ages is call as just Worm but current worm is called as I-Worm (Internet Worm).

Worm of prototype just create so many useless trash files by copy itself continuously and it is not so harmful to system but I-Worm decrease the system speed seriously by attempting copy through the network. During past few years, many different types of I-Worm were created. As a result of it, I-Worms are classified by two different types such as Network Worm and Internet Worm according to its propagation ways. If it is propagated through local network, it is called as Network Worm and if it is propagated through global network like internet, it is called as Internet Worm.

Internet Worm is classified into three categories according to PC infection method. First group of Internet Worm is activated by just reading e-mails. Second group is activated by opening attached files of e-mail. Third group is activated by itself without any PC user's action. Also E-mail Worm is classified as Slow mass-mailers and Fast mass-mailers depending on its dissemination speed. Slow mass-mailers Worm is transferred at the same time when the infected PC users send an e-mail and Fast mass-mailers Worm is disseminated to many e-mail users at once. E-mail Worm use the e-mail client such as Microsoft Outlook and Outlook Express to disseminate the worm to other PC users and it is transferred at the same time to all the users whose e-mail addresses are listed in specific mail client. On that way, if one is infected by e-mail worm then so many other PC users whose e-mail addresses are stored in an infected PC have possibilities of infection. This chain reaction can cause great amount of PC infections and damages in very short time. Current trends of e-mail worm such as Loveletter and Navidad, use very sensitive words which stimulate the PC user or use the title that lewd photos or video files are attached. Moreover, recent e-mail worm disguise that updated virus vaccine files are attached. These methods evoke the PC user's curiosities to open attached files or e-mail without any doubt.

Network Worm is disseminated by local network system and is consist of next three steps.

- Find a Shared Drive
- Mapping Drive
- Copying Worm and Execute

In general, copied worm is not activated immediately and it is stored at starting folder which can be executed automatically with the start of Window. So the copied worm can be activated automatically at the reboot of system. Netlog is one of the Network Worm. Netlog set the IP to search the dissemination target and find out the system which is share the entire C drive in whole subnet system. Then, set the target drive by J drive and copy the worm to Window folder and Window Start folder to make it activated for infection at next start of Windows.

Window Worm is one of the dominating Internet Worm nowadays and there are two types of Window Worm depending on which type of platform they use. Window Worm is activating at Window system and Non-Window Worm is activating at different platform. Window Worm makes use of e-mail, newsletter, IRC, MSN Messenger, Gnutella, IIS and other chatting programs. Most well know Non-Window Worm of love-letter concept is Morris Worm which is activating at Macintosh and Unix system such as Linux and Solaris. Linux Ramen Worm is first Non-Window Worm which produced tremendous amount of damages. Also the Asdmind of Solaris and Simpson of Macintosh is other types of Non-Window Worm which can be found recently.

Worm had been existed at DOS ages. In DOS age, programs and files can be copied by only floppy disk. So even though worms are copied at PC by floppy disk, those worms are copied continuously in only one system itself. There is less risk of dissemination and propagation than now.

In late 1990, development of network system is accelerated by increasing demand of e-mail service and internet boom. Because of weakness of network in 1990, frequent use of e-mail via internet can be considered as major means of worm propagation. Also most of PC operating system was unified by Window and it makes easy to propagate the worm to worldwide. Worm propagation method using network weakness can be grouped in two categories. First method is using the weakness of sharing folder and second method is using the various service weaknesses. Major steps of first method are,

Scan the weakness of sharing folder in the network group which has infected system. If the weakness of sharing folder was detected then copy the worm to that system. Second method is to scan in preset way as many and unspecified systems and check whether the weakness of specific service is exists or not like IIS weakness. Finally copy the worm file to the system which has weakness.

Worm Propagation through e-mail is most effective and powerful method among any other methods. It combined with social and engineering technique. Some of worms are activated by opening the attached files in an e-mail. Recent worm files are downloaded automatically by just opening the e-mail to check. In this manner, malignant code can be spread out very widely in a short time. The macro virus maker tried to make their own malignant code equipped with e-mail function. So the viruses and worms are equipped with e-mail and back door functions. This can be great trial to evolve the simple structure worm to which the worm of fast propagation through the network system is possible.

Sasser Worm also has a combined function of worm and back door. It makes to open the specific port which has weakness and enables to contact as many users as possible. As worm is activating, specific command is executed to download additional worm file.

3 Analysis of Threatening Components by Worm

Sasser.worm.15782.D (Sasser) will be used to analyze the threatening components of worm. Sasser makes use of the LSASS weakness (MS04-011) of Window to scan and attack. MS04-011 is the serial number of security update for Windows system of Microsoft. According to MS04-011(2004. 4. 14), Windows have LSASS weakness, LDAP weakness, PCT weakness, Winlogin weakness, Metafile weakness, Help weakness, Utility Manager weakness, Window Manage weakness, Local explanation table weakness, H.323 weakness, Virtual DOS Machine weakness, Negotiate SSP weakness, SSL weakness, ASN.1 "Double Free" weakness.

Once worm is activated, it copies itself to other system which has same weakness by scan the network. Sasser was made to attack Windows XP systems especially which have LSASS weakness. Under the system which affected by LSASS, buffer overrun weakness which allow the remote code execution can be exist and this is the LSASS weakness. Intruder can install the program by utilizing the buffer overrun weakness and then has a right to view/edit/delete the data. Finally infected system can be remote controlled completely. LSASS offer the interface which can manage the local security, domain certification and Active Directory Process. LSASS handles the certification of both client and server. Also it contains the support function for Active Directory Utility. Hackers can attack the unchecked buffer of the program and then substitute the new malignant code to be activated by executing the changed program. This program can be operated by hacker without any protection. Also it causes the symptoms of program error and this is called as Buffer overrun. Sasser is one of various Netsky's mutations. It copies itself to Windows and then loads the program whose name is "Isasss.exe" to the process. This procedures repeat continuously by loading itself again. Even if the registry is changed to reboot the Windows, it can be executed continuously.

3.1 System Symptoms

- It is generated a file lsasss.exe (15,872byte) in the folder Windows . Then, we should give attention to distinguish a normal file lsass.exe to the worm file lsasss.exe, and we note that it is copied in C:□WinNT under WinNT/2000 series and in C:□Windows under WinXP series.
- It is generated files (Number)_upload.exe in the folder Windows System32 (It is inherited the difference of Windows NT and XP series). But, we can not confirm this phenomenon if the internet is shut off.
- It occurs the system-closed and rebooting caution caused by errors of the lsass code.
- It may not be not infected formally in some computer network system.
    In fact, it may be failed to execute the worm which can be showed the error massage that it is not infected because the code of the file iphlpapi.dll is different (mainly under the Window 2000).
    In the case of the Window XP series, sometimes we can see the error massage 'LSA Shell (Export Version)' which is to close and reboot the system. But it is not the symptoms that the computer is formally infected.
- We register the data lsasss.exe in registry to execute the worm again at the time of rebooting the computer.
- We delete the files ssgrate.exe, drvsys.exe and Drvddll_exe in the registry so that the Bagle worm do not execute at the time of rebooting the Bagle worm infected computer.
- Sufficiently long times later on the infection condition, it's share in CPU is near 100% and so it is difficult to use the system and it can be occurred many processes by one file.
- Some measure of time later since the infection condition is founded, we can view the message in the mutex window which notices the infected fact.

3.2 Network Symptoms

- If the network is infected, it is opened the ports from number 1025 to 5554 and so it is menaced by another attacks. In facts, it enables to confirm by the netstat-na commander in command window in Fig.1.

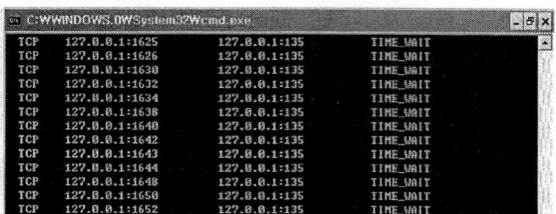

Fig. 1. Netstat Port Monitoring

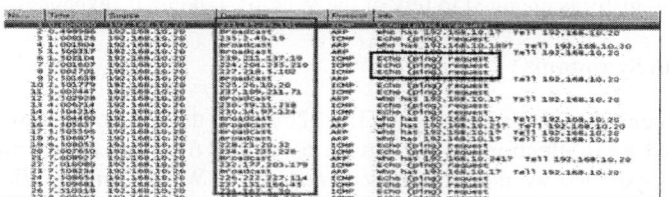

Fig. 2. Search the Next Attack Target

- It becomes to be created and checked several unspecific attack-target addresses. The formation rules as we can confirm, is as follows in Fig.2.
    - For the formal IP addresses (A. B. C .D),
    - A is fixed and the others are random: 40 ~ 50% approx.
    - A and B are fixed, and C and D are random: 20 ~ 30% approx.
    - A, B and C are fixed, and D is random: 30 ~ 40% approx.
- If it is founded the vulnerable points in the process of confirming the attack objects, the infecting action is executed in Fig.3.

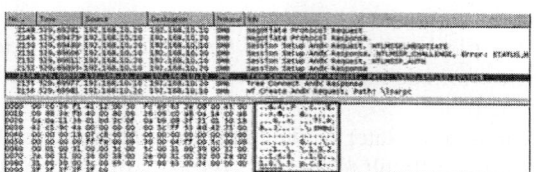

Fig. 3. Execute Script to Propagate (1)

Then, in the process of carrying out the infecting action, a FTP script might be made as the name cmd.ftp. This script contains the commander which can download the worm and can execute this worm. But it can be carried out only if the internet is connected in Fig.4.

Core Technology Analysis and Development for the Virus and Hacking Prevention 875

Fig. 4. Execute Script to Propagate (2)

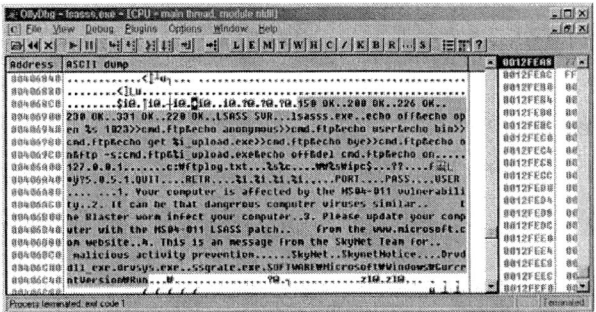

Fig. 5. Mutex Messagees

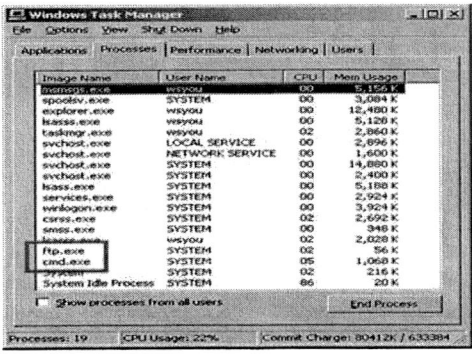

Fig. 6. Commander Execution by Script

* In above dump picture, it contains the picture within the cmd.ftp script and the contents in the mutex window in Fig.5.

* The worm file is downloaded through the open ports which is occurred by executing the files cmd.exe and ftp.exe in the infected Windows in Fig.6.

4 Corresponding Program Modules

As the exclusive Vaccine type for Sasser.worm.15872.D, the basic formula for searching adopts the SLIM (Specific Locate Inspection Method) which keeps the lower wrong diagnosis rate and takes shorter hours for inspection.

Table 1. Comparison the MLIM (Memory Locate Inspection Method)

	Strength	Weakness
SLIM	□Low wrong diagnosis rate □Short hours for inspection □High success rate for Treatment	□Low diagnosis rate for deformation virus □Can Not inspect unless interrupted □Do Not inspect if reside in other program
WLIM	□High diagnosis rate for deformation virus *WLIM(Whole Locate Inspection Method)	□High wrong diagnosis rate □Long hours for inspection □Low success rate for Treatment

4.1 Searching Module

(1) Set up the extensions of files to be searched and searching target directory
(2) Check the extension (designate to execution file) and searching targets if it runs the searching process.
(3) Compare the target execution files with the comparing character strings by dump checking.
(4) Output the object files in view data list if it agrees with the comparing character strings.

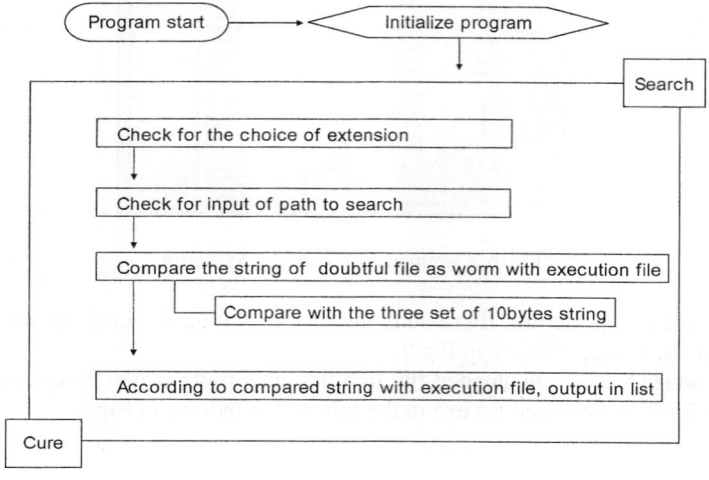

Fig. 7. Vaccine Searching Module

Core Technology Analysis and Development for the Virus and Hacking Prevention 877

In this searching module, we take a notice the two essential points. One is that it must find the object files by searching every subdirectories in the disk before inspecting each file in vaccine program, and the other is that it must be compared with specific character strings for the searched files. In facts even though there are many files in each directory, above all it executes the search for subdirectory. If we describe the directory structure as a tree graph as in fig.8, first after the search the directory A, it accesses the first subdirectory B by recursive call. Once B is searched, again it accesses the first subdirectory, recursively. So in the final first subdirectory R the file searching is executed and then is repeated one after another in ordered R, S, I, J, D, K, E, F, B, L, G, H, C, A.

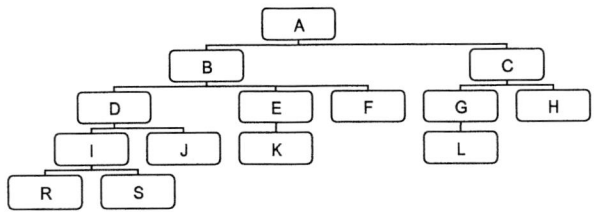

Fig. 8. Directory structure

Next, once a file is accessed, it is compared with specific character strings in file. Here execution files or Worm files are opened to the forms of ASCII codes and we designate these files to several parts of the ASCII codes, and then it executes the comparison these parts (designated character strings) with specific character strings by turns. In this study, we establish the files' designated character strings with three 10 byte character strings in their ASCII codes.

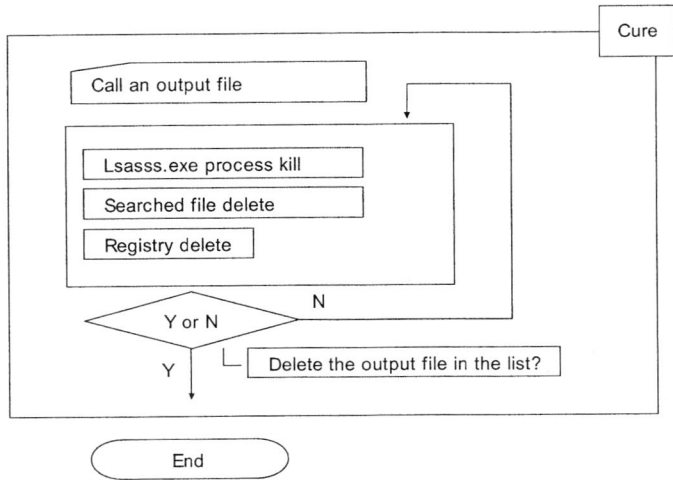

Fig. 9. Vaccine Treatment Module

4.2 Treatment Module

(1) Execute the treatment beginning the first item in the list view if press the treatment button.
(2) End the corresponding process (inspect one after another in the processing lists)
(3) Delete the detected files under the process
(4) Delete the registered key values.

There is no serious problem by Sasser Worm if once one deletes the registry and some its informations. But if the treatment is executed under the Worm action, the worm acting files do not treat. So the treatment process must be executed after closing the programs on the process.

5 Conclusions

Recently according to spread rapidly worldwide of using the internet, there appears highly flourished many malignant programs and it is very various their routes of infection into the world. As the IT infra and internet infrastructure is increasing and changing to the opening network, there is great risk cause by the hacking and viruses. Also it is well-known that the there is a tremendous loss of property by them. There is no doubt that there is a countermeasure to diffuse the fresh version vaccine program into public institutions as well as into each and every person. But we realize the limitations to count the intellectual, high technical and high structured hacking methods and Worm virus in the inside and outside of the country.

In this paper we try to develop a technique to count the malignant programs effectively. We check and analyze the properties of recent hacking methods, reported Worms information and their menace factors. Then Worm virus detection & blocking program is developed which has the functions that it can check the files early to cope the Worm virus attacks by intercepting the value of interrupt vector table of memory structure for checking the action of viruses, and comparing with the character strings of virus. also check the facts that files are deformed or not by generating the statistics sum of encrypting file and check the facts that files are infected or not by the pattern checking of all executable files.

Acknowledgements

This work was supported by a grand No. R12-2003-004-00006-0 from Ministry of Science and Technology.

References

1. Adam, J., "Virus Threats and Countermeasures", IEEE Spectrum, August 1992.
2. Chess, D., "The Future of Viruses on the Internet", Proceedings Virus Bulletin International Conference, October 1997.
3. Denning, P., "Computers Under Attack : Intruders, Worms and Viruses", Addison-Wesley 1990.

4. Hoffman, L., "Rogue Programs : Viruses, Worms and Trojan Horses", New York, Van Nostrand Reinhold, 1990.
5. Nachenberg, C., "Computer Virus-Antivirus Coevolution", Communications of the ACM, January 1997.
6. William, S., "Cryptography and Network Security Principles and Practics", New Jersey, Prence-Hall, Inc., 1999.

Development of Traffic Accidents Prediction Model with Intelligent System Theory

SooBeom Lee[1], TaiSik Lee[2], Hyung Jin Kim[3], and YoungKyun Lee[4]

[1] Assistant Prof., Dept. of Transportation Engineering,
Univ. of Seoul, Korea
mendota@uos.ac.kr
[2] Associate Prof., Dept. of Civil and Environment Engineering,
Hanyang Univ., Korea
cmtsl@hanyang.ac.kr
[3] Associate Prof., Dept of Urban Planning and Engineering,
Yonsei Univ., Korea
hyungkim@yonsei.ac.kr
[4] Director, ITS Policy and Program Division,
Ministry of Construction & Transportation, Kwacheon
ykleefiu@moct.go.kr

Abstract. It is important to clarify the relationship between traffic accidents and various influencing factors in order to reduce the number of traffic accidents. This study developed a traffic accident frequency prediction model using multi-linear regression and quantification theories which are commonly applied in the field of traffic safety to verify the influences of various factors in the traffic accident frequency. The data was collected on the Korean National Highway 17 which shows the highest accident frequency and fatality in Chonbuk Province. In order to minimize the uncertainty of the data, the fuzzy theory and neural network theory were applied. The neural network theory can provide fair learning performance by modeling the human neural system mathematically. In conclusion, this study focused on the practicability of the fuzzy reasoning theory and the neural network theory for traffic safety analysis.

1 Introduction

The highway traffic system is composed of human, vehicle and roadway factors. Traffic accidents can occur because of a single factor, but mostly by a combination and internal reaction of multiple factors. The easiest way to verify the reason of traffic accidents is the analysis of accident data. It should be noted that the reliability of data cannot be guaranteed. Current methods have limitations in overcoming the uncertainty, non-linearity, and variation of time and space which can be included in the process of data collection. It is then concluded that a methodology is necessary to use the collected data while accepting the uncertainty.

In this study, the influences on traffic accidents were analyzed using various factors and considering the possibility of uncertainty.

2 Research Scope and Methodology

The traffic accident data applied in this study was collected on the Korean National Highway 17 which shows the highest accident frequency and fatality in Chonbuk Province. Multiple traffic accident frequency prediction models were developed with the data using multi-linear regression, quantification, fuzzy reasoning, and neural network theories. Among human, vehicle, and roadway factors, this study was limited to those roadway factorr only which included safety appurtenances, geometric features, traffic operation and traffic flow characteristics.

3 Related Works

Zegeer developed a traffic accident frequency prediction model for 2-lane rural highways with independent variables such as vertical alignment, average daily traffic volume, lane width, and shoulder width. The data were collected in 5,000 miles of 2-lane rural highways in seven states of the U.S.A. Kang developed a spatial autoregressive data analysis model to analyze the spatial relationship to traffic accident frequency. The model was based on the factors such as traffic volume by unit section, existence of interchanges and heavy vehicle ratios. Hirofumi performed a study on the development of counter measures for pedestrian-vehicle accidents at intersections based on the q uantification I theory. Kay Fitzpatrick analyzed the relationship between traffic accident frequency and the geometric elements such as lane width, existence of median, horizontal curvature, deflection angle. The study verified that inconsistent lane width is an important cause of accidents.

4 Theories

4.1 Multi-linear Regression Theory

The multi-linear regression analysis is to derive the linearity between multiple independent variables and a dependent variable. It is used to estimate the quantities of relation of independent variables to the dependant variable. The equation (1) is the basic structure of the multi-linear regression model.

$$Y = \alpha + \beta_1 X_1 + \cdots + \beta_l X_l + e \qquad (1)$$

where Y =dependent variable, X =independent variable, α =constant, β =parameter, and e = error.

4.2 Quantification I Theory

The quantification I theory estimates the production of external standard, Y by the variations of explaining characteristics $(X_1, X_2, \cdots, X_m)$. It should be advanced to classify the multiple items related to the phenomenon into several categories. The quantification I theory is to perform the factor analysis for items and categories with the scattered data. It corresponds with multi-linear regression model when the explaining characteristics, x_j are continuous. It can be expressed as a linear relationship as shown in equation (2) when the external variable y_i is the number of accidents and the explaining variable (factor) is the characteristics of the spot on an spot, i.

$$y_i = \sum\sum a_{jk} x_{jki} + \varepsilon_i \qquad (2)$$

where y = external variable, x = independent variable, a = constant, and ε = error.

4.3 Fuzzy Reasoning Theory

The fuzzy reasoning theory is to derive a proposition from multiple fuzzy propositions. It is a way to handle the vagueness of concepts under the conceptual uncertainty of uncertain boundaries. The process is very similar to human reasoning. The rule of the fuzzy reasoning is "IF A_i and B_i THEN C_i". The result of C^* is obtained through the concurrence of i (all the rules including the A_i, B_i) and A^*, B^* (input value). Reasoning structure helps the input value to be applied for each rule and induces an approximate value although the present input value does not match with the standard input value required by the rules. The return value is the compound of all the C_i^*. Figure 1 shows the fuzzy reasoning process.

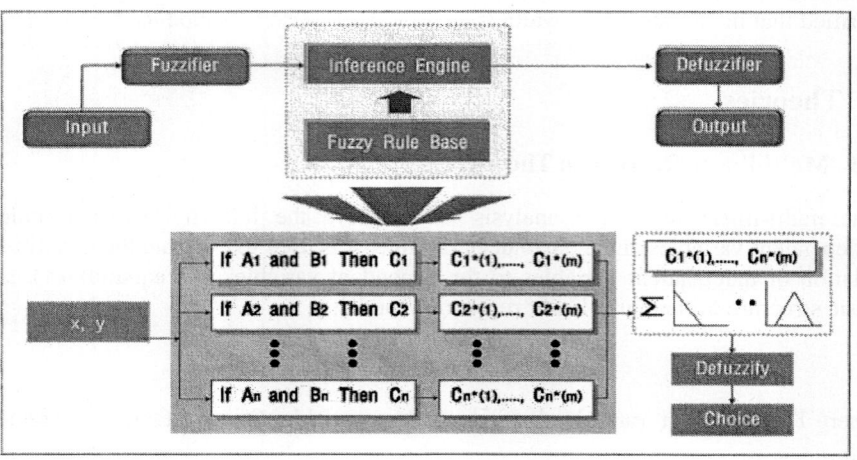

Fig. 1. Fuzzy reasoning process

4.4 Neural Network Theory

The neural network consists of these artificial neurons, as it is the model of human brain's structures for special functions and missions. The neural network performs learning processes through mapping between input and output. A procedure of learning processes is called a learning algorithm and functions to adjust the weight of the neural network to achieve targets. The neural network, a mutual connection of nodes that can perform mathematical operations, can be operated according to proper learning rules. That is, each node performs the mathematical operation with Combination Function and Transfer Function (Activation Function). Inputs to actual nodes are the sum of the values with weights as shown in the equation

$$s_j = \sum_{i=0}^{m} (w_{ji} \cdot x_i) \qquad (3)$$

where w_{ji} = connection weight and x_i = input value.

In the equation (3), s_j, actual input, passes through non-linear function called Transfer Function or Activation Function. The most widely used non-linear function is Sigmoid Function. The equation (4) provides the y_j, the results of s_j input. Figure 2 shows the multi-layer neural network.

$$y_j = f(s_j) = \frac{1}{1+\exp(-s_j)} \qquad (4)$$

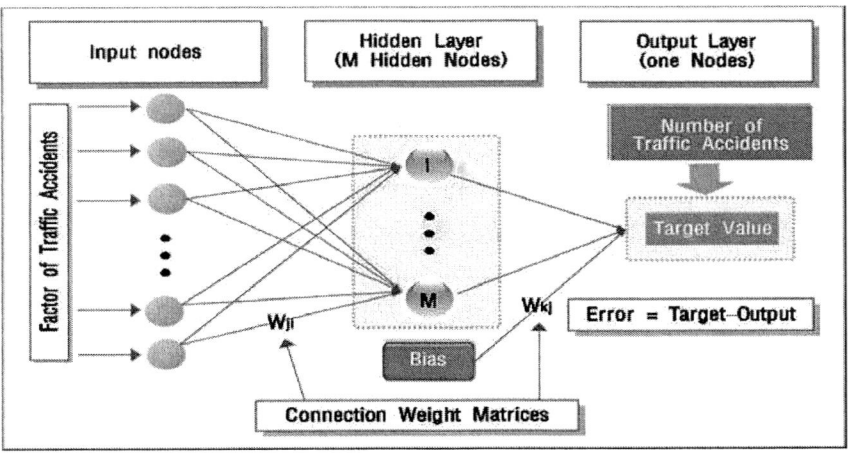

Fig. 2. Multi-layer neural network

5 Development of Traffic Accident Frequency Prediction Model

This study was based on a microscope approach. The factors of the traffic accidents selected were traffic volume, vertical alignment, cross slope, horizontal curvature and roadway width. The factors were arranged by the magnitude of the influence to traffic accidents. The correlations between the variables were also reviewed.

5.1 Multi-linear Regression Model

A multi-linear regression model was developed to estimate the influences of geometric elements such as vertical alignment, horizontal curvature, cross slope, and roadway width as shown on the Table 1. The fitness of the model was evaluated with R^2-value and F-value. The R^2 value with more independent variables showed a higher value, than when the evaluation was performed with the $Ajusted\ R^2$. In addition, a t-test was also performed to evaluate the contribution of independent variable to the dependent variable. According to the result, R^2 was 0.239 and it seemed relatively low. However, compared with other multi-linear models for traffic accident prediction models, the value is acceptable. The reason for the low value is that the human factor is the more influencing factor than the vehicle or roadway factors. It also included the limitation of reliability on the data. The result of the F-test also showed that the model was valid with 95% of confidence level.

Table 1. Result of multi-linear regression model

Variable	Parameter	Standard Error	β	t-Value	Level of Significance
Constant	51.095	9.548		5.351	0.000
Traffic Volume	0.138	0.109	0.085	1.272	0.205
Vertical Alignment	1.353	0.402	0.228	3.367	0.001
Horizontal Curvature	-0.133	0.087	-0.105	-1.522	0.130
Cross Slope	-48.93	8.081	-0.376	-5.552	0.000
Roadway Width	-1.590	0.501	-0.217	-3.173	0.002
$R^2 = 0.239 (Ajusted\ R^2 = 0.217)$					

5.2 Quantification I Model

Another traffic accident frequency prediction model was developed based on the quantification I theory with the same independent variables for the multi-linear regression model. In this process, several performance measures were derived such as the respective contribution of variable, the amount of items and category, the range of items, multi-correlation factors, and average estimation errors. The explaining variable set was prepared by category for each item after combining cases. The result is shown on the Table 2.

According to the result, it is possible to derive the range of items to judge the level of explanation. It should be noted that the contribution of a factor increased as the range of items widened. The range of traffic volume was the widest, and it could then be concluded that the traffic volume was the most influential. The explanation was evaluated with the multi-correlation factor of 63.56%, and the error of estimation was

4.765. The R^2 value was 0.400 and it was higher than that from the multi-linear regression model.

Table 2. Result of quantification I model

Item	Category	Parameter	Range
Traffic Volume	Less than 18,000	-3.28090	6.35349
	More then 18,000	3.07259	
Vertical Alignment	Less than 1%	-0.91310	1.42211
	More then 1%	0.50900	
Horizontal Curvature	Less than 650m	-1.74912	3.46041
	More then 650m	1.71130	
Cross Slope	Less than 3%	1.45299	3.71885
	More then 3%	-2.26585	
Roadway Width	Less than 17.70m	0.13223	0.37518
	More then 17.70m	-0.24294	
Correlation Coefficient : 0.6356		$R^2 = 0.400$	

5.3 Fuzzy Reasoning Model

1) Preparation of Membership Function for the Application of Fuzzy Reasoning Theory
In the development of the fuzzy reasoning model, the selected input variables were the same as the previous models, and the output variable was provide in the form of the membership function with the traffic accident frequency. The traffic volume was classified into three membership functions: high, medium, and low. And, the vertical grade, cross slope, horizontal curvature and roadway width were classified into two functions: high/low or wide/narrow. Figure 3 shows the membership functions.

2) Calculation
The applied fuzzy rule was the IF-THEN rule which corresponds to common sense. The fuzzy reasoning was based on the Min-Max Centroid method, and the range and level of overlapping of the membership function by each variable were decided by the mean of the data and 50% of overlapping. The result is shown on Table 3. The average PI of the fuzzy reasoning model is a measure of fitness, and calculated by equation (5).

$$PI(average) = \frac{\sum_{i}^{n}|R_i - P_i|}{n} \quad (5)$$

where n = number of data points, R_i = actually measured value, and PI = predicted value.

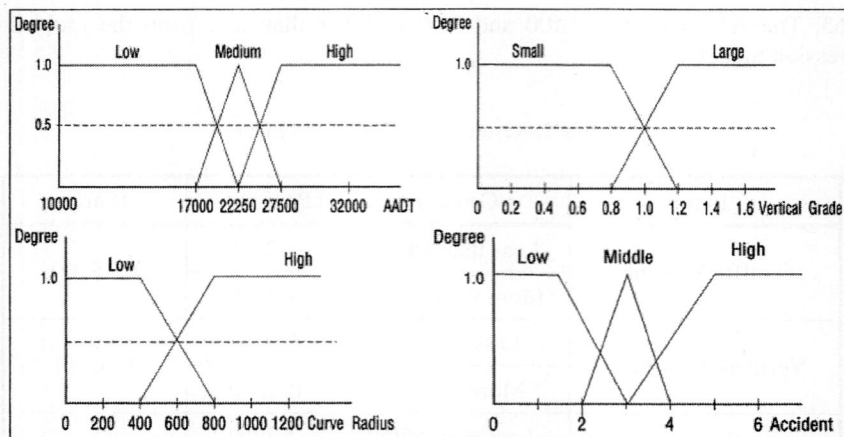

Fig. 3. Membership functions

Table 3. Result of fuzzy reasoning model

Variable	
Traffic Volume (Low, Medium, High) Vertical Alignment and Cross Slope (Small, Large) Horizontal Curvature and Roadway (Low, High)	
PI	0.421
MSE	0.523
Hit ratio (%)	62.35
R^2	0.612

The PI value and MSE were 0.421 and 0.523 respectively. The R^2 value by the quantification I theory was 0.612and it showed higher value than those from the advanced process.

5.4 Neural Network Model

The learning data used to make the neural network model was the same as the input variables in the previous models. Whole learning data were normalized as the pre-process. It means that the data was transformed and that the average and the standard deviations were [0,1]. These values were transformed to the actual values by post-process after the development of the model. The Error Back Propagation Algorithm was applied. The algorithm is commonly applied to make the neural network model. The neural network model was developed for each variable. When the number of nodes on the input layer was n, the number of nodes on the hidden layer varied n to 6n, and the optimal neural network model could be developed. The maximum number of the feed-back was determined to be 10,000 because it is known as the optimal number in terms of the efficiency. The learning method of practice for the adjustment of the neural network in this study was the momentum-adaptive learning rate method. It could avoid the error converging to the local minima and increase the efficiency of learning. The constant of momentum and the initial learning ratio were acquired from the sensitivity analysis to be 0.1. The performance of the model could be evaluated by the *MSE* as shown in the equation (6). When *MSE* is 0, the observed and the estimated values are perfectly the same.

$$MSE = \frac{\sum_{i=1}^{n}(obs_i - \exp_i)}{n} \qquad (6)$$

where n = number of data points, obs_i = target value, and $\exp_i$ = output value.

Table 4. Result of neural network model

Variable	
Traffic Volume (Low, Medium, High) Vertical Alignment and Cross Slope (Small, Large) Horizontal Curvature and Roadway (Low, High)	
PI	0.353
MSE	0.491
Hit ratio (%)	65.23
R^2	0.672

The result is shown on the Table 4, and *PI* and *MSE* were 0.353 and 0.491 respectively. The R^2 value was 0.672, and it is the highest compared with the results from the other models.

6 Conclusions

This study developed traffic accident frequency prediction models by multi-linear regression analysis, quantification I theory, the fuzzy reasoning theory, and the neural network theory. By comparing the outputs from the models with the actual data, the performances of the models were evaluated. According to the result of the comparison, the models of the fuzzy reasoning theory and the neural network theory were superior to those of the multi-linear regression analysis and the quantification I theory in terms of R^2 and the estimated standard error. Each model hires 5 geometry elements as input variables. Models are aimed to evaluate how each input variable affects the occurrence of traffic accidents. The multi-linear regression model suggested that cross slope, roadway width and vertical alignment influence traffic accidents. The quantification I model showed that traffic volume, roadway width and cross slope affects traffic accidents. It could then be concluded that the fuzzy reasoning analysis and the neural network analysis could provide more effective methodology in traffic safety analysis. In addition, the priority of countermeasure implementation could be decided by the level of contribution of each factor.

References

1. Clark, C.T. and Schkade, L.L., Statistical Analysis for Administrative Decisions, South-Western Publishing Co., Cincinnati, 1974.
2. C. V. Zegeer, J. Hummer, L. Herf, D. Reinfurt, and W. Hunter, Safety Effects of Cross-Section Design for Two-Lane Roads, Report No. FHWA-RD-87-008, Federal Highway Administration, Washington, D. C., 1986.
3. Faghri, A. and Hua, J., Evaluation of Artificial Neural Network Applications in Transportation Engineering, Transportation Research Board 1358, 1991.
4. Fitzpatrick, K. et al., Speed prediction rot two lane rural highways, Research Report, FHWA-RD-99-171, 2000.
5. Haykin. S., Neural Networks-A Comprehensive Foundation. Prentice Hall, New Jersey, 1999.
6. Laurence Capus and Nicole Tourigny Road Safety Analysis: A Case-Based Reasoning Approach, Transportation Research Board, January 1998.
7. Marir F and I. Watson, Case-based Reasoning, A Categorized Bibliography, The Knowledge Engineering Review, Vol. 9, N0. 3, 1994.

Prefetching Scheme Considering Mobile User's Preference in Mobile Networks

Jin Ah Yoo[1], In Seon Choi[2], and Dong Chun Lee[3]

[1] Dept. of Public Opinion Polls, Jeonju Univ., Korea
 `gina@jj.ac.kr`
[2] Div. of Information and Comm. Eng., Chonbuk National Univ., Korea
 `ischoi@dcs.chonbuk.ac.kr`
[3] Dept. of Computer Science Howon Univ., Korea
 `ldch@sunny.howon.ac.kr`

Abstract. This paper proposes a mobile computing prefetching method considering the user's interest and the common popularity. For mobile computing environments, there exist restrictions as like bandwidth, latency and traffic. Since, in the present, we prefer the united multimedia mobile information service to the voice-based one, these obstacles are regarded as big problems. To solve those problems, a variety of techniques as well as caching or prefetching have been studied. However most of them are not sufficient in providing the amount of data that the user wants. We suggest a prefetching method to bring information early by using information about the user's former interest and the popularity. Comparing to the previous methods in numerical results, the proposed method improves the prefetching performance to give the maximum effectiveness and reduces the failure rate of information searching.

1 Introduction

The cellular communication environment is currently based on the voice communication, but the new environment is required gradually to ease data communication by increasing the bandwidth. As the information communication industry develops rapidly and the number of Internet user increases, the mobile computing environment is converted from wire Internet service to wireless Internet service in the technology. The application services that are general in wire Internet should be offered to a similar extent in wireless Internet. However technologies applied in existent wire environment have a lot of restrictions to apply directly to wireless network environment [1]. Namely mobile information service requires quick context-aware conversion in movement, so the mobile user must get new information when moved to new location. The low bandwidth, high latency, traffic and frequent connection due to the characteristics of mobile environment are remained the obstacles to users. We should find the solution by utilizing the existing bandwidth instead of increasing bandwidth that would cause additional expense. That is a prefetching method. The basic idea is that we use prefetching information in advance and use them again to accommodate wireless network property. But prefetching data expected to be referred in the near future has the flaw that needs lots of memory and spends much computing time. Therefore to improve the performance of mobile information service we need to set suitable prefetching zone which gives maximum effectiveness.

We propose the most natural strategy applied prefetching with context-aware service using the user mobility model of mobile device. The prefetching zone is set by applying popularity of general publics, the frequency that the user have visit a location and time that the user have stayed at the location. That is to improve the effectiveness of memory in the limited memory of mobile computing by maximizing the utilization rate of prefetching information. Also, we applied the user's interest and popularity to minimize missing rate in prefetching zone.

2 Related Work

In mobile information systems servers can no longer be conscious about the data available on a client or even about the clients connected to the network. It is thus the clients' responsibility to initiate the validity procedure.

In [20], an adaptive network prefetching scheme is proposed. This scheme predicts the files' future access probabilities based on the access history and the network condition. The scheme allows the prefetching of a file only if the access probability of the file is greater than a function of the system bandwidth, delay and retrieval time.

Prefetching method is a well established technique to improve performance in tradition distributed systems based on fixed nodes, and several papers exist about this topic [11, 15]. Some papers have also considered the utility of this technique in the framework of mobile computing, in general from the viewpoint of improving the access to remote file systems; the use of mobility prediction has been also considered for this purpose [6, 7, 8, 10].

Dar [3] proposed to invalidate the set of data that is semantically furthest away from the current user context. This includes the current location, but also moving behaviors like speed, direction of the user.[19] make use of predefined routes to detect the regions of interest for which data is required. In such a way they have location information for the whole ongoing trip and do not have to compute the target areas while on the move.

Cho[1] provides an interesting approach by considering the speed and moving direction of the mobile user. These two aspects are important elements of the movement pattern. The speed provides about the velocity with which a user changes locations. Moreover, the size of the user's area is largely dependent on the speed. Whenever the user crosses the borders of the current zone, new prefetching zones is computed. Depending on the speed in the moment that the user leaves the scope of a zone, the new one considers more or less adjacent network cells. Frequency prefetching method [2] analyzes mobility pattern of user accumulated during fixed period. Informations that is worth being used to the future with this are prefetched. Frequency is based on the speed. If predict with data that is accumulated during given period, there is problem to itself. To solve this problem, some factors should be added, and we propose Frequency, Interest and Popularity (FIP) scheme that mobile users have preference.

3 FIP Prefetcing Scheme

We present mobile environment structure for FIP algorithm in Figure 1 (a). When a mobile user demands information, the acquiring process is composed of level (1)-(7). The request information links to a server from the mobile device through the base station. The server is transmitting data items stored in Database to the mobile client terminal. Prefetching in Figure 1 (a) is based on Virtual Prefetching Table (VPT)and Actual Prefetching Table (APT)to get data in advance which a user is going to need. VPT is a virtual location derived from velocity-based prefetching, which is applied to accumulated frequency.

For extensive FIP prefetching a user's move pattern is assumed in the following. First, a user visits a location in which the user interested repeatedly in mobile environment.

Second, revisit rate would be high on the location many users visited. If the above assumptions are satisfied, there exists a parameter showing the relation between the user and the location. Not only User's interest must be considered if the user visits a location frequently but popularity obtained from other users. So we use popularity as well as user's interest in prefetching algorithm. Note that those variables are independent of location attribute. We can get APT of prefetching from frequency, interest and popularity.

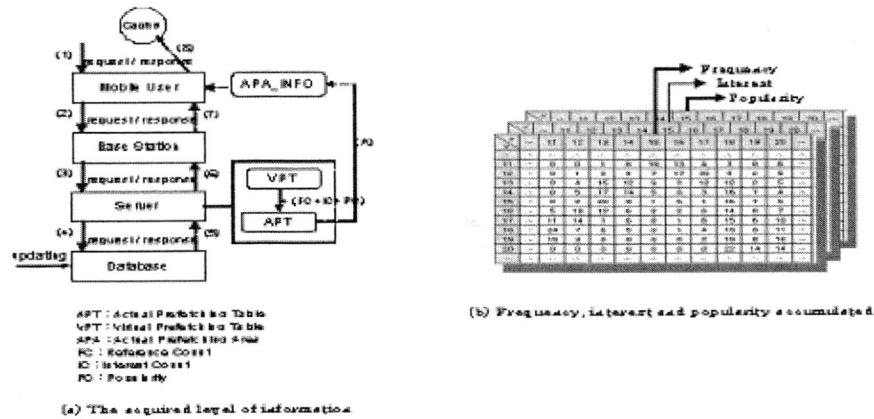

Fig. 1. The general plan to apply FIP prefetching scheme

However, to reach to the target object, if observe object visit process by user's mobile pattern, object that pass is existed. Therefore, we use model that consider time to distinguish target object. In addition, user's mobile pattern is explained by user's interest for object. That is, access probability may be high in object that user has some interest degree or popularity that do not visit all objects. Interest or popularity depends on time, and access object is also changed. Therefore, interest degree and popularity in Figure 1(b) apply each access probability value by time. We use exponential distribution to obtain interest degree and popularity. Also it follows in popular degree and the user access pattern uses a Zipf distribution. When most from the object which is popularity decreasing a ranking, it is a distribution which shows the probability which will be approached from the object which has a ranking. Given that a user is moving from x to y, we can obtain the following prefetching model equations,

$$FIP(X,Y) = F(X,Y) + I(X,Y) + P(Y,W).$$

where is frequency score based on velocity, represents interest probability value, and is popularity probability value.

When user moves, mobile place has interrelation between user and object. Also, because mobile pattern has regional attribute, use modulation parameters that can control regional intensive training in model. Environment that popularity should be applied first than interest degree of specification information can happen. One is when user entered first in object, and there will be the other in object that user is updated newly. However, all of this two environment are excluding for conveniences sake and suppose that information given was accumulated like Figure 1(b). We can get the most satisfied data to user.

Begin
/* Step 1. Initialization */

User's moving range coordinate is allocated;
Allocate a,b value to modulation parameters of Frequency, Interest and popularity;

/* Step 2. Accumulation */
Frequency, Interest and Popularity are accumulated for 30 days;

/* Step3. FIP scheme */
while(x and y coordinate aren't the end)
Obtain value from accumulated information;
Compare value of Interest and Popularity;
if(choose big value in Interest and Popularity), them
Add after multiply b , (1-b) to value;
Add Frequency in result;
/* Step4. Prefetching to apply FIP scheme */
End.

Fig. 2. FIP scheme

4 Performance Evaluation

The FIP scheme is considered spatial locality and temporal locality. It is improved prefetching method. In order to analyze efficiency, the typical move scenario is presented. In the velocity-based mobility model, a user moves in a two-dimensional portion's area with the constant speed and direction during any given unit time period. In this simulation, a mobile user is assumed to move around 25 by 25 portions. The user repeats process that move to position preserve of following destination 20 times according to given coordinate value beforehand. We establish virtual prefetching area with the different velocity in each move. A simulation has been done among the prefetching strategies with Velocity Prefetching(VP), Frequency Prefetching(FP) and FIP.

The numerical result has been measured in three aspects,

- the amount of prefetched portion information,
- the utilization rate of prefetched portion information,
- missing rate of information retrieval.

The simulator has been implemented in C using event-based simulator CSIM [7]. In our experiment, we were assumed for accumulating the mobility reference count for 30 days in simulation time.

4.1 The Amount of Prefetched Portion Information

With given the user mobility scenario, the number of prefetched portion information has been evaluated to show the communication and storage overhead of three prefetcing strategies. Table 1 shows the simulation result. It reports the total number of prefetched information, and the mean number of prefetched information. Proposed strategy FIP was shown improvement of performance about 0.32 when compared with FP.

4.2 The Utilization Rate of Prefetched Portion Information

Now, it is meaningful to figure out the utilization rate. The utilization rate can be achieved with the number of prefetched information actually participated for the user's location-aware service out of all the number of prefetched information. It just reflects the predictability degree for the given prefetching strategy. With the given user mobility model, the utilization rate with FP is much better than that with VP; the former shows that over 0.737 of prefetched portion information is utilized for real service, while the later figures around 0.576. With FIP, the rate is getting to 0.897. Table 2 summarizes the results of this experiment.

4.3 Missing Rate of Information Retrieval

In section 4.1 and 4.2, improved result of the number and the utilization rate of prefetched portion information was seen. This is very important element as service methodology to recognize current situation of mobile information service. In spite of these advantages, it cannot overlook vulnerable point that is information retrieval failure rate. Consequently, we considered user's interest and everybody's popularity. As a result, the missing rate of the information retrieval with FIP was shown improvement of performance of 0.15 than that with FP. Table 2 summarizes the results of this experiment.

Table 1. The number of prefetched portion information

Test Case	The total number	The mean number
Velocity prefetch scheme	201	10.05
Frequency count prefetch scheme	109	5.45
Preposed prefetch scheme	82	4.01

Table 2. Utilization rate of the prefetched portion information

Test Case	The mean Utilization rate
Velocity prefetch scheme	0.576
Frequency count prefetch scheme	0.737
Preposed prefetch scheme	0.897

Table 3. Missing rate of the information retrieval

Test Case	The mean Meaning rate
Frequency count prefetch scheme	0.311
Preposed prefetch scheme	0.160

5 Conclusions

In this paper we suggested prefetching method that consider space locality and time locality at the same time applying frequency, interest, population, and time in user's mobile pattern in mobile networks. This is method that improves existent speed prefetching and frequency prefetching much more, and prefetching done information maximum use can, and prefetching in done information actual capacity effective.

Based on mobile scenario, about number and information actual capacity of information division, analyzed comparison. Also, analyzed information retrieval failure rate by information actual capacity.

Through this, FIP prefetching method proposed that present is overcoming restriction items of mobile environment effectively.

References

1. Cho, G. (2002). Using Predictive Prefetching to Improve Location Awareness of Mobile Information Service, Lecture Notes in Computer Science Vol. 2331, 2002 pp. 1128-1136.
2. Choi. I. "Applying Mobility Pattern to Location-aware Mobile Information Services," a Master's Thesis, Feb. 2003
3. Dar, S., M. J. Franklin, B. T. Jónsson, D. Srivastava & M. Tan (1996). *Semantic Data Caching and Replacement*. Proc. 22nd VLDB Conf. Mumbai, Bombay, India: 330-341.
4. Feiertag, R. J. & E. I. Organick (1972). *The Multics Input/Output System*. Proc. Third Symp. Operating Systems Principles, Palo Alto, CA, USA: 35-41.
5. Gitzenis, S., & N. Bambos (2002). *Power-Controlled Data Prefetching/Caching in Wireless Packet Networks*. Proc. Of IEEE INFOCOM 2002, 21st Ann. Joint Conf. IEEE Computer and Communications Societies, New York, USA: 1405-1414.
6. G. Liu. "Exploitation of location-dependent caching and prefetching techniques for supporting mobile computing and communications". Proc. Of WIRELESS-94, 1994
7. G.Y. Liu, G.Q. Maguire. "A predictive mobility management scheme for supporting wireless mobile computing". Proc. Int. Conf. ICUP-95, Tokyo, Japan, Nov. 1995.

8. George Liu , Gerald Maguire, Jr., A class of mobile motion prediction algorithms for wireless mobile computing and communication, Mobile Networks and Applications, v.1 n.2, p.113-121, Oct. 1996
9. Imai, N., H. Morikawa & T. Aoyama, *Prefetching Architecture for Hot-Spotted Network.* Proc. IEEE ICC2001, 2001
10. James J. Kistler, M. Satyanarayanan, Disconnected operation in the Coda File System, ACM Transactions on Computer Systems v.10 n.1, p.3-25, Feb. 1992
11. K. Korner. "Intelligent caching for remote file setvice". Proc. of the ICDC1990, 1990
12. Kirchner, H., B. Mahleko, M. Kelly, R. Krummenacher & Z. Wang (2004). *eureauweb – An Architecture for a European Waterways Networked Information System.* Proc. Of ENTER'04, 2004
13. Kubach, U. & K. Rothermel (2001). *Exploiting Location Information for Infostation-Based Hoarding.* Proc. Of MobiCom'01, 2001
14. Personè, V. d. N., V. Grassi & A. Morlupi (1998). *Modeling and Evaluation of Prefetching Policies for Context-Aware Information Services.* Proc. Of MobiCom'98, 1998
15. R. Hugo Patterson , Garth A. Gibson , M. Satyanarayanan, A status report on research in transparent informed prefetching, ACM SIGOPS Operating Systems Review, v.27 n.2, p.21-34, April 1993
16. Ren Q. & M.H. Dunham (2000). *Using Semantic Caching to Manage Location Dependent Data in Mobile Computing.* Proc. Of MobiCom'00), 2000
17. S.M. Park, D.Y. Kim, G.H. Cho, Improving prediction level of prefetching for location-aware mobile information service, Future Generation Computer Systems 20, 2004, pp. 197-203
18. Tait, C., H. Lei, S. Acharya & H. Chang, *Intelligent file hoarding for mobile computers.* Proc. Of MobiCom'95, 1995
19. Ye, T., H.-A. Jacobsen & R. Katz (1998). *Mobile awareness in a wide area wireless network of info-stations* .Proc. of MobiCom'98, 1998
20. Z. Jiang and L. Kleinrock. An Adaptive Network Prefetch Scheme. *IEEE Journal on Selected Areas in Communications*, 16(3):1–11, April 1998.

System Development of Security Vulnerability Diagnosis in Wireless Internet Networks

Byoung-Muk Min[1], Sok-Pal Cho[2], Hong-jin Kim[3], and Dong Chun Lee[4]

[1] School of Computing Soongsil Univ., Korea
ceo@nanoware21.com
[2] Dept. of C&C Eng. Sungkyul Univ., Korea
[3] Dept. of Computer Information, KyungWon College, Korea
[4] Dept. of Computer Science Howon Univ., Korea

Abstract. In this paper we design and implement the system that diagnoses security vulnerability in wireless Internet networks. The aim of the paper is to interpret the communication network related to security vulnerability reporting process, and focuses on how the information of a vulnerability is received and processed and how the information is managed after the reception in wireless Internet networks.

1 Introduction

The challenges in the communication process have been discussed widely, for example, on different mailing lists during the past few years. Many difficulties may occur in the vulnerability reporting process. No consensus exists between groups that take part in the reporting process about the ethically correct disclosure. However, to our knowledge, the communication related to the disclosure of the vulnerabilities has not been studied before [1, 2].

Wireless Internet networks security has become a primary concern in order to provide protected communication between mobile nodes in a hostile environment. Unlike the wire-line networks, the unique characteristics of wireless Internet networks pose a number of nontrivial challenges to security design, such as open peer-to-peer network architecture, shared wireless medium, stringent resource constraints, and highly dynamic network topology. These challenges clearly make a case for building multifence security solutions that achieve both broad protection and desirable network performance.

The unreliability of wireless links between nodes, constantly changing topology due to the movement of nodes in and out of the networks, and lack of incorporation of security features in statically configured wireless routing protocols not meant for wireless Internet environments all lead to increased vulnerability and exposure to attacks. Security in wireless Internet networks is particularly difficult to achieve, notably because of the limited physical protection of each node, the sporadic nature of connectivity, the absence of a certification authority, and the lack of a centralized monitoring or management unit [3].

The current draft for the IEEE 802.11 security architecture recommends that this authentication process be completed using Extensible authentication Protocol-

Transparent Layer Security (EAP-TLS), which has been included as the default authentication method in Window XP. Unfortunately, a complete EAP-TLS handshake, including RADIUS messages, requires on the order if 1 s -a number far too large to support ant form of streaming media. To answer this question, the IEEE included "Pre-authentication" in the draft, which permits a mobile station to "per-authenticate" itself to the next AP. Unfortunately, pre-authentication has several shortcomings [7, 8, 9]. First, a station can only pre-authenticate to another Access Point (AP) on the same LAN (i.e., the station cannot authenticate beyond the first access router as a single administrative domain might have multiple access routers). Second, a full EAP-TLS authentication to all potential next APs is not a scalable solution in terms of the number of stations and the APs as most networks use a centralized authentication server (RADIUS) that can quickly become a bottleneck. This obviously prevents WiFi networks from reaching much of the previously discussed vision [4, 5, 6].

The above considerations raise the issue of how to better secure wireless networks. This will be as critical as securing fixed-line Internet systems in the emerging markets as highlighted above. Each of these security breaches and associate risks can be minimized or negated with the proper use of security policy and practices, network design, system security applications, and the correct configuration of security controls.

In this paper we develop system that diagnoses security vulnerability in wireless Internet networks.

2 Security Vulnerabilities in Wireless Internet Networks

There are various reasons why wireless Internet networks are at risk, from a security point of view. In traditional wireless networks, mobile devices associate themselves with an access point, which is in turn connected to other wire-line machinery such as a gateway or name server that manages the network management functions. Wireless Internet networks, on the other hand, do not have a centralized piece of machinery such as a name server, which if present as a single node can be a single point of failure. The absence of infrastructure and the subsequent absence of authorization facilities impede the usual practice of establishing a line of defense, distinguishing nodes as trusted and nontrusted. There may be no ground for an a priori classification, since all nodes are required to cooperate in supporting the network operation, while no prior Security Association (SA) can be assumed for all the network nodes. Freely roaming nodes form transient associations with their neighbors, joining and leaving subdomains independently with and without notice.

An additional problem related to the compromised nodes is the potential Byzantine failures encountered within wireless Internet networks routing protocols where in a set of nodes could be compromised in such a way that incorrect and malicious behavior cannot be directly noted at all. Such malicious nodes can also create new routing messages and advertise nonexistent links, provide incorrect link state information, and flood other nodes with routing traffic, thus inflicting Byzantine failures on the system.

The wireless links between nodes are highly susceptible to link attacks, which include passive eavesdropping, active interfering, leakage of secret information, data tampering, impersonation, message replay, message distortion, and Denial of Service

(DoS). Eavesdropping might give an adversary access to secret information, violating confidentiality. Active attacks might allow the adversary to delete messages, inject erroneous messages, modify messages, and impersonate a node, thus violating availability, integrity, authentication, and nonrepudiation [6, 8,10].

The presence of even a small number of adversarial nodes could result in repeatedly compromised routes; as a result, the network nodes would have to rely on cycles of timeout and new route discoveries to communicate. This would incur arbitrary delays before the establishment of a non-corrupted path, while successive broadcasts of route requests would impose excessive transmission overhead. In particular, intentionally falsified routing messages would result in DoS experienced by the end nodes.

Moreover, the battery-powered operation of wireless Internet networks gives attackers ample opportunity to launch a DoS attack by creating additional transmissions or expensive computations to be carried out by a node in an attempt to exhaust its batteries.

Attacks against wireless Internet network's can be divided into two groups: Passive attacks typically involve only eavesdropping of data whereas active attacks involve actions performed by adversaries, for instance the replication, modification and deletion of exchanged data. External attacks are typically active attacks that are targeted to prevent services from working properly or shut them down completely. Intrusion prevention measures like encryption and authentication can only prevent external nodes from disrupting traffic, but can do little when compromised nodes internal to the network begin to disrupt traffic. Internal attacks are typically more severe attacks, since malicious insider nodes already belong to the network as an authorized party and are thus protected with the security mechanisms the network and its services offer. Thus, such compromised nodes, which may even operate in a group, may use the standard security means to actually protect their attacks [10, 11, 12].

In summary, a malicious node can disrupt the routing mechanism employed by several routing protocols in the following ways.

Attack the Route Discovery Process by:
- Changing the contents of a discovered route.
- Modifying a route reply message, causing the packet to be dropped as an invalid packet.
- Invalidating the route cache in other nodes by advertising incorrect paths.
- Refusing to participate in the route discovery process.

Attack the Routing Mechanism By:
- Modifying the contents of a data packet or the route via which that data packet is supposed to travel.
- Behaving normally during the route discovery process but drop data packets causing a loss in throughput.

Launch DoS Attacks By:
- Sending a large number of route requests. Due to the mobility aspect of MANET's, other nodes cannot make out whether the large number of route requests are a consequence of a DoS attack or due to a large number of broken links because of high mobility.
- Spoofing its IP and sending route requests with a fake ID to the same destination, causing a DoS at that destination.

The above discussion makes it clear that wireless networks are inherently insecure, more so than their wire-line counterparts, and need vulnerability diagnosis schemes before it is too late to counter an attack. If there are attacks on a system, one would like to detect them as soon as possible (ideally in real time) and take appropriate action.

3 Design and Implementation of Vulnerability Diagnosis Systems

Vulnerability diagnosis systems extend to previous vulnerability diagnosis tool so that this system diagnoses wireless networks as possible, and to vulnerability between mobile hosts and APs. After diagnosis to vulnerability, mobile host transmit diagnosis results to vulnerability diagnosis manager while connecting on online state. This vulnerability diagnosis follow as: (1) ESSID broadcasting, (2) Open connection authentication, (3)Vulnerability diagnosis of useless WEP, (4) WEP Key generation using RC4 algorithm vulnerability, (5)Vulnerability diagnosis through challenge-response pair collection, and (6)Possibility of attack between wireless clients in Fig. 2.

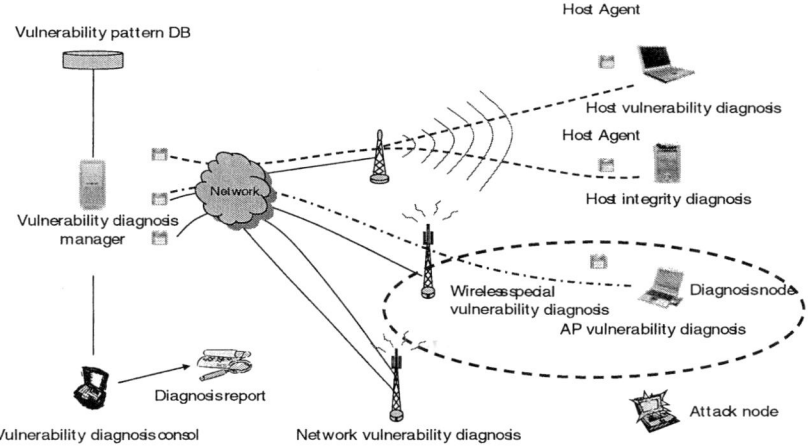

Fig. 1. Vulnerability diagnosis configuration

(1) ESSID broadcasting
Mobile host must know ESSID of AP for connecting AP. All most AP open to Ap name and ESSID doesn't broadcast. But current security production broadcasts ESSID. AP decides to ESSID broadcasting through the active proving.

(2) Open connection Authentication
Open connection authentication is method that doesn't use authentication to connect mobile host to AP and obtain mode to use connection authentication through the active proving. This authentication approve of access authority to mobile host, and can do easily packet sniffing.

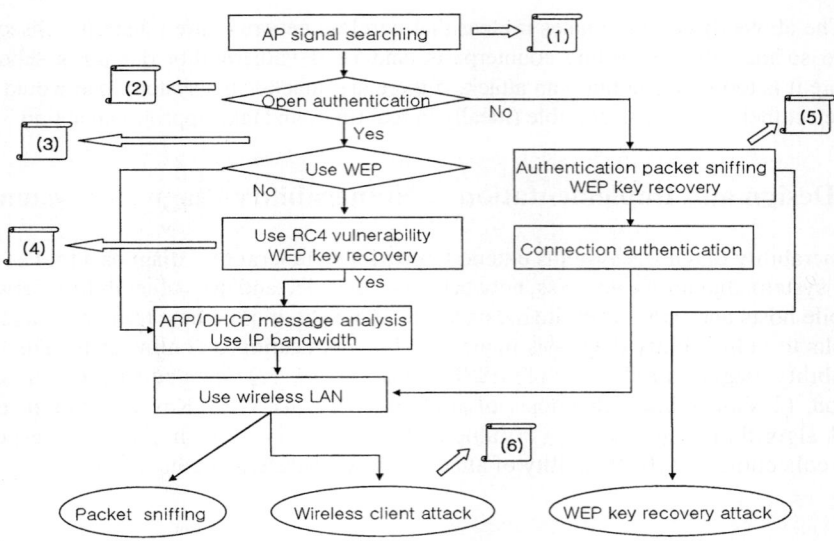

Fig. 2. Procedure of vulnerability diagnosis

(3) Vulnerability diagnosis of useless WEP
If mobile host makes useless of WEP encrypted key, vulnerability diagnosis provides attacker with plain text, user ID, and password through packet collections without hacking in wireless networks. ESSID that obtain from active proving may penetrate to network through opened AP profile, and is be collected to attacker all messages which transmit on wireless networks. Also, it provides attacker with mobile host's information that has been made use of wireless Internet networks. We will know to use of WEP key considering active proving method to make use of obtaining ESSID broadcasting AP profile and when it approach to AP, it may obtain to WEP key using RC4 algorithm vulnerability through packet collections.

(4) WEP Key generation using RC4 algorithm vulnerability
WEP key make use of method that seek to analyze through packet collection using powerful PC or Notebook PC. WEP key encryption is based on RC4 stream encrypting algorithm, and test to decryption environment beyond characteristic condition

(5) Vulnerability diagnosis through challenge-response pair collection
It is security vulnerability to show from authentication method through public key. Vulnerability diagnosis that approaches through challenge-response pair collection collect challenge value that request challenge continuously to AP and response value that encrypted to public key, and when it processes challenge-response using same random numbers, it makes possible response immediately and generation of WEP key using generating packets in initial authentication

In implement environment, proposed system develops vulnerability diagnosis of wireless Internet networks to use easily, and makes use of IPAQ5550 with PDA package including wireless Internet functions which can easily show in popular environment. Also, we use development tool with Microsoft Visual Studio .NET 2003 Professional, Microsoft .NET Compact Framework 1.1 Library, and Microsoft PocketPC 2003 SDK, running HP iPAQ5550 with IEEE 802.11b wireless interface.

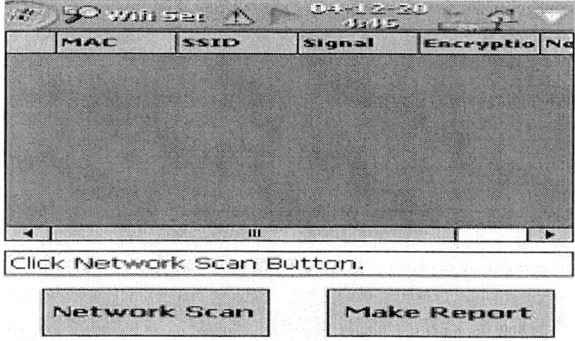

Fig. 3. Initial Screen

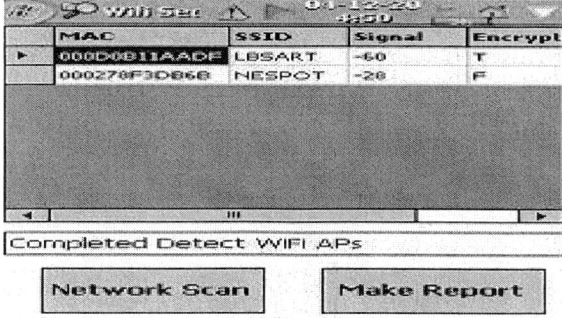

Fig. 4. Screen after scanning network

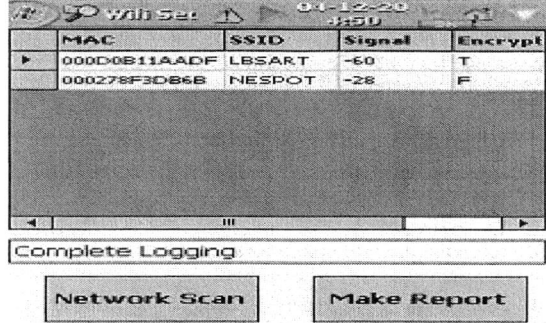

Fig. 5. Screen after make report

In Fig. 3 and Fig.4, initial execution screen consist of button of making report, upper part of printing data , and button of process scan instruction, and middle text box output to current state. After processing network scan, scan test is discovered two APs, and when it process connection test about each AP, scan test is completed.

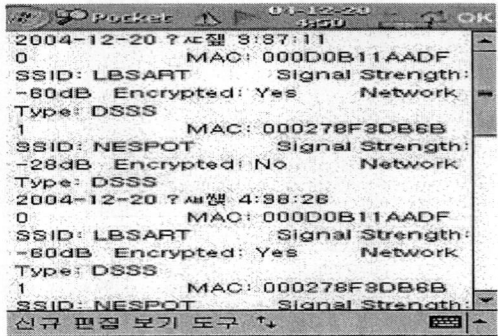

Fig. 6. Screen of report file

Fig. 7 shows report file that processed diagnosis result. In report file, make report instruction process instruction that outputs diagnosis result as report file, and it write down test time and current state situation. In diagnosis result, this shows vulnerability in wireless Internet networks.

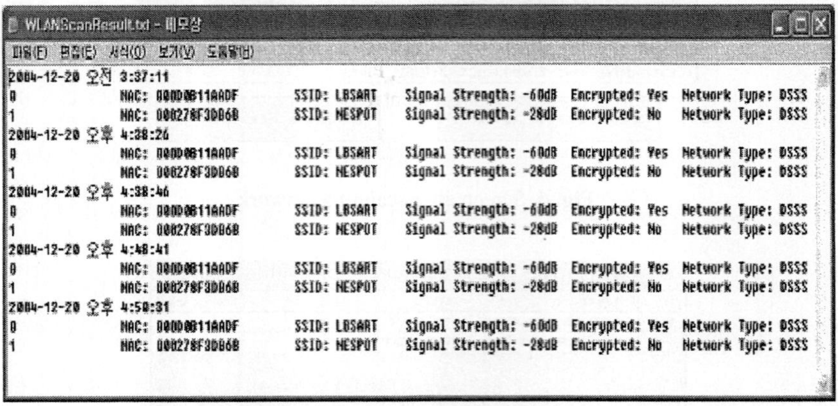

Fig. 7. Generated report file

4 Conclusions

The paper is to interpret the communication network related to security vulnerability reporting process, and focuses on how the information of a vulnerability is received and processed and how the information is managed after the reception in wireless Internet networks. The proposed system can diagnose vulnerability patterns that can generate through wire-line Internet and wireless Internet networks, and can report diagnosis result to network manager rapidly.

Acknowledgements

This work was supported by a grand No. R12-2003-004-00007-0 from Ministry of Science and Technology.

References

1. M. Balazinska and P. Castro, "Characterizing Mobility and network usage in a Corporate Wireless Local Area Network, "Int'l. Conf. Mobile Systems, Apps, and Services, May 2003.
2. A. Mishra, M. Shin, and W. Arbaugh, "An Empirical Analysis of the IEEE 802. 11 Mac Layer Handoff Process, "ACM SIGCOMM Comp. Commun. Rev., vol. 33, Apr. 2003.
3. R. Koodli and C. Perkins, "Fast Handover and context Relocation in Mobile Networks, "ACM SIGCOMM Comp. Commun. Rev., vol. 31, Oct. 2001.
4. IEEE Std. P802. 1X, "Standards for Local and Metropolitan Area Networks: Standard for Port Based Network Access Control," Oct. 2001.
5. J. Edney and W. A. Arbaugh, Real 802.11 Security, Addison Wesley, 2003.
6. IEEE Std. 802.11i, "Draft Amendment to Standard for Telecommunications and Information Exchange between Systems-lan/man Specific Requirements, Part11: Wireless Medium Access Control and Physical Layer (phy) Specifications: MAC Security Enhancements.," May 2003.
7. A. Mishra, M. Shin, and W. Arbaugh, "Context Caching Using Neighbor Graphs for Fast Haridoffs in a Wireless Network," to appear, Proc. IEEE INFOCOM 2004.
8. W. A. Arbaugh and B. Aboba, "Experimental Handoff Extension to RADIUS," Internet draft, May 2003.
9. S. Pack and Y. Choi, "Fast Inter-AP Handoff Using Predictive-Authentication Scheme in a Public Wireless LAN," IEEE Networks, Aug. 2002.
10. S. Pack and Y. Choi, "Pre-Authenticated Fast Handoff in a public Wireless LAN based on IEEE 802. 1x Model," IFIP TC6 Pers. Wireless Commun., Oct. 2002.
11. M. Nakhjiri, C. Perkins, and R. Koodli, "Context Transfer Protocol," Internet Draft: draft-ietf-seamoby-ctp01.txt, Mar. 2003.
12. R. Perlman, "An Algorithm for Distributed Computation of a Spanning Tree in an Extended LAN," 1985, pp. 44-53.
13. R. Perlman, Interconnections, 2nd Edition: Bridges, Routers, Switches and Internetworking Protocols, Pearson Education, Sept. 1999.

An Active Node Management System for Secure Active Networks

Jin-Mook Kim, In-sung Han, and Hwang-bin Ryou

Dept. of Computer Science, Kwangwoon Univ., Korea
447-1 Wolgye-Dong, Nowon-Gu, Seoul, Korea
{jmkim, ishan, ryou}@netlab.kw.ac.kr

Abstract. While conventional networks have functions such as queuing and delivering data packets only, active networks have been introduced and researched since 1990s, where they have additional functions such as performing operations on the packets being transmitted in the networks. Because of such versatile functions, active networks are obviously more complex than conventional networks, but raise considerable security issues at the same time. In this paper, we propose an active node management system for secure active networks that is based on a discrete approach which resolves the weak points of active networks. The proposed system provides the functions of node and user management in the active networks, and improves the security of packet transmission by packet cryptography and session management. We implement the proposed system and show numerical results on performance.

1 Introduction

New concepts and technologies usually take long to be adapted in conventional networks because of the characteristics of heterogeneity among the devices in the networks. In addition, in order for a new technology to be adapted in the networks, it is required to be standardized, which usually takes significant amount of efforts and time [3].

Active networks, proposed by DARPA, are a novel approach to network architecture in which customized programs are executed within the networks. They were first described in [8], where the authors postulated that this approach would provide two key benefits: it would enable a range of new applications that leveraged computation within the network; and it would accelerate the pace of innovation by decoupling services from the underlying infrastructure. However, an active network allows its nodes to be controlled dynamically by program code. It enables users to build their own application-specific protocols to communicate on the network without any limitations of communication standard of protocol suite. The flexibility achieved from active network also produces higher network security risks because program code, which control behavior of the nodes, is also traversed in the network. Therefore, the network security is one of the most important mechanisms that must be provided to active network.

In this paper, an active node management system for secure active networks, based on a discrete approach, is proposed. The discrete approach may also be called a programmable node (switch/router) approach, where programs are injected into programmable active nodes separately from the actual data packets that traverse through the network. A user can send a program to a network node, where it would be stored and later executed when the data arrives at the node, processing the data. The data can have some information that would let the node decide how to handle it or what program to execute. So, the proposed system in this paper basically authenticates the user who will inject a program code at an active node, and manages the authentications of the active node. Also, packet cryptography and session techniques are used for secure transmission of packets and programming codes, and their executions. We compare the existing active network structure and the proposed system model, so the system can provide the effective safety and minimize the loss of efficiency.

2 Related Work

2.1 Active Networks Compositions

Various research results were announced in DARPA in research about an active network to be proposed. To fulfill the proper operations at active node, many research methods are discovered with comparing the existing passive network structure, such as the authentication of middle node and the harmony of running environment and the special programming language for run time.

According to them, a variety of researches are processed, but we concentrate on the transmission method that transmits program code and data at active node.

ANTS(active network Transfer System) as the early-stage research creates the structure of active network and the composite research results that makes data packet include programming code and installs the necessary functions to the active node. Also, SwitchWare that strengthen the flexible programming for the safety of network structure and security is suggested, too. ABone(active network Backbone) which figures out the difficulty of preparing the realistic structure of active network designs packet structure and support a variety running environment. These existing researches become a basis for the further studies of active network.

This paper classifies two transmission methods for the composition of the safer active network. The first one is the discrete approach that firstly divides program code and data, and transmits them. The second one is the capsule method that integrates program code and data as the active packet and transmits them.

The capsule method creates "active packet (capsule)" that contains program code and data without saving program code at active node, and transmits it to the network. Secondly, active node divides the program code and data from the received active packet. The third procedure is loading program code to the runtime environment in active node, and process data by program code. Finally, they recombine program code and processed data and creates active packet and transmits it to the next active node.

ANTS project in MIT and PLANet in Pennsylvania are using this method, but when the program code is very large, the capsule method has many problems, such as traffic overhead, if packet is lost then packet re-transmission, so the efficiency can be reduced. The discrete approach divides program code and data before transmission. It means the program code is installed at active node before the execution. The active node user transmits data with program code identifier. Secondly active node which receives packet checks the identifier and run the proper program code at the active node. Thirdly, it uses running program to process data, and finally it creates packet from the processed data and transmits it. ActiveIP and SwitchWare researched active network with this method.

The Discrete approach can be adapted to the only already-installed program code, and the only network manager can add program code, so it is impossible to add the new program that the generic active hosts want to add. This paper uses the Discrete Approach, and resolves its weak point.

2.2 The Security of Active Network

The active network should provide the solution for authentication, authorization and integrity to support the basic security service. In the Discrete Approach, the authentication of program code sender and the secret and integrity of the program code itself are the essential security points. If the program code is modified on bad purpose or it has the potential problem, it will become the unexpected error, so not only low performance of the entire active nodes but also a big security problem will be raised. In addition, if the authentication of program code is not performed, the hacker will modify the program code, and it will be a serious security problem.

Now many projects of active network security, such as SANE, Seraphim, PLAN and Safety-Net are ongoing, but they cannot assure the basis of safety in the active network. Therefore, new security system that removes weak points is strongly necessary.

3 System Proposal

As we reviewed the existing studies, the security model that provides the basic security solutions such as authentication, authority, integrity is necessary. If the basic security problem is ignored, the performance of the entire active network node will be lowered, and the privacy violation and network congestion will be caused.

To resolve these security threats, we should authenticate active node users on the Discrete Approach. To authenticate active node users, we can restrict the access of hacker who tends to transmit the offensive program code, and block the forgery of program code. Also, we can reduce the deterioration of performance that the program reinstallation in active node causes through the management of frequently-using program code. The active network structure that we propose is shown at Fig. 1.

The proposed system focuses on the authentication of middle node and the safe transmission of program code in active node. In our system, the active node management server provides the certification that means the certified middle node is trustable. It is similar with PKI technique that certification organization certifies internet services. The node management server which is reliance in the domain manages and authenticates clients in the active network structure. Also, active node management server in each domain can authenticates each other.

The active node management server authenticates and manages program codes, too. In order to use it, the client requests authentication service for program code and client, itself and receives the acceptance of them. The authenticated program code can install and run at active node.

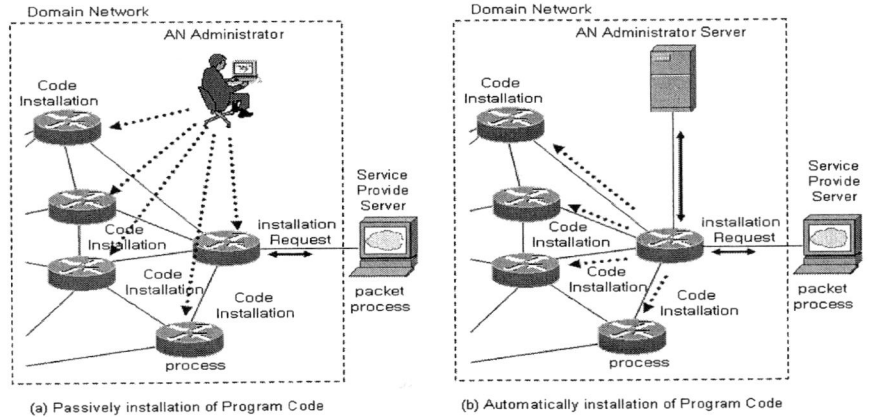

Fig. 1. The safer structure in active network

The existing discrete approach installs the program code at active node, so only network manager can manage and install the program code. In our system, when the client requests it, the active node management server authenticates clients and the client registers the program code at the active node management server. Then, the active node management server check the program code whether it is safe or not, and then if it is safe, the server install the code at active node. Therefore, if the client requests the operation job, the program will run with only authentication code, so it makes active node efficient and safe. In addition, the management of middle nodes can prevent the forgery of program code through the analysis of program code.

4 System Design and Implementation

Our system has two major modules, which are the active node management server and middle node agent in active node. The active node management server processes the preparation of program code registration, and provides and manages information that middle node agent needs.

The active node management server which receives the request of middle node creates the couple of keys and distributes them. It transmits the active node agent to active node by using the key and active node interacts with active node through received active node agent. Therefore, safe and efficient active network is working on our system.

First of all, the node management server is classified with functional modules. It has four modules, middle node key management module, middle node control module, program code management module and middle node run-time environment repository. Each functional module is classified with the specific modules depends on the function and role.

4.1 Active Node Management Server

4.1.1 Key Management Module

When active node key management module receive the request of program code registration, it creates a pair of public key and private key from the IP information of middle node, and send the private key to active node , and store the public key at active node run-time environment repository.

When the number of middle nodes in the domain is too big, active node key manager can be specialized as two modules, one stores only public keys and the other stores the information about relationship of two keys.

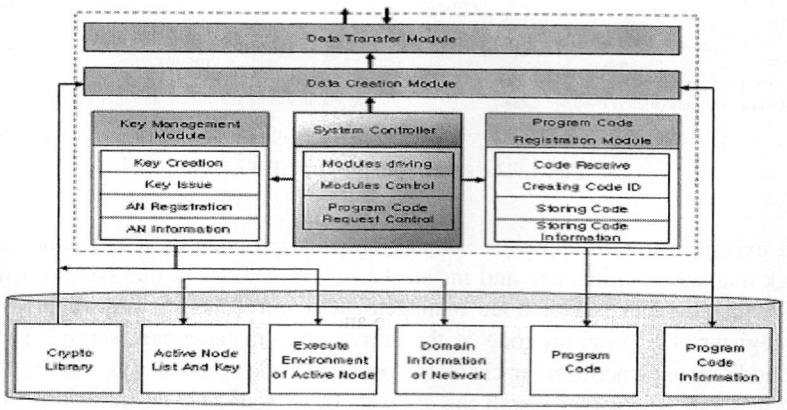

Fig. 2. Architecture of Active Node Administration System

4.1.2 System Controller

This module manages and controls many problems and troubles when modules of active node management server and active node agent work together. If we break down this module, it will become two parts. The first one is data transmission module that is in charge of the connection session for the data transmission, cryptography and creation of secret key. The second one is the data creation module that is in charge of the creation of active packet header, the collection of program code and capsulation of data.

4.1.3 Program Code Registration Module

The program code management module manages and stores the program code that will be sent to active node and executed. It checks the availability of program code, and if it's available, the module will transmit it to active node and install it.

To accomplish this procedure, we need many modules, the module that collects and stores program code, the authentication module for the program provider, the module that manages the run-time environment and the module that stores the specification of program code. To manage and store the received program code at the repository, this module creates program code repository key that is consists of 10 digits serial number. It increases the efficiency of managing program code upon the request of client.

4.1.4 Middle Node Run-Time Environment Repository

Active node run-time environment repository has simple jobs. It stores the program code key that has 10 digit serial key, public key for data transmission and the secret key for session. Also it organizes directories for each information part and makes database to manage them. In addition, it stores the information of the run-time environment for each program code and manages it, so the communication with the active node agent becomes more efficient.

4.2 Active Node Agent

The active node agent is in charge of the communication of active node management server at active node. The major function is collecting the status of middle node and transmitting them to the active node management server. The active node management server can check the environment from this information.

Also, it receives a pair of public key for the security communication with the node management server, it stores and manages them safety. It exchanges the session key for the transmission of cryptic program code and data.

5 Experiments and Discussion

In this paper, we describe the experimental active network configuration used for this research and discuss the results. For the experiment, we inject program codes to the active nodes by using the proposed active node management system and then evaluate how the performance of the active nodes is affected by that.

5.1 Environments of Experiment

For the experimental network configuration where a node management server injects program codes to active nodes, Transit-Stub structure created by GT-ITM is used, which models the current internet configuration. The tool used in the experiment is ns-2 that is developed by LBNL (Lawrence Berkeley National Laboratory) [15]. Performance metric is delay time at active nodes for hop counts. The protocols for performance comparison are TCP, UDP and ANEP (Active Node Encapsulation Protocol). Figure 3 shows the network topology for the experiment.

The experimental network consists of clients and servers which exist in different networks. In order to do so, we assume that there are connections between routers that are located in different networks. When the service-providing server requests, an active node management server installs program codes that has caching function in the active nodes in the managing domain. When a client sends a service request message to a service-providing server, both the installed active node and the

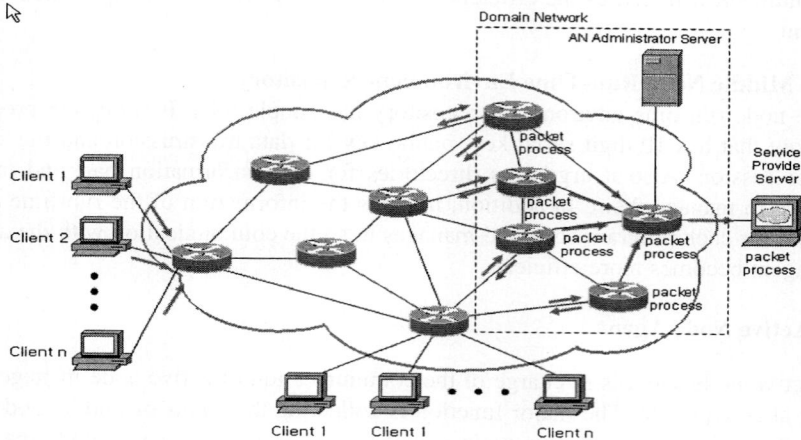

Fig. 3. Active network topology for Experimental

service providing server process the request of the client. TCP and UDP are used for performance comparison of the active networks with caching function. TCP is a reliable end-to-end transport protocol, but it is heavy because it has many functions for achieving reliability. UDP is an unreliable end-to-end protocol and used in the casewhere there is no need to decapsule ANEP packets. It is lightweight comparing to TCP, because it does not have the functions for reliable packet delivery. For the correctness of the experiment, the amount of data and the length of the data packets are same for those protocols. The senders and the receivers are randomly chosen.

5.2 Experimental Results

We assume that there are the traffics only for the experiment in the network, so that there is no situation with packet loss. In such assumption, the lost packet recovery function is not activated and reflected in the result of the performance. Therefore, we can test the pure performance of an active network, because no packet loss due to congestion is excluded. The experimental program code is a caching function that transmits data after copying it. The function of program code has two types. When an active node has no cached data, it saves data at the active node, and when an active node has cached data, it compares the size of received packet, and the node decides whether it transmits or sends the packet back.

The program code for this experiment is installed at the middle active node which is located between client and Service Provide Server. This code saves the result from Service Provide Server to the active node when the packet requested data is not existed, and it runs the process of caching data when the data is existed. The client side caching technique is already popular, but we took the additional experiment which compares the active node using this program code and the generic network.

Figure 4 is the result to measure the execution duration of Service Provide Server upon the number of clients. The Service Provide Server can be in a heavy traffic, so we show the graph which describes the transmission delay time of Service Provide Server as the increasement of the number of clients. As shown at Figure 5, the duration of service provide server is decreased, because many duplicated result is generated from different clients.

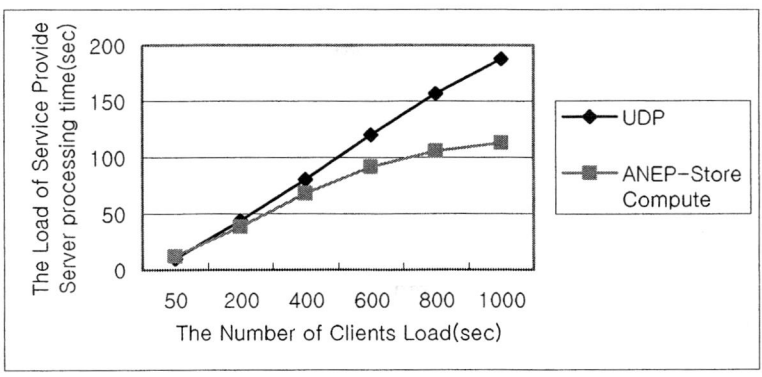

Fig. 4. Protocol latency of Service Provide Server Load

The result that compares the generic network using UDP and caching equipped active network having program code is shown by this experiment. However, we cannot be sure the active network which has the caching function at the end node is the most efficient, because we just use the simple caching technique. Therefore, we need to adopt it to the generic network for clearness.

Figure 5 is the result that shows response delay time when the clients transmit service requesting packet to Service Provide Server as increasing the number of active nodes. The number of clients in this experiment is 500. We compare some protocols, TCP and UDP for the end point transmission and ANEP that can check the packet at the active node and transmit the caching data after operation.

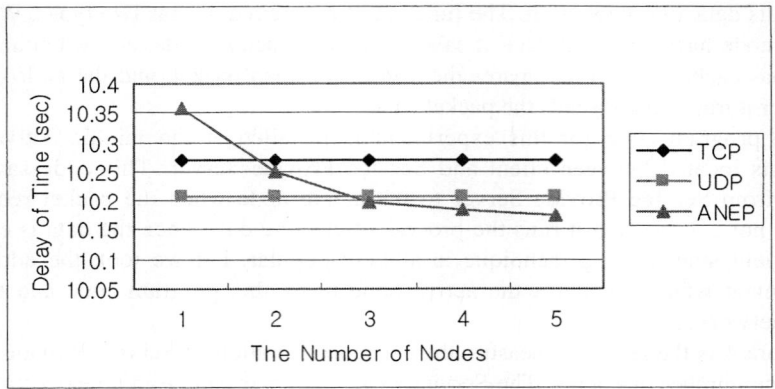

Fig. 5. Client request-response time

As we can see at Figure 5, TCP and UDP protocol have the same value when the number of middle nodes is changed, but ANEP packet reduces the response delay time for the client's request. This result means the ANEP packet is handled quickly as the number of active nodes is increased. It means that the response time of packet processed result is decreased as the number of active nodes is increased.

6 Conclusions

In this paper, an active node management system is proposed and implemented in order to protect networks against setting malicious program code to vulnerable active node and forging program code during transmission. It efficiently manages active nodes in a manageable domain and transmits program code using encryption. We implemented the system and gave numerical results on the performance of the active networks with the system in terms of setting delay time of program code and processing delay time of each protocol. The results show that the performance is not so degraded by having such management system in active networks.

References

1. K. Calvert, et.al., "Direction in Active Networks", IEEE Comm. Mag.,October 1998.
2. Danny Raz and Yuval Shavitt, "An Active Network Approach to Efficient Network Management",IWAN'99,1999.
3. A. B. Kulkarni," Implementation of a Prototype Active Network.", In OPENARCH '98, 1998.
4. AN Security Working Group, "security Architecture for Active Nets", Nov, 2001.
5. Konstantinos Psounis, "Active Networks: Applications, Security, Safety and Architectures",IEEE Communications Surveys, First Quarter 1999.
6. R. H. Campbell, et al., "Seraphim: Dynamic Interoperable Security Architecture for Active Networks", IEEE OPENARCH 2000, Tel-Aviv,Israel, Mar. 2000

7. DARPA AN Node OS Working Group, "NodeOS Interface Specification," http://www.cs.princeton.edu/nsg/papers/nodeos99.ps, January, 2000.
8. D. Tennenhouse and D. Wetherall, "Towards an Active Network Architecture," Computer Communication Review 26(2), April 1996.
9. D. Tennenhouse et al, "A Survey of Active Network Research," IEEE Communications Magazine, January 1997.
10. D. Wetherall and U. Legedza and J. Guttag,"Introducing new internet services: Why and how",IEEE Network Magazine, 1998.
11. D. Wetherall, et al., "ANTS: A Toolkit for Building and Dynamically Deploying Network Protocols" IEEE OPENARCH'98 Proc., San Francisco, Apr. 1998.
12. D. J. Wetherall, "Service Introduction in an Active Network", Ph.D. Thesis Submitted to the Department of Electrical Engineering and Computer Science, M.I.T., Feb. 1999.
13. D. Decasper and Plattner, B., "DAN: Distibuted Code Caching for Active Networks",INFOCOM'99, New York, 1999.
14. Michael Hicks, Pankaj Kakkar, Jonathan T. Moore,Carl A. Gunter and Scott Nettles, "PLAN : A Packet Language for Active Networks", ICFP,1998.
15. NS Network Simulator, http://www.isi.edu/nsnam/ns.

A Systematic Design Approach for XML-View Driven Web Document Warehouses

Vicky Nassis[1], Rajugan R.[2], Tharam S. Dillon[2], and Wenny Rahayu[1]

[1] Dept. of CS-CE, La Trobe University, Melbourne, Australia
{vnassis, wenny}@cs.latrobe.edu.au
[2] eXel Lab, Faculty of IT, University of Technology, Sydney, Australia
{rajugan, tharam}@it.uts.edu.au

Abstract. EXtensible Markup Language (XML) has emerged as the dominant standard in describing and exchanging data among heterogeneous data sources. The ever increasing presence of XML web contents in large volumes creates the need to investigate Web Document Warehouses (WDW) and Web Document Marts, as a means of archiving and analysing large web contents for context-aware web/business intelligence. To address such an issue, in this paper, we focus on intuitively adopting our pervious work on XML-view based XML Document Warehouse design for building a Web Document Warehouse (WDW). To demonstrate this, here, we carryout a systematic approach to conceptual modelling and transformation of the warehouse conceptual model into a logical/schema (XML Schema) model and an in-depth analysis of deriving and querying context-aware WDW dimensions.

Keywords: OO conceptual models, web document warehouse, XML views.

1 Introduction

Data Warehousing (DW) has been an approach adopted for handling large volumes of historical data for detailed analysis and management support. Transactional data in different databases is cleaned, aligned and combined to produce good data warehouses. Since its introduction in 1996, eXtensible Markup Language (XML) [1] has become the *defacto* standard for storing and manipulating self-describing information (meta-data), which creates vocabularies in assisting information exchange between heterogenous data sources over the web [2] [3, 4]. Due to this, there is considerable work to be achieved in order to allow electronic document handling, electronic storage, retrieval and exchange. It is envisaged that XML will also be used for logically encoding documents for many domains. Hence it is likely that a large number of XML documents will populate the would-be repository and several disparate transactional databases.

The concern of managing large amounts of XML document data raises the need to explore the web warehouse approach through the use of XML document marts and XML document warehouses. One of the major challenges in dealing with web related technologies such as web warehouse is the content of the web itself. Typically web contents may range from a combination of static, un/semi-structured textual data to binary multimedia streams and dynamic on-the-fly hypermedia contents. Another dimension to web content is that, they are distributed and hosted by multiple geographically distributed servers and databases. In direct contrast to DBMS managed structured data (relational or Object-Relational or OO), the web content do not conformed to traditional data models which assume a fixed data model/schema for a given set of data domain. Therefore since the introduction of Internet and the World-Wide-Web, researchers, standard organizations and the Industries rallied around to adopt a web data language that is semantically rich and descriptive yet conforms to open standard schema, which can support and describe all types of data and content on the web, including the traditional structured data. To this date, XML together with its Schema language adequately fill the void for a uniform web data model/language.

One of the early XML data warehouse implementations for web data includes the Xyleme Project [3-5]. The Xyleme project was successful and it was made into a commercial product in 2002. It has well defined implementation architecture and proven techniques to collect and archive web XML documents into an XML warehouse for further analysis. Another approach by Fankhauser & Klements [6] explores some of the changes and challenges of a document centric XML warehouse. We argue that, coupling these approaches with a well defined conceptual and logical design methodology will help future design of such XML warehouse for large-scale XML systems.

In our pervious work [7, 8], we proposed a design methodology for building native XML Document Warehouses (XDW). In this paper, we mainly focus on intuitively adopting the XDW design for building a Web Document Warehouse (WDW) for web documents. To demonstrate this, we carryout a systematic approach to conceptual modelling and transformation of the warehouse conceptual model into a logical/schema (XML Schema) model and an in-depth analysis for deriving context-aware WDW dimensions illustrated using a walk-through with a case study example.

As an example to illustrate our concepts, we investigate a possible web document warehouse for web based, conference publishing system (CPSys), for managing and distributing conference proceedings for various International conferences held in different cities throughout the year. The main component of a conference publication comprises of a collection of papers (past and present), stored in various geographically distributed conference databases/systems, in varying proceedings format such as ACM, LNCS or IEEE. Logically, we treat all the different conferences and their proceedings as one big (logical) conference proceeding on the web (similar to the concept of a "global view" in enterprise systems).

The rest of this paper is organized as follows. Section 2 presents a brief overview of our web warehouse design methodology followed by section 3, which provides a detailed discussion on design and transformation of one of the major components of the web warehouse model namely, WebFACT. Section 4 describes in detail about context-aware warehouse dimensions including a systematic querying approach and query algorithm specifications for such dimensions. Section 5 concludes the paper with some discussion on future research directions.

2 Brief Overview of Our Proposed Web Warehouse Conceptual Design Methodology

Our XDW model in [7, 8], to our knowledge, is unique in its kind and it involves three levels: **(1)** *User Requirement Level*; which includes the warehouse user requirement document and OO requirement model, *(2) XML Warehouse Conceptual Level;* composed of an XML based web FACT repository (WebFACT Section 3) and a collection of logically grouped **conceptual views** which are context-aware dimensions that satisfy captured warehouse user requirements, and *(3) XML Schema Level;* involves transformation of the conceptual model into XML Schema.

The WebFACT is meaningful warehouse FACT and provides a snapshot of the underlying web content for a given *context*. A **context** is more than a measure [9, 10] but instead is an item that is of interest for the organization as a whole. The role of conceptual views is to provide perspectives of the document hierarchy stored in WebFACT repository. These can be grouped into logical groups, where each one is very similar to that of a **subject area** [11, 12] appearing in Object-Oriented conceptual modeling techniques. Each subject-area in the WDW model is referred to as *Virtual Dimension* (VDim) to keep in accordance with dimensional models. Here, it is refer to as *context-aware* VDim as a VDim is derived based on a given *context* and exists only in that given context. For example, in our case study example, a VDim called Monthly_CFP_Posting may exist only in the context Conferences. VDim is called *virtual* since it is modeled using an XML *conceptual view* [13-15] (which is an *imaginary* XML document) behaving as a dimension to a given perspective from the WebFACT.

3 Web FACT Repository (WebFACT)

In building the WebFACT, we note that it differs from traditional data warehouse flat FACT table, which is normally modeled as an ID packed FACT table with its associated data perspectives as dimensions. But, in regards to XML, a *context* refers to the involvement of embedded semantics, such as ordering and homogeneous compositions as well as non-relational constructs such as set, list, and bag (Figure 1). Therefore, we argue that, a *meaningless* FACT does not provide semantic constructs that are needed to accommodate an XML *context*.

Here, concentrating most importantly on the interaction amongst the objects will help to determine their relationship configuration and cardinality, which will be reflected in the conceptual design. An examination of our complete UML conceptual model (Figure 1) emphasizes the structural complexity of the WebFACT in which real world objects can be hierarchically decomposed into several sub-elements where each of these can include further embedded elements. Such decompositions are necessary to provide granularity to a real-world object and if needed, additional semantics are added at different levels of the hierarchy. For example, the `Publ_Conference` class hierarchy is decomposed with additional semantics such as `Publ_Region`, `Publ_Country` and `Publ_City`. It is important to make the WebFACT as expressive as possible whilst retaining the overriding objective of relevance.

At the logical level we utilize XML schema definition language to capture the WebFACT semantics (classes, relationships, ordering and constraints). Therefore, modeling of the WebFACT is constrained only by the availability of XML schema elements and constructs. Below is a description of our case study example and a demonstration, which carries out the steps involved to lead to the formation of our WebFACT conceptual model.

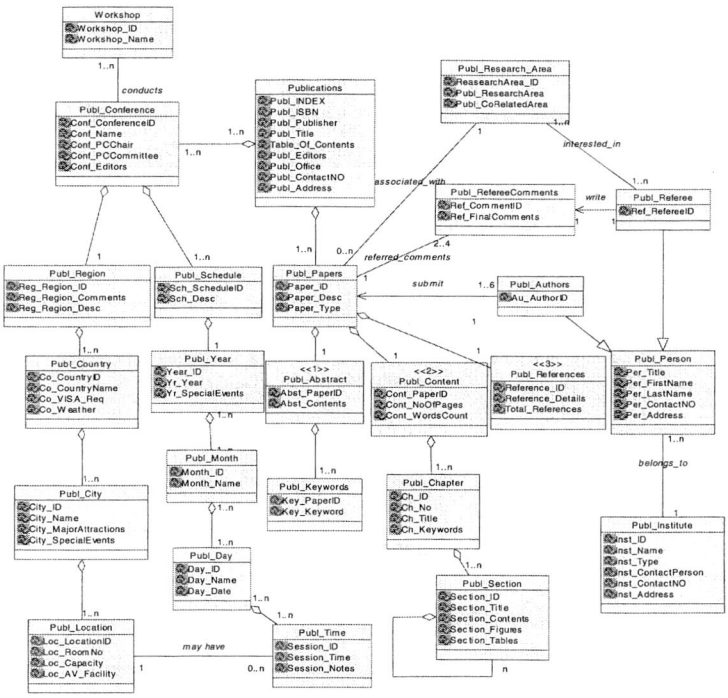

Fig. 1. The complete WebFACT of the Conference Publication System (CPSys) case study

3.1 Transformation of WebFACT OO Conceptual Model to XML Schema

Using the generic rules [16-18] we are able to accomplish the complete transformation of our OO conceptual model into XML Schema. Initially we envisage that the WebFACT table will be an entire, major document of its own containing elements, further embedded elements, and relationships amongst these, which would translate into smaller complex or simple structured components.

We assume that class C, corresponds to the composite document (WebFACT) containing additional classes $C_1,...C_n$. The steps that follow direct the way to form an XML Schema segment.

Step 1. Create an element named C with a ComplexType, *CType*
```
<xs:element name="C" type="CType"/>
```
Step 2. For each of the embedded class C_i of C, create a sub-element C_i with a ComplexType, C_iType. `<xs:element name="Cᵢ" type="CᵢType"/>` Elements can also have a simple structure when they carry single valued attributes from the in-build data types of XML Schema (eg. string, integer). To apply these rules and be able to transform the OO diagrams to XML Schema at this point we will focus on two significant cases that appear in the complete WebFACT conceptual model (Figure 1) namely; Ordered and Homogeneous compositions.

3.1.1 Ordered Composition

The composite element `Publ_Papers` is an aggregation of the sub-elements `Publ_Title`, `Publ_Abstract` and `Publ_References` occurring in a specified order.

Step 1.
```
<xs:element name="Publ_Papers" type=" Publ_PaperType"/>
```
Step 2.
```
<xs:complexType name="Publ_PaperType">
    <xs:sequence>
        <xs:element name="Paper_ID" type="xs:ID"/>
        <xs:element name="Paper_Desc" type="xs:string"/>
        <xs:element name="Paper_Type" type="xs:string"/>
            <xs:sequence>
                <xs:element name="Publ_Abstract" type="Publ_AbstractType"/>
                <xs:element name="Publ_Content" type="Publ_ContentType"/>
                <xs:element name="Publ_References" type="Publ_ReferenceType"/>
            <xs:sequence>
        </xs:sequence>
</xs:complexType>
```

3.1.2 Homogeneous Composition

In a homogeneous aggregation, one "whole" object consists of "part" objects, which are of the same type [19], such as the object `Publ_Chapter` can consist of one or more `Publ_Sections`.

Step 1.
```
<xs:element name="Publ_Chapter" type="Publ_ChapterType"/>
```

Step 2.
```
<xs:complexType name=" Publ_ChapterType">
    <xs:sequence>
        <xs:element name="Ch_ID" type="xs:ID"/>
        <xs:element name="Ch_No" type="xs:short"/>
        <xs:element name="Ch_Title" type="xs:string"/>
        <xs:element name="Ch_Keywords" type="xs:string"/>
        </xs:sequence>
            <xs:element name="Publ_Section" type="Publ_SectionType"/>
        </xs:sequence>
    </xs:sequence>
</xs:complexType>
```

4 Context-Aware Virtual Dimensions

A user requirement, which is captured in the OO Requirement Model, is transformed into one or more *Conceptual Views* [13-15] also referred to as *Virtual Dimension/s*, in association with the WebFACT. These are typically *aggregate views* or *perspectives* of the underlying stored documents of the transaction system/s. A valid user requirement is such that, it can be satisfied by one or more XML conceptual views for a given context (i.e. WebFACT). But in the case where for a given user requirement there is no transactional document or data fragment to satisfy it, further enhancements are necessary to make the requirement feasible to model with a certain WebFACT. Therefore modeling of VDims is an iterative process where user requirements are validated against the WebFACT in conjunction with the transactional system.

Definition 1: *One Virtual Dimension composes of one (or more logically grouped) conceptual view, thus satisfying one (or more logically related) user document warehouse requirement/s.*

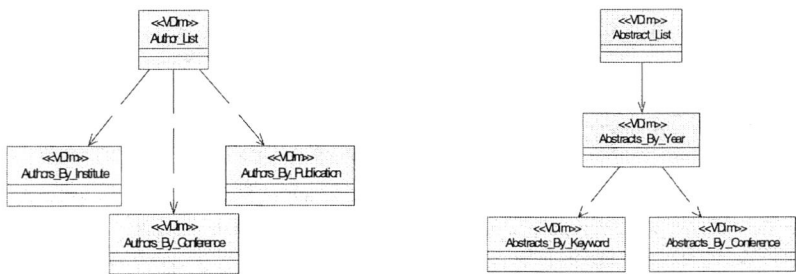

Fig. 2. A conceptual view hierarchy with <<construct>> stereotype

Fig. 3. VDim "Abstract_List" Package Contents

We introduced a new UML stereotype called <<VDim>> to model the virtual dimensions at the XDW conceptual level using This stereotype is similar to a UML class notation with a defined set of attributes and methods. The method set can have either; constructors (to construct a VDim) or manipulators (to manipulate the VDim attribute set). As shown in Figures 2 & 3, the relationship between the VDims is modeled with a dashed, directed line, denoting the <<construct>> stereotype. Though VDims can have additional semantic relationships such as generalization, aggregation, association [13, 15], these can be shown using standard UML notations. In addition to this, two VDims can also have <<construct>> relationships with dependencies such as those shown in Figure 3, between VDims Abstracts_By_Year and Abstracts_By_Keyword.

The following section deals with the concept of *Virtual Dimensions* in more detail. We will show their development, based on the derived user requirements as well as formally defining the four different types of virtual dimensions. For demonstration purposes our case study example of a *Conference Publication System (CPSys)* will be applied.

4.1 Warehouse Requirements and the Systematic Querying Approach for Virtual Dimension/s (VDim/s)

Previously for the conversion of WebFACT into XML Schema we used the transformations discussed in the papers [16-18]. In the case of VDims, this is actually the transformation between the *conceptual views* into *XML View* [13, 15] schemas.

In [13], the XML-View definition indicates that for each *conceptual view* there is a corresponding XML document, which may contain more embedded XML documents. Querying the resulting XML document using any query language for XML with language specific syntax performs extracting the information needed in order to meet all the conditions of each user requirement. The query process is well suited for simple, straightforward requirements but it tends to get very complex especially when dealing with user requirements having a wide-ranging context.

4.2 Logical Implementation of the Virtual Dimension

Considering the above facts regarding the query process, in the segment of queries that follow, the preferred query language is XQuery [20-22]. However XQuery at this stage proves limited regarding group by and aggregate functions for data warehouse operations (need to use nested loops which are difficult to operate) and are still in progress for future development [6]. Due to this, and in order to illustrate the purpose of querying XML documents in relation to VDim, using the notion of XQuery, in this paper we also propose a *generic query algorithm,* which will be the foundation to build smaller algorithmic segments to suit the structure of the case queries from common cases and hence enable full capture of each user requirement.

We use the terms defined in W3C namely; `Query Context` and `Return` from the presentation of use case queries to guide us in deriving and defining the initial parameters to be used in our proposed algorithm. We consider that each class involved in the search query can be of simple or complex structure, meaning that it can contain; only attributes, only elements and sometimes a combination of both. What follows is a list of the conditions and explanations of the *keywords* that will be used to design our *generic query algorithm*.

Keyword 1: `Query Context`: This is the full path of classes used to locate the attributes and elements to be queried. Classes can have further embedded attributes and/or elements.

Keyword 2: `Query Context Value`: This is the parameter value (specified by the user) used to execute the query and is located in the last occurring attribute or element from the class specified in the `query context` path. The value can be; a precise word, phrase, number or include a group of values denoted in the query by the word `All`.

Keyword 3: `Return Context`: The results obtained with all the query's conditions met. The returned class's attributes or elements do not have to correspond to the ones being specified in the query context. While the search is conducted within the class/s specified in the `query context` it is possible that the required result attribute/s or element/s might originate from a different class. Also in some cases the outcome may be comprised of new assembled attribute/s and/or element/s.

Keyword 4: Sort: This function is applicable for the queries performing the ordering of end result records. We are able to sort the final entries based on a **subject factor,** which is directly related to the user's interest and preference regarding the presentation of results. A **subject factor value** can be defined by a name; a number or it can be of any value provided it exists within the class' attribute/s and/or element/s involved in the query.

Keyword 5: Merge: The merging function facilitates in combining separate entry listings together. The **subject factor** as previously stated is used to sort records based on a value **type**. When records are ordered, it is likely that there may be more than one entry which belongs under the same subject factor value. In order to present the results in a more uniform layout, we **merge** the records occurring under the common subject factor value and remove duplicates. This approach of displaying results eliminates repetition while still maintaining all the listed entries.

Regardless of what each VDim aims to fulfil, we categorize four different types of VDims. What follows is a formal description and diagram illustration of the different types of VDims which are; Selection, Sorting, Implicit Join and Explicit Join.

4.2.1 Selection Virtual Dimension

This consists of selecting and extracting instance documents within one or more classes. The required instance documents correspond to what has been specified in the `query context value,` which can be an exact term or involve a group of values. The records obtained from the search class, constructs a new "partial" class meaning that only a part of the original class may be required and extracted. This is shown in Figure 4.

4.2.2 Sorting Virtual Dimension

In the Selection VDim the resultant document list would be displayed in random order, but in the Sorting VDim it is required for these to appear in a certain order. Common instance cases highlight the fact that sorting is to be done in alphabetical or chronological order. Figure 5 demonstrates that the resulting class A1 is now sorted.

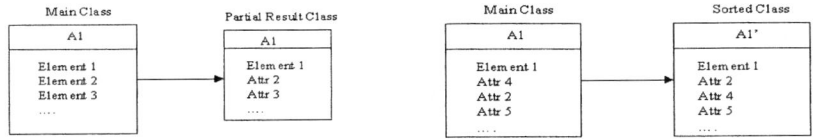

Fig. 4. Selection VDim   **Fig. 5.** Sorting VDim

4.2.3 Implicit Join Virtual Dimension

The concept of class hierarchical decomposition proves necessary to provide granularity to a real world object. Therefore it is likely that a VDim will involve joining of two or more classes/elements, which may appear at different levels within a hierarchy. The case of an *Implicit Join VDim* (Figure 6) is applicable where the class join occurs following hierarchical paths in connection with a common root class, in other words the process carried out is analogous to that of *Path Traversing* [23].

4.2.4 Explicit Join Virtual Dimension

This type of VDim denotes that joining is not limited to occurring within the hierarchical paths originating from the same class. Now it can also emerge from classes belonging to different source components, which are not directly related. For instance in Figure 7 we are able to join the components B and G.

Using the *keywords* presented and the four types of *Virtual Dimensions* presented, we able to construct a *generic query algorithm* as shown below, encompassing all these existing components. *Note that "[ ]" indicates that the contained function is optional.

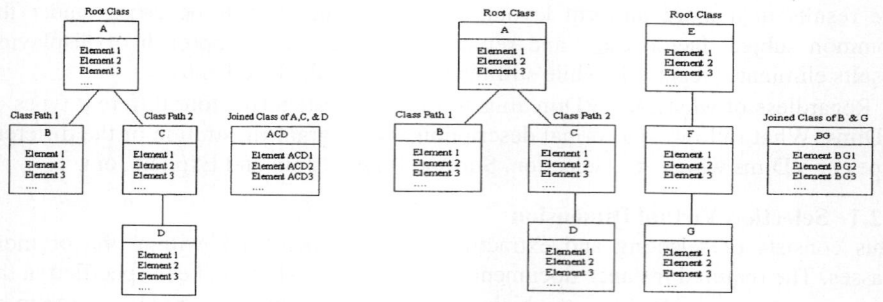

Fig. 6. Implicit Join VDim   **Fig. 7.** Explicit Join VDim

Generic Query Algorithm
```
Get Query Context Value
For [All | Each] value/s within the attribute/s or element/s of the Query Context
path
    If the value is empty
    Go to the next value
[Check for exact match
    If there is no match
    Go to the next value]
Return Context attribute/s and/or element/s value/s
[Sort by Subject Factor Value]
[If the Subject Factor Value is a duplicate
    Merge entries under one common Subject Factor Value
    Delete duplicated Subject Factor Value
ELSE ( )]
```

Due to page limitations, in this paper, we do not provide a complete illustration of the algorithm's purpose by conveying its main capabilities and how it has been applied to a sample set of queries based on examples. A full such illustration, including the design and implementation of the most suited algorithm for each individual case discussed above can be found in our work [8].

5 Conclusion and Future Work

In this paper, we proposed an intuitively approach for designing Web Document Warehouse (WDW) for web documents. In addition, we carried out a systematic approach to conceptual modelling and transformation of the warehouse conceptual model into a logical/schema (XML Schema) model and an in-depth analysis for deriving context-aware WDW dimensions (using user requirements), illustrated using a walk-through with a case study example. In doing so, we highlighted some of the challenges faced and the benefits of using native XML technologies such as XML schema and XML views in the warehouse design methodology.

For future work, the following issues deserve investigation. a) develop formal semantics to automate mapping between web content and WDW repositories and the performance issues associated with such a demanding task, b) incremental update of warehouse dimensions (materialized views) and issues related to supporting web OLAP queries over such dimensions and c) validation and derivation of warehouse user requirements [24] using requirement engineering techniques such as use-case and goal-oriented approaches.

References

1. W3C-XML, *Extensible Markup Language (XML) 1.0, (http://www.w3.org/XML/)*. 2004, The World Wide Web Consortium (W3C).
2. Pokorn'y, J. *XML Data Warehouse: Modelling and Querying*. in *Proc. of the Baltic Conf. (BalticDB-IS '02)*. 2002: Institute of Cybernetics at Tallin Technical University.
3. Lucie-Xyleme. *Xyleme: A Dynamic Warehouse for XML Data of the Web*. in *IDEAS '01*. 2001. Grenoble, France: IEEE Computer Society 2001.
4. Lucie-Xyleme, *Lucie Xyleme: A dynamic warehouse for XML Data of the Web*. IEEE Data Engineering Bulletin, 2001. **24, No 2**: p. 40-47.
5. Xyleme, *Xyleme Project (http://www.xyleme.com/)*. 2001.
6. Fankhauser, P. and T. Klement. *XML for Data Warehousing Chances and Challenges (Extended Abstract)*. in *Int. Conf. on DaWaK '03*. 2003. Prague: Springer-Verlag GmbH.
7. Nassis, V., et al. *XML Document Warehouse Design*. in *6th Int. Conf. DaWaK '04*. 2004. Zaragoza, Spain: Springer.
8. Nassis, V., et al., *Conceptual and Systematic Design Approach for XML Document Warehouses*. Int. J.of Data Warehousing and Mining, 2005. **1, No 3**.
9. Golfarelli, M., D. Maio, and S. Rizzi, *The Dimensional Fact Model: A Conceptual Model for Data Warehouses*. Int. J.of Cooperative Information Systems, 1998. **7**(2-3): p. 215-247.
10. Trujillo, J., et al., *Designing Data Warehouses with OO Conceptual Models*, in *IEEE Computer Society, "Computer"*. December, 2001. p. 66-75.
11. Dillon, T.S. and P.L. Tan, *Object-Oriented Conceptual Modeling*. 1993: Prentice Hall, Aus.
12. Coad, P. and E. Yourdon, *Object-oriented analysis*. 2nd ed. Yourdon Press computing series. 1991, London: Prentice Hall. xiv, 233 p.
13. Rajugan R., et al. *XML Views: Part 1*. in *14th Int. Conf. on Database and Expert Systems Applications (DEXA '03)*. 2003. Prague, Czech Republic: Springer 2003.
14. Rajugan R., et al. *XML Views, Part II: Modeling Conceptual Views Using XSemantic Nets*. in *Workshop & SS in 30th IEEE, IECON '04*. 2004. S.Korea: IEEE.
15. Rajugan R., et al. *XML Views, Part III: Modeling XML Conceptual Views Using UML*. in *7th Int. Conf. on Enterprise Information Systems (ICEIS '05)*. 2005. Miami, USA.
16. Feng, L., E. Chang, and T.S. Dillon, *Schemata Transformation of Object-Oriented Conceptual Models to XML*. Int. J.of Com Sys. Sci. & Eng., 2003. **18, No. 1**(1): p. 45-60.
17. Xiaou, R., et al. *Modeling and Transformation of Object-Oriented Conceptual Models into XML Schema*. in *12th Int. Conf. on DEXA '01*. 2001: Springer.

18. Xiaou, R., et al. *Mapping Object Relationships into XML Schema.* in *Proc. of OOPSLA Workshop on Objects, XML and Databases.* 2001.
19. Rahayu, W., et al. *Aggregation versus Association in Object Modeling and Databases.* in *Proc. of the 7th Australasian Conf. on IS.* 1996. Australia: ACS Inc.
20. W3C-XQuery, *XQuery 1.0: An XML Query Language,* in *XQuery,* 2004.
21. W3C-XPath, *XML Path Language (XPath) Version 1.0,* in *XML Path Language.* 1999,
22. Chamberlin, D.D. and H. Katz, *XQuery from the experts : a guide to the W3C XML query language.* 2003, Boston: Addison-Wesley. 1 v.
23. Bertino, E., *A survey on indexing techniques for object-oriented databases,* in *Query Processing for Advanced Database Systems.* 1994, Morgan Kaufmann.
24. Nassis, V., et al., *Goal-Oriented Requirement Engineering for XML Document Warehouses,* in *Processing and Managing Complex Data for Decision Support.* 2005, Idea Group Publishing (submitted chapter).

Clustering and Retrieval of XML Documents by Structure*

Jeong Hee Hwang and Keun Ho Ryu

Database Laboratory, Chungbuk National University, Korea
{jhhwang, khryu}@dblab.chungbuk.ac.kr

Abstract. We not only propose a method for XML document clustering using common structures but also show the application of our technique to XML retrieval. Our approach first extracts the frequent structures from XML documents by the decomposed method of tree. And then, we perform a new XML document clustering algorithm using common structures, which does not use measure of pairwise similarity between XML documents. The high speed and cluster cohesion of our clustering algorithm are shown in our experiment results.

1 Introduction

XML(eXtensible Markup Language) is a language for specifying semistructured data and a standard for data representation and exchange on the web. XML includes arbitrary tags for representing document elements, and allows the elements to be organized in a nested structure. Therefore, it has become crucial to address the question of how we can efficiently query and search XML documents[1,2,3].

Grouping XML documents according to their structural homogeneity can help in devising indexing techniques for such documents and improving the construction of query plans[4]. Several methods for detecting the similarity of XML documents are more concerned with the common structure in tree collection[5,6,7].

Although several XML document clustering methods[4,8,9,10,11] and related works[7,12] have been carried out, there remain some problems as follows. First, the application of some approaches is limited to DTDs within similar domain. Second, most approaches use hierarchical agglomerative clustering, denoted HAC in this paper, or k-means algorithms[13]. However, they need in general the number of clusters, and also it is based on a measure of pairwise similarity between documents. Therefore, it is difficult to determine both correct parameter and similarity computation method among XML documents. Third, the existing works are not scalable for large schemaless XML document collections because of the computing time of pairwise similarity matching among diverse tree structures.

Therefore, in order to address these problems, in this paper, we use a clustering algorithm for transactional data, CLOPE[14], which is very fast and scalable, while

---

* This work was supported by Ubiquitous Bio-Information Technology Research Institute in Korea.

being quite effective to the large volume of data. And also, for efficient clustering process, we add the notion of large items[15] which are popular items in a cluster.

This paper proceeds as follows. Section 2 presents a method of extracting the representative structures from XML document. Section 3 defines a novel clustering criterion method using common structures. Section 4 describes how to apply our clustering method to XML retrieval. The proposed method is tested and evaluated through experiments in section 5. Section 6 concludes the paper with summary.

2 Discovering the Frequent Structures of XML Documents

An XML document can be modeled as an ordered labeled tree[4,16]. We decompose an XML document trees into structures including path information, called Mine X_path(*mX_path*). It is similar to [9] but there are some differences in that [9] defines paths considering both attribute node and content node in an XML document, so that any tuples representing paths include null value if there exists only either attribute node or content node in a path. However, we focus on paths of elements with content value, not considering attribute in an XML document because attributes in an XML document do not have much influence on structure information. Therefore we do not admit null value in representing *mX_path*. We formally define Mine X_path(*mX_path*) as follows.

Definition 1 (Mine *X_path(mX_path)*). Given a node n_i with content value in an XML document tree, *mX_path* of node n_i is defined as a 2-tuple ((PrefixPath(n_i), ContentNode(n_i)).

PrefixPath(n_i) is an ordered sequence of tag name from root to node n_{i-1} which includes hierarchical structure. ContentNode(n_i) is a sequence of tag names with content value under the same PrefixPath(n_i) sequence and the order among the content sequence is ignored, but duplicate element names are not omitted.

An XML document(X_Doc_i) is composed of many *mX_path* sequences, and the order of *mX_paths* is ignored because we consider each *mX_path* as individual item in an XML document.

```
<book>
   <preface number = "0">
    <author>
       <name>J. Philips</name>
       <name>W. Moore</name>
    </author>
   </preface>
   <chapter number = "1">
    <content> .. </content>
   </chapter>
   <chapter number = "2">
    <author>
       <name> Frank Allen</name>
    </author>
    <content> .. </content>
   </chapter>
</book>
```

Fig. 1. An XML document

An XML document example is shown in Figure 1, and its corresponding tree is shown in Figure 2. We decompose each XML document tree to *mX_paths*. And then, as shown in Figure 3, we replace similar elements in *mX_paths* with same integer by referring element mapping table that we construct. We use WordNet Java API[17] to determine whether two elements are synonyms. We finally extract frequent structures from each document on the basis of the transformed *mX_path* sequences in Figure 3, using PrefixSpan algorithm[18]. The extraction process of frequent structures consists of two stages.

First stage is to extract frequent paths from *PrefixPath* sequence of the *mX_path* by

PrefixSpan algorithm. We regard a *PrefixPath* of each *mX_path* as a sequence and each element of *PrefixPath* as an item. Second stage is to find frequent full paths containing *ContentNode* to frequent *PrefixPath* extracted by first stage. The complete algorithm is given in Algorithm mX_path_FrequentSearch.

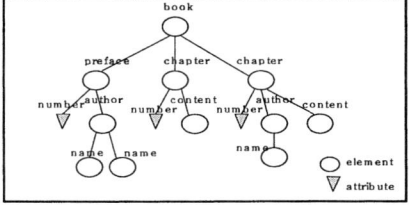

mX_path	Original paths	Transformed paths
mX_path₁	<{book/preface/author}, {name, name}>	<{1/2/3}, {4, 4}>
mX_path₂	<{book/chapter}, {content}>	<{1/5}, {6}>
mX_path₃	<{book/chapter/author}, {name}>	<{1/5/3}, {4}>
mX_path₄	<{book/chapter}, {content}>	<{1/5}, {6}>

Fig. 2. An XML document tree   **Fig. 3.** Transformed *mX_path* sequence

Algorithm mX_path_FrequentSearch
Input: *mX_path* sequences in an XML document (X_Doc_i),
  the minimum support threshold *min_sup*(the total number of *mX_paths* in a document
  * minimum_support $\theta(0< \theta \leq 1)$)

Output: the complete set of frequent path Fp_i

1. find the set of frequent paths Fp_i in the set of *PrefixPath* sequence in each *mX_path*
2.   call PrefixSpan algorithm (refer to [18])
3. find the set of frequent paths Fp_i considering the *ContentNode* of each *mX_path*
4.   **for** each *ContentNode* in each *mX_path*
5.   **if** the number of *ContentNode* ≥ min_sup
6.   **if** the number of the same name of *ContentNode* ≥ min_sup
7.   the *mX_path* including *ContentNode* is included in the set of Fp_i
8.   **else**
9.   the only *PrefixPath* ifself, *mX_path* excluding *ContentNode*, is included in
     the set of Fp_i if there is already no itself in Fp_i

In order to perform XML document clustering, we input not just maximal length path but all the frequent paths of length over min_length(maximal frequent structure length * length_rate σ (0< σ ≤1)) into clustering algorithm, because it is to avoid missing frequent structures of different length.

3 Clustering XML Documents Using Common Structures

We assume an XML document as a transaction, the frequent structures extracted from each document as the items of the transaction, and then we perform the document clustering using common structures. The CLOPE method[14] uses following concept and we also apply it to criterion for cluster allocation.

Figure 4 shows the quality difference of clustering result by histograms that represent common items and their occurrence frequency as width and height. Clustering (a) is better than (b) because it includes more common items.

The item set included in all the transaction is defined as I = {$i_1, i_2, ..., i_n$}, cluster set as C = {$C_1, C_2, ..., C_m$}, and transaction set as T = {$t_1, t_2, ..., t_k$}. As a criterion to allocate a transaction to appropriate cluster, we define the cluster allocation gain.

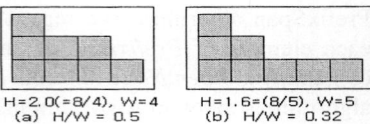

Fig. 4. Clustering Histograms

Definition 2 (Cluster Allocation Gain). The cluster allocation gain is the sum of the rate of the total occurrences of the individual items in every cluster according to cluster allocation. The following equation expresses this.

$$Gain(C) = \frac{\sum_{i=1}^{m} G(C_i) \times |C_i|}{\sum_{i=1}^{m} |C_i|} = \frac{\sum_{i=1}^{m} \frac{T(C_i)}{W(C_i)^2} \times |C_i|}{\sum_{i=1}^{m} |C_i|}$$

Where G is the occurrence rate(H) to individual item(W) in a cluster, H = T (the total occurrence of the individual items) / W (the numbers of the individual items), and G = T/W^2. $|C_i|$ is the number of transaction in cluster C_i.

Gain is a criterion function for cluster allocation of the transaction, and the higher the rate of the common items, the more the cluster allocation gain. Therefore we allocate a transaction to the cluster to be the largest *Gain*.

However if we use only the rate of the common items, not considering the individual items like the previous method, CLOPE, it causes some problems that the allocation gain about a new cluster is considerably high, because *Gain* about a new cluster includes $H(=W)/W^2$, and also it causes not only cluster to produce over the regular size but also cluster cohesion to be reduced. In order to improve this problem, we use the notion of large items as follows and apply it to cluster participation.

An item support in a cluster is defined as the number of the transactions including the item in a cluster. An item is popular in the cluster if the number of the transactions including the item in cluster C_i is over support, Sup = $\theta * |C_i|$, about the user specified minimum support, θ (0 < θ <= 1), and we called it large item in cluster C_i.

Definition 3 (Cluster Participation). It is the rate of the common items between the items of transaction t_k composed of the frequent structures and the large items $C_j(L)$ in the cluster C_j. And it means the probability of transaction t_k to be assigned to cluster C_j. We represent it as follows.

$$P\_Allo(t_k \Rightarrow C_j) = \frac{|t_k \cap C_j(L)|}{|t_k|} \geq \omega$$

(0 < ω < 1 : minimum participation)

$|t_k|$ is the number of the items of the transaction t_k. We use cluster participation two times in the clustering procedure. The first is used for appraising clusters to be needed to compute the *Gain*, which means that we compute *Gain* about the only allocatable clusters that satisfy the given minimum participation ω_1. The second is that if there is any cluster that satisfies the given minimum participation ω_2 about insertion of a transaction, our approach does not produce a new cluster, but allocates the transaction to the existing cluster with maximum participation. Therefore, ω_2 is used for determining the creation of new cluster, so that cluster participation can control the number of cluster. When ω_2 is small, the production of the cluster is suppressed.

To easily find the cluster for allocating a transaction, we define the difference operation, considering a transaction at each step, as follows.

Definition 4 (Difference Operation). The difference operation is the different *Gain* of the inserted transaction to the existing cluster. We use inserted difference ($diff\_Gain(\Delta^+)$) which is formally defined as follows.

$$diff\_Gain(\Delta^+) = New\_Gain(C_i) - Old\_Gain(C_i)$$

$$= \frac{T'(C_i)}{W'(C_i)^2} \times (|C_i|+1) - \frac{T(C_i)}{W(C_i)^2} \times |C_i|$$

W' is the number of the individual items and T' is the total occurrence frequency of individual items when the transaction is inserted. We allocate the transaction to the cluster of the largest $diff\_Gain(\Delta^+)$. Figure 5 shows the flow chart of XML document clustering.

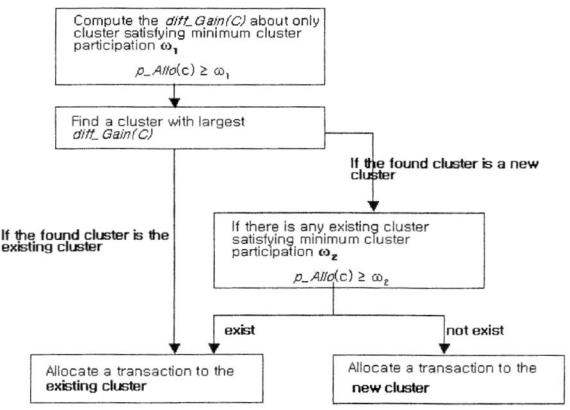

Fig. 5. XML document clustering flow chart

For quality measure of clustering result, we also adopted the cluster cohesion and the inter-cluster similarity, and we defined as follows.

Definition 5 (Cluster Cohesion). The cluster cohesion($Coh(C_i)$) is the ratio of the occurrence frequency of large items, $|L(C_i)|$, to the total occurrence frequency of items, $T(C_i)$, in the cluster C_i. This is calculated by the following formula, and if it is near 1, it is a good quality cluster.

$$Coh(C_i) = \frac{|L(C_i)|}{T(C_i)}$$

Definition 6 (Inter-cluster Similarity). The inter-cluster similarity based on the large items is the rate of the common large items of the cluster C_i and C_j. We calculate the inter-cluster similarity by the following formula, and if it is near 0, it is a good clustering.

$$Sim(C_i, C_j) = \frac{|C_i(L) \cap C_j(L)| \times \frac{|L(C_i \cap C_j)|}{|L(C_i + C_j)|}}{L(C_i) + L(C_j)}$$

Where $L(C_i)+L(C_j)$ is the total number of large items in the cluster C_i and C_j, $|C_i(L) \cap C_j(L)|$ is the occurrence frequency of common large items in the cluster C_i and C_j, $|L(C_i \cap C_j)|$ is the total occurrence frequency of the common large items, and $|L(C_i+C_j)|$ is that of all the large items in cluster C_i and C_j.

4 XML Document Retrieval

The retrieval process of XML Document based on the cluster with similar structures consists of three phases as follows.

1. Simplify query into simple structure by using the element mapping table.
2. Reduce the search space to the cluster of similar structure, finding the most similar cluster by comparing structure of the large item in each cluster with the query.
3. Display the ranked XML documents by computing the similarity between query and documents in the similar cluster.

For XML document retrieval, it is necessary to compute the structural similarity between query tree and XML documents. We consider both edges and paths in the tree structure. To do this, we formulate edge similarity and path similarity as follows.

Definition 7 (Edge Similarity). Given an ordered labeled tree T and Query tree Q, edge similarity is defined by the ratio of the number of common edge to the total number of edge between T and Q. if there is an edge u→v and u'→v' in the T and Q respectively, and also u=u' and v=v', we say that u→v and u'→v' are matched. We can get the edge similarity by following formula.

$$EdgeSim(Q, T) = \frac{|E_Q \cap E_T|}{|E_Q \cup E_T|}$$

Example 1. Figure 6 denotes two XML document trees to show how to compute edge similarity.

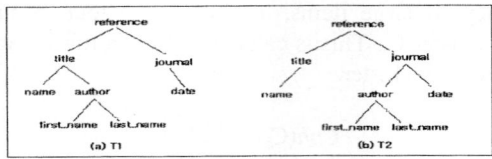

Fig. 6. Example of similar XML document trees

We can obtain the following edge sets in tree T_1 and T_2 (the element name is replaced by the first alphabet).

$$E_{T1} = \{r \to t, r \to j, t \to n, t \to a, j \to d, a \to f, a \to l\}$$
$$E_{T2} = \{r \to t, r \to j, t \to n, j \to a, j \to d, a \to f, a \to l\}$$

So the edge similarity between T_1 and T_2 by definition 7 is computed as follows.

$$EdgeSim(T_1, T_2) = \frac{|E_{T_1} \cap E_{T_2}|}{|E_{T_1} \cup E_{T_2}|} = \frac{6}{8}$$

Definition 8 (Path Similarity). Given an ordered labeled tree T and Query tree Q, a path is denoted by consecutive edge from the root node to the particular node with the different depth(i.e., $v1 \to v2$, $v1 \to v2 \to v3$, $v1 \to v2 \to ... \to vn$) and path similarity is defined by the ratio of the number of common path to the total number of path between T and Q. If there are paths $v1 \to v2 \to v3$ and $v1' \to v2' \to v3'$ in the T and Q respectively, and also $v1 = v1'$, $v2 = v2'$, and $v3 = v3'$, we say that $v1 \to v2 \to v3$ and $v1' \to v2' \to v3'$ are matched. We can get the path similarity by following formula.

$$PathSim(Q, T) = \frac{Max_{com} |P_Q \cap P_T|}{Max_{path} |P_Q, P_T|}$$

where $Max_{path}|P_Q, P_T|$ is the largest path length among paths in the T, Q and $Max_{com}|P_Q \cap P_T|$ is the largest common path length in the T, Q.

We can obtain the following path sets in the tree T_1, T_2 in Figure 6.

$$PathSim(T_1, T_2) = \frac{Max_{com} |P_{T_1} \cap P_{T_2}|}{Max_{path} |P_{T_1}, P_{T_2}|} = \frac{3}{4}$$

According to above definition, the formula for computing similarity between Q and T that considers both edge similarity and path similarity is defined as following.

$$Sim(Q,T) = \alpha * EdgeSim(Q, T) + \beta * PathSim(Q, T) \quad (1)$$
$$(\alpha + \beta = 1, \alpha \geq 0, \beta \geq 0)$$

where $\alpha > \beta$ gives more emphasis on the edge similarity, and $\alpha < \beta$ gives more emphasis on the path similarity, by default $\alpha = 0.5$, $\beta = 0.5$. Therefore, the similarity of tree T_1 and T_2 in Figure 6 by equation (1) is $(0.5*6/8 + 0.5*3/4 = 0.75)$.

To show the application of our approach to XML retrieval, we have implemented the user interface in which user can input at most three elements of ordered hierarchical structure in the left part of window. Figure 7 shows a search result of the ranked similar structured XML documents about query 'book/title' in the right window. And also the contents of XML document selected by user are displayed under right corner of window.

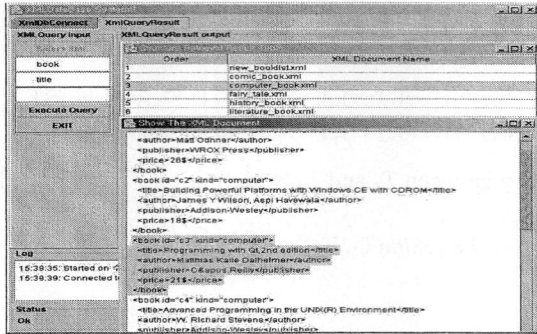

Fig. 7. A query and search result for retrieving XML documents

5 Experimental Results

In this section, we evaluate the performance of *mX_path* decomposition and our clustering method in comparison with the previous methods by doing some experiments. The data used are 300 XML documents, taken from the Wisconsin's XML data bank [19].

5.1 Evaluating Tree Decomposition

To decompose the path from the XML tree structure is significant because of a basis o f clustering algorithm. Therefore, we evaluate our path decomposition method, XML_ C, by comparing the previous method[9], denoted S_C in this paper.

We conducted two experiments. In the first experiment, we compare the execution t ime of the XML_C with that of S_C by increasing document sets. The result of the ex periment is shown in Figure 8. We find that the execution time of XML_C is faster th an that of S_C. In the second experiment, we measure the ratio of the number of deco mposed path to the total number of nodes in a tree structure, and compare it with that of S_C by increasing document sets. The result of the experiment is shown in Figure 9. We see that XML_C is less than S_C at the rate of decomposed path. We can also n otice that the more the rate of attribute included in a tree, the more difference of rate b etween XML_C and S_C, with increasing the documents.

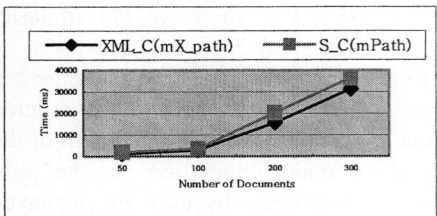

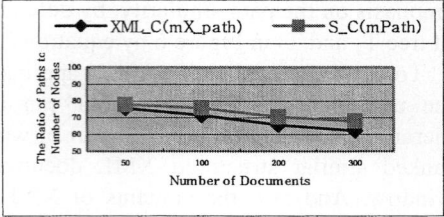

Fig. 8. Decomposition execution time

Fig. 9. The ratio of path to the total number of nodes in an XML document

5.2 Clustering Efficiency

We conducted intensive experiments about clustering algorithm, comparing XML_C with CLOPE and HAC. In order to extract the representative structure of each document, we set the minimum_support to 0.5 and the length_rate to 0.7, considering the best parameter value found by repeated experiments. The average length and the number of frequent path structure extracted in each document are 4.3 and 6.2 respectively. To perform HAC algorithm, we generated the similarity matrix, computing similarity among the mX_path sequences, and then we performed HAC[13] on the basis of the matrix.

Figure 10 illustrates the execution time of the algorithms as the number of document increases. The important result in this experiment is that the performance of XML_C is comparable to CLOPE, while being much better than HAC. This is because it takes much time to calculate the similarities between documents in HAC. On the other hand, XML_C has slightly better performance than CLOPE. This means that XML_C using cluster participation comes into effect on the performance in contrast to CLOPE.

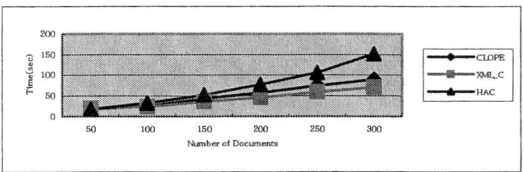

Fig. 10. Execution time

We also experiment the cluster cohesion and inter-cluster similarity based on the large items, under the circumstances of different support. But CLOPE and HAC don't use the notion of large item. Therefore, we extract the large items according to support in the same manner of XML_C from the clustering results, after running both CLOPE and HAC respectively. The result of the cluster cohesion and the inter-cluster similarity according to the minimum support is shown in Figure 11 and Figure 12.

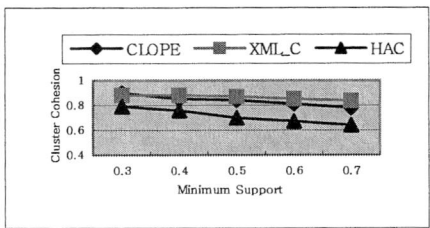 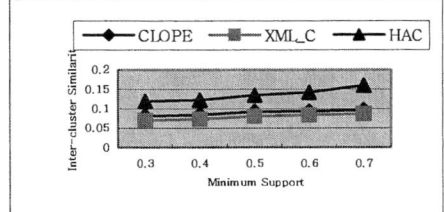

Fig. 11. Cluster cohesion   **Fig. 12.** Inter-cluster similarity

As we expected, the XML_C exhibits the highest cluster cohesion among these three algorithms in Figure 11. On the other hand, The cluster cohesion of HAC has the lowest performance. We see that XML_C keeps it well with good quality approximating 1. The main reason is that it uses the global criterion, *Gain*, increasing common items in each cluster during the cluster assignment.

Figure 12 shows inter-cluster similarity measurement, and the performance ranking of three algorithms is XML_C > CLOPE > HAC (">" means better). This results means that XML_C maintains the similar documents with common structures better than others.

In summary, we can determinate that XML_C is better overall at cluster cohesion and inter-cluster similarity. This means that our algorithm groups similar structured XML documents together and gathers dissimilar structured XML documents apart.

6 Conclusion

In this paper, we presented a novel approach to XML document clustering technique using common structures, which does not use any measure of pairwise similarity between XML documents. We first extracted the frequent structures from XML documents using decomposition mechanism. And then we performed the XML document clustering by common structures, considering that an XML document as a transaction and the extracted frequent structures from documents as the items of the transaction. Our experiment results showed that our approach reduced the execution time cost, while maintaining the higher clustering quality.

References

[1] P. Kotasek and J. Zendulka. An XML Framework Proposal for Knowledge Discovery in Database. *The 4th European Conference on Principles and Practice Knowledge Discovery in Databases,* 2000.
[2] J. Widom. Data Management for XML: Research Directions. *IEEE Computer Society Technical Committee on Data Engineering,* 1999.
[3] R. Nayak, R. Witt and A. Tonev. Data Mining and XML Documents. *International Conference on Internet Computing,* 2002.
[4] F. D. Francesca, G. Gordano, G. Manco, R. Ortale and A. Tagarelli. A General Framework for XML Document Clustering. *Technical report, n(8), ICAR-CNR,* 2003.
[5] K. Wang and H. Liu. Discovery Typical Structures of Documents: A Road Map Approach. *In ACM SIGIR,* 1998.
[6] T. Asai, K. Abe, S. Kawasoe, H. Arimura and H. Sakamoto. Efficient Substructure Discovery from Large Semi-structured Data. *In the proceedings of the Second SIAM international conference on Data Mining,* 2002.
[7] A. Termier, M. C. Rouster and M. Sebag. TreeFinder: A First Step towards XML Data Mining. *IEEE international conference on Data Mining (ICDM),* 2002.
[8] M. L. Lee, L. H. Yang, W. Hsu and X. Yang. XClust: Clustering XML Schemas for Effective Integration. *Proc. 11th ACM international conference on Information and Knowledge Management,* 2002
[9] Y. Shen and B. Wang. Clustering Schemaless XML Document. *In the proceedings of the 11th international conference on Cooperative Information System,* 2003

[10] J. Yoon, V. Raghavan and V. Chakilam. BitCube: Clustering and Statistical Analysis for XML Documents. *In the proceedings of the 13th international conference on Scientific and Statistical Database Management*, 2001.
[11] A. Doucet and H. A. Myka. Naïve Clustering of a Large XML Document Collection. *The Proceedings of the 1st INEX*, Germany, 2002.
[12] J. W. Lee, K. Lee and W. Kim. Preparation for Semantics-Based XML Mining. *IEEE International Conference on Data Mining(ICDM)*, 2001.
[13] A.K. Jain, M.N. Murty and P.J. Flynn. Data Clustering: a review. *ACM Computing Surveys*, Vol. 31, 1999.
[14] Y. Yang, X. Guan and J. You. CLOPE: A Fast and Effective Clustering Algorithm for Transaction Data. *In the proceedings of the 8th ACM SIGKDD International Conference on Knowledge Discovery and Data Mining*, 2002.
[15] K. Wang and C. Xu. Clustering Transactions Using Large Items. *In the proceedings of ACM CIKM-99*, 1999.
[16] L. Mignet, D. Barbosa and P. Veltri. The XML web: a first study. *In the proceedings of the twelfth international conference on World Wide Web*, 2003.
[17] http://sourceforge.net/projects/javawn
[18] J. Pei, J. Han, B. M. Asi and H. Pinto. PrefixSpan: Mining Sequential Pattern Efficiently by Prefix-Projected Pattern Growth. *In the proceedings of the International Conference on Data Engineering(ICDE)*, 2001.
[19] NIAGARA query engine. http://www.cs.wisc.edu/niagara/data.html.

A New Method for Mining Association Rules from a Collection of XML Documents*

Juryon Paik, Hee Yong Youn, and Ungmo Kim

Department of Computer Engineering, Sungkyunkwan University,
300 Chunchun-dong, Jangan-gu, Suwon,
Gyeonggi-do 440-746, Republic of Korea
quasa277@hotmail.com, {youn, umkim}@ece.skku.ac.kr

Abstract. With the sheer amount of data stored, presented and exchanged using XML nowadays, the ability to extract *interesting* knowledge from XML data sources becomes increasingly important and desirable. In support of this trend, several encouraging attempts at developing methods for mining XML data have been proposed. However, efficiency and simplicity are still barrier for further development. In this paper, we show that any XML document can be mined for association rules using only a specially devised hierarchical data structure called HoPS without multiple XML data scans. It is flexible and powerful enough to represent both simple and complex structured association relationships inherent in XML data.

1 Introduction

Data mining is traditionally used to extract *interesting* knowledge from large amounts of data stored in databases or data warehouses. This knowledge can be represented in many different ways such as clusters, decision trees, decision rules, etc. Among them, association rules [1] have been proved effective in discovering interesting relations in massive amounts of data.

Currently, XML [12] is penetrating virtually all areas of Internet application programming and is bringing about huge amount of data encoded in XML. With the continuous growth in XML data sources, the ability to extract knowledge from them for decision support becomes increasingly important and desirable. There are some proposals to exploit XML within the knowledge discovery tasks, but most of them still rely on the traditional relational framework with an XML interface. Due to the inherent flexibilities of XML, in both structure and semantics, mining knowledge in the XML Era is faced with more challenges than in the traditional well-structured world. Hence, compared to the fruitful achievements in well-structured data, mining association rules in the semistructured XML world still remains at a preliminary stage.

---

* This work was supported in part by Ubiquitous computing Technology Research Institute and by the University IT Research Center Project, funded by the Korean Ministry of Information and Communication.

In this paper, we present *HoPS (Hierarchical structure of PairSet)*, a data structure that can be used to extract association rules from native XML documents, shortly "XML association rules", which was first introduced by Braga et al. [3]. The proposed structure not only reduces significantly the number of rounds for candidate tree items pruning, but also simplifies greatly each round by avoiding time-consuming tree item join operations.

The remainder of this paper is organized as follows. We review some closely related works in Section 2. Section 3 defines association rules in the context of relational databases and discusses the notions of association rules for XML. In Section 4 we introduce some basic concepts needed to discuss implementation details and how XML association rules are extracted using by HoPS from a set of XML documents. We conclude this paper and discuss the future direction of our research in Section 5.

2 Related Works

Since the problem of mining association rules was first introduced [1], a large amount of work has been done in various directions. The famous Apriori algorithm for extracting association rules was published independently in [2] and in [10]. Although the first algorithms assumed that transactions were represented as sets of binary attributes, along the years many algorithms for the extraction of association rules from multivariate data have been proposed. This includes the use of quantitative attributes [9] as well as the integration of concept hierarchies to support the mining of association rules from different levels of abstractions [5,8]. Singh et al. [7] proposed to mine association rules that relate structural data values to concepts extracted from unstructured and/or semistructured data. However, as far as we know, they are not suitable for mining association rules from *native* XML documents. Under the traditional association framework, the basic unit of data to look at is database *record*, and the construct unit of a discovered association rule is *item*. Since the structure of XML is a directed acyclic graph, but not static tabular structure, it is required to find the counterparts of record and item in mining association relationships from XML data. Recently, tools for extracting association rules from XML documents have been proposed in [4,11], but both of them are approaching from the view point of a XML query language. This causes the problem of language-dependent association rules mining.

In this paper, we focus on rule detection from a collection of XML documents describing the same type of information. Hence, each of XML documents corresponds to a database record, and possesses a tree structure. Accordingly, we extend the notion of associated item to an XML tree, and build up associations among trees rather than items. Compared to previous works, the work reported in this study aims to provide 1) a data structure model for mining XML association rules, which can deal with associations among both contents and structures of XML data; 2) techniques for the data structure-guided mining of association rules from large XML data; and 3) query language-neutral association rule mining.

3 Background Concepts

3.1 Association Rules for Relational Data

Association rules were first introduced by Agrawal et al. [1] to analyze customer habits in retail databases. Association rule is an implication of the form $X \Rightarrow Y$, where the rule *body* X and *head* Y are subsets of the set $\mathcal{I}$ of *items* ($\mathcal{I} = \{I_1, I_2, \ldots, I_n\}$) within a set of *transactions* $\mathcal{D}$ and $X \cap Y = \phi$. A rule $X \Rightarrow Y$ states that the transactions T ($T \in \mathcal{D}$) that contain the items in X ($X \subset T$) are *likely* to contain also the items in Y ($Y \subset T$). Association rules are characterized by two measures: the *support*, which measures the percentage of transactions in $\mathcal{D}$ that contain both items X and Y ($X \cup Y$); the *confidence*, which measures the percentage of transactions in $\mathcal{D}$ containing the items X that also contain the items Y. More formally, given the function $freq(X, \mathcal{D})$, which denotes the percentage of transactions in $\mathcal{D}$ containing X, we define:

$$support(X \Rightarrow Y) = freq(X \cup Y, \mathcal{D})$$

$$confidence(X \Rightarrow Y) = \frac{freq(X \cup Y, \mathcal{D})}{freq(X, \mathcal{D})}$$

Suppose there is an association rule "*bread, butter* $\Rightarrow$ *milk*" with confidence 0.9 and support 0.05. The rule states that customers who buy *bread* and *butter*, also buy *milk* in 90% of the cases and that this rule holds in 5% of the transactions. The problem of mining association rules from a set of transactions $\mathcal{D}$ consists of generating all the association rules that have support and confidence greater than two user-defined thresholds: minimum support (min_sup) and minimum confidence (min_conf). To help the specification of complex association rule mining tasks, a number of query languages have been proposed. In particular, [6] introduced the MINE RULE operator, an extension of SQL specifically designed for modeling the problem of mining association rules from relational data.

3.2 Association Rules for XML Data

In this subsection, we briefly review some notations of tree model and describe the basic concepts of association rules for XML data, shortly "XAR".

Definition 1 (Labeled Tree). *A labeled tree is a tree where each node of the tree is associated with a label.*

Every XML document is represented by a labeled tree. For brevity, we call a labeled tree as simply a tree in the rest of this paper.

Definition 2 (Fragment). *We say that a tree $F = (N_F, E_F)$ is a **fragment** of a tree $T = (N, E)$, denoted by $F \preceq T$, if and only if $N_F \subseteq N$ and for all edges $(u, v) \in E_F$, u is an ancestor of v in T.*

Note that in the definition of fragment, we require that for any edge (u, v) occurring in a fragment F, there be a path from node u to node v in tree T. Definition 2 preserves the ancestor relation but not necessarily the parent relation.

Definition 3 (Titem). *The basic construct unit in XAR is referred to as **titem** (tree-structured item). Since any part of an XML document is fragment, any fragment can be a titem.*

With the above definitions, we formally define XML association rules and related measurements. Let $\mathcal{D} = \{T_1, T_2, \ldots, T_n\}$ be a collection of XML documents and let $|\mathcal{D}|$ be the number of documents in $\mathcal{D}$. Let $\mathcal{F} = \{F_i, i > 0 \mid F_i \subset T_j, \text{ for some } j \in [1, n]\}$.

Definition 4 (XML Association Rule). *An **XML association rule** is an implication of the form $X \Rightarrow Y$, which satisfies the following two conditions:*

i) $X \in \mathcal{F}, Y \in \mathcal{F}$;
ii) $(X \not\subseteq Y) \wedge (Y \not\subseteq X)$.

Different from the traditional association rules where associated items are usually denoted using simple structured data, the items in XAR have hierarchical tree structures, as indicated by the first clause of the definition. It is worth pointing out that each of the fragments contains only one basic root node. The second clause of the definition requires that in an XAR, titems are independent each other.

Definition 5 (Support and Confidence). *Let $\mathcal{I} = \{I_1, I_2, \ldots, I_m\}$ be a set of titems. Given a set of XML documents $\mathcal{D}$, the **support** and **confidence** of an XML association rule $X \Rightarrow Y$ are defined as:*

$$support(X \Rightarrow Y) = freq(X \cup Y, \mathcal{D}) = \frac{|\mathcal{D}_{X \cup Y}|}{|\mathcal{D}|}$$

$$confidence(X \Rightarrow Y) = \frac{freq(X \cup Y, \mathcal{D})}{freq(X, \mathcal{D})} = \frac{|\mathcal{D}_{X \cup Y}|}{|\mathcal{D}_X|},$$

where $\mathcal{D}_{X \cup Y} = \{T_i \mid \forall I_j \in (X \cup Y), I_j \subset T_i, \text{ for some } i \in [1, n], j \in [1, m]\}$, and $\mathcal{D}_X = \{T_i \mid \forall I_j \in X, I_j \subset T_i, \text{ for some } i \in [1, n], j \in [1, m]\}$.

Example 1. Let us consider three XML documents in Fig. 1(a) where various information about computer products are represented, e.g., the ordering information (identified by the node label `Order`) and the detailed information of a product (`Details`). Consider the problem of mining frequent association rule that if a customer purchases `HP Desktop`, usually s/he is likely to buy `HP Printer`. Fig. 1(b) shows this XML association rule mined from $\mathcal{D}$.

Example 2. In the example depicted in Fig. 1(a), the *support* of the association rule $freq(X \cup Y, \mathcal{D})$ is 0.66 since the titems X and Y appear together in two documents T_1 and T_3 out of three documents, i.e., in 66% of the cases. Likewise, $freq(X, \mathcal{D})$ is 1 since the fragment X appears in all of three documents, i.e., in 100% of the cases. Therefore, we can compute the *confidence* of the XML association rule Fig. 1(b) as:

$$\frac{freq(X \cup Y, \mathcal{D})}{freq(X, \mathcal{D})} = \frac{0.66}{1}$$

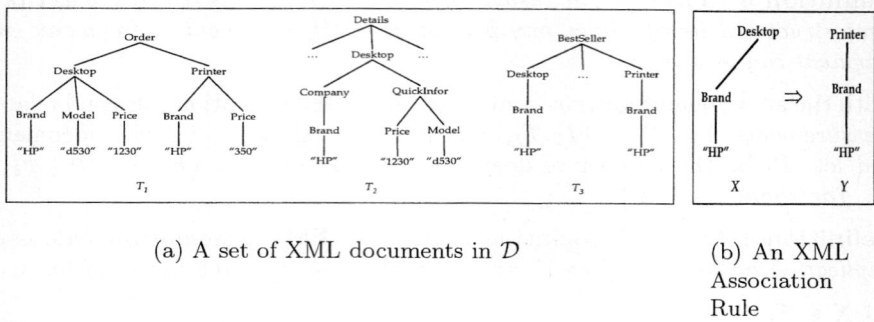

(a) A set of XML documents in $\mathcal{D}$

(b) An XML Association Rule

Fig. 1. Association Rule Mined from $\mathcal{D}$

which returns 0.66, i.e., in 66% of computer products purchased by `HP Desktop` together with `HP Printer` or, in the XML documents, in 66% of the fragments in $\mathcal{D}$ in which appears the titem X appear also the titem Y.

4 Overview of HoPS

Like the association rules for relational data, the problem of mining association rules for XML data consists of generating all XARs greater than min_sup and min_conf. We devised *HoPS* as a tool to mine simple/complex association rules from XML documents. HoPS allows the specification of complex mining task compactly and intuitively. Towards this goal, there is an important component. That is a specially devised data structure called *PairSet*. In this section, we describe the techniques used to mine data structure-guided XML association rules. Our mining process proceeds in three phases; The first phase is to transform each tree-structured data into hierarchical structure PairSets. The second phase is to manipulate PairSets and to reflect the min_sup. The third phase is to mine XML association rules from the PairSets by reflecting the two association measures.

4.1 Transforming Tree-Structured XML Document into PairSets

Definition 6 (Key). *Let K_d be a collection of node labels assigned on the nodes at depth d in every tree in $\mathcal{D}$. We assume that depth of root node is 0. We call each member in K_d by a **key**.*

At this point, note that there may exist some nodes labeled with the same names in $\mathcal{D}$. Thus, for each key we need to identify the list of trees, *tids*, in which the key belongs. Note that we employ hash technique to keep all the necessary information of each key's parent node into a hash table to facilitate the mining of association rules in later.

Definition 7 (PairSet). *A PairSet, $[P]^d$, is defined as a set of pairs (k_d, t_{id}) where k_d is a key in K_d and t_{id} is a list of tree indexes in which k_d belongs.*

The time complexity for generating PairSets is $O(n)$, where n is the total number of nodes in trees. According to a minimum support, a collection of PairSets can be classified as two classes.

Definition 8 (Frequent Key). *Given some user-defined min_sup and a pair (k_d, t_{id}), the key k_d is called **frequent** if $|t_{id}| \geq min\_sup \times |\mathcal{D}|$.*

Definition 9 (Frequent Fragment Set). *Given a PairSet, $[P]^d$, a pair in $[P]^d$ is called **frequent fragment set** if its key is frequent. Otherwise, it is called **candidate fragment set**. We denote frequent fragment set and candidate fragment set by $[F]^d$ and $[C]^d$, respectively.*

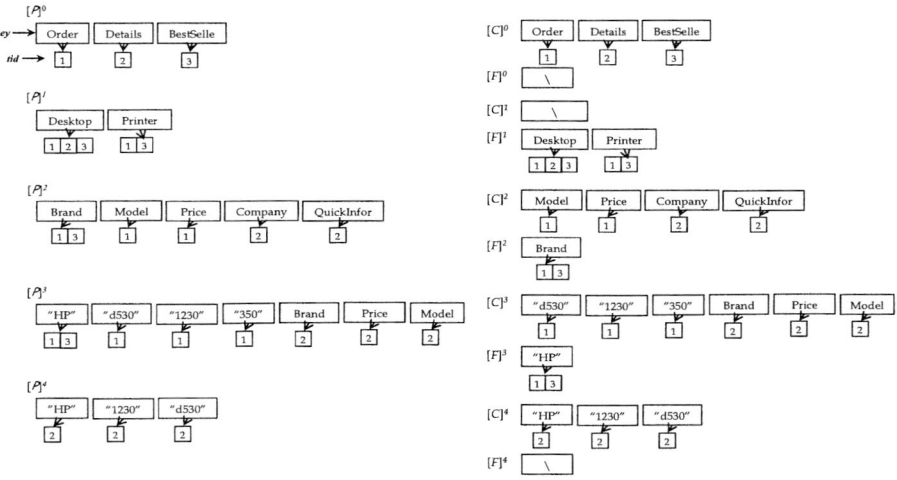

Fig. 2. PairSets Generation from a set of XML trees in $\mathcal{D}$

Example 3. The left side of Fig. 2 depicts a collection of PairSets transformed from $\mathcal{D}$ in Fig. 1(a). For instance, we know that the key Desktop in $[P]^1$ belongs to all of three trees T_1, T_2 and T_3. The right side of Fig. 2 illustrates the two classes, frequent fragment set and candidate fragment set, derived from PairSets in the left side. The given min_sup is 0.66.

4.2 Manipulating the Data Stored in PairSet Structure

The initial frequent fragment sets (depicted in the right side of Fig. 2) are derived solely by considering a min_sup over the absolute depth of node labels. These frequent fragment sets do not have competent fragments to mine XML association rules because both tree characteristic and fragment property (Definition 2) are not reflected in those initial frequent fragment sets. The tree characteristic stems from the fact that same labels can be placed several times throughout a XML tree. To handle those features, we introduce an operation called cross-filtering.

Cross-Filtering: Let $\mathcal{FS} = \{[F]^0, [F]^1, \ldots, [F]^d\}$ be a set of initial frequent fragment sets and $\mathcal{CS} = \{[C]^0, [C]^1, \ldots, [C]^d\}$ be a set of initial candidate fragment sets. The ith round (for $i = 1, 2, \ldots d$) of cross-filtering consists of the following two steps:

- Step 1 (Pruning phase). $Difference$ ($[C]^{i-1}$ vs. $[F]^i$) and
 ($[C]^i$ vs. $[F]^{i-1}$ through $[F]^0$)

 Given the current candidate fragment sets, this step is to eliminate the pairs which are already included in the previous frequent fragment sets. Due to this step, we can reduce significantly the number of pairs in candidate fragment sets.

- Step 2 (Merging phase). $Union$ ($[C]^i$ vs. $[C]^{i-1}$)

 Given the current candidate fragment sets, this step is to obtain a new frequent fragment set. Since our PairSet is a hierarchical structure, it is not necessary to generate additional candidate fragments by using join operations. It is enough to search only the candidate fragment sets at previous depths. Always only two candidate fragment sets are computed to find new pairs of frequent fragment sets without using join operation.

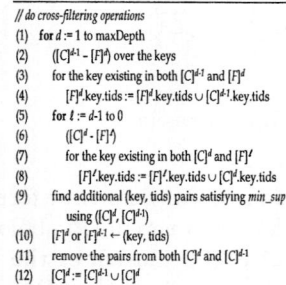

(a) Pseudo Code of Cross-Filtering Operation

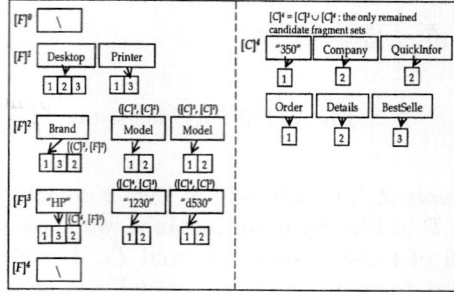

(b) Final Frequent Fragment Sets Corresponding to Fig. 2

Fig. 3. Cross-Filtering Operation

Fig. 3(a) shows the pseudo code of cross-filtering operation. In line 10, it depends on the information of parent node stored in hash table which frequent fragment set has to include a newly discovered pair. Fig. 3(b) depicts the final frequent and candidate fragment sets after applying cross-filtering over the two sets in the right side of Fig. 2.

4.3 Correlating Concrete Contents with Structures

For node labels obtained after the second phase in subsection 4.2, we only uses the node labels in final frequent fragment sets to mine XARs. There are three considerations during this correlating phase. Firstly, we instantiate every symbol \ in the frequent fragment sets with a '*'. Because the meaning of an empty frequent fragment set at depth d is that there is no any other node label satisfying the min_sup at depth d, conversely speaking, whatever a node label is, it can be placed at depth d without any restriction. Typically, a symbol '*' indicates such a meaning.

Secondly, although all pairs in each frequent fragment set satisfy a min_sup, not all of fragments being generated from the sets satisfy the min_sup. This stems from the fact that association relationships between keys have to conform the min_sup. The support has to be computed at the relationships, called edges, between keys of different frequent fragment sets. Given a set of XML documents $\mathcal{D}$ and a collection of frequent fragment sets $\mathcal{FS}$, let (k_i, t_{k_i}) be a pair in $[F]^i$ and (k_j, t_{k_j}) pair in $[F]^j$. Assume that both $|t_{k_i}|$ and $|t_{k_j}|$ are greater than $|\mathcal{D}| \times min\_sup$. We denote this feasible edge by $e_{ij} = (k_i, k_j)$ where k_i is an ancestor of k_j by Definition 2. The support of an edge is computed as:

$$support(e_{ij}) = support((k_i, k_j)) = freq((k_i, k_j), \mathcal{D})$$
$$= \frac{|t_{k_i} \cap t_{k_j}|}{|\mathcal{D}|} = \begin{cases} \text{associated} & \text{if } support(e_{ij}) \geq min\_sup, \\ \text{not associated} & \text{if } support(e_{ij}) < min\_sup. \end{cases} \quad (1)$$

Finally, all of titems satisfying the min_sup are generated by incrementally appending edges and hierarchically calculating the supports (in other words, calculating the supports of edges).

Thirdly, we need to compute the confidence of between titems. Then, this confidence is compared with a min_conf to mine all the XML association rules that have support and confidence greater than or equal to two user-defined thresholds. Let $\mathcal{I} = \{I_1, \ldots, I_m\}$ be a set of titems. Assume that $I_1 = e_{ij} = (k_i, k_j)$ and $I_2 = e_{pq} = (k_p, k_q)$ be titems consisting of 1 edge. The confidence of $I_1 \Rightarrow I_2$ is computed as:

$$confidence(I_1 \Rightarrow I_2) = \frac{freq(I_1 \cup I_2, \mathcal{D})}{freq(I_1, \mathcal{D})} = \frac{|t_{k_i} \cap t_{k_j} \cap t_{k_p} \cap t_{k_q}|}{|t_{k_i} \cap t_{k_j}|} \quad (2)$$

All the XML association rules that have support and confidence greater than or equal to two thresholds, min_sup and min_conf, are mined from the both equation (1) and (2). Following Theorem 1 is derived.

Theorem 1. *Given a set of XML documents $\mathcal{D}$ and a collection of frequent fragment sets $\mathcal{FS}$, all of XML association rules conforming to **any min_conf** are mined from the $\mathcal{FS}$.*

Proof. Let $[F]^i$ and $[F]^j$ be arbitrary nonempty frequent fragment sets in $\mathcal{FS}$, for $i < j$. Let (k_i, t_{k_i}) be a pair in $[F]^i$ and $(k_j, t_{k_j}), (e_j, t_{e_j})$ pairs in $[F]^j$. Assume that $|t_{k_i} \cap t_{k_j}| = \alpha$, $|t_{k_i} \cap t_{e_j}| = \beta$, and $|t_{k_i} \cap t_{k_j} \cap t_{e_j}| = \gamma$. Given min_sup and min_conf are s_m and c_m, respectively, where $0 < s_m, c_m \leq 1$. Let x and y be a tree-id in t_{k_j} and t_{e_j}, respectively, for $x \neq y$.

- **titem**$_a$ ($x \in t_{k_i} \wedge y \notin t_{k_i}$): The k_i and k_j forms a titem. Then,

$$support(titem_a) = freq((k_i, k_j), \mathcal{D}) = \frac{|t_{k_i} \cap t_{k_j}|}{|\mathcal{D}|} = \frac{\alpha}{|\mathcal{D}|}$$

Assume that the support of titem$_a$ is greater than or equal to s_m.
- **titem**$_b$ ($y \in t_{k_i} \wedge x \notin t_{k_i}$): The k_i and e_j forms a titem. Then,

$$support(titem_b) = freq((k_i, e_j), \mathcal{D}) = \frac{|t_{k_i} \cap t_{e_j}|}{|\mathcal{D}|} = \frac{\beta}{|\mathcal{D}|}$$

Assume that the support of titem$_b$ is greater than or equal to s_m.
- **Association Rule.** The confidence of an implication of the form $titem_a \Rightarrow titem_b$ is computed by using the equations (1) and (2):

$$confidence(titem_a \Rightarrow titem_b) = \frac{\gamma}{\beta}$$

$$= \begin{cases} \text{rules with } support \geq s_m \text{ and } confidence \geq c_m \text{ if } \gamma \geq \beta \times c_m, \\ \text{rules with } support \geq s_m \text{ and } confidence < c_m \text{ if } \gamma < \beta \times c_m. \end{cases}$$
(3)

Once the frequent fragment sets $\mathcal{FS}$ satisfying min_sup are derived from a set of XML documents $\mathcal{D}$, it is straightforward to generate both simple and complex XML association rules from $\mathcal{FS}$. There is no any other additional burdensome process to reflect the varying min_conf.

5 Conclusion

This paper presents a new data structure for extended XML association rules, with the aim to discover associations inherent in massive amounts of XML data. The newly devised data structure PairSet makes the process of mining complex XML association rules significantly simplified comparing to previous approaches. This is because the multiple scans of the database are reduced due to the list of tree indexes and the frequent fragment sets help to avoid lots of candidate fragments generations. We discuss language-neutral techniques used in mining data structure-guided XML association rules.

We are currently developing the practical algorithms for mining both simple and complex structured association relationships, and evaluating the performance on synthetic XML data.

References

1. R. Agrawal, T. Imielinski, and A. N. Swami. Mining association rules between sets of items in large databases. In Proc. of the ACM SIGMOD International Conference on Management of Data, pp.207–216, 1993.
2. R. Agrawal and R. Srikant. Fast algorithms for mining association rules. In Proc. of the 20th International Conference on Very Large Data Bases, pp.478–499, 1994.
3. D. Braga, A. Campi, M. Klemettinen, and P. L. Lanzi. Mining association rules from xml data. In Proc. of the 4th International Conference on Data Warehousing and Knowledge Discovery (DaWaK'02), volume 2454 of LNCS. Springer, pp.21–30, 2002.
4. D. Braga, A. Campi, S. Ceri, M. Klemettinen, and P. L. Lanzi. A tool for extracing XML association rules. In Proc. of the 14th IEEE International Conference on Tools with Artificial Intelligence (ICTAI'02), pp.57–64, 2002.
5. J. Han and Y. Fu. Discovery of multiple-level association rules from large databases. In Proc. of the 21st International Conference on Very Large Data Bases, pp.420–431, 1995.
6. R. Meo, G. Pasila, and S. Ceri. An extension to SQL for mining association rules. Data Mining and Knowledge Discovery, 2(2), pp.195–224, 1998.
7. L. Singh, P. Scheuermann, and B. Chen. Generating association rules from semi-structured documents using an extended concept hierarchy. In Proc. of the 6th International Conference on Information and Knowledge Management (CIKM'97), pp.193–200, 1997.
8. R. Srikant and R. Agrawal. Mining generalized association rules. In Proc. of the 21st International Conference on Very Large Data Bases, pp.409–419, 1995.
9. R. Srikant and R. Agrawal. Mining quantitative association rules in large relational tables. In Proc. of the 1996 ACM SIGMOD International Conference on Management of Data, pp.1–12, 1996.
10. H. Toivonen. Sampling large databases for association rules. In Proc. of the 22th International Conference on Very Large Data Bases, pp.43–52, 1996.
11. J. W. W. Wan and G. Dobbie. Extracting association rules from XML documents using XQuery. In Proc. of the 5th ACM International Workshop on Web Information and Data Management (WIDM'03), pp.94–97, 2003.
12. The World Wide Web Consortium (W3C). Extensible Markup Language (XML) 1.0 (Third Edition) W3C Recommendation, http://www.w3.org/TR/2004/REC-xml-20040204/, 2004.

Content-Based Recommendation in E-Commerce

Bing Xu[1], Mingmin Zhang[1], Zhigeng Pan[1,2], and Hongwei Yang[1]

[1] College of Computer Science, Zhejiang University,
310027 Hangzhou, P.R.China
[2] Institute of VR and Multimedia, HZIEE,
310037 Hangzhou, P.R.China
{xubin, zmm, zgpan, yanghongwei}@cad.zju.edu.cn

Abstract. Recommendation system is one of the most important techniques in some E-commerce systems such as virtual shopping mall. With the prosperity of E-commerce, more and more people are willing to perform Internet shopping, which resulted in an overwhelming array of products. Traditional similarity measure methods make the quality of recommendation system decreased dramatically in this situation. To address this issue, we present a novel method that combines the clustering which is based on apriori-knowledge and content-based technique to calculate the customer's nearest neighbor, and then provide the most appropriate products to meet his/her needs. Experimental results show efficiency of our method.

1 Introduction

E-commerce provides customers with ways to access necessary information without any restriction. Recommendation systems using all kinds of tools to respond to customer needs, understand customer behavior, and best use the limited available customer attention.

The nearest collaborative filtering recommendation system is one of the most successful techniques in E-commerce. It produces the recommendation for the customer based on the item rating of the nearest neighbor. With the prosperity of the E-commerce, the data of customer and item increasing dramatically resulted in the extreme sparsity of rating data. The traditional similarity measure methods have their deficiency in this situation. All of them cannot find the nearest neighbor accurately leads to the quality of recommender system decreasing dramatically. In this situation, our paper presents a novel method that combines the clustering which is based on apriori-knowledge and content-based technique to calculate the customer's nearest neighbor and then provide the most appropriate products to meet his/her needs. Our experimental results are showing that the method has a bright future.

The paper is organized as follows: in Section 2 we briefly review the previous work related to our research. In Section 3 deficiency of the traditional similarity measure in sparse dataset is analyzed. Content-based and clustering recommendation algorithm is described in Section 4. We show the experiment results in Section 5, and make concluding remarks in Section 6.

2   Related Work

Various learning approaches have been applied to construct customer profiles and to discover customer preferences to make recommendation.

The earliest approach used nearest-neighbor collaborative filtering algorithms [2][3]. Nearest neighbor algorithm is a rather lazy algorithm. A new customer is generally associated to a target customer, the algorithm chooses the nearest customers from computing the distances between different customers and then recommends the products to the customers.

Another method, the Bayesian networks create a model based on a training set with a decision tree at each node and edges representing customer information. The model can be built off-line very quickly, from a few hours to a few days. The resulting model is very economic, fast, and essentially as accurate as the nearest neighbor methods [4]. Identifying groups of customers appearing to have similar preferences also uses clustering techniques. In some cases, clustering techniques usually less accuracy than the nearest neighbor algorithms [4]. For this reason, clustering techniques can be applied as a "first step" of nearest neighbor algorithms.

Other methods such as classifiers are general computational models for assigning a category to an input. Classifiers have been quite successful in a variety of domains ranging from the identification of fraud and credit risks in financial transactions to medical diagnosis and intrusion detection. Association rules have also been used to analyze patterns of preference across products, and to recommend products to customers based on other products they have selected [1]. Horting is a graph-based technique in which nodes are customers, and edges between nodes indicate degree of similarity between two customers, this method searches for the nearest neighbor node in the graph, and then synthesizes ratings of the neighbor node to produce the recommendation [5].

Content-based recommendation approaches have also been applied to the basic problem of making accurate and efficient products recommendation in E-commerce. The text categorization methods adopted by Mooney and Roy [6] in their LIBRA system that makes content-based book recommendations exploiting the product descriptions found in Amazon.com, use a naive Bayes text classifier as in [7]. A reinforcement learning method is applied by Personal Web Watcher [8], a content-based system that recommends web-page hyperlinks by comparing them with a history of previous pages visited by the customer.

The new generation of web personalization recommender tools is attempting to incorporate techniques for pattern discovery in Web usage data. Web usage systems run a number of data mining algorithms on usage or click stream data gathered from web sites in order to discover customer profiles. A paper by Schafer [9] presents a detailed taxonomy and examples of recommender systems in E-commerce applications and how they can provide one-to-one personalization and capture customer loyalty at the same time.

3 The Deficiency of Traditional Similarity Measure

3.1 The Traditional Similarity Measure

Collaborative filtering recommender produces the target customer's recommended list according to other customer's rating. It accepts such assumption, if some customers have similar ratings on certain items; they have the same rating for other items. Collaborative filtering recommender system uses statistics method to search for the target customer's some nearest neighbors and then, according to the nearest neighbors' item rating, predicts the item rating of the target customer and produces the related recommender list.

To find the nearest neighbor of the target customer, we firstly use similarity measure method to calculate the similarity of the customers, and then select some nearest neighbors who have the closest similarity to the target customer. The accuracy of finding the nearest neighbor of the target customer affects the quality of the recommendation system directly and also plays the important role in the overall collaborative filtering algorithm.

The similarity measure methods include cosine-based similarity, correlation-based similarity and adjusted cosine similarity.

Cosine-Based Similarity: In this case, two customers are thought of as two vectors in the dimensional item-space. If the customer does not rate an item, the item will be looked as zero. The similarity between them is measured by computing the cosine of the angle between these two vectors. Formally, in the $m \times n$ rating matrix, similarity between customer i and j, denoted by $sim(i,j)$ is given by $sim(i,j) = \cos(i,j) = \frac{i \cdot j}{\|i\| \|j\|}$ Where $\cdot$ denotes the dot-product of the two customers.

Correlation-Based Similarity: In this case, similarity between two customers i and j is measured by computing the person r correlation $corr(i,j)$ lets the set of items which are both rated by customers i and j be denoted by I_{ij}, then the correlation similarity is given by

$$sim(i,j) = corr_{i,j} = \frac{\sum_{u \in U}\{R_{u,i} - \bar{R}_i\}\{R_{u,j} - \bar{R}_j\}}{\sqrt{\sum_{u \in U}\{R_{u,i} - \bar{R}_i\}^2}\sqrt{\sum_{u \in U}\{R_{u,j} - \bar{R}_j\}^2}} \quad (1)$$

Here $R_{u,i}$ denotes the rating of customer i on item u, $\bar{R}_i$ is the average of the $i-th$ customer's rating.

Adjusted Cosine Similarity: The cosine-based similarity does not consider the rating scale between different customers. Adjusted cosine similarity adopts a method by subtracting the average rating of the customer to mend its deficiency. Lets the set of items that are both rated by customers i and j be denoted by I_{ij}; I_i and I_j are the item set rated by customer i and customer j separately. Then the correlation similarity is given by

$$sim(i,j) = \frac{\sum_{c \in I_{ij}}\{R_{i,c} - \bar{R}_i\}\{R_{j,c} - \bar{R}_j\}}{\sqrt{\sum_{c \in I_i}\{R_{i,c} - \bar{R}_i\}^2}\sqrt{\sum_{c \in I_j}\{R_{j,c} - \bar{R}_j\}^2}} \quad (2)$$

Here $R_{i,c}$ denotes the rating of customer i on item c, R_i is the average of the $i-th$ customer's rating.

3.2 Analysis of the Traditional Similarity Measure

In commercial recommendation system, many approaches based on nearest neighbor algorithm have been very successful. But the widespread use revealed some potential challenges, such as:

Sparsity: in practice, many commercial recommendation systems are used to evaluate large item sets in these systems, even active customers may have purchased well under 1% of the items. Accordingly, a recommendation system based nearest neighbor selected from the customers may be unable to make any item recommendations for a particular customer. As a result, the accuracy of recommendations may be poor.

Scalability: finding the nearest neighbor requires computation that shows with both the number of customers and the number of items. With millions of customers and items, a typical item-based recommender system will suffer serious scalability problems.

For example, in the cosine-based similarity method, the items that the customers do not evaluate are zero. Lets the rating item of the customer is denoted by R_{ij}. Then

$$R_{ij} = \begin{cases} r_{ij} & r_{ij} \neq \Phi \\ 0 & otherwise \end{cases} \qquad (3)$$

Where r_{ij} is the rating of customer i on item j. If the customer i rates the item j, then R_{ij} equals to r_{ij}, else R_{ij} equals to zero.

This settlement can enhance computational performance, however, in the case of the extreme sparsity of the items and the greatness of the quantity, the reliability of the assumption is poor. Because, in practice, the preference of the customers is different for un-rating items and they cannot be the same rating, i.e., zero. Adjusted cosine similarity also has this problem.

In the method of correlation-based similarity, let u_i denote the item set rated by the customer i, u_j is the item set rated by the customer j. The intersection of items rated both by customer i and customer j is $u_i \Lambda u_j$.

In common sense, they can only get the higher similarity when there exists many items whose ratings are very adjacent for the two customers. When the rating items are very sparse, the item set that both rated by the two customers are very small, only one or two items. In this situation, even the two customers have very high similarity; we cannot say they are similar actually. This method also has some deficiency.

From the above, we can say that the traditional similarity measure cannot measure the similarity between the customers effectively when the rating data are extremely sparse. This resulted in the inaccurate neighbors and the decrease of the recommender accuracy.

4 Content-Based and Clustering Recommendation Algorithm

The weakness of the similarity measure for large and sparse database leads us to explore alternative recommending algorithm. One simple method is to set up the unevaluated items as a constant, and the middle rating is often used in general. The experiment shows that this modified method can improve the quality of the recommendation system. However, it is impossible that the unrated items have the same value, so the modified method cannot solve the traditional similarity measure for sparse database radically. Therefore, we put forward a novel method. Firstly, split the unvalued items into several parts using the apriori-knowledge clustering, and then predict the unvalued items, finally, find out the nearest neighbors using these rating items and accomplish the recommendation. This method can let unvalued items get a reasonable value and accordingly it can provide more rating items for the nearest neighbor algorithm.

In the following, we introduce content-based collaborative filtering recommendation algorithm in detail. The algorithm is divided into two steps: find the nearest neighbor and process recommendation.

4.1 Obtaining Nearest Neighbor

We must find the union of the rating items before calculating the similarity between the customers. In the union, the unvalued items are predicted through the apriori-knowledge clustering method and then the neighbors can be found by calculating the rating items belong to the union. This method can not only deal with the problem that the unvalued items are equal to a constant in cosine-based similarity effectively but also treat with the shortcoming that the workable rating items are very few in correlation-based similarity, which resulted in obtaining the exactness of the nearest neighbor and improving the quality of recommendation.

In the traditional clustering method, there is no supervise nor some apriori-knowledge. The clustering result is laid on the given data and the certain selected clustering algorithm, which to a degree resulted in the poor result. The reason is that unclassified data do not offer any information in the process of the clustering. In practice, there are some known knowledge in the related domain, and we can obtain some classified examples known as apriori-knowledge. The knowledge can be used in the process of the clustering to guide clustering which will increase the accuracy of the clustering.

The algorithm of Clustering based on apriori-knowledge is show as below:
Step 1: each example whose classification has been known can be as the single element for each s_i and construct the initial set $\{s_i\}$, $1 \cdots n$.
Step 2: calculate the similarity of every two sets.

$$\frac{1}{|C_l| \bullet |C_k|} \sum_{\substack{x_l \in c_l \\ x_k \in c_k}} s(x_l, x_k) \tag{4}$$

Where c_l, c_k are two different classification set.

Step 3: suppose s_m and s_n are the two sets that have the greatest similarity. If the examples in c_m and c_n belong to the same class, we will unit the two sets and go to step 2. If not, we turn to step 4.

Step 4: output the k value.

Step 5: use the k value and select $c_1, c_2 \cdots c_k$ randomly from the dataset as the initial training factor, then use SOM algorithm [13] to obtain the clustering result.

For example, suppose $E = \{e_1, e_2 \cdots e_{15}\}$ are composed of two classification. $c_1 = \{e_1, e_2, e_3, e_4, e_5, e_{12}, e_{13}, e_{14}, e_{15}\}$, $c_2 = \{e_6, e_7, e_8, e_9, e_{10}, e_{11}\}$. Here, $\{c_1, c_2\}$ represent two classifications individually. Table 1 shows the dataset.

Table 1. The Dataset

Attribute Example	Attr1	Attr2	Attr3	Attr4	Attr5	Attr6	Attr7	Attr8	Attr9	Attr10
1	1	1	1	0	1	0	0	0	1	0
2	1	1	1	0	0	1	0	0	1	0
3	1	1	1	1	1	0	0	0	1	0
4	1	1	1	0	0	0	0	0	1	0
5	1	1	1	1	0	0	0	0	1	0
6	0	1	0	1	1	1	0	1	0	1
7	0	0	1	1	1	1	0	1	1	0
8	0	1	0	1	1	1	0	1	0	1
9	0	1	0	1	1	1	0	1	0	1
10	0	0	1	1	1	1	0	1	0	1
11	1	0	0	1	1	1	0	1	0	1
12	0	0	0	1	0	0	1	0	1	0
13	0	0	0	1	0	1	1	0	1	0
14	0	0	0	1	1	1	1	0	1	0
15	0	0	0	1	1	0	1	0	1	0

Assume the examples $\{e_1, e_2, e_9, e_{15}\}$ have known their attributes and classifications, others only know: their attributes. The description of the algorithm is as follows:

(1) In the $\{e_1, e_2, e_9, e_{15}\}$, each example can be looked as an independent subset, $s_1 = \{e_1\}, s_2 = \{e_2\}, s_3 = \{e_9\}, s_4 = \{e_{15}\}$;

(2) Calculate each subset's average similarity according to Equation (1), we can conclude that the similarity between and are the most.

Then, estimate whether they belong to the same class or not.

If yes, we will unit s_1 and s_2 and produce new subset $s_1 = \{e_1, e_2\}, s_2 = \{e_9\}, s_3 = \{e_{15}\}$. Repeat this step until the subset cannot unit.

If not, go to step (3).

(3) Output the result $s_1 = \{e_1, e_2\}, s_2 = \{e_9\}, s_3 = \{e_{15}\}$.

(4) Select the original centers randomly: e_2, e_4, e_{10}.

Use the self-organizing map to calculate the clustering results:
$s_1 = \{e_1, e_2, e_3, e_4, e_5\}$, $s_2 = \{e_6, e_7, e_8, e_9, e_{10}, e_{11}\}$, $s_3 = \{e_{12}, e_{13}, e_{14}, e_{15}\}$

The clustering centers are: $s_1 = \{1, 1, 1, 0, 0, 0, 0, 0, 1, 0\}$,
$s_2 = \{0, x, 0, 1, 1, 1, 0, 1, 0, 1\}$, $s_3 = \{0, 0, 0, 1, x, x, 1, 0, 1, 0\}$; Where x represent any value belongs to $\{0, 1\}$.

Then estimate the class for each subset. Examples e_1, e_2 belong to subset s_1 and the ex-ample e_9 belongs to s_2. From step (3) and step (4), we conclude that $\{e_3, e_4, e_5, e_{12}, e_{13}, e_{14}\}$ belong to class c_1, the example e_{15} belongs to s_3, so $\{e_6, e_7, e_8, e_{10}, e_{11}\}$ belong to class c_2.

Here, we use UCI Machine Learning database to verify the algorithm. Table 2 shows the result. Note that, in the column of selected examples, 2*5 express that

Table 2. The Experiment Result

Database name	Attribute character	Example numbers	Class number	Selected examples	Accuracy1	Accuracy2
Voting-record	symbol	435	2	2*5	89.4%	86.2%
Zoo	symbol	101	7	7*2	94.8%	90%
Soybean-large	symbol	683	19	19*2	75.1%	62.5%
Thyroid-disease	symbol, numeric	3772	3	3*10	90.0%	74.8%
Iris	numeric	150	3	3*3	91.1%	87.7%

voting-record database has two classes. In each class, we select 5 examples as the known classification examples. Column 6 showed the accuracy obtained by using our method and column 7 showed the accuracy obtained by ordinary clustering methods. From Table 1, we can see that amendatory clustering techniques can acquire better results.

The further research work about the relationship between classified data, the precision of the clustering algorithm and the number of the clustering center are showed in [10].

The unvalued items can get a rating after the above algorithm. The cosine-based similarity or correlation-based similarity is then used in the union to calculate the similarity of the customers.

4.2 Producing Recommendation

After the neighbors acquired, $P_{u,i}$ is the final rating of the customer u to product i, which can be obtained from the product i by the neighbors. The equation is as follows [4]:

$$P_{u,i} = \bar{R}_u + \frac{\Sigma_{n \in neighbor} sim(u, n) \times (R_{n,i} - \bar{R}_n)}{\Sigma_{n \in neighbor}(|sim(u, n)|)} \quad (5)$$

Where $sim(u, n)$ is the similarity measure between customer u and n, and where $R_{n,i}$ is the rating of customer n to product i. $\bar{R}_u$ is the average rating of customer n, so $\bar{R}_u$ denotes the average rating of customer u.

5 Experiment Results and Analysis

Here, the experiment dataset comes from [12]. There is a brief description of the data. This data set consists of:
  * 100,000 ratings (1-5) from 943 customers on 1682 movies.
  * Each customer has rated at least 20 movies.
  * Simple demographic info for the customers (age, gender, occupation, zip)
the most important files in the data set are:
  u.data – The full u data set, 100000 ratings by 943 customers on 1682 items.
  u.genre – A list of the genres.
  u.item – Information about the items (movies); The last 19 fields are the genres, a 1 indicates the movie is of that genre, a 0 indicates it is not; movies can be in several genres at once.

In the experiment, items whose rating are 1 can be seen as a class, items whose rating are 2 or 3 can be seen as another class, items whose rating are 4 or 5 can be a class too. Using the u.item files to predict the items for the unevaluated items based on the apriori-knowledge clustering algorithm and then use u.data to implement the similarity of the customers as well as the products recommendation.

5.1 Experiment Results

We adopt MAE (Mean Absolute Error) as the evaluated standard.
The definition is:

$$MAE = \frac{\Sigma_{i=1}^{N}|p_i - q_i|}{N} \quad (6)$$

Where p_i represents the degree of satisfaction that the customer assess the product, q_i represents the degree of the satisfaction that recommendation algorithm assess the product, and N represents the total customers. MAE represents the mean absolute error between the real ratings items and the predicable rating items. The more decreased the MAE is, the more the quality of recommendation is increased.

According to [11], cosine-based similarity method can obtain better result than correlation-based similarity method. So, the cosine-based similarity method is chosen as the measure for our experiment.

The experiment result is shows in Fig.1.

5.2 Analysis of the Experiments

The most difference between the traditional collaborative filtering and the combination of content-based and the apriori-knowledge is how to get the validate neighbors. In the traditional collaborative filtering recommender algorithms, the correlation-based similarity method only employs the rating items that the customers are both rated. In the sparse dataset, since the intersection contains few items, so the nearest neighbor calculated may not be the actual neighbor. The experiment results also show that the quality of the recommendation based on the

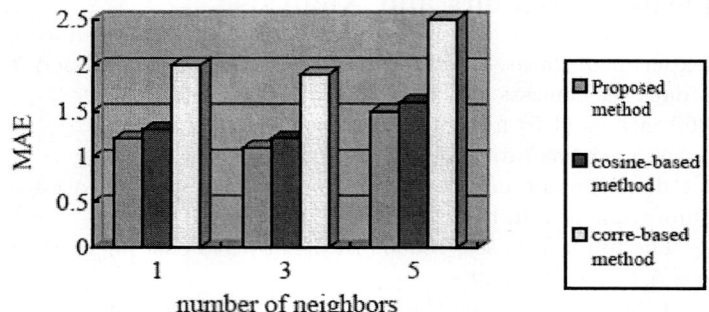

Fig. 1. The comparison of accuracy of recommendation algorithms

correlation-based similarity method is poor. Another method, cosine-based similarity method employs the union of the rating items. In the union, the un-valued items are endued with the same value. Though the quality of the recommendation are improved, the unvalued items have the same value are unreasonable. A content-based clustering method uses the apriori-knowledge clustering to rate the unevaluated items and then find out the nearest neighbor of the target customer. The experiment results show this method has the highest accuracy of the recommendation.

6 Conclusion

This paper begins with the analysis of the deficiency in the traditional similarity measure method for the greatly sparse rating data. To deal with the problem, content-based recommendation method is proposed. This method uses the characters of the products as well as the rated products that can be regarded as the classified data. These apriori-knowledge can be used to supervise the clustering and accordingly the rating of the unevaluated products can be predicted. This method can solve the problem that exists in traditional similarity measure method effectively and get the more accurate neighbors of the target customers. The experiment shows that our method can greatly increase the quality of the recommendation.

Acknowledgements

This research work is supported by 973 project (grant no: 2002CB312100), and TRAPOYT Program in Higher Education Institution of MOE, PRC.

References

1. Lee, C.-H., Kim, Y.-H., Rhee, P.-K. Web Personalization expert with combining collaborative filtering and association rule mining technique. Expert Systems with Applications (2001) 21(3) 131-137

2. Resnick, P.,Iacovou, N., Suchak, M., Bergstrom, P., and Riedl, J. Grouplens: An open architecture for collaborative filtering of netnews. In Proceedings of ACM CSCW'94 Conference on Computer-Supported Cooperative Work, (1994) 175-186
3. Shardanand, U. and Maes, P. Social information filtering: Algorithms for automating "word of mouth". In Proceedings of ACM CHI'95 Conference on Human Fac-tors in Computing Systems, (1995) 210-217
4. Breese, J., Heckerman, D., and Kadie, C. Empirical analysis of predictive algorithms for collaborative filtering. In Proceedings of the 14th Conference on Uncertainty in Artificial Intelligence (UAI-98), (1998) 43-52
5. Wolf, J., Aggarwal, C., Wu, K-L., and Yu, P. Horting Hatches an Egg: A New Graph-Theoretic Approach to Collaborative Filtering. In Proceedings of ACM SIGKDD International Conference on Knowledge Discovery & Data Mining, San Diego, CA. (1999) 201-212
6. Mooney, R. J. and Roy, L. Content-based book recommending using learning for text categorization, Proceedings of the VACM Conference on Digital Libraries, San Antonio, USA, (2000) 195-204
7. Abbattista, F., Degemmis, M., Fanizzi, N., Licchelli, O., Lopes, P., Semeraro, G., Zambetta F. Learning customer profiles for content-based filtering in e-commerce. Italian Artificial Intelligence Conference. (2002)
8. Thorsten Joachims, Dayne Freitag, Tom Mitchell. WebWatcher: A tour guide for the world wide web, Proceedings of the XV International Joint Conference on Ar-tificial Intelligence, Nagoya, Japan (1997) 770-775
9. [9] Schafer J. B., Konstan J. Electronic commerce recommender applications, Journal of Data Mining and Knowledge Discovery, (2001) vol. 5 num. 1-2. 115-152
10. WANG XINGQI the doctor thesis. Research on algorithm of machine learning and its application, Zhejiang University.(2002)
11. Deng Ai-lin, Zhu Yang-yong, Shi Bai-le. A collaborative filtering recommendation algorithm based on item rating prediction. Journal of Software. (2003) 14(9): 1621-1628
12. http://www.research.compaq.com/src/eachmovie
13. Kohonen, T. The self-organizing map. Proceedings of the IEEE, (2001) 78(9):1464-1480

A *P*ersonalized *M*ultilingual Web Content *M*iner: *PM*Web*M*iner

Rowena Chau[1], Chung-Hsing Yeh[2], and Kate A. Smith[3]

School of Business Systems, Faculty of Information Technology,
Monash University, Clayton, Victoria 3800, Australia
{Rowena.Chau, ChungHsing.Yeh,
Kate.Smith}@infotech.monash.edu.au

Abstract. This paper presents the development of a novel personal concept-based multilingual Web content mining system. Multilingual linguistic knowledge required by multilingual Web content mining is made available by encoding all multilingual concept-term relationships within a multilingual concept space using self-organising map. With this linguistic knowledge base, a personal space of interest is generated to reveal the conceptual content of a user's multiple topics of interest using the user's bookmark file. To personalise the multilingual Web content mining process, a concept-based Web crawler is developed to automatically gather multilingual web documents that are relevant to the user's topics of interest As such, user-oriented concept-focused knowledge discovery in the multilingual Web is facilitated.

1 Introduction

The rapid expansion of the World Wide Web throughout the globe means electronically accessible information is now available in an ever-increasing number of languages. With majority of this Web data being unstructured text [2], Web content mining technology capable of discovering useful knowledge from multilingual Web documents thus holds the key to exploit the vast human knowledge hidden beneath this largely untapped multilingual text.

Web content mining has attracted much research attention in recent years [6]. It has emerged as an area of text mining specific to Web documents focusing on analysing and deriving meaning from textual collection on the Internet [3]. Currently, Web content mining technology is still limited to processing monolingual Web documents. The challenge of discovering knowledge from textual data which are significantly linguistically diverse has been well recognised by text mining research [13]. In a monolingual environment, the conceptual content of documents can be discovered by directly detecting patterns of frequent features (i.e. terms) without precedential knowledge of the concept-term relationship. Documents containing an identical known term pattern thus share the same concept. However, in a multilingual environment, *vocabulary mismatch* among diverse languages implies that documents exhibiting similar concept will not contain identical term patterns. This *feature incompatibility* problem thus makes the inference of conceptual contents using term pattern matching inapplicable. To enable multilingual Web content mining, linguistic knowledge of concept-term relationships is essential to exploit any knowledge relevant to the domain of a multilingual document collection. Without such linguistic knowledge, no text or Web mining algorithm can effectively infer the conceptual content of the multilingual documents.

In addition, in the multilingual WWW, a user's motive of information seeking is global knowledge discovery. This is particularly important among users, such as knowledge workers, researchers, government officials and business executives, who need to stay competent by keeping track of the global knowledge development within his/her domain of interest. Instead of looking for specific documents that can be characterised by a few query terms in a specific language, the user is interested in all documents written in all languages that are relevant to his/her personal topics of interest. In such cases, the user usually does not have any ideal documents in mind. Any documents that are conceptually relevant to the user's personal information interest can be useful. As such, multilingual Web content mining facilitating *user-oriented concept-focused knowledge discovery* that focuses on knowledge relevant to the user's personal topics of interest becomes highly desirable. To support user-oriented concept-focused knowledge discovery in the multilingual Web, it is thus a challenge to automatically gather relevant multilingual Web documents based on the conceptual content of the user's information interest.

To address these issues, a personal concept-based multilingual Web content mining system is developed. This is achieved by constructing a multilingual concept space as the linguistic knowledge base. The concept space encodes all multilingual concept-term relationships from parallel corpus using fuzzy clustering. Given this concept space, a personal space of interests modelling the user's multiple topics of interest is generated using the user's bookmark file. Based on the conceptual content of the user's topics of interest as represented in his/her personal space of interest, a concept-based Web crawler is used to automatically gather relevant multilingual documents from the Web. As such, user-oriented concept-focused knowledge discovery in the multilingual Web is facilitated and, thus, personal multilingual Web content mining is realised.

In subsequent sections, we first present the architecture of the personal concept-based multilingual Web content mining system. Technical details about the development of the multilingual concept space for encoding the multilingual linguistic knowledge, the generation of the user's personal space of interest and the design of the concept-based Web crawler are discussed in Section 3, 4, and 5 respectively. Finally, a conclusive remark is given in Section 6.

2 The System Architecture

A personal multilingual Web content mining system called *PMWebMiner* (*P*ersonal *M*ultilingual *Web* Content *Miner*) is developed using a concept-based approach. This concept-based approach to multilingual Web content mining is due to a notion that while languages are culture bound, concepts expressed by these languages are universal [12]. Moreover, conceptual relationships among terms are inferable from the way that terms are set down in the text. Therefore, the domain-specific multilingual concept-term relationship can be discovered by analysing relevant multilingual training documents.

Given such multilingual concept-term relationship, conceptual content of all multilingual text can then be revealed. In our system, this concept-term relationship is used as the linguistic knowledge base for extracting the conceptual content of both the multilingual Web documents gathered by a Web crawler and the user's topics of information interest as embedded in the user's bookmarked documents. Figure 1 shows the architecture of *PMWebMiner*.

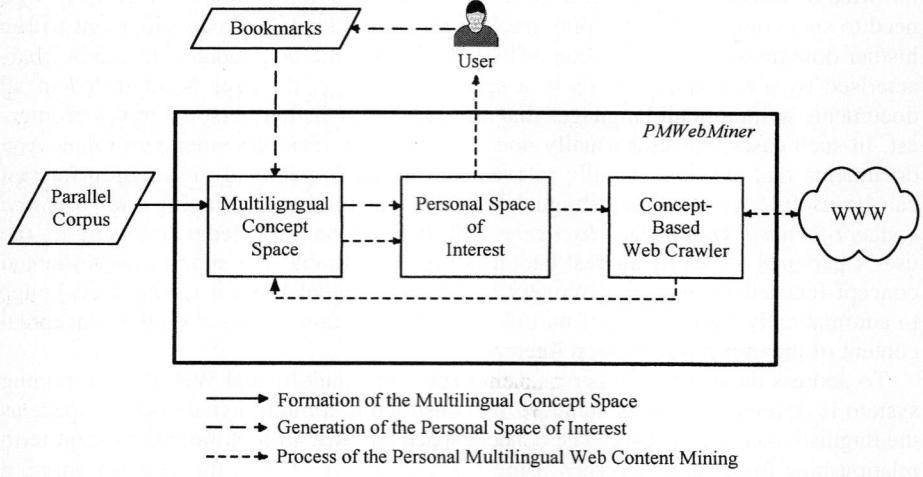

Fig. 1. The *P*ersonal *M*ultilingual *Web* Content *Miner* (*PMWebMiner*)

This system consists of three modules, namely the multilingual concept space, the personal space of interest and the concept-based Web crawler. First, a parallel corpus, which is a collection of documents and their translations, is used as the training documents for constructing a concept space using a self-organising map [5]. The concept space encodes all multilingual concept-term relationships by clustering semantically related multilingual terms into concept classes. As the linguistic knowledge base for multilingual Web content mining, this concept space is used to generate a personal space of interest reflecting the conceptual content of a user's multiple topics of information interest using the user's bookmark file. Based on these personal topics of interest, a concept-based Web crawler traverses the Web to gather relevant multilingual Web documents. Finally, to present the retrieved documents in accordance with the user's topics of interest, the concept space is applied again to reveal the conceptual content of these documents, and the personal space of interest is used as a text filter to facilitate personalized concept-based text filtering. As such, personal multilingual Web content mining aiming for user-oriented concept-focused global knowledgediscovery is realised.

3 Formation of the Multilingual Concept Space

From the viewpoint of automatic text processing, the relationships between terms' meanings are inferable from the way that the terms are set down in the text. Natural language is used to encode and transmit concepts. A sufficiently comprehensive sample of natural language text, such as a well-balanced corpus, may offer a fairly complete representation of the concepts and conceptual relationship applicable within specific areas of discourse. Given corpus statistics of term occurrence, the associations among terms become measurable, and sets of semantically/conceptually related terms are detected.

To construct multilingual linguistic knowledge base encoding lexical relationships among multilingual terms, parallel corpora containing sets of documents and their translations in multiple languages are ideal sources of multilingual lexical information. Parallel documents basically contain identical concepts expressed by different sets of terms. Therefore, multilingual terms used to describe the same concept tend to occur with very similar inter- and intra-document frequencies across a parallel corpus. An analysis of paired documents has been used to infer the most likely translation of terms between languages in the corpus [1],[4],[7]. As such, co-occurrence statistics of multilingual terms across a parallel corpus can be used to determine clusters of conceptually related multilingual terms.

Given a parallel corpus D consisting N pairs of parallel documents, meaningful terms from every languages covered by the corpus are extracted. They form the set of multilingual terms for constructing the multilingual concept map. Each term is represented by an n-dimensional term vector. Each feature value of the term vector corresponds to the weight of the nth document indicating the significance of that document in characterising the meaning of the term. Parallel documents which are translated versions of one another within the corpus, are considered as the same feature. To determine the significance of each document in characterising the contextual content of a term based on the term's occurrences, the TF.IDF weighting scheme is used. When contextual contents of every multilingual term are well represented, they are used as the input into the self-organising algorithm for constructing the multilingual concept map.

Let $\mathbf{x}_i \in R^N$ ($1 \leq i \leq M$) be the term vector of the ith multilingual term, where N is the number of documents in the parallel corpus for a single language (i.e. the total number of documents in the parallel corpus divided by the number of languages supported by the corpus) and M is the total number of multilingual terms. The self-organising map algorithm is applied to form a multilingual concept map, using these term vectors as the training input to the map. The map consists of a regular grid of nodes. Each node is associated with an N-dimensional model vector. Let $\mathbf{m}_j = [m_{jn} | 1 \leq n \leq N]$ ($1 \leq j \leq G$) be the model vector of the jth node on the map. The algorithm for forming the multilingual concept map is given below.

Step 1: Select a training multilingual term vector $\mathbf{x}_i$ at random.
Step 2: Find the winning node s on the map with the vector $\mathbf{m}_s$ which is closest to $\mathbf{x}_i$ such that

$$\|\mathbf{x}_i - \mathbf{m}_s\| = \min_j \|\mathbf{x}_i - \mathbf{m}_j\| \tag{1}$$

Step 3: Update the weight of every node in the neighbourhood of node s by

$$\mathbf{m}_t^{new} = \mathbf{m}_t^{old} + \alpha(t)(\mathbf{x}_i - \mathbf{m}_t^{old}) \qquad (2)$$

where $\alpha(t)$ is the gain term at time t ($0 \leq \alpha(t) \leq 1$) that decreases in time and converges to 0.

Step 4: Increase the time stamp t and repeat the training process until it converges.

After the training process is completed, each multilingual term is mapped to a grid node closest to it on the self-organising map. A multilingual concept space is thus formed. This process corresponds to a projection of the multi-dimensional term vectors onto an orderly two-dimensional concept space where the proximity of the multilingual terms is preserved as faithfully as possible. Consequently, conceptual similarities among multilingual terms are explicitly revealed by their locations and neighbourhood relationships on the map. Multilingual terms that are synonymous are associated to the same node. In this way, conceptual related multilingual terms are organised into term clusters within a common semantic space. The problem of feature incompatibility among multiple languages is thus overcome.

4 Generation of a Personal Space of Interest

With the overwhelming amount of information in the multilingual WWW, not every piece of information is of interest to a user. In such circumstances, a user profile, which models the user's information interests, is required to filter out information that the user is not interested in.

Common approaches to user profiling [8],[9],[10] build a representation of the user's information interest based on the distribution of terms found in some previously seen documents which the user has found interesting. However, such representation has difficulties in handling situations where a user is interested in more than one topic. In addition, in a multilingual environment, the feature incompatibility problem resulted from the vocabulary mismatch phenomenon across languages makes a language-specific term-based user profile insufficient for representing the user's information interest that spans multiple languages. To overcome these problems, we propose a concept-based representation of the user's information interest by generating a personal space of interest. To reflect the conceptual content of a user's multiple topics of information interest using language-independent concepts rather than language-specific terms implies that the resulting user profile is not only more semantically comprehensive but also independent from the language of the documents to be filtered. This is particularly important for multilingual Web content mining where knowledge relevant to a concept in significantly diverse languages has to be identified.

To understand the user's information interests for personalising multilingual Web content mining, the user's preference on the WWW is used. Indicators of these preferences can be obtained from the user's bookmark file. To generate a concept-based personal space of interest from a user bookmark file, Web documents pointed by the bookmarks are first retrieved. Applying the multilingual concept space as the linguistic knowledge base, each Web document is then converted into a concept-based document vectors. To do so, the multilingual concept space is applied.

On the multilingual concept space, conceptually related multilingual terms are organised into term clusters. These term clusters, denoting language-independent concepts, are thus used to index multilingual documents in place of the documents' original language-specific index terms. As such, a concept-based document vector that explicitly expresses the conceptual context of a document regardless of its language is obtained.

To achieve this, each document pointed to by the bookmark file is indexed by mapping its text, term by term, onto the multilingual concept space whereby statistics of its 'hits' on each multilingual term cluster (i.e. concept) are recorded. This is done by counting the occurrence of each term on the multilingual concept space at the node to which that term is associated. This statistics of term cluster occurrences can be interpreted as a kind of transformed 'index' of the multilingual document. The concept-based personal space of interest is generated with the application of the self-organising map algorithm, using the transformed concept-based document vectors as inputs.

Let $\mathbf{y}_i \in R^G$ ($1 \leq i \leq H$) be the concept-based document vector of the ith bookmarked multilingual document, where G is the number of nodes existing in the multilingual concept space and H is the total number of documents pointed to by the bookmark file. In addition, let $\mathbf{m}_j = [m_{jn} | 1 \leq n \leq G]$ ($1 \leq j \leq J$), be the G-dimensional model vector of the jth node on the map. The algorithm for forming the personal space of interest is given below.

Step 1: Select a training concept-based document vector $\mathbf{y}_i$ at random.
Step 2: Find the winning node s on the map with the vector $\mathbf{m}_s$ which is closest to document $\mathbf{y}_i$ such that

$$\|\mathbf{y}_i - \mathbf{m}_s\| = \min_j \|\mathbf{y}_i - \mathbf{m}_j\| \tag{3}$$

Step 3: Update the weight of every node in the neighbourhood of node s by

$$\mathbf{m}_t^{new} = \mathbf{m}_t^{old} + \alpha(t)(\mathbf{y}_i - \mathbf{m}_t^{old}) \tag{4}$$

where $\alpha(t)$ is the gain term at time t ($0 \leq \alpha(t) \leq 1$) that decreases in time and converges to 0.
Step 4: Increase the time stamp t and repeat the training process until it converges.

After the training process, multilingual documents from the bookmark file that describe similar concepts are mapped onto the same node forming document clusters on the self-organising map. Each node thus defines a topic of interest of the user. A personal space of interest reflecting the conceptual content of the user's multiple topics of interest in the multilingual Web is formed. This personal space of interest is then applied to realise personal multilingual Web content mining in two ways. First, it is used to guide the Web crawler in gathering user-specific conceptually relevant multilingual documents from the Web. Second, when potentially relevant documents are collected, it is used as a multilingual text classifier to filter, categorise and organise the documents in accordance with the user's topics of interest.

5 Development of the Concept-Based Web Crawler

The major task of the concept-based Web crawler in our personal multilingual Web content mining system is to gather related Web documents in multiple languages based on the conceptual content of the user's information interests. To achieve this, two novel techniques are developed. First, *concept-based multilingual Web crawling* is proposed to retrieve related multilingual Web documents based on their conceptual content. This is enabled with the generation of a set of concept-focused multilingual seed URLs. Second, *personal concept-based multilingual text filtering* is introduced to personalised the mined Web content. From the documents collected by the Web crawler, it screens out irrelevant documents based on the user's topics of interest, before they are presented to the user.

5.1 Concept-Based Multilingual Web Crawling

Related Web documents are often connected by their in- and out- links. Web crawlers traverse the Web to collect related documents by following links. To start crawling the Web for documents relevant to a topic, a set of seed URLs are required by the Web crawler. In a monolingual dimension, it is sufficient to use any documents previously considered as relevant to a topic by the user as the seeds for gathering other related documents written in the same language. However, related documents written in different languages are rarely linked. To make a Web crawler capable of retrieving relevant documents in various languages, a set of seed URLs in every language must be made available.

To address the above issue, a novel approach to generate multilingual seed URL set is proposed. The idea is: for every language, we use the term that is most prototypical to the conceptual content of a user's interest in a topic to run a Web search. Top-ranked Web documents returned by the search engine are then used as the seed URLs to initialise the concept-based Web crawling. To realise this, identification of the topic-related multilingual terms is the major challenge. Obviously, asking the user to nominate a set of terms that best describe his/her interest in each topic in every language is impractical if not impossible. To solve this problem, the personal space of interest and the multilingual concept space are applied. On one hand, the personal space of interest that capture the conceptual content of the user's multiple topics of interest has revealed concepts covered by every topic of the user's interest. On the other hand, the multilingual concept space that encodes the multilingual concept-term relationships has associated all multilingual terms to the concepts to which they belong. Hence, they provide the essential clues for effectively inferring a set of user-oriented topic-related terms for every language to form an initial query for generating the multilingual seed URLs.

As indicated on the personal space of interest, the weight vector of each output node on the self-organising map has already revealed the significance of each concept with respect to a topic of the user's interest. Now, the task is to identify, for every topic, a set of N terms, representing this topic's N most significant concepts in each language. The number of concepts, N, is determined based on the assumption that all concepts are evenly distributed among the topics when the personal space of interest is formed. So, we obtain the average number of concepts for every topic by dividing the total number of concepts as identified in the multilingual concept space by the total number of topics existing on the personal space. Then, the top N concepts exhibiting the highest concept weights in the weight vector of a topic (an output node on the personal space) are selected to represent a topic. Hence, each topic is now represented by its N most significant concepts. To identify a term that is prototypical with respect to a concept of a topic in every language, the multilingual concept space comes in. For all terms of the same language, the one that is closest to the output node of a concept in the multilingual concept space is considered to be most relevant to that concept. Therefore, this term is selected as the prototypical term to represent this concept in that topic. As such, for every language, each topic will be represented by a set of terms where each term represents a concept. In other words, one set of terms for one language is identified for every topic.

The objective of the concept-based Web crawler is to gather multilingual documents that are relevant to a user's various topics of interest. It needs a set of seed URLs that is relevant to the concepts of such topic to start from. To generate this set of seed URLs for the concept-based Web crawler, we submit a query to the search engine using each set of terms that represent the concepts of a topic in every language to retrieve a set of the most relevant (e.g. top 5) Web document in each language. These documents then become the seed URLs for the Web crawler to collect other related documents describing similar concepts in various languages on the Web. This approach is to ensure the Web crawler will explore the Web context diversely and yet still remained conceptually focused. Multilingual Web document collected by the Web crawler are then passed on to a concept-based multilingual text categorization system for organising into the taxonomy to facilitate global knowledge discovery.

5.2 Personal Concept-Based Multilingual Text Filtering

Web crawlers traverse the Web to collect related documents by following links. There is no guarantee that documents connected by links should all be highly relevant to a certain topic. To ensure that only documents that are relevant to the user's information interest are mined by our Web mining system, conceptual content of these documents must be assessed. Towards this end, a process of personalized concept-based multilingual text filtering is incorporated. The purpose is: from what the Web crawler has collected, we screen out irrelevant documents, and present only the relevant ones to the user for examination.

The objective of concept-based multilingual text filtering is to sift through the arbitrary multilingual Web documents in accordance with their conceptual content. Analyzing documents by content depends heavily on the document representation scheme. To compare multilingual documents by the concepts they describe, contextual contents of documents need to be expressed explicitly with a suitable indexing scheme. In information retrieval, the goal of indexing is to extract a set of features that represent the contents, or the 'meaning' of a document. Among several approaches suggested for document indexing and representation, the vector space model [11] represents documents conveniently as vectors in a multi-dimensional space defined by a set of language-specific index terms. Each element of a document vector corresponds to the weight (or occurrence) of one index term. However, in a multilingual environment, the direct application of the vector space model is infeasible due to the feature incompatibility problem. Multilingual index terms characterising documents of different languages exist in separate vector spaces.

To overcome the problem, a better representation of document contents incorporating information about semantic/conceptual relationships among multilingual index terms is desirable. Towards this end, the multilingual concept space obtained in Section 3 is applied again. On the multilingual concept space, conceptually related multilingual terms are organised into term clusters. These term clusters, denoting language-independent concepts, are thus used to index multilingual documents in place of the documents' original language-specific index terms. As such, a concept-based document vector that explicitly expresses the conceptual context of a document regardless of its language is obtained. The term-based document vector of the vector space model, which suffers from the feature incompatibility problem, can now be replaced with the language-independent concept-based document vector. The transformed concept-based document vectors are then analyzed using the personal space of interest as a personal concept-based multilingual text classifier/filter to screen out the irrelevant ones.

To achieve this, each document collected by the Web crawler is indexed in the same ways as we previously did on the bookmarked Web documents during the generation of the personal space of interest. By mapping the text of every document, term by term, against concepts onto the multilingual concept space, we count the frequency of every concept (i.e. a node on the self-organising map) where a term of that document has been associated. This statistics of concept frequencies is then interpreted as a kind of transformed 'index' of the multilingual document. This transformed concept-based document vectors has now made all multilingual Web documents comparable in terms of their conceptual content. Hence, to sift through the documents by their relevance to the user's information interest, every concept-based document vector is mapped to the personal space of interest to find the best matching unit. Then, a similarity score between the two vectors representing the best matching unit and the documents, respectively, are calculated. Documents with a similarity score higher than or equal to the average similarity score between the bookmarked documents that belong to the same topic are presented to the user. Otherwise, it will be discarded.

6 Conclusion

This paper has presented the architecture of a personal concept-based multilingual Web content mining system, called *PMWebMiner*. A multilingual concept space is constructed to enable an automatic and unsupervised discovery of the multilingual linguistic knowledge from a parallel corpus. A personal space of interest is generated from the user's bookmark file to model a user's multilingual information interests comprising multiple topics. A concept-based Web crawler is developed to gather related Web documents in multiple languages based on the conceptual content of the user's topics of interest. This approach overcomes the vocabulary mismatch phenomenon in the multilingual environment by developing the multilingual concept space, which serves as a universal semantic space accommodating all languages. Personalisation in the context of multilingual Web content mining is facilitated by two novel techniques including concept-based multilingual Web crawling and personal concept-based multilingual text filtering. As such, personal multilingual Web content mining aiming for user-oriented concept-focused global knowledge discovery is realised. This personal multilingual Web mining system is particularly useful for knowledge workers who need to keep track of the global knowledge development in his/her personal domain of interest in order to stay competent.

References

1. Carbonell, J. G., Yang, Y., Frederking, R. E., Brown, R. D., Geng, Y. and Lee, D (1997) Translingual information retrieval: a comparative evaluation. In Pollack, M. E. (ed.) *IJCAI-97 Proceedings of the 15th International Joint Conference on Artificial Intelligence,* pp. 708-714.
2. Chakrabarti, S. (2000) Data mining for hypertext: a tutorial survey. *ACM SIGKDD Exploration,* 1(2), pp. 1–11.
3. Chang, C., Healey, M. J., McHugh, J. A. M. and Wang, J. T. L. (2001) *Mining the World Wide Web: an information search approach.* Kluwer Academic Publishers.
4. Davis, M., (1996) New experiments in cross-language text retrieval at nmsu's computing research lab. In Proceedings of the Fifth Retrieval Conference (TREC-5) Gaithersburg, MD: National Institute of Standards and Technology.
5. Kohonen, T. (1995) *Self-Organising Maps.* Springer-Verlag, Berlin.
6. Kosala, R. and Blockeel, H. (2000) Web mining research: a survey. *ACM SIGKDD Exploration,* 2(1), pp. 1–15.
7. Landauer, T. K. and Littman, M. L. (1990) Fully automatic cross-language document retrieval. In *Proceedings of the Sixth Conference on Electronic Text Research,* pp. 31-38.
8. Lang, K. (1995) NewsWeeder: Learning to filter news. In *Proceeding on the 12th International Conference on Machine Learning, Lake Tahoe, CA, Morgan Kaufmann,* pp. 331-339.
9. Lieberman, H., Van Dyke, N. W. and Vivacqua, A. S. (1999) Let's browse: A collaborative browsing agent. In *Proceedings of the 1999 International Conference on Intelligent User Interfaces, Collaborative Filtering and Collaborative Interfaces,* pp. 65-68.
10. Mukhopadhyay, S., Mostafa, J., Palakal, M., Lam, W., Xue, L. and Hudli, A. (1996) An adaptive multi-level information filtering system. In *Proceedings of The Fifth International Conference on User Modelling,* pp. 21-28.
11. Salton, G. (1989) Automatic Text Processing: The Transformation, analysis, and Retrieval of Information by Computer. Addison-Wesley, Reading. MA.
12. Soergel, D. (1997) Multilingual thesauri in cross-language text and speech retrieval. In *Working Notes of AAAI Spring Symposium on Cross-Language Text and Speech Retrieval,* Stanford, CA, pp. 164-170.
13. Tan, A-H. (1999) Text Mining: The state of the art and the challenges. In *Proceedings of PAKDD'99 workshop on Knowledge Disocovery from Advanced Databases,* Beijing, pp. 65-70.

Context-Based Recommendation Service in Ubiquitous Commerce*

Jeong Hee Hwang, Mi Sug Gu, and Keun Ho Ryu

Database Laboratory, Chungbuk National University, Korea
{jhhwang, gumisug, khryu}@dblab.chungbuk.ac.kr

Abstract. As Ubiquitous commerce is coming, personalization service is getting interested. And also the recommendation method that offers useful information to the customers becomes more important. However, the previous methods depend on specific method and are restricted to the E-commerce. For applying these recommendation methods into U-commerce, we propose a modeling technique of context information related to personal activation in commercial transaction and show incremental preference analysis method, using preference tree which is closely connected to recommendation method in each step. And also, we use an XML indexing technique to efficiently extract the recommendation information from a preference tree.

1 Introduction

Recently, u-commerce which is connected to the existing e-commerce is developed. So it is interesting to provide the personalization services suitable to the requirements and the activities of the customers in ubiquitous space based on the electronic and physical space. However, it has a deficiency that does not consider a personal preference analysis[1,2,3].

The current recommendation methods are divided into a demographic technique [12,13], content based filtering[4,12], collaborative filtering[2,4], and case based filtering[2,3]. Each method has different efficiency and precision about recommenddation according to the applying areas and the activation levels[2]. Therefore, it is necessary to develop the mechanism that merges and connects the characteristics of each recommendation method rather than the one dependent on the special recommenddation method.

In u-commerce environment, to support personalization services, a system should recognize preferred items through preference analysis such as personal profile or case of personal commerce. Based on this, the system recommends the items that are expected for the customers to prefer according to the situation, so it can help commerce easily. However, the current recommendation method based on personal preference analysis is restricted to e-commerce. So it can't support the concepts of physical and electronic space in ubiquitous space[7,8,9].

Accordingly, in this paper we propose how to model the context information[10,11] in u-commerce focused on the definition of context entities that are customers in

---

* This work was supported by University IT Research Center Project in Korea.

commerce. And based on this, we also propose the mechanism of preference analysis which represents heterogeneous personal preference analysis method using preference tree.To extract recommendation information from the preference tree, we apply XML indexing method. Also, to show the efficiency of the proposed method, we present the example of the recommendation method based on the context modeling, and verify the effect of XML indexing through experiments.

The remainder of this paper is structured as follows. Section 2 shows how to model the context information in the u-commerce. Section 3 describes the incremental preference analysis method using the preference tree and section 4 shows the example of recommendation service based on the proposed context modeling. Finally, section 5 concludes the paper.

2 Context Modeling in Ubiquitous Commerce

For context modeling in u-commerce environment, we follow the definition of context described in Merriam-Webster's Collegiate Dictionary. It means that the context entities are the customers who are doing commerce, the situation context is the information occurring according to the context entities, and the environment context is defined as the information of the commerce or shopping mall to extract the context information of context entities.

2.1 U-Environment Context Modeling

The context is represented as the context entities changing not only the position but also the individual activity over time. And the context entities, CE, are composed of time attribute, activity attribute, object attribute, and space attribute, and they are represented as CE=$<T_A, A_A, O_A, G_A>$. Time attribute of CE consists of $T_A=<ct_s,ct_e>$, where ct_s is the starting time, and ct_e is the end time. Activity attribute of customers is "selected" and "visited". In ubiquitous environment all objects and entities are divided into entities of the smallest unit, and the object entities are O_A of CE. G_A classifies the physical and electronic space in u-commerce.

The context information has to be transformed to the generalized space concept. For example, the customer's location(x, y) coordinates in the physical space or directory information in the electronic space itself has no meaning information to provide services. Table 1 and Table 2 show a context information example of the non-generalized personal location and generalized spatial location respectively.

Table 1. Non-generalized individual location context

CE ID	Action	Object	note
356583455	visited	abc	directory
356573455	visited	X35221 y25235	coordinates

Table 2. Generalized individual location context

CE ID	Action	Object	note
356583455	visited	Store 2030	Electronic Shopping mall
356573455	visited	Aisle 2535	Physical Shopping mall

U-commerce based on the context provides the services according to the individual visited location and the selected objects. Therefore if someone visits the cyber shop or section on the web, the concept space is represented through hierarchical relation of items which have the semantic sub-concept.

Environment context modeling like Figure 1 has a category modeling of the concept hierarchy and a boundary grouping modeling based on the nearest entities. Environment context such as the structure of the shopping mall, has a semantic similarity structure such as shopping mall with the concept hierarchy, section, market, and objects in electronic space.

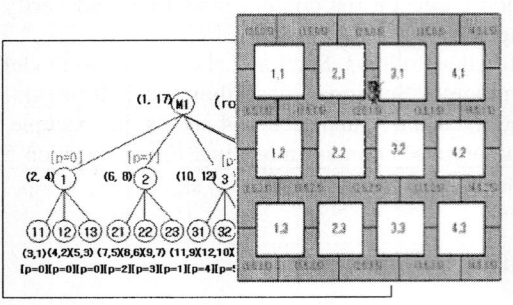

Fig. 1. Environment context modeling

And also it has a structure based on the physical distance in physical space. Therefore, through connecting between physical space and electronic space in u-commerce, we can be provided the information of the physical and electronic space.

2.2 Database Schema

To utilize the context information for the personalization services, we need the generalized context entity attribute, the current situation context, the situation context history and the situation context database. First, the database of the generalized context entity attribute information is composed of the customer's profile information. Table 3 is the example of the generalized context entity attribute. The current situation context is the context entity, the information of the customer's current activity, and it is used for the context recognition service. Table 4 is the example of the situation context information.

Table 3. Generalized context entity attribute

CE ID	Name	Gender	...
356583455	Hong	M	...
356573455	Lee	F	...

Table 4. Current situation context information

CE ID	Time	Action	Object	note
356583455	2004-12-01-20-12	visited	Store 2030	Electronic space
356573455	2004-12-02-18-10	selected	Aisle 2535	Physical space

And the context entity history information is used for the recommendation of the objects through analyzing the customer's preferences. Table 5 is the example of the context entity history information. The environment context information in u-commerce is provided by the host of the shopping mall to get the information of the context entity which is the customer's location or the selected items. Table 6 is the example of the environment context information.

Table 5. Context entity history information

CEID	Start time	End time	Visit Duration	Action	Object	note
356583455	2004-12-01-20-12	2004-12-01-20-22	10(min)	visited	Store 2030	Electronic space
356673455	2004-12-02-18-10	2004-12-02-18-11	1(min)	visited	Aisle 2535	Physical space

Table 6. Environment context information

EID	Section No.	Store No.	Name	Physical Location Info	Object	note
ST-583455	2	3	Store 2030	Polygon(75, 75,125,75..)	210,115 167.23	Shopping gmall
AL-583454	2.5	3.5	Aisle 2535	Polygon(50, 50,100,0..)	-	passage

3 Incremental Preference Analysis by Preference Tree

To efficiently support recommendation technique to the every commercial activetion level, we propose the incremental preference analysis based on the preference tree using the organic recommendation method. And to extract the information from the preference tree, we use the XML indexing method.

3.1 Incremental Preference Analysis

We show the rate of suitability of the existing recommendation method according to the commercial activity level in Figure 2. In the beginning level, profile based method is high, but in the middle, collaborative filtering and case based method are higher. And in the end level, collaborative filtering is higher because of the preference analysis about the items that have never been selected. In this paper, we efficiently use features of these techniques according to the commerce activity level of users.

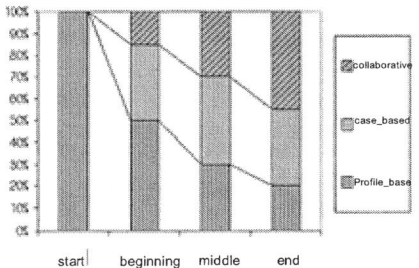
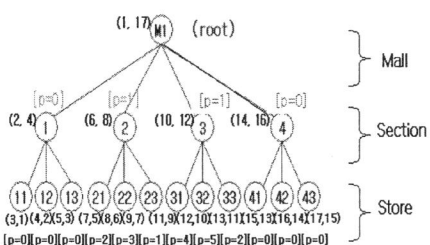

Fig. 2. Suitability of each recommendation method according to the commercial activation level

Fig. 3. Preference tree

To analyze the preference, we construct the preference tree based on the concept hierarchy modeling u-shopping mall tree which is the environment context in this paper. We put the preference, p, as each node of the tree which is the item attribute value, and arrange the shopping mall index structure and preference information. Figure 3 shows the preference tree structure, in which the number in the brace is index number and the value in the square bracket is the initial value of the preference, p. Each node has initialized preference, p.

In the beginning level of commerce or for the customers who visit the sites first time, there's no information for the preference analysis. So it is impossible to apply collaborative filtering method or individual case based method.

Accordingly, we give the preference to the coincident one by mapping the personal profile and the item profile. For example, if male Ken of age 32 is mapped to Section 2 which is for male over 18. Therefore, the preference of Ken in Section 2 is put to 1 and the preference of Section 1 which is not mapped is put to 0.

Figure 4 is the example of initialization using the preference, and the tree on the right hand side shows that the evaluation value about the preference is inherited to the sub items.

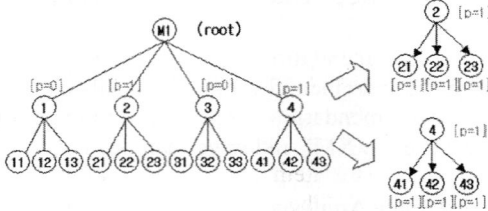

Fig. 4. Individual preference tree initialization and preference inheritance

The preference tree is initialized through mapping the profile. Whenever the user's commercial transaction occurs, we use the following analysis methods. Through the individual case or collaborative preference analysis based on the other user's information, we accumulate the preference value about the mall incrementally. And using these, we can provide the suitable recommendation information to the users.

Individual Case Based Analysis
In the existing e-commerce the individual case based information was collected using click stream or log file. In u-commerce which contains the existing commercial transaction, it is possible to analyze the activities of context entity, connecting the activities in the different access space, that is physical and electronic space.

The activity attribute of the context entity is divided into "selected" and "visited". Analyzing the generalized context information such as the time length of the visited mall or the selected items, we can compute the preference coefficient.

Based on the preference coefficient we accumulate the evaluation value of the preference according to the number of visit. It is defined as the following equation.

$$P_A = \sum_{i=1}^{n} f_i * prr$$

p : preference evaluation, f : visited number of mall (selected number of items), prr: preference coefficient

Generally the visiting time in the mall and the preference coefficient of the items are determined differently according to the inclusion or hierarchy structure. Calculating the preference coefficient is determined through real experiment or verification.

User Based Collaborative Preference Analysis

Because the case based analysis cannot grasp the individual preference about the items in the mall which is not visited, to predict the user's preference we compute the correlation coefficient with others using the equation (1). And using the Pearson correlation coefficient, equation (2), the prediction value of the preference is calculated substituting the correlation coefficient of others and the average of the user's preference. The correlation coefficient value is from -1 to 1. If the value is 1, it means the perfect positive relationship, and if -1, then it means the perfect negative relationship, and if 0, it means that there's no correlation [3,4,5,17].

$$corr_{AB} = \frac{Cov(A,B)}{\delta_A \delta_B} = \frac{\sum_i (A_i - \bar{A})(B_i - \bar{B})}{\sqrt{\sum_i (A_i - \bar{A})^2} \sqrt{\sum_i (B_i - \bar{B})^2}} \quad (1) \quad P_{a,i} = \bar{r}_a + \frac{\sum_{u=1}^{n}(r_{u,i} - \bar{r}_u) * s_{a,u}}{\sqrt{\sum_{u=1}^{n} s_{a,u}}} \quad (2)$$

Item Based Collaborative Preference Analysis

User based collaborative preference analysis predicts the individual preference about particular items which are not known, using the information of the other's preference about the items. But item based collaborative preference is analyzed through the similarity of items. The equation (1) calculates the similarity among items, substituting the preference value of other users about particular items. As a result we can get the similarity among items in Table 7.

Table 7. Correlation among items

	Item A	Item B	Item C	Item D
Item A	1	0.54	-0.53	0.31
Item B	0.54	1	-0.32	0.54
Item C	-0.53	-0.32	1	-0.53
Item D	0.31	0.540	-0.53	1

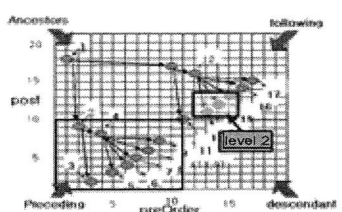

Fig. 5. Preference map based on the index

Using the similarity table we can construct the similarity tree about each item. The preference tree represents the preference of items about the users and the similarity tree represents the similarity about an item against other items.

3.2 Extracting the Information from the Preference Tree

According to the situation of the context entity, in case of "visited", the items based on the preference are recommended. And in case of "selected", the items based on the similarity are recommended. To recommend efficiently according to the recommendation situation, the method is required which extracts fast the information about preferred or similar items over the given level from the preference or similarity tree. Therefore in this paper, we used the XML indexing method to get access to the tree structure fast.

XML indexing aims not only a cut of search time but also the maintenance of hierarchy structure of XML using the least data. We give the preorder and postorder in each node and partition the field through comparing the index number by using XML index tree suggested in [19]. Then we can construct the XML database based on RDBMS.

To recommend based on the preference, first we construct the u-shopping mall structure designed by context modeling, using XML indexing, and then give the index value and the preference to the nodes representing items. Then the documents are partitioned by axes and we can extract the information based on the preference. It is the application of the extended preference map based on the index tree.

Figure 5 is an example of the preference map based on the index querying about the "following" area and preference is 2 by a standard of the current node (preorder=11, postorder=9).

Table 8. Preference map base on the index

Field record	Pre Order	Post Order	Parent	EID	Preference
18	18	16	2	P1020-005	7
19	19	17	2	P1020-006	6

Table 9. Similarity map information based on the index

field record	Pre Order	Post Order	parent	EID	Correlation
46	46	55	1	P2020	-
47	47	45	2	P1020-001	0.14

Similar to the preference map based on XML indexing in Figure 5, we construct index based similarity map, adding the attributes to index table, and then extract the similarity information. Then it is applied to the recommendation based on the similarity. Table 8 and Table 9 is database schema of the index based preference map and the index based similarity map respectively.

4 Implementation and Experiment

In this section, to verify the efficiency of the recommendation method based on the context modeling, we show the example of the recommendation service using the randomly generated experiment data and experiment to extract the information fast from the preference tree using XML indexing method.

To explain the example of the personalization recommendation service, we assume that the customer's information of the preference and the similarity among items is registered. And the information of items according to the given context is generated into XML documents.

The implementation environment is Windows 2000 server, Visual C# .Net, and Ms-SQL 8.0 of RDBMS. The used experiment data are the individual preference data sets of items and the similarity data sets of particular items.

4.1 The Application of the Recommendation Service

In this section, we input the context entities, the context information such as the activity attributes of the customers and the object attributes in section 2.1 as parameters. And then we show the example of the recommendation services providing the personalized recommendation information in XML documents according to the given situation.

To the users who have the attribute of 'visited', we provide the services of items based on the preference. While to the users who have the attribute of 'selected', we provide the services of items based on the similarity. In the physical space the distance is preferred, whereas in the electronic space the category is preferred.

In the physical space, the service system moving along the passage recommends the nearest mall and items to the customers. At that time, the information of the items is extracted according to the preference of the customers. Figure 6 shows the execution result about the query Q1; "recommend the mall over 6 of the preference and the items in the mall, to the customer who is located in the passage EID='2525'".

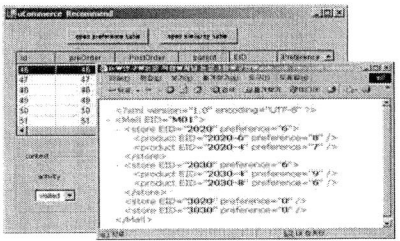
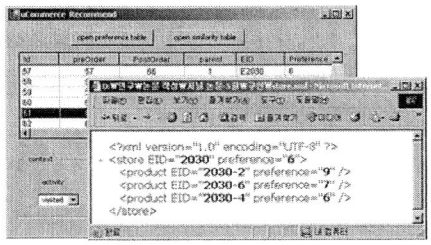

Fig. 6. Recommendation according to the visited passage

Fig. 7. Recommendation according to visited mall

If the customer visits the mall, he is recommended the preferred items in the mall. The items are equally recommended in the physical and electronic space, according to the preference. And through the predicted preference the items are arranged and extracted. Figure 7 shows the execution result to the query Q2; "Recommend the items over 6 of the preference according to the visited mall EID='2030'".

When recommending the items to the customers, we use the preference to provide the predicted similar items to the customers. So we can arrange the items by the number of the blocks which move according to the distance in the physical space. And we arrange the items by the similarity and group them by each section. Figure 8 shows the execution result to the query Q3; "In the electronic space, recommend the items over "0.43" of the similarity in selecting the item '2030-6'".

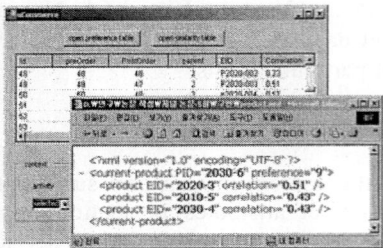 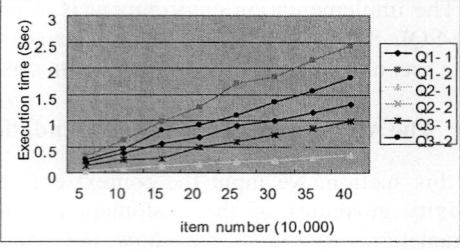

Fig. 8. Recommendation of similar items by selecting the items

Fig. 9. The query performance time according to the number of items

4.2 Experiment and Performance Evaluation

To evaluate the performance of XML indexing method applied to the context and the similarity information modeling for the efficient recommendation, we compare and evaluate the performance between the recommendation query with index and the recommendation query without index.

There are three kinds of queries used in the experiment. First, we set the queries in Figure 6, Figure 7 and Figure 8 as Q1, Q2, Q3. Second, the recommendation queries using XML index are Q1-1, Q2-1, Q3-1. Third, the recommendation queries using all the XML documents are Q1-2, Q2-2, Q3-2. And we evaluate the query performance time about the general items with increasing the number of items from 50000 to 400000. The results are shown in Figure 9. The order of the query performance time is Q1-2 > Q1-1 > Q2-2 > Q2-1 > Q3-2 > Q3-1.

The queries about each situation are set to Q1, Q2, Q3. And the query applying XML index (Q1-1, Q2-1, Q3-1) is more efficient than the query about all the XML documents (Q1-2, Q2-2, Q3-2) as the number of items is increasing. So the XML indexing method is more efficient in the query performance time, because it conducts the query about the data extracted from the selected tree.

Of the queries used in the experiment, Q1 requires the longest query performance time. It is because Q1 executes the complicated calculation such as searching for the nearest mall and extracting the preferred items in each mall, comparing with other queries. On the contrary, because the query Q3 requires the simple calculation which searches only one time when visiting the mall, the least performance time is required.

5 Conclusion

As u-commerce is coming, it is becoming important to provide the personalized information based on the context more and more. However, the recommendation method which is crucial for personalization services is just restricted to the e-commerce, and it is dependent on the particular method rather than flexible to the commerce activity level in analyzing the preference. And the method which analyzes the individual commerce in e-commerce can't apply to the ubiquitous commerce directly.

In this paper, we designed the context information modeling, such as the situation of the commercial activity and the environment information of users which are the context entities in u-commerce. And we proposed the way analyzing the preference incrementally through connecting the preference tree and each method organically. In addition, to verify the efficiency of the context based modeling and the proposed method, we implemented the example of the recommendation method through query, and we experimented on the effect of the XML indexing method.

Future research includes to design the recommendation system specifically and to study the algorithm to manage the changes of the preference tree and the data efficiently.

References

[1] P. J. Denny. Co-evolution in uCommerce: Emerging Business Strategies and technologies. *Telecommunications Review*, Vol.13, No. 1, pp.48~56, 2003.
[2] R. Schank. Dynamic Memory: A Theory of Learning in Computers and People. *Cambridge University Press*, New York, 1982.
[3] D. M. Nichols. Implicit Ratings and Filtering. *Proceedings of the 5's DELOS Workshop on Filtering and Collaborative Filtering*, Budapaest, Hungary, ERCIM, 1997.
[4] B. Sarwar, G. Karypis, J. Konstan and J. Riedl. Analysis of Recommendation Algorithms for E-Commerce. *ACM Conference on Electronic Commerce*, 2000.
[5] D. A. Gregory, A. Dey, R. Orr and J. Brotherton. Computer-awareness in Wearable an d Ubiquitous Computing. *Georgia Institute of Technology*, 1997.
[6] K. Henricksen, J. Indulska, and A. Rakotonirainy. Generating Context Management Infrastructure from High-Level Context Models. *In 4th International Conference on Mobile Data Management - Industry Track*, 21-24 January, 2003.
[7] A. Rakotonirainy. Context-Oriented Programming for Pervaisve Environments. *University of Queensland Technical Report* , September 2003.
[8] R. Robinson. Context Management in Mobile Environments. *PhD. Honours Thesis, School of Information Technology and Electrical Engineering*, University of Queensl, 2000.
[9] G. CHEN, and D. KOTZ. A Survey of Context-Aware Mobile Computing Research. *Dartmouth Computer Science Technical Report TR2000-381*, Department of Computer Science - Dartmouth College.
[10] A. Rakotonirainy, S. W. Loke, and G. Fitzpatrick. Context-Awareness for the Mobile Environment. *dstc.edu.au*, 2000.
[11] M. Bamshad, D. Honghua, L. Tao and N. Miki. Discovery and Evaluation of Aggregate Usage Profiles for Web Personalization. *WEBKDD-Special Issue for the Data Mining and Knowledge Discovery journal*, 2000.
[12] M. J. Pazzani. A Framework for Collaborative, Content-Based and Demographic Filtering. *Artificial Intelligent Review*, pp.394 -408, 1999.
[13] B. Krulwich. LIFESTYLE FINDER: Intelligent User Profiling Using Large-Scale Demographic Data. *Artificial Intelligence Magazine*, Vol.18, No.2, pp.37-45, 1997.
[14] J. F. Allen. Maintaining Knowledge about Temporal Intervals. *Communication of the Association of Computing Machinery*, Vol.26, No.11, 1983.

[15] B. Mobasher, H. Dai, T. Luo, Y. Sung and J. Zhu. Integrating Web Usage and Content Mining for More Effective Personalization. *In Proc. of First Intl. Conf. On E-Commerce and Web Technologies (ECWeb 2000)*, pp. 165-176, 2000.
[16] S. M. Shin, V. T. Nguyen, J. H. Hwang, K. H. Ryu. A New XML Indexing Approach to Support Structure-based Queries Processing of XML Document. *Proceedings of the International Conference on the Computer and Information Science (ICIS)*, pp.356-361, 2004.
[17] P. Resnic, N. Iacocou, M. Sushak, P. Bergstrom and J. Riedl. GroupLens: An Open Architecture for Collaborative Filtering of Netnews. *Proceedings of the Computer Supported Collaborative Work Conference*, 1994.
[18] B. Sarwar, G. Karypis, J. Konstain and J. Reidl. Item-Based Collaborative Filtering Recommendation Algorithms. 2000.
[19] T. Grust. Accelerating Xpath Location Steps. *In SIGMOD Conference*, 2002.

A New Continuous Nearest Neighbor Technique for Query Processing on Mobile Environments

Jeong Hee Chi, Sang Ho Kim, and Keun Ho Ryu[*]

Database Laboratory, Chungbuk National University, Korea
{jhchi, kimsh, khryu}@dblab.chungbuk.ac.kr

Abstract. Recently, as growing of interest for LBS(location-based services) techniques, researches for NN(nearest neighbor) query which has often been used in LBS, are progressed variously. However, the results of conventional NN query processing techniques may be invalidated as the query and data objects move. To solve these problems, in this paper we propose a new nearest neighbor query processing technique, called CTNN, which is possible to meet continuous query processing for mobile objects. In order to evaluate the proposed techniques, we experimented with various datasets and experimental results showed that the proposed techniques can find accurately NN objects. The proposed techniques can be applied to navigation system, traffic control system, distribution information system, etc., and specially are most suitable when both data and query are mobile objects.

1 Introduction

Recently, with the advance of mobile computing technologies and wireless internet technologies, the development of technologies on location based services has been actively performed. These services aim at providing the information that users want by applying various query processing techniques and efficient data storage management technologies to mobile objects. Nearest neighbor query is one of the most frequently used query processing techniques, which returns objects that exist nearest to a given query as results, and can be applied in location based services like following examples: 'A' searches for a pizza house that is located nearest to his home by using mobile phone then orders a pizza; a sailing vessel 'B' wishes to ask for supply of oil by searching for a filling ship that is sailing nearby.

As explained the above examples, location based services process queries for various types of objects and many techniques have been proposed for efficient performance of such services. Since, unlike spatial objects, mobile objects are continuously changing their positions over time, The results of NN query that are calculated at a certain time point may not be invalid in different times. However, the existing techniques for NN queries that are used by most systems ignore such characteristics. Therefore it may not provide users with valid information on total query time.

In order to solve the problems happened on the existing works, in this paper we propose continuous trajectory nearest neighbor (CTNN) query processing techniques.

---

[*] This work was supported by University IT Research Center Project in Korea.

Our proposed techniques are new NN query processing techniques that can find mobile objects which maintain the closest distance with queries by comparing trajectory information.

The rest of the paper is organized as follows. Related works are discussed in section 2. Section 3 describes the basic models and assumptions used in this paper. The proposed techniques are introduced in section 4. Section 5 shows some of our experimental results. Finally, conclusions are given in section 6.

2 Related Works

NN query processing techniques have been studied in broad fields including multimedia database, data mining, and moving object database[1]. Also an index for processing NN query and approximate query processing technique for reducing response time has been proposed[2,3,4].

Kollios in [5] proposes a technique using duality transform technique which transforms moving object trajectory segments on (x, y) plane into the points on (v, a') plane where v is speed and a' is y coordinate value. This enables us to apply to all kinds of query and data object types on just two dimensional planes, but it cannot be applied to continuous NN search which base on moving object trajectory, expressed as segments on 3 dimensional (x, y, t) plane. Benetis in [5] proposes a permanent NN query processing technique using TPR-tree [6]. The technique calculates the distance between node rectangle and query point by using differentiation function according to DF search method[1], and retrieve an object that has the shortest distance for each interval [now, now+Δ]. Since the distance calculation function very much depends on the location and speed of the object, whenever its values change, the function has to be recalculated and it can cause much overhead. Tao in [8] proposes TPNN (time-parameterized NN) and CNN (continuous NN) query. These are continuous NN queries that select NN objects by calculating TPNN or CNN time point, whose object information may change on overall query time, storing each NN object information and valid time interval in time list TL and returning at the last time of query as a result. Therefore, the result consists of several continuous and divided interval like TL = {<a, [t_1, t_2]>, <b, [t_2, t_3]>}, and NN object information valid within each interval. TPNN query firstly calculates in advance the close future time that is closest to query about all objects, checks whether the calculated time value becomes actual NN object on time point that is calculated in sequence of calculated time value, and then selects a result. CNN query is proposed to reduce calculation overhead in TPNN query. CNN query determines NN object and its valid time interval by using neighboring vicinity and vertical bisection plane. There will be bigger the overhead as more the number of objects and more frequent the direction change of object movement since a number of trajectory segments increase.

3 Preliminaries

In this section, we describe the basic data structure of CTNN query, and introduce a few assumptions and some notations used in this paper.

3.1 Data Model

We assume that mobile objects are point objects and change their positions continuously over time such as car, person, and aircraft. Location information of objects is regularly extracted through equipments like GPS, and they are stored in database when updates such as insertion, deletion of objects, change of speed and direction occur. In other words, mobile objects are stored in form of <id_i, (x, y), t> at each time t when updates occur, and information about speed and direction is not stored. The movement path of objects is called a trajectory in geographical or geometrical terms, and generally the trajectory of mobile object is expressed as polyline, groupings of line segments that are continuous and separately divided as Fig.1.

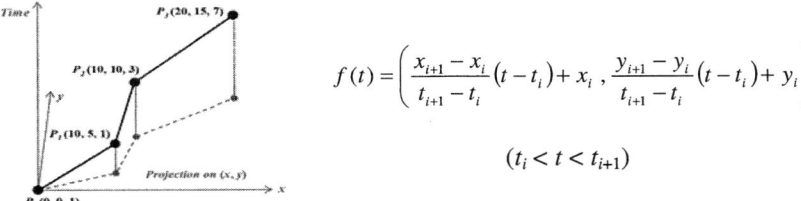

$$f(t) = \left(\frac{x_{i+1} - x_i}{t_{i+1} - t_i}(t - t_i) + x_i , \frac{y_{i+1} - y_i}{t_{i+1} - t_i}(t - t_i) + y_i \right)$$

$$(t_i < t < t_{i+1})$$

Fig. 1. Trajectory of mobile object   **Formula 1.** Function of location estimation

Spatial coordinates at non-stored time in database can be presumed by using function of linear location estimation, as shown formula 1. As the formula uses coordinate values at time t_i and t_{i+1} stored in database, and calculates the object location at time t that exists in that interval, spatial location of the object on one line segment can be deduced with just the information about the two end points on one line segment.

3.2 Assumptions and Notations

CTNN query is processed based on following assumptions.

1. If a certain object p is changed in less than 1 minute after it is selected as NN object, then p is ignored and the valid time interval is included in former or latter interval. This time interval can be used meaninglessly in practice since it is too short to express a result. But it can be adjusted depending on application.
2. If the distance between two objects is less than 1 meter, then these objects are considered to have met or intersected. In real world, objects having the same coordinates at the same time mean collision, and this situation almost never happens. Therefore two objects that exist within 1 meter from a certain time point are regarded as objects that exist on the same place at the same time. But it can be also adjusted depending on application.

And following notations are used.

Table 1. Notations

$\Box a_i$	i^{th} segment of object a ($i \geq 1$, integer)
(x_a^t, y_a^t)	Spatial coordinate of object a at time t
(x_{ai}, y_{ai})	i^{th} stored coordinates of object a
$\Delta x (\Delta y)$	Movement distance on x(y) axis per unit time intervals
$x_{ab} (y_{ab})$	Absolute of difference value between a and b on x(y) axis
$|a - b|^t$	Spatial distance between a and b from time t = $\sqrt{(x_{ab}^t)^2 + (y_{ab}^t)^2}$
lp	Slope of segment $p = \Delta x / \Delta y$
∂p	Movement distance per unit time interval : Displacement of p

4 Continuous Trajectory NN Search

CTNN query, first of all, searches all the data objects that exist in given temporal range query, then time interval of these segments is clipped into query time interval. Then segments that maintain the closest distance with the query trajectory through NN query processing are selected as a result. Process of selecting NN objects is largely made up of evaluation time setup stage for continuous query and object trajectory information comparison stage for trajectory query.

4.1 Evaluation Time Setup

Basic evaluation time consists of two end point times of every segment that includes data objects and query object. As shown in Fig.2(a), Searching for NN objects simply from the start point of segments will yield result, which is TL={<a, [t_s, t_e]>}. If we, like Fig.2(b), search for NN object on displacement point and intersection point, the result will be TL={<a, [t_s, t_4]>, <b, [t_4, t_6]>, <a, [t_6, t_e]>}. In other words, the object 'a' that is searched first is NN object until time t_4. And NN object 'b' after time t_4 intersects with object 'a' at time t_6, so the NN information changes to 'a' again. This may lead to more correct result than simple methods, but it still is not enough.

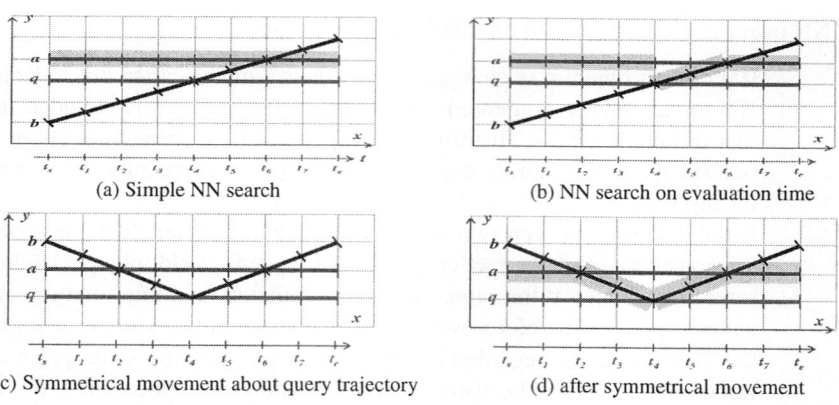

(a) Simple NN search (b) NN search on evaluation time

(c) Symmetrical movement about query trajectory (d) after symmetrical movement

Fig. 2. Simple example of CTNN query

Therefore we use symmetrical movement of data segment to get more detailed information on intersection points between segments, as shown in Fig.2(c). By making query segment as criterions, and gaining intersection information after gathering object segments toward one corner, we can get correct NN objects as shown in Fig.2(d). Overall estimation time is set through algorithm 1 and 2 as shown below.

Input: Data object segment collection *seg* , query segment *q*
Output: intersection point list *IL* , displacement point list *EL*

1   *End points of segments of seg , q* → *EL*
2   **Find_Intersection( )** → *IL*
3   *Return time point within EL , IL*

Algorithm 1. Set_EvaluationTime

Input: Data object segment collection *seg* , query segment *q*
Output: Intersection point list *IL*

1   Search intersection point
    1.1   *Consider seg as line segment on 2 dimensional space*
    1.2   **Sweep Line Algorithm**[9] → extract intersection point *ip*
2   *Set up t, intersection time of ip* // when assuming that objects A and B reach ip each at t_A, t_B
    2.1 $t = t_B$ *If* $t_A > t_B$
    2.2   Otherwise $t = t_A$
3   *ip , t* → *IL*
4   *IL* return

Algorithm 2. Find_Intersection

Input: Data object segment collection *seg* , query object *q*
Output: time list *TL*

// In performing stages, when there exists many candidate objects, always apply **Calculate_Displacement( )**

1   *With q as criterions, segments in one side space are symmetrically moved to another side of space*
2   **Set_EvaluationTime( )**
3   ⌊$seg - q|^{ts}$ // start time of *q*
    3.1   *Select NN object p that has mindist*
    3.2   *TL* = <*p* , [t_s , t_{el}]>
4   **Until** t_{el} // end time of □q_l
    // TL updated if conditions are satisfied in the order of phase
    4.1   *When there exists intersection i_l inside IL in time t_i,*
        4.1.1   *If □q intersects with □p'* = {<*p* , [t_s , t_i]>, <*p'* , [t_i , t_{el}]>}
        4.1.2   *If □p intersects with □p"* = {<*p* , [t_s , t_i]>, <*p"* , [t_i , t_{el}]>}
        4.1.3   *Otherwise TL* = <*p* , [t_s , t_{el}]>
    4.2   *When there exists formation point p' inside EL in time t_c,*
        4.2.1   *If |p' - q|^{tc} < |p - q|^{tc} then TL* = {<*p* , [t_s , t_c]>, <*p'* , [t_c , t_{el}]>}
        4.2.2   *Otherwise TL* = <*p* , [t_s , t_{el}]>
    4.3   *When there exists termination point p' inside EL in time t_d,*
        4.3.1   *If p' == p then repeat 2 stages and select NN object p', TL* = {<*p* , [t_s , t_d]>, <*p'* , [t_d , t_{el}]>}
        4.3.2   *Otherwise TL* = <*p* , [t_s , t_{el}]>
5   **Until** t_e // End time of □*q*
    5.1   *Repeat 4 stages per each query segment unit, with NN object p in t_{si} as criterions, to t_{ei}*
6   *Return TL*

Algorithm 3. Find_NNObject

4.2 NN Object Search

NN object searching step of CTNN query is processed basically like the following algorithm 3. Here we assume the process of NN query that selects one NN object.

CTNN query selects NN objects in the next time interval only with additional calculation process, after performing distance calculation between the whole data object and query point at the start time. Only when a certain object is inserted or deleted within valid time interval of p, it is checked whether valid time interval is changed due to this update or other NN objects are selected, and minimum distance calculation is performed. On other evaluation times, NN object information can be changed to p' when there is a segment p' that intersects with query or p' that intersects with current p.

4.3 Calculation of Displacement

CTNN query selects objects that move while keeping close distance with query as much as possible, as NN objects. At the same time, it decides whether candidate objects satisfy trajectory query conditions on each estimation time point according to k values. For example, in case of NN query, if the number of candidate object that has the nearest distance with query from one estimation time point t is 1, then we can choose the object as result without further comparisons. On the other hand, if there are many candidates, we select one NN object by comparing slop, direction, and displacement values. But since it will take a long processing time if we take into account slope, direction, and displacement values of all segments, we select one NN object through the following algorithm 4, and it is called displacement calculation.

Input: $\Box A$, $\Box B$ // segment of object A and B
Output: integer T

1   If $/A == /B$ and $\partial A == \partial B$,
   1.1   If $x_{ab} == y_{ab} == 0$ then $T = 1$ // identical
   1.2   Otherwise $T = 0$ // parallel
2   If $|/A| == 1/|/B|$,
   2.1   If sign of $/A ==$ sign of $/B$,
      2.1.1   If $x_{ab} == y_{ab}$ then $T = x_{ab} / \|\Delta y|-|\Delta x\|$
      2.1.2   Otherwise $T = 0$
   2.2   If sign of $/A \,!=$ sign of $/B$,
      2.2.1   If $(|\Delta y|+|\Delta x|) / \|\Delta y|-|\Delta x\| == x_{ab} / y_{ab}$ then $T = x_{ab} / (|\Delta y|+|\Delta x|)$
      2.2.2   Otherwise $T = 0$
3   If $/A == -/B$,
   3.1   If $x_{ab} == 0$ and $y_{ab} \,!= 0$ then $T = y_{ab} / (|\Delta y|+|\Delta x|)$
   3.2   If $y_{ab} == 0$ and $x_{ab} \,!= 0$ then $T = x_{ab} / (|\Delta y|+|\Delta x|)$
   3.3   Otherwise $T = 0$
4   Otherwise, $X = x_{ab} / |\partial A - \partial B|$ and $Y = y_{ab} / |\partial A - \partial B|$
   4.1   If $/A > 1$ and $/B > 1$,
      4.1.1   If $X > Y$ then $T = X$
      4.1.2   Otherwise $T = Y$
   4.2   If $(/A$ and $/B) > 0$ and $(/A$ or $/B) < 1$ then, $T = \lfloor x_{ab} - y_{ab}\rfloor / |\partial A - \partial B|$ (rounded off)
   4.3   Otherwise $T = X + Y$

Algorithm 4. Calculate_Displacement

Displacement calculation calculates intersection time point that is generated between two segments. If the result of displacement calculation between object 'a' and 'q' from time t is 1, then it means that 'a' and 'q' intersect at t+1 time. When displacement calculation is carried out between candidate object and query, objects that have the smallest value represent the object that meets the query first, and it represents the object that moves while keeping the closest distance to query at the same time. Therefore we can gain correct NN object result by performing displacement calculation between each object and query when there are many candidate objects. In this case, negative number means that two segments intersect at the past time rather than t. Segments with 0 mean segments that do not meet q after time t, in other words, segments that move parallel with each other or move apart from each other.

4.4 Example of CTNN Query Processing

Fig.3 is shown the result of detecting intersection points through the sweep line algorithm. At start time of a given query, t_{qs}, Object a and b belong to lower side, list U, and c belong to upper side, list H, of the given query. The first intersection time t_1 is contained in the list U, so at t_{qs}, calculate distance between a, b and q, and then a becomes candidate NN object in time interval $[t_{qs}, t_1]$. Next, as c which is contained in H, has less distance than distance between a and q at $t1$, so move segment of c to list U during time interval $[t_{qs}, t_1]$. In list U, c has nearer segment than a on the query since $|c-q|^{tqs} < |a-q|^{tqs}$. So, NN object in $[t_{qs}, t_1]$ becomes c as shown in Fig.4(a).

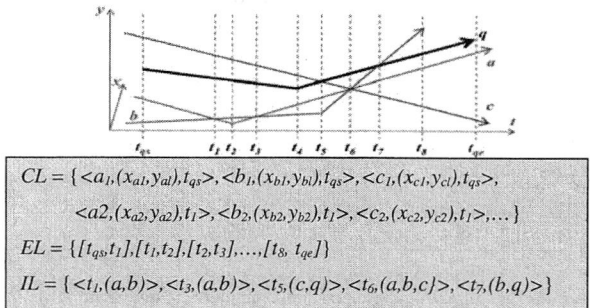

Fig. 3. Result of detecting intersection point

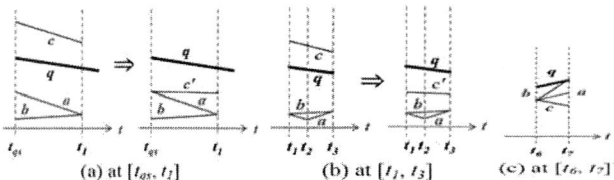

Fig. 4. Calculation of NN Object in Each Subinterval

Then, grasp whether c intersects with a or b. Since there is no intersection with c until t_1, so c remains candidate NN object until next intersection time t_3. And also c is kept as NN object until t_6 as Fig.4(b). At t_6, intersection occur a, b, and c. From there,

b is the NN object as Fig.4(c). Continuing on this way, we can calculate NN object until t_{qe}, and can gain the result from *TL* finally.

5 Experiment and Performance Evaluation

To evaluate the performance of the proposed CTNN technique, we compare it with CNN technique. Since CNN technique on volatile data in [8] is valid only for k-NN query, in our experiment we just compare the proposed method with the step of deciding split timestamp which occurred the new nearest neighbors among CkNN query processing steps.

5.1 Experimental Environment

All experiments were run on Pentium III PC with 256MB main memory. We generated various synthetic datasets of moving objects in form of <ID, time, minutes, spatial coordinate(x,y)> through GSTD(Generator of SpatioTemporal Datasets). And we used trajectory segment number of moving object, spatial range of object movement and time interval of segment, that is, time difference of two end points of segment, as experimental parameters shown in table 2.

Table 2. Experimental parameters

Experimental parameters	Value
Number of object segments to evaluate query response time	50, 100, 150, 200, 250, 300, 350, 400, 450, 500
Number of object segments to evaluate correctness of result	5, 10, 15, 20, 25, 30, 50, 100
Spatial distribution range	50,000m^2, 10,000m^2, 5,000m^2, 1,000m^2
Time Interval	3 hour, 5 hour, 10 hour

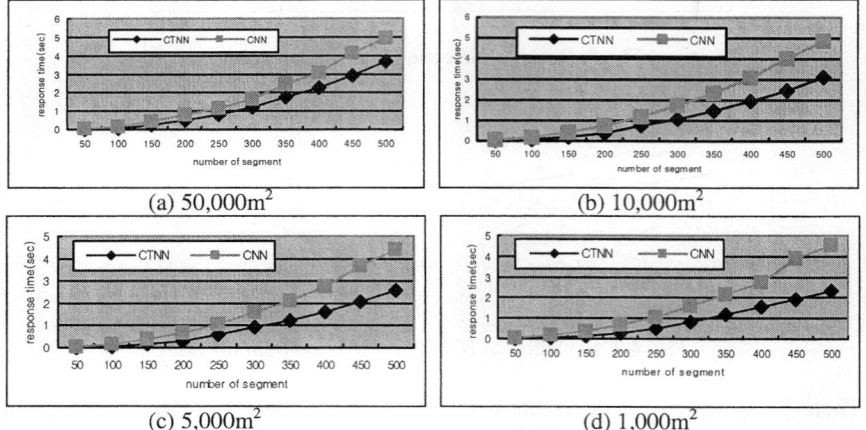

Fig. 5. Query Response Time under Varying Spatial Distribution Range

5.2 Effect of Spatial Distribution Range

We firstly experimented response time of the proposed technique with respect to spatial distribution range. Fig.5 shows the results of the proposed technique and CNN technique according to different spatial distribution range. As shown in Fig.5, the general trend of the result is that the response time increases as the number of segment increases, that is, each technique is not largely affected in query response time according to spatial distribution range.

In case CTNN technique, this is because the number of intersection points extracted from CTNN technique does not change much according to spatial distribution ranges as shown in Fig.6.

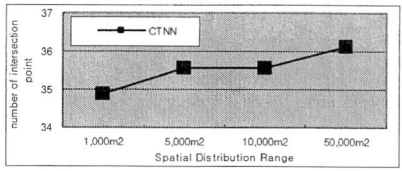

Fig. 6. Number of Intersection Point

5.3 Effect of Time Interval

We experimented how query response time of each CTNN technique differs per number of segments when random segments are generated in time interval (3, 5, 10 hours).

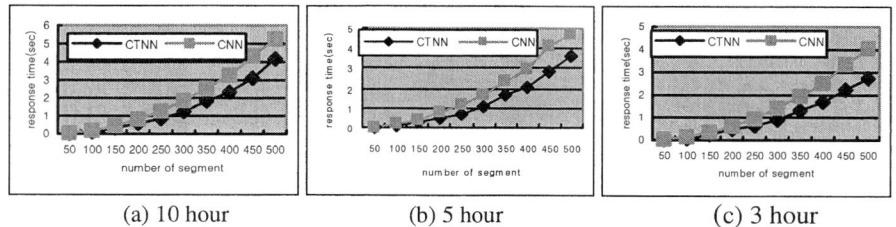

(a) 10 hour (b) 5 hour (c) 3 hour

Fig. 7. Query Response Time according to Time Interval

As shown in Fig.7, both CTNN and CNN techniques are affected by query response time according to the time difference between two end points of each segment. This is because when two end times of each segment are set, if segments are randomly generated, then speed of each segment changes, therefore intersection points are also affected while affecting the length of segment or distribution density of segments.

5.4 Comparison of Correctness

In this section, we describe the correctness of CTNN technique, with randomly generated segments that have consistent proportion of spatial and time range.

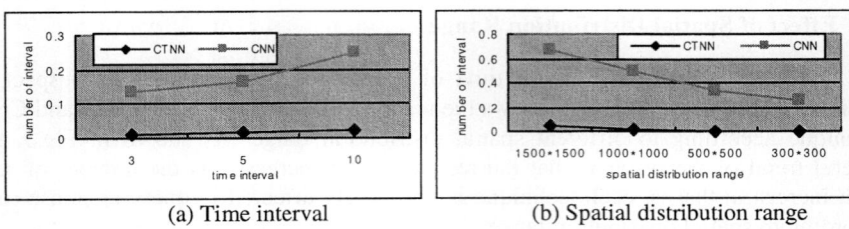

(a) Time interval (b) Spatial distribution range

Fig. 8. Correctness according to Time Interval and Spatial Distribution Range

We compare the correct result values with the number of time intervals that are searched by each technique. The reason is because the difference between time points can be correctly discerned only when the number of interval is the same. Therefore, a comparison is made in the case when NN objects are searched from each dataset with simple technique and the result searched by CTNN technique. If the number of interval is the same as the actual result, it is considered to match when the time difference between start and end is within 1 second, otherwise the interval difference value is set to 1. And difference in the number of intervals between the result searched by CTNN technique and the correct result is generated as correctness value. As shown in Fig.8, CTNN technique can retrieve more accurate NN object than CNN technique.

6 Conclusion

In this paper, we proposed new NN query processing techniques that can be most valuably used when both query and data objects are all mobile objects. Since previous works select NN objects without considering trajectory information of data and query, it can lead to incorrect result. Also because selected NN object information can change rapidly, it need frequent updates. In order to solve these problems, CTNN technique used trajectory NN query concept, and selected mobile objects which maintain the closest distance with query trajectory through displacement calculation, as NN objects. Also in order to explore continuous time during which NN objects are changed, it is checked whether the result is changed at displacement and intersection points of object segments. Through experiment, we showed that what difference the proposed technique and existing technique shows in response time according to spatial range of segments and time range, whether they show any difference in correctness.

In future studies, we will carry out studies on efficient CTNN technique processing by using index structure, and studies on CTNN technique about movement range objects.

References

1. Nick Roussopoulos, Stephen Kelley, Fredeic Vincent, "Nearest Neighbor Queries", SIGMOD Conference 1995, pp.71~79
2. Cui Yu, Beng Chin Ooi, Kian-Lee Tan, H.V. Jagadish, "Indexing the Distance : An Efficient Method to KNN Processing", VLDB 2001, pp.421~430

3. King-Ip, Congjun Yang, "The ANN-tree : An Index for Efficient Approximate Nearest Neighbor Search", DASFAA 2001, pp.174~181
4. Stephan Volmer, "Fast Approximate Nearest Neighbor Queries in Metric Feature Spaces by Buoy-Indexing", VISUAL 2002, pp.36~49
5. .George Kollios, Dimitrios Gunopulos, Vassilis J. Tsotras, "Nearest Neighbor Queries in a Mobile Environment", Spatio-Temporal Database Management 1999, pp.119~134
6. Rimantas Benetis, Christian S. Jensen, Gytis Karciauskas, Simonas Saltenis, "Nearest Neighbor and Reverse Nearest Neighbor Queries for Moving Objects", IDEAS 2002, pp.44~53
7. Simonas Saltenis, Christian S. Jensen, "Indexing the Position of Continuously Moving Object", SIGMOD Conference 2000, pp.331~342
8. Yufei Tao, Dimitris Papadias, "Spatial Queries in Dynamic Environments", TODS 2003
9. M. de Berg, M. van Kreveld, M. Overmars and O. Schwarzkopf, "Computational Geometry Algorithms and Applications", Springer-Verlag, March 1997.

Semantic Web Enabled Information Systems: Personalized Views on Web Data

Robert Baumgartner[1], Christian Enzi[1], Nicola Henze[2], Marc Herrlich[2], Marcus Herzog[1], Matthias Kriesell[3], and Kai Tomaschewski[2]

[1] DBAI, Institute of Information Systems, Vienna University of Technology, Favoritenstrasse 9-11, 1040 Vienna, Austria
{baumgart, enzi, herzog}@dbai.tuwien.ac.at
[2] ISI- Semantic Web Group, University of Hannover, Appelstr. 4, D-30167 Hannover, Germany
{henze, herrlich, tomaschewski}@kbs.uni-hannover.de
[3] Institute of Mathematics (A), University of Hannover, Welfengarten 1, D-30167 Hannover, Germany
kriesell@math.uni-hannover.de

Abstract. In this paper a methodology and a framework for personalized views on data available on the World Wide Web are proposed. We describe its main two ingredients, Web data extraction and ontology-based personalized content presentation. We exemplify the usage of these methodologies with a sample application for personalized publication browsing[1].

Keywords: personalized information management, semantic web, web intelligence, web data extraction.

1 Introduction

The vision of a next generation Web, a *Semantic Web*, in which machines are enabled to understand the meaning of information in order to better inter-operate and better support humans in carrying out their tasks, is very appealing and fosters the imagination of smarter applications that can retrieve, process and present information in enhanced ways. In this vision, a particular attention should be devoted to *personalization*: By bringing the user's needs into the center of interaction processes, personalized Web systems overcome the one-size-fits-all paradigm and provide individually optimized access to Web data and information.

We claim that a huge class of Semantic Web-enabled information systems should be able to extract relevant information from the Web, and to process and combine pieces of distributed information in such a way that the content selection and presentation fits to the current and individual needs of the user.

---

[1] This research has been partially supported by REWERSE - Reasoning on the Web (rewerse.net), Network of Excellence, 6th European Framework Program.

From this viewpoint, such systems need to focus especially on the *information extraction process*, and the *personalized content syndication process*. The actual authoring process of information, and the information management processes, are important aspects, too, if we consider portal-like applications. However, there is a sustainable need of systems which can detect and process already existing Web information.

In this paper, we describe the Web data extraction task (Section 2), and an approach for personalized content presentation (Section 3). Section 4 finally exemplifies our vision of Semantic Web-enabled information systems with an example scenario: browsing publication data with personalized support. We realized this scenario in the *Personal Publication Reader (PPR)* application. The paper ends with conclusions and outlook on future work.

2 Web Data Extraction and Integration

Today the Semantic Web is still a vision. In contrary, the unstructured Web nowadays contains millions of documents which are not queryable as a database and heavily mix layout and structure. Moreover, they are not annotated at all. There is a huge gap between Web information and the qualified, structured data as usually required in corporate information systems. According to the vision of the Semantic Web, all information available on the Web will be suitably structured, annotated, and qualified in the future. However, until this goal is reached, and also, towards a faster achievement of this goal, it is absolutely necessary to (semi-)automatically extract relevant data from HTML document and automatically translate this data into a structured format, e.g., XML. Once transformed, data can be used by applications, stored into databases or populate ontologies.

Whereas information retrieval targets to analyze and categorize documents, information extraction collects and structures entities inside of documents. For Web information extraction languages and tools for accessing, extracting, transforming, and syndicating the Data on the Web are required. The Web should be useful not merely for human consumption but additionally for machine communication. A program that automatically extracts data and transforms it into another format or markups the content with semantic information is usually referred to as *wrapper*. Wrappers bridge the gap between unstructured information on the Web and structured databases. A number of classification taxonomies for wrapper development languages and environments have been introduced in various survey papers [3, 9, 10]. High-level programming languages, machine learning approaches and interactive approaches are distinguished.

2.1 Extracting Web Data with Lixto

Lixto [1] is a methodology and tool for visual and interactive wrapper generation developed at the University of Technology in Vienna. It allows wrapper designers to create so-called "XML companions" to HTML pages in a supervised way. As internal language, Lixto relies on *Elog*. Elog is a datalog-like language especially designed for wrapper generation. The Elog language operates on Web objects, that are HTML elements, lists of HTML elements, and strings. Elog rules can be specified fully visually without knowledge of the Elog language. Web objects can be identified based on internal, contextual, and range conditions and are extracted as so-called "pattern instances".

In [4], the expressive power of a kernel fragment of *Elog* has been studied, and it has been shown that this fragment captures monadic second order logic, hence is very expressive while at the same time easy to use due to visual specification.

Besides expressiveness of a wrapping language, robustness is one of the most important criteria. Information on frequently changing Web pages needs to be correctly discovered, even if e.g. a banner is introduced. Visual Wrapper offers robust mechanisms of data extraction based on the two paradigms of tree and string extraction. Moreover, it is possible to navigate to further documents during the wrapping process. Predefined concepts such such as "is a weekday" and "is a city" can be used. The latter is established by connecting to an ontological database. Validation alerts can be imposed that give warnings in case user-defined criteria are no longer satisfied on a page.

Visually, the process of wrapping is comprised of two steps: First, the identification phase, where relevant fragments of Web pages are extracted. Such extraction rules are semi-automatically and visually specified by a wrapper designer in an iterative approach. This step is succeeded by the structuring phase, where the extracted data is mapped to some destination format, e.g. enriching it with XML tags. With respect to populating ontologies with Web data instances, another phase is required: Each information unit needs to be put into relation with other pieces of information.

2.2 Visual Data Processing with Lixto

Heterogeneous environments such as integration and mediation systems require a conceptual information flow model. The usual setting for the creation of services based on Web wrappers is that information is obtained from multiple wrapped sources and has to be integrated; often source sites have to be monitored for changes, and changed information has to be automatically extracted and processed. Thus, push-based information system architectures in which wrappers are connected to pipelines of postprocessors and integration engines which process streams of data are a natural scenario, which is supported by the Lixto Transformation Server [2]. The overall task of information processing is composed into stages that can be used as building blocks for assembling an information processing pipeline. The stages are to acquire the required content from the source locations, to integrate and transform content from a number of input channels and tasks such as finding differences, and format and deliver results in various formats and channels and connectivity to other systems.

The actual data flow within the Transformation Server is realized by handing over XML documents. Each stage within the Transformation Server accepts XML documents (except for the wrapper component, which accepts HTML), performs its specific task (most components support visual generation of mappings), and produces an XML document as result. This result is put to the

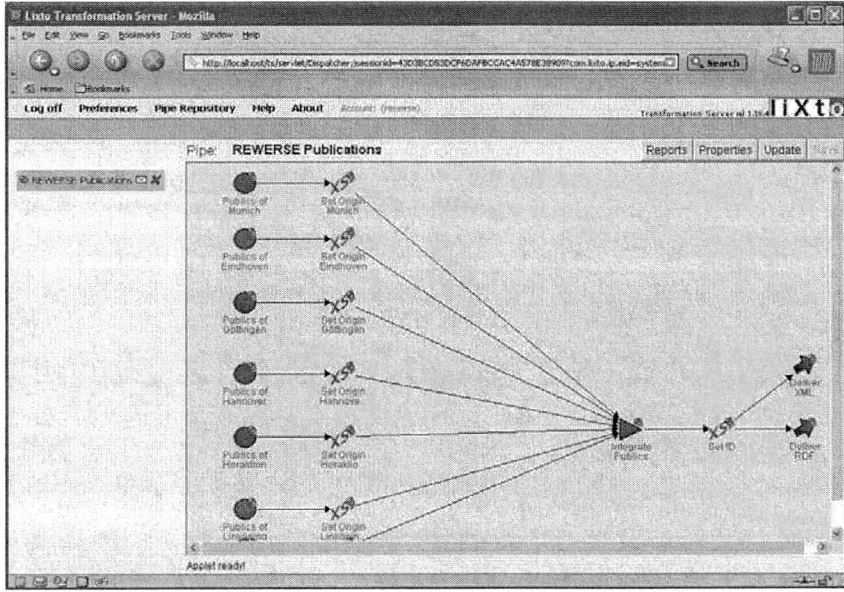

Fig. 1. Lixto Transformation Server: REWERSE Publication Data Flow

successor components. Boundary components have the ability to activate themselves according to a user-specified strategy and trigger the information processing on behalf of the user. From an architectural point of view, the Lixto Transformation Server may be conceived as a container-like environment of visually configured information agents. The pipe flow can model very complex unidirectional information flows (see Figure 1). Information services may be controlled and customized from outside of the server environment by various types of communication media such as Web Services.

3 Personal Readers – Personalization Services for the Semantic Web

Flexible information systems which need to be capable of adjusting to different application domains require a different architecture: not a monolithic approach, but several, independent components, each one serving a specific purpose. The recent Web Service-technology focuses on such-like requirements: A Web Service encapsulates a specific functionality, and communicates with other services or software components via interface components (e.g. [16, 11]).

We consider each (personalized) information provision task as the result of a particular service (which itself might be composed of several services, too). The aim of this approach is to construct a Plug & Play - like environment, in which the user can select and combine the kinds of information delivery services he or she prefers. With the Personal Reader Framework[2], we have developed an environment for designing, implementing and maintaining personal Web-content Readers [6, 5]. These personal Web-content Readers allow a user to browse information (the *Reader* part), and to access personal recommendations and contextual information on the currently regarded Web resource (the *Personal* part).

The architecture of the Personal Reader is a rigorous approach for applying Semantic Web technologies. A modular framework of Web Services – for constructing *the user interface*, for *mediating* between user requests and currently available personalization services, for *user modeling*, and for offering *personalization functionality* – forms the basis of a Personal Reader.

The goal of the Personal Reader architecture is to provide the user with the possibility to select services, which provide different or extended functionality, e.g. different visualization or personalization services, and combine them into a Personal Reader instance. The framework features a distributed open architecture designed to be easily extensible. It utilizes standards such as XML[17], RDF[13], etc., and technologies like Java Server Pages (JSP)[8] and XML-based-RPC[18]. The communications between all components / services is syntactically based on RDF descriptions. The architecture is based on different Web Services cooperating with each other to form a specific Personal Reader instance.

4 The Personal Publication Reader

Let us consider the following scenario for describing the idea of the Personal Publication Reader:

> Bob is currently employed as a researcher in a university. Of course, he is interested in making his publications available to his colleagues, for this he publishes all his publications at his institute's Web page. Bob is also enrolled in a research project. From time to time, he is requested to notify the project coordination office about his new publications. Furthermore, the project coordination office maintains a member page where information about the members, their involvement in the project, research experience, etc. is maintained.

From the scenario, we may conclude that most likely the partners of a research project have their own web-sites where they publish their research papers. In addition, information about the role of researchers in the project like "Bob is participating mainly in working group X, and working group X is strongly cooperating with working groups Y and Z" might be available. If we succeed in making this information available to machines to reason about, we can derive new information like: "This research paper of Bob is related to working group X, other papers of working group X on the same research aspects are A, B, and C, etc."

---

[2] www.personal-reader.de

To realize a Personal Publication Reader (PPR), we extract the publication information from the various web-sites of the partners in the REWERSE project: All Web-pages containing information about publications of the REWERSE network are periodically crawled and new information is automatically detected, extracted and indexed in the repository of semantic descriptions of the REWERSE network (see Section 4.1). Information on the project REWERSE, on people involved in the project, their research interests, and on the project organization, is modeled in an ontology for REWERSE (see Section 4.2). Extracted information and ontological knowledge are used to derive a syndicated view on each publication: who has authored it, which research groups are related to this kind of research, which other publications are published by the research group, which other publications of the author are available, which other publications are on the similar research, etc. Information about the current user of the system (such as specific interests of the user, or his membership to the project) is used to individualize the view on the data (see Section 4.3). The realization of the PPR has been carried out in the Personal Reader Framework (see Section 4.4); the prototype of the PPR is accessible via the Web at the URL www.personal-reader.de.

4.1 Gathering Data for the PPR

Each institute and organization offers access to its publication on the Web. However, each presentation is usually different, some use e.g. automatic conversions of *bibtex* or other files, some are manually maintained. Such a presentation is well suited for human consumption, but hardly usable for automatic processing. Consider e.g. the scenario that we are interested in all publications of *REWERSE* project members in the year 2003 which contain the word "personalization" in their title or abstract. To be able to formulate such queries and to generate personalized views on heterogeneously presented publications it is necessary to first have access to the publication data in a more structured form.

In Section 2.1 we discussed data extraction from the Web and the Lixto methodology. Here, we apply Lixto to regularly extract publication data from all *REWERSE* members. As Figure 1 illustrates, the disks are Lixto wrappers that regularly (e.g. once a week) navigate to the page of each member (such as Munich, Hannover, Eindhoven) and apply a wrapper that extracts at least author names, publication titles, publication year and link to the publication (if available).

In the "XSL" components publication data is harmonized to fit into a common structure and an attribute "origin" is added containing the institution's name. The triangle in Figure 1 represents a data integration unit; here data from the various institutions is put together and duplicate entries removed. IDs are assigned to each publication in the next step. Finally, the XML data structure is mapped to a defined RDF structure (this happens in the lower arc symbol in Figure 1) and passed on to the Personal Publication Reader as described below. A second deliverer component delivers the XML publication data additionally.

This Lixto application can be easily enhanced by connecting further Web sources. For instance, abstracts from www.researchindex.com can be queried for each publication lacking this information and joined to each entry, too. Moreover, using text categorization tools one can rate and classify the contents of the abstracts.

4.2 Ontological Knowledge for the PPR: The REWERSE-Ontology

In addition to the extracted information on research papers that we obtain as described in the previous section, we collect the data about the members of the research project from the member's corner of the REWERSE project. We have constructed an ontology for describing researchers and their involvement in REWERSE. This "REWERSE-Ontology" has been built with by aid of the Protege tool [12]. It extends the Semantic Web Research Community Ontology (SWRC) [15]. Like in the SWRC, the REWERSE-Ontology has three subclasses *person*, *organization*, and *project*. Due to the extension of the SWRC, some more subclasses appear in it, e.g. university, department and institute as subclasses of organization.

4.3 Content Syndication and Personalized Views

All the collected information is then used in a personalization service which provides the end user with an interface for browsing publications of the REWERSE project, and having instantly access to further information on authors, the working groups of REWERSE, recommended related publications, etc.

The personalization service of the PPR uses personalization rules for deriving new facts, and for determining recommendations for the user. As an example, the following rule (using the TRIPLE[14] syntax) determines all authors of a publication:

```
FORALL A, P all_authors(A, P) <-
  EXISTS X, R (
  P['http://.../rewerse#':author -> X]@'http:...#':publications
  AND X[R -> 'http://www.../author':A]@'http:...#':publications).
```

Further rules combine information on these authors from the researcher ontology with the author information. E.g. the following rule determines the employer of a project member, which might be a company, or a university, or, more generally, some instance of a subclass of an organization:

```
FORALL A,I works_at(A, I) <-
  EXISTS A_id,X (name(A_id,A)
    AND ont:A_id[ont:involvedIn -> ont:I]@'http:...#':researcher
    AND ont:X[rdfs:subClassOf ->
              ont:Organization]@rdfschema('http:...#':researcher)
    AND ont:I[rdf:type -> ont:X]@'http:...#':researcher).
```

For a user with specific interests, for example "interest in personalized information systems", information on respective research groups in the project, on persons working in this field, on their publications, etc., is syndicated. As an example, the following rule derives all persons working in specific working groups in the project. Personalization is realized by matching the results of this rule with the individual request, e.g `ont:WG[ont:name -> 'WG A3 - Personalized Information Systems']`.

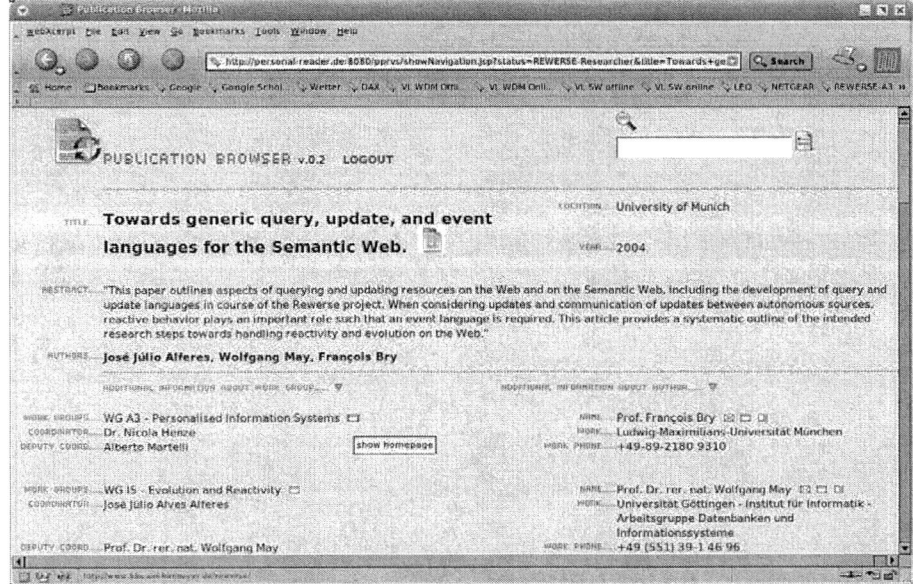

Fig. 2. Data flow of PR

```
FORALL WG,M working_group_members(WG,M) <-
     ont:WG[rdf:type -> ont:WorkingGroup]@'http:..#':researcher
     AND ont:WG[ont:hasMember-> ont:M]@'http://...#':researcher.
```

A screenshot of the PPR application is depicted in fig. 2. The PPR can be accessed via the URL www.personal-reader.de

4.4 Instantiating the Personal Publication Reader

The Personal Publication Reader was implemented using the generic Personal Reader framework. The Personal Publication Reader instance of the Personal Reader consists of the following three components:

- a connector service
- the Personal Publication Reader visualization service
- one or more personalization services

Figure 3 shows the data-flow in the Personal Publication Reader and the services it is composed of:

Step 1: The user logs on to the system and requests information about a publication through the visualization service

Step 2: The visualization service forwards the request to the connector service adding information about where the RDF resource descriptions are located

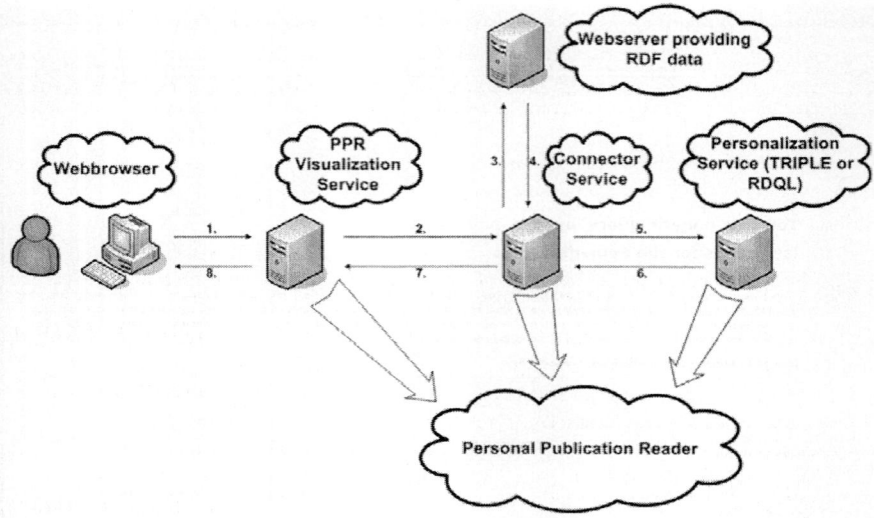

Fig. 3. Data-flow in the Personal Publication Reader

Steps 3 and 4: The connector service retrieves the resource descriptions needed from a web server
  Step 5: The connector service converts - if necessary - the data to a reasoner specific format and forwards it to a personalization service (e.g. based on TRIPLE[14] or Jena's RDF query language RDQL [7])
  Step 6: The personalization service provides the results to the connector service
  Step 7: The connector service converts - if necessary - the results to a specified format and forwards them to the visualization service
  Step 8: The visualization service displays the results to the user in an appropriate manner

5 Conclusion and Future Work

This paper shows an approach for Web data extraction and personalized content syndication for Semantic Web-enabled information systems. For the *Web data extraction process* we use Lixto, an easily accessible technology based on a solid theoretical framework and a visual approach that allows application designers to defined continuously running information agents fetching data from the Web. *Personalized content syndication* has been realized within the Personal Reader Framework, which provides an infrastructure for designing, implementing and maintaining Web content readers. We have demonstrated the realization of our approach in an exemplary application, the Personal Publication Reader. Future research topics in Web data extraction comprise extraction from poorly-structured formats such as PDF, ontology-based wrapping, and techniques for automatic wrapper adaptation. Research on personalized content syndication will explore the application of more complex personalization strategies, and also collaborative approaches for personalization.

References

1. R. Baumgartner, S. Flesca, and G. Gottlob. Visual web information extraction with Lixto. In *Proc. of VLDB*, 2001.
2. R. Baumgartner, M. Herzog, and G. Gottlob. Visual programming of web data aggregation applications. In *Proc. of IIWeb-03*, 2003.
3. S. Flesca, G. Manco, E. Masciari, E. Rende, and A. Tagarelli. Web wrapper induction: a brief survey. *AI Communications Vol.17/2*, 2004.
4. G. Gottlob and C. Koch. Monadic datalog and the expressive power of languages for Web Information Extraction. In *Proc. of PODS*, 2002.
5. N. Henze and M. Herrlich. The Personal Reader: A Framework for Enabling Personalization Services on the Semantic Web. In *Proceedings of the Twelfth GI-Workshop on Adaptation and User Modeling in Interactive Systems (ABIS 04)*, Berlin, Germany, 2004.
6. N. Henze and M. Kriesell. Personalization functionality for the semantic web: Architectural outline and first sample implementation. In *Proccedings of the 1st International Workshop on Engineering the Adaptive Web (EAW 2004), co-located with AH 2004*, Eindhoven, The Netherlands, 2004.
7. Jena - A Semantic Web Framework for Java, 2004. http://jena.sourceforge.net/.
8. SUN - java Server Pages, 2004. http://java.sun.com/products/jsp/.
9. S. Kuhlins and R. Tredwell. Toolkits for generating wrappers. In *Net.ObjectDays*, 2002.
10. A. H. Laender, B. A. Ribeiro-Neto, A. S. da Silva, and J. S. Teixeira. A brief survey of web data extraction tools. In *Sigmod Record 31/2*, 2002.
11. OWL-S: Web Ontology Language for Services, W3C Submission, Nov. 2004. http://www.org/Submission/2004/07/.
12. Protege Ontology Editor and Knowledge Acquisition System, 2004. http://protege.stanford.edu/.
13. RDF Vocabulary Description Language 1.0: RDF S, 2004. http://www.w3.org/TR/2004/REC-rdf-schema-20040210/.
14. M. Sintek and S. Decker. TRIPLE - an RDF Query, Inference, and Transformation Language. In I. Horrocks and J. Hendler, editors, *International Semantic Web Conference (ISWC)*, pages 364–378, Sardinia, Italy, 2002. LNCS 2342.
15. SWRC - Semantic Web Research Community Ontology, 2001. http://ontobroker.semanticweb.org/ontos/swrc.html.
16. WSDL: Web Services Description Language, version 2.0, Aug. 2004. http://www.w3.org/TR/2004/WD-wsdl20-20040803/.
17. XML: extensible Markup Language, 2003. http://www.w3.org/XML/.
18. XML-based RPC: Remote procedure calls based on xml, 2004. http://java.sun.com/xml/jaxrpc/index.jsp.

Design of Vehicle Information Management System for Effective Retrieving of Vehicle Location[*]

Eung Jae Lee and Keun Ho Ryu

Database Laboratory, Chungbuk National University, Korea
{eungjae, khryu}@dblab.chungbuk.ac.kr

Abstract. Vehicle management systems have been developed, which is based on conventional database. However, previous systems cannot efficiently retrieve location data of vehicles, because conventional databases did not take into consideration about property of moving object data such as continuously changing location overtime. In this paper, we design the vehicle information management system that is able to manage and retrieve vehicle locations efficiently in mobile environment. Our proposed system consists of vehicle information collector, vehicle information management server, and mobile clients. The system is able to not only process spatiotemporal queries related to locations of moving vehicles but also provide moving vehicles' locations which are not stored in the system. The system is also able to manage vehicle location data effectively using a moving object index.

1 Introduction

Due to progress in location positioning devices such as GPS(Global Positioning System), mobile users are able to obtain their location more precisely in real time. Therefore mobile user is able to retrieve various information services related to their location at any time. The examples using the location information are transportation vehicle management, air traffic control. Especially vehicle location tracking system(VLTS), which monitors the vehicle's position in a control center in real-time, is the representative research. However previous VLTSs use the traditional commercial DBMS, and cannot efficiently process the query related to the moving object location, which continuously changes over time [1].

In this paper, we propose vehicle information management system which manages location of vehicle in mobile environment. The proposed system consists of three parts: vehicle information collector, vehicle location management server, and mobile client. Vehicle information collector gathers location information of vehicle from the mobile client. Vehicle location management server stores and manages location of vehicle, and provides variety of information to mobile clients related to their location. Mobile client sends their location to server, and displays information retrieved from the server on PDA(Personal Digital Assistant). In the proposed system, we modify TB-Tree[2] for indexing trajectory data of moving object.

---

[*] This work was supported by the RRC program of MOST and KOSEF.

The rest of the paper is as follows. Section 2 describes previous works related to vehicle location tracking system. Section 3 presents system architecture of the proposed method and algorithms. Section 4 and 5 show implemented system, and give a performance evaluation of the modified index. Finally, Section 6 gives a conclusion and shows directions for future works.

2 Previous Works

The vehicle location management system monitors the location and state of the moving vehicle in real-time, and displays to client system using map data for checking operation status of vehicle. The representative vehicle location management systems are Commercial Vehicle Operations(CVO)[3], Advance Public Transport System(APTS)[4], and vehicle management and control system of EuroBus[5].

CVO efficiently manages cargo/cargo vehicles in order to reduce logistic cost, prevent accident, and promote emergency measures. CVO consists of two parts: Freight and Fleet Management System(FFMS), Hazardous Material Monitoring System (HMMS). FFMS traces cargo/cargo vehicles and forwards variety of information to the drivers to prevent empty cargo and to figure out the optimal interval time between vehicles. HMMS administrates vehicles with hazardous cargo and traces their location. APTS supports information related to location of vehicle for public transport in U.S.A. APTS consists of GIS(Geographic Information System), AVL(Automatic Vehicle Location) system, APC(Automated Passenger Counters), TOS(Transit Operations Software). In the APTS, AVL system plays a role to position the location of vehicle equipped with GPS, and transmits location data to central center at regular intervals. In the EuroBus, the whole information related to bus equipped with receiver sends to the central center, and central center supports related information as well as location about bus to the company and bus stop.

There are several researches on vehicle managing system such as DOMINO[1,6] CHOROCHRONOS[2,7], Battle Field Analysis[8,9].

Prototype of DOMINO project is focused on predicting future location of moving objects based on present location of the object, speed, and direction. However this prototype does not support history information of past movement of moving objects. In the CHOROCHRONOS project, researches on data modeling of moving vehicle, indexing method, and vehicle management system which manages the location and trajectory of vehicle ware performed. However this project did not make a prototype like DOMINO until now. The battlefield analysis system defines moving units and tanks as moving vehicle, which have property of moving point object. This system is focused on predicting the motion of moving units in the simulation battlefield. Therefore this system cannot properly deal with real-time received location information in mobile environment.

3 Proposed Methods

In this section, we propose overall system architecture for vehicle information management system, and describe each module in the proposed system. And we present how to modify indexing method and deal with index file for management of moving objects.

3.1 Overall System Architecture

The proposed system consists of Vehicle Information Collector(VIC), Vehicle Location Management Server(VLMS), and Mobile Client(MC) as shown in Fig. 1. VIC converts location information, which is received from GPS into TM coordinate, and transfers location to vehicle location management server. VIC is composed of data receiver, TM coordinate converter, packet data converter, data transceiver, and data store.

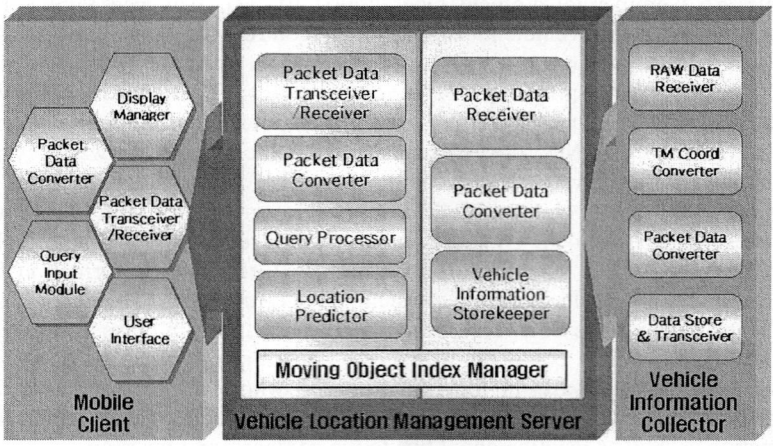

Fig. 1. Architecture of vehicle information management system

VLMS stores and manages vehicle location data, and transfers query result about vehicle information to mobile clients. VLMS consists of packet data transceiver/receiver, packet data converter, query processor, location predictor, vehicle information storekeeper, and moving object index manage. MC requests information about vehicle to the VLMS in real-time, and displays information received from server on PDA.

In the VLMS, packet data receiver gathers vehicle information transferred from the VIC periodically. Packet data converter transforms packet data into moving object data that is able to manage in index manager. Because moving object always changes its position, all locations generated by vehicle cannot be stored into the database. Therefore the location of moving object is sampling over time using distance or time-interval criteria. When the user requests query about location that is not stored in the system, the system cannot support properly query result. In the proposed system, location predictor estimates past and future location data, which is not stored in the system.

3.2 Management of Moving Vehicle Location

The proposed method uses moving object indexing method for managing enormous volume of vehicle location data as shown in Fig.2. Moving object index manager is composed of past location index and current location index. Current location index is implemented using R*-Tree[10]. When the location of vehicle is changed, previous location data is converted into line segment for inserting past location index.

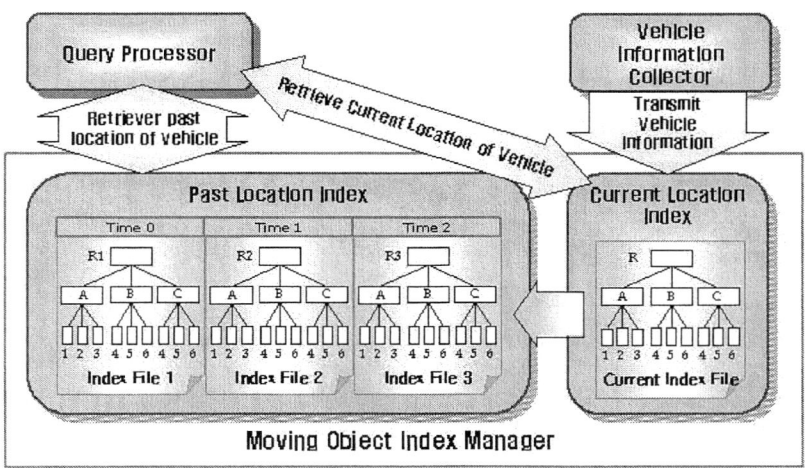

Fig. 2. Structure of Moving Object Index Manager

For managing trajectory of moving vehicle efficiently, we modify the TB-Tree which is able to support concurrency control and multi-version index file. TB-Tree is proposed for strict preservation of trajectories so that the leaf node only contain line segments that belong to the same trajectory. Since spatial proximity is not considered, spatial discrimination decreases and the classical range query cost increases. However trajectory preserver helps in efficient answering of pure spatiotemporal queries.

In the mobile environment, multi-user accesses data in server simultaneously. So VLMS needs to support concurrency control. The proposed system adopts TDIM(Top Down Index Modification)[11] for supporting concurrency control proposed by Ravi Kanth et. al.

Due to the property of continuously moving, vehicle generates a great quantity of location data over time. Increasing data causes to grow the index size and deteriorates query performance. In this paper, we use multi-version framework for managing index file. Index manager stores locations of moving vehicle into new index file day by day. If time interval of new data is between two days, new data splits into two data by time interval, and stores into two pages. For example, if new data is reported between Feb. 10 and Feb. 11, the data is separated into two data that time stamps are Feb. 10 and Feb. 11, and then inserted into index file. Multi-version framework of index file causes to deteriorate query performance in the case of query for time interval more than two days. However, because amount of location data is rapidly increased over time in the mobile environment, it makes it easy to manage and migrate index file. Multi-version framework also has a merit to increase performance of query which time interval is within one day, because tree-height becomes lower in each page file.

3.3 Query Processing

Query processor extracts date from the query, and calls the search function for each date. Algorithm1 shows how to process the range query for retrieving vehicle located in specified range at specified time interval. Time interval of query is extracted from input parameter. If time interval is within one day, index retrieves vehicle location from corresponding single page file. If time interval is more than two days, index searches data from all page files overlapped with time interval. The past location of moving vehicle is append-only data. There is no delete operation, and each node does not need to coalesce with other node.

```
Algorithm 1. RangeQuery(VehicleID, st, sx, sy, et, ex, ey)
INPUT   VehicleID is time stamp for query,
        (st,sx,sy)~(et,ex,ey) is range for retrieving trajectory
OUTPUT  Result: Retrieved data
BEGIN
   Extract start and end date from (st, et)
   IF st == et THEN
      PageFile is page file corresponding to time interval
      Result = retrieve data from PageFile
   ELSE
      WHILE (st != et)
         PageFile is page file corresponding to st
         Result = retrieve data from PageFile
         Increase st
      END WHILE
   Return Result
END
```

3.4 Moving Object Index Manager

If new trajectory data is reported, trajectory information is changed into MBB(Minimum Bounding Box). Then new data is inserted into index. Algorithm2 shows insertion algorithm of the proposed system. As shown in Algorithm2, when new data is inserted into the system, index manager locks only one node in TB-Tree for processing concurrency control.

```
Algorithm 2. Insert(VehicleID, st, sx, sy, et, ex, ey)
INPUT   VehicleID is time stamp for query,
        (st,sx,sy)~(et,ex,ey) is trajectory of vehicle
BEGIN
    IF time interval of new data is more than two days THEN
       Split new data into two new-data by each day
    END IF
    Choose leaf path for inserting new data
    Node = Root node
    WHILE (Node is not leaf node)
       Obtain shared lock
       IF MBB needs to be modified THEN
          Release shared lock
          Re-obtain the exclusive lock
          Update MBB
       END IF
       Release the lock on the Node
       Set Node to its child node
```

```
END WHILE
   Obtain exclusive lock
   Add new data into leaf node
   Release the lock on the leaf node
END
```

4 Implementation and Analysis

In this section, we present the user interface of the proposed system. The vehicle information management system is implemented using JDK 1.3(JAVA), MS-SQL 2000.

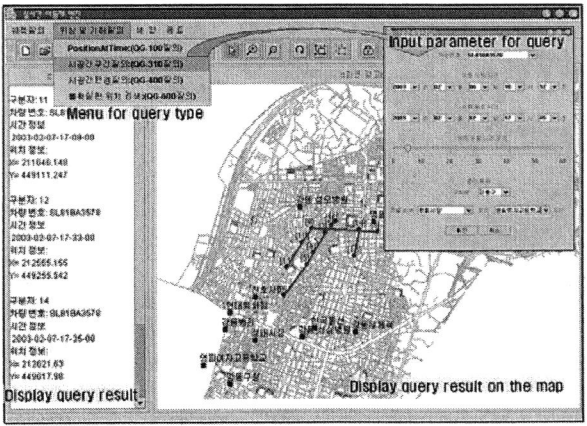

Fig. 3. Display result of range query

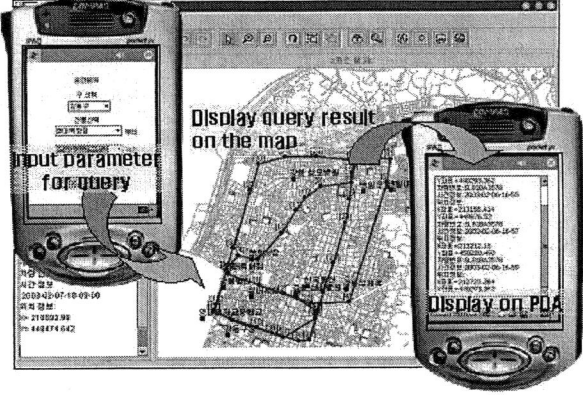

Fig. 4. Display query result on PDA

4.1 Query for Vehicle Information

Query is requested from both VLMS and MC. Fig. 3 shows the result of spatiotemporal range query. If user request query, VLMS search vehicle information, and then display query result with a text and map data.

User also request query from mobile client such as PDA. Fig. 4 shows execution of query from the mobile client. Mobile client sends parameter for query such as vehicle unique ID, query range to the VLMS, then VLMS executes query processing. And VLMS sends query result to mobile client. Finally, mobile client displays query result on PDA.

4.2 Comparison with Previous Works

Prototype of DOMINO project deals with only uncertain future location of vehicle and does not store history information of moving object. So prototype of this project cannot support properly services related past location. CHOROCHRONOS project presented application scenario applied in transport management system based on GPS. However they didn't make a prototype until now. The battlefield analysis system is focused on predicting the motion of moving units and tanks for decision-making in the simulation field. However the system cannot process real-time location data generated in mobile environment. The proposed system is able to deal with trajectory data as well as current location of moving object. And the indexing method used in the system supports concurrency control for managing real-time data.

5 Experimental Results for Modified Moving Object Index

In this section, we present experimental result of modified moving object index. Due to the lack of real dataset, we use the synthetic datasets using CitySimulator[12]. City simulator is a scalable, three-dimensional city that enables generating dynamic spatial data simulating the motion of up to 1 million people by modeling a scalable, three-dimensional city. The parameters of the generator are given the default value. And we generate datasets by simulation that 25, 50, 100, 250, 500, 1000 mobile users report each their location during 500 iteration. Each set of queries consists of 100 individual queries, and each experimental result accumulates for 100 queries.

5.1 Index size and Creation Time

Fig. 5 shows insertion cost of the modified index as the various page sizes. As shown in the figure, the cost of insertion operation is smaller as the page size is larger. If page size is larger, then fewer nodes is need for the same dataset. Fewer number of nodes cause to less height of tree. Therefore overall insertion and query cost is reduced.

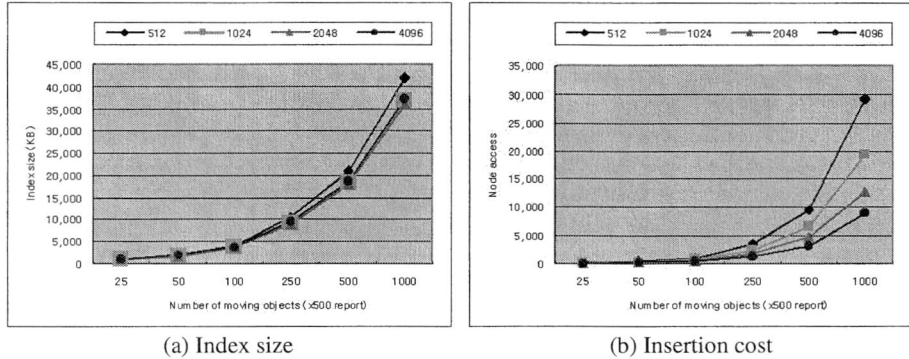

(a) Index size   (b) Insertion cost

Fig. 5. Index size and insertion performance

5.2 Range Query

Range query is one of the important query types for the mobile environment. Fig. 6 shows the result of the range query. Dataset used for range query is set to 10% range of overall data domain. We experiment on various page sizes: 512Bytes, 1KB, 2KB, 4KB. Fig. 6(a) shows the result of range query at the user specified time stamp. And Fig. 6(b) shows the result of spatiotemporal range query. As shown in the figure, query performance is influenced by page size.

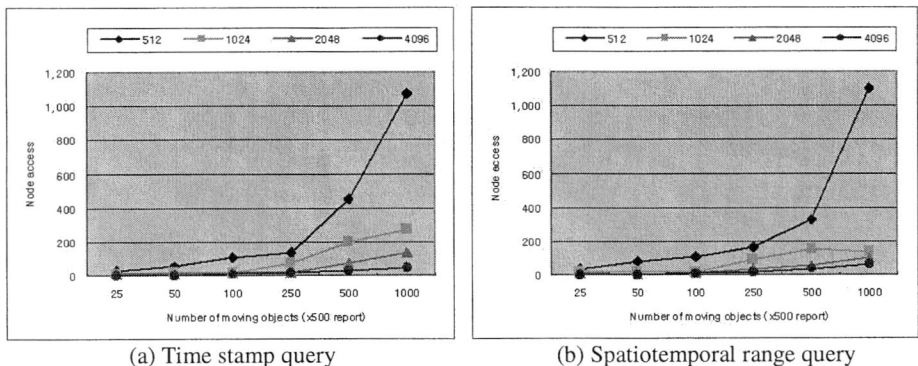

(a) Time stamp query   (b) Spatiotemporal range query

Fig. 6. Performance of range query

5.3 Multi-version Page File

In the proposed system, index file splits into multi-version file. For the performance evaluation of the multi-version file, we generate test dataset assumed that 100 vehicles report their location for three days, and 200 times in every day. Fig. 7 and Fig.8 show the performance of index between single version file and multi-version file. As shown in Fig. 7, multi-version file has lower tree-height than single version file. Therefore query performance of multi-version file is better than single-version file for the query within one day. However, when the time interval of query is more than two days, the performance of single version file is better than multi-version file. Fig. 8(a) shows the insertion performance between two kinds of methods. As shown in the figure, insertion performance of multi-version file is better than single version file. This result is caused by lower height of index tree in each index file of multi-version file than single version file.

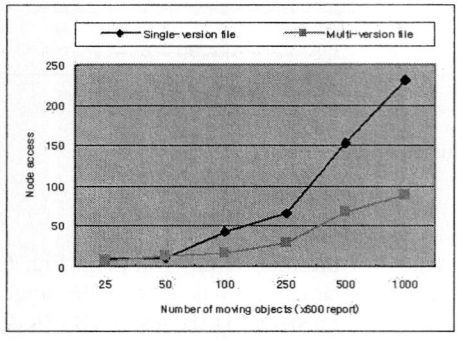
(a) Query with in one day

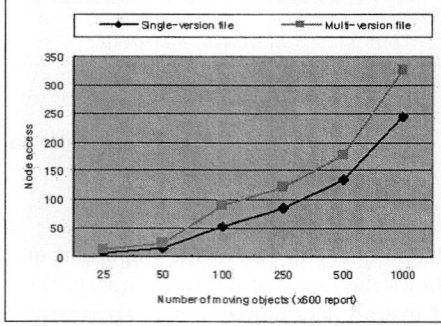
(b) Time interval for more than two days

Fig. 7. Performance of range query

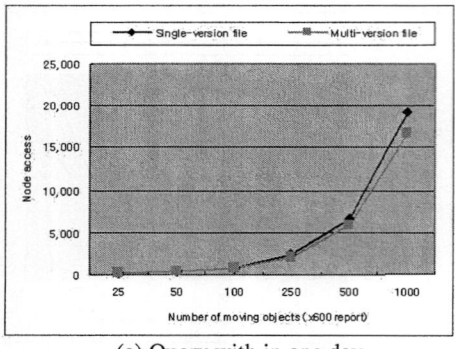
(a) Query with in one day

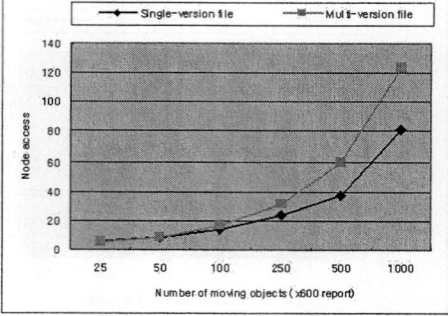
(b) Time interval for more than two days

Fig. 8. Performance of range query

6 Conclusions and Future Work

Recently, it is rapidly increasing interest on location-based services in mobile environment. In this paper, we proposed vehicle information management system which is capable of managing location of vehicle in real-time. For efficiently managing location of moving vehicle, we expand TB-Tree to support concurrency control. And the expanded indexing method adopts multi-version framework for managing index file. The experimental result showed that the proposed indexing method using multi-version framework outperformed original TB-Tree.

Currently we are focusing on the improving of query performance by increasing the space discrimination ability of the moving object indexing method. And we are developing enhanced mobile client, which supports displaying map data on PDA.

References

1. O. Wolfson, B. Xu, S. Chamberlain, and L. Jiang, "Moving Objects Databases: Issues and Solutions", *Proc. Of the 10th Intl. Conf. on Scientific and Statistical Database Management(SSDBM'98)*, Capri, Italy, 1998.
2. D. Pfoser, C. S. Jensen, and Y. Theodoridis, "Novel Approaches in Query Processing for Moving Objects", *Proc. of the Intl. Conf. on VLDB*, 2000.
3. IVHS America, Strategic Plan for Intelligent Vehicle-Highway Systems, Report No:IVHS-AMER-92-3, U.S. DOT, 1992.
4. Federal Transit Administration, "Advanced Public Transportation Systems: The State of the Art Update '96", U.S. Department of Transportation FTA-MA-26-7007-96-1, January 1996.
5. EUROBUS, "Case Study on Public Transport Contribution to Solving Traffic Problems", EUROBUS Project, Deliverable 18(version 2.0), 1994.
6. P. Sistla, O. Wolfson, S. Chamberlain, and S. Dao, "Modeling and Querying Moving Objects", *Proc. Of the 13th Intl. Conf. on Data Engineering(ICDE'97)*, Birmingham, UK, April 1997.
7. S. Saltenis, C. S. Jensen, S. Leutenegger, and M. Lopez, "Indexing the Positions of Continuously Moving Objects", *Proc. of the ACM SIGMOD Conf.*, 2000.
8. K. H. Ryu, and Y. A. Ahn, "Application of Moving Objects and Spatiotemporal Reasoning", *Time Center TR-58*, 2001.
9. Y. A. Ahn, J. S. Park, and K. H. Ryu, "Moving Object Inference Engine and Applying to Battlefield", *Proc. Of the Intl. Conf. on Artificial Intelligence*, Las Vegas, Nevada, USA, June 2003.
10. N. Beckmann, H. P. Kriegel, R. Schneider, and B. Seeger, "The R*-tree: An Efficient and Robust Access Method for Points and Rectangles", *Proc. of the ACM SIGMOD Conf.*, Atlantic City, New Jersey, USA, May 1990.
11. K. V. Ravi Kanth, D. Serena, and A. K. Singh, "Improved Concurrency Control Techniques for Multi-dimensional Index Structures", *Proc. of the 1st Merged Intl. and Symposium on Parallel and Distributed processing(IPPS/SPDP)*, 1998.
12. http://www.alphaworks.ibm.com/tech/citysimulator, 2001

Context-Aware Workflow Language Based on Web Services for Ubiquitous Computing

Joohyun Han, Yongyun Cho, and Jaeyoung Choi

School of Computing, Soongsil University,
1-1 Sangdo-dong, Dongjak-gu, Seoul 156–743, Korea
{jhhan, yycho, choi}@ss.ssu.ac.kr

Abstract. The services for a ubiquitous computing environment have to automatically provide users with adaptive services according to dynamically changing context information, which is obtained from both the users and their environment. Workflows used in business processes and distributed computing environments have supported service automation by connecting many tasks with rules and/or orderings. To adapt these workflows to ubiquitous computing, we must specify the context information on their transition conditions. In this paper, we propose uWDL, Ubiquitous Workflow Description Language, to specify the context information on the transition constraints of a workflow in order to support adaptive services. And it is designed based on Web services, which are standardized and independent of heterogeneous and various platforms, protocols, and languages. In order to verify the effectiveness of uWDL, we designed and implemented a scenario described with uWDL. And we demonstrated that the uWDL system provides users with autonomic services in ubiquitous computing environments.

1 Introduction

Ubiquitous is a Latin word, which means "Existing or being everywhere at the same time." The concept of ubiquitous computing was introduced by Mark Weiser in 1988. He described ubiquitous computing as "The method of enhancing computer use by making many computers available throughout the physical environment, but making them effectively invisible to the user [1]." This means that the computing environments are absorbed into the physical world and integrated in everyday life [2]. In such a ubiquitous environment, more sensibility is required than in traditional computing environments. Deriving context information such as the location, status, and actions, of users, devices, and network by sensing physical environment is necessary to provide users with context-aware services based on this context information.

The workflow model in [3] supports service automation through a sequence of rules in processing tasks. It has been successively applied to traditional computing environments such as business processes and distributed computing in order to perform service composition, flow management, parallel execution, and time-driven services. It is also the task of the workflow to provide a service to the right person or the right application at the right time so that the service for a specific task can be carried out [4]. Users of ubiquitous computing environments may want to receive services in an appropriate form, at the appropriate time, and without user intervention on dynamically changing environments [5, 6]. In order to support automatic service in these environments, we need to adapt workflows to ubiquitous computing. However it must satisfy at least the following two requirements: first, a context-aware state transition is required to provide users with appropriate services according to a user's current situation. Second, a platform- and language-independent standard service interface needs to integrate, manage, and execute ubiquitous applications, which consist of heterogeneous protocols on various platforms [7].

In this paper, we propose uWDL (Ubiquitous Workflow Description Language), which specifies the context information on the transition conditions of workflow services to provide users with an adaptive service for a user's current situation. It specifies context information as the perspective of rule-based reasoning which can effectively infer his/her current situation in a simple and flexible way. Furthermore it is designed based on Web services interfaces, which are not only already specified and widely used, but independent of various platforms, protocols, and languages. By interpreting a scenario described with uWDL and executing the scenario, the uWDL system can effectively provide users with context-aware and autonomic services.

2 Related Work

There have been many studies to enable applications to become capable of context-awareness in ubiquitous computing environments. Context-aware applications such as PARCTAB [8], Aura [9], and One.world [10] define context-triggered action as 'if-then' rules. They support context-aware services through context-triggered actions in execution time. However, these studies focus on recognizing a user's situation, within only one service not in a flow of services. In a ubiquitous computing environment, there exist many services for various purposes, and the specific services which have the flow of tasks needed to accomplish their purposes must be selected based on the user's current situation. Because uWDL is a workflow language, it can easily represent a flow of adequate services and select the service according to the user's situation.

Gaia [11] supports a service environment in which ubiquitous applications can communicate context information with each other. But because it is based on CORBA middleware, it depends on a specific protocol which is not widely used. LuaOrb, that is Gaia's script language, can instantiate applications and interact with execution nodes to create components and easily glue them together, but it can't express dependency or parallelism among the services because it describes only a sequential flow of specific services. Because uWDL is a workflow language based on Web services which are platform- and language-independent standard service interfaces, it can express dependency and parallel execution among the services in the heterogeneous ubiquitous computing environments.

BPEL4WS [12], WSFL [13], and XLANG [14] are Web service-based workflow languages for business processes and distributed computing environments. They use the results of previous services and event information for a current service's transition conditions. They support service transition, and they use XML-typed messages defined in other services using XPath. However, they have some problems adapting a ubiquitous computing environment. First, in ubiquitous computing, workflow services must not simply use the results of specific service or event information but instead context information which can be obtained from both users and their environment as a service's transition conditions. And context information is a complex data set, which includes data types, values, and relations among the data types. XPath cannot sufficiently describe such diverse context information as this, because it can use only condition and relation operators to decide transition conditions. uWDL uses rule-based triplet elements - subject, verb, and object - in order to express high-level context information as transition conditions, which existing workflow languages are unable to support. uWDL can efficiently express the high-level context information using the triplets, and offer appropriate services according to the current situation.

3 Requirements of Workflow in Ubiquitous Computing

3.1 Context-Aware Workflow

WfMC (Workflow Management Coalition) states that a workflow expresses flows of subtasks until a process is completed using standardized methods [3]. Between the subtasks in a workflow, there exist various relationships such as dependency, ordering, and concurrency. Workflows express flows of subtasks in a standardized method using a workflow language. A workflow management system manages and controls flows of subtasks using state-transition constraints specified in the workflow.

Figure 1 shows the evolution of workflow. Workflow was initially applied to the automation of business processes. A business process is very static, and its requirements and functionalities are relatively simple. Currently, the workflow has been applied to service automation in business processes, distributed computing environments, and especially Grid services. In the near future, the workflow

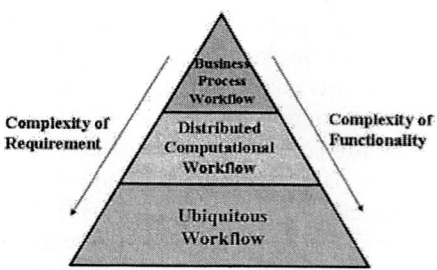

Fig. 1. Evolution of workflow

may be evolved to support service automation in ubiquitous computing, since the workflow is the best means to resolve automation problems in dynamically changing environments. As the complexity of requirements and functionalities increase, the workflow must be evolved.

Workflow in ubiquitous computing environments depends on context information obtained from physical environments and provides context-aware services automatically based on the information retrieved. Ubiquitous workflow needs to specify ubiquitous context information as state-transition constraints.

3.2 The Need of Web Services as a Middleware for Ubiquitous Computing

Web services are middleware, which allow us to make remote procedure calls (RPCs) for an object over the Internet or network. Web services aren't the first technology to allow us to do this, but they differ from previous technologies in that they use platform-neutral standards such as HTTP, XML, and SOAP which make the services able to be used in any language, with any protocol, and on any platform. A workflow in ubiquitous computing environments has to integrate, manage, and execute many applications implemented by different languages on various platforms and protocols. Therefore, it requires Web services interfaces which are standardized and independent of heterogeneous and various platforms, protocols, and languages. The characteristics of Web services' platform-neutral standards satisfy workflow requirements in ubiquitous environments.

4 A Ubiquitous Workflow Description Language

Although current workflow languages such as BPEL4WS, WSFL, and XLANG can specify the data flow among services based on Web services, these workflow languages do not support the ability to select services using context, profile, or event information in ubiquitous computing environments. Therefore, it is difficult to express relationships among the services using traditional workflow languages in ubiquitous environments.

uWDL (Ubiquitous Workflow Description Language) is a Web service-based workflow language that describes service flows and provides the functionalities to select an appropriate service based on high-level contexts, profiles, and events information, which are obtained from various sources and structured by Ontology [15]. To provide these functionalities, uWDL specifies the context and profile information as a triplet of subject, verb, and object for rule-based reasoning, which can effectively represent the situation in a simple and flexible way. Figure 2 shows the schema structure of uWDL.

4.1 <node> Element

The <node> element points to an operation that provides a functionality of Web services in ubiquitous environments. Web services use WSDL (Web Services Definition Language) to describe the port types and operations of specific

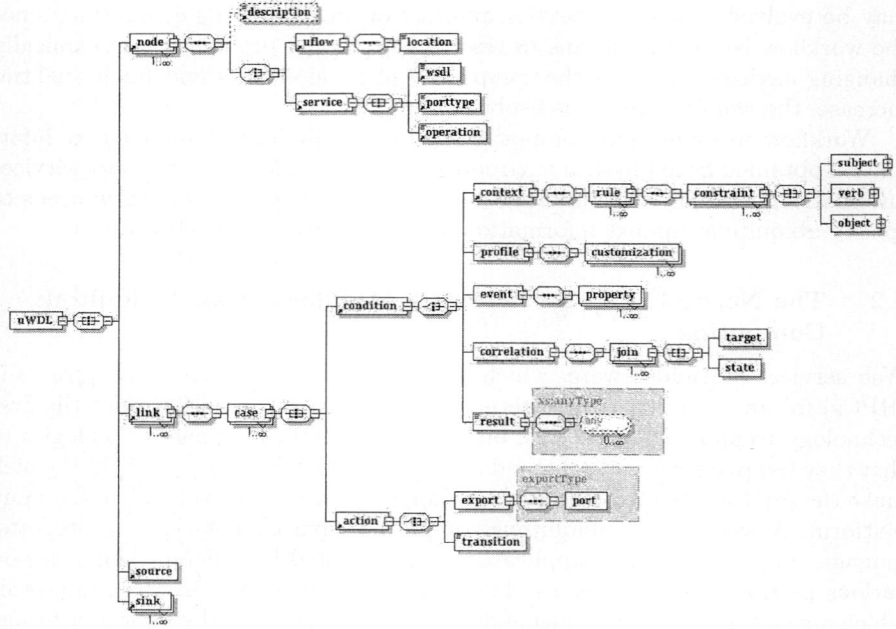

Fig. 2. uWDL Schema

Web services. So, uWDL uses the <service> element and subelements of the <service> - <wsdl>, <porttype>, and <operation> - to describe the service location, type, and operation of a specific Web service. And the <uflow> element directs a reference to another uWDL document.

4.2 <link> Element

The <link> element is the most important part of the uWDL. It specifies context, profile, and event, aggregated from ubiquitous environments, and defines the flow of services. The <link> element is composed of <condition> and <action> elements. The <condition> element uses <context>, <profile>, and <event> subelements to specify the context, profile, and event status of a specific node, respectively. If the calculated value of the status satisfies a given condition, the action described in the <action> element is performed. The <action> element consists of <export> and <transition> elements, where <export> has a control link and a data link according to its attribute, and <transition> specifies the state change of the current node.

The <condition> element is responsible for selecting the appropriate service based on the context, profile, and event information. The important element is <context>, which contains <constraint> to specify context information standardized by Ontology. The <constraint> element has the subelements of <subject>, <verb>, and <object>. The information of subject, verb, and object are provided by entities [16], which represent abstract information such as location, computing device, user activity, and the social situation of diverse domain in ubiquitous computing environments. The <constraint> element expresses a context based on the relationship of the subject and the object, which are instances of the entity. The <subject>, <verb>, and <object> elements also have an attribute of 'type'. The type attribute represents a property of an entity in a domain. The composite attribute of the <constraint> element has an attribute value of 'and', 'or', and 'not'. By using these values of the attribute, it is possible to express the relationship among the simple contexts and describe a high-level complex context. The <rule> element means a set of the <constraint> elements, and represents the high-level expression to infer a social situation.

5 The Architecture for Handling Contexts in uWDL

A uWDL document designed for a specific scenario must be translated and executed to provide an adaptive service for a user's situation. To achieve this purpose, we need a process to manipulate contexts aggregated from a sensor network. Figure 3 shows the architecture for handling the contexts expressed in uWDL.

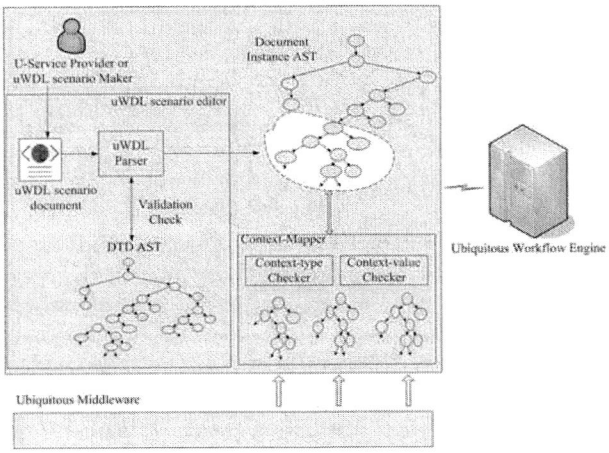

Fig. 3. The architecture for handling the context in uWDL

Any information obtained from ubiquitous middleware is objectified with an entity. An entity represents a person, a place or an object relevant to the interaction between a human and the corresponding virtual computer world. And a context means information used to characterize the situation of entities. An entity categorized into a specific domain has a type and a value. The uWDL parser parses an uWDL scenario document and produces a DIAST (Document Instance Abstract Syntax Tree) [17, 18] as a result. A DIAST represents a syntax of a scenario document. A DIAST is used to compare contexts expressed in a scenario with entities aggregated from a sensor network to verify their coincidence.

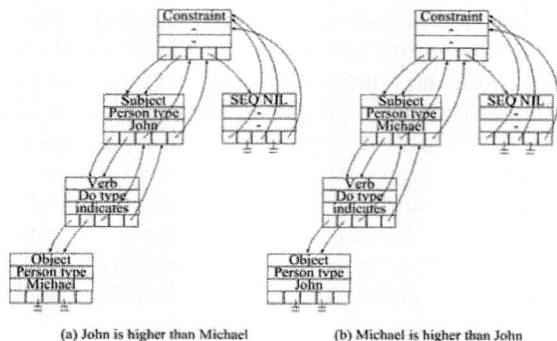

Fig. 4. Construction of an ambiguous subtree for "John indicates Michael"

A context consists of a triplet of {subject, verb, and object} in sequence. A context is described with one or more constraint elements, and each constraint is represented by a triplet. They are used as nodes to construct a DIAST's subtree. In Figure 3, the partial subtree in dotted lines indicates a subtree that makes up context constraints in the scenario.

The context mapper extracts types and values from objectified entities aggregated from the sensor network, and composes a subtree which consists of subject, verb, and object information. It then compares a type and a value of an entity with those of the constraint element in the DIAST subtree, respectively. If a type in the entity matches with its counterpart in the constraint element, the context mapper regards it as a correct subelement of the constraint element. If each entity has the same type, it may be ambiguous to decide a context's constraint according to its entity type only. The problem can be resolved by comparing the value of the objectified entity with that of the constraint element in the DIAST subtree.

For example, assume that there is a scenario, "John indicates Michael." The context mapper may receive entities objectified as (PersonType, John) and (PersonType, Michael). By comparing entity types with the DIAST's subtree for the scenario, the subtree can be drawn as Figure 4(a). Because the entity value of (PersonType, Michael) in Figure 4(b) differs from that of subject element in the DIAST's subtree for the scenario, it is not possible for Figure 4(b) to exist as a subject element. Consequently, the context mapper is designed to consider not only entity type but also its value to resolve the ambiguity and suggest a proper service.

6 Experiments

For testing, we developed a uWDL scenario editor with which we created a scenario of an office meeting. In this section, we show a process to decide the state transition according to the context information. The purpose is "Implementing a service which prepares an office meeting automatically according to a user's schedule." Figure 5 conceptualizes the scenario.

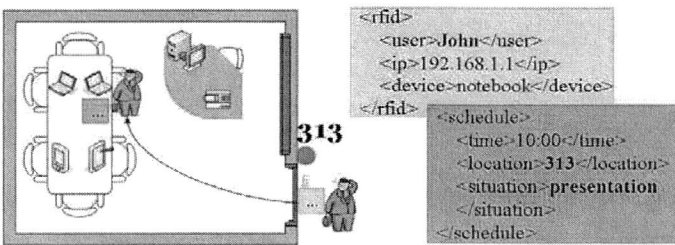

Fig. 5. Office meeting preparation scenario

The scenario designed by using the uWDL scenario editor is as follows: "John records an appointment on his notebook computer that there is a presentation in Room 313 at 10:00 AM. John moved to Room 313 to participate in the meeting at 9:40 AM. There is a RFID sensor above room 313's door, and John's basic context information (such as name and notebook's IP address) is transmitted to a server. If the conditions, such as user location, situation, and current time, are satisfied, then the server automatically downloads his presentation file and executes a presentation program."

Figure 6 shows a uWDL instance document created with the uWDL scenario editor for the above scenario. The uWDL scenario editor is composed of available services, constraint info, and so forth. The available services show a list of the Web services available in the current environment. The structure window shows the structure of DIAST's subtree for the constraint element highlighted in the editing window. The constraint information presents a list of the entity types aggregated from a sensor network within the current environment, and a list of the entity values registered in each entity type. If the context mapper receives entities objectified as (SituationType, presentation), (UserType, Michael), (UserType, John), and (LocationType, 313) from the sensor network during processing of the scenario, it compares the entities' types and values with those of the constraint element highlighted in the DIAST subtree of Figure 6. At the moment, because entity (UserType, Michael) is not suitable for anyone of the subtree's elements, it is removed. As a result, the entity (UserType, John) is selected as a context for the DIAST's subtree.

7 Conclusion

In this paper, we proposed uWDL (Ubiquitous Workflow Description Language), which can describe service flows for a ubiquitous computing environment. Because uWDL is based on Web services, it is able to integrate, manage, and execute various and heterogeneous services in ubiquitous environments. In a

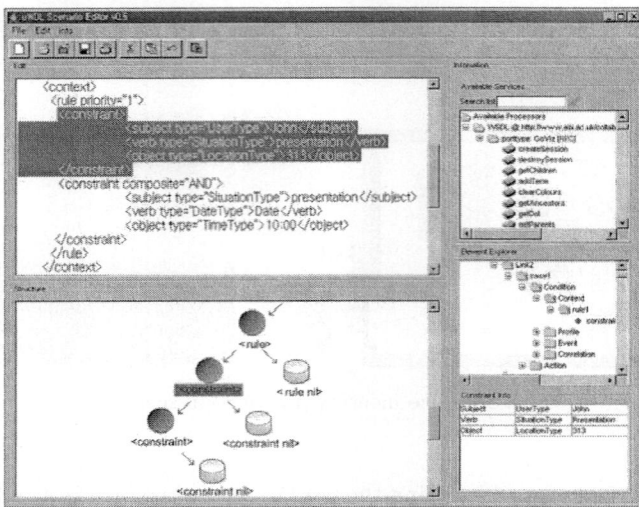

Fig. 6. uWDL Scenario Editor

ubiquitous computing environment, a service engine needs a method to recognize a user's context and situation information to decide the state transitions of a service flow. As explained in section 2, current workflow languages are for business processes and distributed computational workflows, and they are not suitable for describing the context and situation information needed for service transitions in a service flow of ubiquitous computing environments.

We designed uWDL so that it can specify the context information on transition constraints of a service workflow in ubiquitous computing environments. So, users are provided with an appropriate service according to the user's context information. We implemented an uWDL scenario editor and developed a sample scenario described with the uWDL. And we demonstrated that the uWDL system provides users with autonomic services in ubiquitous computing environments. In the near future, we will expand the uWDL schema to express more detailed situations by assigning semantic information to Web services.

Acknowledgements

This research is supported by the Ubiquitous Autonomic Computing and Network Project, the Ministry of Information and Communication (MIC) 21st Century Frontier R&D Program in Korea.

References

1. M., Weiser: Some Computer Science Issues in Ubiquitous Computing. Communications of the ACM. Vol.36. No.7 (1993) 75–84
2. M., Weiser: The Computer for the 21st Century. Sci. Amer. (1991)

3. D., Hollingsworth: The Workflow Reference Model. Technical Report. TC00–1003. Workflow Management Coalition (1994)
4. Wil, van, der, Aalst, Kees, van, Hee: Workflow Management, Models, Methods, and Systems. The MIT Press. pp.147 (2002)
5. D., Saha, A., Mukherjee: Pervasive Computing: A Paradigm for the 21st Century. IEEE Computer. IEEE Computer Society Press (2003) 25–31
6. Guanling, Chen, David, Kotz: A Survey of Context-Aware Mobile Computing Research. Technical Report. TR200381. Dartmouth College (2000)
7. Mack, Hendricks, Ben, Galbraith, Romin, Irani, et al.: Professional Java Web Services. WROX Press. (2002) 1–16
8. R., Want, B., N., Schilit, N., I., Adams, et al.: The ParcTab Ubiquitous Computing Experiment. Technical Report CSL–95–1. Xerox Palo Alto Research Center (1995)
9. D., Garlan, D., Siewiorek, A., Smailagic, P., Steenkiste: Project Aura–Towards Distraction Free Pervasive Computing. IEEE Pervasive Computing (2002)
10. Robert, Grimm: System support for pervasive applications. Phd thesis. University of Washington (2002)
11. Manuel, Roman, Christopher, K. : Gaia–A Middleware Infrastructure to Enable Active Spaces. IEEE Pervasive Computing (2002) 74–83
12. Tony, Andrews, Francisco, Curbera, et al.: Business Process Execution Language for Web Services. BEA Systems. Microsoft Corp. IBM Corp., Version 1.1 (2003)
13. Frank, Leymann: Web Services Flow Language (WSFL 1.0). IBM (2001)
14. Satish, Thatte: XLANG Web Services for Business Process Design. Microsoft Corp. (2001)
15. Deborah, L., McGuinness, Frank, van, Harmelen, (eds.): OWL Web Ontology Language Overview. W3C Recommendation (2004)
16. Anind, k., Dey: Understanding and Using Context, Personal and Ubiquitous Computing. Vol 5. Issue 1. (2001)
17. Aho, A., V., Sethi, R., Ullman, J., D.: Compilers: Principles, Techniques and Tools. Addison-Wesley (1986)
18. Bates, J., Lavie, A.: Recognizing Substring of LR(K) Languages in Linear Time. ACM TOPLAS. Vol.16. No.3. pp.1051–1077 (1994)

A Ubiquitous Approach for Visualizing Back Pain Data

T. Serif[1], G. Ghinea[1], and A.O. Frank[2]

[1] School of Information Systems, Computing and Mathematics,
Brunel University, Uxbridge, Middlesex, UB8 3PH, UK
{Tacha.Serif, George.Ghinea}@brunel.ac.uk
[2] Department of Rehabilitation Medicine and Rheumatology,
Northwick Park Hospital, Harrow, HA1 3UJ, UK
andrew.frank1@btinternet.com

Abstract. We describe a wireless enabled solution for the vizualisation of back pain data. Our approach uses *pain drawings* to record spatial location and type of pain and enables data collection with appropriate time stamping, thus providing a means for the seldom-recorded (but often attested) time-varying nature of pain, with consequential impact on monitoring the effectiveness of patient treatment regimes. Moreover, since the implementation platform of our solution is that of a Personal Digital Assistant, data collection takes place ubiquitously, providing back pain sufferers with mobility problems (such as wheelchair users) with a convenient means of logging their pain data and of seamlessly uploading it to a hospital server using WiFi technology. Stakeholder results show that our approach is generally perceived to be an easy to use and convenient solution to the challenges of anywhere/anytime data collection.

1 Introduction

The integrated use of telecommunications and information technology in the health sector leads to new challenges in organizing, storing, transmitting and presenting health information in both a timely and efficient manner for effective health-related decision-making. Innovations range from routine hospital information systems [1] to sophisticated AI-based clinical decision support systems [2].

Moreover, in today's information intensive society, consumers of health care want to be better informed of their health options and are, therefore, demanding easy access to relevant health information. Simultaneously, clinicians are eager to exploit advances in telecommunication technology in order to put in practice new methods of data gathering and patient monitoring. Whilst the use of the Internet in this respect is by now traditional [3], it is only recently that wireless technologies have been harnessed to act as tools coming to the aid of patients and clinicians alike.

To this end work has focused on patient monitoring systems and context-aware hospitals. Thus, a patient-monitoring system that utilizes WAP-enabled devices as mobile access terminals is described in [4]. Using this system, authorized users, hospital personnel and patients' relatives, can access a patient's physiological data stored on the hospital's computer. On the other hand, a context-aware hospital mobile prescription system that can identify and react according to the location of tagged items

(PDAs, beds, hospital trolleys), prescribing the correct medication to patients based on their bed identification number is detailed in [5], while a context-aware messaging system, which can download the appropriate data to a doctors' PDA according to its location was depicted in [6]. From a different perspective [7] examined the use of small-screened mobile devices for healthcare services and showed no significant difference between the use of PDAs and laptops when they are used for nursing documentation.

In this paper, we present the implementation and experiences of a wireless-enabled monitoring system for back pain patients. The motivation behind our work lies in the fact that, whilst back pain is a worldwide problem with considerable implications on countries' health-care budgets and national economies, there is a relative paucity of tools for the collection and digitization of back pain data. Moreover, the disabling pain experienced by back pain sufferers means that in many cases such data collection cannot take place unless medical personnel is present at the patient's domicile, a situation which in most cases is both unrealistic and impractical. The consequence of this state of affairs is that there is under-reporting of back pain data, as well as an almost total lack of available, continuously-polled back pain data, notwithstanding the evidence in support of the fact that, for chronic back pain sufferers, pain has a time-dependent nature [8], and that this relation is as of yet still not completely understood. Accordingly, the structure of this paper is as follows: Section 2 presents an overview of the area of back pain, while Section 3 reviews work done on the visualization of back pain data. Such work provides the foundation for our project, which is described in detail in Section 4. Lastly, Section 5 presents the results of an evaluative study of our back-pain tool, while the implications of our work are elaborated upon in Section 6, where conclusions and possibilities for future work are identified.

2 Back Pain

Back pain is a worldwide experience. Disabling back pain appears to be a problem for western and industrialized societies, possibly related to the development of welfare states. Thus, according to a Department of Health survey, in Britain back pain affects 40% of the adult population, 5% of which have to take time off to recover [9]. This causes a large strain on the health system, with some 40% of back pain sufferers consulting a GP for help and 10% seeking alternative medicine therapy [9]. Due to the large number of people affected, back pain alone cost industry £9090 million in 1997/8, with between 90 and 100 million days of sickness and invalidity benefit paid out per year for back pain complaints [10]. Back pain is not confined to the UK alone, but is a worldwide problem: in the USA, for instance, 19% of all workers' compensation claims are made with regard to back pain. Although this is a lot less than the percentage of people affected by back pain in the UK, it should be noted that not all workers in the USA are covered by insurance and not all workers will make a claim for back pain [11]. Moreover, back pain does not affect solely the adult population: studies across Europe [12] show that back pain is very common in children, with around 50% experiencing back pain at some time.

Like most types of pain, back pain is difficult to analyze, as the only information that can be used is suggestive descriptions from the patient. However, these patients may have developed psychological and emotional problems, due to having to deal with the pain. Because of these problems, patients can have difficulty describing their pain, which can lead to problems during the treatment. In some patients, the psychological problems may have aided the cause of the back pain, by adding stress to the body, or the stress of the back pain may have caused psychological problems [13]. It is because of this factor that patients suffering from back pain are usually asked to fill out questionnaires of different types in order to help the medical staff, not only to know where the pain is located, but also to identify the patient's mental state before treatment begins. In addition, the patient is usually required to mark on a diagram, usually of a human body, where the pain is located, and the type of pain. This type of diagram is known as a 'pain drawing' and forms the primary focus of our paper.

3 Pain Drawings and Visualization

Pain drawings, as depicted in Figure 1, have been successfully used in pain centers for over 50 years [17] and act as a simple self-assessment technique, originally designed to enable the recording of the spatial location and type of pain that a patient is suffering from [18]. They have a number of advantages including being economic and simple to complete, and can also be used to monitor the change in a patient's pain situation [18].

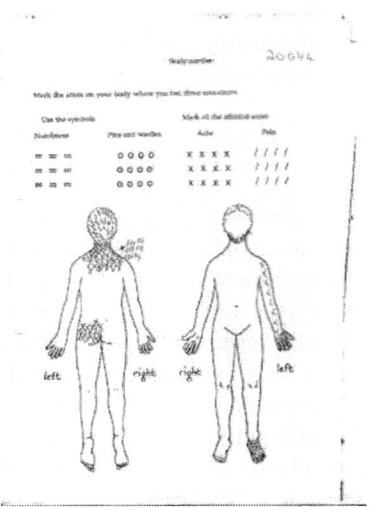

Fig. 1. Example Pain Drawing

Pain drawings have proven to be a versatile tool for recording information as diverse as psychological distress, type of pain, and disability [15]. In order to link the pain drawing to either psychological, emotional or causes of pain; several scoring systems have been developed and described in the literature. These broadly fall into four categories: *grid methods, body region methods, penalty point system* and *visual inspection methods*. Whilst the first two record the presence or absence of pain within defined regions, the last two do require subjective interpretation.

With the grid method [19] an overlay of a grid is placed over the pain drawing. The grid is designed so that each cell is approximately the same size. By using the grid, unskilled testers could calculate the amount of surface area that was in pain. Body region methods, on the other hand, break down the surface of the human body in very simple regions, in order to indicate areas that are in pain. Thus, in a study exploring lumbar discogenic pain, Ohnmeiss et al. [20] used five general regions: low back and buttocks, posterior thigh, posterior leg, anterior thigh and anterior leg. Other ways of regionalising the human body can also be used, such as based on dermatomes have also been employed [21].

Penalty point systems, such as the one described by Ransford et al. [16], work by awarding points for every un-natural placement of pain on a pain drawing. Different areas and rules are made so that there is a weighting depending on the irregularities in the drawing. If more points are scored than normal, then that person may have a psychological problem that needs addressing. In this particular case, pain drawings are used not only as a recorder of pain location, but also as an economical psychological screening instrument to see if a patient would react well to back pain treatment [16], for, as previously mentioned, back pain can be caused by psychological and emotional problems, as well as occupational factors, and hence medical treatment itself may not remove the cause of the pain. Whilst psychological screening for back pain treatment usually entails patients completing costly, time-consuming and difficult to understand questionnaires, by using a penalty point scoring method, it was found that pain drawings could predict 93% of the patients that needed further psychological evaluation just by looking at their completed pain drawing, a conclusion later corroborated in [13].

Visual inspection methods use trained evaluators, who look at the pain drawings and from their experience are able to say what they believe to be wrong with the patient, or if psychological testing is needed [22]. Thus, Uden and Landin [23] have used this method for to identifying patients with lumbar disc herniation. In their approach, drawings were classified as indicative or non-indicative of symptomatic disc disease. If pain was primarily in a radicular pattern from the back into one or both lower extremities, the drawing was classified as indicative. The drawing was classified as non-indicative if pain was indicated to be restricted to the low back only, was indicated to be widespread in a sporadic pattern, or was indicated by extraneous marks made inside and outside of the body to show pain or other sensations.

Most of the methods described can be and are used in practice in conjunction with *sensation type* approaches, which allow not only the placement of pain to be noted but also the particular type of pain encountered. This is done using a key, therefore allowing more information to be collected and acts as an aid to the clinic as to what the cause of the pain is.

The consensus of the literature seems to be that the pain diagram is a powerful tool in the role that it is designed for, namely to record the spatial location and pain type. However, pain drawings are usually stored in a paper format, which allows no further evaluation of the data that is stored upon it and makes searching through the data somewhat an arduous task. To compound the issue, when information from the pain drawings is digitized, it invariably results in loss of information, since current systems that are used for analysis of the pain drawings and the associated questionnaires revolve around statistical packages, such as Excel and SPSS, incapable of handling diagrammatic data. Thus, although diagrammatic data is collected, it is not used as the key component to the data analysis tools. This is somewhat a problem, as people will find it easier to show through a diagram the way that they feel, instead of answering closed questions in questionnaires. Such data cannot therefore be used to its full potential and, in particular, cannot be used in helping with queries within the dataset. Lastly, the paper-based solution of existing methods makes it impractical to record pain variations over time, in spite of the time-dependent nature of pain in chronic sufferers [8].

4 Implementation

4.1 Aim

In our work, we have sought to alleviate the problems identified above and have developed a wireless-enabled, ubiquitous solution that uses the pain drawing as an actual user-friendly visual aid to the input and analysis of back pain datasets. Whilst our solution is generic and applicable to all back pain sufferers which have access to wireless technology, we have specifically targeted wheelchair users due to their severe mobility limitations (which might mean that they might not, for instance, easily have access to a desktop-based computer) and their dynamic pain patterns, which are now easily logged by the developed application. In so doing, we specifically address the issue of pain variability in time, as identified by Gibson and Andrew [8], and our application can thus also be used as a data gathering tool for this still incompletely understood phenomenon, the solution of which has potentially important implications in the monitoring of the effectiveness of back pain treatment and medication.

4.2 Data Collection

In order to function as an effective data gathering tool, the developed application, in keeping with previously identified best practice [14] incorporates a questionnaire complemented by visual input of pain location and type, via a pain drawing.

The questionnaire was elaborated in consultation with clinicians from Northwick Park Hospital in London and representatives of the UK National Forum of Wheelchair User Groups. Clinicians were interested in recording data pertaining to a patient's medical background as well as that which captured the variation of pain patterns with the time of day. On the other hand, the wheelchair users were interested in the usability, flexibility and privacy aspects of the application. Both stakeholder groups agreed that a wireless solution would be beneficial for the added versatility that it offers.

It was agreed that the pain drawing should incorporate four different pain types, namely *numbness*, *pins & needles*, *pain* and *ache* and that grid-scoring should be used. As opposed to traditional methods [19], in which transparencies of the grid are made, and the drawings are scored by placing the grid over each and counting the number of squares in which the patient indicated symptoms, our approach conceptually slices the body contour into squares. The advantage brought with this approach was that we were able to code the pain location with its coordinates from an image to a database, and vice versa.

4.3 Application Structure

The underlying structure of our application is based on a three-tier wireless system model where the three main components are: a mobile, wireless-enabled device, a Web server with scripting capability, and a backend database.

In this model, the patient inputs on a wireless-enabled device (in our case, a PDA), pain information. This is done at specific time intervals, as requested by clinicians, and the information is saved to a local backend database. Whenever the user is within a wireless-enabled zone, s/he then connects to a Web/Database server via a wireless access point, using the Hypertext Transfer Protocol over Secure Socket Layer (HTTPS). Moreover, the connection between the PDA and the wireless access point is itself secured through the use of 128-bit Wired Equivalent Privacy (WEP) encryption.

Upon receiving such requests, the server responds back and asks for appropriate authorisation. After this has been successfully completed, the data is then uploaded to the hospital server. The clinician then uses his/her computer to logon to the Web server and downloads information regarding any specific patient and their pain pattern from the database for further analysis.

4.4 Application Architecture

The developed Back Pain Application is designed and implemented using Microsoft Embedded Visual Basic, a language specifically geared to help developers build applications for the next generation of communication and information-access devices running Windows CE.

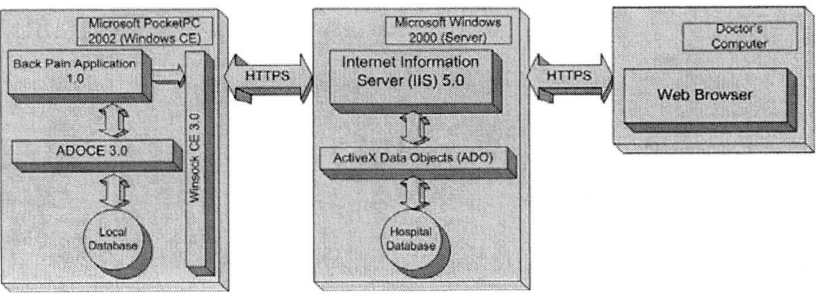

Fig. 2. System architecture diagram

The system architecture diagram (Figure 2) shows the main components that make the wireless model system model work. Accordingly, the Back Pain Application was implemented on an HP iPAQ 5450 PDA with 16-bit touch-sensitive transflective thin film translator (TFT) liquid crystal display (LCD) that supports 65,536 colour. The display pixel pitch of the device is 0.24 mm and its viewable image size is 2.26 inch wide and 3.02 inch tall. It runs Microsoft Windows for Pocket PC 2002 (Windows CE) operating system on an Intel 400Mhz XSCALE processor and contains 64MB standard memory as well as 48MB internal flash ROM. The Web server was implemented on an

Intel Pentium III running at 1 GHz, with 512MB RAM and a 50GB hard disk. In our work, a 10Mbps D-Link DWL-700AP wireless access point was used.

The application reads the coordinates of the pain locations from the touch-sensitive screen and using ADOCE 3.0 (Active Data Objects for CE) connects to a local Microsoft Pocket Access database file. Through this connection, the application saves the pain coordinates and patient questionnaire data to the database. When the user is within wireless Internet coverage, the application uses Winsock CE 3.0 (Windows CE Sockets) to send a connection request to the server (Figure 3). The server runs Windows 2000 operating system and Internet Information Server (IIS) 5.0, which Open Database Connectivity (ODBC) to connect to the hospital database.

The doctor's interface is made of dynamically created Active Server Pages (ASP), which can be accessed using any conventional web browser running on a computer connected to the Internet. Thus, after successful authorization, medical personnel can download a particular patient's data to their personal computer. This is achieved through the ASP code dynamically creating an SQL query to the database, the results of which are presented dynamically on the viewed Web page.

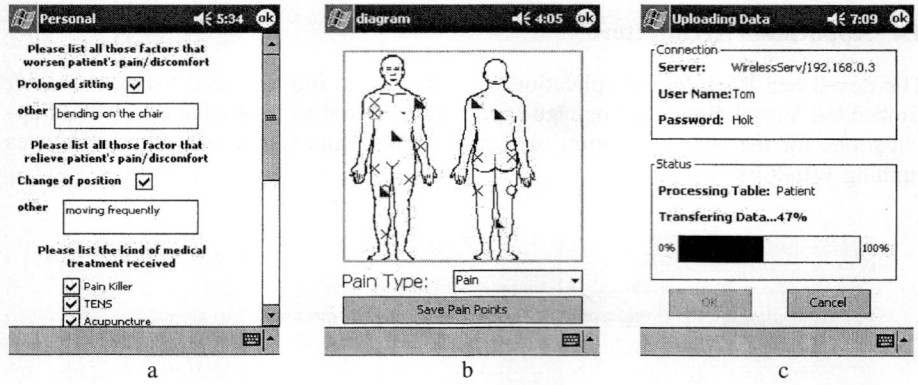

Fig. 3. a) Patient treatment details b) Diagram with pain points c) Upload screen

5 Functionality and Evaluation

5.1 Pilot Evaluation

The first version of the developed application was given out, together with a brief user manual, to three wheelchair users from the collaborating group for a 5-day pilot evaluation. The feedback provided could be broadly categorized into two groups. The first concerned ways through which any potential misunderstandings of the questionnaire content could be clarified. The second grouped issues such as font size (too small in our initial prototype) and color schemes used by the application (which had to take into account users' potential color blindness). The users did not encounter navigation problems, nor were there any problems raised with regard to clarity of the pain diagram, or indeed with the saving and transferring of recorded data. All the concerns identified by the participants of the pilot study were addressed in the subsequent version of our application.

5.2 User Evaluation

The developed application was evaluated with a sample of 25 wheelchair users, members of the UK National Forum of Wheelchair User Groups. There were 13 females and 12 males in the sample, aged between 27-64 years old, each of which had varying degrees of daily wheelchair use. Each participant was given 5 days in which to evaluate the application, as well as a short (3-page) user manual, and instructions at which times of the day they should record their pain measurements.

Table 1. Evaluation Questionnaire

	Strongly Disagree	Disagree	Slightly Disagree	Neither	Slightly Agree	Agree	Strongly Agree
It is important to be able to record my pain on a PDA.							
It is useful to be able to log pain data across time.							
I find the process of inputting pain data on a PDA easy.							
Process of transferring data from PDA to the main database is easy.							

Input of data took place mainly at the user's domiciles (or wherever they happened to be when the recording of data had to take place), with no personnel being on hand to offer help in this respect, save for the information contained in the manual. While the degree of local connectivity of each patient varied (three of which had wireless LANs already installed in their homes), they were told to use their own means and resources in order to upload the collected data to the hospital server. At the end of the evaluation period, participants were requested to complete a questionnaire (Table 1), in which they recorded their opinions on a Likert scale of 1-7 about the usability and feasibility of using such an application in practice. Patients were also asked to note down any other observations that they might wish to make.

The results of the evaluation are given in Figure 4. Analysis of the results highlighted a general consensus that wheelchair users had in respect of the ability to record pain data on a mobile device being beneficial to their lifestyles (an observation also confirmed through informal and formal, written feedback, at the end of the questionnaires). Although some users, especially those suffering from arthritis and/or poorer eyesight did encounter difficulties in using the relatively small interface of the PDA, overall the participants agreed that the processes of recording pain data was a relatively easy one and that the ability to record data across time, irrespective of the particular location in which users found themselves, was indeed beneficial. Lastly, although participants did encounter barriers in respect of their attempts to upload data, with some of them using ingenious resources (such as using WiFi-enabled cafes or local shopping malls) to accomplish this task, nonetheless participants felt generally positive about ubiquitous data collection and transmission capabilities, with the feeling that proliferation of WiFi hotspots would remove such barriers in the future.

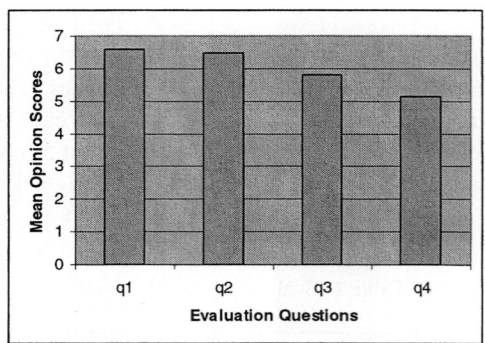

Fig. 4. Evaluation results

6 Conclusions

In this paper we have described the design, implementation and evaluation of a wireless-enabled solution for back pain data collection and wireless transmission to a remote clinical database. Employing a user-friendly visual approach to data input, in our solution such activities are carried out in a ubiquitous fashion. The fact that the collected data, including pain drawings, are digitized makes it easier for it to be collected, time-stamped and analyzed, while the fact that such input takes place on a PDA means that this can happen irrespective of the location of the user without clinical supervision. Finally, recognizing the mobility problems that many back pain patients endure, our solution is WiFi enabled, thus facilitating remote, ubiquitous, data access and management and absolving patients of the need to actually physically hand in their completed questionnaires.

However, our proof-of-concept study has not addressed issues such as scalability, and security will need to be updated in line with future developments in the area. Nonetheless, our experience has shown that the provision of such goals are worth pursuing, for only then will the potential of true anytime/anywhere data collection become be realized.

References

[1] Chan, T.: A web-enabled framework for smart card applications in health services. Communications of the ACM, Vol. 44, No. 9 (2000) 77 – 82
[2] Hernando M.E. *et al.*: Evaluation of DIABNET: A decision support system for therapy planning in gestational diabetes. Computer Methods and Programs in Biomedicine, Vol. 62 (2000) 235-248

[3] Hooda J. S. *et al.*: Health Level-7 compliant clinical patient records system. Proc. of the ACM 2004 symposium of Applied computing. Cyprus, (2004) 259- 263

[4] Hung, K., Zhang, Y.: Implementation of a WAP-Based Telemedicine System for Patient Monitoring. IEEE Transactions on Information Technology in Biomedicine, Vol. 7 No. 2, (2003)

[5] Bardram, J. E.: Applications of context-aware computing in hospital work: examples and design principles, Proceedings of the 2004 ACM symposium of Applied Computing. Cyprus (2004) 1574-1579

[6] Munoz M. A. *et al.*: Context-Aware Mobile Communication in Hospitals. IEEE Computer, Vol. 36 No. 9 (2003) 38-46

[7] Rodriguez N. J. *et al.*: PDA vs. Laptop: A Comparison of Two Versions of a Nursing Documentation Application. Proc. of the 16th IEEE Symposium on Computer-Based Medical Systems (CBMS'03)

[8] Gibson, J., Frank, A. O.: Pain experienced by Electric Powered Indoor/Outdoor Chairs (EPIOC) users: a pilot exploration using pain drawings. Proc. of the 2nd meeting of the European Federation of Physical and Rehabilitation Medicine. Vienna (2004) 114

[9] Boucher, A.: The Prevalence of Back pain in Great Britain in 1998. Department of Health, (1999) Available at http://www.doh.gov.uk/public/back pain.htm

[10] Frank, A., De Souza, L. H.: Conservative Management of Low back pain. International Journal of Clinical Practice, Vol. 55, No. 1, (2000) 21 – 31

[11] Jefferson, J. R., McGrath, P. J.: Back pain and peripheral joint pain in an industrial setting, Arch. Phys. Med. Rehabil., Vol. 77 (1996) 385-390

[12] Balague F. *et al.*: Non-specific low back pain in children and adolescents: risk factors. European Spine Journal, Vol. 8 (1999) 429-438

[13] Von Baeyer C. L. *et al.*: Invalid Use of Pain Drawings in Psychological Screening of Back pain Patients. Pain, Vol. 16, (1983) 103-107

[14] Mann III N.H *et al.*: Initial-Impression Diagnosis Using Low-Back Pain Patient Pain Drawings. Spine, Vol. 18 No. 1, (1992) 41-53

[15] Parker H. *et al.*: The Use of the Pain Drawing as a Screening Measure to Predict Psychological Distress in Chronic Low Back pain. Spine, Vol. 20 No. 2 (1995) 236-243

[16] Ransford A. O. *et al.*: The Pain Drawing as an Aid to Psychologic Evaluation of Patients With Low-Back pain. Spine, Vol. 1 No. 2 (1976) 127-134

[17] Palmer, H.: Pain charts: A Description of a Technique whereby Functional Pain may be Diagnosed from Organic Pain. New Zealand Medical Journal, Vol. 48 No. 264 (1949) 187-213

[18] Rankine J. J. *et al.*: Pain Drawings in the Assessment of Nerve Root Compression: A Comparative Study With Lumbar Spine Magnetic Resonance Imaging. Spine, Vol. 23 No.15 (1998) 1668-1676

[19] Gatchel R.J. *et al.*: Qualifications of Lumbar Function: Part 6: The use of Psychological Measures in Guiding Physical Function Restoration. Spine, Vol. 11 (1996) 36-42

[20] Ohnmeiss *et al.*: The Association between Pain Drawings and CT/discographic pain responses. Spine, Vol. 20 (1995) 729-733

[21] Ghinea G. *et al.*: Using Geographical Information Systems for Management of Back-Pain Data. Journal of Management in Medicine, Vol. 2/3 No. 16 (2002) 219-237

[22] Chan C. W. *et al.*: The Pain Drawing and Waddell's Nonorganic Physical Signs in Chronic Low-Back pain. Spine, Vol. 18 No. 3 (1993) 1717-1722

[23] Uden, A., Landin, L.A.: Pain drawing and myelography in sciatic pain. Clin. Orthop., Vol. 216 (1987) 124-130

Prototype Design of Mobile Emergency Telemedicine System

Sun K. Yoo[1,3], S.M. Jung[2], B.S. Kim[2], H.Y. Yun[2], S.R. Kim[4], and D.K. Kim[2]

[1] Dept. of Medical Engineering, Yonsei Univ. College of Medicine,
134 Shinchon-dong Seodaemun-ku, Seoul, Korea
sunkyoo@yumc.yonsei.ac.kr
[2] Graduate School of Biomedical Engineering, Yonsei Univ.,
134 Shinchon-dong Seodaemun-ku Seoul, Korea
[3] Center for Emergency Medical Informatics, Human Identification Research Institute,
Seoul, Korea
[4] Dept. of Computer Science, Dongduk Woman's Univ., Seoul, Korea

Abstract. High bit rate wireless cellular service using CDMA 1X-EVDO is now popular in Korea, particularly in urban areas, since it was launched commercially in 2002. This cellular service allows the real-time transmission of patient images and vital sign signals simultaneously in a moving ambulance application. In this paper, we designed a prototype emergency telemedicine system that can transfer both biological signal and patient motion video from a moving vehicle using a CDMA 1X-EVDO reverse link. To cope with the limited bandwidth of the reverse link (transmission bandwidth of cellular device) relative to the the forward link (receiving bandwidth), priority control between the vital sign and video images, frame rate control using MPEG-4 compression, and error control using automatic repeat request were incorporated into the application layer protocol of the designed prototype system. Many on-road experiments have been performed to evaluate the actual performance and to demonstrate the applicability in a real situation. In most cases, the biological signal and patient video images with reduced frame rate were successfully transmitted from the moving vehicle in urban areas

1 Introduction

The mobile emergency telemedicine system can provide an efficinet means of patient care from a medical specialist in an urgent situation, since cellular communication with high data rate enables the transmission of video information and vital signs of the patient instantly from the scene ubiquitously [1,2]. In accordance with the advances in telecommunication, there has been a constant development of the emergency telemedicine system using cellular communication for which ambulances are the most important means of transportation [3,4]. Since 2002, a 2.5 G mobile telecommunication system, the CDMA2000 1X-EVDO (Evolution-Data Only), has been in use. The CDMA2000 1X-EVDO system has an asymmetric data rate structure between uplink bandwidth and downlink bandwidth. The maximum forward bandwidth is 2.4576 Mbps and the maximum backward bandwidth is 153.6 Kbps.

The purpose of this research is to design a mobile emergency telemedicine system using the reverse (backward) channel of the CDMA2000 1X-EVDO system that can transmit multimedia data from a moving vehicle [5]. The information transmitted from the mobile emergency telemedicine system includes electrocardiography (ECG), SpO2, non-invasive blood pressure (NIBP) monitoring , respiration, and patient video, which helps a remote medical specialist to understand the accident scene and to diagnosis patients with external injury via teleconsultation. Throughout field measurements, we analyzed the actual transmission bandwidth of the CDMA2000 1X-EVDO reverse channel in terms of transmission speed and vehicle location. Rather than focus on the thoretical performance,, we determined an appropriate transmission strategy to improve the efficiency of the designed mobile emergency system under the conditions of a mobile ambulance.

2 Materials and Methods

As shown in Fig. 1, the mobile emergency telemedicine unit transmitted data through the reverse link with a maximum theoretical bandwidth of 153.6 Kbps from an ambulance. The transmitted bitstream over air interface passed to a wired IP network via PDSN at the base station, and finally was destined to the emergency center receiver unit.

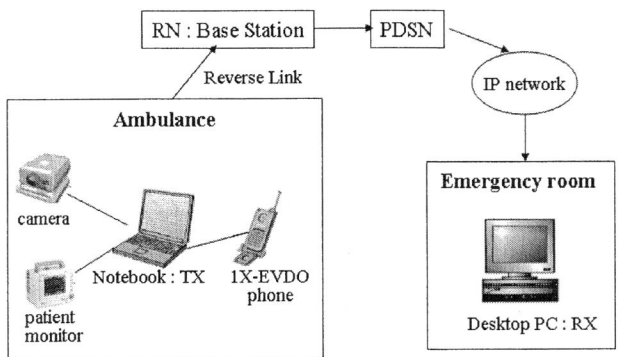

Fig. 1. The configuration of the mobile emergency telemedicine system

The CDMA 2000 1X-EVDO employed the radio link protocol (RLP) version 3, with a limited number of automatic repeat requests as the device level communication protocol. In the case of a high error environment, the RLP discards some packets. Hence, we devised a special header having a sequence number (segment number) to manage both the biological signal and patient video information efficiently to limit packet loss. A data unit is the standard unit that processes the biological and video signal, where as the segment unit is a real data unit transmitted by the UDP protocol. The data number (DN) distinguishes the type of data, e.g. vital sign or patient video.

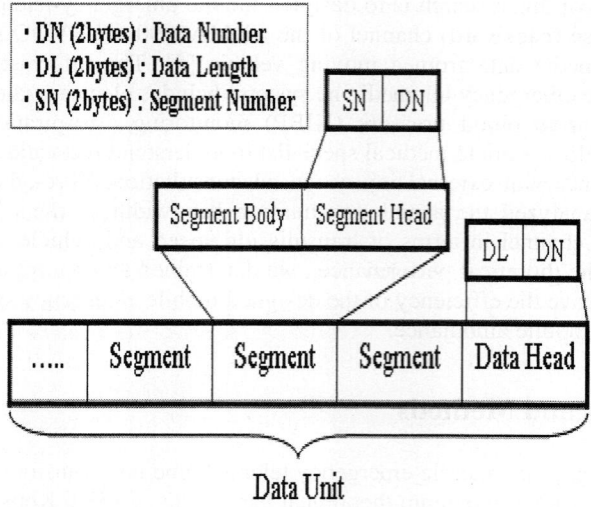

Fig. 2. Data structure of the mobile emergency telemedicine system

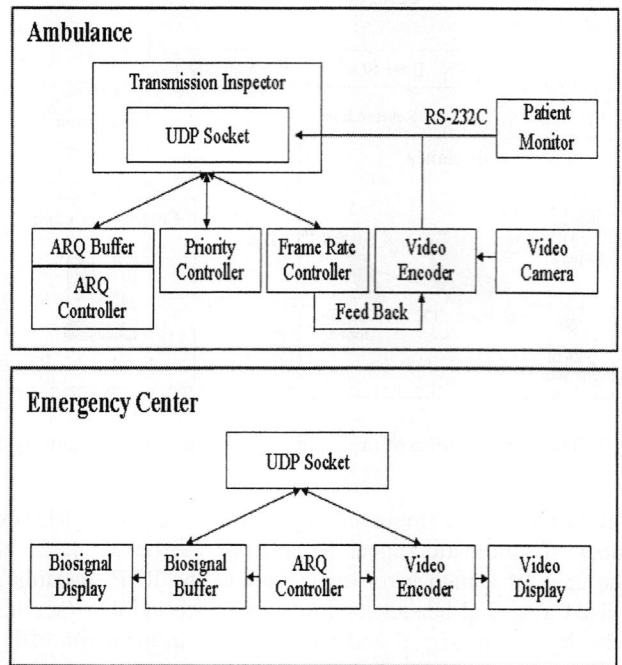

Fig. 3. Dataflow daigram for the mobile emergency telemedicine system

Figure 3 shows data flow diagrams for the mobile emergency telemedicine system, which consist of an ambulance unit with a reverse CDMA2000 1X-EVDO cellular link (transmiting patient's biological signal and video information), and an emergency center unit with a wireline connection (displaying the received data for remote consultation). The ambulance unit consists of data input and transmission control with UDP protocol parts. Biological signals, measured by the patient monitor, are inputted to a laptop computer through an RS-232C interface. Patient video images are compressed by MPEG4 format (spatial resolution 640x480). Both streams are stored at the ARQ controller buffer of the application layer to supplement lost packets due to the ARQ of the RLP protocol. The transmission inspector inspects failures of the RLP protocol corresponding to the lost packet (checking each packet's sequence number and ARQ timing), and requests the re-transmission up to once for video and twice for vital signs at 2-second intervals (maximum permissible delay) to make mobile communication more reliable.

We also employed the following control strategies to compensate for the varing bandwidth depending on the number of users and power control at the radio station.

‼ Priority control: We assumed that biological signals are more important than the patient's video information [1]. Hence, if the bandwidth available is unable to handle the biological signal and patient video with the minimum frame rate and maximum quantization scale, the priority controller assigns higher priority to the biological signal than the patient video to secure biological signal transmission. As the bandwidth approaches 8 Kbps, which is the minimum required bandwidth for reliable ECG transmission, the priority controller starts to discard lower priority data including compressed patient video.

‼ Frame rate control: The transmitter controls the video frame rate according to the buffer state of the transmission unit. Before priority control, the frame rate controller increses the quantization level and reduces the frame rate.

‼ Error control: The receiver recognizes error occurrence using the data head and sequence number. If the received packet is lost, the receiver's ARQ controller sends an ARQ signal requesting re-transmission to the transmission unit. If the transmitter receives an ARQ signal from the receiver, it re-transmits relevant data in the ARQ Buffer.

3 Results

Based on the designed prototype system, we performed several field tests to design an emergency telemedicine system fitted to the reverse channel of CDMA2000 1X-EVDO in a real situation. By using the user datagram protocol (UDP), we measured actual transmission bandwidth at the receiver unit by chaning the speed of the ambulance. The error probability (P_E) related to packet data loss, was then measured.

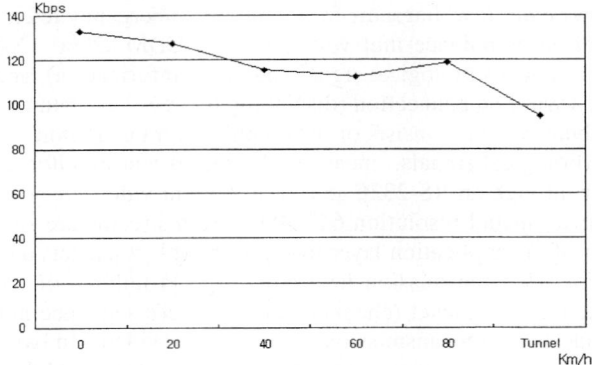

Fig. 4. Measured mean bandwidth in terms of vehicle speed and location

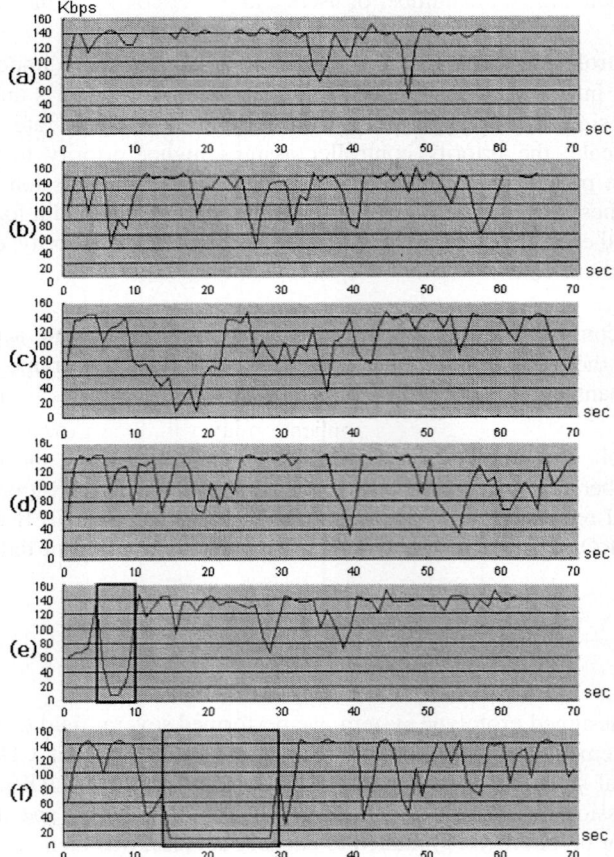

Fig. 5. Transmission Patterns in terms of vehicle speed
(a) stop (b) 20km/h (c) 40km/h (d) 60km/h (e) 80km/h (f) tunnel

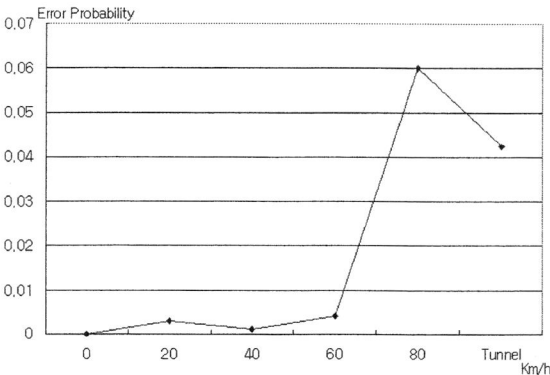

Fig. 6. Error probability in terms of vehicle speed

Bandwidth measurements of the reverse channel of CDMA 1x EVDO were conducted 20 times each in the morning, afternoon, and evening. We measured bandwidth and P_E by changing vehicle speed. Figure 4 shows the measured mean bit rate. A maximum bandwidth of 132 Kbps was observed at suspension state, which is similar to the theoretical maximum bandwidth of 153.6 Kbps for the reverse link of CDMA2000 1X-EVDO. As the vehical speed increased, transmission speed decreased gradually. However, bandwidths higher than 100 Kbps were always maintained except during tunnel measurement.

Figure 5 shows the transmission pattern of UDP packets for 70 seconds associated with the designated transmission speed. Bandwidthes higher than 140 Kbps were mostly maintained regardless of vehicle speed. But, the duration and frequency of burst error incresed as the vehicle speed increased. Particularly, in cases of 80 km/h and tunnel, relatively longer disruption of burst error was measured, which necessatiated an efficient ARQ scheme at the application layer to cope with packet loss due to a limited number of re-transmissions at the RPL protocol layer.

Figure 6 shows error probability in terms of vehicle speed, corresponding to lost packets due to unreliable RLP protocol. Lower error probability is maintained if the speed is lower than 60 km/h. However, when the speed reaches 80 km/h, error rate increases rapidly. Hence, the mobile telemedicine system can be reliably operated in urban areas, where vehicle speed is generally lower than 60 km/h.

The following was observed in moving vehicle experimentation. The average time delay for biological signal and patient video transmissions were 2.5 and 3 seconds, respectively. Frame rates for patient video were 0.5 ~ 5 frames per second, with an average of 2.5 frames per second. In the case of an unmoving state, the biological signal was continuously transmitted, while patient video was infrequently disrupted. Regarding vehicle speed from 20 to 60 km/h, the biological signal was continuously transmitted, while patient video was suspended 3 times per minute. Delay (4 ~ 5 seconds) occurred in biological signal and patient video transmissions when vehicle speed was higher than 80 km/h. Finally, when the ambulance shook due to a corner or rugged road surface, the frame rate for patient video dropped to less than 1 frame/sec regardless of speed or tunnel.

4 Discussion and Conclusions

High bit rate wireless cellular service using CDMA 1X-EVDO is now common in Korea, particularly in urban areas, since it was launched commercially in 2002. This cellular service allows real-time transmission of patient images and vital sign signals simultaneously in a moving ambulance application. Future work will focus in comparing network performance with general packet radio service(GPRS) and in the provision of a mobile emergency telemedicine system covering the needs of the whole areas to apply to the practical emergency situations[6, 7].

In this paper, we designed a prototype emergency telemedicine system that can transfer both biological signals and patient motion video from a moving vehicle using a CDMA 1X-EVDO reverse link. To cope with the limited bandwidth of the reverse link (transmission bandwidth of cellular device) compared to the forward link (receiving bandwidth), priority control between vital signs and video images, frame rate control using MPEG-4 compression, and error control using automatic repeat request, was incorporated into the application layer protocol of the designed prototype system.

On-road experiments have been performed to evaluate the actual performance and to demonstrate applicability in a real situation. In most cases, the biological signal and patient video images with a reduced frame rate were successfully transmitted from a moving vehicle in urban areas.

Acknowledgements

This study was supported by a grant of the Korea Health 21 R & D Project, Ministry of Health & Welfare, Republic of Korea (02-PJ3-PG6-EV08-0001).

References

1. Pattichis CS. et al:Wireless Telemedicine Systems. An Overview, *IEEE Antenna's and Propagation Magazine*, 44(2), (2002)pp. 143-153
2. Hung HK., et al.: 'Implementation of a WAP-based telemedicine system for patient monitoring', *IEEE Transactions on Information Technology in Biomedicine*, 7(2), (2003) pp.101 -107
3. Kyriacou E., et al.: 'Multi-purpose HealthCare Telemedicine Systems with mobile communication link support', BioMedical. *Engineering OnLine* (2003)
4. Woodward B. et al.: 'Design of a telemedicine system using a mobile telephone', *IEEE Transactions on Information Technology in Biomedicine*, 5(1), (2001) pp. 13-15
5. C.S0001-A, *'Introduction to cdma2000 Standards for Spread Spectrum System'*(2000)
6. Istepanian, R. S. H. 'Guest Editorial Introduction to the Special Section on M-Health: Beyond Seamless Mobility and Global Wireless Health-Care Connectivity', *IEEE TRANSACTIONS ON INFORMATION TECHNOLOGY IN BIOMEDICINE, VOL. 8, NO. 4*, DECEMBER (2004) p.p 405-411
7. Alesanco, A., Olmos, S., Salcedo, J., Istepanian, R., Garcìa, J., 'Mobile Telecardiology System', *International Congress on Computational Bioengineering, c13A, EspaOa* (2003)

An Intermediate Target for Quick-Relay of Remote Storage to Mobile Devices

Daegeun Kim, MinHwan Ok[*], and Myong-soon Park

Dept. of Computer Science and Engineering, Korea University,
Seoul, 136-701, Korea
{vicroot, myongsp}@ilab.korea.ac.kr, [*]panflute@korea.ac.kr

Abstract. Requests for application services that require large data space such as multimedia, game and database[1] have greatly increased. Nowadays those requests locomote for various services using mobile devices. However, mobile devices have difficulty in sustaining various services as in a wired environment, due to the storage shortage of the mobile device. The research[5] which provides remote storage service for mobile appliances using iSCSI has been conducted to overcome the storage shortage in mobile appliances. In research we found that when iSCSI was applied to mobile appliances, iSCSI I/O performance dropped rapidly if a iSCSI client had moved from the server to a far away location. It occurred due to the specific character of iSCSI, which is very sensitive to delay time. In this paper, we suggest an intermediate target server that localizes iSCSI target to achieve a breakthrough against the shortcomings of iSCSI performance dropping sharply as latency increases when mobile appliances recede from a storage server.

1 Introduction

The explosive growth of the mobile appliance market has made a lot of demands in mobile-related services. Efforts to apply wired network environment services, which need large amount of storage space, such as multimedia and databases, to mobile appliances with a wireless network environment, were performed. However, mobile appliances should be small and light to support mobility, so that they use a small flash memory instead of a hard-disk having large data space. In the case of PDAs, these usually have memory space of 32 ~ 64M. Cell-phones or smart phones permit smaller memory space. Therefore, there is difficulty saving multimedia data such as mpg, mp3, etc. and installing large software such as database engines. Limited storage space in mobile appliances has been a barrier for applying various services in a wired environment. As a result, the necessity for remote storage services in mobile appliances has overcome limited storage space of mobile devices, and store large amounts of data, or provide various application services [5] [8] [9].

---

[*] Corresponding Author.

Remote storage services have two classifications. One is file-level I/O and the other is block-level I/O. File-level I/O service is provided by communication between a server and client file systems. In the case of file I/O, because client's I/O request is passed to a device via the server file system, data share is possible between other clients through a locking mechanism which the server file system offers. However, file system overhead drops I/O performance. In block-level service, client's I/O request is directly passed to a storage device without a server file system, in the form of a block I/O command. Therefore, it is not able to provide data sharing services by itself, but I/O performance is better than file-based I/O services [9].

The bandwidth of wireless networks, which mobile devices use, is usually lower than those of wired networks. If the purpose of the remote storage service is to extend individual storage space for mobile appliances without data sharing, block-level I/O service is more suitable. iSCSI is a standard protocol that transports SCSI command. This is a representative block I/O through TCP/IP network. However, iSCSI has a problem that I/O performance drops sharply if network latency increases between iSCSI initiator and target.

For the environment iSCSI-based remote storage service is applied to mobile appliances, we have achieved a breakthrough against the problem of iSCSI performance falling rapidly accordingly the mobile client recedes from the storage server and network latency increases. This paper comprises as follows. In section 2, we present the iSCSI basic operations to explain the reason why iSCSI performance drops down when distance between iSCSI initiator and target is long. We also depict the research related with improving iSCSI performance using a client's local cache. In section 3, we propose ways to improve iSCSI performance using an intermediate server, and we talk about system architecture and algorithms in section 4. We analyze the result of simulation with NS2. Finally we shall conclude and describe our future work in section 5.

2 Background

2.1 iSCSI

The iSCSI (Internet Small Computer System Interface) is an emerging standard storage protocol that can transfer a SCSI command over IP network [5]. Since the iSCSI protocol can make clients access the SCSI I/O devices of server host over an IP Network, client can use the storage of another host transparently without the need to pass through a server host's file system [6].

In iSCSI layer, which is on top of TCP layer, common SCSI commands and data are encapsulated in the form of iSCSI PDU (Protocol Data Unit). The iSCSI PDU is sent to the TCP layer for the IP network transport. Through this procedure, a client who wants to use storage of the remote host, can use, because the encapsulation and the decapsulation of SCSI I/O commands over TCP/IP enable the storage user to access a remote storage device of the remote host directly [3]. Likewise, if we build a remote storage system for mobile appliances using the iSCSI protocol, mobile clients can use the storage of a server host directly, like their own local storage. It enables mobile appliances to overcome the limitation of storage capacity, as well as the ability to adapt various application services of wired environment in need of mass scale data. Different from the traditional remote storage system, based on file-level I/O, iSCSI protocol provides block unit I/O. Therefore it can make more efficient transmission throughput than the traditional remote storage systems, like CIFS and NFS.

Another characteristic of the iSCSI protocol is that it operates on a standard and commonly used network component like Ethernet. Since iSCSI protocol can be plugged directly into an Ethernet environment, it is easier to manage than another storage data transmission protocol, such as Fibre Channel. Moreover, iSCSI can reduce the costs to build a storage system due to the use of infra network without additional adjustment. The iSCSI protocol was defined as the standard SCSI transmission protocol recently by IETF and a lot of related research is being conducted with the development of Gigabit Ethernet technology.

Fig. 1 shows the exchanged control and data packets' sequence in the read and write operation of iSCSI protocol [10].

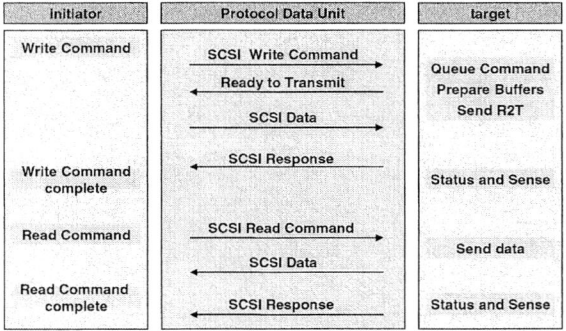

Fig. 1. iSCSI Basic Operation

Like Fig. 1, iSCSI exchanges the control packets (SCSI Command, Ready to Transmit, SCSI Response) and data packet (SCSI Data) to process one R/W operation. Three control packets and one data packet are used in a write operation and spend 2 x RTT. It takes 2 control packets and 1 x RTT for a read operation. Because the control packet, including header informations, is no more than 48 byte, bandwidth waste becomes serious when the initiator recedes a little from the target. Therefore, the distance between iSCSI initiator and target, and the network latency influence iSCSI I/O performance.

2.2 iCache

iCache is a research to improve iSCSI performance using local cache of a client system. Initiator's systems have specific cache space for iSCSI data, and iSCSI block data is cached to minimize network block I/O. Therefore, iSCSI does not send I/O requests through the network every time the disk I/O happens. Instead it reads cached blocks or send blocks cached in LogDisk at once to the server for improving iSCSI performance. iCache's buffer space consists of two hierarchical caches comprising NVRAM and LogDisk. Data is stored sequentially in NVRAM. When enough data is gathered, iCache process moves data from NVRAM to LogDisk. Blocks which are frequently accessed, are kept in NVRAM where access speed is fast. iCache stores less accessed data in the LogDisk. Furthermore, destage operation is achieved by kernel thread called a LogDestage. In a Destage Operation, NVRAM's data is moved to the LogDisk, or the LogDisk's data is sent to remote storage using iSCSI protocol.Cache techniques used in iCache are based on DCD technology, [11] proposed to improve Disk I/O performance.

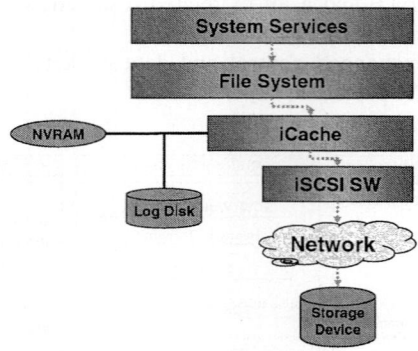

Fig. 2. iCache Architecture

However it is difficult to apply iCache to mobile devices which lack memory, because iCache needs additional memory and hard-disk space to embody the local cache, NVRAM and LogDisk.

3 Quick-Relay of Remote Storage

In this section, we discuss how localizing the iSCSI target with an intermediate server, called intermediate target, and solve the problem of iSCSI performance dropping when iSCSI mobile clients recede from the storage server. iSCSI response time for read operation is minimized with the nearest intermediate target. This prefetches read blocks for the client to use, or response beforehand packet to iSCSI initiator for the written blocks.

In the proposed design, we assume the following three factors to simplify the problem.

- Supposing a distributed storage server environment, iSCSI intermediate target server spread over wide areas. A mobile client is connected with the nearest iSCSI intermediate target through an iSNS (Internet Storage Name Server) [3].
- As transmission distance increases, propagation delay of physical media, and the sum of the queuing delay of intermediate routers increase.

- iSCSI latency includes propagation delay by distance, queuing delay and transmission delay by bandwidth, and ignores processing delay of end nodes.

3.1 iSCSI Intermediate Target

In an iSCSI based remote storage service for mobile nodes, when mobile node serviced in area A moves to area C, the distance between iSCSI initiator and target is prolonged and packet transfer time increases. In section 3, we showed that the SCSI command is processed sequentially in SAM-3 [7] and iSCSI basic operations need an exchange of several control packets to process one command. Because small control packets of the iSCSI protocol influence iSCSI response time, if the distance between the server and the client is long, the link utilization drops sharply and iSCSI performance becomes low.

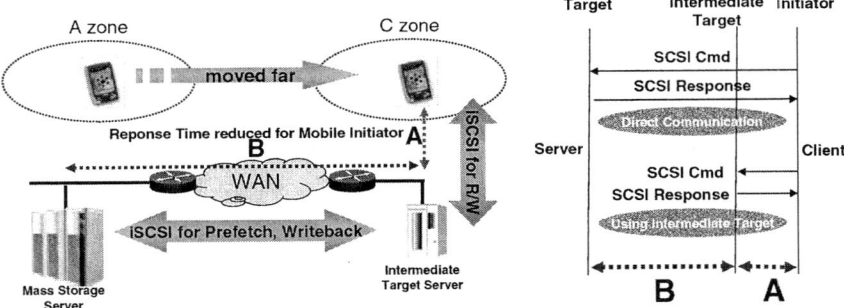

Fig. 3. iSCSI Target Localization with Intermediate Server

Fig. 4. SCSI Response Time Reduction

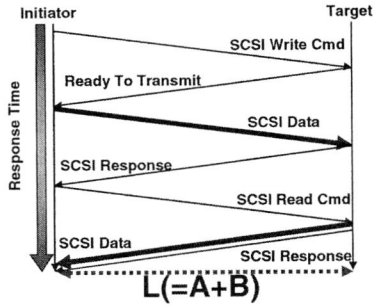
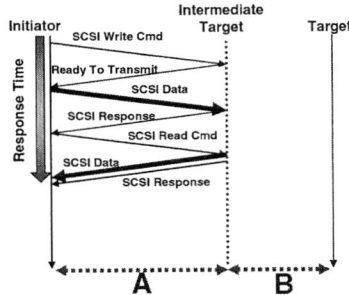

Fig. 5. Direct Communication between iSCSI Initiator and Target

Fig. 6. Response Time by Intermediate Target

We introduced an intermediate server (iSCSI Intermediate Target) to reduce the packet transfer delay time between server and client, and heighten practical utilization of the network bandwidth. Intermediate target prefetches the next block to be used by the client for iSCSI read operations and can give quick responses for iSCSI write operations. The nearest intermediate server from mobile client is selected by iSNS and the client can run iSCSI protocol with the intermediate server, which also has an iSCSI connection with a remote storage server. Therefore, response delay time of an I/O request shortens because the client has an iSCSI connection with a nearby intermediate server, instead of a long-distance storage.

Fig. 5 and Fig. 6 show that response delay times are different when the iSCSI Initiator and Target communicate directly with each other, and when put with an iSCSI intermediate server. Three control packets should be exchanged to process a write command and two control packets for a read command. If the distance between initiator and target is short, iSCSI I/O response time is reduced because the control packet size is small, but it has same propagation and queuing delay. If the end nodes' processing delay of an iSCSI packet is ignored, response delay time is reduced by A/L ratio when introducing an intermediate server as shown in Fig. 6.

3.2 System Architecture

Fig. 7 shows a module diagram of an intermediate target system. An intermediate target system consists of iSCSI initiator, target, and block management module. An iSCSI intermediate server has two iSCSI connections. One is a connection between the intermediate target server and mobile client and the other is a connection with the storage server. The target module has an iSCSI session with a mobile client's iSCSI initiator and the initiator module has one with the iSCSI target module of the storage server. Two modules perform the same role, such as general iSCSI Target/Initiator module. However, the first iSCSI connection between mobile client and intermediate

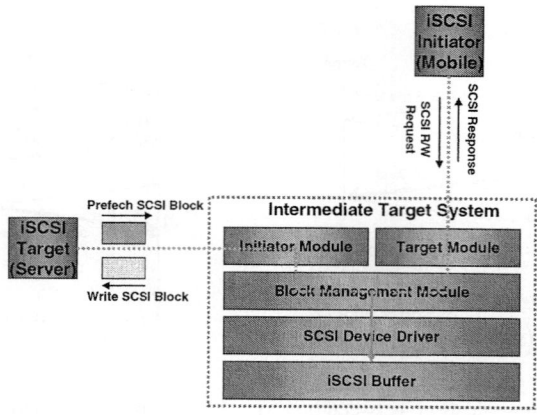

Fig. 7. Intermediate Target System

target is used for I/O requests of the client, and the latter is used to prefetch next blocks used by the client from the storage server or to deliver write blocks to a storage server. The target/initiator module of an intermediate server is controlled by a block management module. ISCSI Buffer is managed by FIFO as a way to leave blocks used most recently.

Fig. 8 shows processing algorithm for client I/O requests by block management module of intermediate server. In the case of a read request, a block management module searches requested blocks in the iSCSI buffer to service the requested block with a target module. If the requested block exists in the iSCSI buffer, target module of intermediate target server will reply to the client. But when there is a non requested block in the buffer, the initiator module of the intermediate target server, which has a connection with the storage server, sends the client's request to storage server and receives the block to send to the client. At this time, the block management module sends a read request to the next logical address of the block that the client required. When the intermediate server receives a write I/O request, the intermediate server sends a reply message to the client and sends the write block to the storage server in no time. Therefore, a mobile client's iSCSI initiator has an effect of an iSCSI connection with a nearby storage server.

```
When Intermediate Target Receives I/O Request
i : Requested Block
{
 if(Read Op)
 {//Client's I/O Request is Read Operation
    if(i exists in iSCSI Buffer)
    {
       Send i and Response to Initiator
    }
    else
    {
       Send Read Request of i to Target
    }
    Send Read Request of i+1 from Target
    and Put at Head of iSCSI Buffer
    //Prefetch next block from Storage Server
 }
 else (Write Op)
 {//Client's I/O Request is Write Operation
    Put i at Head of iSCSI Buffer
    Send Response to Initiator
    Send Write Request of i to Target
 }
}
```

Fig. 8. Block Management Algorithm

4 Simulation

4.1 Simulation Environment

In Fig. 9, the network of an iSCSI initiator, target and intermediate target is simulated using a network simulator 2.27. The bandwidth of the link between node1 and node2 is limited by 14400bps to sumulate wireless network (CDMA 2000 1x). Performance is measured by changing the delay time from 1ms to 64ms to analyze the change of iSCSI performance according to the distance between iSCSI initiator and target under the supposition that delay is proportional to distance. We selected 512 byte data size which is SCSI block size used in PDA of Windows CE base, which is one of the representative mobile appliances.

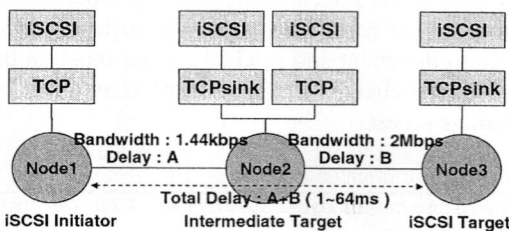

Fig. 9. Network Configuration with NS2

4.2 Experimental Results and Analysis

Fig. 10 shows the difference in iSCSI performance, by data type, when an intermediate target is used or not. In the case of the ratio of delay time A and B specified by 7:3, we can see that the difference in iSCSI performance is very small regardless of data type when total delay is less than 2ms. In a simulation using NS2, iSCSI throughput is influenced by a propagation delay, which depends on physical media, and transmission delay that depends on data size and bandwidth. In cases where iSCSI has a short delay, when the distance of iSCSI initiator from target is short, transmission delay is much bigger than the difference of propagation delay by introducing intermediate target. Therefore it has little influence on iSCSI performance. However, when the iSCSI initiator is distant from the target, the difference of propagation delay time is much longer than transmission delay. And, the intermediate target greatly effects on the iSCSI throughput. According to Fig. 11, we know that iSCSI performance differences appeared greatest when propagation delay of the iSCSI initiator and target was prolonged. In the case of multimedia, text and application data, each has improved throughput by 30%, 27% and 17% when the latency is 64ms. Performance differences according to data type are due to the fact that the iSCSI buffer hit rate for a read block is different. We suppose that the hit ratio of multimedia data, which is accessed sequentially, is 90 percent and hit ratios are 80% and 50% in the case of text and application data.

An Intermediate Target for Quick-Relay of Remote Storage to Mobile Devices   1043

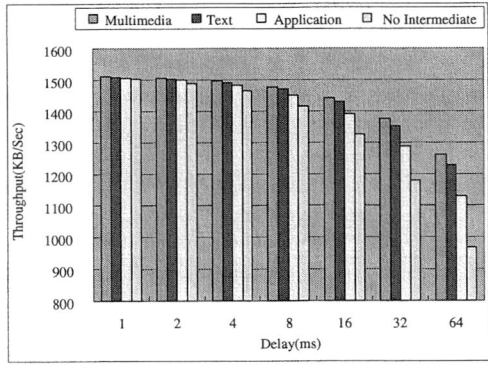

Fig. 10. Performance Result by Data Types (Read Operation)

Fig. 11 shows throughput change by distance ratio of an intermediate target for iSCSI write operations. In cases of a write operation, throughput does not change regardless of data type, so we experimented how the performance of iSCSI write operation changes according to the intermediate target's position. Same as a read operation, iSCSi write operation has a small difference in performance when total distance is short, but the difference is larger if the distance is longer. Furthermore the results show that performance appears high when delay time of the iSCSI initiator and intermediate target is smaller. iSCSI performance difference was biggest at the point where total delay time is 64ms, as is the case of read operations. And in the case of the ratios, A:B = 5:5, A:B = 3:7, A:B = 9:1, performance elevation of each 25%, 40%, 59% occured. Therefore, we know that proposed method show high performance when the iSCSI intermediate target is close to the initiator.

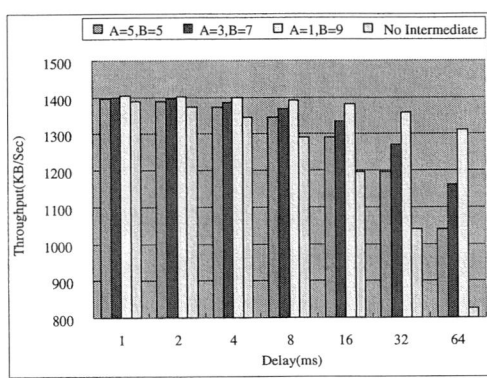

Fig. 11. Throughput Change by Distance Ratio of Intermediate Server (Write Operation)

5 Conclusion and Future Work

In this paper, we described efficient ways introducing iSCSI intermediate target, to achieve a breakthrough against the problem of iSCSI performance falling when applying iSCSI-based remote storage services for mobile appliances. In the proposed method, localizing the iSCSI target improved low link utilization and iSCSI performance degradation problems when the initiator receded from the target. Through simulation results we know that iSCSI performance falls rapidly if the distance (latency) of the initiator and target is long, and could see that iSCSI performance is high if the intermediate target is close to the initiator.

In future work, we will apply various prefetch algorithms for data type, or improve present FIFO iSCSI buffer management algorithm to heighten buffer hit rate. To accomplish this future work, we have to research how data type is classified and methods of sending information, and other buffer management algorithms.

References

1. S. J. Shepard: Embedded Databases Can Power Emerging World of Information to Go. IEEE Distributed Systems Online.
2. Block Device Driver Architecture, http://msdn.microsoft.com/library/en-us/wceddk40/html/_wceddk_system_architecture_for_block_devices.asp
3. T. Clark: IP SANs : A Guide to iSCSI, iFCP, and FCIP Protocols for Storage Area Networks. Addison-Wesley (2002)
4. X. He, Q. Yang, M. Zhang: A caching strategy to improve iSCSI performance. Proc. of IEEE LCN 2002
5. S. Park, B.-S. Moon: Design and Implementation of iSCSI-based Remote Storage System for Mobile Appliance. Data Storage System Research Group (2003) 236-240
6. Y. Lu and D. H. C. Du: Performance Study of iSCSI-Based Storage Subsystems. IEEE Communication Magazine. Aug. (2003)
7. SAM-3 : Information Technology - SCSI Architecture Model 3. Working Draft. T10 Project 1561-D. Revision7 (2003)
8. S. Park, B.-S. Moon, M.-S. Park: Design, Implementation, and Performance Analysis of the Remote Storage System in Mobile Environment. ICITA 2004
9. B.-S. Moon, M.-S. Park: A Performance Analysis of the CIFS, NBD, iSCSI on the Wireless Network by using CDMA-2000 1x. STORAEG 2003. Data Storage System Research Group
10. J. Satran: iSCSI Draft20, http://www.ietf.org/internet-draft/draft-ietf-ips-iscsi-20.txt
11. Y. Hu and Q. Yang: DCD-disk caching disk : A New Approach for Boosting I/O Performance. Proc. of ISCA'96 (1996)
12. K. Z. Mesh, J. Satran: Design of the iSCSI Protocol. Proc. of NASA Goddard Conference on MSS'03

Reflective Middleware for Location-Aware Application Adaptation

Uzair Ahmad[1], S.Y. Lee[1], Mahrin Iqbal[2], Uzma Nasir[2], A. Ali[2], and Mudeem Iqbal[3]

[1] Computer Engineering Dept. Kyung Hee University, 449-701 Suwon, Republic of Korea
[2] NUST Institute of Information Technology
[3] FAST, Pakistan
{uzair, sylee}@oslab.khu.ac.kr,
{54mahrin, 54uzma, arshad.ali}@niit.edu.pk

Abstract. Today mobile computing is pervasively taking over the traditional desktop computing. Mobile devices are characterized by abrupt and unannounced changes in execution context. The applications running on these devices need to be autonomous and thus dynamically adapt according to the changing context. Existing middleware support for the typical distributed applications is strictly based on component technology. Future mobile applications require highly dynamic and adaptive services from the middleware components i.e. context-aware autonomic adaptation. Traditional middleware do not address this emerging need of wide ranges of mobile applications mainly because of their monolithic and inflexible nature. It is hypothesized that such application adaptation can be achieved through meta-level protocols that can reflectively change the state and behaviors of the system. We integrate Component Technology with our active Meta Object Protocols for enabling mobile applications to become adaptive for different contexts. This paper implements the application adaptation service of this middleware. It specializes the concept of Meta Object Protocols for autonomic adaptation of mobile applications. ActiveMOP provide robust and highly flexible framework for autonomic component development for mobile applications. It proved to be a very simple and powerful way to programmatically develop location-driven applications based on autonomic components.

1 Introduction

Adaptivity is a distinguishing characteristic of emerging Ubiquitous Computing environments that clearly separates it from desktop computing. Software are supposed to be adaptive towards changing environments, ad hoc networks, multitude of communication protocols and most importantly user's current context.

Desktop computing is no longer sufficient to meet the ever-increasing information needs of the highly mobile world. Future computing environments promise to free the user from the constraints of this stationary desktop Computing. But mobile devices are characterized by abrupt and un-announced changes in execution context. The applications running on these devices need to be autonomous and thus dynamically reconfigure according to the changing context [1].

The conventional Middleware do not provide appropriate support for dealing with the dynamic aspects of this new mobile computational infrastructure. Due to this reason today the research focus is shifting from Static Design time services to runtime adaptive services. [2], [3], [4] and [5]. Application adaptation can be achieved through meta-level protocols that can reflect internal state and behaviors of the system.

1.1 Application Adaptation and Meta Object Protocols

Reflection is a discipline to achieve inspection and adaptability [4], [6], [5] and the protocols designed to achieve reflection are known as the *"Meta Object Protocols"*. MOPs can enable software to change itself e.g. alter its behaviors, reconfigure its settings, recover the damages, optimize its structures, and install new security mechanisms. As shown in Fig. 1, the meta object can be changed at runtime to get adaptive services.

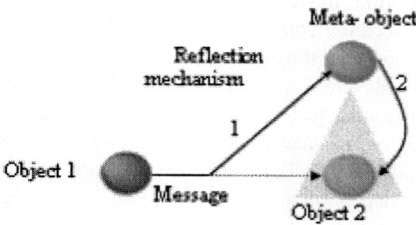

Fig. 1. MOPs at work

Meta Object Protocols are an abstraction of the computational process where the protocols governing the execution of the program are exposed. A *Meta Object* is bound to the base object and controls the execution of the object. The method calls going to base object are intercepted and diverted to the Meta Object. By changing the implementation of the Meta Object the object's execution can be adjusted in a principled way.

1.2 A Middleware Using MOPs

A major advantage of using middleware to develop software is that it hides the details of the underlying layers and operating system specific interfaces. Developers of distributed applications can write code that looks similar to code for centralized applications; the middleware takes care of networking, method dispatching, scheduling etc. The code running on top of the middleware is easily portable and the programmer need not to worry about the internals of the operating system and of the middleware.

System and application code may use meta-interfaces to inspect the internal configuration of the middleware and when needed, reconfigure it to adapt to changes in the environment. Hence it becomes possible to choose networking protocols, security policies, encoding algorithms [3], and various other components to provide personalized service for different contexts and locations.

Our focus is on building a fully dynamic middleware thus providing development APIs for the development of Location-Driven applications so that at runtime behavior adaptive services can be provided depending upon the location of the mobile client. This research specializes the concept of Meta Object Protocols and introduces ActiveMOP that has been used in our Location Driven Adaptive Middleware (LDAM) for robust highly flexible application adaptation.

2 Existing Work

Dynamic TAO [7] is an extension of the C++ TAO ORB [8], enabling runtime reconfiguration of the ORB internal engine and of applications running on top of it. But being built on top of the static code of TAO it does not provide much flexibility for offering adaptability. The Open ORB project [9] aims at the design of highly configurable and dynamically reconfigurable middleware platforms to support applications with dynamic requirements. It is built using components, component frameworks and reflection. But using many component frameworks increases the size of the middleware implementation; extra management functionality for managing reconfiguration exhausts the constrained resources of the mobile device. Moreover these existing systems like Dynamic TAO and Open ORB are built for application domains, such as multimedia and real-time only and do not address many other issues related to middleware in mobile computing. [5] Identifies that the key property in supporting mobile computing is the ability to seamlessly interoperate with the range of ubiquitous devices that are encountered by the mobile device as it changes location. Therefore, the Universal Interoperable Core (UIC) [5] has been developed; this reflective middleware is loosely based on the reconfiguration techniques of Dynamic TAO. The platform can change between different middleware personalities. But the implementation of UIC concentrates on synchronous middleware styles and does not implement all paradigm types that could be encountered in a ubiquitous environment. [10]

ReMMoC [10] also examines the use of reflection and component technology to overcome the problems of heterogeneous middleware technology in the mobile environment as in OpenORB. But, ReMMoC consists of two key component frameworks: (1) a binding framework for interoperation with mobile services implemented upon different middleware types, and (2) a service discovery framework for discovering services advertised by a range of service discovery protocols [10]. Hence it over comes the problem of exhaustion of the mobile devices resources. It uses OpenCOM [2] as its underlying component technology. OpenCOM uses a subset of Microsoft COM technology, hence is not an open source middleware. It needs an Active Space (a physical space where mobile devices communicate via certain set mechanisms) called Gia for its execution.

To our best knowledge MOP are not used to specifically address the Location-driven Adaptation of applications for handheld devices and mobile clients. This problem faces a number of challenges in order to be operative for really interactive mobile applications. There is a need to extend component technology to take the benefit of MOP in order to support rich behavior adaptation. They should be supporting heterogeneous client handheld devices and independent of a particular wireless operating environment. Furthermore unlike existing MOP implementations, it should be self managing to support autonomic component development.

We propose a design pattern called the Autonomist Design Pattern that can be used to design Meta Object Protocols for dynamic adaptation of mobile applications. ActiveMOP has been built using Autonomist DP that supports behavior as well as application adaptation. Location Driven Adaptive Middleware (LDAM) is a middleware that implements ActiveMOP to provide development as well as run time adaptive services to location driven mobile applications across different environments in a principled manner.

3 Autonomist DP

As part of this research we have developed a design pattern, known as the Autonomist DP that supports design of Meta Object Protocols for application adaptation and construction of components that are Autonomic in nature. The design pattern offers adaptation at the behavioral as well as the application level. Its participants are shown in Fig 2. The Autonomic Component is at the base level, while the Autonomic Service and Service Template are at the meta level. The Meta Controller acts as a mediator between the base and the meta level and intercepts all the calls coming to the base level, sets the meta level with new behaviors and calls the base level that experiences new changed behaviors.

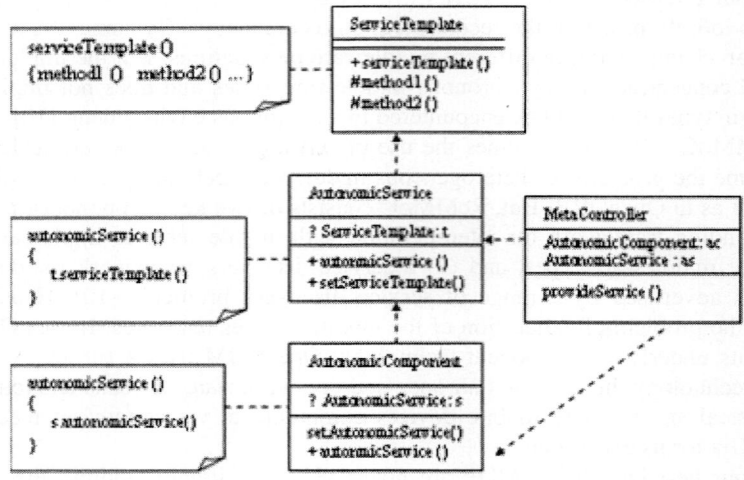

Fig. 2. Autonomist DP Participants

4 ActiveMOP

To show the strength of Autonomist DP, ActiveMOP was constructed. Fig. 3 shows ActiveMOP working. When a mobile client invokes the base level, the Meta Controller intercepts this call. The Meta Controller registers this client with the location service (which is peer research involving location sensing MOPs). After registration the Meta Controller is constantly updated about the clients location and sets the Autonomic Service with each update. In turn the Autonomic Service sets the Behavior Template based on the location value. Since location updation and behavior switching is done actively, that is why this MOP is called the ActiveMOP.

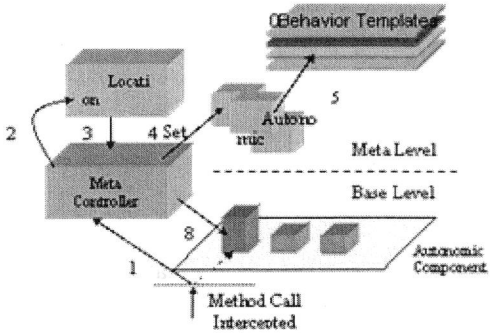

Fig. 3. ActiveMOP working

5 LDAM

Location Driven Adaptive Middleware (LDAM) is a middleware that provides development as well as run time adaptive services for location driven mobile applications across different environments. ActiveMOP has been used to make LDAM capable of behavior level as well as application level adaptation specifically based on changing location.

5.1 System Architecture of LDAM

In the LDAM architecture the behavior adaptive services are realized through the components of ActiveMOP. All the components are part of a centralized Container that pools different components for instant execution.

- Container

Container is the main holder of all the components. It controls all the activity of the reflective middleware. When the container is started up, it initializes all the Autonomic Components in a pool. It also initializes a Meta space that contains a pool of various template behaviors, the autonomic service object pool and the location service object pool.

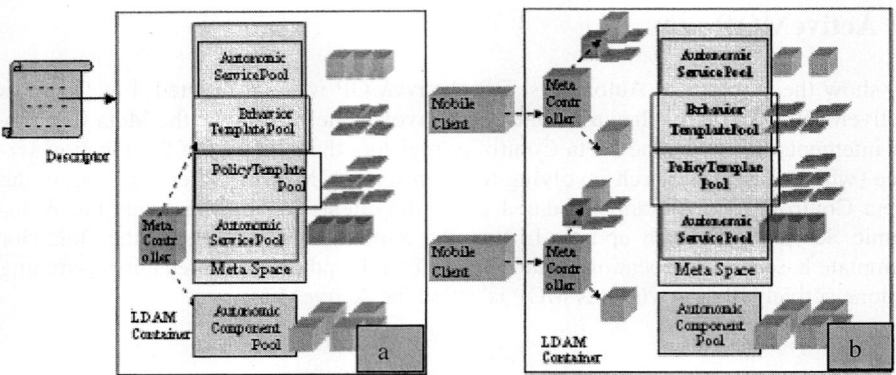

Fig. 4. (a) Descriptor defines LDAM container; (b) Dedicated Meta Controller for each mobile client

At the time of arrival of a client, the container fetches a behavior adaptive service (Autonomic Service) object and a location service object from the respective pools and binds it with a dedicated Meta Controller.

- Meta Controller

The Meta Controller acts as a mediator and interceptor of calls coming to the base level. It sets up the Meta level based upon the current location of the client and calls the base level method to execute the new behavior. The Meta Controller offers network transparency and hides the implementation details from the client.

- Autonomic Component

The Autonomic Component is a representative of the client application. It is the base level component containing an Autonomic Method, the execution of which is controlled by the Meta Controller.

- Behavioral Templates

The behavioral templates are different template applications that are actually the adaptive behaviors for each particular location. Service Template is an interface that defines these behavior templates. These different BTs are the multiple implementations identifying a complete working component present against each Autonomic Component.

- Autonomic Service

Autonomic service is setup by the Meta controller for each client. The location attribute is passed to the autonomic service by the Meta controller whenever behavior adaptation is required. The task of the Autonomic Service is to match the behavior with its corresponding Location. The resulting Template Behavior is set for execution in the service residing in the Autonomic Component. The Meta Controller invokes the Autonomic Method to execute the new template behavior through this service.

- Location Service

The second part of this research is to provide Meta Object Protocols (MOPs) for location monitoring, acquisition and updation. It facilitates the development of location sensitive applications by providing standard vocabulary in the form of APIs that encapsulate in them the description of location awareness thus shielding the programmer from many tedious and error prone aspects of location determination of an object representing a mobile device.

6 Implementation Results

The initial size of the client is very small, but without behavior adaptation the increase in behaviors causes a drastic increase in client size.

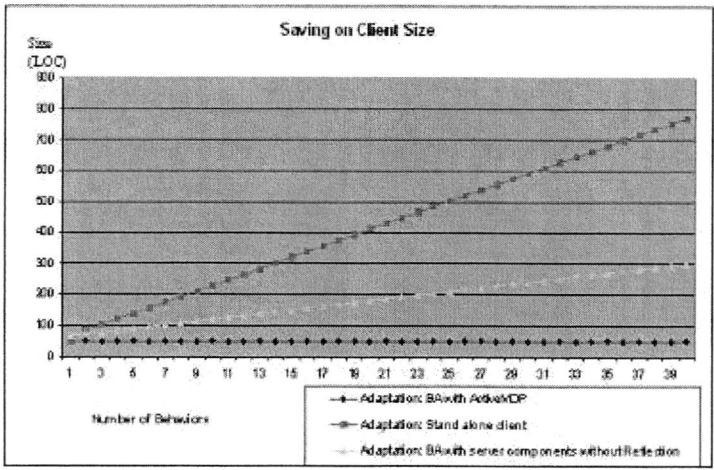

Fig. 5. Behavior Adaptation (BA) service size comparison

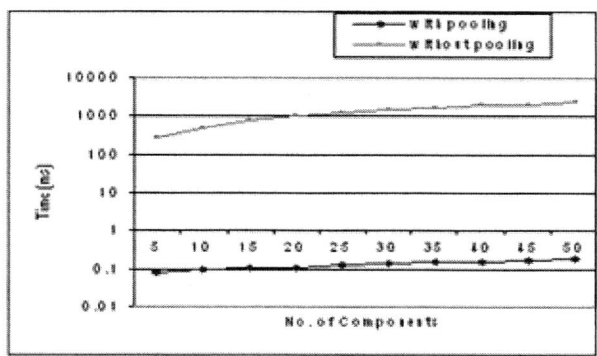

Fig. 6. Pooling vs. no pooling

Through behavior adaptation in LDAM though, no code for the behavioral templates is present on the client. For as many behaviors there is no change in client size. This is shown in Fig.5.

Pooling was incorporated in LDAM after seeing the remarkable results found for component loading when they were already pooled. These results have been shown in Fig. 6. As the number of components increases the time for initializing them increases rapidly.

The startup time for the client was noted to vary between 2ms to 20ms. This was the case when the Container initializes the Meta Controller by fetching components from each pool. Incase there is no pooling the components have to be instantiated each time a new client comes. This was noted to vary in range of 20 to 37 milliseconds.

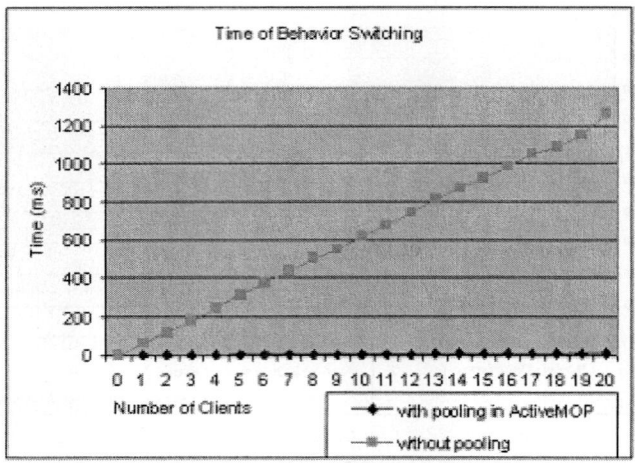

Fig. 7. Behavior switching time comparison

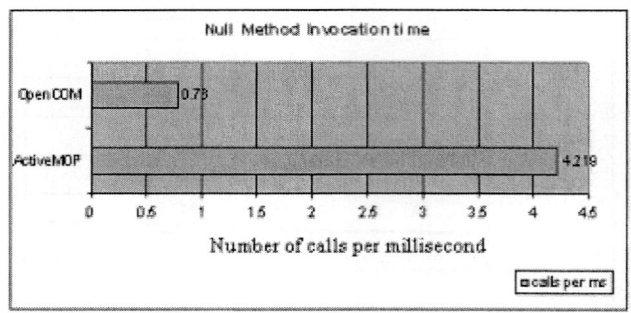

Fig. 8. ActiveMOP comparison with OpenCOM

Pooling also resulted in fast execution of behavior adaptation process. At location change time the behavior switching was transparent for the client, i.e. it took the Meta controller nearly 0 ms to change the behavior. Whereas without pooling, when each time instantiation of components is required, it took 15 to 64 ms to switch a behavior. Pooled components save both startup as well as execution time of the client.

Fig. 8. Shows ActiveMOP comparison with OpenCOM for, the number of null method (method with no code) invocations possible in one millisecond. Through ActiveMOP 4.219 calls can be made in one millisecond while in OpenCOM only 0.78 calls can be made in one millisecond.

All the tests for LDAM were conducted on Pentium III 800 M-Hz machine with 256MB RAM. The tests for OpenCOM had been performed on Dell Precision 410MT workstation with 256Mb RAM and Intel Pentium III processor 550Mhz. The operating system was Microsoft's Windows2000.

7 Conclusion

In this paper a generic architecture for a behavioral adaptive middleware targeting location driven applications is presented where emphasis has been laid upon dynamic behavior adaptation of the client device keeping in view the fact that mobile devices have a small memory footprint and the behavior adaptive services need to be lightweight and flexible. This middleware was named LDAM, Location-Driven Adaptive Middleware. It is a combination of component technology and Meta object protocols to be capable of providing adaptive services to the mobile client.

The Autonomist Design Pattern opens up new avenues for application programmers who want to design systems to implement adaptation at different granularity levels i.e. application, behaviors. ActiveMOP provides complete supports for location-based adaptation at the behavior level as well as the application level. This adaptation is transparent for the client since pooling the components makes behavior-switching time reduce to being negligible. The LDAM client enjoys a vast range of behavioral services with no burden on the memory.

References

1. Research.IBM.com/autonomous (2002)
2. Clarke, M., Blair, G., Coulson, G. and Parlavantzas, N. "An Efficient Component Model for the Construction of Adaptive Middleware". In Proceedings of Middleware 2001, Heidelberg, Germany (November 2001)
3. Fabio Kon et al, The case for reflective middleware Communications of the ACM Volume 45, Issue 6 (June 2002)
4. L. Capra et al "Exploiting Reflection in Mobile Computing Middleware". In ACM, SIGMOBILE Mobile Computing and Communications Review.
5. Roman, M., Kon, F. and Campbell, R. H. "Reflective Middleware: From Your Desk to Your Hand". IEEE DS Online, Special Issue on Reflective Middleware, 2001.
6. Thomas Ledoux. OpenCORBA: A Reflective Open Broker. In Proceedings of Reflection'99, number 1616 in LNCS, pages 197–214, St. Malo, France, July 1999. Springer Verlar.

7. Fabio Kon, Manuel Rom´an, Ping Liu, Jina Mao, Tomonori Yamane, Luiz Claudio Magalhaes, and Roy H. Campbell. Monitoring, Security, and Dynamic Configuration with the dynamicTAO Reflective ORB. In Proceedings of the IFIP/ACM International Conference on Distributed Systems Platforms and Open Distributed Processing (Middleware'2000), number 1795 in LNCS, pages 121–143, New York, April 2000. Springer-Verlag.
8. Douglas C. Schmidt and Chris Cleeland. Applying Patterns to Develop Extensible ORB Middleware. IEEE Communications Magazine Special Issue on De-sign Patterns, 37(4):54–63, May 1999.
9. Gordon S. Blair, Geoff Coulson, An-ders Andersen, Lynne Blair, Michael Clarke, F´abio Costa, Hector Duran-Limon, Tom Fitzpatrick, Lee Johnston, Rui Moreira, Nikos Parlavantzas, and Katia Saikoski. The Design and Implementation of Open ORB 2. IEEE Distributed Systems Online, 2(6), 2001.
10. Paul Grace, Gordon S. Blair and Sam Samuel. Interoperating with Services in a Mobile Environment. Distributed Multimedia Research Group, Computing Department. Lancaster University, Lancaster, LA1 4YR, UK., Global Wireless Systems Research, Bell Laboratories, Lucent Technologies, Quadrant, Stonehill Green, Westlea, Swindon, SN5 7DJ. lsamuel@lucent.com, p.grace@lancaster.ac.uk, gordon@comp.lancs.ac.uk

Efficient Approach for Interactively Mining Web Traversal Patterns

Yue-Shi Lee, Min-Chi Hsieh, and Show-Jane Yen

Department of Computer Science and Information Engineering, Ming Chuan University,
5 The-Ming Rd., Gwei Shan District, Taoyuan County 333, Taiwan, R.O.C.
leeys@mcu.edu.tw

Abstract. Web mining is one of the mining technologies, which applies data mining technique in large amount of web data to improve the web services. Web traversal pattern mining discovers most of users' access patterns from web logs. When we understand the users' behaviors, we can make some appropriate actions for different purposes. However, it is considerably difficult to select a perfect minimum support threshold during the mining procedure to find the interesting rules. Even though the experienced experts, they also cannot determine the appropriate minimum support to find the interesting rules. Thus, we must constantly adjust the minimum support until the satisfactory mining results can be found. This will waste a lot of time on these repeating mining processes with the same data. Therefore, many researchers pay attention to the interactive data mining in recent years. The essence of interactive data mining is that we can use the previous mining results to reduce the unnecessary processes when the minimum support is changed. In this paper, we propose an efficient interactive web traversal pattern mining algorithm to reduce the mining time and make the mining results to satisfy the users' requirements.

1 Introduction

With the trend of the information technology, huge amounts of data would be easily produced and collected from the *electronic commerce* environment every day. It causes the web data in the database to grow up at amazing speed. Hence, how should we obtain the useful information and knowledge efficiently based on the huge amounts of web data has already been the important issue at present.

Web mining [1] refers to extracting useful information and knowledge from large amounts of web data, which can be used to improve the *web services*. Mining *web traversal patterns* [2] is to discover most of users' access patterns from web logs. These patterns can not only be used to improve the website design, e.g. provide efficient access between highly correlated objects, and better authoring design for web pages, etc., but also be able to lead to better marketing decisions, e.g. putting advertisements in proper places, better customer classification, and behavior analysis, etc.

In the following, we describe the definitions about web traversal patterns: Let $I = \{x_1, x_2, ..., x_n\}$ be a set of all web pages in a website. A *web traversal sequence* $S = <w_1, w_2, ..., w_m>$ $(w_i \in I, 1 \leq i \leq m)$ is a list of web pages which is ordered by traversal time, and the web page can repeatedly appear in a web traversal sequence. The *length* of the web traversal sequence S, which is denoted as $|S|$, is the total number of web pages in S. A web traversal sequence with length l is called an l-sequence. Suppose that there are two web traversal sequences α and β, if $\alpha \subseteq \beta$, α is the *web traversal sub-sequence* of β, and β is the *web traversal super-sequence* of α. For instance, if there are two web traversal sequences $\alpha = <ACD>$ and $\beta = <ABCBD>$, then α is a web traversal sub-sequence of β and β is a web traversal super-sequence of α.

A *web traversal sequence database D*, as shown in Fig. 1, contains a set of records. Each record includes *traversal identifier (TID)* and a *traversal sequence*. A traversal sequence is a web traversal sequence, which stands for a complete browsing behavior by a user. The *support* of a web traversal sequence α is the ratio of traversal sequences which contains α to the total number of traversal sequences in D. It is usually denoted as *Support (α)*. A web traversal sequence α is a *web traversal pattern* if *Support (α) $\geq$ min_sup*, in which the *min_sup* is the user specified *minimum support* threshold. For instance, in Fig. 1, if we set *min_sup* to 80%, then *Support (<CA>)* = 4/5 = 80% $\geq$ *min_sup* = 80%. Hence, *<CA>* is a web traversal pattern. If the length of a web traversal pattern is l, then it can be called *l-web traversal pattern*.

TID	Traversal Sequence
3	CDEAD
4	CDEAB
5	CDAB
6	ABDC
7	ABCEA

Fig. 1. Web Traversal Sequence Database

Based on the *min_sup*, all the web traversal patterns can be found. Thus, it is very important to set an appropriate *min_sup*. If the *min_sup* is set too high, it will not find enough information for us. On the contrary, if the *min_sup* is set too low, it will find excess noise and waste our time. However, it is very difficult to select a perfect minimum support threshold in the mining procedure to find the interesting rules. Even though experienced experts, they also cannot determine the appropriate minimum support threshold. Therefore, we must constantly adjust the minimum support until the satisfactory results can be found. This will waste our time on these repeat mining processes. Since it is difficult to know exactly what can be discovered within a database, an interactive scheme is needed. So, many researchers pay attention to the *interactive mining* [3] in the recent years.

In this paper, we propose a novel *interactive web traversal pattern mining* algorithm and a storage method to reduce the mining time and make the mining results satisfied the real requirements. The rest of the paper is organized as follows. Section 2 introduces the most recent researches related to this work. Section 3 describes our web traversal pattern mining algorithm and storage structure. Before concluding, we compare our method with the methods without interactive mining.

2 Related Work

Path traversal pattern mining [4, 5, 6, 7, 8, 9] is the technique that find navigation behaviors for most of the customers in the web environment. The web site designer can use this information to improve the web site design, and to increase the web site performance. Besides, this information can also provide the navigation suggestions to customers. Many researches focused on this field, e.g., *FS (Full Scan)* algorithm, *SS (Selective Scan)* algorithm [4], *MFTP (Mining Frequent Traversal Patterns)* algorithm [5], and other algorithms [6, 7]. Nevertheless, these algorithms have the limitations that they can only discover the *simple path traversal pattern*, i.e., the page cannot repeat in the pattern. *Non-simple path traversal pattern, i.e., web traversal pattern,* contains more information. It can be used to predict customer's behaviors more accurately. The related research is *IPA (Integrating Path traversal patterns and Association rules)* algorithm [8, 9].

Interactive mining has been proposed for mining *sequential patterns*. [3] proposed a *KISP (Knowledge base assisted Incremental Sequential Pattern)* algorithm for interactively finding sequential patterns. KISP algorithm extends the *hash tree* in *GSP* algorithm to count the *candidate sequence* and constructs a *KB (Knowledge Base)* structure in hard disk to minimize the response time for iterative mining. KISP algorithm used the previous information in KB and extends the content for further mining. Based on the KB, KISP algorithm can mine sequential patterns on different minimum support thresholds without using original database.

KISP differs from our proposed algorithm in several ways. First, our algorithm is proposed to interactively discover web traversal patterns. KISP algorithm is proposed for the sequential patterns. KISP does not consider the web site structure to increase the mining performance. Second, we cannot directly apply KISP algorithm on mining the web traversal patterns. The combination method in mining sequential patterns is different from mining web traversal patterns. Third, KISP algorithm cannot obtain the *longest sequential patterns* immediately. Our algorithm can easily obtain the longest web traversal patterns. In next section, we will describe our algorithm in details.

3 Algorithm for Interactive Web Traversal Pattern Mining

For interactive mining, we use the previous mining results to discover new patterns and reduce mining time. Therefore, how to choose a well storage structure to store previous mining results becomes very important. In this paper, lattice structure is selected as our storage structure. Fig. 2 shows the original lattice structure O for the database described in Fig.1, when min_sup is set to 50%. In O, only web traversal patterns are stored in this structure. To interactively mine the web traversal patterns and speed up the mining processes, we extend O to record more information. The extended lattice structure E is shown in Fig. 3. In Fig.3, each node stands for one web traversal sequence. We append the support information into the upper part of each node. This information can help us to calculate and accumulate the support when the interactive mining is proceeding.

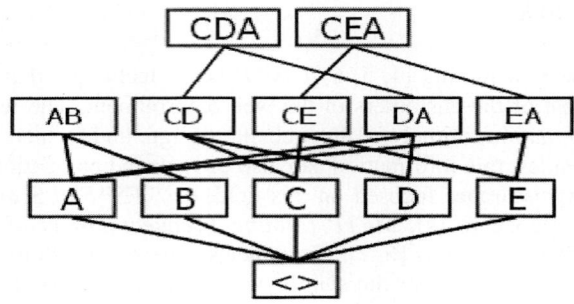

Fig. 2. Original Lattice Structure

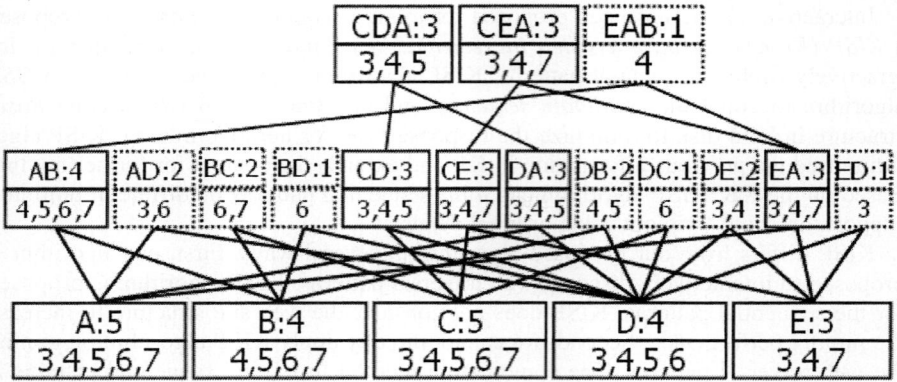

Fig. 3. Extended Lattice Structure

Moreover, we also append the TID information, which the web traversal sequence occurs, into the lower part of each node. This information can help us to reduce the unnecessary database scans. According to TID information, we only need to scan some records of web traversal sequence database. If the mining system does not have enough memory space to store the TID information, then we need to scan all records of database such that our algorithm can also work. Different from original lattice, we put all web traversal sequences, which their support count are greater than or equal to one, into the lattice structure. The reason for putting all web traversal sequences into the lattice structure is that it is more convenient for us in interactive mining. The lattice structure is saved in hard disk level-by-level. The total size of the lattice structure (including TID information) is about 5 times larger than the original database in average. Because our method is to mine the patterns level-by-level, it will not cause the memory be broken when we just load one level of lattice structure into memory.

The lattice structure is a well storage structure. It can quickly find the relationships between patterns. For example, if we want to search for the patterns related to "A" web page, we just traverse the lattice structure from the node represented as "A" web page. Moreover, if we want to find the longest web traversal patterns, we traverse the lattice structure once and output the patterns in leaf nodes, whose supports are greater than or equal to *min_sup*.

The following shows an example for mining web traversal patterns. The database is shown in Fig.1. The web site structure shown in Fig. 4 is also used in this example. Fig. 3 is the mining result, when *min_sup* is set to 50%. If the sequence is a web traversal pattern, it is denoted by the single line. Otherwise, it is denoted by the dotted line.

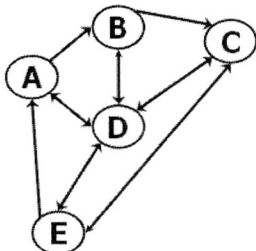

Fig. 4. Web Site Structure

The reason for using the web site structure is that we want to avoid the unqualified web traversal sequences to be generated in the mining processes. For example, assume that our web site has 300 web pages and all of them are all 1-traversal patterns. If we do not refer to the web site structure, then 299×300=89,700 2-sequences are generated in next step. But, in most situations, most of them are unqualified. Assume that the average number of links between web pages is 10. If we refer to the web site structure, then just 300×10=3,000 2-sequences are generated.

Different from the combination method in sequential pattern mining [10], we use the middle term to join the candidates. For example, <ABCDE> is joined by <ABCD> and <BCDE>. BCD is the middle term in this example. Besides, we also check all of the qualified web traversal sub-sequences with length l-1 to reduce some unnecessary combinations. In this example, we need to check <ABDE> and <ABCE>. If one of them is not a web traversal pattern, <ABCDE> is also not a web traversal pattern. We do not need to check <ACDE>, because <ACDE> is an unqualified web traversal sequence (no connections between web pages A and C).

The following shows an example for interactive mining. First, we change the *min_sup* from 50% to 70%. Because we increase the minimum support threshold, we just traverse the lattice structure once and output the web traversal patterns, which the supports are greater than or equal to 70% (occurs at least four records). In this example, the outputs are <CDA>, <CEA>, and <AB>. Second, we change the *min_sup* from 50% to 40% (occurs at least two records). Because we decrease the minimum support threshold, we should scan the database again and update the lattice structure. First of all, we scan the first level of the lattice structure. Because no new 1-web traversal patterns are generated (no updates in this level) in this case, we scan the second level of the lattice structure. Then, we find that <AD>, <BC>, <DB> and <DE> become new 2-web traversal patterns. It is shown in Fig. 5. The double line represents these four new patterns.

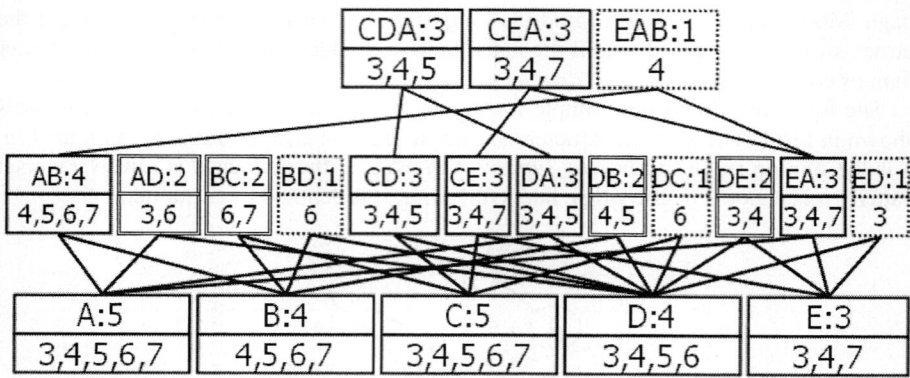

Fig. 5. Update on Second Level of Lattice

According to the web site structure, we join these four new patterns with the original web traversal patterns (denoted by single line). The results are shown in Fig. 6. In this example, although <BCD> can be joined by <BC> and <CD>, we also need to check the web traversal sub-sequences. In this case, we need to check that whether <BD> is a 2-web traversal pattern or not, because there is a direct link between web pages B and D in the web site structure. In this case, <BD> is not a 2-web traversal pattern. Hence, <BCD> are not 3-web traversal pattern and eliminated from the candidate set. If there is not a direct link between B and D in web site structure, the <BCD> also can be a candidate, because we can not decide whether it is a web traversal pattern or not in current situation.

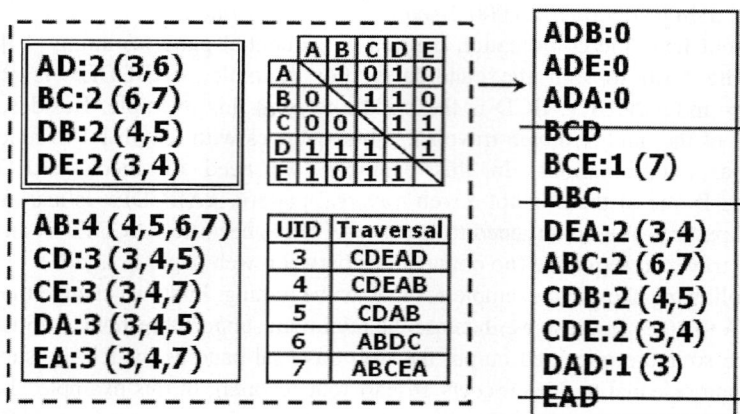

Fig. 6. The Results of Joining New-Generated Patterns with Original Web Traversal Patterns

We recursively join the new-generated patterns with the original web traversal patterns and count the candidates, until no candidates can be generated. Fig. 7 is the final result of the lattice structure when *min_sup* is set to 40%. The single line represents the web traversal patterns. The c++ like algorithm is listed below.

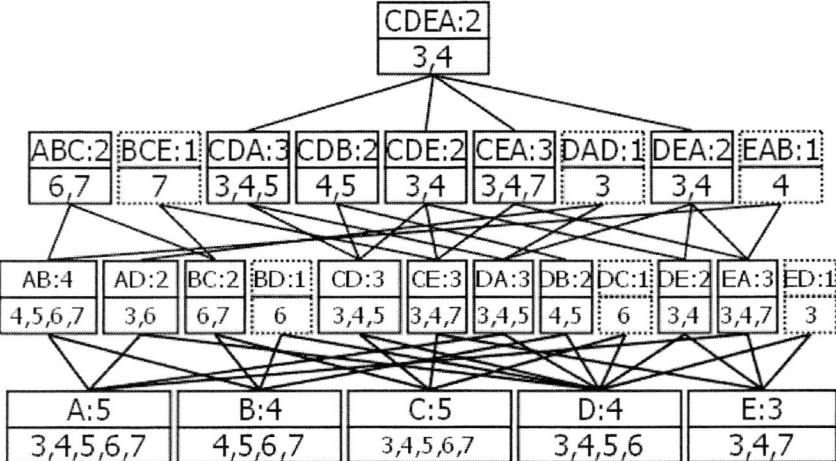

Fig. 7. The Final Result

```
Void Interactive_Mining(D, s, s', W, L)
{
   if(s'≤s)
   {
      Level=1;
      Lattice_Top_Level=Get_Top_Level(L);
      while(Level≠Lattice_Top_Level)
      {
         L=Interactive_Update_Lattice(D, s, s', W, L, Level);
         Level++;
         if(Have_New_Frequent_in_Top_Level(s, s', L))
            Ori_Mining(D, s', W, L, Level);
      }
   }
   Traverse_Lattice(L, s');
}
```

```
Lattice Interactive_Update_Lattice(D, s, s', W, L, Level)
{
   Level_k=Read_Lattice_Into_Memory(L,Level);
   A=Find_Frequent_Pattern(Level_k, s);
   B=Find_Frequent_Pattern(Level_k, s');
   New_Freq=B-A;
   New_Cand=Interactive_Generate_Candidate(A, New_Freq, W);
   L=Count_Candidate(New_Cand, D, L, Level);
   return L;
}
```

```
Candidate Interactive_Generate_Candidate(Ori_Freq, New_Freq, W)
{
  New_Cand=∅;
  for each frequent $F_i$ in New_Freq
  {
    for each frequent $F_j$ in New_Freq
    {
      if(i≠j)
      {
        Cand_temp= $F_i \otimes F_j$;
        if(Subset_Check(Cand_temp, W, Ori_Freq, New_Freq))
          New_Cand = New_Cand ∪ Cand_temp;
      }
    }
    for each frequent Fj in Ori_Freq
    {
      Cand_temp= $F_i \otimes F_j$;
      if(Subset_Check(Cand_temp, W, Ori_Freq, New_Freq))
        New_Cand = New_Cand ∪ Cand_temp;
    }
  }
  return New_Cand;
}
```

In the above three algorithms, D denotes the web traversal sequence database, W denotes the web site structure, L denotes the lattice structure, s denotes the current min_sup, and s' denotes the new min_sup. The longest web traversal patterns will be outputted as the results. The *Interactive_Mining* is the major algorithm. If s' greater than s, we just traverse the lattice. Otherwise, we update the lattice level-by-level. In *Interactive_Update_Lattice* algorithm, we read the lattice structure level-by-level. In each level, we generate new candidates *New_Cand* by using *Interactive_Generate_Candidate* algorithm and then count these new candidates by using *Count_Candidate* algorithm. Finally, the updated lattice is returned as the results. The *Interactive_Generate_Candidate* algorithm generate new candidates by self-joining new web traversal patterns *New_Freq* and join new web traversal patterns *New_Freq* and original Ori_Freq web traversal patterns.

4 Experimental Results

In the following experiments, we use a real web traversal sequence database and several simulated datasets to test the execution performance of our proposed method. This real database is a networked database. It stores information for renting DVD movies. There are 82 web pages in the web site. We collect the user traversing data from 02/18/2001 to 02/24/2001 (seven days). There are 428,596 log entries in this original database. Before mining the web traversal patterns, we need to transform these web logs into the web traversal sequence database. The steps are listed as follows. Because we want to get most of users' traversal patterns, the log entries referred to images are not important. Thus, all log entries with access filename suffix like .JPG, .GIF, .SME, .CDF are removed. Then, we organize the log entries according to the user's IP address and time limit. After these processes, we can acquire the web traversal sequence database like Fig. 1.

According to these steps, we organize the original log entries into 12,157 traversal sequences. The comparison of using interactive mining and without interactive mining is depicted in Fig. 8. It is based on different *min_sup*. The initial *min_sup* is set to 20%. Then, we continually decrease the *min_sup*.

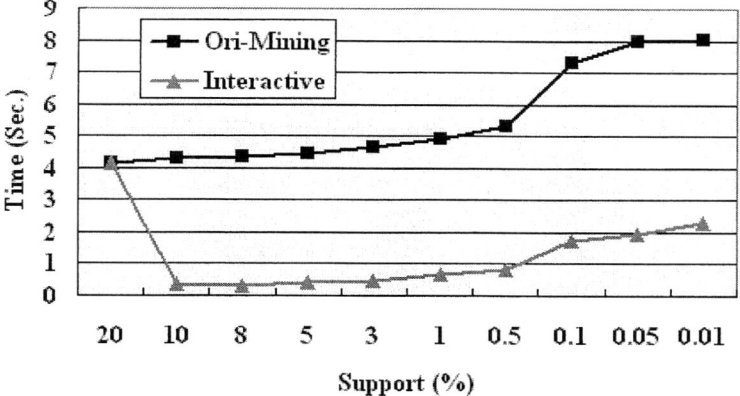

Fig. 8. Execution Time in Real Database

Table 1. The Ratio of Using Interactive Mining and Without Interactive Mining

Dataset \ min_sup	20%	10%	5%	1%	0.5%	0.1%	0.05%	0.01%
10K	1.00	3.25	5.27	3.37	6.74	7.62	7.99	8.28
30K	1.00	29.25	4.27	5.76	22.98	25.51	26.74	27.75
50K	1.00	12.11	3.87	6.55	38.59	43.39	46.11	47.39
100K	1.00	7.54	3.46	7.01	64.42	88.26	92.36	97.24

In the simulation datasets, we set the number of web pages to 300. Then, we generate 4 datasets with 10K, 30K, 50K and 100K traversal sequences, respectively. The experiment result is shown in Table 1. The initial *min_sup* is also set to 20%. Then, we continually decrease the *min_sup* as the previous experiment. Table1 lists the ratio of using interactive mining and without interactive mining, i.e., the execution time of original mining / execution time of interactive mining. Obviously, the mining time can be reduced dramatically. Our algorithm speeds up 3.25 to 97.24 times, according to the dataset size and the minimum support threshold.

5 Concluding Remarks

The interactive web traversal pattern mining was not discussed in the past. We consider that it is necessary to have interactive mining, because the user's behavior will change and we usually try various thresholds for the final desirable patterns. Thus, if we integrate the concept of interactive mining into the web traversal pattern mining, we will not waste our time on these repeating mining processes with the same data. The experimental results show that the interactive mining is very efficient. It can help us to quickly find a perfect minimum support.

Acknowledgement

Research on this paper was partially supported by National Science Council grant NSC93-2213-E-130-006, NSC93-2213-E-030-002.

References

1. Cooley, R., Mobasher, B., Srivastava, J.: Web Mining: Information and Pattern Discovery on the World Wide Web. Proceedings of the IEEE International Conference on Tools with Artificial Intelligence, (1997).
2. Srivastava, J., et al.: Web Usage Mining: Discovery and Applications of Usage Patterns from Web Data. SIGKDD Explorations, (2000) 12-23.
3. Lin, M.Y., Lee, S.Y.: Improving the Efficiency of Interactive Sequential Pattern Mining by Incremental Pattern Discovery.", Proceedings of the 36th Hawaii International Conference on System Sciences, (2002).
4. Chen, M.S., Park, J.S., Yu, P.S.: Efficient Data Mining for Path Traversal Patterns in a Web Environment. IEEE Transaction on Knowledge and Data Engineering, Vol. 10, No. 2, (1998) 209-221.
5. Yen, S.J.: An Efficient Approach for Analyzing User Behaviors in a Web-Based Training Environment. International Journal of Distance Education Technologies, Vol. 1, No. 4, (2003) 55-71.
6. Chen, M.S., Huang, X.M., Lin, I.Y.: Capturing User Access Patterns in the Web for Data Mining. Proceedings of the IEEE International Conference on Tools with Artificial Intelligence, (1999) 345-348.
7. Pei, J., Han, J., Mortazavi-Asl, B., Zhu, H.: Mining Access Patterns Efficiently from Web Logs. Proceedings of the Pacific-Asia Conference on Knowledge Discovery and Data Mining, (2000) 396-407.
8. Lee, Y.S., Yen, S.J., Tu, G.H., Hsieh, M.C.: Web Usage Mining: Integrating Path Traversal Patterns and Association Rules. Proceedings of International Conference on Informatics, Cybernetics, and Systems, (2003) 1464-1469.

9. Lee, Y.S., Yen, S.J., Tu, G.H., Hsieh, M.C.: Mining Traveling and Purchasing Behaviors of Customers in Electronic Commerce Environment. Proceedings of IEEE International Conference on e-Technology, e-Commerce and e-Service, (2004) 227-230.
10. 10.Agrawal, R., Srikant, R.: Mining Sequential Patterns. Proceedings of International Conference on Data Engineering, (1995) 3-14.

Query Decomposition Using the XML Declarative Description Language

Le Thi Thu Thuy[1] and Doan Dai Duong[2]

[1] Information Technology Department – Hue University of Sciences
[2] Informatics Department – Hue University of Education
{thuthuy, ddduong.sp}@hueuni.edu.vn

Abstract. Query decomposition is one of the most important phases of query processing in an integrated database system. A global query is decomposed into several sub-queries conforming to local formats, which can be used to extract data from distributed databases. In this paper a new query decomposition methodology for integrated XML databases is introduced. A special construction of mappings is also introduced, which provides information for query decomposition while efficiently avoiding data redundancy. Based on a set of given mappings, a global query is simultaneously decomposed into n sub-queries in one step, thus reducing time complexity. XML Declarative Description (XDD) - an extensible XML language incorporating a new variable class - provides the means to model as well as build the algorithms for the proposed system.

1 Introduction

The last few years have witnessed a great increase in the number of web sites on the world-wide computer network. A web site can be considered as a collection of semi-structured data, hence the quest for a structured standard language for the web is on-going. XML[1] - eXtensible Markup Language - has been proposed to fill this gap. It permits a new class of databases - XML databases. In order to support users during information access and utilization, distributed XML databases can be integrated and provide a unified representation of all participating XML databases. In an integrated XML database system, each participating XML database source can follow its own schema, which typically differs from the integrated schema. Users can pose their queries based on the integrated schema. These queries cannot be used to query the local sources directly due to the different formats of the global schema and the local ones. In order to extract data from these sources for further processing, the global XML query must be decomposed into all possible XML sub-queries. Each sub-query conforms to a schema of local sources; thus it can be executed to get the relevant data. The most recent database integration systems are: MIX [3, 4, 12], ADDSIA [5], DIXSE [6], Tukwila [8, 9]. Even though these systems have achieved certain results, their query decomposition algorithms still contain limitations caused by the heterogeneous formats of the query, metadata and programming language. None of them can simultaneously produce sub-queries for local sources in one step, thus they unnecessarily increase complexity and computer usage.

---

[1] http://www.w3.org/TR/REC-xml

Towards a solution to these existing problems, we introduce a new algorithm for query decomposition in integrated XML databases. In our approach, the global query is simultaneously decomposed into n sub-queries conforming to local source formats. Each of them can then be used to query the local data source directly. Moreover, all components of query decomposition are modeled as XML clauses, hence they can be flexibly processed in a uniform environment. We also construct a data dictionary (*mappings*) that can be used for both query decomposition and data conversion.

Section 2 gives a brief overview of XDD. Section 3 describes the proposed query decomposition framework, its steps and algorithms. Section 4 demonstrates system prototype. Section 5 concludes the paper.

2 XML Declarative Description

XML Declarative Description (XDD) [2, 11, 14, 15] is an XML based modeling language with well-defined declarative semantics and supports for computation and inference mechanisms. In order to obtain this expressive power, it incorporates variables with the conventional definition of XML elements.

Ordinary XML elements, that is XML expressions without variables are called *ground XML expressions*. Those containing variables are called *non-ground XML expressions*. Table 1 lists all types of variables in *non-ground XML expressions*. This representation facilitates expressing a set of *ground XML expressions* with similar characteristics by a *non-ground XML expression*.

Table 1. Types of XDD variable

Variable type	Begin with	Instantiation to
N-variables: Name-variables	$N	Element type or attribute names
S-variables: String-variables	$S	Strings
P-variables: Attribute-value-pair-variables	$P	Sequences of zero or more attribute value pairs
E-variables: XML expression-variables	$E	Sequences of zero or more XML expressions
I-variables: Intermediate-expression-variables	$I	Part of XML expressions

Moreover, an important concept of XDD is the XML clause [15] of the form:
$$H \leftarrow B_1, \ldots, B_m, \beta_1, \ldots, \beta_n$$
where $m, n \geq 0$; H and B_i ($i=1..m$) are XML expressions. Each β_j ($j=1..n$) is a predefined *XML constraint* - useful for defining restrictions on XML expressions or their components. The XML expression H is called the *head* of the clause while the set $\{B_1, \ldots, B_m, \beta_1, \ldots, \beta_n\}$ is the *body*. When the *body* is empty, such a clause is referred to as an *XML unit clause*; otherwise it is a *non-unit clause* with both *head* and *body*. A unit clause (H←.) is often denoted simply by H and referred to as a *fact*.

3 Query Decomposition

Fig. 1 shows an overview of the XML query decomposition. The input is an XML query followed by the integrated schema. Its outputs are XML sub-queries conformed to local schemas. In order to achieve the results, mappings and algorithms modeled by XDD are used.

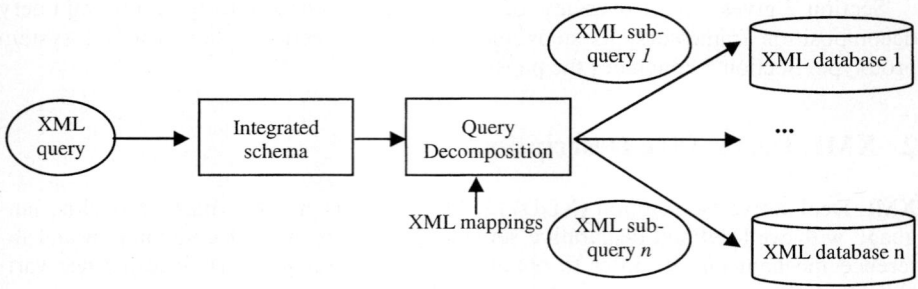

Fig. 1. Query decomposition overview

3.1 Query

User's queries can be expressed in terms of XQuery, which will be transformed into the XDD format. However, it is assumed that user queries are in the XDD format and comprise single XML clauses.

A query has three parts: *constructor*, *pattern* and *filter* which are *head*, *body* and *constraints* of the XML clause, respectively [11]. The following example shows an XML query modeled by XDD.

`<Answer>`
    `<name>$S:name</name>`
    `<nationality>$S:nation</nationality>`
`</Answer>`
←
    `<Student>`
        `<name>$S:name</name>`
        `<nationality>$S:nation</nationality>`
        `<GPA>$S:gpa</GPA>`
        `$E:properties`
    `</Student>`
    `[ $S:gpa>3.5]`
	% List *name* and *nationality*
	% of all *student* elements
	% which contain *GPA* sub-
	% element with a value of
	% more than 3.5.

When the *pattern* matches with a *ground XML expression* of the given XML document source and the *filter* is also satisfied, the result is returned in the form of the *constructor* with specifications (variables are replaced by *ground XML elements*). In order to decompose a global query into sub-queries, the query decomposition uses mappings to find corresponding elements between the integrated and the local schemas.

3.2 Mappings

The mappings introduced are composed in XML format and modeled by XDD as XML expressions. In order to describe the correspondence between an object (*XML attribute, XML element*) in the integrated XML schema and others in the local ones, the mapping contains two sub-XML expressions. The first represents the object in the integrated schema, while the second describes all corresponding objects in schemas of the local sources. Because XDD rules can be applied to manipulate XML expressions, mappings containing those sub-XML expressions can be processed to generate the expected results. The general form of the mappings is shown in Fig. 2.

```
<Mapping>                                          % This mapping specifies that
    <$N:globalTag>                                 % the element $E:exp in the
        $E:exp                                     % integrated schema corre-
    </$N:globalTag>                                % sponds to elements
    <local>                                        % $E:exp1,    $E:exp2    in
        <$N:tagName1 source=$S:source1>            % sources $S:source1 and
            $E:exp1                                % $S:source2, respectively.
        </$N:tagName1>
        <$N:tagName2 source=$S:source2>
            $E:exp2
        </$N:tagName2>
    </local>
</Mapping>
```

Fig. 2. General form of mapping

The following example is a mapping between the integrated schema and schemas of sources A and B.

```
<Mapping>                                          % This mapping speci-
    <student>                                      % fies that the country
        <country>$S:country</country>              % element which is child
    </student>                                     % of the student element
    <local>                                        % in the integrated
        <SATstudent source="A">                    % schema has two corre-
            <country>$S:country</country>          % sponding elements
        </SATstudent>                              % that are country and
        <SOMstudent source="B">                    % nationality in local
            <nationality>$S:country</nationality>  % sources A and B.
        </SOMstudent>
    </local>
</Mapping>
```

With such mappings, when the elements in the integrated schema or in the local ones are defined as *XML expressions*, they are very useful. They can be embedded themselves in XML clauses or XML rules that can apply for logic model in executing. We use that efficiently for query decomposition. In detail, if an element in the integrated schema is exactly known, its corresponding elements in local sources will be derived naturally by using XML rules. The general rule applying for mappings in query decomposition is shown.

```
$E:expression
 ←
   <Mapping>
      <globalTag>$E:exp</globalTag>
      <local>$E:expression</local>
   </Mapping>
```

% This rule specifies that if an ele-
% ment *$E:exp* in the integrated
% schema is explicitly known, its cor-
% responding elements in the local
% sources contained in *$E:expression*
% are yielded via the mapping.

An application for this rule is shown in Fig. 3. When this rule is executed, the body of the rule will match with one mapping and the variables ($E:exp,$E:expression) are replaced by explicit instances *(ground XML expressions)*, thus the *head* of rule will contain the expected results.

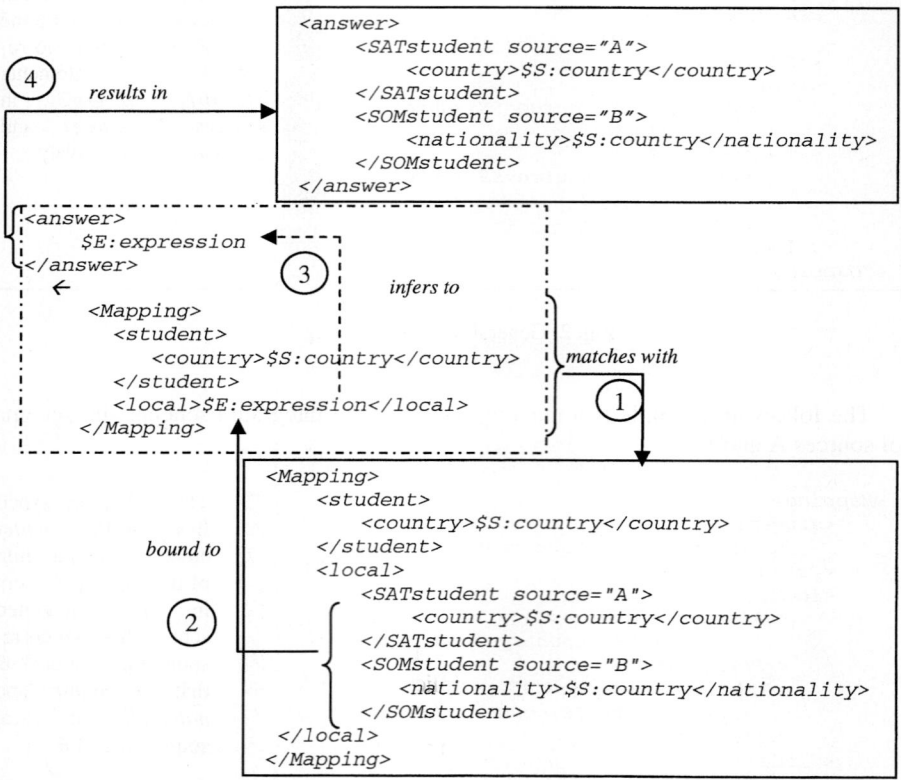

Fig. 3. Applying to the mapping generation rule

Another phase of the integration system, after querying is to convert the data format following global schema. Those mappings can, once again be used. They find the correspondence between one element in an explicit local source and an element in the integrated schema. These mappings (*facts*) are created once in the schema integrating phase but can be used for both query decomposition and data conversion. Thus, this way of building such mappings and their attached rules is extremely useful. It saves effort on the part of whoever creates the mappings and omits data redundancies in the system. Other applications of mappings will be expressed in detail in the next section.

3.3 Query Decomposition Algorithm

In order to succeed with query decomposition, we model all related algorithms based on XDD theory. In other words, we build them by using a set of XML rules. While input, output queries and mappings are modeled by XDD, the algorithms for query decomposition are also modeled by XDD. We believe that this will create a harmonious combination among internal parts of the system. An XML rule for query decomposition is:

```
<$N:LocalTag source=$S:source>          %   This rule specifies that if the
    $E:exp1                              %   query contains the $N:tag1
    <$N:tag2>$E:content</$N:tag2>        %   element, which is mapped to
    $E:exp2                              %   the $N:tag2 element in the
</$N:LocalTag>                           %   $S:source source, the sub-
←                                        %   query will contain the
    <Query>                              %   $N:tag2 element in spite of
        $E:exp1                          %   $N:tag1.
        <$N:tag1>$E:content</$N:tag1>
        $E:exp2
    </Query >
    <Mapping>
        <$N:GlobalTag>
            <$N:tag1>$E:content</$N:tag1>
        </$N:GlobalTag >
        <local>
            $E:exp3
            <$N:LocalTag source=$S:source>
                <$N:tag2>$E:content</$N:tag2>
            </$N:LocalTag>
            $E:exp4
        </local>
    </Mapping>
```

Fig. 4 shows an example of the integrated schema and an XML query. Fig. 5.a and 5.b represent two schemas of local sources *A*, *B* and their sub-queries after decomposition, respectively. In this paper, for simplicity, we use a tree to express the instance of an XML schema.

The *local* XML expressions of mappings (see Fig. 2) contain many elements belonging to different sources. The name of each source is the value of *source* attribute. In the XML rule for query decomposition, at each value of *$N:tag1*, two variables *$E:exp3* and *$E:exp4* would change automatically, depending on the value of *$S:source* in the mappings. Thanks to this feature, all possible values of *$N:tag2* can be treated simultaneously. By repeating this rule recursively, the output sub-queries are concurrently yielded. The number of sub-queries depends on the number of *$S:source* value.

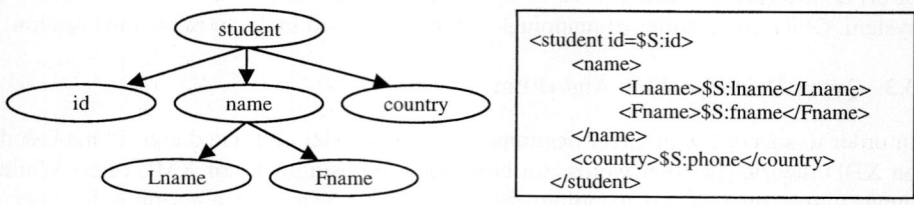

Fig. 4. An example of integrated schema and input query

Altogether, this step processes an XML query to obtain XML sub-queries in the form of the local sources. Low-level metadata processing, which could yield mappings for both query decomposition and data conversion, helps to save human effort and to avoid redundant data.

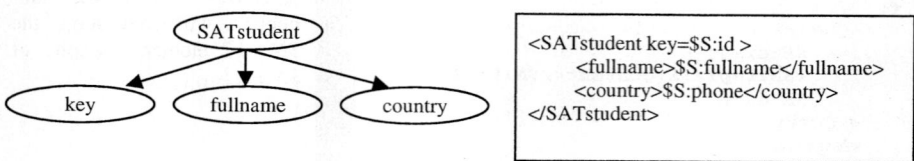

5.a Schema of source *A* and the query after decomposing

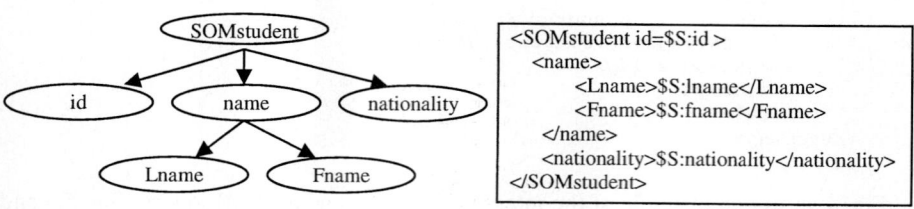

5.b Schema for source *B* and the query after decomposing

Fig. 5. An example of output sub-queries

4 System Prototype

The prototype of the proposed framework for query decomposition of integrated XML databases is implemented successfully. After careful analysis of XDD theory, all components of the framework are changed semantically into XET [1] for separate implementation, while Java is used to combine them into one big system as well as to support building of a system interface by which users can easily contact the system and collect their data.

4.1 Overview of XET

XML Equivalent Transformation (XET) is a rule based language modeled by XDD. Similar to XDD, XET can manipulate and represent XML elements with variables. Its five types of variables: *Nvar, Evar, Svar, Ivar, Pvar* are similar to XDD's variables: *$N, $E, $S, $I, $P*, respectively. A program in XET is a well formed XML document which contains the three main parts: *fact*, i.e., ground XML elements; *XML rules*, i.e., head, body and variable, and *rule priority* - the order of the rules performed. There are two kinds of XET engine: on ETC[2], another enhanced version on ETI[3]. For the sake of its flexibility, implementation in the second one is chosen.

4.2 System Functions

The prototype is built with three different database sources - three different XML documents which sample information of students in *SAT school*, *SOM school* and *Student Union*. These documents are attached with different schemas which contain conflicts about structure and format. These heterogeneous documents are harmonized by the available integrated XML schema and *two-direction mapping*. Besides, there exist certain data conflicts between the three documents and certain metadata are assumed to have been given to resolving them.

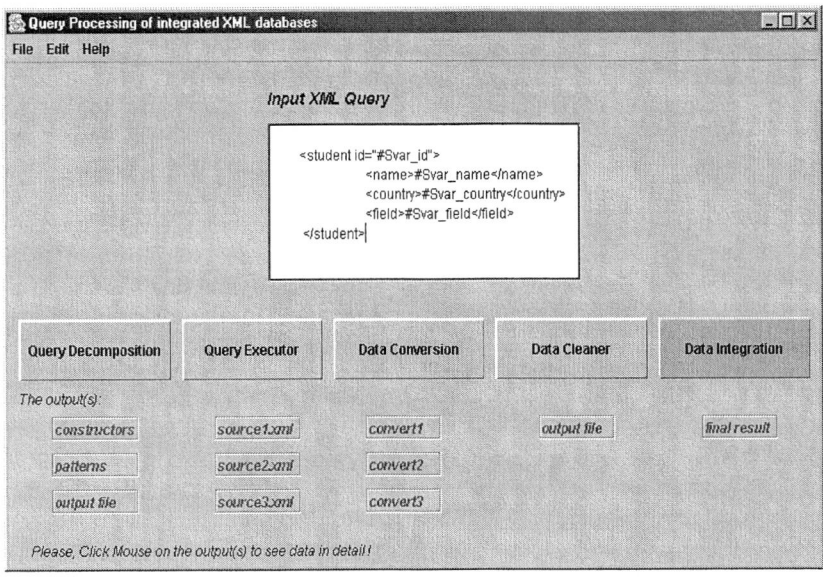

Fig. 6. System interface

---

[2] http://kr.cs.ait.ac.th/et
[3] http://assam.cims.hokudai.ac.jp/eti

In order to allow users to get quickly integrated data from local sources, a system interface has been developed - an important client component. It consists of a small window by which users pose queries; five buttons for executing the steps of the prototype: Q*uery decomposition, query execution, data conversion, data cleaner* and *data integration*, respectively; during run time; below each component is its outputs, kept in the XML document. Users click mouse on each output button to see the output of each component by *XML Spy application*. Besides, if users like to see only the final result and ignore all mediate results, they just need to click the *Data integration* button. The result will be shown in the XML file placed below it. Figure 6 shows the system interface.

Assuming that users already know the integrated schema, XML query modeled by XDD is again followed by the format of the integrated schema. The prototype starts work whenever users click their mouse on the buttons of the interface.

5 Conclusions

The prototype of the proposed framework for query decomposition of integrated XML databases was successfully implemented. After careful analysis of XDD theory, all components of the framework are transformed into XET programming language [1] for execution. The system is running simply, flexibly; and free of syntax errors. Java is also used to support building a system interface by which users can easily interact and collect their data. This proposal is the first query decomposition system of integrated XML databases that uses XDD as underlying model. It is one of a few systems that are capable of decomposing a global query to obtain n sub-queries in local format simultaneously. Since these sub-queries can be used to query directly in local sources, the extracted data is structural and semantic which is easier for the further integrating processes.

For the present, we are concentrating on XML queries in simple forms. Complex queries are being built and tested. Its power can be appreciated when it is considered as a complete database integration system [13] where every component is modeled by using XDD such as: XML schema integration, query processing, data integration, conflict resolving, etc. Besides, the proposed framework should be extended in a network environment, where XML databases are distributed for Ecommerce purposes. A typical application that could be seen is query processing of integrated web pages on the internet where XML web pages share a global schema.

References

1. Anutariya, C., Wuwongse, V., and Wattanapailin, V.: An Equivalent-Transformation-Based XML Rule Language. Proceedings of the International Workshop on Rule Markup Languages for Business Rules in the Semantic Web, Sardinia, Italy (2002)
2. Anutariya, C., Wuwongse, V., Nantajeewarawat, E., and Akama, K.: Towards a Foundation for XML Document Databases. Proceedings of 1st International Conference on Electronic Commerce and Web Technologies (EC-Web 2000), London, UK. Lecture Notes in Computer Science, Springer Verlag, Vol. 1875 (2000) 324-333

3. Baru, C., Gupta, A., Ludaescher, B., Marciano, R., Papakonstantinou, Y., and Velikhov, P.: XML-Based Information Mediation with MIX. In Demo Session. ACM-SIGMOD'99, Philadelphia, PA (1999)
4. Baru, C., Ludaescher, B., Papakonstantinou, Y., Velikhov, P. and Vianu, V.: Features and Requirements for an XML View Definition Language: Lessons from XML Information Mediation. Position paper in W3C's QueryLanguage Workshop (1998)
5. Bi, Y., Lamb, J.: Facilitating Integration of Distributed Statistical Databases Using Metadata and XML (2001).
   Available online: http://webfarm.jrc.cec.eu.int/ETKNTTS/Papers/final_pa- pers/en187.pdf
6. Gianolli, P., Mylopoulos, J.: A semantic approach to XML based data integration. Proceedings of the 20th. International Conference on Conceptual Modeling (ER), Yokohama, Japan (2001)
7. Ives, Z. G., Florescu, D., Friedman, M. A., Levy, A. Y., and Weld, D. S.: An adaptive query execution system for data integration. Proceedings of ACM SIGMOD International Conference on Management of Data (SIGMOD) (1999) 299-310
8. Ives, Z., G., Halevy, A., I., Weld, D., S.: Integrating Network-Bound XML Data. IEEE Data Engineering Bulletin. Volume 24 (2001) 20-26
9. Ives, Z., G., Levy, Y., and Weld, D., S.: X-scan: a Foundation for XML Data Integration (2002). Available online: http://data.cs.washington.edu/integration/x-scan.
10. Jakobovits, R.: Integrating Heterogeneous Autonomous Information Sources. University of Washington Technical Report, UW-CSE-971205 (1997)
11. Kiyoshi, A., Anutariya, C., Wuwongse, V., and Nantajeewarawat, E.: Query Formulation and Evaluation for XML Databases. Proceedings of the 1st IFIP Workshop on Internet Technologies, Applications, and Societal Impact (WITASI'02), Wroclaw, Poland (2002)
12. The MIX (Mediator of Information using XML) (1999). Available online: http://www.database.ucsd.edu/project/MIX.
13. Thuy, L., T., T., and Wuwongse, V.: Query Processing of Integrated XML Databases. Proceedings of the 5th International Conference on Information Integration and Web-based Applications & Services. Jakarta, Indonesia (2003) 335-344.
14. Wuwongse, V., Akama, K., Anutariya, C., and Nantajeewarawat, E.: A Data Model for XML Databases. Journal of Intelligent Information Systems. Vol. 20, No. 1 (2003) 63-80
15. Wuwongse, V., Anutariya, C., Akama, K., and Nantajeewarawat, E.: XML Declarative Description (XDD): A Language for the Semantic Web. IEEE Intelligent Systems, Vol. 16, No. 3 (2001) 54-65

On URL Normalization[†]

Sang Ho Lee[1], Sung Jin Kim[2], and Seok Hoo Hong[1]

[1] School of Computing, Soongsil University, Seoul, Korea
shlee@comp.ssu.ac.kr, vonsaki@korea.com
[2] School of Computer Science and Engineering,
Seoul National University, Seoul, Korea
sjkim@oopsla.snu.ac.kr

Abstract. Since syntactically different URLs could represent the same resource in WWW, there are on-going efforts to define the URL normalization in the standard communities. This paper considers the three additional URL normalization steps beyond ones specified in the standard URL normalization. The idea behind our work is that in the URL normalization we want to minimize false negatives further while allowing false positives in a limited level. Two metrics are defined to analyze the effect of each step in the URL normalization. Over 170 million URLs that were collected in the real web pages, we did an experiment, and interesting statistical results are reported in this paper.

1 Introduction

A Uniform Resource Locator (URL) is a string that represents a web resource (here, a web page). A URL is composed of five components: scheme, authority, path, query and fragment [1]. Given a URL, we can identify a corresponding web page in the World Wide Web (WWW). If some URLs locate the same web page in the WWW, we call them equivalent URLs in this paper. Syntactically identical URLs are certainly equivalent, but there are cases in which syntactically different URLs represent the same web page (hence, are equivalent).

One of the most common operations on URLs is simple comparison to determine if two URLs are equivalent without using the URLs to access their web pages. For example, a web crawler (web robot) encounters millions of URLs during collecting web pages in the WWW, and it needs to determine if a newly found URL is equivalent with ones of the already-crawled URLs to avoid duplication of request actions. Syntactically different URLs that are indeed equivalent give rise to a large amount of processing overhead; a web crawler requests, downloads, and stores the same page repeatedly, consuming unnecessary network bandwidth, disk I/Os, disk space, and so on. Extensive normalization [2][5] prior to comparison of URLs is often used by many web crawlers to prune a search space or reduce duplication of request actions.

[†] This work was supported by Korea Research Foundation Grant. (KRF-2004-005-D00172)

The details of URL normalization methods that have been used (or are currently used) in real web crawlers have not been described in the literature yet. The URL normalization issues have not drawn technical interests successfully in the research communities yet, in spite of their importance in web application developments. Moreover, the existing URL normalization methods have been developed rather on the basis of developers' personal heuristics, not from an extensive analysis of experimental simulation. We need to approach this matter in an analytic and systematic way.

It is possible to determine that two URLs are equivalent, but it is never possible to be sure that two URLs represent different web pages. For example, an owner of two different domain names could decide to serve the same resource from both, resulting in two different URLs. Any URL normalization methods could cause occurrences of so-called false negative (determining equivalent URLs to be not equivalent, hence not being able to transform equivalent URLs into a syntactically identical string). People want to use URL normalization methods that transform equivalent URLs into an identical string as much as possible (hence reduce the occurrence rate of false negatives as low as possible).

There are on-going efforts to define the URL normalization in the standard body [1]. The standard URL normalization defines a number of steps to transform URLs into the canonical form of URLs, so that it helps determine whether given two URLs are equivalent. The standard URL normalization transforms URLs that are determined to be equivalent, into the syntactically identical canonical form. The standard URL normalization is designed to minimize false negatives while strictly avoiding false positives (determining non-equivalent URLs to be equivalent); it never transforms non-equivalent URLs into a syntactically identical string. However, we notice that in reality there are cases in which we can minimize false negatives significantly while allowing false positives in a very limited way.

This paper considers the three URL normalization steps that are beyond the standard URL normalization. Discussed three steps are the case sensitivity at the path component of a URL, the last slash symbol in the path component of URLs, and the designation of a default page. The idea behind our approach is that in the URL process we want to minimize false negatives further while allowing false positives in a limited level. We believe that our approach is worthy of investigation even though after all we may not collect some web pages due to false positives.

We define two metrics: the redundancy rate and the coverage loss rate. The redundancy rate shows how many web pages are duplicated due to false negatives. The coverage loss rate shows how many web pages are lost due to allowing false positives. Over 170 million URLs that were collected in the real web sites in Korea, we report statistical information on the redundancy and loss of web pages for each method of the URL normalization. The investigation shows the cases in which our approach works well in practice. Analyzing the statistical data on the effect of the URL normalization, we propose the new methods for URL normalization that reduces the rate of false negatives significantly while allowing false positives just a little.

Our paper is organized as follows. In section 2, the syntax of a URL and the standard URL normalization are presented. Section 3 describes the two metrics, and section 4 presents our transformation options for the URL normalization along with experimental results. Section 5 contains the closing remarks.

2 URLs and Their Normalization

A URL is composed of five components: the scheme, authority, path, query, and fragment components. The scheme component contains a protocol (here, Hypertext Transfer Protocol) that is used for communicating between a web server and a client. The authority component has three subcomponents: user information, host, port. The user information may consist of a user name and, optionally, scheme-specific information about how to gain authorization to access the resource. The user information, if present, is followed by a commercial at-sign ("@") that delimits it from the host. The host component contains a location of a web server. The location can be described as either a domain name or IP (Internet Protocol) address. A port number can be specified in the component. The colon symbol (":") should be prefixed prior to the port number. The path component contains directories including a web page and a file name of the page. A directory and a file are separated by the slash symbol ("/"). The query component contains parameter names and values that may to be supplied to web applications. The query string starts with the question symbol ("?"). A parameter name and a parameter value are separated by the equal symbol ("="). A pair of parameter name and value is separated each other by the ampersand symbol ("&"). The fragment component is used for indicating a particular part of a document. The fragment string starts with the sharp symbol ("#"). Fig. 1 shows all the components of a URL.

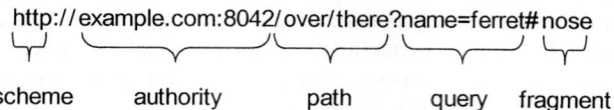

Fig. 1. URL Example

The URL normalization is a process that transforms a URL into a canonical form. During the URL normalization, syntactically different URLs that are equivalent should be transformed into a syntactically identical URL, and URLs that are not equivalent should not be transformed into a syntactically identical URL. The standard document [1] describes the syntax-based standard URL normalization as below.

First, a URL with a default port number (80 for the HTTP protocol) and a URL without the port number represent the same page. For instance, "http://example.com:80/" and "http://example.com/" represent the same page. During the normalization, the default port number is truncated from a URL.

Second, the scheme and authority components are case insensitive. For example, "http://EXAMPLE.com" and "http://example.com" represent the same page. During the normalization, all the letters in the components are changed into lower-case letters.

Third, a URL with path string null and a URL with path string "/" represent the same page. For instance, "http://example.com" and "http://example.com/" represent the same page. If a path string is null during the normalization, then the path string is transformed into "/".

Fourth, unreserved characters (namely, uppercase and lowercase letters, decimal digits, hyphen, period, underscore, and tilde) can be encoded into a three-digit string, where the percent symbol ("%") should be located at the first position, and the last two digits are a hexadecimal number representing an ASCII code of the character under consideration. For instance, "http://example.com/~smith" and "http://example.com/%7Esmith" represent the same page. During the normalization, encoded unreserved characters are decoded.

Fifth, a URL with the fragment and a URL without the fragment represent the same page. For instance, "http://example.com/list.htm#chap1" and "http://example.com/list.htm" represent the same page. During the normalization, the fragment in the URL is truncated.

3 Two Metrics

This section presents two metrics (namely the redundancy rate and the coverage loss rate) to analyze the effect of the URL normalization. These metrics will be used to evaluate a normalization method later.

A set of equivalent URL candidates is defined as a set of URLs that may possibly represent the same web page under a particular normalization method. The set of all URLs, U, is then represented as $U_1 \cup U_2 \ldots \cup U_n$, where U_i denotes the ith set of equivalent URL candidates and $U_i \cap U_j = \varnothing$ for $i \neq j$. It should be noted that the decomposition of U into U_i is fully dependent on a normalization method in question. Further, we do not need to consider a set of equivalent URL candidates in which the number of elements is exactly one, because there is no equivalent URL candidate to compare. For the purpose of computation of the redundancy rate and coverage loss rate, we drop off all U_i's with only one element without loss of generality.

Suppose that we have N sets of equivalent URL candidates in which the number of elements is at least two. D_i is the set of successfully downloaded documents (or pages) from all the URLs in U_i. $n(D_i)$ is the number of documents in D_i, $un(D_i)$ is the number of unique documents in D_i (simply, we count the number of unique contents in D_i). The redundancy rate is defined as equation (1).

$$Redundancy\ rate\ =\ \frac{\sum_{i=1}^{N}(n(D_i) - un(D_i))}{\sum_{i=1}^{N} n(D_i)} \qquad (1)$$

The redundancy rate shows how many web pages are duplicated due to equivalent URLs. If all the equivalent URL candidates in U represent different web pages, then the redundancy rate becomes 0.

Suppose that we transform all the URLs in U_i into a syntactically identical URL, hoping that all the URLs in U_i are equivalent. During the transformation, a normalization method under consideration could transform a non-equivalent URL into a syntactically identical one (i.e., false positive occurs). This implies that some valid web pages may be lost (not collected) during the transformation. For example, if U_i is composed of a number of URLs representing different web pages, then $(un(D_i) - 1)$

documents will be lost. The coverage loss rate is defined in equation (2), where $D_{downloaded}$ denotes the number of documents downloaded successfully with the transformed URLs.

$$Coverage\ loss\ rate\ =\ 1 - \frac{D_{downloaded}}{\sum_{i=1}^{N} un(D_i)} \qquad (2)$$

In essence, the redundancy rate shows how much web pages are redundantly collected due to false negatives, and the coverage loss rate shows how much valid web pages are lost due to false positives. Our approach toward the URL normalization is to reduce the redundancy rate significantly while limiting the coverage loss rate to be within a (user-definable) threshold.

Example 1.
Assume that we have 10 URLs (u1 to u10) as below. For convenience, the scheme component (i.e. "http://") is omitted.
  U = { www.microsoft.com/ (u1), www.microsoft.com/default.asp (u2),
    www.microsoft.com/index.htm (u3), www.microsoft.com/index.html (u4),
    nasdaq.com/asp/ownership.asp (u5), nasdaq.com/ASP/ownership.asp (u6),
    ietf.org/tao.html (u7), ietf.org/TAO.html (u8), ietf.org/Tao.html (u9),
    www.acm.org/ (u10) }

We can think of several criteria to decompose U into sets of equivalent URL candidates. Suppose that we decompose U under the designation of default pages of web servers. Then we have seven sets of equivalent URL candidates:
  U_1 = {u1, u2, u3, u4}, U_2 = {u5}, U_3 = {u6}, U_4 = {u7},
  U_5 = {u8}, U_6 = {u9}, U_7 = {u10}.

The sets U_2, ..., U_7 have only one member, so those sets are not considered any more. Suppose that the downloaded pages from URLs u1, u2, u3, and u4 are ⓐ, ⓐ, ⓑ, and ⓑ, respectively. Assume that u1 is the normalized URL for U_1 (i.e., u2, u3 and u4 are transformed to u1). Then we have following computation:
  The redundancy rate = (4 − 2)/4 = 0.5,
  The coverage loss rate = 1 − (1/2) = 0.5 (note that page ⓑ is lost).

Now consider that we decompose U under other criterion (such as the case sensitivity at the path component). Then we have seven sets of equivalent URL candidates:
  U_1 = {u1}, U_2 = {u2}, U_3 = {u3}, U_4 = {u4},
  U_5 = {u5, u6}, U_6 = {u7, u8, u9}, U_7 = {u10}.

Suppose that the downloaded pages from URLs u5, u6, u7, u8, and u9 are ⓒ, ⓒ, ⓓ, ●, and ●, respectively, where ● denotes the downloading failure. Assume that u5 and u7 are the normalized URLs for U_5 and U_6, respectively. Then we have following computation:
  The redundancy rate = ((2 − 1) + (1 − 1)) / (2+1) = 0.33,
  The coverage loss rate = 1 − (2/2) = 0.

It should be noted that the set of all URLs can be decomposed differently, depending on the decomposition criteria. ∎

4 New URL Normalization Methods

This section considers additional normalization steps beyond the ones specified in the standard document [1]. In our experiment, the web robot [4] crawled approximately the 350,000 Korean sites in the breadth-first fashion during one week in December 2003. The robot requested web pages within nine hops from the root pages of a site. The robot was not allowed to download dynamically-generated pages, whose URLs contain the question symbol inside. Consequently, 50 million web pages were downloaded successfully. We extracted 170 million URLs from the collected web pages, and normalized the URLs according to the standard URL normalization. Using the real URLs we collected, we did an experiment to get the effectiveness of normalization options for URLs in each case.

4.1 Case Sensitivity at the Path Component

The Windows operating system manages names of directories and files in a case-insensitive fashion. In a web server working on the Windows operating system, URLs can be composed with various combinations of the upper-case and lower-case letters. For instance, "http://www.nasdaq.com/asp/ownership.asp" and "http://www.nasdaq.com/ASP/ownership.asp" are equivalent. On the other hands, since the Unix and Linux operating systems manage names of directories and files in a case-sensitive fashion, two URLs composed with different combinations of the upper-case and lower-case letters are very likely to represent different web pages. For instance, "http://www.acm.org/pubs/journals.html" and "http://www.acm.org/PUBS/journals.html" should be considered not to represent the same web page.

The scheme and host components of a URL are defined to be case-insensitive. The path component of a URL is case-sensitive in principle. Here, we investigate the effects of assuming that the path component is case-insensitive. The set of equivalent URL candidates contains URLs that are syntactically identical except the case sensitivity in the path component. We consider the following three transformation options for the equivalent URL candidates.

   (1) Keep a URL as it is
   (2) Change letters in the path component into the lower-case letters
   (3) Change letters in the path component into the upper-case letters

The first option is the one accepted in the standard URL normalization. The second and third options can cause false positives to occur, because the file and directory names are case-sensitive in the Unix-like operating systems. Analyzing statistical data in our experiment, this paper explores the possibility of accepting the second or third options in the URL normalization.

In our experimental set of URLs, there were 383,444 equivalent URL candidates, which were decomposed into 185,474 sets of equivalent URL candidates. Figure 2 shows the number of requested, downloaded, and unique pages for each option. In the first option, the redundancy rate was 0.52. That is, over half of the successfully downloaded web pages were turned out duplicates.

In the second option, we change all the letters in the path component into the lower-case letters, resulting that each set of equivalent URL candidates has only one URL after normalization. Hence, only one request for the each set is possible. In terms of the number of web requests, there was 48% (= 185,474 / 383,444) reduction after normalization. The coverage loss rate of the second option was 0.01, which means 1% of the contents were lost. As for the third option, the coverage loss rate was 0.11.

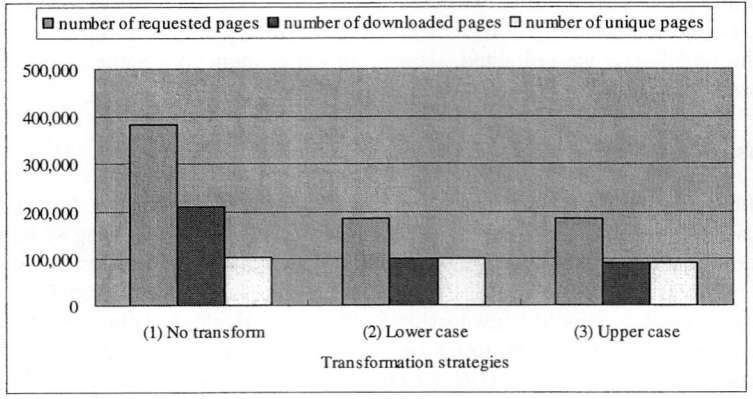

Fig. 2. Results for case sensitivity at the component

In terms of case sensitivity at the path component of a URL, almost all equivalent URL candidates in the same set represent the same page in the real web. We could download more web pages in the second option than those in the third option. In contrast to the first option, the second option (transforming all the letters in the path component into the lower-case letters) not only reduced the number of requests by 48% but also reduced redundant web pages by 52% at the cost of losing 1% of the unique pages that should be downloaded. Taking into account the benefits of the second option, we would say that the coverage loss rate (here, 1%) is negligible.

4.2 The Last Slash Symbol at the Non-empty Path Component

The standard URL normalization defines an empty path of a URL to be equivalent to "/", thus the two URLs are equivalent: "http://example.com" and "http://example.com/". The standard URL normalization, however, does not do anything on the last slash symbol at the non-empty path component at all (i.e., leaves it as it is).

This subsection considers the last slash symbol at the non-empty path component of URL. A URL with the last slash symbol represents a directory. When sending web servers such URL, web clients get either a default page in the requested directory or a temporarily created page showing all files in the directory. Users requesting a directory often omit specifying the last slash symbol in a URL. What really happens in this case is that web servers are likely to redirect the URL into a URL including the last slash symbol. For instance, "http://acm.org/pubs" is redirected into "http://acm.org/pubs/". This redirection allows users not to take care of the last slash symbol when users specify URLs.

Two URLs with and without the last slash symbol in the non-empty path component consist of a set of equivalent URL candidates. Each set of equivalent URL candidates has exactly two URLs that differ only at the last slash symbol. We consider the following three transformation options for the equivalent URL candidates.

(1) Keep a URL as it is
(2) Pick the URL with the last slash symbol as a normalized URL.
(3) Pick the URL without the last slash symbol as a normalized URL.

In our experimental set of URLs, there were 152,924 equivalent URL candidates, which were decomposed into 76,462 sets of equivalent URL candidates. Figure 3 shows the number of requested, downloaded, and unique pages for each option. In the first option, the redundancy rate was 0.49, which means approximately half of the successfully downloaded web pages were duplicates. The coverage loss rates in the second and third options were 0.01 and 0.03, respectively.

Our experiment shows that the possibility that equivalent URL candidates represent different pages is indeed rare. In the second and third options, the coverage loss rates were very small (0.01 and 0.03, respectively). The second option (specifying the last slash symbol at the end of the URL if possible) could reduce 50% redundant web pages, while only losing 1% of the unique pages to be downloaded.

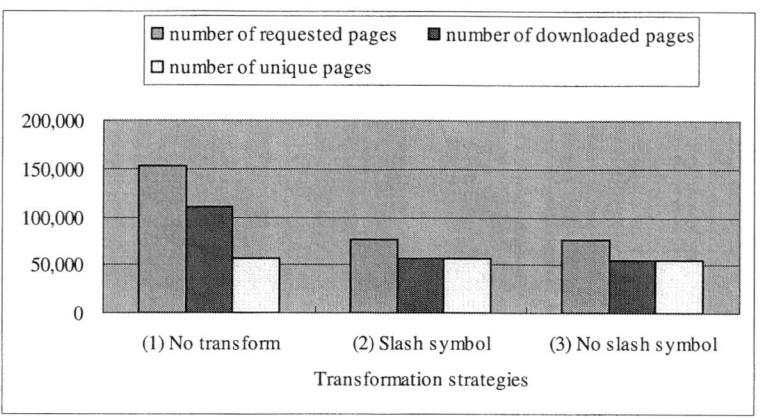

Fig. 3. Results for the last slash symbol

4.3 The Designation of a Default Page

A default page is a file to look for, when a client requests a directory. The default page can be specified in a URL. For instance, "http://www.acm.org/" and "http://www.acm.org/index.htm" represent the same page in the site (acm.org), which sets the designation of a default page as "index.htm". This subsection considers the designation of default pages in a web site.

In reality, any file name could be designated as a default page. It is reported [6] that only two web servers (the Apache web server, the MS IIS (Internet Information Services) web server) comprise over 85% of all the installed web servers in the world. The default pages of those web servers are "index.htm", "index.html", or "default.htm", when they are installed by default. We consider the three file names, i.e., "index.htm", "index.html", and "default.htm", as default pages in our experiment.

The URLs with and without the default page names consist of a set of equivalent URL candidates. We consider the following two transformation options for the equivalent URL candidates.

(1) Keep a URL as it is.
(2) Eliminate default page names in the path component. This means that a URL with a default page specified in the path component is normalized by eliminating the default page name.

In our experimental set of URLs, there were 147,266 sets of equivalent URL candidates and 305,835 equivalent URL candidates. Figure 5 shows the number of requested, downloaded, and unique pages for each option. The redundancy rate of the first option was 0.42. The coverage loss rate of the second option was 0.19. The second option could reduce 42% of redundant pages, but lost 19% of the unique web pages to be downloaded.

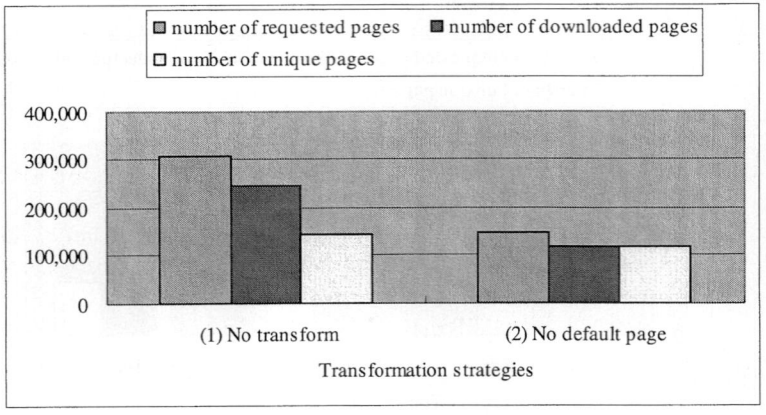

Fig. 4. Results for the designation of a default page

5 Conclusion and Future Work

We have considered three additional normalization steps beyond the ones specified in the standard document. Using two metrics this paper defined, we have analyzed the effects of each normalization step on the set of real URLs we collected in the all the Korean web pages. Our experimental results are summarized in Table 1. Changing all letters in the path component into low-case letters and specifying the last slash symbol could reduce approximately 50% redundant web pages with the small coverage loss rates (i.e., 0.01). Despite the reductions of the redundancy rates, eliminating a default page name caused a large portion (19%) of the web pages to be lost.

Table 1. Summary of experimental results

Type	Normalization options	Redundancy rate	Coverage loss rate
Case sensitivity at the path component	Leave as it is	0.52	0
	Change into lower cases	0	0.01
	Change into upper case	0	0.11
Last slash symbol at the path component	Leave as it is	0.49	0
	Slash symbol	0	0.01
	No slash symbol	0	0.03
Default pages	Leave as it is	0.42	0
	No page default specification	0	0.19

Under the two cases (case sensitivity at the path component and last slash symbol), does work out the idea that we can reduce the redundant web pages significantly at the cost of allowing false positives at a limited threshold. Adoption of our approach, which allows false positives to take place as well as reduces false negatives considerably, is completely up to the users. If efficiency is more important than correctness in URL process of web applications, our approach is an excellent choice. But if users think that correctness is more important than efficiency, they will stick to only the standard URL normalization.

The URL normalization is composed of the several steps to transform a URL into another. The order of transformation steps could lead to different URLs. As a future work, we plan to investigate how to effectively establish the normalization steps that consider not only the steps in the standard community but also the steps described in this paper.

References

1. Berners-Lee, T., Fielding, R., and Masinter, L.: Uniform Resource Identifiers (URI): Generic Syntax, http://gbiv.com/protocols/uri/rev-2002/rfc2396bis.html (2004)
2. Burner, M.: Crawling Towards Eternity: Building an Archive of the World Wide Web, Web Techniques Magazine, Vol. 2, No. 5. (1997) 37-40
3. Heydon, A. and Najork, M.: Mercator: A Scalable, Extensible Web Crawler, International Journal of WWW, Vol. 2, No. 4. (1999) 219-229
4. Kim, S.J. and Lee, S.H.: Implementation of a Web Robot and Statistics on the Korean Web, Springer-Verlag Lecture Notes in Computer Science, Vol. 2713, (2003) 341-350
5. Kim, S.J. and Lee, S.H.: How Web Pages Change: An Empirical Study, submitted for publications (2004)

Clustering-Based Schema Matching of Web Data for Constructing Digital Library

Hui Song[1,2], Fanyuan Ma[2], and Chen Wang[2]

[1] Department of Computer Information Technology,
Donghua University, 200051 Shanghai, China
[2] Department of Computer Science and Engineering,
Shanghai Jiao Tong University, 200030 Shanghai, China
{songhui_17, fanyuan ma, wangchen}@ sjtu.edu.cn

Abstract. The abundant information on the web attracts many researches on reusing the valuable web data in other information applications, for example, digital libraries. Web information published by various contributors in different ways, schema matching is a basic problem for the heterogeneous data sources integration. Web information integration arises new challenges from the following ways: web data are short of intact schema definition; and the schema matching between web data can not be simplified as 1-1 mapping problem. In this paper we propose an algorithm, COSM, to automatic the web data schema matching process. The matching process is transformed into a clustering problem: the data elements clustered into one cluster are viewed as mapping ones. COSM is mainly instance-level matching approach, also combined with a partial name matcher in calculating the elements distance metrics. A pretreatment for data is carried out to give rational distance metrics between elements before clustering step. The experiment of algorithm testing and application (applied in the Chinese folk music digital library construction) proves the algorithm's efficiency.

1 Introduction

The web contains a huge volume of information contributed by various organizations, groups and individuals, which provides abundant raw material for constructing digital library [4]. Many researches try to develop tools to facilitate constructing data collections of digital library from web [2,4] through web data integration. Though these tools have automated the first step of the task: extracting data from the web pages, the second step, mapping heterogeneous web data into the target data collection of DL, i.e. schema matching, are most done manually.

Schema matching is a fundamental problem in many database application domains, such as information integration, digital library, data warehousing and semantic query processing, also including the web-oriented data integration [1]. However the automatic schema matching between web information is more difficult [3]. New challenges arise in many ways: The data obtained from web pages with automatic extracting algorithms [14,15,16] mostly are flat or nested data lists, lack of data type and data illustration. Also, attached with descriptive content, extracted data are not as clean as the ones in rational databases. In general, the web data schema lacks the intact definition in the traditional schema-matching problem. Another side, there are great diversity of web data sets in the same application domain for different purpose of web sites. Just finding pairwise correspondences becomes impractical.

This paper considers the specific problem of schema matching – determining the correspondences elements in heterogeneous data sets obtained from web sites. The data sets with associated schema are all extracted from web pages, i.e. no additional information is added into the data set manually. Because of the opaque character of the schema and data, we call it the opaque schema-matching problem.

Solutions for automatic schema matching have received steady attention over the years [5,11,12,13]. The key conclusion from this body of research is that an effective schema matching tool requires a principled combination of many techniques, such as linguistic matching of names of schema elements, comparison of their data instances, considering structural similarities between schemas, and using domain knowledge and user feedback.

Our approach for the opaque schema-matching problem is constructed on the former works, combining the instance matching with schema name matching. Since not all elements having names, the instance-level matching is more valuable to gain an insight into the contents of data sets. In this paper we describe the COSM (Clustering-based Opaque Schema Matching) algorithm to automatically create semantic mappings between web data sets. With little information about the correspondence cardinality of web data schema, we attempt to consider the schema mapping with a new paradigm, into a clustering problem. The elements in different schemas clustered into one cluster are similar ones, and the instances of each element are viewed as an object to process the clustering iteration. A heuristic pre-step is taken to identify the likely data type of the pure text data so that the distance of data objects can be exactly measured.

The experiments on several web data sets obtained from real web sites validate the effective of COSM. It also has been successfully used in the construction of Chinese folk music digital library, for mapping the web data into the folk music collections.

The rest of paper is organized as follows. We give the formal representation of web data schema matching problem in Section 2, and the details of our algorithm COSM are described in Section 3. Section 4 reports on experiments result and section 5 reviews related work. Section 6 is the conclusion.

2 Opaque Schema Matching

Automatic information extracted technology is an important method to access the structured data from web pages for reusing. The data sets obtained have following character:

1) Data schema is deduced by wrapper production program, it is always represented as a table, flat or nested table.

2) No schema definitions such as data type, constraint and etc of each element, some of data elements may have name extracted from web pages.
3) No matter what the content, the data are stored as text string. Some of them are dirty for meaningless characters.

Example 1. From two web sites of bookstore, we get the book detail information with the following structure: $S_a = \{C_{a1}, C_{a2}, C_{a3}, C_{a4}, C_{a5}\}\}$, $S_b = \{C_{b1}, C_{b2}, C_{b3}, C_{b4}, C_{b5}, C_{b6}\}$, and the corresponding elements name list are $N_a = \{$**U, U, U, by, our price**$\}$, $N_b = \{$**U, author, U, price, member price, book description**$\}$ respectively, where "U" represents no name is found for this element from the web page.

Table 1. Data insxtances for schema S_a and S_b

C_{a1}	C_{a2}	C_{a3}	C_{a4}	C_{a5}	
4685.	Data Mining	This book described...	Jiawei Han	42.1	
6480.	Introduction to Algorithm	Since first edition ...	Thomas H.Cormen	57.8	
C_{b1}	C_{b2}	C_{b3}	C_{b4}	C_{b5}	C_{b6}
Principles of Data Mining	David Hand	08/01/2003	$48.00	$38.40	Many subjects are in face of ...
Data System Concepts	Abraham Silberschatz	12/10/2002	$69.00	$55.20	Database system has become ...

Our approach rephrases the problem of spatial n-m mapping as a clustering problem [10]: for the elements in different schemas, we want to find the relation between these elements. The elements with strong correspondence relationship are clustered into one cluster. Some clusters may have more than two elements and some clusters may keep one's. For example 1, the two schemas can be clustered into 6 clusters: $\{\{C_{a1}\}, \{C_{a2}, C_{b1}\}, \{C_{a3}, C_{b6}\}, \{C_{a4}, C_{b2}\}, \{C_{b3}\}, \{C_{a5}, C_{b4}, C_{b5}\}\}$.

The opaque schema-matching problem can be formal represented as following.

Definition 1. The opaque schema is defined as the data schema of the structure data extracted from web pages, which can be express as a nest set of string:

1) $S = \{C_1, C_2, ..., C_n\}$ is a schema, C denotes the basic type of set. $C \in \Sigma$, where Σ is an alphabet of symbols.
2) IF S is a schema, then $\{S\}$ is also a schema.
3) IF $S_1, S_2, ..., S_n$ is a schema, then $\{S_1, S_2, ..., S_n\}$ is also a schema.
4) All data schemas are symbol string by using 1) ~ 3) finite times.

Definition 2. For two opaque schemas S_a and S_b, the corresponding elements name are expressed as $N_a = \{N_{a1}, N_{a2}, ..., N_{ama}\}$, $N_b = \{N_{b1}, N_{b2}, ..., N_{bmb}\}$ respectively, where m_a and m_b are the element number of S_a and S_b. Some values of N_i may be null. $V_a:m_a \times q = \{v_{a1}, v_{a2}, ..., v_{aq}\}$ and $V_b:m_b \times q = \{v_{b1}, v_{b2}, ..., v_{bq}\}$ give the data instances of S_a and S_b, where q is the instances number. The schema matching algorithm outputs a cluster set of the elements contained in two schemas: $P = \{P_1, P_2, ..., P_n\}$, $P_i = \cup S_i$, $S_i \in S_a \cup S_b$.

3 Instances Cluster Analysis

The spatial n-m mapping problem of web data schema requires the clustering algorithm has ability to deal with different types of attributes and to be insensitive to noisy data. Many algorithms are designed to cluster interval-based (numerical) data. However, the web data are all extracted as text, we introduce a data pretreatment step to meet the requirement.

3.1 Data Pretreatment

For clustering analysis, the fundamental problem is measuring the gap between two objects [10]. In general clustering application, we have full knowledge about each attribute's data type, text, numeric or date etc. So it's not very hard to find the suitable distances metrics.

For web data, the situation is complex. No data type information is explicit labeled. They are just text string. The distance metric may produce bias, if we simply deal implicit numerical and date type data as text string.

A pretreatment must be taken to reduce the bias. The purpose of this step tries to identify the implicit data type from the data themselves, and then we can choice proper distances metrics for each data type. Through research on various data sets from real web sites, we classify the data type as text, numeric and date (including time) categories, for each of them having similar distance metric. Following heuristics are used to distinguish the data type of instances:

1) If the string *str* is a numerical character list, no other type character, i.e. $str = \cup \{i\}$, $i \in \{0, 1, 2, \ldots, 9\}$, then *str* is numeric type data;
2) If the string *str* contains the normal numerical format string (e.g. -, e), i.e. there exists substr(str)∈ *NFE*(Numeric Format Expression), then *str* may be numerical type;
3) If the string *str* contains the normal date and time format string (e.g. - -, //, pm), i.e. there exists substr(str)∈ *DFE*(Date Format Expression), then *str* may be date and time type;
4) There exists one and only one substr(str)∈ *DFE* or *NFE* in the data of numeric or date type;
5) Subtrings can prefix and suffix to numerical or date type expression, the percentage of the sum of the length of prefix and suffix cannot exceed th_d;
6) Every data instances of an element must be one data type.

The numeric data type contains general numeric, currency, accountant, percentage, fraction and scientific notation. We use the electronic table, Excel, to give the normal numeric, date and time expressions list.

The pretreatment algorithm checks each instance of one element $\{v_i\}$, $i=1,\ldots,q$ with heurist rules1)~5) r_j, $j=1,\ldots,5$. The check result $sa(i,j)$ can be four cases: numeric, date, numeric or date and text. For a instance string v_i, if $\exists$ j, $sa(i,j)$ = text, then v_i is labeled as text data, otherwise it calculates k_{nu}, k_{da}, the occurrence of numeric and date type giving by every rule. If $k_{nu}>k_{da}$, then the data type of v_i is numeric, otherwise is date. For the element, we calculate ke_{nu}, ke_{da} and ke_{te}, the occurrence of numeric, date and text type giving by every instance. The data type corresponding to $max(ke_{nu}, ke_{da}, ke_{te})$ is assigned to the elements.

3.2 Object Distance Metrics

For accurately measure the distance between the object instances, we introduce two different metrics for the text data and the rest data separately.

Definition 3. For two text elements S_v, S_w, N_v, N_w are the corresponding names, and V, W are the documents composed of the correspondence instances. The similarity distance of these two elements is defined as the weighted mean of three components: *isim*, the similarity of the text content of V, W; the normalization difference of the length of V, W and *nsi*, the name similarity of N_v, N_w.

$$dist(V,W) = w_{l1} \times (1 - isim(V,W)) + w_{l2} \times (1 - nsi(N_v, N_w))$$
$$+ w_{l3} \times \frac{|\lg L(V) - \lg L(W)|}{\underset{i}{Max}(\lg L(X_i)) - \underset{i}{Min}(\lg L(X_i))}$$

where $\sum_{i=1}^{3} w_{li} = 1$. If *nsi* is zero, then $w_{l2} = 0$. $L(x)$ gives the length of text document, and $\{X_i\}$ is the document set of all text elements.

The similarity distance between two text instances, $isim(V,W)$, uses VSM (Vector Space Model), a mathematic mode to calculate the text-matching degree, is widely applied in traditional IR and search engine [17]. VSM works well on long textual elements, such as book description. It is not good at short text elements, such as title, name. To revise the *dist* value, we add the weighted term of the data length difference.

The name of elements is also useful for measure the distances of them, though not all of them have a certain name. Element names are broken up into sets of name tokens $\{u_i\}$, $N = \cup \{ u_i \}$. For Chinese language text, the tokens are determined by a thesaurus lookup. The similarity of two names, *nsi*, is the average of the similarity of each token with a token in the other set. The similarity or two name token u_1 and u_2, $si(u_1, u_2,)$ is looked up in a synonym and hyponym thesaurus (Wordnet [18] for English and Hownet [19] for Chinese). Each thesaurus entry is annotated with a coefficient in the range of [0,1] that indicates the strength of the relationship. In the absence of such entries, we match sub-strings of the words u_1 and u_2, to identify common prefixes or suffixes. The similarity of two names N_1 and N_2 is calculated as follows:

$$nsi(N_1, N_2) = \frac{\sum_{u_1 \in n_1} \left(\sum_{u_2 \in n_2} si(u_1, u_2) \right)}{|N_1| \times |N_2|}$$

For two elements named N_1 and N_2, if $N_1 =$ **null** or $N_2 =$ **null,** then the name similarity $nsi(N_1, N_2)$ is zero.

Definition 4. For two numeric (or date) type elements S_v, S_w, N_v, N_w are the corresponding names, V, W are the corresponding data vectors. The distance of S_v, S_w is measured as weighted mean of normalization absolute deviation of V, W and the name similarity of N_v, N_w.

First, we calculate the mean of all instances $\{v_i\},\{w_i\}$ of the elements S_v, S_w respectively, and then use the normalization difference between the mean values to express the distance between two elements:

$$disd(V,W) = w_d \times \frac{\left|\sum_{i=1}^{q} v_i/q - \sum_{i=1}^{q} w_i/q\right|}{Max(x_i) - Min(x_i)} + (1-w_d) \times nsi(N_v, N_w))$$

where $\{x_i\}$ is the data set of all instances in numeric (or date) type. The date type instances are transformed into absolute time format to calculate the difference.

3.3 Hierarchical Clustering Algorithm

A hierarchical clustering method works by grouping the data objects into a tree of clusters. For this problem, a cluster containing a lot of elements is lower probability case. The agglomerative hierarchical clustering method is chosen. At first, every element of the two schemas are considered as an object and placed into their own atomic cluster. The clusters are then merged step-by-step according to some criterion, until the distance of two closest clusters exceeds the threshold. Because the distance metrics between three data type greatly differ with each other, the different types of elements are dealt separately. The cluster merging only works between the elements of the same data type, the thresholds for three data type are also set separately.

For schema S_a and S_b, the instances vector of elements v_{ai} and v_{bi} form the objects of the elements. The merging rule is: the clusters P_x and P_y will be merged if an object in P_x and an object in P_y form the minimum distance between any two objects from different clusters. The clusters containing elements of different schemas merge first, i.e. only when one cluster containing elements come from different schemas can merge elements randomly. The termination criterion is the mean distances of the objects in one cluster larger than the giving threshold.

$$Disc(P) = \frac{1}{n_i \times n_i} \sum_{p \in P} \sum_{p' \in P} |p - p'|, \; n_i \text{ is the element number of } P.$$

The detail algorithm is giving as following:

Algorithm: *Instance-based Clustering (Instance V_a, Instance V_b)*
Input: The instances of two schemas V_a, and V_b, the schema name mapping table
Output: a cluster set P_i
Method:
    for each elements set of data type dt
        $P_i = \{v_i\}, v_i \in V_a \cup V_b, i = 1, \ldots, m_a + m_b$
        Repeat
            $SelMin(P_x, P_y)$ // Function: give the data type of one element
            $P_x = P_x \cup P_y$
        Until $Disc(P_i) > threshold(dt)$, $i = 1, \ldots, k$
    return P_i

Fig. 1. Hierarchical clustering algorithm

4 Experimental Result

To evaluate the COSM algorithm, we test it with four domains of sources from real web sites. Our goals were to evaluate the matching accuracy of COSM, and the contribution of our data pretreatment process and name matcher. We also applied the algorithm in WGaDL, a web data gather tool for digital library, to automatically map web data extracted from different web sites into the target schema of DL collection.

4.1 Experiment Setup

We report the evaluation of COSM on four domains, whose characteristics are stated in table 2. The test data are obtained by extracting structure data sets and schemas from web pages. Automatic tool helps this work [16]. For each web source, we get the schema and a lot of instances. In preparing the data, we performed only trivial schema cleaning on discard elements irrelevant to the domain concepts, such as page ID of web pages. Table 2 shows the elements number of one schema and indicates the number of elements with name.

Table 2. Domains and the characteristics of sources

Domains	Sources	Extracted data instances for each source	Schema elements	Elements with name
Books	5	350	8~17	5~15
Faculty staff	5	50	6~10	2~6
Real Estate	5	300	9~22	8~20
Movies	5	300	12~15	9~10

For each domain, the schemas are mapped two by two in the same domain. So with 5 sources of a domain, we can get 10 mapping pairs. Firstly, we created a mapping result manually as answer, and then compare the COSM' mapping result with the answer to measure the algorithm's accuracy. Next, we measure the contribution of the data pretreatment process and the name matcher. Third, we investigated how sensitive it is to the amount of instances available from each source. *The matching accuracy of two schemas* is defined as the percentage of the clusters grouped by matchable elements that are output correctly by COSM. *The average matching accuracy of a domain* is the accuracy average over all ten pairs in the domain.

4.2 Matching Result Analysis

Figure 2 shows the average matching accuracy for different domains running with 300 data instances for each source. The contribution of data pretreatment process and name matcher are depicted. For each domain, the three bars (from left to right) represent the average accuracy produced respectively by the complete algorithm COSM, COSM without name matcher and COSM without data pretreatment process, i.e. all the data are treated as text data in hierarchical clustering step.

The results show that COSM achieves high accuracy across four domains, ranging from 83.8% to 92.3%. In contrast, if COSM without data pretreatment, the accuracy decrease to 56.2%~74.2%. The contribution of data pretreatment varied greatly between different domains because of the difference of the percentage of text elements. For movies domain, most elements are text data, so pretreatment step only improves the accuracy 10.3%. Name matcher also improves the accuracy in some extend. Especially notable in the movies domain, most data elements are given and some elements' content are similar to each other, so the accuracy increases 18.4% by name matcher.

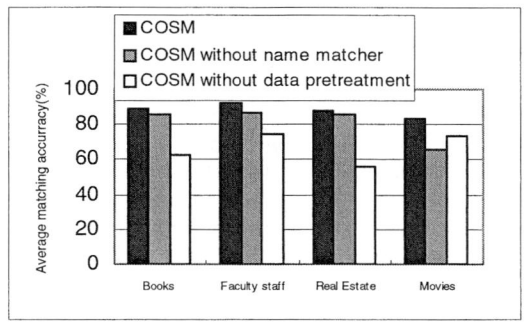

Fig. 2. Average matching accuracy of CMOS

Figure 2 shows the variation of the average domain accuracy as a function of the number of data instances available from each source. The results show that on these domains the performance of COSM stabilizes fairly quickly. From 5~30, the accuracy climbs steeply, and after 100, it levels off. This result indicates that COSM appears to be robust and the instances needed for running the algorithm can be controlled.

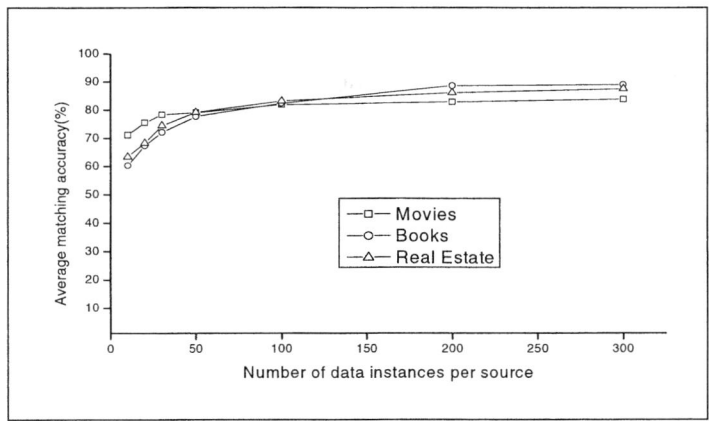

Fig. 3. Matching accuracy with variety of data instancesMatching result analysis

Figure 3 shows the variation of the average domain accuracy as a function of the number of data instances available from each source. The results show that on these domains the performance of COSM stabilizes fairly quickly. From 5~30, the accuracy climbs steeply, and after 100, it levels off. This result indicates that COSM appears to be robust and the instances needed for running the algorithm can be controlled.

5 Related Work

Schema matching is a critical step for data integration [1]. We relate our work to existing works in following aspects.

The previous works [5,7,8,9] assume their input as either relational or structured schemas, and those schemas are designed internally for developers. As a consequence, the attributes of the schemas are explicit defined, also as the data type. We focus on a newcome problem, web extracted data integration. The data schema and instances are all extracted from web pages by machine automatically. The web site changing quickly and the huge web data make manual work unpractical.

The Cupid and related method [5] are based on schema-level matching combined with structure-level matching. This approach does not suit the opaque situation for incomplete data schema definition.

Our work is mainly based instance mapping, related to the LSD system [8]. LSD focuses on 1-1 mapping and maps source to a fix mediated schema so it can view the mapping problem as classification problem. The web data schema cannot be controlled and simple treated as a 1-1 mapping problem. This motivates this paper's clustering idea, for it need no having the classification knowledge before. Clustering approach also avoids the work of collecting and labeling training data for each application. Embley's work [7] aimed to the n-m mapping problem and applied machine-learning algorithms to terminological relationships and data-value characteristics, which are just not possessed by web data schema.

The work described in [9] talks about the opaque column name and data values problem. It's a just supplement to other automated schema mapping tools, can not be used independently.

6 Conclusion

In this paper, we study a special schema-matching problem, matching the data schema extracted from web pages, when automatically constructing digital library collections from web data. The web data schema matching has more challenge than traditional work since we lost the control of the data schema and instances. We present a novel view to resolve the problem: transforming the spatial n-m mapping problem into clustering problem. It avoids the traditional approach's requirement of complete schema definition and the 1-1 mapping limitation. The algorithm is mainly based on instance-level matching but also combined with schema name matching. It takes a data pretreatment step to deal with the special representation of web data. The experiment results across four domains prove the efficiency of our algorithm. In WGaDL tools, it has been successfully used to automate the process of mapping heterogeneous web data schemas into the target one.

We see a number of promising avenues for future work. Now we just consider the elements in the web schema as equal level. In fact, they may be organized in nested structure. Combined the structure-level matching into the clustering algorithm may improve the matching accuracy. On the other side, when we apply this algorithm to a special domain, the domain constraints are existed. The constraints-based clustering algorithm is more complex but more valuable for real application.

References

1. Erhard Rahm, Philip A. Bernstein. On Matching Schemas Automatically. VLDB Journal, 10(4), 2001.
2. S. Lawrence, C. L. Giles, and K. Bollacker. Digital libraries and Autonomous Citation Indexing. *IEEE Computer*, 32(6):67–71, June 1999.
3. Bin He, Kevin Chen-Chuan Chang. Statistical Schema Matching across Web Query Interfaces. *ACM SIGMOD 2003*, San Diego, CA
4. P. P. Calado, M. A. Goncalves and etc. The Web-DL Environment for Building Digital Libraries from the Web. *In Proceedings of the 3th ACM/IEEE Joint Conference on Digital Libraries, JCDL 2003*, May 27 - 31, 2003 Houston, Texas USA
5. J. Madhavan, P.Bernstein, and E.Rahm. Generic Schema Matching with Cupid. *The Proceeding s of VLDB*, 2001
6. Naveen Ashish and Craig Knoblock. Wrapper Generation for Semi-Structured Internet Sources. *Proc. of the ACM SIGMOD Workshop on Management of Semistructured Data*, Tucson, Arizona, May 1997
7. Li Xu, David W. Embley. Discovering Direct and Indirect Matches for Schema Elements. *The IEEE conference of DASFAA '03*, Japan, 2003
8. A. Doan, P. Domingos, and A. Halevy. Reconciling schemas of Disparate Data Sources: A machine Learning Approach. *SIGMOD 2001*, Santa Barbara, California, USA
9. Jaewoo Kang, Jeffrey F. Naughton. On Schema Matching with Opaque Column Names and Data Values. *ACM SIGMOD 2003*, San Diego, CA
10. L. Kaufman and P.J.Rousseeuw. Finding Groups in Data: An introduction to Cluster Analysis. New York: John Wiley & Sons, 1990.
11. Hong-hai and Erharm Rahm. COMA – A System for Flexible Combination of Schema Matching Approaches. *In Proc. of the 28th VLDB*, 2002.
12. Anhai Doan, Jayant madhavanm Pedro Domin-gos, and Alon Y. Halevy. Learning to Map between Ontologies on the Semantic Web. *In Proc. of the 11th WWW*, 2002.
13. Sergey Melnik, Hector Garcia-monina, and Erhard Rahm. Similarity Flooding: A Versatile Graph Matching Algorithm. *In Proce. of the 18th ICDE*, 2002.
14. V. Crescenzi, G. Mecca, and P. Merialdo. ROADRUNNER: Towards automatic data extraction from large web sites. *In Proc. of the 2001 Intl. Conf. on Very Large Data Bases, pages 109–118*, 2001.
15. Arvind Arasu, Hector Garcia-Monina. Extracting structured data from web pages. *ACM SIGMOD 2003*, San Diego, CA.
16. Hui Song, Fanyuan Ma, and Giri Suraj. Data Extraction and Annotation for Dynamic Web Pages. *In proceeding of IEEE conference EEE'04*, Taibei, 2004
17. W. Cohen and H. Hirsh. Joins that generalize: Text classification using whirl. *In Proc. of the fourth Int. Conf. on KDD*, 1998.
18. http://www.cogsci.princeton.edu/~wn/
19. http://www.keenage.com/

Bringing Handhelds to the Grid Resourcefully: A Surrogate Middleware Approach*

Maria Riaz, Saad Liaquat Kiani, Anjum Shehzad, and Sungyoung Lee

Computer Engineering Dept., Kyung Hee University,
Yongin-Si, Gyeonggi-Do, 449-701, Republic of Korea
{maria, saad, anjum, sylee}@oslab.khu.ac.kr

Abstract. This paper presents the design of a middleware approach that aims at assisting handheld devices in accessing Grid services by wrapping the computational and resource intensive tasks in a surrogate and shifting them to a capable machine for execution. The performance of the surrogate approach is evaluated with the help of a test scenario. The reduction in computational intensity at the handheld device, achieved through task delegation, is examined and the optimization of communication mechanisms, that reduce the load on a resource constrained handheld device, is presented[1].

1 Introduction

With ever decreasing costs and increasing functionality in small sized chips, mobile handheld devices e.g., Personal Digital Assistants (PDA) and smart phones are becoming mainstream now. While mobile elements will improve in absolute ability, they will always be resource-deprived relative to their static counterparts (desktops/workstations). In [1], the author argues that for a given cost and level of technology, considerations of weight, power, size and ergonomics will exact a penalty in computational resources such as processor speed, memory size, and disk capacity. These devices do not have enough resources in effect to utilize the Grid services comprehensively.

Owing to monotonically increasing mobility of users and greater adoption of handheld devices, job submission to Grid through handheld devices comes up as a viable option for maximizing usability of devices and efficiency of services. Constraints that hinder handheld devices from such interactions include limited network bandwidth, CPU power, memory (small network buffers) and intermittent connectivity. Keeping such limitations in mind, we aim to define a middleware approach that will allow handheld devices, e.g. PDA units, to interact with Grid services while inducing minimal burden on the device itself. We demonstrate a solution based on Jini Network Technology's [2] Surrogate Architecture [3] which provides a network framework in which a device can deploy a client or a service on a device other than itself.

---

* This work is partially supported by Ministry of Commerce, Industry and Energy, Republic of Korea.

Since we are stepping in a new realm of Grid access through handheld devices, many design and performance challenges need to be considered and countered. In the domain of Grid infrastructure, where services and data resources are replicated across geographical boundaries [4], [5], communication costs can be minimized by careful selection of intermediate network. The communication mechanisms involved in job submission, execution and resource access are optimized at three levels: 1) Selection of the host to which the device will submit the job/task for execution, 2) Resource access by the surrogate during execution and 3) filtering and optimization of intermediate results that are to be transferred to the device from the remote machine.

One possible approach for facilitating handheld device interaction with the Grid is to narrow down the criteria for Grid access and make it less resource hungry; but doing so will also take away several benefits. How can a resource constrained device be configured and supplemented with software based techniques to make it Grid-interaction capable? A handheld device wishing to host a service and unable to do so can be allowed to delegate this task to a relatively powerful machine (desktop, server). Conversely, if the interaction with remote Grid services proves too much for limited local resources of a handheld device, it can deploy the actual client functionality at an intermediate machine and receive the results in a form that is in keeping with its hardware resources. This second scenario has a greater probability of being used in real world applications and is the focus of our research.

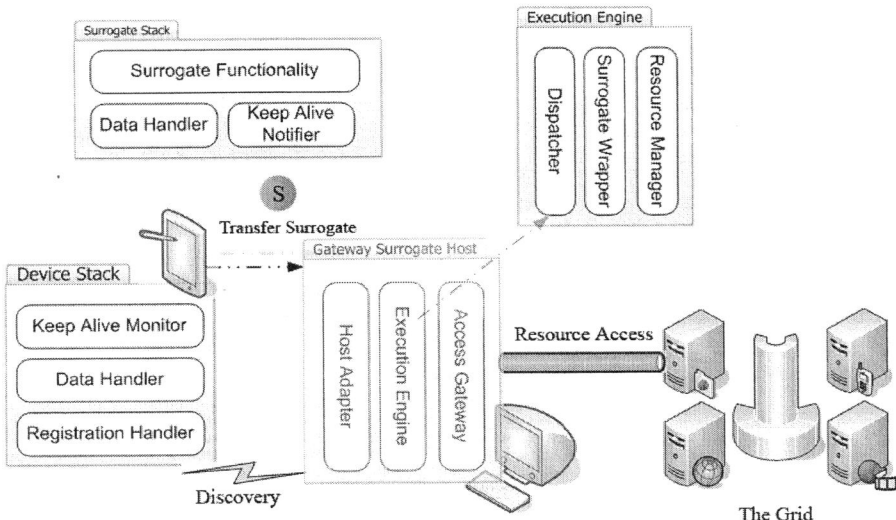

Fig. 1. Interaction between a mobile handheld device with Grid services through middleware deployed at a 'in-between' gateway

The 'client' process, transferred from the device, is called a *'surrogate'* (The term 'surrogate' is used to describe an entity that performs some action on behalf of another entity). The middleware component at intermediate machine, which provides the execution environment and access to extensive resources for the handheld device's surrogate, is called the 'Gateway *Surrogate Host*' or simply '*Host*'. An interconnect mechanism, defined as "logical and physical connection between the surrogate host and a device" [6], also needs to exist. A handheld device that can communicate on IP (wirelessly or wired) can be programmed to shift its processing to a host capable machine.

An overview of our middleware approach is presented in Sect. 2. Section 3 deals with the communication mechanisms and the proposed optimizations in the middleware. Prototype implementation and test results are presented in Sect. 4. We conclude our discussion in Sect. 5 and also list relevant related work.

2   3 – Tier Architecture

The main concept driving our approach is to shift the 1) access to generic Grid services and 2) intensive task processing, from a resource constrained handheld device to a resource rich system (i.e. the *Surrogate Host*). This is to be achieved by wrapping the access and processing mechanisms in a 'surrogate' module. Consider the example of a physicist who needs to see graph plots, on his PDA, of data produced as a result of high energy collisions between atomic particles. The amount of information stored in data-stores from which graphs are to be generated will be in the range of several gigabytes or even terabytes. The processing of such data for the purpose of plotting graphs is not a job to be handled by a handheld device. Moreover, the handheld device may have low network bandwidth, further diminishing the prospects of a successful remote analysis by a user. By utilizing the Jini Surrogate Architecture based middleware support, one can 'pack' the functionality of access mechanisms for data-stores and graph plotting routines in a *surrogate* and transfer the *surrogate* to a *host* machine. The *host* machine will provide the *surrogate* with the necessary resource rich execution environment and network connectivity. The *surrogate* is able to communicate back to the device (PDA) through available interconnects e.g. IP, USB, Bluetooth etc. The aforementioned tasks of service access and intensive processing can be shifted from the handheld device to a more appropriate host machine, with the device only managing less intensive tasks of displaying the tailored results returned from its surrogate. The middleware framework consists of three distinct tiers namely Device, Surrogate and Gateway. These are discussed one by one in the subsequent paragraphs.

2.1   Tier 1 – Gateway

Gateway Surrogate Host is the middleware component that aids the Device to overcome resource and computational limitations by accepting tasks, packed as surrogates, for execution. The middleware provided at these hosts consists of three main sub-modules. *Host Adapter* sub-module offers an interface to client devices for accessing the *Gateway Surrogate Host*. It enables the initial communication between a device and the host so that both can agree on the transfer of the *surrogate* after authenticating the device and its related surrogate. Once the *surrogate* is available at the host, it is delivered to the *Execution Engine* sub-module. It consists of a *Surrogate Wrapper* that exposes the functionality of the surrogate that is required to facilitate surrogate's execution at the host. *Dispatcher* allocates a separate thread for the execution of the surrogate from a thread pool, and then activates the surrogate. Resources required for surrogates' execution are resolved and handled by the *Resource Manager* module. These resources include memory and disk space, processor, JVM (for Java based surrogates, as is the case with our implementation), network resources etc. The *Access Gateway* sub-module provides interface to the external resources e.g. discovery of available Grid services and resources.

A *Gateway Surrogate Host* announces attributes relevant to its properties and capabilities including, but not restricted to:

- ID, Location, Currently hosted surrogates etc
- Network address and Discovery/Listening port for incoming Device/Client requests
- Available/Allocated Resources e.g. CPU, Memory, Storage, Throughput
- Environment e.g. Java VM availability and version, SOAP/WSDL [7], [8] XML parser etc
- Grid services available through this Surrogate Host

Advertising these attributes allows clients to identify appropriate hosts based on their location, network proximity and other desired features. Administrator of a host can restrict the number of *surrogates* that are allowed to execute, restrict memory, bandwidth allocation etc on per *surrogate* basis. Security policies can be configured based on public/private key pairs and digital certificates.

2.2 Tier 2 – Surrogate

A generic surrogate for Grid service access contains the following features: client authentication based on public/private key pairs; generic functionality to communicate and interact using WSDL/SOAP for web service based Grid services; persistency safe i.e. to be put to persistent storage if its functionality is periodic; migration – to be able to stop and save current execution, mark restore points and migrate to a different Surrogate Host. The functionality of the generic surrogate, as shown in surrogate stack in Figure 1, is incorporated at the top layer along with the specific logic of the extended Surrogate. Moreover, the surrogate has complementary modules for communicating with the middleware stack at the Device. Surrogate can be hosted in the file system of the Device or it can be stored at a URL accessible store e.g. a web server or FTP server. Some clients may be void of any Surrogates. These sorts of clients/devices are still able to use other deployed surrogates if they can provide valid credentials as their rightful owner or users.

2.3 Tier 3 – Device

At the Device, a lightweight middleware stack is provided to facilitate the coordination with its exported surrogate. The stack consists of a *Surrogate Handler* module which has three sub modules for providing services complementary to the middleware at the *Gateway Surrogate Host*. *Registration Handler* discovers and selects the *Host* and registers and transfers the surrogate. Once the surrogate is transferred, *Keep Alive Monitor* keeps track of the status of the surrogate. *Data*

Handler retrieves the results from the surrogate-side corresponding module, and makes them available for the application executing at the device. *Surrogate* to be transferred can be stored at the *Device* or at a URL accessible store e.g. a web server or FTP server.

3 Optimizations

There is a critical requirement of clients/devices being able to discover available *Gateway Surrogate Hosts*. A good discovery mechanism is required to avoid single points of failure in the system. For reasons of efficiency and fault tolerance, multiple discovery techniques are provided in the architecture.

The foremost method of discovery is multicast announcements from *Gateway Surrogate Hosts*. This automatically provides for finding 'nearby' hosts by the devices (as multicast is geographically contained within a limited network boundary by most administrators). HTTP based discovery is provided as a supplement. All available *Gateway Surrogate Hosts* register with a web service hosted on a known location. Client devices/applications can query for a particular host by submitting appropriate parameters to this service over HTTP.

The surrogate paradigm will function most efficiently when the network delays between the device/client side and *surrogate* are minimal. Moreover, efficiency also depends on the proximity of *surrogate* to the actual service being accessed. Since a mobile user may be in motion with respect to the *Gateway Surrogate Host* as well as the Grid resources it wants to access, support is needed in the architecture to optimize both proximity based parameters. Each *Gateway Surrogate Host* will keep track of its access quality towards available Grid service hosts/networks. On the other hand, before deploying a *surrogate*, client side application can determine its network connectivity and temporal efficiency with a specific *host*. This procedure poses a certain one time per start-up burden, but improves runtime performance relative to a scenario where such optimizations are left to good luck.

Table 1. Attributes published by a surrogate host

Name	Description
Host Identification	ID, Location, Network address and Discovery/Listening port for incoming Device/Client requests
Host Resources	Currently Hosted Surrogates, Available/Allocated Resources e.g. CPU, Memory, Storage, Throughput
Host Environment	JVM availability, version; SOAP/WSDL etc
Network Resources	Grid services available through this Host, Proximity to service and client side (in terms of network access)

Table 1 lists the attributes computed and advertised by each host allowing clients to select hosts based on location, proximity and other desired features. The following pseudo-code describes a selection approach for 1) the device to choose a host and 2) the surrogate to select resources:

1) Discover available Surrogate Hosts
        Listen for Multicast Announcements from Hosts
        Query Web Service W for available Hosts
   Select Optimal Gateway Surrogate Host
        For all discovered Hosts
             Retrieve attributes
             Choose best host through function 'f'
   Transfer Surrogate

2) Retrieve Resource List from Gateway Surrogate Host
        For all known Resources
             Retrieve Resource attributes
   Choose optimal resource

In order to elaborate the given algorithms, let D be a set of *Devices* willing to transfer surrogates and let G be a set of available *Gateway Surrogate Hosts*:

$$D = \{d_1, d_2, d_3, ..., d_n\} \quad (1)$$

$$G = \{g_1, g_2, g_3, ..., g_n\} \quad (2)$$

Let R be the resources known to a particular Gateway Surrogate Host g_i that might be of interest to arriving surrogates:

$$R = \{r_1, r_2, r_3, ..., r_n\} \quad (3)$$

where R_{gi} will a subset of resources R known to host g_i.

The set A_{gi}, of attributes associated with a Gateway Surrogate Host g_i is listed as follows:

$$A_{g_i} = \{T_{g_i, d_j}, M_{g_i}, C_{g_i}, N_{g_i}\} \quad (4)$$

where $T_{gi,dj}$ represents the network throughput available between the device d_j and a host g_i, M_{gi} represents the available memory resources and C_{gi} represents the average idle CPU availability. Basing on the type of surrogate, a subset of these parameters is chosen to decide the most suitable host for the surrogate of the device.

A device with a CPU intensive surrogate task can choose a *Gateway Surrogate Host*, as follows:

$$g_{sel} = \max_{g_i \in G} \{f(C_{g_i}, N_{g_i})\} \quad (5)$$

where g_{sel} is the Gateway Surrogate Host selected as a function of processing power and number of surrogates hosted to avoid contention for CPU.

Similarly, a number of attributes can be retrieved from job schedulers and resource managers in generic grid infrastructures such as approximate wait time (AWT), network throughput, CPU availability, wait queue length; [9] describes a 'resource utilization status' (RUS) being maintained by a grid computing facilities that indicates resource availability. Attributes associated with each resource r_i include:

$$Ar_i = \{T_{r_k}, C_{r_k}, RUS_{r_k}, \ldots\} \tag{6}$$

where T_{rk} is the network throughput [10] available between the resource and the Gateway Surrogate Host and C_{rk} is the CPU availability at the resource host.

The surrogate can select the resource to access basing on these attributes.

$$r_{sel} = g\{A_{r_i}\} \tag{7}$$

where r_{sel} is the Resource selected as a function over attributes of available resources.

The attributes of a host and resource along with corresponding selection functions, as shown in (5) and (7), help in optimizing access to the resources.

4 Implementation Overview

The authors have provided a bare-bones implementation of the proposed architecture so that before this design is tested for actual Grid service interaction, its viability can be validated in a general scenario. The scenario of choice should involve considerable CPU, memory and network utilization. Simple Network Management Protocol [11] is a widely accepted and utilized way of monitoring network entities. We have chosen to verify our approach by monitoring a remote server for 14 system statistics periodically, through a handheld device. Handheld device has network connectivity through a wireless LAN interface. A desktop machine is configured to act as a *Gateway Surrogate Host*. A *Surrogate* has been coded for the handheld device with the functionality of monitoring the remote server through SNMP queries and adjusting the results to be sent back to the *Device*. The results of these queries are to be displayed on the handheld device in the forms of dynamic line, bar and pie charts/graphs. Performance of the device and the impact of the running system will be measured and the benefits and shortcomings of the approach will be highlighted.

The *Gateway Surrogate Host* module has been implemented by modifying and extending the *Surrogate Host* provided with the reference implementation of Jini Surrogate Architecture specification. The extensions include addition of useful attributes to be announced, additional discovery mechanism and addition of an SNMP agent. IBM's J9 VM for java is used to implement the *surrogate* for the handheld device and contains classes which implement the functionality of the task that the *Device* wishes to execute i.e., monitoring. Moreover, it contains the 'device-to-surrogate' interconnect implementation which, in the case of this scenario, is based on IP Interconnect Specification.

4.1 Analysis

Measurements were taken to analyze the performance of the *Device* during the course of execution. The client application on the PDA consumes fewer than 6 MB of memory at maximum. This also includes the foot print of the J9 JVM and Java AWT classes. Delay during the transmission of results from the su*rrogate* and their display in the form of graphs on the Device were found to be negligible (quite less than 1 second) owing to 100 % signal strength of the wireless connection and CPU availability to client application on the PDA The size of result object depends on the type of values stored in the fields. The 14 statistical values are received in 5 'Result' objects and amount to, on average, 62 bytes of results per 5 seconds with additional 44 bytes after every minute. An interesting comparison is made by considering the number of result parameters and their size as retrieved by the *surrogate* (executing at the *Gateway Surrogate Host*) with the corresponding values at the *Device*. A significant amount of information can be condensed by applying intermediate calculations and filtration of values at the *surrogate* module.

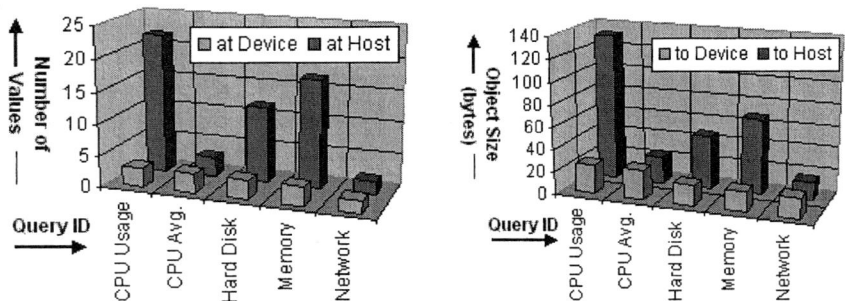

Fig. 2. Left: Comparison between number of values at the *Host* and values sent to the *Device*; **Right:** Comparison between size of intermediate results at the *Host* and size of results at *Device*

It can be observed that the number of parameters is reduced by 75% (4 times reduction) when transferring results to the *Device*. Similarly, more than 64% of the data has been filtered out in intermediate calculations and trimming at the *surrogate*. This performance markup is in addition to the communication and delay reduction achieved by careful selection of host machine and resource access mechanisms during surrogate lifetime, as explained earlier. The burden on PDA has been reduced to a few hundred bytes of data and graph formation.

5 Conclusion and Related Work

Research and development in enabling handheld held devices to interact with Grid services is in its early stages. Signal [12] proposes a mobile proxy-based architecture that can execute jobs submitted to mobile devices, in-effect making a grid of mobile devices, but this approach may affect the fault tolerance of the system as the mobile device hosting the proxy also has to deal with the adverse effects of a mobile/wireless environment. Also, the proxy schedules the jobs submitted to it by other mobile devices, but in our case as the middleware has far more resources at its disposal, so there is no particular need for scheduling. In [13] mobile agent paradigm is used to develop a middleware to allow mobile users' access to the Grid. It focus's on providing this access transparently and keeping the mobile host connected to the service. GridBlocks [14] builds a Grid application framework with standardized interfaces facilitating the creation of end user services. They state that SOAP usage on mobile devices may be 2-3 times slower than a proprietary communication protocol, but the advantages of using SOAP (such as overcoming device heterogeneity) maybe far more profitable than the effects of this limitation

A solution based on Jini Surrogate Architecture, to access Grid services, is demonstrated in this paper. In the proposed approach, a resource constraint device wishing to access a resource-demanding service is allowed to delegate this task to a relatively powerful machine (desktop, server). Specifically, CPU intensive, network oriented tasks can be efficiently delegated to such systems when network connectivity is available. In case of intermittent connectivity, applications and services requiring on demand or periodic network access can benefit from this approach.

Optimization of the overhead caused by an additional layer between the source service and the destination device, location based dynamic scalability, and multi-protocol discovery services, are the main focus of the research. The implementation has been tested for a moderately intensive task. We intend to extend and implement the architecture to interact with existing Grid services and analyze the working of our framework incorporating HTTP discovery, client authentication, and surrogate migration support. A notable constraints suffered by our approach include the requirement of Java virtual machine at the device. Furthermore, at present we have not addressed the notions of client/surrogate authentication and authorization and are the focus of our future work.

References

1. Satyanarayanan, M.: Fundamental Challenges in Mobile Computing. In: Proceedings of the 15th Annual ACM Symposium on Principles of Distributed Computing, Philadelphia (1996)
2. Sun Microsystems, Inc.: Jini^{TM} Architecture specification. http://www.sun.com/jini/specs/
3. Sun Microsystems, Inc.: JiniTM Technology Surrogate Architecture Specification. http://surrogate.jini.org/sa.pdf (2003)
4. S. Vazhkudai, S., Tuecke, S., Foster, I.,:Replica Selectionin the Globus Data Grid. Proceedings of the first IEEE/ACM International Conference on Cluster Computing and the Grid (CCGRID 2001), IEEE Computer Society Press,(2001) 106-113,
5. Lee, B., Weissman, J.B.: Dynamic Replica Management in the Service Grid. In: High Performance Distributed Computing 2001 (HPDC-10''01), San Francisco, California (2001) p. 0433
6. Sun Microsystems, Inc.: Jini^{TM} Technology IP Interconnect Specification. http://ipsurrogate.jini.org (2001)
7. Curbera, F., Duftler, M., Khalaf, R., Nagy, W., Mukhi, N., Weerawarana, S.: Unraveling the Web Services Web – An Introduction to SOAP,WSDL, and UDDI. In: IEEE Internet Computing, vol. 6, no. 2,(2002) 86–93
8. Box D., et al. Simple Object Access Protocol 1.1. Technical report, W3C.,http://www.w3.org/TR/2000/NOTESOAP-20000508/ (2000)

9. Shan, H., Oliker,L, Biswas, R.: Job Superscheduler Architecture and Performance in Computational Grid Environments. In: Super Computing Conference 2003 (SC2003), Phoenix, Arizona (2003) 15-21
10. Wolski, R.: Dynamically Forecasting Network Performance Using the Network Weather Service. In: Journal of Cluster Computing, (1998)
11. Stallings W.: SNMP, SNMPv2, SNMPv3, and RMON1 and RMON2. 3rd Edition Addison-Wesley, California (1999) 71-82
12. Hwang, P. Aravamudham Middleware Services for P2P Computing in Wireless Grid Networks. IEEE Internet Computing vol. 8, no. 4, July/August 2004, pp. 40-46
13. Bruneo, M. Scarpa, A. Zaia, A. Puliafito, Communication Paradigms for Mobile Grid Users. Proceedings 10th IEEE International Symposium in High-Performance Distributed Computing, (2001)
14. Gridblocks project (CERN) http://gridblocks.sourceforge.net/docs.htm

Mobile Mini-payment Scheme Using SMS-Credit

Simon Fong[1] and Edison Lai[2]

[1] Faculty of Science and Technology, University of Macau, Macao
ccfong@umac.mo
[2] Faculty of Science and Technology, University of Macau, Macao
edisonr@macau.ctm.net

Abstract. Mobile Phone has been widely used in the world and adopted in different application areas involved with SMS. SMS has been evolved for downloading logo, ring-tones, advertisement, security notes, and location based system. This paper describes an extension to current functionalities of SMS. "SMS Credit" is a prototype, which allows customers to perform electronic payment using their mobile phones via SMS without the need to modify existing devices or to acquire new equipment. Users simply send SMS to connect to the background GSM system and key-in the charged amount, their account will then be automatically credited once the transaction is authorized. Specific wireless terminal is installed at participating merchants and service providers, which allow customers to identify and authorize a transaction. Example scenarios on actual application on telecommunication and electronic commerce field will be discussed.

1 Introduction

Mobile phones become an essential communication tool nowadays that almost every person owns one. They are so popular nowadays, as wallets or watches, that almost every citizen in an urbanized city will carry one along with them.

GSM (Global System for Mobile communications) networks are highly developed and its coverage extends to almost every part of the city. Since any person with suitable equipment can capture information sent over the radio medium, the network itself is designed to provide a secure channel for transferring digital information through the air. To achieve this, a connection must first be established and both the user and the background system have to mutually authenticate themselves before information is exchanged. Data is transmitted through a secure private connection afterwards and all the information is encrypted using cryptographic algorithms.

With its popularity and ease-of-use, application of this secure wireless network can be further extended from a simple voice communication to other commercial areas. Mobile phones with integrated electronic purse applications are already available in the market. However, in order to access these new functionalities, customers usually have to purchase new equipments. In this paper, a prototype will be developed that allows customers to perform electronic payment using their mobile phones, without the need of modification of their mobile device or acquiring any new equipment.

Section 2 reviews some existing mobile payment models that allow customers to perform electronic transactions using their mobile phone. Section 3 presents the framework for mini-payments using mobile phones. In section 4, the software prototype of this payment model will be discussed. Finally, it concludes the benefit of this mini payment for the GSM operator and the customer.

2 Literature Review

Mobile payment systems are nowadays readily available as a commercialized product. Wireless devices can be used either as an Internet surfing tool, allowing users to perform different types of electronic transaction or as an electronic wallet that allows users to obtain resources and services from traditional distribution medium.

Various online mobile business transactions models are developed and discussed [1]. However, its development is constrained as it faces the traditional limitation of mobile devices: limited bandwidth, low-resolution display and lack of a proper user interface.

Current mobile devices and technology might be more suitable for conducting electronic transactions over the traditional retail medium. Nokia 6510 first introduces the concept of an electronic wallet, which made use of the advance WAP technologies to perform electronic transactions. Siemens have introduced the concept of pay@once [2] that enables users to conduct micro-payment at registered merchants and services providers. Network operators act as a payment gateway that charges their subscribers on behalf of internal or external merchants. It makes use of SMS (Short Message Service) for transferring payments from one account to another.

A mobile wallet model is proposed, allowing existing credentials, such as bank and credit cards, be easily included within the architecture without any change to this legacy system and their functionality, thus realizing payment using off-line digital cash. The authors proposed the usage of a two-slot "Subscriber Identity Module" (SIM) in the implementation of the mobile payment system.

Money can quickly and safely be transferred from one account to another via mobile phone. It is as simple as sending an SMS: just type in the recipient's mobile phone number, the desired amount of money and confirm with your PIN [3].

Telemoney [4] offers a similar payment which allows customers to perform their payment on registered merchants and service providers using their mobile phone. The main idea of *Telemoney* is to use the client's mobile as a digital identity tool to authorize online transactions performed on the Internet. Compared to the payment model in this paper, it offers a similar service called *TeleCab* which allows customers to pay their taxi fees by simply inputting a set of parameters into their mobile. *Telepay* is another service that allows users to pay by their mobile phones by beaming their *Telemoney* ID to the service terminal.

In summary, the model and prototype were discussed on the online payment area. For electronic payment over the traditional retail medium, most of these solutions require new devices or modification of existing equipments.

3 Mini Payment Model

3.1 Design Requirement

The main requirement of this project is to develop a mini-payment model using mobile phones without the need of modification of existing wireless devices. The model is designed to allow consumers to conduct electronic payment by simply key-in several codes via Short Message Services (SMS) to specific MSISDN (Mobile Station International ISDN) using the 12-keys input pad available in most standard mobile phones.

Furthermore, customers expect an efficient payment solution when dealing with low value transactions. Standard macro-payment or electronic cash protocols make use of cryptographic algorithms to achieve security and privacy. Transactions must be authorized by a central independent third party. This results in a central bottleneck on most of the on-line payment schemes, a single point of failure that increases payment latency. It also raises the cost of the transaction, and imposes a minimum cost per transaction, as the bank is faced with the real cost of authorizing each transaction [5].

In this payment model, Network Operator (NO) will act as a central authority that handles all the transactions. They will charge on behalf of the merchant and vendors and credit the customer's account once the transaction is completed. GSM network architecture provides mechanism for the authentication of the user and the encryption of the transmitted messages. As it is dealing with low value transactions, payment information will be encrypted using the standard cryptographic algorithms offered by the GSM Network. To minimize the risk of payment using stolen or loss mobile devices, users have to authenticate themselves using a PIN code for authorizing each transaction.

Privacy cannot be achieved in this model and users are expecting to reveal their identity. As in credit card payment models, users have to identify themselves using a card and the corresponding account is credited. In this payment model, users are identified by their MSISDN.

3.2 System Overview

Assumptions

1. The billing architecture of the Network operator could be extended to handle external financial transactions.
2. Security of messages transmitted through radio communication network is achieved by using cryptographic services offered by the GSM Network.
3. Users are identified by their MSISDN.
4. Background operation of the billing system will be out of scope of this paper.

This model requires network operators to offer additional services more than the existing voice-call system. Traditionally, NOs provide basic telecommunication services (voice-calls, SMS, etc.) that charge their subscribers according to their usage. In this payment model, NOs have a new role: they act as a Payment Service Provider (PSP) that charges their consumer on behalf of an external merchant for the purchase of a product via their billing infrastructure.

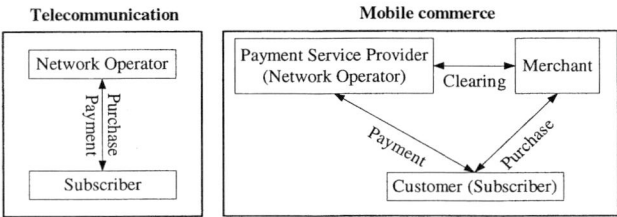

Fig. 1. Role of network operator

Each subscriber's account does not only represent the amount charged for the usage of the telecommunication services offered by the NO, it also accounts and credits the consumer for any m-commerce purchases and services initiated using their mobile devices. As a consequence, the NO who is willing to participate in this business has to modify his payment infrastructure in order to accept new internal and external merchants.

Authorization terminals can be installed on the participating merchant and service providers. These are special GSM devices that are designed to allow consumers to confirm and authorize a particular transaction.

In order to access this payment service, the customer must be a valid mobile subscriber of the network operator, and he/she must have applied short code messaging service and has been registered as a *Mobile payment user"*.

The following figure shows how the model works:

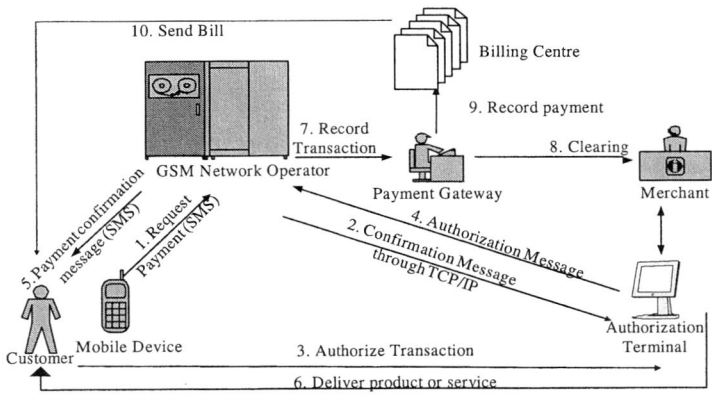

Fig. 2. Mini-payment model with authorization terminal

1. Consumer sends SMS to the Payment Gateway. A set of parameters is key-in using their mobile device, separated with the special key "#": vendor's code, their secret PIN and the corresponding amount to be charged or key-in using their mobile device, separated with the special key "*": Product short code, their secret PIN.

2. A confirmation message through TCP/IP is sent to the merchant's authorization terminal. Consumer's mobile number and amount to be charged are shown on the terminal for authentication.
3. Consumer authorizes the transaction by key-in a second PIN code in the through authorization terminal.
4. Authorization terminal sends a message back to the Network Operator along with the secret PIN code to confirm that the customer has authorized the transaction.
5. Consumer receives a SMS message, which notifies the payment completion.
6. Merchant or service provider delivers the product or service to the customer.
7. The transaction is recorded on the Payment Gateway server.
8. Merchant can request for clearing through the Payment Gateway.
9. Payment Gateway sends transaction records to the billing center.
10. The corresponding credited amount is recorded and included in the monthly bill and sent to the subscriber.

3.3 Detailed Payment Model

The proposed payment model has just been discussed in a simplified form. The actual payment model involves interaction with the GSM background system. Details on how users are authenticated and messages are encrypted by using the services offered by the GSM network has not been discussed.

In summary, the process is shown as follows:

1. Customer chooses a valid short code of a merchant, keys in the short code and PIN code in SMS and sends out. The request will be transferred to the BTS through radio, and finally transferred to MSC through the BSC.
2. The GSM network authenticates customer by a service table, which is located in the HLR (Home Location Registers) related to the MSC. It will check the request from the customer (mobile phone short code) with its short code table and finds the matching code in the service located in the HLR.
3. If this short code and service belongs to the mini payment service, it will route the request to SMS credit system, which has been integrated as part of the Value Added System (VAS).
4. After the VAS received the request, the short code and PIN will be decoded according to the content in short message and the system in return will either accept the customer's request by sending a confirmation "short message" or denied the customer with a rejection "short message".
5. Customer request will be routed to the payment gateway, which will authenticate the request with both the merchant and the customer billing center.
6. The VAS will inform the merchant for requesting final authorization from the customer.
7. The merchant terminal will display the authorization requested message for final confirmation.
8. If the customer agrees with the transaction, a message will be sent back to the VAS, which will route the message to the payment gateway.
9. The billing account of the customer will be credited, and the merchant account will be debited if the customer agreed with the transaction.

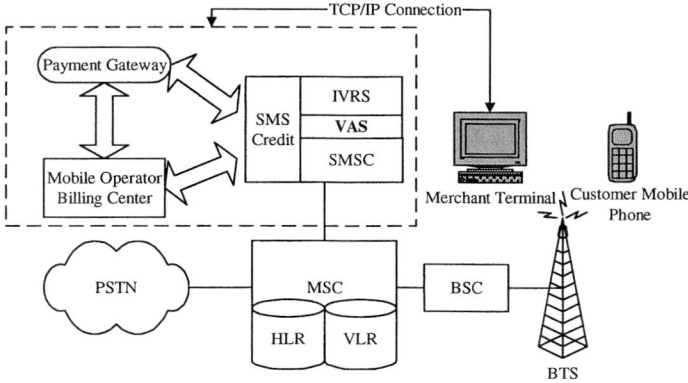

Fig. 3. GSM system overview

4 Software Prototype

4.1 Implementation

In this software prototype, there are three components:

1. The phone Client is only a mobile phone or a mobile device, which emulates the basic functionalities of a mobile client acting as a wireless payment device.
2. GSM Server emulates the background system that serves as a payment gateway recording each transaction. This GSM server is composed with a mobile phone or a SMS modem which is used to retrieve the SMS of Phone Client Request
3. Merchant terminal emulates the device installed on the merchant's sites for transaction verification and authorization.

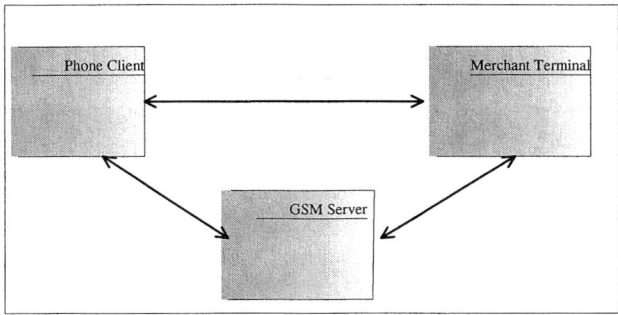

Fig. 4. Component diagram

4.2 Phone Client

The phone client mainly emulates the basic functionalities of a mobile device used as an electronic payment tool on the model proposed in the last section. By using the 12-keys input pad offered on most mobile phones, user input a set of string via SMS for the initialization of a payment.

In order to be correctly interpreted by the GSM server, the string must have a standardized format. In this mini-payment, there are two payment services, which offer credit transfer and short code payment

1. Credit Transfer-The billing account of customer will be credited with this specific payment amount, and the merchant account will be debited if customer agreed with the transaction.

#[Merchant ID]#[User PIN]#[Payment Amount]#

The decimal point is represented by *. In the following example, the amount of $88.80 is represented by 88*8.

2. Short Code Payment-Customer only needs to key in the Short Code for requesting the product or service. The payment amount charged based on Merchant Service or Product Code Table.

* [Product or Service Short Code]*[User PIN]*

In this model, after the user inputs the whole payment string, user must click on the SEND button in order to transfer the SMS message to the GSM Server. TCP/IP is used to simulate the communication protocol between the merchant terminal and the background GSM server. The server is listening to the port 1007 to receive messages transmitted from the merchant terminal.

As discussed before, information transmitted through the GSM network is secured and messages transmitted through the air are encrypted when a session is established in the background system. The unique PIN number contained in the message cannot easily be eavesdropped by attackers.

4.3 GSM Server

GSM server primarily emulates the background system connected to the GSM network, the billing system and the payment gateway. The basic functionalities are:

1. Listen to client payment requests.
2. Identify and authenticate the client.
3. Send authorization confirmation messages to merchant terminals.
4. Record transactions and transfer it to the billing system.

When the server receives the payment message, the interpreter will decode it and perform the following checks:

1. Locate the customer from the database by its Customer ID. An error message is returned if the corresponding ID does not exist on the database.

2. Locate the merchant from the database by its Merchant ID. An error message is returned if the corresponding ID does not exist on the database. If a match is found, the corresponding port number is retrieved from the database.
3. Finally, the PIN of the customer is retrieved from the database and compared with the one received from the user. If they do not match, an error message is returned.

The transaction proceeds if the correct PIN is matched. The message is sent to the corresponding merchant terminal according to their port number.

In the payment system, the merchant service query is updated through TCP/IP connection or a SMS message is sent to the merchant terminal, which is a specialized designed similar to the GSM client. This device is designed to retrieve merchant service query or received SMS messages transmitted from the GSM server and act as an authorization terminal where customers are able to confirm their transaction.

4.4 Merchant Terminal

Merchant terminal models the special designed GSM client used to authorize a transaction. It will listen to payment confirmation messages by TCP/IP sent from Server or from GSM servers in SMS format.

When a message arrives, it will display the corresponding client information in the terminal screen, along with the amount to be paid. User can select to accept the transaction or reject it.

If the user chooses to accept the transaction, a confirmation message is sent back to the GSM server. When the message arrives to the payment gateway, the transaction is recorded in the database and later will be sent to the billing system for processing. The server will finally send a payment receipt in SMS message format to the client's mobile to acknowledge the transaction.

If the user chooses to reject the transaction, a reject message is sent back to the server and the transaction will be aborted.

5 Conclusion

GSM network can provide a strong security for each mobile subscriber, and it is found to be a suitable medium for performing off-line transaction of small amount.

A model has been developed allowing GSM subscribers to effectuate payments by their mobile phones without need of modification of existing equipment or acquiring new equipments. Users simply key in a set of parameters to initialize a payment to registered merchants or service providers, and they account will be charged accordingly. Network operators will act as acquirers to credit customer's account to pay-in-advance to merchants they deal with.

This value-added service offered by network operators helps them to increase their market share as they are offering a much more convenient payment option to their subscribers.

6 Future Work

The future work for this mobile payment scheme is to prepare a simulation analysis on this payment protocol and evaluate its performance. Moreover, fraud control and cost analysis are major study work.

References

1. Upkar Varsney, Ron Vetter: A Framework for the Emerging Mobile Commerce Applications, 34th Hawaii International Conference on System Sciences (2001),2
2. Web site for Siemens Mobile Payment:
   http://www.siemens.ie/mobile/mobile_commerce/payment_solution_arch.htm
3. Web site for Siemens Mobile Payment:
   http://www.siemens.ie/mobile/mobile_commerce/mobile_beaming.htm
4. Web site for Telemoney Mobile Payments:
   http://www.telemoneyworld.com/how_it_works.htm
5. Manfred, Lilge: Evolution of Prepaid Service towards a Real-time Payment System, IEEE Intelligent Workshop (2001).

Context Summarization and Garbage Collecting Context[1]

Faraz Rasheed, Yong-Koo Lee, and Sungyoung Lee

Computer Engineering Dept. Kyung Hee University,
449-701 Suwon, Republic of Korea
faraz@oslab.khu.ac.kr, {yklee, sylee}@khu.ac.kr

Abstract. Typical ubiquitous computing environments contain a large number of data sources, in the form of sensors and infrastructure elements, emitting a huge amount of contextual data (called context) continuously that need to be processed and stored in some context repository. Usually, this data is for software system's internal use to provide proactive services. Hence, it makes sense not to store this entire huge amount of data but to identify and remove some irrelevant data (garbage collecting context), summarize the left over and only store this summarized and more meaningful data. We believe that such a summarization will result in improved performance in query processing, data retrieval, knowledge reasoning and machine learning. Besides, it will also save the storage space required to store context repository. In this paper, we will present the idea and motivation behind context summarization and garbage collecting context and some possible techniques to achieve this.

1 Introduction

The idea of Ubiquitous Computing [1] is gaining popularity with every passing day. Several research groups are developing their own ubiquitous computing projects [2], [3]. Ubiquitous (or pervasive) computing provides computing environment where computing resources are spread through out, present everywhere in the environment and providing services to users seamlessly & invisibly without any explicit user intervention. A ubiquitous computing environment, thus, contains a number of devices, sensors, and software systems.

Context awareness is among the foremost features of any ubiquitous computing environment. In order to provide appropriate services, an application needs to be aware of the user and environmental context. Similarly at lower levels of abstraction, an application (or middleware) is required to be aware of the computational context including device and network state. So, what is 'context' itself? We take context as the 'implicit situational understanding' and consider all the information that defines a situation as context. So, location, temperature, network bandwidth, device profile, user identity can all be taken as the context information or simply context.

Since a Ubiquitous Computing system needs to deal with such huge and diversified information (context), there should be an appropriate context model to define,

---

[1] This work is supported by grant No. R01-000-00357-0 from Korea Science and Engineering Foundation (KOSEF).

represent, and store context efficiently in some context repository. The management of context information in ubiquitous computing imposes lots of issues and challenges. M. J. Franklin [4] has identified a number of issues in ubiquitous data management such as those posed by adaptivity, ubiquity, mobility and context awareness

We approach the context (or data) management in ubiquitous computing from a different perspective. We are working on to identify the relevance and significance of information that a Ubicomp system receives from sensors and its surrounding. We believe that identifying and removing the irrelevant context (we call it 'garbage collecting context') and summarizing the available or incoming context (which we call the 'context summarization') will result in the improved performance of knowledge reasoning, inference making, machine learning and efficient use of computing resource including the storage space required by the Context Repository.

The rest of the paper is organized as follows. Section 2 contains related work. Shortcomings of existing systems are in Section 3. Our proposed solution is in Section 4. Four techniques of Context Summarization are in Section 5 and our proposed model is in Section 6. Section 7 contains issues and challenges. Finally we mention about risk factors involved in Section 8 and conclude the paper in Section 9.

2 Related Work

Unfortunately, Garbage Collecting Context (GCC) and Context Summarization (CS) have not yet got the attention of researchers. One primary reason is majority of Ubicomp systems are academic projects and have not been deployed in real environment and used for elongated periods. The issues identified in this work come in one's attention when actual system is deployed and run for considerable time. The general focus of research community in ubiquitous computing is not towards context data management; but how to make ubiquitous computing operational in first place?

Several existing ubiquitous computing systems support features like noise filtering, privacy control, feature extraction [6], [3], [10] but we believe that using separate components for GCC and CS with clearly defining the responsibility of each component will produce better results; mainly because of the separation of concerns.

In Database Systems, data mining and data ware housing [7] use the concept of histogram [8] and multidimensional views of database and work on the aggregate, consolidated data instead of raw data to support higher level decision making and to identify the hidden patterns in data. Hence, when we extract underlying meaning from context data, it can be considered as something like 'Context Mining' where we extract higher level context from the lower level context. Online Analytical Processing (OLAP) and data mining are not done on actual data but on the historical, consolidated and aggregate data while we are performing the context summarization on actual context. Contrary to data mining and OLAP, we want to transform the raw context to summarized form taking less storage space and provide improved and efficient reasoning and machine learning. Anyhow, the concepts explored in the field of data mining and OLAP are highly useful for the Context Summarization.

Researchers in DBMS have also analyzed time series data streams for very large databases [9]. Here, they analyze the data coming in continuous streams with time. They have proposed solutions on how to manage, represent and store the time series data streams. This is also highly related to the context summarization.

In traditional DBMS, data is seldom removed. But in our context summarizer, we do replace raw context with summarized information. Why? The answer lies in why, in first place, we are storing the context? We are storing the context and maintaining context history so as to reason on context, draw inferences from the context and make the machine learn. If the context is summarized properly, the application can reason, infer and learn about activities more efficiently; because, they need the history and consolidated data which we are providing as a result of context summarization.

3 Problem Definition

A ubiquitous computing environment comprises of a number of different sensors providing context information like Environmental context (temperature, pressure, light), Audio / Video, Location context, Computational context (network bandwidth, underlying operating system, hardware specification), the list goes on and on...

Context information comes in a continuous stream with each sensor emitting data regularly (at least during some interesting activity). We are heading towards flood of context data. Such a huge amount of data requires proper management. At this point, we need to answer what to do with such a huge amount of data? Do we need to store all of this? More importantly do we really need such a large amount of data?

Several data items sensed from the environment are required for some instant processing and reasoning, e.g., presence of a person can be used to trigger the activity of turning lights on or caching data related to a particular user. But, we also need to store context for later use; knowledge reasoning, inference making & machine learning. For instance, we may need to keep the context of user presence for some on going (near future) activity or to infer what she might be up to.

But storing all such context information imposes several issues. First, it requires considerable amount of storage space. Ubiquitous computing systems are essentially distributed, therefore, migrating larger amount of data puts significant burden over network traffic. Secondly, query processing and data retrieval on large context repository requires significant computing resources decreasing the overall throughput of the system. Thirdly, several contexts needs to be discarded and should not be stored permanently. For example, the data with low precision, because of noise, needs to be filtered out before sensitive operations (e.g., heartbeat rate of a patient). Privacy control also prevents us from storing all information, e.g., the information that user is in washroom. Lastly, efficiency of techniques such as knowledge reasoning, inference making and machine learning depends heavily on the size of supplied data.

4 Proposed Solution

First we need to identify low precision, irrelevant and redundant context; the one that is no longer useful and remove such context information. We call this process as Garbage Collecting Context (GCC).

Secondly, we need to summarize the actual (raw) context in such a way that it is more meaningful, can be used more efficiently for reasoning, etc and takes up less storage space. We name this process as the Context Summarization (CS).

A simple analogy is human behavior towards received news. Every day, we read a lot of news in newspaper, on internet and television. But do we (need to) remember all the words and information that make up a particular activity or event? What we actually (need to) remember is some compact information about a particular event that what has actually happened. For example, Bob watches a soccer match for 70 minutes but after the match is over, he does not remember exactly what had happened in the 14th minute of the game. What he actually remembers is a summary of the match like who has won the match, few ups and down, and how many goals were scored and by whom. This is very close to what we mean by *Context Summarization* that instead of storing each and every raw information, only keep summarized and meaningful context information. Coming back to scenario, after the match is over, Bob tends to forget some information, e.g., how far did the ball go when Player X kicked it and who received it. Also, as time goes by, he also tends to forget more details like a spectator had broken in to the field. This act of discarding irrelevant information is analogous to the concept behind *Garbage Collecting Context*.

4.1 Garbage Collecting Context (GCC)

GCC is analogous to the garbage collection in programming languages [5] where we identify the memory areas no longer needed by a program and free it.

GCC can be used to filter out the noise in the data, i.e., the data with low precision so that it does not affect sensitive operations. Some systems [3] provide the precision value or the probability of the correctness of sensed value which can be employed.

GCC can be used to identify and remove the context no longer needed by an application. For example, if an application is storing temperature values after every 5 minutes then it may not require raw history forever. But generally, discarding information is not considered as a good idea; therefore, here we can employ the idea of context summarization and replace the raw history with this summarized history.

Privacy control can also be dealt using GCC. In this case, certain privacy policies determine which context should not be stored and included in the system processing. For example, the location of user in private places (like washrooms) and activities during the lunch break should not be processed and stored permanently in the system.

4.2 Context Summarization (CS)

Where Garbage Collecting Context (GCC) identifies and removes the irrelevant and less significant context, Context Summarizer (CS) operates on incoming and existing context to extract useful context from the original data so that it consumes less storage space, improves the efficiency of query processing, reasoning and machine learning. Consider a temperature sensor emitting temperature value after every 5 minutes.

Table 1. Temperature values stored after every 5 minute

Time	Temp.
12:05	23 °C
12:10	21 °C
...	
15:35	15 °C
...	

Using Context Summarization, e.g., we can group (average) this on the daily basis. Another possible way could be when a day is divided into periods like morning, afternoon, evening and context information is kept for each such period (See Table 2).

Table 2. Temperature values stored for different periods of day

Date	Period	Avg. Temp	Max. Temp	Min. Temp
12/01	Morning	5 °C	8 °C	3 °C
12/01	Afternoon	10 °C	14 °C	8 °C
12/01	Evening	9 °C	11 °C	7 °C
12/01	Night	7 °C	8 °C	5 °C
12/02	Morning	4 °C	8 °C	1 °C
...				

The above example demonstrates the summarization of historical data. CS can also be applied as data is received from sensors. For example, when receiving data from audio/video sensors we can extract useful information from it. With audio, we can extract Intensity and Audio type (music, talk, telephone ring). From video sensor, we can extract Pixel percent change, Motion pattern, etc. As a result, instead of storing actual audio & video context, we can summarize and only store relevant information.

One of the benefits of performing context summarization is reduced storage space, which will result in the faster query execution and data retrieval. It will also make the data migration in distributed environment more efficient with fewer burdens on network traffic. Another important motivation behind CS is to store only the relevant context information in such a way that it is more useful for context consumers.

Reasoning about context and drawing inferences is the primary reason for keeping context in context repository in first place, and is primary tool for providing context aware services. For example, if a ubiquitous computing system knows that when Bob comes to his office in morning, he likes to check his emails, then a system can start downloading his emails when Bob enters the room in the morning. The amount and quality of input data makes reasoning engine perform more efficiently. We believe that if context summarization is done properly then it will result in less data; optimized for reasoning and inference making and machine learning.

Context Summarization can either be '*Active*' or '*Passive*'. In *Active Context Summarization*, the context is summarized as it is received from the context sources; sometimes, even before it being stored in the Context Repository, e.g., the summarization of audio and video context. Active Context Summarization is usually irregular and event-based and is performed more frequently.

Passive Context Summarization is usually performed on the context already stored in the repository. The summarization of temperature (as discussed earlier) and other numerical valued contexts comes in this category. It is regular and periodic, i.e., performed in background after a regular interval or at some pre-specified time. It is performed less frequently and may consume considerable computing resources.

5 Context Summarization Techniques

We have identified several categories of context information based on the similarities and nature of context. Each technique is designed for a particular category of context.

5.1 Aggregation

In aggregation, the history of context information is aggregated to generate compact and consolidated context. Numerical context types like temperature, light intensity, humidity, available network bandwidth can be summarized using this technique. In section 4.2, we have demonstrated how this technique.

Aggregation is a passive, regular and periodic type of context summarization, which works in background periodically and is performed less frequently. It removes the original (raw) context after the context summarization has been performed.

5.2 Categorization

Here we categorize different context entities, e.g., user and device profile can be categorized into user and device groups. In this way, we can track the network bandwidth utilization by some particular user group (say doctors) or by some particular device group (say PDAs) during office hours.

Categorization is passive and static type of context summarization, i.e., it is not performed frequently. It can be performed at system startup by some human or the system can learn itself to define categories. In any case, categorization supports machine learning and higher level reasoning. Unlike other techniques, it does *not* remove original context information such as existing user or device profiles.

5.3 Context Extraction

In Context Extraction, useful interesting context is extracted from continuous streams like audio and video. For example, it can be applied to video stream received from Camera, Webcam to extract features like pixel percent change, picture motion pattern (like stable, regular, irregular), etc. Similarly, audio context can also be summarized.

It is an active, irregular and event based context summarization. It can start any time with an interesting activity. Unlike other techniques, it can be triggered even before the context is stored in the repository and even discard it before storage. It results in saving a lot of storage space but may take considerable time in doing so.

5.4 Pattern Identification

Context can be summarized by identifying existing patterns in the context repository or history of activities. Location context can be summarized using this technique. Consider the location context history in the context repository (See Table 3).

Table 3. Location Context History of Users and Rooms

Time	User	Room
09:05	1	1
09:02	2	1
10:08	1	2
11:26	3	3
11:44	3	3
...		

Using pattern identification, a system may deduce the pattern of user's location during week days and come out with something as presented in Table 4.

Table 4. Pattern Identification for User Location

Time Period		User	Room	Probability
From	To			
09:00	12:00	1	1	0.76
13:00	17:00	1	1	0.83
09:00	12:00	2	2	0.67
13:00	17:00	2	1	0.89
...				

Similarly, a system can identify pattern of room occupants at various time periods. Using categorization with pattern identification, a system may also infer which user group (doctors, programmers, etc) occupies which room at different periods of time.

Pattern Identification is again passive, regular and periodic class of summarization that works in background. It is resource intensive and thus performed less frequently. On the positive side, it results in reducing considerable amount of storage space and also supports higher level inference making, machine learning and in predicting future intentions of a user or device in the specific situation. Pattern identification works on the history of context and replaces larger history with patterns of activities.

6 Proposed Model for GCC and CS

The first question, while designing and developing the GCC and CS, is should these components be part of middleware or not? We believe that making these components part of a middleware will yield us the re-usability of design and code.

We prefer designing these components (GCC and CS) as frameworks [11] so that applications only need to provide the *hotspots* (areas of specification). Hence, *Garbage Collecting Context (GCC)* can be developed in such a way that application specific techniques for Noise Filtering and Privacy Policies can be induced even while the application is operational. For example, an application can specify, through XML, that from 1 pm to 2 pm, there is a lunch time at room X, so the location and other activities of users over there should not be monitored. Figure 1 presents the proposed architecture of Garbage Collecting Context (GCC) module.

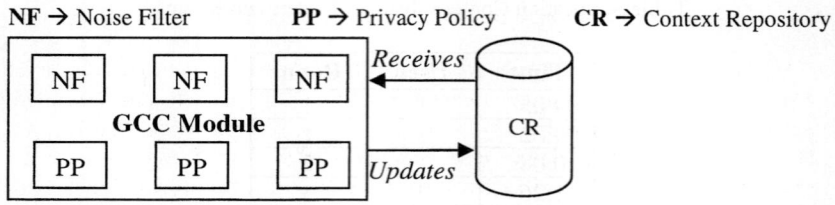

Fig. 1. Garbage Collecting Context Module

GCC retrieves context data from Context Repository (CR), identifies noise (corrupted) context using Noise Filters (NF), applies Privacy Policies (PP) to remove privacy sensitive context and updates the context repository. In a particular implementation, GCC may not delete the context as it identifies it as garbage but only mark it or move such context to some other repository for some later analysis.

Context Summarizer (CS) can also be developed with framework technique. There are various context summarizer sub-modules for each different category of context, called Context Category Summarizer (CCS). Thus temperature, network bandwidth, etc can all be summarized using a single CCS. Context Summarizer (CS) is supplied context information along with Context Meta-Data (CMD). This context meta-data, usually represented through XML, specifies the category of supplied data, so that CS may decide which Context Category Summarize (CCS) should be used to summarize this context. All CCS sub-modules implement a particular interface so that CS can access each of them uniformly. Because of the framework based design of the CS, new CCS can be added and the existing CCS can be updated while the application is operational. Figure 2 shows the architecture of Context Summarizer (CS).

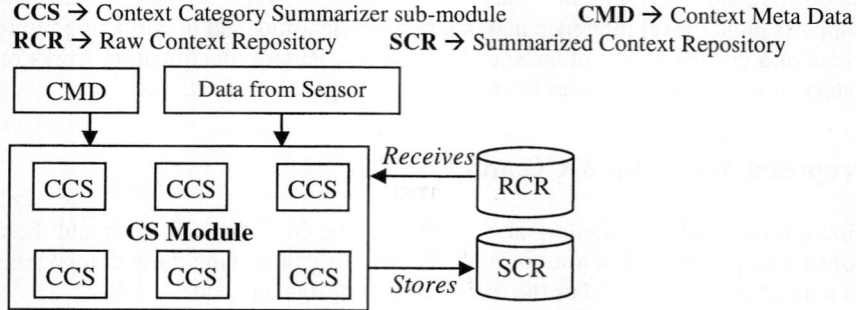

Fig. 2. Context Summarization Module

7   Issues and Challenges in GCC and CS

Garbage Collecting Context (GCC) and Context Summarization (CS) have their own research issues and challenges both at conceptual and implementation level. Here, we identify several such issues and wherever possible identify few applicable solutions.

7.1 What Can Be Summarized and When the Context Should Be Summarized?

Can we summarize each and every type of context? We do not think so; we should only apply context summarization where our application specific analysis identifies some performance or storage optimization. Also, different kind of context information can not be efficiently summarized using a single method. Rather, we can form several categories according to similarities in the nature of context information and define mechanisms to summarize each different category of context data.

Another important question asks what is the appropriate time to trigger context summarization? Should we start context summarization just as sensors provide contexts? (*may be useful for audio and video sensors*) or as the application processes it to some more useful form and stores in the context repository (*may be applicable for location context*) or once the context has become the history and is not directly useful for application (*applicable for temperature and similar type of contexts*) or periodically after some regular interval of time or some pre-specified time?

7.2 Performance Overhead and Other Challenges

Perhaps the foremost concern to apply these techniques is the performance cost. Do the benefits achieved by these methods justify the computing resource consumption? We believe that a proper application of GCC and CS (like those discussed in section 5) will yield the performance improvement and will not eat up many resources. In any case, the overall system should not be ceased or hung-up during the execution of GCC and CS modules, the resources (like context repository) should not be locked for noticeable period of time. But the problem is how to achieve this? We need GCC and CS only when there is considerable amount of context information, a considerable amount of context means a considerable amount of processing and resource consumption to produce useful output.

Other issues and challenges include how are we going to deal with distributed and ubiquitous nature of middleware, data repository and applications? What are security, trust and service level guarantees required for systems using GCC and CS techniques?

8 Risks Involved

Garbage Collecting Context (GCC) and Context Summarization (CS) are sensitive in nature as they directly access context information and modify it. Information is always one of the most important assets of any system and organization. In this section, we will briefly mention about some risk factors that should be considered while developing and implementing GCC and CS techniques.

Both GCC and CS result in some data and precision loss. Failing to compensate this precision lost may affect performance and overall throughput of the system badly.

Improper context summarization may make reasoning and machine learning more difficult, complicated, inefficient, incorrect and misleading instead of improving it.

GCC and CS modify Context Repository (CR). Several modules of middleware and application access CR simultaneously. Such a sudden modification may be unexpected and make these modules produce unexpected results and must be avoided.

9 Conclusion

Garbage Collecting Context (GCC) and Context Summarization (CS) are new, interesting and useful research areas with a number of interesting research issues. We have presented both the benefits and risk factors involved in using these techniques and have also identified four techniques for implementing Context Summarization (CS). We have also proposed a model for implementing these concepts and identified certain research issues and challenges expected to be faced. We conclude with the fact that they are sensitive operations which must be handled carefully and applied after rigorous testing. Finally, 'to summarize and how to summarize?' that is the question!

References

1. M. Weiser, The computer for the 21st century. ACM SIGMOBILE 1999 Review
2. Dey, A.K., et al.: A Conceptual Framework and Toolkit for Supporting Rapid Prototyping of Context-Aware Applications. Human-Computer Interaction (HCI) Journal, Vol. 16. (2001)
3. Hung Q. Ngo et al: Developing Context-Aware Ubiquitous Computing Systems with a Unified Middleware Framework. EUC 2004: 672-681
4. Michael J. Franklin, Challenges in Ubiquitous Data Management. . Informatics: 10 Years Back, 10 Years Ahead, LNCS #2000, R. Wilhiem (ed)., Springer-Verlag 2001
5. Richard Jones, The Garbage Collection, http://www.cs.ukc.ac.uk/people/staff/rej/gc.html
6. Jason I. Hong, James A. Landay, Support for location: An architecture for privacy-sensitive ubiquitous computing, Proceedings of the 2nd international conference on Mobile systems, applications, and services, June 2004
7. Alex Berson , Stephen J. Smith, Data Warehousing, Data Mining, and OLAP, McGraw-Hill, Inc., New York, NY, 1997
8. D. Barbara et al., The New Jersey Data Reduction Report, Bulletin of the IEEE Technical Committee on Data Engineering December 1997 Vol. 20
9. Lin Qiao et al, Data streams and time-series: RHist: adaptive summarization over continuous data streams, Proceedings of the eleventh international conference on Information and knowledge management, Nov 2002
10. Moore, D., I. Essa, and M. Hayes, Exploiting Human Actions and Object Context for Recognition Tasks, In Proceedings of IEEE International Conference on Computer Vision 1999 (ICCV'99), Corfu, Greece, March 1999
11. Mohamed Fayad, Douglas C. Schmidt, Object-Oriented Application Frameworks, Communications of the ACM, Volume 40 Issue 10, Oct 1997

EXtensible Web (*xWeb*): An XML-View Based Web Engineering Methodology

Rajugan R.[1], William Gardner[1], Elizabeth Chang[2], and Tharam S. Dillon[1]

[1] eXeL Lab, Faculty of IT, University of Technology, Sydney
{rajugan, wgardner, tharam}@it.uts.edu.au
[2] School of Information Systems, Curtin University of Technology, Australia
ChangE@cbs.curtin.edu.au

Abstract. XML is becoming increasingly a popular medium in industrial informatics for (a) storing and representing unstructured and semi-structured information such as web content and (b) messaging between heterogeneous data sources. For both these purposes it is important to provide a high level, model driven solution to design and implement websites that are capable of handling heterogonous schemas and documents. For this, we need a methodology, that provides higher level of abstraction of the domain in question (here the web) with rigorously defined standards that are to be more widely understood by all stakeholders of the system. To achieve this, in this paper, we propose an XML-view driven design and architecture solution (methodology) for web engineering called *xWeb*. This methodology uses Object-Oriented (OO) conceptual modeling techniques in combination with proven higher-level web user interface engineering and data engineering techniques that provides a comprehensive, yet generic web design methodology. Also, *xWeb* architecture utilizes XML technologies without requirements for a middleware and provides support for web portals.

1 Introduction

Web engineering and website development have evolved from coding simple HTML based static pages into complex a software engineering discipline. The traditional web engineering techniques, which were based around textual file based structures, provided only limited or no facilities for modeling higher-level design concepts that go beyond the granularity of file-based textual information [1]. But, today's websites not just deliver static contents, but also support (web) application driven data transactions to complex multimedia web contents. Thus with such complex increasing complex web contents such as interactive and hypermedia based web sites, designers went beyond traditional HTML files and turned towards middleware and scripting technologies, from Common Gateway Interface (CGI) scripts to advanced SOAP messages. While there are speedy advancement in the implementation and deployment level technologies, there is a lack of sufficient techniques with modeling and design capabilities that are available to classical (non-web based) software solutions. One of the most common architecture for the deployment of such web system is the 3-tier architecture, which consist of the presentation layer, application layer, and data layer. Many web engineering solutions (for simple web site design to e-Commerce engineering) are focused mainly on building, designing and maintaining web pages (using various styles and techniques) to support user-requests (from displaying simple text information to B2B and B2C transactions) without consideration for all four combined aspects of web engineering namely; (1) a well defined design methodology (such as OO) using standard modeling language (such as OMG's UML™) to capture and model (a) domain requirements, (b) user requirements, (c) user-interface requirements (user-interface engineering), (d) data requirements (data engineering) and (e) platform/architecture requirements, (2) a generic architecture adoptable to various implementation techniques (3) well defined generic data standard (such as XML) to describe and represent web data and (4) process to address post-development maintenance and expansion. Though there exists tools that can perform one or two of the above stated aspects in regard to web engineering, to our knowledge no tool or techniques provide all four aspects. In this paper we propose web design/engineering methodology (called eXtensible Web or *xWeb*) that we argue will provide all four aspects of the software engineering process for a comprehensive web engineering methodology.

xWeb incorporates proven software engineering techniques such as; (1) OO conceptual modeling [2, 3], (2) data modeling and engineering using XML views [4-6] and (3) web user interface engineering [7, 8]. In related work, the authors have shown that, *XML view* based design methodology is well suited for modeling and building architectural solutions for the complex domains such as web portals [9] and XML data warehouses [10-12]. Also, in work [13], the authors proposed an XML-view based middleware solution to modeling and implement a role-based user access control (UAC) mechanism. In the case of capturing web user interface requirements in *xWeb*, we adopt *Web User Interaction Analysis Model* (WUiAM) [7, 8], which is used in requirement capture and design of a large e-Logistics/e-Commerce solution [14-16] and taught as part of the user interface engineering course in a leading Australian University (in 2001/'02).

In the following sections, we discuss the uniqueness and the application of *xWeb* methodology using a simple case-study example (Section 1.2). Though we like to provide detail discussion on all aspects of the *xWeb* methodology and its components, due to page limitation, we provide detailed discussion on unique aspects of *xWeb* and only a brief discussions on established/common software engineering aspects, as *xWeb* methodology derives/utilizes many well proven software engineering techniques such as OO conceptual modeling [2, 3] (using UML).

The rest of this paper is organized as follows. In Section 1.1, we briefly look at some early work done in relation to web engineering (and web portal engineering) followed by an outline of an example case-study used to illustrate the concepts presented in this paper. Section 1.3 briefly introduces our own work; *xWeb*, followed by Sections 2 - 2.2 where we look into the *xWeb* modeling techniques and gradually build our case-study website. Section 2.3 provides an in-depth discussion on the *xWeb* components that form the basic *xWeb* implementation architecture. Section 3 concludes the paper which includes some discussion on our future research directions.

1.1 Related Work

Here we look at some work done in both areas of web engineering as well as web portal design, as our work incorporates both. There exists many work and tools in dealing with one or more aspects of general web engineering principle such as [1, 17-20], but none that address the whole spectrum of web engineering issues. Also, there are many existing works that deal with the possibility of application of portal in different areas of use [21], and the classification and discussion of different type of portal [22, 23]. Only few have look into the issue of the actual design and development of a portals. One of the most interesting work includes [24], where the authors looks at the development of portal from a software engineering perspective. In [25], usability issues are taken into account and the importance of evaluating these on customizable portals is also discussed. Yet most of these works do not provide a comprehensive design and technological solution for addressing both web data and web user interface design issues under one design methodology. We argue that, such a combined design methodology is a must for any web system development such as portals.

In [26], the authors have argued that there are two aspects of technical architecture that a web modeling language must possess for it to be used effectively on the development of web systems, namely information architecture and functional architecture. In the related literature, there is a lack of consideration to the idea that the implementation of a web user interface (WUI) is quite difference to that of a traditional software system, as traditional software GUI is mainly constructed through the use of GUI widgets. Also, the kinds of device that are used for the display of WUI are much diverse, such as PDA's, mobile phone, etc. Over the years, several techniques have been introduced in the literature for the modeling and design of web-based systems. There is a heavy concentration in the earlier methods to be; (1) hypertext oriented [27, 28], or (2) data centric or data driven [20]. While some of the more recent methods have its base on (3) object oriented paradigm [Con99]. These models were found to not pay sufficient attention to users, who are central in web systems. These systems, hypertext, or data centered approaches need to be contrasted with the (4) User-Centered approach [7, 18, 29].

In general, most of the abovementioned methods have navigational design addressed in the process. However, the navigational model is often a by-product of the underlining domain model, which does not always provide the user view required as the user would like to perceive the information. Rather, it had only map this data model that are a suitable representation of the data for storage and efficient for system manipulation directly onto the presentation layer. It can be observed there is the assumption that all data source come form the internal system. However, with the swift advent of technologies such as web services, agent-base system, the final contexts that are presented to user on client device may include content from a number of different data sources. This will certainly have fundamental effect on the way how the whole system is to be built.

1.2 Motivating Case Study

To demonstrate our work, we developed our research group's website [30] using *xWeb* methodology and as an *xPortal* system. It is a simple web site used as information source (research collaboration, news, publication list etc.), for public

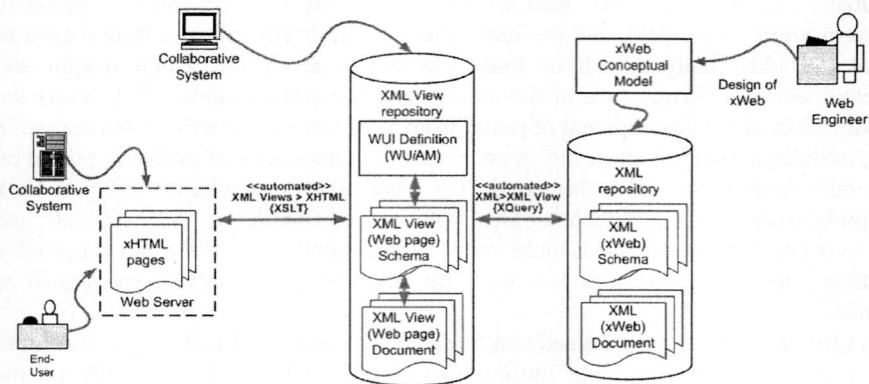

Fig. 1. The *xWeb* and its Components (Context Diagram)

relations (PR), references (for members and their students/collaborators) and for collaborative work (with other research and industry partners/entities). The site has four main user groups, namely; (1) visitors (any user/agent visiting the site), (2) collaborative partners, (3) members and (4) administrators. Each group has some predefined privileges in regards to accessing the web site, visitor being the least privilege group and the administrator being the highest privilege group.

1.3 Our Approach

Our approach to web engineering is an XML-view driven design methodology and architecture (shown in Figure 1), with extensive support for web user analysis and design using *Web User Interaction Analysis Model* (WUiAM), for modelling and building user-centric, content-based websites and web portals. The design methodology provides three levels of abstraction namely; (1) conceptual, (2) logical or schema and (3) document levels. Intuitively, *xWeb* methodology provides a 3-step design process to engineer web contents.

At the conceptual level we; (1) design the web site and its semantics (such as site layout, structure, data, user access control) using a generic UML model (shown in Figure 2), which serves as the XML repository for the web site, (2) develop abstract user interface definitions [2] using abstract (web) user interface (AUI) objects or user perspectives [19] and (3) we derive conceptual views [4] to build the web pages (i.e. a web page construct corresponds to one (or more) conceptual view/(s)) and web portals (view of a view/ aggregate view).

At the logical level we; (1) transform the XML repository captured in UML model to XML (schema and document) using the website data (page contents, layouts, resources etc) as shown in Figure 4, (2) transform AUI objects to web user interface (WUI) definitions such as stylesheet definitions and (3) transform conceptual views to XML view [4, 6] schema definitions (schema & constructor or XQuery [31] definitions) using transformation rules described in [32-34].

And finally at the document level, the transformation is threefold; (1) fill XML repository with web data, (2) transform XML view definitions to (imaginary) XML (view) documents (materialized views) and (3) generate XHTML documents using materialized XML (view) documents and UI definitions (an XSLT transformation).

The main advantage of the *xWeb/xPortal* design is that, the user does not feel any difference between a classical (HTTP) website and a *xWeb* driven website, as there is no loss of performance traditionally related to XML technologies (XQuery) as the user access directly access (X)HTML pages and not native XML documents. As shown in Figure 1, the HTTP web server serves as the front end to users and the XML repository serves as the data source for both *xWeb* and *xPortal* servers.

2 The *xWeb* Methodology

To understand *xWeb*, it is important to understand both the inter-related *xWeb* architecture and the design methodology itself. The design-time methodology has the following steps;

Step 1. Development of conceptual model of the web domain in question: At the conceptual level, the website domain and data requirements are analysed and captured (using UML) using typical OO conceptual modelling techniques with extensive use-case analysis (discussed in Section 2.1) and WUiAM analysis (discussed in Section 2.2). Also identified are the abstract (web) user interface (AUI) objects [19], such as screens, their components and the navigation links.

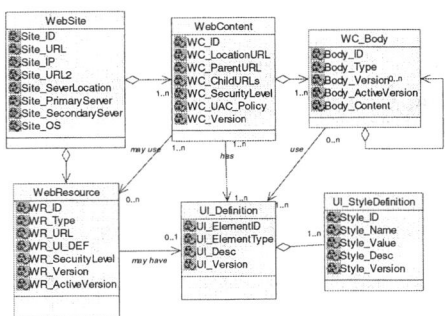

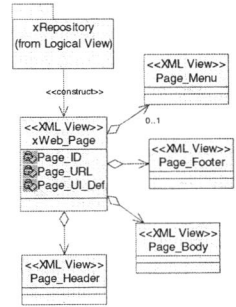

Fig. 2. XML (web) Repository (conceptual level)   **Fig. 3.** XML (view) Repository (conceptual level)

Step 2. Development of conceptual model of the XML (web) repository: Based on the conceptual model development in Step 1, the conceptual model is mapped to the XML (web) repository model under three categories namely; (1) web contents, (2) web resources (hypermedia objects host at the website)and (3) user interface definitions (such as AUI). The model developed for the example case study is shown in Figure 2.

Step 3. Development of website conceptual model: Here, based on XML (web) repository, conceptual views (XML views at the conceptual level) [4, 6] are constructed (discussed in Section 2.3.2) to satisfy one "screen" which in turn satisfy one or more web user requirement/(s). The conceptual views constructed form the XML (view) repository. Figure 3 shows this model, which is developed for the case study example.

Step 4. Development of the XML (web) repository logical model: Here, the conceptual model of the XML (web) repository is mapped to XML schema using schemata transformation rules defined by Ling Feng et al. [32-34].

Step 5. Development of the XML (view) repository logical model: The conceptual views defined in Step 3 are mapped to XML views, where XML view schemas are constructed.

Step 6. Development of the XML (web) repository document model: At this stage, the XML (web) repository is populated with the web contents. Later the populated repository (XML) document/(s) are validated against the XML (web) repository schema generated during Step 4.

Step 7. Development of the XML (view) repository document model: Here, the XML views defined are materialised and validated against the (XML) view schema constructed in Step 5 above. The view materialization can be done using automated tools (which uses XQuery or SQL 2003) or manually using simple XPath or XSLT queries.

Step 8. Development of the *xWeb* (XHTML) Pages: Here the materialised XML views are transformed into XHTML pages using XSLT and style-sheet definitions, thus enabling classical web servers to host *xWeb* pages.

2.1 *xWeb*: Use-Case Analysis

Though we do not intend to address the issue of user access control design for web sites, this paper do not provide In regards to privilege and access to *xWeb* based systems, we look back at work done in [14-16], where we discussed in detail about communities (open, closed and locked) for web based systems based on their access privileges. Based on that, in our case study example, we identified four main groups of actors who interact with system. They are (1) members, (2) administrators, (3) collaborators and (4) visitors. The member and administrator group belongs to the locked community while collaborative partner group belong to closed community. The visitors (and other web users) belong to open community where they mainly have view privilege only. The user access hierarchy (actors) is shown in Figure 4, while the complete use-case analyses together with use-cases are shown in Figures 5-6.

2.2 *xWeb*: User Interaction Analysis Using WUiAM

For analysing user interaction, we use our own Web User Interaction Analysis Model (WUiAM), which we argue captures the user interaction for WUI at a higher level of abstraction. The proposed WUiAM, which is a modelling method for representing the possible user-system interaction requirements, is a systematic approach that allows the specification of an analysis model based on a task/activity oriented approach. The information captured in WUiAM should be isolated from any specific visual or graphical design concerns; it gives a logical view of the WUI that is under consideration. It does not mean to replace some of the currently available conceptual and design modelling methods for Web systems, but as an added set of models targeting the area of user interface of a Web system, which can be integrated and complemented with other domain modelling methods, hence providing a comprehensive system

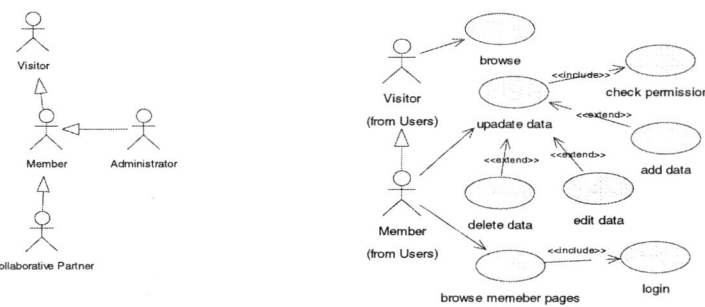

Fig. 4. Use-Case analysis (Actor hierarchy)

Fig. 5. Use-Case analysis model (member/ visitor interaction)

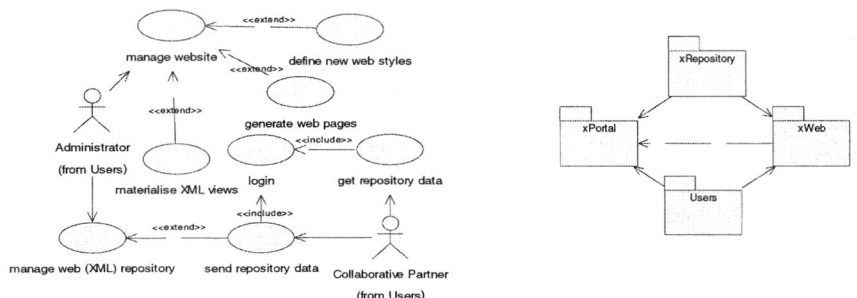

Fig. 6. Use-Case analysis (Administrator/Collaborative partner interaction)

Fig. 7. Relationship between xWeb components

development process. We note here that in addition to domain analysis, which leads to a domain conceptual model, we are here proposing a method of user interface analysis leading to a logical user interface model. Due to page limitation, we do not show a complete WUiAM model for our case-study example here.

2.3 *xWeb* Components

In this section, we briefly discuss the major components of the *xWeb*, its purpose and the implementation model of the components. *xWeb* composes of three logical components namely the XML repository, XML view repository and the *xWeb* page server. These individual components are discussed in the following sections. Figure 7 shows the basic relationships that exists between some these components (shown using UML packages).

2.3.1 XML (Web) Repository

The *xWeb* repository hosts the website contents (site meta-data, data and user interface definitions) in a generic XML encoded textual format. The repository composes of a descriptive, semantically rich repository (XML) schema and an XML document that store the web contents. Therefore the XML repository serves as organized web information source for building, hosting and distributing web data for intend use. Simply said, it replicates a simple yet generic XML based database management system (DBMS) hosting web contents.

Storing and maintaining web content in such an XML repository reduces structural ambiguity among web content, yet maintain semantic richness of each individual web object (this in contrast to hosting web content as relational data). Some of the perceived benefit of such repository include and not limited to; (1) generic yet semantically descriptive web content repository (a direct result of using XML and XML Schema for content description, (2) semantically richer than relational and/or HTML counter parts where schema descriptions are limited and/or optional, (3) since it is in native XML format, the web content dissemination (such as publication data in our motivating example website) among collaborators and/or stakeholders are easier than using propriety messaging formats, (4) based on native XML technology, thus data is descriptive and support heterogeneous web structures and (5) since the web content is independent from WUI objects and/or constraints (in contrast to HTML data), the data along with the WUI definitions (not technology specific presentation layers) are readily distributable and re-usable in other applications such as web portal generation and collaboration. In the case of collaborative website engineering, using XML repository helps in; (1) keeping the content generic yet semantically descriptive, (2) XML (view) Schema driven, therefore no need of additional schema mapping at the source and the target in the content distribution chain, (3) text based data (XML) data thus support for Unicode & multilingual support and (4) keeping the captured web user interface (WUI) definitions independent and free of the web content structure and format.

For physical storage model for the repository, it is readily implemented and hosted in any data storage technology that provides support for native XML documents (and schema) manipulations (from XML-enabled high end DB servers such as Oracle 9i or above to simple custom made XML DBs).

2.3.2 XML (View) Repository

The XML (view) repository is only logically (not physically) different from its web repository counterpart. Here, instead hosting the original web content and its associated structures, it hosts maintain (XML) views. An XML view represents a web "user" screen (with web contestants and associated WUI definitions) in XML. The description and the semantics of the user screen is provides by the XML view Schema, which in addition provide validity the user screen. Typically, the XML views in the repository organized in a hierarchical manner that loosely reflect a web document tree structure (such root view/node closely resembles the classical index.html page in a classical web server documents and so on).

Physically, the view repository is part of the XML (web) repository storage model (similar to that of external schema in the relational databases) and implemented as part of the XML (web) repository (external schema). Typically the XML views and the schemas are persistence in the view repository together with their view definitions as they serve as the middleware (thus avoiding overhead of additional XML-aware middleware) between the presentation oriented XHTML pages and the core XML repository based web contents. Here, the middleware support is provided in the form of (XML) view updates and re-writes.

2.3.3 xWeb (XHTML) Page Server

The *xWeb* Page server is a typically a web server serving clients of X/HTML pages. The XHTML pages are generated (preferably not in real-time) using the XML views stored in the XML (view) repository (in batch mode, depending on the web content type) and the WUI definitions using XSLT transformation. A detail discussion of such transformation (XML view to XHTML) can be found in [9] in the context of web portals. The main advantage of using XHTML based pages to build screens is performance and compatibility. Other perceived advances of using HTML based screens include; (1) no propriety standards (classical HTML server pages) or browsers needed to view the pages (though originally the web content is based in an XML encoded format), (2) easy to implement and maintain (server/client technologies), (3) no new scripts and/or web languages embedded into the web pages and (4) no middleware and/or servers are needed.

3 Conclusion and Future Work

Since the introduction of XML and later with its Schema language, the web has been revolutionized. Today XML is considered as the language of the web. Here, in this paper, we proposed an XML view based web engineering methodology focusing on user-centric web design.

For future work, a lot of issues deserve investigation. First, the process of automation, where the schemata transformation, view construction and model mapping between the levels (conceptual, logical and document) are automated. Secondly incorporating XForms as (web) user interface objects deserve further investigation. Another area that needs refinement is handling interactive (web) objects (to support web applications), where web contents are updated in the XML (web) repository via XML (view) repository in real-time. Also investigation into formally modeling user access control (UAC) mechanisms as part of the web engineering process, as web contents need to be access controlled.

References

1. Gaedke, M., C. Segor, and H.-W. Gellersen. *WCML: paving the way for reuse in object-oriented Web engineering*. in *Proc. of the ACM SIGAPP '00*. 2000. Italy: ACM Press.
2. Chang, E.J., *Object Oriented User Interface Design and Usability Evaluation*, in *Department of CS-CE*. 1996, La Trobe University, Melbourne, Australia.
3. Dillon, T.S. and P.L. Tan, *Object-Oriented Conceptual Modeling*. 1993: Prentice Hall, Australia.
4. Rajugan R., et al. *XML Views: Part 1*. in *14th DEXA '03*. 2003. Prague: Springer 2003.
5. Rajugan R., et al. *XML Views, Part II: Modeling Conceptual Views Using XSemantic Nets*. in *Workshop & SS on Ind. Info., 30th IEEE IECON '04*. 2004. S.Korea: IEEE.
6. Rajugan R., et al. *XML Views, Part III: Modeling XML Conceptual Views Using UML*. in *7th Int. Conf. on Enterprise Information Systems (ICEIS '05)*. 2005. Miami, USA.
7. Gardner, W., E. Chang, and T.S. Dillon. *Analysis Model of Web User Interface for Web Applications*. in *16th Int. Conf. on ICSSEA 2003*. 2003. France.
8. Gardner, W., E. Chang, and T.S. Dillon. *Two Layer Web User Interface Analysis Framework Using SNN and iFIN*. in *OTM 2003 Workshops HCI-SWWA*. 2003.
9. Gardner, W., et al. *xPortal: XML View Based Web Portal Design*. in *17th Int. Conf. on Software & Systems Engineering and their Applications (ICSSEA '04)*. 2004. Paris, France.
10. Nassis, V., et al. *A Systematic Design Approach for XML-View Driven Web Document Warehouses*. in *Int. Wrks. on Ubi .Web Sys.& Inte. (UWSI '05)*. 2005. Singapore.
11. Nassis, V., et al., *Conceptual and Systematic Design Approach for XML Document Warehouses*. Int. Journal of Data Warehousing and Mining, 2005. **1, No 3**.
12. Nassis, V., et al. *XML Document Warehouse Design*. in *Data Warehousing and Knowledge Discovery, 6th Int. Conf. (DaWaK '04)*. 2004. Zaragoza, Spain: Springer.
13. Steele, R., et al. *Design of an XML View Based User Access Control (UAC) Middleware*. in *IEEE Int. Conf. on e-Technology, e-Commerce and e-Service (EEE-05)*. 2005. HK: IEEE.
14. Chang, E., et al. *Virtual Logistics and Partner to Partner Information Exchange System*. in *5th SCI 2001 and the 7th ISAS 2001*. 2001. Orlando, USA.
15. Chang, E., et al. *A Virtual Logistics Network and an e-Hub as a Competitive Approach for Small to Medium Size Companies*. in *Int. Human.Society@Internet Conf.* 2003. SKorea.
16. *iPower Logistics (http://www.logistics.cbs.curtin.edu.au/)*. 2004.
17. Conallen, J., *Building Web Application with UML*. 2 ed. 2003: Addison-Wesley.
18. Troyer, O.M.F.D. and C.J. Leune. *WSDM: a User Centered Design Method for Web Sites*. in *7th Int. World Wide Web Conf.*. 1998.
19. Chang, E. and T.S. Dillon. *Integration of User Interfaces with Application Software and Databases Through the Use of Perspectives*. in *1st Int. Conf. on ORM '94*. 1994. Australia.
20. Ceri, S., P. Fraternali, and A. Bongio. *Web Modeling Language (WebML): a Modeling Language for Designing Web Site*. in *9th Int. World Wide Web Conf., WWW2000*. 2000.
21. Gant, J.P. and D.B. Gant. *Web portal functionality and State government E-service*. in *Proc.of the 35th Annual Hawaii Int. Conf. on System Sciences (HICSS '02)*. 2002. Hawaii.
22. Tatnall, A., *Web Portals: The New Gateways to Internet Information and Services*, ed. A. Tatnall. 2004: Idea Group Publishing.

23. Reynolds, D., P. Shabajee, and S. Cayzer. *Semantic Information Portals.* in *Proc. of the 13th Int. World Wide Web Conf. (WWW '04).* 2004. USA.
24. Bellas, F., D. Fernández, and A. Muiño. *A Flexible Framework for Engineering "My" Portals.* in *13th Int. World Wide Web Conf., WWW2004.* 2004.
25. A. Aragones and W. Hart-Davidson. *Why, When and How do Users Customize Web Portels?* in *Proc. of IEEE Int. Professional Communication Conf. (IPCC '02).* 2002.
26. Gu, A., B. Henderson-Sellers, and D. Lowe. *Web Modeling Languages: the gap between requirements and current exemplars.* in *8th AUSWEB'02.* 2002.
27. Garzotto, F., P. Paolini, and D. Schwabe, *HDM - A Model-Based Approach to Hypertext Application Design.* ACM Transactions on IS (TOIS), 1993. **11**(1): p. 1-26.
28. Schwabe, D., G. Rossi, and S.D.J. Barbosa. *Systematic hypermedia application design with OOHDM.* in *The 7th ACM Conf. on Hypertext and Hypermedia.* 1996: ACM Press.
29. Chang, E. and T. Dillon. *Software Development Methodology and Structure of a Class of Multimedia Web Based Systems.* in *2nd, APWeb 99.* 1999. Hong Kong.
30. *Extended Enterprises & Business Intelligent Laboratory.* 2004, http://exel.it.uts.edu.au/.
31. W3C-XQuery, *XQuery 1.0: An XML Query Language.* 2004, W3C Consortium .
32. Xiaou, R., et al. *Modeling and Transformation of Object-Oriented Conceptual Models into XML Schema.* in *12th Int. Conf. on DEXA '01.* 2001: Springer.
33. Xiaou, R., et al. *Mapping Object Relationships into XML Schema.* in *Proc. of OOPSLA Workshop on Objects, XML and Databases.* 2001.
34. Feng, L., E. Chang, and T.S. Dillon, *Schemata Transformation of Object-Oriented Conceptual Models to XML.* Int. J. of Comp Sys. Sci. & Eng., 2003. **18, No. 1**(1): p. 45-60.

A Web Services Framework for Integrated Geospatial Coverage Data

Eunkyu Lee, Minsoo Kim, Mijeong Kim, and Inhak Joo

Telematics·USN Research Division, ETRI,
161, Gajeong-dong, Yuseong-gu, Daejon, Korea (ROK)
{ekyulee, minsoo, kmj63341, ihjoo}@etri.re.kr
http://www.etri.re.kr/e-etri/

Abstract. This paper refers Web Coverage Service which provides geospatial data as Coverages that are geospatial information representing space-varying phenomena. It also takes care of Coverage Portrayal Service that defines a standard operation for producing visual pictures from coverage data. This paper proposes an interoperable and integrated coverage service. It takes international recommendations of coverage operations, WCS and CPS, on web environments and implements them in a single server system. For interoperability, it takes XML Web Services enabling cooperation of separate systems regardless of their platforms and languages. Through this paper, we are able to propose a prototype model for real geospatial services from an implementation viewpoint. It is expected that this endeavor can develop existing HTTP-based geospatial web services into XML Web Services, which enhances the Internet GIS services to give users fundamental benefits of ubiquitous accesses.

1 Introduction

The use of Internet and related web technologies for accessing spatial data as well as for performing basic spatial query and analysis has opened the world of GIS to the masses. Web GIS means an access to GIS applications/functionality and data via a web browser. Web-based mapping is the fastest growing segment of the spatial-software market. For some applications, web-based applications can provide map visualization and analysis at the lowest possible cost per user. The result is more rapid and informed decision-making, which translates into higher revenue growth and greater cost reduction.

With the semantics of information and information processing, the *OpenGIS Information Framework* from *Open Geospatial Consortium (OGC)* defines conceptual schemas for describing models for any geospatial information and associated communities. A geographic feature is an abstraction of a real world phenomenon with a location relative to the Earth. It may be represented as a discrete phenomenon with its geographic and temporal coordinates. Feature types may have properties, feature operations, and feature associations expressed in a meta model, and attributes. Such geographic information has been treated based on their types that classified into two parts. Vector data deals with discrete phenomena that are recognizable objects having well-defined boundaries or extents such as streams and buildings. Raster data are about continuous phenomena varying over space and having no specific extents. In order to represent one feature or a collection of features to model spatial relationships between, and the spatial distribution of, earth phenomena, a raster image, or a polygon overlay (defined coverages) can be used.

This paper refers and implements Web Coverage Service, which supports the networked interchange of geospatial data as Coverages that are geospatial information representing space-varying phenomena. We also take care of Coverage Portrayal Service that defines a standard operation for producing visual pictures from coverage data. Communication interfaces of the basic service are advanced to allow different types of client systems to be accessed through. In addition to designs of SOAP protocols, web operations for geospatial coverages are represented with WSDL documents.

The rest of this paper is organized as follows. Section 2 gives an overview of the Internet GIS and related web specifications derived from OGC. Section 3 describes basics on XML Web Services technologies and steps of geospatial data toward interoperable web services. Design, implementation, and discussion of the proposed service system are presented in Section 4. Finally, we conclude this paper in Section 5.

2 Geospatial Web Services

For international recommendations of the Internet GIS, OGC has researched a suite of web service specifications that have explicit bindings for HTTP in order to easily provide geo-spatial data on browser.

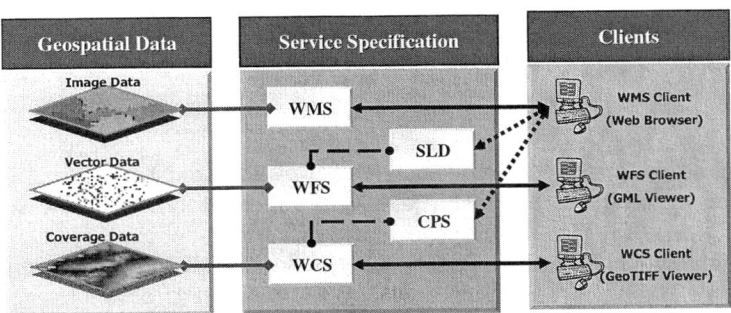

Fig. 1. This figure shows specifications of geospatial web services derived from OGC

Figure 1 shows major specifications and their relations. *"Web Map Services Implementation Specification"* (WMS) produces maps of georeferenced data, where the maps are generally rendered in a simple image format such as PNG, GIF, or JPEG [4]. *"Web Feature Services Implementation Specification"* (WFS) delivers vector representations of simple geospatial features; exactly a GML which is shown in the next section [5]. *"Web Coverage Services Implementation Specification"* (WCS) provides geospatial data as 'Coverages' in digital information [6]. Coverage represents satellite imagery which is currently encoded as GeoTIFF, HDF-EOS, DTED, and NITF. Due to raw data formats that are not browsed on a client easily, WFS and WCS require a client-side rendering. For a thin client, however, *"Styled Layer Descriptor Implementation Specification"* (SLD) [8] and *"Coverage Portrayal Service"* (CPS) could help to display those raw data on its web browser; the two specifications convert raw-typed data such as GML or GeoTIFF data into a JPEG or GIF.

3 Advance of Geospatial Data Toward XML Web Services

3.1 XML Web Services

XML Web Services are the fundamental building blocks in the move to distributed computing on the Internet. XML Web Services expose useful functionality to web users through a standard protocol. *"Simple Object Access Protocol"* (SOAP) defines the XML format for message protocols and describes how to represent program data as XML [10]. *"Web Service Description Language"* (WSDL) describes how the messages are exchanged and specifies what a request message must contain and what a response message will look like [11]. *"Universal Discovery Description and Integration"* (UDDI) describes businesses and services it offers; it contains a company offering the service, industrial categories, and interface descriptions of the service [12]. In order to provide an interoperable architecture of XML Web Services, *Web Service Architecture* (WSA) from *World Wide Web Consortium* (W3C) identifies global elements required to ensure interoperability between web services [13]. WSA consists of three primary roles such as service provider, service consumer, and service broker, and operates with interactions of roles to perform three basic operations: *Publish, Find,* and *Bind*.

3.2 Geospatial Specifications for Interoperable Web Services

OGC has proposed service specifications such as WMS, WFS, and WCS referred to as *OGC Web Services* (OWS) [2] allowing representation of geospatial information with appropriate XML schemas. Recently, OGC has announced new specifications for supporting interoperable functionality of web services: *"OpenGIS Reference Model"* (ORM) [3], *"Web Registry Server Specification"* (WRS) [14], and *"Geography Markup Language Implementation Specification"* (GML) [9]. WRS supports 'one stop shopping' for the registration, metadata harvesting and descriptor ingest, push and pull update of descriptors, and discovery of OWS services using HTTP. Even though WRS does not consider SOAP protocols currently, its conceptual service architecture follows up that of service broker in WSA; WRS interfaces allow service objects to populate a database of service descriptions and query the description database to discover service location. GML is an XML encoding for the transport and storage of geographic information including the geometry and properties of geographic features. GML could make OWS services interoperable for data exchanges by encoding the semantics, syntax and schema of geospatial and geoprocessing-related information resources. GIS industries have researched and proposed many ideas, algorithms, and extensions related with GML since its first appearance [21-23].

3.3 OWS 1.2 SOAP Experiments

With efforts for communication interoperability, OGC does a SOAP experiment by porting OWS services to XML Web services, which might offer several benefits including easier distribution over heterogeneous environments, integration of geospatial functionality and data, and utilization of the advantage of huge amount of infrastructure [1]. In the experiment report, it is discussed how OWS services can be ported into XML Web services and what are issues and problems discovered and needed for future discussion. A target for the SOAP experiment is WMS producing maps of georeferenced data, where a map may be generally rendered in a pictorial format such as GIF or JPG. WMS defines four web operations: *GetCapabilities, GetMap, GetFeatureInfo, and DescribeLayer*. At the beginning of the experiment, a platform independent model of WMS operations is defined using *Unified Modelling Language* (UML). UML models for HTTP Get/Post interfaces in the existing WMS specifications are discussed. Then it proposes a new UML model of WMS for SOAP protocol based on the two existing UML models with several class changes such as MapType, ImageType, etc. The next step in the experiment is to create WMS web services based on the proposed UML model. The procedure includes a creation of the appropriate XML schemas, WSDL documents, and a definition of the bindings where clients can use to communicate with a server. The last step in the experiment is a test for interoperability on a broad range of XML Web Services toolkit, .NET, Axis, and XML spy, by invoking WMS operations. The experiment identifies issues involved with using XML Web Services standards and the toolkits to call WMS server. The issues are almost related to structures inside XML schemas, WSDL documents, or SOAP messages: import of multiple XML schema files, name/namespace conflicts, top-level elements in SOAP messages, enumerations, etc. Then the experiment proposes WMS XML schema and WSDL documents.

4 Proposed System: An Interoperable Web Coverage Service for Integrated Geospatial Imagery

4.1 Coverage Service Description

Coverage is a feature that associates positions within a bounded space to feature attribute values, a function from a spatiotemporal domain to an attribute domain, including a digital elevation matrix and a raster image. As a type of coverage, imagery is a common way of collecting information where the value of a continuous phenomenon is sampled at regular and discrete locations. WCS promoted by OGC provides access to potentially detailed and rich sets of geospatial information. It returns representations of space-varying phenomena that relate a spatio-temporal domain to a range of properties. Portrayal services provide visualization of geographic information, rendered maps such as perspective views of terrain and portrayed maps. They are usually sequenced into a value-chain of services to support production decisions and workflows. CPS may be chained to WCS to convert data to a map viewable by a thin client, and defines a standard operation for producing visual pictures from coverage data.

Even though WCS and CPS give fine ideas together, it is conceptually hard to relate them into a single workflow. If a thin client wants to get a geographic image on its web browser, it, at first, connect to CPS. CPS, then, communicate with WCS get corresponding coverage data, which is completely from the coverage concept. The acquired coverage is transformed into an appropriate format, and then transmitted to the clients. In order to reduce the complex steps, integrating approach should be explored.

4.2 System Architecture of Proposed WCS Server

Figure 2 shows overall architecture of the proposed WCS server including a service consumer page. Raw format of satellite imagery data stored in main-memory based database system is provided into the server. Basic functionalities and operations for services from WCS provide appropriate coverage data, GeoTIFF format, to clients and middleware system. Additional extension of WCS is able to provide imagery data, JPG format, to thin clients by taking CPS operations into current WCS specification. Services on the server can be provided through SOAP protocols based on XML Web Services environments. The portrayal part includes a service consumer page that can communicate with the server to receive coverage and imagery data and display them on a specified view or a normal web browser.

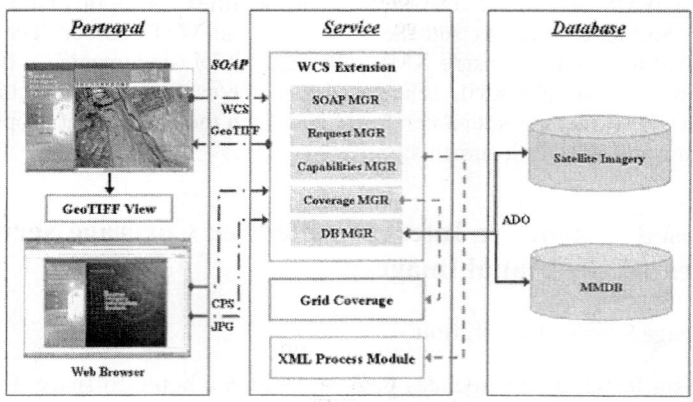

Fig. 2. This figure shows overall architecture of the proposed WCS server

4.3 Web Operations for GeoTIFF and JPG

In order to implement coverage service on web environments, we take WCS operations providing GeoTIFF format of output coverage data. GeoTIFF is a data interchange standard for raster graphic images and extends the TIFF format to support a raster data georeferencing capability. In addition to WCS specification, the server takes CPS concept for portrayal services for a thin client, which provides JPG output format. Due to none of any specification regulating operations of CPS on detail, the basic design idea for implementation of the server providing coverage data is on an extension of the existing WCS specification in order to support fundamental functionalities of CPS. Figure 3 shows designs of web operations of the server.

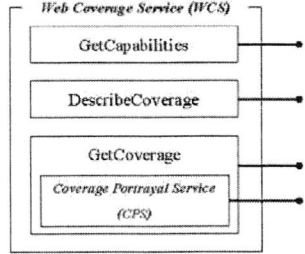

Fig. 3. Web operations, including WCS and CPS together, provide GeoTIFF and JPG data

Fundamentally, the server supports three operations of WCS: GetCapabilities, DescribeCoverage, and GetCoverage. In the GetCapabilities operation, the server take only required requesting parameters: REQUEST and SERVICE and the values are 'GetCapabilities' and 'WCS'. The response contains a capability document encoded by GML. The DescribeCoverage operation also supports two requesting parameters: REQUEST and SERVICE and responds with descriptions of coverage all data also encoded by GML. The GetCoverage operation has an extension, and supports required requesting parameters: SERVICE, VERSION, COVERAGE, CRS, BBOX, WIDTH, HEIGHT, DEPTH, and FORMAT. In order to provide JPEG coverage data, we update the limit of output data format specified in WCS specification. Clients can request JPEG coverage data to the server for displaying simple portrayal format.

4.4 SOAP Communications

For extensible interoperability, the server takes concepts of XML Web Services. From WSA, the server plays the role of service provider. Therefore, it does care about SOAP protocols and WSDL documents. The below listing shows a SOAP messages for service request of a GetCoverage operation. Requesting parameters listed are corresponding to those described in previous subsection and use data types of character and integer. The response data, GeoTIFF and JPG, are transmitted in forms of binary that are configured by binary character in the message.

Figure 4 shows a part of WSDL on the server: GetCoverage operation again. It defines message types consisting of a few elements of requesting parameters and response data. At the beginning, an administrator configures the server system. He selects target data to be served that is stored in database system. In our system, main-memory based database storage is used in order to reduce data access time. The format of imagery data stored may be any type, for example, a raw format of satellite imagery. Instead, the server takes advantage of Grid Coverage component that can process and transform satellite imagery data into GeoTIFF coverage data. In order to support CPS service, GeoTIFF coverage data from the Grid Coverage component can be transformed into JPG format data. Or the server can directly receive JPG data from database storage, if it contains data in the same format. On receiving a request of

GetCoverage operation, the server can provide GeoTIFF coverage data or JPG data based upon the requested output format. Additionally, the server includes a processing module for XML data. On receiving a request of GetCapabilities or DescribeCoverage operation, the server provides appropriate XML documents, exactly GML format except for an error situation, based upon the request.

Example of a SOAP message in the system

```xml
<?xml version="1.0" encoding="utf-8"?>
<soap:Envelope xmlns:xsi="http://www.w3.org/2001/XMLSchema-instance"
xmlns:xsd="http://www.w3.org/2001/XMLSchema"
xmlns:soap="http://schemas.xmlsoap.org/soap/envelope/">
  <soap:Body>
    <GetCoverage xmlns="http://hostserver">
      <SERVICE>string</SERVICE>
      <VERSION>string</VERSION>
      <COVERAGE>string</COVERAGE>
      <CRS>string</CRS>
      <BBOX>string</BBOX>
      <WIDTH>int</WIDTH>
      <HEIGHT>int</HEIGHT>
      <DEPTH>int</DEPTH>
      <FORMAT>string</FORMAT>
    </GetCoverage>
  </soap:Body>
</soap:Envelope>
```

Fig. 4. This figure shows a part of WSDL documents: details on a *GetCoverage* operation

4.5 Service Consumer Page

An additional implementation in this paper includes a service consumer page. From the view point of WSA, it plays a role of service consumer receiving coverage data from WCS as a service provider and displaying the received image data on a user web browser. It can call the operations of the server and receive/process corresponding data. A viewer is nested into the page based on OCX technology and uses public libraries for processing TIFF, GeoTIFF, and JPG data. It can support basic navigation operation such as loading, saving, enlarging, and moving of coverage data displayed.

Figure 5(a) shows the page displaying coverage data. When the consumer calls the GetCoverage operation of WCS server, the page receives coverage data of binary

A Web Services Framework for Integrated Geospatial Coverage Data 1143

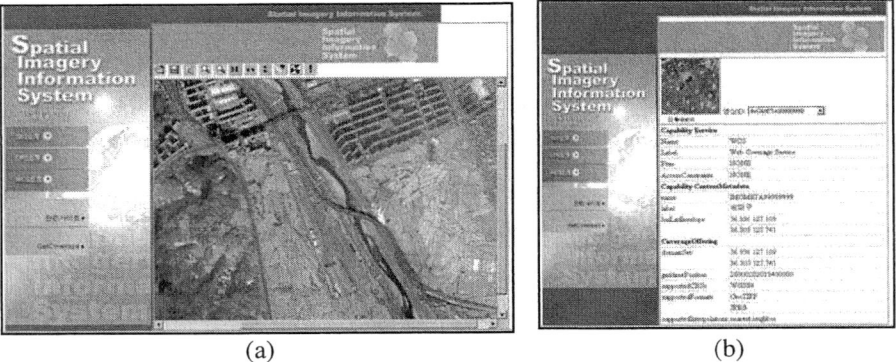

(a)   (b)

Fig. 5. The figure shows a part of the consumer page displaying selected coverage data received from the proposed service system, and detailed descriptions on coverage information

format that may be a format of GeoTIFF or JPG. It is displayed on the OCX viewer nested on the page. Figure 5(b) shows the service consumer page displaying information of coverage data. When the consumer selects coverage and request coverage information, the page calls the DescribeCoverage operation of WCS server and receives corresponding response of a GML document describing the requested coverage. The content list of the GML document may be entirely depends on information regarding the coverage stored in database system. It may include cost information as well as name, label, and supported format of coverage.

4.6 Discussion

The proposed server includes CPS operations in WCS specification, which is a quiet different approach from that by OGC. That means a distinction between coverage including geographic imagery and normal images. With advance of ubiquitous technologies, coverage has become interesting research target due to its specific characteristics and relationship to phenomena in the real world. Therefore, more extensive attention has to be on it and this paper proposes a prototype model at a starting point. The unification of coverages is expected to provide easier accesses to coverages with various types of data and efficient methods for data management and analysis with countless ubiquitous information. With regarding to data distribution, the server uses XML Web Services that are expected to provide interconnected and complicated services, which may make us more familiar. The server is able to provide more applicable services utilizing pre-existing coverage and imagery data with SOAP protocols and WSDL documents. Additional cooperation of the server with systems in other industry, for example traditional GIS systems, CAD visualization systems, traffic management systems, is also expected to create real applications and services with outstanding quality.

On comparison the proposed server in this paper with SOAP experiment in OGC, objectives of the server are interoperability, cooperation, and efficiency, and OGC experiment has distribution, integration, and infrastructure. Even though they notify their goals in quiet different terms, the final target in common is developing the HTTP based geospatial web services into XML Web Services and proposing advanced geospatial services with fundamental benefits of XML Web Services. In order to achieve the goals, OGC experiment proposes a basic structure model based on analyses of inner architectures of XML schemas and WSDL documents. On the other hand, the proposed server gives an implementation prototype based on actual service architectures and integration processes.

It is expected that two endeavors are able to propose the first step into the next geospatial vision. However, it is needed to do more research with intensive design and experiment model for actual services at the same time. Moreover, untouched research areas may include transaction models for integration and cooperation of multi-level workflows and liaison approach for distribution and interoperability, which maximizes benefits from concepts, architectures, and applications of XML Web Services.

5 Conclusion and Further Works

This paper proposes an interoperable and integrated web coverage service. It takes international recommendations of coverage operations, WCS and CPS, on web environments and implements them in a single server system. For interoperability support, it takes XML Web services enabling cooperation of separate systems regardless of their platforms and languages. With regarding to the proposed server, contributory points and discussing speech are described. This paper also gives analyses and comparisons about approach of an international organization. Through this paper, we propose a prototype model for real geospatial services from an implementation viewpoint. Current research works give basic approaches, and they remain a huge room to be discussed and explored. They may include system environments and message architectures as well as transaction models and liaison approaches. Therefore, we are going to make more efforts on those research areas, especially on integrated geospatial services with information systems in other industries.

References

1. Jerome. S., Charles S., OWS 1.2 SOAP Experiment Report, Open Geospatial Consortium (2003)
2. Lieberman, J., OWS1 Web Service Architecture v0.3, Open Geospatial Consortium (2003)
3. Kurt B., OpenGIS Reference Model, Open Geospatial Consortium (2003)
4. J. L. Beaujardiere, Web Map Service (WMS) Implementation Specification v1.1.1, Open Geospatial Consortium (2002)
5. Panagiotis V., Web Feature Service (WFS) Implementation Specification v1.0.0, Open Geospatial Consortium (2002)
6. John. E., Web Coverage Service (WCS) v1.0.0, Open Geospatial Consortium (2003)
7. D. Nebert, OpenGIS Catalog Services Specification v1.1.1, Open Geospatial Consortium (2002)
8. Lalonde, W., Styled Layer Descriptor (SLD) Implementation Specification v1.0.0, Open Geospatial Consortium (2002)

9. Simon. C., Paul D., Ron L., Clemens P., Arliss W., OpenGIS Geography Markup Language (GML) Implementation Specification version 3.0, Open Geospatial Consortium (2003)
10. SOAP, Simple Object Access Protocol, World Wide Web Consortium (W3C) (2003)
11. WSDL, Web Service Description Language, World Wide Web Consortium (W3C) (2004)
12. UDDI, Universal Discovery Description and Integration, http://www.uddi.org/
13. WSA, Web Service Architecture, World Wide Web Consortium (W3C) (2004)
14. L. Reich, Web Registry Server Specification v0.0.2, Open Geospatial Consortium (2001)
15. Michael W. and Matt D., Integrating Spatio-thematic Information, Proc. GIScience'2002, LNCS 2478 (2002) PP.346-361
16. Barbara P. Buttenfield, Transmitting Vector Geospatial Data across the Internet, Proc. GIScience'2002, LNCS 2478 (2002) PP.51-64
17. E. Lee, M. Kim, B. Oh, and B. Jang, System Comparisons for GML Services, Proc. ACRS2003ISRS (2003)
18. Guoray Cai, GeoVSM: An Integrated Retrieval Model for Geographic Information, Proc. GIScience'2002, LNCS 2478 (2002) PP.65-79
19. Buttenfield B., Sharing Vector Geospatial Data on the Internet, Conf. the international Cartographic Association (1999) PP.35-44
20. Laurini, R. and D. Thompson, Fundamentals of Spatial Information Systems, London, Academic Press (1992)
21. Y. Ahn, S. Park, S. Yoo, and H. Bae, Extension of Geography Markup Language (GML) for Mobile and Location-Based Applications, Proc. International Conference on Computational Science and Its Applications, LNCS 3044 (2004) PP.1079-1088
22. Z. Guo, S. Zhou, Z. Xu, and A. Zhou, G2ST: A novel Method to Transform GML to SVG, Proc. 11th ACM International Symposium on Advances in Geographic Information Systems 2003 (2003) PP.161-168
23. J. Guan and S. Zhou, Distributed Geo-Referenced Information Accessing and Integrating based on Mobile Agents and GML, Proc. IEEE International Conference on Web Information Systems Engineering 2004 (2004) PP.54-62

Open Location-Based Service Using Secure Middleware Infrastructure in Web Services

Namje Park[1], Howon Kim[1], Seungjoo Kim[2], and Dongho Won[2]

[1] Information Security Research Division, ETRI,
161 Gajeong-dong, Yuseong-gu, Daejeon, 305-350, Korea
{namjepark,khw}@etri.re.kr
[2] School of Information and Communication Engineering, Sungkyunkwan University,
300 Chunchun-dong, Jangan-gu, Suwon-si, Gyeonggi-do, 440-746, Korea
skim@ece.skku.ac.kr, dhwon@dosan.skku.ac.kr

Abstract. Location-based services or LBS refers to value-added service by processing information utilizing mobile user location. For this kind of LBS, the role of security service is very important in the LBS that store and manage the location information of mobile devices and support various application services using those location information. And in all phases of these functions that include acquisition of location information, storage and management of location information, user management including authentication and information security, and management of the large-capacity location information database, safe security service must be provided. We show the security methods for open LBS in this paper.

1 Introduction

Given the recent advancement of mobile telecommunications technology and rapid diffusion of mobile devices, the importance of wired and wireless Internet services utilizing the past and present location information of users carrying mobile terminals with location tracking function is growing. LBS refer to value-added services that detect the location of the users using location detection technology and related applications. LBS is expected to play an essential role in creating value-added that utilizes wired and wireless Internet applications and location information, since these are very useful in various fields.

In view of the current controversy on the information-collecting practices of certain online sites concerning their members particularly with regard to the disclosure of personal information, it is only natural that there is heightened concern on the disclosure of personal information regarding the user's present location, given the unique characteristics of LBS. Easily disclosed information through certain online sites include member information, i.e., name, resident registration number, and address. Moreover, there is a growing concern that such personal information are leaked for purposes other than what has been originally intended. Such concern is even more serious since location information on customers and possibility of tracking their movements can constitute a direct encroachment of other people's privacy by themselves. Hence, there is a growing need to conduct research on LBS security both in korea and abroad to prevent disclosure of personal information of individuals especially in the areas of authentication and security.

Furthermore, an open LBS service infrastructure will extend the use of the LBS technology or services to business areas using web service technology. Therefore, differential resource access is a necessary operation for users to enable them to share their resources securely and willingly.

This paper describes a novel security approach to open LBS to validate certificates based on the current LBS environment using the web services security mechanism, presents a location-based platform that can block information leak and provide safe LBS, and analyzes authentication and security service between service systems and presents relevant application methods.

2 Framework Model for Providing Secure Services

2.1 Security Service Framework

Web services can be used to provide mobile security solutions by standardizing and integrating leading security solutions using XML messaging. XML messaging is considered the leading choice for a wireless communication protocol. In fact, there are security protocols for mobile applications that are based on XML messaging. Some of these include SAML, which is a protocol for transporting authentication and authorization information in an XML message. It can be used to provide single sign-on web services. On the other hand, XML signatures define how to sign part or all of an XML document digitally to guarantee data integrity. The public key distributed with XML signatures can be wrapped in XKMS (XML Key Management Specification) formats. In addition, XML encryption enables applications to encrypt part or all of an XML document using references to pre-agreed symmetric keys. Endorsed by IBM and microsoft, WS-security is a complete solution to providing security to web services. It is based on XML signatures, XML encryption, and same authentication and authorization scheme as SAML (Security Assertion Markup Language).

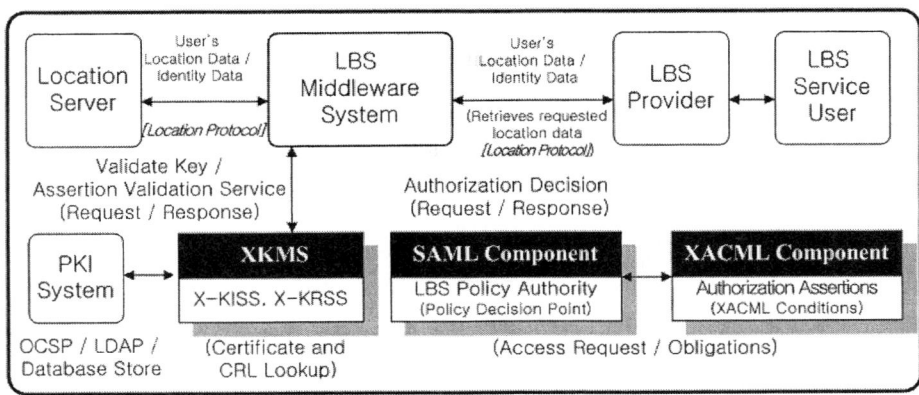

Fig. 1. Proposed secure LBS middleware service model

When a LBS-mobile device client requests access to a back-end application, it sends authentication information to the issuing authority. Depending on the credentials presented by the LBS-mobile device client, the issuing authority can then send a positive or negative authentication assertion. While the user still has a session with the mobile applications, the issuing authority can use the earlier reference to send an authentication assertion stating that the user was in fact authenticated by a particular method at a specific time. As mentioned earlier, location-based authentication can be done at regular time intervals. This means that the issuing authority gives location-based assertions periodically as long as the user credentials enable positive authentication.

Security technology for LBS is currently based on KLP (Korea Location Protocol). Communication between the LBS platform and Application Service Providers should be examined from the safety viewpoint vis-à-vis XML security technology. As shown in the security service model of the LBS platform in figure 1, the platform should have an internal interface module that carries out authentication and security functions to provide the LBS application service safely to the users[2].

2.2 Structure of Mobile Location Protocol in Korea

The protocol used for data exchange between the LBS server and terminals operates based on the MLP (Mobile Location Protocol) protocol established by LIF (Location Inter-Operability Forum). KLP is korea location protocol. The application of the authentication and security factors to KLP should be configured considering the following points:

The KLP is an application-level protocol for querying the position of mobile stations independent of underlying network technology. The KLP serves as the interface between a location server and a location-based application. The details of the location information security structure between the location information providers and the LBS providers should be defined in terms of confidentiality, integrity, and authentication and access control element as shown in figure 2.

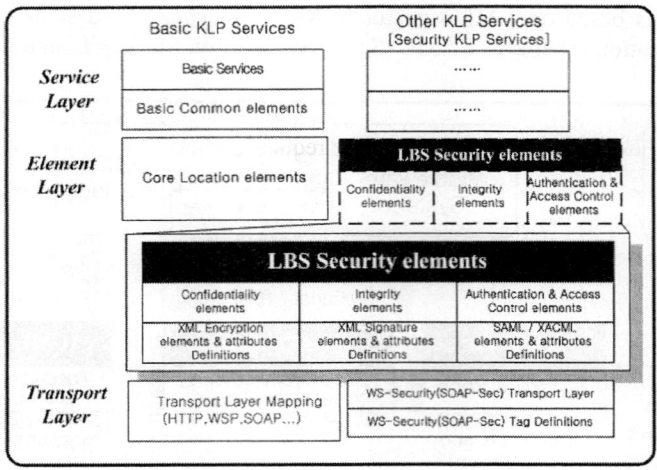

Fig. 2. Security protocol for secure open LBS services

These elements should be configured into a standard reference system for LBS security and authentication on WS-security-based transport layers. Based on this structure, the LBS security service transmits location information safely between the LBS platform and the Service Providers in accordance with the XML-based request and response model.

2.3 Standard Element Layer Definition for LBS Security

For LBS provided on the service layers of KLP, KLP service protocols can be divided into five different services, whereas message transmission can be defined in three types as request, answer, and report messages.

Security functions that should be provided to LBS service layers should be defined as LBS security elements. For element layers, there are seven main definitions: subscriber identification element, functional element, location element, configuration element, location accuracy element, network element, and context element. For the seven types of element definition DTD (Data Type Definition), the attribute parameters requiring security are displayed in bold font.

① Subscriber Identification Elements Definitions

Among subscriber identification elements, there are three elements requiring security: 'msid', which represents the identification information of mobile telecommunications subscribers; 'codeword', which is the access code defined in each subscriber terminal, and; 'session', which is the information on the session of the LBS client with the subscriber terminal. There is a need to encrypt data using XML encryption tag elements.

② Functional Element Definitions

As an element requiring security among functional elements, the 'url' represents the necessary address information to send an answer to the report. 'url' is part of the 'pushaddr' item that can contain the ID and password. Thus, it is necessary to encrypt data using XML encryption tag elements.

③ Location Element Definitions

As the element requiring security among location elements, 'time' represents the time when the service was carried out upon the request for location information. It is therefore necessary to encrypt data using XML encryption tag elements.

④ Shape Elements Definitions

'X', 'Y', and 'Z' are elements requiring security among configuration elements. As coordinate values, 'X', 'Y', and 'Z' are the basic units of location information. Thus, these data have to be encrypted using XML encryption tag elements.

⑤ Quality of Position Definitions

Since location accuracy elements represent accuracy based on the location information for which security has already been dealt with, there is no need for separate security. Nonetheless, it is necessary to examine security elements that can be applied to the necessary attribute parameters for enhancing the quality of the future LBS service.

⑥ Network Parameter Element Definitions

CDMA (Code Division Multiple Access), GSM (Group Special Mobile), CDMA-2000, and WCDMA (Wideband CDMA) are defined in network elements. As such, transport layer security should be dealt with on the transport layer of KLP. Likewise, the security function depending on the network infrastructure should be examined separately.

⑦ Context Element Definitions

As an element requiring security among context elements, an identifier can be used in an element for the provisioning of the privacy structure. This includes ID that allows the use of location information service, 'sessionid' that can substitute for 'pwd', 'pwd' as the password for a registered user implementing the location service, and 'serviceid' as an identifier for distinguishing services and applications that access the network. For 'sessionid' and 'pwd', it is necessary to encrypt data using XML encryption tag elements. In addition, since 'serviceid' requires security to access service, such security should be based on authentication using the PKI (Public Key Infrastructure) interface.

3 Security Protocol for Secure Open LBS Middleware Services

Three types of principals are involved in the proposed protocol: LBS application (server/client), SAML processor, and XKMS server (including PKI). The proposed invocation process for the secure LBS security service consists of two parts: initialization protocol and invocation protocol.

The initialization protocol is a prerequisite for invoking LBS web services securely. Through the initialization protocol, all principals in the proposed protocol set the security environments for their web services (Fig. 3). The following is the flow of setting the security environments:

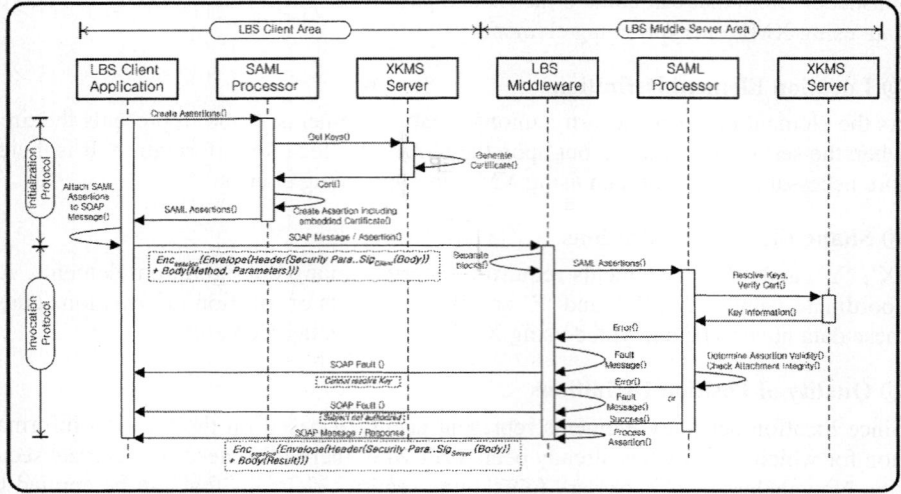

Fig. 3. Security protocol for secure open LBS services

The client first registers information for using web services. It then gets its id/password, which will be used for verifying its identity when it calls web services via a secure channel. The client gets SAML assertions and installs a security module to configure its security environments and to make a secure SOAP message. It then generates a key pair for digital signature and registers its public key to a CA.

The client creates a SOAP message containing authentication information, method information, and XML signature. XML then encrypts and sends to a server such message. The message is in the following form: $Enc_{session}(Envelope\ (Header\ (SecurityParameters, Sig_{client}(Body))+Body(Method,\ Parameters))))$, where $Sig_x(y)$ denotes the result of applying x' s private key function (i.e., the signature generation function) to y. The protocol shown in Fig. 3 shows the use of end-to-end bulk encryption [3,7,9]. Security handlers in the server receive, decrypt, and translate the message by referencing security parameters in the SOAP header. To verify the validity of the SOAP message and authenticity of the client, the server first examines the validity of the client's public key using XKMS. If the public key is valid, the server receives it from a CA and verifies the signature. The server invokes web services upon completion of the assessment of the security of the SOAP message. It then creates a SOAP message that contains the result, signature, and other security parameters. The server encrypts the message using a session key and sends it back to the client. Finally, the client evaluates the validity of the SOAP message and server and receives the result.

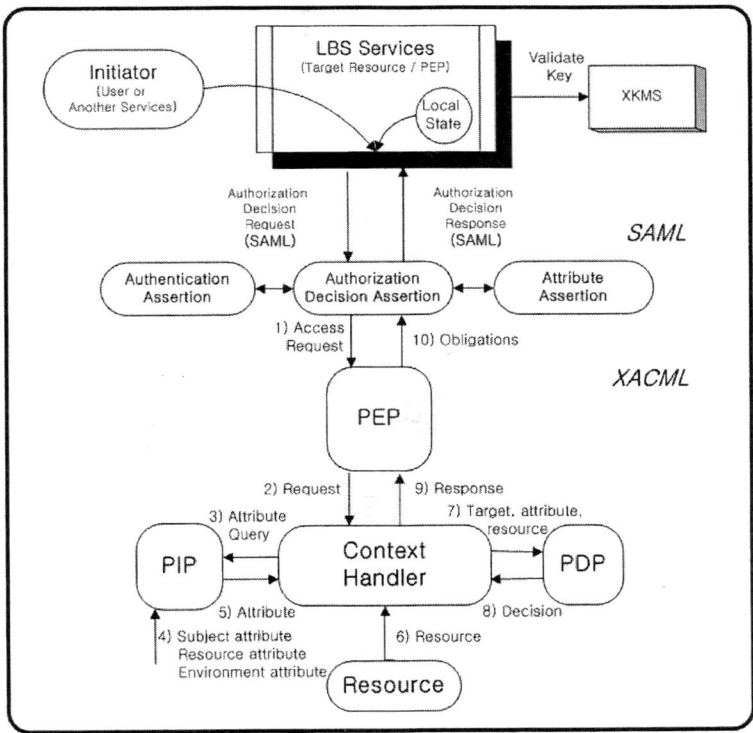

Fig. 4. SAML/XACML message flow using XKMS in open LBS

In the existing LBS service, there is no mechanism for differential resource access. To establish such security system, a standardized policy mechanism is required. The XACML specification is employed to establish the resource policy mechanism that assigns a differential policy to each resource (or service)[2,3]. SAML also has such policy mechanism, whereas XACML provides a very flexible policy mechanism that is applicable to any resource type. For the proposed implementing model, SAML provides a standardized method of exchanging authentication and authorization information securely by creating assertions from the output of XKMS (e.g., assertion validation service in XKMS). XACML replaces the policy part of SAML (Fig. 4). Once the three assertions are created and sent to the protected resource, verification of authentication and authorization at the visiting site is no longer necessary. SSO (Single Sign-On) is a main contribution of SAML in distributed security systems[1].

Figure 4 shows the flow of SAML and XACML integration for differential resource access. Once assertions are created from the secure identification of the PKI-trusted service, the access request is sent to the policy enforcement point (PEP) server (or agent) and to the context handler. The context handler then parses and sends to the PIP (policy information point) agent the attribute query. The PIP gathers subject, resource, and environment attributes from the local policy file, with the context handler giving the required target resource value, attribute, and resource value to the PDP (policy decision point) agent. Finally, the PDP determines and sends to the context handler the access possibility to enable the PEP agent to allow or deny the request [4].

4 Simulations

We have modeled our architecture as a closed queuing system as in figure 5, and we analyzed of approximate Mean Value Analysis (MVA) as described in [5,6,8]. In the scenario of figure 5, the secure LBS procedure has two job classes, initial secure location update step and secure LBS roaming step. $r_{im,jn}$ means the probability that a class m job moves to class n at node j after completing service at node i. And *ratio* represents a ratio of total users to secure LBS roaming users[8]. Analyze steps of class switching closed queuing system are following.

Step1: Calculate the number of visits in original network by using (1)

$$e_{ir} = \sum_{j=1}^{K} \sum_{s=1}^{C} e_{js} r_{js,ir} \qquad (1)$$

where K = total number of queues, C = total number of classes.

Step 2: Transform the queuing system to chain.
Step 3: Calculate the number of visits $e*_{iq}$ for each chain by using (2)

$$e_{iq}^{a} = \frac{\sum_{r \in \pi_q} e_{ir}}{\sum_{r \in \pi_q} e_{1r}} \qquad (2)$$

where r = queue number in chain q, $q\,\square$ = total queue number

Step 4: Calculate the scale factor α_{ir} and service times s_{iq} by using (3) with (1).

$$s_{iq} = \sum_{r \in \pi_q} s_{ir} \alpha_{ir} , \quad \alpha_{ir} = \frac{e_{ir}}{\sum_{s \in \pi_q} e_{is}} \tag{3}$$

Step 5: Calculate the performance parameters for each chain using MVA.

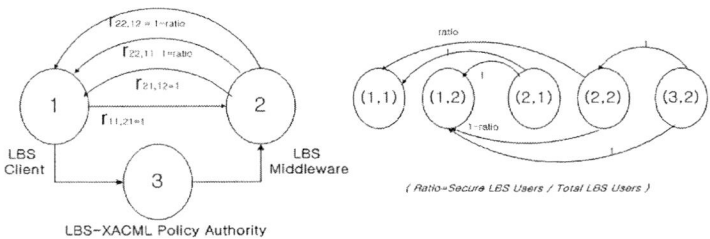

Fig. 5. Multiple class queuing system in the Secure LBS push scenario

XKMS has been implemented based on the design described in previous section. Package library architecture of XKMS based on CAPI (Cryptographic Application Programming Interface) is illustrated in figure 6. Components of the XKMS are xml security library, service components api, application program. Although XKMS service component is intended to support xml applications, it can also be used in order environments where the same management and deployment benefits are achievable. XKMS has been implemented in java and it runs on JDK ver. 1.4 or more.

The manner in which the various XKMS service builds upon each other and consumes each other's services is shown in the following diagram.

The figure for representing testbed architecture of XKMS service component is as follows figure 6.

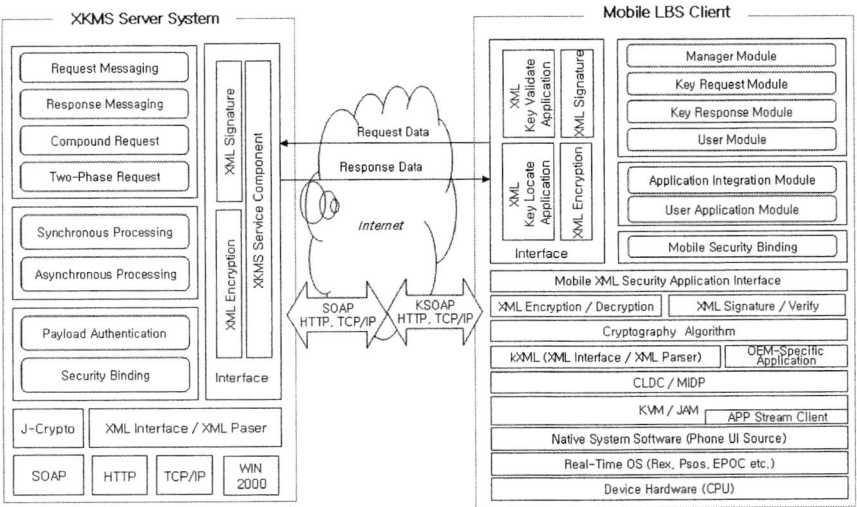

Fig. 6. Testbed Architecture of XKMS component for Open LBS

Figure 7 showed difference for 0.2 seconds that compare average transfer time between client and server of XML encryption &decryption by XML Signature base on XML security library. According as increase client number on the whole, showed phenomenon that increase until 0.3 seconds.

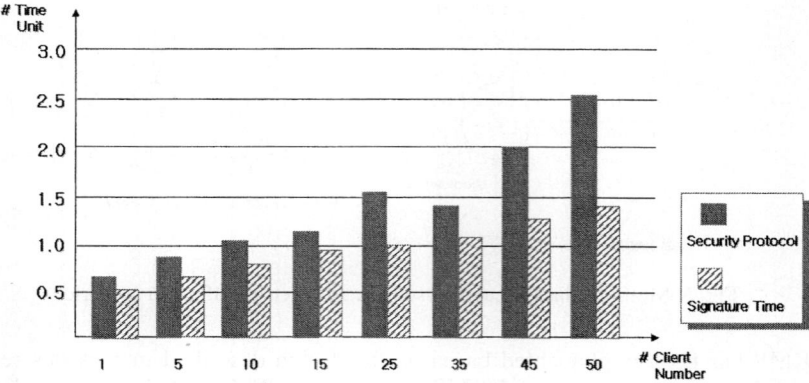

Fig. 7. Simulation Result of XKMS Protocol

Figure 7 is change of average transmission time according as increase client number in whole protocol environment. If client number increases, we can see that average transfer time increases on the whole. And average transfer time increases rapidly in case of client number is more than 45. Therefore, client number that can process stably in computer on testbed environment grasped about 40(At the same time). When compare difference of signature time and protocol time, time of xml signature module is occupying and shows the importance of signature module about 60% of whole protocol time.

5 Conclusion

This paper sought to present a location-based platform that can block information leak and provide safe LBS as well as to establish methods for authentication and security application between service systems for presentation. Toward this end, LBS security requirements were examined and analyzed. In particular, the trend of technology and standard was analyzed to provide safe LBS. To formulate an authentication method as well as a security technology application method for LBS on MLP, MLP security elements were identified based on LBS security requirements by defining the MLP security structure, which serves as the basis for KLP.

A novel security approach to open LBS was proposed to validate certificates based on the current LBS security environment using XKMS and SAML and XACML in xml security. This service model allows a client to offload certificate handling to the server and to enable the central administration of XKMS polices. To obtain timely certificate status information, the server uses several methods such as CRL (Certificate Revocation List), OCSP, etc. The proposed approach is expected to be a model for the future security system that offers open LBS security.

References

1. M. Myers, R. Ankney, A. Malpani, S. Galperin, and C. Adams: X.509 Internet Public Key Infrastructure Online Certificate Status Protocol – OCSP, RFC 2560 (1999)
2. Namje Park, et. Al.: The Security Consideration and Guideline for Open LBS using XML Security Mechanism, ASTAP 04/FR08/EG.IS/06. (2004)
3. M. Naor and K. Nissim: Certificate Revocation and Certificate Update, IEEE Journal on Selected Areas in Communications, 18 (4) (2000)
4. Chanho Lee , et. Al.: A Scalable Structure for a Multiplier and an Inversion Unit in GF(2m), ETRI Journal, V.25, No.5 (2003) 315-320
5. Yuichi Nakamur, et. Al.: Toward the Integration of web services security on enterprise environments, IEEE SAINT 02. (2002)
6. Boudewijn R. Haverkort John: Performance of Computer Communication Systems : A Model-Based Approach, Wiley & Sons (1999)
7. Sungmin Lee et. Al.: TY*SecureWS:An integrated Web Service Security Solution based on java, LNCS 2738 (2003) 186-195
8. Minsoo Lee, et. Al: A Secure Web Services for Location based Services in Wireless Networks. Networking2004 (2004)
9. Mi-Jung Choi, et. Al.: XML-Based Network Management for IP Networks, ETRI Journal, Vol.25, No.6 (2005) 445-463

ANNEX A. Security Elements & Attributes in DTD

```
<!--// Subscriber Identification Element Definitions //-->
    1 Line) <!ELEMENT msid     (#PCDATA)>
    5 Line) <!ELEMENT codeword (#PCDATA)>
    10 Line) <!ELEMENT session (#PCDATA)>
    12 Line) <!ELEMENT start_msid (msid)>
    13 Line) <!ELEMENT stop_msid (msid)>
<!-- // Function Element Definitions // -->
    11 Line) <!ELEMENT pushaddr (url, id?, pwd?)>
    17 Line) <!ELEMENT url (#PCDATA)>
<!--// Shape Element Definitions //-->
    5 Line) <!ELEMENT coord    (X, Y?, Z?)>
    6 Line) <!ELEMENT X (#PCDATA)>
    7 Line) <!ELEMENT Y (#PCDATA)>
    8 Line) <!ELEMENT Z (#PCDATA)>
<!--// Location Element Definitions //-->
    1 Line) <!ELEMENT pos     (msid, (pd | poserr), net_param?)>
    2 Line) <!ELEMENT eme_pos (msid, (pd | poserr), esrd?, esrk?)>
    3 Line) <!ELEMENT trl_pos (msid, (pd | poserr))>
    5 Line) <!ELEMENT pd (time, shape, (alt, alt_acc?)?, speed?, direction?, lev_conf?)>
    10 Line )<!ELEMENT time (#PCDATA)>
<!--// Context Element Definitions //-->
    2 Line) <!ELEMENT sessionid (#PCDATA)>
    4 Line) <!ELEMENT requestor (id, serviceid?)>
    5 Line) <!ELEMENT pwd (#PCDATA)>
    6 Line) <!ELEMENT serviceid (#PCDATA)>
    9 Line) <!ELEMENT subclient (id, pwd?, serviceid?)>
<!--// Quality of Position Definitions //-->
    None
<!--// Network Parameter Element Definitions //-->
    None
```

Ubiquitous Systems and Petri Nets

David de Frutos Escrig*, Olga Marroquín Alonso**,
and Fernando Rosa Velardo*

Departamento de Sistemas Informáticos y Programación,
Universidad Complutense de Madrid, E-28040 Madrid, Spain
{defrutos, alonso, fernandorosa}@sip.ucm.es

Abstract. Several years before the popularization of the Internet, Mark Weiser proposed the concept of ubiquitous computing with the purpose of enhancing the use of computers by making many computers available throughout the physical environment, but making them effectively invisible to the user. Nowadays, such idea affects all areas of computing science, including both hardware and software. In this paper, a formal model for ubiquitous systems based on Petri nets is introduced and motivated with examples and applications. This simple model allows the definition of two-level ubiquitous systems, composed of a collection of processor nets providing services, and a collection of process nets requesting those services. The modeled systems abstract from middleware details, such as service discovery protocols, and security infrastructures, such as PKI's or trust policies, but not from mobility or component compatibility.

1 Introduction

The term *ubiquitous computing* was coined by Mark Weiser [11, 12] in order to describe environments full of devices that compute and communicate with its surrounding context and, furthermore, interact with it in a highly distributed but pervasive way. By pervasive we mean that users will not be aware of the existence of such environment, much in the same way as they pay little attention to other technologies, already fully integrated in their everyday life. Thus, ubiquitous computing is a vast field that involves not only many areas of computer science, including hardware components, network protocols, and computational methods, but also social sciences.

Since Weiser's vision [11], a great deal has been achieved, mainly because of advances in micro-electronics, that make possible the design of smaller embedded devices. However, the state of the art is probably not as developed as expected. One of the reasons may be the lack of widely accepted formal models, needed at various levels of abstraction, in order to understand "the probably largest *engineered artifact* in human history" (see [7]).

---

* Work partially supported by the MCYT project MIDAS, TIC2003-01000.
** Work partially supported by the MCYT project MASTER, TIC2003-07848-C02-01.

Nevertheless, some recently developed formal models can be applied in fields related to ubiquitous computing like workflow, flexible manufacturing or agent-oriented approaches (mobile agents or intelligent agents as in AI research). Among these models we are here interested in those based on Petri nets for their amenable graphical representation and their solid theoretical basis. For instance, the interest of *Elementary Object Systems* [8, 9] has been illustrated in numerous case studies [8]. Elementary Object Systems are composed of a system net and one or more object nets that move along the former, like ordinary tokens of it. Such tokens are able to change their marking, but not their structure, either when lying on a place or when being moved by a transition of the system net. In this way, the change of the object net marking can be either independent from the system net or triggered by it.

In contrast with Elementary Object Systems and their reference semantics, which allow to access a net token from many places at the same time, in *Nested Petri nets* [5, 6] each token is located at a single place at each time. Net tokens may be produced, copied and removed during a system run, as expressed by labels on arcs. The number of those tokens, as well as the level of nestedness, is unlimited, thus obtaining multi-level nested systems, whose behaviour consists of three kinds of steps: An autonomous step in a given level of a Nested Petri net follows the ordinary firing rule for high-level Petri nets; horizontal synchronization is defined as the simultaneous firing of two element nets located in the same place of a system net; and vertical synchronization is the firing of a system net together with the firing of its token nets that are involved.

In earlier papers [3, 4], we have introduced another multi-level extension of the Elementary Object Systems called *Ambient Petri nets*, which allows the arbitrary nesting of ambients permitted in the Ambient Calculus [1]. As a consequence, it is possible to find in the places of an Ambient Petri net both ordinary and high-level tokens. The latter move along the net due to the firing of ambient transitions, labeled by capabilities that are obtained from names: Given a name n of a component net, that is, a bounded place where computation happens, the capability $in\,n$ allows to enter into n, the capability $out\,n$ allows to exit out of n, and the capability $open\,n$ allows to open n. Besides, ordinary transitions consume and produce only ordinary tokens by following the firing rule from ordinary Petri nets. In [4] the basic model of Ambient Petri nets has been extended with the aim of supporting the replication operator from the Ambient Calculus, $!P$, which generates an unbounded number of parallel replicas of P. By combining these elements, together with concepts such as limitation of access to locations, Ambient Petri nets provide a framework to describe wide area network mobility.

Although the described models were not originally conceived in the framework of ubiquitous computing, they can be used for handling some of its most important aspects, specially mobility. Nevertheless, many features of ubiquitous computing, such as the supply and demand of resources between processors and processes, context awareness, and ad-hoc nets, are not naturally modeled. In order to formalize such features, in this paper we define a new model based on

Petri nets called *Ubiquitous nets*, whose basic version relies on ordinary Petri nets, though it can be easily enhanced by considering coloured Petri nets.

Ubiquitous nets allow to model both devices (processors) and software components located in processors (processes), in such a way that processes change their location due to the firing of special *movement transitions*. Besides, we abstract from middleware details, such as those dealing with service discovery or transport protocols, so transitions offering or requesting a service are detected by others just by its mere existence. Then, a service is supplied whenever its offer and request are co-located, that is, whenever the firing of the corresponding *service-supply* and *service-request* transitions can be done at the same time. The synchronization criteria is merely syntactical, that is, two transitions can synchronize whenever the corresponding labels match. Nevertheless, in real open systems we could always make use of specific-domain ontologies, in order to avoid this lack of flexibility.

The paper is structured as follows. Section 2 gives the formal definition of ubiquitous nets by considering the simplest possible model in which processes move from location to location with no processor interaction. Section 3 illustrates those definitions with a simple example composed of three processor nets and a process net that is required to follow an authentication protocol in order to obtain a specific service. Extensions of the basic model are introduced and motivated in Section 4. Finally, conclusions and areas for further study are discussed in Section 5.

2 Formal Definitions

Ubiquitous systems are defined in order to model the supply and demand of services/resources of both processors and processes, respectively. To define the exchange of such services, we consider a countable alphabet of labels $\mathcal{S}$, which will denote available resources. Moreover, we assume the existence of two bijections $!\colon \mathcal{S} \to \mathcal{S}^!$ and $?\colon \mathcal{S} \to \mathcal{S}^?$, by means of which we associate to each label $s \in \mathcal{S}$ two synchronizing actions, $s! \in \mathcal{S}^!$ and $s? \in \mathcal{S}^?$, respectively. The countable alphabet of labels $\mathcal{S}^!$ will denote resources provided by processor nets, while the countable alphabet of labels $\mathcal{S}^?$ will denote resources requested by process nets. Besides, we consider a countable alphabet of labels $\mathcal{A}$, which will denote autonomous actions performed by either processors or processes.

In order to identify the different components of a ubiquitous system, we start with two given sets of processor names, $\mathcal{N}_r$, and process names, $\mathcal{N}_s$, thus taking $\mathcal{N} = \mathcal{N}_r \cup \mathcal{N}_s$. Then we have:

Definition 1. *A **processor net** is a labeled Petri net $L = (P, T, F, \lambda)$ where:*

- *P and T are disjoint sets of places and transitions.*
- *$F \subseteq (P \times T) \cup (T \times P)$ is the set of arcs of the net.*
- *λ is a function from T to the set $\mathcal{A} \cup \mathcal{S}^!$.*

Definition 2. *A **process net** is a labeled Petri net* $A = (P, T, F, \lambda)$ *where:*

- *P and T are disjoint sets of places and transitions.*
- $F \subseteq (P \times T) \cup (T \times P)$ *is the set of arcs of the net.*
- λ *is a function from T to the set* $\mathcal{A} \cup \mathcal{S}^? \cup \mathcal{M}_s$, *where* $\mathcal{M}_s = \{go_l \mid l \in \mathcal{N}_r\}$.

As stated above, ubiquitous systems are composed of a collection of processor nets which provide services to a collection of process nets. Therefore, a processor net has two types of transitions: *autonomous transitions* (those with $\lambda(t) \in \mathcal{A}$), and *service-supply transitions* (those with $\lambda(t) \in \mathcal{S}^!$). Those services are requested by process nets, which migrate from processor to processor due to the execution of *go* transitions. Therefore, a process net has three types of transitions: *autonomous transitions* (those with $\lambda(t) \in \mathcal{A}$), *service-request transitions* (those with $\lambda(t) \in \mathcal{S}^?$), and *movement transitions* (those with $\lambda(t) \in \mathcal{M}_s$).

Definition 3. *A **plain ubiquitous system** is a pair of the form* $\mathbb{U} = \langle \mathbb{R}, \mathbb{S} \rangle$ *where* $\mathbb{R} = \{l^1 : L^1, \ldots, l^m : L^m\}$ *is a finite collection of named processor nets or locations and* $\mathbb{S} = \{a^1 : A^1, \ldots, a^n : A^n\}$ *is a finite collection of named process nets or agents, with* $m > 0$, $n \geq 0$, $\forall k \in \{1, \ldots, m\}$ $l^k \in \mathcal{N}_r$ *and* $L^k = (P_l^k, T_l^k, F_l^k, \lambda_l^k)$, *and* $\forall k \in \{1, \ldots, n\}$ $a^k \in \mathcal{N}_s$ *and* $A^k = (P_a^k, T_a^k, F_a^k, \lambda_a^k)$.

Note that, in contrast with Ambient Petri nets, ubiquitous nets define two-level systems with no distinguished root net, so we consider that all processors and process nets have the same significance. The static nature of processor nets, whose location is fixed, highlights the static character of some devices such as operating systems. On the other hand, the dynamic nature of process nets, that move from processor to processor in order to request the execution of services, reflects the behaviour of processes performed in a distributed way by the whole distributed system.

The current location of processes is described by means of a location function *loc*, which, given a process net, returns its communicating processor. In the following, we assume that in an ubiquitous system each component net has a different name. $\mathcal{N}_r(\mathbb{U}) = \{l^1, \ldots, l^m\}$ will denote the set of processor names in $\mathbb{U}$, and $\mathcal{N}_s(\mathbb{U}) = \{a^1, \ldots, a^n\}$ will denote the set of process names in $\mathbb{U}$.

Definition 4. *Given a plain ubiquitous system* $\mathbb{U}$, *we define a **location function** for* $\mathbb{U}$ *as a function* $loc : \mathcal{N}_s(\mathbb{U}) \to \mathcal{N}_r(\mathbb{U})$. *A **located ubiquitous system** is a plain ubiquitous system for which we have defined a location function.*

A located ubiquitous system describes the structure of both processors and processes. In order to suitably represent their state, we use the usual concept of marking, in such a way that places are occupied by ordinary tokens that move along the system by following the ordinary firing rule.

Definition 5. *A **dynamic located ubiquitous system** is a located ubiquitous system for which we have defined an ordinary marking* $M : P \to \mathbb{N}$, *where P is the full set of places of the ubiquitous system, that is,* $P = (\bigcup_{k=1}^{m} P_l^k) \cup (\bigcup_{k=1}^{n} P_a^k)$.

Similarly, we define the full sets of transitions and arcs of the ubiquitous system as the sets $T = (\bigcup_{k=1}^{m} T_l^k) \cup (\bigcup_{k=1}^{n} T_a^k)$ and $F = (\bigcup_{k=1}^{m} F_l^k) \cup (\bigcup_{k=1}^{n} F_a^k)$, respectively, and the full labelling function of the system as $\lambda(t) = \lambda_c^k(t)$ if $t \in T_c^k$ with $c \in \{a, l\}$.

As stated before, both processors and processes can perform *autonomous transitions* that model independent actions, that is, actions whose execution does not depend on the surrounding context of the evolving net. The corresponding firing rule is the one for ordinary Petri nets, since the location function does not change. Besides, a process can move from its current location to any other due to the execution of *movement transitions*, labeled by go_l with $l \in \mathcal{N}_r$, which specify the new destination l. Note that in this basic model we disregard security issues, so we consider that the full set of processor names is known to the full collection of processes of the system. Finally, the supply of a service s is modelled by means of the synchronized firing of two transitions, $t_1 \in T_a^k$ and $t_2 \in T_l^{k'}$ with $\lambda_a^k(t_1) = s?$ and $\lambda_l^{k'}(t_2) = s!$, corresponding respectively to the request of the service by the process net A^k, and its offering by the processor net $L^{k'}$. Those transitions can only be fired simultaneously, each of them following the firing rule for ordinary Petri nets.

Definition 6. *Let $\langle \mathbb{U}, loc \rangle$ be a located ubiquitous system and M be a marking of it. An **autonomous** transition $t \in T$, with $\lambda(t) \in \mathcal{A}$, is enabled at marking M if $\forall p \in {}^\bullet t \; M(p) > 0$. The reachable state of $\mathbb{U}$ after the firing of t is that described as follows:*

- *The reachable marking M' is defined by $M'(p) = M(p) - F(p,t) + F(t,p)$ for all $p \in P$.*
- *The location function loc does not change.*

Definition 7. *Let $\langle \mathbb{U}, loc \rangle$ be a located ubiquitous system and M be a marking of it. A **movement** transition $t \in T_a^k$, with $\lambda_a^k(t) = go_l \in \mathcal{M}_s$, is enabled at marking M if $\forall p \in {}^\bullet t \; M(p) > 0$. The reachable state of $\mathbb{U}$ after the firing of t is that described by:*

- *The reachable marking M' is defined by $M'(p) = M(p) - F(p,t) + F(t,p)$ for all $p \in P$.*
- *The current location of process net a^k changes, getting $loc(a^k) = l$. The location of the rest of the process nets remains the same.*

Definition 8. *Let $\langle \mathbb{U}, loc \rangle$ be a located ubiquitous system and M be a marking of it. A pair of **service-supply/service-request** transitions (t_1, t_2), with $t_1 \in T_a^k$ and $t_2 \in T_l^{k'}$, $\lambda(t_1) = s? \in \mathcal{S}^?$ and $\lambda(t_2) = s! \in \mathcal{S}^!$, and $loc(a^k) = l^{k'}$, is enabled at marking M if $\forall p \in {}^\bullet t_1 \cup {}^\bullet t_2 \; M(p) > 0$. The reachable state of $\mathbb{U}$ after the firing of (t_1, t_2) is that described as follows:*

- *The reachable marking M' is defined by*

$$M'(p) = M(p) - \sum_{v \in \{t_1, t_2\}} F(p, v) + \sum_{v \in \{t_1, t_2\}} F(v, p) \quad \forall p \in P$$

- *The location function loc does not change.*

Due to the firing rule in Definition 8, services are provided in mutual exclusion with the purpose of avoiding their concurrent use, since they are considered unshareable resources. In this way, whenever a process requests a service s to its processor, both nets must synchronize the firing of the corresponding transitions in order to satisfy such demand. At that moment, service s becomes unavailable for any process requesting the same resource. Note that if there exist more than one process net demanding such service, the choice among them is made in a non-deterministic way.

3   A Simple Application

In order to illustrate the behaviour of ubiquitous nets, in this section we present an example that models a system composed of three processor nets, L^1, L^2 and L^3, and a process net, A, initially located in L^3 (Figure 1). Processor L^1 can be interpreted as an electronic notes system [2] that requires authentication to view its contents (action identified as service $s2$), and L^2 can be seen as an electronic thermometer [10] in which the action of consulting the temperature is denoted by $s3$. Both processors can also give the local time, which is denoted as service

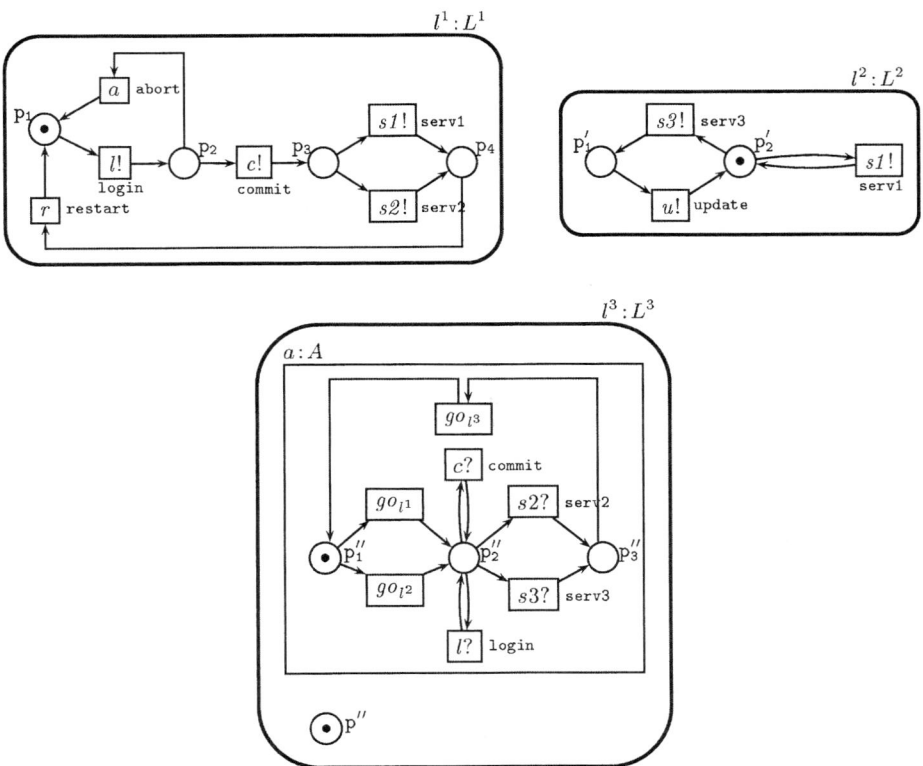

Fig. 1. An ubiquitous system modeled by Petri nets

$s1$. Processor net L^3, composed of a single place and no transitions, can neither evolve in an independent way nor interact with any process, since it is just a container that allows to store agent nets.

Therefore, process A can move either to L^1 or L^2, where it demands services $s2$ or $s3$ (trying to view the contents of the notes system or asking for the temperature, respectively). In order to supply service $s2$, processor L^1 requires first to log in and if it is successfully done then to proceed with a commitment (otherwise it aborts), whereas processor L^2 does not demand any such authentication to offer its services. On the other hand, after the firing of a movement transition, process A may not only log in if asked, but also try to commit with no previous logging, thus trying to force the authentication protocol.

In this scenario, it is clear that after the firing of transition go_{l^2}, process A obtains service $s3$ from its new location L^2, which must be updated before offering again its services (for instance, checking the temperature again). Then, process A needs to restart before going back to its initial state, where now it may choose to execute transition go_{l^1}. As a consequence of this firing, process A moves to L^1, that demands it to log in before proceeding with a commitment, which is needed to supply service $s2$.

4 Extensions of the Basic Model

Ubiquitous nets allow to define two-level systems focusing on both the supply and demand of services and the mobility of processes. Nevertheless, their simplicity produces some drawbacks, that can be easily removed by introducing some extensions in the defined basic model.

In the first place, real systems constrain both mobility of processes and access to their resources as a general rule. In particular, processes do not migrate by themselves, but are moved by processors. As a consequence, it is reasonable to limit the access to processors depending on their current availability to receive processes. Moreover, processes would need to obtain the permission of their present location to move away to the desired processor.

Therefore, in general it is necessary to model a three-way synchronization among the moved process, its current location and its new destination. In order to do it we introduce processor transitions labelled by lgo_l and lin. Their intended meaning is that the processor firing a transition labelled by lgo_l allows any process that can fire transition go_l to exit out of it. This migration can only take place when the destination processor, l, executes the admission transition lin at the same time.

Definition 9. *A **go-processor net** is a Petri net $L_{go} = (P, T, F, \lambda)$ where:*

- *P and T are disjoint sets of places and transitions.*
- *$F \subseteq (P \times T) \cup (T \times P)$ is the set of arcs of the net.*
- *λ is a function from T to $\mathcal{A} \cup \mathcal{S}^l \cup \mathcal{M}_r$, where $\mathcal{M}_r = \{lin\} \cup \{lgo_l \mid l \in \mathcal{N}_r\}$.*

Definition 10. *Let $\langle \mathbb{U}, loc \rangle$ be a located ubiquitous system with go-processor nets and M be a marking of it. A tuple of **movement** transitions (t_1, t_2, t_3), with*

$t_1 \in T_l^i$, $t_2 \in T_a^k$ and $t_3 \in T_l^{i'}$, $\lambda(t_1) = lgo_{l^{i'}}$, $\lambda(t_2) = go_{l^{i'}}$ and $\lambda(t_3) = lin$, and $loc(a^k) = l^i$, is enabled at marking M if $\forall p \in {}^\bullet t_1 \cup {}^\bullet t_2 \cup {}^\bullet t_3$ $M(p) > 0$. The reached state of $\mathbb{U}$ after the firing of (t_1, t_2, t_3) is that described by:

- The reachable marking M' is defined by

$$M'(p) = M(p) - \sum_{v \in \{t_1, t_2, t_3\}} F(p,v) + \sum_{v \in \{t_1, t_2, t_3\}} F(v,p) \quad \forall p \in P$$

- The current location of the process net a^k changes, getting $loc(a^k) = l^{i'}$. The location of the rest of the process nets remains the same.

In this extended model, authentication could be simply modeled: We only have to replace lin transitions by others labeled by lin_a, where a is the name of the incoming process, that is, the one the destination location is ready to receive. In this way, it is easy to limit the access to some services by taking into account the names of processes, and hence a processor net will only admit those agents whose name appears in labels of the form lin_a.

However, in this model processes must include in their code concrete information about their desired movements, although in some cases such movements are performed in a non-deterministic way (this happens, for instance, whenever there exists a choice between movement transitions). Furthermore, processors must have a static knowledge of the names of those processes whose entry is allowed. Nevertheless, this assumption is a bit coarse, and more flexible mechanisms to control authentication and mobility are desirable. Regarding the latter, processor names will appear as token values. Then, a synchronized firing of the movement transitions, that in this new version of the model would be labeled by lgo, go and lin, respectively, produces the migration of the involved process to the location indicated by the consumed token. In this way, labels of tokens represent a permission or a capability to enter into the corresponding processors.

Definition 11. *A **ggo-processor net** is a Petri net $L_{ggo} = (P, T, F, \lambda)$ where:*

- *P and T are disjoint sets of places and transitions.*
- *$F \subseteq (P \times T) \cup (T \times P)$ is the set of arcs of the net.*
- *λ is a function from T to the set $\mathcal{A} \cup \mathcal{S}^! \cup \{lin, lgo\}$.*

Definition 12. *A **ggo-process net** is a Petri net $A_{ggo} = (P, T, F, \lambda)$ where:*

- *P and T are disjoint sets of places and transitions.*
- *$F \subseteq (P \times T) \cup (T \times P)$ is the set of arcs of the net.*
- *λ is a function from T to the set $\mathcal{A} \cup \mathcal{S}^? \cup \{go\}$.*
- *Each $t \in T$ such that $\lambda(t) = go$ has a distinguished precondition, where $_t \in P$, whose tokens should be labelled with processor names.*

In order to define the behaviour of ggo-systems, we have to separate ordinary places from those distinguished ones storing processor names. This could be easily formalized using a simple version of coloured Petri nets, though its definition could be rather cumbersome. Here we propose a less formal but clearer definition.

Definition 13. *A **dynamic located ubiquitous ggo-system** is a located ubiquitous system with ggo-processor and ggo-process nets, for which we have defined a marking $M: P \to \mathbb{N} \cup \mathcal{M}(\mathcal{N}_r(\mathbb{U}))$, where P is the set of places of the ubiquitous system, and all the tokens are ordinary ones except from those in distinguished places, that is, $M(p) \in \mathbb{N} \ \forall p \in P \setminus P_{go}$ and $M(p) \subseteq \mathcal{M}(\mathcal{N}_r(\mathbb{U})) \ \forall p \in P_{go}$, where P_{go} is the set of distinguished places of the ubiquitous system $P_{go} = \{where_t \mid t \in T, \lambda(t) = go\}$.*

Therefore, each process in a ggo-system has some distinguished places in P_{go} connected as preconditions of its *go* transitions. In this way, whenever a three-way synchronization is performed, the name of the new location is taken from the distinguished place $where_t$, taking as t the corresponding transition labeled by *go*. Then, it is simple to model a mechanism for the transmission of processor names to processes by means of special agents that move from location to location providing the corresponding process nets with their labeled tokens. Such mechanism is not formalized here due to lack of space.

Definition 14. *Let $\langle \mathbb{U}, loc \rangle$ be a located ubiquitous ggo-system and M be a marking of it. A tuple of **movement transitions** (t_1, t_2, t_3), with $t_1 \in T_l^i$, $t_2 \in T_a^k$ and $t_3 \in T_l^{i'}$, $\lambda(t_1) = lgo$, $\lambda(t_2) = go$ and $\lambda(t_3) = lin$, $loc(a^k) = l^i$ and $l^{i'}$ is a processor name stored in the distinguished precondition $where_{t_2}$ of t_2, is enabled at marking M if $\forall p \in {}^\bullet t_1 \cup {}^\bullet t_2 \cup {}^\bullet t_3 \ M(p) > 0$. The reachable state of $\mathbb{U}$ after the firing of (t_1, t_2, t_3) is that described as follows:*

- *The reachable marking M' is defined by*

$$M'(p) = M(p) - \sum_{v \in \{t_1, t_2, t_3\}} F(p, v) + \sum_{v \in \{t_1, t_2, t_3\}} F(v, p) \quad \forall p \in P \setminus P_{go}$$
$$M'(p) = M(p) \quad \forall p \in P_{go}$$

- *The current location of the process net a^k changes, getting $loc(a^k) = l^{i'}$. The location of the rest of the process nets remains the same.*

Following the above definition, the marking of places in P_{go} does not change due to the firing of transitions, since the distinguished input place of a *go* transition is just a container for the names of the available destinations. The coloured formal version of ggo-systems would be more flexible, allowing us to indicate how the tokens annotated with processor names are transmitted and consumed, in such a way that processes can have a dynamic knowledge of their possible destinations.

5 Conclusions and Future Work

We have introduced a model for two-level ubiquitous systems, in which a collection of processors provide services to a collection of processes that request those services. In the simplest version of the model, processors remain fixed in their locations, whereas processes move from processor to processor in order to obtain the resources they need.

Supply and demand of services is modeled by the synchronized firing of two transitions: a *service-demand transition* located in the involved process and a *service-supply transition* located in the corresponding processor. Besides, mobility is formalized by the execution of *movement transitions*, by means of which each process sets its destinations.

This simple model is then enhanced with some extensions that include, first a three-way synchronization mechanism that limits resource access (a process moves if and only if its current processor lets it go and its new location lets it in), and then introduces the colouring of some tokens, that allow to dynamically determine the destinations of the moving processes.

As work in progress, we are currently introducing new features in our model to cover the most of the characteristic properties of ubiquitous computing, mainly a procedure to dynamically transmit private processor names to processes, in such a way that access to locations can be adequately constrained. We will do that by using a simple version of coloured Petri nets. Certainly, this will lead us to the analysis of security properties by means of, for example, typing mechanisms to suitably restricting the contents of net places. In addition to this, we will generalize the described framework in order to encompass the dynamic generation of new processes during a system run. With this purpose, we will introduce a set of process types $\{A_1, A_2, \ldots, A_k\}$, which are ordinary process nets initialized in such a way that the firing of a special transition $create_i$ will generate a new copy of the process of type i.

References

1. L. Cardelli. *Mobility and Security*. Proceedings of the NATO Advanced Study Institute on Foundations of Secure Computation, pp.3-37. IOS Press, 2000.
2. K. Cheverst, A. Dix, D. Fitton and M. Rouncefield. *'Out To Lunch': Exploring the Sharing of Personal Context through Office Door Displays*. Proceedings of the Australasian Computer-Human Conference-OzCHI 2003, pp.74-83. 2003.
3. D. Frutos Escrig and O. Marroquín Alonso. *Ambient Petri Nets*. Foundations of Global Computing 2003, ENTCS vol.85, 27 pp. Elsevier Science, 2003.
4. D. Frutos Escrig and O. Marroquín Alonso. *Replicated Ambient Petri Nets*. Computational Science-ICCS 2003, LNCS vol.2658, pp.774-783. Springer-Verlag, 2003.
5. I.A. Lomazova. *Nested Petri Nets; Multi-level and Recursive Systems*. Fundamenta Informaticae vol.47, pp.283-293. IOS Press, 2002.
6. I.A. Lomazova. *Modeling Dynamic Objects in Distributed Systems with Nested Petri Nets*. Fundamenta Informaticae vol.51, pp.121-133. IOS Press, 2002.

7. R. Milner. *Theories for the Global Ubiquitous Computer.* Foundations of Software Science and Computation Structures-FoSSaCS 2004, LNCS vol.2987, pp.5-11. Springer-Verlag, 2004.
8. R. Valk. *Petri Nets as Token Objects: An Introduction to Elementary Object Nets.* Applications and Theory of Petri Nets 1998, LNCS vol.1420, pp.1-25. Springer-Verlag, 1998.
9. R. Valk. *Concurrency in Communicating Object Petri Nets.* Concurrent Object-Oriented Programming and Petri Nets, LNCS vol.2001, pp.164-195. Springer-Verlag, 2001.
10. R. Want. *Enabling Ubiquitous Sensing with RFID.* Computer vol.37(4), pp.84-86. IEEE Computer Society Press, 2004.
11. M. Weiser. *Some Computer Science Issues in Ubiquitous Computing.* Communications of the ACM vol.36(7), pp.74-84. ACM Press, 1993.
12. M. Weiser. *The Computer for the 21st Century.* Proceedings of Human-computer Interaction: Toward the Year 2000, pp.933-940. Morgan Kaufmann Publishers Inc, 1995.

Virtual Lab Dashboard: Ubiquitous Monitoring and Control in a Smart Bio-laboratory

XiaoMing Bao, See-Kiong Ng, Eng-Huat Chua,
and Wei-Khing For

Institute for Infocomm Research, 21 Heng Mui Keng Terrace,
Singapore 119613
{baoxm, skng, stuwkf}@i2r.a-star.edu.sg

Abstract. Our Smart Bio-Laboratory project seeks to deploy smart technologies pervasively within the wet laboratory to facilitate bio-scientists in their daily experimental activities. In this work, we develop a "Virtual Lab Dashboard" (VLD) platform for ubiquitous local and remote monitoring and control in a smart bio-laboratory. We have implemented a prototype VLD system that employs wireless mobile computing, Java, and LAMP (Linux, Apache, MySQL, PHP/Perl/Python) technologies to enable lab users to access different bio-equipments at different locations in different modes, showing that current wireless and embedded technologies can be effectively utilized to create a unique smart work space for scientific experimentation and research in the wet laboratories.

1 Introduction

With the recent successful completion of the Human Genome Project[7, 9], technologically sophisticated bio-instruments designed for large-scale, genome-level experimental interrogations of whole biological systems have become commonplace in today's bio-laboratories. Increasingly complex experimental procedures are now being routinely carried out by the scientists in the bio-laboratories. Biology is becoming highly computerized in response to the mountains of data being generated and the computationally grand challenges of managing and analyzing the experimental data.

In the meantime, a phenomenal proliferation of wireless networks and mobile computing has taken place in the computing world. Together with the technological advances in embedding computing power in non-computing devices, appliances and equipment, pervasive computing has emerged as a key technological theme of the post-PC era. Already, numerous pervasive computing implementations of smart spaces have been reported in terms of smart homes[6], smart museums[3], smart towns[8],and so on. We conceive that the modern *wet laboratory*[1], with its dispersed heterogeneous experimentation devices and its increasing needs for computational power, makes a perfect candidate for smart technology deployment.

One key characteristic of the wet laboratory environment is that lab scientists often need to move from one workstation to another in order to carry out their tasks. At the same time, each job that they run on a workstation often lasts for hours and requires periodic monitoring. This causes unnecessary reduction in productivity in the laboratories, as the lab scientists are "tied" to one workstation physically (or at least need to stay in the vicinity) in order to monitor progress and control adjustments. To free the lab users from this constraint, it is necessary to provide a means to remotely access, monitor, and control anytime, anywhere so that the lab scientists can do other productive tasks after they have started a lengthy experimental run at a workstation. In our Smart Laboratory Project, we envisage an intelligent laboratory environment in which the various lab appliances at each of the work stations can be accessed, monitored, and even controlled remotely from anywhere anytime via a "virtual lab dashboard" on another computer, PDA, or even mobile phone. This not only frees the lab scientists from having to be physically near the work station, but it can also enable new applications, for example in remote collaboration, as another scientist in a different lab somewhere else can share access to a local work station running an experiment.

There are also many other potential usages for such virtual lab dashboards. For example, in military situations when the sending over of biomedical samples for local analysis may be infeasible, biomedical experts outside the (field) lab can help access and control the lab appliance in the proper running of the experiments remotely. Another potential use for a virtual lab dashboard is to enable lab scientists to set-up, monitor, or control multiple lab equipment simultaneously. For example, a farm of thermocyclers (a common bio-equipment for the amplification of DNAs) is often deployed in a high throughput genomic laboratory. Lab scientists can use the virtual lab dashboard to set up the desired parameters and apply to each thermocyclers in the farm concurrently instead of having to set up and monitor each individual thermocycler manually with the same set of settings.

The rest of this paper is structured as follows. In Section 2, we provide the background on our Smart Laboratory project, as well as some related work. We then present the system architecture of our VLD platform in Section 3, and we show how VLD can be deployed to provide ubiquitous local and remote access and control in various wet-lab scenarios in Section 4. Finally, we conclude in Section 5 with discussions on various "concept-to-reality" research challenges associated with the Smart Bio-laboratory.

---

[1] The "wet laboratories" in this paper refer to the traditional bio-laboratories where biological ("wet") experiments are carried at the lab benches. The "dry laboratories", on the other hand, are a new kind of bio-laboratories in the form of computer labs where *bioinformatics*—a new computationally-based cross-discipline between computer science and biology—is the primary tool used to analyze the massive biological data generated in the wet laboratories.

2 Background

The grand objective of our Smart Laboratory Project is to convert the traditional wet bio-laboratory into an intelligent and productive work place for the scientists. We hope to create a *pervasive LIMS* (Laboratory Information Management System) that provides ubiquitous online accessing, monitoring, controlling, and capturing of experiments while allowing maximal freedom of mobility for the laboratory scientists and requiring minimal manual computing efforts from them.

A recent pioneering attempt in the transformation of the bio-laboratory environment into a smart space is the Labscape project[1]. Labscape focuses on deploying smart technologies for meta-data capturing to create an intelligent lab system through automatically acquiring knowledge about the lab users' work processes by the computer. In this work, our Virtual Lab Dashboard (VLD)'s focus is on addressing the user-oriented issues of providing ubiquitous monitoring and control in a smart bio-laboratory. There are also other related works such as virtual or online laboratories[4], as well as telemetric applications such as remote RES (Renewable Energy Sources) monitoring systems[5]. These works focused on the provision of *remote* monitoring of equipment, typically in an engineering environment. In our project, our VLD is to be a smart laboratory platform that provides for borderless man-machine interface *both* in and out of the bio-lab environment. In other words, we focus on designing a smart platform that can provide for both ubiquitous monitoring and control services *remotely* over the internet or other public network, and *locally* within the bio-lab environment. Our VLD addresses the following two key issues:

1. *Heterogeneous lab instruments.* It is typical for a biology laboratory to house multiple lab instruments from different vendors for similar or different experimental procedures. A common platform for integrating this heterogeneous environment has to be provided.
2. *Mobile lab scientists.* Frequent physical movements of the scientists in the laboratory are necessary for the coordination and progress monitoring of multiple heterogeneous laboratory devices. Ubiquitous access of the equipment must be provided in a smart laboratory.

In the next section, we describe our VLD platform that (i) integrates diverse bio-instruments using a re-configurable plug-and-play middleware; and (ii) provides of remote multi-modal access for diverse bio-instruments via a common online lab dashboard.

3 System Architecture

We have designed and implemented a prototype VLD platform that links lab equipment together wirelessly and makes these equipment remotely accessible by lab users using current wireless and embedded technologies. Figure 1 shows the architecture of our current system.

As mentioned previously, a bio-laboratory usually houses multiple lab instruments from different vendors for similar or different experimental procedures.

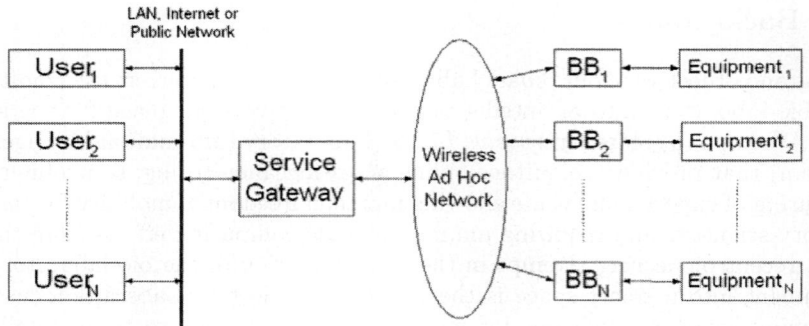

Fig. 1. Smart Lab Prototype VLD System Architecture

These lab instruments have different vendor-specific communication protocols for remote monitoring and control purpose, creating a big challenge for effective system integration. Although there are some well-known enterprise middlewares like CORBA, DCOM, JINI that are developed for supporting application software on different platforms to communicate with each other over the network, these technologies are not applicable in the smart bio-laboratory because the firmware in the various lab instruments are fixed and not re-programmable. To deal with this heterogeneity issue, we have included a component called the Bio-Bridge (BB) Device in our VLD platform which is a embedded middleware that links the heterogeneous lab instruments to the network in a plug-and-play fashion while providing a unified communication interface for remote access to the lab instruments. In the current manifestation of VLD, all the BB Devices are connected to the Service Gateway wirelessly via 802.11b ad hoc network, using a window controlling scheduling scheme among the BB Devices and the Service Gateway for peer-to-peer communication. Various web-based Smart Lab services can then provided to the remote users via a VLD Service Gateway (more details on this in Section 4).

3.1 BB Device

The BB Device is a low cost, small form-factored embedded system. It uses UbiCom IP2022, a high speed (120MIPS) dedicated network processor, which is specially designed to deliver high performance software I/O solutions over a wide range of communication protocols. It is physically connected to the lab equipment through RS232/485 or GPIB, which are common interfaces for almost all of advanced lab equipment. Figure 2 shows a prototype BB Device and its connection to a thermocycler. With the on board 802.11b PCMCIA card, it links the connected lab equipment to the end user through the wireless network and the VLD Service Gateway.

The BB Device plays an essential role in our VLD platform for creating a smart workspace where all the lab equipment can be plugged into the system "on-the-fly" to be ubiquitously accessible by the scientists in a smart bio-lab. For this, it must understand the language that the connected lab equipment

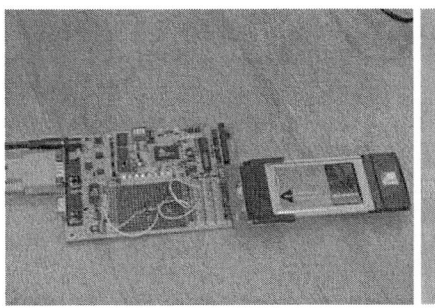

Fig. 2. A BB Device and its connection to a bio-equipment

speaks and speaks the language that the equipment can understand. This can be achieved by a combination of off-line learning and in-system reconfiguration procedure. Additionally, with the RSSI (Received Signal Strength Index) in its wireless LAN module, it can support smart laboratory features such as the proximity service provided by our Smart Lab system: When a user (scientist) is identified within the lab premise, the system performs a real-time relative positioning scheme together with the application program on the user side (either running on a laptop or a PDA carried by the mobile user). A direct wireless link between the user and the connected lab equipment is established when the user is identified closest to it and this virtual linking is maintained dynamically by the system without intervention from the user. In so doing, the system provides on-the-go access to different lab equipment, which is highly preferable by scientists conducting multiple experiments in a bio-laboratory and have to move about frequently in (and out of) the lab to monitor the various experimental procedures. More details on this are provided in Section 4.1. In the case when a user is not physically within the wet lab, the BB Device then serves as a "bridge" in providing a virtual remote link between the user and the lab equipment he/she wants to access. The BB Device receives any commands sent from the Service Gateway and forwards them to its connected lab equipment, collects and buffers real-time experimental data (if any) from the lab equipment, and updates the respective database on the Service Gateway continuously. In this way, the BB device helps the VLD system provide remote monitoring and control services for the authorized users (see also Section 4.2).

3.2 VLD Service Gateway

The second component of the VLD platform is the Service Gateway. In our current prototype system, it is physically composed of a WaveCom FastTrack GSM Modem, a LinkSys WUSB11 802.11b adapter and a Dell Optiplex Desktop PC. Figure 3 shows the architectural block diagram of the gateway.

The Apache web server provides client with access to dynamic web pages that are made up of PHP scripts and Java applets. Two different levels of access are provided in our current system: Administrative access and User access. The core

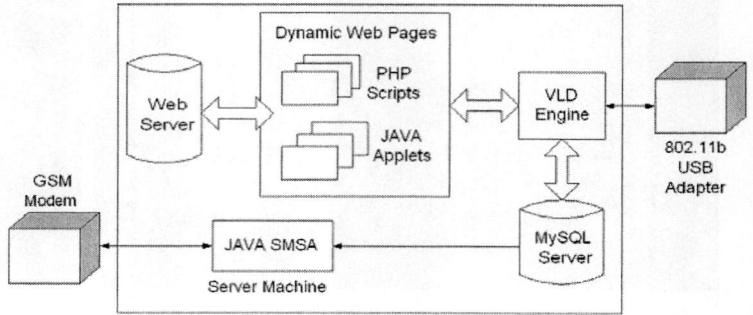

Fig. 3. VLD Service Gateway Block Diagram

program is depicted as the VLD Engine; it keeps tracking of active user status and active BB Device (lab-equipment) status, controls the access to database, schedules the peer-to-peer wireless communication between the Service Gateway and the BB Devices. A specific communication protocol is designed for inter PHP-VLD communication.

The VLD platform stores its system information and experimental data in a MySQL DBMS server. All VLD databases are updated through the core VLD Engine. While the system supports multiple connections to the DBMS server by different PHP scripts for database queries, the granting of permission by VLD Engine is always required before the direct connection can be created. This is essential in the case of multiple users accessing multiple lab equipment through the web server.

Our current system design also includes Java SMSA (SMS Alerting) which is an independent Java program running on the Service Gateway. It listens for incoming SMS from the GSM Modem, parses the SMS command and sender information, performs both the sender and the command authentication, and once authenticated, queries the database and sends the appropriate SMS about (say) lab-equipment status to the user in the way that is specified in the user's SMS request commands.

4 Ubiquitous Multi-modal Access to Bio-equipment

In this section, we describe how the VLD platform supports ubiquitous access to bio-equipment in a smart bio-laboratory. We consider three different use case scenarios:

1. *Smart on-the-go access within the lab.* Bio-scientists frequently need to move from one work station to another in the laboratory to perform various experimental procedures, monitor the experimental status of lab equipment, record experimental data and set experimental parameters when necessary. In fact, a busy biologist can change locations as high as a total of 76 times

in the span of 60 minutes (i.e. more than once per minute), as reported in [1]. It is thus desirable that the linking of a bio-instrument to a scientist can be automatically done when they approach an equipment. With the help of BB Devices and a novel relative positioning algorithm running on the mobile computing device (e.g. a laptop or a palmtop) carried by the scientist, a wireless communication link is dynamically established and maintained between the scientist and the nearest lab equipment.
2. *Internet access through a web browser.* As most of the biological experiments are time consuming, lab scientists would rather not be tethered within the lab for hours before the completion of their experiments. The VLD delivers web-based monitoring and control services so that lab scientists can go back to their offices to work on other productive tasks while logging into the VLD Service Gateway via a web browser for monitoring their experiments remotely.
3. *Mobile access through public network.* In the event when the lab scientist is not within the reach of internet (e.g., during commute), the VLD system provides an alternate way to access the lab equipment through mobile phones. The GSM SMS alerting service is developed for this purpose. While service content adaptation to the mobile user remains a research issue, we focus here on simple text alerting services that would be helpful to the scientist in efficient time management for monitoring lengthy (oftentimes overnight) experiments.

4.1 Intelligent On-the-Go Access Within the Lab

When a lab scientist parks his mobile computing equipment (a laptop, PDA, or even a wearable computer) next to a lab equipment, he expects to work on this equipment of his interest. This means that a virtual link has to be set up automatically and wirelessly between his personal computing device and the lab instrument of his interest without any explicit input from the user. For a Smart Lab to achieve this intelligently, the key contextual information to establish the virtual link is the relative proximity of the user and the various equipment in the lab.

To provide such intelligent virtual linking between user and machine, the VLD platform must embed solutions for relative proximity detection. In the VLD platform, the BB devices are configured to broadcast a beacon which consists of Service Set Identity (SSID) periodically with the IEEE 802.11b wireless LAN API provided in BB device software stack. This allows the users' mobile devices such as laptop, PDA which are within its purview to detect its beacon so that relative proximity amongst multiple BB-device enabled bio-instruments can be determined using the relative Received Signal Strength Indications (RSSIs)[10, 2]. Similarly, the BB devices can be used to determined whether a user is in the lab or outside the lab by comparing the RSSI with a pre-set threshold value (The threshold value is pre-set as the pre-calibrated minimum RSSI of a user's mobile device when he is in the lab). The BB device can then use this information to detect that a user has exited the lab when the RSSI is smaller than this pre-set threshold. This allows the VLD system to perform intelligent context switch from the local wireless virtual link between man and machine to the remote communication mode between server and machine where the data are transmitted to the server for the user to retrieve remotely. Such smart features are made feasible by exploiting the BB device as a common middleware platform to embed context sensitive features for the Smart Lab.

4.2 Remote Internet Access Though a Web Browser

Our VLD platform provides web-based ubiquitous remote monitoring and control services so that a user only needs to have a general web browser to access the lab equipment through the internet without any requirement to install additional user/application specific programs on the user side. Figure 4 shows the access diagram in terms of the various PHP scripts on the web server.

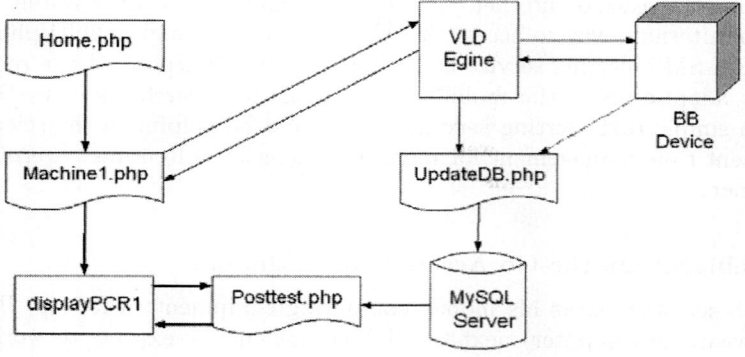

Fig. 4. Lab-equipment Web Access Diagram

Let us consider the following scenario: a user is in his office and wishes to access Smart Lab via his desktop PC. He invokes an internet browser on his desktop PC and enters Smart Lab's URL to access the VLD. Standard log in procedures ensue—upon successful authentication, the Home.php webpage is presented to user, in which a list of user-accessible bio-instruments in the lab is presented for remote access and/or control over LAN. After the user has selected a bio-instrument for access (say MACHINE1), the PHP script machine1.php is invoked to manage the communication between the user and the MACHINE1 via the VLD Engine. At this point, the VLD Engine creates a TCP connection to BB device associated with MACHINE1 and adds it into the scheduling list for the peer-to-peer communication within the backend wireless ad hoc network. It will also establish a direct database access link for the MACHINE1 to update the database with its on-going experimental data (through UpdateDB.php).

On the user side, machine1.php downloads a Java applet displayPCR1 to the user browser for real-time graphic display of experimental data. The PHP

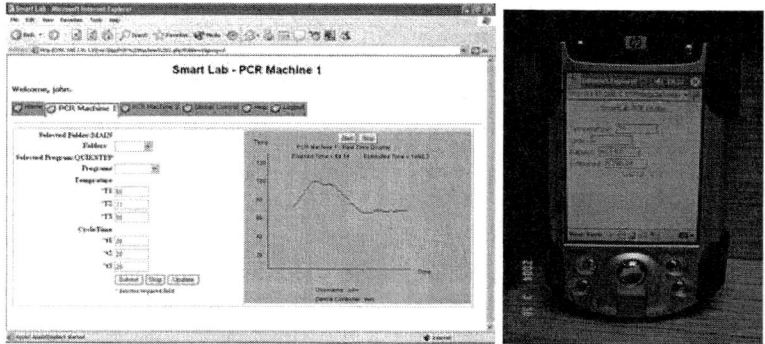

Fig. 5. Multi-modal remote access on a user's laptop and PDA

script `Posttest.php` periodically queries the database and updates the applet. Since in VLD, we provide web-based remote monitoring and control services, the same smart lab service is accessible on the users' PDAs and even web-enabled mobile phones. For the latter mode of remote access, the VLD system provides alternative text-based web pages, since some PDAs or web-enabled mobile phones may have problems displaying graphically intensive pages. Figure 5 shows a screen shot of the dynamic web pages displayed by the web browser on a user's laptop and PDA.

4.3 Remote Mobile Access Through Public Network

An SMS text-based command interface has also been designed in VLD for simple but interactive access to lab equipment via mobile phones that are not web-enabled or connected to the internet. In this case, a user can send an SMS message to the Java SMSA on the VLD Service Gateway, for example: "`SHOW ACTIVE MID*;`" to ask for the machine IDs of all the bio-instruments on which that he is currently running an experimental procedure. The Java SMSA on the Service Gateway will then query the registered user database to authenticate the user using pre-registered phone numbers. Once authenticated, the user will receive an SMS from VLD's Java SMSA with such content: "`YOUR ACTIVE MIDS ARE PCR1 PCR2 SEQ1`". After that, the user may send another SMS to program an alert on PCR2: "`SET ALERT PCR2 -I30MIN -DTEMPERATURE -DCYCLE`", requesting VLD's Java SMSA to send an SMS Alert with the temperature and cycle data of `PCR2` every 30 minutes. In this way, the VLD provides remote access to the Smart Lab even if the user is out of the lab's campus or on the road.

5 Conclusion and Discussions

In this work, we address the fundamental need for the integration of diverse bio-instruments, as well as the increasing demand for ubiquitous monitoring and controlling of experimental procedures in the bio-laboratories. We have implemented a prototype VLD platform that employs wireless mobile computing, Java, and LAMP (Linux, Apache, MySQL, PHP/Perl/Python) technologies to enable lab users to access different bio-equipments at different locations in different modes.

There are still numerous "concept-to-reality" issues that need to be addressed before a full-fledged enterprise system can be deployed into a real bio-laboratory. These include various further research issues in pervasive computing such as plug-and-play middleware design, automatic service content adaptation to multimodal users, intelligent scheduling of wireless ad hoc network, and so on, as well as research issues that are beyond the scope of pervasive computing—for example, distributed data management and data security. While such practical issues need to be addressed before the realization of a full-fledged enterprise system, we have demonstrated here that current wireless and embedded technologies can be utilized to create a unique smart work space for scientific experimentation and research in the wet laboratories. We also surmise that the wet laboratory forms a unique "sand-box" for further research and development in pervasive computing and other smart space technologies, providing many inimitable challenges to drive the necessary advances in these current emerging technological domains.

References

1. L. Arnstein, G. Borriello, S. Consolvo, et al. Labscape: A smart environment for the cell biology laboratory. IEEE Pervasive Computing Magazine, 1(3):1421, July-September 2002.
2. Paramvir Bahl and Venkata N. Padmanabhan. Radar: An in-building rf-based user location and tracking system. IEEE Infocom 2000, 2:775784, March 2000.
3. Juan Carlos Cano, David Ferrndez Bell, and Pietro Manzoni. A context-aware guide for a museum using bluetooth wireless technology. The Fifth European Wireless Conference: Mobile and Wireless Systems beyond 3G, 2004.
4. Bing Duan, Keck-Voon Ling, and Habib Mir M. Hosseini. Developing and implementing online laboratory for control engineering education. In The Eighth International Conference on Control, Automation, Robotics and Vision, 2004.
5. Kostas Kalaitzakis, Eftichios Koutroulis, and Vassilios Vlachos. Development of a data acquisition system for remote monitoring of renewable energy systems. Measurement (Elsevier Science), 34(2):7583, 2003.
6. Cory D. Kidd, Robert Orr, Gregory D. Abowd, et al. The aware home: A living laboratory for ubiquitous computing research. Proceeding of 2nd International Workshop on Cooperative Buildings, pages 191198, 1999.
7. E. S. Lander, L. M. Linton, B. Birren, et al. Initial sequencing and analysis of the human genome. Nature, 409(6822):860921, 2001.
8. Salil Pradhan, Cyril Brignone, Jun-Hong Cui, et al. Websigns: Hyperlinking physical locations to the web. IEEE Computer, 34(8):4248, August 2001.
9. J. C. Venter, M. D. Adams, E. W. Myers, et al. The sequence of the human genome. Science, 291(5507):130451, 2001.
10. Moustafa Youssef, Ashok Agrawala, and A. Udaya Shankar. Wlan location estimation determination via clustering and probability distributions. IEEE PerCom 2003, March 2003.

On Discovering Concept Entities from Web Sites

Ming Yin[1], Dion Hoe-Lian Goh[1], and Ee-Peng Lim[2]

[1] Division of Information Studies,
School of Communication and Information,
Nanyang Technological University, Singapore 639798
{asmyin, ashlgoh}@ntu.edu.sg
[2] Centre for Advanced Information Systems,
School of Computer Engineering,
Nanyang Technological University, Singapore 639798
aseplim@ntu.edu.sg

Abstract. A web site usually contains a large number of concept entities, each consisting of one or more web pages connected by hyperlinks. In order to discover these concept entities for more expressive web site queries and other applications, the *web unit mining* problem has been proposed. Web unit mining aims to determine web pages that constitute a concept entity and classify concept entities into categories. Nevertheless, the performance of an existing web unit mining algorithm, iWUM, suffers as it may create more than one web unit (*incomplete* web units) from a single concept entity. This paper presents a new web unit mining algorithm, kWUM, which incorporates site-specific knowledge to discover and handle incomplete web units by merging them together and assigning correct labels. Experiments show that the overall accuracy has been significantly improved.

1 Introduction

A web site usually contains a large number of concept entities. Each concept entity consists of one or more web pages. For example, in a university web site, a professor's homepage together with web pages about his/her research interests, teaching activities, or curriculum vitae constitute one concept entity. It is clear that a single web page is not sufficient to represent a concept entity. Thus the concept of *web unit* is introduced [6]. A web unit is a set of web pages that jointly provide information about a concept entity. The web unit level is the granularity more suitable for representing concept entities. Web units are useful for indexing and organizing web information since they have richer and more complete content than any individual web page.

In order to construct and classify web units, Sun and Lim [6] proposed *web unit mining*. It is related to the web classification research. Unlike previous web classification efforts that conduct classification either at the web page level [1, 2, 4, 5] or at the web site level [3, 8, 7], web unit mining conducts classification at the web unit level. An iterative web unit mining algorithm, iWUM, has been developed [6]. Experiments show that concept entities can be better determined by iWUM compared to the standard web page classification approach and report 20% improvement in overall accuracies.

Nevertheless, there is some room for improving the iWUM algorithm. Experiments showed that iWUM might create more than one web units (*incomplete* web units) from a single concept entity. Each *incomplete* web unit contains incomplete information about a concept entity. We therefore propose a new web unit mining algorithm, kWUM, which finds web folders in a given web site containing potential web units and handles incomplete web units by merging them together and assigning correct labels. We also created a new evaluation dataset called UniKB from two university web sites. Our experiments show that overall accuracy of kWUM has been significantly improved compared to that of iWUM.

The rest of the paper is organized as follows. In Section 2, we present the web unit mining problem. In Section 3, we propose our new web unit mining algorithm, kWUM. Experiment with UniKB are given in Section 4. Finally, we conclude this paper in Section 5.

2 Web Unit Mining

A web unit consists of exactly one *key page* and zero or more *support pages* that jointly provides information about a concept entity [6]. Consider the *course* web unit shown in Figure 1. It consists of eight web pages. The first page is the course's (CS100) homepage and the others provide supplementary information of the course. The homepage is the entry point to all information about the course and thus the key page; the others are support pages. Similarly, for a *faculty* web unit, the key page is a faculty's homepage; support pages include those pages about his/her research interests, teaching activities, and so on.

```
http://course-path/CS100/CS100.html              → Key Page
http://course-path/CS100/lecture-programs.html
http://course-path/CS100/instructors.html
http://course-path/CS100/officehours.html
http://course-path/CS100/exams/final.html        } Support Pages
http://course-path/CS100/exams/prelim.html
http://course-path/CS100/programs/program1.html
http://course-path/CS100/programs/program2.html
```

Fig. 1. An example web unit for course CS100

Web unit mining consists of two sub-problems, namely *web unit construction* and *web unit classification*. In the former, web pages representing a single concept entity are identified so as to form a web unit. The latter involves assigning web units correct concept labels.

Sun and Lim [6] proposed an algorithm called iterative Web Unit Mining (iWUM). iWUM carries out web unit construction and web unit classification in an iterative manner. It first groups closely-related web pages (based on hyperlink connectivity) into small units, which are then classified and merged with one another to form large units. This classifying-merging procedure repeats until there is no change of category labels assigned to the web units. The rules of merging allow a single *labelled* web unit to merge with neighboring *unlabelled* web units, but not *labelled* web units. Note that a *labelled* web unit refers to one that has been assigned a category label; otherwise it is called an *unlabelled* web unit. In this method, an error in web unit classification will lead to errors in web unit construction, resulting in *incomplete* web units. An *incomplete* web unit is one that covers only a subset of web pages forming a concept entity. For example, a concept entity consists of web pages $p_1, p_2, \ldots, p_n$. If more than one web unit is created from them, say $u_1 = \{p_1, p_2\}$, $u_2 = \{p_3, p_4\}$ and $u_3 = \{p_5, p_6, \ldots, p_n\}$, they are *incomplete* web units. The larger the number of web pages of a concept entity, the more likely *incomplete* web units are created. Our initial experiment showed that among the web units mined by iWUM, 15% of them were *incomplete* web units.

3 Knowledge-Based Web Unit Mining (kWUM)

In order to address the issue of *incomplete* web units, we propose a new web unit mining algorithm called Knowledge-based Web Unit Mining (kWUM). kWUM first takes the web units produced by iWUM and then attempts to merge *incomplete* web units and assign correct category labels. It utilizes site-specific knowledge about the constraints of web unit distribution in a web site to discover and handle *incomplete* web units.

3.1 Web Directory and Web Unit Distribution

We are interested in analyzing the distribution of web units among web folders in a web site. The location of a web unit can be determined by its key page URL. Given a set of web units from a web site, we can construct a *web directory*, which is a tree structure with web folders as internal nodes and web units as leaf nodes. Web folders are extracted from the locations of web units, in

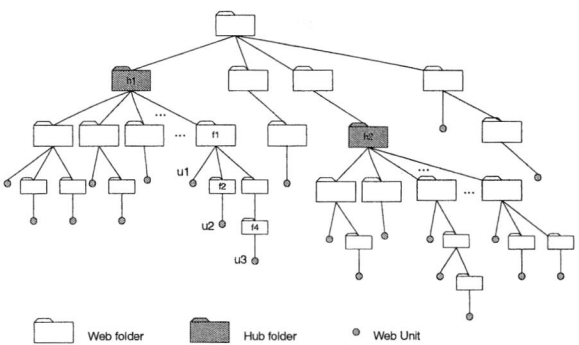

Fig. 2. Web directory and web units

particular, from the *path* components of their URLs. Note that a typical URL follows the syntax: *type://hostname[:port number][/path][filename]*. An example web directory is shown in Figure 2.

We have observed from the web directory in our experiments that *incomplete* web units are either located in the same web folder or form *invalid* parent-child pairs. In kWUM, this key observation has been utilized in finding *incomplete* web units. Two web units, e.g. u_i and u_j, can form a parent-child pair if the web folder containing u_j is a sub-folder or descendent folder of the web folder containing u_i. This parent-child web unit pair is denoted as $pc(u_i, u_j)$. In Figure 2, $pc(u_1, u_2)$ and $pc(u_1, u_3)$ are parent-child web unit pairs. Based on the web units' category labels, we have different types of parent-child web unit pairs, e.g. *faculty-course* web unit pairs, *faculty-faculty* web unit pairs, and so on. We then divide them into two groups: *valid* or *invalid*, based on site-specific knowledge described in Section 3.2.

3.2 Site-Specific Knowledge

Given a web site and a set of concepts relevant to the web site, there are some constraints on how the concept entities are located in the web directory. We model these constraints as invalid parent-child concept pairs and call them *site-specific knowledge*. Each invalid parent-child concept pair suggests that it is highly unlikely to have an entity of the child concept be found in the sub-folder or descendent folder of the web folder containing an entity of the parent concept. The set of invalid parent-child concept pairs varies from web site to web site. Consider university web sites that contain three categories of web units, *faculty*, *student*, and *course*. *Faculty-faculty* and *faculty-student* concept pairs are invalid concept pairs as it is unlikely that a faculty or a student puts his/her own home page under another faculty's home page. For some specific university web sites, more invalid parent-child concept pairs may be specified. If a university web site W_1 has dedicated web space for course materials, it is unlikely that course web pages are stored under the web folder assigned to faculty. As a result, *faculty-course* concept pairs is invalid for W_1. We assume that this site-specific knowledge about invalid parent-child concept pairs of a web site is provided by a domain expert who is familiar with that web site.

3.3 kWUM Algorithm

The details of kWUM are given in Algorithm 1. A few notations are listed in Table 1. kWUM requires three inputs, *a collection of web pages from a web site*, *a set of category labels*, and the *parent-child pair validity table* that represents the site-specific knowledge.

We first utilize iWUM to create a preliminary set of web units (Algorithm 1, Line 1). We then discover and handle *incomplete* web units based on web structure and site-specific knowledge. A web directory is built with web units produced by iWUM (Algorithm 1, Line 2). Web folders whose sub trees contain many web units, or *hub folders*, are discovered by $FindHubFolders$ (Algorithm 1, Line 3).

Algorithm 1 kWUM Algorithm

Input: A collection of web pages from a web site,
        A set of category labels,
        Parent-child pair validity table
Output: Web units

1: generate an initial set of web units by applying iWUM, $\{u\}$
2: build a web directory with web units in $\{u\}$, $root$ as the root folder
3: hfl =FindHubFolders($root$)
4: **for** each hub folder $hf \in hfl$ **do**
5:    **for** each sub-folder $sh_i \in sub(hf)$ **do**
6:       merge multiple web units within the same web folders under $subT(sh_i)$
7:       RemoveInvalidPair(sh_i)
8:    **end for**
9: **end for**

Table 1. Notations and their meanings

Notation	Description
$sub(f)$	sub-folders of web folder f
$subT(f)$	subtree rooted at web folder f, including f, its sub-folders and descendent folders
$N(f)$	number of labelled web units under $subT(f)$

A hub folder is one that has many sub-folders possibly containing *incomplete* web units. We then discover and handle *incomplete* web units in each sub-folder of the hub folders (Algorithm 1, Line 4-9). For each sub-folder of hub folders, sh_i, we first merge web units within the same web folder under $subT(sh_i)$ (Algorithm 1, Line 6). The key page of each merged web unit is identified based on filenames and hyperlinks. Then we carry the removal of invalid parent-child web unit pairs (Algorithm 1, Line 7).

As shown in Algorithm 2, hub folders are found by travelling the web directory in a breadth-first manner, where web folder $root$ is examined first, followed by its sub-folders and descendent folders. For each web folder f, $E(f)$ is calculated as defined in Equation 1 (Algorithm 2, Line 6). We set a minimum threshold E_{min}

Algorithm 2 FindHubFolders

Input: web folder r
Output: list of hub folder

1: create an empty key folder list kfl
2: create an empty folder list fl
3: $fl.push\_back(root)$
4: **while** $|fl| > 0$ **do**
5:    $f=fl.pop\_front()$
6:    calculate $E(f)$
7:    **if** $E(f) > E_{min}$ **then**
8:       $kfl.push\_back(f)$
9:    **else**
10:       **for** each sub-folder $cf_i \in sub(f)$ **do**
11:          $fl.push\_back(cf_i)$
12:       **end for**
13:    **end if**
14: **end while**
15: return kfl

such that web folders satisfying $E(f) \geq E_{min}$ will be determined as hub folders. If f is a hub folder, it is inserted into a hub folder list, hfl (Algorithm 2, Line 8); otherwise, all sub-folders of f are inserted into a queue, fl, in order to find hub folders among them later (Algorithm 2, Line 10–12). Note that sub-folders of a hub folder will not be further examined. This is because we assume that no hub folders should be found under another hub folder. Finally, when fl is empty, all hub folders are stored in hfl.

$$E(f) = \begin{cases} -\sum_{cf_i \in sub(f)} \frac{N(cf_i)}{N(f)} * \log_2 \frac{N(cf_i)}{N(f)} & \text{if } f \text{ has sub-folders} \\ 0 & \text{otherwise} \end{cases} \quad (1)$$

The detailed removal procedure is shown in Algorithm 3. We take a top-down approach to examine each web folder f in the sub-tree of sh_i, $subT(sh_i)$. Let u be the web unit in f, all parent-child web unit pairs with u as the parent web unit are examined (Algorithm 3, Line 5–7). We conduct a majority voting between the number of valid pairs and invalid pairs, num_v and num_i. If $num_v \geq num_i$, the category label of u is assumed to be correct. Otherwise, we try to assign another category cat for u such that after re-computing the number of valid and invalid parent-child pairs, $num_v - num_i$ is maximized and non-negative. If such a cat is not found, u's category label is not changed (Algorithm 3, Line 8–12). For each child web unit of u, cu_j, if it forms an invalid parent-child pair with u, it is merged into u, u's key page unchanged (Algorithm 3, Line 15); otherwise the web folder containing cu_j, is inserted into a queue, fl (Algorithm 3, Line 17–18). In this way, all invalid parent-child pairs with u as the parent web unit are removed. We then pop up a web folder from fl and repeat the same processing. Finally, when fl is empty, all invalid pairs under $subT(sh_i)$ are removed.

Algorithm 3 RemoveInvalidPair

Input: Web folder sh_i

1: create an empty list fl
2: $fl.push\_back(sh_i)$
3: **while** $|fl| > 0$ **do**
4:     $f = fl.pop\_front()$
5:     get web unit u contained in f
6:     get the set of web units $\{cu\}$ such that $\forall cu_j \in \{cu\}, pc(u, cu_j)$
7:     count number of valid and invalid parent-child pairs, num_v and num_i
8:     **if** $num_v < num_i$ **then**
9:         **if** exists category cat for web unit u such that $num_v >= num_i$ **then**
10:             assign u to category cat that maximizes $num_v - num_i$
11:         **end if**
12:     **end if**
13:     **for** each web unit $cu_j \in \{cu\}$ **do**
14:         **if** $pc(u, cu_j)$ is invalid pair **then**
15:             merge cu_j and its descendent web units with u
16:         **else**
17:             get web folder f_j that contains cu_j
18:             $fl.push\_back(f_j)$
19:         **end if**
20:     **end for**
21: **end while**

4 Experiments

In this section, we conduct experiments of iWUM and kWUM on a dataset called UniKB. The main objective is to compare the performance of kWUM and iWUM in terms of precision, recall, and F1 measures, as defined in [6] (illustrated in Appendix A).

4.1 UniKB

Web unit mining was originally evaluated using the WebKB dataset [3]. However, WebKB is a small and relatively old dataset, containing only 4159 pages from four university web sites collected in 1997. The small number of pages in each web site cannot reflect present web sites of complex structures. Therefore, we create a new dataset, *UniKB*. In early April 2004, we downloaded web pages (HTML, HTM, SHTML) from *www.cs.washington.edu* and *www.cs.utexas.edu* (These two web sites are also used in WebKB). After removing the following web pages: 1) pages that require login authentication, 2) dynamic pages, 3) PowerPoint slides, email archives in HTML format, and 4) reference documentations, we call this collection *UniKB*. We then manually identified web units corresponding to concept entities. Three categories of concepts are used, namely *course, faculty, and student*. The statistics of UniKB are shown in Table 2, where u and p refer to the number of web units and web pages respectively; and *other* refers to those pages not belonging to any web unit.

Table 2. UniKB dataset overview

Concept	course		faculty		student		other
University	u	p	u	p	u	p	p
Washington	1233	19346	59	1689	220	1865	2930
Texas	132	1534	51	1104	155	1358	4202

Table 3. Parent-child validity table

Washington

Parent unit	Child unit		
	course	faculty	student
course	1	0	0
faculty	0	0	0
student	0	0	0

Texas

Parent unit	Child unit		
	course	faculty	student
course	1	0	0
faculty	1	0	0
student	0	0	0

We also manually identified *valid* and *invalid* parent-child concept pairs for each web site and created parent-child validity tables, as shown given in Table 3.

---

[3] http://www-2.cs.cmu.edu/ webkb/

In Washington, there is only one valid parent-child concept pair, *course-course*. In Texas, there are two valid parent-child concept pairs, *course-course* and *faculty-course*. Note that in Table 3, 1 denotes a valid parent-child concept pair and 0 denotes an invalid one.

4.2 Results and Discussion

The performance of iWUM and kWUM are shown in Table 4. We measure the precision (Pr), recall (Re), and $F1$ [4] for each category. Note that $<Cat>(W)$ and $<Cat>(T)$ denote the category Cat in Washington and Texas respectively. In the rightmost column, we measure the percentage of improvement kWUM achieves over iWUM.

Table 4. iWUM and kWUM results ($\alpha = 1$)

Concept	iWUM results			kWUM results			Comparison (in %)		
	Pr	Re	$F1$	Pr	Re	$F1$	Pr	Re	$F1$
Course(W)	0.867	0.509	0.642	0.887	0.604	0.719	+2.3	+18.7	+12.0
Faculty(W)	0.548	0.727	0.625	0.652	0.818	0.726	+19.0	+12.5	+16.2
Student(W)	0.481	0.518	0.499	0.595	0.668	0.630	+23.7	+29.0	+26.3
Course(T)	0.267	0.578	0.366	0.336	0.538	0.414	+25.8	-6.9	+13.1
Faculty(T)	0.475	0.569	0.518	0.533	0.784	0.635	+12.2	+37.8	+22.6
Student(T)	0.394	0.735	0.514	0.544	0.684	0.606	+38.1	-6.9	+17.9

In general, web units are extracted with reasonable accuracies using iWUM and kWUM, although the performance can be quite poor for some categories. kWUM significantly outperforms iWUM. kWUM has better $F1$ values for all three categories from both universities (ranging from 12.0% to 26.3%). kWUM is also consistently better in Pr scores. This is because kWUM merges some *incomplete* web units into larger web units and thus reduces the number of web units that should not exist. For *course* and *student* in Texas, the Re scores of kWUM are slightly worse than iWUM by 6.9%, although this does not affect the improvement of $F1$ values. In other words, kWUM sometimes trades off Re scores for improving Pr scores.

Another interesting finding is that in both kWUM and iWUM results, the performance of different categories in different universities varies. *Course(W)* has very high precision scores (0.887/0.867) while *Course(T)* has very low scores (0.336/0.267). Washington has much better results in *course* and *faculty* categories than Texas. We notice that Washington is better-structured compared to Texas, especially for its *course* and *faculty* web units. *Course* web units in Washington has their own dedicated web space while those in Texas are often embedded in faculties' web folders. We also observe that key pages of course units in Washington follow similar styles or templates in their content. On the

---
[4] $F1 = 2*Pr*Re/(Pr+Re)$.

contrary, key pages of *course* units in Texas have no common style, some of which even contain little information. As a result, web site structure and key page quality are the two major factors that affect the performance of both kWUM and iWUM.

5   Conclusion and Future Work

Web unit mining aims to discover concept entities in a web site. An existing web unit mining algorithm, iWUM, suffers as it creates *incomplete* web units, with each covering only a subset of web pages of a single concept entity. By observing the distribution of web units within a web directory, we find that *incomplete* web units often form *invalid* parent-child pairs. Thus we propose a new web unit mining algorithm, kWUM, which analyzes the distribution of web units within a web directory, and discover and handle *incomplete* web units by removing *invalid* parent-child pairs.

We conducted experiments on a dataset called UniKB, which consists of web pages from two large university web sites. The results show that concept entities can be extracted with a reasonable accuracy and kWUM significantly outperforms iWUM in terms of precision and F1 measures. The performance of both iWUM and kWUM for different concept categories in different universities varies, depending on web structure and key page content.

Web unit mining is a new research field. There is much room for further improvement, e.g. improving performance for web sites that are not well-constructed. In our next phase of work, we plan to apply web units to other applications, e.g. developing more expressive queries for web site search.

References

1. S. Chakrabarti, B. Dom, and P. Indyk. Enhanced hypertext categorization using hyperlinks. In *Proceedings of the 1998 ACM SIGMOD*, Seattle, Washington, USA, June 2-4, 1998, pages 307–318.
2. M. Craven and S. Slattery. Relational learning with statistical predicate invention: Better models for hypertext. *Journal of Machine Learning*, 43(1-2):97–119, 2001.
3. M. Ester, H.-P. Kriegel, and M. Schubert. Web site mining: a new way to spot competitors, customers and suppliers in the world wide web. In *Proceedings of the 8th ACM SIGKDD*, Edmonton, Alberta, Canada, July 23 - 26, 2002, pages 249–258.
4. J. Furnkranz. Hyperlink ensembles: A case study in hypertext classification. *Journal of Information Fusion*, 1:299–312, 2001.
5. H.-J. Oh, S. H. Myaeng, and M.-H. Lee. A practical hypertext catergorization method using links and incrementally available class information. In *Proceedings of the 23rd ACM SIGIR*, Athens, Greece, July 24 - 28, 2000, pages 264–271.
6. A. Sun and E.-P. Lim. Web unit mining: finding and classifying subgraphs of web pages. In *Proceedings of the 12th CIKM*, McLean, Virginia, USA, November 4-9 2002, pages 108–115.

7. L. Terveen, W. Hill, and B. Amento. Constructing, organizing, and visualizing collections of topically related web resources. *ACM Transactions on Computer-Human Interaction*, 6(1):67–94, 1999.
8. Y. Tian, T. Huang, W. Gao, J. Cheng, and P. Kang. Two-phase web site classification based on hidden markov tree models. In *Proceedings of IEEE/WIC Web Intelligence*, Beijing, China, October 13-17, 2003.

Appendix A – Performance Metrics for Web Unit Mining

It is difficult to use the standard *precision* and *recall* measures to evaluate web unit mining performance. A web unit is set of web pages. A mined web unit could only partially match a labelled web units. Furthermore, the key page and support pages may have different importance. Given a web unit u_i constructed by a web

Table 5. Contingency table for web unit u_i

Web unit evaluation		Perfect web unit u'_i		
		$u'_i.k$	$u'_i.s$	NU
Constructed web unit u_i	$u_i.k$	TK_i	SK_i	–
	$u_i.s$	KS_i	TS_i	FS_i
	NU	NK_i	NS_i	–

unit mining method, we must first match it with an appropriate labelled web unit u'_i, also known as the *perfect web unit*. We define u'_i to be the labelled web unit containing $u_i.k$ and u'_i has the same label as u_i; $u_i.k$ can be either the key page or a support page of u'_i. The contingency table for matching a web unit u_i with its perfect web unit u'_i is shown in Table 5. Each table entry represents the overlapping web pages between the key/support pages of u_i and u'_i. For example $TK_i = \{u_i.k\} \cap \{u'_i.k\}$ and $TS_i = u_i.s \cap u'_i.s$. The entries in the last column and row account for pages that appear either in u_i or u'_i, but not both. $FS_i = u_i.s - (u'_i.s \cup \{u'_i.k\})$. $NK_i = \{u'_i.k\} - \{u_i.k\} - u_i.s$. $NS_i = u'_i.s - \{u_i.k\} - u_i.s$. Note that $|TK_i| + |KS_i| + |NK_i| = 1$ and $|TK_i| + |SK_i| = 1$. If the perfect web unit for u_i does not exist, u_i is considered invalid and will be assigned zero precision and recall values. Otherwise, the *precision* and *recall* of a web unit, u_i, are defined as follows.

$$Pr_{u_i} = \frac{\alpha \cdot |TK_i| + (1-\alpha) \cdot |TS_i|}{\alpha + (1-\alpha) \cdot (|KS_i| + |TS_i| + |FS_i|)} \quad (2)$$

$$Re_{u_i} = \frac{\alpha \cdot |TK_i| + (1-\alpha) \cdot |TS_i|}{\alpha + (1-\alpha) \cdot (|SK_i| + |TS_i| + |NS_i|)} \quad (3)$$

To account for the importance of key pages, a *weight factor* α to represent the degree of importance is introduced. More information about the performance metrics for web unit mining can be found in [6].

Towards a Realistic Microscopic Traffic Simulation at an Unsignalised Intersection

Mingzhe Liu, Ruili Wang, and Ray Kemp

Institute of Information Sciences and Technology, Massey University,
Private Bag 11222, Palmerston North, New Zealand
{m.z.liu, r.wang, r.kemp}@massey.ac.nz

Abstract. In this paper we propose a microscopic traffic flow model to simulate the flow at an unsignalised intersection. The model is built on fine grid cellular automata (CA), and is able to simulate actual traffic flow. Several important novel features are employed in our model. Firstly, the average car-following headway (=distance /velocity) 1.5 seconds has been observed in local urban networks and this *1.5-second rule* is built to our model. Secondly, vehicle movement on urban streets is simulated, based on the assumption of velocity following a Gaussian (normal) distribution and is calibrated by field data. Thirdly, driver behaviour is modelled, using a truncated Gaussian distribution. Finally, the limited priority mechanism is involved in this paper. The model has been validated against real data for its several components.

1 Introduction

Unsignalised intersections, regarded as complex subsystems, are important components of complex urban networks. There are two main types of unsignalised intersections: two-way stop-controlled (TWSC) intersections and all-way stop-controlled (AWSC) intersections [1]. AWSC are typical in North America. In New Zealand, there are two-way give-way-controlled (TWGWC) intersections which are similar to TWSC intersections.

At a TWSC/TWGWC intersection, vehicles from major streams have priority over vehicles from minor streams. Delays at an intersection are expected by both minor-stream vehicles and major-stream non-straight ahead vehicles, whereas major-stream straight-ahead vehicles suffer no delays [2]. Recently, a limited priority mechanism has been used for merging of roundabouts [3] and freeway [4] with high flows. This is based on the assumption that vehicles from major streams may encounter slightly delay in order to accommodate vehicles from minor streams. Our simulation shows that appropriate limited priority can slightly enhance the capacity of minor streams. However, this kind of driver behaviour is only suitable for turning-left vehicles from minor streams for left-side driving, e.g., in New Zealand.

Interactions between drivers have attracted attention from many disciplines, such as physics, mathematics and computer science. Various interaction models have been developed, for example, gap-acceptance models [5] and cellular automata (CA) models [1, 6, 10, 13]. The limitations of gap-acceptance models have been analysed and detailed in [1]. Thus, in this paper we focus on using a CA model to simulate traffic flow at a TWSC intersection.

The employment of CA modelling traffic flow at intersections has become popular in the last few years [1, 6-9]. For an unsignalised intersection, vehicle manoeuvres may include driving on the road and on the intersection.

Many models such as in [6, 7, 10] have been developed to deal with driving on a straight urban road. To our knowledge, previous models normally implicitly assume that the headways (= distance/velocity) are 1 second, that is, the 2-second rule is not considered in those models [11]. Theoretically, it should be observed by all drivers for safety reason, although the headways that drivers use are shorter than 2 seconds [12], but are normally longer than 1 second in the real world. In our research we have recorded 10 hours of traffic data between 16 August 2004 and 27 August 2004. The average car-following headway of 1.5 seconds has been observed in local urban networks and this 1.5-second rule is built into our model.

Ruskin and Wang [1] proposed a *Minimal Acceptable sPace* (MAP) method to simulate interactions between drivers at single-lane intersections. The method is able to simulate heterogeneous driver behaviour and inconsistent driver behaviour. In their model, driver behaviour is randomly classified into four categories: *conservative, rational, urgent and reckless*, and each group has its own MAP. Meanwhile, inconsistent driver behaviour is simulated by reassignment of categories with given probabilities at each time step. Although the assumption to categorise driver behaviour into four groups is coarse, this approach, to our knowledge, is the first model to reveal the impact of driver behaviour on traffic flow at unsignalised intersections.

This paper is organized as follows. In Section 2, two important novel features employed in this paper are described. Firstly, the headway of 1.5 seconds is built into our model. The second is that driver behaviour and vehicle movement are modelled using a (truncated) Gaussian distribution. In Section 3, vehicle movement on urban roads is calibrated by field data and interaction models are also calibrated using field data provided in [18]. Furthermore, we applied our model to an unsignalised intersection with limited priority mechanism. Finally, the conclusion is given in Section 4.

2 Methodology

In order to realistically describe microscopic highway traffic, a fine grid CA model has been proposed [13], where the length of each cell is 1.5 m, corresponding to a unit velocity of 5.4 km/h. To our knowledge, 1.5 m is the shortest length of cell that has been used in traffic flow modelling.

In this paper the finer discretization of cells in our CA model is used, that is, the length of each cell is equal to 1 m in a real road, which provides a better resolution modelling actual traffic flow than other models, in which the length of each cell is larger than 1 m. A unit of velocity is therefore equal to 3.6 km/h and each time step is 1 second. Since 1 unit of acceleration is 1 m/s^2, this corresponds to a 'comfortable acceleration' [14].

In urban networks, a lower velocity should be considered due to speed constraints. Normally, the legal maximum velocity in urban networks is 50 km/h, however some people will drive at speeds about 58 km/h, which is just below the limit (61 km/h) of being apprehended. Therefore, in our model, we assume the maximum speed of each vehicle is in the range of 50.4 km/h – 57.6km/h, which corresponds to the number of cells that a vehicle can move forward 14-16 cells in 1 second.

Different cells are required for different vehicle configuration. Following are average values based on recording 10-hour data sets at morning peak hour and these are adopted in this paper.

Table 1. Vehicle components and required cells

Vehicle types	Occupied cells	Percentage (%)
Motorcycles (M)	3	2
Personal Vehicles (P)	5	78
Vans and minibuses (V)	7	11
Buses (B)	10	6
Other large vehicles (O)	13	3

As mentioned, driver behaviour is inconsistent, so that even under similar conditions a driver may behave differently with time. With regard to a driver n, the value $d(n)$ can be viewed as a velocity or distance, therefore, it can be one of a set of data sets in the data records, namely, $d(n) \in \{d_1(n), d_2(n), d_3(n),..., d_l(n)\}$, $l = 1, 2, 3, ...$, L. The value $d(n)$ consist of two independent parts:

$$d(n) = d + \tau_n \quad (1)$$

That is, $d(n)$ is a sum of a constant part d plus a random and independent *error* component τ_n. Moreover, the *error* portion can itself be thought of as a sum of m components [15]:

$$\tau_n = g(e_1 + e_2 + e_3 + \cdots + e_m) \quad (2)$$

Here, e_i is a random variable that can take on only two values:
$e_i = 1$, when factor i is valid
$e_i = -1$, when factor i is not valid

In reality, the basic value d stands for the driver's habits under normal driving circumstance. τ_n is influenced by many stochastic factors such as temporary road construction, pedestrian crossing, etc. but normally impacts little on the distance d. So $d(n)$ fluctuates with the basic value d. According to the central limit theorem [15], τ_n can be referred to as Gaussian distribution approximately, i.e., $\tau_n \sim N(0, \sigma_n^2)$. Therefore, $d(n)$ also follows Gaussian distribution approximately, i.e., $d(n) \sim N(d, \sigma_n^2)$.

From the point of view of statistics, every vehicle entering an intersection can be viewed as an independent event. According to the joint distribution theorem [17], if driver A follows Gaussian distribution $N(\mu_1, \sigma_1^2)$, driver B follows $N(\mu_2, \sigma_2^2)$,......, driver M follows $N(\mu_m, \sigma_m^2)$, then for independent driver A, B,......, M., the joint distribution of driver A, B,..., M follows Gaussian distribution $N(\mu, \sigma^2)$, namely,

$$A + B + \ldots\ldots + M \sim N(\mu, \sigma^2) \tag{3}$$

Thus vehicle velocity and interaction space can be modelled based on Gaussian distribution. Normal Acceptable Space (NAS) is used in this paper to describe driver behaviour under normal conditions. The value of NAS is the number of required cells on a major stream for a vehicle from a minor stream to enter the intersection. As such the heterogeneous driver behaviour and inconsistent driver behaviour can be incorporated by NAS and its deviations from it.

2.1 Modelling Vehicles Manoeuvre on Urban Streets

Two types of traffic flow *free flow* and *following flow* can be observed in urban networks. In free flow driving, speed changes of all vehicles can be assumed to follow Gaussian distribution. This assumption is based on a fact that each individual vehicle speed change can be seen as a Gaussian distribution.

For a vehicle driving between intersections, velocity changes can be illustrated in Fig. 1, where five stages are involved. In initial stage A, acceleration is delayed due to physical reasons. In stage B, vehicles move with a higher speed and lead to the desired velocity in stage C. In stage C, speeds randomly fluctuate within a comfortable acceleration/deceleration range. According to the distance between current position and the downstream intersection, vehicles are again decelerated differently in stages D and E.

In car-following states, drivers adjust their velocity depending upon velocity changes of the preceding vehicle. Thus, the acceleration and deceleration of all vehicles can be modelled based on Gaussian distribution. The update rules are performed for the nth vehicle, where $x_n(t)$ is its position, $v_n(t)$ is velocity, and $g_n(t)$ is the number of free cells in front of the vehicle:

1. Free flow mode

Velocity is adjusted as illustrated in Fig. 1.

A: $v_n(t + 1) \rightarrow v_n(t) + 1$
B: $v_n(t + 1) \rightarrow v_n(t) + 2$
C: $v_n(t + 1) \rightarrow v_n(t) - 1$ with probability p_1 or,
$v_n(t + 1) \rightarrow v_n(t) + 1$ with probability p_2 or,
unchanged with probability p_3
D: $v_n(t + 1) \rightarrow v_n(t) - 2$
E: $v_n(t + 1) \rightarrow v_n(t) - 1$

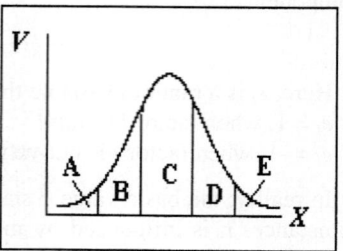

Fig. 1. Velocity changes of vehicles in terms of the current position and the distance to the downstream junction. V and X denote the current velocity and position, respectively

2. Following flow mode

If $g_n(t) < v_n(t)$ then: $v_n(t + 1) \rightarrow g_n(t + 2/3)$

This rule is based on the 1.5-second rule. In other words, the vehicle can only drive up to 2/3 of the total distance between the vehicle and the vehicle in front.

3. Randomization

If $v_n(t) > 0$, then the velocity of the n-th vehicle is decelerated randomly with probability p_b, i.e. $v_n(t+1) \rightarrow \max\{0, v_n(t) - 1\}$

4. Vehicle movement

$$x_n(t+1) \rightarrow x_n(t) + v_n(t+1)$$

2.2 Interaction Rules for Entry Intersections

In general, a driver may accept a value which is shorter than the NAS due to long waiting time or other urgent conditions, while a driver may also accept a value which is larger than the NAS due to bad weather, night visibility or other factors. Let x_{min} represent the number of minimum acceptable cells and x_{max} stand for the number of maximum acceptable cells for a driver to interact with other drivers. If $x > x_{max}$, a vehicle can pass the intersection without delay and, there is no interaction. Values less than x_{min} are rejected due to safety factors and values larger than x_{max} do not need to be considered due to no interaction involved (free flow). Therefore, the resulting model can be viewed as a *truncated* Gaussian distribution [16], where the left and right parts have been cut off. The *truncated* Gaussian distribution can be written as follows:

$$f(x) = \frac{1}{\sigma\sqrt{2\pi}} e^{\frac{(x-\mu)^2}{2\sigma^2}} \qquad x_{min} \leq x \leq x_{max} \qquad (4)$$

At unsignalized intersections there is a hierarchy of streams. Fig.2 (a) suggests a classification of four levels of priority in terms of the twelve arrows produced for traffic driving on the right. However, this method cannot be directly applied to driving on the left, especially when special traffic rules are dominating the traffic flow. For instance, in New Zealand, turning-left vehicles give way to turning-right vehicles travelling in the opposite direction. Westbound and eastbound are denoted as major streams, while southbound and northbound as minor streams in the model.

For simplicity, we introduce a concept of *conflicting points*, where vehicles have the higher priority and potentially conflict with the object. For instance, for a vehicle on the stop line in lane 7, its *conflicting points* are in lane 5 and 12.

Simulation conditions for vehicle j, k, and n are described here. The vehicle n has passed the intersection and vehicle j is approaching the entrance. Vehicle k is at the entrance and is preparing for entry. Let l_k denote the length of vehicle k; $m_k(t)$ denote NAS of vehicle k; $p_{k,j}(t)$ denote the number of free cells between vehicle k and j; $s_{k,n}(t)$ denote spacing between vehicle k and n; and $v_j(t)$ denote velocity of vehicle n at time step t. The following entry rules are performed by driver k and other drivers in parallel.

1. Assigning NAS of driver k in terms of the truncated Gaussian distribution.
2. Calculating individual conflicting point $p_{k,j}(t)$, $j = 1, 2, \ldots, q$. q is the number of conflicting points. If $m_k(t) \leq \min\{p_{k,j}(t), j = 1, \ldots, q\}$ and $l_k(t) \leq s_{k,n}(t)$, the waiting vehicle k can enter the unsignalised intersection, otherwise vehicle k is rejected to enter.
3. If vehicle k is waiting for entry, the update rule at each time step is as follows:
$m_k(t) = m_k(t) - \sigma_k$ if random() $< p$, p is the predefined number within $[0, 1]$
$m_k(t) = m_k(t) + \sigma_k$ otherwise
where $m_k(t)$ and σ_k are NAS (mean) and its deviation of vehicle k.

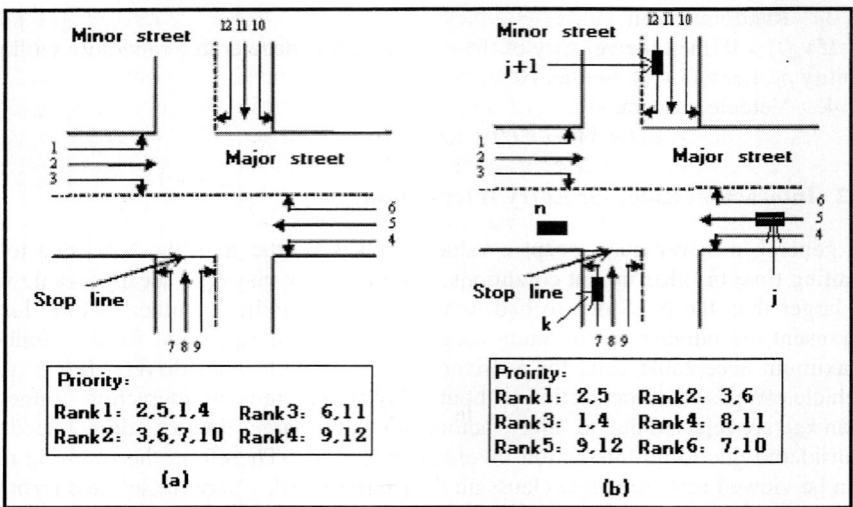

Fig. 2. The illustration of roads and an unsignalised intersection. (a) is given by [2]. (b) shows the stream ranking in New Zealand, according to the road rules

3 Numerical Results

3.1 Calibration and Validation

As preliminary work, we have calibrated vehicle movement on a straight lane. Fig. 3 shows observed single-vehicle movement and its simulation by using the proposed method in Section 2.1, where $p_1 = p_2 = 0.3$, $p_3 = 0.4$, $p_b = 0.1$. Probability density of each part is assumed to follow the standard Gaussian distribution. The dual-regime of acceleration and deceleration means that the model maintains in good agreement with the real world, especially in the initial acceleration and final deceleration phases.

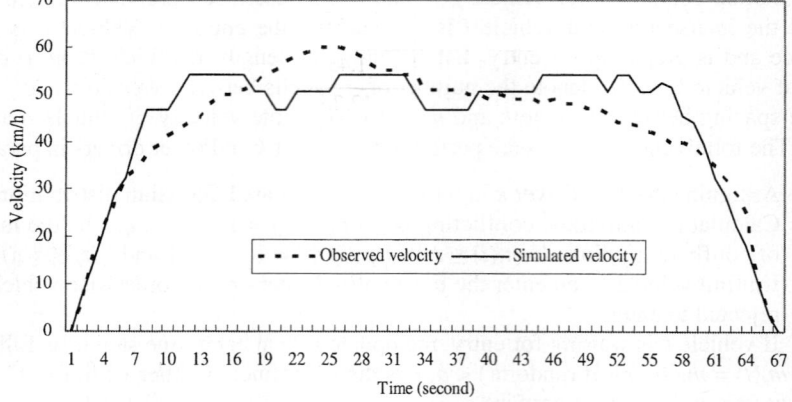

Fig. 3. Simulation of single-vehicle velocity between two intersections

Our model has been calibrated using field data provided in Table 2 [18]. Input parameters, such as turning rate, volume, Peak Hour Factor (= the ratio of the hourly volume to the peak 15-minute flow rate) [19], and percentage of vehicle types are the same as in the literature [18]. The length of normal vehicles, like cars, is set as 5 cells and heavy vehicles like buses or trucks are set as 10 cells. In [18], westbound and eastbound are minor streams, which are different from Section 2.2, where there are major streams.

The model then is validated by comparing results of capacity, delay, and queue length illustrated in Table 3. The 95% queue-length is calculated in terms of the queue equation in [20]. Our model appears to agree well with the field data at the calibration stage.

Table 2. Field data provided in [18]. VOL = Volume, PHF = Peak-Hour Factor, HFR = Hourly Flow Rate, PHV = Percent Heavy Vehicles, LT = Left Turn, ST = Straight ahead, RT = Right Turn

	Major Street						Minor Street					
	Northbound			Southbound			Westbound			Eastbound		
	LT	ST	RT	LT	ST	RT	LT	ST	RT	LT	ST	RT
VOL	8	184	5	165	273	10	10	35	116	1	67	11
PHF	0.95	0.95	0.95	0.95	0.95	0.95	0.95	0.95	0.95	0.95	0.95	0.95
HFR	8	193	5	173	287	10	10	36	122	1	70	11
PHV	3	0	0	3	0	0	3	3	3	3	3	3

Table 3. Comparison results of our model and field data [18]

	Westbound		Eastbound	
	[18]	Our model	[18]	Our model
Capacity	485	506	272	301
V/C	0.35	0.33	0.3	0.27
95% Queue Length	1.53	1.43	1.23	1.07
Delay	16.3	12.25	23.8	18.18

3.2 Application of the Model

For the purpose of a realistic microscopic simulation, vehicles arrival rate, turning rate, vehicles type, driver behaviour, special interaction (limited priority mechanism), well-designed division of speed, etc. have been considered and built in our model. By calibrating the model, we can implement a realistic simulation.

The rate of limited priority [3] of vehicles from minor streams is arbitrarily set as 1%; that is, one of every 100 minor-stream vehicles would enter the intersection in terms of a limited priority rule.

Vehicles arrive is assumes to follow a Poisson distribution. If all arriving vehicles pass the intersection without queuing, the flow rate $\lambda = 0.1, 0.2, 0.3...$ equivalents are 360vph (vehicle per hour), 720vph, 1080vph... respectively. In the simulations, the arrival rate of the major-stream vehicles vary between 0.1 and 0.2, while the arrival rate of the minor-stream vehicles is fixed to be 0.1. The turning rates of LT, ST, and RT are 0.2, 0.6, and 0.2 on all approaches.

Capacity and performance (delay and queue length) of a TWSC intersection have been investigated and detailed under different arrival rate (traffic volume) and turning rate (turning proportion) in [1]. Here, capacity is studied based on different vehicle configuration and arrival rate (=westbound: eastbound). Experiments were carried out for 3600 time steps (equivalent to 1 hour) for a street-length of 100 cells on all approaches. The NAS of all drivers ranges within $[x_{min}, x_{max}]$, where x_{min}, x_{max} are taken as 16 and 26 cells and the deviation of the truncated Gaussian distribution is 2 cells.

Table 4. Capacity of a minor stream for various vehicles component, arrival rate, and limited priority. Vehicle types see Table 1

Vehicles type (%)	With limited priority				Without limited priority			
	Arrival rate on major streams (westbound: eastbound)				Arrival rate on major streams (westbound: eastbound)			
M:P:V:B:O	0.1:0.1	0.1:0.2	0.2:0.1	0.2:0.2	0.1:0.1	0.1:0.2	0.2:0.1	0.2:0.2
0:100:0:0:0	358	336	322	286	356	331	317	276
0:80:11:9:0	350	328	310	267	348	323	304	255
2:78:11:6:3	341	320	301	243	335	311	287	234

Table 4 shows that vehicle components, arrival rate and limited priority rule can affect capacity. We find that the capacity of the minor-stream decreases in general when arrival rate of major-stream vehicles increases. The capacity of the minor-stream also decreases when percentage of private cars decreases and percentage of other vehicles increases. However, this effect differs as vehicles types vary. We also find that capacity has a lower increase with limited priority than that with no limited priority under the same conditions, especially for congested traffic.

4 Conclusion

In this paper, we have described a prototype CA model which attempts to simulate traffic flow at a TWSC intersection. Several important novel features are employed in our model. Firstly, on average car-following headway of 1.5 seconds has been observed in local urban networks and this *1.5-second rule* has been used in the car-following process. Secondly, vehicle movement along urban streets is simulated based on the assumption that velocity change follows a Gaussian distribution. Thirdly, driver behaviour is modelled using the truncated Gaussian distribution.

Vehicle manoeuvres on urban streets has been calibrated using field data. The simulation results show that the dual-regime of acceleration and deceleration provide good agreement of the model formulation with the real world, especially in the initial acceleration and final deceleration phases.

The numerical results indicate that the performance (delay and queue length) of TWSC intersections can be described well, based on the assumption of driver behaviour following a Gaussian distribution. In order to model a realistic microscopic simulation, vehicle arrival rate, turning rate, vehicle type, driver behaviour, special interaction (limited priority mechanism), and categorisation of speed, etc. are built into our model.

Acknowledgement

The support of the Massey University Research Fund (MURF) and the ASIA 2000 Foundation High Education Exchange Programme (HEEP) is a gratefully acknowledged.

References

1. Ruskin, H.J. and Wang, R.: Modelling Traffic Flow at an Urban Unsignalised Intersection, Proceedings of International Conference on Computational Science, (April 21-24, Amsterdam, Netherland), Lecture Notes in Computer Science (LNCS), Vol. 2329 (2002) 381-390
2. Troutbeck, R.J.: Average delay at an unsignalized intersection with two major streams each having a dichotomized headway distribution. Transportation Science, (1986)
3. Troutbeck, R. and Kako, S.: Limited priority merge at unsignalised intersections. Transportation Research Part, A 33 (1999) 291-304
4. Bunker, J. and Troutbeck, R.: Prediction of minor stream delays at a limited priority freeway merge. Transportation Research part B 37 (2003) 719-735
5. Tian, Z., Vandehey, M., Robinson, B.W., Kittelson, W., Kyte, M., Troutbeck, R., Brilon, W. and Wu, N.: Implementing the maximum likelihood methodology to measure a driver's critical gap. Transportation Research Part A. Vol. 33, Issues 3-4, April (1999) 187-197
6. Simon, P.M., Nagel, K.: Simplified cellular automata model for city traffic. Physical Review E, Vol 58 (1998)
7. Chowdhury, D., Santen, L. and Schadschneider, A.: Phys. Rep. 329 (2000) 199
8. Dupuis, A., Chopard, B.: Cellular Automata Simulation of Traffic: A Model for the City of Geneva. Networks and Economics, Vol. 3 (2003) 9-21
9. Wang, R., Ruskin, H.J.: Modelling Traffic Flow at a Multilane Intersection, Proceedings of International Conference on Computational Science and its Applications, (May 18 - 21, 2003 Montreal, Canada), Lecture Notes in Computer Science (LNCS), Springer-Verlag
10. Barlovic, R., Brockfeld, E., Schreckenberg, M., Schadschneider, A.: Optimal traffic states in a cellular automaton model for city traffic. Traffic and Granular Flow, 2001.10.15 - 2001.10.17, Nagoya University, Japan
11. Wang, R.: Modelling Unsignalised traffic Flow with Reference to Urban and Interurban Networks. Doctorate Thesis. Dublin City University. (2003)
12. Neubert, L., Santen, L., Schadschneider, A. and Schreckenberg, M.: Single-vehicle data of highway traffic: A statistical analysis. Phys. Rev. E 60 (1999) 6480
13. Knospe, W., Santen, L., Schadschneider, A. and Schreckenberg, M.: Towards a realistic microscopic description of highway traffic. J.Phys. A: Math. Gen. 33 (2000) L477-L485.
14. Institute of Transportation Engineers. Traffic Engineering Handbook. (1992)

15. Hays, W.L.: Statistics, the 5th Edition, University of Texas as Austin, Harcourt Brace College Publishers (1994)
16. Barr, D., Sherrill, E.: Mean and Variance of Truncated Normal Distributions, The American Statistician, Vol. 53 (1999) 357-361
17. Bhattacharyya, G.K., Johnson, R.A.: Statistical Concepts and Methods. John Wiley & Sons, Inc. (1977) Printed in the United States of America
18. http://www.co.st-johns.fl.us/BCC/gmsvcs/Planning/DRI/Transportation.Appendix/HCS.2-4-03.pdf
19. Transportation Research Board, Highway Capacity Manual. National Research Council, Washington, D.C., U.S.A. (2000)
20. http://www.rpi.edu/dept/cits/files/ops.ppt

Complex Systems: Particles, Chains, and Sheets

R.B. Pandey

Department of Physics and Astronomy,
University of Southern Mississippi, Hattiesburg,
MS 39406-5046, USA

Abstract. Particles, chains, sheets etc. are primary constituents to address complex issues in complex systems with appropriate coarse grained models. Flow in a driven immscible fluid mixture, film growth, protein relaxation, and dynamics of a self-avoiding sheet are used to illustrate specific issues as examples.

Keywords: complex system, computer simulation, self-organized structures, multi-scale dynamics.

1 Introduction

The range of topics and issues to be covered in this meeting reflect increasing importance of computers, computing, and computer simulations. Biological and physical systems (besides computational finance, traffic flow, etc.) span almost all areas of science and technology with a variety of problems over a range of scale (atomic, nano, micro, meso, macro). Most analytical theories are severely limited for complex systems which exhibit linear and non-linear response properties on different spatial and temporal scales. Computer simulation remains the main choice to probe multi-scale phenomena from microscopic details of constituents to macroscopic observables in such complex systems. Most real systems are still too complex to be fully addressed by computing and computer simulations alone. Coarse grained descriptions are almost un-avoidable in developing models for many of these model systems. In this talk I will constrain to a very small set of systems described by coarse grained models with basic units such as particles, chains, sheets, and possibly globules. I apologize for not citing the huge list of literature and constraining to some of our recent computer simulations on discrete lattices.

Studying the motion of a single particle in various environments is helpful in understanding physical properties of laboratory systems such as the dynamics of colloidal particles in solvent, conductivity of a conductor-insulator alloy etc. Analysis of a random walk motion of a particle provides useful insight into the overall characteristics of the medium/environment and its global pathways for propagating physical properties. For example, the variation of the mean square displacement (R^2) with the time (t) step of a particle executing its random walk on percolating clusters exhibit diffusive behavior, $R^2 = Dt$ above percolation threshold (p_c) and becomes anomalous at p_c, $R^2 \simeq t^{2k}$ with $k \sim 0.2$ on a cubic lattice ($d = 3$) and $k \sim 0.3$ on a square lattice ($d = 2$). Effect of an external bias (B) on the mobility of the particle adds further complexity as the bias competes with the pore barriers. Thus, the transport behavior of a mobile particle depend on the the characteristics of mobile particles, media, external parameters, time scale, correlations etc.

Understanding the many interacting particles such as complex fluid is more complex where cooperative dynamics and competing factors (parameters) lead to a variety of self-organized structures and non-linear response. In this talk, we consider steady-state morphology and flow [2, 3] in a multi-component driven system [4]–[7], film growth and roughness [8, 9], relaxation of a HP protein chain model [10], conformation and dynamics of sheet, a model for clay platelet and tethered membrane [11].

2 Self-organizing Flow

We consider the mixture of two components A and B (each with molecular weight M_A and M_B) on a cubic lattice of size L^3 [3]. Particles A and B are randomly distributed initially at about 50% of the lattice sites with one particle at a site (hard-core interaction). A nearest neighbor interaction between particles (A, B) and empty (pore) sites (O) is described by,

$$E = \sum_i \sum_n J(i, n) \quad (1)$$

where index i runs over all sites occupied by particles and n over all sites within a range r_i of i; $r_i = 1$ (nearest neighbor). The interaction matrix elements,

$$J(A, A) = J(B, B) = -J(A, B) = -J(B, A) = -\epsilon; J(A, O) = J(B, O) = -1. \quad (2)$$

$\epsilon(= 1, 2, ...)$ describes the miscibility gap. The gravitational energy U at the height z is given by, $U = M_{A/B} z$. A source of particles is connected to the bottom plane ($z = 1$) to release particles into the lattice. This concentration gradient causes an upward driving force. Additionally, we consider a hydrostatic pressure bias (H) along the vertical ($\pm z$) direction with probabilities,

$$H_{+z} = (1 + H)/6, \quad H_{-z} = (1 - H)/6; \quad 0 \leq H \leq 1; H_{\pm x} = H_{\pm y} = 1/6 \quad (4)$$

The Metropolis algorithm is used to move randomly selected particles (A and B) to their neighboring empty sites with probability $exp\{-\beta(\Delta E + \Delta U)\}$, $\beta = 1/k_B T$. Particles from the source are injected into the bottom plane ($z = 1$) with a probability which depends on their relative concentrations (p_A, p_B). A particle can drop out from the bottom or escape from the top ($z = L$). Periodic boundary conditions are implemented along the transverse (x, y) directions. An attempt to move each particle once is defined as one Monte Carlo step (MCS) time. Most data are generated on 100^3 samples although different sample sizes ($L = 30-300$

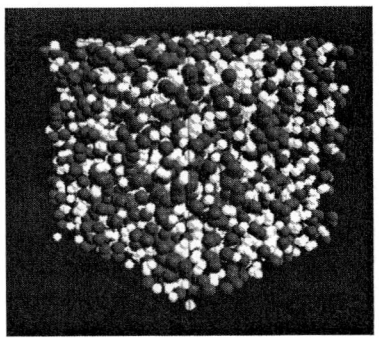 

Fig. 1. Snap shots, A (red) and B (white) on a 30^3 lattice with $H = 0.1, M_A = 0.1, M_B = 0.3, \epsilon = 1$ at time step $t = 1$ (Left), $t = 10^3$ (Right)

are used to check for the finite size effects. There are many variables including the molecular weight weight ratio $\alpha = M_B/M_A$.

Figure 1 shows a typical snap to illustrate the model, the self organizing constituents and corresponding phases. Note that the number of particles are not conserved, but the steady-state morphology is reached with a constant mass flux. For a immiscble system ($\epsilon = 1$), one can see distinct phases with high density at the bottom, dilute gas phase toward the top separated by bi-continuous interface (i.e., figure 1). The phase separation can also be identified by analyzing the density profiles (figure 2). The flux density (j) exhibits linear (low H) and volcanic (high H) responses to the pressure bias H and depends on the molecular weight (figure 3). A range of structural morphologies with various non-linear response and transport emerge as we vary the parameters (H, α, T, ϵ) in such a multi-component driven fluid [2, 3]. Some of these findings are also shared by driven granular materials [7].

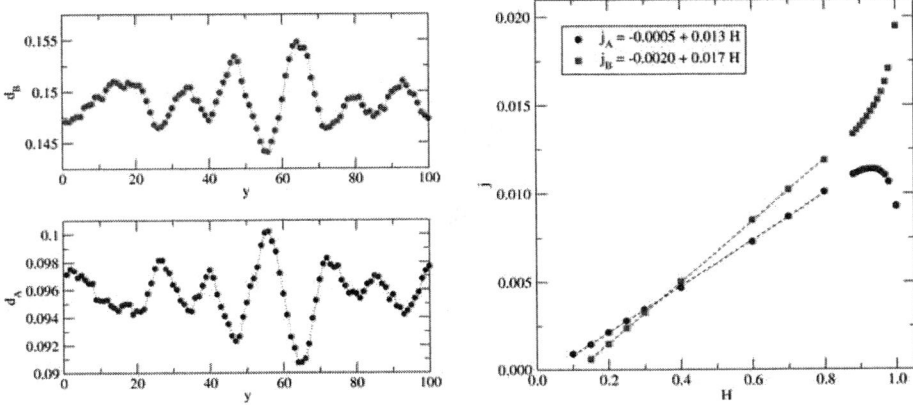

Fig. 2. Left: Transverse density profile of A and B on a 100^3 lattice with 128 independent samples, $\epsilon = 1, M_A = 0.1, M_B = 0.3, H = 0.1$. Right: Flux density ($j_A, j_B$) of A and B versus bias H

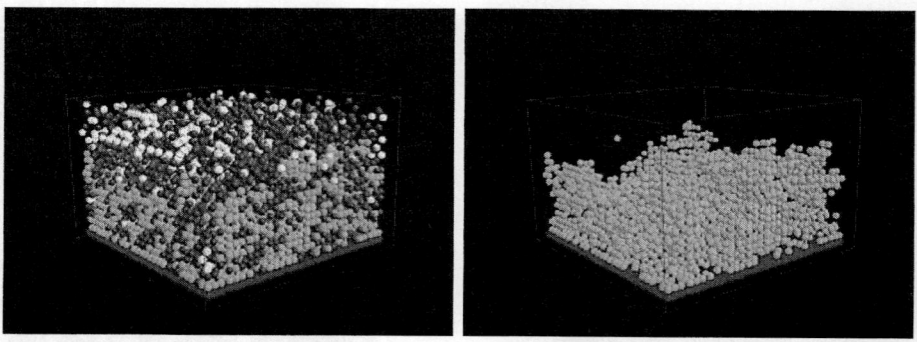

Fig. 3. Snapshots of the film growth. Left: substrate (blue), polar (red), non-polar (white), water (green), reacted component (gold) on a $40 \times 40 \times 30$, with $p_P + p_H + p_W = 0.6$, initial water concentration $p_W = 0.3$ and $p_P = p_H = 0.15$ for rate of reaction $r = 0.05$ at temperature $T = 1$ in arbitrary unit. Interaction $\epsilon_1 = 2, \epsilon_2 = 1$ is used. Right: covalently bonded sites of the film

3 Film Growth and Roughness

Polar (P) and non-polar (H) mixture with concentrations p_P and p_H respectively in an aqueous (W) solution are used to model film formation on an adsorbing substrate [8]. Particles with appropriate interactions ($J(P,P) = J(H,H) = -\epsilon_1; J(H,P) = 0; J(P,W) = -J(H,W) = -\epsilon_2$) and molecular weights ($M_P = M_H = 1, M_W = 0.1$) move and equilibrate (as in previous section) on a discrete lattice. The kinetic reactions are implemented for covalent bonding among these reacting elements and the substrate. Aqueous component may evaporate during the film formation while the fraction of polar and non-polar components is kept constant. Bonded elements become immobile. The interacting components include mobile constituents polar, non-polar, water, and the fixed substrate. The kinetic reaction may arrest these elements as a part of the film grows while the mixture equilibrate and water continue to evaporate. Because of the higher molecular weight, polar and non-polar groups deposit on the substrate where they become reactive. Figure 3 shows the snapshot of this multi-component system.

The parameters include water initial concentration (p_W), fraction of polar and non polar groups (p_P, p_H), rate of reaction (r), and temperature (T). We study the phase partitioning as the film grows and water evaporates. Density profile, growth of the interface width (W), and surface roughness of the film are examined as a function of these parameters. Figure 4 shows the variation of the interface width (w_b) with the time steps. It is easy to note that the interface width grows with the power law and saturated in the asymptotic time regime. At low values of the water concentration and low reaction rate, the initial power-law growth is followed by a secondary power-law growth before the saturation. The saturated interface width is a measure of roughness of the film. Variation

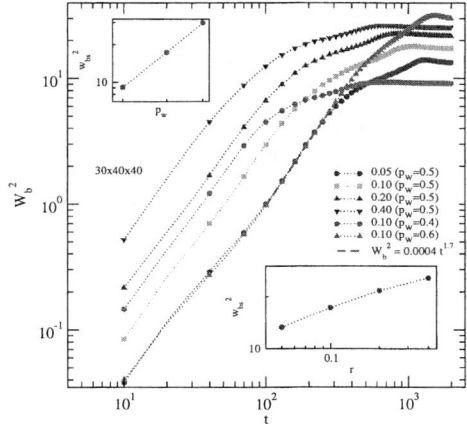

Fig. 4. Variation of W_b^2 with the time (t) steps at different water concentrations (p_W) and rate of reactions. Interaction parameters are same as above figure with 10 independent runs. Inset figures show variation of saturated interface width (W_{bs}) with r and p_W

of the saturated interface width with the initial water concentration and rate of reaction show that the roughness increases on increasing both p_W and r. This model [9] is designed to understand the phase partitioning during the film growth of s waterborne two-component polyurethanes (WB 2K-PUR) which consists of a hydrophobic polyisocyanate cross-linker, a hydrophobic polyester polyol, and water. Internal reflection infrared (IRIR) measurements and atomic force microscopic (AFM) images have shown that the roughness of the film increases with the humidity. We see from figure 4 that the increase of the interface width (roughness) with the initial water concentration is consistent with the experimental observations.

4 HP Protein Relaxation

The structural stability of protein chain has been a subject of immense interest in recent years. How the protein relaxes to its native structure (conformation) is one of the primary question addressed by computer simulation models [10]. A protein is a large polymer chain of twenty amino acid groups (characterized by their hydrophobic(H), polar (P), and electrostatic (E) interactions) in specific sequence [12]. Since majority of amino acids fall into hydrophobic and polar groups, HP chains (H and P nodes tethered together by covalent bonds in a chain) have become the norm for many coarse grained protein chain models for describing its general structure and dynamics. We select a bond-fluctuation model (over constant bond chain on-lattice and bead-spring chain off-lattice) for optimizing the computational efficiency and degrees of freedom. We consider protein chains of length N with hydrophobic (H) and polar (P) nodes connected by fluctuating bonds on a cubic lattice. Unlike particles (on a lattice site) considered in above sections, a node occupies a cube (i.e., eight lattice sites). The bond length l (between consecutive nodes) can vary, $l^2 = 4 - 10$ excluding 8 in unit of lattice constant. In addition to excluded volume effect, we consider a short range interaction (eq. 1). The range (r_i) of interaction $r_i^2 = 4, 5, 10$ includes nearest neighbor, next nearest neighbor sites, etc. For a range, the interaction matrix elements

$$J(H,S) = -J(P,S) = \epsilon_1; J(H,P) = \epsilon_2; J(H,H) = J(P,P) = \epsilon_3. \qquad (5)$$

Typically, $\epsilon_1 = \epsilon_2 = 0, 1, 2, 3, ...$; $\epsilon_3 = 0$. As before the energy is measured in units of $k_B T$. The Metropolis algorithm is used to move chain nodes randomly to their neighboring sites (i.e. cubes). An attempt to move each node once defines one Monte Carlo step (MCS) as time unit. Physical quantities of our interests are, the radius of gyration R_g, mean square displacements of each node ($< R_n^2 >$) and that of their center of mass ($< R_c^2 >$). We have explored a range of random and ordered sequences but we will constrain to random sequences here.

Figure 5 shows snapshots at various time steps to illustrate an approach to a globular conformation. Figure 6 shows the relaxation of the radius of gyration of the HP chain for various interaction. The protein chain relaxes well with weak (nearest neighbor) interaction but remain fairly extended. A substantial decay of R_g occurs only with stronger and longer range interactions. Thus in order to for an HP chain to collapse to its native globular conformation, it must be in solvent and that the interaction should be relatively strong. The problem with such interaction is that it takes very long time for the chain to relax. A conformational histogram presented in figure 3 shows how the population inversion occurs, i.e., the population of the native structures increases from their initial random conformations with the time steps. The scaling of the radius of gyration of such native conformation shows a trend toward a globular conformation.

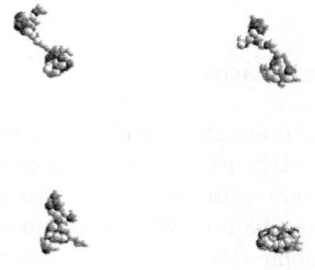

Fig. 5. Snapshots of HP protein conformations at time steps $t = 5, 10, 15, 20$ in unit of 10^5 for $N = 100$ with interaction $\epsilon_1 = 5$, $r_i^2 = 5$ on a 200^3 lattice

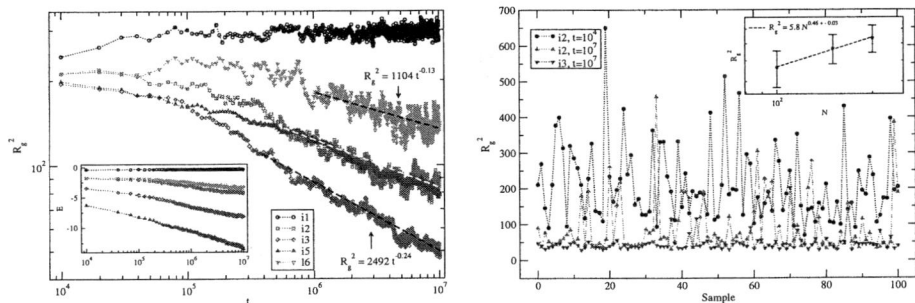

Fig. 6. Left: R_g^2 versus t on a log-log scale for $N = 100$ with $r_i^2 = 5$ and strength $\epsilon_1 = 1(i1), 2(i2), 3(i3), 5(i5)$ and $r_i^2 = 10(l6), \epsilon_1 = 1(i1)$ (down triangles). The inset figure shows the variation of corresponding energy on a semi-log scale. Sample size 200^3 is used with 100 independent runs. Right: Variation of R_g^2 with independent runs for $N = 100$ with $r_i^2 = 5$ at $t = 10^4$, strength $\epsilon_1 = 2(i2)$, and $t = 10^7$, $\epsilon_1 = 2(i2), 3(i3)$. Inset figure is R_g^2 versus N plot on a log-log scale for $\epsilon_1 = 2(i2)$ with each data point generated for $t = 10^7$ MCSs with a selective sampling of 100 runs; y-axis range is $35 - 80$. Sample size 200^3. (Ref. [10])

5 Self-avoiding Sheet

Studying the conformation and dynamics of semi-flexible sheets is useful in understanding the physical properties of planar molecules such as ex-foliated and dispersed silicate platelets, β-sheets of protein and tethered membranes [11]. Dynamics of a polymer chain (a model protein in previous sections) is governed by the collective segmental motion which in turn depends on the movement of underlying nodes. As an extension, it would be interesting to know how the segments of a flexible sheet move, conform, and dictates its global properties. For a coarse-grained sheet (i.e., a mesh of nodes connected by flexible bonds) ample degrees of freedom are necessary to capture the various structural relaxation modes. Connected nodes on a discrete lattice with fluctuating bond lengths can accelerate the relaxation dynamics while preserving the pertinent structural details via ample degrees of freedom - the choice used here.

We consider sheet of size L_s^2 (initially a square configuration) on a cubic lattice (L^3). Each node of the sheet occupies a unit cube, i.e., eight lattice sites. The excluded volume constraints similar to bond-fluctuating chain makes it a self-avoiding sheet (SAS). Further interaction among the nodes, between nodes and solvent, and constraints on the bonds connecting the neighboring nodes may be considered. Similar to movement of a node in a bond-fluctuating chain, each node of SAS execute its stochastic moves to its neighboring (26) sites selected randomly subject to excluded volume and bond-length constraints. Figure 7 shows snapshots of the self-avoiding sheet to illustrate its conformational changes. Variation of the mean square displacement of the central node with the time step is presented in Figure 8. Since a node is connected with four neighboring nodes, its

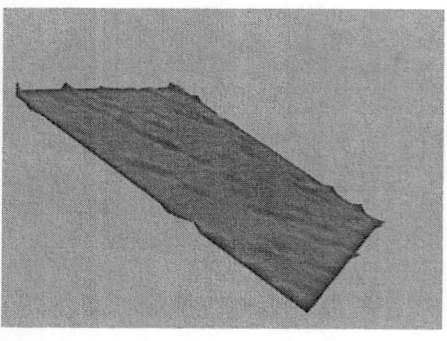 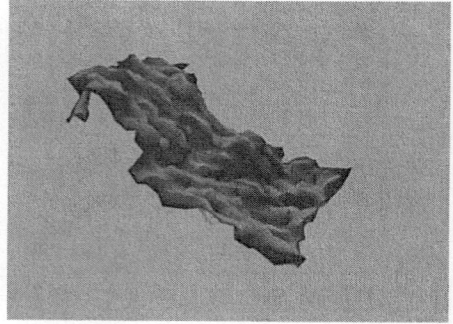

Fig. 7. Snapshot of sheet of size $L_s^2 = 32^2$ at time step $t = 1$ (left) and $t = 10^6$ (right) on a 100^3 lattice

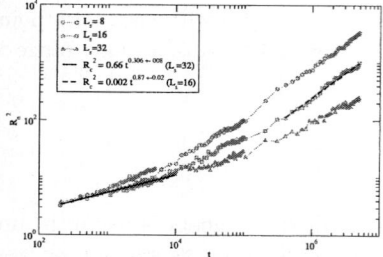 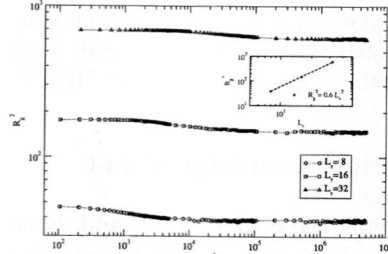

Fig. 8. Left: Mean square displacement of the center node of sheets with the time steps with $L_s = 8, 16, 32$ on a 100^3 lattice. Inset figures show the short time dynamics of the center node: $L_s = 32$ on 100^3 (triangle), $L_s = 64$ on 200^3 (square). The inset of the inset (circle) are short time data ($t = 2000 - 5000$) MCS after equilibrating the $L_s = 32$ sheet for $t = 10^6$ time steps. 10-500 independent runs are used. Right: $\langle R_g^2 \rangle$ versus t. Inset figure shows R_g^2 versus L_s plot on a log-log scale. (Ref. [11])

movement is more constrained than that in a linear chain. Accordingly, the short time dynamics is expected to differ from the Rouse dynamics of the node in a chain. Note that the larger the sheet, the longer it will take to reach asymptotic diffusive behavior for the mean square displacement of the node, i.e., $R_n^2 \propto t$. In short time regime, $R_n^2 \propto t^{0.26}$. Since the connectivity (four) of the interior nodes of the sheet is larger (twice) than that of linear chain nodes (two), the smaller power law exponent (nearly half) is consistent with intuition. Figure 8 shows the variations of the radius of gyration with time steps. The inset shows scaling of gyration radius with the linear size of the sheets, $R_g^2 \propto L_s^2 (= N)$. This implies that the self-avoiding sheet remains nearly flat consistent with the previous simulation on self-avoiding tethered membranes.

6 Conclusion

Particles with appropriate interactions can be used to model complex multi-component systems. For a driven system, simulations show how these components organize, mix, and segregate and how they respond to external bias. Evolution of various phases and linear and non-linear responses for a range of bias values are successfully predicted. Increase in roughness of the film growth with humidity due to phase partitioning of polar and non-polar group observed in laboratory experiments has been confirmed by simulations. Using HP model of protein, we have shown how a protein chain relaxes in solvent and that a relatively strong interaction is necessary to reach its native conformation. Short time dynamics of a self-avoiding sheet shows power-law behavior. SAS remains relatively flat consistent with recent work on tethered membranes. Thus, with particles, interactions, and their covalent bonding, it is feasible to address complex issues in complex system with appropriate coarse graining.

Acknowledgment. This work spans over several research projects which are in part supported by NSF-EPSCoR, Air Force Research Laboratory, Naval Research Laboratory, and NSF-MRSEC (Materials Science and Engineering Research Center).

References

[1] Stauffer, D., Aharony, A.: Introduction to Percolation Theory, Second Edition, Taylor and Francis (1994)

[2] Pandey, R.B., Gettrust, J.F., Seyfarth, R., Cueva-Parra, Luis A.: Self-organized phase segregation in a driven flow of dissimilar particles mixtures, International Journal of Modern Physics C, vol. 14, No. 7(2003) 955-962

[3] Pandey, R.B., Gettrust, J.F.: Structural response in steady-state flow of a multi-component driven system: Interacting lattice gas simulation, Physica A, Vol. 345, Elsevier (2005) 555-564

[4] B. Schmittmann, B., Zia, R.K.P.: Statistical Mechanics of Driven Diffusive Systems, Academic Press (1995)

[5] Marro, J, Dickman, R.: Nonequilibrium Phase Transitions in Lattice Models, Cambridge University Press (1999)

[6] Mehta, A (Ed.): Granular Matter: An Interdisciplinary Approach, Springer, New York (1994)

[7] Coniglio, A (Ed.): Unifying Concepts In Granular Media And Glasses, Elsevier (2004)

[8] Pandey, R.B., Urban, M.W.: Surface phase partitioning in film formation of waterborne polyurethanes: Monte Carlo simulations and internal-reflection IR imaging, Langmuir, Vol. 20, No. 20(2004) 2970-2974

[9] Otts,D., Cueva-Parra, Luis, Pandey, R.B., Urban, M.W.: Film formation from aqueous polyurethane dispersions of reactive hydrophobic and hydrophilic components: Spectroscopic studies and Monte Carlo simulations, reprint (submitted to Langmuir, 2005)

[10] Bjursell, J., Pandey, R.B.: Relaxation to native conformation of a bond-fluctuating protein chain with hydrophobic and polar nodes, Physcal Review E, Vol. 70, (2004) 052904-1-052904-4
[11] Pandey, R.B., Anderson, K., Heinz, H., Farmer, B: Conformation and Dynamics of a Self-Avoiding Sheet: Bond-Fluctuation Computer Simulation, Journal of Polymer Science Part B (in press, 2005)
[12] Branden, C., Tooze, J.: Introduction to Protein Structure, 2nd edition, Garland, New York (1999)

Discretization of Delayed Multi-input Nonlinear System via Taylor Series and Scaling and Squaring Technique

Zhang Yuanliang, Hyung Jo Choi, and Kil To Chong

Electronics and Information Engineering,
Chonbuk National University, Chonju, South Korea 561-756
kitchong@chonbuk.ac.kr

Abstract. A new discretization method for the calculation of a sampled-data representation of nonlinear continuous-time system is proposed. The suggested method is based on the well-known Taylor-series expansion and zero-order hold (ZOH) assumption. The mathematical structure of the new discretization method is analyzed. On the basis of this structure the sampled-data representation of nonlinear system with time-delayed multi-input is derived. First the new approach is applied to nonlinear systems with two inputs. And then the delayed multi-input general equation has been derived. In particular, the effect of the time-discretization method on key properties of nonlinear control systems, such as equilibrium properties and asymptotic stability, is examined. And 'hybrid' discretization schemes that result from a combination of the 'scaling and squaring' technique with the Taylor method are also proposed, especially under conditions of very low sampling rates. Practical issues associated with the selection of the method's parameters to meet CPU time and accuracy requirements, are examined as well. A performance of the proposed method is evaluated using a nonlinear system with time-delay: maneuvering an automobile.

1 Introduction

Time delay is often encountered in various engineering systems, such as chemical process, hydraulic, and rolling mill systems and its existence is frequently a source of instability. Many of these models are also significantly nonlinear which motivates the research in control of nonlinear systems with time delay. In the field of the discretization, for the original continuous-time systems with time free case the traditional numerical techniques such as the Euler and Runge-Kutta methods have been used for getting the sampled-data representations [1]. But these methods need a small sampling time interval. Because it is necessary to meet the desired accuracy and they cannot be applied to the large sampling period case. But due to the physical and technical limitation slow sampling is becoming inevitable. A time-discretization method which expands the well-known time-discretization of the linear time-delay system to nonlinear continuous-time control system with time-delay [2] can solve this problem. And this method is applied to the nonlinear control systems with delayed multi-input and the nonlinear control systems with non-affine delayed input [3,4]. The effect of this approach on system-theoretic properties of nonlinear systems, such as equilibrium properties, relative order, stability, zero dynamics and minimum-phase characteristics revealing the natural and transparent way in which Taylor methods permeate the relevant theoretical aspects is also studied [5]. Nowadays, modern nonlinear control strategies are usually implemented on a microcontroller or digital signal processor. As a direct consequence, the control algorithm has to work in discrete-time. For such digital control algorithms, one of the following time discretization approaches is typically used: i) time-discretization of a continuous time control law designed on the basis of a continuous time system, ii) time-discretization of a continuous time system resulting in a discrete-time system and control law design in discrete-time. It is apparent that the second approach is an attractive feature for dealing directly with the issue of sampling. Indeed, the effect of sampling on system-theoretic properties of the continuous-time system is very important because they are associated with the attainment of the design objectives. It should be emphasized that in both design approaches time discretization of either the controller or the system model is necessary. Furthermore, notice that in the controller design for time-delay systems, the first approach is troublesome because of the infinite dimensional nature of the underlying system dynamics. As a result the second approach becomes more desirable and will be pursued in the present study. This paper proposed the time discretization method of the nonlinear control systems with multiple time-delays in control [3]. The proposed discretization scheme applies the Taylor series expansion according to the mathematical structure developed for the delay-free nonlinear system [5] and delayed single-input nonlinear system [2]. The effect of sampling on system-theoretic properties of nonlinear systems with time-delayed multi-input, such as equilibrium properties and stability is examined. And the well known "scaling and squaring" technique which is widely used for computing the matrix exponential [6] is expanded to the nonlinear case when the sampling period is too large.

2   Nonlinear System with Delayed Multi-input

As shown in [2], a discrete-time nonlinear time-delayed input system can be obtained using Taylor series, and it has been shown that the expansion of single dimensional system to n dimensional system. Similarly the single input case can be expanded to multi-input case. The discretization method of general nonlinear system with multi-input delay is developed using Taylor series expansion. The general multi-input nonlinear system in state space form without time delay can be represented as follows.

$$\frac{dx(t)}{dt} = f(x(t)) + \sum_{i=1}^{n} g_i(x(t))u_i(t) \qquad (1)$$
$$= f(x(t)) + u_1(t)g_1(x(t)) + u_2(t)g_2(x(t)) + \ldots + u_n(t)g_n(x(t))$$

The values of $A^{[l]}(x, u)$ are evaluated recursively as follows.

$$A^{[1]}(x, u) = f(x) + u_1 g_1(x) + u_2 g_2(x) + \ldots + u_n(t) g_n(t)$$

$$A^{[2]}(x, u) = f'(x)x' + u_1 g_1'(x)x' + u_2 g_2'(x)x' + \ldots + u_n g_n'(x)x' = \frac{\partial A^{[1]}(x, u)}{\partial x} x'$$

$$\vdots$$

$$A^{[l+1]}(x, u) = \frac{\partial A^{[l]}(x, u)}{\partial x}(f(x) + u_1 g_1(x) + u_2 g_2(x) + \ldots + u_n(t) g_n(x))$$

(2)

A system with only two time-delayed inputs will be considered for simplicity in this section. A time-delayed two-input nonlinear continuous-time control system can be expressed with the following state-space form.

$$\frac{dx(t)}{dt} = f(x(t)) + u_1(t - D_1)g_1 x(t)) + u_2(t - D_2)g_2(x(t)) \quad (3)$$

The delays of the inputs are as in Eq. (3),

$$\begin{aligned} u_1(t - D_1) &\to (D_1 = q_1 T + \gamma_1) \\ u_2(t - D_2) &\to (D_2 = q_2 T + \gamma_2) \end{aligned} \quad (4)$$

So the inputs are as follows;

$$u_1(t - D_1) = \begin{cases} u(kT - q_1 T - T) \equiv u(k - q_1 - 1) & \text{if } kT \leq t < kT + \gamma_1 \\ u(kT - q_1 T) \equiv u(k - q_1) & \text{if } kT + \gamma + 1 \leq t < kT + T \end{cases}$$

$$u_2(t - D_2) = \begin{cases} u(kT - q_2 T - T) \equiv u(k - q_2 - 1) & \text{if } kT \leq t < kT + \gamma_2 \\ u(kT - q_2 T) \equiv u(k - q_2) & \text{if } kT + \gamma_2 \leq t < kT + T \end{cases}$$

(5)

It is necessary to specify how the sampled-data representation is affected if the delayed two-input are applied in the system.

Case. 1. ($\gamma_1 < \gamma_2$)
i) if $kT \leq t < kT + \gamma_1$

$$u_1 = u_1(k - q_1 - 1), \quad u_2 = u_2(k - q_2 - 1)$$

$$x(kT + \gamma_1) = x(kT) + \sum_{l=1}^{\infty} A^l(x(kT), u_1(k - q_1 - 1), u_2(k - q_2 - 1))\frac{\gamma_1^l}{l!} \quad (6)$$

ii) if $kT + \gamma_1 \leq t < kT + \gamma_2$

$$u_1 = u_1(k - q_1), \quad u_2 = u_2(k - q_2 - 1)$$

$$x(kT+\gamma_2) = x(kT+\gamma_1) + \sum_{l=1}^{\infty} A^l(x(kT + \gamma_1), u_1(k - q_1), u_2(k - q_2 - 1))\frac{(\gamma_2 - \gamma_1)^l}{l!}$$

(7)

iii) if $kT + \gamma_2 \leq t < kT + T$

$$u_1 = u_1(k - q_1), \quad u_2 = u_2(k - q_2)$$

$$x(kT + T) = x(kT + \gamma_2) + \sum_{l=1}^{\infty} A^l(x(kT + \gamma_2), u_1(k - q_1), u_2(k - q_2))\frac{(T - \gamma_2)^l}{l!} \tag{8}$$

where $k = 0, 1, 2, 3, \ldots$

Case 2. ($\gamma_1 > \gamma_2$)
i) if $kT \leq t < kT + \gamma_2$

$$u_1 = u_1(k - q_1 - 1), \quad u_2 = u_2(k - q_2 - 1)$$

$$x(kT + \gamma_2) = x(kT) + \sum_{l=1}^{\infty} A^l(x(kT), u_1(k - q_1 - 1), u_2(k - q_2 - 1))\frac{\gamma_2^l}{l!} \tag{9}$$

ii) if $kT + \gamma_2 \leq t < kT + \gamma_1$

$$u_1 = u_1(k - q_1 - 1), \quad u_2 = u_2(k - q_2)$$

$$x(kT+\gamma_1) = x(kT+\gamma_2) + \sum_{l=1}^{\infty} A^l(x(kT+\gamma_2), u_1(k - q_1 - 1), u_2(k - q_2))\frac{(\gamma_1 - \gamma_2)^l}{l!} \tag{10}$$

iii) if $kT + \gamma_1 \leq t < kT + T$

$$u_1 + u_1(k - q_1), \quad u_2 = u_2(k - q_2)$$

$$x(kT+T) = x(kT+\gamma_1) + \sum_{l=1}^{\infty} \infty A^l(x(kT+\gamma_1), u_1(k - q_1), u_2(k - q_2))\frac{(T - \gamma_1)^l}{l!} \tag{11}$$

where $k = 0, 1, 2, 3, \ldots$

3 'Scaling and Squaring' Technique (Extrapolation to the Limit)

When the sampling period T is large, the order N of the Taylor method should be large, so that the necessary accuracy can be achieved. This is mathematically reflected upon the asymptotic behavior of $\frac{T^N}{(N+1)!} \to 0$ as $N \to \infty$. If the computational capabilities of the available computer system are adequate, Taylor's method can be applied and provide a fairly accurate discrete-time model. However when T is large enough, then $A^{[l]}T^l/l!$ might become extremely large (due to the finite-precision arithmetic) before it becomes small at higher powers, where convergence takes over.

According to the 'scaling and squaring technique', if T is too large, one can subdivide the sampling interval into two subintervals of equal lenght $T/2$, approximate the exponential matrix series for $T/2$ and square the result. This is validated by the following property:

$$\exp(AT) = (\exp(A\frac{T}{2}))^2 \tag{12}$$

The 'scaling' procedure terminates until one finds an appropriate positive integer m such that $T/2^m$ is small enough for the given approximation scheme. The sampling interval is thus subdivided into 2^m equally spaced subintervals of length $T/2^m$ and the exponential matrix is approximately evaluated for $T/2^m$. Finally, 'squaring' the matrix $\exp(AT/2^m)$ m times performs the computation of $\exp(AT)$:

$$\exp(AT) = (((\exp A\frac{T}{2^m}))^2 \cdots)^2 \tag{13}$$

The "scaling and squaring" technique may be easily extended to the nonlinear case within the context of the Taylor method, and a nonlinear analogue of the linear result may be readily derived. If T is large, instead of applying the single-step Taylor method for a large order $tildeN$ (in order to obtain the value $x(k+1)$ when the value $x(k)$ is known), one may divide the interval $[t_k, t_{k+1})$ to 2^m equally spaced subintervals, and then apply a computationally 'lighter' Taylor expansion of order $N < \tilde{N}$ with a time step $T/2^m$ for the 2^m intermediate subintervals.

Suppose now that $\Omega(\tilde{N}, T) : R^n \to R^n$ is the operator that corresponds to the Taylor expansion of order $\tilde{N}$ with a time step T, and when it acts on $x(kT)$ the outcome is:

$$x(kT + T) = \Omega(\tilde{N}, T) x(kT) \tag{14}$$

where $\Omega(\tilde{N}, T)(\cdot) = I + \sum_{l=1}^{\tilde{N}} A^{[l]}(x(k), u(k)) \frac{T^l}{l!}$

Using operator notation the resulting discrete-time system may now be written as follows[5]:

$$x(kT + T) = \left[\Omega\left(N, \frac{T}{2^m}\right)\right]^{2^m} x(kT) \tag{15}$$

The above ASDR may be viewed as the direct result of the combination of Taylor's method with the 'scaling and squaring' technique.

One major implementation issue concerning both the single-step Taylor discretization and the 'scaling and squaring' discretization scheme is the criteria that can be used for the selection of the positive integers N and m. It is evident, that different criteria reflecting different requirements on the proposed discretization algorithm's performance, result in different values for N and m. In our search for a desirable pair of (m, N), we would restrict our requirements for the proposed discretizatio scheme, to the three elements: (i) simplicity and CPU time considerations; (ii) numerical convergence and accuracy requirements; and (iii) numerical stability. Practically, a natural point of departure for the selection of an appropriate (m, N), pair is a comparison of the magnitude of the sampling period T with the fastest time constant $1/\rho$ of the original continuous-time system. If T is small compared to $2/\rho$, then we can set $m = 0$ and we apply the single-step Taylor discretization method. Since T is small, a low order N single-step Taylor discretization method is usually sufficient to meet the expected accuracy requirement. Whenever T is larger than the fastest time constant $1/\rho$, we apply the 'scaling and squaring' discretization technique. The sampling interval is therefore subdivided into 2^m subintervals, and a low-order N single step Taylor discretization method is applied for each subinterval. So it requests that the following inequality should hold:

$$\frac{T}{2^m} < \frac{2}{\rho} \tag{16}$$

Since the requirements for numerical convergence and stability are also met. The preceding inequality may be rewritten as follows:

$$m > \log_2(\frac{T\rho}{2}) \tag{17}$$

The positive integer m is now selected to be:

$$m = \max\Big(\big[\log_2\big(\frac{T}{\theta}\big)\big] + 1, 0\Big) \tag{18}$$

where $\theta < 2/\rho$ is arbitrarily chosen and $[x]$ denotes the integer part of the number x. It is evident, that smaller values of the arbitrarily selected number θ would result to more stringent bounds on $T/2^m$. Note, that the maximum between $\log_2(T/\theta) + 1$ and 0 is considered, because T/θ might be already so small, yielding a negative $\log_2(T/\theta)$. In this case m is chosen to be 0, in accordance to what was previously stated.

And the 'scaling and squaring' technique can be applied to the nonlinear control systems with time-delayed multi-input. In this case we do not consider the single sampling interval T but the subintervals of $\gamma_1, \gamma_2 - \gamma_1, \ldots, T - \gamma_n$. And the method to choose m can also be used by changing T of that preceding equality into these subintervals of $\gamma_1, \gamma_2 - \gamma_1, \ldots, T - \gamma_n$. That is,

$$\begin{aligned} m_{\gamma_1} &= \max\Big(\big[\log_2(\frac{\gamma_1}{\theta})\big] + 1, 0\Big), \\ m_{\gamma_2-\gamma_1} &= \max\Big(\big[\log_2(\frac{\gamma_2-\gamma_1}{\theta})\big] + 1, 0\Big), \ldots, \\ m_{T-\gamma_n} &= \max\Big(\big[\log_2(\frac{T-\gamma_n}{\theta})\big] + 1, 0\Big), \end{aligned} \tag{19}$$

4 Simulation

One example is considered in the computer simulations. The example is a simplified model of maneuvering an automobile. Exact solutions for the systems are required in order to validate the proposed discretization method of nonlinear systems with the delayed multi-input. In this paper the continuous Matlab

Table 1. The x_1 response of system when $(T = 0.5 sec)$ with SS technique

Time step	Matlab (x_1)	Taylor(x_1) $(l=1, m=0, n=1)$	Taylor(x_1) $(l=4, m=3, n=4)$	Taylor(x_1) $(l=8, m=6, n=8)$
4	1.1224	1.125	1.1224	1.1244
8	0.6667	0.6668	0.6666	0.6666
12	0.3968	0.3966	0.3968	0.3968
16	0.6995	0.6991	0.6994	0.6995
20	1.1410	1.1408	1.1410	1.1410
24	1.0890	1.0890	1.0899	1.0899
28	0.6180	0.6178	0.6177	0.6177
32	0.4028	0.4024	0.4026	0.4026
36	0.7519	0.7513	0.7517	0.7517
40	1.1650	1.1647	1.1649	1.1649

ODE solver is used as an exact solution. In the simulation the discrete values obtained using the Taylor series expansion method are compared with the values obtained through the continuous Matlab ODE solver at the corresponding sampled period. The front axle of a simplified automobile maneuvering system is shown in Fig 1. The middle of the axles linking the front wheels has position $(x_1, x_2) \in R^2$, while the rotation of this axis is given by the angle x_3. The states x_1, x_2 related with rolling are directly controlled by input u_1 and the state x_3 related with rotation is directly controlled by u_2, thus the governing nonlinear differential equation can be obtained as followings;

$$\frac{d}{dt}\begin{bmatrix} x_1 \\ x_2 \\ x_3 \end{bmatrix} = \begin{bmatrix} \sin x_3 \\ \cos x_3 \\ 0 \end{bmatrix} u_1(t - D_1) + \begin{bmatrix} 0 \\ 0 \\ 1 \end{bmatrix} u_2(t - D_2) \quad (20)$$

The eigenvalue of the linear approximation Eq. (20) are small, thus $2/\rho$ is large. At first we choose a small sampling period and small time delay to verify the discretization method proposed in this paper. The inputs of u_1 and u_2 are assumed to be step functions respectively whose magnitudes are $u_1 = 1$ and $u_2 = 2.5$.

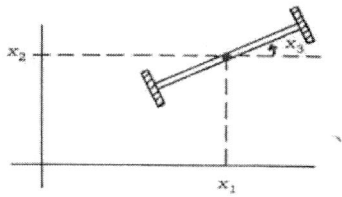

The front axis of a car.

Fig. 1. The front axis of automobile

Table 2. The x_2 response of system when $(T = 0.5sec)$ with SS technique

Time step	Matlab (x_2)	Taylor(x_2) ($l=1, m=0, n=1$)	Taylor(x_2) ($l=4, m=3, n=4$)	Taylor(x_2) ($l=8, m=6, n=8$)
4	0.8112	0.8107	0.8112	0.8112
8	0.9578	0.9577	0.9578	0.9578
12	0.5623	0.5623	0.5623	0.5623
16	0.1914	0.1912	0.1914	0.1913
20	0.3764	0.3759	0.3764	0.3764
24	0.8523	0.8519	0.8523	0.8523
28	0.9373	0.9372	0.9373	0.9373
32	0.5096	0.5096	0.5096	0.5095
36	0.1820	0.1817	0.1819	0.1819
40	0.4238	0.4233	0.4238	0.4238

The initial conditions are assumed to be $x_1(0) = 0, x_2(0) = 0, x_3(0) = 30°$ and the sampling period is 0.002sec. The inputs delays are 0.0015 for u_1 and 0.0036 for u_2. Thus we can use a single-step Taylor method and choose $N = 3$. The numerical differences between the Matlab ODE solver and the proposed method for state x_1 are from 1.9×10^{-10} to -3.296×10^{-8} and those for state x_2 are from -4.636×10^{-8} to -5.823×10^{-8}.

From the above definition we know that $\gamma_1 = 0.3$, $\gamma_2 = 0.2$. Assuming $l = m_{\gamma_2}, m = m_{\gamma_1 - \gamma_2}$ and $n = T - m_{\gamma_1}$ are the scaling and squaring coefficients m of the three time intervals of $[kT, kT+\gamma_2), [kT+\gamma_2, kT+\gamma_1)$ and $[kT+\gamma_1, kT+T)$. We choose $N = 3$ and $l = 1, m = 0, n = 1; l = 4, m = 3, n = 4$ and $l = 8$,

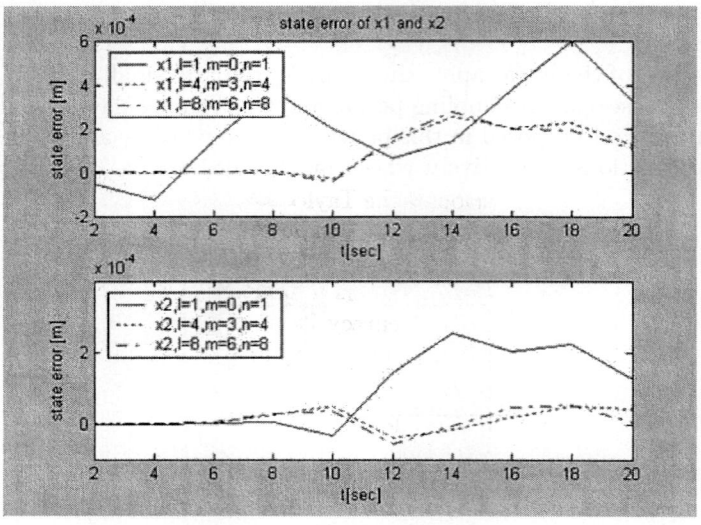

Fig. 2. State error responses of the simplified automobile for the case 2

Table 3. The computing time data for various (l, m, n) $(T = 0.5sec)$

Extrapolation to the limit parameters $(l, m.n)$	Order of the Taylor method N	Computing time (sec)
(1,0,1)	3	7.18
(4,3,4)	3	21.21
(8,6,8)	3	386.85

Table 4. The computing time data for various (l, m, n) $(T = 20sec)$

Extrapolation to the limit parameters (l, m, n)	Order of the Taylor method N	Computing time (sec)
(0,0,0)	$300(\epsilon = o(10^{-1}))$	81.03
(1,0,1)	$100(\epsilon = o(10^{-2}))$	26.70
(2,1,2)	$40(\epsilon = o(10^{-2}))$	21.35
(3,2,3)	$15(\epsilon = o(10^{-2}))$	17.63
(4,3,4)	$9(\epsilon = o(10^{-2}))$	25.01
(4,3,4)	$5(\epsilon = o(10^{-2}))$	38.19
(4,3,4)	$3(\epsilon = o(10^{-2}))$	71.46

$m = 6, n = 8$ respectively. The numerical differences between the Matlab ODE solver and the proposed method for state x_1 and x_2 are shown in Table 1. and Table 2. In the case of $l = 1, m = 0, n = 1$ the numerical differences between the Matlab ODE solver and the proposed method for state x_1 range from -1.277×10^{-4} to 5.967×10^{-4} and those for state x_2 range from -2.307×10^{-5} to 5.250×10^{-4}; in the case of $l = 4, m = 3, n = 4$ the numerical differences between the Matlab ODE solver and the proposed method for state x_1 range from -3.301×10^{-5} to 2.529×10^{-4} and those for state x_2 range from -4.040×10^{-5} to 5.067×10^{-5}; and in the case of $l = 8, m = 6, n = 8$ the numerical differences between the Matlab ODE solver and the proposed method for state x_1 range from -4.106×10^{-5} to 2.767×10^{-4} and those for state x_2 range from -5.671×10^{-5} to 5.490×10^{-5}.

The differences in the responses the Taylor method and the Matlab solver are shown in Fig. 2. The time used for the computing by these three cases is shown in Table 3. From these data we can conclude that it will be cautious to choose the extrapolation to the limit parameters (l, m, n). When some small parameters (l, m, n) can satisfy the desired accuracy it is not useful to choose larger ones because it will aggravate the computing task highly. Then we continue enlarging the sampling interval. And when $(T = 20, D_1 = 8, D_2 = 32)$ the computing time as the Extrapolation to the limit parameters (l, m, n) change is shown in the Table 4. As the sampling period becomes larger and larger we have to enlarge N to satisfy the desired accuracy that will aggravate the computing pressure highly if we do not use the "scaling and squaring" method. And if T is large enough, then $A^{[l]}T^l/l!$ might become extremely large due to the finite-precision arithmetic before it becomes small at higher powers, where convergence takes over. So that in this case we cannot get the accurate results using only the single-step Taylor method, especially in the case of the system whose fastest time constant is small. The "scaling and squaring" technique can overcome this problem.

5 Conclusions

A new approach for discrete-time representation of nonlinear control system with delayed multi-input in control is proposed. It is based on the ZOH assumption and the Taylor-series expansion that is obtained as a solution of continuous-time system. The effect of sampling on system-theoretic properties of nonlinear systems with time-delayed multi-input, such as equilibrium properties and stability is examined. And the well known "scaling and squaring" technique is expanded to the nonlinear case when the sampling period is too large.

References

1. Franklin, G. F., Powell, J. D. and Workman, M. L.: Digital Control of Dynamic Systems, Addison-Wesley, New York, 1998
2. Kazantzis, N., Chong, K. T., Park, J. H. and Parlos, A. G.: Control-relevant Discretization of Nonlinear Systems with Time-Delay Using Taylor-Lie Series, American Control Conference, pp. 149-154, 2003
3. Ji Hyang Park and Kil To Chong : Time-Discretization of Nonlinear Systems wit Delayed Multi-Input Using Taylor Series, KSME International Journal, Vol. 18, no. 7, pp. 1107-1120, 2004
4. Ji Hyang Park and Kil To Chong : Time-Discretization of Non-affine Nonlinear System with Delayed Input Using Taylor-Series, KSME International Journal, Vol. 18, no. 8, pp. 1297-1305, 2004
5. Kazantzis, N. and Kravaris, C. : System-Theoretic Properties of Sampled-data Representations of Nonlinear Systems Obtained via Taylor-Lie Series, Int. J. Control, Vol. 67, pp. 997-1020, 1997
6. Nicholas J. Higham : The scaling and squaring method for the matrix exponential revisited, Numerical Analysis Report 452, Manchester Center for Computational Mathematics, July 2004.

On the Scale-Free Intersection Graphs

Xin Yao[1], Changshui Zhang[1], Jinwen Chen[2], and Yanda Li[1]

[1] Department of Automation, Tsinghua University,
State Key Laboratory of Intelligence Technology and System,
Beijing 100084, P.R. China
yaoxin99@mails.tsinghua.edu.cn*
[2] Department of Mathematics, Tsinghua University,
Beijing 100084, P.R. China

Abstract. In this paper we study a network model called scale-free intersection graphs, in which there are two types of vertices, terminal vertices and hinge vertices. Each terminal vertex selects some hinge vertices to link, according to their attractions, and two terminal vertices are connected if their selections intersect each other. We obtain analytically the relation between the vertices attractions and the degree distribution of the terminal vertices and numerical results agree with it well. We demonstrated that the degree distribution of terminal vertices are decided only by the attractions decay of the terminal vertices. In addition, a real world scale-free intersection graphs, BBS discussing networks is considered. We study its dynamic mechanism and obtain its degree distribution based on the former results of scale-free intersection graphs.

1 Introduction

In many real world networks, such as authors collaboration networks [New1, New2], actor collaboration networks [ASBS], etc., the connections between two vertices are formed by their common selections. The authors connect with each other because they have cooperated on the same paper and the actors connect with each other because they have performed in the same movies. The construction of such networks is illustrated in Fig. 1. Newman has studied the construction and properties of the author collaboration networks [New1, New2] and some analysis of such networks are given in [New3]. Random intersection graph, which is a random graph model describing these phenomena, has been presented and studied with rigorous mathematical methods [KSS, FSS, Sta]. This model is a non-scale-free graph and this make it different from many real world networks. In this paper, we present the model call scale-free intersection graphs

---

* This work was funded in part by the the Fundamental Research Foundation of Tsinghua University under Grant No.JC2003031, National Science Foundation of China under Grant No.60234020, the Ministry of Science and Technology of China under Grant No.2001CCA01400 and the National Science Foundation of China under Grant No.10371063.

to describe these real world networks and study the BBS discussion networks as an example to support our model.

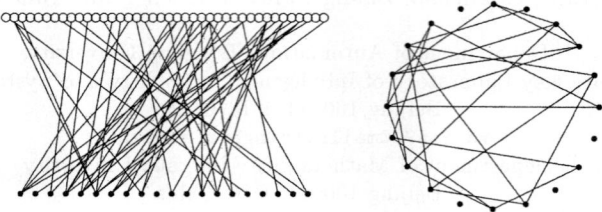

Fig. 1. The illustration of intersection graphs. In the left of the figure, each solid dot selects some of the small circles to link. Any two solid dots connect each other if their selections are intersect and then the intersection graphs are formed as shown in the right of the figure

We distinguish two types of vertices, hinge vertices and terminal vertices. The terminal vertices select some of the hinge vertices to link and two terminal vertices are connected if they have common selections. The set of terminal vertices and the edges among them constitute the intersection graph. For example, in the coauthorship networks, the authors are terminal vertices and the articles are hinge vertices.

The mechanism of the terminal vertices linking to hinge vertices decides the property of the intersection graphs. In this paper, we define the *attractions* of the hinge vertices and terminal vertices. The probability that there exists a connection between a hinge vertex and a terminal vertex is proportional to the attractions of both these vertices. We obtain the relation between the vertices attractions and the decay of the vertex degree sequence. As a consequence the degree distribution of the scale-free intersection graphs can be obtained from the scaling relations.

In the real world networks, the vertices may be added at different time. The early added vertices may have more links because they stay in the network longer than the new added vertices and the preferential attachment mechanism often make the high degree vertices gain more links. If we define that the vertices with more links have higher attraction, the scale-free intersection graphs may describe many real world networks. We present the BBS discussing networks of Tsinghua university in China as an example. We describe its formation mechanism and give the dynamical analysis. The results demonstrate the BBS discussing networks agree with our model.

2 The Model and the Analytical Results

Let $\mathscr{M} = \{m : 1 \leq m \leq M\}$ be the set of hinge vertices and $\mathscr{N} = \{n : 1 \leq n \leq N\}$ be the set of terminal vertices. Define the attraction of the hinge vertex

$m \in \mathcal{M}$ as $m^{-\alpha}$ and the attraction of the terminal vertex $n \in \mathcal{N}$ as $n^{-\beta}$. Let the terminal vertex n link to a hinge vertex m with probability

$$P(n,m) = n^{-\beta} m^{-\alpha}. \tag{1}$$

Therefore, the probability that both terminal vertices n_1 and n_2 connect with hinge vertex m

$$P(n_1, n_2; m) = \frac{1}{n_1^\beta n_2^\beta} \frac{1}{m^{2\alpha}}. \tag{2}$$

Since n_1 will connect with n_2 if there exists a hinge vertex m such that both n_1 and n_2 link to it, the probability of n_1 connects to n_2 is

$$P(n_1, n_2) = 1 - \prod_{m}^{M} (1 - P(n_1, n_2; m)) \tag{3}$$

$$= 1 - \exp\left[\sum_{m=1}^{M} \log\left(1 - P(n_1, n_2; m)\right)\right], \tag{4}$$

where

$$\sum_{m=1}^{M} \log(1 - P(n_1, n_2; m))$$

$$\simeq \log\left[\frac{(n_1 n_2)^\beta - 1/M^{2\alpha}}{(n_1 n_2)^\beta - 1}\right] + \frac{2\alpha}{2\alpha - 1} \frac{1}{(n_1 n_2)^\beta} \left[M^{-2\alpha+1} - 1\right]. \tag{5}$$

Substituting Eq.(5) to Eq.(4), we have

$$P(n_1, n_2) \simeq 1 - \frac{(n_1 n_2)^\beta - 1/M^{2\alpha}}{(n_1 n_2)^\beta - 1} \exp\left[\frac{2\alpha}{2\alpha - 1} \frac{M^{-2\alpha+1} - 1}{(n_1 n_2)^\beta}\right]. \tag{6}$$

It can be seen from Eq.(6) that the asymptotic property of the probability $P(n_1, n_2)$ will be different for the different case of α.

2.1 The Case: $\alpha > \frac{1}{2}$ and $\alpha = \frac{1}{2}$

If $\alpha > \frac{1}{2}$, then $-2\alpha + 1 < 0$ and $M^{-2\alpha+1} \to 0$. When n_2 is large, we have

$$P(n_1, n_2) \simeq 1 - \frac{(n_1 n_2)^\beta}{(n_1 n_2)^\beta - 1}\left(1 - \frac{2\alpha}{2\alpha - 1} \frac{1}{(n_1 n_2)^\beta}\right)$$

$$\simeq \frac{1}{2\alpha - 1} \frac{1}{(n_1 n_2)^\beta}. \tag{7}$$

It can be seen from Eq.(7) that the connecting probability of n_1 and n_2 is proportional to the product of their attractions. For the given terminal vertex n_1, the probability of linking to n_2 decays with the increase of n_2 by $n_2^{-\beta}$. Therefore, the mean degree of the terminal vertex can be obtained as follows,

$$k(n_1) \simeq \int_1^N P(n_1, n_2) \mathrm{d}n_2$$
$$\simeq \frac{N^{1-2\beta}}{(2\alpha - 1)(1 - \beta)} \left(\frac{N}{n_1}\right)^\beta. \tag{8}$$

If $\beta \neq 1/2$, the degree distribution of the terminal vertices is size dependent, i.e., their degree distribution depends on N. It can be seen that if $\beta < 1/2$, we get the accelerating growth networks in which the density of the links become higher when N increases, and if $\beta > 1/2$, we get the decelerating growth networks in which the density of the links become lower when N increases. Hence, the exponent of the degree distribution of the scale-free intersection network can be obtained by the methods in [DM3].

The analysis of the $\alpha = 1/2$ case is similar to above and we can also obtain that $k(n_1) \sim N^{1-2\beta}(N/n_1)^\beta$.

2.2 The Case: $\alpha < \frac{1}{2}$

For the case that $\alpha < \frac{1}{2}$, it is difficult to give the exact estimation of Eq.(6). However, we can obtain the following inequalities directly from Eq.(3),

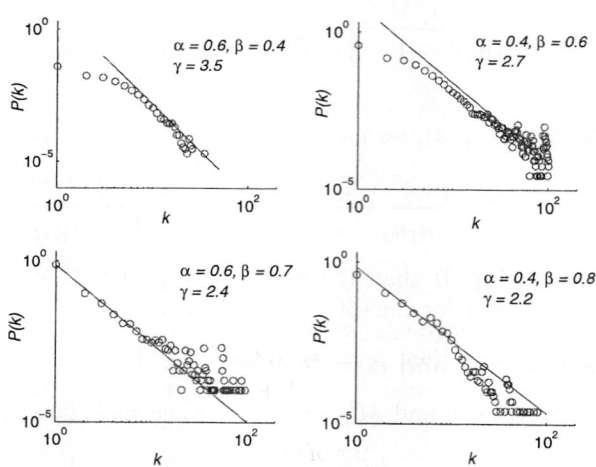

Fig. 2. The degree distribution of scale-free intersection graphs. $M = 20000, N = 20000$. We take $P(n, m) = Cn^{-\beta}m^{-\alpha}$, where $C < 1$ is a small constant in order to avoid a completely connected network and the character of the degree distribution holds. The circles are the simulating results for the corresponding parameter α, β on the figures. The solid lines in the figure are the theoretical value for the exponents of degree distributions obtained from scaling relation $\gamma = 1 + (1/\beta)$, which are $3.5, 2.6, 2.4, 2.2$, respectively

$$\frac{1}{(n_1 n_2)^\beta} \leq P(n_1, n_2) \leq \frac{M^{1-2\alpha}}{1-2\alpha} \frac{1}{(n_1 n_2)^\beta} \qquad (9)$$

in which the right hand side inequality comes from $\prod_{i=1}^{n} a_i - \prod_{i=1}^{n} b_i \leq \sum_{i=1}^{n}(a_i - b_i)$ where $0 \leq a_i, b_i \leq 1$. Integrating by n_2 from 1 to N in Eq.(9), we have

$$\frac{1}{(1-\beta)N^{1-2\beta}} \left(\frac{N}{n_1}\right)^\beta \leq k(n_1) \leq \frac{M^{1-2\alpha}}{(1-\beta)(1-2\alpha)N^{1-2\beta}} \left(\frac{N}{n_1}\right)^\beta \qquad (10)$$

Since the N and M is given as the size of system, we may demonstrate that

$$k(n_1) \sim C_{M,N} \left(\frac{N}{n_1}\right)^\beta \qquad (11)$$

Therefore, the degree distribution can be also obtained by the scaling relation. The simulating results of the scale-free intersection network are illustrated in Fig. 2 in which the numerically results agree with above theoretical analysis.

3 The BBS Discussing Networks and the Dynamic Model

The BBS(*Bulletin Board System*) is a popular networks system for users to discuss [KZ]. Each user in the system post the articles in the board. The articles can be classified as initial articles and reply articles. The initial article is the first post of a topic and may initiate a discussion about this topic. The reply articles are the articles which reply to some other articles. If we regard a topic, which is constituted by an initial article and some other reply articles, as a hinge vertex and regard a user as a terminal vertex, we obtain an example of scale-free intersection graph — the BBS discussing networks.

In the following we describe the detailed dynamical property of the BBS discussing networks. At each time step a new user(terminal vertex) is added and about $t^\delta (0 \leq \delta \leq 1)$ new articles are posted, which results in an accelerating growth network. The probability of a new article being an initial article is p. Hence, about pt^δ new topics(hinge vertices) are initiated and about $(1-p)t^\delta$ reply articles(links from the terminal vertices to the hinge vertices) are generated. It can be seen that there are about $t^{1+\delta}/(1+\delta)$ links between hinge vertices and terminal vertices at time t. The probability that the new reply articles belongs to a topic is proportional to the number of articles of this topic, i.e., the probability of the terminal vertices linking to a hinge vertex is proportional to the number of links of that hinge vertex. Let $\ell_H(m,t)$ denote the number of links of the hinge vertex m at time t, we have the following rate equation,

$$\begin{aligned}\frac{\partial \ell_H(m,t)}{\partial t} &= \frac{(1-p)t^\delta \ell_H(m,t)}{\sum_{m'} \ell_H(m',t)} \\ &= \frac{(1-p)(1+\delta)\ell_H(m,t)}{t},\end{aligned} \qquad (12)$$

which gives

$$\ell_H(m,t) = \left(\frac{t}{t_m}\right)^{(1-p)(1+\delta)}, \qquad (13)$$

where t_m is the time that the hinge vertex m is added to network. Since pt^δ new hinge vertices are added at each time step, there are about $pt^{1+\delta}/(1+\delta)$ hinge vertices at time t. Therefore we have $t_m = [m(1+\delta)/p]^{1/(1+\delta)}$. It can be derived from Eq.(13) that the distribution of the links number of hinge vertices follows scale-free property and the exponent is $1 + \frac{1}{(1-p)(1+\gamma)}$. It means that in the BBS discussing network, the distribution of the topic size is $P(\ell_H) \sim \ell_H^{-1-\frac{1}{(1-p)(1+\gamma)}}$. It can be seen that if more new topics(initial articles) generated at each time step (p is large), the tail of the topic size distribution will be light, if more reply articles generated at each time step(p is small), the tail of the topic size distribution will be heavy.

In the BBS discussion networks, the user who have posted more articles will be replied by more users and he is more apt to post new articles for continuing the discussion. Therefore, for any terminal vertex n, the increase of its links' number follows mechanism of preferential attachment, too. Let $\ell_T(n,t)$ denote the number of links of terminal vertex n at time t, we have the following rate equation,

$$\begin{aligned}\frac{\partial \ell_T(n,t)}{\partial t} &= \frac{t^\delta \ell_T(n,t)}{\sum_{n'} \ell_T(n',t)} \\ &= \frac{(1+\delta)\ell_T(n,t)}{t},\end{aligned} \qquad (14)$$

which gives

$$\ell_T(n,t) = \left(\frac{t}{n}\right)^{1+\delta}, \qquad (15)$$

and the distribution of the users' articles posting number of the users(the links number of the terminal vertices) is $P(\ell_T) \sim \ell_T^{-1-\frac{1}{1+\delta}}$ by the scaling relations.

The probability of terminal vertex n linking to hinge vertex m at time t is

$$\begin{aligned}f_{n,m}(t) &= \frac{t^\delta \ell_T(n,t)}{\sum_{n'} \ell_T(n',t)} \frac{(1-p)t^\delta \ell_H(m,t)}{\sum_{m'} \ell_H(m',t)} \\ &= \frac{1-p}{t^2}\left(\frac{t}{n}\right)^{1+\delta}\left(\frac{t}{t_m}\right)^{(1-p)(1+\delta)}.\end{aligned} \qquad (16)$$

When the network evolves to time N, the probability that there exists at least one connection between n and m is

$$P_{n,m}(N) = 1 - \prod_{t=n \wedge t_m}^{N}(1 - f_{n,m}(t)). \qquad (17)$$

It follows that

$$f_{n,m}(N) \leq P_{n,m}(N) \leq \sum_{t=n \wedge t_m}^{N} f_{n,m}(t), \qquad (18)$$

which gives

$$\frac{1-p}{N^{2-(2-p)(1+\delta)}} \left(\frac{1}{n}\right)^{1+\delta} \left(\frac{1}{t_m}\right)^{(1-p)(1+\delta)} \leq P_{n,m}(N)$$
$$\leq N^{(2-p)(1+\delta)-1} \left(\frac{1}{n}\right)^{1+\delta} \left(\frac{1}{t_m}\right)^{(1-p)(1+\delta)}.$$
(19)

It can be seen from Eq.(19) that the BBS discussion graphs exhibit the similar property of the scale-free intersection networks as Eq.(1), if we regard $n^{-(1+\delta)}$ and $t_m^{-(1-p)(1+\delta)}$ as the attractions the the terminal vertices n and the hinge vertex m. Therefore, the mean degree of the terminal vertices can be obtained by Eq.(8) or Eq.(11) and the degree distribution follows from the scaling relation discussed in [AB].

We take the real data from BBS system of Tsinghua University and the statistical results agree with above analysis, see Fig. 3.

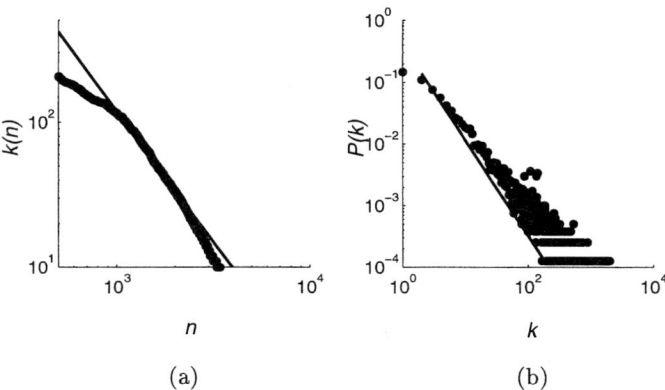

Fig. 3. BBS discussing network. The number of users is 7883 and the topic number is 37190. (a) The decay of the links number of users. The solid line is the function $h(x) \sim x^{-1.75}$ and the dots are the links number of the BBS users. (b) The degree distribution of the scale-free intersection network constituted by the users. The solid line is the function $h(x) \sim x^{-1.57}$ and the dots are the degree distribution of the users' degree

4 Discussion and Summation

We present the scale-free intersection networks model for describing and analyzing a large class of real world networks. The analytical results of the connectivity properties are obtained. It is demonstrated that the scaling exponent of the degree distribution of scale-free intersection networks is decided by the links distribution of the terminal vertices. The case for $\beta \leq \frac{1}{2}$ describes the accelerating growth and $\beta > \frac{1}{2}$ describes the decelerating growth. We study the BBS discussing network, a real world scale-free intersection network, obtain the distribution of the topic size and the distribution of article-posting number of the users. We show that the BBS discussing network can be analyzed by the method presented in Section 2 and therefore demonstrate the validation of the scale-free intersection network model.

The prevalent existence of scale-free intersection network make it important to study its variety of properties, such as clustering, shortest path, etc. We leave these for future work.

References

[ASBS] Amaral, L. A. N., Scala, A., Barthélémy, M., and Stanley, H. E.,: Classes of small-world networks. Proc. Natl. Acad. Sci. USA **97** (2000) 11149-11152

[AB] Albert, R., Barabási, A.-L.: Statistical mechanics of complex networks. Rev. Mod. Phys. **74** (2002) 47-97

[DM3] Dorogovtsev, S.N., Mendes, J.F.F.: Effect of the accelerating growth of communications networks on their struction. Phys. Rev. E. **63** (2001) 025101

[Sta] Stark, D: The vertex degree distribution of random intersection graphs, Random Structures and Algorithms. **24** (2004) 249-258

[FSS] Fill, J. A., Scheinerman, E. R., and Singer-Cohen, K. B.: Random intersection graphs when $m = \omega(n)$: An equivalence theorem relating the evolutio of $G(m, n, p)$ and $G(n, p)$ models, Random Structures and Algorithms. **16** (2000) 156-176

[KSS] Karoński, M., Scheinerman, E. R., and Singer-Cohen, K. B.: On random intersection graphs: the subgraph problem. Combin. Probab. Comput. **8** (1999) 131-159

[KZ] Kou, Z. B., Zhang, C. S.: Reply networks on a bulletin board system. Phys. Rev. E. **67** (2003) 036117

[New1] Newman, M. E. J.: The Structure of scientific collaboration networks. Proc. Nat. Acad. Sci. U. S. A. **98** (2001) 404

[New2] Newman, M. E. J.: Scientific collaboration networks. I. Network construction and fundamental results. Phys. Rev. E. **64** (2001) 6131

[New3] Newman, M. E. J.: Properties of highly clustered networks. Phys. Rev. E. **68** (2003) 026121

A Stochastic Viewpoint on the Generation of Spatiotemporal Datasets*

MoonBae Song, KwangJin Park, Ki-Sik Kong, and SangKeun Lee

Department of Computer Science and Engineering, Korea University,
5-1, Anam-dong, Seongbuk-Ku, Seoul 136-701, Korea
{mbsong, kjpark, kskong, yalphy}@disys.korea.ac.kr

Abstract. The issue of standardized generation scheme of spatiotemporal datasets is a research area of growing importance. In case of the lack of large real datasets, especially, benchmarking spatio-temporal database requires the generation of synthetic datasets simulating the real-word behavior of spatial objects that move and evolve over time. Recently, a few studies have been conducted on the generation of artificial datasets from a different point of view. For more realistic datasets, this paper proposes a novel framework, called *state-based movement framework* (SMF) to provide more generalized framework for both describing and generating the movement of complexly moving objects which simulate the movement of real-life objects. Based on Markov chain model, a well-known stochastic model, the proposed model classifies the whole trajectory of a moving object into a set of movement state. From some illustrative examples, we show that the proposed scheme is able to generate various realistic datasets with respect to the given input parameters.

1 Introduction

Movement is ubiquitous in physical world. Everything and every physical entities are on the move. For example, people, clouds, and airplanes are always moving around in a dynamic environment. Geographical objects (e.g., mountain, river, and continent) can also be understood as moving objects in the viewpoint of a very long period of time. The movement can exist even in a logical world. For instance, the stock price moves along a two-dimensional space (*time, price*). Understanding the movement is a fundamental issue for various research area such as robotics, computer vision, physical simulation, biology, spatio-temporal databases, and mobile wireless networks [1].

Generating spatio-tmporal datasets is a process of describing, representing, and generating the temporal behavior of spatial objects (so-called *evolving phenomena*). These objects are a huge number of real-life objects such as mobile

---

* This work was done as a part of Information and Communication Fundamental Technology Research Program, supported by Ministry of Information and Communication in Republic of Korea.

clients, soldiers in battlefield, a forest fire, a plane, and so on. The most important reason for this is that the real datasets of spatio-temporal patterns of real-life mobile users are very hard to obtain. The issue of standardized generation scheme of spatio-temporal datasets is a research area of growing importance. In case of the lack of large real datasets, especially, benchmarking spatio-temporal database requires the generation of synthetic datasets simulating the real-word behavior of spatial objects that move and evolve over time. Recently, a few studies have been conducted on the generation of artificial datasets from a different point of view.

There are several examples of spatiotemporal data generators such as Oporto [7], a network-based generator [3], City Simulator [4], GSTD [9], and G-TERD [10]. And these approaches categorized into the following three classes.

Random number generator-based. GSTD [9] and G-TERD [10] are based on a rich set of random distribution to compute the size, shape, and location of the data. The GSTD (generating spatio-temporal datasets) algorithm was proposed by Theodoridis *et al.* [9]. A generator of time-evolving regional data (G-TERD) was proposed in Tzouramanis et al. [10]. The basic concepts that determine the function of G-TERD are the structure of complex regional objects, their color, maximum speed, zoom and rotation angle per time slot, the influence of other moving or static objects on the speed and on the moving direction of an object, the position and movement of scene-observer, the statistical distribution of each changing factor and, finally, time.

Infrastructure-based traffic generator. In network-based moving objects generator proposed by Brinkhoff [3], the driving application is the field of traffic telematics. Important concepts of the generator are the maximum speed and the maximum edge capacity, the maximum speed of the object classes, the interaction between objects, etc. City Simulator of IBM [4] is a scalable, three-dimensional model city that enables creation of dynamic spatial data simulating the motion of up to 1 million people moving along streets, buildings, and between building floors in three dimensions.

Realistic scenario-based generator. The Oporto generator [7] uses a realistic application, the modeling of fishing ships. Ships are attracted by shoals of fish, while at the same time they are repulsed by storm areas. Fishes themselves are attracted by plankton areas.

Based on the concept of state transition between various movement states, this work is aimed to a flexible framework for an application-specific purpose. In the remainder of the paper, due to the notability and adequateness of the name of the well-known generator GSTD, we will denote the general process of the "generating spatio-temporal datasets" as $\mathcal{GSTD}$.

2 Observation: Movement Behavior of Real-Life Objects

In realistic movement, diverse movement patterns are mixed: slow moving objects, fast moving objects, and stationary objects. Even a single moving object has a different movement type over time. The "realistic" is a quite relative concept because of the difficulty of proving the realistic. No one can prove the reality of a mobility pattern in a formal manner. Everybody have just insisted on an *aspect* of real movements. This is also caused by the great diversity of real-life objects and its movements. Surely, this aspect serves as a basis for the research area of mobility modeling. Depending on the viewpoint on the real movements, the corresponding mobility model will be changed.

Basically, our scheme was motivated by the observation that the behavior of real-life moving objects can be interpreted as a set of movement component (repeatedly) such as linear, pause, and random. As we already mentioned, a great diversity of mobility patterns of real-life objects is quite natural. But, there are some specific repeated patterns in the movements. In our work, we will classify the whole trajectory of a user into 'pause', 'linear movement', and 'random movement' in the rough. In this paper, these movements are called *basic movement states*. And, of course, we have to consider the temporal pattern of movements as Markovian process.

There exists definitely a *pattern* (similar to the above mentioned "aspect"). For example, average speed, direction, destination, and so on. The spatio-temporal pattern in real-life applications can be summarized as following characteristics [3, 8, 9, 11]:

- A moving object is not always on the move. Rather, the most-time consuming part of our daylife is a static situation (e.g., in his/her home, office, meeting room for a long time). A mobile subscriber will mainly switch between two states: *stop* and *move* [8].
- The majority of objects in the real world do not move according to statistical parameters but, rather, move intentionally [8, 9].
- Moving objects belong to a class (e.g., pedestrian, vehicles, trains, ...). This class restricts the maximum speed of the object [3]. Different groups of moving objects exhibit different kinds of behavior [9].
- Motion has a random part and a regular part, and the regular part has a periodic pattern [11].

3 State-Based Movement Framework (SMF)

3.1 Basic Definitions

In this section, we propose the *state-based movement framework* (SMF) considering a complex mobility pattern as a set of simple movement states using a finite state Markov chain based on the classification discussed in Section 2.

Definition 1 (Movement State). *A movement state s is a 3-tuple form of* $(\bar{\mathbf{v}}, \sigma, \mathbf{\Phi}, \mathbf{t})$, *where* $\bar{\mathbf{v}} = (\bar{v}_0, \bar{v}_1, \ldots, \bar{v}_{d-1})$ *and* $\sigma = (\sigma_0, \sigma_1, \ldots, \sigma_{d-1})$ *are d-dimensional vector for mean and standard deviation of velocity for each dimension. And,* $\mathbf{\Phi} = (\phi_0, \phi_1, \ldots, \phi_{d-1})$. ϕ_i *is a movement generating function over* $(\bar{v}_i, \sigma_i)$ *which is either probabilistic or non-probabilistic. And, time function* $\mathbf{t}$ *is continuous or discrete, regular or irregular.* $\mathbf{\Sigma}$ *is a finite set of movement states.*

Definition 2 (State-based Movement Framework). *The state-based movement framework (SMF) describes a user mobility patterns using a finite state Markov Chain* $\{state_n\}$*, where* $state_n$ *denotes the movement state at step* n*,* $state_n \in \Sigma$*. And, the chain can be described completely by its* transition probability *as* $p_{ij} \equiv Pr\{state_{n+1} = j | state_n = i\}$ *for all* $i, j \in \Sigma$*. These probabilities can be grouped together into a* transition matrix *as* $\mathbf{P} \equiv (p_{ij})_{i,j \in \Sigma}$.

An important question is, as stated in [2], why such a general mobility model is not as popular as the restrictive models so abundant in the literature. The most important reason is that the generalized model has nothing to be assumed to start the analysis. Therefore, it is need to instantiate a useful instance for analytic approaches. We think that a common framework can be used to generate a realistic and expressive instance for various fields of research.

Definition 3 (Self-Transition Probability Vector). *The self-transition probability vector* $\tilde{\pi}$ *of a transition probability matrix is defined as* $\tilde{\pi} = (p_{ii})_{i \in \Sigma}$.

Definition 4 (Temporal Locality). *The* temporal locality *(or locality)* τ *is defined as*

$$\tau = \left(\prod_{i \in \Sigma} \left(\begin{cases} 1, & \text{if } p_{ii} = 0; \\ p_{ii}, & \text{otherwise.} \end{cases} \right) \right)^{1/|\Sigma|}. \tag{1}$$

3.2 The Instantiation of the SMF Model

As we mentioned before, the complex mobility patterns in the real world can be interpreted as a set of basic movement states. The most definite movement states are *pause* (P), *linear movement* (L), and *random movement* (R). These states $\{P, L, R\}$ are called basic states in the paper. In this section, we present two practical instances of the proposed SMF model based on the three states described above, the state space $\Sigma_0 = \{L, R\}$ and $\Sigma_1 = \{P, L, R\}$. It is need to describe a correlation between these probabilities $p_{ij}, i, j \in \Sigma$ because of the difficulty of finding a general rule for the matrix analysis. Let us define a measurement that estimates how much each state has an influence on the whole movement patterns in this simplified model.

Definition 5 (Dominancy). *The dominancy* η_s *of state* $s \in \Sigma$ *is defined as* $\eta_s = \frac{\sum_{i \in \Sigma} p_{is}}{\sum_{i,j \in \Sigma, j \neq s} p_{ij}}$. *As a special case of* dominancy, η_P, η_L, *and* η_R *are called* stationarity(ρ), linearity(ℓ), *and* randomness(γ) *respectively.*

For practical purpose, the above parameters is quite important to describe the various feature of mobility patterns. In the experiment we performed, a more realistic mobility patterns could produced with varying these parameters. These parameters will be used in the generation of spatio-temporal datasets.

3.3 Determining the State Transition Matrix

One of the most important things in the SMF model is to determine the transition matrix **P**. The matrix can be determined by the user profile, the spatiotemporal data mining process, or in an ad hoc manner. Determining the state transition matrix is a summarization process from the past user movements. Given a sequence of optimal states $\mathcal{O} = o_1 o_2 ... o_{|\mathcal{O}|}$, there are several variations depending on how to set every N_{ij}. $\hat{p}_{ij} = N_{ij} / \sum_{k \in \Sigma} N_{ik}$. Based on these assumptions, we propose three schemes to estimate the transition matrix: (1) **Markov**, (2) **Window**, and (3) **EWMA** (exponential weighted moving average).

1. **Markov:** The **Markov** scheme uses the number of state transition of first-order Markov chain. The total number of station transition on this account is $|\mathcal{O}| - 1$, e.g., $\sum_{i,j \in \Sigma} N_{ij} = |\mathcal{O}| - 1$.

$$N_{ij} = \sum_{k=1}^{|\mathcal{O}|-1} \left(\begin{cases} 1, & \text{if } o_k = i \text{ and } o_{k+1} = j \\ 0, & \text{otherwise.} \end{cases} \right), \quad (2)$$

where a sequence of optimal states $\mathcal{O} = o_1 o_2 ... o_{|\mathcal{O}|}$.

2. **Window:** In **Window** scheme, only at most W recent transitions are allowed to affect the estimated values. Thus,

$$N_{ij} = \sum_{k=|\mathcal{O}|-1-W}^{|\mathcal{O}|-1} \left(\begin{cases} 1, & \text{if } o_k = i \text{ and } o_{k+1} = j \\ 0, & \text{otherwise.} \end{cases} \right), \quad (3)$$

where W is window size for the computation of N_{ij}.

3. **EWMA** (exponential weighted moving average): The final approach that we consider is use of exponential weighted moving average (**EWMA**) estimators. The EWMA estimator is a low pass filter that has been used widely for estimation of various time series data (e.g. TCP round trip time). The **EWMA** scheme assigns exponentially decreasing weights to previous transition so that recent transition bear higher weights. With an adjustable parameter ω $(0 < \omega \leq 1)$, the most recent transition will receive a weight of ω and the next one a weight of ω^2, and so on.

$$N_{ij} = \sum_{k=1}^{|\mathcal{O}|-1} \omega^{|\mathcal{O}|-k-1} \left(\begin{cases} 1, & \text{if } o_k = i \text{ and } o_{k+1} = j \\ 0, & \text{otherwise.} \end{cases} \right), \quad (4)$$

where $0 < \omega \leq 1$.

By means of this statistical technique, the reasonable estimation of **P** can be estimated from the user's past movement. Obviously, it is possible to dynamically instantiate a SMF model corresponding to the realistic movement.

Example 1. *Let's consider a sequence of optimal states $\mathcal{O}$ for 51 time units (i.e, $|\mathcal{O}|=51$) $\mathcal{O}$ = PPPPRRLLLLRRLPLLPPRRLLLLRLLLLRPPRRRRRPLLLLRRPPRRLLL. Using **EWMA** scheme, we have*

$$N_{EWMA} = \begin{matrix} & P & L & R \\ P \\ L \\ R \end{matrix} \begin{pmatrix} 0.698 & 0.275 & 0.764 \\ 0.043 & 3.092 & 0.557 \\ 0.817 & 0.857 & 1.852 \end{pmatrix} \text{ and } \widehat{P}_{EWMA} = \begin{matrix} & P & L & R \\ P \\ L \\ R \end{matrix} \begin{pmatrix} 0.402 & 0.158 & 0.440 \\ 0.012 & 0.837 & 0.151 \\ 0.232 & 0.243 & 0.525 \end{pmatrix},$$

where $\omega = 0.9$. From the estimated matrix $\widehat{P}_{EWMA}$, we can easily calculate the mobility parameters such as: $\ell = 0.70313$, $\gamma = 0.59239$, $\rho = 0.27411$, $\tau = 0.56132$, and $\pi = (0.1275, 0.5724, 0.3001)$. □

3.4 Definition of Movement State from Real Movements

Based on the concept of *basic movement states* = {P, L, R}, we described the SMF model which provides a general framework for describing and generating the realistic moving behavior. In a real environment, however, a more specific definition of the movement state is needed to determine the movement state from the real movement-trace in the form of triples (x, y, t). These traces are usually called *trajectory* (see Figure 1(a)).

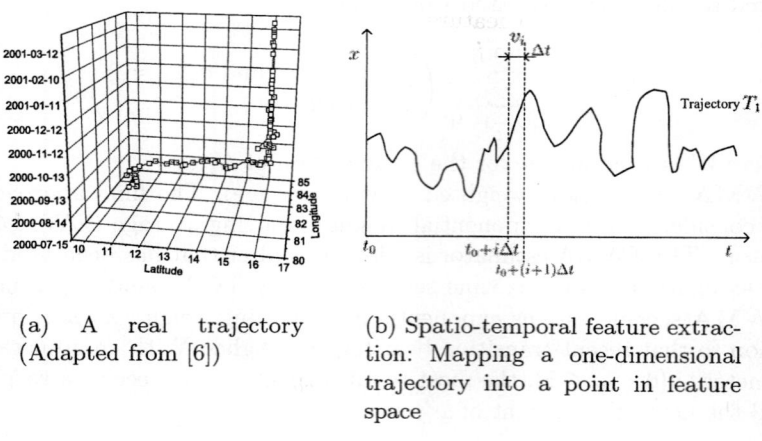

(a) A real trajectory (Adapted from [6])

(b) Spatio-temporal feature extraction: Mapping a one-dimensional trajectory into a point in feature space

Fig. 1. An example trajectory and feature extraction scheme for trajectory

Definition 6 (Trajectory). *A trajectory T is a kind of time series data* [1] *in the form of triples (x, y, t), where (x, y) is the position of the moving object at time t. $T = \langle (x_0, y_0, t_0), (x_1, y_1, t_1), \ldots, (x_{|T|-1}, y_{|T|-1}, t_{|T|-1}) \rangle$.*

In the viewpoint of trajectory, the movement state can be viewed as a variation of velocity vector and location in a given time interval. Then, the corresponding movement feature vector can be generated by a process so called

---

[1] The definition of time series is observations of a variable made over time.

feature extraction. A trajectory, by the process, can be mapped into a point in the feature space $(\bar{v}, \sigma_v)$. In this aspect, the definition of movement state is derived from the result of a clustering algorithm on a point dataset in the feature space $(\bar{v}, \sigma_v)$. Figure 1(b) shows a feature extraction process of a one-dimensional trajectory T_1. First, the trajectory is evenly sampled with a sampling interval Δt. Given a trajectory T, the location within the trajectory T at time t is denoted by $T(t)$. Therefore, the i-th velocity v_i can be computed by $v_i = T(t_0 + (i+1)\Delta t) - T(t_0 + i\Delta t)$ where t_0 is the initial timestamp when the trajectory T was stored in a location database. And V is a finite set of v_i for $0 \leq i \leq |V| - 1$. So, $V = \{v_0, v_1, ..., v_{|V|-1}\}$. Thus, $\bar{v}$ and σ_v are defined as follows.

$$\bar{v} = \frac{1}{|V|} \sum_{i=0}^{|V|-1} v_i, \text{ and } \sigma_v = \varepsilon \sqrt{\frac{1}{|V|-1} \sum_{i=0}^{|V|-1} (v_i - \bar{v})^2}, \qquad (5)$$

where ε is a space expansion parameter because of the difference of $\bar{v}$-axis and σ_v-axis.

After generating the feature vectors of each trajectory, a clustering algorithm is used to define the movement states. Clustering is the unsupervised classification of patterns (or feature vectors) into groups (clusters). In our work, k-Means clustering algorithm is used. The distance between two points (nothing but the trajectories in our work) in measured in the *Euclidean distance* $\mathcal{D}_2$ which is a special case $(p = 2)$ of the *Minkowski metric* $\mathcal{D}_p(\mathbf{x}_i, \mathbf{x}_j) = \left(\sum_{k=1}^{d}(x_{i,k} - x_{j,k})^p\right)^{1/p} = ||\mathbf{x}_i - \mathbf{x}_j||_p$, where d is the dimensionality of the data. The *Manhattan metric* is a special case where $p = 1$ (i.e., $\mathcal{D}_1(\mathbf{x}_i, \mathbf{x}_j) = \sum_{k=1}^{d}|x_{i,k} - x_{j,k}|$), while the *Supermum metric* or dominance metric has $p = \infty$ (i.e., $\mathcal{D}_\infty(\mathbf{x}, \mathbf{y}) = \lim_{p \to \infty} L_p(\mathbf{x}, \mathbf{y}) = \max_{k=1}^{d}|x_{i,k} - x_{j,k}|$).

In k-Means clustering algorithm, the value of k is assumed to be predefined. Unfortunately there is no general theoretical solution to find the optimal number of clusters k_{opt} for any given data set. A simple approach is to compare the results of multiple runs with different k classes and choose the best one according to a given criterion for $k \geq 2$.

3.5 The SMF-Based $\mathcal{GSTD}$ Algorithm

The $\mathcal{GSTD}$ process, that is to say, is the process of describing the temporal behavior of spatial objects. The previous approaches to $\mathcal{GSTD}$ is a quite application-specific and model-based approach. These perceptional differences in understanding the realistic movements have eliminated diversity the $\mathcal{GSTD}$ process. As a result, the realistic movements can hardly be generated by previous approaches. The uniqueness of this work is based on the generalization. By the definition of appropriate movement states and the determination of state transition matrix, we can easily generate trajectories of moving objects which are simulating the realistic movements.

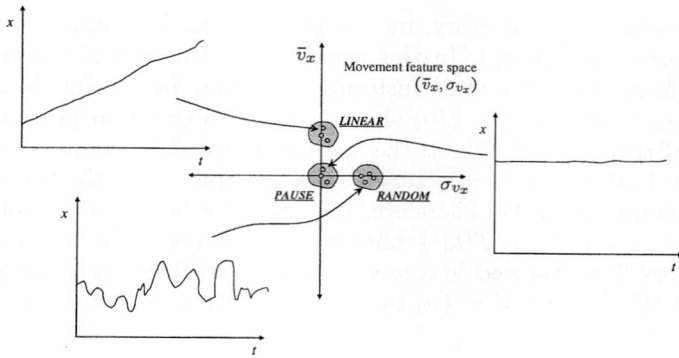

Fig. 2. Definition of movement states using *clustering* in the movement feature space $(\bar{v}_x, \sigma_{v_x})$

In SMF-based $\mathcal{GSTD}$ process, already mentioned in previous section, two basic procedures are need to generate an instance for the SMF model.

- *Define the movement states*: Based on Definition 2, appropriate movement states are need to be defined. Σ is a set of movement states.
- *Determine the state transition matrix*: For a probabilistic mixing of movement states, the state transition matrix **P** has to be determined. The **P** is $|\Sigma| \times |\Sigma|$ matrix. For this purpose, the above mentioned dominancy and temporal locality are used.

Define the movement states: Basically, there are two distinct approaches for defining the movement states: *realistic* and *synthetic* way. First, as described in Section 3.4, the movement states can be extracted from real trajectories (*realistic* way). Otherwise, we may consider variations which are based on the basic movement states (P, L, R) (*synthetic* way).

Determine the transition matrix: Determining the transition matrix also has two distinct approaches: *realistic* and *synthetic* way. First, as described in Section 3.3, the state transition matrix can be formulated as a statistical inference problem by performing maximum-likelihood estimation for the optimal state sequence $\mathcal{O}$ (*realistic* way). Otherwise, the matrix can be determined by using the mobility parameters of dominancy and temporal locality (*synthetic* way).

The former is a more realistic model for describing the realistic movements, while the latter is suitable for performance evaluation in terms of the impacts of specific mobility parameters. The former was already mentioned in Section 3.3, so this section will describe the latter. First of all, we only consider the states L and R among the basic movement states. Then, we have

$$\mathbf{T}(\tau) = \begin{matrix} \\ \mathsf{L} \\ \mathsf{R} \end{matrix} \begin{matrix} \mathsf{L} & \mathsf{R} \\ \begin{pmatrix} \tau & 1-\tau \\ 1-\tau & \tau \end{pmatrix} \end{matrix} \text{ and } \mathbf{L}(\ell) = \begin{matrix} \\ \mathsf{L} \\ \mathsf{R} \end{matrix} \begin{matrix} \mathsf{L} & \mathsf{R} \\ \begin{pmatrix} \frac{\ell}{\ell+1} & \frac{1}{\ell+1} \\ \frac{\ell}{\ell+1} & \frac{1}{\ell+1} \end{pmatrix} \end{matrix}, \qquad (6)$$

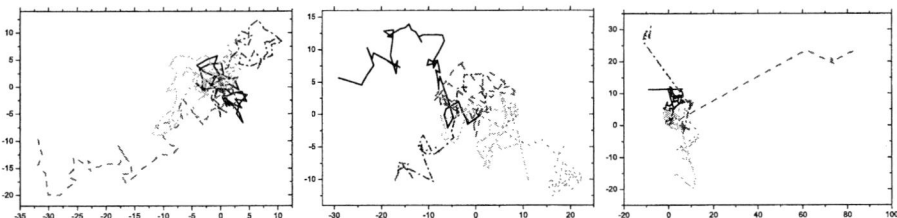

Fig. 3. Five synthesized trajectories from the matrix $\mathbf{T}(\tau)$ based on $\tau = 0, 0.5, 0.95$

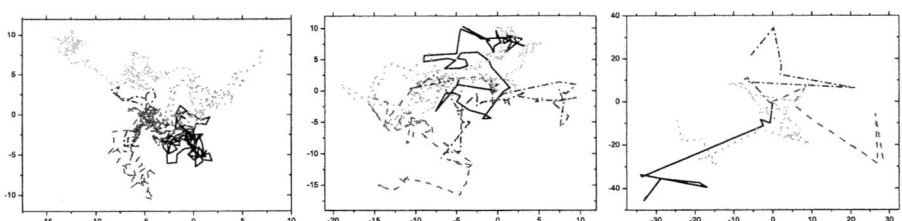

Fig. 4. Five synthesized trajectories from the matrix $\mathbf{L}(\ell)$ based on $\ell = 0, 1.0, 10$

where $0 \leq \tau \leq 1$ and $0 \leq \ell \leq \infty$. The matrices $\mathbf{T}(\tau)$ and $\mathbf{L}(\ell)$ are for describing the effect in term of the variations of temporal locality τ and linearity ℓ respectively. The matrix $\mathbf{T}(\tau)$, namely, is a variation of $\mathbf{P}_0$, which has half of linearity and half of randomness on a probabilistic viewpoint. The matrix $\mathbf{L}(\ell)$, namely, describes how much the linearity ℓ has an influence on the whole movement patterns in a probabilistic viewpoint.

In contrary to $\mathbf{T}(\tau)$ and $\mathbf{L}(\ell)$, the matrix $\mathbf{S}(\rho)$ considers the stationary state P and this is for describing the effect in term of the variations of stationarity ρ.

$$\mathbf{S}(\rho) = \begin{array}{c} \\ \mathsf{P} \\ \mathsf{L} \\ \mathsf{R} \end{array} \begin{pmatrix} \mathsf{P} & \mathsf{L} & \mathsf{R} \\ \frac{\rho}{\rho+1} & \frac{1/2}{\rho+1} & \frac{1/2}{\rho+1} \\ \frac{\rho}{\rho+1} & \frac{1/2}{\rho+1} & \frac{1/2}{\rho+1} \\ \frac{\rho}{\rho+1} & \frac{1/2}{\rho+1} & \frac{1/2}{\rho+1} \end{pmatrix}, \tag{7}$$

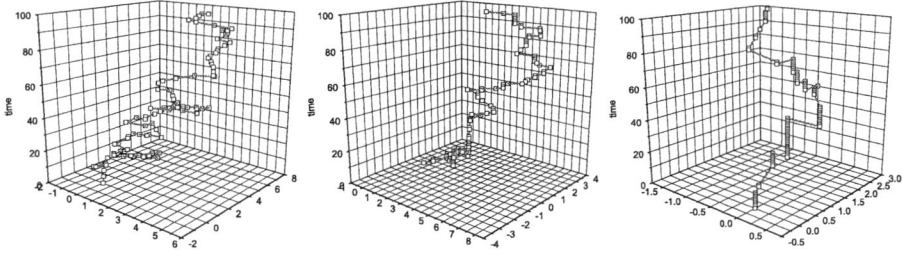

Fig. 5. A synthesized trajectory from the matrix $\mathbf{S}(\rho)$ based on $\rho = 0, 1.0, 10$

where $0 \leq \rho \leq \infty$. And the temporal locality is $\tau = \sqrt[3]{\frac{\rho}{4(\rho+1)^3}}$. The matrix $\mathbf{S}(\rho)$, namely, describes how much the stationarity ρ has an influence on the whole movement patterns in a probabilistic viewpoint.

Due to the lack of space, we do not describe the generation algorithms in details. Figures 3, 4, and 5 are the visualization of examples datasets generated from the state transition matrix $\mathbf{T}$, $\mathbf{L}$, and $\mathbf{S}$ described.

4 Concluding Remarks and Future Work

In this paper, we considered the problem of modeling, describing, and simulating the realistic movements of real-life objects. We argue that a generalized movement framework for spatio-temporal databases is a crucial for various research area such as mobile computing, robotics, computer vision, and so on. Future works of the SMF model may be summarized as follows.

- A more elegant feature extraction technique for the real trajectories.
- Support for regional objects such as earthquake, typhoon, wild fire and a shoal of fish.
- Support for clustered movements and infrastructure(network)-based movement in SMF.
- Development of portable software API for the SMF-based $\mathcal{GSTD}$ process.

References

1. Pankaj K. Agarwal, et al., "Algorithmic Issues in Modeling Motion", *ACM Computing Surveys*, 34(4), December 2002.
2. Amiya Bhattacharya and Sajal K. Das. "LeZi-Update: An Information-Theoretic Approach to Track Mobile Users in PCS Networks," In *Proc. of MobiCom*, 1999.
3. T. Brinkhoff. "A Framework for Generating Network-Based Moving Objects," *GeoInformatica*, 2002.
4. IBM alphaWorks: City Simulator, *alphaWorks Emerging Technologies*, November 2001. http://www.alphaworks.ibm.com/tech/citysimulator.
5. D.L. Minh. *Applied Probability Models*, Brooks/Cole, 2001.
6. Real datasets from Caribbean Conservation Corporation & Sea Turtle Survival League. http://www.cccturtle.org/sat3.htm
7. J.-M Saglio and J. Moreira. "Oporto: A Realistic Scenario Generator for Moving Objects," In *Proc. of DEXA Workshop*, 1999.
8. Y.-C. Tseng, L.-W. Chen, M.-H. Yang, and J.-J. Wu. "A Stop-or-Move mobility model for PCS networks and its location-tracking strategies," *Computer Communications*, 26:1288–1301, 2003.
9. Y. Theodoridis, J.R.O. Silva, and M.A. Nascimento. "On the Generation of Spatiotemporal Datasets," In *Proc. of SSD*, 1999.
10. T. Tzouramanis, M. Vassilakopoulos and Y. Manolopoulos. "On the Generation of Time-Evolving Regional Data," *GeoInformatica*, Vol.6, No.3, pp. 207-231, 2002.
11. Ouri Wolfson. "Moving Objects Information Management: The Database Challenge," *Proc. of NGITS*, 2002.

A Formal Approach to the Design of Distributed Data Warehouses

Jane Zhao

Massey University, Department of Information Systems &
Information Science Research Centre,
Private Bag 11 222, Palmerston North, New Zealand
j.zhao@massey.ac.nz

Abstract. Data warehouses provide data for on-line analytical processing (OLAP) systems, which deal with analytical tasks in businesses. As these tasks do not depend on the latest updates by transactions, the input from operational databases is separated from the outputs to dialogue interfaces for OLAP. In this paper a layered formal specification for data warehouses and OLAP systems using Abstract State Machines (ASMs) is presented. The approach explicitly exploits the fundamental idea of separating input from operational databases and output to OLAP systems. Then it will be shown how this specification can be extended to distributed data warehouses.

1 Introduction

Data Warehouses are data-intensive systems that are used for analytical tasks in businesses such as analysing sales/profits statistics, cost/benefit relation statistics, customer preferences statistics, etc. The term used for these tasks is "on-line analytical processing" (OLAP) in order to distinguish them from operational data-intensive systems, for which the term "on-line transaction processing" (OLTP) has become common.

The idea of a data warehouse [2, 3] is to extract data from operational databases and to store them separately. The justification for this approach is that OLAP largely deals with condensed data, thus does not depend on the latest updates by transactions. Furthermore, OLAP requires only read-access to the data, so the separation of the data for OLAP from OLTP allows time-consuming transaction management to be dispensed with.

The main idea of data warehouses implies a separation of input from operational databases and output to views that contain the data for particular OLAP tasks. In addition, the work in [4] shows that each data mart together with the OLAP functions working on it defines a so-called "dialogue object". According to this view we obtain a data warehouse architecture as shown in Figure 1, a three-tier model, consisting of operational database, the data warehouse, and the data marts with OLAP functions.

Usually the design of data warehouses and OLAP systems uses rather informal methods as e.g. in [2, 9]. Such approaches have been criticised in [1] for

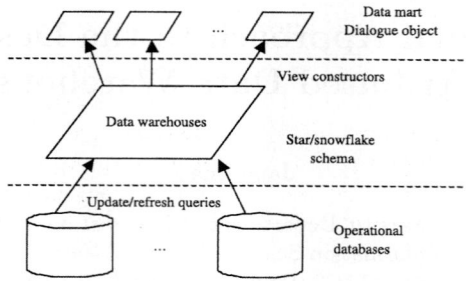

Fig. 1. The general architecture of a data warehouse and OLAP

their lack of solid foundations and semantics and the inappropriateness of layers of abstraction. Our work aims at showing that the use of the ASM method [1] solves these problems, as Abstract State Machines (ASMs) provide a flexible high-level language with a clear mathematical semantics.

In [11] we started developing an ASM ground model for data warehouses and OLAP system based on the fundamental idea of separating input from operational databases from output to OLAP systems. According to this idea we obtain a model of three interconnected ASMs, one for the operational database(s), one for the data warehouse, and one for the OLAP system.

This idea was extended in in [8] focussing on cost-efficient distribution of data warehouses. This last idea exploits fragmentation techniques from [6] and the recombination of fragments, but still remains on rather informal grounds. In this paper we start from the layered ASM specification for data warehouses and OLAP systems and show, how this specification can be extended to distributed data warehouses.

We start with a presentation of the three-tiered data warehouse ground model in Section 2, accompanied by a simple concrete example. The we approach the distribution design in Section 3. We conclude with a short summary.

2 An ASM Ground Model for Data Warehouses

In order to develop an ASM ground model we consider again the three-tier model from Figure 1. At the bottom tier, we have the operational database model, which has the control of the updates to data warehouse. The updates are abstracted as data extraction from the operational database to maintain the freshness of the data warehouse. In the middle tier, we have the data warehouse which defines the data structure from a business point of view. It can either be modelled in multidimensional database[9], or in relational as in our case, as star or snowflake schemata [3]. At the top tier, we have the data marts, which are constructed out of dialogue objects with OLAP operations. Based on this three-tier architecture, we end up with three linked ASMs, the DB-ASM, the DW-ASM, and the OLAP-ASM in our ground model.

2.1 The Operational Database ASM

Since we are only interested in functions for the data warehouse, the operational database model is abstracted to have the data extraction functions only. Basically the data warehouse presents a consolidated view of the operational database. The consolidation is achieved through aggregation of the transaction data, which will be realised by the data extraction rules defined in DB-ASM. As data source types other than relational ones can be wrapped as relational, we assume for simplicity that all data sources are relational, thus the signature in DB-ASM would just describe the relation schemata. As relations can be seen as boolean-valued functions, each relation with n attributes in one of the operational databases will define an n-ary function in the signature of DB-ASM. The rules on DB-ASM correspond to the extraction of data for each of the relations in the data warehouse schema. DB-ASM will export these rules to DW-ASM for refreshing the data in the data warehouse and import the data warehouse signatures from DW-ASM, which are needed when data extraction is initiated from DB-ASM.

Example 1. Let us give a more concrete model by looking at an example taken from [4–p.358]. In this case we have a single operational database with five relation schemata as illustrated in the left hand HERM diagram in Figure 2. Since the difference between the data extraction rules is not significant, we will only present one rule definition as an example in the following DB-ASM model.

```
ASM DB-ASM
IMPORT DW-ASM(Shop, Product, Customer, Time, Purchase)
EXPORT extract
SIGNATURE
   Store(2)(controlled),
   Part(3) (controlled),
```
$\quad\quad\text{Customer}_{db}(3)$ (controlled),
```
   Buys(4) (controlled),
   Offer(5) (controlled),
   Schm_sel(1) (monitored)
BODY
```
$\quad\quad main =$ **if** $\text{selected}(extract)$ **then** $extract(Schm\_sel)$ **endif**
$\quad\quad extract(Schm\_sel) =$
$\quad\quad\quad$ **case** Schm_sel **of**
$\quad\quad\quad\quad$ Purchase: $extract\_purchase$
$\quad\quad\quad\quad$ Customer: $extract\_customer$
$\quad\quad\quad\quad$ Shop: $extract\_shop$
$\quad\quad\quad\quad$ Product: $extract\_product$
$\quad\quad\quad\quad$ Time: $extract\_time$
$\quad\quad\quad$ **endcase**
$\quad\quad extract\_purchase =$
$\quad\quad\quad$ **forall** i, p, s, t, p', c **with** $\exists q.\text{Buys}(i,p,q,t) \neq \bot \wedge$
$\quad\quad\quad\quad \exists n, a.\text{Customer}_{db}(i,n,a) \neq \bot \wedge \exists k, d.\text{Part}(p,k,d) \neq \bot \wedge$

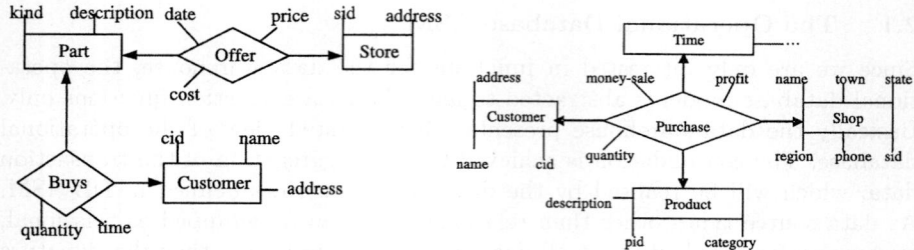

Fig. 2. The operational database and data warehouse schemata

$\exists a'.\text{Store}(s, a') \neq \bot \land \exists d.(\text{Offer}(p, s, p', c, d) \neq \bot \land date(t) = d)$
  do let $Q = sum(q \mid \text{Buys}(i, p, q, t) \neq \bot), S = Q*p', P = Q*(p'-c)$
  in $\text{Purchase}(i, p, s, t, Q, S, P) := 1$
  enddo
$extract\_customer = \ldots$

2.2 The Data Warehouse ASM

For the data warehouse ASM we follow the same line of abstraction as for the model of operational databases, i.e. the signature for DW-ASM will contain functions that correspond to the relation schemata used in the data warehouse star (or snowflake) schema. DW-ASM will import the data extraction rules from DB-ASM and the signatures of the OLAP views from OLAP-ASM. On the other hand it will export signaturs of star schema to DB-ASM and view creation rules to OLAP-ASM. These rules will be defined in the same way as the rules for the extraction operations in DB-ASM. As the basic function of the data warehouse is to maintain the data in both data warehouse and OLAP, we define two more rules, *refresh_schm*, and *refresh_view*. The latter one is required when views are materialised. We also need two monitored functions, S and V. The former contains the name of the selected schema to be refreshed, the latter has the name of the selected view to be maintained.

Note that DW-ASM does not look significantly different from DB-ASM. The reason for this is that both ASMs specify simple relational databases and view creation operations on them. Of course, we would like to have efficient refresh-operations. In particular, we would reduce the data warehouse updates to incremental changes. The specification of corresponding rules will be left as a refinement task.

Example 2. Let us continue the grocery store example using a simple star schema for the data warehouse as illustrated by the right-hand side HERM diagram in Figure 2. According to this we must specify five simple data extraction rules. In addition, we will only present one sample view creation rule.

  ASM DW-ASM
  IMPORT DB-ASM(*extract*), OLAP-ASM(V_sales)

EXPORT *create_view*
SIGNATURE Schm_sel(1) (monitored), View_sel(1) (monitored),
  Shop(6) (controlled), Product(3) (controlled), Customer(3) (controlled),
  Time(4) (controlled), Purchase(7) (controlled)
BODY
  main =
    if selected(refresh-schm) then *extract(Schm_sel)*
    elsif selected(refresh-view) then *refresh_view(View_sel)*
    endif
  refresh_view(View_sel) = case View_sel of Sales: *refresh_V_sales* endcase
  create_view(View_sel) = case View_sel of Sales: *create_V_sales* endcase
  create_V_sales =
    forall s, r, st, m, q, y with
    $\exists n, t, ph.\text{Shop}(s, n, t, r, st, ph) \neq \bot \wedge \exists \ldots .\text{Time}(\ldots, m, q, y, \ldots) \neq \bot$
    do let $Q = sum(q' \mid \exists c, p, t, s', p'. \text{Purchase}(c, p, s, t, q', s', p') \neq \bot$
         $\wedge \text{month}(t) = m \wedge \text{quarter}(t) = q \wedge \text{year}(t) = y)$,
       $S = sum(s' \mid \exists c, p, t, q', p'. \text{Purchase}(c, p, s, t, q', s', p') \neq \bot$
         $\wedge \text{month}(t) = m \wedge \text{quarter}(t) = q \wedge \text{year}(t) = y)$
    in V_sales$(s, r, st, m, q, y, Q, S) := 1$
    enddo
  refresh_V_sales = *create_V_sales*

2.3 The OLAP ASM

The top-level ASM dealing with OLAP is a bit more complicated, as it realises the idea of using dialogue objects for this purposes. The general idea from [7] is that each user has a collection of open dialogue objects, i.e. data marts for our purposes here. At any time we may get new users, or the users may create new dialogue objects without closing the opened ones, or they may close some of the dialogue objects, or quit when they finish their work with the system.

Thus, we have a shared function req-queue to keep all the requests from the users of the OLAP system. If the user logs onto the OLAP system, we keep his user id *usr* to a controlled function user, with user(*usr*) $\neq \bot$. Similarly, we use a controlled function datamart with datamart(*dm*) $\neq \bot$ iff *dm* is the identifier of a data mart in the system. In addition, we need another controlled function owner with owner(*dm*) = *u* indicating that user *u* owns the data mart *dm*.

Then part of the functionality of OLAP-ASM deals with adding and removing users and data marts. In particular, when a user leaves the system, all data marts owned by him must be removed as well.

The major functionality, however, deals with performing OLAP operations on existing data marts or opening new data marts. In the latter case we use the imported rules, create_view from DW-ASM. Besides, we have to create a new identifier for the data mart and associate an owner to it.

If the user from a dialogue object issues an operation other than quit, i.e. the user does not want to leave the system, or close, i.e. the user does not want to finish work on the current data mart, or open, i.e. no new data mart is to be created, then we may request to receive additional input, from the user before the selected operation can be executed. For each user requests, we use five more monitored functions in, usr, dm, view and op.

Example 3. The OLAP ASM for the grocery store application looks as follows:

    ASM OLAP-ASM
    IMPORT DW-ASM(*create_view*)
    EXPORT V_sales
    SIGNATURE
      V_sales(8) (controlled), req_queue(1) (shared),
      user(1) (controlled), datamart(1) (controlled), owner(1) (controlled),
      in(1) (monitored), dm(1) (monitored), usr(1) (monitored),
      op(1) (monitored), view(1) (monitored),
      the_view(1) (controlled), the_arity(1) (monitored)
    BODY
      main =
        forall req : req_queue
        do if user($req.usr$) = $\bot \land req.op$ = login
        then user($req.usr$) := 1
        elsif user($req.usr$) = 1
        then case req.op **of** quit: *quit* close: *close* open: *open*
        else *OLAP_process* **endcase**
        req_queue := req_queue \ req **enddo**
      quit =
        forall dm' **with** datamart(dm') $\neq \bot \land$ owner(dm') = usr
          do datamart (dm') := $\bot$; the_view(dm') := $\bot$ **enddo** ;
        user(req.usr) := $\bot$
      close =
      owner(req.dm) := $\bot$; datamart(req.dm) := $\bot$; the_view(req.dm) := $\bot$
      open =
       choose dm' **with** datamart(dm') = $\bot$
       do datamart(dm') := 1 **enddo**
       seq create_view(req.view)
       seq the_view(dm') :=req.view; r :=the_arity(req.view)
       seq forall $x_1, \ldots, x_r$ **with** req.view($x_1, \ldots, x_r$) $\neq \bot$
        do $dm'(x_1, \ldots, x_r)$:= 1 **enddo**
      OLAP_process = *get(request.in)*;
        OLAP_op(req.usr,req.dm,req.op,req.in)

3   Distribution Design

The ASM method assumes that we first set up a ground model as presented in the previous section. In particular, we have assumed separate ASMs for the database, the data warehouse and the OLAP tier. Each of these ASMs uses

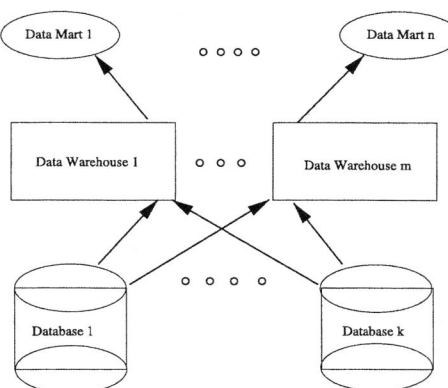

Fig. 3. Distributed Data Warehouse Architecture

separate *controlled functions* to model states of the system by logical structures and *rules* expressing transitions between these states. The ASMs are then linked together via queries that are expressed by these transitions.

3.1 The Approach to Distribution Design in General

In order to develop a distributed model of a data warehouse we proceed as follows:

1. Replicate the data warehouse and the OLAP ASMs: For each node in the network assume the same copy of the data warehouse ASM and the OLAP ASM. This is illustrated in Figure 3.
2. Remove controlled functions and rules in local OLAP ASMs: If the needed OLAP functionality is different at different network nodes, then this step will simply reduce the corresponding OLAP ASM.
3. Fragment controlled functions in local data warehouse ASMs: This will reorganise and reduce a local data warehouse ASM, if the corresponding OLAP ASM does not need all of the replicated data warehouse. The refresh rules are then adapted accordingly.
4. Recombine fragments in local data warehouse ASMs: This rules will reorganise a local data warehouse ASM according to query cost considerations. The refresh rules are then adapted accordingly. A cost model and pragmatics for choosing which fragments to recombine was presented in [5] and will not be repeated here.

3.2 Details of the Distribution Approach

Replicating the data warehouse and ASM machines has no impact to the machines as such. However, it replaces DW-ASM and OLAP-ASM by several identical machines OLAP-ASM$_i$ and DW-ASM$_i$. Also, removing functions and rules that are not used by a site does not cause any problems. We simply simplify OLAP-ASM$_i$ and DW-ASM$_i$ for some i.

For instance, when a replicated OLAP-ASM at node i has no interest in the sales in north region, then the view total_sales_north can be set to undefined. Consequently, when total_sales_north is set to undefined, the corresponding view creation should be removed as well.

For fragmentation we distinguish between horizontal and vertical fragmentation [5]. If we fragment a controlled function S_i with arity ar_i horizontally by using some selection predicate ψ, we replace it by new controlled functions S_{i1} and S_{i2}, both keeping the arity ar_i subject to the following condition:

$$S_{i1}(x_1,\ldots,x_{ar_i}) \neq \bot \Leftrightarrow S_i(x_1,\ldots,x_{ar_i}) \neq \bot \wedge \psi(x_1,\ldots,x_{ar_i}) \text{ and}$$
$$S_{i2}(x_1,\ldots,x_{ar_i}) \neq \bot \Leftrightarrow S_i(x_1,\ldots,x_{ar_i}) \neq \bot \wedge \neg\psi(x_1,\ldots,x_{ar_i}).$$

Example 4. Assume ψ is defined as: net_price_product ≥ 1000. The we replace purchase of arity 7 by purchase_exp and puchase_nonexp, both having arity 7, provided the following condition is satisfied:

$$\text{p\_exp}(c,s,p,t,Q,S,P) \neq \bot \Leftrightarrow \text{purchase}(c,s,p,t,Q,S,P) \neq \bot \wedge S/Q \geq 1000.$$
$$\text{p\_nexp}(c,s,p,t,Q,S,P) \neq \bot \Leftrightarrow \text{purchase}(c,s,p,t,Q,S,P) \neq \bot \wedge S/Q < 1000$$

where c is customer id, s is shop id, p is product id, t is time, Q is quantity, S is money sale, and P is profit.

If we use a horizontal fragmentation, then we have to add refresh rules for the new functions S_{i1} and S_{i2}:

extract_S_{i1} :[extract_S_{i1} = extract_S_i;
    forall $x_1,\ldots,x_{ar_i}$ with $S_i(x_1,\ldots,x_{ar_i}) = 1 \wedge \psi(x_1,\ldots,x_{ar_i}) = 1$
    do $S_{i1}(x_1,\ldots,x_{ar_i}) := 1$ enddo]

and

extract_S_{i2} :[extract_S_{i2} = extract_S_i;
    forall $x_1,\ldots,x_{ar_i}$ with $S_i(x_1,\ldots,x_{ar_i}) = 1 \wedge \psi(x_1,\ldots,x_{ar_i}) = \bot$
    do $S_{i2}(x_1,\ldots,x_{ar_i}) := 1$ enddo]

Example 5. Continuing the previous example we obtain two new refresh rules as follows:

extract_p_exp :[extract_p_exp = extract_purchase;
    forall c,s,p,t,Q,S,P with purchase$(c,s,p,t,Q,S,P) = 1 \wedge S/Q \geq 1000$
    do p_exp$(c,s,p,t,Q,S,P) := 1$ enddo]

and

extract_p_nexp :[extract_p_nexp = extract_purchase;
    forall c,s,p,t,Q,S,P with purchase$(c,s,p,t,Q,S,P) = 1 \wedge S/Q < 1000$
    do p_nexp$(c,s,p,t,Q,S,P) := 1$ enddo]

Rule 1. Fragment controlled function vertically by σ:

$$\frac{\text{DW-ASM} \triangleright S_i(ar_i)}{\text{DW-ASM}' \triangleright S_{i1}(ar_{i1}), \ldots, S_{in}(ar_{in})} \varphi$$

where φ is defined by:

$$S_{ij}(x_1, \ldots, x_{ar_{ij}}) \neq \bot \Leftrightarrow \exists y_1, \ldots, y_{ar_i}.S_i(y_1, \ldots, y_{ar_i}) \wedge \bigwedge_{1 \leq p \leq ar_{ij}} x_p = y_{\sigma_j(p)}$$

and $\bigcup_{j=1}^{n} \{\sigma_j(1), \ldots, \sigma_j(ar_{ij})\} = \{1, \ldots, ar_i\}$.

Rule 2. Add refresh rules according to vertical fragmentation σ:

$$\overline{\begin{array}{l}\text{DB-ASM}' \triangleright \text{extract\_}S_{ij} : [\text{ extract\_}S_{ij} = \text{extract\_}S_i;\\ \quad \texttt{forall } x_1, \ldots, x_{ar_i} \texttt{with } S_i(x_1, \ldots, x_{ar_i}) = 1\\ \quad \texttt{do } S_{ij}(x_{\sigma_j(1)}, \ldots, x_{\sigma_j(ar_{ij})}) := 1 \texttt{ enddo}]\end{array}}$$

Rule 3. Recombine fragments according to query cost as introduced in [5,8]:

$$\frac{\text{DW-ASM} \triangleright S_{i1}(ar_{i1}), S_{i2}(ar_{i2})}{\text{DW-ASM}' \triangleright S_{i1} \oplus S_{i2}(ar_{new})}$$

The combined fragment $S_{i1} \oplus S_{i2}$ can be obtained from S_i by using selection with $\varphi \vee \psi$ followed by projection using $\sigma + \tau$, which is defined by

$$(\sigma + \tau)(j) = \begin{cases} \sigma(j) & \text{for } j = 1, \ldots, k \\ \tau(j - x) & \text{else} \end{cases}$$

assuming σ is defined on $\{1, \ldots, k\}$ and τ on $\{1, \ldots, \ell\}$. Alternatively, we may re-combine S_{i1} and S_{i2} by an outer-join, which would produce a result different from $S_{i1} \oplus S_{i2}$, in which irrelevant values are replaced by $\bot$. This is described in the following rule of adapting refresh rule.

Rule 4. Adapt refresh rule according to fragment recombination:

$$\overline{\begin{array}{l}\text{DB-ASM}' \triangleright \text{extract\_}S_{i1} \oplus S_{i2} :\\ [\text{ extract\_}S_{i1} \oplus S_{i2} = \text{extract\_}S_{i1}; \text{extract\_}S_{i2};\\ \quad \texttt{forall } x_1, \ldots, x_{ar_{i1}}, y_1, \ldots, y_{ar_{i2}}\\ \quad \texttt{with } S_{i1}(x_1, \ldots, x_{ar_{i1}}) = 1 \vee S_{i2}(y_1, \ldots, y_{ar_{i2}}) = 1\\ \quad \texttt{do if } \bigwedge_{1 \leq j \leq m} x_{ar_{i1}-m+j} = y_j\\ \quad\quad \texttt{then } x(x_1, \ldots, x_{ar_{i1}-m}, y_1, \ldots, y_{ar_{i2}}) := 1\\ \quad\quad \texttt{else } x(x_1, \ldots, x_{ar_{i1}}, \bot, \ldots, \bot) := 1; x(\bot, \ldots, \bot, y_1, \ldots, y_{ar_{i2}}) := 1\\ \quad\quad \texttt{endif}\\ \quad \texttt{enddo}]\end{array}}$$

4 Conclusion

In this paper we presented a continuation of the work in [11] dealing with the use of Abstract State Machines (ASMs) for the design of data warehouses and OLAP systems. Based on the general idea of data warehouses separating input from operational databases from output to dialogue-based on-line analytical processing (OLAP) systems, we can model such systems by three interleaved high-level ASMs. In this article we extended this specification to capture distributed data warehouses.

The approach covers requirements at a very high level of abstraction, but nevertheless provides the rigorous mathematical semantics of ASMs from the very beginning of systems development. There is no switch in terminology during the design, which will ease the validation of requirements, and enable even formal verification when this becomes suitable.

The work in [10] contains an intensive overview of our complete ASM-based formal approach t5o the design of data warehouses and OLAP systems including a first version of a refinement calculus. In our future research we will extend the calculus.

References

1. BÖRGER, E., AND STÄRK, R. *Abstract State Machines.* Springer-Verlag, Berlin Heidelberg New York, 2003.
2. INMON, W. *Building the Data Warehouse.* Wiley & Sons, New York, 1996.
3. KIMBALL, R. *The Data Warehouse Toolkit.* John Wiley & Sons, 1996.
4. LEWERENZ, J., SCHEWE, K.-D., AND THALHEIM, B. Modelling data warehouses and OLAP applications using dialogue objects. In *Conceptual Modeling – ER'99* (1999), J. Akoka, M. Bouzeghoub, I. Comyn-Wattiau, and E. Métais, Eds., vol. 1728 of *LNCS*, Springer-Verlag, pp. 354–368.
5. MA, H., SCHEWE, K.-D., AND ZHAO, J. Cost optimisation for distributed data warehouses. In *Proc. HICSS 2005* (2005).
6. ÖZSU, T., AND VALDURIEZ, P. *Principles of Distributed Database Systems.* Prentice-Hall, 1999.
7. SCHEWE, K.-D., AND SCHEWE, B. Integrating database and dialogue design. *Knowledge and Information Systems 2*, 1 (2000), 1–32.
8. SCHEWE, K.-D., AND ZHAO, J. Balancing redundancy and query costs in distributed data warehouses – an approach based on abstract state machines. In *Conceptual Modelling 2005 – Second Asia-Pacific Conference on Conceptual Modelling*, S. Hartmann and M. Stumptner, Eds., vol. 43 of *CRPIT*. Australian Computer Society, 2005, pp. 97–105.
9. THOMSON, E. *OLAP Solutions: Building Multidimensional Information Systems.* John Wiley & Sons, New York, 2002.
10. ZHAO, J., AND MA, H. ASM-based design of data warehouses and on-line analytical processing systems. Tech. Rep. 10/2004, Massey University, Department of Information Systems, 2004. submitted for publication.
11. ZHAO, J., AND SCHEWE, K.-D. Using abstract state machines for distributed data warehouse design. In *Conceptual Modelling 2004 – First Asia-Pacific Conference on Conceptual Modelling* (Dunedin, New Zealand, 2004), S. Hartmann and J. Roddick, Eds., vol. 31 of *CRPIT*, Australian Computer Society, pp. 49–58.

A Mathematical Model for Genetic Regulation of the Lactose Operon

Tianhai Tian and Kevin Burrage

Advanced Computational Modelling Centre,
University of Queensland,
Brisbane, Queensland 4072, Australia
{tian, kb}@maths.uq.edu.au

Abstract. In this paper we construct a mathematical model for the genetic regulatory network of the lactose operon. This mathematical model contains transcription and translation of the lactose permease (LacY) and a reporter gene GFP. The probability of transcription of LacY is determined by 14 binding states out of all 50 possible binding states of the lactose operon based on the quasi-steady-state assumption for the binding reactions, while we calculate the probability of transcription for the reporter gene GFP based on 5 binding states out of 19 possible binding states because the binding site O_2 is missing for this reporter gene. We have tested different mechanisms for the transport of thiomethylgalactoside (TMG) and the effect of different Hill coefficients on the simulated LacY expression levels. Using this mathematical model we have realized one of the experimental results with different LacY concentrations, which are induced by different concentrations of TMG.

1 Introduction

The ultimate goal of molecular cell biology is to understand the physiology of living cells in terms of information that is encoded in the genome of cell. After the completion of the sequencing of the human genome, a great challenge in understanding biological complexity is to reconstruct the regulatory networks governing the dynamics of cellular processes. Much of the complexity lies in regulatory networks that couple DNA replication, transcription, translation, environmental stimuli, cell metabolism, cell division etc. Dissecting the complexities of regulatory networks is essential for understanding regulatory functions of biological pathways, such as transcriptional pathways, metabolic pathways and signal transduction pathways. During the past two decades, substantial progress has been made in understanding the biochemical properties of genetic regulatory networks by mathematical modelling [2], [3], [4], [11], [12]. Because of the sheer number of components and regulatory interactions, in recent years, it has become increasingly clear that sophisticated mathematical models and computational methods will be needed to manage, interpret and understand the complexity of biological information [9].

Many systems of gene regulation in humans are quite complex and it is difficult to model these complicated biological systems due to the lack of quantitatively regulatory information and experimental data. By studying simple models of gene regulation, we hope to gain insight into how more complex gene regulatory systems work. The lactose operon has been considered as a model system for understanding the molecular biology of gene expression and its regulation. The lactose operon encodes the genes in the pathway for the import of lactose or lactose analogue into the cell and its transformation to glucose and galactose. The basic components of this network have been well characterized and a large amount of information has been acquired, making it an ideal candidate for global analysis in the emerging field of genetic network analysis. For this model system, people are very interested in the bistability properties, which is the capacity to achieve multiple internal states in response to a single set of external inputs, and have designed different types of mathematical models for realizing bistability properties [7], [10], [13], [14].

Recently Ozbudak *et al.* [6] have studied the multistability property in the lactose utilization network. The thio-methylgalactoside (TMG) is used to induce the cell and a large amount of information has been obtained for the bistability properties with respect to different concentrations of glucose and TMG. In addition, a conceptual mathematical model has been developed to find out the conditions for realizing bistability in the network of the lactose operon. Based on the regulatory mechanisms proposed in [6], we develop a more detailed mathematical model in this paper to simulate the genetic regulation in the lactose operon. In section 2 we discuss the construction of mathematical model and numerical simulations are presented in section 3. Discussions are given in section 4.

2 Mathematical Model

Genetic regulatory mechanisms in the lactose operon are presented in Fig. 1. This network comprises three genes: *lacZ*, *lacY* and *lacA*. *lacZ* codes for β-galactosidase, an enzyme responsible for the conversion of lactose into allolactose and subsequent metabolic intermediates. *lacY* codes for the lactose permease (LacY), which facilitates the uptake of lactose and similar molecules, including TMG, a non-metabolizable lactose analogue. *lacA* codes for an acetyltransferase, which is involved in sugar metabolism [6].

Transcription of the lactose operon is inhibited by a *lac* repressor (LacI) but is enhanced by the cyclic AMP receptor protein (CRP). The activity of repressor LacI is decreased by the binding of intracellular TMG. Thus uptake of TMG induces the synthesis of LacY, which in turn promotes further TMG uptake. This mechanism leads to a positive feedback loop that creates potential for bistability. On the other hand, cAMP accumulates inside the cell in the absence of glucose. cAMP can bind to and trigger activation by CRP, which will enhance transcription of the *lac* operon. Glucose in the network can inhibit *lac* expression by two negative regulatory functions: to decrease the synthesis of cAMP and to interfere with LacY activity. As cAMP levels are not affected by TMG uptake, the extracellular concentrations of TMG and glucose can be used independently to regulate the activities of LacI and CRP, the two *cis*-regulatory inputs of the lac operon.

In this network, a single copy of the green fluorescent protein gene (*gfp*) is incorporated into the chromosome of *E. coli* in order to report the activities of LacY. This report gene has the similar regulatory mechanisms as those of LacY but the binding site O_2 is missing in the reporter gene [6]. Another report gene *HcRed* in Fig. 1 will not be included in the mathematical model.

The first mechanism discussed here is the TMG transport into the cell, which is assumed to be the same as that of lactose transport [7], [13]. The transport of TMG is enhanced by the concentration of LacY but is inhibited by the concentration of glucose

$$\text{TMG}_{\text{in}} = k_{\text{in}} \left[\frac{[\text{TMG}_{\text{ext}}]}{[\text{TMG}_{\text{ext}}] + K_{\text{in}}} \right] \left[\frac{K_{\text{Glu}}}{K_{\text{Glu}} + [\text{Glu}]} \right] [\text{LacY}] \qquad (1)$$

$$= k_{\text{in}} \left[\frac{[\text{TMG}_{\text{T}}] - [\text{TMG}_{\text{int}}]}{[\text{TMG}_{\text{T}}] - [\text{TMG}_{\text{int}}] + K_{\text{in}}} \right] \left[\frac{K_{\text{Glu}}}{K_{\text{Glu}} + [\text{Glu}]} \right] [\text{LacY}].$$

where $[\text{TMG}_{\text{ext}}]$ and $[\text{TMG}_{\text{int}}]$ are concentrations of extracellular and intracellular TMG, respectively. $[\text{TMG}_{\text{T}}]$ is the total TMG concentration and degradation of TMG is not considered here. $k_{\text{in}} = 1080 \text{ min}^{-1}$ is the TMG uptake rate, $K_{\text{in}} = 0.05 \ \mu M$ is the saturation constant for TMG uptake, $K_{\text{Glu}} = 271 \ \mu M$

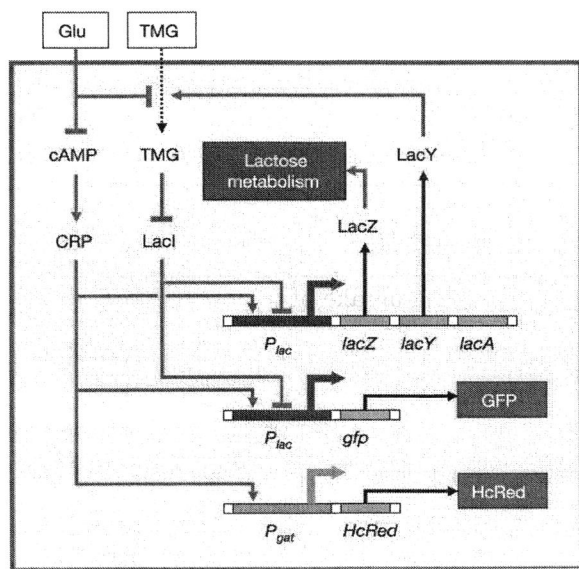

Fig. 1. The lactose utilization network. Reproduced with permission from Nature [6], copyright 2004, Macmillan Magazines Ltd

is the TMG transport inhibition constant by glucose [7]. [LacY] and [Glu] are concentrations of LacY and glucose, respectively.

For the reversible TMG transport, it is assumed that TMG efflux is dependent on the internal TMG concentration but independent on the extracellular glucose [7], namely

$$\text{TMG}_{\text{out}} = k_{\text{in}} \left[\frac{[\text{TMG}_{\text{int}}]}{[\text{TMG}_{\text{int}}] + K_{\text{in}}} \right] [\text{LacY}]. \quad (2)$$

The same transport constants k_{in} and K_{in} are used here for TMG efflux.

Two possible mechanisms have been discussed for the function of glucose in cAMP synthesis [13]. This synthesis process is inhibited either by extracellular glucose or by internal glucose that has been transported into the cell. When the concentration of extracellular glucose is low, there is little qualitative difference between these two inhibition mechanisms [13]. For simplicity, we do not discuss the transport of glucose and assume that the cAMP synthesis is regulated by extracellular glucose

$$s(\text{cAMP}) = k_{\text{cAMP}} \frac{K_{\text{cAMP}}}{[\text{Glu}] + K_{\text{cAMP}}}, \quad (3)$$

where $K_{\text{cAMP}} = 40$ μM is the inhibition constant for the effect of glucose on cAMP synthesis, and $k_{\text{cAMP}} = 5.5$ μM min^{-1} is the cAMP synthesis rate.

One or two cAMP molecules can bind to CRP to form cAMP-CRP complex CAP or CAP$_2$, then only complex CAP will bind to the specific DNA binding site to enhance transcriptional initiation of the *lac* operon. As CRP has constant concentration [CRP] = 2.6μM, mass balances on cell complex involving CRP and cAMP are considered. The concentration of CAP can be calculated by [13]

$$[\text{CAP}] = \frac{2K_{\text{CAP}}[\text{cAMP}]_f}{(K_{\text{CAP}} + [\text{cAMP}]_f)^2} [\text{CRP}], \quad (4)$$

where $K_{\text{CAP}} = 3.0$ μM, and [cAMP$_f$] is the concentration of free cAMP

$$[\text{cAMP}]_f = \frac{[\text{cAMP}] - K_{\text{CAP}} - 2[\text{CRP}]}{2}$$
$$+ \frac{1}{2}\sqrt{([\text{cAMP}] - K_{\text{CAP}} - 2[\text{CRP}])^2 + 4[\text{cAMP}]K_{\text{CAP}}}.$$

The affinity of the repressor LacI to its specific DNA binding sites is decreased by the binding of TMG to the repressor. Similar to the discussion in [6], it is assumed that the concentration of free LacI in cell is

$$[R] = \frac{K_R^n}{K_R^n + [\text{TMG}_{\text{int}}]^n} [R_T], \quad (5)$$

where $R_T = 0.029$ μM is the total concentration of repressor in cell [7]. K_R is the half-saturation concentration, and it is used as a free parameter to match the experimental data. The Hill coefficient $n = 2$ is used in [6] but we will test different values in simulation.

The transcriptional initiation rate is determined by the probabilities of the binding states in which a mRNA polymerase (mRNAP) is bound to site 2 while site 3 is either empty or bound by a complex CAP. Santillán and Mackey [7] recently have given a full list of the 50 possible binding states of the *lac* operon. Detailed information can be found in Tables 1 and 2 in [7]. Here we use a function

$$s(LacY) = k_1 f([\text{mRNAP}], [\text{CAP}], [R]) \quad (6)$$

to represent the transcriptional rate. Here it is assumed that the mRNAP concentration is constant ([mRNAP] = $3.0\mu M$) [7]. The rate of transcriptional initiation is $k_1 = 0.18$ min^{-1}.

As the O_2 binding site is missing in the reporter gene GFP [6], there are only 19 possible binding states and 5 favourable binding states for transcription, namely they are epee, epce, rpee, cpee and cpce by using the notation in Table 1 in [7]. Based on the energies of these 19 binding states, the transcriptional rate of GFP is given by

$$s(gfp) = k_1 \frac{4.5*10^{-8} + 19.64c + 15.16r + 0.021c^2}{1 + 25.54R + 900.44c + 13646.79r^2 + 17334.44rc + 0.0544c^2}, \quad (7)$$

where $c = $ [CAP] and $r = $ [R].

Based on the above considerations, we can construct a mathematical model for describing the genetic regulation of the lactose operon, given by

$$\begin{aligned}
\frac{d[\text{TMG}_{\text{int}}]}{dt} &= \text{TMG}_{\text{in}} - \text{TMG}_{\text{out}} \\
\frac{d[\text{cAMP}]}{dt} &= s(\text{cAMP}) - d_1[\text{cAMP}] \\
\frac{d[\text{mRNA}_{\text{LacY}}]}{dt} &= s(LacY) - d_2[\text{mRNA}_{\text{LacY}}] \\
\frac{d[\text{LacY}]}{dt} &= k_2 \text{mRNA}_{\text{LacY}} - d_3[\text{LacY}] \\
\frac{d[\text{mRNA}_{\text{GFP}}]}{dt} &= s(gfp) - d_4[\text{mRNA}_{\text{GFP}}] \\
\frac{d[\text{GFP}]}{dt} &= k_2[\text{mRNA}_{\text{GFP}}] - d_5[\text{GFP}],
\end{aligned} \quad (8)$$

where $k_2 = 18.8$ min^{-1} is the translation initiation rate [7], $d_1 = 2.1 + 0.02$ min^{-1}, $d_2 = 0.01$ min^{-1}, $d_3 = 0.01$ min^{-1} [7], $d_4 = 0.0677$ min^{-1} and $d_5 = 0.04$ min^{-1} [1] are degradation rates of cAMP, mRNA and protein of LacY, mRNA and protein of GFP, respectively. The initial condition of the above system can be zero or very low concentrations for each variable.

3 Simulation Results

When simulating system (8), we found that the ratios of the expression levels of LacY and GFP are nearly the same with regard to different total TMG concentrations. The transcriptional rates of these two genes, defined by (6) and (7),

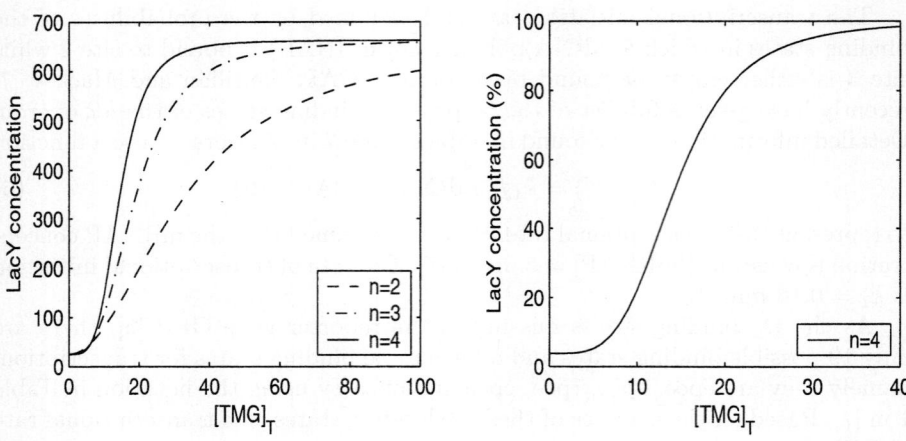

Fig. 2. Concentrations of LacY at equilibrium states based on different Hill coefficients n and different total TMG concentrations $[TMG]_T$

are nearly the same, although the values of transcriptional probabilities for all possible binding states and for the favourable binding states are different. The difference in degradation rates determines the different expression levels of these two genes. Thus we will only report the expression levels of LacY in the following study.

We first discuss the effect of Hill coefficient n in (5) on the expression levels of LacY. It has been assumed that $n = 2$ in (5) to determine the activity of the repressor [6]. As the lac repressor is a tetramer, four TMG proteins may bind to repressor LacI and only free repressor has a high affinity for its specific DNA

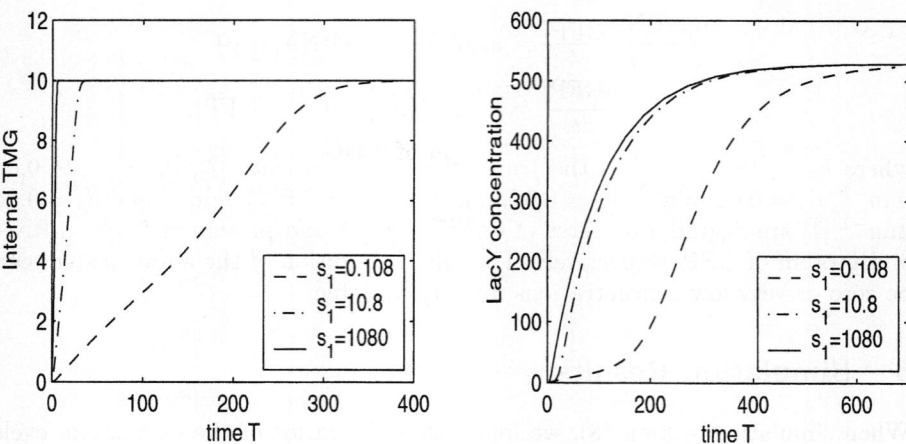

Fig. 3. Concentrations of LacY and internal TMG based on different TMG transport rate s_1

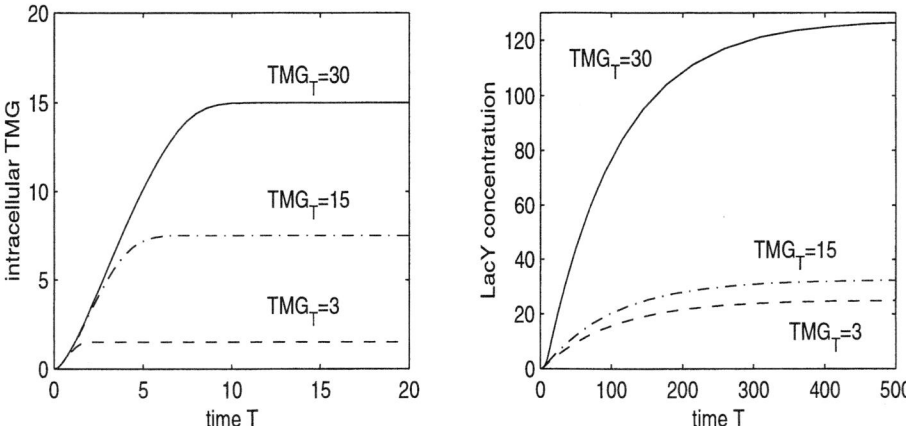

Fig. 4. Concentrations of intracellular TMG (TMG$_{int}$) and LacY based on different total TMG concentrations (TMG$_T$)

binding sites. Thus we first test the concentration of LacY in steady state with different n. For doing this we simulated the system in the time interval $[0, 1000]$ with different initial total TMG concentration [TMG]$_T$ ranging from 0 to 100. The left figure in Fig. 2 gives the expression levels of LacY at steady-states (the concentration of LacY at $t = 1000$) with $K_R = \sqrt{10}$, $n = 2, 3, 4$ and different [TMG]$_T$ concentrations. Compared with the steady state expression level [LacY] = 662.00 that is obtained when [TMG]$_T$ = 100, the relative expression levels of LacY when [TMG]$_T$ = 30 are 0.4797, 0.8089 and 0.9525 for $n = 2, 3$ and 4, respectively. Simulation results with $n = 4$, presented in the right figure of Fig. 2, are consistent with the experimental results that were presented in the lower panel of Fig. 2 (b) in [6]. It is suggested to use the Hill coefficient $n = 4$, and this is consistent with the fact that the repressor is a tetramer [7] and the analysis in [7]. On the other hand, coefficient K_R can be used to adjust the slope of curves in Fig. 2. Here we choose $K_R = \sqrt{10}$ in order to match experimental results. Thus all of the following simulations in this paper are based on $n = 4$ and $K_R = \sqrt{10}$.

Next we test the transport mechanism of TMG. As there is little information on TMG transport, we use the transport mechanism of lactose directly in this paper. We simulated system (8) with different transport rate s_1. Numerical results indicates that transport rate s_1 has little influence on the expression levels of LacY at steady states. Concentrations of LacY and internal TMG will reach the steady state quickly when a larger s_1 is used. Thus a large transport rate such as $s_1 = 1080$ should be used in mathematical modelling because the activity of *lac* operon increases quickly in experiment.

Finally the dynamics of system (8) is reported. Here we only give the concentrations of internal TMG and LacY with different total TMG concentrations [TMG$_T$], because the concentration of cAMP is nearly a constant and the mRNA concentration of LacY is proportional to that of LacY. The intracellular TMG will reach the steady state quickly, at which about half of TMG is inside the cell. Compared with the internal TMG concentration, LacY concentration reaches the steady state slowly and more time is needed if the total TMG concentration is larger.

4 Discussions

In this paper we have developed and numerically analyzed a mathematical model for the genetic regulation of the lactose operon in *E. coli*. A system with six differential equations is used to model transcription and translation processes of LacY and the reporter gene GFP. Based on numerical simulations we have discussed transport mechanisms of TMG and the effect of different Hill coefficients on the simulated expression levels of LacY. Numerical simulations have given good approximation to one of the experimental results.

All of the simulations in this paper are based on the initial conditions with low LacY concentrations, which correspond to the initially uninduced cell in the lower panel of Fig. 2 (b) in [6]. We also simulated system (8) based on initial conditions with high LacY concentrations. However, we did not realize the experimental results with fully induced *lac* expressions in the upper panel of Fig. 2 (b) in [6]. This suggests that there are some mechanisms in the *lac* operon for maintaining the high expression state but these mechanisms have not been included in the mathematical model. More detailed mathematical models should address this issue. In addition, we have presented the dynamics of the *lac* operon when no glucose exists in the media. This mathematical model can get good simulation results only when the glucose concentration is very low in the media. This is consistent with the discussion by Wong *et al.*, [13]. As it has been indicated that the glucose concentration which inhibit lactose transport is critically important, we should discuss the transport of glucose and inhibitation mechanism by the internal glucose when the glucose concentration is large in the media.

Experimental results indicate that the *lac* operon is a stochastic system [5], [6]. When TMG with different concentrations was added into the media, different proportions of cells were induced and bimodel population distributions were observed [6]. By using the deterministic model here we can only realize different average LacY concentrations. Stochastic models, based on either detailed biochemical reactions [5] or stochastic differential equations [8], should be constructed in order to realize the single cell experimental results, which is a challenging and exciting future area for research.

References

1. Basu, S., Mehreja, R., Thiberge, S., Chen, M., Weiss, R.: Spatiotemporal control of gene expression with pulse-generating networks. Proc. Natl. Acad. Sci. **101** (2004) 6355-6360
2. D'haeseleer, P., Liang, S., Somogyi, R.: Genetic network inference: from co-expression clustering to reverse engineering. Bioinformatics **16** (2000) 707-726

3. Hasty, J., McMillen, D., Isaacs, F., Collins, J.J.: Computational studies of gene regulatory networks: in numero molecular biology. Nat. Rev. Genet. **2** (2001) 268-279
4. de Jong, H.: Modelling and simulation of genetic regulatory systems. J. Comput. Biol. **9** (2002) 67-103
5. Kierzek, A.M.: STOCKS: stochastic kinetic simulations of biochemical systems with Gillespie algorithm. Bioinformatics **18** (2002) 470–481
6. Ozbudak, E.M., Thattal, M., Lim, H.N., Shraiman, B.I., van Oudenaarden, A.: Multistability in the lactose utilization network of *Escherichia coli*. Nature **427** (2004) 737-740
7. Santillán, M., Mackey, M.C.: Influence of catabolite repression and inducer exclusion on the bistable behaviour of the *lac* operon. Biophysical J. **86** (2004) 1282-1292
8. Tian, T., Burrage, K.: Bistability and switching in the lysis/lysogeny genetic regulatory network of Bacteriophage lambda. J. Theor. Biol. **227** (2004) 229-237
9. Tyson, J.J., Chen, K., Novak, B.: Network dynamics and cell physiology. Nature Rev. Mol. Cell. Bio. **2** (2002) 908-916
10. Vilar, J.M.G., Guit, C.C., Leibler, S.: Modeling network dynamics: the *lac* operon, a case study. J. Cell Biol. **161** (2003) 471-476
11. Wahde, M., Hertz, J.: Modeling genetic regulatory dynamics in neural development. J. Comput. Biol. **8** (2001) 429-442
12. Wessels, L.F.A., van Someren, E.P., Reinders, M.J.T.: A comparison of genetic network models. Pacific Symp. Biocomput. **6** (2001) 508-519
13. Wong, P., Gladney, S., Keasling, J.D.: Mathematical model of the *lac* operon: inducer exclusion, catabolite repression, and diauxic growth on glucose and lactose. Biotechnol. Prog. **13** (1997) 132-143
14. Yildirim, N., Mackey, M.C.: Feedback regulation in the lactose operon: a mathematical modeling study and comparison with experimental data. Biophysical J. **84** (2003) 2841-2851

Network Emergence in Immune System Shape Space

Heather J. Ruskin[1] and John Burns[1,2]

[1] School of Computing, Dublin City University, Dublin 9, Ireland
[2] Institute of Technology Tallaght, Dublin 24, Ireland
{jburns, hruskin}@computing.dcu.ie

Abstract. We present a model which enables us to study emergent principles of immune system T-cell repertoire self-organisation, based on a stochastic cellular automata model of a simplified lymphatic compartment. An extension of the immune system shape space formalism is developed such that each activated effector T-cell clonotype and viral epitope are represented as nodes, and edges between nodes models the affinity or clearance pressure applied to the antigen presenting cell bearing the target epitope. When the model is repeatedly exposed to infection by heterologous or mutating viruses, a distinct topology of the network space emerges which parallels recent biological experimental results in the area of cytotoxic T-cell activation, apoptosis, crossreactivity, and memory - especially with respect to repeated reinfection. The model presented here is a stochastic agent-based approach, which allows a broad distribution of results to be studied by tuning crucial T-cell life-cycle probabilities.

1 Introduction

In this paper, we propose a new approach which we feel supports well current biological experimentation in the area of cytotoxic T-cell activation, apoptosis, crossreactivity, and memory - with particular emphasis on repeated reinfection by heterologous antigens. Furthermore, the model presented here is a stochastic agent-based approach, which allows a broad distribution of results to be studied by tuning crucial T-cell life-cycle probabilities. It is our objective to study the means by which organising principles emerge which may govern the macroscopic behaviour of the immune system, based on a direct microscopic model. This approach is taken because reductionist techniques are likely to fail when confronted with global or system wide phenomena of the immune system.

This paper is organised as follows: Section 2 explains the principles of the extended model used throughout, Section 3 presents some of the important results from the model, while Section 4 discusses the importance of the findings.

2 The Model

The dynamics governing affinity between antigen and lymphocyte is developed directly from the *shape space* formalism [1]. In shape space, each unique antigen epitope and CTL clone is represented as a point within the two dimensional ($N = 2$) discrete space, of size 50×50. Surrounding each CTL clone is a disc of radius ρ [1]. Any antigen epitope located within the disc will subject to clearance pressure with a force inversely proportional to the distance between the two. In this model, we examine the immune response to primary and secondary virus challenge by both conserved and variable epitopes borne by APC. Specifically, we examine the emergence of cross-reactive responses between viruses as diverse as lymphocytic choriomeningitis (LCMV), Pichinde virus (PV) and vaccinia [3, 4, 5].

A shape space network. We make the following refinement to shape space: recent work [6, 3] has shown that effector CTL memory cells are capable of recognising diverse epitopes of unrelated viruses, and the authors conclude that cross-reactivity between heterologous viruses may be a key factor in influencing the hierarchy of CD8$^+$ T-cell responses and the shape of the memory T-cell pools. Therefore, we propose that shape space can be used to model both homogeneous viruses with conserved and mutated epitopes [7], *as well as* heterologous (derived from a separate genetic source) viruses with cross-reactive epitopes. A network model of shape space emerges naturally from the real space model as follows: each immunogenic epitope e_k and activated CTL clonotype c_j are considered as nodes in the space.

Clearance pressure applied between two nodes c_k and e_k is represented as a *directed* edge between them ($c_j e_k$). Each edge carries an implicit weight, representing the distance between the two nodes in shape space and therefore, a measure of the affinity between the nodes[2]. After initial infection, most c_j undergo programmed apoptosis (a crucial regulator of immune system homeostaisis). In our network model, this means that edges between such nodes are deleted. However, recruitment to the memory pool consumes around $5 - 10\%$ of activated CTL [8, 9, 10], therefore, these nodes remain active in shape space, preserving the edge connected to the stimulatory epitope e_k. In Section 3 we discuss how the topology of this network emerges and evolves during runtime, and how such the network topology can explain phenomena of cross-reactive memory, epitope mutation and escape, and how different immune responses arise when the same stimulus is applied to two different networks.

In the particular example shown in Fig. 1, and antigen presenting cell has two cytotoxic lymphocyte cells as nearest neighbours. When *apc* and *ctl*$^-$ are adjacent in real space, shape space affinity rules determine whether a bind between the two will arise. Whereas shape space is used to model the affinity between CTL response and viral peptide epitope, the life-cycle of the real space effector T cells is the same as that presented in [11].

Recirculation. Within the real space model of the lymphatic compartment, CTL cells constantly recirculate, sampling their nearest neighbours in order to detect

---

[1] Other models [2] have placed the antigen epitope at the locus of the disc of influence, without loss of generality.
[2] Conversely, a measure of the vigour which c_j mounts against, or suppresses e_k.

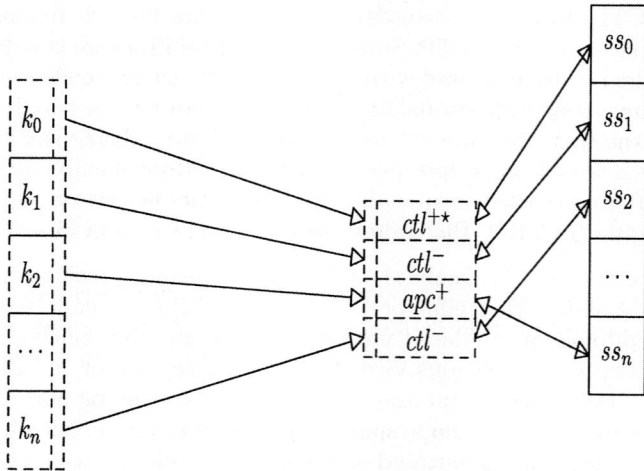

Fig. 1. On the left, the array of lattice sites $k_0, k_1, \ldots, k_n$ represents the real space data structures in memory. Each k_i holds a pointer to a structure describing the site occupant for the current time-step τ. Pointers are used in order to implement fast movement about the lattice: to move occupants around, a pointer swap is all that is required. On the right, the lattice of sites $ss_0, ss_1, \ldots, ss_n$ is the array holding shape space information (such as clonotype density levels). Information flows bi-directionally from shape space: local space entities are assigned their clonotype or epitope co-ordinate, and local space population changes are updated in shape space

the presence of infected antigen-presenting cells. On the lattice, each immune cell c_i has two neighbourhoods: an *inner* and an *outer*, denoted R_i and R_o, respectively, with $|R_i| = 8$ and $|R_o| = 16$ (with radius 1 and 2, respectively). As part of the update of the lattice, each k_i recirculates within the real space, implemented as follows: first, R_i is examined in order to locate an unoccupied position into which the immune cell may move. If an empty cell is located within R_i, k_i will move into it with probability $P(inner) = 0.9$. If no space is available within R_i, R_o is searched for a free space. If a free space is located in R_o, k_i will move into it with probability $P(outer) = 0.7$. If both R_i and R_o are occupied, then no movement of k_i will occur in this time-step. If a free location is found in R_i or R_o, the new coordinates of k_i are calculated and the cell is moved. The values chosen for $P(inner)$ and $P(outer)$ are subjective and reflect the concept of cell motion into *proximate* and *nearby* space, respectively.

2.1 Model Parameters

Crossreactivity: In this model, the number of different clonotypes which respond to the same (randomly selected) epitope is a ratio of the clonal cutoff parameter to the length of shape space, or $\rho/L = \hat{\rho}$. This is crucial parameter is known as *crossreactivity*. Some research [4], has suggested this figure to be as high as

50 – 111. In the original work of Perelson and Oster [1], the fraction of clonotypes which bind a randomly selected antigen was estimated conservatively at 10^{-3}, therefore, a shape space of size 2.5×10^3 (used here), the number of different clonotypes which respond to the same epitope would be ~ 2.5. In previous work [12], we found that in a model of healthy primary response, the number of clonotypes responding to a randomly selected epitope was on the average 25 (with $\hat{\rho} = 0.14$), and this value is also used in the model we present here. This value is lies roughly in between the low estimate of [1] and the high estimate of [4].

Experiment. *Objective:* To analyse the critical nature of crossreactive memory in response to epitopes from heterologous antigens. In particular, we study the emergent network in shape space in an effort to understand principles of self-organisation.

Method: At time $\tau = 0$, 5000 antigen are placed at random locations on the real space lattice, to simulate antigen-presenting cells entering the lymphatic compartment with marker epitopes displayed on the cell surface. At this stage, the immune system will be in a virgin state without having been exposed to any previous antigen. A primary response will ensue, and a low level fraction of cells will enter the long-lived memory pool, and the remainder of the effector cells will undergo apoptosis. At time $\tau = 1500, \tau = 3000, \tau = 4500$, further infections are introduced. At each infection point, all antigen presenting-cells carry the same epitope, but the epitope is different to the one seen at the previous infection point [3]. In this way, we simulate four heterologous infections challenging the immune system, and this enables us to study the activation network, as well to study the critical nature of crossreactive memory. The simulation executes for 6000 time-steps, simulating some 125 days of real time. This scenario is referred to as $\mathcal{E}_1$ in the rest of this paper.

Assumptions. In $\mathcal{E}_1$ we assume a case of non-proliferating antigen, and therefore the clearance rate of infected cells from the real space is a function of recognition and stimulation only (ie, *affinity*). In both experiments, effector T-cells have a 2% chance of entering the memory pool once the infection has been cleared. We assume that shape space is completely covered. That is, for n clonotypes in 2-dimensional shape space:

$$n\pi r^2 \gg L_s^2 \qquad (1)$$

Eqn. 1 is a realistic assumption in that an *escape* mutation does not mean no further clearance pressure is ever brought to bear on the mutated antigen. On the contrary, in this model, escape mutation means that no active effector cells (either memory or primary response) can apply clearance pressure at the time of mutation. In such a case, the immune response is synthesised as a *de novo* primary response - characterised by relatively slow precursor cell activation and population growth rates, with a consequent elongated antigen clearance profile,

---

[3] In all, four distinct epitopes e_1, e_2, e_3, e_4, will be presented to the real space model.

typically extended over eight days or more. We replicate the on-going thymus generation of cytotoxic T cell precursors by injecting into the lattice some 5×10^3 precursor CTL cells (ctl^-) at the start of each infection event.

3 Results

The ouput of the first experiment ($\mathcal{E}_1$) is summarised in Fig. 2 and Fig. 3. The network topology of shape space is shown in Fig. 2, at the end of four infections, introduced at $\tau = \{0, 1500, 3000, 4500\}$. To monitor the state of the CTL memory pool, the lattice was sampled at $\tau = \{1500, 3000, 4500, 6000\}$, and at each sample, only CTL memory clonotypes and immunogenic epitopes were recorded. As such, CTL cells which had not undergone apoptosis, or become memory cells, are not shown. Clearly, most immune challenges do not present

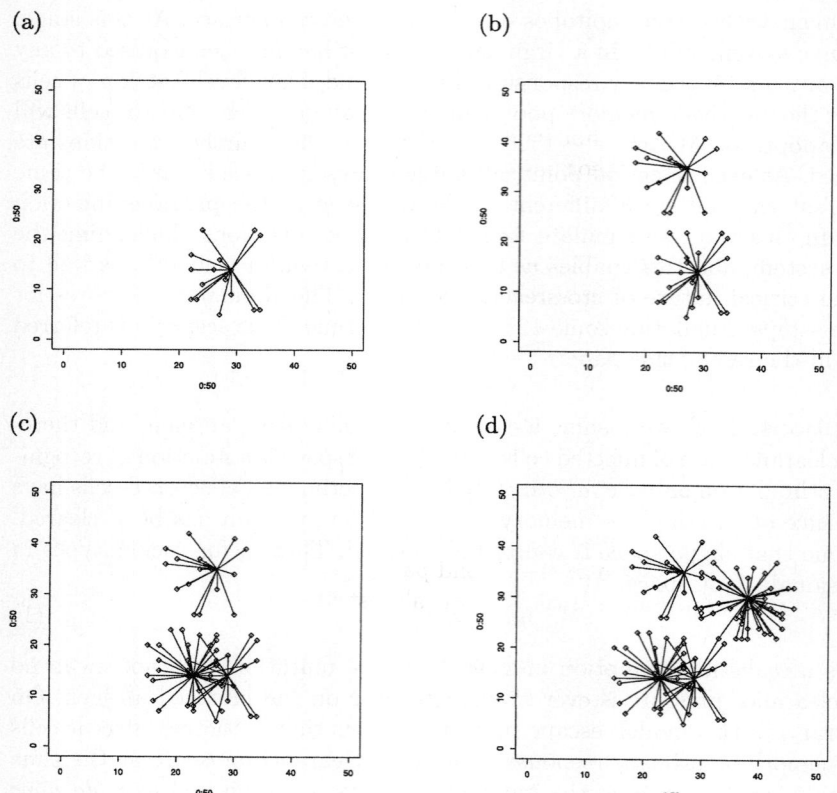

Fig. 2. The development of a four-epitope network in shape space represented over 6000 time-steps. Only CTL memory clonotypes are shown. At the centre of each cluster is the immunogenic epitope, and each node connected to the cluster centre is a stimulated CTL

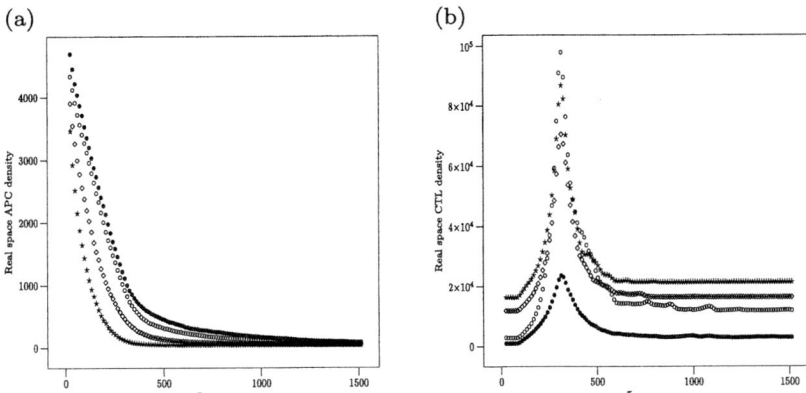

Fig. 3. Real space density levels of antigen presenting cell (a) and CTL cell (b) respectively in the model of the lymphatic compartment over the course of each infection. Plots are superimposed in order to convey cell density levels. In both panels, cell levels are shown by •, ○, ⋄, ⋆, to represent first, second, third and fourth infections respectively

themselves at equal-dose and equally-spaced time intervals, so our choice of infection time and sample time is somewhat contrived. However, as it serves to illustrate the underlying theory, we will continue with this configuration. After the first infection has been cleared, the CTL memory cells are arranged in a cluster formation around the immunogenic epitope (Fig. 2(a)). As discussed in Section 2.1, there is a spread of memory within a disc of radius $\hat{\rho}$ from the epitope. As this model is stochastic, the shape space activation spread is "irregular" in nature. Real space clearance rates of the infected cells, and CTL density levels are shown in Fig. 3 (a) and (b), respectively.

The first infection (•) is cleared with an infected cell half-life of about 3.2 days ($\tau = 156$), with 5% remaining after 10.2 days, and this is broadly in keeping with clearance profiles expected during primary response. When the second infection has been cleared, the shape space network has developed further (Fig. 2(b)): two unconnected clusters emerge. Clearance rate associated with second infection (Fig. 3 (a), (○)) indicates that no advantage was conferred on the immune response during elimination of the second pathogen: a normal primary response was required - and the clearance rate was almost identical the to first infection (•), with a half life of $132 \leq \tau \leq 144$, and a 95% clearance obtained at 10.25 days. As can be seen from Fig. 3 (a), both first and second infections are similar in clearance profile indicating that no previously primed memory cells participated in cell removal, and this is borne out by the two-cluster configuration in shape space in Fig. 2 (b). The randomly chosen epitope location in shape space for the third infection places it in a position such that $|\epsilon_1 - \epsilon_3| \leq 2\hat{\rho}$, and some crossreaction between memory CTL arises. Between the two clusters, some 8 CTL have $deg(2)$ [4], and the two clusters are fused into one. The advantage con-

---

[4] These nodes were known as α nodes in [13].

ferred on the immune system when memory cells respond to a challenge is clear: because they are primed from the point of a previous infection they produce armed effector cells without spending time in clonal expansion. Having already been primed by a previous encounter with the specific pathogen, they undergo expansion with lower death rates than during primary response: they therefore accumulate more quickly [14].

From Fig. 3 (a), the half life of the third infected cell population ($\diamond$) is ~ 72, which is a efficiency improvement of around 50% compared to the previous two. When the final infection is analysed, an important condition has arisen in the shape space network: all small clusters have joined together merged into one large cluster, due to the critical influence of crossreactive memory clonotypes. Only one clonotype is responsible for connecting the two clusters of Fig. 3 (c) to into one large cluster. The benefit of this single cluster network shape space is demonstrated by the clearance dynamics of 3 (a): the final infection ($\star$) is cleared so rapidly (half-life ~ 36) that it would probably be asymptomatic. Analysis of the CTL density levels in real space (Fig. 3 (b)) explains how clearance of infected cells is so rapid: each consecutive challenge *with the same level dose of infected cells*, is met with increasingly rapid effector population growth, and a gradually increasing memory pool.

As each infection is introduced and cleared, the average degree of the network in shape space increases. The conditions under which immunity to one virus can reduce the effects of challenge by a second virus are clear from Fig. 2: one or more nodes in the first cluster must also have an edge to the second cluster. Thus, damage to or suppression of these critical nodes will have a significantly greater impact than damage to leaf nodes. Crossreactive memory nodes are called α nodes, and non-crossreactive leaf nodes are β nodes (following the terminology presented in [13]). Also from Fig. 2, the clinical phenomena of two identical

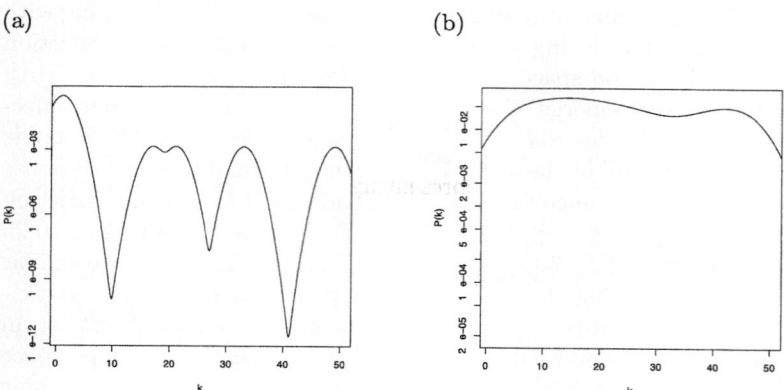

Fig. 4. Log plots of the edge degree distribution in shape space network (a) and random network (b). Clearly, (a) has a central mode around $deg(1)$ and statistically significant modes around $deg(20)$, $deg(35)$ and $deg(49)$. For comparison eacn node in (b) has degree probability drawn from a uniform distribution in the range $[0, 50]$

infections have a different disease outcome has a possible explanation in shape space network topology: the density and distribution of α nodes will play a crucial role in determining if the ensuing immune response is primary or secondary. The degree $deg(k)$ of a node is the number of edges associated with that node.

In Fig. 4, we compare the degree distribution of the shape space network at time $\tau = 6000$ (a), with that of a random network (b) in which each the degree of each node is same probability drawn from a uniform distribution in the range $[0, 50]$. Clearly, Fig. 4 (a) is not a random distribution and is typical of the density distributions observed over many trials: it has a clear mode at 1 (from Fig. 2 (d), the majority of nodes of $deg(1)$) and three minor modes centred around 20, 35 and 49, which represents the cluster "centres" of the immunogenic epitopes. This non-random degree distribution clearly demonstrates some form of consistent structure emerging from a highly stochastic model of the real space, where the emergent network topology after four infections is arranged around immunogenic epitopes connected in hub-and-spoke formation to memory CTL cells, and in turn, cross-reactive memory CTL clonotypes connect cluster formations to each other such that the overall network is connected at the end of the simulation.

4 Discussion and Conclusions

Most discrete computational models of immune response to viral infections have used real space or shape space formalisms. In this study, however, we have presented a model based on a combination of the two, with the objective of demonstrating how emergent behaviour and principles of self organisation may arise from a many-particle microscopic system. This is achieved by using a stochastic model of the lymphatic system as stimulus to shape space differentiation and distribution. We have developed an extension to shape space which goes beyond restrictive early network models, to demonstrate a mechanism by which early and protective immunity can be mediated by memory T cells generated by a previous heterologous viral infection. This important feature emerges in the shape space network as a memory T cell node v of $deg(v) \geq 2$. An edge between two nodes $c_k e_k$ is added whenever one exerts clearance pressure against the other. The pressure applied between two nodes c_k and e_k is represented as a directed edge between them. Each edge carries an implicit weight, representing the distance between the two nodes in shape space and therefore, a measure of the affinity between the nodes.

Of course, the *degree* of protective immunity offered by cross-reacting memory cells is dependent on the distance between the memory T cell clonotype and the immunogenic epitope, with optimal immunity arising when reinfection is by the same antigenic epitope. In our results, we have seen increasingly effective clearance dynamics as the memory pool increases, and each T cell clone has a 2% chance of becoming a long-lived memory cell. Although we have not modelled it here, clearly, the immune system cannot continually increase the size of the memory pool, and [3] has suggested that some (as yet undefined) process probably exists to purge non-crossreactive (or at the very least, the relatively weaker cross-reactive) memory cells in order that the pool does not grow beyond the physical limits imposed by the restrictive spaces of the lymphatic system. However, this purging method, if it exists, is as yet unknown.

Individuals vary considerably in their responses to viral infections, ranging from subclinical to severe. There are many factors that contribute to this variation in responsiveness, including the dose and route of infection, as well as the physiological state and genetic background of the host. In this study, we show that memory T cells in immune system shape space network which are specific to unrelated viruses may also contribute to the host's primary response to a second virus. The beneficial effect of these early protective memory T cells is to slow down the spread of infection, much like a natural immune-mediated response, allowing time for more suitable high affinity antigen-specific T and B cell responses to develop. It is reasonable to expect that this level of resistance may be the difference between clinical and subclinical infections or lethal and nonlethal infections, and we propose that such clinical outcome may be explained, at least in part, by the varying toplogy of the immune system shape space network of memory CTL cells.

References

1. Perelson, A.S., Oster, G.F.: Theoretical Studies of Clonal Selection: Minimal Antibody Repertoire Size and Reliability of Self-Non-Self Discrimination. J. Theor. Biol. **81(4)** (1979) 645–70
2. Smith, D., Forrest, S., Ackley, D., Perelson, A.: Using lazy evaluation to simulate realistic-size repertoires in models of the immune system. Bull. Math. Biol. **60(4)** (1998) 647–658
3. Brehm, M., Pinto, A., Daniels, K., Schneck, J., Welsh, R., Selin, L.: T cell immunodominance and maintenance of memory regulated by unexpectedly cross-reactive pathogens. Nat. Immunol. **3(7)** (2002) 627–634
4. Mason, D.: A very high level of crossreactivity is an essential feature of the T-cell receptor. Immunology Today **19(9)** (1998) 395–404
5. Borghans, J.A., De Boer, R.J.: Crossreactivity of the T-cell receptor. Immunology Today **19(9)** (1998) 428–429
6. Selin, L.K., Varga, S.M., Wong, I.C., Welsh, R.M.: Protective Heterologous Antiviral Immunity and Enhanced Immunopathogenesis Mediated by Memory T Cell Populations. J. Exp. Med. **188** (1998) 1705–1715
7. Nowak, M.A., May, R.M.: 14. In: Virus Dynamics. Oxford Univ. Press, Oxford New York (2000) 149–181
8. Murali-Krishna, K., Lau, L.L., Sambhara, S., Lemonnier, F., Altman, J., Ahmed, R.: Persistence of Memory CD8 T Cells in MHC Class I-Deficient Mice. Science **286** (1999) 1377–1381
9. De Boer, R.J., Oprea, M., Kaja, R.A., Murali-Krishna, Ahmed, R., Perelson, A.S.: Recruitment Times, Proliferation, and Apoptosis Rates during the $CD8^+$ T-Cell Response to Lymphocytic Choriomeningitis Virus. J. Virol. **75(22)** (2001) 10663–10669
10. Badovinac, V.P., Porter, B.B., Harty, J.T.: CD8+ T cell contraction is controlled by early inflammation. Nat. Immunol. **5** (2004) 809–817

11. Burns, J., Ruskin, H.J.: A Stochastic Model of the Effector T Cell Lifecycle. In P.M.A. Sloot and Bastien Chopard and Alfons G. Hoekstra, ed.: Cellular Automata. Volume 3305 of Lecture Notes in Computer Science., Berlin Heidelberg, Springer-Verlag (2004) 454–463
12. Burns, J., Ruskin, H.J.: Diversity Emergence and Dynamics During Primary Immune Response: A Shape Space, Physical Space Model. Theor. in Biosci. **123(2)** (2004) 183–194
13. Burns, J., Ruskin, H.J.: Network Topology in Immune System Shape Space. In Sloot, P., Gorbachev, Y., eds.: Computational Science - ICCS 2004. Volume 3038 of Lecture Notes in Computer Science., Berlin Heidelberg, Springer-Verlag (2004) 1094–1101
14. Grayson, J., Harrington, L., Lanier, J., Wherry, E., Ahmed, R.: Differential Sensitivity of Naive and Memory CD8+ T cells to Apoptosis in Vivo. J. Immunol. **169(7)** (2002) 3760–3770

A Multi-agent System for Modelling Carbohydrate Oxidation in Cell[*]

Flavio Corradini, Emanuela Merelli, and Marco Vita

Università di Camerino, Dipartimento di Matematica e Informatica,
Camerino 62032, Italy
{flavio.corradini, emanuela.merelli, marco.vita}@unicam.it

Abstract. A cell consists of a large number of components interacting in a dynamic environment. The complexity of interaction among cell components and functions makes design of cell simulations a challenging task for biologists. We posit that the paradigm of agent-oriented software engineering (AOSE), in which complex systems are organized as autonomous software entities (agents) situated in an environment and communicating via high-level languages and protocols (ontologies), may be a natural approach for such models. To evaluate this approach, we constructed a model of cell components involved in the metabolic pathway of carbohydrate oxidation. The agent-oriented organization proved natural and useful in representing three different views of the cell system (functional, dynamic, and static structural) and in supporting bioscientists querying a system very close to their mental model.

1 Introduction

A biological cell is a complex system. It consists of a large number of components interacting with each other to perform cell functions. Each cell is a self-contained and self-maintaining entity: It can take in nutrients, convert these nutrients into energy, carry out specialized functions, and reproduce itself. The modelling and simulation of cell behaviour belongs to an emerging research area, Systems Biology [7], whose aim is to understand how biological systems function, by studying at different abstraction levels the relationships and interactions between various parts of a biological system, e.g. organelles, cells, physiological systems, organisms etc., and producing a model as close as possible to the biological reality. The modelling of complex systems, as suggested in Kitano [7], implies a deep understanding of the biological system both in terms of its structure and its behaviour. Cells grow through the functioning of cell metabolism. Cell metabolism is the process by which individual cells process nutrient molecules.

---

[*] This work was supported by Center of Excellence for Research 'DEWS: Architectures and Design Methodologies for Embedded Controllers, Wireless Interconnect and System-on-chip' and Italian CIPE project 'Sistemi Cooperativi Multiagente'. Work completed while the second author was on Fulbright leave at University of Oregon.

In particular, carbohydrate oxidation is the process by which a cell produces energy through chemical transformation of carbohydrates [5]. In biochemistry, those cell processes, which imply a series of chemical reactions within a cell, are called metabolic pathways.

In this paper, we propose a multi-agent system (MAS) [6] to model carbohydrate oxidation cell metabolism, to evaluate how agent-based computing paradigm in modelling three different views of the cell system: functional, dynamic and structural. The first view shows the cell's functionalities and *who* are the components (actors) that perform cell functions, i.e the agent society and agents' roles. The second view shows *when* and *how* any components, within its role, participates to cell functions, i.e. the agents communication protocols and tasks flow description. The third view shows *what* concepts describe the conceptual model of cell system behaviour. A set of variables and a set of relationships, comprise the agents' domain ontology knowledge.

The multi-agent approach allows one to design systems very close to the mental model of its user. The resulting multi-agent system is very flexible. It supports changes to the model during simulation time, by adding and removing components, by changing the behaviour of components, and by moving, cloning and consuming components as happens naturally in the cell. Furthermore, the agents' autonomy and pro-activity allow one to easily represent the non-deterministic behavioural model of the cell as in nature. Thus, one is allowed to progressively refine components and develop the simulation system.

We have designed the cell simulation system using the PASSI ToolKit [1] and we have implemented it on the Hermes platform [3]. For lack of space, many details are omitted. The full description of the system can be found in [4].

In the next section of the paper, we summarize the cell process of carbohydrate oxidation. In Section 3, we introduce AOSE and describe the PASSI methodology. In Section 4, we present our MAS for modelling carbohydrate oxidation. Section 5 concludes.

2 Introduction to the Cell System Structure and Behaviour

The cell consists of a large number of components interacting in a dynamic environment. We study the process of carbohydrate oxidation (CO), i.e the energy production process, that happens inside a cell through the transformation of carbohydrates [5]. We have considered the case of fructose, glucose, mannose, maltose, lactose, saccharose, glycogen and starch. Each kind of carbohydrate has its metabolic pathway, i.e. a series of chemical reactions, that can happen in presence of oxygen (aerobic respiration in the mitochondria) or anaerobic metabolism (fermentation). The CO process is performed by two main cell components: Cytoplasm and the Mitochondria.

---

[1] http://mozart.csai.unipa.it/passi

Cytoplasm is the viscid, semifluid component between the cell membrane and the nuclear envelope, where the first stage of carbohydrate molecule transformation takes place. It consists primarily of water. It also contains various organelles (e.g. mitochondria) as well as salts, dissolved gasses and metabolites. During the CO process, the following metabolic pathways take place.

Glycolysis: The first stage of the CO process, glycolysis converts one molecule of *Glucose* into two molecules of *Pyruvate* [2] w.r.t.

$$Glucose + 2ATP + 4ADP + 2Pi + 2NADox \rightarrow 2Pyruvate + 2ADP + 4ATP + 2NADrid + 2H_2O$$

where ATP, ADP, Pi, $NADox$ and $NADrid$ are energy packets. Depending on the cell typology and under anaerobic conditions the *Pyruvate* is processed differently by Lactic Fermentation or by Alcoholic Fermentation.

Lactic Fermentation: In absence of oxygen, lactic fermentation reduces the Pyruvate to Lactate. It occurs in anaerobic microorganisms (sour milk) and animal cells (muscle pain).

Alcoholic Fermentation: In the absence of oxygen, alcoholic fermentation reduces Pyruvate to Ethanol. It occurs in yeast (Saccharomyces) causing the transformation of carbohydrates present in grapes and malted barley into Ethanol.

Mitochondria consists of two sub-components, the Inner Mitochondrial Membrane (IMM) and the Mitochondrial Matrix (MM), where the aerobic respiration takes place via four processes.

Transportation: In IMM, transportation process transports Pyruvate from the Cytoplasm to MM.

Electron Transport Chain: In IMM, electron transport chain transfers electrons from reduced NAD and FAD (energy packets) to the final electron acceptor, molecular oxygen. The function of this chain is to permit the controlled release of free energy to drive the synthesis of ATP (oxidative phosphorylation).

Partial oxidation of Pyruvate: In MM, partial oxidation of Pyruvate degrades the Pyruvate w.r.t.

$$2Pyruvate + 2NADox + 2CoA \rightarrow 2acetylCoA + 2NADrid + 2CO2$$

Kreb's Cycle: In MM, Kreb's cycle forms part of the break down of carbohydrates, fats and proteins into carbon dioxide and water in order to generate energy, w.r.t.

$$Acetil - CoA + 3NADox + FADox + GDP + Pi \rightarrow CoA + 3NADrid + FADrid + GTP + 2CO_2$$

Following the Kitano suggestion [7], to model a biological system, we need to *identify* (1) the cell structure, to *analyse* (2) the cell behaviour, to *control* (3) the system simulation and to *design* (4) systems. In the next section, we introduce the PASSI methodology used to instantiate the Kitano's approach.

---

[2] http://www.genome.jp/kegg/pathway/map/map00010.html

3 Agent Oriented Software Engineering

methodologies and modelling techniques to support multi-agent systems (MAS) engineering. Based on the agent computing paradigm, it represents a guide for software engineers whose aim is to analyze, to model and to implement a complex system. In fact, by a multi-agent system we can simulate complex systems whose modelling becomes almost impossible without a rigorous guide. Several methodologies have been proposed in literature, among these, we have used the PASSI methodology because it is made up of many models, concerning different design levels. It supports the identification of a huge amount of variables, different system's components behaviours, and the UML definition of different communications protocols by means of domain ontologies. Further, the PASSI ToolKit provides all the system specification by UML diagrams very helpful for the implementation. For our purposes, we have considered two design models:

The *System Requirements Model* is an anthropomorphic model of the system requirements in terms of system functionalities (target) and system actors (agency). It is specified through four steps: 1. *Domain Description:* a functional description of the system, using conventional use-case diagrams. 2. *Agent Identification:* separation of responsibilities among actors (agents), represented by stereotyped packages. 3. *Role Identification:* a functional description of agent roles respect to the system functionalities and agents responsibilities, using sequence diagrams. 4. *Task Specification:* specification of the agent behaviour, using activity diagrams.

The *Agent Society Model* is the social organization in terms of interactions and dependencies among agents. It is specified by three more steps: 5. *Ontology Description:* a description of the knowledge given to individual agents and the high-level language for their interactions, using of class diagrams and OCL constraints. 6. *Role Description:* a description of agent roles respect to the system functionalities, and the agent tasks, using class diagrams. 7. *Protocol Description:* a grammar of each pragmatic communication protocol in terms of performative speech-act, using sequence diagrams.

4 Modelling Carbohydrate Oxidation with a MAS

The PASSI methodology naturally leads to the identification of structure and behaviour of the cell and its components, in modelling the carbohydrate oxidation. In fact, the first two steps (1-2) of the system requirements model provide the functional view of the cell system; the next two steps (3-4) the dynamic view of the cell system and the last three steps (5-7) the structural view of the cell system. The model is summarized below; a full detailed description can be fund in [4].

Functional Model. It defines the cell functionalities, the cell components and their roles by performing the first two UML diagrams.

Domain Description Diagram: in Section 2, we have listed seven functions the cell system during the carbohydrate oxidation, known as metabolic pathways. Thus, it is natural to define the cell functional description by a UML diagram with seven use cases: *Glycolysis, Lactic Fermentation, Alcoholic Fermentation, Transportation, Partial oxidation of Pyruvate, Krebs Cycle, Electron Transport Chain*. To the final diagram, we have added four more functionalities, to support the execution environment and user interaction: *Data Updating, Initialization, Quantitative Analysis* and *User Interface*.

Agent Identification Diagram: the agent-based simulation system aims to be close as much as possible to the mental model of its user, in our case the bioscientist, we have decided that every reactive cell component has been represented as an agent. In Section 2, we have identified that Cytoplasm and the Mitochondria with its two sub-components Inner Mitochondrial Membrane (IMM) and Mitochondrial Matrix (MM) are involved in the seven cell functionalities. Since every functionality of the cell is associated with one of its components, it is natural to associate agents with their roles. Figure 1[3] shows the UML diagram where the cell components are agents. Note that component has different roles consistently with their cell functionalities. The agents can communicate with each other through a high-level communication protocol, and can change the environment state by modifying the environmental variables. Beside the three cell agent components we have defined two more agents as it follows:

AgCytoplasm: an agent that simulates Cytoplasm. It plays the *AlcholicFermentation, LacticFermentation, Glycolysis* roles.
AgMitocondrialInnerMembran: an agent that simulates the IMM. It plays the *Transport, Electron Transport Chain* roles.
AgMitocondrialMatrix: an agent that simulates the MM. It plays the *Oxidative Decarboxilation Of Pyruvate, Citric Acid Cycle* i.e. the *Kreb's cycle* roles.
AgEnvironment: an agent that simulates the execution environment in which any cell process occurs.
AgInterface: an user assistant agent. It embodies the interface between the user and the cell simulation system.

Dynamic Model. A dynamic model describes the cell components behaviours during the OC process, by means of third and fourth type of PASSI diagrams.

Role Identification Diagrams: through these diagrams we describe the temporal order by which the agents activities are executed w.r.t. the communication

---

[3] The diagram is part of the real implementation and thus the labels are in Italian the native language of the designers. Glicolisi=Glycolysis; Fermentazione Lattica=Lactic Fermentation; Fermentazione Alcolica=Alcoholic Fermentation; Trasporto=Transportation; Catena Respiratoria=Electron Transport Chain; Acido Citrico=Kreb's cycle; INIZIALIZZAZIONE=Initialization; AggiornaDatiCondivisi=updating; MostraATPCtosol=ATP; MostraQuantitaAttuali=Environment; Utente=User.

A Multi-agent System for Modelling Carbohydrate Oxidation in Cell   1269

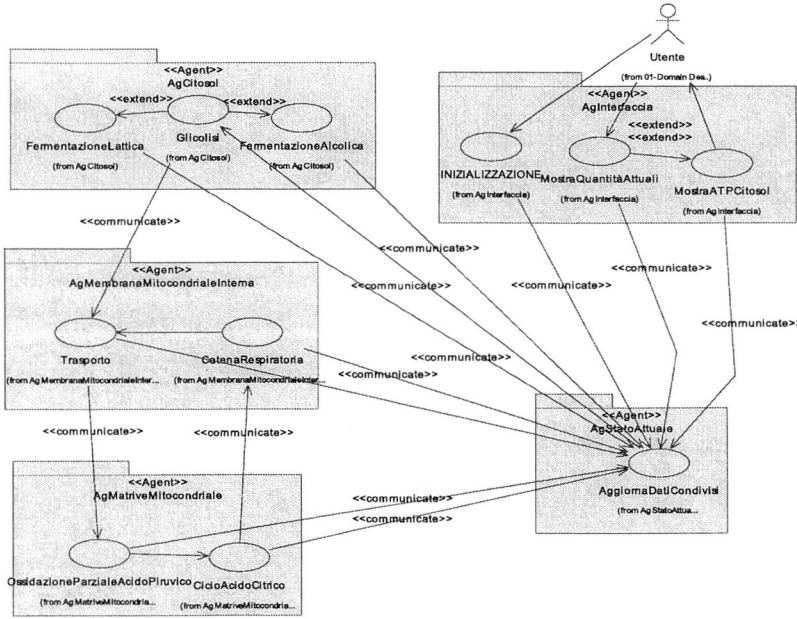

Fig. 1. Agent Identification: the Cell Components Stereotypes

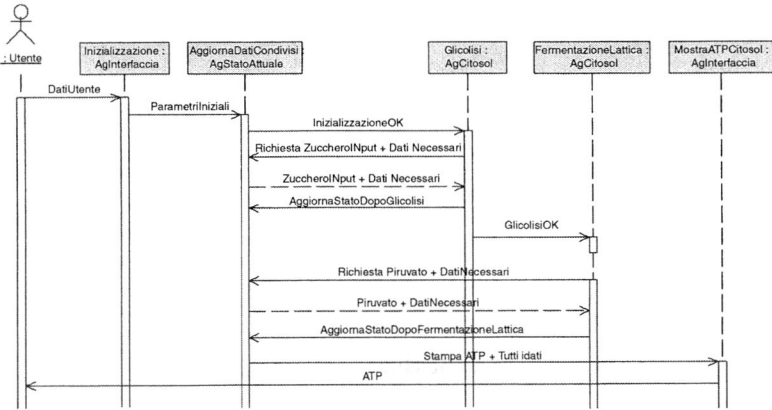

Fig. 2. Roles Identification Diagram: the Lactic Fermentation Sequence Diagram

protocol. For any metabolic pathway we have a sequences diagram which describes the temporal order of the communication acts occurring among agents undertaking a role in that pathway. Figure 2 shows only the diagram related to the Lactic Fermentation pathway, all others are described in [4]. Note that the AgCitosol agent undertakes two different roles in this pathway, that related to the Glycolise functionality, and that of the Lactic Fermentation. Any invoked action primitive that, like "Richiesta di Zucchero" (Sugar Request), belongs to the coordination protocol is described within the Ontology Description Diagram.

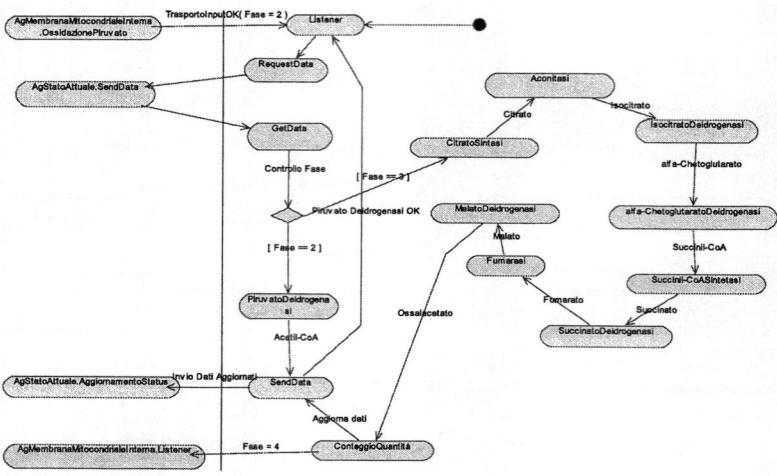

Fig. 3. Task Specification: the Mitochondrial Matrix Activity Diagram

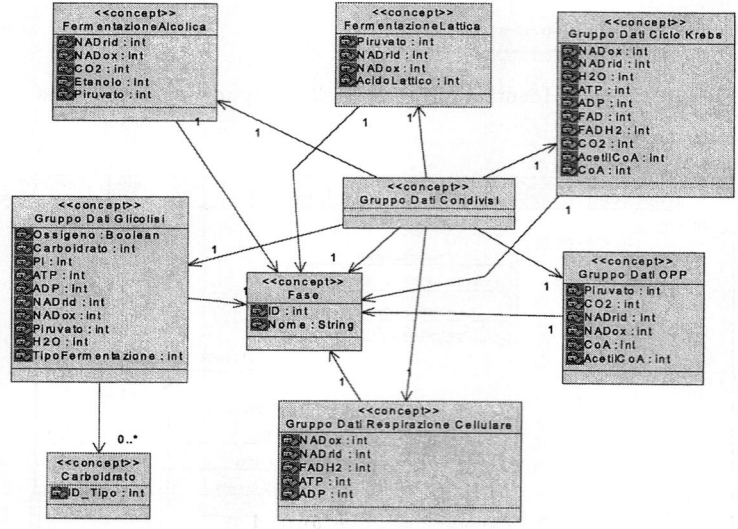

Fig. 4. Ontology Description: the Cell Conceptual Model Class Diagram

Task Specification Diagrams: In the cell system, each component has its own behaviour for each of its roles. Thus, every agent of the system has its own activity diagram that specifies its behaviour within any of its roles. Figure 3 shows the behaviour of the *AgMitocondrialInnerMembrane* agent, which simulates the Mitochondrial Matrix component. On the right hand side of the figure we can see the tasks comprising the Kreb's Cycle, while the other activities are related to Partial oxidation of Pyruvate. The diagrams are fully described in [4].

A Multi-agent System for Modelling Carbohydrate Oxidation in Cell   1271

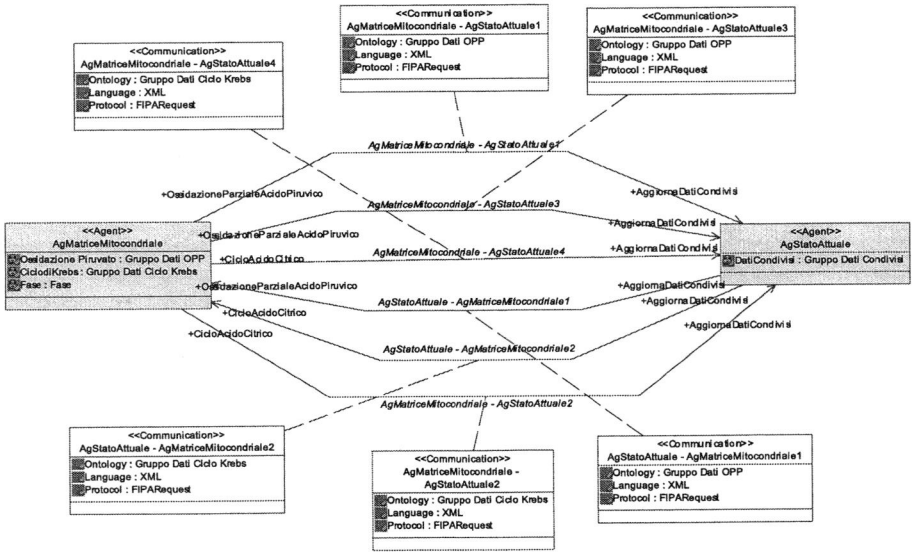

Fig. 5. Communication Ontology: the Cell Relationship Model Class Diagram

Static Structural Model. The static structural model is a conceptual model of the cell system. Following the PASSI methodology, it corresponds to the ontological description of the cell components and their interaction language, which gives rise to the agents' knowledge and agents' communication language. In this first version of the work, we have defined our cell ontology in terms of concepts, attributes (variables), relations, actions and predicates. In the future, we aim to include the standard notation provides by KEGG and SBML specification [4].

Ontology Description Diagram: Our domain is the cell. Thus, the ontology description relates to each metabolic pathway and substance used within the OC process. A metabolic pathway is described in terms of variables of the chemical reactions, by its relations with other pathways, and by actions, all those allowed in the pathway. A substance, as the carbohydrate, is described both in terms of its attributes and its predicates. Figure 4 shows only part of the ontology, the complete description is available in [4]. Beside, the description of cell components, the Communication Ontology Diagram describes the components interaction i.e. the agents communication language. The communication language consists of a set of primitives. Each primitive specifies the sender and its role, the receiver and its role and the type of message. Figure 5 shows part of the communication language, that occurs between two agents, *AgStatoAttuale* (i.e. environment) and *AgMitocondrialMatrix*. The communication primitive identified by *AgMatriceMitochondriale-AgStatoAttuale*, it is sent by the agent *AgMi-*

---

[4] http://www.genome.jp/kegg/pathway.html, http://sbml.org

tocondrialMatrix in its role *CicloAcitiCitrico* (i.e. Kreb's Cycle), to the agent *AgStatoAttuale* in its role of *AggiornaDatiCondivisi* (i.e. environment); the message is described w.r.t the *GruppoDatiCicloKreb* ontology (see Figure4, it is coded by XML and it is sent by the FIPA communication protocol[5].

Roles Description Diagram: it allows to define, for every agent w.r.t. its roles, the set of tasks and communications primitives the agent can use.

Protocol Description Diagram: it describes the communication protocol shared among agents during the ontological communication at higher level. The system has been implemented by using the FIPA Request protocol.

The simulation system, we have implemented, takes in input a carbohydrate *fructose, glucose, mannose, maltose, lactose, saccharose, glycogenous and starch* and behaves as a cellulae process, either aerobic (aerobic respiration in the mitochondria) or anaerobic (fermentation). It supports query as "Which is the amount of ATP formed when a determined enzyme is inhibited or lack in the system?", "Monitoring the metabolic pathway (the level of intermediate metabolites during the pathways)", "Establish the best carbohydrate source in terms of ATP production", "Monitoring the ratio of NADrid/NADox and FADrid/Fadox during the oxidation process". Table 1 summarizes some experimental cases in presence or in absence of oxygen; it is important to observe as *Maltose, Lactose* and *Saccharose* are very more energetic than the other carbohydrates; Also

Table 1. CellMAS simulation results for the oxidation of carbohydrate. For each listed Carbohydrate,note, that there is always equilibrium among ATP and ADT because the use of one provokes the creation of the other

Carbohydrate	Quantity	Oxygen	Fermentation	ATP	ADP	NA Drid	NA Dox	FA Drid	FA Dox	Acetil -CoA	Pyruvate	CO2	Ethanol	Lactic Acid
Glucose	1	yes	/	38	-38	0	0	0	0	0	0	6	0	0
Fructose	1	yes	/	38	-38	0	0	0	0	0	0	6	0	0
Mannose	1	yes	/	38	-38	0	0	0	0	0	0	6	0	0
Maltose	1	yes	/	76	-76	0	0	0	0	0	0	12	0	0
Lactose	1	yes	/	77	-77	0	0	0	0	0	0	12	0	0
Saccharose	1	yes	/	76	-76	0	0	0	0	0	0	12	0	0
Glycogen	1(1)	yes	/	39	-39	0	0	0	0	0	0	6	0	0
Starch	1(1)	yes	/	39	-39	0	0	0	0	0	0	6	0	0
Glycogen	3(2)	yes	/	234	-234	0	0	0	0	0	0	36	0	0
Glucose	3	No	Lactic	6	-6	0	0	0	0	0	0	0	0	6
Lactose	1	No	Lactic	5	-5	0	0	0	0	0	0	0	0	4
Glucose	1	No	Alcoholic	2	-2	0	0	0	0	0	0	2	2	0

Glycogen and *Starch* should be considered because they are polysaccharides; they are quantified in table with the quantity of *Polysaccharides* (length of each polysaccharides). It is important to notice that there is always equilibrium among ATP and ADP; NADox and NADrid, FADox and FADrid because the use of one, provokes the creation of the other and vice versa. The *Pyruvate* and *Acetyl − CoA* are important intermediates substances that are created and used during the process. The value zero of that substances meaning that they are completely consumed.

---

[5] http://www.fipa.org

5 Conclusion and Future Work

We found the close correspondence of system views to models natural and helpful in producing a simulation system, in contrast to a conventional mathematical model which lacks this modular structure. We expect the transparency and modularity of the system to be helpful in accommodating elaboration with new components and functionality, in collaboration with domain experts.

The simulation studies have encouraged us to exploit the agent paradigm to model the cell system at a finer grain. We will exploit mobility to simulate cell components movement [1], which will be possible because the system has been implemented with Hermes middleware for mobile computing [3]. We also aim to use biological ontologies, like those in KEGG, to specify the domain description model, and use a formal notation like SBML for system behavioral analysis.

Acknowledgements. We would like to thank Michal Young and Nicola Cannata for valuable comments on a preliminary version of this paper.

References

[1] L. Cardelli. Abstract machines of systems biology. In *Transaction on Computation System Biology*, LNCS. Springer-Verlag, 2005. to appear.
[2] P. Ciancarini and M. Wooldridge. Agent-oriented software engineering. In S. Verlag, editor, *1st International Workshop on Agent-Oriented Software Engineering*, volume 1957 of *LNCS*, pages 1–24, 2001.
[3] F. Corradini and E. Merelli. Hermes: agent-based middleware for mobile computing. In *Tutorial Book of SFM-05*, LNCS. Springer-Verlag, 2005. to appear.
[4] F. Corradini, E. Merelli, and M. Vita. Cell-mas: ATP production from carbohydrates oxydation. Technical Report TR05-2004, Università di Camerino, Dicember 2004.
[5] R. Garret and C. Grisham. *Biochemistry*. Sunder College Publishing, 1995.
[6] N. Jennings. An agent-based approach for building complex software systems. *Communication of ACM*, 44(6):35–41, Apr. 2001.
[7] H. Kitano. *Foundations of Systems Biology*. MIT Press, 2002.
[8] F. Zambonelli and A. Omicini. Challenges and research directions in agent-oriented software engineering. In *JAAMAS*, 2004.

Characterizing Complex Behavior in (Self-organizing) Multi-agent Systems

Bingcheng Hu and Jiming Liu

Department of Computer Science, Hong Kong Baptist University,
Kowloon Tong, Hong Kong
{bchu, jiming}@comp.hkbu.edu.hk

Abstract. In this paper, we show our work on characterizing complex behavior in self-organizing (SOMAS) and non-self-organizing (NonSOMAS) multi-agent systems. Through experiments and analysis, we investigate how Self-Organized Criticality (SOC) phenomena arise in SOMAS rather than in NonSOMAS. Furthermore, we compare the order of agent performance in the two types of systems and explain its implications.

1 Introduction

In RoboNBA [9][10] and RoboCup [7], a player needs to cooperate with its teammates such that its team can perform better. For example, a player tends to pass the ball to a teammate that is in a good position, or tries to run to a position where it is not only ready to attack but also safe to catch the ball. Players exhibiting this cooperative behavior interact with each other and produce aggregated effects on the team.

Generally speaking, there are two types of multi-agent systems: self-organized (SOMAS) and non-self-organized (NonSOMAS). Under certain conditions, the aggregated behavior of multi-agent systems (MAS) cannot be directly understood from the individual agent behavior due to complex interactions among agents, and thus they are considered as SOMAS exhibiting emergent and complex behavior [13][17][20]. The aforementioned cooperative player team is a good example of SOMAS. On the other hand, if the behavior of agents cannot aggregate together, the MAS is considered as non-self-organized.

In order to characterize complex behavior in MAS, we need to evaluate the following properties of MAS:

1. Is there any Self-Organized Criticality (SOC) phenomenon in the system? SOC, as described by Bak [3][4], manifests in a wide range of natural and synthetic systems. Examples of natural systems are living systems [1], national GDP [19], and cities [22]. On the other hand, random boolean networks [5][6], sandpile models [4], and cyberspace [2][11][21] are relevant synthetic systems. Basically, SOC refers to the phenomena that power law distributions appear in natural or synthetic systems. In addition, Darly [8] attempted to use the idea of "emergence" to explain SOC. However, the problem of under what conditions SOC will appear in MAS has generally been overlooked by researchers in the MAS area.

2. We can check how the order of individual agents and MAS evolve. The order of a system refers to how accurately we can predict its future behavior. The order of a system is defined by entropy in statistical physics. The higher the entropy, the lower the order. However, occasionally it is not appropriate to use the entropy to define the order, such as when time series data are present. On the other hand, the second law of thermodynamics states that the entropy of an isolated system always increases, which mean the order of the system always decreases. Our findings should obey this law.

Now the question that remains is how to induce different complex behavior of MAS. First, we need to specify the behavior of individual agents. These agents must be able to aggregate at both individual-level (low-level) and system-level (high-level). Furthermore, the aggregation at the two levels must interact collectively, which means the low-level aggregation can influence the high-level aggregation and the high-level can also have an impact on the low-level [18]. Finally, we should also characterize MAS under different configurations of individual agents, such as different behavior of agents and different interactions among them.

1.1 Organization

The paper is organized as follows. In Section 2, we present and explain the mathematical formulation of our MAS. Section 3 defines and discusses the measurements that we adopt. Experiments and discussions are included in Section 4, followed by Section 5, which concludes the paper.

2 Formulation

We study two types of MAS to characterize different behavior generated by them.

2.1 Non-self-organizing Multi-agent Systems (NonSOMAS)

In NonSOMAS, there are no critical interactions between agents, and therefore agents can be approximately viewed as independent of each other. Hence, one single agent is able to simulate the whole system. For example, in the computer simulated football match RoboCup, if players (a player can be consider as an agent) do not cooperate with their teammates, one player is representative enough for the whole team.

The formulation in Subsection 2.1 only considers one player, which competes with the external environment. For each clock cycle, if the player beats the external environment, its performance will be increased; otherwise it will be deducted. This is designed according to the spirit "only the fittest will survive" or positive feedback [17], in short. The update function of player performance is defined as:

$$\alpha(t+1) = \begin{cases} \alpha(t) + \beta \ Q(\alpha(t)) = 1 \\ \alpha(t) - \beta \ Q(\alpha(t)) = 0 \end{cases} \quad (1)$$

- $\alpha(t)$ is the performance of the player at time t.
- $Q(x)$ denotes the process that the player competes with the external environment:

$$Q(x) = \begin{cases} 1 & x > rand(\gamma) \\ 0 & x \leq rand(\gamma) \end{cases} \quad (2)$$

- β is the step size for α.
- $rand(x)$, $x > 0$, generates a random number within the range of [-x, x].

2.2 Self-organizing Multi-agent Systems (SOMAS)

The formulation in Subsection 2.2 takes into consideration competitive scenarios of MAS. Specifically, it attempts to capture the dynamic process of RoboNBA or RoboCup game: each match is composed of a number of attacks, which themselves consist of a number of clock cycles. At each clock cycle, one player is selected from each of the two teams and competes with each other. Note that we use some approximations, such as constant intervals of attacks and clock cycles.

The Process. A complete competition process is divided into a number of attacks, and an attack is composed of several clock cycles. The dynamics of SOMAS can be described by a set of variables and update functions (By dynamics, we mean the different behavior of an agent or MAS as time evolves):

- $\alpha_{ij}(t)$ denotes the performance of player j from team i at time t, where $i \in \{0, 1\}$ and $j \in \{1, 2, \cdots, 5\}$
- The update function of $\alpha_{ij}(t)$ is defined as:

$$\begin{aligned}
\alpha_{ij}(t+s) &= F(\alpha_{01}(t+s-1), \cdots, \alpha_{05}(t+s-1), \alpha_{11}(t+s-1), \cdots, \\
&\quad \alpha_{15}(t+s-1)), where, \; mod(t,m) = 0, 1 \leq s < m. \\
\alpha_{ij}(t+m) &= G(\alpha_{01}(t+m-1), \cdots, \alpha_{05}(t+m-1), \alpha_{11}(t+m-1), \cdots, \\
&\quad \alpha_{15}(t+m-1), R_i(t), R_j(t), M_i(t/m), M_j(t/m)), \\
&\quad where, \; mod(t,m) = 0.
\end{aligned} \quad (3)$$

- m is the number of clock cycles of an attack.
- $mod(t, m)$ denotes the operation modulus after division.

Low-Level Aggregation. At each clock cycle, one player is selected from each team and competes with one another. The performance of players is adjusted accordingly in favor of positive feedback. In addition, the results of competitions are recorded. We consider the dynamics of performance as low-level aggregation, since it is defined on players, which are the lowest-level components in SOMAS. The following describe the dynamics of low-level aggregation:

- F updates the performance at all clock cycles except the last one of an attack.

$$F = \begin{cases} \alpha'_{ij} & H(t+s,i) \neq j \\ \alpha'_{ij} + \beta & H(t+s,i) = j \; \&\& \; Y(\alpha'_{ij}, \alpha'_{nz}) = 1 \\ \alpha'_{ij} - \beta & otherwise \end{cases} \quad (4)$$

- $\alpha'_{ij} = \alpha_{ij}(t+s-1)$ and $\alpha'_{nz} = \alpha_{nz}(t+s-1)$.
- $n = V(i), z = H(t+s, n)$.
- $V(i)$ returns the ID of team i's opponent team.

$$V(i) = \begin{cases} 1 & i = 0 \\ 0 & i = 1 \end{cases} \tag{5}$$

- $H(t+s, i)$ selects a player from team i and returns the player ID at time $t+s$. By default, $H(x)$ uses random selection.
- $Y(\alpha_{ij}(t+s-1), \alpha_{nz}(t+s-1))$ denotes the process that player j from team i competes with player z from team n, and decides which one wins.

$$Y(\alpha_1, \alpha_2) = \begin{cases} 1 & \alpha_1 + rand(k) > \alpha_2 + rand(k) \\ 0 & otherwise \end{cases} \tag{6}$$

- k denotes the influential power of the external environment.

High-Level Aggregation. At the last clock cycle of an attack, the competition records and morale of two teams determine the result of the attack. In addition, the morale of the each team is modified accordingly in favor of positive feedback. We consider the dynamics of morale as high-level aggregation, since it is defined on a team, which is a higher hierarchical organization than players in SOMAS. The morale is influenced by the player performance and can change it in turn. The following describe the dynamics of the high-level aggregation:

- $M_i(t/m)$ is the morale of team i at attack number t/m. $M_i((t+m)/m)$ is updated by function P.

$$P = \begin{cases} M_i(t/m) + \delta \, O(R_i(t+m), R_n(t+m), M_i(t/m), \\ \qquad M_n(t/m)) = 1, where \, n = V(i) \\ M_i(t/m) - \delta \; otherwise \end{cases} \tag{7}$$

- O determines which team wins at an attack.

$$O = \begin{cases} 1 & R_i(t+m) - R_n(t+m) + \chi(M_i(t/m) \\ & -M_n(t)) > 0, where \, n = V(i) \\ 0 & otherwise \end{cases} \tag{8}$$

- $R_i(t)$ denotes the number of winning clock cycles of team i at time t. It is only valid within an attack interval.

$$R_i(t+s) = \begin{cases} R'_i + 1 & 1 \leq s < m \; \&\& \; Y(\alpha'_{ij}, \alpha'_{nz})) = 1 \\ R'_i & otherwise \end{cases} \tag{9}$$

- $n = V(i), z = H(t+s, n)$.

$$R'_i = \begin{cases} 0 & mod(t+s-1, m) = 0 \\ R_i(t+s-1) & mod(t+s-1, m) > 0 \end{cases} \tag{10}$$

Collective Feedback. The process as defined in P and O refers to how low-level aggregation influences high-level aggregation. In addition, one player is selected from each team and its performance is modified accordingly. The process as defined in G illustrates how high-level aggregation influences low-level aggregation:

- G updates the performance at the last clock cycle of an attack.

$$G = \begin{cases} \alpha_{ij}^2 & I(t+m,i) \neq j \\ \alpha_{ij}^2 + \beta & I(t+m,i) = j \text{ \&\& } O(R_i(t+m), R_n(t+m), \\ & M_i(t/m), M_n(t/m)) = 1, \text{ where } n = V(i) \\ \alpha_{ij}^2 - \beta & \text{otherwise} \end{cases} \quad (11)$$

where,
- $\alpha_{ij}^2 = \alpha_{ij}(t+m-1)$.
- $I(t+m,i)$ selects a player from team i at time $t+m$ and returns the player ID. It is similar to $F(t+s,i)$. By default, it uses random selection.

Table 1 specifies parameters used in SOMAS.

Table 1. Parameters used in SOMAS

	Initial value	Range	Remark
α	5000	0-10000	performance for a player
M	5.0	0-10	morale for a team
γ	10000	a constant	range of a random number
m	6	a constant	no. of clock cycles per attack
β	1	a constant	step size for performance
δ	0.005	a constant	step size for morale
χ	0.1	a constant	an coefficient in O

3 Measurements

In order to characterize the aforementioned complex behavior, we have designed the following measurements.

- *The order of low-level and high-level aggregation.* The order can be defined by the autocorrelation function: $C(\tau) = lim_{N \to \infty} \frac{1}{N} \sum_{i=1}^{N} c(t_i) c(t_i + \tau)$ for discrete systems. We divide the sum by the number of $N - \tau$ to ensure fairness for each $C(\tau)$. If $C(\tau)$ decreases relatively slowly, $c(t)$ is considered relatively ordered. If $C(\tau)$ decreases rapidly (e.g., an exponential decay), $c(t)$ is considered as less ordered.
- *"Avalanches".* It refers to the various degree of changes in player performance or team morale. The size of an avalanche is defined as the number of steps of continuous increases or decreases. We call them avalanches because they are similar to those described in [4].
- *Distribution of avalanches.* It can be an exponential distribution, a power law distribution, or even a random one. The type of distribution makes a profound difference.

4 Experiments

4.1 Experiment 1

This experiment is carried out to examine the dynamics generated by NonSOMAS. Figure 1 is a typical output for extensive experiments. Figure 1 (a) illustrates the performance of the player as a function of time. From Figure 1 (b), we can see that the performance avalanche follows an exponential distribution, which indicates that a random variable is dominating the dynamics. Let the fixed probability for an increase in performance be p and then the probability for a corresponding decrease will be $1 - p$. So the probability for a n step continuous increase or decrease will be $(1-p)p^n$ or $p(1-p)^n$, respectively, given changes are independent of one another. Note we use 5000 for the initial value of α and it is compared with $rand(10000)$. Therefore, the probabilities to increase or decrease the performance are equally 0.5 and the probabilities for a n step increase or decrease are 0.5^{n+1}. Consequently, the performance avalanche follows an exponential distribution if the random variable dominates the dynamics and proper initial values are used.

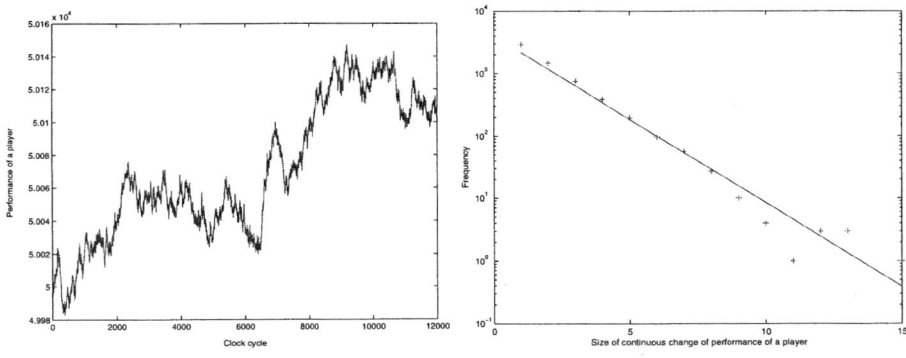

(a) Dynamics of performance of a player in NonSOMAS.

(b) An exponential distribution of performance avalanche.

Fig. 1. Experiment 1: Dynamics and some regularities of player performance in NonSOMAS

4.2 Experiment 2

This experiment is carried out based on SOMAS. Through comparison on Experiment 1 and Experiment 2, we try to have some basic understanding on the Self-Organized Criticality phenomena appearing in SOMAS.

Figure 2 is a typical output for extensive experiments. Figure 2 (a) illustrates the performance of a player as a function of time. From Figure 2 (b), we can see that the performance avalanche follows a power law distribution, which is significantly different from the exponential distribution we observed in Experiment 1. This result is quite robust since it appears under different conditions. For example, we use a wide range of k values in $Y(x, y)$ function, and the power law distribution is still there. We are quite

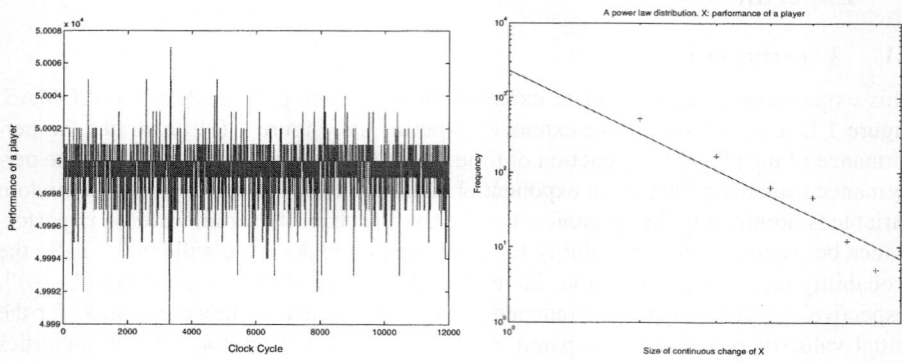

(a) Dynamics of performance of a player in SOMAS.

(b) A power law distribution of performance avalanche.

Fig. 2. Experiment 1: Dynamics and some regularities of player performance in SOMAS

confident at the moment the dynamics is not dominated by a random variable otherwise an exponential distribution will appear. What are the underlying mechanisms that generate this power law distribution? We would like to refer to SOC proposed by Bak [3] to explain the underlying mechanisms: under particular conditions, the system organizes by driving itself slowly and eventually it comes to a critical point, where it exhibits complex behavior, such as power law distribution (scale free in space). However, Bak did not point out specific conditions under which SOC would appear.

For our general competitive multi-agent model, SOMAS, we consider the SOC phenomena appearing in it is due to the following reasons:

1. Critical interactions between players are important. In NonSOMAS, there is no critical interaction between players and their behavior is relatively ordered and predictable. On the other hand, in SOMAS, there are a total of ten players interacting together. Thus player behavior is relatively harder to predict and thus less ordered. From Figure 3, we can observe the remarkable difference between the order of player performance in the two models. From Figure 3 (a), we can see that the autocorrelation function value is decreasing quite slowly for NonSOMAS. However, in Figure 3 (b), there is an exponential decay as τ increases, which indicates lower order and predictability for player performance in SOMAS.
2. A higher level organization, team morale (M) in SOMAS, is essential. Due to the existence of team morale and its collective interactions with player performances through functions G, O, and P, we have a collective design of the model rather than more traditional ones, such as top-down or bottom-up design. Thanks to this collective design and backward causation embedded inside it, complex behavior of the model becomes possible [12][15][18]. Figure 4 illustrates a typical dynamics for team morale. From Figure 5 (a), interestingly, we can observe a power law distribution on the team morale avalanche. Furthermore, we can observe a slow decay in Figure 5 (b).

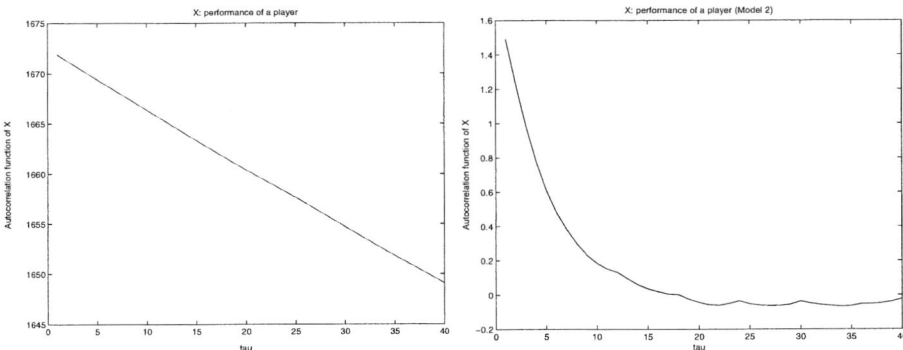

(a) Autocorrelation as a function of τ on performance of a player in NonSOMAS (x axis: τ, y axis: autocorrelation function)

(b) Autocorrelation as a function of τ on performance of a player in SOMAS (x axis: τ, y axis: autocorrelation function)

Fig. 3. Comparison on the speed of autocorrelation function decay for player performance as τ increases

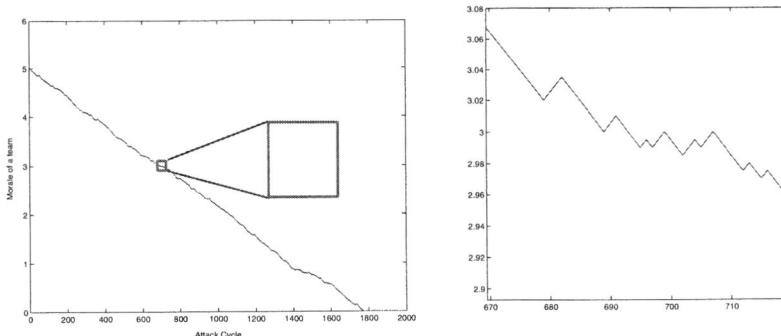

(a) Dynamics of morale of a team (x axis: morale, x unit is 0.005, y axis: attack cycle). The rectangle is enlarged in Figure 4 (b)

(b) A sub plot of Figure 4 (a) (x axis: morale [670-730], y axis: attack cycle [2.90-3.08]).

Fig. 4. Experiment 2: Morale dynamics of a team in SOMAS

In summary, we can observe SOC phenomena in both low-level (performance) and high-level (team morale) aggregation in SOMAS. Furthermore, thanks to the collective design specified in Subsection 2.2, the SOC phenomena exhibit by decreasing the order on the lower-level aggregation and preserving relatively high order on the higher-level aggregation in the model. This result is consistent with the properties of self-organization process in MAS [14].

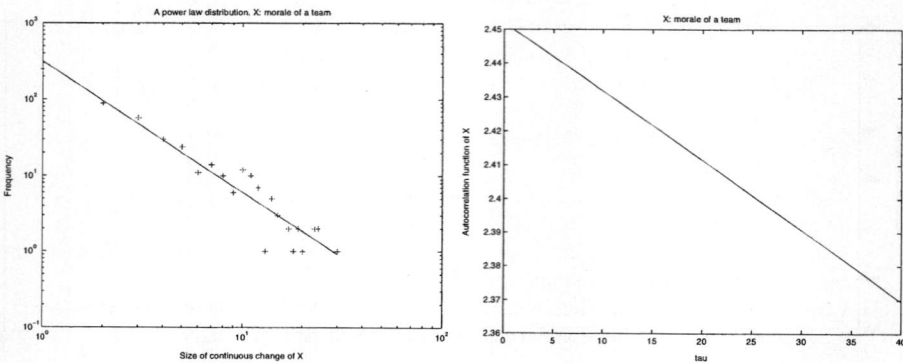

(a) A power law distribution of morale avalanche of a team (x axis: size of morale avalanche, y axis: frequency)

(b) Autocorrelation as a function of τ on team morale in SOMAS (x axis: τ, y axis: autocorrelation function)

Fig. 5. Experiment 2: Regularities of morale of a team in SOMAS

5 Conclusions

In this paper, we built models for two types of MAS (NonSOMAS and SOMAS) in order to induce and characterize complex behavior in MAS. Based on the two models, we carried out three experiments. Comparing and analyzing the experimental results, we gained some understanding of the Self-Organized Criticality phenomena that occurred in SOMAS. Furthermore, we discovered that the avalanches of player performance and team morale followed a power law distribution.

References

1. C. Adami. Self-organized criticality in living systems. Physics Letter A, 203:2932, 2 1995.
2. R. Albert, H. Jeong, and A. Barabasi. Diameter of the world-wide web. Nature, 401:130, 1999.
3. P. Bak. How Nature Works. Springer-Verlag, 1996.
4. P. Bak, C. Tang, and K.Wiesenfeld. Self-organized criticality. Physical Review A, 38(1):364 374.
5. A. Barabasi and R. Albert. Emergence of scaling in random networks. Science, 286:509 512, October 1999.
6. A. Barabasi, R. Albert, and H. Jeong. Scale-free characteristics of random networks: The topology of the world-wide web. Physica A, 281:7077, 2000.
7. M. Chen, K. Dorer, E. Foroughi, F. Heintz, Z. X. Huang, S. Kapetanakis, K. Kostiadis, J. Kummeneje, J. Murray, I. Noda, O. Obst, P. Riley, T. Steffens, Y. Wang, and X. Yin. RoboCup server manual. RoboCup Federation, February 2003.
8. V. Darley. Emergent phenomena and complexity. Artificial Life, 1994.
9. B. Hu, J. Liu, and X. Jin. From local behaviors to global performance in a multi-agent RoboNBA system. In IEEE/WIC/ACM Intelligent Agent Technology (IAT04), pages 309 314, Beijing, 2004.

10. B. Hu, J. Liu, and X. Jin. Multi-agent RoboNBA simulation: From local behaviors to global characteristics. Special Issue: Agent-Directed Simulation of the Simulation: Transactions of the Society for Modeling and Simulation International, 2005, to appear.
11. B. Huberman and L. Adamic. Growth dynamics of the world-wide web. Nature, 40:450457, 1999.
12. P. Marcenac and S. Calderoni. Self-organisation in agent-based simulation. IREMIA, University of La Runion.
13. M. Mitchell and M. Newman. Complex systems theory and evolution. Sante Fe Institue Working Paper, 4 2001.
14. H. Parunak and S. Brueckner. Entropy and self-organization in multi-agent systems. In International Conference on Autonomous Agents, 2001.
15. H. Parunak, S. Brueckner, M. Fleischer, and J. Odell. Co-x: Defining what agents do together. Workshop on Team and Coalition Formation, AAMAS02, Bologna, Italy.
16. C.W. Reynolds. Flocks, herds, and schools: A distributed behavioral. In Computer Graphics: Proceedings of SIGGRAPH 87, volume 21(4), pages 2534, July 1987.
17. J. Rosnay. Feedback. http://pespmc1.vub.ac.be/FEED-BACK.html.
18. R. Sawyer. Simulating emergence and downward causation in small groups. In: S. Moss and P. Davidsson (Eds.) Multi-Agent-based Simulation. Second InternationalWorkshop, MABS, 2000.
19. N. Shiode and M. Batty. Power law distributions in real and virtual worlds. University College London, UK.
20. R. V. Sole, E. Bonabeau, J. Delgado, P. Fernandez, , and J. Marin. Pattern formation and optimization in army ant raids. Artificial Life, 6:219227, 2001.
21. S. Valverde and R. Sole. Self-organized critical traffic in parallel computer network. Sante Fe Institue Working Paper, 2001.
22. G. Zipf. Human Behavior and The Principle of Least Effort. Addison Wesley, Cambridge, MA, 1949.

Protein Structure Abstraction and Automatic Clustering Using Secondary Structure Element Sequences

Sung Hee Park[1], Chan Yong Park[1], Dae Hee Kim[1], Seon Hee Park[1], and Jeong Seop Sim[2]

[1] Bioinformatics Team, Electronics and Telecommunications Research Institute,
161 Gajung, Yusung, Daejeon, Korea
{sunghee, cypark, dhkim98, shp}@etri.re.kr
http://www.etri.re.kr
[2] School of Computer Science and Engineering, Inha University, Korea
jssim@inha.ac.kr

Abstract. To study protein clustering is very important in diverse fields such as drug design and environmental industry. For a meaningful clustering, protein structure must be considered. But, protein structures are very complicated and have so much information such as angles, 3-dimensional coordinates. Thus, it is not easy to efficiently compute their relations. In this paper, we present a method to efficiently abstract and cluster protein structures using secondary structure element sequences. Since a secondary structure element sequence is an abstract representation of protein structure, it can be regarded as a useful descriptor to cluster a set of proteins at the abstraction level. Using secondary structure element sequences and their distances, we implemented an automatic protein clustering system and verify their efficiency by experimental results.

1 Introduction

During recent years, many efforts have been made to analyze the relation between structure and function. Most previous research work focused on classifying protein families based on homology [1][2][3][4][5]. A major assumption of previous works is that the protein families or functional categories are known in advance and the protein features like sequence or structural features used to make the classification model are labeled with the corresponding families or categories. As a well known technique in statistics and computer science, clustering has been proven very useful in detecting unknown object categories and revealing hidden correlations and pattern among objects. In this paper, we are involved in the problem of automatic clustering of protein structure.

Protein clustering is very important and has applications in such diverse fields as drug design, molecular biology, and environmental industry. By recently, protein clustering has been carried out by protein sequence [6][7][8]. To cluster proteins effectively, we must take into account structures of proteins. But due to the complex features of proteins, it is not easy to effectively and efficiently figure out their similarities in the aspects of structures and functions. To resolve these difficulties, most research on clustering focused on defining efficient similarity between proteins [9][10][11][12][13]. Holm and Sander tried to calculate similarity by alignment of residue-residue (C^a-C^a) distance matrices in [9]. But this approach is computationally very complex and sensitive to errors. Recently, some efficient similarities were proposed. Schwarzer and Lotan proposed a fast similarity that is calculated using segments-segment distance matrices in [10] instead of residue-residue distance matrices in [9], where segment consists of several residues. Another approach, by Singh and Brutlag, was proposed that calculate similarities with the vector representations represent vectors of secondary structures [11]. We found a trend of abstraction of protein structure from previous work.

In this paper, we present a method to automatically cluster proteins using a well known abstract descriptor.

We construct our paper as follows. In section 2, we define and explain some notations. In section 3, we describe the algorithm of our clustering system. In section 4, we explain our implemented system and result analysis. Section 5, we conclude.

2 Preliminaries

2.1 Secondary Structure Element Sequence

Functionally related protein domains consist of some combinations of secondary structures. These secondary structure representations are more abstract representations of protein structures than the primary structure representations. The secondary structure element sequence (SSES for short) is a descriptor which describes a protein structure as follows: first, analyze a protein and extract secondary structures, next label each extracted secondary structure as alpha, and beta from N-terminal direction to C-terminal direction of protein amino acid peptide chains. For example, Table 1 shows

Table 1. The primary and secondary structure sequence of protein 1a3x:a

Representation type	Representation(protein 1a3x:a)
Protein Sequence	(1) MSRLERLTSL NVVAGSDLRR TSIIGTIGPK TNNPETLVAL RKAGLNIVRM (51) NFSHGSYEYH KSVIDNARKS EELYPGRPLA IALDTKGPEI RTGTTTNDVD (101) YPIPPNHEMI FTTDDKYAKA CDDKIMYVDY KNITKVISAG RIIYVDDGVL (151) SFQVLEVVDD KTLKVKALNA GKICSHKGVN LPGTDVDLPA LSEKDKEDLR (201) FGVKNGVHMV FASFIRTAND VLTIREVLGE QGKDVKIIVK IENQQGVNNF (251) DEILKVTDGV MVARGDLGIE IPAPEVLAVQ KKLIAKSNLA GKPVICATQM (301) LESMTYNPRP TRAEVSDVGN AILDGADCVM LSGETAKGNY PINAVTTMAE (351) TAVIAEQAIA YLPNYDDMRN CTPKPTSTTE TVAASAVAAV FEQKAKAIIV (401) LSTSGTTPRL VSKYRPNCPI ILVTRCPRAA RFSHLYRGVF PFVFEKEPVS (451) DWTDDVEARI NFGIEKAKEF GILKKGDTYV SIQGFKAGAG HSNTLQVSTV (500 residues)
Secondary Structure Element Sequence	αβααβαβαββββαααβααβααβαααααααβαβα (21 alpha helices and 12 Beta sheets)

the primary structure sequence and the SSES of the protein whose PDB code is 1a3x. The "A" chain of PDB code 1a3x is composed of quite long (about 500) amino acid peptide bonds. We call this descriptor a primary structure amino acid sequence. See second row in Table 1. Also, we can represent a protein by an SSES as shown in third row of Table 1. As we can see, if we represent a protein with an SSES, we can reduce the amount of data required to abstract the protein to 7 percent of data with a primary structure sequence.

2.2 Similarity (or Distance) Measure Between Two Proteins

In this paper, we use the distance (or error) to measure the similarity between two proteins. That is, if two proteins are very similar, their distance is small (or near) and if they are very different, their distance is large (or far). The distance of two proteins is defined to be the distance between each secondary structure sequence of the two proteins. Thus, we can compare distances between proteins with generic sequence comparing algorithms, such as the Hamming distance, the edit distance (also called Levenshtein distance), and weighted edit distances [14]. The Hamming distance is the number of mismatched characters in the two sequences but defined only when the lengths of two sequences are the same. The edit distance is the minimum number of required operations (such as insert, delete, and change operations) to transform one sequence into another. The weighted edit distance is a generalization of the edit distance. All these distances can be computed by dynamic programming.

2.3 Clustering Method and Cluster Validity Technique Used in Our System

General clustering algorithms can be used such as maximum-distance algorithm, K-means algorithm, and ISODATA (Iterative Self-Organizing Data analysis technique) algorithm. For our automatic clustering system, K-means clustering algorithm was used.

K-Means clustering algorithm. K-means algorithm can be defined as follows. This nonhierarchical method initially takes the number of clusters among the population equal to the final desired number of clusters. In this step itself the final required number of clusters is chosen such that the points are mutually farthest apart. Next, it examines each object in the population and assigns it to one of the clusters depending on the minimum distance. The position of centroid is recalculated every time an object is added to the cluster and this continues until all the components are grouped into the final required number of clusters.

K-means clustering algorithm finds clusters in a set of unlabeled data as much as the desired number of, K, initial cluster. Automatic K-means clustering should provide best cluster partition. The clustering partition that optimizes the validation measure under consideration is chosen as the best partition [15]. We use cluster validation measures so that we may select the optimal number of cluster center. A cluster validity index indicates the quality of a resulting clustering process. There are diverse cluster validation measures such as silhouette method [16], Dunn's method [17], Davies-Bouldin method [18], and so on [6][19][20]. We used silhouette method as validity index for our system. Silhouette method is described below.

Silhouette method. For a given cluster, X_j ($j = 1, ..., c$), the silhouette technique assigns to the i-th sample of X_j a quality measure, s(i) (I = 1, ..., m), known as the silhouette width. This value is a confidence indicator on the membership of the ith sample in cluster X_j and it is defined as:

$$S(i) = \frac{(b(i) - a(i))}{\max\{a(i), b(i)\}},\qquad(1)$$

where $a(i)$ is the average distance between the i-th sample and all of the samples included in X_j; and $b(i)$ is the minimum average distance between the i-th sample and all of the samples clustered in X_k ($k=1, \ldots, c$; $k \neq j$).

When $S(i)$ is close to 1, one may infer that the ith sample has been assigned to an appropriate cluster. When $S(i)$ is close to zero, it suggests the sample lies equally far away from both clusters, in other words, has been "misclassified". For a given cluster, Xj it is possible to calculate a cluster silhouette S_j, which characterizes the heterogeneity and isolation properties of such a cluster. It is calculated as the sum of all samples' silhouette widths in Xj. Moreover, for any partition, a global silhouette values or silhouette index, GS_u, can be used as an effective validity index for a partition U.

$$GS_u = \frac{1}{c}\sum_{j=1}^{c} S_j,\qquad(2)$$

In this case the partition with the maximum silhouette index value is taken as the optimal partition.

3 Clustering Algorithms Using SSES

Now we explain our clustering algorithm using SSES. Our system consists of following three modules: i) preprocessing module, ii) distance matrix computing module, and iii) clustering module. Details are shown below. See Fig. 1.

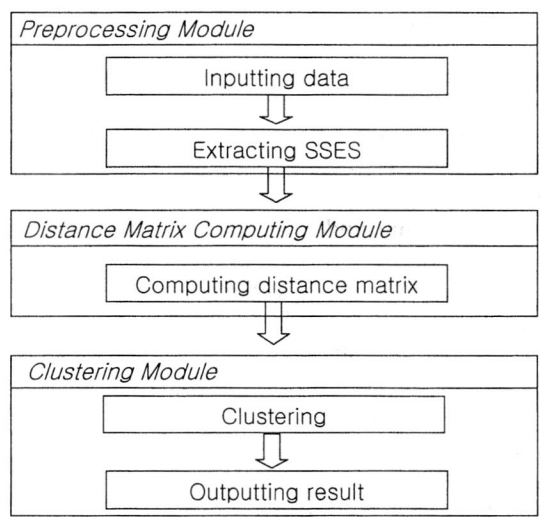

Fig. 1. Processes flow of automatic clustering system

3.1 Preprocessing Module

In this module, we preprocess the input and make SSES. The input of the protein structure data input module can either be a file which consists of 3 dimensional coordinates of atoms that compose proteins or be a secondary structure sequence. Input file format is Protein Data Bank (PDB for short) format. When the input to the module is a PDB file, there are two cases. In case when there is a SSE field, then we parse the PDB file for the protein and the SSES can be read directly. In case when there is no SSE filed, STRIDE program [21] is used to extract SSES from the PDB file. When the input is a secondary structure sequence, we can directly use this sequence to measure the similarity between proteins.

3.2 Distance Matrix Computing Module

This module compares two proteins using the distance between each SSES of two proteins. That is, we compare two proteins by computing the distance between two secondary structure sequences. In this paper, we use the edit distance for the distance measure. The edit distance can be computed easily using dynamic programming. Let $D(X,Y)$ be the edit distance between two sequences X and Y. Let $X[i]$ denote the i-th symbol (or character) of X and $X[i..j]$ denote $X[i]X[i+1]...X[j]$. Then we can compute $D(X[1..i],Y[1..j])$ easily with three values $D(X[1..i-1],Y[1..j])$, $D(X[1..i],Y[1..j-1])$, and $D(X[1..i-1],Y[1..j-1])$ by following recurrence.

$D(X[1..i],Y[1..j])$ = min { $D(X[1..i-1],Y[1..j])+1$, $D(X[1..i],Y[1..j-1])+1$, $D(X[1..i-1],Y[1..j-1])+Z$}, where Z is 0 when $X[i]=Y[j]$ and otherwise, Z is 1.

By the above recurrence, we can compute all the distances between all secondary structure sequences of proteins and build a distance matrix. This distance matrix will be used several times in the clustering module. See Fig. 2.

3.3 Clustering Module

We use K-means clustering method. This method has been applied to analyze expression profiles in several biomedical and systems biology studies [22].

First, let the number of cluster K = 2. and then assign two initial sample to two initial cluster center $Za(1)$, $Zb(1)$ ($Za \neq Zb$), each other. Next, it examines each object in the population and assigns it to one of two clusters depending on the minimum distance. At this time, we use distance matrix. Distance matrix has all distance information for all samples. Ties are resolved arbitrarily. For example, in the case of Fig. 2 (a), for initial cluster center $Za(1)$ = A, $Zb(1)$ = B ($A \neq B$), C is assigned to Za closer than A. All other samples D,..., I are also assigned to closer cluster center. For the two Cluster Sa, Sb, find new cluster centers $Za(2)$, $Zb(2)$. New cluster center is given by a sample with a minimum radius from samples in the cluster, where radius of a sample is defined to maximum distance from it to all other sample, $Ri = \max(D(i,j))$ for I, j $\in$ S, where $D(i,j)$ is distance from sequence i to j. In other words, a sample with minimum radius is located near cluster center, as the radius of center of circle is minimum in comparison with any other point inside the circle.

For all other cluster, finding new cluster center is same. If $Z_i(n+1) = Z_i(n)$, for i = 1, 2, the algorithm has converged. Otherwise, it examines each object in the population and assigns it to one of two new clusters depending on the minimum distance.

For example, in Fig. 2, the new cluster centers are protein C and E.

For a validity technique, Silhouette method is used. Equation (2) can be applied to estimate the "correct" number of clusters for partition where the number of cluster k, (in case of our system, 2 < k < 20) with maximum silhouette index value GS_u in equation (2) is taken as the optimal number of clusters.

	A	B	C	D	E	F	G	H	I
A	0	3	4	5	4	3	5	7	2
B	3	0	2	3	5	5	4	3	3
C	4	2	0	2.5	2	4	3	4	4
D	5	3	2.5	0	7	8	7	5	3
E	7	5	2	7	0	2	3	7	2
F	3	5	4	8	2	0	4	4	6
G	5	6	3	7	3	4	0	6	4
H	7	3	4	5	7	4	6	0	4
I	2	3	4	3	2	6	4	4	0

(a)

	A	B	C	D	E	F	G	H	I
A	0	3	4	5	4	3	5	7	2
B	3	0	2	3	5	5	4	3	3
C	4	2	0	2.5	2	4	3	4	4
D	5	3	2.5	0	7	8	7	5	3
E	7	5	2	7	0	2	3	7	2
F	3	5	4	8	2	0	4	4	6
G	5	6	3	7	3	4	0	6	4
H	7	3	4	5	7	4	6	0	4
I	2	3	4	3	2	6	4	4	0

(b)

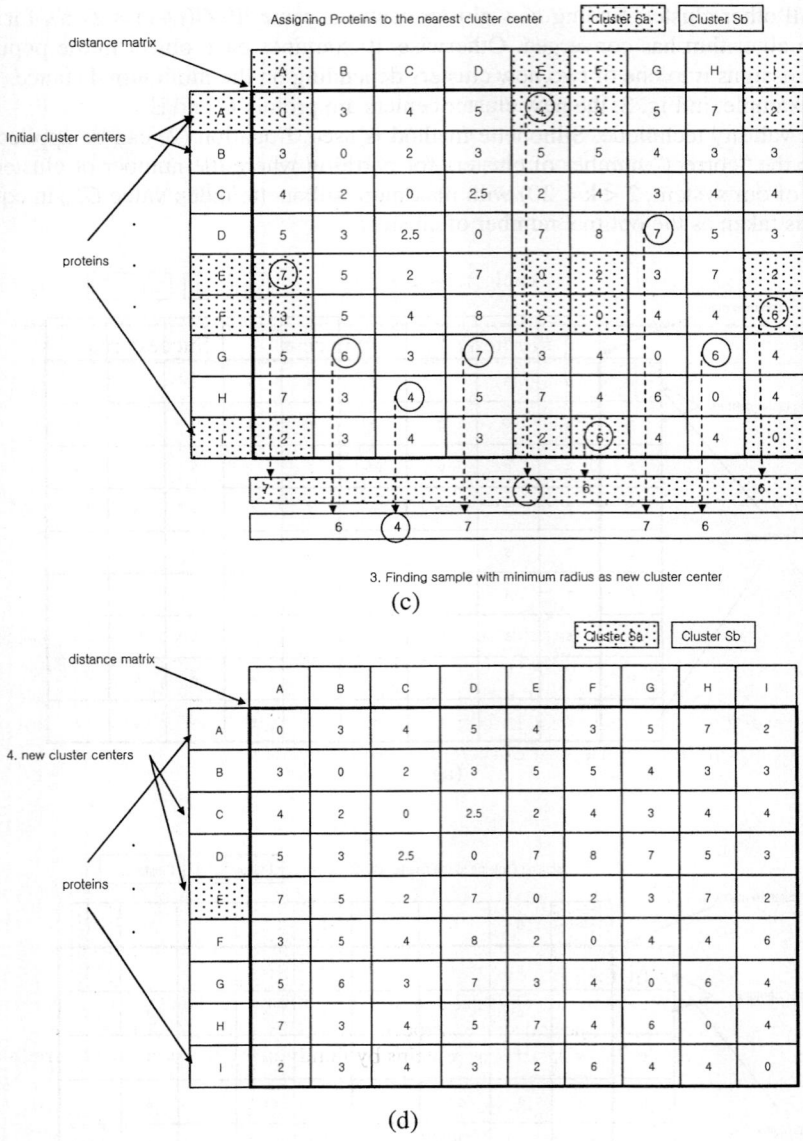

Fig. 2. An example K-means clustering with distance matrix: (a) initial cluster centers when k = 2 (b) assigning proteins to their nearest cluster center (c) finding sample with minimum radius as new cluster center (d) new cluster center

4 Computer Experiments and Results

Our clustering system was implemented in C++ programming language on Windows operating system. We used silhouette validation measure as a cluster validation measure. We tested our system for the dataset used in [5]. There are four clusters in the protein data used in [5]. We iteratively clustered the proteins with the initial number of clusters from 2 to 9, and applied the silhouette validation measure. The suggested number of clusters by silhouette validation measure was 4. The results are shown at Table 1. Note that we use only the SSES, that is, we don't use any other information such as angles, types and lengths. Nevertheless, our system clusters the input proteins quite well. That is, the number of clusters is the same as [5] and the success ratio is comparable to the results in [5].

Table 2. Success rate

Class	#Proteins	#Success	Success rate
All alpha	18	13	72%
All beta	32	28	87%
Alpha/beta	35	11	31%
Alpha+beta	6	0	0%

5 Conclusion

In this paper, we used SSES to represent protein structure. Since we just use the sequence data (in the aspect of data type, actually, including structural information) to cluster proteins with complex structure, our system is very simple but accurate. We highly believe that if we use other information such as angles and/or types as well as SSES information, the accuracy of our system would be better.

References

1. A.G. Murzin and S.E. Brenner and T. Hubbard and C. Chotia.: J. Mol.Biol. 247 (1995) 536-540
2. J.F. Gibrat, T. Madej, and S. H. Bryant: Surprising similarities in structure comparison. Curr. Opin. Struct. Biol. 6(3) (1996) 377-385
3. Apostolico, A. and Bejerano, G.: Optimal amnesic probabilistic automata or how learn and classify proteins in linear time and space.: Proc. of ACM RECOMB (2000) 25-32
4. Bailey, T. and Grundy, W.: Classifying proteins by family using the product of correlated p-values. Proc. of ACM RECOMB (1999) 10-14
5. Dorohonceanu, B. and Nevill-Manning, C.: Accelerating protein classification using suffix trees. Proc. of Intelligent Systems for Molecular Biology (2000)
6. N. Bolshakova, F. Azuaje: Improving expression data mining through cluster validation. Fourth Annual IEEE EMBS Special Topic Conference on Information Technology Applications in Biomedicine (2003)
7. Jiong Yang, Wei Wang: Towards Automatic Clustering of Protein Sequences. CSB 2002 (2002) 175-186
8. Dubey, A., S. Hwang, C. Rangel, C. E. Rasmussen, Z. Ghahramani and D. L. Wild: Clustering Protein Sequence and Structure Space with Infinite Gaussian Mixture Models. Pacific Symposium on Biocomputing 2004 (2003)

9. L. Holm and C.Sander: Protein Structure Comparison by alignment of distance matrices. Journal of Molecular Biology, Vol. 233 (1993) 123-138
10. Rabian Schwarzer and Itay Lotan: Approximation of Protein Structure for Fast Similarity Measures. Proc. 7th Annual International Conference on Research in Computational Molecular Biology(RECOMB) (2003) 267-276
11. Amit P. Singh and Douglas L. Brutlag: Hierarchical Protein Structure Superposition using both Secondary Structure and Atomic Representation. Proc. Intelligent Systems for Molecular Biology (1993)
12. S.H. Park, S.J. Park, S.H. Park: A Protein Structure Retrieval System Using 3D Edge Histogram. Key Engineering Materials, Vols. 277-279 (2005) 324-330.
13. T. Ohkawa, S. Hirayama, and H. Nakamura: A method of comparing protein structures based on matrix representation of secondary structure pairwise topology. In 4th IEEE Symposium on Intelligence in Neural and Biological Systems (2001) 10-15
14. Gusfield, D.: Algorithms on Strings, Trees, and Sequence. Cambridge University Press. (1997)
15. F. Azuaje, N. Bolshakova: Clustering genome expression data: design and evaluation principles. In: D. Berrar, W. Dubitzky, M.Granzow, Ed.: Understanding and using microarray analysis techniques: A practical guide. London: Springer Verlag (2002)
16. P.J. Rousseeuw: Silhouettes: a graphical aid to the interpretation and validation of cluster analysis. J. Comp App. Math, vol. 20. (1987) 53-65
17. J. Dunn: Well separated clusters and optimal fuzzy partitions. J. Cybernetics. Vol. 4. (1974) 95-104
18. D.L. Davies, D.V. Bouldin: A cluster separation measure. IEEE Transactions on Pattern Recognition and Machine Intelligence. Vol. 1. No. 2. (1979) 95-104
19. J.C. Bezdek, N.R. Pal: Some new indexes of cluster validity. IEEE Transactions on Systems, Man and Cybernetics. vol. 28, part B. (1998)301-315,
20. N. Bolshakova, F. Azuaje, "Cluster validation techniques for genome expression data classification", Signal Processing, 2002, in press.
21. Frishman,D & Argos,P: Knowledge-based secondary structure assignment. Proteins: structure, function and genetics. 23 (1995) 566-579.
22. J. Quackenbush: Computational analysis of microarray data. Nature Reviews Genetics. Vol. 2. (2001) 418-427

A Neural Network Method for Induction Machine Fault Detection with Vibration Signal

Hua Su[1], Kil To Chong[1], and A.G. Parlos[2]

[1] Department of Electrical and Computer Engineering, Chonbuk National University,
Jeonju, Korea 561-756
{hua_su, kitchong}@chonbuk.ac.kr
[2] Department of Mechanical Engineering, Texas A&M University,
College Station, TX 77843-3123, USA
a-parlos@tamu.edu

Abstract. Early detection and diagnosis of induction machine incipient faults are desirable for online condition monitoring, product quality assurance, and improved operational efficiency. However, conventional methods have to work with explicit motor models and cannot be used for vibration signal case because of their non-adaptation and the random nature of vibration signal. In this paper, a neural network method is developed for induction machine fault detection, using FFT. The neural network model is trained with vibration spectra and faults are detected from changes in the expectation of vibration spectra modeling error. The effectiveness and accuracy of the proposed approach in detecting a wide range of mechanical faults is demonstrated through staged motor faults, and it is shown that a robust and reliable induction machine fault detection system has been produced.

1 Introduction

Induction machines play an important role in the safe and efficient operation of industrial plants because they are considered inherently reliable due to their robust and relatively simple design [1]. However, many electric machine components are especially susceptible to failures. In general, fault detection of induction motors has concentrated on sensing failures in one of the three major components, the stator, the rotor, and the bearings [2]. Because of the potential savings provided by motor fault detection system, many methods have been developed for the induction motors. Among most existing methods, the processing of machine vibration signals is the most popular choice for extracting diagnostic features, as vibration analysis techniques are quite effective in assessing a machine's health [3]. It is claimed that vibration monitoring is the most reliable method of assessing the overall health of a rotor system [4].

Among the methods analyzing the vibration signal, spectrum analysis in frequency domain is the most popular and computational approach, as machine defects produce vibrations with distinctive characteristic frequencies. The difficulty in fault detection is the problem of sorting through the enormous number of frequency lines present in vibration spectra to extract useful information associated with the health of induction motors. In order to overcome this problem, some studies have been reported recently [6-7]. But since there is no reliable way of predicting which kind of fault will happen, many potential defect frequencies must be monitored. Vibration spectra also often contain bewildering mixtures of extraneous frequencies that provide little or no pertinent diagnostic information about the health of motors.

The recent success of neural network for modeling highly complex system offers the potential for minimizing the above problems and realizing model-based fault detection [8]. In this research, a neural network method is developed in combination with FFT to extract fault spectra features used in the detection of machine faults. Steady-state vibration signals are detected first and a neural network motor model is trained with vibration spectra to generate residuals to compute fault indicators. In this study, the developed fault detection system is demonstrated with experimental results on real induction motor. The paper emphasizes the high-performance of the proposed system in detecting the most widely encountered mechanical faults.

Following this introduction, in Section II, the proposed neural network method is briefly described. Section III presents the detail procedures used in developing the fault indicator and Section IV presents the experimental results obtained from the faults staged on induction motor tested. Finally, in Section V, the summary and conclusions drawn from this study are presented.

2 Proposed Neural Network Fault Detection System

The underlying principle of residual-based methods, summarized in Fig. 1, is to compare the outputs of the machine to the prediction of neural network model. Residual-based methods are based on the use of analytical, rather than physical redundancy. In contrast to physical redundancy, in which measurements from different sensors are compared, sensory measurements are compared with computa- tionally obtained values of the corresponding variables [9]. The comparison between computationally obtained quantities and measurements results in the socalled residuals.

The overwhelming majority of currently used motor fault detection systems are based on the processing and analysis of raw motor measurements, such as vibration signals used in the proposed method. In general, the measured motor vibration signals are non-stationary. The steady-state signal can be accounted for by using Fourier-based technique. In the proposed fault detection system, a neural network prediction

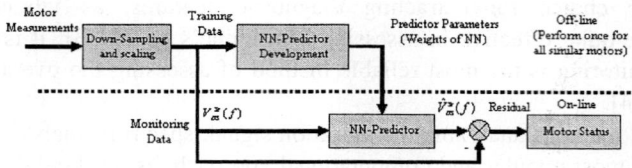

Fig. 1. Principle of Residual-based Fault Detection Method

model is used to generate residuals. The residuals are then processed to extract fault information by computing appropriate indicators. The proposed fault detection system combines elements from model-based and signal-based approaches. The overall system is schematized as shown in Fig. 2, where all time dependence is in the discrete-time domain.

The data acquisition system allows sampling of vibration signal $V_{raw}(k)$, and it is converted to $V_{DS}(k)$ by the low pass filtering and down-sampling to filter out the noise of the motor and support bearings. Then the steady-state data $V_{DS}^{St}(k)$ is extracted from the transient data by signal segmentation, where DS denotes the signal on down sampling and St denotes the signal on steady state. The Spectrum $V_{DS}^{St}(f)$ is converted using FFT signal processing. The neural network model $\hat{V}_{DS}^{St}(f|i)$ is trained and validated in spectra to generate residual $r_{DS}^{St}(f|i)$. Further details regarding the development of the motor model and residual generation are given in the following section.

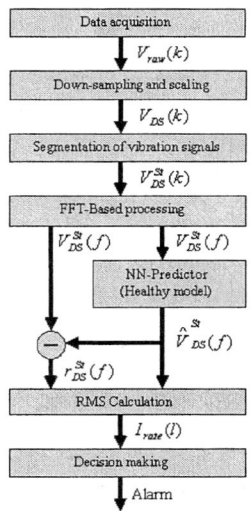

Fig. 2. Overview of the NN Fault Detection System

The residuals and the output of the NN model are used to generate the fault indicator $I_{rate}(l)$. By placing appropriate thresholds on the indicator magnitudes, a decision is made regarding the presence of a fault.

3 Signal Processing Procedures and Model Development

3.1 Steady-State Segmentation

Most features used in fault detection assume the presence of a stationary signal from which faults features, such as mean, variance or spectral estimates, are extracted [10]. However, the vibration signals of a motor are non-stationary signals, which contain both the transient signals and the steady-state signals.

Therefore, to obtain high-performance motor model that are not influenced by time-varying machine characteristics, the motor signatures must be extracted from steady-state signals of the motor vibration signals. In a recent paper [9], the authors derived a segmentation algorithm applied to the current measurements of the stator. The idea underlying signal segmentation is that for a signal to be considered stationary, its fundamental and harmonics must remain constant over time. Since the transient signals result in changes in the motor vibration signal harmonics, which are significantly smaller than fundamental, a statistical method is used for processing the vibration signal in the time-domain. The RMS values for the vibration signals are calculated over the window, which has a defined interval. Choosing a window size depends on the size of the used data. If the RMS value at successive windows does not vary, then the signal is considered stationary. The equations are shown as follows

$$V_{rms}(i) = \sqrt{\frac{1}{N_W}\sum_{j=1}^{N_W} V_{DS}^2(j)} \tag{1}$$

$$|V_{rms}(i+1) - V_{rms}(i)| < \beta \quad i = 1,2,...,n \tag{2}$$

where $V_{rms}(i)$ is the RMS value of vibration signal in each window, N_w is the window size, β is a user-defined threshold, and n is the total number of windows in the signal. The comparison is carried throughout the entire signal. If this algorithm does not result in the selection of the steady-state segments, then the threshold can be increased to allow for relaxation of the signal stationary.

3.2 Neural Network Based Model Development

As the random nature of the vibration signal, explicit motor models cannot be developed with conventional methods. Therefore, a neural network based model can be achieved more efficiently due to the facts that neural network is nonlinear empirical model which can capture the nonlinear system dynamics and do not require knowledge of specific system parameters.

Neural Model Formulation. In this work, we use a feedforward back-propagation (BP) network that undergoes supervised learning to model the output of the motor system in vibration spectra. Each of the processing elements of a BP network is governed by the following equation,

$$x_{[l,i]} = \sigma_{[l,i]}(\sum_{j=1}^{N_{[l-1]}} \omega_{[l-1,j][l,i]} x_{[l-1,j]} + b_{[l,i]}) \tag{3}$$

for $i = 1,..., N_{[l]}$ (the node index), and $l = 1,..., \ell$ (the layer index), where $x_{[l,i]}$ is the i th node output of the l th layer for sample t, $\omega_{[l-1,j][l,i]}$ is the weight, the adjustable parameter, connecting the j th node of the $(l-1)$ th layer to the i th node of the l th layer, $b_{[l,i]}$ is the bias, also an adjustable parameter, of the i th node in the l th layer, and $\sigma_{[l,i]}(\cdot)$ is the discriminatory function of the i th node in the l th layer.

The relation between inputs and outputs in BP network can be expressed using general nonlinear input-output models, as follows,

$$\hat{y}(k;W) = f(u(k);W) \tag{4}$$

where W is weight matrix which is to be determined by the learning algorithm, f represents the nonlinear transformation of the input approximated by a BP network, and in here hyperbolic tangent function is used. The input vector $u(k)$ is defined as,

$$u(k) = [y(k),..., y(k-n_y+1)] \tag{5}$$

where n_y is the maximum number of lags in the output.

Learning Algorithms. Using the structure of Eq. 4, the neural network model is trained using Levenberg-Marquardt (LM) algorithm. In this training phase, the error function to be minimized is given by,

$$\varepsilon = \sum_{t=0}^{NP-1} \sum_{k=1}^{n} [\hat{y}_k(t) - y_k(t)]^2 \qquad (6)$$

where n is the number of outputs included in the training, and NP is the number of training samples. The LM algorithm is designed to approach second-order training speed without having to compute the Hessian matrix. When the performance function has the form of a sum of squares, then the Hessian matrix can be approximated as,

$$H = J^T J \qquad (7)$$

and the gradient can be computed as,

$$g = J^T \varepsilon \qquad (8)$$

where J is the Jacobian matrix that contains first derivatives of the network errors with respect to the weights and biases, and ε is the network error. Then the processing element is updated by,

$$x_{k+1} = x_k - [J^T J + \mu I]^{-1} J^T \varepsilon \qquad (9)$$

The detailed computation of the gradients involved in LM learning algorithm can be found in many neural network references, such as [11].

Motor Vibration Spectra Model Development. The windowed healthy condition vibration signal is used for neural network training and validation. The vibration spectrum of each window after FFT processing is expressed as $V_{DS}^{NF}(f \mid i)$ and is used as inputs of NN model shown in Fig. 3, where i is the index of windows in the training set and NF (no fault) means the vibration spectra in healthy condition. The designed BP network has three layers; one hidden layer with 9 nodes, one input layer and one output layer both with N_f nodes, which is equal to the number of amplitude in vibration spectrum. The model structure of NN is decided after various experiments while the pruning method is used also. The general structure of this feedforward BP network is shown in Fig. 4. Based on the discussion about NN predictor above, the motor vibration spectra model $\hat{V}_{DS}(f \mid i)$ can be obtained as,

$$V_{DS}^{NF}(f \mid i) = FFT\{V_{DS}^{NF}(k)\} \qquad (10)$$

$$\hat{V}_{DS}(f \mid i) = NN\{V_{DS}^{NF}(f \mid i)\} \qquad (11)$$

the size of window is set user-defined, but should be the same as the window used for monitoring measurements.

Initially the motor predictor is developed for small machine, with training data representing the high-load level. After developing this baseline model, additional models valid at lower load levels are developed by incrementally tuning the baseline high-load level model. The training data set consists of 2800 samples for estimation, and 1400 samples for validation. The validation data set is used to determine the best stopping point in the predictor training to prevent over training, and select the predictor structure.

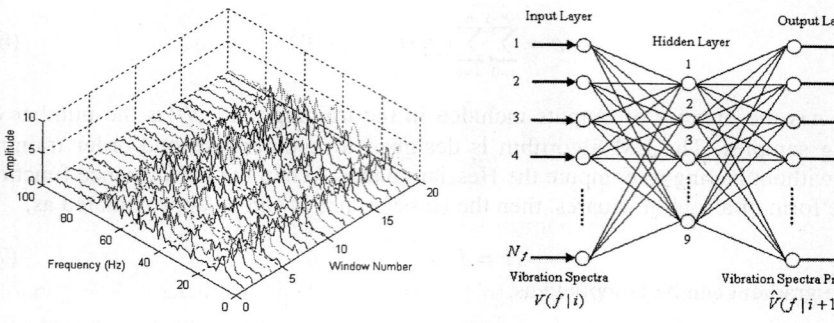

Fig. 3. Windowed Vibration Spectra   **Fig. 4.** General Structure of BP Network

3.3 Residual Generation

Residual is one of the most important elements in the residual-based method. A residual-based method should ideally decouple such parasitic effects from the effects of incipient faults as they are observed on the outputs. Theoretically, the residuals should be small values in normal operation since the trained neural predictor is capable of giving a satisfactory prediction of the spectrum. However, the residual generated by faults will significantly deviate from the nominal value. In other words, faults can be easily analyzed by this value.

In this research, the residuals are generated in vibration spectra. Consider one data window having the time interval $[t_1, t_2]$, the residual $r_{DS}(f \mid i)$ of the i th window is expressed as

$$r_{DS}(f \mid i) = V_{DS}^T(f \mid i) - \hat{V}_{DS}(f \mid i) \tag{12}$$

where superscript T means the measurements in spectra is also after denoising processing.

3.4 Description of the Fault Indicator

The fault indicator proposed here for detecting faults is based on the observation that the vibration signals, and as a result the residuals, are distorted in the presence of such faults. Consequently, in the presence of such faults the harmonic components in the residuals increase when compared to a baseline. Therefore, vibration harmonics variations provide some clues for detecting the presence of faults, whereas tracking variations in the vibration fundamental might result in false alarms. Relative changes in the harmonics, as seen through the processing of the residuals, appear promising for the detection of changes in motor condition.

In a monitoring time interval, let the size of a moving window be $p = t_2 - t_1$, which is the same as the size of the window for residual generation. And consider that the moving window moves by p at a time. The following moving window RMS values are computed for the model prediction and residual

$$r_{rms}(i) = \sqrt{\frac{1}{N_f}\sum_{f=1}^{N_f} r_{DS}^2(f \mid i)} \qquad i = 1,2,\ldots,m \tag{13}$$

$$\hat{V}_{rms}(i) = \sqrt{\frac{1}{N_f}\sum_{f=1}^{N_f} \hat{V}_{DS}^2(f \mid i)} \qquad i = 1,2,\ldots,m \tag{14}$$

where $m = t_N - t_1$ is the total number of moving windows, N_f is the number of data in the spectrum.

The relative change in the harmonic component of the residuals can be quantified by the ratio $r_{rms}(i)/\hat{V}_{rms}(i)$. In this study, the normalized harmonic content of the residuals is used as an indicator for detecting faults as follows

$$I_{rate}(i) = r_{rms}(i)/\hat{V}_{rms}(i) \tag{15}$$

By placing appropriate threshold on the indicator magnitude, faults can be detected when the value of the indicator is higher than the threshold value. The primary limitation of the indicator is reflected in the accuracy of the motor reference model. The performance of the proposed approach can be improved by regulating the window size and the threshold value.

4 Experiments and Analysis

4.1 Experimental Settings and Staged Motor Faults

The experiment system is setup to collect the data needed for testing the neural network fault detection system. An off-site industrial scale testbed is utilized for data acquisition from larger motors. In acquiring the necessary digital data, various anomalies are introduced to the motors, and also motor faults are staged.

The staged incipient faults include several mechanical faults. The results of a few of these anomalies and staged faults are presented here. A 3 - ϕ, eight pole, 597 kW

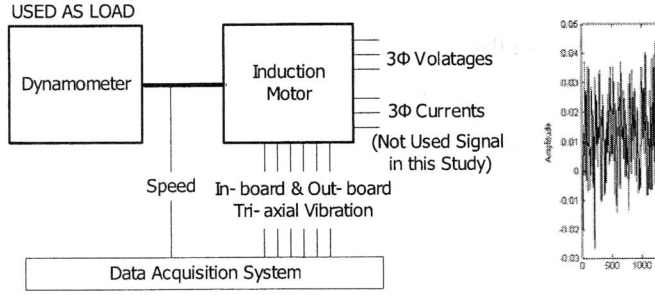

Fig. 5. Configurations for the motor system

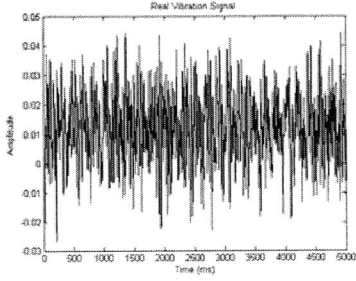

Fig. 6. Motor Vibration Signal

Allis Chalmers motor is run directly from the power supply mains. The motor is connected to the dynamometers used to load them. A simplified schematic of the experiment system is shown in Fig. 5.

A 13-channel IOTech data-acquisition system is used to record the six vibration signals, the three line voltages, the three line currents, and encoder speed signal at 40 kHz sampling frequency. Channel 1-6 are to collect vibration signals, and only the vibration signals are used in this study, which are down-sampled to 4000 Hz for further processing. A wide range of case studies for the motor are collected. There include healthy cases, and cases with operational anomalies. In Fig 6, the vibration signal in normal case is illustrated to show the complex random nature from real motor instead of simulated system.

4.2 Results and Analysis of the Experiment

The experiment results are collected from the 597 kW motor. The window size used in these experiments is 1 s, and the amount of collected data for one test is 60,000. The vibration spectra of the output of the machine and the neural network model is compared in Fig. 7.

Two air-gap eccentricity tests are performed using the 597 kW motor. The first case consists of moving the rotating center at the end of the inboard shaft 25% upward, whereas the second case moving the rotating center at the end of the out board shaft 20% downward and 10% to the right. Following data collection, down-sampling and scaling is performed. The vibration signals are processed through the data segmentation stage, revealing the steady-state signals of the motor operation. The residuals are then generated by subtracting the measurements from the NN model in spectra over the window. The indicator values for the healthy motor response are considered as baseline to set the threshold. The air-gap vibration spectra and the network model spectra are depicted in Fig. 8. Detection of air-gap eccentricity faults using the proposed indicator and threshold is shown in Fig. 9. The normal condition runs for 400s, and then the first case is switched on for 200s and the second case for 200s also. The threshold is set to 0.55 according to the experiment and the results show that the fault conditions can be easily classified.

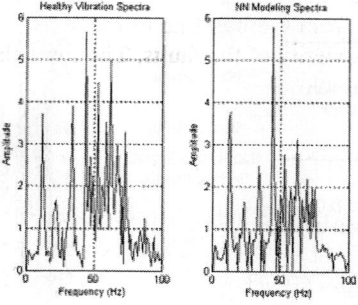

Fig. 7. Healthy Vibration Spectra and its Modeling

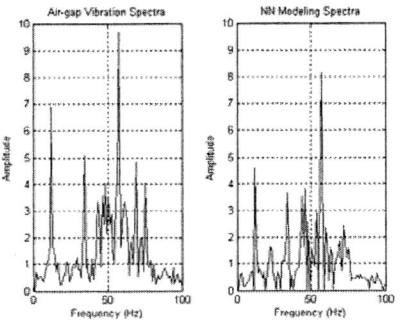

Fig. 8. Vibration Spectra of Air-gap

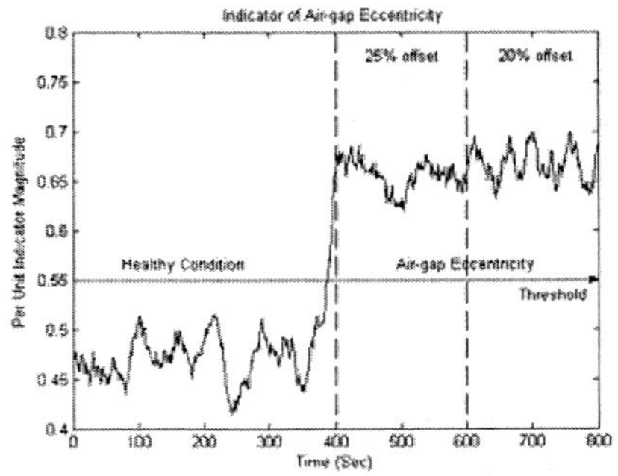

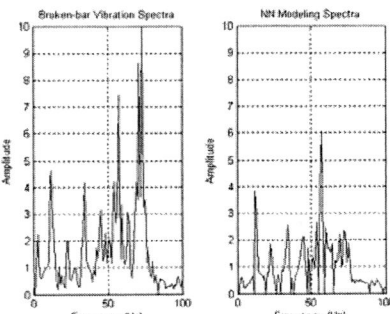

Fig. 10. Vibration Spectra of Broken Bar

Fig. 9. Indicator of Air-gap Eccentricity

Fig. 11. Indicator of Broken Rotor Bar

Another mechanical fault is broken rotor bars. Experiments are performed to obtain motor measurements with four cases of broken bars using the 597 kW motor. The four cases are a half broken bar, one broken bar, two broken bars, and four broken bars. The measurements are further processed as in the case of the air-gap eccentricity, and the fault indicators are obtained. The compared vibration Spectra and the fault indicator are given in Fig. 10 and 11, respectively. The proposed indicator clearly shows the changes from the baseline to broken bar faults, and the magnitude change increases with the severity of the faults. The same threshold is used to classify the healthy and fault condition.

5 Conclusion

In this paper, the development and testing of a residual-based fault detection and diagnosis system for induction machine is presented. The proposed system uses a vibration spectra predictor developed using back-propagation neural network. The investigations are based on the notion that neural network can capture the nonlinear system dynamics and do not require knowledge of specific system parameters. FFT has been used to convert the steady-state vibration signal into its spectra. The resulting motor vibration residuals are stationary and the RMS values are used to compute the fault indicator.

A high-performance fault detection approach exhibits high probability of fault detection and low probability of false alarm. The results of the proposed method are promising and demonstrate the scalability of the proposed method for induction motor condition monitoring and fault detection. The proposed system is shown to be quite effective in detecting the early stages of many frequently encountered motor faults.

References

1. N. Arthur and J. penman, "Inverter fed induction machine condition monitoring using the bispectrum," *Proc. IEEE Signal Processing Workshop Higher-Order Statistics*, Banff, AB, Canada, July 21-23, pp.67-71, 1997.
2. P. J. Tavner and J. Penman, *Condition Monitoring of Electrical Machines*, Letchworth, England, Research Studies Press, 1987.
3. C. M. Riley, B. K. Lin, T.G. Habetler, and R. R. Schoen, "A method for sensorless on-line vibration monitoring of induction machines," *IEEE Trans. Ind. Applicat.*, vol. 34, pp. 1240-1245, Dec. 1998.
4. P. A. Lagan, "Vibration Monitoring," *Proc. IEE Colloquium on Understanding your Condition Monitoring*, pp. 1-11, 1999.
5. N. Arthur and J. Penman, "Induction machine condition monitoring and with higher order spectra," *IEEE Trans. Industrial Electronics*, vol. 47, no. 5, pp. 1031-1041, Oct. 2000.
6. "Effective Machinery Measurement Using Dynamic Signal Analyzers," Hewlett Packard, Applicat. Note 243-1, 1990.
7. G. Betta, C. Liguori, A. Paolillo, and A. Pietrosanto, "A DSP-based FFT-analyzer for the fault diagnosis of rotating machine based on vibration analysis," *IEEE Trans. Instrumentation and Measurement*, vol. 51, no. 6, Dec. 2002.
8. A. F. Atiya and A. G. Palos, "New results on recurrent network training: Unifying the algorithms and accelerating convergence," *IEEE Trans. Neural Networks*, vol. 13, pp. 765-786, May 2000.
9. K. Kim and A. G. Parlos, "Model-based fault diagnosis of induction motors using non-stationary signal segmentation," *Mechanical System and Signal Processing*, Vol. 16, pp. 223-253, 2002.
10. K. Kim and A. G. Parlos, "Reducing the impact of false alarms in induction motor fault diagnosis," *ASME Journal of Dynamic Systems*, Measurement, and Control, vol. 125, pp. 80-95, 2003.
11. M. Norgaard, O. Ravn, N.K. Poulsen, and L.K. Hansen, *Neural Networks for Modeling and Control of Dynamic Systems*, London, Springer-Verlag, 2000

Author Index

Abawajy, J.H.   III-60, IV-1272
Ahiska, S. Sebnem   IV-301
Åhlander, Krister   I-657
Ahmad, Uzair   II-1045
Ahn, Beumjun   IV-448
Ahn, Byeong Seok   IV-360
Ahn, Chang-Beom   I-166
Ahn, EunYoung   I-1122
Ahn, Hyo Cheol   IV-916
Ahn, Jaewoo   I-223
Ahn, Joung Chul   II-741
Ahn, Kwang-Il   IV-662
Ahn, Seongjin   I-137, I-242, I-398, II-676, II-696, II-848, IV-1036
Ahn, Yonghak   II-732
Akbari, Mohammad K.   IV-1262
Akyol, Derya Eren   IV-596
Alcaide, Almudena   III-729, IV-1309
Alexander, Phillip J.   IV-1180
Ali, A.   II-1045
Alkassar, Ammar   II-634
Aloisio, Giovanni   III-1
Alonso, Olga Marroquin   II-1156
Altas, Irfan   III-463
An, Sunshin   I-261
Anikeenko, A.V.   I-816
Anton, François   I-669, I-683
Aquino, Adélia J.A.   I-1004
Araújo, Madalena M.   IV-632
Aranda, Gabriela N.   I-1064
Arteconi, Leonardo   I-1093
Aylett, Ruth   IV-30

Bacak, Goksen   III-522
Baciu, George   I-737
Bae, Hae-Young   IV-812
Bae, Hanseok   I-388
Bae, Hyeon   IV-1075, IV-1085
Bae, Hyerim   III-1259
Bae, Ihn Han   II-169
Bae, Jongho   IV-232
Bae, Kyoung Yul   I-204
Baek, Dong-Hyun   IV-222
Baek, Jang Hyun   IV-528

Baek, Jang-Mi   III-964
Baek, Jun-Geol   IV-148
Baek, Sunkyoung   I-37
Baig, Meerja Humayun   I-806
Baik, MaengSoon   III-89, IV-936
Baker, Robert G.V.   III-143
Bang, Young-Cheol   IV-989
Bannai, Hideo   III-349
Bao, Hujun   III-215
Bao, XiaoMing   II-1167
Barbatti, Mario   I-1004
Barco, Raquel   IV-958
Barlow, Jesse   IV-843
Baumgartner, Robert   II-988
Bayhan, G. Mirac   IV-596
Bekker, Henk   IV-397
Bénédet, Vincent   I-838
Bernholdt, D.E.   III-29
Bernholt, Thorsten   I-697
Bertazzon, Stefania   III-118, III-152
Bhat, M.S.   IV-548
Bhatia, Davinder   IV-1190
Bhattacharyya, Chiranjib   IV-548
Bierbaum, Aron   III-1119
Borruso, Giuseppe   III-126
Bozer, Yavuz A.   IV-437
Braad, Eelco P.   IV-397
Brucker, Peter   IV-182
Brunstrom, Anna   IV-1331
Burns, John   II-1254
Burrage, Kevin   II-1245
Byun, Sang-Yong   III-788
Byun, Yung-Cheol   III-788

Caballero-Gil, Pino   III-719
Cafaro, Massimo   III-1
Cai, Guoyin   III-173
Çakar, Tarık   IV-1241
Caminati, Walther   I-1046
Castro, Julio César Hernández   IV-1292
Catanzani, Riccardo   I-921
Cattani, Carlo   III-604
Cechich, Alejandra   I-1064
Cha, ByungRae   II-254

Author Index

Cha, Jae-Sang II-332, II-341, II-373, II-411, II-429, II-449, IV-1319
Cha, Jeon-Hee I-11
Cha, Jung-Eun III-896
Cha, Kyungup III-1269
Chae, Jongwoo II-147
Chae, Kijoon I-591
Chae, Oksam II-732, IV-20
Chae, Soo-young II-458
Chan, Choong Wah II-657
Chanchio K. III-29
Chang, Chun Young I-1204
Chang, Dong Shang IV-577
Chang, Elizabeth II-1125
Chang, Hangbae IV-128
Chang, Hung-Yi IV-1007
Chang, Jae Sik IV-999
Chang, Jae-Woo I-77
Chang, Ok-Bae III-758, III-836, III-878, III-945
Chang, Pei-Chann IV-172, IV-417
Chang, Soo Ho I-46
Chau, Rowena II-956
Che, Ming III-284
Che, Yinghui III-225
Chen, M.L. III-29
Chen, Chia-Ho IV-417
Chen, Chun IV-826
Chen, J.C. IV-333
Chen, Jianwei IV-519
Chen, Jinwen II-1217
Chen, Ling III-338
Chen, Shih-Chang IV-1017
Chen, Taiyi I-967
Chen, Tse-Shih I-19
Chen, Tung-Shou IV-1007
Chen, Weidong I-865
Chen, Wen II-806
Chen, Yefang IV-1
Chen, Yun-Shiow IV-172
Chen, Zhiping P. IV-733
Cheng, Xiangguo IV-1046
Cheon, Seong-Pyo IV-1075
Chernetsov, Nikita III-133
Chi, Jeong Hee II-977
Ching, Wai-Ki IV-342, IV-843
Cho, Byung Rae IV-212
Cho, Cheol-Hyung I-707, III-993
Cho, Chiwoon III-1297
Cho, Dongyoung I-232
Cho, Eun-Sook III-778, III-868
Cho, Hyeon Seob II-832
Cho, Kyung Dal II-474
Cho, Miyoung I-37
Cho, Nam Wook III-1297
Cho, Seokhyang I-498
Cho, Sok-Pal II-781, II-896
Cho, Sung-Keun IV-48
Cho, SungHo I-204
Cho, Wanhyun IV-867
Cho, Yongju III-1289
Cho, Yongsun I-1
Cho, Yongyun II-1008
Cho, Yookun II-353
Cho, You-Ze I-378
Cho, Youngsong I-707, I-716, III-993
Cho, YoungTak IV-20
Choi, In Seon II-889
Choi, DeaWoo IV-103
Choi, Deokjai I-195
Choi, Dong-seong IV-62
Choi, Eunmi II-187, III-858
Choi, Gyunghyun IV-261
Choi, Hee-Chul III-938
Choi, Honzong III-1289
Choi, Hyang-Chang II-82
Choi, Hyung Jo II-1207
Choi, Ilhoon III-1229
Choi, Jaemin II-567
Choi, Jaeyoung II-1008, II-1018, IV-10
Choi, Jong Hwa IV-86
Choi, Jong-In III-1148
Choi, Jonghyoun I-271
Choi, Kee-Hyun III-99
Choi, Kun Myon I-448
Choi, Mi-Sook III-778
Choi, Sang-soo II-458
Choi, Sang-Yule II-341, II-429, IV-1319
Choi, Sung-ja II-215
Choi, SungJin III-89, IV-936
Choi, Wonwoo I-137
Choi, WoongChul IV-1231
Choi, Yeon-Sung II-71
Choi, YoungSik I-186
Chong, Kil To II-1207, II-1293
Chong, Kiwon I-1
Choo, Hyunseung I-291, I-448, I-468, I-529, I-540, IV-989
Choudhary, Alok Kumar IV-680
Chow, K.P. III-651

Chow, Sherman S.M.   III-651
Choy, Yoon-Chu   I-847
Chua, Eng-Huat   II-1167
Chun, Junchul   I-1135
Chun, Kilsoo   II-381
Chun, Kwang Ho   II-749
Chun, Kwang-ho   II-723
Chung, Chin Hyun   I-638, I-1213
Chung, Hyun-Sook   III-788
Chung, Jinwook   I-137
Chung, Kwangsue   IV-1231
Chung, Min Young   I-348, I-448, I-529
Chung, Mokdong   II-147
Chung, Tae-sun   IV-72
Chung, Tae-Woong   IV-836
Chung, Tai-Myung   I-146, I-468
Chung, YoonJung   II-92, II-274
Chung, Youn-Ky   III-769
Chunyan, Yu   I-875, I-974
Cornejo, Oscar   IV-712
Corradini, Flavio   II-1264
Costantini, Alessandro   I-1046
Cotrina, Josep   II-527, II-624
Couloigner, Isabelle   III-181
Croce, Federico Della   IV-202
Cruz-Neira, Carolina   III-1070, III-1119
Cui, Kebin   I-214
Cui, Shi   II-657

Das, Amitabha   I-994
Das, Sandip   I-827
Dashora, Yogesh   IV-680
Datta, Amitava   I-87, II-686, III-206
Da-xin, Liu   IV-753
Debels, Dieter   IV-378
de Frutos Escrig, David   II-1156
Deris, M. Mat   III-60
Dévai, Frank   I-726
Dew, Robert   III-49
Díez, Luis   IV-958
Dillon, Tharam S.   II-914, II-1125
Ding, Jintai   II-595
Ding, Yongsheng   III-69
Djemame, Karim   IV-1282
Doboga, Flavia   III-563
Dol'nikov, Vladimir   III-628
Dongyi, Ye   I-875, I-974
Du, Tianbao   I-1040
Duan, Pu   II-657

Dumas, Laurent   IV-948
Duong, Doan Dai   II-1066

Emiris, Ioannis   I-683
Enzi, Christian   II-988
Eong, Gu-Beom   II-42
Epicoco, Italo   III-1
Ercan, M. Fikret   III-445
Ergenc, Tanil   III-463
Escoffier, Bruno   IV-192, IV-202
Espírito-Santo, Isabel A.C.P.   IV-632
Estévez-Tapiador, Juan M.   IV-1292, IV-1309
Eun, He-Jue   II-10

Faudot, Dominique   I-838
Feng, Jieqing   III-1023
Fernandes, Edite M.G.P   IV-488, IV-632
Fernandez, Marcel   II-527, II-624
Ferreira, Eugenio C.   IV-632
Fiore, Sandro   III-1
Fong, Simon   II-1106
For, Wei-Khing   II-1167
Frank, A.O.   II-1018
Froeklich, Johannes   I-905, I-938
Fu, Haoying   IV-843
Fúster-Sabater, Amparo   III-719

Gaglio, Salvatore   III-39
Gálvez, Akemi   III-472, III-482, III-502
Gao, Chaohui   I-1040
Gao, Lei   III-69
Garcia, Ernesto   I-1083
Gardner, William   II-1125
Gatani, Luca   III-39
Gaur, Daya Ram   IV-670
Gavrilova, M.L.   I-816
Gavrilova, Marina L.   I-748
Geist A.   III-29
Gerardo, Bobby   II-71, II-205
Gervasi, Osvaldo   I-905, I-921, I-938
Ghelmez, Mihaela   III-563
Ghinea, G.   II-1018
Ghose, Debasish   IV-548
Gil, JoonMin   IV-936
Gimenez, Xavi   I-1083
Goetschalckx, Marc   IV-322
Goff, Raal   I-87
Goh, Dion Hoe-Lian   II-1177
Goh, John   IV-1203

Goh, Li Ping IV-906
Goi, Bok-Min I-488, IV-1065
Gold, Christopher M. I-737
Goldengorin, Boris IV-397
Goscinski, Andrzej III-49
Goswami, Partha P. I-827
Gower, Jason II-595
Grinnemo, Karl-Johan IV-1331
Großschädl, Johann II-665
Grząślewicz, Ryszard II-517
Gu, Mi Sug II-966
Guan, Xiucui IV-161
Guan, Yanning III-173
Guo, Heqing III-691, IV-1028
Guo, X.C. IV-1040
Gupta, Pankaj IV-1190

Ha, JaeCheol II-245
Haji, Mohammed IV-1282
Han, Chang Hee IV-222, IV-360
Han, In-sung II-904
Han, Joohyun II-1008
Han, Jung-Soo III-748, III-886
Han, Kyuho I-261
Han, SangHoon I-1122
Han, Young-Ju I-146
Harding, Jenny A. IV-680
Hartling, Patrick III-1070, III-1119
He, Ping III-338
He, Qi III-691, IV-1028
He, Yuanjun III-1099
Hedgecock, Ian M. I-1054
Heng, Swee-Huay II-603
Henze, Nicola II-988
Heo, Hoon IV-20
Herbert, Vincent IV-948
Hernández, Julio C. IV-1301
Herrlich, Marc II-988
Herzog, Marcus II-988
Higuchi, Tomoyuki III-381, III-389
Hirose, Osamu III-349
Hong, Changho IV-138
Hong, Choong Seon I-195, I-339
Hong, Chun Pyo I-508
Hong, Helen IV-1111
Hong, In-Sik III-964
Hong, Jung-Hun II-1
Hong, Jungman IV-642
Hong, Kicheon I-1154
Hong, Kiwon I-195

Hong, Maria I-242, IV-1036
Hong, Seok Hoo II-1076
Hong, Xianlong IV-896
Hou, Jia II-749
Hsieh, Min-Chi II-1055
Hsieh, Ying-Jiun IV-437
Hsu, Ching-Hsien IV-1017
Hu, Bingcheng II-1274
Hu, Guofei I-758
Hu, Xiaohua III-374
Hu, Xiaoyan III-235
Hu, Yifeng I-985
Hu, Yincui III-173
Hua, Wei III-215
Huang, Zhong III-374
Huettmann, Falk III-133, III-152
Huh, Eui-Nam I-311, I-628, I-1144
Hui, Lucas C.K. III-651
Hung, Terence I-769, IV-906
Hur, Nam-Young II-341, IV-1319
Hur, Sun II-714, IV-606
Huynh, Trong Thua I-339
Hwang, Chong-Sun III-89, IV-936, IV-1169
Hwang, Gi Yean II-749
Hwang, Ha Jin II-304, III-798
Hwang, Hyun-Suk II-127
Hwang, Jae-Jeong II-205
Hwang, Jeong Hee II-925, II-966
Hwang, Jun I-1170, I-1204
Hwang, Seok-Hyung II-1
Hwang, Suk-Hyung III-827, III-938
Hwang, Sun-Myung II-21, III-846
Hwang, Yoo Mi I-1129
Hwang, Young Ju I-619
Hwang, Yumi I-1129
Hyun, Chang-Moon III-927
Hyun, Chung Chin I-1177
Hyuncheol, Kim II-676

Iglesias, Andrés III-502, III-472, III-482, III-492, III-547, III-1157
Im, Chae-Tae I-368
Im, Dong-Ju II-420, II-474
Imoto, Seiya III-349, III-389
In, Hoh Peter II-274
Inceoglu, Mustafa Murat III-538, IV-56
Iordache, Dan III-614
Ipanaqué, Ruben III-492
Iqbal, Mahrin II-1045

Author Index

Iqbal, Mudeem II-1045
Izquierdo, Antonio III-729, IV-1309

Jackson, Steven Glenn III-512
Jahwan, Koo II-696
Jalili-Kharaajoo, Mahdi I-1030
Jang, Dong-Sik IV-743
Jang, Injoo II-102, II-111
Jang, Jongsu I-609
Jang, Sehoon I-569
Jang, Sung Man II-754
Jansen, A.P.J. I-1020
Javadi, Bahman IV-1262
Jeon, Hoseong I-529
Jeon, Hyong-Bae IV-538
Jeon, Nam Joo IV-86
Jeong, Bongju IV-566
Jeong, Chang Sung I-601
Jeong, Eun-Hee II-322, II-585
Jeong, Eunjoo I-118
Jeong, Gu-Beom II-42
Jeong, Hwa-Young I-928
Jeong, In-Jae IV-222, IV-312
Jeong, JaeYong II-353
Jeong, Jong-Youl I-311
Jeong, Jongpil I-291
Jeong, Kugsang I-195
Jeong, KwangChul I-540
Jeong, Seung-Ju IV-566
Ji, Joon-Yong III-1139
Ji, Junfeng III-1167
Ji, Yong Gu III-1249
Jia, Zhaoqing III-10
Jiang, Chaojun I-1040
Jiang, Xinhua H. IV-733
Jiao, Xiangmin IV-1180
Jin, Biao IV-1102
Jin, Bo III-299
Jin, Guiyue IV-1095
Jin, Jing III-416
Jin, YoungTaek III-846
Jin, Zhou III-435
Jo, Geun-Sik IV-1131
Jo, Hea Suk I-519
Joo, Inhak II-1136
Joung, Bong Jo I-1196, I-1213
Ju, Hak Soo II-381
Ju, Jaeyoung III-1259
Jun, Woochun IV-48
Jung, Changho II-537

Jung, Ho-Sung II-332
Jung, Hoe Sang III-1177
Jung, Hye-Jung III-739
Jung, Jason J. IV-1131
Jung, Jin Chul I-252
Jung, Jung Woo IV-467
Jung, KeeChul IV-999
Jung, Kwang Hoon I-1177
Jung, SM. II-1028

Kang, Euisun I-242
Kang, HeeJo II-420, II-483
Kang, Kyung Hwan IV-350
Kang, Kyung-Woo I-29
Kang, MunSu I-186
Kang, Oh-Hyung II-195, II-284, II-295
Kang, Seo-Il II-177
Kang, Suk-Hoon I-320
Kang, Yeon-hee II-215
Kang, Yu-Kyung III-938
Karsak, E. Ertugrul IV-301
Kasprzak, Andrzej IV-772
Kemp, Ray II-1187
Khachoyan, Avet A. IV-1012
Khorsandi, Siavash IV-1262
Kiani, Saad Liaquat II-1096
Kim, B.S. II-1028
Kim, Byunggi I-118
Kim, Byung Wan III-1306
Kim, Chang Han II-647
Kim, Chang Hoon I-508
Kim, Chang Ouk IV-148
Kim, Chang-Hun III-1080, III-1129, III-1139, III-1148
Kim, Chang-Min I-176, III-817, IV-38
Kim, Chang-Soo II-127
Kim, Chul-Hong III-896
Kim, Chulyeon IV-261
Kim, Dae Hee II-1284
Kim, Dae Sung I-1111
Kim, Dae Youb II-381
Kim, Daegeun II-1035
Kim, Deok-Soo I-707, I-716, III-993, III-1060, IV-652
Kim, D.K. II-1028
Kim, Do-Hyeon I-378
Kim, Do-Hyung II-401
Kim, Dong-Soon III-938
Kim, Donghyun IV-877
Kim, Dongkeun I-857

Kim, Dongkyun   I-388
Kim, Dongsoo   III-1249
Kim, Donguk   I-716, III-993
Kim, Dounguk   I-707
Kim, Eun Ju   I-127
Kim, Eun Suk   IV-558
Kim, Eun Yi   IV-999
Kim, Eunah   I-591
Kim, Gi-Hong   II-771
Kim, Gui-Jung   III-748, III-886
Kim, Guk-Boh   II-42
Kim, Gye-Young   I-11
Kim, Gyoung-Bae   IV-812
Kim, Hae Geun   II-295
Kim, Hae-Sun   II-157
Kim, Haeng-Kon   II-1, II-52, II-62, II-137, III-769, III-906, III-916
Kim, Hak-Keun   I-847
Kim, Hang Joon   IV-999
Kim, Hee Sook   II-483, II-798
Kim, Hong-Gee   III-827
Kim, Hong-jin   II-781, II-896
Kim, HongSoo   III-89
Kim, Hoontae   III-1249
Kim, Howon   II-1146
Kim, Hwa-Joong   IV-538, IV-722
Kim, Hwankoo   II-245
Kim, HyoungJoong   IV-269
Kim, Hyun Cheol   I-281
Kim, Hyun-Ah   I-427, III-426, IV-38
Kim, Hyun-Ki   IV-887
Kim, Hyuncheol   I-137, II-676
Kim, Hyung Jin   II-789, II-880
Kim, InJung   II-92, II-274
Kim, Jae-Gon   IV-280, IV-322
Kim, Jae-Sung   II-401
Kim, Jae-Yearn   IV-662
Kim, Jae-Yeon   IV-743
Kim, Jang-Sub   I-348
Kim, Jee-In   I-886
Kim, Jeom-Goo   II-762
Kim, Jeong Ah   III-846
Kim, Jeong Kee   II-714
Kim, Jin Ok   I-638, I-1187
Kim, Jin Soo   I-638, I-1187
Kim, Jin-Geol   IV-782
Kim, Jin-Mook   II-904
Kim, Jin-Sung   II-31, II-567
Kim, Jong-Boo   II-341, IV-1319
Kim, Jong Hwa   III-1033

Kim, Jong-Nam   I-67
Kim, Jongsung   II-567
Kim, Jong-Woo   I-1177, II-127
Kim, Ju-Yeon   II-127
Kim, Jun-Gyu   IV-538
Kim, Jung-Min   III-788
Kim, Jungchul   I-1154
Kim, Juwan   I-857
Kim, Kap Sik   III-798
Kim, Kibum   IV-566
Kim, KiJoo   I-186
Kim, Kwan-Joong   I-118
Kim, Kwang-Baek   IV-1075
Kim, Kwang-Hoon   I-176, III-817, IV-38
Kim, Kwang-Ki   III-806
Kim, Kyung-kyu   IV-128
Kim, Mihui   I-591
Kim, Mijeong   II-1136
Kim, Minsoo   II-225, II-1136, III-1249, III-1259
Kim, Misun   I-550, I-559
Kim, Miyoung   I-550, I-559
Kim, Moonseong   IV-989
Kim, Myoung Soo   III-916
Kim, Myuhng-Joo   I-156
Kim, Myung Ho   I-223
Kim, Myung Won   I-127
Kim, Myung-Joon   IV-812
Kim, Nam Chul   I-1111
Kim, Pankoo   I-37
Kim, Sang Ho   II-977, IV-79
Kim, Sang-Bok   I-628
Kim, Sangjin   IV-877
Kim, Sangkyun   III-1229, III-1239, IV-122
Kim, Seungjoo   I-498, II-1146
Kim, Soo Dong   I-46, I-57
Kim, Soo-Kyun   III-1080, III-1129, III-1139
Kim, Soung Won   III-916
Kim, S.R.   II-1028
Kim, Sung Jin   II-1076
Kim, Sung Jo   III-79
Kim, Sung Ki   I-252
Kim, Sung-il   IV-62
Kim, Sung-Ryul   I-359
Kim, Sungshin   IV-1075, IV-1085
Kim, Tae Hoon   IV-509
Kim, Tae Joong   III-1279
Kim, Tae-Eun   II-474

Kim, Taeho   IV-280
Kim, Taewan   II-863
Kim, Tai-Hoon   II-341, II-429, II-468, II-491, IV-1319
Kim, Ungmo   II-936
Kim, Won-sik   IV-62
Kim, Wooju   III-1289, IV-103
Kim, Y.H.   III-1089
Kim, Yon Tae   IV-1085
Kim, Yong-Kah   IV-858
Kim, Yong-Soo   I-320, I-1162
Kim, Yong-Sung   II-10, II-31, III-954
Kim, Yongtae   II-647
Kim, Young Jin   IV-212, IV-232
Kim, Young-Chan   I-1170
Kim, Young-Chul   IV-10
Kim, Young-Shin   I-311
Kim, Young-Tak   II-157
Kim, Youngchul   I-107
Ko, Eun-Jung   III-945
Ko, Hoon   II-442
Ko, Jaeseon   II-205
Ko, S.L.   III-1089
Koh, Jae Young   II-741
Komijan, Alireza Rashidi   IV-388
Kong, Jung-Shik   IV-782
Kong, Ki-Sik   II-1225, IV-1169
Koo, Jahwan   II-696, II-848
Koo, Yun-Mo   III-1187
Koszalka, Leszek   IV-692
Kravchenko, Svetlana A.   IV-182
Kriesell, Matthias   II-988
Krishnamurti, Ramesh   IV-670
Kuo, Yi Chun   IV-577
Kurosawa, Kaoru   II-603
Kutyłowski, Jarosław   II-517
Kutyłowski, Mirosław   II-517
Kwag, Sujin   I-418
Kwak, Byeong Heui   IV-48
Kwak, Kyungsup   II-373, II-429
Kwak, NoYoon   I-1122
Kwon, Dong-Hee   I-368
Kwon, Gihwon   III-973
Kwon, Hyuck Moo   IV-212, IV-232
Kwon, Jungkyu   II-147
Kwon, Ki-Ryong   II-557
Kwon, Ki-Ryoung   III-1209
Kwon, Oh Hyun   II-137
Kwon, Soo-Tae   IV-624
Kwon, Soonhak   I-508

Kwon, Taekyoung   I-577, I-584
Kwon, Yong-Moo   I-913

La, Hyun Jung   I-46
Lægreid, Astrid   III-327
Laganà, Antonio   I-905, I-921, I-938, I-1046, I-1083, I-1093
Lago, Noelia Faginas   I-1083
Lai, Edison   II-1106
Lai, K.K.   IV-250
Lamarque, Loïc   I-838
Lázaro, Pedro   IV-958
Ledoux, Hugo   I-737
Lee, Bo-Hee   IV-782
Lee, Bong-Hwan   I-320
Lee, Byoungcheon   II-245
Lee, Byung Ki   IV-350
Lee, Byung-Gook   III-1209
Lee, Byung-Kwan   II-322, II-585
Lee, Chang-Mog   III-758
Lee, Chong Hyun   II-373, II-411, II-429, II-449
Lee, Chun-Liang   IV-1007
Lee, Dong Chun   II-714, II-741, II-762, II-889, II-896
Lee, Dong Hoon   I-619, II-381
Lee, Dong-Ho   IV-538, IV-722
Lee, DongWoo   I-232
Lee, Eun-Ser   II-363, II-483
Lee, Eung Jae   II-998
Lee, Eung Young   III-1279
Lee, Eunkyu   II-1136
Lee, Eunseok   I-291
Lee, Gang-soo   II-215, II-458
Lee, Geuk   II-754
Lee, Gi-Sung   II-839
Lee, Hakjoo   III-1269
Lee, Ho Woo   IV-509
Lee, Hong Joo   III-1239, IV-113, IV-122
Lee, Hoonjung   IV-877
Lee, Hyewon K.   I-97, I-118
Lee, Hyoung-Gon   III-1219
Lee, Hyun Chan   III-993
Lee, HyunChan   III-1060
Lee, Hyung-Hyo   II-82
Lee, Hyung-Woo   II-391, II-401, IV-62
Lee, Im-Yeong   II-117, II-177
Lee, Insup   I-156
Lee, Jae-deuk   II-420
Lee, Jaeho   III-1060

Lee, Jae-Wan   II-71, II-205, II-474
Lee, Jee-Hyong   IV-1149
Lee, Jeongheon   IV-20
Lee, Jeongjin   IV-1111
Lee, Jeoung-Gwen   IV-1055
Lee, Ji-Hyen   III-878
Lee, Ji-Hyun   III-836
Lee, Jongchan   I-107
Lee, Jong chan   II-781
Lee, Jong Hee   II-856
Lee, Jong-Hyouk   I-146, I-468
Lee, Joon-Jae   III-1209
Lee, Joong-Jae   I-11
Lee, Joungho   II-111
Lee, Ju-Il   I-427
Lee, Jun   I-886
Lee, Jun-Won   III-426
Lee, Jung   III-1080, III-1129, III-1139
Lee, Jung-Bae   III-938
Lee, Jung-Hoon   I-176
Lee, Jungmin   IV-1231
Lee, Jungwoo   IV-96
Lee, Kang-Won   I-378
Lee, Keon-Myung   IV-1149
Lee, Keun Kwang   II-474, II-420
Lee, Keun Wang   II-798, II-832, II-856
Lee, Key Seo   I-1213
Lee, Ki Dong   IV-1095
Lee, Ki-Kwang   IV-427
Lee, Kwang Hyoung   II-798
Lee, Kwangsoo   II-537
Lee, Kyunghye   I-408
Lee, Malrey   II-71, II-363, II-420, II-474, II-483
Lee, Man-Hee   IV-743
Lee, Mi-Kyung   II-31
Lee, Min Koo   IV-212, IV-232
Lee, Moon Ho   II-749
Lee, Mun-Kyu   II-314
Lee, Myung-jin   IV-62
Lee, Myungeun   IV-867
Lee, Myungho   IV-72
Lee, NamHoon   II-274
Lee, Pill-Woo   I-1144
Lee, S.Y.   II-1045
Lee, Sang Ho   II-1076
Lee, Sang Hyo   I-1213
Lee, Sangsun   I-418
Lee, Sang Won   III-1279
Lee, Sang-Hyuk   IV-1085
Lee, Sang-Young   II-762, III-945
Lee, Sangjin   II-537, II-567
Lee, SangKeun   II-1225
Lee, Sangsoo   II-816
Lee, Se-Yul   I-320, I-1162
Lee, SeongHoon   I-232
Lee, Seoung Soo   III-1033, IV-652
Lee, Seung-Yeon   I-628, II-332
Lee, Seung-Yong   II-225
Lee, Seung-youn   II-468, II-491, II-499
Lee, SiHun   IV-1149
Lee, SooBeom   II-789, II-880
Lee, Su Mi   I-619
Lee, Suk-Hwan   II-557
Lee, Sungchang   I-540
Lee, Sunghwan   IV-96
Lee, SungKyu   IV-103
Lee, Sungyoung   II-1096, II-1106, II-1115
Lee, Sunhun   IV-1231
Lee, Suwon   II-420, II-474
Lee, Tae Dong   I-601
Lee, Tae-Jin   I-448
Lee, Taek   II-274
Lee, TaiSik   II-880
Lee, Tong-Yee   III-1043, III-1050
Lee, Wonchan   I-1154
Lee, Woojin   I-1
Lee, Woongjae   I-1162, I-1196
Lee, Yi-Shiun   III-309
Lee, Yong-Koo   II-1115
Lee, Yonghwan   II-187, III-858
Lee, Yongjae   II-863
Lee, Young Hae   IV-467
Lee, Young Hoon   IV-350
Lee, Young Keun   II-420, II-474
Lee, YoungGyo   II-92
Lee, YoungKyun   II-880
Lee, Yue-Shi   II-1055
Lee, Yung-Hyeon   II-762
Leem, Choon Seong   III-1269, III-1289, III-1306, IV-79, IV-86, IV-113
Lei, Feiyu   II-806
Leon, V. Jorge   IV-312
Leung, Stephen C.H.   IV-250
Lezzi, Daniele   III-1
Li, Huaqing   IV-1140
Li, Jin-Tao   II-547
Li, JuanZi   IV-1222
Li, Kuan-Ching   IV-1017

Author Index   1311

Li, Li   III-190
Li, Minglu   III-10
Li, Peng   III-292
Li, Sheng   III-1167
Li, Tsai-Yen   I-957
Li, Weishi   I-769, IV-906
Li, Xiao-Li   III-318
Li, Xiaotu   III-416
Li, Xiaowei   III-266
Li, Yanda   II-1217
Li, Yun   III-374
Li, Zhanhuai   I-214
Li, Zhuowei   I-994
Liao, Mao-Yung   I-957
Liang, Xiaohui   III-225
Liang, Y.C.   I-1040
Lim, Cheol-Su   III-1080, III-1148
Lim, Ee-Peng   II-1177
Lim, Heui Seok   I-1129
Lim, Hyung-Jin   I-146
Lim, In-Taek   I-438
Lim, Jongin   II-381, II-537, II-567, II-647
Lim, Jong In   I-619
Lim, Jongtae   IV-138
Lim, Myoung-seob   II-723
Lim, Seungkil   IV-642
Lim, Si-Yeong   IV-606
Lim, Soon-Bum   I-847
Lim, YoungHwan   I-242, IV-1036
Lim, Younghwan   I-398, II-676, II-848
Lin, Huaizhong   IV-826
Lin, Jenn-Rong   IV-499
Lin, Manshan   III-691, IV-1028
Lin, Ping-Hsien   III-1050
Lindskog, Stefan   IV-1331
Lischka, Hans   I-1004
Liu, Bin   II-508
Liu, Dongquan   IV-968
Liu, Fenlin   II-508
Liu, Jiming   II-1274
Liu, Jingmei   IV-1046
Liu, Joseph K.   II-614
Liu, Ming   IV-1102
Liu, Mingzhe   II-1187
Liu, Xuehui   III-1167
Liu, Yue   III-266
Lopez, Javier   III-681
Lu, Chung-Dar   III-299
Lu, Dongming   I-865, I-985

Lu, Jiahui   I-1040
Lu, Xiaolin   III-256
Luengo, Francisco   III-1157
Luo, Lijuan   IV-896
Luo, Xiangyang   II-508
Luo, Ying   III-173
Luo, Yingwei   I-301, II-822

Ma, Fanyuan   II-1086
Ma, Liang   III-292
Ma, Lizhuang   I-776
Mackay, Troy D.   III-143
Małafiejski, Michal   I-647
Manera, Jaime   IV-1301
Mani, Venkataraman   IV-269
Manzanares, Antonio Izquierdo   IV-1292
Mao, Zhihong   I-776
Maris, Assimo   I-1046
Markowski, Marcin   IV-772
Márquez, Joaquín Torres   IV-1292
Martoyan, Gagik A.   I-1012
Medvedev, N.N.   I-816
Meng, Qingfan   I-1040
Merelli, Emanuela   II-1264
Miao, Lanfang   I-758
Miao, Yongwei   III-1023
Michelot, Christian   IV-712
Mielikäinen, Taneli   IV-1251
Mijangos, Eugenio   IV-477
Million, D.L.   III-29
Min, Byoung Joon   I-252
Min, Byoung-Muk   II-896
Min, Dugki   II-187, III-858
Min, Hyun Gi   I-57
Min, Jihong   I-1154
Min, Kyongpil   I-1135
Min, Seung-hyun   II-723
Min, Sung-Hwan   IV-458
Minasyan, Seyran H.   I-1012
Minghui, Wu   I-875, I-974
Minhas, Mahmood R.   IV-587
Mirto, Maria   III-1
Miyano, Satoru   III-349
Mnaouer, Adel Ben   IV-1212
Mo, Jianzhong   I-967
Mocavero, Silvia   III-1
Moon, Hyeonjoon   I-584
Moon, Kiyoung   I-609

Morarescu, Cristian   III-556, III-563
Moreland, Terry   IV-1120
Morillo, Pedro   III-1119
Moriya, Kentaro   IV-978
Mourrain, Bernard   I-683
Mun, Ki-Young   I-311
Mun, Young-Song   I-97, I-118, I-242, I-271, I-398, I-408, I-459, I-550, I-559, I-569, I-628, II-676, II-848, IV-1036
Murat, Cécile   IV-202
Muyl, Frédérique   IV-948

Nait-Sidi-Moh, Ahmed   IV-792
Nakamura, Yasuaki   III-1013
Nam, Junghyun   I-498
Nam, Kichun   I-1129
Nam, Kyung-Won   I-1170
Nandy, Subhas C.   I-827
Nariai, Naoki   III-349
Nasir, Uzma   II-1045
Nassis, Vicky   II-914
Ng, Michael Kwok   IV-843
Ng, See-Kiong   II-1167, III-318
Nicolay, Thomas   II-634
Nie, Weifang   III-284, III-292, III-416
Nikolova, Mila   IV-843
Ninulescu, Valerică   III-635, III-643
Nodera, Takashi   IV-978
Noël, Alfred G.   III-512
Noh, Angela Song-Ie   I-1144
Noh, Bong-Nam   II-82, II-225
Noh, Hye-Min   III-836, III-878, III-945
Noh, Seung J.   IV-615
Nozick, Linda K.   IV-499
Nugraheni, Cecilia E.   III-453

Offermans, W.K.   I-1020
Ogiela, Lidia   IV-852
Ogiela, Marek R.   IV-852
Oh, Am-Sok   II-322, 585
Oh, Heekuck   IV-877
Oh, Nam-Ho   II-401
Oh, Sei-Chang   II-816
Oh, Seoung-Jun   I-166
Oh, Sun-Jin   II-169
Oh, Sung-Kwun   IV-858, IV-887
Ok, MinHwan   II-1035
Olmes, Zhanna   I-448
Omar M.   III-60

Ong, Eng Teo   I-769
Onyeahialam, Anthonia   III-152

Padgett, James   IV-1282
Páez, Antonio   III-162
Paik, Juryon   II-936
Pan, Hailang   I-896
Pan, Xuezeng   I-329, II-704
Pan, Yi   III-338
Pan, Yunhe   I-865
Pan, Zhigeng   II-946, III-190, III-245
Pandey, R.B   II-1197
Pang, Mingyong   III-245
Park, Myong-soon   II-1035
Park, Bongjoo   II-245
Park, Byoung-Jun   IV-887
Park, Byungchul   I-468
Park, Chan Yong   II-1284
Park, Chankwon   III-1219
Park, Cheol-Min   I-1170
Park, Choon-Sik   II-225
Park, Daehee   IV-858
Park, Dea-Woo   II-235
Park, DaeHyuck   IV-1036
Park, DongGook   II-245
Park, Eung-Ki   II-225
Park, Gyung-Leen   I-478
Park, Hayoung   II-442
Park, Hee Jun   IV-122
Park, Hee-Dong   I-378
Park, Hee-Un   II-117
Park, Heejun   III-1316
Park, Jesang   I-418
Park, Jin-Woo   I-913, III-1219
Park, Jonghyun   IV-867
Park, Joon Young   III-993, III-1060
Park, Joowon   II-789
Park, KwangJin   II-1225
Park, Kyeongmo   II-264
Park, KyungWoo   II-254
Park, Mi-Og   II-235
Park, Namje   I-609, II-1146
Park, Sachoun   III-973
Park, Sang-Min   IV-652
Park, Sang-Sung   IV-743
Park, Sangjoon   I-107, I-118
Park, Seon Hee   II-1284
Park, Seoung Kyu   III-1306
Park, Si Hyung   III-1033
Park, Soonyoung   IV-867

Park, Sung Hee   II-1284
Park, Sung-gi   II-127
Park, Sung-Ho   I-11
Park, Sungjun   I-886
Park, Sung-Seok   II-127
Park, Woojin   I-261
Park, Yongsu   II-353
Park, Youngho   II-647
Park, Yunsun   IV-148
Parlos, A.G   II-1293
Paschos, Vangelis Th.   IV-192, IV-202
Pedrycz, Witold   IV-887
Pei, Bingzhen   III-10
Peng, Jiming   IV-290
Peng, Qunsheng   I-758, III-1023
Penubarthi, Chaitanya   I-156
Pérez, María S.   III-109
Phan, Raphael C.-W.   I-488, III-661, IV-1065
Piattini, Mario   I-1064
Pietkiewicz, Wojciech   II-517
Ping, Lingdi   I-329, II-704
Pirani, Fernando   I-1046
Pirrone, Nicola   I-1054
Podoleanu, Adrian   III-556
Ponce, Eva   IV-1301
Porschen, Stefan   I-796
Prasanna, H.M.   IV-548
Przewoźniczek, Michał   IV-802
Pusca, Stefan   III-563, III-569, III-614

Qi, Feihu   IV-1140
Qing, Sihan   III-711
Qu, Na   III-225

Rahayu, Wenny   II-914, II-925
Rajugan, R.,   II-914, II-1125
Ramadan, Omar   IV-926
Rasheed, Faraz   II-1115
Ravantti, Janne   IV-1251
Raza, Syed Arshad   I-806
Re, Giuseppe Lo   III-39
Ren, Lifeng   IV-30
Rhee, Seung Hyong   IV-1231
Rhee, Seung-Hyun   III-1259
Rhew, Sung Yul   I-57
Riaz, Maria   II-1096
Ribagorda, Arturo   III-729
Riganelli, Antonio   I-905, I-921, I-938
Rim, Suk-Chul   IV-615

Rob, Seok-Beom   IV-858
Rocha, Ana Maria A.C.   IV-488
Roh, Sung-Ju   IV-1169
Rohe, Markus   II-634
Roman, Rodrigo   III-681
Rosi, Marzio   I-1101
Ruskin, Heather J.   II-1254
Ryou, Hwang-bin   II-904
Ryu, Keun Ho   II-925, II-977
Ryu, Han-Kyu   I-378
Ryu, Joonghyun   III-993
Ryu, Joung Woo   I-127
Ryu, Keun Ho   II-966
Ryu, Keun Hos   II-998
Ryu, Seonggeun   I-398
Ryu, Yeonseung   IV-72

Sadjadi, Seyed Jafar   IV-388
Sætre, Rune   III-327
Sait, Sadiq M.   IV-587
Salzer, Reiner   I-938
Sánchez, Alberto   III-109
Sarfraz, Muhammad   I-806
Saxena, Amitabh   III-672
Seo, Dae-Hee   II-117
Seo, Jae Young   IV-528
Seo, Jae-Hyun   II-82, II-225, II-254
Seo, Jeong-Yeon   IV-652
Seo, Jung-Taek   II-225
Seo, Kwang-Kyu   IV-448, IV-458
Seo, Kyung-Sik   IV-836
Seo, Young-Jun   I-928
Seong, Myoung-ho   II-723
Seongjin, Ahn   II-676, II-696
Serif, T.   II-1018
Shang, Yanfeng   IV-1102
Shao, Min-Hua   III-701
Shehzad, Anjum   II-1096, II-1106
Shen, Lianguan G.   III-1003
Shen, Yonghang   IV-1159
Sheng, Yu   I-985
Shi, Lie   III-190
Shi, Lei   I-896
Shi, Xifan   IV-1159
Shim, Bo-Yeon   III-806
Shim, Donghee   I-232
Shim, Young-Chul   I-427, III-426
Shimizu, Mayumi   III-1013
Shin, Byeong-Seok   III-1177, III-1187
Shin, Chungsoo   I-459

Shin, Dong-Ryeol   I-348, III-99
Shin, Ho-Jin   III-99
Shin, Ho-Jun   III-806
Shin, Hyo Young   II-741
Shin, Hyoun Gyu   IV-86
Shin, Hyun-Ho   II-157
Shin, In-Hye   I-478
Shin, Kitae   III-1219
Shin, Myong-Chul   II-332, II-341, II-499, IV-1319
Shin, Seong-Yoon   II-195, II-284
Shin, Yeong Gil   IV-1111
Shin, Yongtae   II-442
Shu, Jiwu   IV-762
Shumilina, Anna   I-1075
Sicker, Douglas C.   IV-528
Siddiqi, Mohammad Umar   III-661
Sierra, José M.   IV-1301, IV-1309
Sim, Jeong Seop   II-1284
Sim, Terence   III-1197
Simeonidis, Minas   III-569
Sinclair, Brett   III-49
Singh, Sanjeet   IV-1190
Sivakumar, K.C.   IV-1341
Skworcow, Piotr   IV-692
Smith, Kate A.   II-956
So, Yeon-hee   IV-62
Soares, João L.C.   IV-488
Soh, Ben   III-672
Soh, Jin   II-754
Sohn, Bangyong   II-442
Sohn, Chae-Bong   I-166
Sohn, Hong-Gyoo   II-771
Sohn, Sungwon   I-609
Sokolov, B.V.   IV-407
Solimannejad, Mohammad   I-1004
Son, Bongsoo   II-789, II-816, II-863
Song, Hoseong   III-1259
Song, Hui   II-1086
Song, Il-Yeol   III-402
Song, MoonBae   II-1225
Song, Teuk-Seob   I-847
Song, Yeong-Sun   II-771
Song, Young-Jae   I-928, III-886
Soriano, Miguel   II-527, II-624
Sourin, Alexei   III-983
Sourina, Olga   IV-968
Srinivas   IV-680
Steigedal, Tonje Stroemmen   III-327
Sterian, Andreea   III-585, III-643

Sterian, Andreea-Rodica   III-635
Sterian, Rodica   III-592, III-598
Strelkov, Nikolay   III-621, III-628
Su, Hua   II-1293
Suh, Young-Joo   I-368
Sun, Dong Guk   III-79
Sun, Jizhou   III-284, III-292, III-416, III-435
Sun, Weitao   IV-762
Sung, Jaechul   II-567
Sung, Ji-Yeon   I-1144
Suresh, Sundaram   IV-269
Swarna, J. Mercy   IV-1341

Ta, Duong Nguyen Binh   I-947
Tadeusiewicz, Ryszard   IV-852
Tae, Kang Soo   I-478
Tai, Allen H.   IV-342
Tamada, Yoshinori   III-349
Tan, Chew Lim   III-1197
Tan, Kenneth Chih Jeng   IV-1120
Tan, Soon-Heng   III-318
Tan, Wuzheng   I-776
Tang, Jiakui   III-173
Tang, Jie   IV-1222
Tang, Sheng   II-547
Tang, Yuchun   III-299
Taniar, David   IV-1203
Tao, Pai-Cheng   I-957
Tavadyan, Levon A.   I-1012
Techapichetvanich, Kesaraporn   III-206
Teillaud, Monique   I-683
Teng, Lirong   I-1040
Thuy, Le Thi Thu   II-1066
Tian, Haishan   III-1099
Tian, Tianhai   II-1245
Tillich, Stefan   II-665
Ting, Ching-Jung   IV-417
Tiwari, Manoj Kumar   IV-680
Toi, Yutaka   IV-1055
Toma, Alexandru   III-556, III-569
Toma, Cristian   III-556, III-592, III-598
Toma, Ghiocel   III-563, III-569, III-576, III-585, III-614
Toma, Theodora   III-556, III-569
Tomaschewski, Kai   II-988
Torres, Joaquin   III-729
Trunfio, Giuseppe A.   I-1054
Turnquist, Mark A.   IV-499
Tveit, Amund   III-327

Ufuktepe, Ünal   III-522, III-529
Umakant, J.   IV-548
Urbina, Ruben T.   III-547

Vanhoucke, Mario   IV-378
Velardo, Fernando Rosa   II-1156
Varella, E.   I-938
Vita, Marco   II-1264
Vizcaíno, Aurora   I-1064

Wack, Maxime   IV-792
Wahala, Kristiina   I-938
Walkowiak, Krzysztof   IV-802
Wan, Zheng   I-329, II-704
Wang, Chen   II-1086
Wang, Chengfeng   I-748
Wang, Chuanpeng   III-225
Wang, Gi-Nam   IV-702
Wang, Guilin   III-701, III-711
Wang, Hao   III-691, IV-1028
Wang, Hei-Chia   III-309
Wang, Hui-Mei   IV-172
Wang, Jianqin   III-173
Wang, Jiening   III-284
Wang, K.J.   IV-333
Wang, Lei   IV-733
Wang, Pi-Chung   IV-1007
Wang, Ruili   II-1187
Wang, Shaoyu   IV-1140
Wang, Shu   IV-1
Wang, S.M.   IV-333
Wang, Weinong   II-806
Wang, Xiaolin   II-822
Wang, Xinmei   IV-1046
Wang, Xiuhui   III-215
Wang, Yanguang   III-173
Wang, Yongtian   III-266
Weber, Irene   III-299
Wee, H.M.   IV-333
Wee, Hyun-Wook   III-938, IV-333
Wei, Sun   IV-753
Weng, Dongdong   III-266
Wenjun, Wang   I-301
Wille, Volker   IV-958
Wirt, Kai   II-577
Won, Chung In   I-707
Won, Dongho   I-498, I-609, II-92, II-1146
Won, Hyung Jun   III-1259
Won, Jae-Kang   III-817

Wong, Duncan S.   II-614
Woo, Gyun   I-29
Woo, Seon-Mi   III-954
Woo, Sinam   I-261
Wu, C.G.   I-1040
Wu, Chaolin   III-173
Wu, Enhua   III-1167
Wu, Hulin   IV-519
Wu, Yong   III-1099
Wu, Yue   IV-250
Wu, Zhiping   II-595

Xia, Yu   IV-290
Xiaolin, Wang   I-301
Xinpeng, Lin   I-301
Xiong, Guomin   II-822
Xirouchakis, Paul   IV-538, IV-722
Xu, Bing   II-946
Xu, Dan   III-274
Xu, Guilin   IV-30
Xu, Jie   I-758
Xu, Qing   III-292
Xu, Shuhong   I-769, IV-906
Xu, Xiaohua   III-338
Xu, Zhuoqun   II-822
Xue, Yong   III-173

Yamaguchi, Rui   III-381
Yamamoto, Osami   I-786
Yamashita, Satoru   III-381
Yan, Chung-Ren   III-1043
Yan, Dayuan   III-266
Yan, Hong   III-357
Yang, Byounghak   IV-241
Yang, Chao-Tung   IV-1017
Yang, Ching-Nung   I-19
Yang, Dong Jin   II-647
Yang, Hae-Sool   II-1, II-52, III-739, III-827, III-938
Yang, Hongwei   II-946
Yang, Jie   III-416
Yang, KwonWoo   III-89
Yang, Tao   III-266
Yang, X.S.   III-1109
Yang, Xin   I-896, IV-1102
Yang, Yoo-Kil   III-1129
Yantır, Ahmet   III-529
Yao, Xin   II-1217
Yates, Paul   I-938
Yazici, Ali   III-463

Ye, Dingfeng   II-595
Ye, Lu   III-190, IV-30
Yeh, Chung-Hsing   II-956
Yen, Show-Jane   II-1055
Yi, Yong-Hoon   II-82
Yim, Wha Young   I-1213
Yin, Jianfei   III-691, IV-1028
Yin, Ming   II-1177
Yingwei, Luo   I-301
Yiu, S.M.   III-651
Yoo, Cheol-Jung   III-758, III-836, III-878, III-945
Yoo, Chun-Sik   II-31, III-954
Yoo, Hun-Woo   IV-458, IV-743
Yoo, Hyeong Seon   II-102, II-111
Yoo, Jin Ah   II-889
Yoo, Seung Hwan   I-252
Yoo, Seung-Jae   II-870
Yoo, Sun K.   II-1028
Yoon, Chang-Dae   II-332, II-373, II-429
Yoon, Mi-sun   IV-62
Yoon, Yeo Bong   III-19
Yoshida, Ryo   III-389
You, L.H.   III-197
You, Jinyuan   III-10
You, Peng-Sheng   IV-368
Youn, Chan-Hyun   I-320
Youn, Hee Yong   I-519, II-936, III-19, IV-916, IV-1149
Youn, Hyunsang   I-291
Youn, Ju-In   II-10
Younghwan, Lim   II-676
Youngsong, Mun   II-676
Yu, Eun Jung   III-1306
Yu, HeonChang   III-89, IV-936
Yu, Jiangying   III-225
Yu, Sang-Jun   I-166
Yuan, Qingshu   I-865, I-985
Yun, HY.   II-1028
Yun, Sung-Hyun   II-391, II-401, IV-62
Yun, Won Young   IV-558
Yunhe, Pan   I-875, I-974
Yusupov, R.M.   IV-407

Zantidis, Dimitri   III-672
Zaychik, E.M.   IV-407
Zhai, Jia   IV-702
Zhai, Qi   III-284, III-435
Zhang, Changshui   II-1217
Zhang, Fuyan   III-245
Zhang, Jian J.   III-197, III-1003, III-1109
Zhang, Jianzhong   IV-161
Zhang, Jiawan   III-292, III-416, III-435
Zhang, Jin-Ting   IV-519
Zhang, Jun   III-691, IV-1028
Zhang, Kuo   IV-1222
Zhang, Mingmin   II-946, III-190, III-245
Zhang, Mingming   IV-1, IV-30
Zhang, Qiaoping   III-181
Zhang, Qiong   I-967
Zhang, Shen   II-686
Zhang, Ya-Ping   III-274
Zhang, Yan-Qing   III-299
Zhang, Yi   III-435
Zhang, Yong-Dong   II-547
Zhang, Yu   III-1197
Zhang, Yang   I-214
Zhang, Yuanliang   II-1207
Zhao, Jane   II-1235
Zhao, Qinping   III-235
Zhao, Weizhong   IV-1159
Zhao, Yang   III-274
Zhao, Yiming   IV-1
Zheng, Jin Jin   III-1003
Zheng, Weimin   IV-762
Zheng, Zengwei   IV-826
Zhong, Shaobo   III-173
Zhou, Hanbin   IV-896
Zhou, Hong Jun   III-1003
Zhou, Jianying   III-681, III-701
Zhou, Qiang   IV-896
Zhou, Suiping   I-947
Zhou, Xiaohua   III-402
Zhu, Jiejie   IV-30
Zhu, Ming   IV-1102
Zhuoqun, Xu   I-301
Żyliński, Pawel   I-647

Lecture Notes in Computer Science

For information about Vols. 1–3392
please contact your bookseller or Springer

Vol. 3525: A.E. Abdallah, C.B. Jones, J.W. Sanders (Eds.), Communicating Sequential Processes. XIV, 321 pages. 2005.

Vol. 3517: H.S. Baird, D.P. Lopresti (Eds.), Human Interactive Proofs. IX, 143 pages. 2005.

Vol. 3510: T. Braun, G. Carle, Y. Koucheryavy, V. Tsaoussidis (Eds.), Wired/Wireless Internet Communications. XIV, 366 pages. 2005.

Vol. 3508: P. Bresciani, P. Giorgini, B. Henderson-Sellers, G. Low, M. Winikoff (Eds.), Agent-Oriented Information Systems II. X, 227 pages. 2005. (Subseries LNAI).

Vol. 3503: S.E. Nikoletseas (Ed.), Experimental and Efficient Algorithms. XV, 624 pages. 2005.

Vol. 3501: B. Kégl, G. Lapalme (Eds.), Advances in Artificial Intelligence. XV, 458 pages. 2005. (Subseries LNAI).

Vol. 3500: S. Miyano, J. Mesirov, S. Kasif, S. Istrail, P. Pevzner, M. Waterman (Eds.), Research in Computational Molecular Biology. XVII, 632 pages. 2005. (Subseries LNBI).

Vol. 3498: J. Wang, X. Liao, Z. Yi (Eds.), Advances in Neural Networks – ISNN 2005, Part III. L, 1077 pages. 2005.

Vol. 3497: J. Wang, X. Liao, Z. Yi (Eds.), Advances in Neural Networks – ISNN 2005, Part II. L, 947 pages. 2005.

Vol. 3496: J. Wang, X. Liao, Z. Yi (Eds.), Advances in Neural Networks – ISNN 2005, Part II. L, 1055 pages. 2005.

Vol. 3495: P. Kantor, G. Muresan, F. Roberts, D.D. Zeng, F.-Y. Wang, H. Chen, R.C. Merkle (Eds.), Intelligence and Security Informatics. XVIII, 674 pages. 2005.

Vol. 3494: R. Cramer (Ed.), Advances in Cryptology – EUROCRYPT 2005. XIV, 576 pages. 2005.

Vol. 3492: P. Blache, E. Stabler, J. Busquets, R. Moot (Eds.), Logical Aspects of Computational Linguistics. X, 363 pages. 2005. (Subseries LNAI).

Vol. 3489: G.T. Heineman, J.A. Stafford, H.W. Schmidt, K. Wallnau, C. Szyperski, I. Crnkovic (Eds.), Component-Based Software Engineering. XI, 358 pages. 2005.

Vol. 3488: M.-S. Hacid, N.V. Murray, Z.W. Raś, S. Tsumoto (Eds.), Foundations of Intelligent Systems. XIII, 700 pages. 2005. (Subseries LNAI).

Vol. 3483: O. Gervasi, M.L. Gavrilova, V. Kumar, A. Laganà, H.P. Lee, Y. Mun, D. Taniar, C.J.K. Tan (Eds.), Computational Science and Its Applications – ICCSA 2005, Part IV. LXV, 1362 pages. 2005.

Vol. 3482: O. Gervasi, M.L. Gavrilova, V. Kumar, A. Laganà, H.P. Lee, Y. Mun, D. Taniar, C.J.K. Tan (Eds.), Computational Science and Its Applications – ICCSA 2005, Part III. LXV, 1340 pages. 2005.

Vol. 3481: O. Gervasi, M.L. Gavrilova, V. Kumar, A. Laganà, H.P. Lee, Y. Mun, D. Taniar, C.J.K. Tan (Eds.), Computational Science and Its Applications – ICCSA 2005, Part II. LV, 1316 pages. 2005.

Vol. 3480: O. Gervasi, M.L. Gavrilova, V. Kumar, A. Laganà, H.P. Lee, Y. Mun, D. Taniar, C.J.K. Tan (Eds.), Computational Science and Its Applications – ICCSA 2005, Part I. LXV, 1234 pages. 2005.

Vol. 3479: T. Strang, C. Linnhoff-Popien (Eds.), Location- and Context-Awareness. XII, 378 pages. 2005.

Vol. 3477: P. Herrmann, V. Issarny (Eds.), Trust Management. XII, 426 pages. 2005.

Vol. 3475: N. Guelfi (Ed.), Rapid Integration of Software Engineering Techniques. X, 145 pages. 2005.

Vol. 3468: H.W. Gellersen, R. Want, A. Schmidt (Eds.), Pervasive Computing. XIII, 347 pages. 2005.

Vol. 3467: J. Giesl (Ed.), Term Rewriting and Applications. XIII, 517 pages. 2005.

Vol. 3465: M. Bernardo, A. Bogliolo (Eds.), Formal Methods for Mobile Computing. VII, 271 pages. 2005.

Vol. 3463: M. Dal Cin, M. Kaâniche, A. Pataricza (Eds.), Dependable Computing - EDCC 2005. XVI, 472 pages. 2005.

Vol. 3462: R. Boutaba, K. Almeroth, R. Puigjaner, S. Shen, J.P. Black (Eds.), NETWORKING 2005. XXX, 1483 pages. 2005.

Vol. 3461: P. Urzyczyn (Ed.), Typed Lambda Calculi and Applications. XI, 433 pages. 2005.

Vol. 3460: Ö. Babaoglu, M. Jelasity, A. Montresor, C. Fetzer, S. Leonardi, A. van Moorsel, M. van Steen (Eds.), Self-star Properties in Complex Information Systems. IX, 447 pages. 2005.

Vol. 3459: R. Kimmel, N.A. Sochen, J. Weickert (Eds.), Scale Space and PDE Methods in Computer Vision. XI, 634 pages. 2005.

Vol. 3456: H. Rust, Operational Semantics for Timed Systems. XII, 223 pages. 2005.

Vol. 3455: H. Treharne, S. King, M. Henson, S. Schneider (Eds.), ZB 2005: Formal Specification and Development in Z and B. XV, 493 pages. 2005.

Vol. 3454: J.-M. Jacquet, G.P. Picco (Eds.), Coordination Models and Languages. X, 299 pages. 2005.

Vol. 3453: L. Zhou, B.C. Ooi, X. Meng (Eds.), Database Systems for Advanced Applications. XXVII, 929 pages. 2005.

Vol. 3452: F. Baader, A. Voronkov (Eds.), Logic for Programming, Artificial Intelligence, and Reasoning. XI, 562 pages. 2005. (Subseries LNAI).

Vol. 3450: D. Hutter, M. Ullmann (Eds.), Security in Pervasive Computing. XI, 239 pages. 2005.

Vol. 3449: F. Rothlauf, J. Branke, S. Cagnoni, D.W. Corne, R. Drechsler, Y. Jin, P. Machado, E. Marchiori, J. Romero, G.D. Smith, G. Squillero (Eds.), Applications of Evolutionary Computing. XX, 631 pages. 2005.

Vol. 3448: G.R. Raidl, J. Gottlieb (Eds.), Evolutionary Computation in Combinatorial Optimization. XI, 271 pages. 2005.

Vol. 3447: M. Keijzer, A. Tettamanzi, P. Collet, J.v. Hemert, M. Tomassini (Eds.), Genetic Programming. XIII, 382 pages. 2005.

Vol. 3444: M. Sagiv (Ed.), Programming Languages and Systems. XIII, 439 pages. 2005.

Vol. 3443: R. Bodik (Ed.), Compiler Construction. XI, 305 pages. 2005.

Vol. 3442: M. Cerioli (Ed.), Fundamental Approaches to Software Engineering. XIII, 373 pages. 2005.

Vol. 3441: V. Sassone (Ed.), Foundations of Software Science and Computational Structures. XVIII, 521 pages. 2005.

Vol. 3440: N. Halbwachs, L.D. Zuck (Eds.), Tools and Algorithms for the Construction and Analysis of Systems. XVII, 588 pages. 2005.

Vol. 3439: R.H. Deng, F. Bao, H. Pang, J. Zhou (Eds.), Information Security Practice and Experience. XII, 424 pages. 2005.

Vol. 3437: T. Gschwind, C. Mascolo (Eds.), Software Engineering and Middleware. X, 245 pages. 2005.

Vol. 3436: B. Bouyssounouse, J. Sifakis (Eds.), Embedded Systems Design. XV, 492 pages. 2005.

Vol. 3434: L. Brun, M. Vento (Eds.), Graph-Based Representations in Pattern Recognition. XII, 384 pages. 2005.

Vol. 3433: S. Bhalla (Ed.), Databases in Networked Information Systems. VII, 319 pages. 2005.

Vol. 3432: M. Beigl, P. Lukowicz (Eds.), Systems Aspects in Organic and Pervasive Computing - ARCS 2005. X, 265 pages. 2005.

Vol. 3431: C. Dovrolis (Ed.), Passive and Active Network Measurement. XII, 374 pages. 2005.

Vol. 3429: E. Andres, G. Damiand, P. Lienhardt (Eds.), Discrete Geometry for Computer Imagery. X, 428 pages. 2005.

Vol. 3427: G. Kotsis, O. Spaniol (Eds.), Wireless Systems and Mobility in Next Generation Internet. VIII, 249 pages. 2005.

Vol. 3423: J.L. Fiadeiro, P.D. Mosses, F. Orejas (Eds.), Recent Trends in Algebraic Development Techniques. VIII, 271 pages. 2005.

Vol. 3422: R.T. Mittermeir (Ed.), From Computer Literacy to Informatics Fundamentals. X, 203 pages. 2005.

Vol. 3421: P. Lorenz, P. Dini (Eds.), Networking - ICN 2005, Part II. XXXV, 1153 pages. 2005.

Vol. 3420: P. Lorenz, P. Dini (Eds.), Networking - ICN 2005, Part I. XXXV, 933 pages. 2005.

Vol. 3419: B. Faltings, A. Petcu, F. Fages, F. Rossi (Eds.), Constraint Satisfaction and Constraint Logic Programming. X, 217 pages. 2005. (Subseries LNAI).

Vol. 3418: U. Brandes, T. Erlebach (Eds.), Network Analysis. XII, 471 pages. 2005.

Vol. 3416: M. Böhlen, J. Gamper, W. Polasek, M.A. Wimmer (Eds.), E-Government: Towards Electronic Democracy. XIII, 311 pages. 2005. (Subseries LNAI).

Vol. 3415: P. Davidsson, B. Logan, K. Takadama (Eds.), Multi-Agent and Multi-Agent-Based Simulation. X, 265 pages. 2005. (Subseries LNAI).

Vol. 3414: M. Morari, L. Thiele (Eds.), Hybrid Systems: Computation and Control. XII, 684 pages. 2005.

Vol. 3412: X. Franch, D. Port (Eds.), COTS-Based Software Systems. XVI, 312 pages. 2005.

Vol. 3411: S.H. Myaeng, M. Zhou, K.-F. Wong, H.-J. Zhang (Eds.), Information Retrieval Technology. XIII, 337 pages. 2005.

Vol. 3410: C.A. Coello Coello, A. Hernández Aguirre, E. Zitzler (Eds.), Evolutionary Multi-Criterion Optimization. XVI, 912 pages. 2005.

Vol. 3409: N. Guelfi, G. Reggio, A. Romanovsky (Eds.), Scientific Engineering of Distributed Java Applications. X, 127 pages. 2005.

Vol. 3408: D.E. Losada, J.M. Fernández-Luna (Eds.), Advances in Information Retrieval. XVII, 572 pages. 2005.

Vol. 3407: Z. Liu, K. Araki (Eds.), Theoretical Aspects of Computing - ICTAC 2004. XIV, 562 pages. 2005.

Vol. 3406: A. Gelbukh (Ed.), Computational Linguistics and Intelligent Text Processing. XVII, 829 pages. 2005.

Vol. 3404: V. Diekert, B. Durand (Eds.), STACS 2005. XVI, 706 pages. 2005.

Vol. 3403: B. Ganter, R. Godin (Eds.), Formal Concept Analysis. XI, 419 pages. 2005. (Subseries LNAI).

Vol. 3402: M. Daydé, J.J. Dongarra, V. Hernández, J.M.L.M. Palma (Eds.), High Performance Computing for Computational Science - VECPAR 2004. XI, 732 pages. 2005.

Vol. 3401: Z. Li, L.G. Vulkov, J. Waśniewski (Eds.), Numerical Analysis and Its Applications. XIII, 630 pages. 2005.

Vol. 3400: J.F. Peters, A. Skowron (Eds.), Transactions on Rough Sets III. IX, 461 pages. 2005.

Vol. 3399: Y. Zhang, K. Tanaka, J.X. Yu, S. Wang, M. Li (Eds.), Web Technologies Research and Development - APWeb 2005. XXII, 1082 pages. 2005.

Vol. 3398: D.-K. Baik (Ed.), Systems Modeling and Simulation: Theory and Applications. XIV, 733 pages. 2005. (Subseries LNAI).

Vol. 3397: T.G. Kim (Ed.), Artificial Intelligence and Simulation. XV, 711 pages. 2005. (Subseries LNAI).

Vol. 3396: R.M. van Eijk, M.-P. Huget, F. Dignum (Eds.), Agent Communication. X, 261 pages. 2005. (Subseries LNAI).

Vol. 3395: J. Grabowski, B. Nielsen (Eds.), Formal Approaches to Software Testing. X, 225 pages. 2005.

Vol. 3394: D. Kudenko, D. Kazakov, E. Alonso (Eds.), Adaptive Agents and Multi-Agent Systems II. VIII, 313 pages. 2005. (Subseries LNAI).

Vol. 3393: H.-J. Kreowski, U. Montanari, F. Orejas, G. Rozenberg, G. Taentzer (Eds.), Formal Methods in Software and Systems Modeling. XXVII, 413 pages. 2005.

Lecture Notes in Computer Science 3481

Commenced Publication in 1973
Founding and Former Series Editors:
Gerhard Goos, Juris Hartmanis, and Jan van Leeuwen

Editorial Board

David Hutchison
  Lancaster University, UK
Takeo Kanade
  Carnegie Mellon University, Pittsburgh, PA, USA
Josef Kittler
  University of Surrey, Guildford, UK
Jon M. Kleinberg
  Cornell University, Ithaca, NY, USA
Friedemann Mattern
  ETH Zurich, Switzerland
John C. Mitchell
  Stanford University, CA, USA
Moni Naor
  Weizmann Institute of Science, Rehovot, Israel
Oscar Nierstrasz
  University of Bern, Switzerland
C. Pandu Rangan
  Indian Institute of Technology, Madras, India
Bernhard Steffen
  University of Dortmund, Germany
Madhu Sudan
  Massachusetts Institute of Technology, MA, USA
Demetri Terzopoulos
  New York University, NY, USA
Doug Tygar
  University of California, Berkeley, CA, USA
Moshe Y. Vardi
  Rice University, Houston, TX, USA
Gerhard Weikum
  Max-Planck Institute of Computer Science, Saarbruecken, Germany